Human Anatomy & Physiology Laboratory Manual

MAKING CONNECTIONS

CAT VERSION

Catharine C. Whiting
University of North Georgia

With contributions by

Karen L. Keller
Frostburg State University

PEARSON

Editor-in-Chief: Serina Beauparlant
Senior Acquisitions Editor: Gretchen Puttkamer Roethle
Project Manager: Caroline Ayres
Project Editor: Kari Hopperstead
Development Editor: Alan Titche
Art Development Editors: Kelly Murphy and Elisheva Marcus
Editorial Assistant: Arielle Grant
Director of Development: Barbara Yien
Art Development Manager: Laura Southworth
Program Management Team Lead: Mike Early
Project Management Team Lead: Nancy Tabor
Production Management: S4Carlisle Publishing Services
Copyeditor: Lorretta Palaji
Compositor: S4Carlisle Publishing Services
Design Manager: Marilyn Perry
Interior Designer: tani hasegawa
Cover Designer: Side By Side Design
Illustrators: Imagineering
Rights & Permissions Project Manager: Donna Kalal
Rights & Permissions Management: Rachel Youdelman
Photo Researcher: Maureen Spuhler
Senior Manufacturing Buyer: Stacey Weinberger
Senior Marketing Manager: Allison Rona
Senior Anatomy & Physiology Specialist: Derek Perrigo
Media Content Producer: Nicole Tache

Cover Photo Credit: Peathegee, Inc./Blend Images/Corbis

Library of Congress Control Number
Library of Congress Cataloging-in-Publication Data is available upon request.

1 2 3 4 5 6 7 8 9 10—V364—18 17 16 15 14

0-321-78700-5 (Student edition)
978-0-321-78700-2 (Student edition)
0-13-397580-0 (Instructor's Review Copy)
978-013-397580-2 (Instructor's Review Copy)

www.pearsonhighered.com

About the Author

Catharine C. Whiting, *University of North Georgia*

Cathy Whiting began her college career at Waycross Junior College before transferring to the University of Georgia and earning a B.S. in biology. She earned both an M.S.T. and a Ph.D. at the University of Florida, training under an extraordinary mentor, Dr. Louis J. Guillette, a brilliant researcher, author, and educator who taught her how to do science and, more importantly, how to teach. With 20 years of college teaching experience, Whiting seeks to engage her students through active learning in order to facilitate the development of critical-thinking and problem-solving skills. She has discovered that passionate teaching leads to passionate learning. The recipient of several teaching awards including Faculty Member of the Year, Advisor of the Year, and Master Teacher, she considers her greatest reward to be the privilege of teaching and impacting the lives of students.

Contributor

Karen L. Keller, *Frostburg State University*

Karen Keller earned both her B.S. and M.S. degrees in biology from Frostburg State University and her Ph.D. in physiology from the University of Georgia, College of Veterinary Medicine. She has taught at community college and four-year college levels and has extensive experience teaching introductory biology, anatomy and physiology, musculoskeletal anatomy, microbiology, comparative vertebrate anatomy, histology, and parasitology courses. In addition, she advises students interested in pursuing careers in the health professions and is a member of the American Association of Anatomists, the Human Anatomy and Physiology Society, and the Northeast Association of Advisors for the Health Professions.

About the Author

Catherine C. Whiting, University of North Georgia

Preface

Why Did I Write This Lab Manual?

I have been teaching in a wide variety of settings since I graduated from the University of Georgia—as a laboratory assistant, as a high school teacher, as a graduate assistant, as a tutor/mentor for college athletes, as an assistant professor of biology at a small liberal arts university, and, currently, as a professor of biology at the University of North Georgia. Regardless of the setting, I have always regarded teaching as an incredible opportunity and a great privilege. Through the years, I have learned that effective teaching requires much hard work, dedication, and enthusiasm. It involves a life-long pursuit of both content knowledge and understanding how students learn. It involves challenging students to develop critical-thinking and problem-solving skills. Most importantly, it involves building relationships with students and investing in their lives. As a matter of fact, it was a late afternoon conversation with a group of students after lab in the fall of 2009 that inspired me to pursue writing a lab manual.

I set out to write a lab manual that was first and foremost a tool of engagement. In my experience, engaging students in an active learning environment is the key to student success in both the lecture and laboratory settings. When students are engaged, exciting things happen. Attendance improves. Students enjoy being in class. Grades soar! Students begin to focus on learning instead of worrying about what is going to be on the test. My hope is that instructors will be able to use and adapt the activities in this manual to cultivate their own active learning environment and to experience the joy of watching students fully engage in the learning process. Imagine having to run students out of the lab so that the next lab can get started. You will be amazed at what your students can accomplish when they are engaged, challenged, and inspired!

How Is This Lab Manual Different?

Human Anatomy & Physiology Laboratory Manual: Making Connections distinguishes itself from other A&P lab manuals by focusing heavily on addressing the **three biggest teaching challenges** for A&P lab instructors: getting students to **engage** in the lab, to **prepare** for the lab, and to **apply** concepts in the lab.

Getting Students Engaged in the Lab

For many instructors this is the #1 teaching problem in the lab course. The whole active-learning approach of *Human Anatomy & Physiology Laboratory Manual: Making Connections* is centered around getting students engaged in the lab and asking questions. We achieve this by including a **rich variety of hands-on activities** that use **different learning modes** including labeling, sketching, touching, dissecting, observing, conducting experiments, interacting with groups, and making predictions.

This lab manual includes many tried and true lab activities but also has some unique activities to help facilitate active learning, including those listed in the table below.

Examples of Active Learning in This Lab Manual

Unit	Activity	How it facilitates active learning
Unit 2 Introduction to Organ Systems	Activity 3—Studying Homeostasis and Organ System Interactions	Students work together to research and explain how organ systems interact during the patellar reflex; high engagement factor; challenging task that requires students to think critically and discuss their ideas with lab group members
Unit 6 Histology	Activity 4—Tissue Identification Concept Map	Students must interact (discuss, question, argue, etc.) to determine the best set of questions to identify the assigned tissue types; encourages students to think about tissues rather than to just memorize them; high engagement and high energy; demands critical-thinking and problem-solving skills
Unit 10 The Appendicular Skeleton	Activity 2—Identifying Bones-in-a-Bag	Students identify bones and their features by touch only; high engagement and interaction as students discuss and review the assigned features of each bone as it is pulled out of the bag
Unit 13 Gross Anatomy of the Muscular System	Activity 1—Determining How Skeletal Muscles Are Named	Students complete an interactive overview activity that helps them understand how skeletal muscles are named; this activity teaches students a very useful approach to learning specific skeletal muscles (origin, insertion, innervation, and action) and prepares them for the remaining activities in the unit; actively engages students as they perform various muscle actions and locate muscles on different anatomical models throughout the lab

(continued)

Examples of Active Learning in This Lab Manual *(continued)*

Unit	Activity	How it facilitates active learning
Unit 15 The Central Nervous System: Brain and Spinal Cord	Activity 3—Identifying the Meninges/Ventricles and Tracing the Flow of Cerebrospinal Fluid	Students engage in a high-energy, interactive cerebrospinal fluid "dance" as they learn about the production, flow, and return of CSF to venous circulation
Unit 19 The Endocrine System	Activity 3—Investigating Endocrine Case Studies: Clinician's Corner	Mini case studies encourage students to apply the information that they have learned in Activity 1 and Activity 2; builds critical-thinking and problem-solving skills
Unit 24 Blood Vessel Physiology	Activity 1—Tracing Blood Flow—General Systemic Pathways	Students use their knowledge of heart and blood vessel anatomy obtained in previous units along with anatomical models to trace the pathway of blood from the left ventricle to four peripheral sites (eye, forearm, abdomen, and leg) and back to the right atrium; they work together to diagram, label, and explain the exchange of materials at the capillary bed
Unit 25 The Lymphatic System	Activity 4—Using a Pregnancy Test to Demonstrate Antigen–Antibody Reactions	An interactive "wet lab" that engages students as they perform an enzyme-linked immunosorbent assay (ELISA) to detect the presence of an antigen (human chorionic gonadotropin) in unknown samples
Unit 28 Anatomy of the Digestive System	Activity 3—Examining the Histology of Selected Digestive Organs	Interactive question set encourages student engagement and challenges students to make predictions and draw conclusions concerning the relationship between structure and function at the histological level
Unit 31 Physiology of the Urinary System	Activity 2—Simulating the Events of Urine Production and Urine Concentration	Hands-on activity using beads to simulate renal function; a question set takes students through a step-by-step process with increasingly challenging questions to help them better understand the role of the kidneys in maintaining homeostasis, as well as to further identify structure/function relationships

Key features of *Human Anatomy & Physiology Laboratory Manual: Making Connections* that help facilitate active learning include:

- **Lab Boosts** invite students to do hands-on demonstrations of key concepts.

LabBOOST >>>

Anatomy of the Renal Corpuscle
Understanding the anatomy of the renal corpuscle can be confusing. Here is a trick to help you learn the anatomy of the visceral layer of the glomerular capsule. Draw or tape a "nucleus" to the back of each of your hands. Your hands represent podocytes. Now, wiggle your fingers. Your fingers represent pedicels which are foot-like processes of the podocytes. Bring your fingers together so that they interdigitate (palms facing you). Note the slit-like openings between your fingers. These openings represent filtration slits. This visceral layer of the glomerular capsule overlies the glomerulus and its fenestrations to form the renal corpuscle.

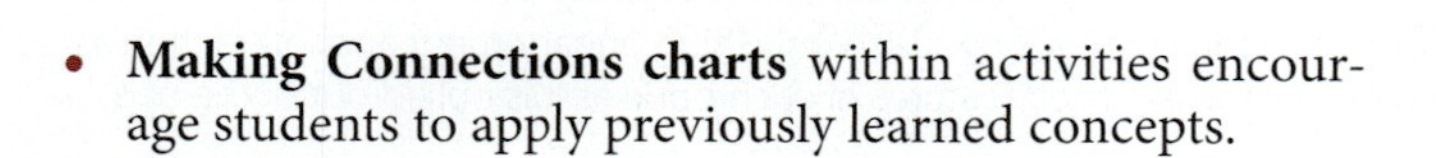

- **Making Connections charts** within activities encourage students to apply previously learned concepts.
- **Guided questions** within activities help students think about the relevant concepts and how they apply to the activity.
- **Quick Tips** provide hints for performing activities or mnemonics for remembering key terms.

QUICK TIP The following mnemonic device can help you remember the relative abundance of each leukocyte type, from most abundant to least: ***n***ever ***l***et ***m***onkeys ***e***at ***b***ananas (***n***eutrophils, ***l***ymphocytes, ***m***onocytes, ***e***osinophils, ***b***asophils).

- **Clinical Connection boxes** highlight relevant diseases or conditions and help reinforce learning of key concepts.

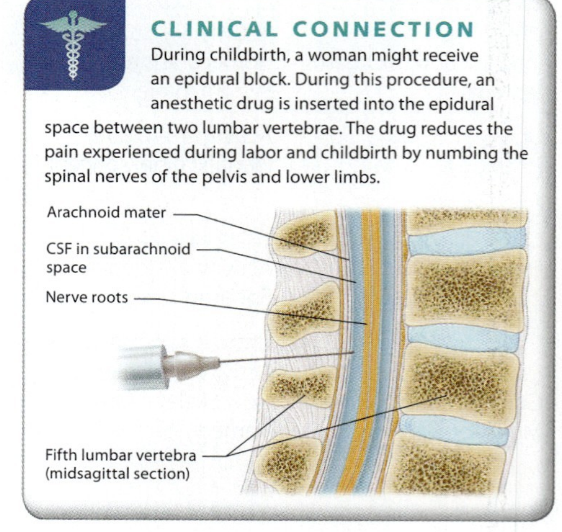

CLINICAL CONNECTION
During childbirth, a woman might receive an epidural block. During this procedure, an anesthetic drug is inserted into the epidural space between two lumbar vertebrae. The drug reduces the pain experienced during labor and childbirth by numbing the spinal nerves of the pelvis and lower limbs.
Arachnoid mater
CSF in subarachnoid space
Nerve roots
Fifth lumbar vertebra (midsagittal section)

Getting Students to Prepare for Lab

This manual helps address this problem by providing extensive **pre-lab assignments** that include pre-lab activity questions *for each activity* in the unit. These pre-lab questions are intended to get the student to peruse the lab activities *before* lab. Assignable pre-lab assessments are also available in MasteringA&P.

Getting Students to Apply Concepts

A third challenge and goal in the lab course is to get students to see the connections between concepts learned in lecture and their application in the lab. This manual fosters students' ability to make these connections with unique **Think About It** questions that begin each unit and **Making Connections** charts within activities. **Post-lab Assignments** also include **Bloom's Level II Review Questions** and **Concept Mapping.**

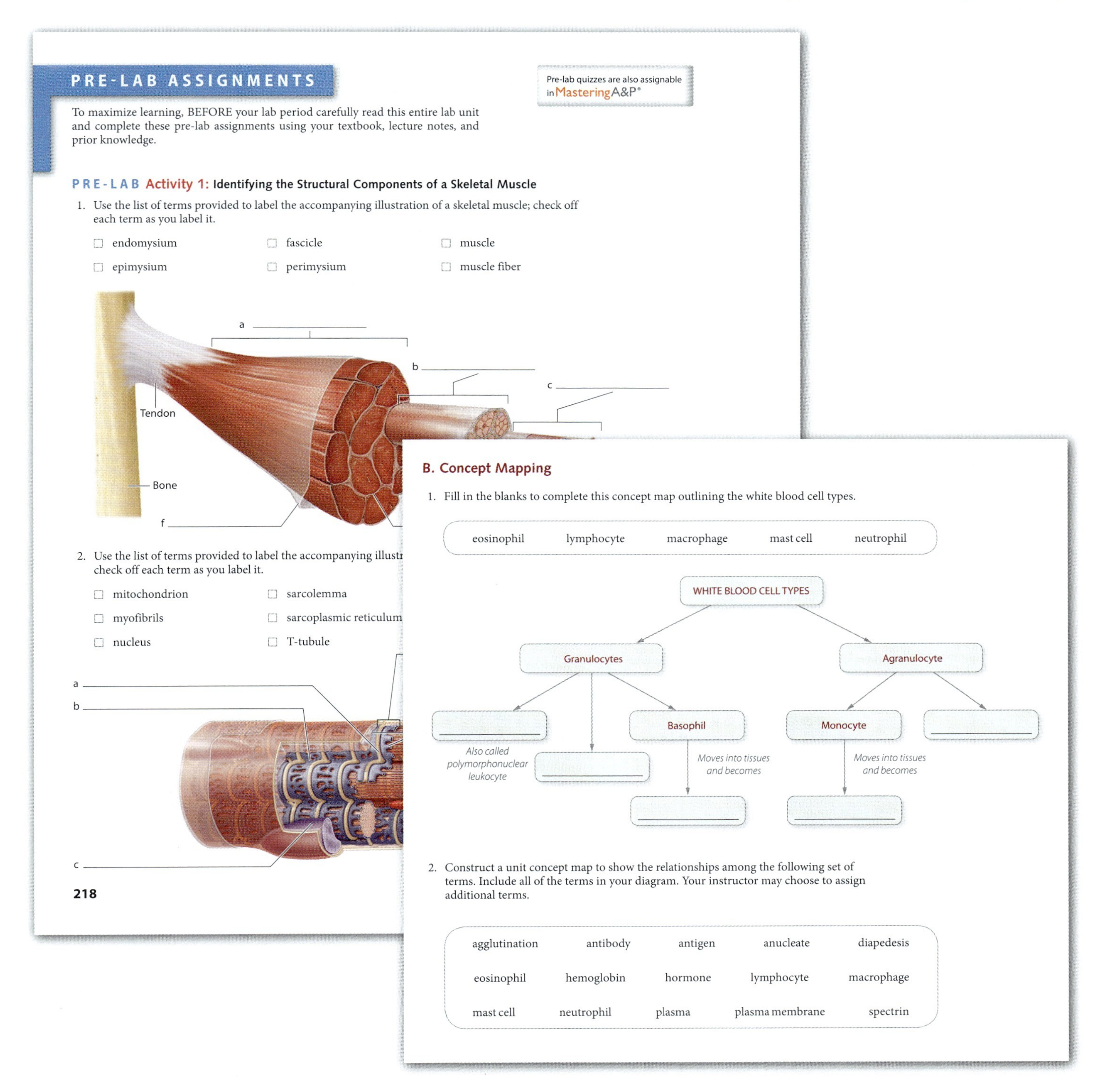

Other Key Features

Human Anatomy & Physiology Laboratory Manual: Making Connections features a rich and varied art program and integration of key media and equipment used in the lab.

Companion Lab Manual to Erin Amerman's *Human Anatomy & Physiology*

This lab manual reflects the terminology and explanations found in the Amerman textbook.

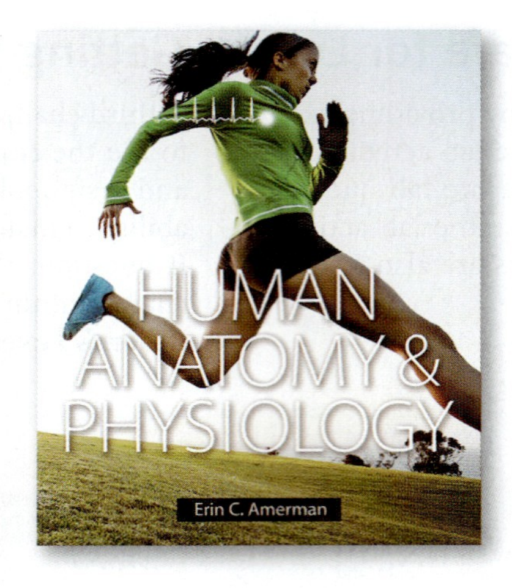

Superb Art from Amerman Textbook

The art from the Amerman textbook includes anatomical illustrations, photos, histology photomicrographs, and physiology sequence figures.

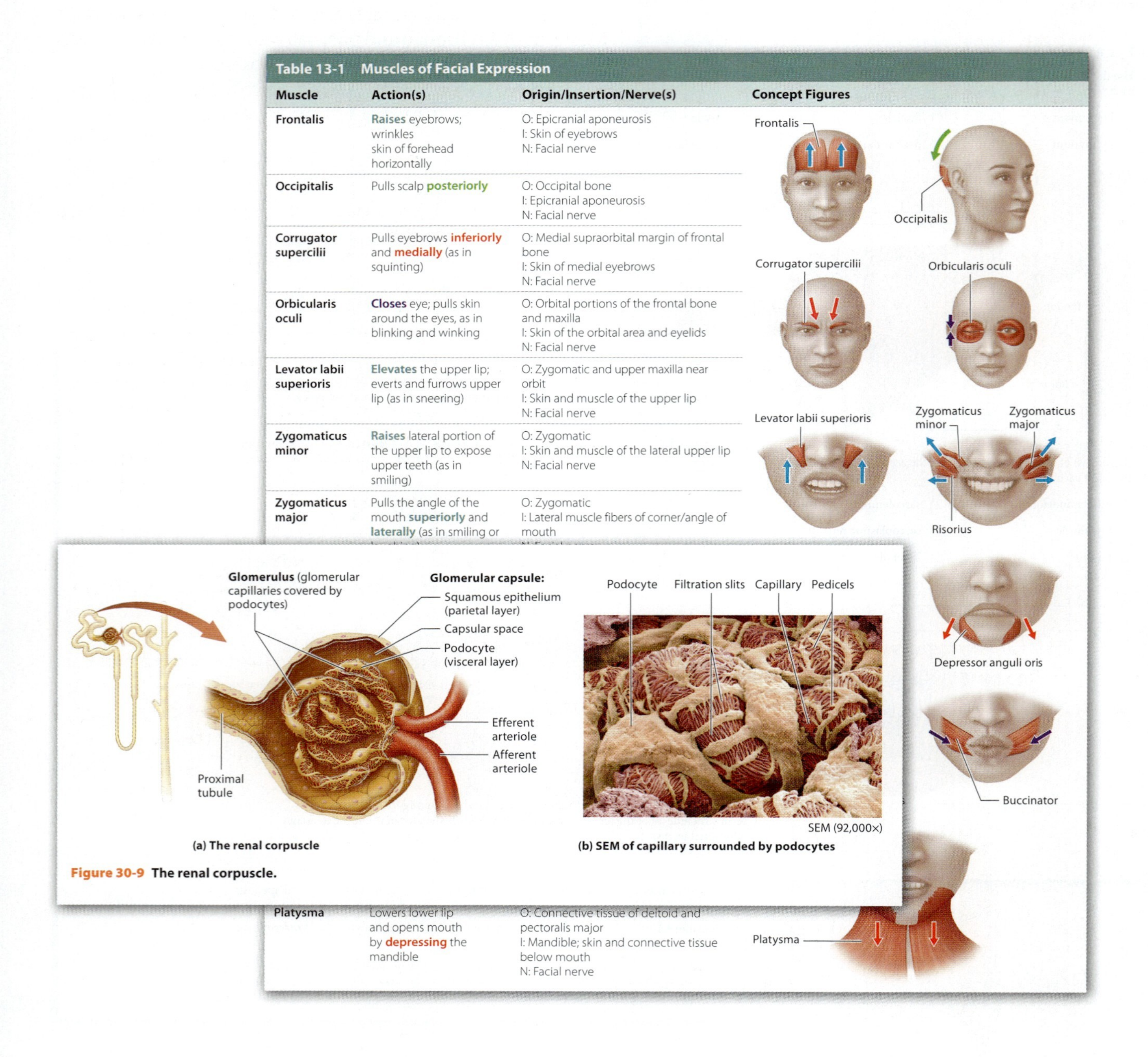

Table 13-1 Muscles of Facial Expression

Muscle	Action(s)	Origin/Insertion/Nerve(s)
Frontalis	**Raises** eyebrows; wrinkles skin of forehead horizontally	O: Epicranial aponeurosis I: Skin of eyebrows N: Facial nerve
Occipitalis	Pulls scalp **posteriorly**	O: Occipital bone I: Epicranial aponeurosis N: Facial nerve
Corrugator supercilii	Pulls eyebrows **inferiorly** and **medially** (as in squinting)	O: Medial supraorbital margin of frontal bone I: Skin of medial eyebrows N: Facial nerve
Orbicularis oculi	**Closes** eye; pulls skin around the eyes, as in blinking and winking	O: Orbital portions of the frontal bone and maxilla I: Skin of the orbital area and eyelids N: Facial nerve
Levator labii superioris	**Elevates** the upper lip; everts and furrows upper lip (as in sneering)	O: Zygomatic and upper maxilla near orbit I: Skin and muscle of the upper lip N: Facial nerve
Zygomaticus minor	**Raises** lateral portion of the upper lip to expose upper teeth (as in smiling)	O: Zygomatic I: Skin and muscle of the lateral upper lip N: Facial nerve
Zygomaticus major	Pulls the angle of the mouth **superiorly** and **laterally** (as in smiling or	O: Zygomatic I: Lateral muscle fibers of corner/angle of mouth
Platysma	Lowers lower lip and opens mouth by **depressing** the mandible	O: Connective tissue of deltoid and pectoralis major I: Mandible; skin and connective tissue below mouth N: Facial nerve

(a) The renal corpuscle

(b) SEM of capillary surrounded by podocytes

Figure 30-9 The renal corpuscle.

Additional Photos of Lab Specimens

This lab manual contains additional images not found in the Amerman textbook, including photos of **anatomical models, cadaver images,** and **histology photomicrographs**.

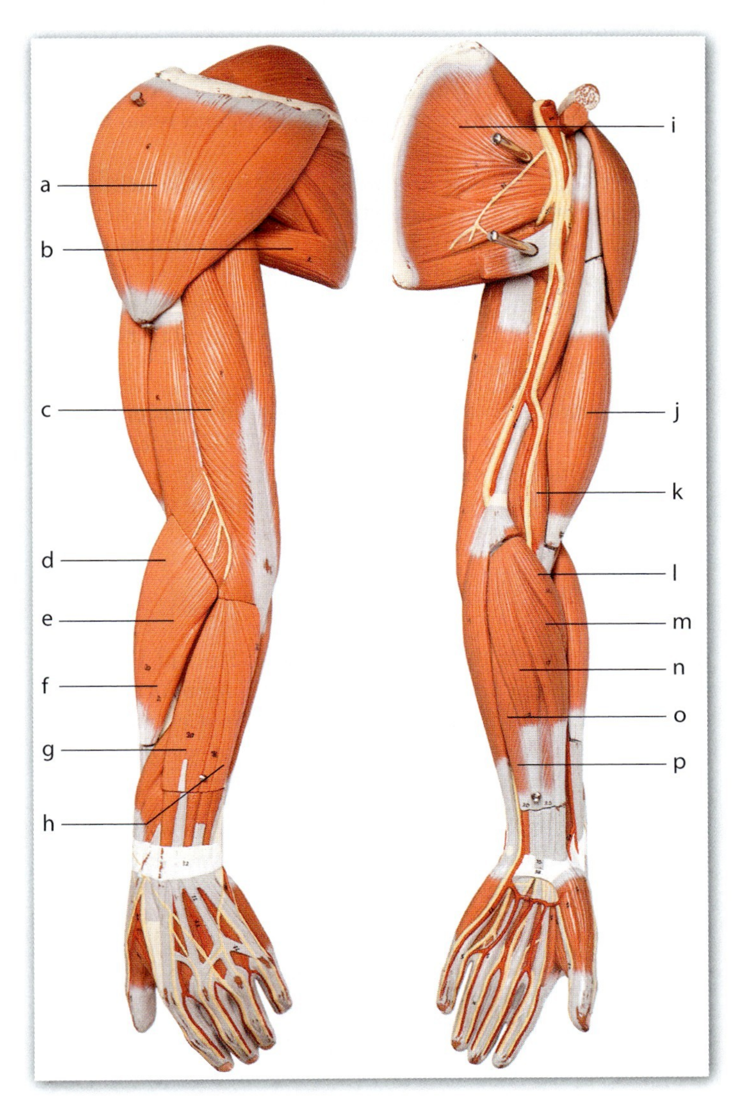

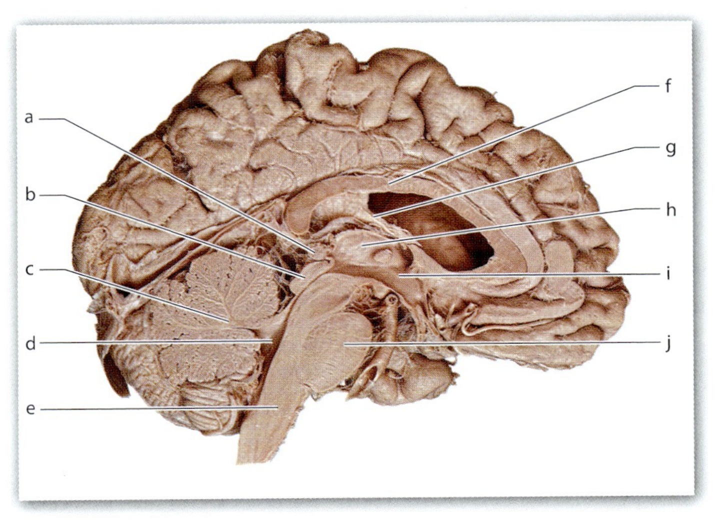

PhysioEx™ 9.1

PhysioEx™ 9.1 is an easy-to-use physiology lab simulation program that allows students to repeat labs as often as they like, perform experiments without animals, and conduct experiments that are difficult to perform in a wet lab environment because of time, cost, or safety concerns. Every exercise includes an overview and every activity includes objectives, an introduction, a pre-lab quiz, the experiment, a post-lab quiz, review sheet questions, and a lab report that students can save as a PDF and print and/or email to their instructor. The online format with easy step-by-step instructions includes everything students need in one convenient place.

Each exercise and activity is referenced in the lab manual where students are directed to access PhysioEx in MasteringA&P. Pre-lab and post-lab quizzes and review sheets for PhysioEx are assignable in MasteringA&P.

PhysioEx 9.1 includes 12 exercises containing a total of 63 physiology lab activities. The program features:

- **Input data variability** allows students to change variables and test various hypotheses for the experiments.
- **Step-by-step instructions** put everything students need to do to complete the lab in one convenient place. Students gather data, analyze results, and check their understanding, all on screen.
- **Stop & Think Questions** and **Predict Questions** help students think about the connections between the activities and the physiological concepts they demonstrate.
- **Greater data variability in the results** reflects more realistically the results that students would encounter in a wet lab experiment.
- **Pre-lab and Post-lab Quizzes** and short-answer **Review Sheets** are offered to help students prepare for and review each activity.
- **Students can save their Lab Report as a PDF,** which they can print and/or email to their instructor.
- **A Test Bank of assignable pre-lab and post-lab quizzes** for use with TestGen or its course management system is provided for instructors.
- **Seven videos of lab experiments** demonstrate the actual experiments simulated on screen, making it easy for students to understand and visualize the content of the simulations. Videos demonstrate the following experiments: Skeletal Muscle, Blood Typing, Cardiovascular Physiology, Use of a Water-Filled Spirometer, Nerve Impulses, BMR Measurement, and Cell Transport.

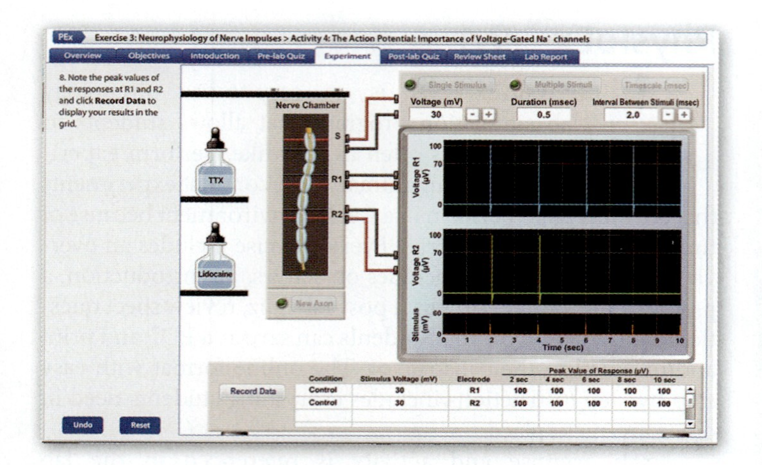

PhysioEx 9.1 topics include:

- Exercise 1: *Cell Transport Mechanisms and Permeability*. Explores how substances cross the cell membranes. Topics include: simple and facilitated diffusion, osmosis, filtration, and active transport.
- Exercise 2: *Skeletal Muscle Physiology*. Provides insights into the complex physiology of skeletal muscle. Topics include: electrical stimulation, isometric contractions, and isotonic contractions.
- Exercise 3: *Neurophysiology of Nerve Impulses*. Investigates stimuli that elicit action potentials, stimuli that inhibit action potentials, and factors affecting the conduction velocity of an action potential.
- Exercise 4: *Endocrine System Physiology*. Investigates the relationship between hormones and metabolism; the effect of estrogen replacement therapy; the diagnosis of diabetes; and the relationship between the levels of cortisol and adenocorticotropic hormone and a variety of endocrine disorders.
- Exercise 5: *Cardiovascular Dynamics*. Allows students to perform experiments that would be difficult if not impossible to do in a traditional laboratory. Topics include: vessel resistance and pump (heart) mechanics.
- Exercise 6: *Cardiovascular Physiology*. Examines variables influencing heart activity. Topics include: setting up and recording baseline heart activity, the refractory period of cardiac muscle, and an investigation of factors that affect heart rate and contractility.
- Exercise 7: *Respiratory System Mechanics*. Investigates physical and chemical aspects of pulmonary function. Students collect data simulating normal lung volumes. Other activities examine factors such as airway resistance and the effect of surfactant on lung function.
- Exercise 8: *Chemical and Physical Processes of Digestion*. Examines factors that affect enzyme activity by manipulating (in compressed time) enzymes, reagents, and incubation conditions.
- Exercise 9: *Renal System Physiology*. Simulates the function of a single nephron. Topics include: factors influencing glomerular filtration, the effect of hormones on urine function, and glucose transport maximum.
- Exercise 10: *Acid-Base Balance*. Topics include: respiratory and metabolic acidosis/alkalosis, and renal and respiratory compensation.
- Exercise 11: *Blood Analysis*. Topics include: hematocrit determination, erythrocyte sedimentation rate determination, hemoglobin determination, blood typing, and total cholesterol determination.
- Exercise 12: *Serological Testing*. Investigates antigen–antibody reactions and their role in clinical tests used to diagnose a disease or an infection.

Note: In addition to being available in MasteringA&P, PhysioEx 9.1 is also available as a CD-ROM packaged with this lab manual for no additional charge. Please contact your Pearson representative for ordering information.

Biopac®

BIOPAC® Activities that utilize the Biopac Student Labs® data acquisition system are included in Unit 12, *Introduction to the Muscular System: Muscle Tissue*; Unit 15, *The Central Nervous System: Brain and Spinal Cord*; Unit 22, *Physiology of the Heart*; and Unit 27, *Physiology of the Respiratory System*.

Instructions for other data acquisitions systems including **iWorx, Intellitool**, and **PowerLab** are available in the Instructor Resources in MasteringA&P.

Practice Anatomy Lab™ (PAL™) 3.0

Practice Anatomy Lab 3.0 (PAL) correlations are indicated by the PAL logo and presented as optional activities. These direct students to related content in the PAL 3.0 software in MasteringA&P.

Note: In addition to being available in MasteringA&P, Practice Anatomy Lab 3.0 is also available as a DVD packaged with this lab manual for no additional charge. Please contact your Pearson representative for ordering information.

Assignable Content in MasteringA&P®

MasteringA&P is an online learning and assessment system proven to help students learn. It helps instructors maximize lab time with customizable, easy-to-assign, automatically graded assessments that motivate students to learn outside of class and arrive prepared for lab. The powerful gradebook provides unique insight into student and class performance. Instructors can easily assign the following:

- **Pre-lab and Post-lab Quizzes for each activity** in the lab manual
- **Clinical Coaching Activities** for select units that include a brief clinical scenario with Bloom's Level II questions with feedback and hints
- **Quizzes and Lab Practicals from PAL 3.0 Test Bank**

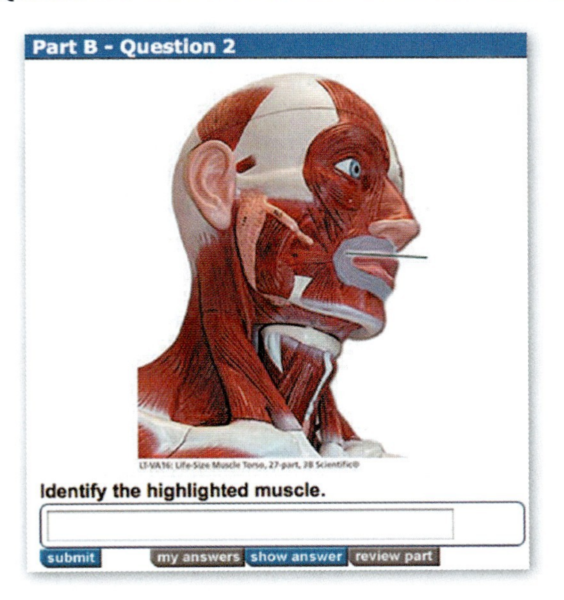

- **Pre-lab and Post-lab Quizzes and Review Sheets for PhysioEx 9.1**

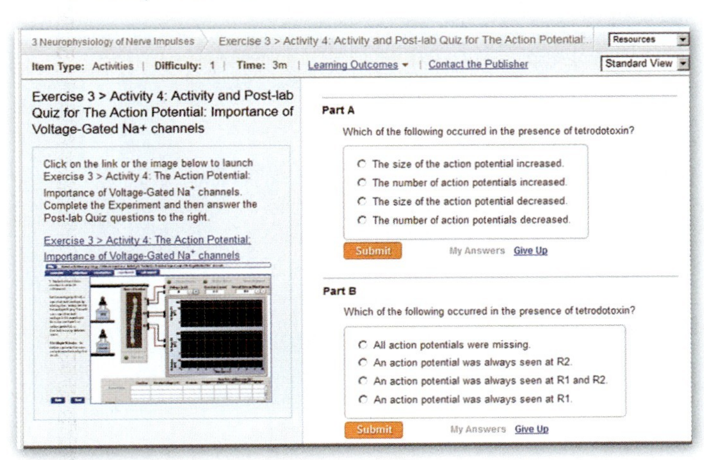

- **Drag-and-Drop Art Labeling Activities and Art Based Questions**

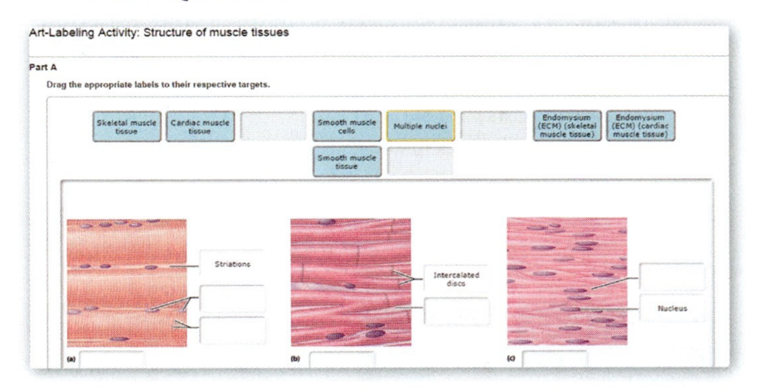

- **Bone and Dissection Video Coaching Activities** help students identify bones and learn how to do organ dissections

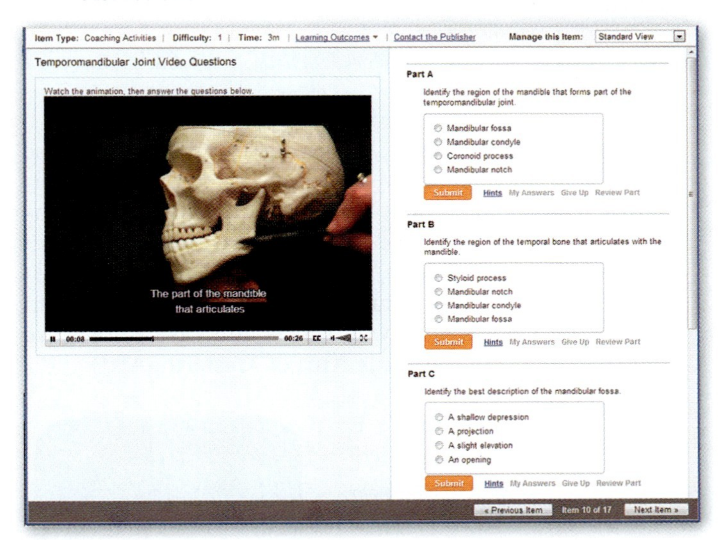

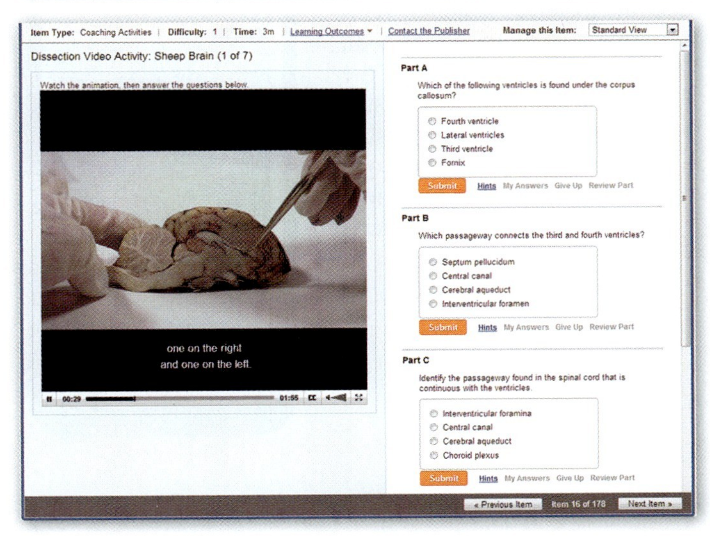

- **A&P Flix™ Animations** are 3D movie-quality anatomy animations that include self-paced tutorials and gradable quizzes. Students learn structures and functions from two sets of anatomy topics:
  - Origins, insertions, actions, and innervations (over 60 animations)
  - Group muscle actions and joints (over 50 animations)

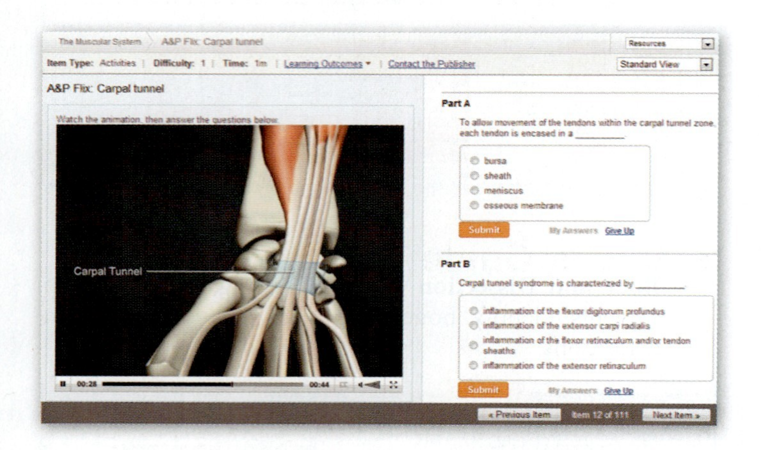

- **Clinical Case Study Coaching Activities** increase problem-solving skills and prepare students for future careers in allied health. Corresponding Teaching Strategies, available in the Instructor Resources in MasteringA&P, enable instructors to "flip" the classroom by providing valuable tips on when and how to use case studies. The worksheets and case studies are also available in the Study Area of MasteringA&P.

- **Learning Catalytics™** is a "bring your own device" student engagement, assessment, and classroom intelligence system. With this classroom lecture tool, instructors can flip the classroom and assess students in real time using open-ended tasks to probe student understanding. Students use their smartphone, tablet, or laptop to respond to questions in class.

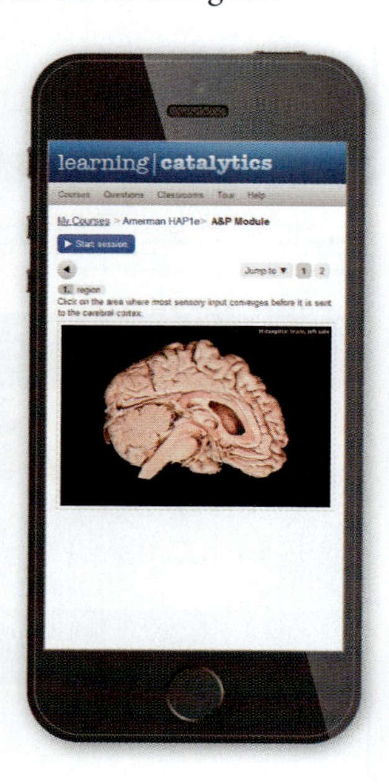

Study Tools in MasteringA&P®

Students get quick access to the following study tools in MasteringA&P:

- **Pre-lab and Post-lab Quizzes** are provided for each activity.
- **Bone and Dissection videos** aid review of key bones and organ dissections found in the lab manual.
- **Dynamic Study Modules** are designed to enable students to study effectively on their own, and to help them quickly access and learn the concepts they need to be more successful on quizzes and exams. These flashcard-style questions adapt to the student's performance and include art and explanations from this lab manual to cement the student's understanding.

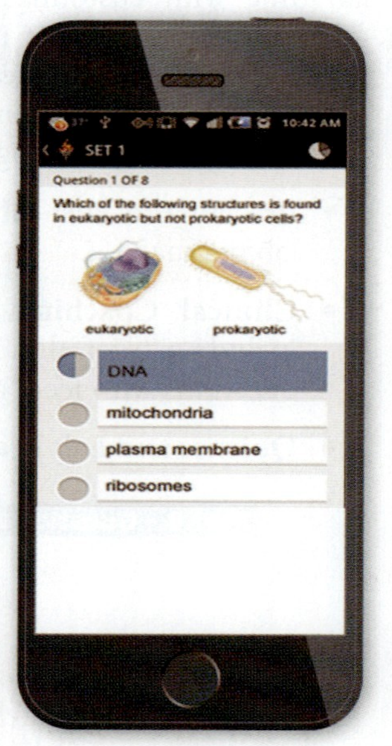

- **Practice Anatomy Lab™ (PAL™) 3.0** is an indispensable virtual anatomy study and practice tool that gives students 24/7 access to the most widely used lab specimens including human cadaver, anatomical models, histology, cat, and fetal pig. PAL 3.0 is easy to use and includes built-in audio pronunciations, rotatable bones, multiple-choice quizzes, and simulated fill-in-the-blank lab practical exams. PAL 3.0 is also accessible on mobile devices.

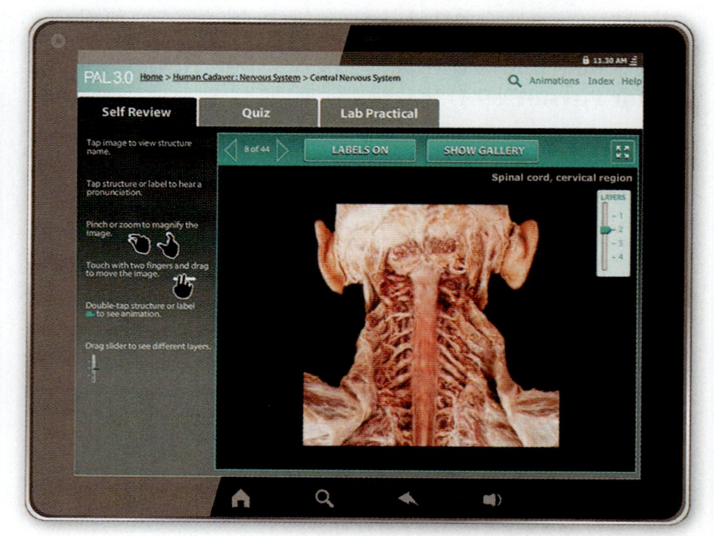

- **PhysioEx™ 9.1** is easy-to-use physiology laboratory simulation software. Every exercise includes an overview and every activity includes objectives, an introduction, a pre-lab quiz, the experiment, a post-lab quiz, review sheet questions, and a lab report that students can save as a PDF and print and/or email to their instructor.

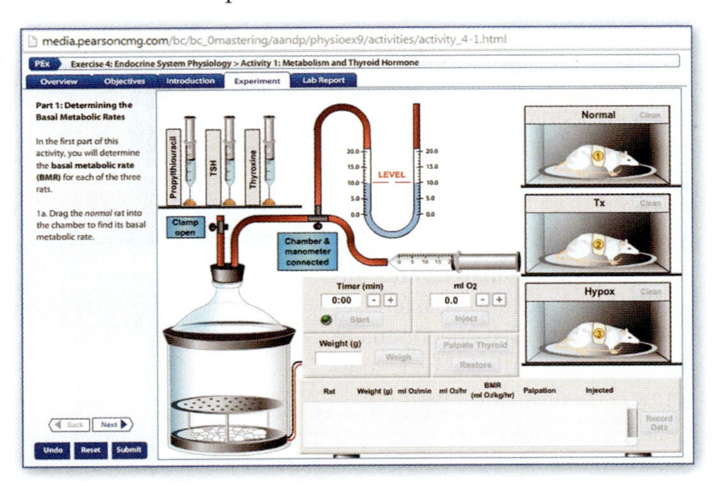

- **Videos of lab experiments**
- **A&P Flix animations**
- **Clinical Case Studies with worksheets**
- **Terminology Challenge worksheets**
- **Histology Atlas**
- **eText also available in *MasteringA&P with eText***

Three Versions

Human Anatomy & Physiology Laboratory Manual: Making Connections is available in three versions for your students: **Main**, **Cat**, and **Fetal Pig**. The Cat and Fetal Pig versions are identical to the Main version except that they include seven additional cat dissection exercises and nine additional fetal pig dissection exercises, respectively, at the back of the lab manual.

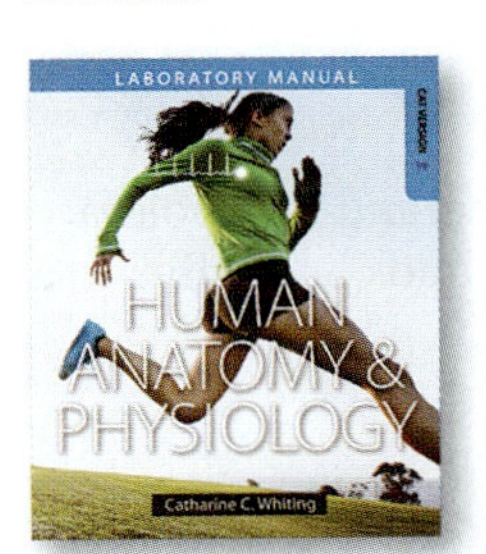

Cat Version
0-321-78700-5 /
978-0-321-78700-2

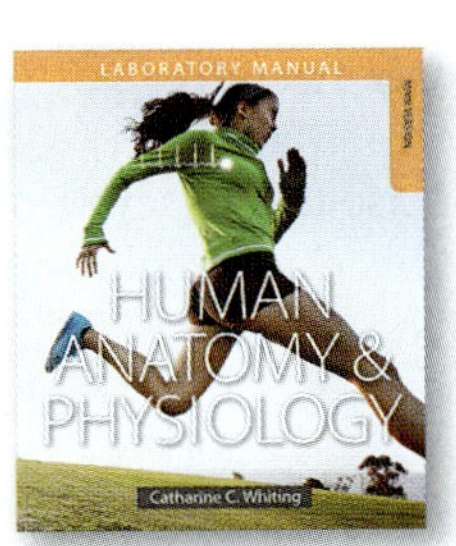

Main Version
0-13-395247-9 /
978-013-395247-6

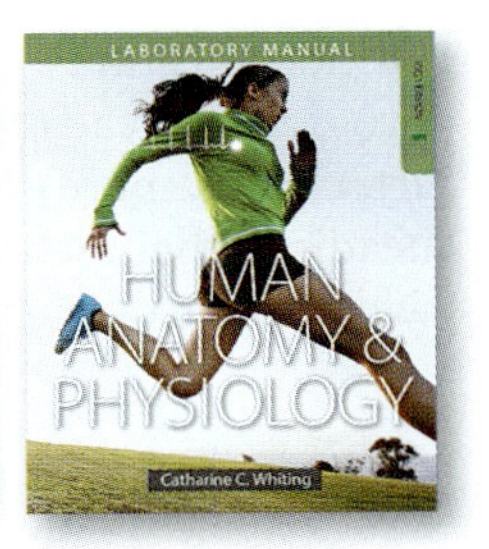

Fetal Pig Version
0-13-399679-4 /
978-013-399679-1

Customization Options

An enhanced custom program allows instructors to pick and choose content to tailor the lab manual *at the activity level*, selecting only those activities they assign. Each activity includes relevant background information, full-color figures, tables, and charts.

For information on creating a custom version of this manual, visit www.pearsonlearningsolutions.com, or contact your Pearson representative for details.

Additional Instructor Resources

Instructor Guide

0-13-405738-4 / 978-013-405738-5

This guide includes detailed instructions for setting up the laboratory, time allotments for each activity, and answers to the pre-lab assignments, activity questions, and post-lab assignments. Additionally, it describes strategies that encourage active learning, including sample concept maps and an overview of using concept mapping to increase student engagement. Finally, it discusses helpful hints for running an effective lab, ways to avoid common pitfalls, and extension activities that can be used to expand activities when time allows.

Instructor Resources in MasteringA&P®

These resources include: editable pre-lab and post-lab quizzes, the Instructor's Guide, instructions for each PhysioEx activity, Terminology Challenge Worksheets, Clinical Case Studies and Teaching Strategies for each case, A&P Flix (anatomy) in PPT, A&P Flix (anatomy) in MPEG, and instructions for other data acquisition systems including iWorx, Intelli-tool, and Powerlab.

Student Supplements

NEW! A Photographic Atlas for Anatomy & Physiology

0-321-86925-7 / 978-0-321-86925-8

by Nora Hebert, Ruth E. Heisler, Jett Chinn, Karen M. Krabbenhoft, Olga Malakhova

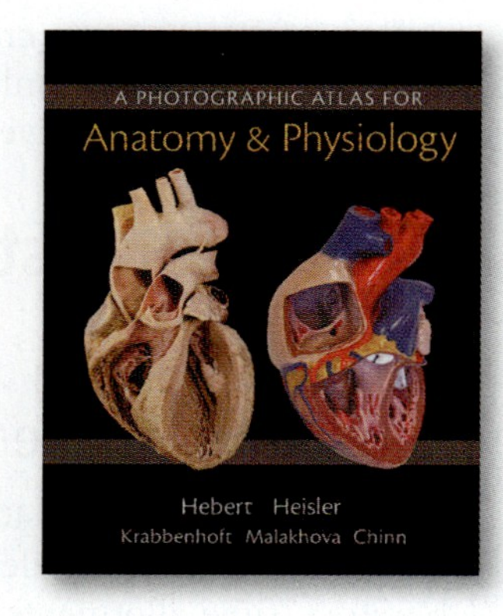

This brand new photo atlas is the perfect lab study tool that helps students learn and identify key anatomical structures. Featuring photos from Practice Anatomy Lab™ 3.0 and other sources, the Atlas includes over 250 cadaver dissection photos, histology photomicrographs, and cat dissection photos plus over 50 photos of anatomical models from leading manufacturers such as 3B Scientific®, SOMSO®, and Denoyer-Geppert Science Company.

Practice Anatomy Lab™ (PAL™) 3.0

0-321-68211-4 / 978-0-321-68211-6 (DVD)

by Nora Hebert, Ruth E. Heisler, Jett Chinn, Karen Krabbenhoft, Olga Malakhova

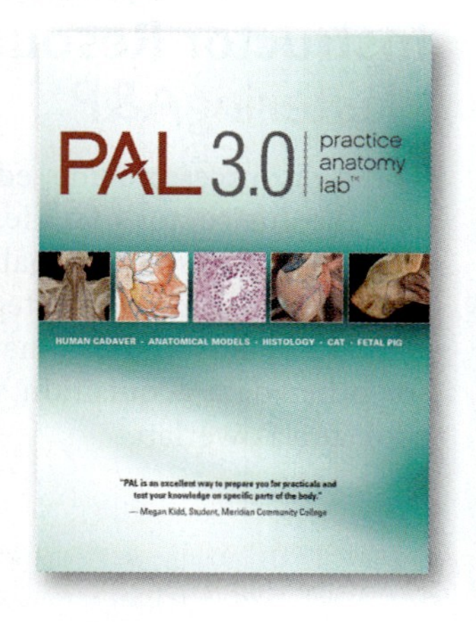

An indispensable virtual anatomy study and practice tool that gives students 24/7 access to the most widely used lab specimens including human cadaver, anatomical models, histology, cat, and fetal pig. PAL 3.0 also includes multiple-choice quizzes and practice fill-in-the-blank lab practicals.

Practice Anatomy Lab 3.0 Lab Guide

0-321-84025-9 / 978-0-321-84025-7 (standalone)
0-321-85767-4 / 978-0-321-85767-5 (with PAL 3.0 DVD)

by Ruth Heisler, Nora Hebert, Jett Chinn, Karen Krabbenhoft, Olga Malakhova

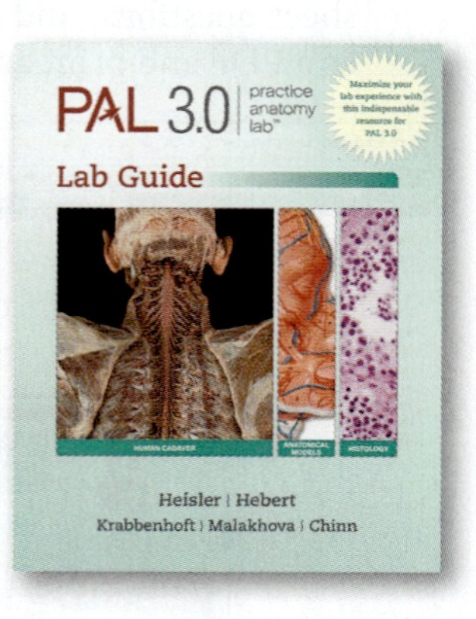

Written to accompany PAL 3.0, the new *Practice Anatomy Lab 3.0 Lab Guide* contains exercises that direct the student to select images and features in PAL 3.0, and then assess their understanding with labeling, matching, short-answer, and fill-in-the-blank questions. Exercises cover three key lab specimens in PAL 3.0—human cadaver, anatomical models, and histology.

The Anatomy Coloring Book, Fourth Edition

0-321-83201-9 / 978-0-321-83201-6

by Wynn Kapit and Lawrence M. Elson

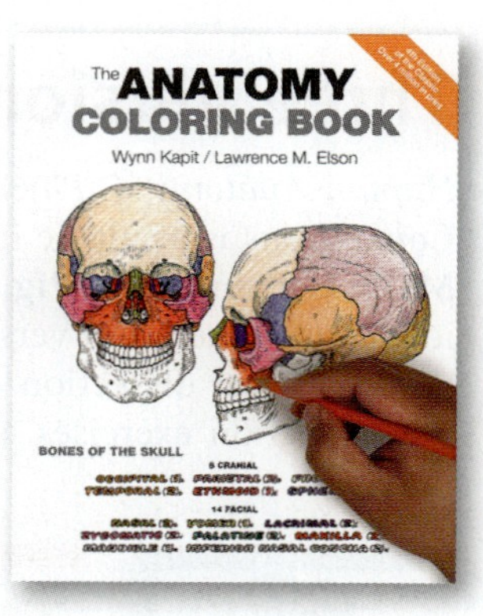

For more than 35 years, *The Anatomy Coloring Book* has been the best-selling human anatomy coloring book! A useful tool for anyone with an interest in learning anatomical structures, this concisely written text features precise, extraordinary hand-drawn figures that were crafted especially for easy coloring and interactive study. The Fourth Edition features user-friendly two-page spreads with enlarged art, clearer, more concise text descriptions, and new boldface headings that make this classic coloring book accessible to a wider range of learners.

Acknowledgments

A project of this magnitude is truly a team effort and I have been a part of an amazing team. I have so many people to thank. I will be forever grateful to Acquisitions Editor Gretchen Puttkamer for bringing me onto the team, for helping me to create a vision for this project, and for having the patience to coach me through those rough beginnings. I owe my deepest gratitude to the outstanding editorial, production, and marketing teams at Pearson. A heartfelt thanks to Serina Beauparlant, Editor-in-Chief, Kari Hopperstead, Project Editor, and Alan Titche, Development Editor, for their unending support, encouragement, and direction. Their hard work and dedication to this project inspired me to give this project my all and to keep my eyes on the finish line. Kudos also to Allison Rona, Marketing Manager, and Derek Perrigo, Senior Anatomy and Physiology Specialist, for their market guidance. I am also grateful to Media Content Producer Nicole Tache for her excellent work spearheading MasteringA&P for this manual.

The production of this book was a herculean task expertly managed by Caroline Ayres, Project Manager at Pearson, and Norine Strang, Senior Project Manager at S4Carlisle Publishing Services. Many thanks to Art Development Editors Kelly Murphy and Elisheva Marcus, and Project Managers Alicia Elliot and Lima Colati, who all provided expert guidance to the amazing team of illustrators at Imagineering Art. Thanks also to Maureen Spuhler for her excellent photo research, and to Lorretta Palagi for her eagle-eyed copyediting.

I want to thank contributor Karen Keller of Frostburg State University for her superb job of writing several units. Many thanks go to Patricia Wilhelm of Johnson & Wales University for her wonderful job of writing the cat dissection unit, and to Kerrie Hoar of University of Wisconsin–La Crosse for the beautiful cat dissection photographs. Thanks also to Sarah Matarese of St. George's School who contributed the Biopac activities and to Wendy Rappazzo of Harford Community College who authored clinical questions for MasteringA&P. I am grateful to Sheri Boyce of Messiah College and Anna Gilletly of Central New Mexico Community College for their assistance in editing and preparing units. A huge thank you to Carolyn Lebsack of Linn-Benton Community College, Steve Leadon of Durham Technical Community College, Kerrie Hoar of University of Wisconsin–La Crosse, Michelle Gaston of Northern Virginia Community College, and Bert Atsma of Union County College for their meticulous accuracy checks. I owe very special thanks to Erin Amerman for writing an outstanding textbook for this lab manual to accompany. Erin is a gifted writer with incredible insight into how students learn.

I would like to thank several of my colleagues at the University of North Georgia for their help, support, and valuable insights. A special thanks to John Hamilton for his expert photography. I owe JB Sharma a debt of gratitude for being an endless source of encouragement and a model of teaching excellence. I want to thank Lynn Berdanier for listening to my wild ideas and for her willingness to try them out in her labs. A special thanks to Mary Mayhew who, despite serving as a biology department head during a challenging time of consolidation, was always available to me and continues to be one of my greatest sources of support, encouragement, and guidance. Finally, to Malynde Weaver, my friend, my colleague, and teacher/advisor extraordinaire—I am blessed to serve alongside you.

To my current and former students, you are the inspiration for this project. Your passion for learning motivates me to be the best teacher that I can be. You are the reason that I teach!

To my mentor, my professor, and my friend, Dr. Louis J. Guillette, thank you for investing in my life and for teaching me the art of being an educator. You believed in me so many years ago and that belief not only changed the direction of my life but it also instilled in me a confidence in my abilities that has taken root and enabled me to pursue my dreams.

I am deeply grateful to my husband, Mark, and to our three incredible children—Jesse, Eli, and Ashton. Mark, this project would have never happened without you. You are the love of my life—an incredible husband and father—and I am blessed beyond measure. You remind me daily with your words and actions what is really important in life and you help me keep my priorities in order. Jesse, Eli, and Ashton, thank you for being so patient (most of the time) when mom needed to write, to talk to Kari, Serina, and Alan, or to take a nap. I know that you never thought this day would come, but it is finished. Mom and Dad owe you an awesome vacation!

Text and Media Reviewers

Pius Aboloye, *North Lake College*
Michele Alexandre, *Durham Technical Community College*
Chris Allen, *College of the Mainland*
Emily Allen, *Rowan College of Gloucester County*
Marcia Anglin, *Miami Dade College–North Campus*
Verona Barr, *Heartland Community College*
Dena Berg, *Tarrant County College–Northwest Campus*
Sheri Boyce, *Messiah College*
Ron Bridges, *Pellissippi State Community College*
Carol Britson, *University of Mississippi*
Geralyn Caplan, *Owensboro Community & Technical College*
Maria Carles, *Northern Essex Community College*
Carol Carr, *John Tyler Community College*
Ellen Carson, *Florida State College–Jacksonville*
Peter Charles, *Durham Technical Community College*
Teresa Cowan, *Baker College*
Ken Crane, *Texarkana College*
Mary Dettman, *Seminole Community College*
Karen Dunbar-Kareiva, *Ivy Tech Community College*
Kathryn Englehart, *Kennebec Valley Community College*
Sondra Evans, *Florida State College–Jacksonville*
Jill Feinstein, *Richland Community College*
Tracy Felton, *Union County Community College*
Christine Foley, *Southwest Texas Junior College–Del Rio Campus*
Lori Frear, *Wake Tech Community College*
Kim Fredricks, *Viterbo University*
Lynn Gargan, *Tarrant County College–Northeast Campus*
Lori Garrett, *Parkland College*
Michelle Gaston, *Northern Virginia Community College*
Carol Gavareski, *Bellingham Technical College*
Anna Gilletly, *Central New Mexico Community College*
Miriam Golbert, *College of the Canyons*
Joanna Greene, *Ivy Tech Community College–Anderson*
Juan Guzman, *Florida Gateway College*
Bill Hanna, *Massasoit Community College*
Lesleigh Hastings, *Wake Tech Community College*
Stephanie Havemann, *Alvin Community College*
Heidi Hawkins, *College of Southern Idaho*
D.J. Hennager, *Kirkwood Community College*
Charmaine Henry, *Baker University*
Julie Huggins, *Arkansas State*
Jody Johnson, *Arapahoe Community College*
Karen Keller, *Frostburg State University*
Suzanne Kempke, *St. Johns River Community College*
Christine Kisiel, *Mount Wachusett Community College*
Ellen Lathrop-Davis, *Community College of Baltimore County*
Steven Leadon, *Durham Technical Community College*
Carolyn Lebsack, *Linn-Benton Community College*
Stephen Lebsack, *Linn-Benton Community College*
Jeffrey Lee, *Essex Community College*
Leona Levitt, *Union County College*
Christine Maney, *Salem State College*
Bruce Maring, *Daytona State College*
Robert Marino, *Capital Community College*
Sarah Matarese, *St. George's School*
Cherie McKeever, *Montana State University–Great Falls College*
Jaime Mergliano, *John Tyler Community College*
Justin Moore, *American River College*
Howard Motoike, *LaGuardia Community College*
Regina Munro, *Chandler Gilbert Community College*
Karen Murch-Shafer, *University of Nebraska–Omaha*
Zvi Ostrin, *Hostos Community College*
Ellen Ott-Reeves, *Blinn College–Bryan Campus*
Debbie Palatinus, *Roane State Community College*
Kevin Ragland, *Nashville State Community College*
Wendy Rappazzo, *Harford Community College*
Jean Revie, *South Mountain Community College*
Travis Robb, *Allen Community College–Burlingame*
Fredy Ruiz, *Miami Dade College*
Tracy Rusco, *East Central College*
Amy Ryan, *Clinton Community College*
Linda Schams, *Viterbo University*
Jeff Schinske, *De Anza College*
Steven Schneider, *South Texas College*
Maureen Scott, *Norfolk State University*
George Steer, *Jefferson College of Health Sciences*
James Stittsworth, *Florida State College–Jacksonville*
Deborah Temperly, *Delta College*
Terry Thompson, *Wor-Wic Community College*
Carlene Tonini-Boutacoff, *College of San Mateo*
Liz Torrano, *American River College*
Lisa Welch, *Weatherford College*
Deb Wiepz, *Madison Area Technical College*
Darrellyn Williams, *Pulaski Technical College*

Brief Contents

UNIT

CAT DISSECTION EXERCISES

Contents

UNIT 24 CIRCULATORY PATHWAYS AND THE PHYSIOLOGY OF BLOOD VESSELS 487

UNIT 25 THE LYMPHATIC SYSTEM 505

UNIT 26 ANATOMY OF THE RESPIRATORY SYSTEM 523

UNIT 27 PHYSIOLOGY OF THE RESPIRATORY SYSTEM 539

UNIT 28 ANATOMY OF THE DIGESTIVE SYSTEM 557

UNIT 29 PHYSIOLOGY OF THE DIGESTIVE SYSTEM 585

UNIT 30 ANATOMY OF THE URINARY SYSTEM 599

1 Introduction to Anatomy and Physiology

UNIT OUTLINE

Anatomy and physiology (A&P) is a fascinating subject, but learning it can be very challenging. Even though learning A&P involves a great deal of memorization, memorizing is only the first step. True learning requires you to apply the knowledge you attain in critical-thinking and problem-solving activities.

True learning requires a tremendous level of discipline, motivation, and determination. It is hard work, and it demands a commitment to daily study. You will encounter a wide variety of study strategies designed to motivate you to become engaged in the learning process. You will be encouraged to participate fully in your lab group, to establish a regular study group, and to be willing to try new study techniques—all so that you can learn more information than you ever thought possible in one or two semesters.

We will begin building this foundation by learning the basic anatomical terminology that you will use throughout the course and when communicating with health care professionals throughout your career.

THINK ABOUT IT *The relationship between structure and function is a key concept in anatomy and physiology. In fact, structure often determines function. How is the structure of the heart related to its function?*

____

____

____

____

____

PRE-LAB ASSIGNMENTS

Pre-lab quizzes are also assignable in MasteringA&P®

To maximize learning, BEFORE your lab period carefully read this entire lab unit and complete these pre-lab assignments using your textbook, lecture notes, and prior knowledge.

PRE-LAB Activity 1: Identifying Body Regions and Exploring Surface Anatomy

1. Which of the following descriptions of the anatomical position is *incorrect*?
   a. arms straight
   b. palms facing posteriorly
   c. toes facing forward
   d. feet slightly apart

2. Match each of the following descriptions with the correct directional term.

a. The sternum is ______________ to the vertebrae.	1. posterior/dorsal
b. The feet are ______________ to the hands.	2. distal
c. The elbows are ______________ to the abdomen.	3. anterior/ventral
d. The skin is ______________ to the skeleton.	4. deep
e. The heart is ______________ to the sternum.	5. lateral
f. The lungs are ______________ to the ribs.	6. superficial
g. The chest is ______________ to the abdomen.	7. inferior
h. The knee is ______________ to the hip.	8. superior
i. The little finger is ______________ to the thumb.	9. proximal
j. The elbow is ______________ to the wrist.	10. medial

3. Use the list of terms provided to label the accompanying illustration; check off each term as you label it.

☐ antecubital
☐ cervical
☐ frontal
☐ patellar
☐ pelvic
☐ femoral

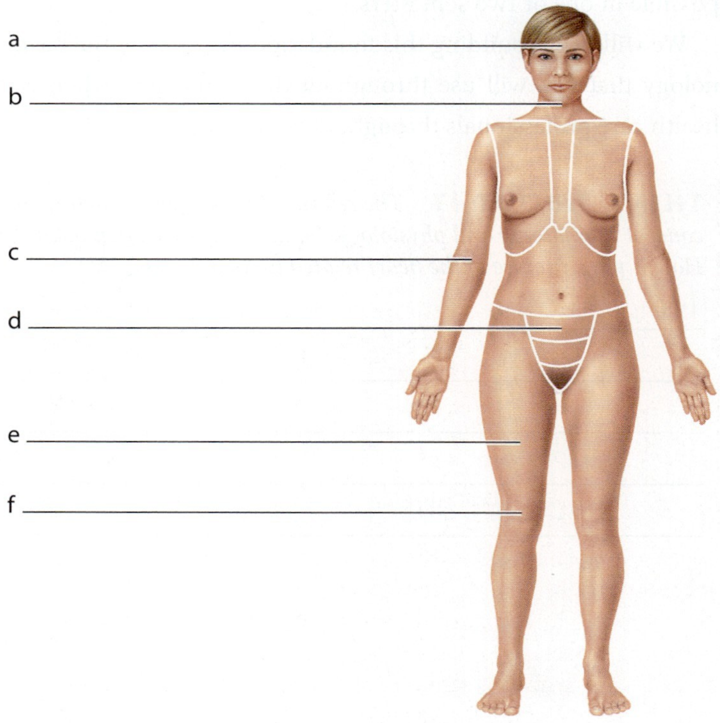

a ______________
b ______________
c ______________
d ______________
e ______________
f ______________

PRE-LAB Activity 2: Identifying Body Cavities and Abdominopelvic Regions

1. The dorsal body cavity is subdivided into the _______________ cavity and the _______________ cavity.
2. The ventral body cavity is subdivided into the _______________ cavity and the _______________ cavity.
3. Which abdominopelvic region(s) is(are) located:
   a. inferior to the umbilical region? _______________
   b. lateral to the epigastric region? _______________
   c. superior to the right iliac region? _______________
   d. inferior to the left hypochondriac region? __________
   e. lateral to the hypogastric region? _______________
4. The innermost layer of the serous membrane surrounding the lungs is called the _______________ pleura.
5. The outermost layer of the serous membrane surrounding the heart is called the _______________ pericardium.

PRE-LAB Activity 3: Demonstrating and Identifying Body Planes of Section

1. Which plane of section divides the body into anterior and posterior parts?_______________
2. Which plane of section divides the body into superior and inferior parts?_______________
3. Which plane of section divides the body into right and left parts?_______________

PRE-LAB Activity 4: Assisting the Coroner

1. The pelvis is _____________________ to the ribs.
   a. superior
   b. lateral
   c. inferior
   d. distal
2. Which of the following descriptions best applies to the term medial?
   a. nose relative to the ears
   b. knee relative to the ankle
   c. lips relative to the nose
   d. ears relative to the eyes
3. Which of the following descriptions best applies to the term superior?
   a. sternum relative to the lungs
   b. wrist relative to the elbow
   c. eyebrows relative to the eyes
   d. fingers relative to the palm
4. The visceral pericardium is _____________________ (superficial/deep) to the parietal pericardium.

Anatomical Terminology

Learning anatomy is sometimes compared to learning a new language because of the vast number of terms that are specific to the study of the human body. However, because these terms are widely used in health care professions, it is essential that you become fluent in them.

Anatomical Position

Anatomical position is the universally accepted standard position that scientists and medical professionals use to communicate information concerning parts of the body. In anatomical position (**Figure 1-1**), the body is erect and facing forward; the arms are straight and at the sides of the body, with the palms facing forward; and the feet are slightly apart, with the toes pointing forward.

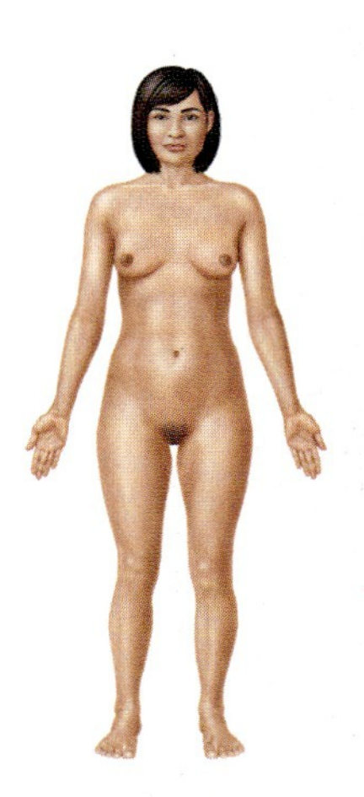

Figure 1-1 Anatomical position.

Directional Terms

Another effective way of communicating in anatomy is to use directional terms, which describe the locations of body structures in relationship to other structures. **Figure 1-2** illustrates the most common directional terms, plus their definitions and some examples of how each might be used.

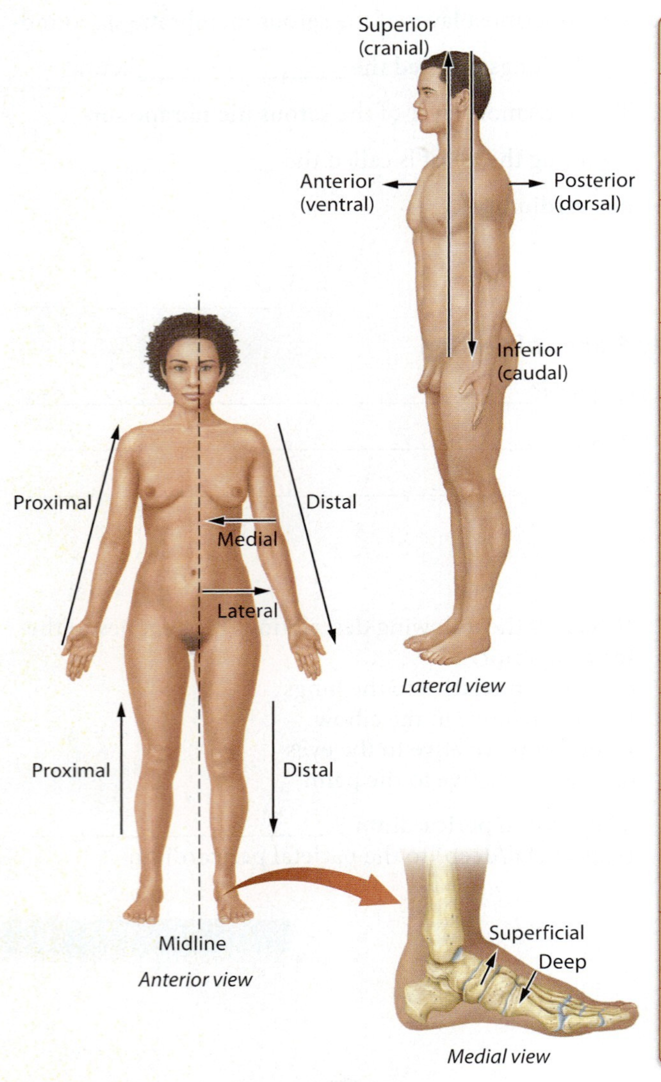

TERM	DEFINITION	EXAMPLES
Anterior (ventral)	Toward the front	• The palms are on the anterior side of the body. • The esophagus is anterior to the spinal cord.
Posterior (dorsal)	Toward the back	• The occipital bone is on the posterior cranium (skull). • The spinal cord is posterior to the esophagus.
Superior (cranial)	Toward the head	• The nose is superior to the mouth. • The neck is superior to the chest.
Inferior (caudal)	Toward the tail	• The nose is inferior to the forehead. • The umbilicus (belly button) is inferior to the chest.
Proximal	Closer to the point of origin (generally the trunk)	• The knee is proximal to the ankle. • The shoulder is proximal to the elbow.
Distal	Farther away from the point of origin (generally the trunk)	• The foot is distal to the hip. • The wrist is distal to the elbow.
Medial	Closer to the midline of the body or a body part; on the inner side of	• The ear is medial to the shoulder. • The index finger is medial to the thumb.
Lateral	Farther away from the midline of the body or a body part; on the outer side of	• The shoulder is lateral to the chest. • The thumb is lateral to the index finger.
Superficial	Closer to the surface	• The skin is superficial to the muscle. • Muscle is superficial to bone.
Deep	Farther below the surface	• Bone is deep to the skin. • Bone is deep to muscle.

Figure 1-2 Common directional terms.

Surface Anatomy/Body Regions

Regional terms are used to identify specific areas on the surface of the body. **Figure 1-3** illustrates the most common terms describing various body regions, and **Table 1-1** provides definitions of each of the regional terms.

In the following lab activity, you will explore the anatomical terms that describe various regions of the surface of the body.

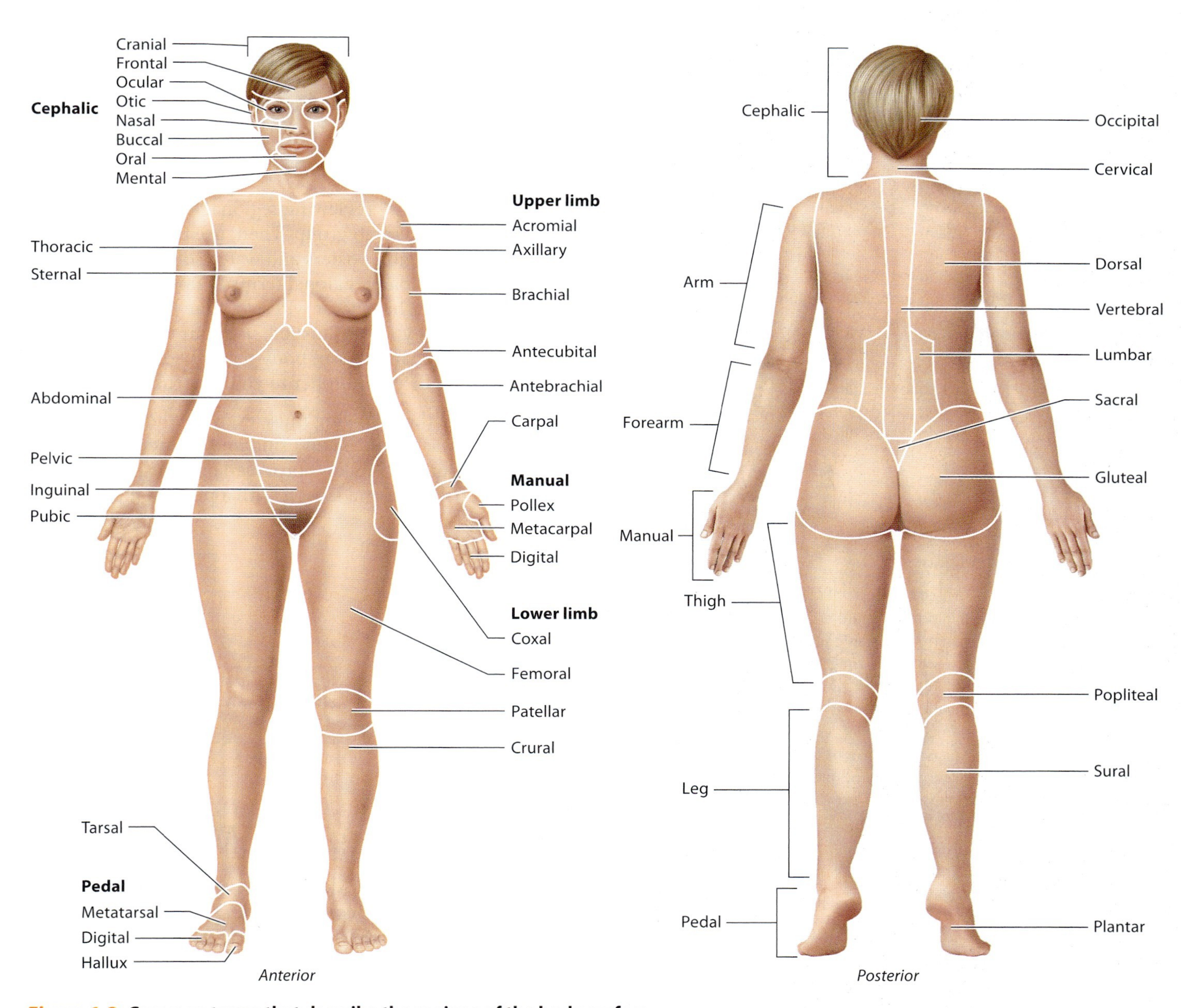

Figure 1-3 Common terms that describe the regions of the body surface.

Table 1-1 Surface Anatomy/Regional Terms

Term	Meaning
Anterior and Posterior Regions	
Abdominal	pertaining to the abdomen
Cephalic	pertaining to the head
Cervical	pertaining to the neck
Dorsal	pertaining to the back of the body
Gluteal	pertaining to the buttocks
Inguinal	pertaining to the groin
Lumbar	pertaining to the lower back
Manual	pertaining to the hand
Occipital	pertaining to the back of the head
Palmar	pertaining to the palm
Pedal	pertaining to the foot
Pelvic	pertaining to the pelvis
Plantar	pertaining to the sole of the foot
Popliteal	pertaining to the posterior surface of the knee
Pubic	pertaining to the pubis
Sacral	pertaining to the sacrum
Sural	pertaining to the posterior surface of the leg
Sternal	pertaining to the sternum
Thoracic	pertaining to the chest
Vertebral	pertaining to the spinal column
Regions of the Head and Face	
Buccal	pertaining to the cheek
Cranial	pertaining to the cranium
Frontal	pertaining to the forehead
Mental	pertaining to the chin
Nasal	pertaining to the nose
Ocular	pertaining to the bony eye socket
Oral	pertaining to the mouth
Otic	pertaining to the ear
Regions of the Upper Limb	
Acromial	pertaining to the point of the shoulder
Antebrachial	pertaining to the forearm
Antecubital	pertaining to the anterior surface of the elbow
Axillary	pertaining to the armpit
Brachial	pertaining to the arm
Carpal	pertaining to the wrist
Digital	pertaining to the fingers
Metacarpal	pertaining to the metacarpals
Pollex	pertaining to the thumb
Regions of the Lower Limb, Anterior View	
Coxal	pertaining to the hip
Crural	pertaining to the anterior surface of the leg
Digital	pertaining to the toes
Femoral	pertaining to the thigh
Hallux	pertaining to the great toe
Metatarsal	pertaining to the metatarsals
Patellar	pertaining to the anterior surface of the knee
Tarsal	pertaining to the ankle

ACTIVITY 1
Identifying Body Regions and Exploring Surface Anatomy

Learning Outcomes

1. Use surface anatomy terms accurately.
2. Use regional terms accurately.

Materials Needed

- ☐ Laminated anterior body region poster
- ☐ Laminated posterior body region poster
- ☐ Water-soluble marking pens
- ☐ Muscle models
- ☐ Labeling tape

Instructions

1. Spend 10 minutes reviewing regional terms with the members of your lab group. Then, use the two laminated body region posters and water-soluble markers to identify as many regional terms as possible from memory. Your instructor will set a time limit for each poster. When you are finished, use your lab manual to determine the number of body regions correctly identified and report the number to your instructor.

2. For each of the following muscles, write the body region in which it is found. Then using muscle models and model identification keys provided by your instructor, find each muscle and label it with a piece of tape.

rectus abdominis m. ______________________

brachialis m. ______________________

biceps femoris m. ______________________

epicranius m. ______________________

mentalis m. ______________________

gluteus maximus m. ______________________

Body Cavities and Membranes

The human body is divided into several fluid-filled cavities, each containing specific organs. The two major body cavities are the **dorsal** (posterior) **cavity** and the **ventral** (anterior) **cavity** (**Figure 1-4**). The dorsal body cavity is subdivided into the **cranial cavity**, which houses the brain, and the **vertebral** (spinal) **cavity**, which houses the spinal cord. The ventral body cavity is divided by the diaphragm into the thoracic cavity and the abdominopelvic cavity.

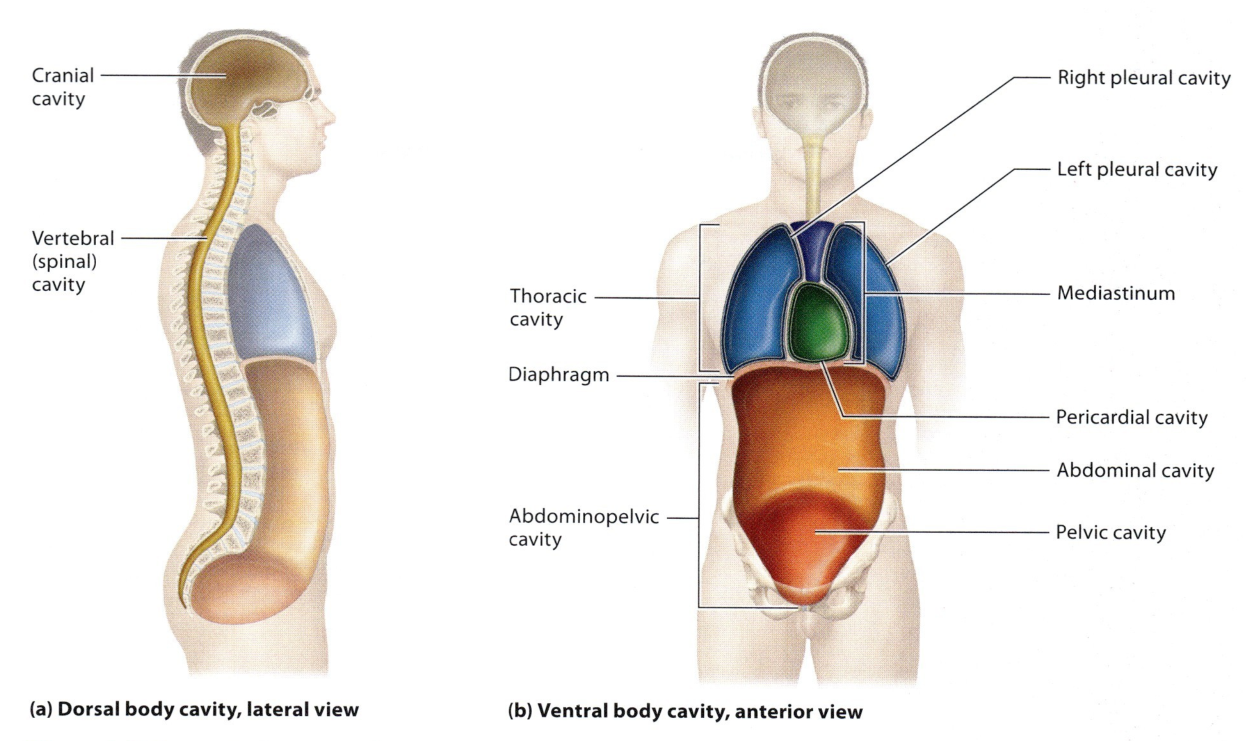

Figure 1-4 The major body cavities.

The **thoracic cavity** can be subdivided into the medial mediastinum and the right and left pleural cavities. The mediastinum contains numerous organs, including the esophagus, trachea, bronchi, and heart, the last of which is enclosed by the **pericardial cavity**. Inferior to the diaphragm is the **abdominopelvic cavity**, consisting of the **abdominal cavity**, which contains the digestive organs, and the **pelvic cavity**, which contains the urinary bladder, reproductive organs, and rectum.

Most of the organs in the ventral body cavity are surrounded by **serous membranes**, which are thin, double-layered sacs. The outer layer of the membrane is the **parietal layer**; the inner layer of the membrane covers the organ and is called the **visceral layer**. The two layers of the serous membranes are separated by a narrow cavity filled with a clear serous (watery) fluid, which is secreted by the membranes and prevents friction as the organs move within the ventral body cavity. The pleura covers the lungs, the pericardium covers the heart, and the peritoneum covers most of the abdominal organs.

The abdominopelvic cavity is typically divided into either four quadrants or nine regions (**Figure 1-5**). Clinicians divide the cavity into four quadrants: the right upper quadrant (RUQ), left upper quadrant (LUQ), right lower quadrant (RLQ), and left lower quadrant (LLQ). Anatomists subdivide the cavity into nine regions: the right hypochondriac, epigastric, left hypochondriac, right lumbar, umbilical, left lumbar, right iliac, hypogastric or pubic, and left iliac regions.

ACTIVITY 2

Identifying Body Cavities and Abdominopelvic Regions

Learning Outcomes

1. Identify the body cavities that make up the dorsal cavity and those that make up the ventral cavity, and list the major organs found in each.
2. Describe the two ways in which the abdominopelvic cavity is commonly subdivided.
3. Explain the structure and function of a serous membrane, and name the serous membranes in the ventral body cavity.

Materials Needed

- ☐ Torso model
- ☐ Quart-sized Ziploc bag
- ☐ Food coloring
- ☐ Miscellaneous anatomical models

Instructions

A. Body Cavities and Abdominopelvic Regions

Identify the body cavities and abdominopelvic regions listed in the following charts on an anatomical model. Then, complete the charts as you name the major organs found in each.

Dorsal Body Cavity	
Subdivision	**Organ(s)**
Cranial cavity	
Vertebral cavity	

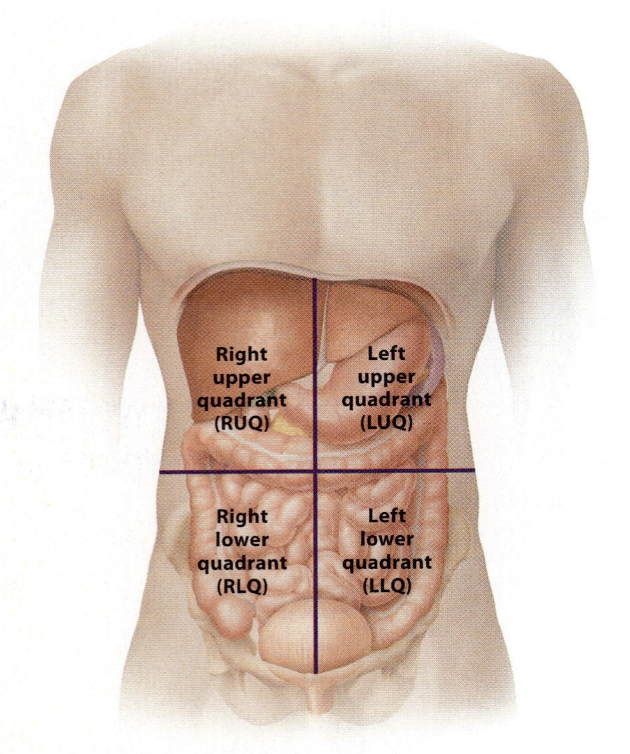

(a) The four abdominopelvic quadrants

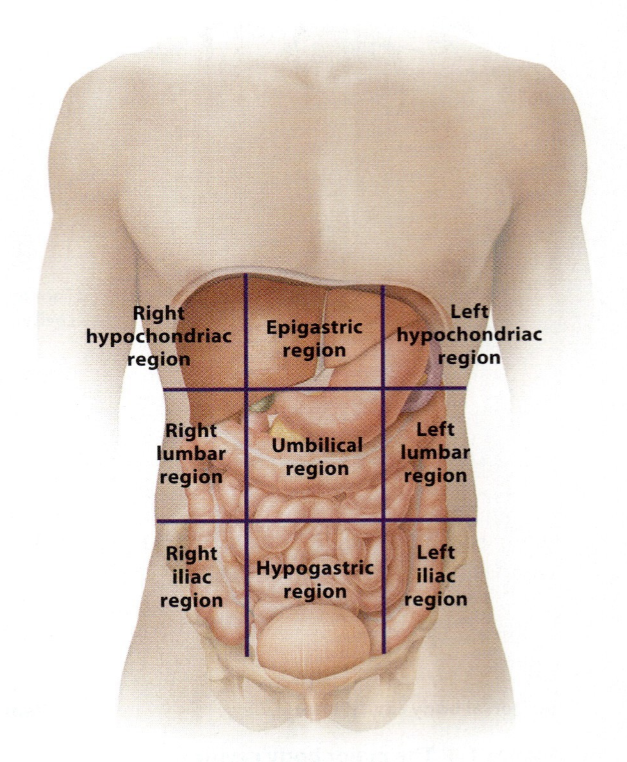

(b) The nine abdominopelvic regions

Figure 1-5 Divisions of the abdominopelvic cavity.

Ventral Body Cavity	
Subdivision	**Organ(s)**
Thoracic cavity	
Abdominopelvic cavity	
• Right hypochondriac region	
• Epigastric region	
• Left hypochondriac region	
• Right lumbar region	
• Umbilical region	
• Left lumbar region	
• Right iliac region	
• Hypogastric region	
• Left iliac region	

B. Serous Membranes

Use a Ziploc bag containing 2 tablespoons of water plus some food coloring as a model for the parietal layer of a serous membrane, the visceral layer of a serous membrane, and a serous fluid-filled cavity.

1. Place the Ziploc bag on top of the heart model.

   The portion of the bag adjacent to the heart represents the ______________________________, the water-filled space represents the ______________________________, and the outermost portion of the bag represents the ______________________________.

2. Next, place the Ziploc bag on top of the lung model.

   The portion of the bag adjacent to the lung represents the ______________________________, the water-filled space represents the ______________________________, and the outermost portion of the bag represents the ______________________________.

3. Finally, place the Ziploc bag on top of the model of the small intestine.

   The portion of the bag adjacent to the intestines represents the ______________________________,
   the water-filled space represents the ______________________________, *and*
   the outermost portion of the bag represents the ______________________________.

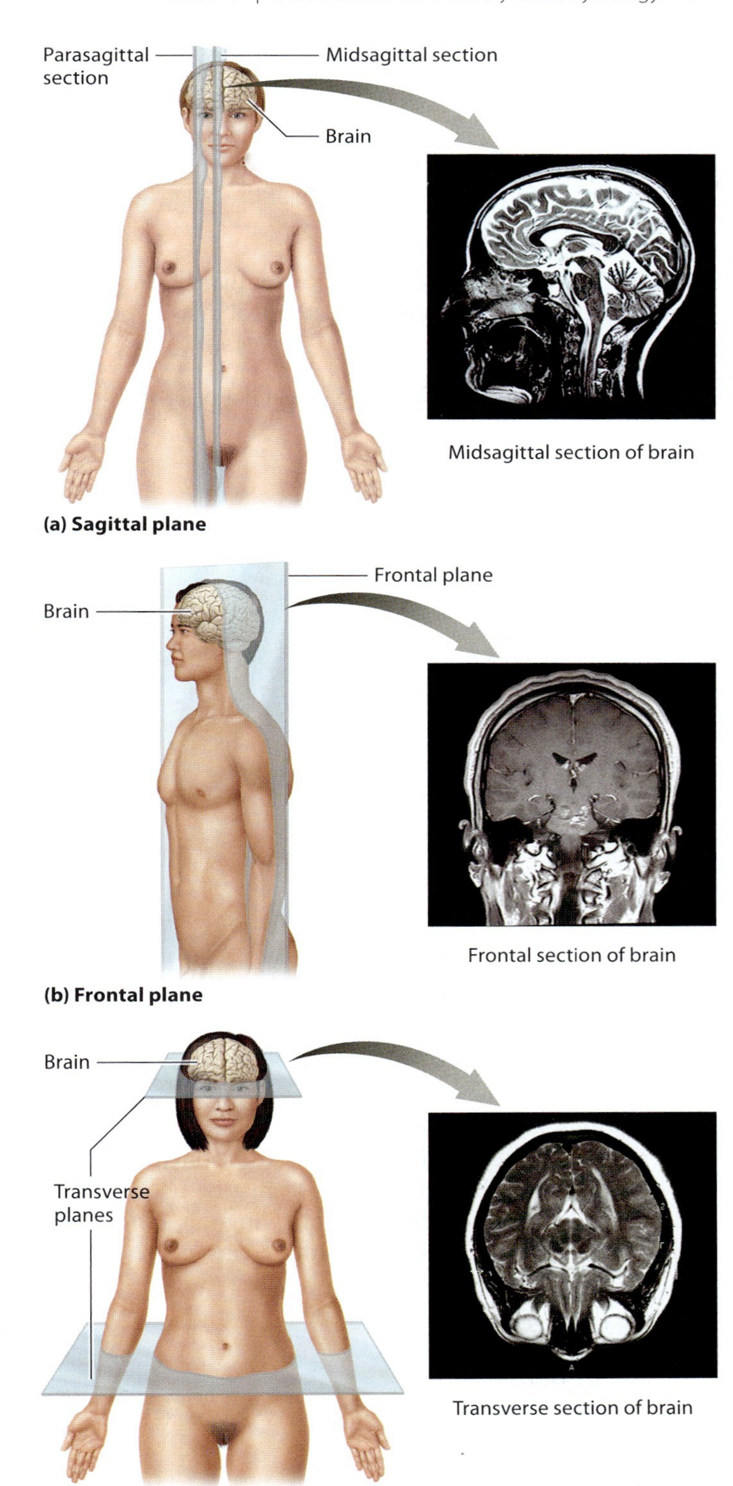

Figure 1-6 Planes of section.

Body Planes of Section

When viewing the internal anatomy of organs on models, on diagrams, and in specimens, it is important to understand the various types of cuts, or sections, that have been made to show the internal structures. In the health professions, these planes of section are also observed in various types of images, such as MRI and CT scans. **Figure 1-6** illustrates the three most commonly used planes:

1. A **sagittal plane** is a section made parallel to the body's longitudinal axis; it divides the body into right and left parts. A **midsagittal** (median) **plane** divides the body into equal right and left parts; a **parasagittal plane** divides the body into unequal right and left parts.
2. A **frontal** (coronal) **plane** is a section made parallel to the body's longitudinal axis; it divides the body into anterior and posterior parts.
3. A **transverse plane** (cross-section) is a section made perpendicular to the body's longitudinal axis; it divides the body into superior and inferior parts.

ACTIVITY 3 Demonstrating and Identifying Body Planes of Section

Learning Outcome

1. Demonstrate and describe anatomical planes of section.

Materials Needed

- ☐ Modeling clay
- ☐ Scalpel
- ☐ Anatomical models

Instructions

A. Modeling Clay Activity

1. Assign each member of your lab group one of the following body planes: a sagittal plane, a coronal plane, or a transverse plane.

2. Each student should then mold a ball of modeling clay into a "head" that has two eyes, a nose, and a mouth.
3. Using a scalpel, each student should demonstrate the assigned body plane by cutting a clay "head" and describing the cut to the other members of the lab group.

B. Identifying Body Planes

Complete the following chart as you identify the anatomical planes of section represented in anatomical models provided by your instructor.

	Anatomical Plane of Section Represented
Model 1	
Model 2	
Model 3	
Model 4	

Applying Anatomical Terminology

The previous lab activities introduced you to commonly used anatomical terms. However, in addition to knowing the meaning of each term, you also must be able to apply them in real-life situations. The following lab activity will give you some practice in using the terms you have just learned.

ACTIVITY 4 Assisting the Coroner

Learning Outcome

1. Apply anatomical terminology in a realistic scenario.

Materials Needed

☐ Torso model with three "wounds"

Instructions

A torso model has been "stabbed" three times. As the coroner on the case, you need to provide the following information for each stab wound: (a) a description of the wound location using at least two directional terms and one regional term (such as "right lumbar"); (b) the body cavity in which the wound is located; and (c) the names of the serous membranes that have been penetrated (if applicable).

1. Stab wound #1:
   a. ______________________________
   ____
   ____
   b. ______________________________
   c. ______________________________
2. Stab wound #2:
   a. ______________________________
   ____
   ____
   b. ______________________________
   c. ______________________________
3. Stab wound #3:
   a. ______________________________
   ____
   ____
   b. ______________________________
   c. ______________________________

Post-lab quizzes are also assignable in MasteringA&P®

POST-LAB ASSIGNMENTS

Name: ______________________ Date: ______________ Lab Section: ________

PART I. Check Your Understanding

Activity 1: Identifying Body Regions and Exploring Surface Anatomy

1. Which of the following terms is correctly matched to its description?
   a. manual—pertaining to the palm
   b. crural—pertaining to the calf
   c. acromial—pertaining to the chest
   d. mental—pertaining to the chin
   e. femoral—pertaining to the leg

2. Another term for the wrist is the:
   a. crural region.
   b. femoral region.
   c. popliteal region.
   d. sural region.
   e. carpal region.

Activity 2: Identifying Body Cavities and Abdominopelvic Regions

1. Identify the three ventral body cavities and the two dorsal body cavities in the following diagram. Then, name one organ found in each cavity.
   a. ______________ ______________
   b. ______________ ______________
   c. ______________ ______________
   d. ______________ ______________
   e. ______________ ______________

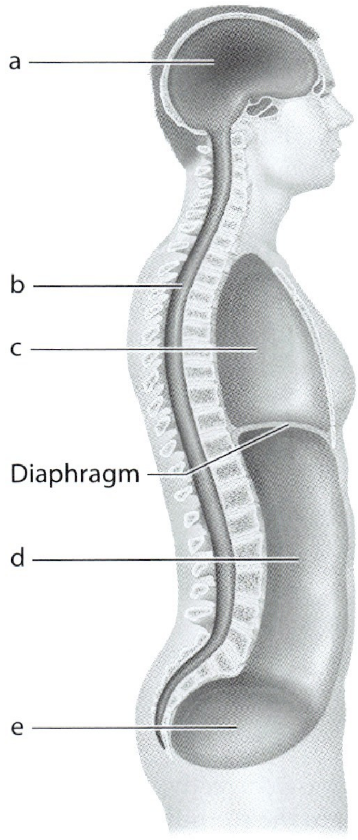

2. The spleen is located in the ______________ abdominopelvic region.
   a. left hypochondriac
   b. umbilical
   c. hypogastric
   d. right hypochondriac
   e. epigastric

3. A bullet that lodges in the heart would:
   a. be located in the ventral body cavity.
   b. penetrate the visceral peritoneum.
   c. be located in the vertebral cavity.
   d. penetrate the parietal pleura.
   e. be located laterally to a bullet that lodges in the lung.

Activity 3: Demonstrating and Identifying Body Planes of Section

1. Identify the planes of section shown in the following diagrams:
   a. ______________________________
   b. ______________________________

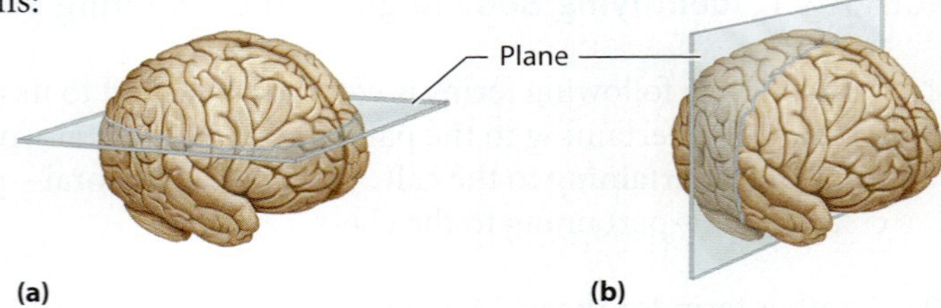

2. Which of the following organs could *not* be viewed in a midsagittal section through the body?
   a. the brain
   b. the heart
   c. the lung
   d. the diaphragm
   e. the pancreas

Activity 4: Assisting the Coroner

1. For each of the wound descriptions below, mark the diagram with an "a," "b," and "c" to represent the location of each wound:
   a. a cut in the medial part of the right femoral region
   b. wound in the left iliac region
   c. bruising in the left thoracic region, midway between the sternal and axillary regions

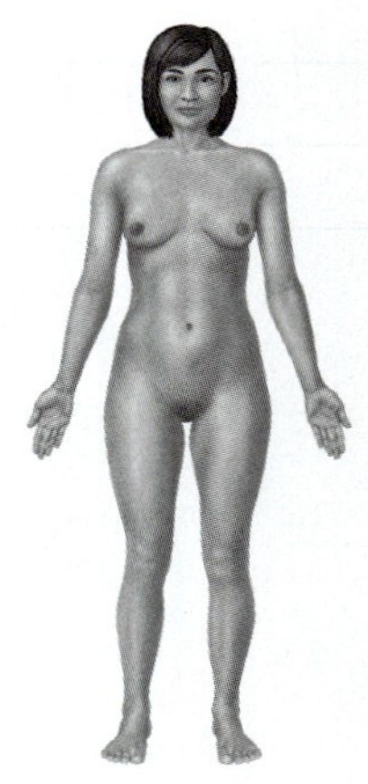

2. Which phrase correctly describes a stab wound that penetrates the anterior liver?
   a. medial to the sternum
   b. inferior to the rib cage
   c. superior to the left inguinal region
   d. in the right lower quadrant

PART II. Putting It All Together

A. Review Questions

Answer the following questions using your lecture notes, your textbook, and your lab notes.

1. Indicate whether each of the following statements is true or false. If the statement is false, correct it so that it is true.
   a. The small intestine is dorsal to the kidneys. ______
   b. The trachea is lateral to the lungs. ______
   c. The urinary bladder is superior to the uterus. ______
   d. The brain is inferior to the skull. ______

2. Assume anatomical position. Is the radius medial or lateral to the ulna? ______

   Explain the importance of using anatomical position as a standard reference point. ______

   ______

3. Use as many directional terms as possible to describe the relationship between:
   a. the antecubital region and the popliteal region. ______

      ______

      ______

   b. the acromial region and the mental region. ______

      ______

      ______

   c. the gluteal region and the sternal region. ______

      ______

      ______

4. Identify the body cavities entered during each of the following medical procedures. Begin with the largest cavity and end with the most specific body cavity. The answer for the first procedure is provided as an example.
   a. spinal tap *dorsal cavity, vertebral cavity*
   b. removal of appendix ______
   c. removal of gallbladder ______
   d. coronary bypass surgery ______

5. Which body plane(s) could provide a view of both:
   a. the spinal cord and the right lung? ______
   b. the trachea and the bladder? ______
   c. the right and left kidneys? ______
   d. the brain and the thyroid gland? ______

B. Concept Mapping

1. Fill in the blanks to complete this concept map outlining the anatomy of the ventral cavity.

abdominopelvic cavity	diaphragm	mediastinum	thoracic cavity	ventral cavity

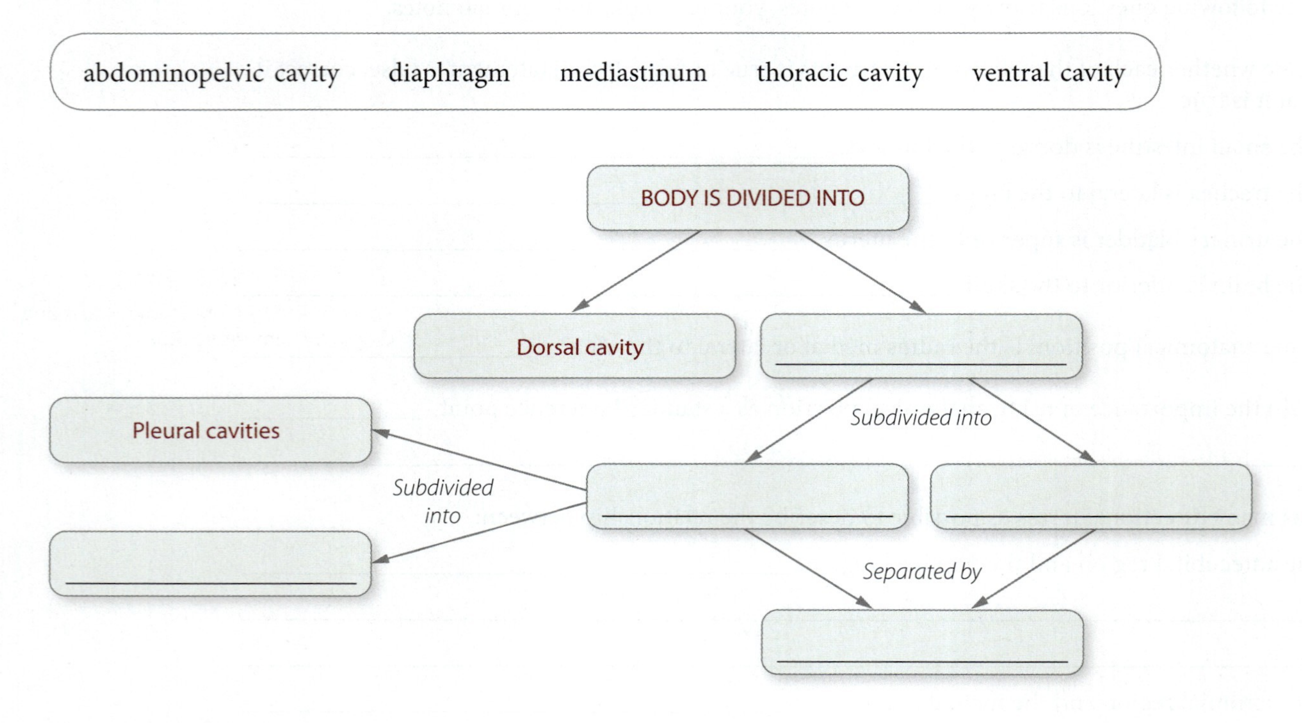

2. Construct a unit concept map to show the relationships among the following set of terms. Include all of the terms in your diagram. Your instructor may choose to assign additional terms.

abdominopelvic cavity	diaphragm	dorsal cavity	heart	hypogastric
lung	medial	mediastinum	pericardial cavity	peritoneum
pleural cavity	stomach	thoracic cavity	transverse	ventral cavity

2 Introduction to the Organ Systems

UNIT OUTLINE

In this unit we examine two topics that are basic to anatomy and physiology: the body's various levels of structural organization—from the chemical level to organ systems and the body as a whole—and how multiple organ systems interact to keep conditions within the body in a relatively stable state.

THINK ABOUT IT *How do the digestive, cardiovascular, and respiratory systems work together to deliver oxygen and nutrients to the cells of the body?*

____

____

____

____

PRE-LAB ASSIGNMENTS

Pre-lab quizzes are also assignable in MasteringA&P®

To maximize learning, BEFORE your lab period carefully read this entire lab unit and complete these pre-lab assignments using your textbook, lecture notes, and prior knowledge.

PRE-LAB Activity 1: Locating and Describing Major Organs and Their Functions

1. Which structural level of organization:

   a. represents the basic unit of life? ______________________________

   b. represents the basic unit of matter? ______________________________

   c. consists of two or more tissue types that work together for a common function? ______________________________

   d. consists of two or more cell types that work together for a common function? ______________________________

2. Match each of the following organ systems with its correct function/description:

_____	a. cardiovascular	1. returns excess tissue fluid to the blood
_____	b. digestive	2. functions in hematopoiesis
_____	c. endocrine	3. transmits electrical impulses
_____	d. integumentary	4. generates heat
_____	e. lymphatic	5. produces gametes
_____	f. muscular	6. rids body of nitrogenous wastes
_____	g. nervous	7. transports oxygen to body cells
_____	h. respiratory	8. absorbs nutrients
_____	i. reproductive	9. functions in gas exchange
_____	j. skeletal	10. produces hormones
_____	k. urinary	11. provides physical barrier

PRE-LAB Activity 2: Using Anatomical Terminology to Describe Organ Locations

1. The dorsal body cavity is subdivided into the _______________ cavity and the _______________ cavity.
2. The ventral body cavity is subdivided into the _______________ cavity and the _______________ cavity.
3. Mark each of the following statements as either true (T) or false (F):

   _____ a. The heart is lateral to the lungs.

   _____ b. The wrist is distal to the elbow.

   _____ c. The ribs are superficial to the skin.

   _____ d. The stomach is posterior to the kidneys.

   _____ e. The thyroid gland is superior to the pancreas.

4. Provide an accurate directional term to complete the following sentences:
   a. The spinal cavity is ______________________ to the cranial cavity.
   b. The thoracic cavity is ______________________ to the abdominopelvic cavity.
   c. The hypogastric region is ______________________ to the umbilical region.
   d. The epigastric region is ______________________ to the right hypochondriac region.
   e. The left lumbar region is ______________________ to left iliac region.

PRE-LAB Activity 3: Studying Homeostasis and Organ System Interactions

1. Define homeostasis. ____.
2. The three components of a feedback mechanism are a ______________, a ______________, and an ______________.
3. *Circle the correct response:* Body temperature regulation is an example of a (positive/negative) feedback mechanism.
4. *Circle the correct response:* Blood clotting is an example of a (positive/negative) feedback mechanism.
5. In this activity you will explore the interactions that occur among four different organ systems during the ______________ reflex.

The Levels of Structural Organization in the Body

Anatomy (the study of structure) and physiology (the study of function) are intimately associated: The function of each of the body's parts stems from the structural adaptations of those parts. Perhaps the best way to begin studying the human body is to consider the various levels at which the body's structures are organized, from the smallest structures to the body as a whole (**Figure 2-1**).

At the lowest level—the chemical level—are **atoms**, the fundamental units, or basic building blocks, of matter. Atoms combine to form **molecules**. In turn, molecules combine to form **macromolecules**, of which there are of four biologically important types: carbohydrates, proteins, lipids, and nucleic acids. These macromolecules are the components of the various parts of **cells**, the functional units of the cellular level of organization and the fundamental units of life. Groups of cells function together to form **tissues**. Figure 2-1 illustrates an example of one general type of tissue: epithelial tissue. (The other three general types of tissue are connective tissue, muscle tissue, and nervous tissue.) Groups of tissues function together to form **organs**, and groups of organs function together to form **organ systems**. Taken together, the body's 11 organ systems constitute the highest level of organization, the **organism** level.

The organ system used to illustrate the hierarchy of the body's structural levels in Figure 2-1 is the digestive system, and the specific organ shown is the esophagus. This long tubular structure conveys food and liquids from the pharynx to the stomach. The structure of the esophagus is well adapted to its function. Its internal lining is composed of a type of epithelial tissue that consists of multiple layers of flattened (squamous) cells, an arrangement that protects the esophagus from abrasion by the passing food materials. Additionally, the walls of the esophagus contain smooth muscle tissue that contracts to propel food toward the stomach. Each squamous epithelial cell and each smooth muscle cell contains specialized organelles that allow the cell to perform its specific functions. These organelles, in turn, are composed of macromolecules, which are composed of molecules, which are composed of atoms.

You will learn much more about the esophagus when you study the digestive system. Here the crucial point is that it is important to understand the hierarchy of the body's structural levels, because the relationship between structure and function occurs at each and every level.

Using this approach to study the organ systems effectively requires you to learn the key principles of cytology (the study of cells) and histology (the study of tissues). Unit 5 (The Cell) and Unit 6 (Histology) will provide a foundation on which you can build your study of the body's organ systems.

THINK ABOUT IT Each time you begin to study an organ system, ask yourself the following questions about the structure and function of each level of the hierarchy:

- Which organs make up this organ system?
- How is each organ structurally adapted to its function(s)?
- Which tissues make up each organ?
- Which specialized cell types are found in each organ?
- How is each type of cell structurally adapted to performing its specific functions? In other words, which cell parts (or organelles) are predominant in each cell type?
- Which types of molecules/macromolecules does the cell contain and/or produce? ■

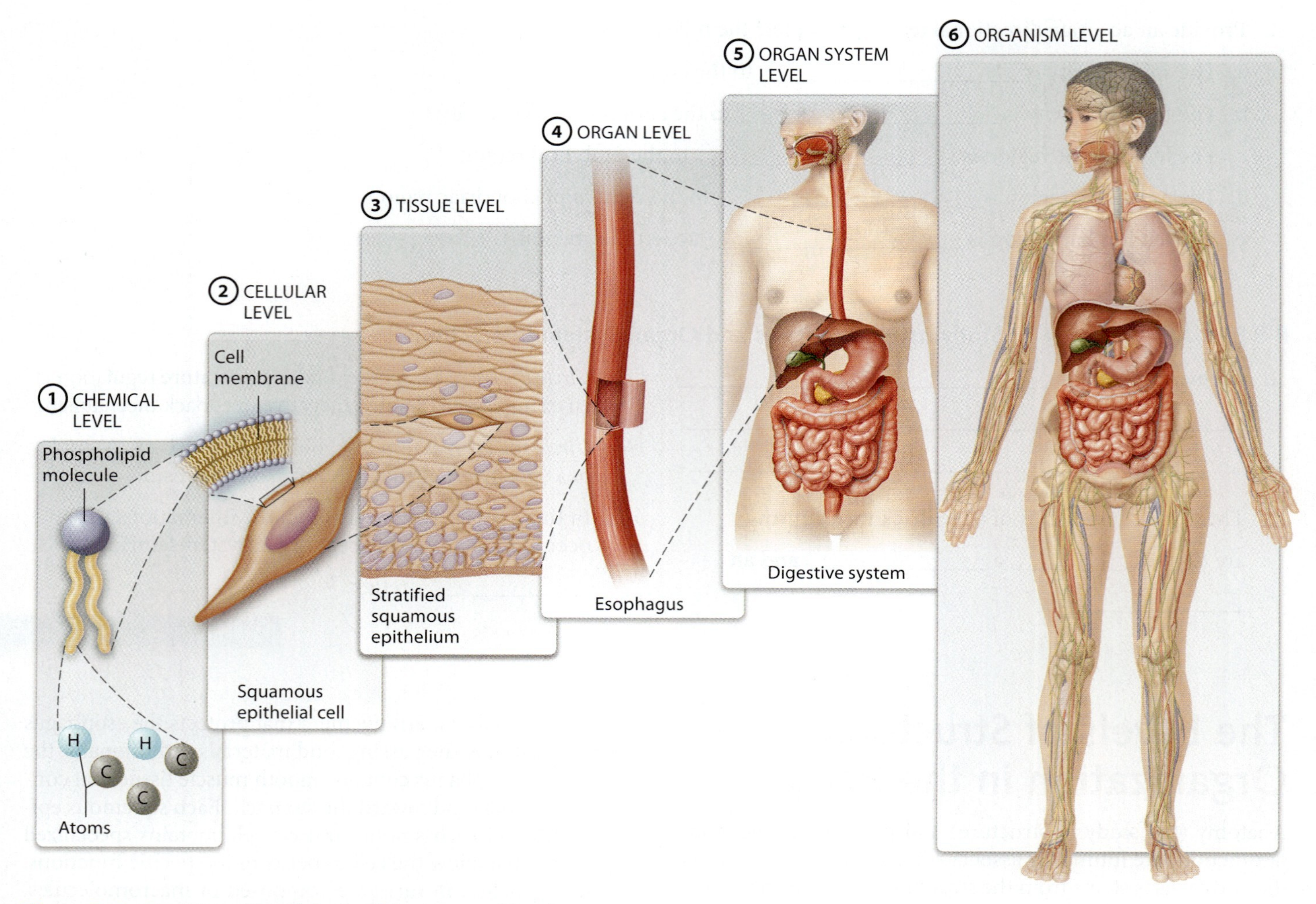

Figure 2-1 The hierarchy of the body's structural levels.

ACTIVITY 1

Locating and Describing Major Organs and Their Functions

Learning Outcomes

1. Locate selected organs on an anatomical model and identify the organ system(s) to which each belongs.
2. Describe the general function(s) of each organ.

Materials Needed

- ☐ Torso model
- ☐ Labeling stickers
- ☐ Markers

Instructions

1. For each of the organs in the organ identification chart that follows, place a sticker on the organ on a torso model, and then label the sticker with the relevant organ system using the following key:

C = cardiovascular	N = nervous
D = digestive	R = respiratory
E = endocrine	S = skeletal
I = integumentary	U = urinary
L = lymphatic	♂ = male reproductive
M = muscular	♀ = female reproductive

If an organ belongs to more than one system, place and label as many stickers as apply.

2. CHART Then complete the following organ identification chart by indicating the organ system(s) to which each listed organ belong(s) and by briefly describing the function of each organ.

Organ Identification Chart

Organ	Organ System(s)	Function
Aorta (artery)		
Bone		
Brain		
Esophagus		
Heart		
Kidney		
Large intestine		
Larynx		
Liver		
Lungs		
Lymph node		
Muscle		
Nerve		
Ovary		
Pancreas		
Pharynx		
Pituitary gland		
Skin		
Small intestine		
Spinal cord		
Spleen		
Stomach		
Testis		
Thyroid gland		
Urinary bladder		
Uterus		
Vagina		
Vena cava (vein)		

ACTIVITY 2

Using Anatomical Terminology to Describe Organ Locations

Learning Outcomes

1. Describe the locations of selected organs using appropriate regional and directional terms.
2. Identify the body cavities in which selected organs are located.

Materials Needed

☐ Torso model used in Activity 1

Instructions

In this activity you will combine what you learned about correct anatomical terminology in Unit 1 with what you are learning about organs and organ systems in this unit.

CHART For each of the following organs, describe in the following chart the organ's location using correct anatomical terminology, including the mention of body cavities and the abdominopelvic regions where applicable. Some answers have been supplied for you.

Organ	Description of Location Using Correct Anatomical Terminology
Brain	*The brain is located in the cranial cavity of the dorsal body cavity. It is superior to the spinal cord.*
Esophagus	
Heart	*The heart is located in the thoracic cavity of the ventral body cavity. It is located medial to the lungs.*
Kidney	
Large intestine	
Larynx	
Liver	
Lungs	
Ovary	
Pancreas	
Pharynx	
Pituitary gland	*Neuroendocrine organ located inferior to the brain. It is located in the cranial cavity of the dorsal body cavity.*
Skin	*The skin is superficial to the internal organs. It is not located in a body cavity.*
Small intestine	
Spinal cord	
Spleen	
Stomach	*The stomach is located predominantly in the epigastric region of the abdominopelvic cavity, a subdivision of the ventral body cavity.*
Testis	
Thyroid gland	
Urinary bladder	
Uterus	

Homeostasis, Feedback Control, and Organ System Interaction

The word **homeostasis**—derived from the Greek words *homios*, meaning "similar," and *stasis*, meaning "standing still"—refers to the body's actions in maintaining internal conditions within a narrow, relatively stable physiological range. Homeostatic control of any physiological event requires the coordinated effort of multiple organ systems. The body maintains this dynamic state of equilibrium through self-regulating control systems, also known as feedback mechanisms or feedback loops. Such control systems contain the following three components:

- A **receptor**, which responds to a particular environmental change or stimulus;
- A **control center**, which receives the information supplied by the receptor, processes the information, and sends out commands; and
- An **effector**, which produces a response (carries out the commands sent from the control center).

The most common control system in the body involves **negative feedback**. In this type of feedback control, a stimulus elicits a physiological response that opposes or counteracts the stimulus. Consider, for example, the regulation of body temperature (**Figure 2-2**). As body temperature falls below the normal range, receptors in the skin (integumentary system) and in certain brain cells (nervous system) send afferent messages to the thermoregulatory center (the control center) in another part of the brain (nervous system). When cells in the control center detect a sufficiently large drop in temperature, they send signals to the system's effectors—muscle cells (muscular system)—stimulating them to shiver, which generates heat. The resulting production of heat (the response) counteracts the original stimulus (a decrease in body temperature) and returns body temperature to within the normal range. Negative feedback then stops the shivering, and homeostasis is restored.

Positive feedback mechanisms, in which the output enhances the original stimulus, are far less common than negative feedback mechanisms. We will learn about positive feedback control in later units when we study the events of blood clotting and the ovarian cycle.

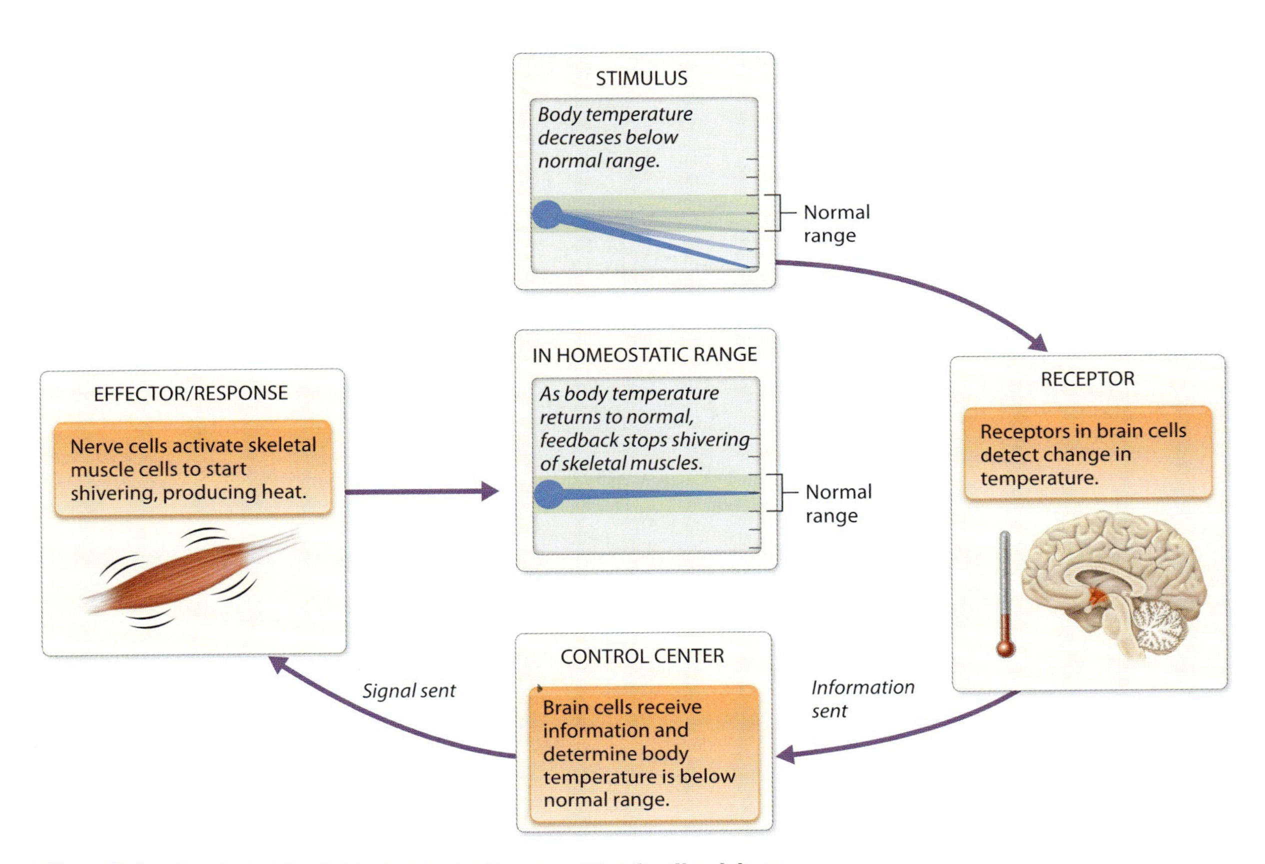

Figure 2-2 Control of body temperature by a negative feedback loop.

ACTIVITY 3

Studying Homeostasis and Organ System Interactions

Learning Outcomes

1. Explain the concept of homeostasis.
2. Distinguish between negative and positive feedback mechanisms.
3. Outline and explain the patellar reflex.
4. Describe the ways in which the nervous, digestive, skeletal, and muscular systems work together during the patellar reflex.

Materials Needed

- ☐ Laminated organ system interactions poster
- ☐ Water-soluble markers

Instructions

In this activity you will explore the interactions among four different organ systems during the patellar (knee-jerk) reflex, which physicians test by tapping on the patellar ligament just inferior to the patella. The tapping of the tendon causes the attached skeletal muscle to stretch slightly (the stimulus), which in turn stimulates specialized sensory receptors called muscle spindles to generate a nerve impulse, which then travels to the spinal cord (the control center). The spinal cord then sends out a motor impulse that triggers the contraction of the slightly stretched muscle (the effector). Thus, the stimulus (muscle stretching) produces an effect (muscle contraction) that counteracts that stimulus.

On the next page is an Organ System Interaction Worksheet that is a miniature version of the poster you will fill out during this activity. Working with the other members of your lab group, use a water-soluble marker to fill out the poster for the patellar reflex according to the following directions:

1. Assign each of the following organ systems to one of the "Organ system" boxes on the organ systems interaction poster: nervous system, cardiovascular system, muscular system, and skeletal system. In each "Organ system" box, list the system's major structures or organs, and describe the system's major functions.

2. In each "Organ system" box, write notes concerning the role of that system in the patellar reflex. In your notes within the "Organ system" boxes, include the following terms:

   rectus femoris muscle, glucose, patellar tendon, bone, mitochondria, myosin, Pacinian corpuscle, brain

   Look up these terms in your textbook to discover their relationship to the patellar reflex.

3. In the large "Interaction circle" in the center of the poster, explain how the components of these four organ systems work together to produce the patellar reflex.

Organ System Interaction Worksheet

(1) Organ system:

Structures:

Functions:

Notes:

(2) Organ system:

Structures:

Functions:

Notes:

Interaction circle

Notes:

Structures:

Functions:

(3) Organ system:

Notes:

Structures:

Functions:

(4) Organ system:

Post-lab quizzes are also assignable in MasteringA&P®

POST-LAB ASSIGNMENTS

Name: ______________________ Date: ____________ Lab Section: ________

PART I. Check Your Understanding

Activity 1: Locating and Describing Major Organs and Their Functions

1. Which of the following organs belong(s) to more than one organ system?
   a. ureter
   b. larynx
   c. esophagus
   d. kidney
   e. pituitary gland

2. Which of the following organ systems is correctly matched with one of its functions?
   a. cardiovascular system—produces blood cells
   b. endocrine system—acts as a fast-acting control system
   c. respiratory system—transports oxygen and carbon dioxide
   d. muscular system—generates heat
   e. urinary system—returns excess tissue fluid to the bloodstream

3. Which of the following pairs of organ systems functions primarily to regulate body functions?
   a. cardiovascular and nervous systems
   b. lymphatic and endocrine systems
   c. endocrine and nervous systems
   d. urinary and endocrine systems
   e. integumentary and nervous systems

4. Immunity is carried out primarily by which of the following pairs of organ systems?
   a. respiratory and cardiovascular systems
   b. cardiovascular and lymphatic systems
   c. skeletal and cardiovascular systems
   d. urinary and integumentary systems
   e. nervous and lymphatic systems

5. Which organ:

   ______________ a. transports blood to the heart?

   ______________ b. is commonly known as the windpipe?

   ______________ c. transports food from the pharynx to the stomach?

   ______________ d. stores urine?

   ______________ e. produces sperm?

Activity 2: Using Anatomical Terminology to Describe Organ Locations

1. Complete the following chart:

Organ	Organ System to Which It Belongs	Dorsal Body Cavity or Ventral Body Cavity?	Specific Body Cavity
Brain			
Gallbladder			
Heart			
Skin			
Urinary bladder			

2. Complete each of the following statements with an accurate directional term:

   a. The thymus is ____________________ to the thyroid gland.

   b. The pituitary gland is ____________________ to the brain.

   c. The liver is ____________________ to the gallbladder.

   d. The sternum is ____________________ to the heart.

   e. The esophagus is ____________________ to the trachea.

3. In which abdominopelvic region is each of the following organs predominantly found?

   ____________________ a. liver

   ____________________ b. urinary bladder

   ____________________ c. spleen

   ____________________ d. stomach

   ____________________ e. right kidney

Activity 3: Studying Homeostasis and Organ System Interactions

1. Distinguish between a negative feedback mechanism and a positive feedback mechanism.

   ____

   ____

2. Does the patellar reflex involve a negative feedback mechanism or a positive feedback mechanism?

   ____________________ Why? ____

3. In the patellar reflex:

   ____________________ a. Which component serves as the receptor?

   ____________________ b. Which component serves as the control center?

   ____________________ c. Which component serves as the effector?

4. Describe one way in which each of the following organ systems contributes to the patellar reflex:

   a. nervous ____

   b. cardiovascular ____

   c. muscular ____

   d. skeletal ____

PART II. Putting It All Together

A. Review Questions

Answer the following questions using your lecture notes, your textbook, and/or your lab notes.

1. Briefly describe one example of how the circulatory, respiratory, and digestive systems work together to maintain homeostasis. ______________________________

   ____

   ____

2. Briefly describe one example of how the nervous, endocrine, and circulatory systems work together to maintain homeostasis. ______________________________

   ____

   ____

3. Calcium ions play a major role in many physiological events. A decline in blood calcium levels is regulated primarily by parathyroid hormone (PTH) released by the parathyroid gland. PTH stimulates certain bone cells to break down bone, the small intestine to absorb more calcium from the diet, and the kidneys to reabsorb more calcium from the blood. As a result of these actions, blood calcium levels rise, and PTH release is then inhibited.

   a. List three organ systems involved in the maintenance of blood calcium levels, and state the role of each organ system in the process. ______________________________

   ____

   ____

   ____

   b. Is blood calcium regulation an example of negative feedback control or positive feedback control? Explain. ______________________________

   ____

   ____

B. Concept Mapping

1. Fill in the blanks to complete this concept map outlining selected organs found in the ventral body cavity.

kidney	lung	spleen	thymus	uterus

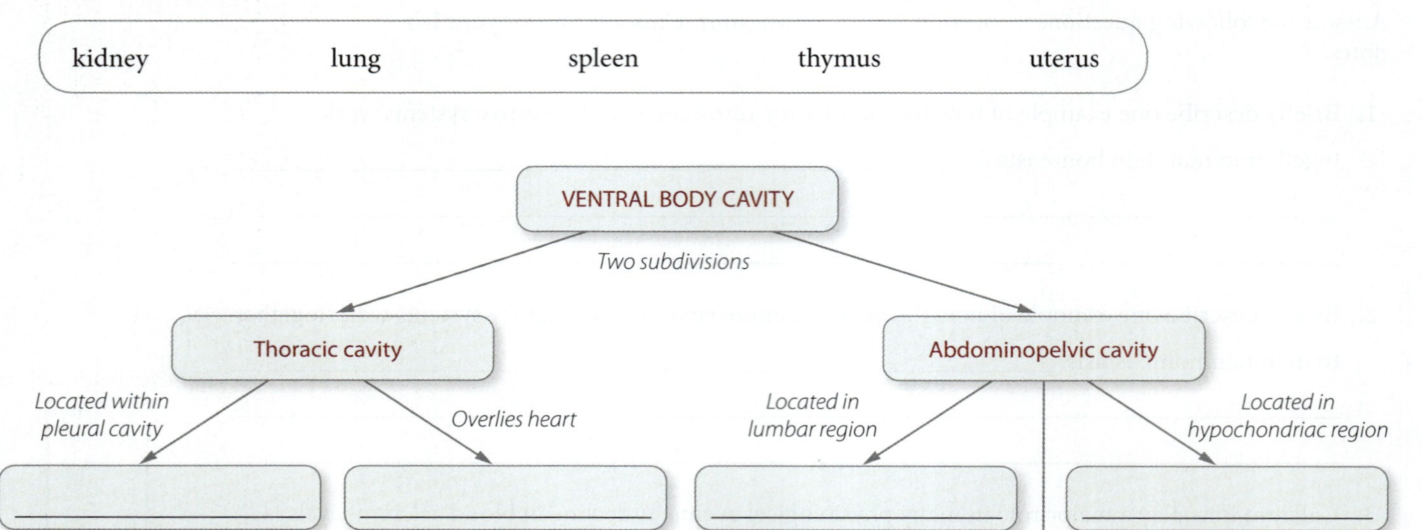

2. Construct a unit concept map to show the relationships among the following set of terms. Include all of the terms in your diagram. Your instructor may choose to assign additional terms.

bone	brain	esophagus	heart	kidney
lung	muscle	pancreas	pharynx	skin
spleen	stomach	testis	thymus	trachea

3 Chemistry

UNIT OUTLINE

Recall from our study of the structural hierarchy of life in Unit 2 that the fundamental unit of matter is the atom. Two or more atoms unite to form molecules. Metabolism—all of the chemical activities that occur in the body—consists of chemical reactions involving interactions between atoms and molecules.

A basic understanding of chemical concepts is critical to your understanding of anatomy and physiology. In this unit, you will learn several important chemical principles as you explore the chemical properties of water, investigate the concept of pH, examine the role of buffers in maintaining homeostasis, and test the function of enzymes.

THINK ABOUT IT *Chemistry is the basis of life. Chemical reactions occur at an incredible rate during physiological events such as the digestion of food, the transport of blood, the transmission of electrical messages, and movement of the body. Metabolic events are classified as anabolic reactions that build molecules or catabolic reactions that degrade molecules. You can remember the difference between anabolic reactions and catabolic reactions by remembering ABCD: Anabolic Builds, Catabolic Degrades.*

State one example of an anabolic event in the body: ______________________

____

State one example of a catabolic event in the body: ______________________

____

PRE-LAB ASSIGNMENTS

Pre-lab quizzes are also assignable in MasteringA&P®

To maximize learning, BEFORE your lab period carefully read this entire lab unit and complete these pre-lab assignments using your textbook, lecture notes, and prior knowledge.

PRE-LAB Activity 1: Exploring the Chemical Properties of Water

1. Rank the following levels of structural organization from 1 (smallest unit) to 5 (largest unit).

   _____ cell _____ organelle

   _____ molecule _____ proton

   _____ atom

2. In terms of chemical bonding, which part(s) of the atom is(are) most important?
   a. Neutrons
   b. Protons
   c. Electrons in the innermost electron shell
   d. Electrons in the outermost electron shell
   e. Nucleus

3. Which of the following types of bonds occurs between water molecules?
   a. Ionic bond
   b. Hydrogen bond
   c. Polar covalent bond
   d. Nonpolar covalent bond
   e. None of these

4. The tendency of water molecules to cling to one another is called _______________ and the tendency of water molecules to cling to another substance is called _______________.

5. Because so many inorganic and organic molecules dissolve in water, water is known as the _______________.

PRE-LAB Activity 2: Determining pH and Interpreting the pH Scale

1. The pH scale:
   a. ranges from 1 to 14.
   b. is based on the hydrogen ion concentration of a solution.
   c. is a logarithmic scale.
   d. was devised in 1909 by a Danish biochemist.
   e. All of these statements are true.

2. Which pH value represents neutral?
   a. 0
   b. 3
   c. 7
   d. 9
   e. 14

3. A solution with a pH of 3.0 is said to be _______________, whereas a solution with a pH of 9.0 is said to be _______________.

4. As blood carbon dioxide levels rise, do blood hydrogen ion levels rise or fall? _______________

PRE-LAB Activity 3: Observing the Role of Buffers

1. Bromocresol purple is a pH indicator that turns _______________ in acidic environments and _______________ in basic environments.

2. When acids and bases are mixed, they react to form a _______________ and _______________.

3. Substances that stabilize the pH of a solution by either releasing H^+ when pH levels rise or binding H^+ when pH levels rise are called _______________.

PRE-LAB Activity 4: Analyzing Enzymatic Activity

1. Enzymes are macromolecules composed of ______________________, and they function as biological ______________________.
2. Enzyme specificity:
   a. occurs because enzymes are proteins.
   b. results because enzymes remain unchanged by the reactions in which they participate.
   c. explains why enzymes speed up chemical reactions.
   d. results from the structure of the enzyme's active site.
3. Define denaturation. ______________________________
4. The basic building block(s) of carbohydrates is(are):
   a. monosaccharides.
   b. amino acids.
   c. glycerol.
   d. fatty acids.
   e. starch.
5. The breaking of chemical bonds by the addition of water is a chemical reaction called ______________________.

The Chemistry of Water

All living organisms require water to survive. Water accounts for up to two-thirds of your body weight. As a result of its chemical structure, water exhibits some unusual properties:

1. **Solubility.** Water is known as the universal solvent because many inorganic and organic molecules will dissolve in H_2O. The medium in which atoms/ions/molecules are dispersed is the **solvent**, and the dispersed substances are called **solutes**. In aqueous solutions, H_2O is the solvent. Water is the body's major transport medium because it is such an excellent solvent.
2. **High heat of vaporization.** The temperature of liquid water must be high before individual water molecules have enough energy to break free and become a gas (water vapor). Furthermore, when water changes phases from a liquid to a gas, it carries a great deal of heat away with it. This is the process by which sweating helps cool the body.
3. **High heat capacity.** Water has the ability to absorb or release large amounts of heat while its temperature changes only slightly.
4. **Reactivity.** Water molecules are participants in many chemical reactions. Hydrolysis reactions use water to split larger molecules into their smaller building blocks. Dehydration synthesis reactions, by contrast, remove water from small molecules in order to combine them into a single larger molecule.
5. **Adhesion/cohesion.** Adhesion is the tendency for water molecules to cling to another substance; cohesion is the tendency for water molecules to cling to each other. Cohesion of water molecules is especially evident at its surface, where the molecules exhibit surface tension.

To begin to understand how these properties of water relate to its chemical structure, consider the structure of the atom shown in **Figure 3-1**. An **atom** consists of a centrally located nucleus containing positively charged **protons** and electrically neutral **neutrons** bound tightly together. The nucleus is orbited by a cloud of negatively charged **electrons**. Figure 3-1 depicts a carbon atom containing six protons, six neutrons, and six electrons. Electrons forming this electron cloud occupy regions called **electron shells**, each of which can hold a certain number of electrons. Only electrons located in the outermost shell are important in terms of chemical bonding. When the outermost shell of an atom is filled, the atom is considered to be stable. Atoms have a tendency to interact with each other in such a way that their outermost shells become filled.

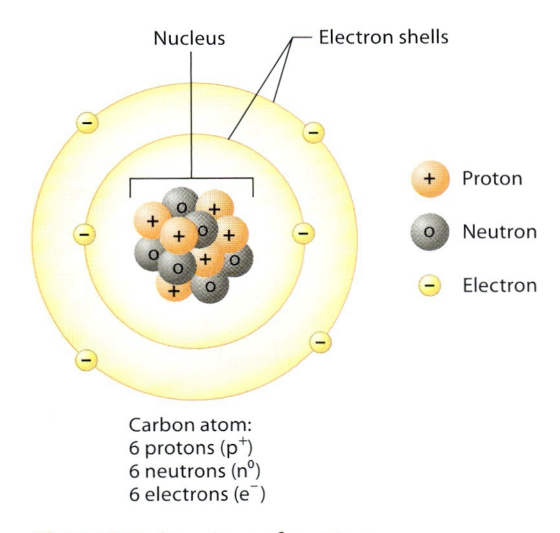

Figure 3-1 Structure of an atom.

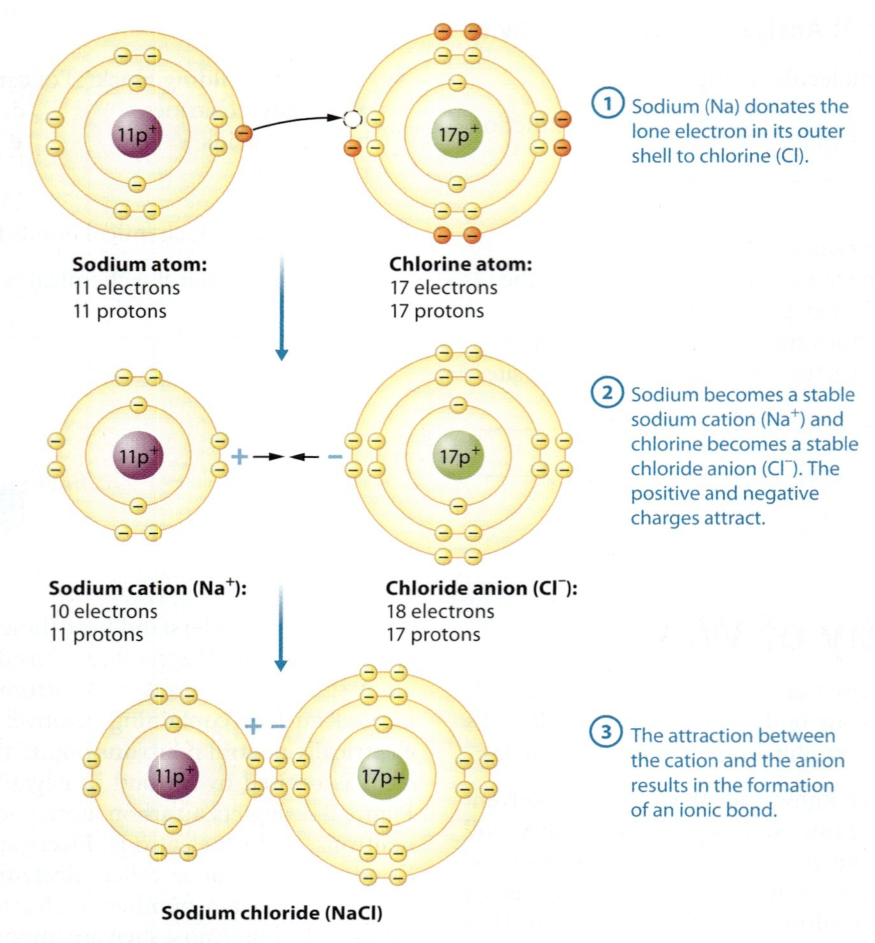

Figure 3-2 **Formation of an ionic bond.**

Three types of chemical bonds are ionic bonds, covalent bonds, and hydrogen bonds. An **ionic bond** results from the attraction between ions, which are atoms that have either a positive or a negative electrical charge. Consider, for example, the formation of an ionic bond to produce the compound sodium chloride (**Figure 3-2**). Sodium (Na) donates the lone electron in its outer shell to chlorine (Cl). Sodium then becomes a stable cation (Na^+) and chlorine becomes a stable anion (Cl^-). The attraction between the cation and the anion results in the formation of an ionic bond.

> **QUICK TIP** You can remember that a cation has a positive charge because a handwritten small letter *t* resembles a plus sign.

A **covalent bond** results from the sharing of electrons. **Figure 3-3** illustrates the formation of a covalent bond between two hydrogen atoms. In this case, the electrons are shared equally between the two hydrogen atoms, forming a hydrogen molecule. Because a hydrogen molecule is

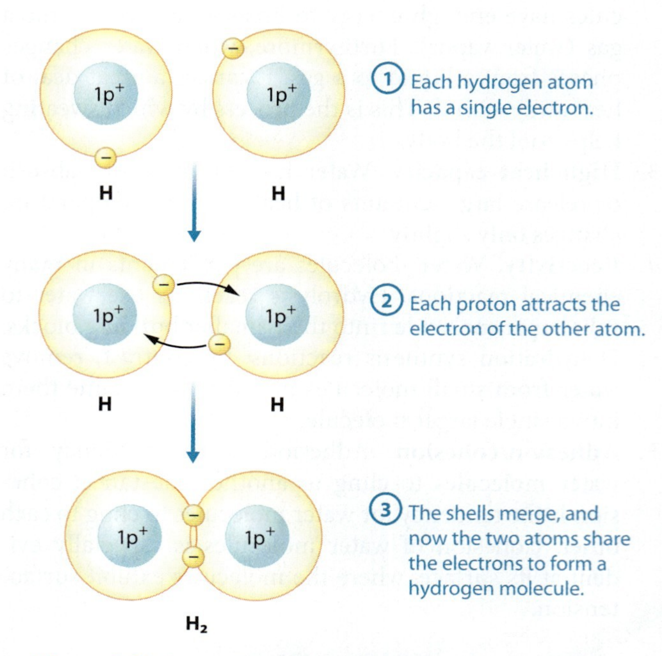

Figure 3-3 **Formation of a covalent bond.**

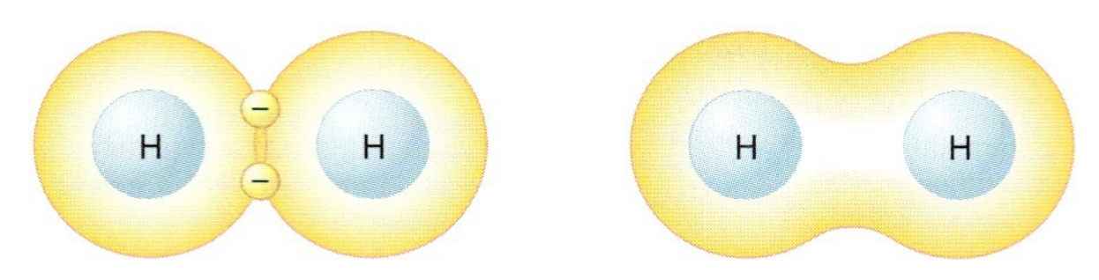

(a) Nonpolar covalent bond—H_2 (hydrogen molecule): Electrons spend equal time around the two hydrogen atoms.

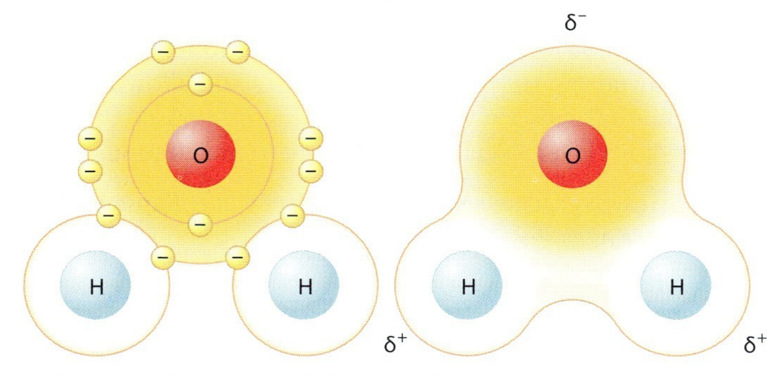

(b) Polar covalent bond—H_2O (water): Electrons spend more time around the more electronegative oxygen atom.

Figure 3-4 Examples of nonpolar and polar covalent bonds.

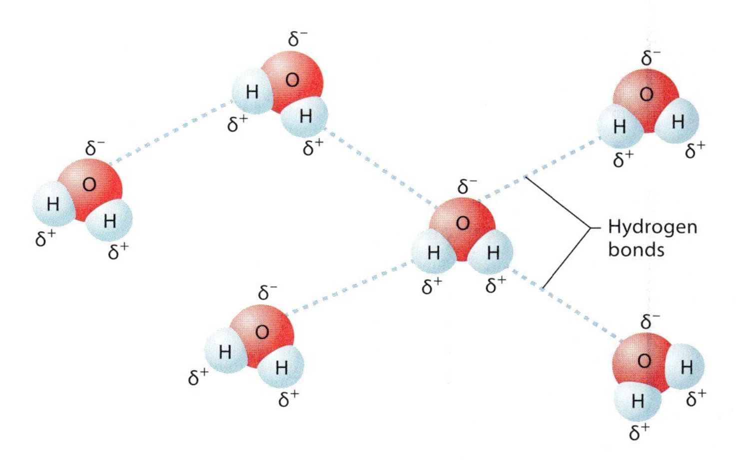

Figure 3-5 Hydrogen bonding between water molecules.

electrically balanced (it has neither positive nor negative poles), it is referred to as a nonpolar molecule and is held together by a **nonpolar covalent bond** (**Figure 3-4a**).

In other cases, however, the electrons are shared unequally between the atoms that make up a molecule. In a water molecule (H_2O), depicted in **Figure 3-4b**, the nucleus of the oxygen atom has a stronger attraction for electrons than the hydrogen nuclei have. As a result, the shared electrons spend more time in the vicinity of the oxygen atom, giving the oxygen a slightly more electronegative charge (indicated by δ^-). The hydrogen atoms have a slightly electropositive charge (indicated by δ^+) because the electrons spend less time in the vicinity of the hydrogen atoms. Because H_2O has both positive and negative poles, water is considered a polar molecule held together by **polar covalent bonds**.

Another type of chemical bond, a **hydrogen bond**, forms when the slightly electropositive charge (δ^+) on a hydrogen atom of one polar molecule is weakly attracted to the slightly electronegative charge (δ^-) on an atom (typically an oxygen atom or a nitrogen atom) of another polar molecule. Hydrogen bonding between water molecules (**Figure 3-5**) involves the attraction of the slightly positive hydrogen atom of one water molecule to the slightly negative oxygen atom of another water molecule. Hydrogen bonding is responsible for many of the chemical properties of water, including its role as the universal solvent.

Figure 3-6 illustrates the behavior of ionic compounds, polar covalent compounds, and nonpolar covalent compounds in water.

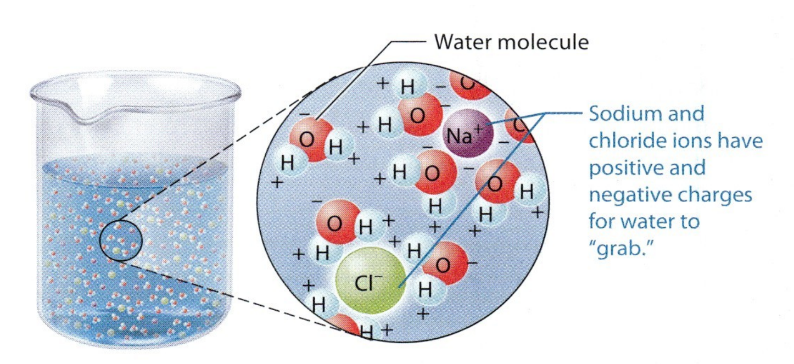

(a) Ionic compounds are hydrophilic.

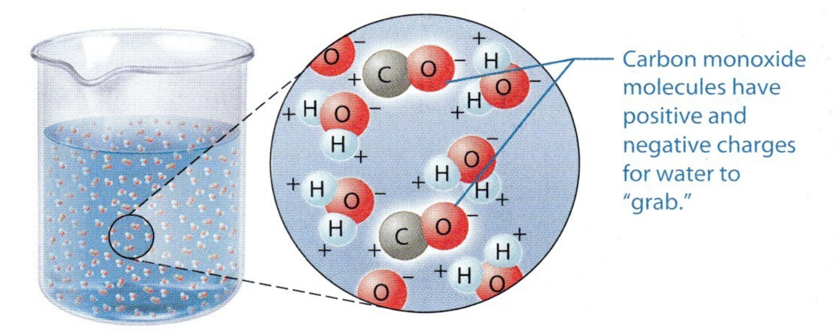

(b) Polar covalent compounds are hydrophilic.

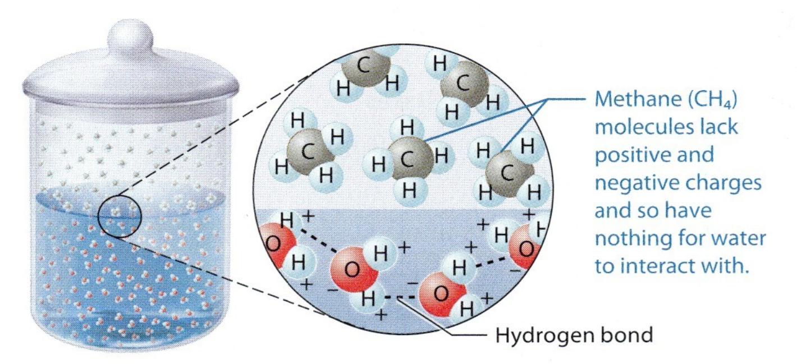

(c) Nonpolar covalent compounds are hydrophobic.

Figure 3-6 The behavior of different types of compounds in water.

ACTIVITY 1

Exploring the Chemical Properties of Water

Learning Outcomes

1. Explain the chemical structure of water.
2. Describe the properties of water.
3. Demonstrate surface tension.
4. Demonstrate water's high heat of vaporization.
5. Analyze the role of surfactant in lung function.

Materials Needed

- ☐ Two pennies
- ☐ Plastic dropper
- ☐ Water
- ☐ Liquid soap
- ☐ Two 50-ml beakers
- ☐ Ethanol
- ☐ Hot plate

Instructions

A. Demonstrating Surface Tension

1. Fill two 50-ml beakers with water. To the second beaker add a few drops of liquid soap and stir to make a soapy solution.
2. Fill a plastic dropper with water from the first beaker. Count the number of drops of water you can fit on a penny before the water begins to spill off the coin.

   Record the number of drops: ______________________

3. Empty the remaining water from the plastic dropper and then fill it with soapy water from the second beaker. Count the number of drops of soapy water you can fit on the penny before the soapy water begins to spill off the coin.

   Record the number of drops: ______________________

   How did the soap affect the cohesive/adhesive abilities of the water molecules?

   ____

   ____

B. Comparing Evaporation Rates

1. Pour 5 ml of water into a 50-ml beaker and label the beaker "water."
2. Pour 5 ml of ethanol into a 50-ml beaker and label the beaker "ethanol."
3. Place both beakers on a hot plate, NOT on an open flame, and slowly warm the water and the ethanol.

   Which fluid (water or ethanol) evaporates more quickly?

   ____

   Provide an explanation for your results.

   ____

   ____

C. Applying the Principles

1. Surfactant is a lipoprotein produced by cells of the air sacs (alveoli) in your lungs. It aids breathing by reducing surface tension within the air sacs and allowing them to remain open. The alveoli are composed of a very thin layer of cells covered by a liquid coating.

   Why would the alveoli have a tendency to collapse in the absence of surfactant?

   ____

   ____

   What is the effect of surfactant on water molecules in the alveolus?

   ____

   ____

   Why does a premature infant often face infant respiratory distress syndrome (IRDS)?

   ____

   ____

   Based on what you've just learned, propose a way that IRDS might be treated.

   ____

   ____

2. Water exhibits a high heat capacity and a high heat of vaporization. Draw a sketch of the chemical structure of water and label it in a way that helps explain why water exhibits these characteristics.

   What is the physiological significance of these two properties of water?

   ____

   ____

   ____

   ____

The pH Scale

The term **pH** is a measure of the concentration of hydrogen ions in a solution. Water dissociates (breaks apart) into hydrogen ions (H^+) and hydroxide ions (OH^-). One liter of pure water contains about 0.0000001 (10^{-7}) moles of hydrogen ions and an equal amount of hydroxide ions. Thus, in 1 liter of water:

Concentration of hydrogen ions $= [H^+] = 1 \times 10^{-7}$ moles/liter

The pH of a solution is defined as the negative logarithm of the concentration of hydrogen ions in that solution:

$$pH = -\log [H^+]$$

Therefore, the pH of pure water $= -(-7) = 7$. A pH of 7 presents a neutral solution in which $[H^+] = [OH^-]$. As the acidity of a solution increases, its pH decreases; as the acidity of a solution decreases, its pH increases.

QUICK TIP You can remember that pH is a measure of the concentration of hydrogen ions by remembering the phrase "the power of Hydrogen."

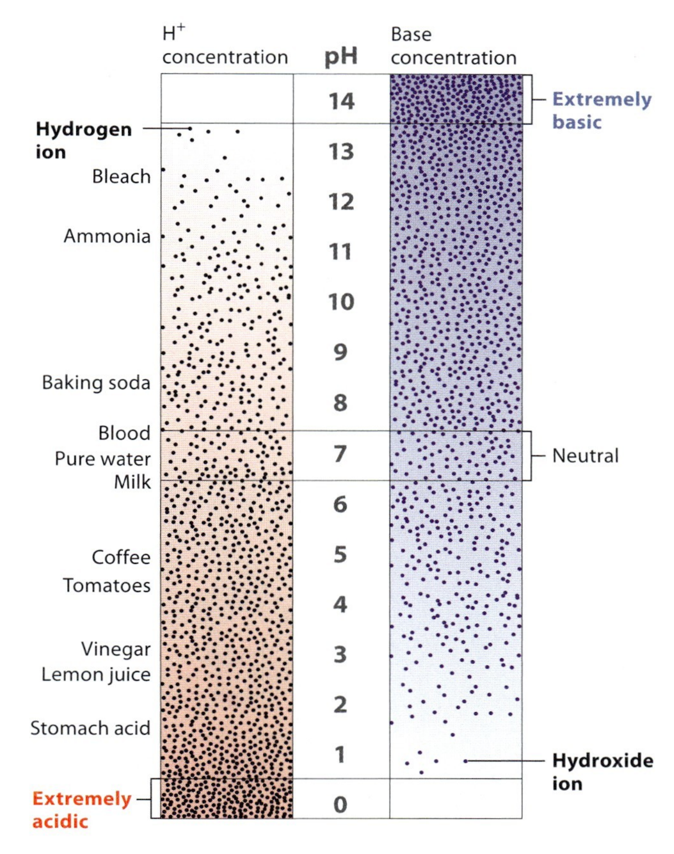

Figure 3-7 The pH scale and pH values for some representative substances.

The pH scale (**Figure 3-7**) was devised in 1909 by a Danish biochemist and brewer, Soren Sorenson, who wanted to ensure that his beer was acidic enough to prevent the growth of bacteria that would spoil the beer. The pH scale ranges from 0 to 14, with 7 representing a neutral solution. Solutions with pH values of less than 7 are acidic, whereas solutions with pH values greater than 7 are basic or alkaline. Note that the pH scale is a logarithmic scale; a solution with a pH of 6.0 has a $[H^+]$ that is 10 times higher than the $[H^+]$ of a solution with a pH of 7.0.

Acids, Bases, and Buffers

Acids, bases, and salts are inorganic compounds called electrolytes because they dissociate in water and are capable of conducting electrical currents. **Acids** release hydrogen ions (H^+) and anions (negatively charged ions). The released H^+ increase the hydrogen ion concentration (pH) of the solution. Because a hydrogen ion consists of only a hydrogen nucleus containing a single proton, an acid is also called a proton donor. A strong acid dissociates completely, whereas a weak acid dissociates very little.

Bases, by contrast, are known as proton acceptors. They take up hydrogen ions (H^+). **Bases** also dissociate when dissolved in water, but they release hydroxide ions (OH^-) and cations (positively charged ions). The released OH^- then accept the H^+ present in solution to produce water, thereby reducing the hydrogen ion concentration (pH) of the solution. **Figure 3-8** illustrates the behavior of acids and bases in water.

When acids and bases are mixed, they react to form a salt and water. **Salts** are ionic compounds containing cations other than H^+ and anions other than OH^-. When dissolved in water, salts dissociate into their component ions. For example, sodium chloride (table salt) dissociates immediately in water, releasing Na^+ and Cl^-. Most salts, like NaCl, are called neutral solutes because their disassociation does not affect the concentrations of either hydrogen ions or hydroxide ions.

Buffers are substances that stabilize the pH of a solution by acting either like an acid (by releasing H^+) when pH levels rise or like a base (by binding H^+) when pH levels fall. Thus, buffers resist changes in pH; in the body, buffers prevent marked changes in the pH of body fluids.

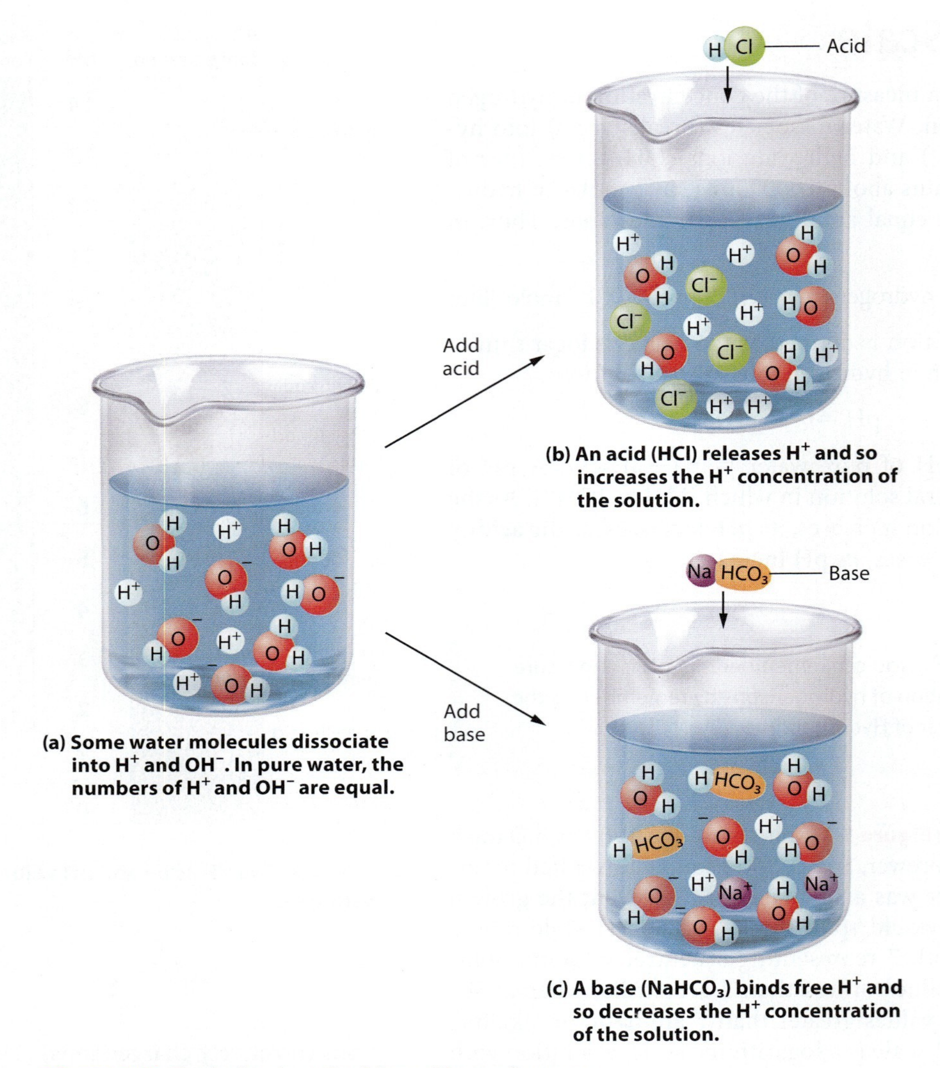

Acid
Add acid
Add base
Base
(a) Some water molecules dissociate into H+ and OH−. In pure water, the numbers of H+ and OH− are equal.
(b) An acid (HCl) releases H+ and so increases the H+ concentration of the solution.
(c) A base (NaHCO3) binds free H+ and so decreases the H+ concentration of the solution.

Figure 3-8 The behavior of acids and bases in water.

ACTIVITY 2

Determining pH and Interpreting the pH Scale

Learning Outcomes

1. Determine the pH of known and unknown solutions.
2. Describe and interpret the meaning of the pH scale.
3. Describe the effects of elevated CO_2 levels on blood pH.

Materials Needed

- ☐ pH indicator strip
- ☐ Various solutions to be tested (distilled water, tap water, black coffee, lemon juice, household bleach, egg white)
- ☐ A plastic dropper for each solution
- ☐ 0.2M NaOH
- ☐ Phenolphthalein
- ☐ Two plastic straws
- ☐ Safety goggles and gloves
- ☐ Two 200-ml beakers
- ☐ Watch or stop watch

Instructions

A. Determining pH Values

Figure 3-9 shows a typical pH indicator strip. You will be using a similar pH strip to determine the pH of several solutions.

1. Put on safety goggles and gloves.
2. Use an unused plastic dropper to place two drops of distilled water onto a pH strip.
3. Compare the color change that occurs with the colors on the side of the pH strip bottle and determine the pH of the solution.

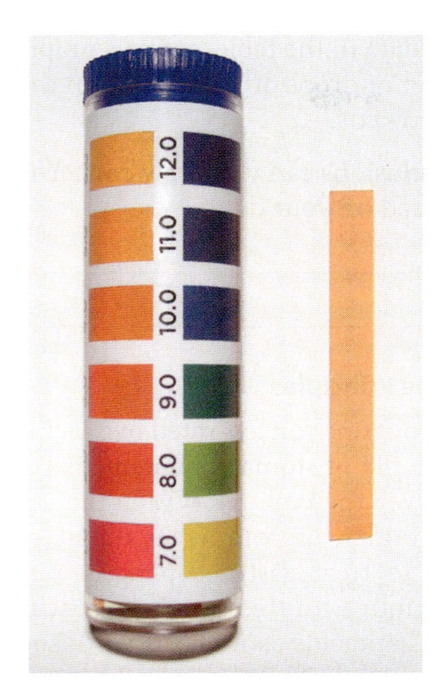

Figure 3-9 **Interpreting pH test strip results.**

4. Record your results in the following table.
5. Repeat steps 2–4 for each solution.

Solution	pH	Acidic, Neutral, or Basic
1. Distilled water		
2. Tap water		
3. Black coffee		
4. Lemon juice		
5. Household bleach		
6. Egg white		

B. Applying the Principles

When oxygen is present, the cells of your body break down the simple sugar glucose ($C_6H_{12}O_6$) to store energy in a molecule called ATP. Carbon dioxide (CO_2) and water are produced as byproducts of this reaction, which is summarized as follows:

$$C_6H_{12}O_6 + 6O_2 \rightarrow 6CO_2 + 6H_2O + ATP$$

As carbon dioxide enters the blood, it combines with water to form carbonic acid (H_2CO_3), which then dissociates into hydrogen ions (H^+) and bicarbonate ions (HCO_3^-):

$$CO_2 + H_2O \leftrightarrow H_2CO_3 \leftrightarrow H^+ + HCO_3^-$$

Therefore, an increase in blood carbon dioxide levels results in an increase in blood hydrogen ion levels.

What effect does an increase in hydrogen ion concentration have on blood pH?

____

____

1. Put 100 ml of distilled water into a 200-ml beaker.
2. Add 1.0 ml of 0.2M NaOH to the beaker and swirl the beaker gently to mix.
3. Add four drops of phenolphthalein to the solution and swirl the beaker gently to mix. Phenolphthalein is a pH indicator that is colorless in acidic solutions and pinkish-red in basic solutions.

   What color is the solution?

   ____

4. Noting the time at which you do so, use a plastic straw to blow bubbles of exhaled air into the solution.

   Record the time it takes for the solution to change color.

   ____

   What does this color change indicate?

   ____

   ____

5. Have the same student who performed step 4 run in place for 5 minutes.
6. While this student is running, prepare another beaker by following the directions in steps 1 and 2.
7. Immediately after the student has run in place for 5 minutes, note the time and have the student exhale through a straw into the solution.

   Record the time it takes for the solution to change color.

   ____

8. Answer the following questions based on your results:

   a. What is the effect of exercise on blood CO_2 levels?

   ____

   ____

   b. How were you able to draw this conclusion?

   ____

   ____

ACTIVITY 3

Observing the Role of Buffers

Learning Outcomes

1. Explain the function of a buffer.
2. Demonstrate the effect of a buffer on pH changes.
3. Explain the role of an antacid in controlling gastroesophageal reflux (heartburn).

Materials Needed

- ☐ Five 200-ml beakers
- ☐ Deionized water
- ☐ Sodium bicarbonate
- ☐ 25 pH indicator sticks
- ☐ 0.1M HCl
- ☐ Mortar and pestle
- ☐ Bromocresol purple
- ☐ Three types of antacid tablets
- ☐ Measuring spoons

Instructions

A. Demonstrating the Role of Buffers

1. Place 100 ml of deionized water into each of two 200-ml beakers labeled "#1" and "#2."
2. To beaker #2, add 0.7 g of sodium bicarbonate (NaH_2CO_3).
3. Use a pH indicator strip or a pH meter (follow instructions provided by your instructor) to determine the pH of the water in beaker #1 and of the solution in beaker #2. Record your results in the first row of the table below.
4. Add two drops of 0.1M HCl to each beaker and swirl gently to mix.
5. Use a pH indicator strip or a pH meter to determine the pH of the solution in each beaker. Record your results in the second row of the table.

Number of Drops of 0.1M HCl Added	pH of Water (Beaker #1)	pH of Water Plus Sodium Bicarbonate (Beaker #2)
0		
2		
4		
6		
8		
10		
12		
14		
16		
18		
20		

6. Continue adding HCl, two drops at a time, and determining pH until you have added a total of 20 drops of HCl to each beaker. Record your results in the table.
7. Using the data in the table, create a graph by plotting the number of drops of HCl added on the *x*-axis and the pH on the *y*-axis.
8. What conclusions can you draw concerning the role of buffers based on your data set?

____

____

B. Applying the Principles

The stomach contains concentrated hydrochloric acid (HCl). As a result, the pH of stomach contents is very low, approximately 2.0. Over-the-counter (OTC) antacids are medications used to buffer, or counteract the acidity of, stomach acid in order to treat gastroesophageal reflux, commonly known as heartburn. In this activity you will test the ability of several OTC antacids as well as that of sodium bicarbonate (baking soda) to buffer acid.

1. Using a mortar and pestle, crush one of the antacid tablets.
2. Add the crushed tablet to 100 ml of deionized water in a beaker and mix until dissolved.
3. Add six drops of Bromocresol purple to the beaker. The solution will now be purple. Bromocresol purple is a pH indicator that changes colors at specific hydrogen ion concentrations. It turns yellow in acidic environments and purple in basic (alkaline) environments.
4. Add 0.1M HCl drop by drop until the solution turns yellow. Record the number of drops needed to turn the solution yellow in the row marked "Antacid #1" in the table below.
5. Repeat steps 1–4 for each of the two remaining types of antacids and record the number of drops added in the appropriate row of the table.
6. Finally, repeat steps 2–4 using ¼ teaspoon of sodium bicarbonate and record the number of drops added in the appropriate row of the table.

Buffering Agent	Number of Drops Added
Antacid #1	
Antacid #2	
Antacid #3	
Sodium bicarbonate	

7. According to your data, which buffering agent worked best?

____

Biologically Important Macromolecules

Recall from our discussion of the structural complexity of life in Unit 2 that there are four types of biologically important macromolecules.

Here you will be learning about two of these types of macromolecules—carbohydrates and proteins—as you prepare to complete the next activity; you will learn more about the other two groups—lipids and nucleic acids—in later units.

Organic (carbon-containing) macromolecules are synthesized via dehydration synthesis reactions, and they are broken down via a reaction called hydrolysis. In dehydration synthesis (**Figure 3-10**), each time monomer (single-unit) building blocks are linked together to form a macromolecule, a molecule of water is produced. Conversely, during hydrolysis, the addition of water molecules breaks a macromolecule into its monomer building blocks.

In the next activity you will be studying the effects of protein catalysts called enzymes on the hydrolysis of the carbohydrate lactose (milk sugar) into its two building blocks: glucose and galactose. Before you start the activity, therefore, you need to understand the basic chemistry of both carbohydrates and proteins.

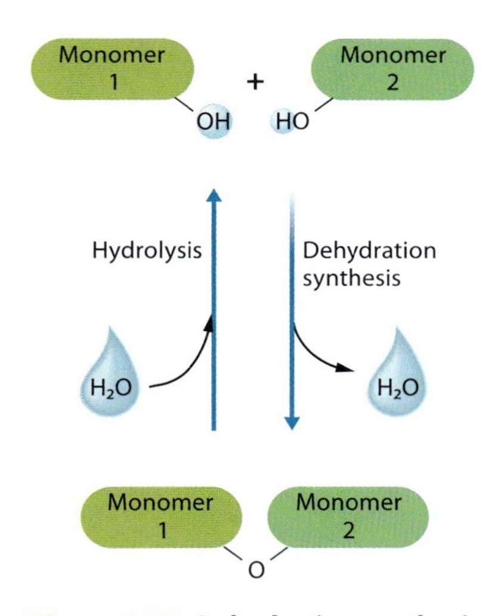

Figure 3-10 Dehydration synthesis and hydrolysis.

Carbohydrates

Carbohydrates contain carbon, hydrogen, and oxygen. In general, the hydrogen and oxygen atoms in carbohydrates occur in a 2:1 ratio, as they do in water—hence the term *carbohydrate,* meaning "hydrated carbon." Carbohydrates can be classified according to size. Monosaccharides ("one sugar") are the monomer building blocks of other carbohydrates. Disaccharides ("two sugars") consist of two monosaccharides joined together via dehydration synthesis; polysaccharides ("many sugars") are polymers containing many monosaccharide units linked together, also via dehydration synthesis. **Table 3-1** summarizes the carbohydrate molecules that are most important to the human body.

A major function of carbohydrates in the human body is as a source of fuel for cells. Only monosaccharides can pass through cell membranes, so polysaccharides and even disaccharides must be broken down into their monosaccharide building blocks before they can enter a cell. As described earlier, this decomposition process involves hydrolysis, the addition of water to break bonds and release simple sugars. Once the simple sugars enter a cell, they are broken down or stored as polysaccharide for future use, and the energy so released becomes available to the cell.

Proteins

The building blocks of proteins are 20 distinct amino acids. When two amino acids are joined by a dehydration synthesis reaction, a peptide bond is formed between the amino acids, producing a molecule called a dipeptide. When three amino acids are joined, a tripeptide is formed, and when 10 or more amino acids are connected, a polypeptide is formed. Although any chain of more than 50 amino acids is called a protein, most proteins are macromolecules containing more than 100 amino acids, and some contain over 10,000 amino acids.

Table 3-1 Biologically Important Carbohydrate Molecules

Monosaccharides (simple sugars; monomers of carbohydrates)	Disaccharides (two linked monosaccharides)	Polysaccharides (polymers or long chains of monomers)
Glucose	Sucrose (table sugar) = glucose + fructose	Starch: a storage carbohydrate in plants; an important component of our diet
Fructose	Lactose (milk sugar) = glucose + galactose	Cellulose: a plant carbohydrate that we are unable to digest; an important source of fiber in our diet
Galactose	Maltose (malt sugar) = glucose + glucose	Glycogen: a storage carbohydrate primarily found in liver cells and skeletal muscle fibers

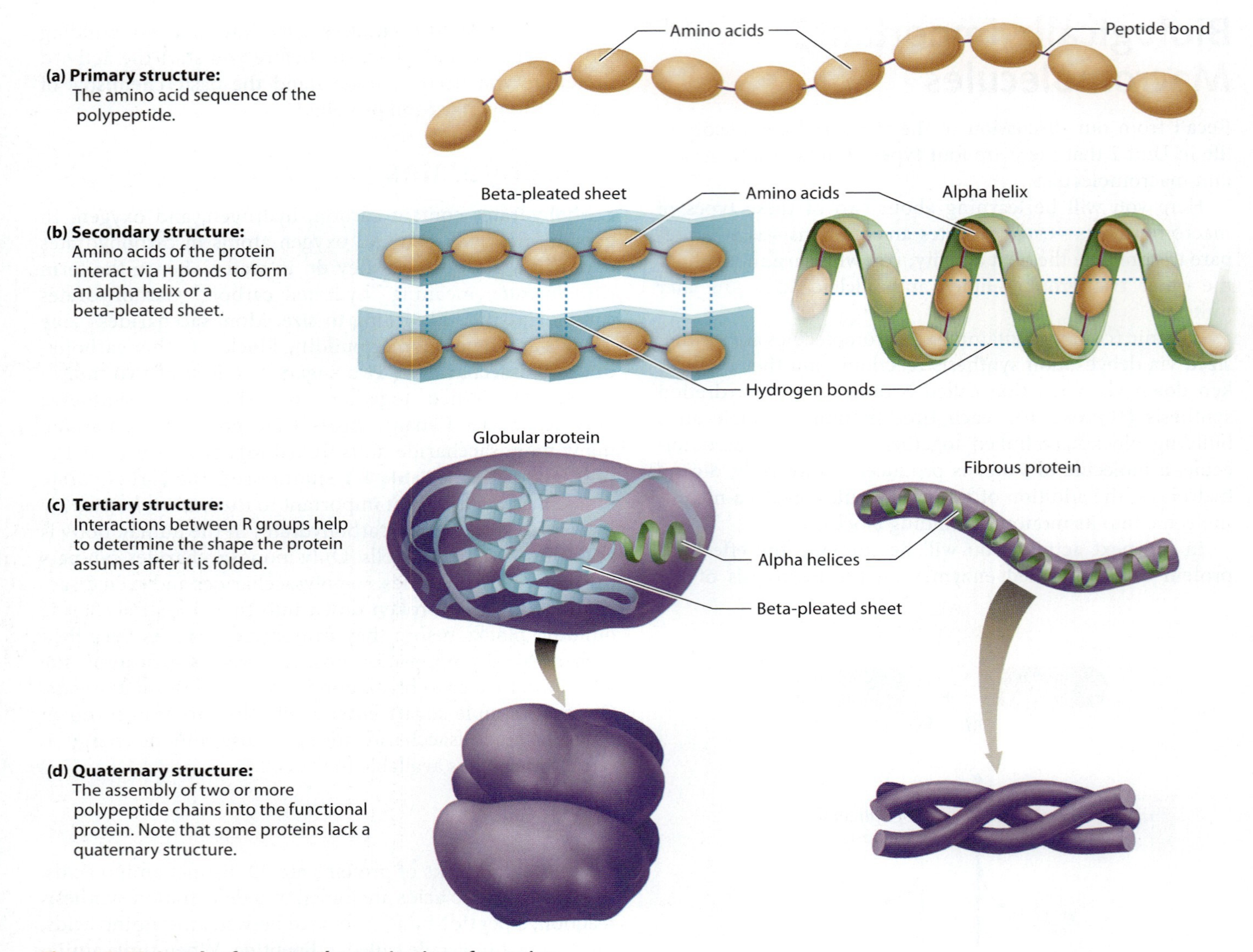

Figure 3-11 Levels of structural organization of proteins.

Proteins can be described as having four levels of structural organization (**Figure 3-11**). The sequence of amino acids in the polypeptide chain is called the primary structure of the protein. In the secondary structure, the polypeptide chain forms either a spring-like coil (an alpha helix), or it folds back and forth on itself (a beta-pleated sheet). Many proteins have a tertiary structure in which further bending and folding of the protein produces either a compact ball-like shape (a globular protein) or a long thin shape (a fibrous protein). A quaternary structure results from the interaction of two or more polypeptide chains. Not all proteins exhibit this fourth level of structural organization. Hemoglobin, the pigment in red blood cells that transports respiratory gases, is an example of a protein with a quaternary structure because it consists of four globular polypeptide chains.

Proteins function in a variety of ways in the body: as structural materials, membrane transport channels, chemical messengers (hormones), and enzymes, and even sometimes as energy sources. The functional activity of any protein depends on its three-dimensional shape. Environmental conditions such as a rise in temperature or a decrease in pH can cause proteins to unfold, altering their three-dimensional shapes. When this process—called **denaturation**—occurs, the protein is no longer able to perform its physiological role.

Enzymes: Biological Catalysts

Enzymes are proteins that serve as biological catalysts—that is, they increase the rate of chemical reactions without being consumed in the process. Each enzyme exhibits specificity, meaning that it performs a particular action on a particular substance, called a substrate. Each enzyme has one or more **active sites**, pockets on the enzyme surface to which substrates bind to form an **enzyme-substrate complex** (**Figure 3-12**). This interaction between the enzyme and substrates changes

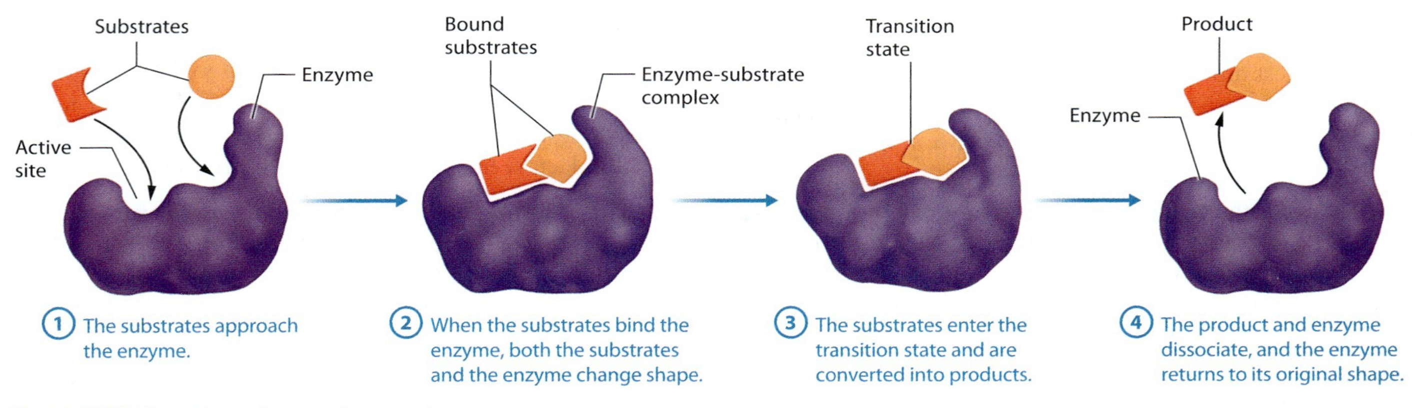

Figure 3-12 Enzyme-substrate interaction.

LabBOOST >>>

Protein Structure Imagine a protein made up of two slinky toys. Straightening the slinky toys all the way out would represent primary structure. Allowing the slinky toys to coil up again would represent secondary structure. Then, wadding up each coiled slinky toy into a ball would demonstrate tertiary structure. Finally, holding the two coiled slinky toys (each of which has been wadded into a ball) together would demonstrate quaternary structure.

the shapes of each, straining chemical bonds in the substrates and making a chemical reaction more likely to occur.

The fit between an enzyme and a substrate is often likened to the fit of a lock and key. Just as a key fits only one particular lock, a particular substrate "fits" into a particular enzyme's active site. The breakdown (hydrolysis) of the disaccharide lactose (milk sugar) into its two monosaccharide building blocks (glucose and galactose) by lactase is explored in Activity 4.

ACTIVITY 4

Analyzing Enzymatic Activity

Learning Outcomes

1. Distinguish among monosaccharides, disaccharides, and polysaccharides.
2. Distinguish between a dehydration synthesis reaction and a hydrolysis reaction.
3. Describe the structural levels of organization of a protein.
4. Explain the function of an enzyme.
5. Demonstrate the chemical breakdown of lactose into its building blocks.
6. Investigate the effects of temperature, pH, and enzyme-substrate specificity on lactase activity.

Materials Needed

- ☐ Twelve test tubes
- ☐ Lactase solution
- ☐ Distilled water
- ☐ 1% lactose solution
- ☐ 1% maltose solution
- ☐ 1% sucrose solution
- ☐ Ten glucose test strips
- ☐ Buffered solution (pH = 2)
- ☐ Buffered solution (pH = 7)
- ☐ Buffered solution (pH = 9)
- ☐ Ice water bath
- ☐ Hot plate for warm water bath
- ☐ Hot plate for boiling water bath
- ☐ Three 500-ml beakers

Instructions

A. The Effect of Temperature on Enzyme Activity

In this activity, you will test the effect of temperature on the ability of lactase (an enzyme) to break down lactose (a disaccharide) into glucose and galactose.

1. Label six test tubes 1–6. Place 1 ml of lactase solution in test tubes 1, 3, and 5. (Lactase is an enzyme that breaks lactose down into its constituent simple sugars, glucose and galactose.) Place 10 ml of 1% lactose solution in test tubes 2, 4, and 6.
2. Place test tubes 1 and 2 in an ice water bath.

3. Place test tubes 3 and 4 in a warm water bath (~45°C).
4. Place test tube 5 only in a boiling water bath. Do NOT put test tube 6 in the boiling water bath.
5. After test tubes 1 and 2 have been in the ice water bath for a minimum of 10 minutes, remove the test tubes and pour the lactase solution (test tube 1) into the lactose solution (test tube 2). Return test tube 2 to the ice water bath. After 15 minutes, use a glucose test strip to measure the amount of glucose in test tube 2, and then record the result in the table that follows.
6. After test tubes 3 and 4 have been in the warm water bath for a minimum of 10 minutes, remove the test tubes and pour the lactase solution (test tube 3) into the lactose solution (test tube 4). Return test tube 4 to the warm water bath. After 15 minutes, use a glucose test strip to measure the amount of glucose in test tube 4, and then record the result in the table that follows.
7. After test tube 5 has been boiling for a minimum of 10 minutes, remove it, allow it to cool for 10 minutes, and then add it to test tube 6. Wait 15 minutes and then use a glucose test strip to measure the amount of glucose in test tube 6, and then record the result in the table that follows.

Test Tube	Amount of Glucose
2	
4	
6	

8. Answer the following questions based on your results:

   a. What was the effect of cooling on lactase activity?

   ____

   b. What was the effect of boiling on lactase activity?

   ____

   c. Predict the optimum temperature for lactase activity.

   ____

B. The Effect of pH on Enzyme Activity

In this activity, you will test the effect of pH on the ability of lactase (an enzyme) to break down lactose (a disaccharide) into glucose and galactose.

1. Label three test tubes 1–3.
2. Place 1 ml of buffered solution (pH = 2.0) in test tube 1.
3. Place 1 ml of buffered solution (pH = 7.0) in test tube 2.
4. Place 1 ml of buffered solution (pH = 9.0) in test tube 3.
5. Add 10 ml of lactose solution to each test tube and swirl each test tube to mix.
6. Add 1 ml of lactase solution to each test tube and swirl each test tube to mix. Begin timing this demonstration after lactase has been added to all three test tubes.
7. After 15 minutes, use a glucose strip to measure the amount of glucose in each test tube, and then record the results in the following table.

Test Tube	Amount of Glucose
1	
2	
3	

8. What is the optimum pH for lactase activity?

   ____

C. The Effect of Enzyme-Substrate Specificity on Enzyme Activity

Now we will demonstrate enzyme-substrate specificity using three substrates—lactose, maltose, and sucrose—and the enzyme lactase.

1. Label three test tubes 1–3.
2. Place 10 ml of 1% lactose solution in test tube 1.
3. Place 10 ml of 1% maltose solution in test tube 2.
4. Place 10 ml of 1% sucrose solution in test tube 3.
5. Place 1 ml of lactase solution in each test tube and mix the contents of each test tube thoroughly. Begin timing this demonstration after lactase has been added to all three test tubes.
6. After 15 minutes, use a glucose strip to measure the amount of glucose in each test tube, and then record the results in the following table.

Test Tube	Amount of Glucose
1	
2	
3	

7. Based on your data, what conclusion can you draw concerning enzyme-substrate specificity?

   ____

   ____

Post-lab quizzes are also assignable in MasteringA&P®

POST-LAB ASSIGNMENTS

Name: ______________________ **Date:** ______________ **Lab Section:** ________

PART I. Check Your Understanding

Activity 1: Exploring the Chemical Properties of Water

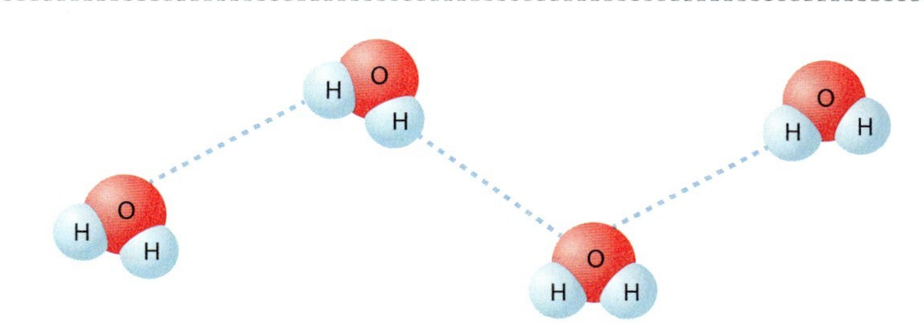

1. On the sketch at right of the chemical structure of water, label the polar covalent bond, hydrogen bond, electronegative charge, and electropositive charge.

2. A weak attraction between a slightly positively charged atom in one polar molecule and a slightly negatively charged atom in another polar molecule is called a ____________________ bond.

3. Cohesion of water molecules at the air/water interface of an alveolus is called ________________ ____________________.

4. In a solution, the medium in which atoms/ions/molecules are dispersed is the ____________________, and the dispersed substances are called the ____________________.

5. Which fluid evaporated more quickly, water or ethanol? Explain why.
   ____
   ____

Activity 2: Determining pH and Interpreting the pH Scale

1. How did you determine the pH of various solutions? ______________________________

2. Indicate the pH that you determined for each of the following solutions and then state whether each solution is acidic, neutral, or basic based on your results:
   a. Black coffee pH = _____ ______________________________
   b. Egg white pH = _____ ______________________________
   c. Tap water pH = _____ ______________________________

3. Which of the following solutions contains the highest concentration of hydrogen ions?
   a. A solution with a pH of 4.0
   b. A solution with a pH of 7.0
   c. A solution with a pH of 9.0

4. Acids:
   a. are proton acceptors.
   b. release hydroxyl ions when dissolved in water.
   c. cause the pH of a solution to rise.
   d. are electrolytes.
   e. None of these statements is correct.

5. An increase in blood carbon dioxide levels causes an increase in blood hydrogen ion levels. Explain.

____

____

6. What effect does an increase in blood hydrogen ion levels have on the pH of the blood?

____

____

Activity 3: Observing the Role of Buffers

1. Define buffer. ____

____

2. What was the role of each of the following substances in the demonstration of the role of buffers?

   a. Sodium bicarbonate ______________________________

   b. HCl ______________________________

   c. Bromocresol purple ______________________________

3. How did the effectiveness of sodium bicarbonate compare to the effectiveness of the three antacids tested in Part B of Activity 3?

____

____

Activity 4: Analyzing Enzymatic Activity

1. Identify each of the molecules as a monosaccharide (M), disaccharide (D), or polysaccharide (P):

   a. Sucrose __________ e. Lactose __________

   b. Glucose __________ f. Glycogen __________

   c. Starch __________ g. Galactose __________

   d. Fructose __________

2. Explain the difference between a dehydration synthesis reaction and a hydrolysis reaction.

____

____

3. Match each of the following levels of structural organization of proteins with its description:

   a. _____ primary 1. compact, globular structure

   b. _____ secondary 2. sequence of amino acids

   c. _____ tertiary 3. alpha helix or beta-pleated

   d. _____ quaternary 4. interaction of two or more polypeptide chains

4. Each enzyme has a pocket on its surface called a(an) ____________________ to which a substrate binds.

5. Distinguish between lactose and lactase and describe what happens to lactose in the presence of lactase.

____

____

6. Environmental conditions such as heat and low pH can cause proteins to lose their three-dimensional shape. This process is called ________________ and results in ____.

PART II. Putting It All Together

A. Review Questions

Answer the following questions using your lecture notes, your textbook, and your lab notes:

1. Why is water such an effective solvent?

____

____

2. Calculate the pH for a solution with $[H^+] = 1 \times 10^{-4}$. pH = ____________

3. A solution with a pH of 2.0 is how many times more acidic than a solution with a pH of 4.0?

____

4. Use a chemical reaction to explain why an increase in blood carbon dioxide levels causes the pH of blood to decrease.

____

____

____

5. Summarize the data presented in the graph that you created for part A of Activity 3.

____

____

____

6. Write a chemical equation that describes the buffering of hydrochloric acid by sodium bicarbonate, and explain how the reaction works.

____

____

____

7. A test tube contains the following: lactase, water, buffer (pH = 2.0), and sucrose. Will digestion of the sucrose occur? Why or why not?

____

____

____

B. Concept Mapping

1. Fill in the blanks to complete this concept map outlining the structure and function of enzymes.

amino	acid	denaturation	enzymes	macromolecules	temperature

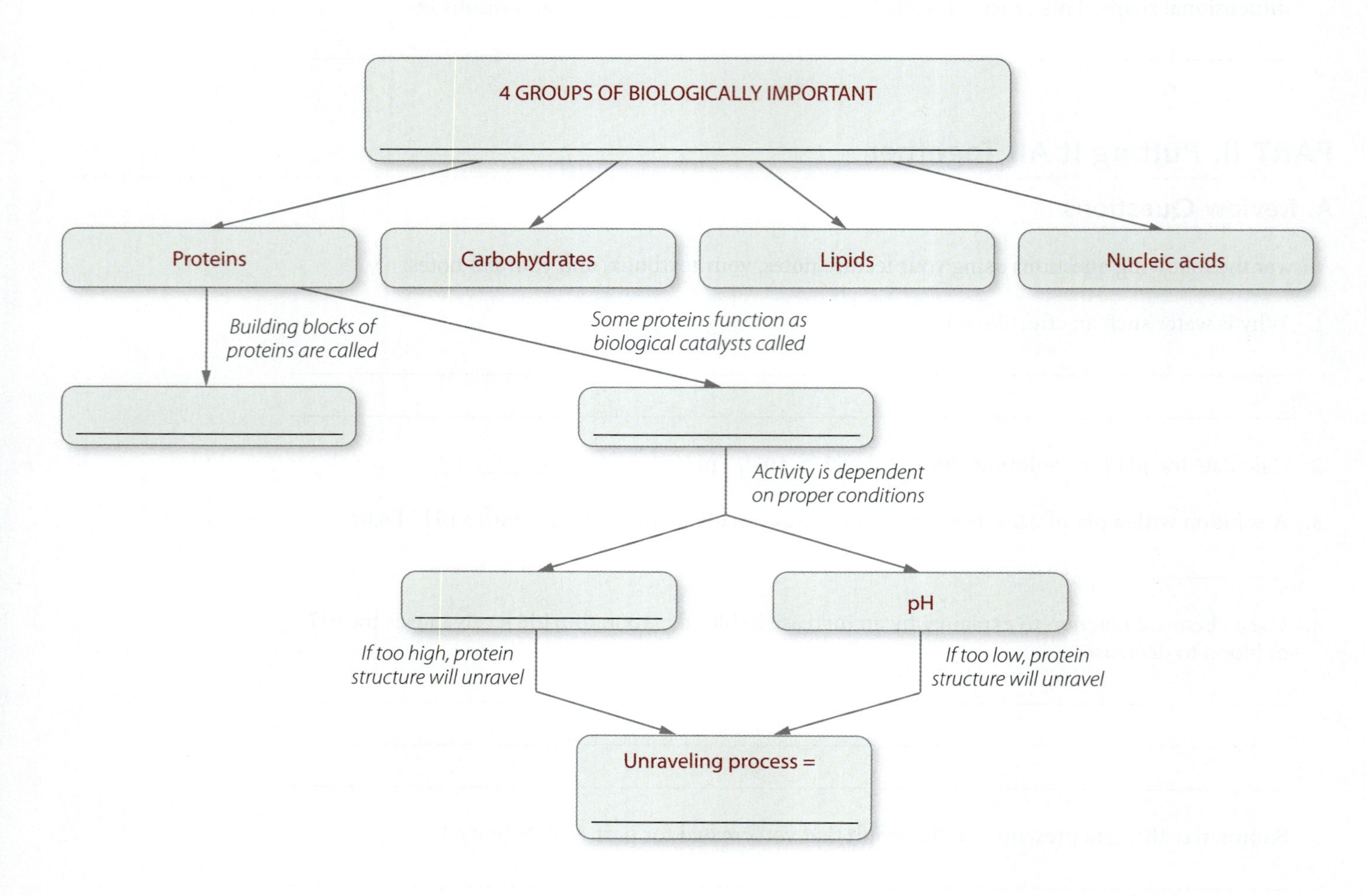

2. Construct a unit concept map to show the relationships among the following set of terms. Include all of the terms in your diagram. Your instructor may choose to assign additional terms.

acid	amino acid	atom	base	covalent bond
denaturation	enzymes	hydrogen bond	ion	ionic bond
macromolecule	molecule	monosaccharide	pH	temperature

4

The Microscope

UNIT OUTLINE

The compound light microscope is an essential tool in the study of human anatomy and physiology. The human eye is not able to perceive objects smaller than 0.1 mm in diameter, so a microscope is needed to study small structures such as a typical cell.

The microscope that you will be using in this unit is called a compound light microscope because it uses two different types of lens: an ocular lens and an objective lens. It is called a light microscope because it requires a light source in order to view an object. Most good laboratory microscopes today are parfocal, meaning that once you focus the specimen using one objective, it will remain "almost" in focus when viewed through another objective. In other words, when you switch to another objective, only a slight adjustment using the fine adjustment knob is required to bring the specimen into sharp focus.

THINK ABOUT IT *The microscope is a very important tool used in the study of anatomy and physiology to examine cells and tissues that are too small to be seen with the naked eye. Microscopes are also invaluable diagnostic tools in the clinical setting. Describe two specific ways in which microscopes are used in the diagnosis and treatment of disease.*

____

____

____

PRE-LAB ASSIGNMENTS

Pre-lab quizzes are also assignable in MasteringA&P®

To maximize learning, BEFORE your lab period carefully read this entire lab unit and complete these pre-lab assignments using your textbook, lecture notes, and prior knowledge.

PRE-LAB Activity 1: Identifying the Parts of the Microscope

1. Use the list of terms provided to label the accompanying photo of a microscope; check off each term as you label it.

- ☐ arm
- ☐ base
- ☐ coarse adjustment knob
- ☐ condenser
- ☐ fine adjustment knob
- ☐ head
- ☐ iris diaphragm lever
- ☐ objective lens
- ☐ ocular lens
- ☐ revolving nosepiece
- ☐ stage
- ☐ substage light

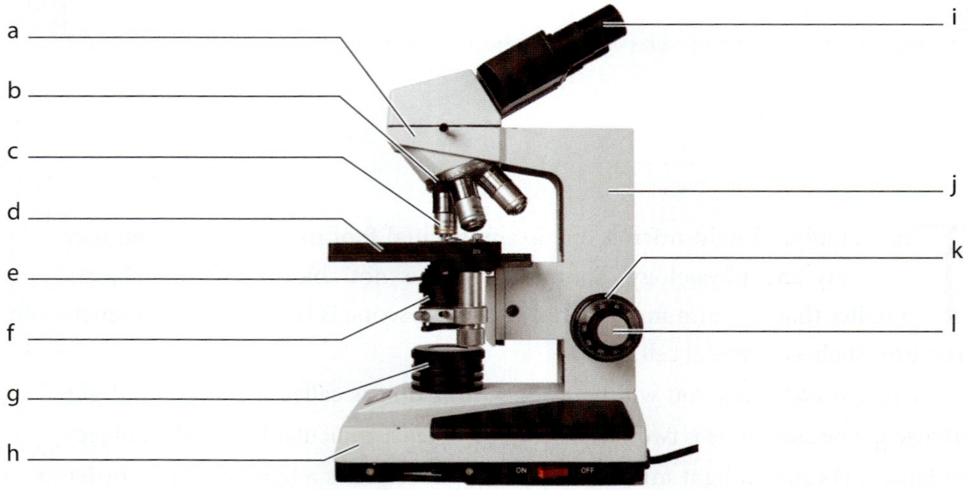

a ______
b ______
c ______
d ______
e ______
f ______
g ______
h ______
i ______
j ______
k ______
l ______

ON OFF

2. Match each of the following microscope parts with its correct description.

_____ a. broad bottom support of microscope
_____ b. controls the amount of light passing through the condenser
_____ c. supports the objective and ocular lenses
_____ d. is used for precise focusing
_____ e. lenses of various powers of magnification
_____ f. serves as a handle for carrying the microscope
_____ g. circular area on stage through which light passes
_____ h. moves the slide on the stage
_____ i. lenses located within the eyepieces
_____ j. moves mechanical stage in large increments

1. arm
2. objective lenses
3. aperture
4. coarse adjustment knob
5. iris diaphragm
6. base
7. ocular lenses
8. head
9. fine adjustment knob
10. mechanical stage controls

PRE-LAB Activity 2: Using the Microscope to View Objects

1. The objective lenses of the compound light microscope are attached to the:
   a. base.
   b. stage.
   c. body tube.
   d. rotating nosepiece.

2. Which objective lens provides the largest total magnification?
   a. Scanning lens
   b. Oil-immersion lens
   c. Low-power lens
   d. High-power lens

3. Which microscope structure concentrates light onto the specimen?
   a. Condenser
   b. Ocular lens
   c. Iris diaphragm lever
   d. Coarse adjustment knob

4. The ________________ is the distance between the specimen and the bottom of the objective lens

PRE-LAB Activity 3: Determining Field Diameter and Estimating the Size of Objects

1. A micrometer equals 1/1000 of a millimeter and is symbolized by μm. Perform the following conversions:
   a. 5 mm = ________________ μm
   b. 0.1 mm = ________________ μm
   c. 1500 μm = ________________ mm
   d. 200 μm = ________________ mm

2. If the field diameter for the high-power objective is 0.5 mm and an object extends halfway across the field of view, estimate the size of the object in micrometers (μm).
   a. 25 μm
   b. 50 μm
   c. 250 μm
   d. 500 μm
   e. 2500 μm

PRE-LAB Activity 4: Perceiving Depth of Field

1. In this procedure, what objects are viewed to investigate depth of field? ________________

2. Depth of field, also known as ________________, is the thickness of a specimen that is in ________________.

PRE-LAB Activity 5: Caring for the Microscope

1. Before putting a microscope away, remove any slide on the stage, rotate the nosepiece to the ________________ objective, and move the microscope stage to its ________________ position.

2. When transporting a microscope you should always use two hands: one hand should support the ________________ and the other hand should hold onto the ________________.

Microscope Basics

It is critical that you learn basic guidelines concerning the proper use of a microscope, because microscopes are very expensive laboratory instruments. Not only will following proper procedures help keep your microscope in proper working order, it will also keep your instructor happy. But first you need to learn the parts of a microscope, and a little about how those parts function.

The parts of a compound light microscope are shown in **Figure 4-1**. The major parts of a microscope—the parts that form its framework—are the base, the arm, and the head (also called the body tube). All other parts attach to these three major parts of the microscope. **Table 4-1** provides an overview of the parts of the microscope and brief descriptions of their functions.

Now read over the basic guidelines for the use of the microscope, which are listed in **Table 4-2**.

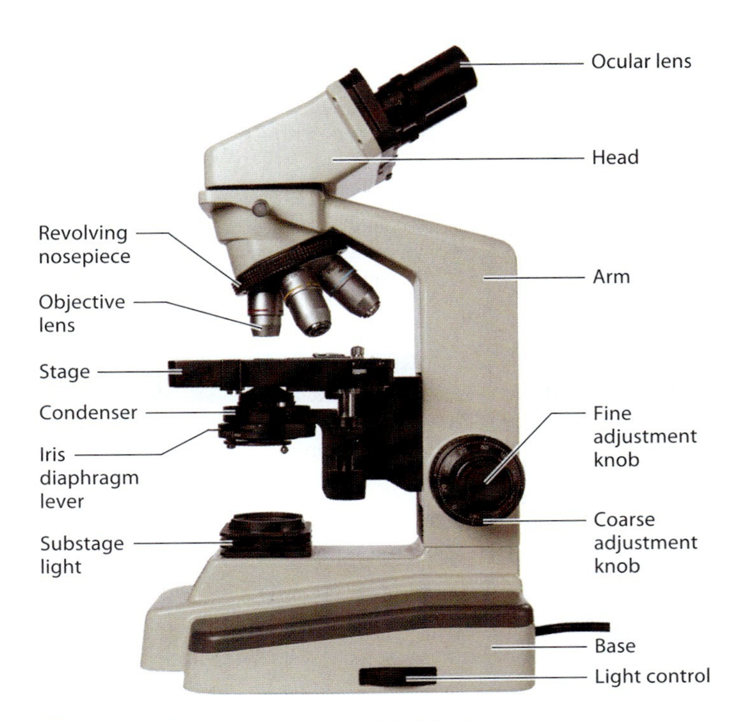

Figure 4-1 Parts of a compound light microscope.

Table 4-1 Parts of the Compound Light Microscope

Microscope Part	Description and Function
Head	The upper part of the microscope that supports the ocular lens(es) and the various objective lenses.
Base	The broad, flat, lower part of the microscope that supports the rest of the instrument.
Arm	The vertical part of the microscope that connects the head to the base.
Ocular lenses	Lens(es) located within the eyepieces. Monocular microscopes have one ocular lens, whereas binocular microscopes have two ocular lenses. Ocular lenses typically magnify an object 10 times.
Objective lenses	Magnifying lenses mounted on a rotating nosepiece. Most microscopes have four objective lenses: scanning (4×), low-power (10×), high-power (40×), and oil-immersion (100×) lenses.
Rotating nosepiece	Connects the objective lenses to the head and allows different objective lenses to be moved into place.
Mechanical stage	Flat, horizontal shelf onto which the slide is placed and typically secured with a spring clamp. Two control knobs can be used to move the stage to position the slide.
Condenser	Small lens located under the stage that concentrates light onto the specimen. A condenser adjustment knob is used to raise and lower the condenser. The condenser should usually be in its uppermost position, just below the aperture (the hole in the stage through which light travels).
Iris diaphragm lever	Located beneath the condenser. Regulates the amount of light that passes through the condenser.
Focus knobs	Located on the arm of the microscope just above the base. The larger coarse adjustment knob moves the stage up and down in large increments and is used to find the specimen and for initial focusing; the smaller fine adjustment knob is used for fine focusing after coarse focusing has been completed.
Substage light	Located in the base. Provides the light that passes through the condenser, the specimen, the lenses, and finally into your eyes. A light control knob located on the base or the arm controls the brightness of the light.

Table 4-2 Basic Guidelines for Microscope Use

1. Always transport the microscope in the upright position with two hands—one hand supporting the base and one hand holding the arm.
2. Place the microscope on the lab table gently to prevent jarring of the instrument.
3. Use only special lens paper and lens-cleaning solution to clean the microscope's lenses.
4. Always begin the focusing process with the lowest power objective, and then move to the higher-power objectives as necessary.
5. Use the coarse adjustment knob only with the lowest-power objective.
6. Never remove any microscope parts.
7. Inform your instructor if you encounter any problems with your microscope.

ACTIVITY 1

Identifying the Parts of the Microscope

Learning Outcomes

1. Demonstrate the proper procedure for transporting a microscope.
2. Identify the parts of a compound light microscope.
3. Describe the function of each microscope part.

Materials Needed

☐ Compound light microscope

Instructions

1. Obtain a compound light microscope. **Figure 4-2** shows the proper technique for carrying a microscope. Always carry a microscope with both hands—one hand holding the arm and one hand supporting the base.
2. Carefully place the microscope on the table.
3. Identify the head, the arm, and the base of the microscope.
4. Now, identify the ocular lens(es).

   Is your microscope a monocular microscope or a binocular microscope? ______________________

   Ocular lenses typically magnify objects __________ *times.*

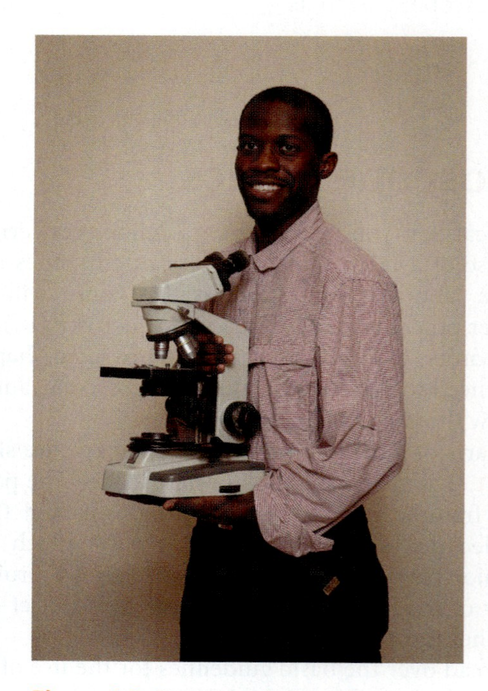

Figure 4-2 **Proper technique for transporting a microscope.**

5. Identify the rotating nosepiece that connects the objective lenses to the head of the microscope. Rotate the nosepiece to demonstrate how the different objective lenses can be locked into place. Each objective lens is labeled with its magnifying power.

   Note the magnifying power of each objective lens and record it below:

   Scanning objective: __________

   Low-power objective: __________

   High-power objective: __________

   Oil-immersion objective: __________

   Total magnification is calculated by multiplying the power of the ocular lens by the power of the objective lens. For example, the total magnification of the scanning objective is 10× (the power of the ocular lens) × 4× (the power of the scanning objective) = 40×.

6. Identify the stage, the platform on which the slide sits. Note the hole in the stage through which light passes.

   What is the name of this opening? ________________

   Does your microscope have a mechanical stage, as shown in Figure 4-2, or does it have spring clips to hold the slide in place?

   ____

7. If your microscope has a mechanical stage, identify the two adjustable mechanical stage knobs that control the movement of the slide.

8. Identify the larger coarse adjustment knob and the smaller fine adjustment knob located on the arm just above the base of the microscope. Rotate the coarse adjustment knob to move the stage up and down. This knob allows for gross focusing of the specimen, whereas the fine adjustment knob allows for fine focusing.

9. Identify the condenser, which concentrates light on the specimen. The iris diaphragm lever regulates the amount of light passing through the condenser. The passage of more light through the condenser decreases contrast between structures; the passage of less light through the condenser increases contrast between structures.

10. Identify the substage light located in the base. The light control knob, often a rheostat dial, controls the brightness of the light.

ACTIVITY 2

Using the Microscope to View Objects

Learning Outcomes

1. Demonstrate the proper use of the microscope to view specimens.
2. Calculate total magnification for each objective lens.
3. Determine the working distance for each objective lens.

Materials Needed

- ☐ Compound light microscope
- ☐ Letter "e" microscope slide
- ☐ Millimeter ruler
- ☐ Immersion oil and dropper
- ☐ Lens paper
- ☐ Lens-cleaning fluid

Instructions

1. Obtain a letter "e" slide. Examine the orientation of the "e" before you place the slide on the stage. Sketch the appearance of the letter "e" in the following rectangle.

   Is the "e" right-side up or upside down? ______________

   Is the "e" oriented left to right or right to left? __________

2. Plug in your microscope and turn it on.

3. Rotate the scanning objective into position, and place the slide on the stage. Record the magnification of the scanning objective in the microscope summary table. Calculate the total magnification when using the scanning objective and record it in the summary table on the next page.

4. Using the stage adjustment knobs, move the slide around until the letter "e" comes into the field of view.

5. Now, bring the letter "e" into focus using the coarse adjustment knob. Once it is mostly in focus, use the fine adjustment knob to bring the letter "e" into sharp focus. Sketch the appearance of the letter "e" in the circle representing the field of view in the summary table.

   Is the "e" right-side up or upside down? ______________

   Is the "e" oriented left to right or right to left? __________

6. Using the mechanical stage control knobs, move the slide away from you.

   In which direction does the image of the letter "e" move as you look through the ocular lens(es)? ______________

7. Now, move the slide to the right.

   In which direction does the image of the letter "e" move as you look through the ocular lens(es)? ______________

Microscope Summary Table

Objective Lens	Sketch of Observed Letter "e"	Magnification of Objective Lens	Total Magnification	Working Distance (mm)	Diameter of Field (mm)
Scanning					
Low-power					
High-power					
Oil-immersion					

8. The **working distance** is the distance between the specimen and the bottom of the objective lens (**Figure 4-3**). Using a millimeter ruler, measure the working distance for the scanning objective, and record the distance in the summary table.

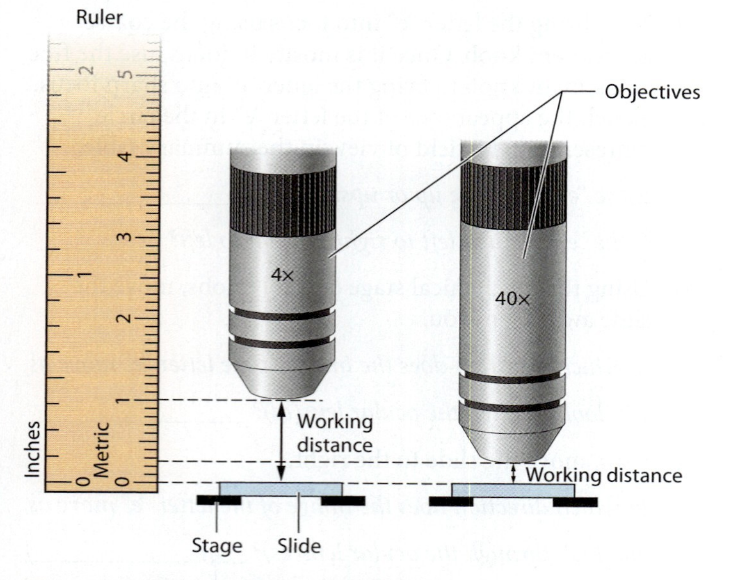

Figure 4-3 Measuring working distance.

9. Rotate the low-power objective into place. Using the fine adjustment knob only, bring the letter "e" into sharp focus.

   Because you only need to use the fine adjustment knob when you switch to a different objective lens, the microscope is said to be ________________.

   What is the total magnification when using the low-power objective? ________________

10. Record the low-power objective's magnification and total magnification in the summary table, and sketch the appearance of the letter "e" in the table as well.

11. Using a millimeter ruler, measure the working distance for the low-power objective and record the distance in the summary table.

12. Adjust the amount of light passing through the specimen using the iris diaphragm lever.

    Do you need to increase or decrease the amount of light when using a higher magnification? ________________

13. Rotate the high-power objective into place. Using only the fine adjustment knob, bring the letter "e" into sharp focus. Sketch the image of the letter "e" in the summary table, and record the high-power objective magnification and total magnification in the table as well.
14. Using a millimeter ruler, measure the working distance for the high-power objective and record the distance in the summary table.
15. Rotate the high-power objective lens out of position, but do not rotate another objective lens into position. With the area above the slide unobstructed, you can now place a single drop of immersion oil on the letter "e." Then you can rotate the oil-immersion lens into position. Make sure that the condenser is as close to the stage as possible and that the diaphragm is completely open to allow the maximum amount of light to pass through the slide. Bring the letter "e" into sharp focus. Sketch the letter "e" in the summary table.
16. Using a millimeter ruler, measure the working distance for the oil-immersion objective and record the distance in the summary table. Record this objective's magnification and total magnification in the table as well.
17. Rotate the oil-immersion objective out of position and remove the slide. Carefully clean both the oil-immersion objective lens and the microscope slide with lens paper and lens-cleaning solution.

ACTIVITY 3

Determining Field Diameter and Estimating Size of Objects

Learning Outcomes

1. Determine the field diameter for each objective lens.
2. Estimate the size of objects using each objective lens.

Materials Needed

- ☐ Compound light microscope
- ☐ Transparent millimeter ruler
- ☐ Small section of millimeter graph paper glued to a microscope slide
- ☐ Microscopic slide of compact bone

Instructions

A. Determining Field Diameter

In this activity you will learn to estimate the size of objects viewed through a microscope. To accomplish this task, you need to determine the diameter of the microscopic field for each of the objective lenses. First you will use a transparent millimeter ruler or a small section of millimeter graph paper glued to a slide to measure the field diameter for the scanning objective.

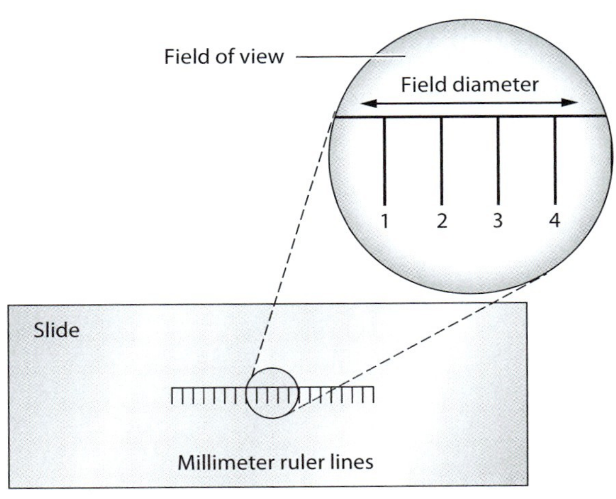

(a) Using a transparent ruler

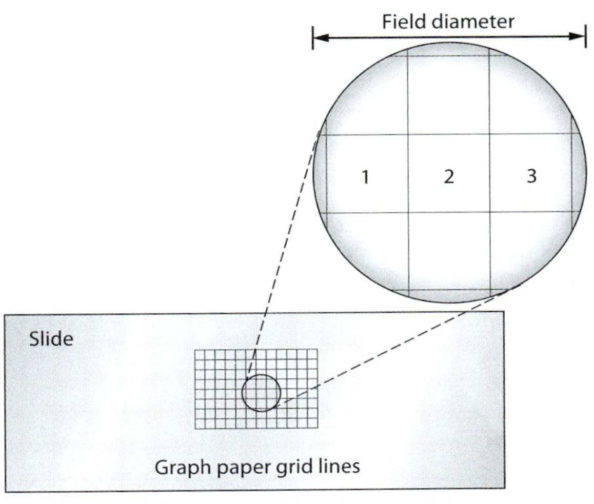

(b) Using a graph paper grid

Figure 4-4 **Two ways to measure field diameter.**

1. Line up a vertical line on the ruler or the graph paper with the left edge of the field of view (as shown in Figure 4-4). Count the number of millimeter squares or lines across the field. Record the field diameter (in mm) for the scanning objective in the summary table in Activity 2.
2. Once you know the field diameter for the scanning objective (lens A), you can use the following formula to calculate the field diameter for the other objective lenses:

$$\text{Field diameter (lens B)} = \frac{\text{field diameter (lens A)} \times \text{total magnification (lens A)}}{\text{total magnification (lens B)}}$$

Use this formula to calculate the field diameter for the low-power, high-power, and oil-immersion objectives, and record your data in the summary table in Activity 2.

Table 4-3 Metric Conversion Table

Metric Unit	Abbreviation	Description	Fraction	Decimal	Scientific Notation
Meter	m	Base unit of length	1/1 m	1.0 m	1.0×10^{1} m
Decimeter	dm	One-tenth of a meter	1/10 m	0.1 m	1.0×10^{-1} m
Centimeter	cm	One-hundredth of a meter	1/100 m	0.01 m	1.0×10^{-2} m
Millimeter	mm	One-thousandth of a meter	1/1000 m	0.001 m	1.0×10^{-3} m
Micrometer	μm	One-millionth of a meter	1/1,000,000 m	0.000001 m	1.0×10^{-6} m
Nanometer	nm	One-billionth of a meter	1/1,000,000,000 m	0.000000001 m	1.0×10^{-9} m

B. Estimating the Size of Objects

Microscopic specimens are typically measured in millimeters (mm) or micrometers (μm), both of which are metric units. Use **Table 4-3** to help you make conversions among metric units as you complete this activity.

1. Based on your results, what is the field diameter for the scanning objective? _________ mm = _________ μm
2. Assume that you viewed and made a sketch of the following object using the scanning objective. Estimate the length of the object in μm.

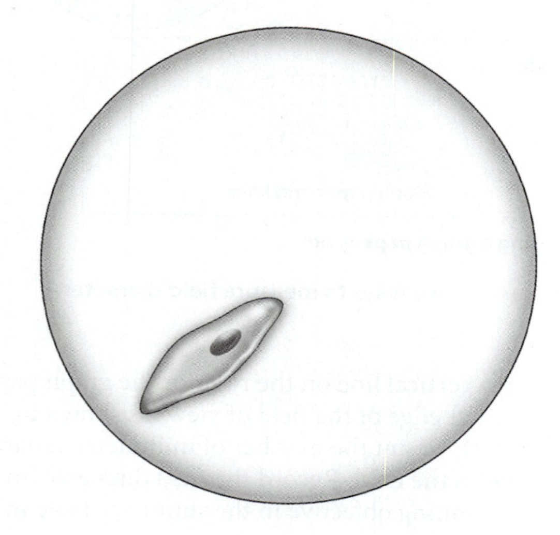

3. Next, obtain a microscopic slide of compact bone. The microscopic unit of compact bone is the osteon. An osteon, shown circled in red in **Figure 4-5**, has a unique histological appearance. It resembles the cross-section of a tree trunk. Place the slide of compact bone on the microscope stage and focus on a single osteon using the scanning objective. Now, switch to the low-power objective and use the fine adjustment knob to bring the slide into sharp focus.

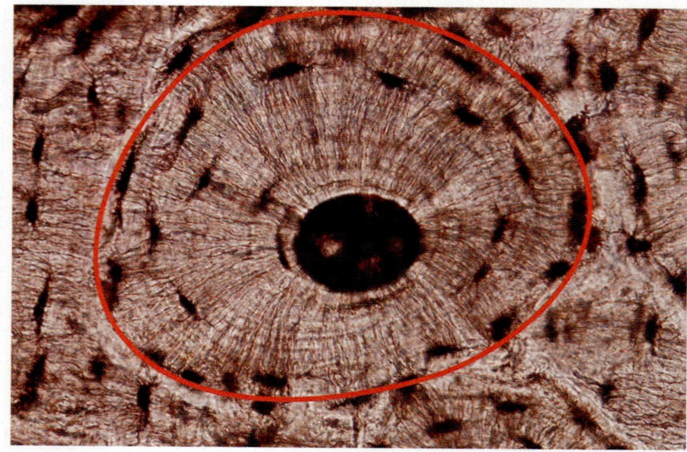

LM (320×)

Figure 4-5 **A histological view of an osteon.**

4. Sketch the osteon:

Total magnification: _____ ×

What is the field diameter (in mm) using the low-power objective? _________________ *mm*

Approximately what percentage of the field diameter is taken up by the osteon? _________________

Estimate the size of the osteon: _________________ *mm*

_________________ *μm*

LabBOOST >>>

Metric Conversions If you are struggling with metric conversions, use this chart to help you convert from one metric unit to another. If you are moving down the chart (from larger units to smaller units), then you will multiply. If you are moving up the chart (from smaller units to larger units), then you will divide. The numbers in parentheses are the factors by which you will multiply or divide. For example, to convert 3 meters to millimeters, you would multiply: 3 m × 10 × 10 × 10 = 3000 mm. To convert 5000 millimeters to centimeters you would divide: 5000 mm ÷ 10 = 500 cm.

multiply ↓		divide ↑
	meters (m)	
	(10)	
	decimeters (dm)	
	(10)	
	centimeters (cm)	
	(10)	
	millimeters (mm)	
	(1000)	
	micrometers (µm)	

ACTIVITY 4
Perceiving Depth of Field

Learning Outcomes

1. Describe the concept of depth of field.
2. Explain the relationship between magnification and depth of field.

Materials Needed

- ☐ Compound light microscope
- ☐ Three colored threads microscope slide

Instructions

When students first begin using a microscope, they usually understand that they need to move the slide side to side and to and fro until they find the specimen. But they often do not realize each specimen has a thickness, and that they can focus on the specimen at many levels of that thickness. **Depth of field**, also known as depth of focus, is the thickness of a specimen that is in sharp focus. To learn to focus up and down in order to observe all levels of a specimen, you will observe a slide with three crossed, colored threads that are located at different levels of the field.

1. Rotate the scanning objective into place, and place the slide with the three colored threads on the stage. Position the slide so that the point of intersection of the three threads is near the center of the stage aperture.
2. Move the slide using the mechanical stage controls until the threads come into the field of view. Position the slide so that the point of intersection of the three threads is in the center of the field.
3. Rotate the low-power objective into place and then focus down until all three threads are out of focus. Slowly focus (moving the stage upward) and note which thread comes into focus first. The first clearly focused thread is the top thread. Note that when you are sharply focused on one colored thread, the other two colored threads will be blurry.
4. Continue to focus upward, noting when the second and third threads come into focus. Record your observations:

   Color of the top thread: ____________________

   Color of the middle thread: ____________________

   Color of the bottom thread: ____________________
5. Predict what happens to the depth of field as magnification increases. ______________________________

ACTIVITY 5
Caring for the Microscope

Learning Outcomes

1. Review the general guidelines for use of a microscope.
2. Demonstrate proper technique for cleaning and storing a microscope.

Materials Needed

- ☐ Compound light microscope

Instructions

Microscopes are very expensive instruments! It is of utmost importance that you take proper care of your microscope. In the final activity of this unit, you will review the important guidelines for using and cleaning a microscope and then learn how to store the microscope safely.

A. Review of Microscope Basics

1. Always transport the microscope in the ____________________ position with two hands—one hand supporting the ____________________ and one hand holding the ____________________.
2. Place the microscope on the lab table gently to prevent ____________________ of the instrument.
3. Use only special ______________________________ and lens-cleaning solution to clean the microscope's lenses.

4. Always begin the focusing process with the ___________________-power objective, and then move to the higher-power objectives as necessary.
5. Use the ___________________ adjustment knob only with the lowest-power objective.
6. Never ___________________ any microscope parts.
7. Inform the ___________________ if you encounter any problems with your microscope.

B. Clean-Up and Storage

1. Turn off the power to the microscope, unplug it, and loosely wrap the cord around the base of the microscope, as shown by your instructor. Allowing the light bulb to cool before moving the microscope will extend the life of the bulb.
2. Remove any slide on the stage, and rotate the nosepiece to the scanning objective.
3. Use lens paper and lens-cleaning fluid to carefully clean the objectives and the ocular lenses. If the oil-immersion lens was used, be sure to clean it thoroughly to remove all of the oil. It is especially important to avoid getting the oil on the other objectives because it will cloud these lenses. You will also need to carefully clean any slides used with the oil and examine the stage area carefully to determine if any oil spilled onto the stage. If so, also clean the stage with lens paper and lens-cleaning fluid.
4. Lower the microscope stage to its lowest position.
5. Put the dust cover on the microscope and return it to its proper storage location, making sure to carry the microscope properly with both hands—one hand under the base and one hand holding the arm of the microscope.

Post-lab quizzes are also assignable in MasteringA&P®

POST-LAB ASSIGNMENTS

Name: ______________________ Date: ______________ Lab Section: __________

PART I. Check Your Understanding

Activity 1: Identifying the Parts of the Microscope

1. Explain the proper procedure for transporting a microscope.

____

2. Label the following parts of the compound light microscope.

a. ______________________

b. ______________________

c. ______________________

d. ______________________

e. ______________________

f. ______________________

g. ______________________

h. ______________________

i. ______________________

j. ______________________

k. ______________________

l. ______________________

3. Which microscope part(s):

a. concentrates light on the specimen? ______________________

b. regulates the amount of light passing through the specimen? ______________________

c. connects the objective lenses to the head of the microscope? ______________________

d. magnifies the specimen? ______________________

Activity 2: Using the Microscope to View Objects

1. Draw the appearance of the letter "R" as viewed under the microscope.

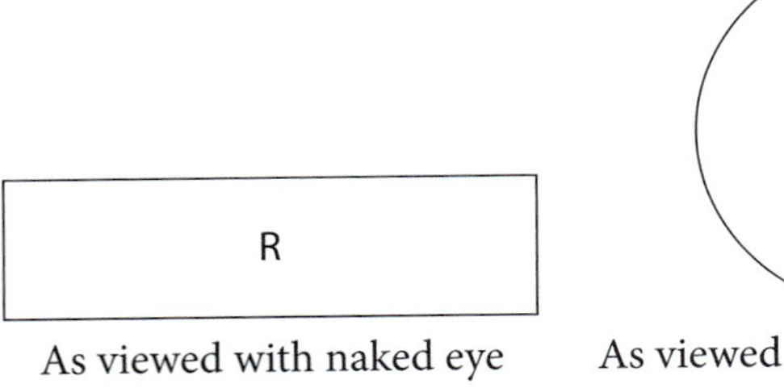

As viewed with naked eye

As viewed under the microscope

2. How is total magnification calculated?

____

What is the total magnification of the low-power objective? ________________

3. Define working distance. ____

____

4. What happens to working distance as total magnification increases? ________________

Activity 3: Determining Field Diameter and Estimating the Size of Objects

1. Perform the following conversions:

a. 3 mm = ________________ μm

c. 350 μm = ________________ mm

b. 0.5 mm = ________________ μm

d. 1 μm = ________________ mm

2. Based on the calculations that you made for your microscope, what is the field diameter for the:

a. scanning objective? ________________ mm = ________________ μm

b. low-power objective? ________________ mm = ________________ μm

c. high-power objective? ________________ mm = ________________ μm

d. oil-immersion objective? ________________ mm = ________________ μm

3. If the field diameter for the 4× objective is 4.5 mm, then what is the field diameter when using the 10× objective? ________________

4. Estimate the length of each of the following objects in μm:

a. ________________

400×

b. ________________

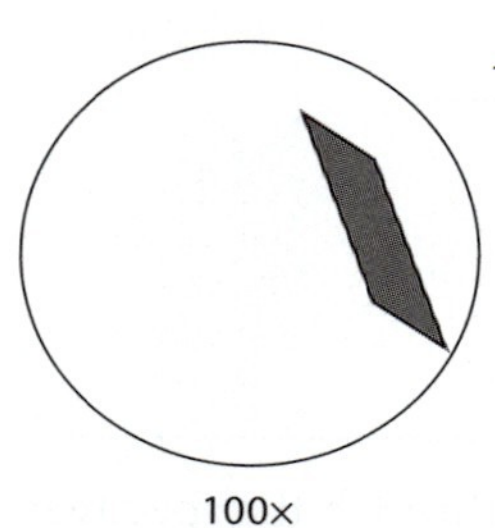

100×

Activity 4: Perceiving Depth of Field

1. Define depth of field.

____

____

2. What happens to the depth of field as total magnification decreases?

____

Activity 5: Caring for the Microscope

1. True/False: If the statement is false, correct it so that it is true.
   a. Begin the focusing process with the high-power objective.
   b. Store the microscope with the oil-immersion objective in position.
   c. Use tissue paper and water to clean the ocular and objective lenses.
   d. The coarse adjustment knob can be used with any objective lens.

2. Why is it important to allow the microscope to cool before putting it away?

____

____

PART II. Putting It All Together

A. Review Questions

1. How does a microscope's iris diaphragm lever work like the iris of an eye?

____

____

2. For each of the following parameters, indicate whether it increases or decreases as you move from lower magnification to higher magnification.

   a. Field diameter ______________

   c. Amount of light required ______________

   b. Working distance ______________

   d. Resolution ______________

3. Assume that you are using the low-power objective and an object takes up a quarter of the field of view. Estimate the diameter of the object in μm. ______________

4. A student is using the high-power lens. His lab partner notices that the working distance is approximately 1 cm. Is the student observing the specimen? ______________

   Explain. ____

   ____

5. Indicate a possible solution for each of the following complaints during microscope usage:

   a. There is not enough light to view a specimen: ______________________

   ____

   b. I can't find the specimen on the slide: ______________________

   ____

   c. I see a dark crescent-shaped structure: ______________________

   ____

B. Concept Mapping

1. Fill in the blanks to complete this concept map concerning microscope structures.

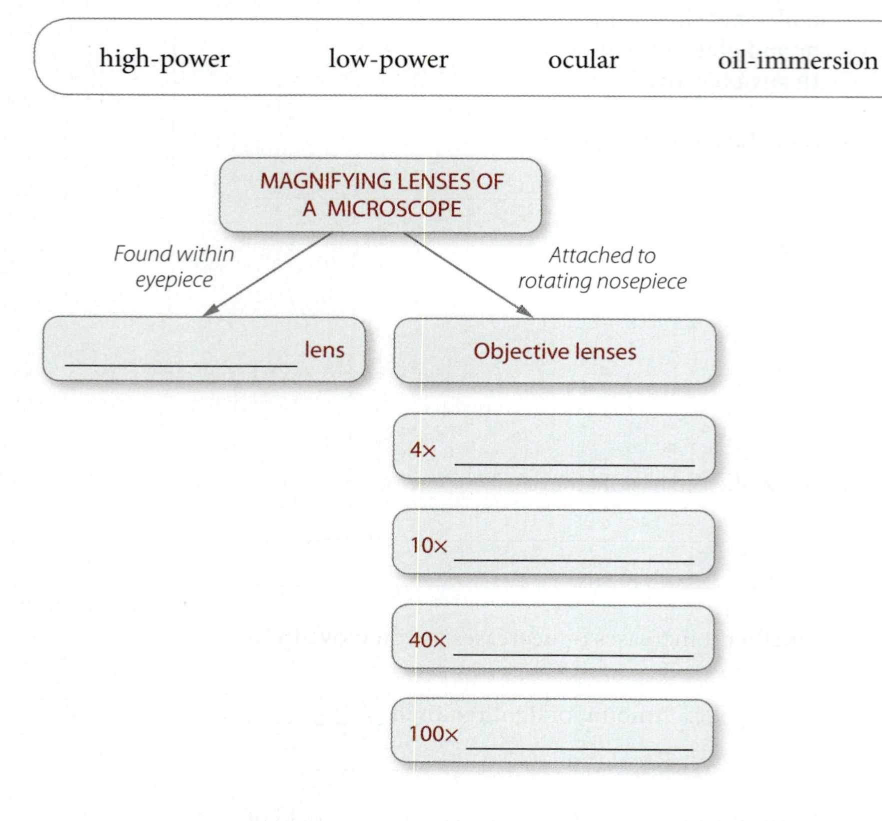

2. Construct a unit concept map to show the relationships among the following set of terms. Include all of the terms in your diagram. Your instructor may choose to assign additional terms.

coarse adjustment knob	condenser	field diameter	high-power objective
iris diaphragm	low-power objective	micrometer	ocular lens
oil-immersion lens	resolving power	scanning objective	stage
substage light	total magnification		working distance

5 The Cell

UNIT OUTLINE

The Functional Anatomy of the Cell

- **Activity 1:** Identifying Cell Components in a Wet Mount
- **Activity 2:** Identifying Cell Structures
- **Activity 3:** Examining the Possible Role of Osmosis in Cystic Fibrosis

Cellular Division

- **Activity 4:** Identifying the Stages of the Cell Cycle

Cellular Adaptations

- **Activity 5:** Exploring Cellular Diversity

PhysioEx™ Exercise 1: Cell Transport Mechanisms and Permeability

- **PEx Activity 1:** Simulating Dialysis (Simple Diffusion)
- **PEx Activity 2:** Simulating Facilitated Diffusion
- **PEx Activity 3:** Simulating Osmotic Pressure
- **PEx Activity 4:** Simulating Filtration
- **PEx Activity 5:** Simulating Active Transport

The **cell** is the fundamental unit of life. Cells of the human body are classified as eukaryotic, meaning that each has a true nucleus during some portion of its life cycle. A typical human body cell also includes cytoplasm containing a variety of organelles (literally, "little organs") and is surrounded by a selectively permeable plasma membrane. Different types of body cells have different appearances because they are structurally adapted (specialized) to perform a wide range of specific functions.

In this unit we will be studying each of these cell types in more detail, but first we will look at the structure and function of a generalized (or composite) cell. A generalized cell contains all the structures that can be found among the body's various cell types, but no cell in the body looks exactly like the generalized cell. Instead, each cell type has a form that is highly adapted to its specific functions.

THINK ABOUT IT *The structural adaptations of specialized cells are related to the cells' functions in the body.*

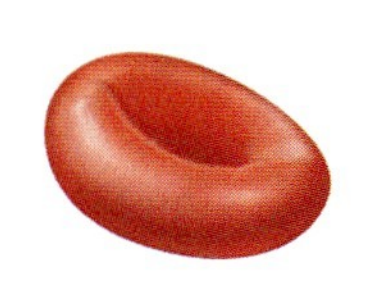

A red blood cell is packed with hemoglobin. How does this adaptation assist its function? ______________________

____

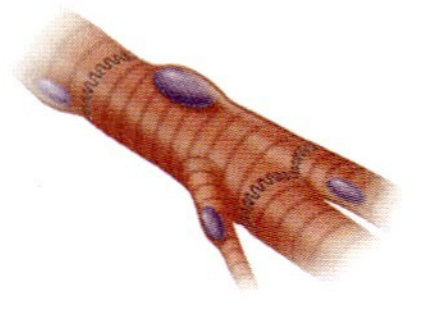

A cardiac muscle cell contains an abundance of contractile proteins. Why is this a necessary adaptation? ______________________

____

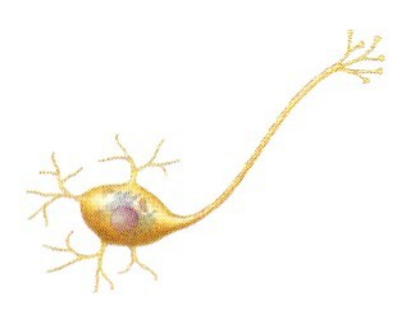

Some nerve cells have a long axon. How does this adaptation assist its function? ______________________

____

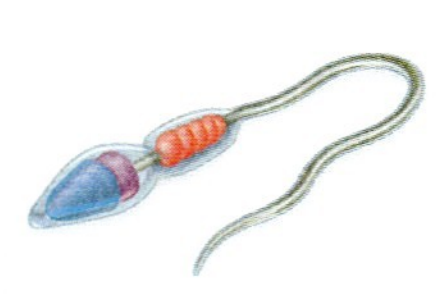

Each sperm cell contains a whip-like flagellum. What function does this adaptation serve? ______________________

____

PRE-LAB ASSIGNMENTS

Pre-lab quizzes are also assignable in MasteringA&P®

To maximize learning, BEFORE your lab period carefully read this entire lab unit and complete these pre-lab assignments using your textbook, lecture notes, and prior knowledge.

PRE-LAB Activity 1: Identifying Cell Components in a Wet Mount

1. Name the three components of a typical cell:

   ____

2. How will you obtain isolated cells for this procedure?

   ____

   ____

PRE-LAB Activity 2: Identifying Cell Structures

1. Use the list of terms provided to label the accompanying diagram of a cell; check off each term as you label it.

- ☐ centrosome
- ☐ cytosol
- ☐ golgi apparatus
- ☐ plasma membrane
- ☐ free ribosomes
- ☐ smooth endoplasmic reticulum
- ☐ lysosome
- ☐ nucleus
- ☐ nucleolus
- ☐ rough endoplasmic reticulum
- ☐ mitochondrion

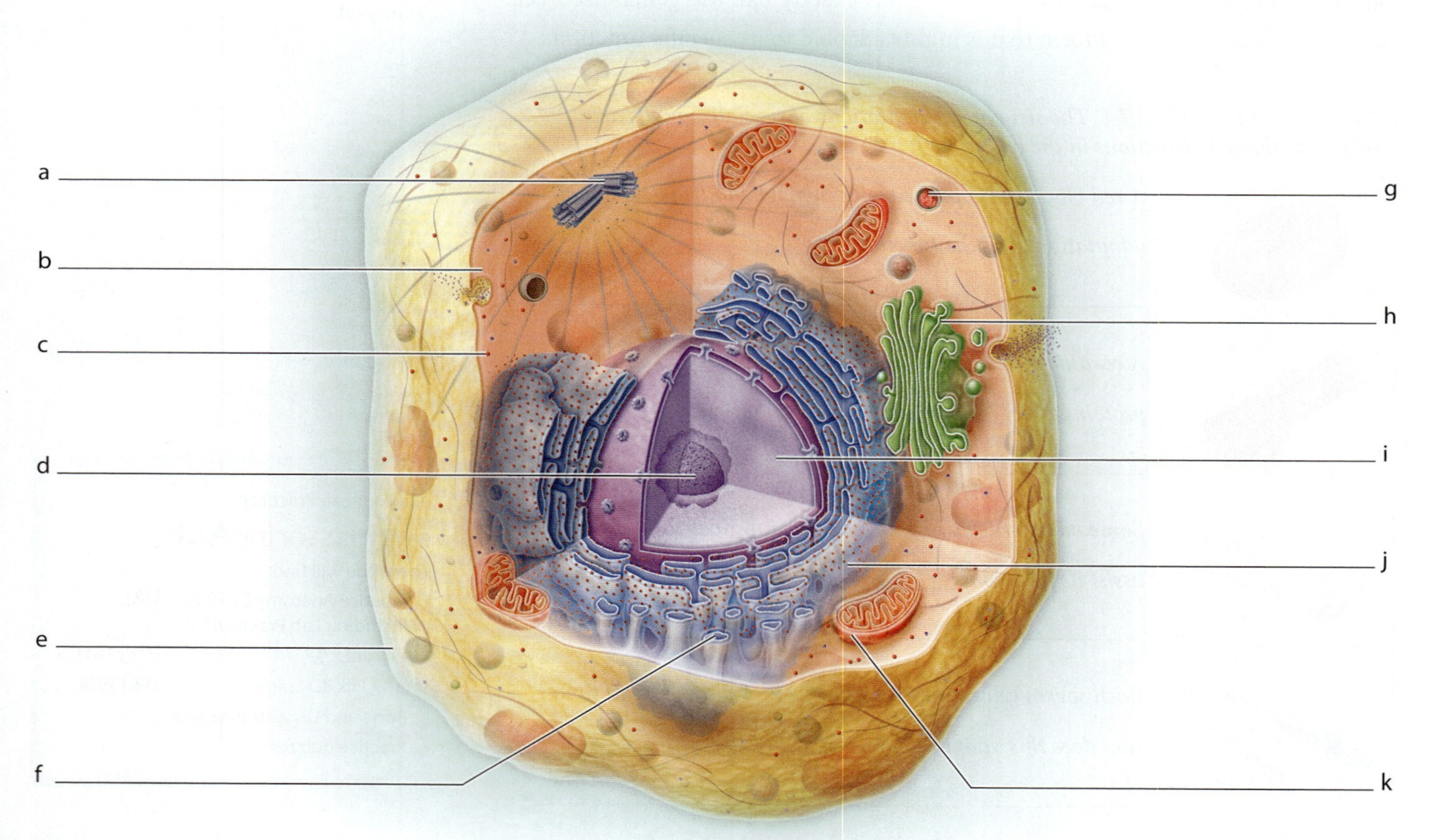

PRE-LAB Activity 3: Examining the Possible Role of Osmosis in Cystic Fibrosis

1. In this procedure, what is used to represent a plasma membrane? ______________________

2. A plasma membrane is ______________________________ (impermeable, semipermeable, freely permeable).

PRE-LAB Activity 4: Identifying the Stages of the Cell Cycle

1. Match each of the following stages of the cell cycle with its correct description:

_____	a. metaphase	1. sister chromatids reach opposite poles
_____	b. prophase	2. DNA is replicated
_____	c. interphase	3. nuclear envelope breaks down
_____	d. anaphase	4. chromosomes line up along the equator
_____	e. telophase	5. sister chromatids split and move to opposite poles

PRE-LAB Activity 5: Exploring Cellular Diversity

1. Match each of the following cell types with its correct description:

_____	a. red blood cell	1. packed with actin and myosin
_____	b. cardiac muscle fiber	2. propelled by a flagellum
_____	c. neuron	3. filled with hemoglobin
_____	d. sperm	4. characterized by an extension called an axon

The Functional Anatomy of the Cell

In this section we will study the structure and function of the cells of the human body. We begin by examining the three major parts of a cell: the plasma membrane, the cytoplasm and its organelles, and the nucleus. As you read about the parts of the cell, refer to **Figure 5-1**, which depicts a generalized body cell.

The Plasma Membrane

The **plasma membrane** is a thin phospholipid bilayer in which various proteins are embedded or attached (**Figure 5-2**). The phospholipids are composed of hydrophilic ("water-loving") phosphate heads and hydrophobic ("water-fearing") lipid tails. Two types of proteins are associated with the phospholipid bilayer: Integral proteins are embedded at least partially within the cell membrane, whereas peripheral proteins are located on the interior or exterior surface of the cell membrane. Some integral proteins, called transmembrane proteins, span the entire width of the phospholipid bilayer.

A plasma membrane is selectively permeable—that is, it allows the passage of some substances but not others. Nonpolar molecules such as oxygen, carbon dioxide, and steroid hormones pass freely through the membrane. However, water-soluble substances, such as amino acids, sugars, proteins, nucleic acids, and many ions, are unable to pass freely through the membrane. Integral proteins can serve as transport channels through which substances move from one side of the membrane to the other. Both integral and peripheral proteins can serve as receptors and enzymes and they also play major roles in cell-to-cell recognition and intercellular joining.

Cells depend on the selectively permeable plasma membrane to import substances such as nutrients, oxygen, and hormones, and to export substances such as carbon dioxide and other waste products. A cell's membrane transport mechanisms are of two types (**Figure 5-3**): passive, in which substances cross the membrane without the cell expending any energy, and active, which requires the cell to expend energy in the form of the high-energy molecule ATP.

Among passive transport mechanisms the most common are diffusion and osmosis (see **Figure 5-3a–d**). Passive transport mechanisms can be either unassisted or assisted (facilitated). The unassisted net movement of molecules from an area of higher concentration to an area of lower concentration is called simple diffusion (see **Figure 5-3a**). When red blood cells carrying oxygen come into proximity of body

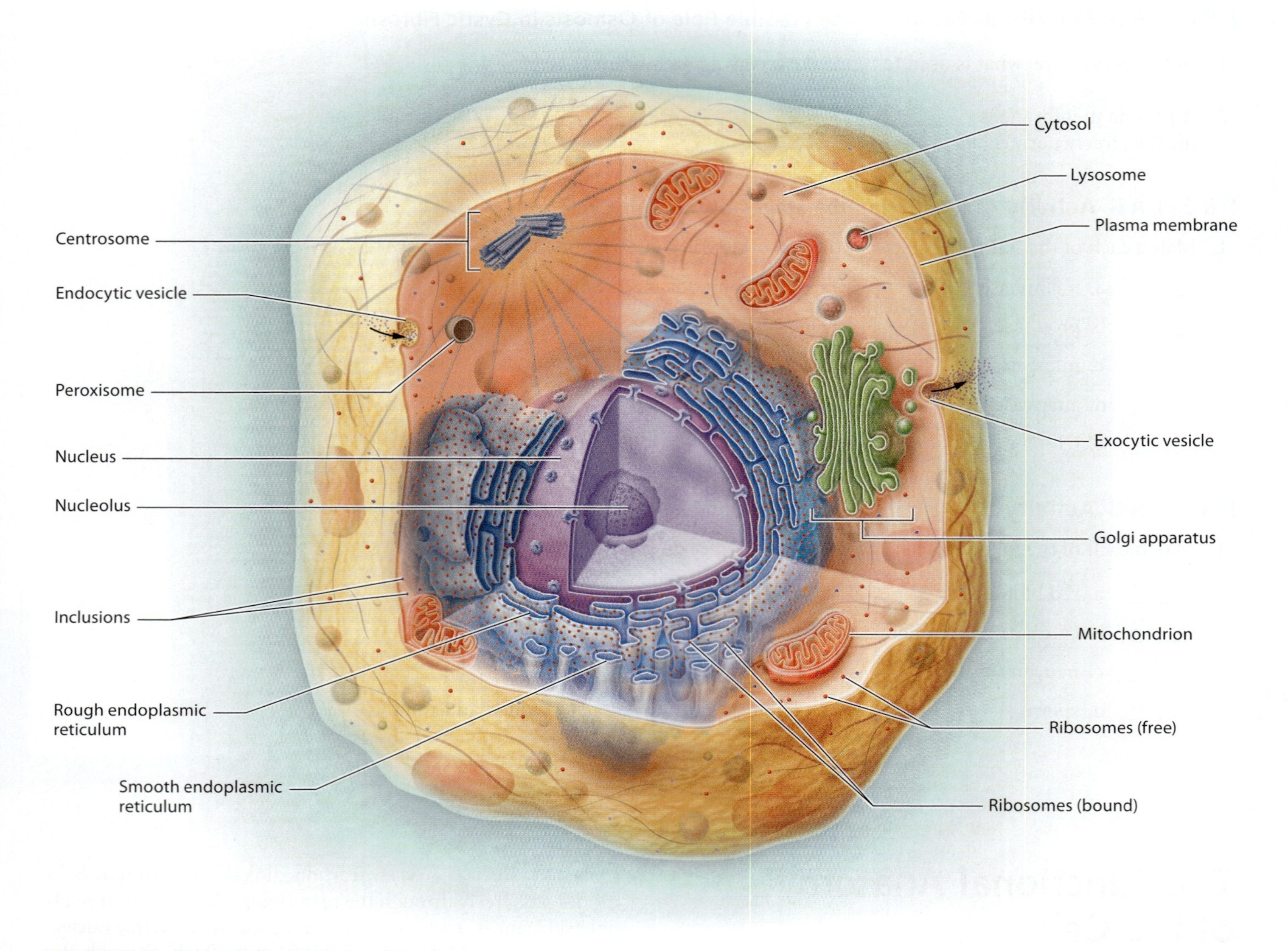

Figure 5-1 A generalized cell showing its various structures and organelles.

cells, oxygen molecules move from the blood (where the oxygen concentration is high) into the body cells (where the oxygen concentration is lower) via simple diffusion. By contrast, the passive process called facilitated diffusion requires the assistance of either a channel protein (channel-mediated facilitated diffusion; see **Figure 5-3b**) or a carrier protein (carrier-mediated facilitated diffusion; see **Figure 5-3c**). **Osmosis** is a special type of diffusion involving the passive movement of a solvent (usually water) from an area of higher water concentration (and thus lower solute concentration) to an area of lower water concentration (and thus higher solute concentration) across a selectively permeable membrane, either through the phospholipid bilayer or through a specific channel protein called an aquaporin (see **Figure 5-3d**).

In another passive process, filtration, substances are driven through a selectively permeable membrane by hydrostatic pressure. (We will study filtration's major role in the formation of urine by the kidneys when we study renal physiology.)

> **CLINICAL CONNECTION**
>
> **Dialysis**
>
> **Dialysis** is a medical procedure used to remove wastes and excess water from the blood of patients experiencing kidney failure. During hemodialysis, one of the main types of dialysis, the patient's blood is cleaned as it passes through a filter (called a dialysis membrane) in a dialysis machine ("artificial kidney"). Dialysis works on the principles of diffusion and filtration.

Two common energy-requiring transport mechanisms are termed active transport and vesicular transport. **Active transport** is the movement of molecules from an area of lower concentration to an area of higher concentration and is accomplished with the expenditure of energy. The sodium-potassium pump is a classic example of an active transport mechanism (see **Figure 5-3e**).

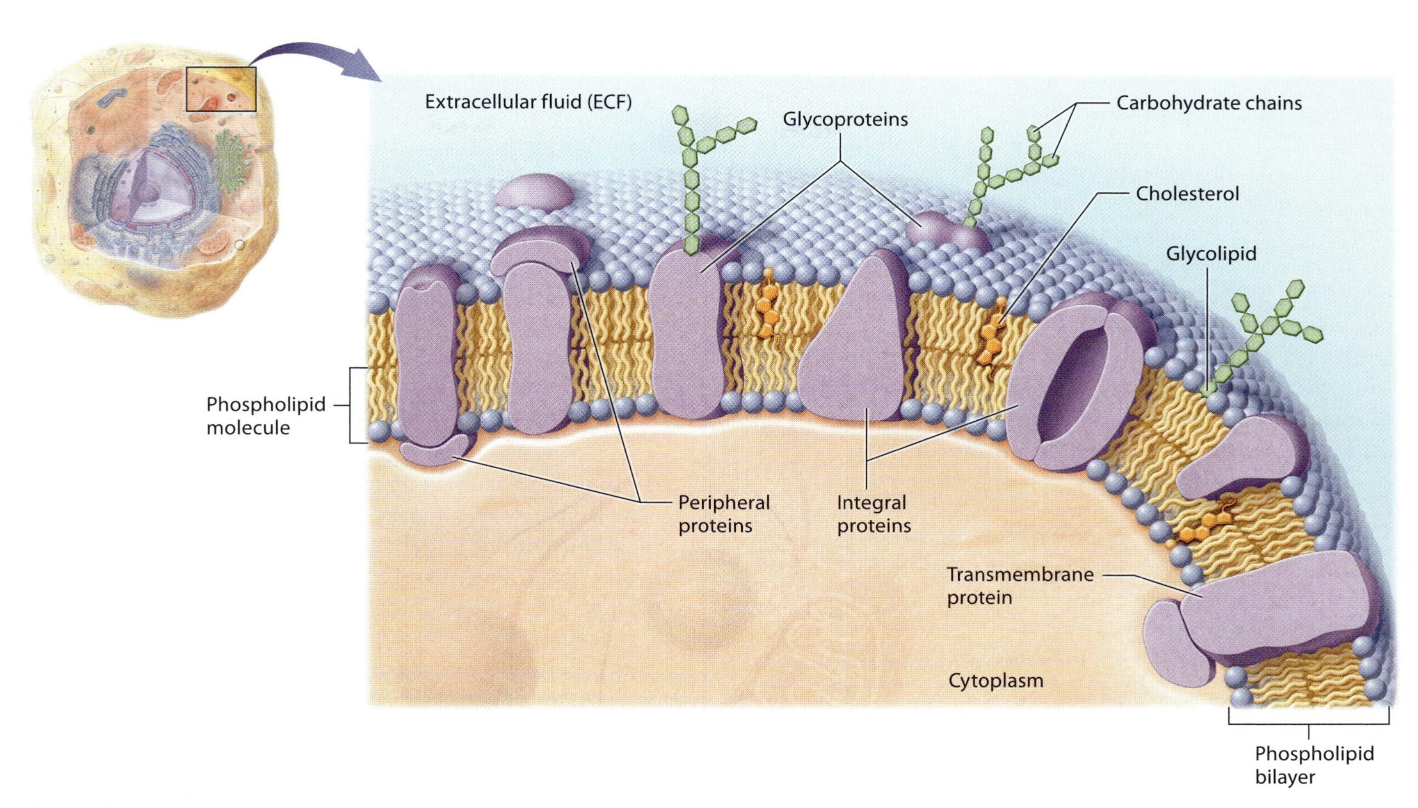

Figure 5-2 The plasma membrane.

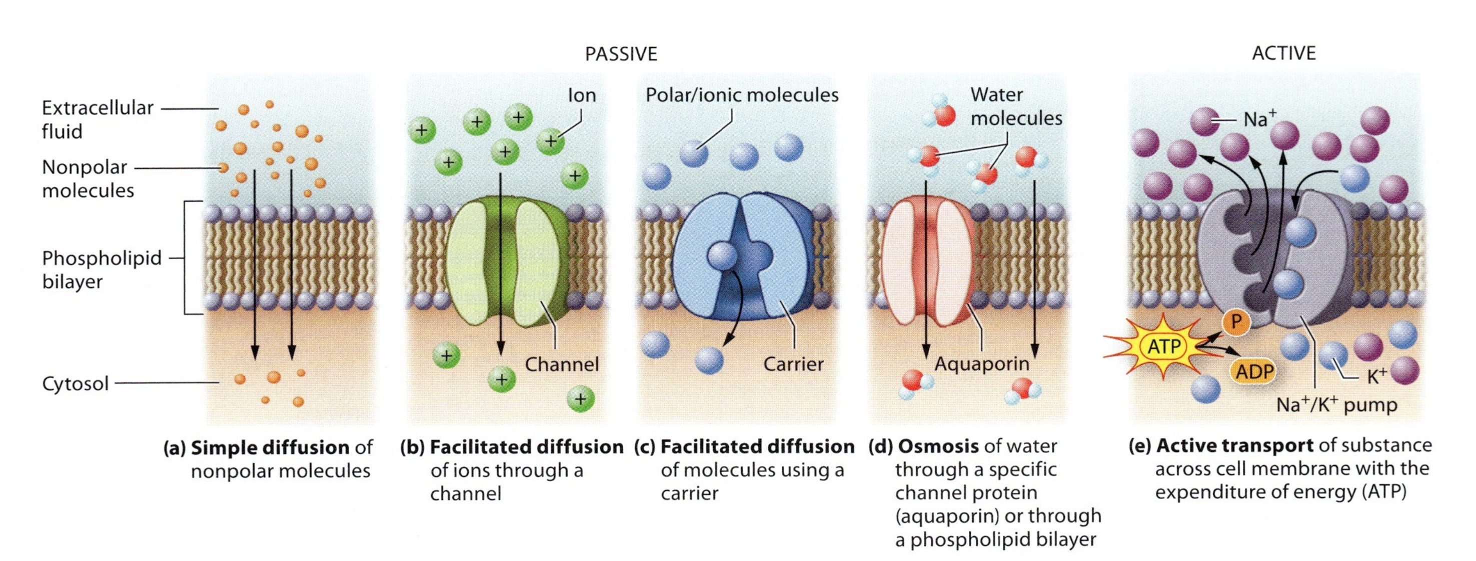

(a) Simple diffusion of nonpolar molecules

(b) Facilitated diffusion of ions through a channel

(c) Facilitated diffusion of molecules using a carrier

(d) Osmosis of water through a specific channel protein (aquaporin) or through a phospholipid bilayer

(e) Active transport of substance across cell membrane with the expenditure of energy (ATP)

Figure 5-3 Mechanisms of membrane transport.

The Cytoplasm and Its Organelles

The **cytoplasm** of a cell consists of a variety of specialized structures called organelles suspended in a fluid portion called the cytosol, which is largely water containing a variety of solutes such as sugars, proteins, and ions. Organelles perform specific functions to maintain homeostasis within the cell.

Note that the cytosol of some cells also contains inclusions, which are stored cellular products such as glycogen granules (in muscle fibers and liver cells), lipid droplets (in adipocytes), and pigment granules (in certain skin cells). The major organelles are depicted in Figure 5-1, and their most important functions are listed in **Table 5-1**.

The Nucleus

The **nucleus** is the control center of the cell. Even though it is actually the largest of a cell's organelles, it is typically considered to be a separate part of the cell. Most cells contain a single nucleus (such cells are termed uninucleate), but some cells such as skeletal muscle fibers and liver cells are multinucleate, and a mature red blood cell is anucleate (lacks a nucleus). The nucleus is surrounded by a double-layered **nuclear envelope** that is penetrated at various points by **nuclear pores** that regulate the movement of substances into and out of the nucleus. Within the nucleus are **nucleoli**, which are round, dark-staining structures that function in ribosome synthesis. Also visible in the nucleus is **chromatin**, which is composed of tightly coiled DNA (the genetic material), proteins, and RNA.

Table 5-1 Functions of the Major Organelles

Organelle	Function
Ribosome	Synthesizes proteins.
Rough endoplasmic reticulum (RER)	Synthesizes and transports proteins.
Smooth endoplasmic reticulum (SER)	Synthesizes lipids and steroids (cholesterol); detoxifies drugs.
Golgi apparatus	Packages and modifies proteins.
Mitochondrion	Synthesizes ATP; is the "powerhouse" of the cell.
Lysosome	Contains enzymes that digest worn-out organelles and substances that have entered the cell.
Peroxisome	Detoxifies toxic substances.
Centrosome with centrioles	Organizes the mitotic spindle during cell division.
Cytoskeleton	
• Microtubules	Support the cell and give it shape; are components of centrioles, cilia, and flagella; form spindle apparatus during mitosis; transport organelles and structures within the cell.
• Intermediate filaments	Strengthen the cell and help maintain its shape; stabilize the position of organelles.
• Microfilaments	Anchor the cytoskeleton to integral proteins of the plasma membrane; enable cellular movements.

LabBOOST >>>

Organelles It is important to know what each organelle looks like and to remember what its function is. But instead of simply memorizing individual structures and their functions, begin training yourself to "tell a story" that explains how the functions of each of the organelles enable the *cell* to achieve its functions.

For example, consider a cell within a salivary gland that produces the digestive enzyme amylase, which is a protein. Many organelles are involved in the process of synthesizing, transporting, packaging, and secreting a protein. First, ribosomes located on the rough endoplasmic reticulum (RER) produce the protein. The RER then transports it to the Golgi apparatus, where it is packaged into a secretory vesicle. The secretory vesicle then moves to the plasma membrane and releases its contents via exocytosis (an example of vesicular transport, an active transport mechanism, which therefore requires the expenditure of energy). The cell's mitochondria provide the ATP needed for all these events to occur. So, remember to practice developing strategies for connecting the individual parts of processes in a way that explains how those parts function together to accomplish some larger function in the body.

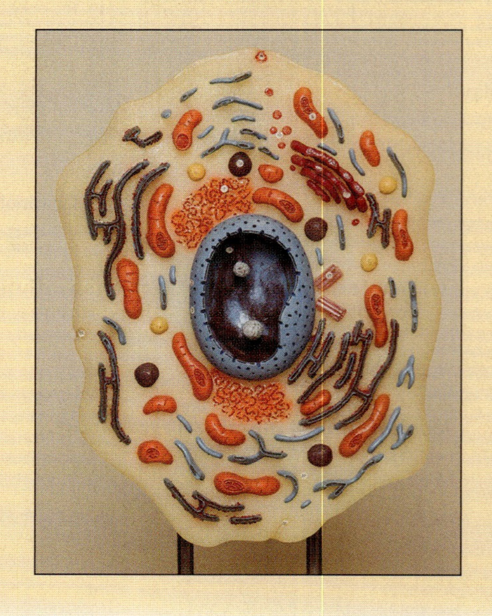

ACTIVITY 1

Identifying Cell Components in a Wet Mount

Learning Outcomes

1. Prepare a wet mount of a cheek cell smear and observe the wet mount under the microscope.
2. Identify the plasma membrane, nucleus, and cytoplasm of a cheek cell.

Materials Needed

- ☐ Microscope slide and coverslips
- ☐ Saline solution
- ☐ Toothpick
- ☐ Methylene blue
- ☐ Filter paper or piece of paper towel
- ☐ Microscope

Instructions

Cheek cells are the major cell type of the epithelial tissue lining the inside of your mouth. This epithelium is composed of many layers of thin, flattened cells. Perform the following steps to make a wet mount of these cells in order to identify the three major components of a typical cell:

1. Place a drop of saline solution on a microscope slide.
2. Carefully scrape the inside of your cheek with the end of a toothpick.
3. Place the cheek scrapings in the drop of saline solution on the slide and stir with the toothpick.
4. Add a drop of methylene blue to the drop of saline and stir again.
5. Place a coverslip over the drop on the slide. Use the folded edge of a piece of filter paper (or paper towel) to absorb any excess fluid.
6. Place the slide on the microscope and observe the scrapings at low power. Now, switch to the high-power objective and examine the scrapings more closely (**Figure 5-4**).
7. Make a sketch of the epithelial cells that you observe. On your sketch label the plasma membrane, the nucleus, and the cytoplasm.

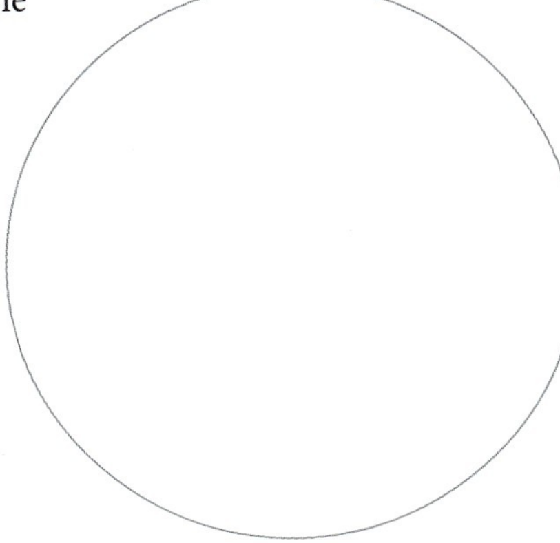

Total magnification: _____ ×

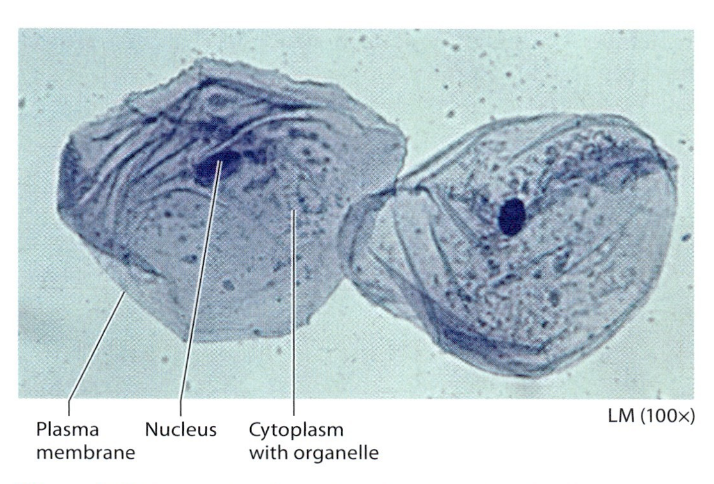

Figure 5-4 Human cheek epithelial cells stained with methylene blue.

8. Dispose of materials and clean up as directed by your instructor.

ACTIVITY 2

Identifying Cell Structures

Learning Outcomes

1. Describe the structure and function of the three major components of a cell (plasma membrane, cytoplasm containing organelles, and nucleus), and identify each on a cell model or anatomical chart.
2. Identify on a cell model or anatomical chart the following cytoplasmic organelles and describe the function of each: ribosome, rough endoplasmic reticulum, smooth endoplasmic reticulum, Golgi apparatus, mitochondrion, lysosome, peroxisome, centrosome, and cytoskeletal elements.
3. Identify on a cell model or diagram the following nuclear structures and describe the function of each: nuclear envelope, nuclear pore, nucleolus, and chromatin.
4. Explain how the components of the cell function together to synthesize, package, transport, and secrete proteins.

Materials Needed

- ☐ Cell model or anatomical chart

Instructions

Identify each of the following cell structures on a cell model, and then write down one key word or phrase to help you remember the function of each structure. The first answer has been provided as an example.

Cell membrane	selectively permeable
Ribosome	
RER	
SER	
Golgi apparatus	
Mitochondrion	
Lysosome	
Peroxisome	
Centrosome	
Cytoskeleton	
Nucleus	
Nuclear envelope	
Nuclear pore	
Nucleolus	
Chromatin	

ACTIVITY 3

Examining the Possible Role of Osmosis in Cystic Fibrosis

Learning Outcomes

1. Demonstrate and explain the process of osmosis.
2. Demonstrate the role that a salt imbalance and osmosis might have in causing the symptoms of cystic fibrosis.

Materials Needed

- ☐ Dialysis tubing
- ☐ String
- ☐ Scissors
- ☐ Wax pencil
- ☐ 0.9% NaCl solution
- ☐ 20% NaCl solution
- ☐ 250-ml beaker
- ☐ Two 500-ml beakers
- ☐ Scale
- ☐ Timer
- ☐ Graduated cylinder
- ☐ Paper towels
- ☐ Graph paper

Cystic fibrosis (CF), the most common lethal genetic disease in the United States, is a progressive disease that causes pulmonary and pancreatic insufficiency as a result of the thickening of secretions produced by the lungs and pancreas. In CF, a gene mutation results in the production of an abnormal protein, which is then degraded so that it cannot perform its usual function in the apical plasma membranes of epithelial cells lining the respiratory and digestive tracts. The absence of the normal protein disrupts normal membrane transport processes.

How this malfunction of these epithelial cells causes the clinical signs and symptoms of cystic fibrosis is not well understood. One hypothesis proposes that chloride ions are unable to leave the cells due to the absence of a normal protein. As a result, water moves into the cell by osmosis, and the mucus on the surface of the epithelial cells becomes dehydrated, thick, and sticky. In normal respiratory tract epithelia, pathogens and foreign material are removed from the respiratory tract through the movement of mucus by ciliary action. In affected cells, however, a thickened, sticky layer of mucus inhibits ciliary action and thus the movement of mucus, thereby increasing the both the frequency and severity of respiratory infections.

The purpose of this activity is twofold: to demonstrate the normal process of osmosis, and to suggest how a salt imbalance can lead to the production of very thick mucus in the respiratory passages. In this activity, fluid-filled bags made of dialysis tubing represent cells, and the dialysis tubing itself represents the plasma membrane of those cells. Like a plasma membrane, the dialysis tubing is "selectively permeable"—it allows some molecules (such as water) to pass freely through it, but prevents the passage of other substances (such as sodium ions and chloride ions). The passage of sodium ions and chloride ions through the plasma membrane or dialysis tubing requires the presence of special transport molecules.

Instructions

Set up and conduct the demonstration as follows:

1. Cut two 6-inch-long strips of dialysis tubing.
2. Pour 100 ml of distilled water into a 250-ml beaker, submerge the two strips of dialysis tubing in the water in the beaker, and soak the strips for 3 minutes.
3. Tie off one end of one of the strips with a piece of string. Rub the other end of the strip between your thumb and first finger to open it.
4. Fill the tubing approximately half full with 0.9% NaCl solution. While being careful not to trap air within the tubing, tie off the other end of the strip with a second piece of string. This bag of fluid represents a "normal cell" of the respiratory passageway. Dry off the outside of the "cell" with a paper towel; then use the scales to weigh the bag, and record the data as "Starting weight of bag" in the chart that follows these instructions.
5. Repeat step 4 with the remaining strip of dialysis tubing using a 20% NaCl solution. This "cell" represents a respiratory passageway cell that has a defective chloride channel

(a "defective cell"). Remember that defective chloride channels cause chloride levels inside the cell to increase because the normal transport of chloride out of the cell is prevented. As a result, water moves into the cell by osmosis and the mucus on the surface of the respiratory passageway cell becomes dehydrated, thick, and sticky.

6. Fill one 500-ml beaker with 400 ml of 0.9% NaCl solution, and label this beaker "Normal Condition."
7. Fill the other 500-ml beaker with 400 ml of 0.9% NaCl solution, and label this beaker "Cystic Fibrosis Patient." The 0.9% NaCl solution in each beaker represents the mucus on the surfaces of cells in the respiratory passageways.
8. Place the "normal cell" in the "Normal Condition" beaker, and place the "defective cell" in the "Cystic Fibrosis Patient" beaker (**Figure 5-5**).

   When the "normal cell" is placed in the beaker of 0.9% NaCl solution, what do you predict will happen? Why?

   ___

   ___

   When the "defective cell" is placed in the beaker of 0.9% NaCl solution, what do you predict will happen? Why?

   ___

   ___

9. Remove the bags from the beakers and weigh them at 5-minute intervals; that is, at 0, 5, 10, 15, and 20 minutes. Each time be sure to wipe excess fluid off the outside of each "cell" before weighing it.
10. Record the resulting weight data in the table below.
11. Plot your data on a piece of graph paper. Using change in weight along the y-axis and time along the x-axis, plot the change in weight over time. Be sure to label both axes and to include units.

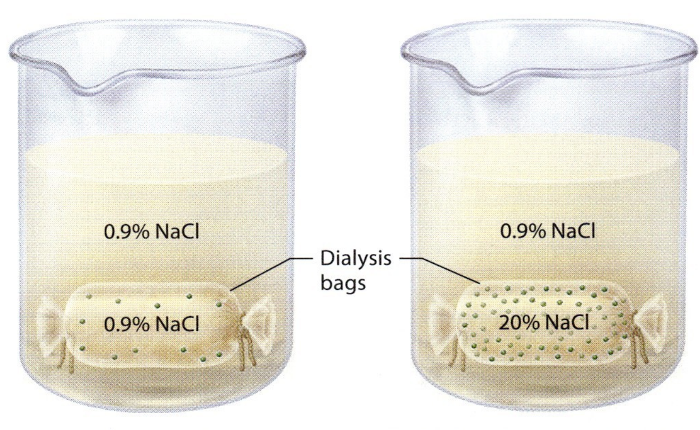

(a) Dialysis bag containing 0.9% NaCl represents a normal cell.

(b) Dialysis bag containing 20% NaCl represents a CF cell.

Figure 5-5 The setup for the osmosis demonstration in Activity 3.

12. Answer the following questions based on your results:
    a. Did a net movement of water occur into the "normal cell" or out of the "normal cell"? Why?

    ___

    ___

    ___

    b. Did a net movement of water occur into the "defective cell" or out of the "defective cell"? Why?

    ___

    ___

    ___

    c. Based on this osmosis demonstration, explain how defective chloride channels in the epithelial cells of the respiratory tract could result in the formation of thickened mucus.

    ___

    ___

	"Normal Cell" Containing 0.9% NaCl			"Defective Cell" Containing 20% NaCl		
Time	Starting weight of bag (in grams)	New weight of bag (in grams)	New weight of bag minus starting weight (in grams)	Starting weight of bag (in grams)	New weight of bag (in grams)	New weight of bag minus starting weight (in grams)
Start: 0 minutes						
5 minutes						
10 minutes						
15 minutes						
20 minutes						

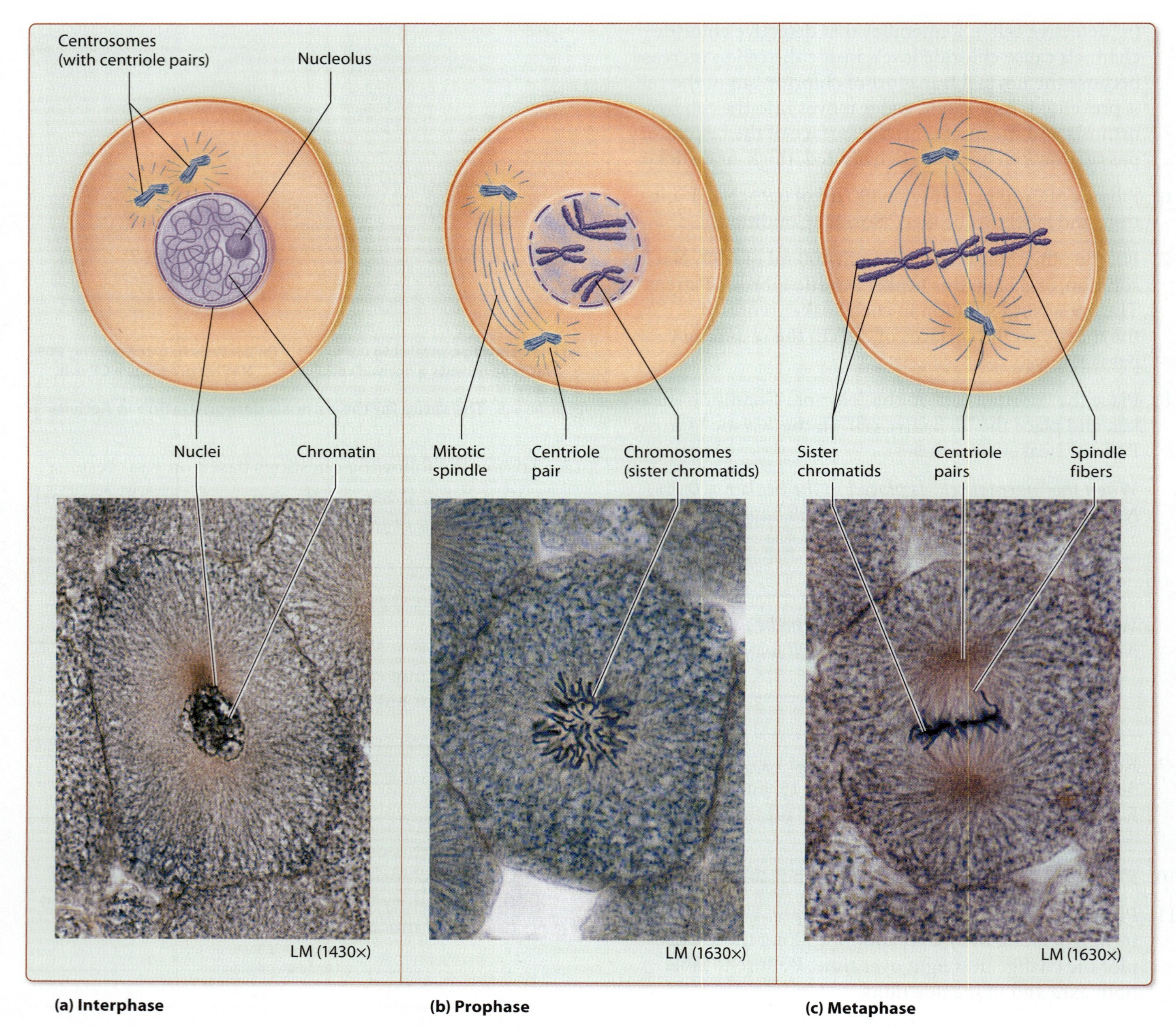

Figure 5-6 The appearance of a cell during (a) interphase and (b–e) the stages of mitosis.

Cellular Division

Next we consider an important aspect of basic cell biology: the cell cycle.

The Cell Cycle

The cell cycle, which consists of the series of events that occur in the life of a cell, is divided into two major parts: interphase (during which the cell grows and is metabolically active) and a mitotic phase (during which the cell divides).

Interphase (**Figure 5-6a**) is further subdivided into three subphases: G_1, S, and G_2. During G_1 ("gap 1"), the cell grows, produces additional organelles, and is metabolically active. Toward the end of G_1, the cell begins to replicate its centrioles in preparation for cell division. During the S phase, replication (DNA synthesis) occurs. During replication, the cell makes an identical copy of its genetic material so that each daughter cell produced during the mitotic phase will have one complete set of chromosomes and will be genetically identical to the parent cell. In other words, each of the 23 chromosomes in humans is "duplicated" and

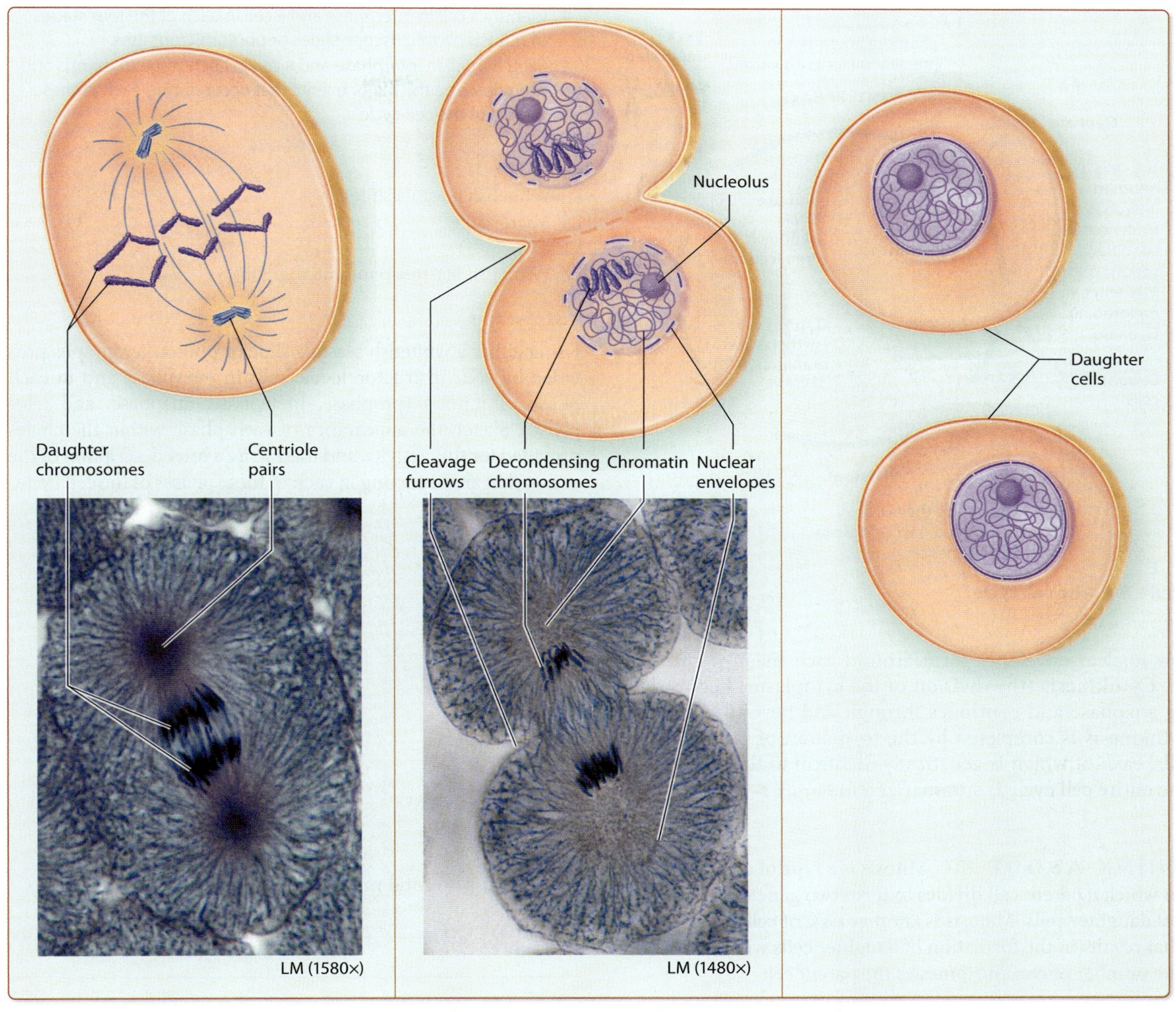

(d) Anaphase **(e) Telophase and cytokinesis** **(f) End result**

consists of two sister chromatids held together by a centromere. During G_2 ("gap 2"), the cell continues to grow, synthesizes the proteins necessary for cell division, and completes the production of new centrioles. The cell is now ready to enter the mitotic phase. Note that some mature, highly specialized cells (such as skeletal muscle fibers and neurons) instead enter an indefinite G_0 phase and never enter the mitotic phase.

The **mitotic phase** of the cell cycle consists of two specific events: **mitosis** (division of the nucleus) and **cytokinesis** (division of the cytoplasm). Mitosis is divided into four phases: prophase, metaphase, anaphase, and telophase.

Remember that interphase precedes the first phase of the mitotic phase: prophase. During **prophase** (see **Figure 5-6b**), the chromosomes (each consisting of two sister chromatids held together by a centromere) coil and become visible, the nuclear envelope breaks down, the centrioles move to opposite poles of the cell, and the mitotic spindle forms. During **metaphase** (see **Figure 5-6c**), the chromosomes line up along the metaphase plate (equator). During **anaphase** (see **Figure 5-6d**), the sister chromatids of each chromosome split apart and move to opposite poles of the cell. During **telophase** (see **Figure 5-6e**), the sister chromatids reach the opposite poles and uncoil to become chromatin once again as

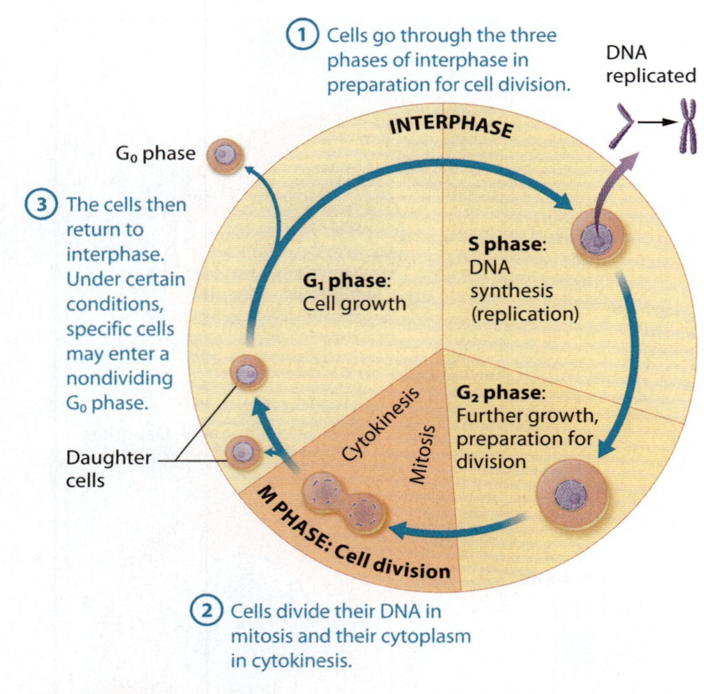

Figure 5-7 **The cell cycle.**

new nuclear envelopes form around each mass of chromatin. **Cytokinesis** (the division of the cytoplasm) begins during anaphase and continues through and beyond telophase. Cytokinesis is completed by the formation of two daughter cells, each of which is genetically identical to the parent cell. The entire cell cycle is summarized in **Figure 5-7**.

THINK ABOUT IT Mitosis is a type of cell division in which a parent cell divides to form two genetically identical daughter cells. Meiosis is another type of cell division that results in the formation of daughter cells with only half the number of chromosomes as the parent cell.

Where does mitosis occur? ______________________________

____

Where does meiosis occur in males? ____________________

Where does meiosis occur in females? __________________

ACTIVITY 4 Identifying the Stages of the Cell Cycle

Learning Outcomes

1. Characterize the stages of the cell cycle.
2. Explain the events that occur during interphase.
3. Outline the events that occur during the four mitotic stages.
4. Identify a cell in interphase and a cell in each of the four stages of mitosis on microscope slides or photomicrographs.
5. Sketch a cell in interphase and a cell in each stage of mitosis and then describe the major events that occur during each of these phases of the cell cycle.

Materials Needed

- ☐ Whitefish blastula slides
- ☐ Microscope

Or

- ☐ Whitefish blastula photomicrographs

Instructions

Using either a whitefish blastula slide or photomicrographs provided by your instructor, locate cells in interphase and in each phase of mitosis (prophase, metaphase, anaphase, and telophase). Sketch the appearance of each phase within the circles provided for this activity, and then write a brief description of the events that are occurring in each of these phases of the cell cycle.

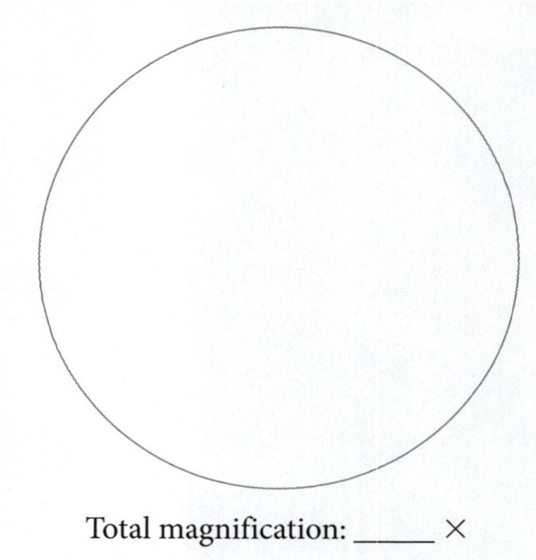

Total magnification: _____ ×

Interphase: ______________________________________

____

Total magnification: _____ ×

Prophase: ______________________________________

____

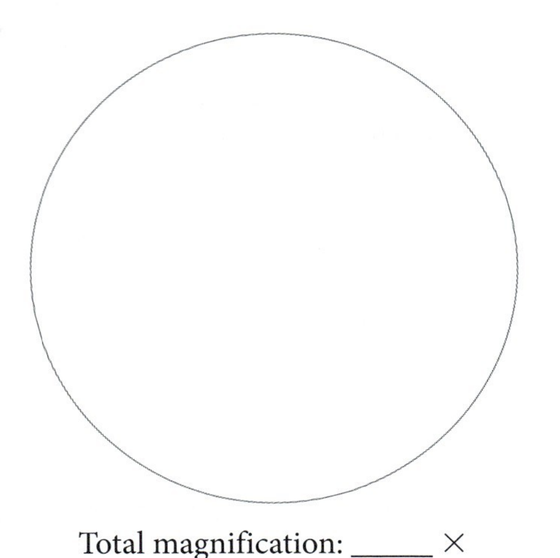

Total magnification: _____ ×

Metaphase: ____

____

Total magnification: _____ ×

Anaphase: ____

____

Total magnification: _____ ×

Telophase: ____

____

Cellular Adaptations

Next we explore how cells that have different functions typically have different structures.

ACTIVITY 5
Exploring Cellular Diversity

Learning Outcomes

1. Sketch the following four specialized cell types: erythrocyte (red blood cell), sperm cell, cardiac muscle fiber, and nerve cell.
2. Compare and contrast the structure and function of each of these four specialized cell types.

Materials Needed

- ☐ Photomicrographs or microscope and slides of a blood smear, sperm cells, cardiac muscle fibers, and a neuron smear

Instructions

Observe each of the following specialized cell types using photomicrographs or microscopic slides provided by your instructor. Make a sketch of each cell type and then state where each of these specialized cell types is found in the body and describe how each is adapted to its function.

Total magnification: _____ ×

Erythrocyte: ____

____

Total magnification: _____ ×

Sperm cell: ____

____

Total magnification: _____ ×

Cardiac muscle fiber: ____

____

Total magnification: _____ ×

Nerve cell: ____

____

Optional Activity

View and label histology slides of cell division at MasteringA&P® > Study Area > Practice Anatomy Lab > Histology > Cytology

PhysioEx EXERCISE 1

Cell Transport Mechanisms and Permeability

The PhysioEx 9.1 computer simulations can be used as an alternative to or as a supplement to the lab activities in this unit. Each simulation allows you the opportunity to investigate important concepts in cellular physiology.

Access the simulations in these activities at MasteringA&P® > Study Area > PhysioEx 9.1.

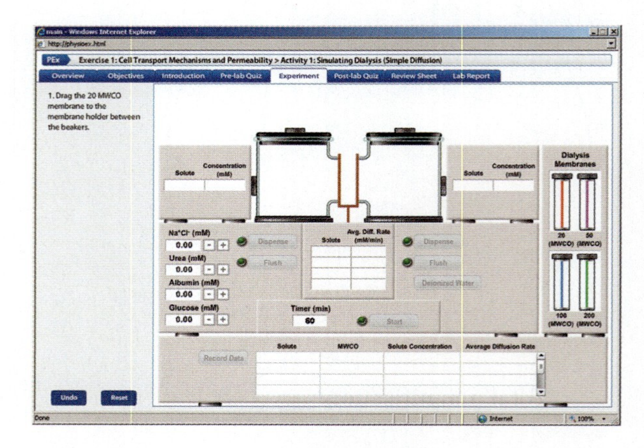

There you will find the following activities:

- **PEx Activity 1:** Simulating Dialysis (Simple Diffusion)
- **PEx Activity 2:** Simulating Facilitated Diffusion
- **PEx Activity 3:** Simulating Osmotic Pressure
- **PEx Activity 4:** Simulating Filtration
- **PEx Activity 5:** Simulating Active Transport

Post-lab quizzes are also assignable in MasteringA&P®

POST-LAB ASSIGNMENTS

Name: ______________________ Date: ______________ Lab Section: ________

PART I. Check Your Understanding

Activity 1: Identifying Cell Components in a Wet Mount

1. Cheek cells:
   a. are anucleate.
   b. are filled with hemoglobin.
   c. are thin, flattened cells.
   d. are classified as connective tissue cells.
   e. contain axons and dendrites.

2. Name the three components of a typical cell.

   ____

   ____

   ____

Activity 2: Identifying Cell Structures

1. Label each of the following components of a typical cell.

   a. ______________________

   b. ______________________

   c. ______________________

   d. ______________________

   e. ______________________

   f. ______________________

   g. ______________________

   h. ______________________

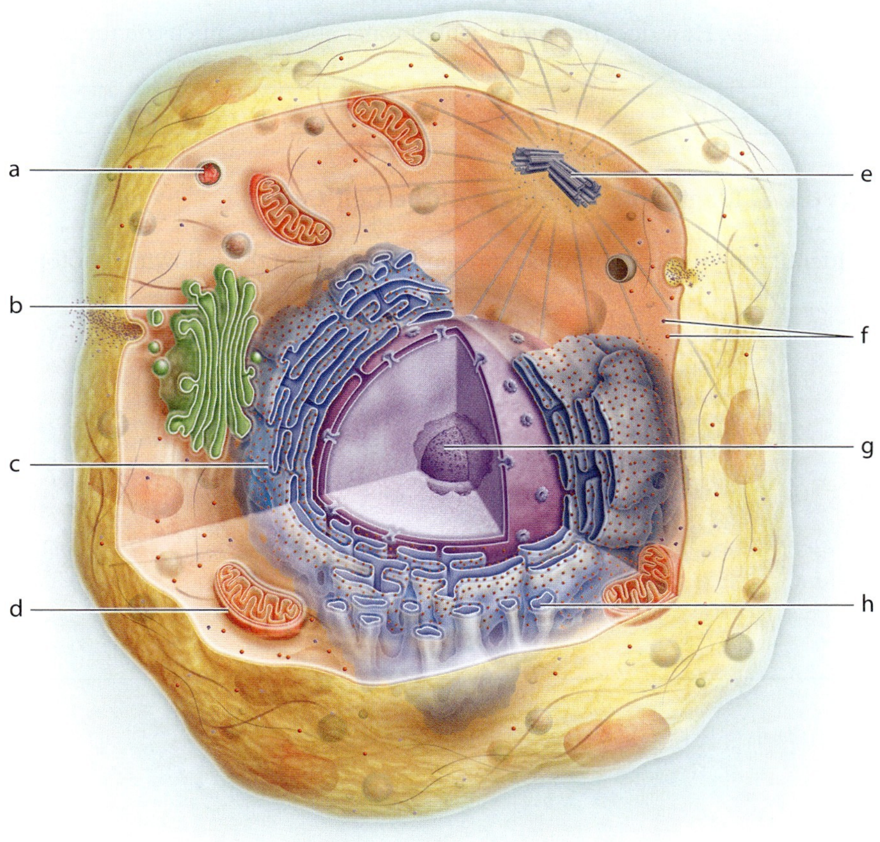

2. Fill in the blank with the appropriate organelle.

   a. Functions in the synthesis of steroid hormones. ____________________

   b. Synthesizes actin and myosin. ____________________

   c. Is the smallest of the cytoskeletal elements. ____________________

   d. Contain digestive enzymes. ____________________

   e. Receives proteins from the rough endoplasmic reticulum. ____________________

Activity 3: Examining the Possible Role of Osmosis in Cystic Fibrosis

1. The movement of a ____________________ (usually water) from an area of ____________ concentration to an area of ____________ concentration through a ____________________ ____________________ membrane is called osmosis.

2. One possible hypothesis explaining the cause of cystic fibrosis involves a defective protein in the plasma membrane. As a result of this defective protein, chloride ions are unable to leave the epithelial cells of the respiratory and digestive tracts and water moves ____________________ the cell by osmosis. How does this movement of water affect the mucus on the surface of the epithelial cells?

   ____

   ____

Activity 4: Identifying the Stages of the Cell Cycle

1. Identify the stages of mitosis represented by each of the following photomicrographs.

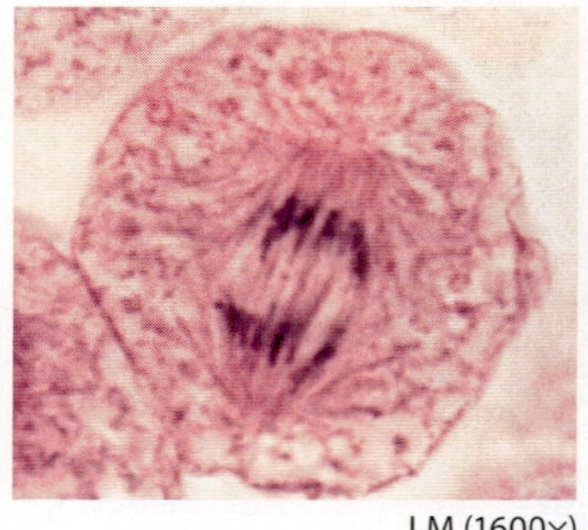

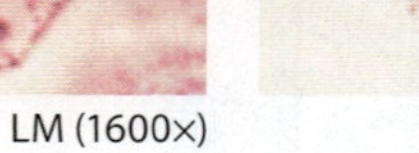

LM (1600×)

a ____________________

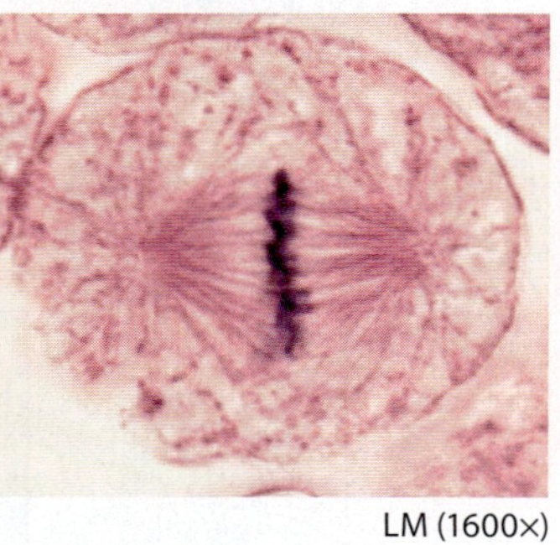

LM (1600×)

b ____________________

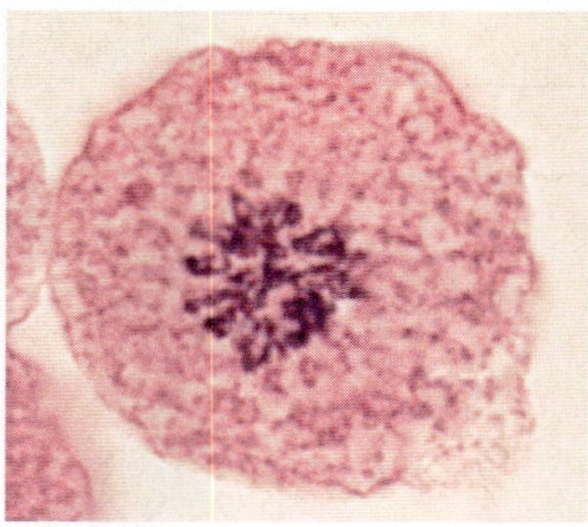

LM (1600×)

c ____________________

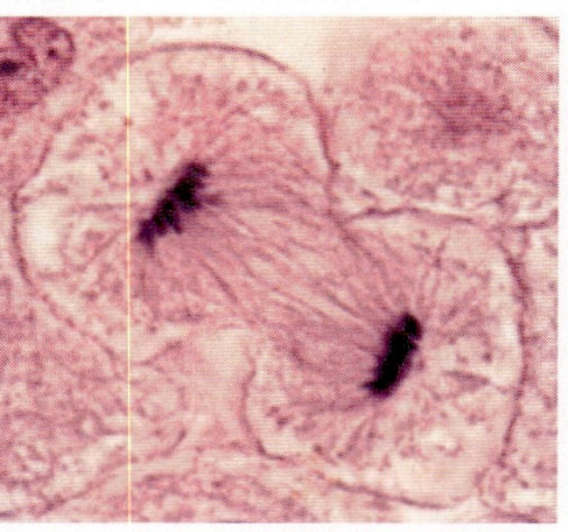

LM (1600×)

d ____________________

Activity 5: Exploring Cellular Diversity

1. Identify the following cell type and label its three parts.

___

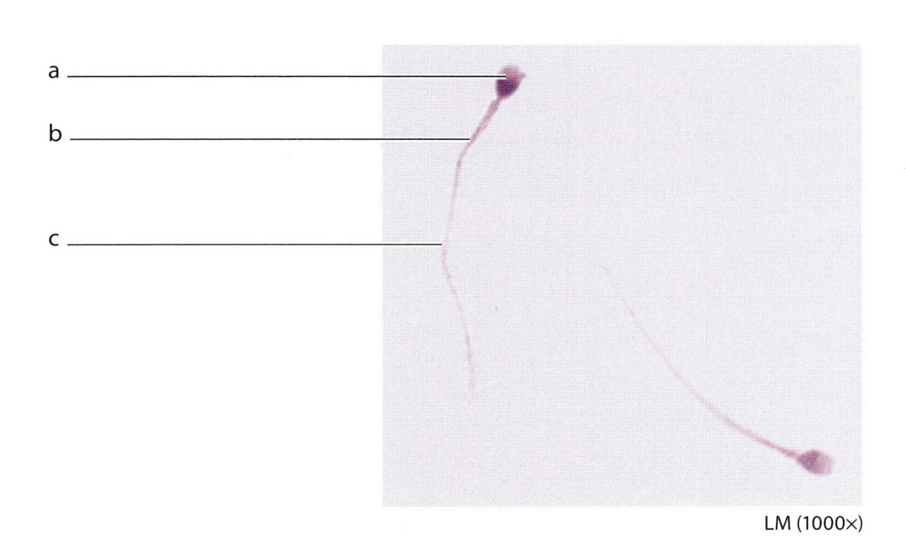

LM (1000×)

2. Identify the indicated cell type and name one unique feature of it.

___

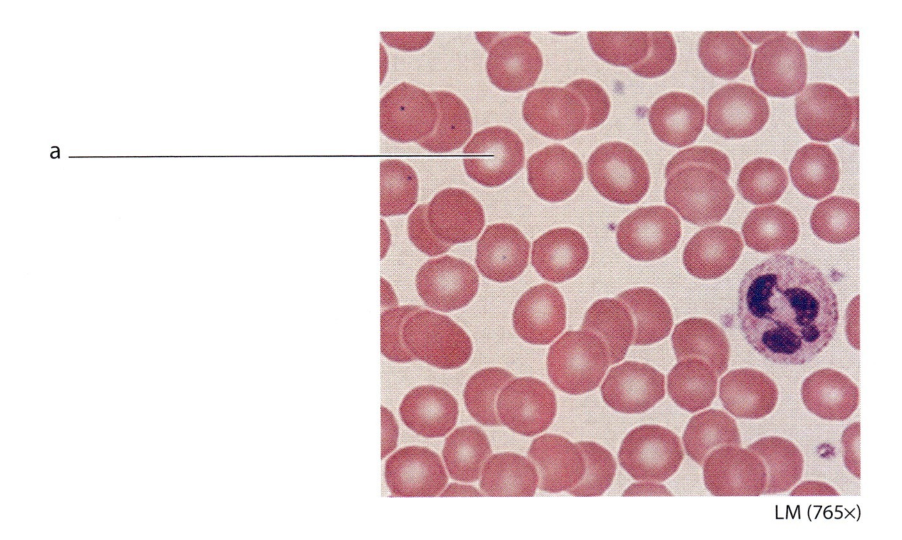

LM (765×)

PART II. Putting It All Together

A. Review Questions

Answer the following questions using your lecture notes, your textbook, and your lab notes:

1. Circle the organelle that you would expect to be most abundant in each of the following cells:
   a. A cell that produces an abundance of steroid hormones
      smooth endoplasmic reticulum; ribosome; rough endoplasmic reticulum
   b. A cell that synthesizes and secretes proteins
      Golgi apparatus; peroxisome; lysosome
   c. A cell that exhibits a high rate of metabolic activity
      ribosome; centrosome; mitochondrion
   d. A cell that detoxifies alcohol
      peroxisome; rough endoplasmic reticulum; inclusion
2. Fill in the blanks as you describe how a pancreatic β-cell produces and secretes insulin: Insulin, a protein, is produced by ribosomes attached to the ____.
   The insulin is then packaged by the ____.
   Secretory vesicles pinch off and the protein is released from the cell via a process called
   ____________________________.
3. In Activity 3, the dialysis tubing containing 0.9% NaCl represented a "normal cell" and the dialysis tubing containing 20% NaCl represented a "defective cell." Each of these "cells" was placed in a 0.9% NaCl solution. Circle the correct term to describe the tonicity of the "cells" and the solution in which each was placed.
   a. The "normal cell" was ________________________ (hypertonic, isotonic, hypotonic) compared to the solution in which it was placed.
   b. The "defective cell" was ________________________ (hypertonic, isotonic, hypotonic) compared to the solution in which it was placed.
4. Determine whether each of the following characteristics is unique to a sperm cell (sperm), unique to a nerve cell (nerve), or common to both cell types (both).
   a. Contains a flagellum. ________________________
   b. Contains mitochondria. ________________________
   c. Is uninucleate. ________________________
   d. Contains an axon. ________________________
   e. Transmits electrical messages. ________________________

5. Determine whether each of the following characteristics is unique to a mature red blood cell (RBC), unique to a cardiac muscle fiber (muscle fiber), or common to both cell types (both).

   a. Is packed with proteins. ________________________

   b. Is anucleate. ________________________

   c. Is surrounded by a cell membrane. ________________________

   d. Is packed with mitochondria. ________________________

   e. Transports oxygen. ________________________

B. Concept Mapping

1. Fill in the blanks to complete this concept map outlining the structure and function of a highly specialized skeletal muscle fiber.

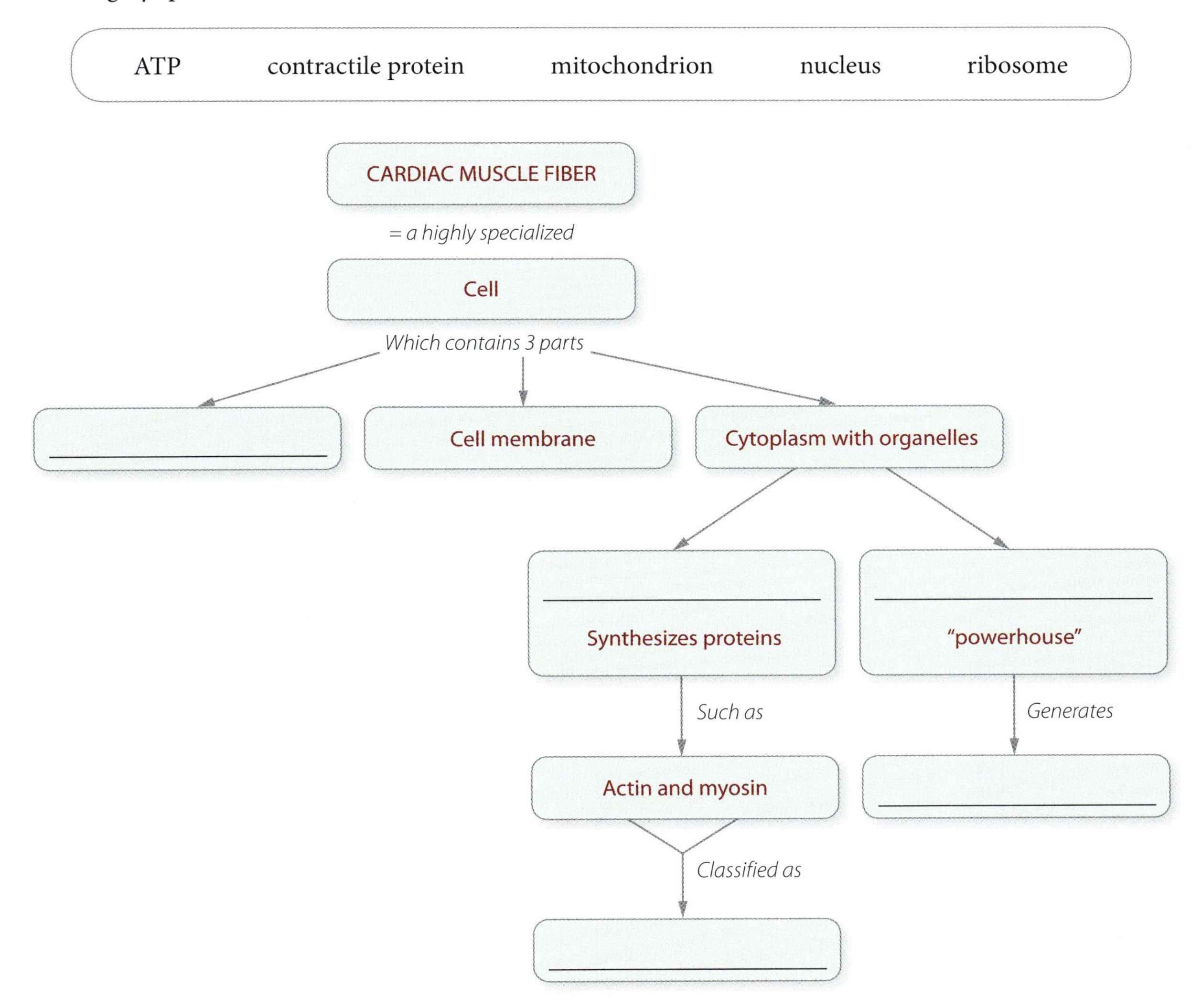

2. Construct a unit concept map to show the relationships among the following set of terms. Include all of the terms in your diagram. Your instructor may choose to assign additional terms.

ATP	cardiac muscle fiber	contractile proteins	erythrocyte	exocytosis
filtration	flagellum	Golgi apparatus	hemoglobin	lysosome
mitochondrion	osmosis	ribosome	secretory vesicle	spectrin

6 Histology

UNIT OUTLINE

A **tissue** is a group of structurally and functionally related cells and their external environment that together perform common functions; histology is the study of the normal structure of tissues. Learning about individual tissue types is much easier if you understand the major locations and basic functions of the four tissue types in the adult human body: **epithelial tissue**, which covers and lines all body surfaces and cavities; **connective tissue**, which is very widespread and performs binding, support, protection, and transport functions; **muscle tissue**, which contracts and generates force; and **nervous tissue**, which generates, sends, and receives electrical signals throughout the body.

Each of the body's organs is composed of two or more tissue types. The small intestine, for example, contains all four tissue types. The innermost layer of the intestinal wall consists largely of epithelial tissue that secretes enzymes and absorbs nutrients. The second layer consists of connective tissue that supports the epithelium and contains a rich supply of blood vessels that carry absorbed nutrients away from the small intestine. The third layer consists of smooth muscle tissue, which contracts to mix the intestinal contents and propel them toward the large intestine. Nerve fibers penetrating all the layers of the intestinal wall transmit sensory information from the small intestine to the CNS, and motor commands from the CNS to the smooth muscle layers.

An understanding of the locations, arrangement, and basic functions of tissues is important in understanding the structure and function of individual organs, but we also need to know the microscopic characteristics of these tissues and the cells that compose them. This information enables us to more fully understand the relationship between structure and function.

THINK ABOUT IT *What shape would you expect a cell to have if its major function is diffusion? Explain.*

____

____

PRE-LAB ASSIGNMENTS

To maximize learning, BEFORE your lab period carefully read this entire lab unit and complete these pre-lab assignments using your textbook, lecture notes, and prior knowledge.

PRE-LAB Activity 1: Examining Epithelial Tissue

1. Epithelial cells can be all of the following shapes *except*:
   a. cuboidal.
   b. squamous.
   c. spherical.
   d. columnar.
2. The term *stratified* means that the epithelial tissue:
   a. has multiple layers of cells.
   b. has cube-shaped cells.
   c. is only found on the external surface of the body.
   d. functions in absorption and secretion.
3. Use the list of terms provided to label the accompanying illustration of tissues and cells. Check off each term as you label it.

☐ cuboidal cell

☐ simple epithelium

☐ columnar cell

☐ stratified epithelium

☐ squamous cell

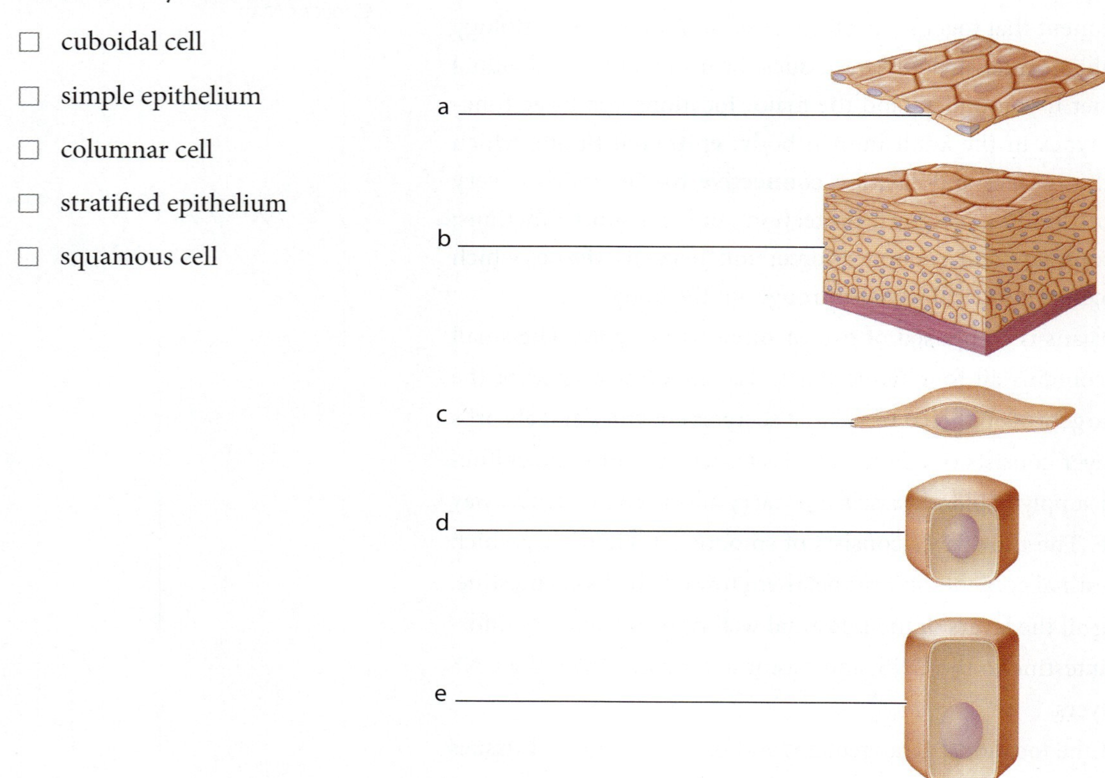

PRE-LAB Activity 2: Characterizing Connective Tissue

1. Name the four subcategories of connective tissue proper: ______________________________

____

2. Match each of the following characteristics with the correct connective tissue:

_____	a. resists unidirectional stress	1. cartilage
_____	b. transports oxygen and hormones	2. dense regular connective tissue
_____	c. contains chondrocytes	3. adipose tissue
_____	d. stores minerals	4. blood
_____	e. resides in thick skin and around joints	5. bone
_____	f. cushions and insulates	6. dense irregular connective tissue

3. Name one way in which the three cartilage types differ from one another. ______________________

____

4. Which of the following cell types is prominent in connective tissue proper?
   a. fibroblast
   b. osteocyte
   c. chondrocyte
   d. erythrocyte
   e. All of these cell types are prominent in connective tissue proper.

PRE-LAB Activity 3: Exploring Muscle and Nervous Tissue

1. Which of the following statements about neurons is true?
   a. They can transmit electrical impulses.
   b. They are only found in the brain.
   c. They offer support and protection to neuroglial cells.
   d. They have a simple, spherical shape.

2. Which types of muscle tissue contain uninucleate cells?
   a. skeletal and cardiac
   b. cardiac and smooth
   c. skeletal and smooth
   d. skeletal, cardiac, and smooth

PRE-LAB Activity 4: Putting It All Together—Tissue Identification Concept Map

1. Match each of the following characteristics with its correct tissue type:

_____	a. propels food through the digestive tract	1. epithelial tissue
_____	b. lines internal and external body surfaces	2. connective tissue
_____	c. conducts electrical impulses	3. muscle tissue
_____	d. transports oxygen to the body tissues	4. nervous tissue
_____	e. functions in transport and secretion	
_____	f. pumps blood	
_____	g. stores fats and minerals	

Epithelial Tissue

Epithelial tissue covers both internal body surfaces (such as the inner lining of the stomach) and external body surfaces (such as the skin). It also forms glands, such as sweat glands. Epithelia perform a wide variety of functions, including transport (diffusion and active transport), secretion, and protection.

An epithelium consists of tightly packed cells sitting on an adhesive, acellular structure called a basement membrane, which attaches it to underlying connective tissue. Epithelial tissue is avascular—it has no blood supply of its own. Thus it depends on the diffusion of gases and nutrients from the underlying connective tissue to obtain the substances it needs.

Each epithelium has two basic characteristics: the number of cell layers and the shape of its cells (**Figure 6-1**). Epithelia that have a single layer of cells are called simple epithelia, whereas those that have two or more layers are called stratified epithelia. Epithelial cells are either flattened (squamous), cube shaped (cuboidal), or column shaped (columnar).

Epithelia are named according to the *combination* of cell shape and number of cell layers (**Figure 6-2**). A single layer of flattened epithelial cells, for example, is called a **simple squamous epithelium** (Figure 6-2a); an example is the air sacs in the lungs. A single layer of cube-shaped epithelial cells and a single layer of column-shaped cells are called a **simple cuboidal epithelium** (Figure 6-2b) and a **simple columnar epithelium** (Figure 6-2c), respectively. The kidney tubules have simple cuboidal epithelia; simple columnar epithelia are found in the small intestine.

A **stratified squamous epithelium** consists of two or more cell layers in which the basal cells (those closest to the basement membrane) are cuboidal or columnar, but the apical cells (those near the apical or free surface) are flattened. (Note that the description of cell shape in a stratified epithelium refers to the shape of the cells on the apical surface.) Stratified squamous epithelia are subdivided into two types: nonkeratinized and keratinized. In a nonkeratinized stratified squamous epithelium (Figure 6-2d), the outermost layer of cells are living, whereas in a keratinized stratified squamous epithelium the outermost cells are dead and filled with the waterproofing protein keratin. Two or more layers of cube-shaped cells form a **stratified cuboidal epithelium** (Figure 6-2e), whereas two or more layers of column-shaped cells form a **stratified columnar epithelium** (Figure 6-2f).

The number of layers directly determines the function of epithelial tissue. The single cell layer of simple epithelia easily permits passage of materials across it, so they are important in transport and secretion. The many cell layers in stratified epithelia provide protection against friction and abrasion.

There are two unique examples of simple and stratified epithelia. **Pseudostratified** (*pseudo* = false) **columnar epithelium** (Figure 6-2g), such as that in the trachea, consists of column-shaped cells of differing heights containing nuclei at different levels of the cells. This epithelium appears to be stratified, but because every cell rests on the basement membrane, it consists of only a single layer of cells. Pseudostratified epithelium can be ciliated (as in Figure 6-2g) or nonciliated. **Transitional epithelium** (Figure 6-2h) is a stratified epithelium found only in the urinary system. The unique dome-shaped cells on its apical surface change shape according to the degree of stretch required of the structure in which they are found. The apical cells are flattened when the structure is distended, and rounded (dome shaped) when the structure is empty.

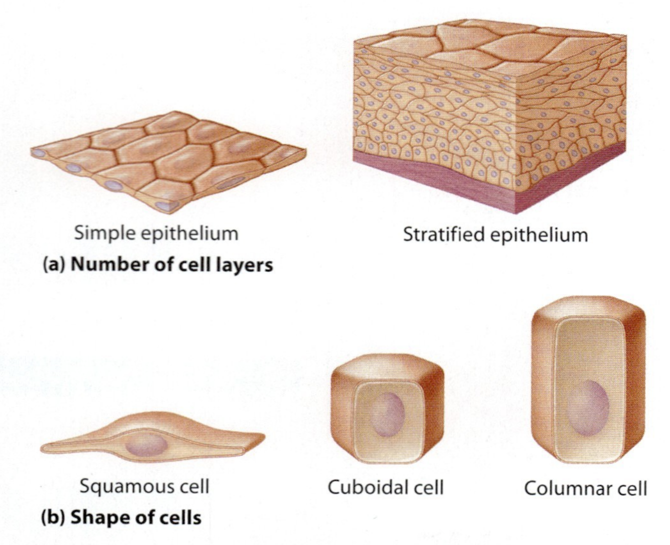

Figure 6-1 Classification of epithelial tissue based on (a) number of cell layers and (b) shapes of the cells.

ACTIVITY 1 Examining Epithelial Tissue

Learning Outcomes

1. Describe the microscopic structure of the following specific types of epithelial tissue: simple squamous, simple cuboidal, simple columnar, nonkeratinized stratified squamous, ciliated pseudostratified columnar, and transitional.
2. Describe the basic functions of the epithelial tissues listed in the previous Learning Outcome.
3. List the major locations where each of the epithelial tissues is found.

Materials Needed

- ☐ Microscope and prepared microscope slides (or photomicrographs) of the following six types of epithelial tissue:
  Simple squamous epithelium
  Simple cuboidal epithelium
  Simple columnar epithelium
  Nonkeratinized stratified squamous epithelium
  Ciliated pseudostratified columnar epithelium
  Transitional epithelium
- ☐ 5″ × 7″ lined index cards
- ☐ Colored pencils

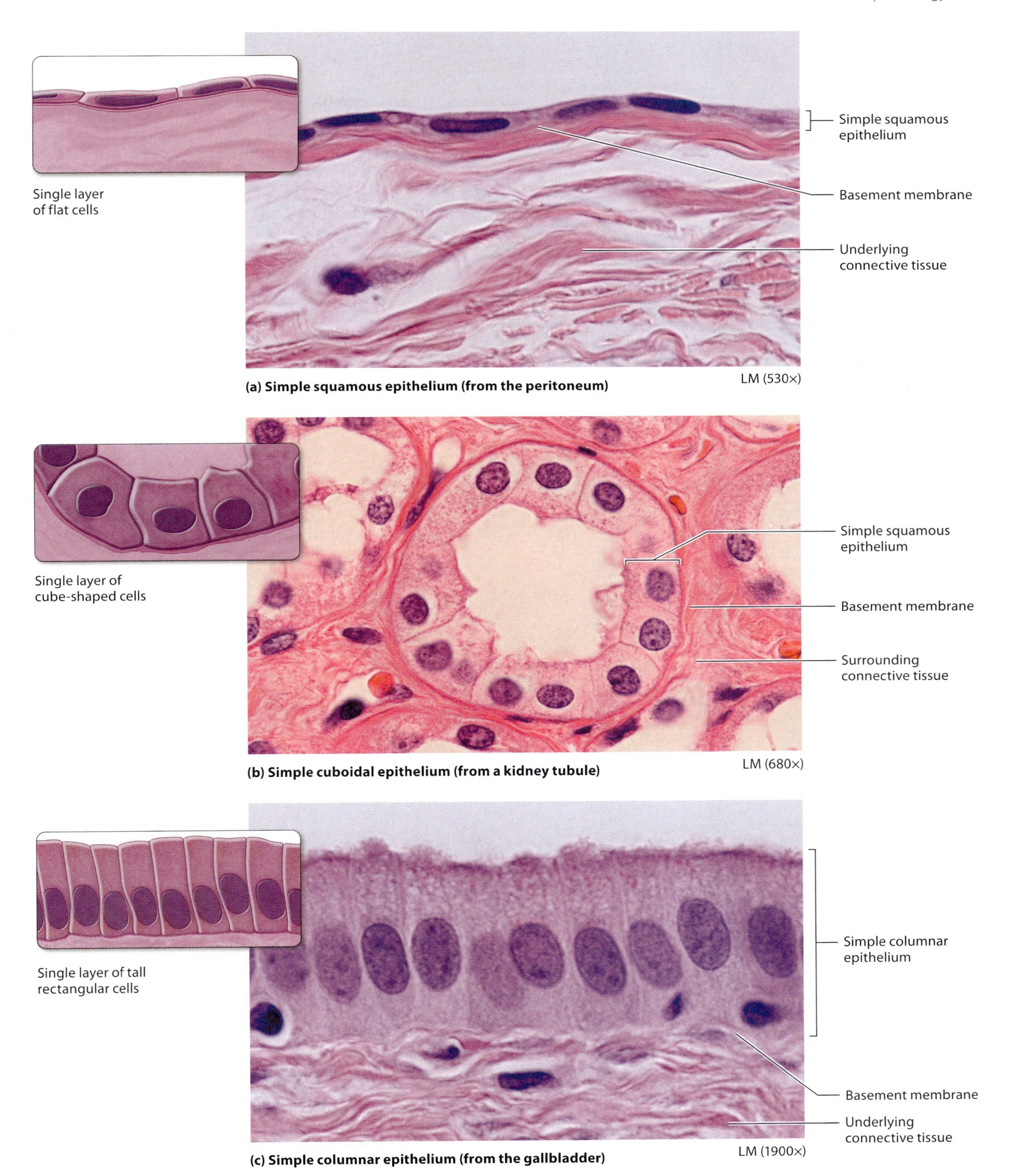

Figure 6-2 The histological appearance of eight types of epithelia *(continues)*.

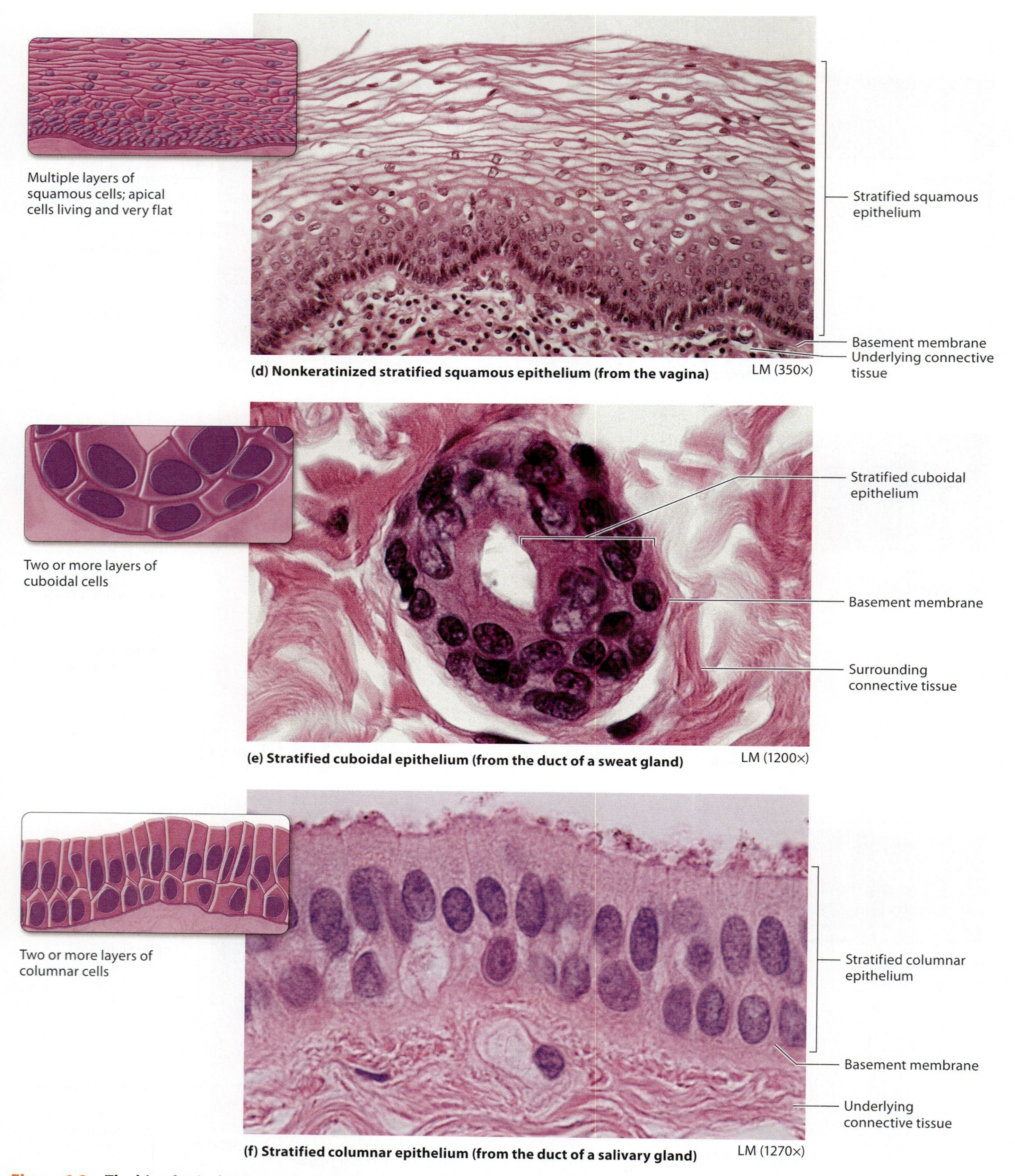

(d) Nonkeratinized stratified squamous epithelium (from the vagina)

(e) Stratified cuboidal epithelium (from the duct of a sweat gland)

(f) Stratified columnar epithelium (from the duct of a salivary gland)

Figure 6-2 The histological appearance of eight types of epithelia *(continued)*.

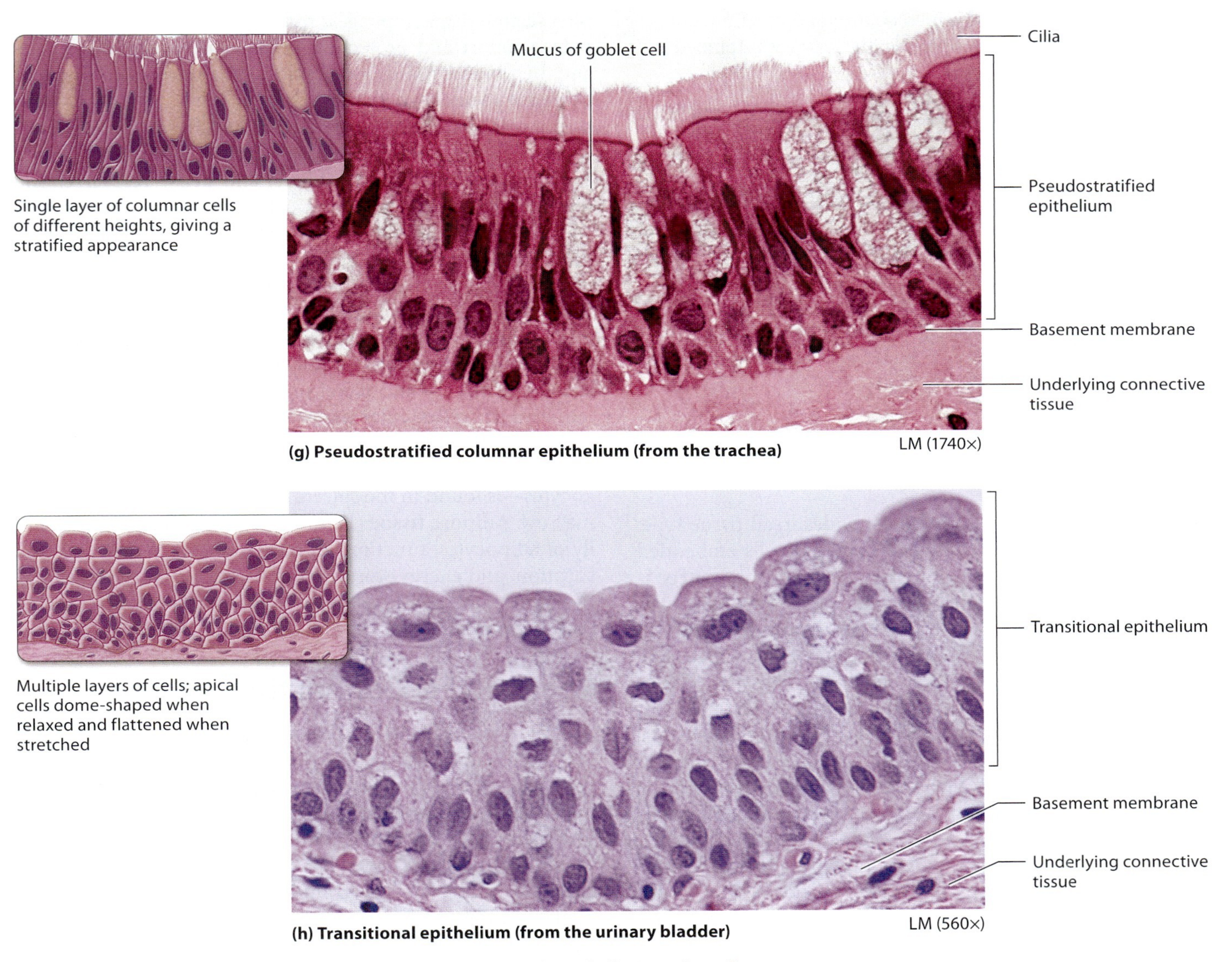

(g) Pseudostratified columnar epithelium (from the trachea)

(h) Transitional epithelium (from the urinary bladder)

Figure 6-2 The histological appearance of eight types of epithelia *(continued)*.

Instructions

Divide the six types of epithelial tissue among the members of your lab group. (Assign each student one or two tissue types.) For each of your assigned tissue types, do the following:

1. Examine the microscope slide or photomicrograph. Then make a colored sketch of the tissue on the unlined side of an index card, labeling the important structures. Make your sketch as large and as clear as possible.
2. On the lined side of the card, use your own words to describe the tissue and list its locations and functions in the body. Create mnemonic devices and/or include any additional information that can help you learn and remember the characteristics of the tissue.
3. After completing steps 1–2, have each group member "teach" the rest of the group about his or her assigned tissue(s).
4. As a group, answer the following questions based on what you have learned about epithelial tissues:
   a. What do all six of these tissues have in common?

   ____

   ____

   ____

   b. Compare and contrast nonkeratinized stratified squamous epithelium and transitional epithelium.

   ____

   ____

   ____

c. Is ciliated pseudostratified columnar epithelium correctly grouped with simple epithelia or with stratified epithelia? Why?

____

____

5. Make copies or take photographs of all index cards so that each student in your lab group will have a complete set of epithelial tissue cards.

6. **Optional Activity**

View and label histology slides of epithelial tissues at MasteringA&P® > Study Area > Practice Anatomy Lab > Histology > Epithelial Tissue

Connective Tissue

Connective tissue is the body's most widespread tissue type. Most connective tissues consist of scattered cells embedded in an abundant extracellular matrix (ECM). The matrix is composed of a ground substance plus various types of protein fibers and ranges from a liquid (as in blood) to a solid (as in bone). Protein fibers include collagen fibers, which give the tissue strength and resist tension and pressure; elastic fibers, which give the tissue flexibility; and reticular fibers, which provide a supporting network for the entire tissue.

The various combinations of ground substance, protein fibers, and cells enable connective tissue to perform a large variety of functions: binding, support, protection, and transport. Connective tissues can be divided into two groups: **connective tissue proper** (**Figure 6-3**), which includes loose connective tissue, dense connective tissue, reticular tissue, and adipose tissue, and **specialized connective tissue** (**Figure 6-4**), which includes cartilage, bone, and blood. The types and subtypes of connective tissue are discussed next.

Connective Tissue Proper

Connective tissue proper (Figure 6-3) contains four major types of cells. The most prominent cells are fibroblasts, which produce the protein fibers of the ECM. Adipocytes (fat cells) are filled with lipid droplets and are found in many different connective tissues. Also present are two specialized types of leukocytes: macrophages (which are phagocytic) and mast cells (which function in inflammation).

Connective tissue proper includes four types of tissue. **Loose connective tissue** (also known as areolar connective tissue; Figure 6-3a) consists of fibroblasts and all three types of protein fibers embedded in a gel-like ground substance, plus scattered macrophages, mast cells, and fat cells. It provides support and protection in the walls of hollow organs and membranes lining body cavities. The primary component of **dense connective tissue** is protein fibers, which provide strength. The three types of dense connective tissue are dense regular (Figure 6-3b) and dense irregular (Figure 6-3c), which primarily contain collagen fibers, and dense regular elastic (also known as elastic; Figure 6-3d), which primarily contains elastic fibers. Dense regular connective tissue is found in tendons and ligaments and resists unidirectional stress because of the parallel arrangement of fibers. Dense irregular connective tissue, which is found in the deep layer of thick skin and around joints, resists stress from every direction because of the haphazard arrangement of fibers. Elastic connective tissue allows stretch and recoil in large blood vessels and certain ligaments. **Reticular tissue** (Figure 6-3e)—named for the fine network of fibers that forms the structure of many organs and supports small structures such as blood vessels and leukocytes—is found in the spleen, liver, lymph nodes, and bone marrow. **Adipose tissue** (Figure 6-3f), which consists primarily of adipocytes, functions in insulation, warmth, shock absorption, and energy storage. Adipose tissue is found deep to the skin; in the abdomen, breasts, hips, buttocks, and thighs; and surrounding the heart and abdominal organs.

Specialized Connective Tissues

As their name suggests, specialized connective tissues (Figure 6-4) perform more specialized functions than does connective tissue proper. **Cartilage** functions in support, maintaining the shape of structures, and shock absorption. The major cells in cartilage are chondrocytes, which are located within cavities called lacunae. The three types of cartilage are hyaline cartilage, found in the trachea and between bones in joints (Figure 6-4a); fibrocartilage, found between intervertebral discs (Figure 6-4b); and elastic cartilage, found in the external ear and the epiglottis (Figure 6-4c). **Bone** functions in support and protection, serves as attachment sites for muscles, produces blood, and stores fat and minerals. The major cells in bones are osteocytes, which (like chondrocytes) are located within lacunae (Figure 6-4d). The extracellular matrix of bone consists of an organic component that gives the tissue flexibility and an inorganic component that gives it strength. **Blood** (Figure 6-4e) contains three so-called "formed elements": erythrocytes (red blood cells), which transport oxygen; leukocytes (white blood cells), which function in immunity; and cell fragments called platelets, which function in blood clotting. The formed elements are suspended in a fluid extracellular matrix called plasma, which transports gases, nutrients, wastes, and hormones throughout the body.

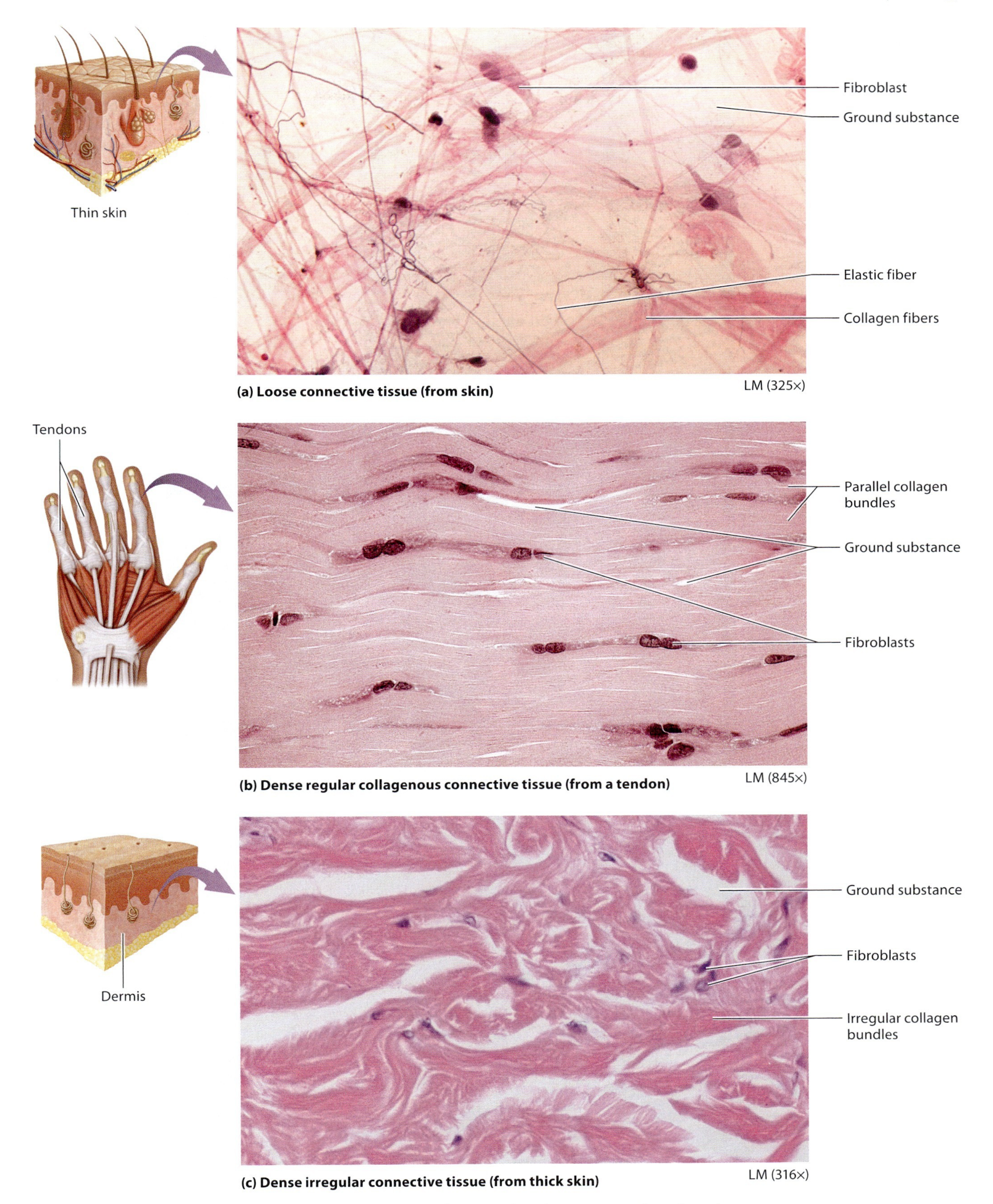

Figure 6-3 The histological appearance of six types of connective tissue proper *(continues)*.

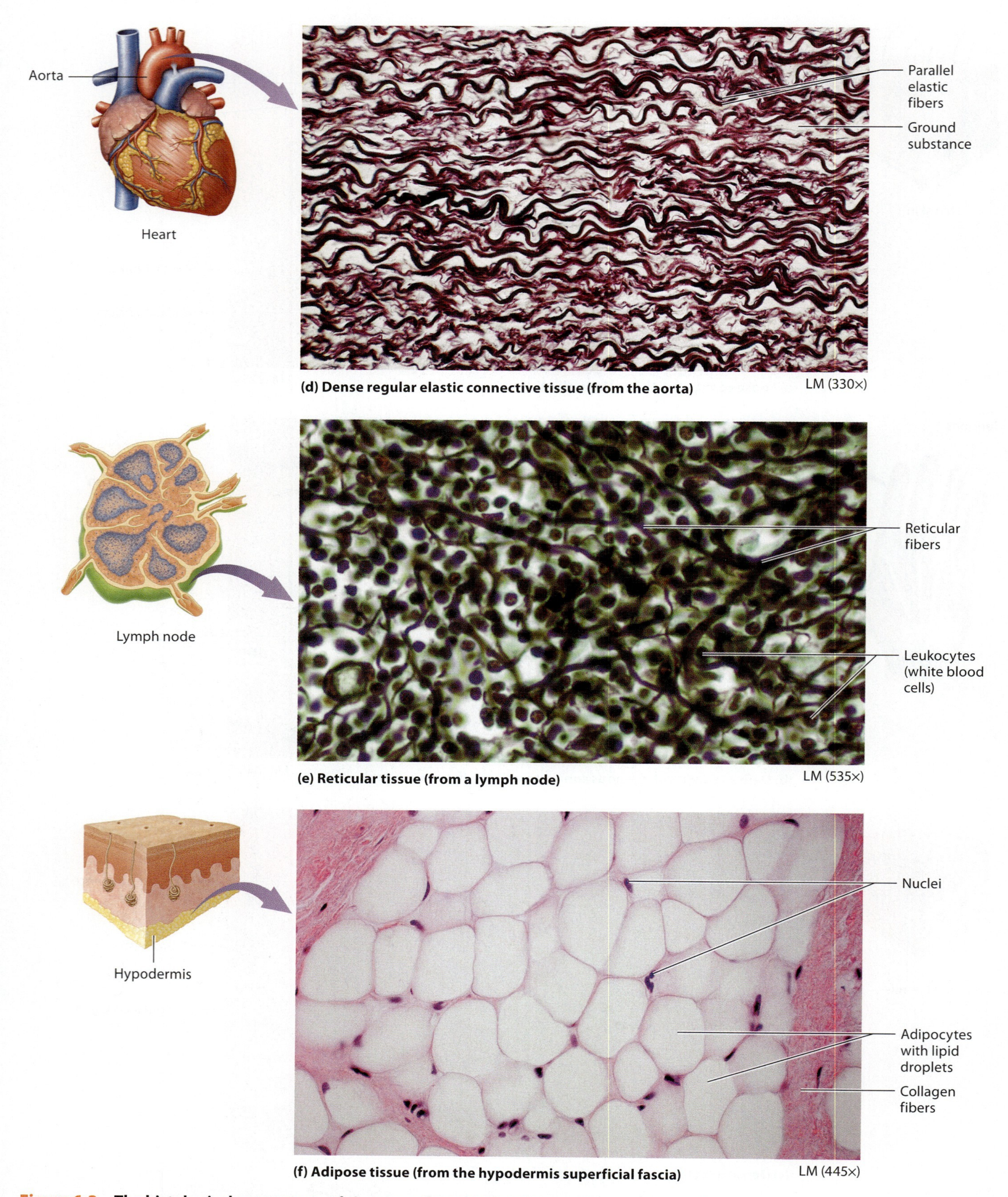

Figure 6-3 **The histological appearance of six types of connective tissue proper *(continued).***

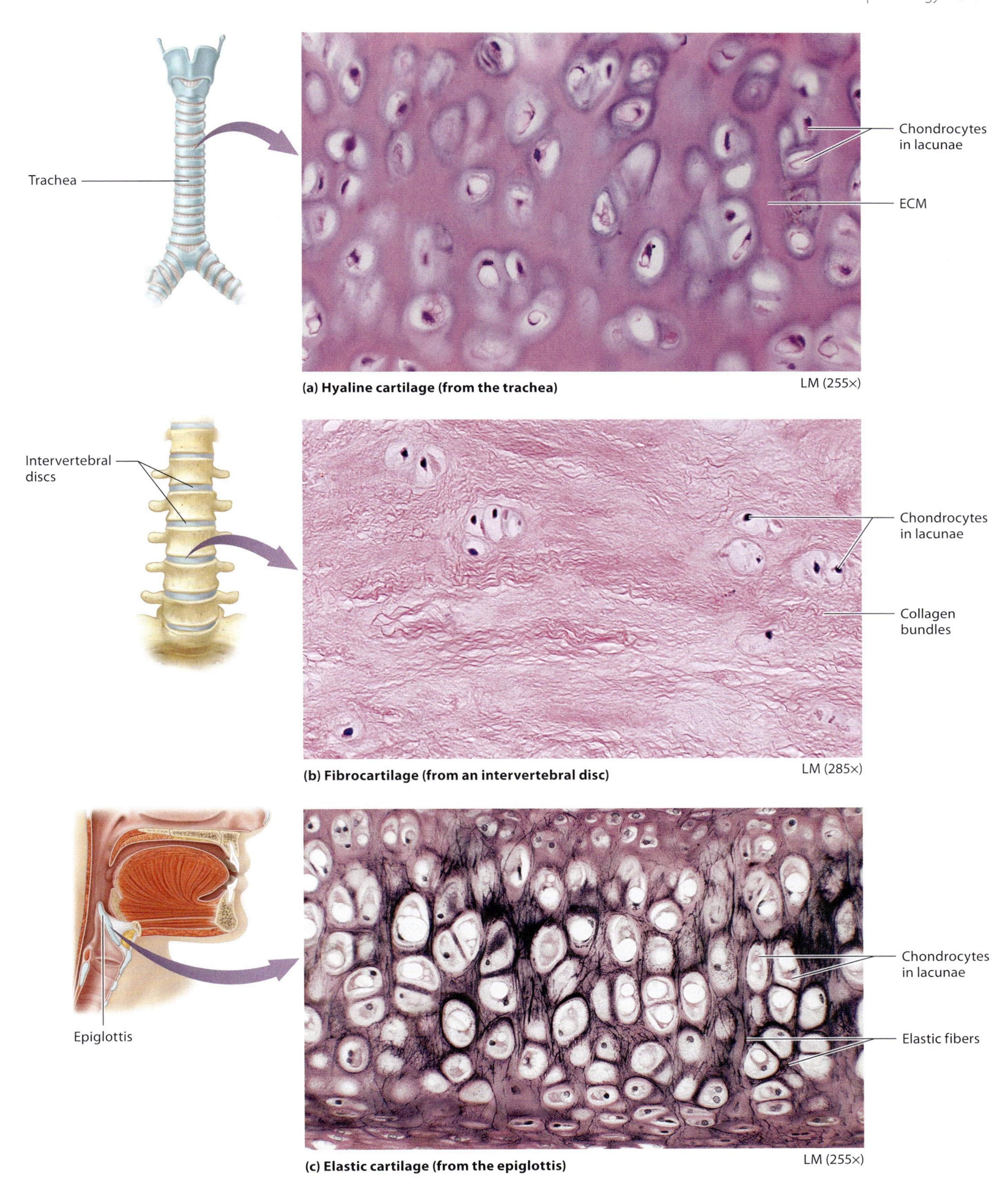

Figure 6-4 **The histological appearance of five types of specialized connective tissue *(continues)*.**

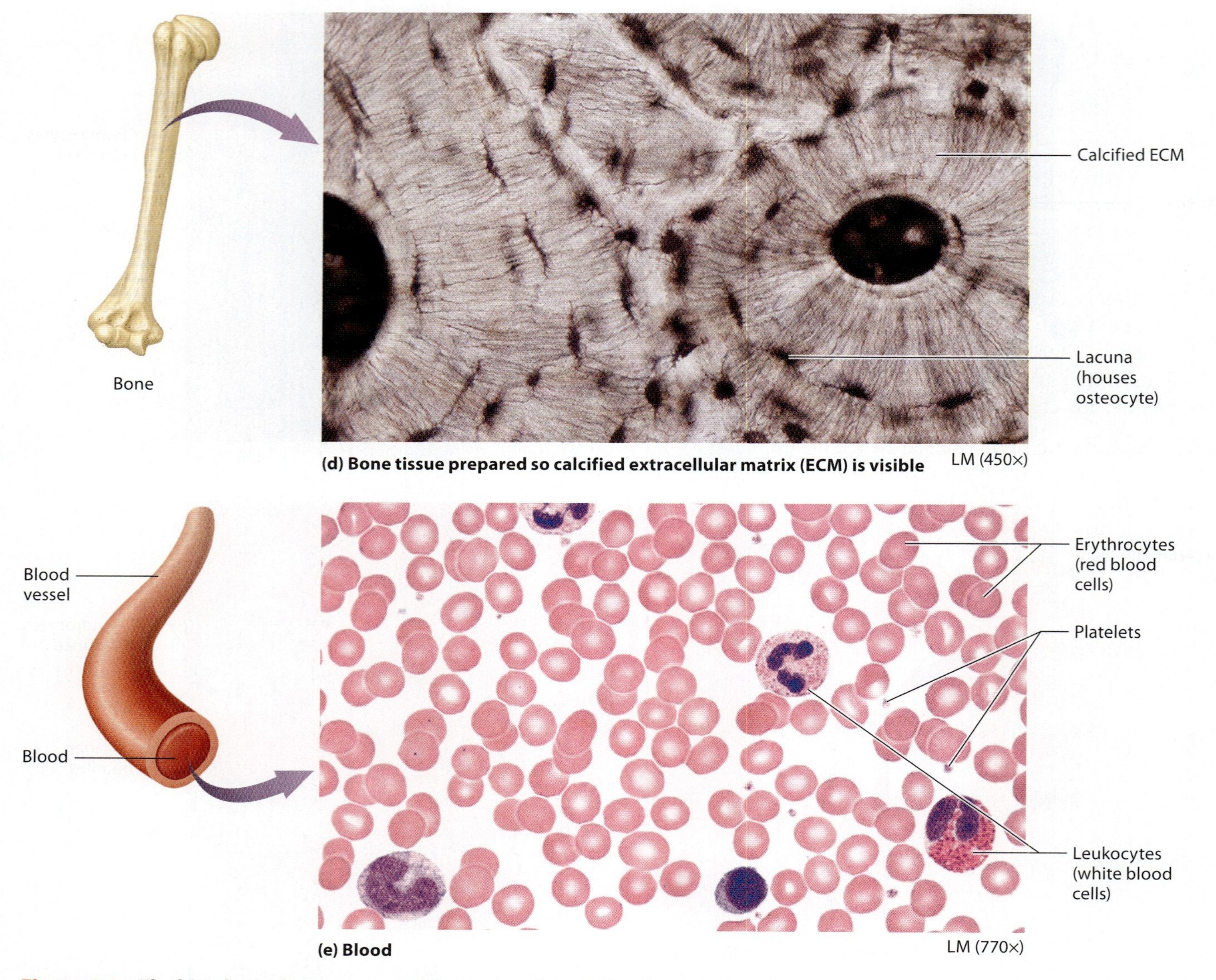

(d) Bone tissue prepared so calcified extracellular matrix (ECM) is visible

(e) Blood

Figure 6-4 **The histological appearance of five types of specialized connective tissue *(continued).***

ACTIVITY 2

Characterizing Connective Tissue

Learning Outcomes

1. Describe the microscopic structure of the following specific types of connective tissue: loose, adipose, dense regular, dense irregular, elastic, reticular, hyaline cartilage, elastic cartilage, fibrocartilage, bone, and blood.
2. Describe the basic functions of the connective tissues listed in the previous Learning Outcome.
3. List the major locations where each of the connective tissues is found.

Materials Needed

- ☐ Microscope and prepared microscope slides (or photomicrographs) of the following types of connective tissue:
  - Loose connective tissue
  - Adipose tissue
  - Dense regular connective tissue
  - Dense irregular connective tissue
  - Elastic
  - Reticular
  - Hyaline cartilage
  - Elastic cartilage
  - Fibrocartilage
  - Bone
  - Blood
- ☐ 5″ × 7″ lined index cards
- ☐ Colored pencils

Instructions

Divide the 11 types of connective tissue among the members of your lab group. (Assign each student two or more tissue types.) For each of your assigned tissue types, do the following:

1. Examine the microscope slide or photomicrograph. Then make a colored sketch of the tissue on the unlined side of an index card, labeling the important structures. Make your sketch as large and as clear as possible.
2. On the lined side of the card, use your own words to describe the tissue and list its locations and functions in the body. Create mnemonic devices and/or include any additional information that can help you learn and remember the characteristics of the tissue.
3. After completing steps 1–2, have each group member "teach" the rest of the group about his or her assigned tissue(s).
4. As a group, answer the following questions based on what you have learned about connective tissues:
   a. What do all 11 of these tissues have in common?

   ____

   ____

   b. Compare and contrast bone and fibrocartilage.

   ____

   ____

   c. How is adipose tissue different from most other connective tissue types?

   ____

   ____

5. Make copies or take photographs of all index cards so that each student in your lab group will have a complete set of connective tissue cards.
6. **Optional Activity**

View and label histology slides of connective tissues at MasteringA&P® > Study Area > Practice Anatomy Lab > Histology > Connective Tissue

Nervous Tissue

Nervous tissue (**Figure 6-5**), which is located in the brain, spinal cord, and nerves, consists of two types of cells: neurons and the more numerous neuroglial cells.

Neurons transmit electrical signals to, from, and within the central nervous system (the brain and spinal cord). Their structure is highly adapted for sending and receiving electrical impulses. In addition to a cell body containing the nucleus, neurons have two types of cell processes: dendrites, which receive and carry impulses toward the cell body, and axons, which carry impulses away from the cell body.

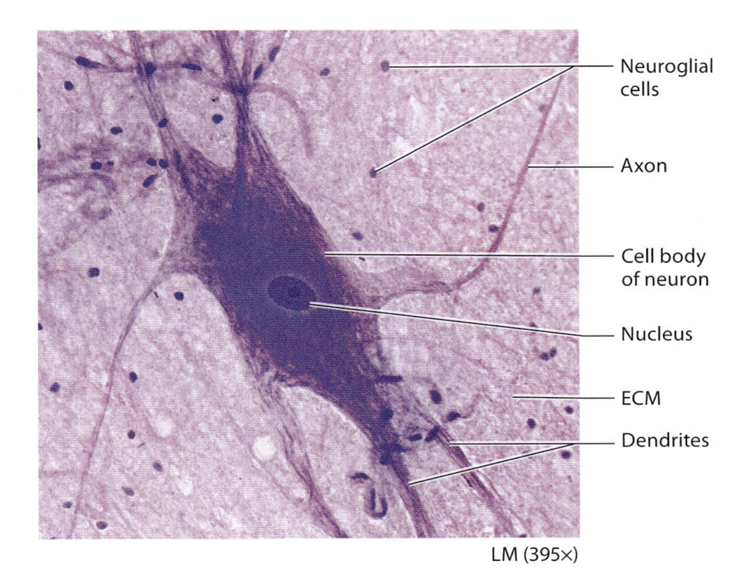

Figure 6-5 **The histological appearance of nervous tissue.**

Neuroglial cells are also referred to as supporting cells because they support, anchor, monitor, nourish, and insulate neurons. Most neuroglial cells are unable to transmit electrical signals and are thus considered "nonirritable."

Muscle Tissue

There are three types of muscle tissue, all of which are capable of contraction: skeletal muscle, which is attached to bone; cardiac muscle, which is found only in the heart; and smooth muscle, which lines hollow organs and blood vessels throughout the body (**Figure 6-6**). Muscle cells are often called muscle fibers.

The muscle fibers of **skeletal muscle** (Figure 6-6a) are long and cylindrical, multinucleate, and surrounded by a thin connective tissue sheath called endomysium. Skeletal muscle fibers appear striated (striped) due to the particular arrangement within them of the contractile myofilaments actin and myosin. Skeletal muscle fibers are under voluntary control. **Cardiac muscle** fibers (Figure 6-6b) are short, branched, typically uninucleate cells that are interconnected by intercalated discs and surrounded by endomysium. Cardiac fibers are also striated but are not under voluntary control. **Smooth muscle** fibers (Figure 6-6c) are thin, uninucleate, tapered cells that lack striations and are not voluntarily controlled.

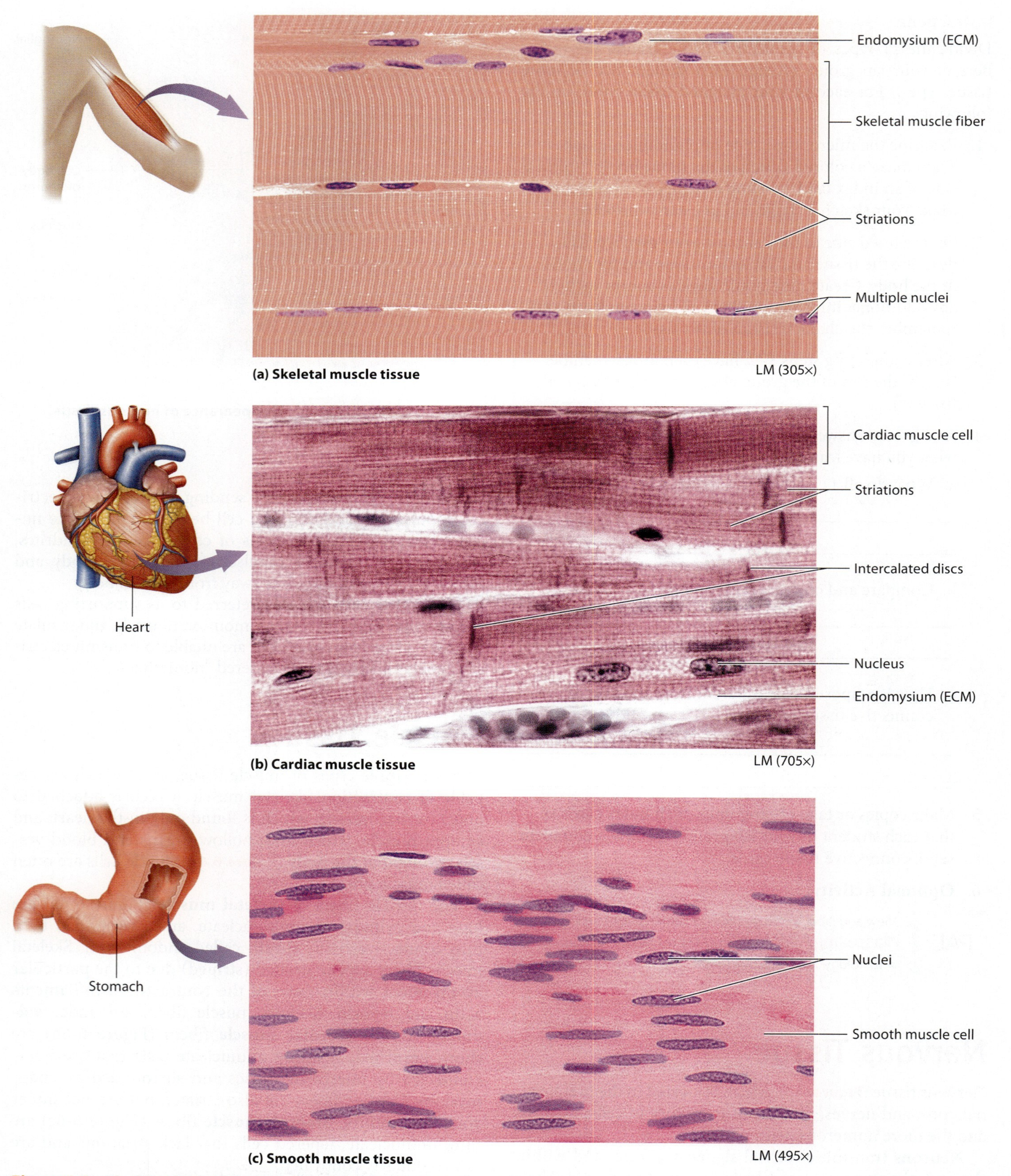

Figure 6-6 The histological appearance of the three types of muscle tissue.

ACTIVITY 3

Exploring Nervous Tissue and Muscle Tissue

Learning Outcomes

1. Identify the microscopic structure, describe the basic functions, and list the major locations of nervous tissue.
2. Describe the microscopic structure of the following specific types of muscle tissue: skeletal, cardiac, and smooth.
3. Describe the basic functions of the muscle tissues listed in the previous Learning Outcome.
4. List the major locations where each of the muscle tissues is found.

Materials Needed

- ☐ Microscope and prepared microscope slides (or photomicrographs) of the following tissue types:
  - Nervous tissue
  - Skeletal muscle tissue
  - Cardiac muscle tissue
  - Smooth muscle tissue
- ☐ 5″ × 7″ lined index cards
- ☐ Colored pencils

Instructions

Divide the four muscle and nervous tissues among the members of your lab group. (Assign each student one or two tissue types.) For each of your assigned tissue types, do the following:

1. Examine the microscope slide or photomicrograph. Then make a colored sketch of the tissue on the unlined side of an index card, labeling the important structures. Make your sketch as large and as clear as possible.
2. On the lined side of the card, use your own words to describe the tissue and list its locations and functions in the body. Create mnemonic devices and/or include any additional information that can help you learn and remember the characteristics of the tissue.
3. After completing steps 1–2, have each group member "teach" the rest of the group about his or her assigned tissue(s).
4. As a group, answer the following questions based on what you have learned about muscle and nervous tissues:
   a. What do all three types of muscle tissues have in common?

   ____

   ____

   b. Compare and contrast smooth muscle tissue and skeletal muscle tissue.

   ____

   ____

   c. In what ways is nervous tissue different from all other tissue types?

   ____

   ____

5. Make copies or take photographs of all index cards so that each student in your lab group will have a complete set of muscle and nervous tissue cards.
6. **Optional Activity**

View and label histology slides of muscle tissues and nervous tissues at MasteringA&P® > Study Area > Practice Anatomy Lab > Histology

Concept Mapping

In the following activity you will use what you have learned in the previous activities to construct a "mental framework" that organizes the material about the body's tissue types in a way that makes sense to you. Health care professionals regularly use so-called "dichotomous keys" to help them identify such things as bacteria; here you will construct a concept map that you can use to recognize and identify the various tissue types.

ACTIVITY 4

Tissue Identification Concept Map

Learning Outcome

1. Construct a concept map using the key terms presented in this unit.

Materials Needed

- ☐ 3″ × 5″ colored photomicrograph of each assigned tissue
- ☐ 3″ × 5″ colored photomicrograph of each unknown tissue
- ☐ Whiteboard or laminated poster board
- ☐ Washable markers
- ☐ Removable tape

Instructions

A. Creating a Tissue Identification Concept Map

In this activity, you and your lab group will create a concept map that can be used to identify tissues by distinguishing among their visible characteristics. The goal of this activity is to develop a set of yes/no questions that will enable you to identify each of the specific tissue types based on *observable* microscopic characteristics. To construct your concept map, follow these directions:

1. To start the process, look at a photomicrograph of an assigned tissue and ask, "Does this tissue have an apical (free) surface and a basal (attached) surface?" If the

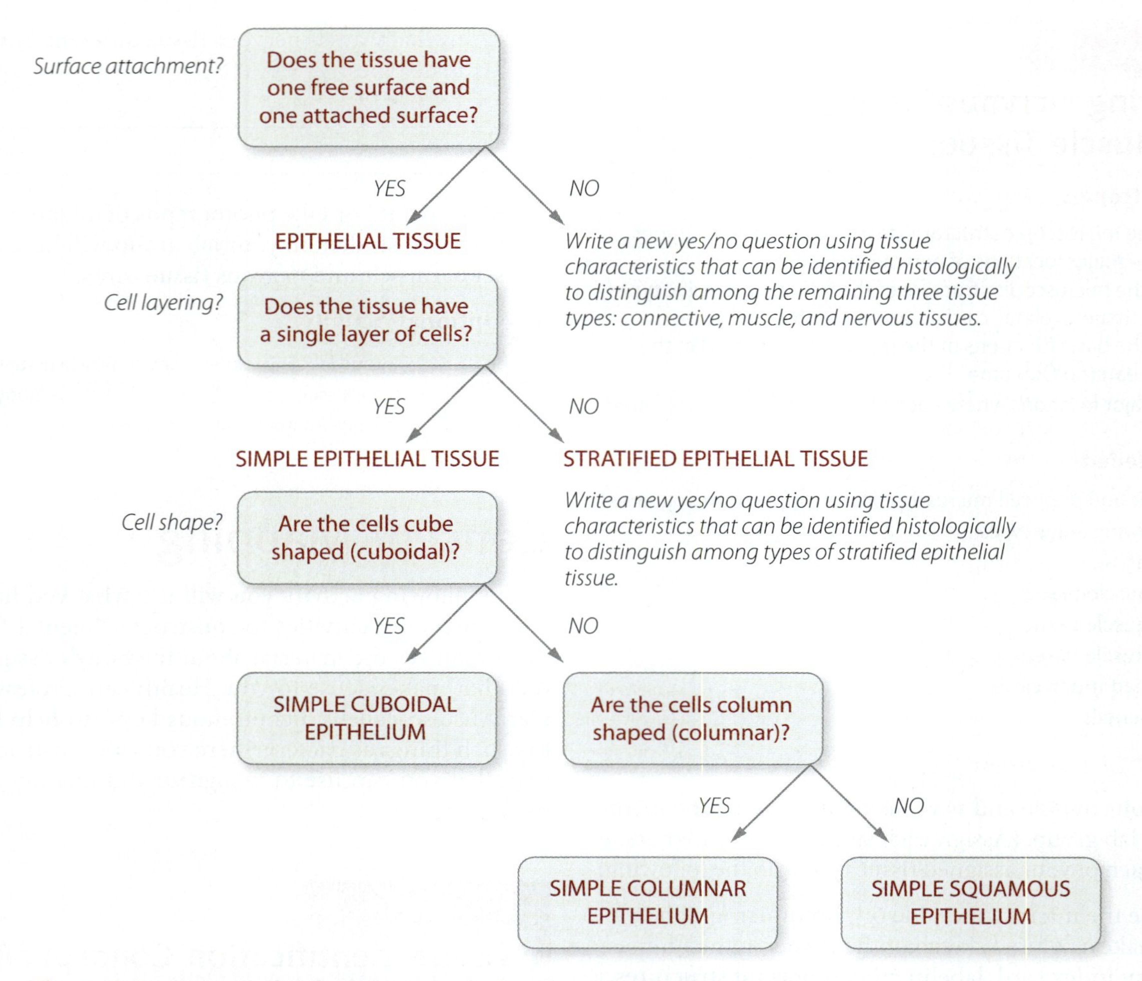

Figure 6-7 An example of one branch of a concept map for identifying tissues.

answer is "yes," then the tissue is a type of epithelial tissue, but you still need to ask more questions to identify which specific type. If the answer is "no," then this tissue is not an epithelial tissue, and you need to ask yourself other yes/no questions that can help you determine whether the tissue is instead connective, muscle, or nervous tissue.

Figure 6-7 illustrates one string of questions and answers that produces a "branch" of the concept map—that for identifying a simple columnar epithelium. Use a marker and a whiteboard or laminated poster board to write the questions and draw the "yes" and "no" arrows for each branch of the concept map.

2. Here are some examples of other yes/no questions that can be helpful in distinguishing among the four major tissue types: *Does the tissue have an abundance of extracellular matrix? Does it consist of tightly packed cells?*

   Remember, all questions must be yes/no and based solely on characteristics you can observe in a photomicrograph. Thus, "Does the tissue contain many elastic fibers?" is a good question, but "Does the tissue contain actin and myosin?" is not because these molecules are too small to be observed in a photomicrograph.

3. Continue writing questions until all the photomicrographs have been completely identified on your concept map. A branch of the concept map is complete when only a single, specific tissue type is alone at the end of the branch (see Figure 6-7). When you reach the end of a branch, tape the photomicrograph to the whiteboard or poster board and write the name of the tissue type under it.

B. Identifying Unknown Tissue Types

Your instructor will provide you with photomicrographs of four unknown tissues (unknowns A–D). Use the concept map you created to identify each of these unknown tissues. You may have already seen these unknowns in the lab or elsewhere in this unit, or you may have never seen them before. In the latter case, use other resources (for example, textbooks or the Internet) to confirm your identification.

Post-lab quizzes are also assignable in MasteringA&P®

POST-LAB ASSIGNMENTS

Name: ______________________ Date: ______________ Lab Section: ________

PART I. Check Your Understanding

Activity 1: Examining Epithelial Tissue

1. Which of the following statements regarding stratified squamous epithelium is true?
   a. It has many layers of cells.
   b. It contains predominantly cube-shaped cells.
   c. It functions in transport.
   d. It is located in the lining of the stomach.
   e. It has the ability to stretch.

2. The tissue through which gases are exchanged between the blood and the air in the lungs is:
   a. transitional epithelium.
   b. simple squamous epithelium.
   c. simple cuboidal epithelium.
   d. stratified columnar epithelium.
   e. pseudostratified columnar epithelium.

Activity 2: Characterizing Connective Tissue

1. Which of the following tissue(s) contain(s) protein fibers? (More than one answer may be correct.)
   a. dense regular connective tissue
   b. hyaline cartilage
   c. bone
   d. loose connective tissue

2. Loose connective tissue:
   a. functions primarily to generate heat.
   b. is classified as a dense connective tissue.
   c. has a mineralized matrix containing all three fiber types.
   d. wraps and cushions organs.
   e. contains fibroblasts, macrophages, and mast cells.

3. The *major* cell type in a tendon is the:
   a. chondrocyte.
   b. osteocyte.
   c. fibroblast.
   d. leukocyte.
   e. mast cell.

Activity 3: Exploring Nervous and Muscle Tissue

1. For each characteristic of muscle tissue in the following chart, place an X under each type of tissue that applies:

Characteristic	Skeletal Muscle	Smooth Muscle	Cardiac Muscle
Has multinucleate cells			
Is striated in appearance			
Is under voluntary control			
Contains intercalated discs			
Contains myofilaments			
Has short branching cells			
Has elongated tapered cells			
Attaches to bones			
Is present in the heart			
Contracts the stomach			

2. Which of the following statements regarding neuroglial cells is *false*?
   a. They send electrical signals to other cells by way of an axon.
   b. They are found in the brain, spinal cord, and nerves.
   c. They support neurons.
   d. They are more numerous than neurons in nervous tissue.

Activity 4: Tissue Identification Concept Map

1. For each of the following images, identify both the general and specific tissue type, and then label the indicated structure. The first image has been identified and labeled as an example.

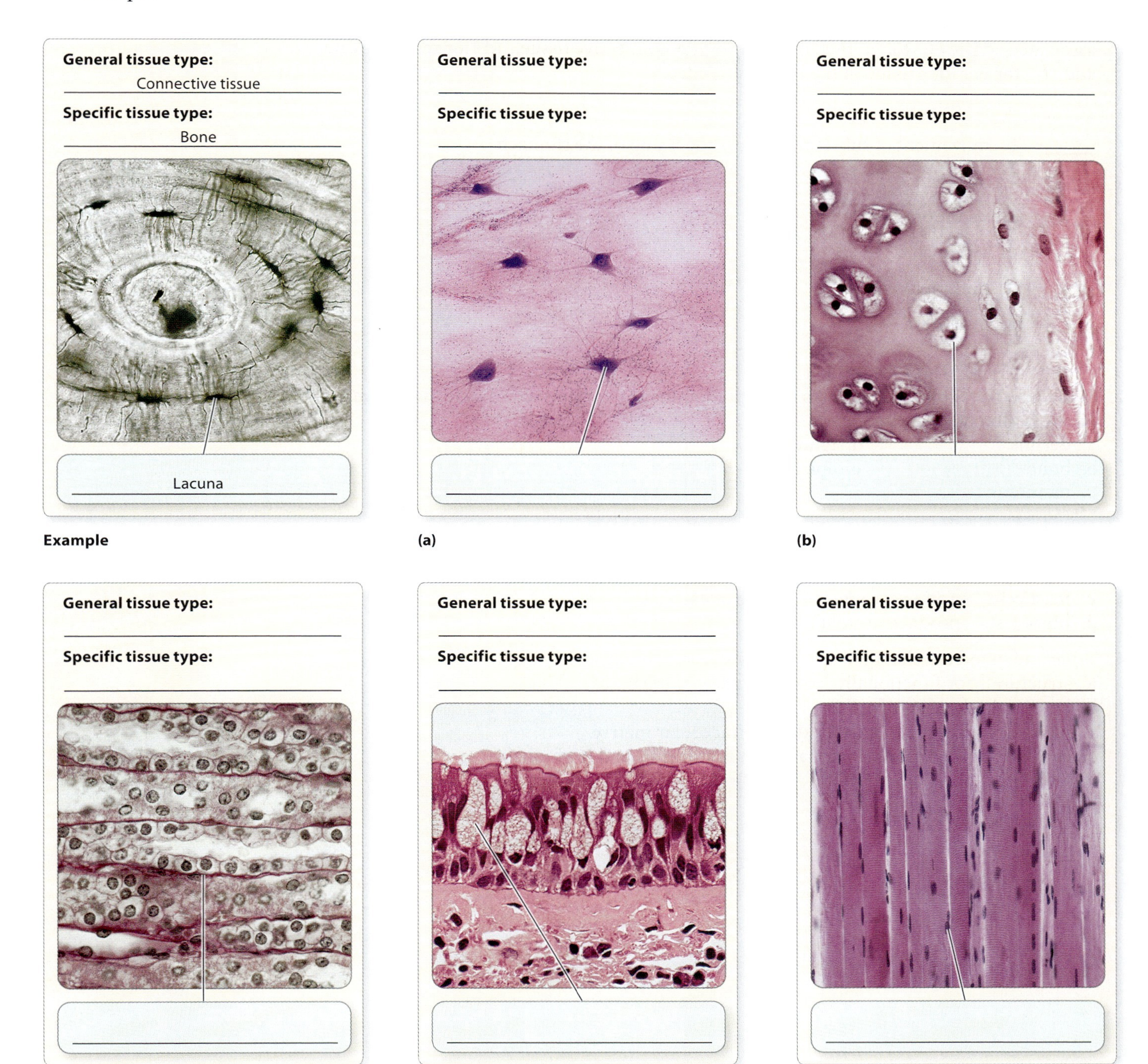

PART II. Putting It All Together

A. Review Questions

Answer the following questions using your lecture notes, your textbook, and your lab notes:

1. For each of the following structures or processes, identify the type of tissue with which it is associated. Write "E" for epithelial tissue, "C" for connective tissue, "M" for muscle tissue, and "N" for nervous tissue in the blanks provided.

   _____ a. erythrocyte

   _____ b. movement of substances across the lining of the stomach

   _____ c. ligament

   _____ d. secretion

   _____ e. spinal cord

   _____ f. multinucleate cells

   _____ g. insulation

   _____ h. actin

2. Which of the following structures is/are predominantly composed of tissue that is *avascular*? (Choose all that are correct.)
   a. dermis
   b. bone
   c. meniscus
   d. tendon
   e. epiglottis

3. Which of the following cell types is/are correctly matched with its function?
   a. neuroglial cells—transmit nerve impulses
   b. leukocytes—function in blood clotting
   c. mast cells—produce ground substance in cartilage
   d. fibroblasts—produce protein fibers

4. Bone and blood are classified as connective tissues because they both:
   a. structurally or functionally "connect" parts of the body.
   b. have multiple types of cells.
   c. have cells scattered in an abundant extracellular matrix.
   d. can store and transport materials such as minerals and hormones.

5. Complete the following chart. The first row has been filled in for you.

Organ	Two Prominent Tissue Types Present and their Functions
Small intestine	*simple columnar epithelium: secretes enzymes, absorbs nutrients; smooth muscle: contracts to churn and propel food being digested*
Trachea	
Urinary bladder	
Heart	
Diaphragm	
Skull	
External ear	

B. Concept Mapping

1. Fill in the blanks to complete this concept map outlining the structure and function of connective tissue.

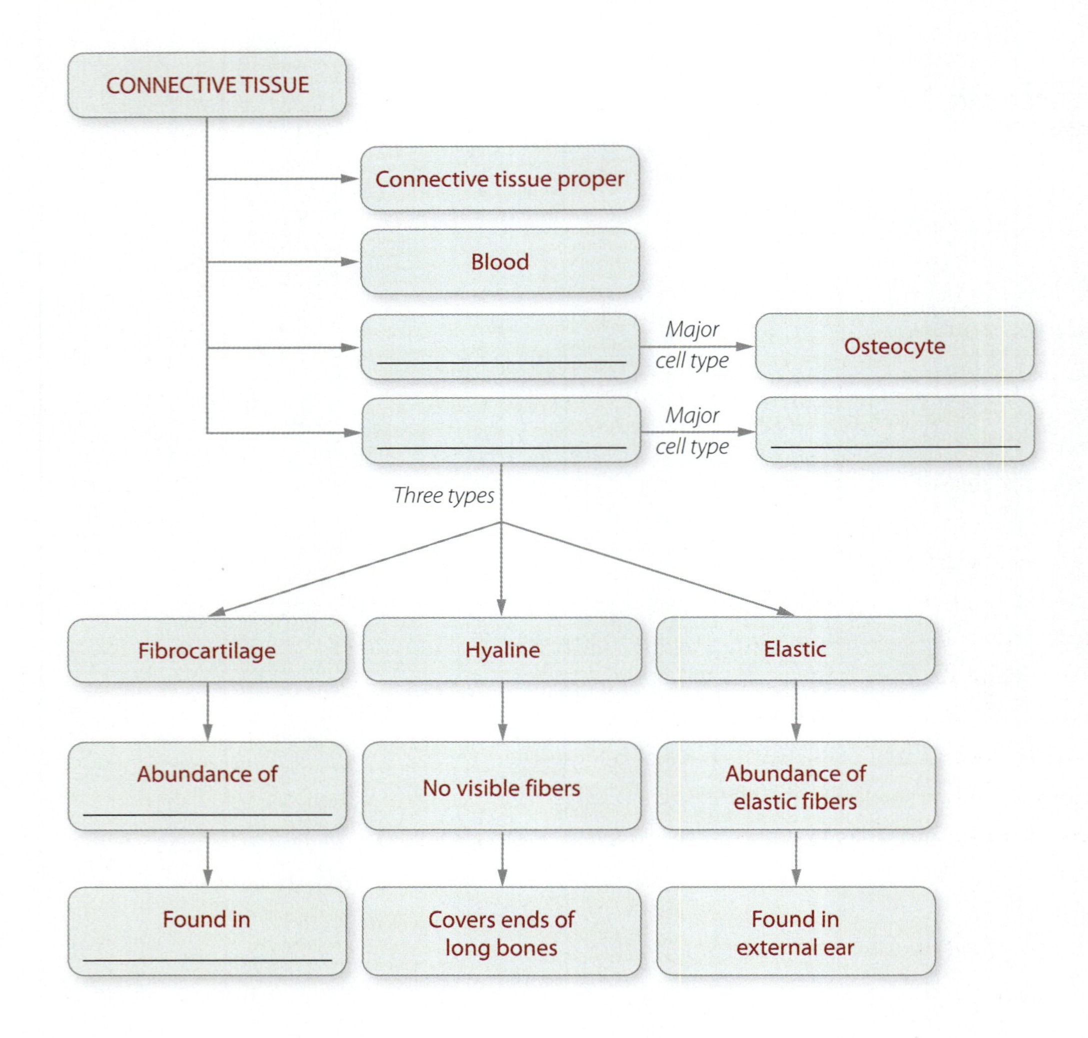

2. Construct a unit concept map to show the relationships among the following set of terms. Include all of the terms in your diagram. Your instructor may choose to assign additional terms.

axon	bone	cartilage	chondrocyte	collagen fibers	dendrite
intervertebral disc	ligament	multinucleate	nervous tissue	secretion	
simple columnar epithelium	skeletal muscle	small intestine	voluntary		

7 The Integumentary System

UNIT OUTLINE

The integumentary system is composed of the skin, or cutaneous membrane, plus its accessory structures: hairs, nails, and glands. The skin, the largest and most visible organ of the body, has numerous functions. The tightly connected cells of the skin provide *protection* by forming an impenetrable barrier to infectious agents. Through the *excretion* of sweat, waste products (including lactic acid and urea) are released onto the skin. Numerous sensory receptors in the skin enable *sensations*, such as touch, pain, temperature, pressure, and vibration. Blood vessels and sweat glands in the skin enable it to act in *thermoregulation* by removing excess heat or preventing heat loss. In response to nerve signals, blood vessels constrict to reduce blood flow and heat loss through the skin, or dilate to increase heat loss. Finally, the skin is the site of the initial steps in the *synthesis of vitamin D,* which is important for the absorption of calcium from the diet.

In this unit we first look at the various parts of each layer of the skin, and then study the histology of skin. Then we will investigate the distribution of sweat glands in skin at various locations of the body.

THINK ABOUT IT *Describe one way in which each of the following organ systems works with the integumentary system to maintain homeostasis:*

Cardiovascular system: ____

____

Nervous system: ____

____

Immune system: ____

____ ■

PRE-LAB ASSIGNMENTS

Pre-lab quizzes are also assignable in MasteringA&P®

To maximize learning, BEFORE your lab period carefully read this entire lab unit and complete these pre-lab assignments using your textbook, lecture notes, and prior knowledge.

PRE-LAB Activity 1: Identifying and Describing Skin Structures

1. Use the list of terms provided to label the accompanying illustration of the skin. Check off each term as you label it.

- ☐ hair
- ☐ dermis
- ☐ adipose tissue
- ☐ hypodermis
- ☐ sweat gland
- ☐ epidermis
- ☐ sebaceous gland

a ______
b ______
c ______
d ______
e ______
f ______
g ______

2. Which layer of the epidermis is most superficial?
   a. stratum corneum
   b. stratum basale
   c. stratum granulosum
   d. stratum spinosum
   e. stratum lucidum

PRE-LAB Activity 2: Examining the Histology of the Skin

1. Which of the following statements correctly describes thick skin?
   a. It is found in the skin of the soles and palms.
   b. The epidermis contains four different layers or strata.
   c. The stratum basale is the thickest stratum.
   d. Thick skin does not contain keratinocytes.

2. For each of the following structures, write in the blank whether it is located in the epidermis (E) or the dermis (D):

   _____ a. melanocyte
   _____ b. collagen
   _____ c. sweat gland
   _____ d. keratinocyte
   _____ e. sebaceous gland
   _____ f. lamellated corpuscle
   _____ g. blood vessels

3. *True or false?*: The shaft of a hair projects above the surface of the skin and is composed of living cells.
   a. true
   b. false

PRE-LAB Activity 3: Determining Sweat Gland Distribution

1. *True or false?*: The skin on the forehead contains both eccrine and apocrine sweat glands.
   a. true
   b. false

2. Which of the following statements regarding eccrine sweat glands is true?
   a. They are only found on the soles of the feet and the palms of the hands.
   b. Their primary function is the regulation of body temperature.
   c. They usually open into hair follicles.
   d. They are found in the deepest part of the epidermis.

The Skin and Its Accessory Structures

The skin is made up of two distinct layers: the superficial **epidermis** and the underlying **dermis** (**Figure 7-1**). A third layer of tissue, the **hypodermis** (or subcutaneous layer) lies deep to the dermis, and although it is not considered a component of the skin, it does help the skin perform its functions. The hypodermis consists of loose connective tissue with many blood vessels and adipose tissue. The right-hand portion of **Figure 7-2** provides a histological view of the epidermis, dermis, and hypodermis.

The Epidermis

The epidermis is avascular and composed of keratinized, stratified squamous epithelium and contains four distinct cell types: keratinocytes, melanocytes, dendritic (Langerhans) cells, and Merkel (tactile) cells. **Keratinocytes** produce the strong waterproofing protein keratin, **melanocytes** produce the protective pigment melanin, and **dendritic cells** are specialized white blood cells that migrate to the epidermis, where they function as phagocytes. **Merkel (tactile) cells** are located at the epidermal–dermal junction and function in light touch reception.

Epidermal cells are arranged in layers (strata) of cells. Thick skin, which is located only in the palms of the hands

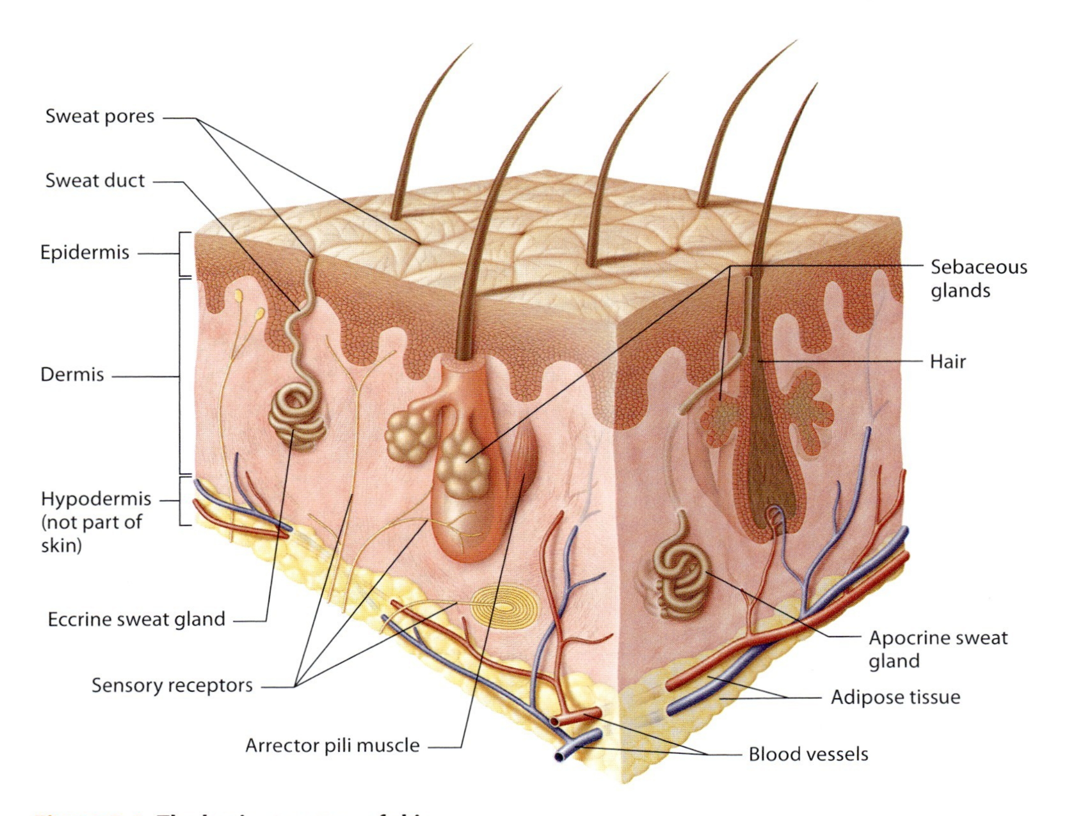

Figure 7-1 The basic structure of skin.

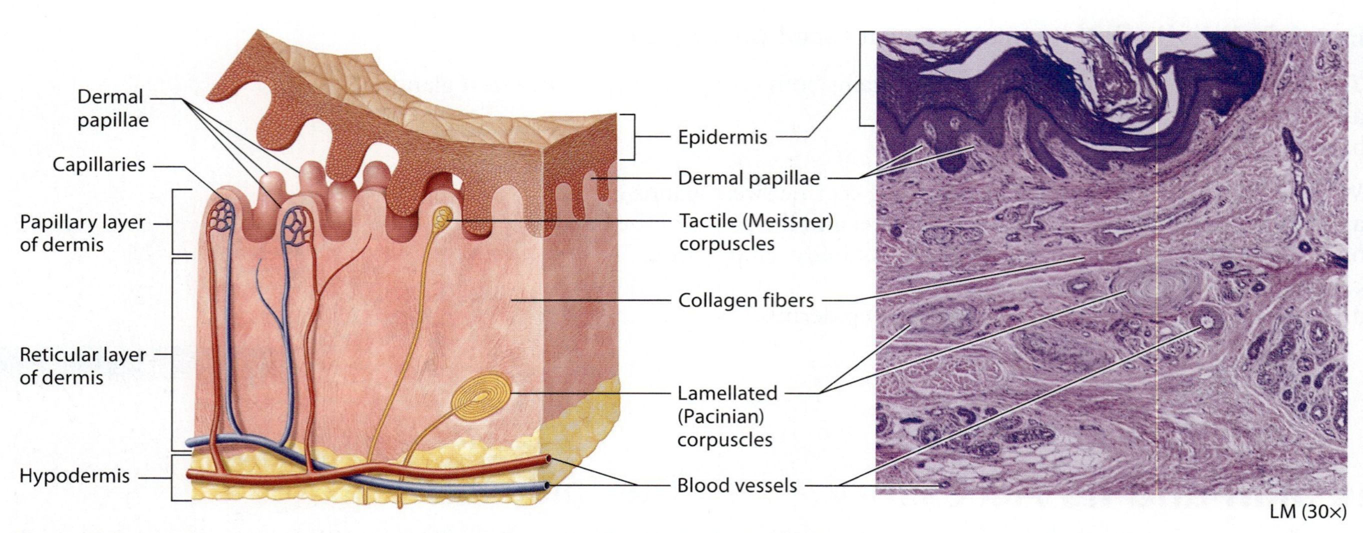

Figure 7-2 A diagrammatic representation (left) and a histological view (right) of the structures of the skin.

and the soles of the feet, contains five strata: the stratum basale, stratum spinosum, stratum granulosum, stratum lucidum, and stratum corneum. Thin skin, which is located everywhere else in the body, contains only four strata (the stratum lucidum is absent). **Figure 7-3** illustrates the histological differences between thick skin and thin skin. **Figure 7-4** provides a diagrammatic representation and a detailed histological view of the epidermis. Refer to Figures 7-3 and 7-4 as you read the following descriptions of the epidermal strata.

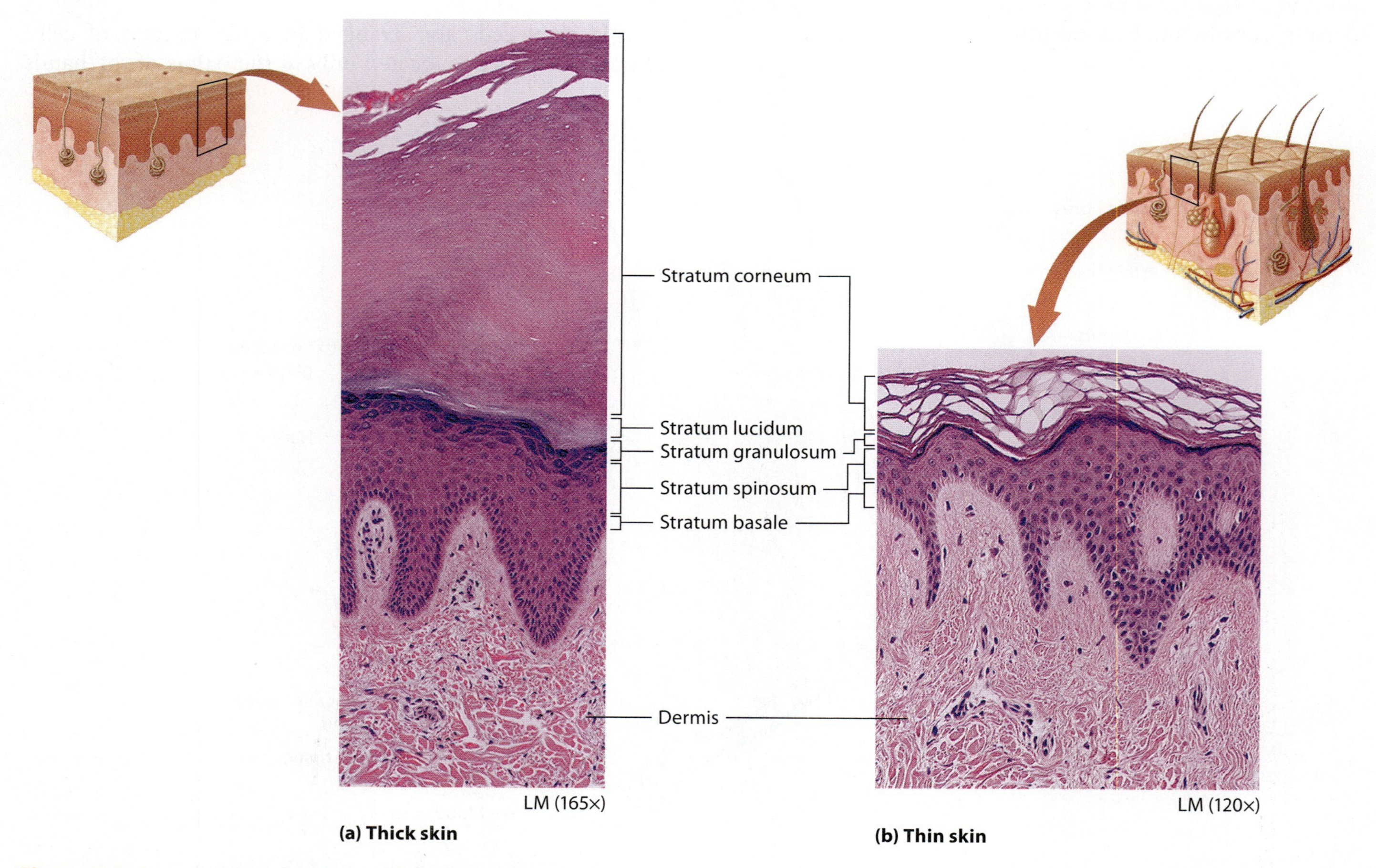

Figure 7-3 A comparison between thick skin and thin skin.

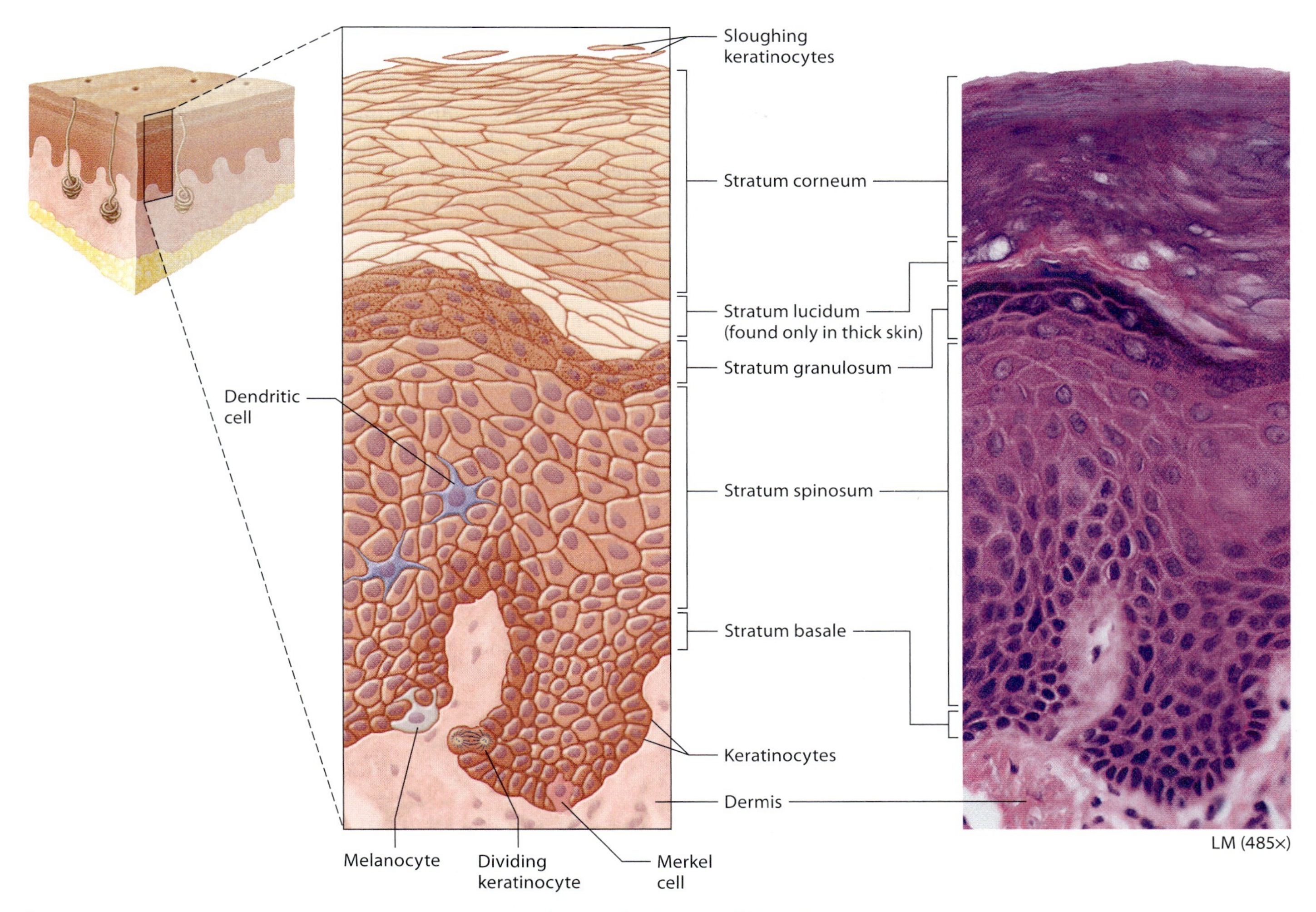

Figure 7-4 A diagrammatic representation (left) and a histological view (right) of the epidermis.

The **stratum basale**, the deepest layer of the epidermis, consists of a single layer of cuboidal epithelial cells (keratinocytes) attached to a basement membrane. These cells are actively mitotic; as they produce more keratinocytes, the newest cells push the older keratinocytes toward the surface of the epidermis. The stratum basale also contains melanocytes, cells that secrete the brown pigment melanin, which is then transferred to neighboring keratinocytes. The accumulation of melanin granules between the keratinocyte's nucleus and its apical surface shields the cell's DNA from the damaging effects of ultraviolet radiation in sunlight.

The **stratum spinosum** contains 8–10 layers of cells, mostly keratinocytes that have a "spiny" appearance in histological preparations. Living stratum spinosum cells do not actually have spines; instead, the spines are artifacts caused by the shrinkage of the cells during the preparation required for viewing them on a microscope slide. In addition to keratinocytes, dendritic cells are prominent in the stratum spinosum.

The thin **stratum granulosum** consists of three to five layers of keratinocytes that undergo a drastic change in appearance as they begin to fill with keratin (a process called keratinization). As this happens, the cells flatten and their organelles, including the nucleus, disintegrate; the cells eventually die.

THINK ABOUT IT Why do cells in the stratum granulosum, the third epidermal layer, begin to die? (Hint: In what way is the epidermis dependent on the underlying dermis?)

____

____

The **stratum lucidum**, which is found only in the thick skin of the palms and soles, consists of two to four translucent layers of flat, dead keratinocytes. The most superficial epidermal stratum, the **stratum corneum**, is the thickest stratum, consisting of 20–30 layers of cells. Its cells are essentially flattened sacs of keratin that are continually shed from the surface of the skin and are replaced by keratinocytes arising from deeper layers.

The Dermis

Underlying the epidermis is the highly vascular dermis, which consists of a papillary layer and a reticular layer (see Figure 7-2). The **papillary layer**, which lies immediately deep to the stratum basale and largely consists of loose connective tissue, is so named because it contains finger-like projections called dermal papillae, many of which contain capillary loops that provide a blood supply to the overlying, avascular epidermis. Other dermal papillae contain tactile (Meissner) corpuscles, which respond to light touch, or free nerve endings, which function as pain or temperature receptors. Dermal papillae are prominent on the hands and feet and form dermal ridges that indent the epidermis and form fingerprints. Touching a surface leaves behind sweat in the shape of your fingerprint pattern (much as an inked rubber stamp does).

The **reticular layer** of the dermis, the deepest layer of the skin, consists largely of dense, irregular connective tissue and accounts for approximately 80% of dermal thickness. The reticular layer contains an abundance of collagen fibers (for strength), elastic fibers (for flexibility), and reticular fibers (to form a supporting network for dermal structures). Lamellated (Pacinian) corpuscles are also located in the reticular layer. Their multilayered capsule is specialized to detect pressure and vibrations.

The accessory structures of the skin include hairs, nails, and glands. Hairs and glands are primarily dermal structures, although portions of them project up into and through the epidermis.

Hairs

Hairs are found all over the body surface—with the exception of the palms, soles, lips, parts of the external genitalia, and nipples. **Hairs** are produced by structures called hair follicles and consist of two basic parts: the shaft, which is composed of dead, keratinized epithelial cells and projects from the skin surface, and the root, which is enclosed by the hair follicle projecting down into the dermis (**Figure 7-5**). Viewed in transverse section, a hair can be seen to consist of three distinct layers of keratinized cells: an inner medulla,

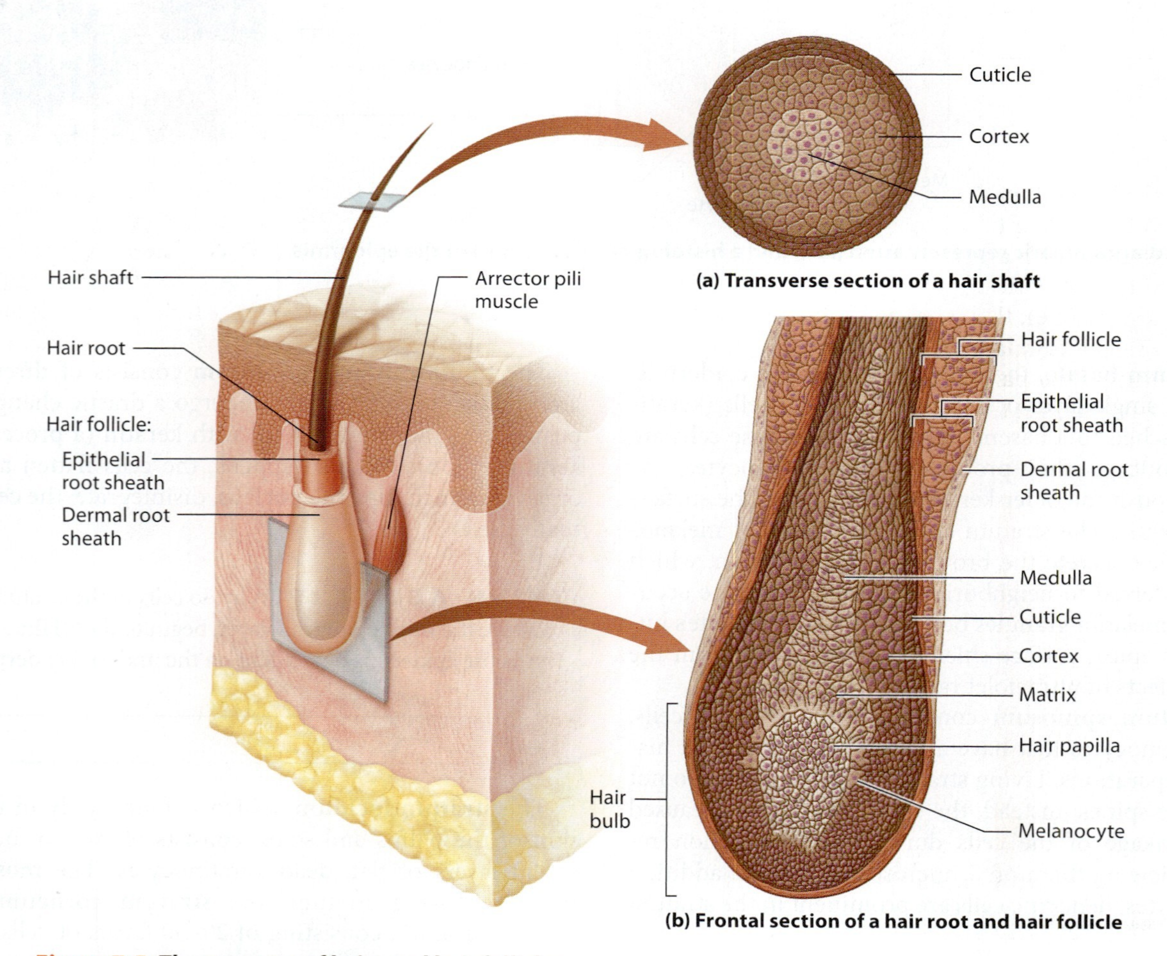

Figure 7-5 **The anatomy of hairs and hair follicles.**

a middle cortex, and an outer cuticle (see Figure 7-5a). Hair color is largely determined by the type and amount of melanin within the cortex.

Each **hair follicle** is formed by layers of epidermal tissue, called the epithelial root sheath, that is surrounded by a dermal root sheath composed of connective tissue. A bundle of smooth muscle fibers, called an arrector pili muscle, attaches to the dermal root sheath. Contracted arrector pili muscles cause goosebumps. The knob-like base of the hair root, called the hair bulb (see Figure 7-5b), contains the matrix, the living, mitotically active part of the hair that adds new hair cells to the base of the hair root. Sensory receptors are connected to the base of the hair bulb. The hair papilla, a small projection of dermal tissue from the dermal root sheath that protrudes into the hair bulb, contains capillaries that supply the growing hair cells with oxygen and nutrients. As mitosis continues within the matrix, hair cells are pushed toward the surface of the skin. These cells undergo keratinization, which is completed by the time the cells approach the skin surface, and then die.

THINK ABOUT IT How are hair papillae similar to dermal papillae? ______________________________

____

Nails

Nails (**Figure 7-6**) are modifications of the epidermis that protect the dorsal ends of the fingers and toes. These structures are associated with a nail: two nail folds, the folds of skin along either side of the nail body (see Figure 7-6a); the eponychium (or cuticle), the thick fold at the proximal end of the nail; and the lunula, the white, crescent-shaped region at the base of the nail. The free edge of the nail (see Figure 7-6b) extends past the end of the digit. The nail body rests on a region of the epidermis called the nail bed. The thickened proximal portion of the nail bed, called the nail matrix, is responsible for nail growth. As the nail matrix produces new nail cells, the older cells become keratinized and die as they move distally across the nail bed.

Sebaceous Glands

Sebaceous glands are exocrine glands—glands that release their secretory products onto external or internal body surfaces. Most sebaceous glands (see Figure 7-1) secrete an oily, acidic substance called **sebum** into hair follicles, although some deposit sebum directly onto the skin surface. Sebum acts as a lubricant that keeps the skin and hair soft and moist and deters the growth of infectious agents. Sebaceous glands are located in skin everywhere except the palms and soles. The histological view of the scalp in **Figure 7-7** shows a sebaceous gland next to a cross-section of a hair follicle.

Sweat Glands

Sweat glands, which are also exocrine glands, consist of a coiled secretory component and a duct composed of simple cuboidal epithelial tissue. Two types of glands produce sweat: eccrine sweat glands and apocrine sweat glands.

Eccrine sweat glands are distributed all over the body, and they secrete sweat that is primarily water but also contains salts and waste products. The coiled secretory portion of the gland is in the dermis, and the duct extends up through the epidermis and opens into a funnel-shaped pore at the skin surface (see Figure 7-1). The primary function of eccrine sweat glands is temperature regulation. When body temperature rises, eccrine sweat glands release sweat that carries heat away as it evaporates from the skin surface.

Apocrine sweat glands are confined to the groin, axillae, and the areola around each nipple. They tend to be larger and deeper in the dermis than eccrine sweat glands, and they have ducts that empty into hair follicles (see Figure 7-1).

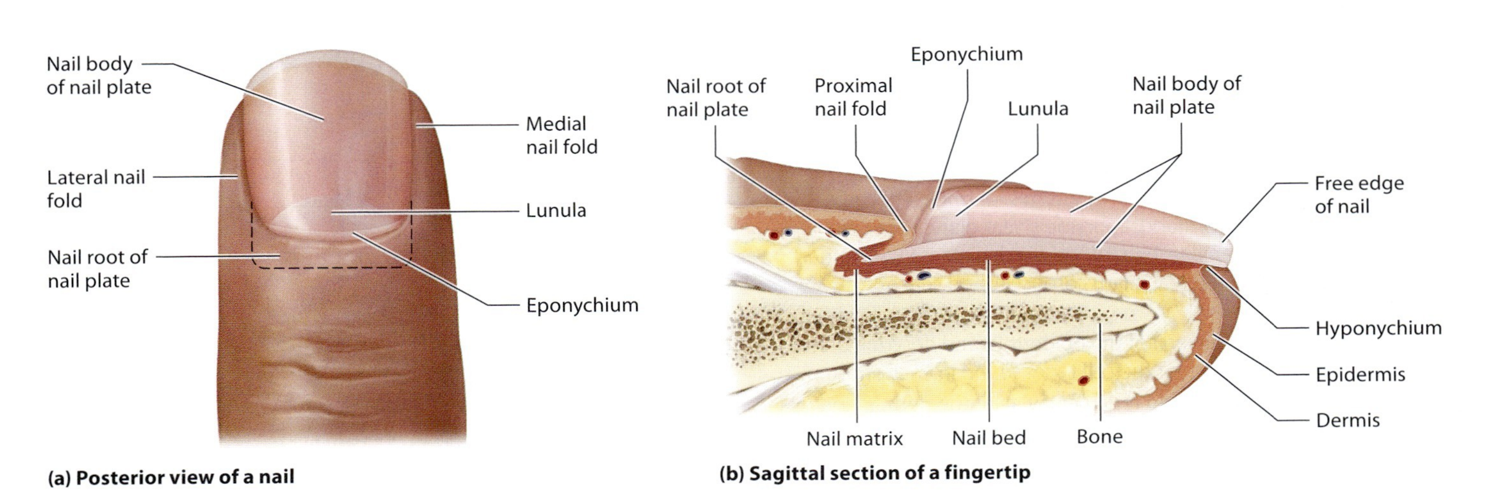

(a) Posterior view of a nail

(b) Sagittal section of a fingertip

Figure 7-6 The structure of a nail.

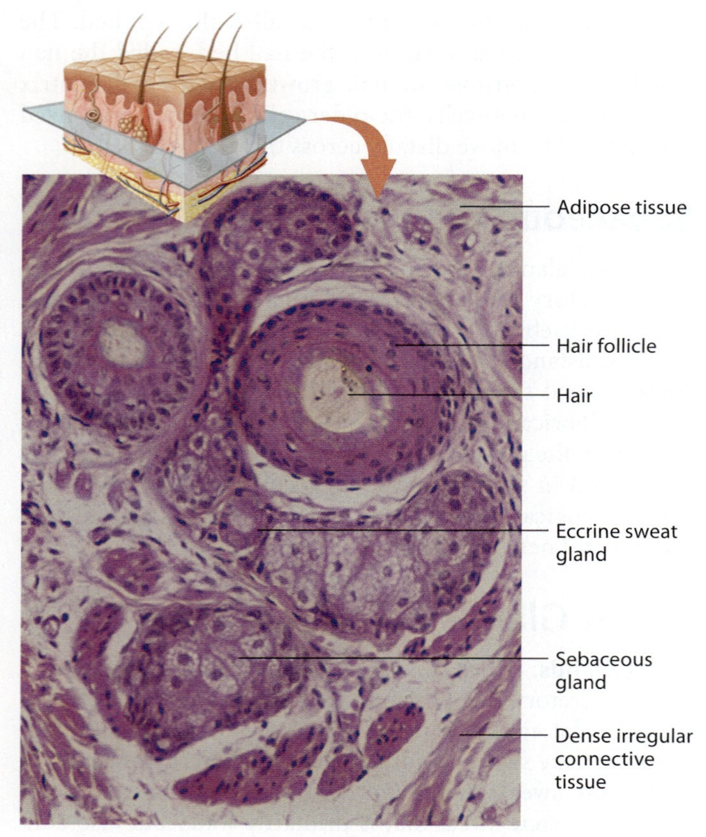

Figure 7-7 A histological view of the scalp.

These sweat glands, which are activated at puberty, respond to pain, emotional stress, and sexual stimulation. In addition to the components found in sweat (water, salts, and wastes), apocrine sweat gland secretions also contain proteins. These secretions are odorless, but when released onto the skin they are metabolized by bacteria, producing body odor.

ACTIVITY 1 Identifying and Describing Skin Structures

Learning Outcomes

1. Identify the following skin structures on a skin model, and describe the function of each: epidermis, dermis, sebaceous gland, eccrine sweat gland, apocrine sweat gland, arrector pili muscle, dermal papillae, hair follicle and hair, tactile (Meissner) corpuscle, lamellated (Pacinian) corpuscle, and sensory nerve fiber.
2. Identify the strata (layers) of the epidermis, and describe the structure and function of the cell types found in each stratum.
3. Describe the papillary and reticular layers of the dermis, and name the specific tissue type that is predominant in each layer.
4. Distinguish between eccrine sweat glands and apocrine sweat glands with respect to structure and function.

Materials Needed

- ☐ Skin model
- ☐ Small stickers for labeling
- ☐ Markers

Instructions

1. Identify the epidermis on the skin model.

   Which specific tissue type makes up the epidermis?

   ____

2. Locate the five strata of the epidermis. For each stratum, list the specific cell types and the approximate number of cell layers. Also, circle the stratum that is found only in thick skin.

   Stratum basale ____________________

   Stratum spinosum ____________________

   Stratum granulosum ____________________

   Stratum lucidum ____________________

   Stratum corneum ____________________

3. Identify the dermis on the skin model.

   Name the two layers of the dermis and list the predominant specific tissue type found in each. ____________________

   ____

4. Using a black marker, label 9 stickers with the letters a–i. As you identify each of the following structures on a skin model, place the corresponding sticker on the structure. Discuss within your lab group the function and anatomical features of each structure, and then agree on and list at least three terms or phrases that best identify or describe the anatomy and functions of each structure. The features of the first structure are given as an example.

   a. Sebaceous gland <u>Produces sebum, is associated with a hair follicle, is an exocrine gland</u>

   b. Eccrine sweat gland ____________________

   ____

   c. Apocrine sweat gland ____________________

   ____

   d. Arrector pili muscle ____________________

   ____

   e. Dermal papillae ____________________

   ____

   f. Hair follicle ____________________

   ____

g. Tactile (Meissner) corpuscle ____________________

____

h. Lamellated (Pacinian) corpuscle ________________

____

i. Sensory receptor ____________________________

____

5. Now, quiz each other about the 9 structures you labeled on the model. First, take turns calling out the three terms or phrases you wrote down for each structure. As the terms/phrases are called out, the other students will name the structure being described and locate it on the model. Then take turns calling out a structure, and have the other students locate it on the model and state the three terms/phrases that describe it.

6. **Optional Activity**

Practice labeling structures of the integument on skin models at MasteringA&P® > Study Area > Practice Anatomy Lab > Anatomical Models > Integumentary System

ACTIVITY 2
Examining the Histology of the Skin

Learning Outcome

1. Identify skin structures on microscope slides or in photomicrographs.

Materials Needed

☐ Microscope and prepared microscope slides (or photomicrographs) of the following structures:
Thin skin with hairs
Thick skin
Pigmented epithelium
Nonpigmented epithelium
Dermis
Skin with a tattoo

☐ Colored pencils

Instructions

1. Observe a slide of a section of thin skin that includes hairs. Use colored pencils to sketch the slide and label the following structures: epidermis, dermis, hair follicle, hair root, and hair shaft.

Total magnification: _____ ×

Which epidermal stratum is absent in this section of skin? ______________________________

Which part of the hair is actively mitotic?

____

2. Observe a slide of a section of thick skin. Use colored pencils to sketch the slide and label the following structures: epidermis (stratum basale, stratum spinosum, stratum granulosum, stratum lucidum, and stratum corneum), epidermal ridge, dermis (papillary layer and reticular layer), and dermal papilla.

Total magnification: _____ ×

Which parts of the skin are responsible for fingerprints?

____

Which epidermal stratum is unique to thick skin?

____

3. Observe a slide of a section of pigmented epithelium and a slide of a section of nonpigmented epithelium. Use colored pencils to sketch each of these slides, and label the following structures: epidermis (stratum basale, stratum spinosum, stratum granulosum, and stratum corneum), melanocyte, and dermis (papillary layer and reticular layer).

Total magnification: _____ ×

Total magnification: _____ ×

Which cells produce melanin? ____________________

Where are these cells located? ____________________

4. Observe a slide of a section of dermis. Use colored pencils to sketch the slide and label all the skin layers and skin structures you see.

Total magnification: _____ ×

Are sweat glands present? __________ *If yes, are these glands eccrine sweat glands or apocrine sweat glands? Explain your answer.*

____

____

Are sebaceous glands present? __________ *How can you distinguish between sebaceous glands and eccrine sweat glands based on structure?*

____

____

5. Observe a slide of a section of skin that contains a tattoo. Use colored pencils to sketch the slide and label all the skin layers and skin structures you see.

Total magnification: _____ ×

In which layer of the skin are the ink granules of a tattoo located? ______________________________

____

Why is a tattoo permanent, whereas a suntan is not?

____

____

6. **Optional Activity**

PAL™ View and label histology slides of the integumentary system at MasteringA&P® > Study Area > Practice Anatomy Lab > Histology > Integumentary System

ACTIVITY 3 Determining Sweat Gland Distribution

Learning Outcome

1. Determine the relative distribution of sweat glands in the skin of the palm, forearm, and forehead.

Materials Needed

- ☐ 1-cm × 1-cm paper squares (three squares per pair of students)
- ☐ Betadine
- ☐ Cotton swabs
- ☐ Surgical tape
- ☐ Scissors
- ☐ Soap and water

Instructions

1. Working in pairs for this activity, formulate and write down a hypothesis concerning the relative distribution of sweat glands in three different sites: palm, forearm, and forehead. ____

____

____

2. Perform the following steps:
   a. Use a cotton swab to apply Betadine to areas of the skin on the palm, forearm, and forehead. Each Betadine-swabbed area should be slightly larger than the paper squares you will be using. Let the Betadine dry thoroughly.
   b. Cover each Betadine-swabbed area with a square of paper and then secure the paper with surgical tape.
   c. After 15 minutes, remove the paper squares and count the number of blue-black dots on the paper. Each dot indicates the presence of an active eccrine sweat gland. Record your data in the following table:

Body Location	No. of Eccrine Sweat Glands/cm^2 of Skin
Palm	
Forearm	
Forehead	

   d. Use soap and water to wash the Betadine from the skin.
   e. Report your data to your instructor, and review the results obtained by the entire class.

3. Was your hypothesis supported by the experimental results? ____

4. What conclusions can you draw from this data set?

____

____

____

____

____

Post-lab quizzes are also assignable in MasteringA&P®

POST-LAB ASSIGNMENTS

Name: ______________________ Date: ______________ Lab Section: ________

PART I. Check Your Understanding

Activity 1: Identifying and Describing Skin Structures

1. Identify the components of skin in the accompanying diagram:

   a. ______________________

   b. ______________________

   c. ______________________

   d. ______________________

   e. ______________________

   f. ______________________

   g. ______________________

   h. ______________________

   i. ______________________

   j. ______________________

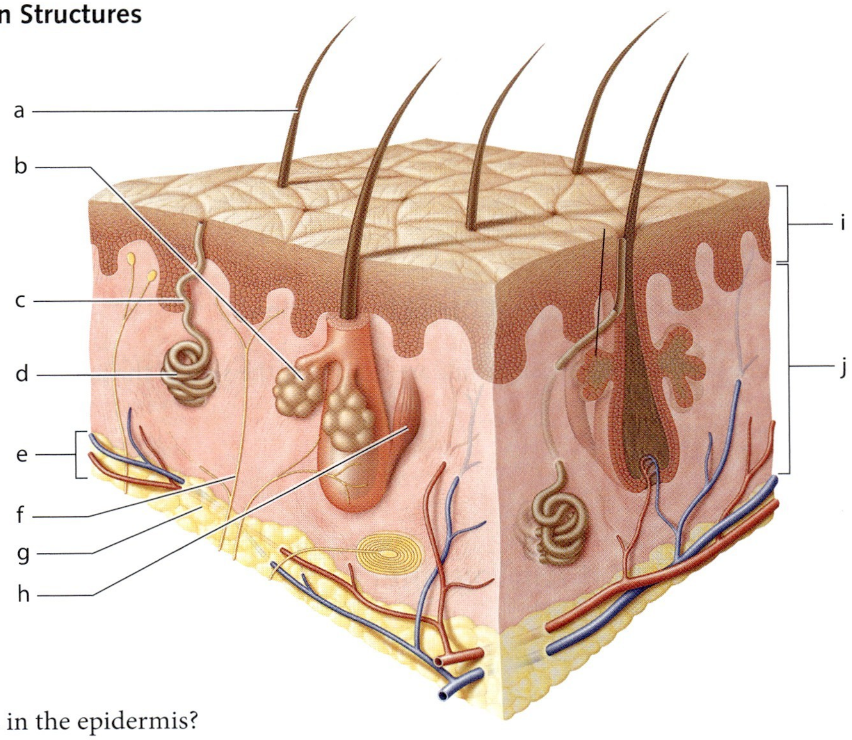

2. Which of the following structures is found in the epidermis?
   a. tactile corpuscle
   b. arrector pili muscle
   c. lamellated corpuscle
   d. eccrine sweat gland
   e. None of these structures are found in the epidermis.

3. The papillary layer of the dermis is composed primarily of which type of tissue?
   a. stratified squamous epithelial tissue
   b. dense irregular connective tissue
   c. simple cuboidal epithelial tissue
   d. loose connective tissue

4. For each of the following characteristics, indicate whether it describes apocrine (A) or eccrine (E) sweat glands:

   _____ a. Are found primarily in the groin and axillae

   _____ b. Function primarily in temperature regulation

   _____ c. Are located deep in the dermis

   _____ d. Have ducts that empty into hair follicles

   _____ e. Open onto the skin surface

   _____ f. Are located across most of the body surface

Activity 2: Examining the Histology of the Skin

1. The narrow epidermal layer that appears clear is the:
   a. stratum granulosum.
   b. stratum corneum.
   c. stratum basale.
   d. stratum spinosum.
   e. stratum lucidum.

2. Which epidermal stratum is missing in thin skin?
   a. stratum corneum
   b. stratum basale
   c. stratum lucidum
   d. stratum spinosum
   e. stratum granulosum

3. *True or false?*: Unlike a suntan, which is lost because of constant renewal of keratinocytes, a tattoo is usually permanent because the pigments are in the dermis, which does not have significant cell turnover.
   a. true
   b. false

Activity 3: Determining Sweat Gland Distribution

1. *True or false?*: Eccrine sweat glands are located in the skin of the palm and forehead, but not in the forearm.
   a. true
   b. false

PART II. Putting It All Together

A. Review Questions

Answer the following questions using your lecture notes, your textbook, and your lab notes:

1. For each skin structure, write one of its functions and the skin layer in which it is found.
   a. hair matrix ______________________
   b. dendritic cell ______________________
   c. keratinocyte ______________________
   d. collagen fibers ______________________
   e. dermal papillae ______________________
   f. stratum basale ______________________

2. If you cut the palm of your hand, but do not bleed:
   a. the cut has penetrated through the epidermis into the dermis.
   b. it is because blood vessels are located in the hypodermis only.
   c. you have injured the epidermis but not the dermis.
   d. it is because blood in the epidermal blood vessels clots very quickly.

3. Which of the following statements about melanocytes and keratinocytes is true?
   a. Melanocytes are in the dermis and keratinocytes are in the epidermis.
   b. Melanocytes produce pigment but keratinocytes do not.
   c. Both cell types produce waterproofing material.
   d. Both cell types divide and die as they move away from their blood supply.

4. *True or false?*: Lamellated corpuscles detect pressure, whereas tactile corpuscles detect light touch.
   a. true
   b. false

B. Concept Mapping

1. Fill in the blanks to complete this concept map outlining components of the skin.

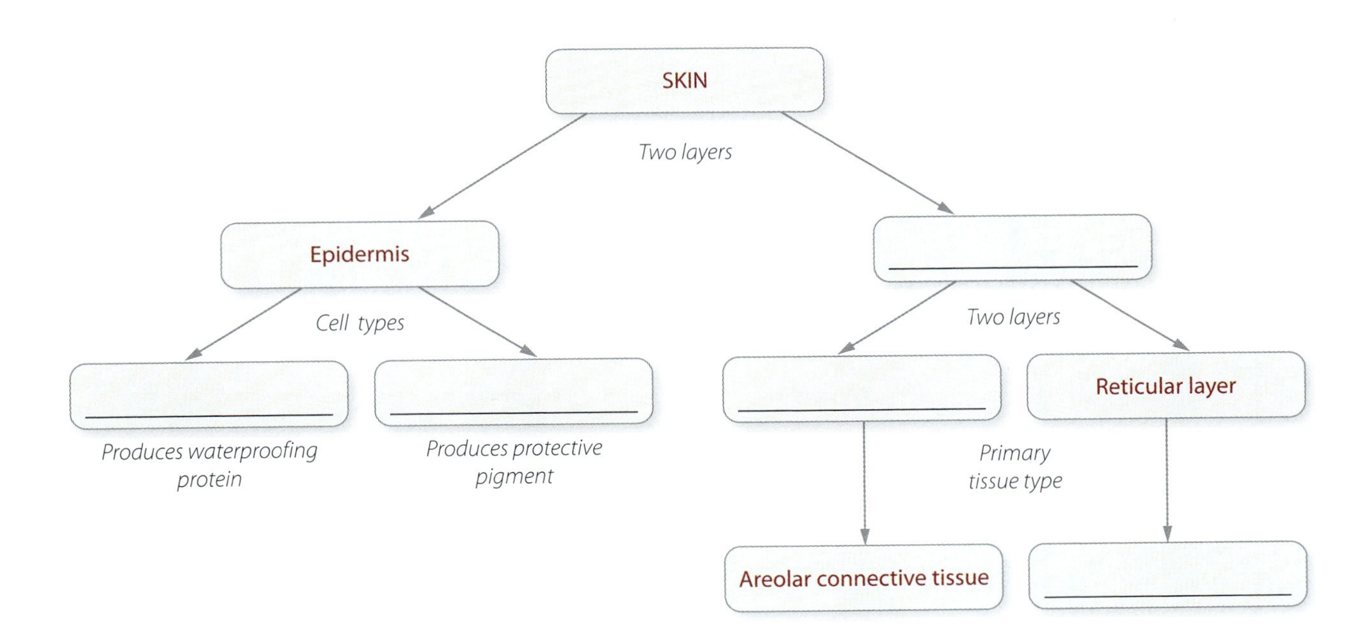

2. Construct a unit concept map to show the relationships among the following set of terms. Include all of the terms in your diagram. Your instructor may choose to assign additional terms.

adipose tissue	apocrine sweat gland		arrector pili muscle
dense irregular connective tissue	dermal papillae	dermis	eccrine sweat gland
epidermis	hypodermis	keratinocyte	lamellated corpuscle
loose connective tissue	melanocyte	sebaceous gland	tactile corpuscle

8

Introduction to the Skeletal System

UNIT OUTLINE

In this unit we begin to explore the skeletal system. We will learn about the functions of this organ system, the gross and microscopic structure of bone, and the chemical composition of bone. We will also review the skeletal cartilages, among the most important accessory structures of the skeletal system. In Units 9 and 10 we will study the two main divisions of the skeleton: the axial and appendicular skeletons. Finally, in Unit 11 we will learn about the structure and function of joints.

THINK ABOUT IT *Describe several functions of bones.* ______________

____

____

____

PRE-LAB ASSIGNMENTS

Pre-lab quizzes are also assignable in MasteringA&P®

To maximize learning, BEFORE your lab period carefully read this entire lab unit and complete these pre-lab assignments using your textbook, lecture notes, and prior knowledge.

PRE-LAB Activity 1: Reviewing Skeletal Cartilages

1. Which type of cartilage cushions the articular surface (joint surface) of bones?
   a. hyaline cartilage
   b. elastic cartilage
   c. fibrocartilage

2. Which of the following cells makes cartilage?
   a. chondrocyte
   b. chondroblast
   c. fibroblast
   d. osteocyte

PRE-LAB Activity 2: Classifying and Identifying the Bones of the Skeleton

1. Match each of the following bone types with its description:

_____	a. short bone	1. thin, plate-like
_____	b. long bone	2. found embedded in a tendon
_____	c. irregular bone	3. box shaped
_____	d. sesamoid bone	4. longer than it is wide
_____	e. flat bone	5. unusually shaped

2. Match each of the following bone types with an example in the body:

_____	a. short bone	1. vertebra
_____	b. long bone	2. cranial bone
_____	c. irregular bone	3. patella
_____	d. sesamoid bone	4. humerus
_____	e. flat bone	5. carpals

PRE-LAB Activity 3: Examining the Gross Anatomy of a Long Bone

1. Match each of the following functions/descriptions with its correct skeletal structure:

_____	a. shaft of a long bone	1. epiphysis
_____	b. bands of dividing hyaline cartilage	2. periosteum
_____	c. filled with adipose tissue	3. spongy bone
_____	d. anchors periosteum to bone	4. epiphyseal plate
_____	e. covers ends of long bones	5. perforating fibers
_____	f. houses red bone marrow	6. diaphysis
_____	g. end of long bone	7. medullary cavity
_____	h. dense outer layer of bone	8. compact bone
_____	i. lines the medullary cavity	9. endosteum
_____	j. houses blood vessels and nerves	10. articular cartilage

PRE-LAB Activity 4: Exploring the Microscopic Anatomy of Compact Bone—The Osteon

1. Osteocytes are primarily "housed" within _____________ in compact bone.
   a. the central canal
   b. canaliculi
   c. lacunae
   d. perforating canals

2. Oxygen reaches individual osteocytes from the blood supply in the central canal via:
   a. perforating canals.
   b. the periosteum.
   c. the endosteum.
   d. canaliculi.

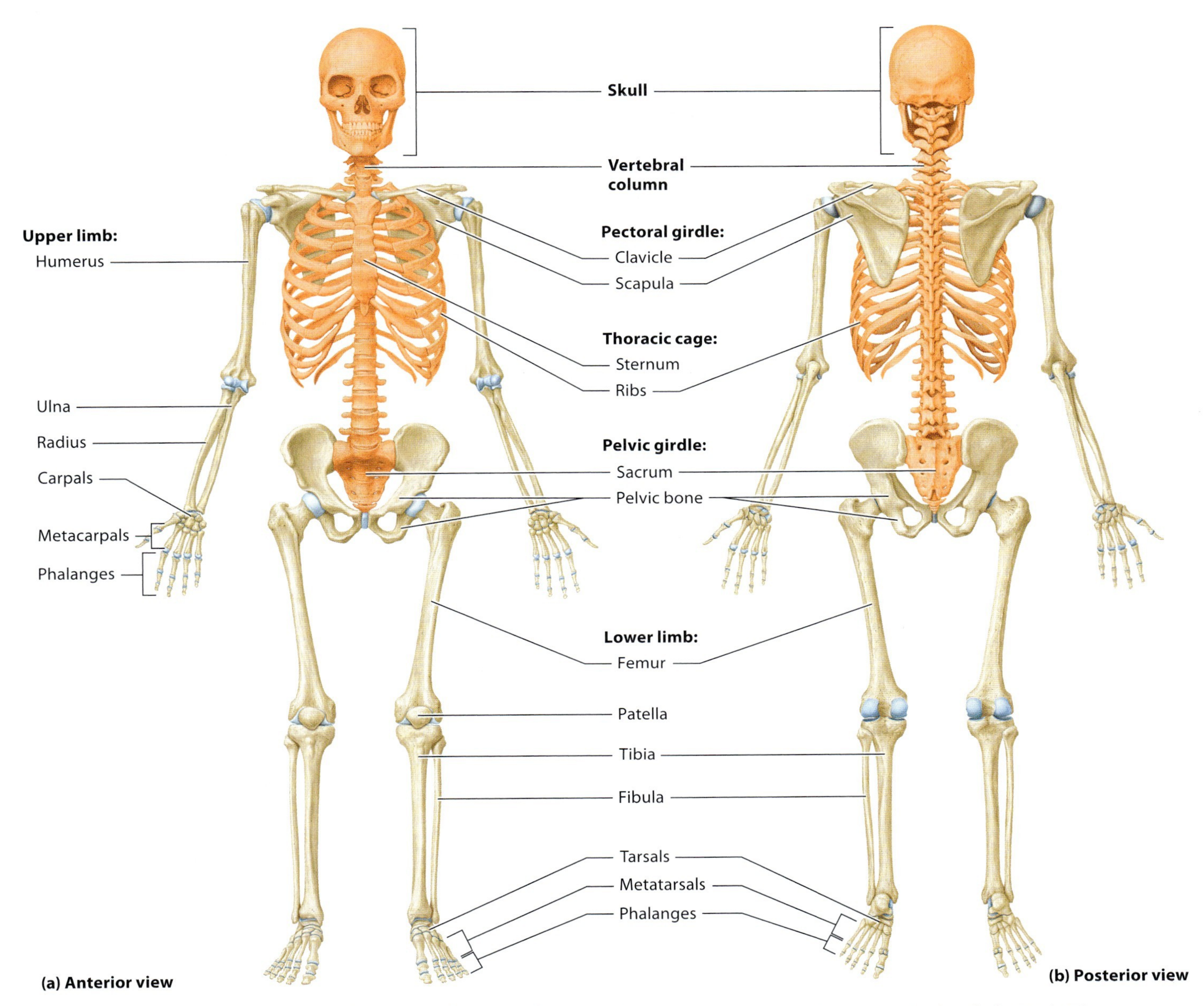

Figure 8-3 **The two subdivisions of the skeletal system: the axial skeleton (orange) and the appendicular skeleton (tan).**

	Features Common to All Cartilage	Features Unique to Hyaline Cartilage	Features Unique to Fibrocartilage	Features Unique to Elastic Cartilage
Structure				
Function				
Location				

Intervertebral discs ______________________________

____

____

Costal cartilage ______________________________

____

____

Meniscus ______________________________

____

____

5. **Optional Activity**

View and label histology slides of cartilage types at MasteringA&P® > Study Area > Practice Anatomy Lab > Histology > Connective Tissue > Images 16–26

Structural Characteristics of Bones

The bones of the human body are classified into five major shapes: long, short, flat, irregular, and sesamoid bones (**Figure 8-4**). Most of the bones of the appendicular skeleton are **long bones**, which are longer than they are wide. **Short bones** are roughly cube shaped; they include the bones of the wrists (the carpals) and ankles (the tarsals). The sternum and most of the skull bones are **flat bones** having thin, plate-like structures. **Sesamoid bones** develop within tendons; the patellas are examples. Bones that do not fit into any of the previous categories are classified as **irregular bones**; the vertebrae, the coxal bones, and the hyoid bone are examples. Additionally, so-called **Wormian** (sutural) **bones** are tiny bones that develop within sutures (joints that connect cranial bones).

Bone markings are projections, holes, ridges, depressions, and other physical characteristics of bone surfaces. Some of these markings serve as attachment sites for muscles, tendons, and ligaments. Others are surfaces where bones meet to form joints with other bones, and still others serve as passageways for blood vessels and nerves. The major bone markings are listed in **Table 8-1**; it is important to learn them because you will rely on this knowledge when you study the specific bones of the axial and appendicular skeletons.

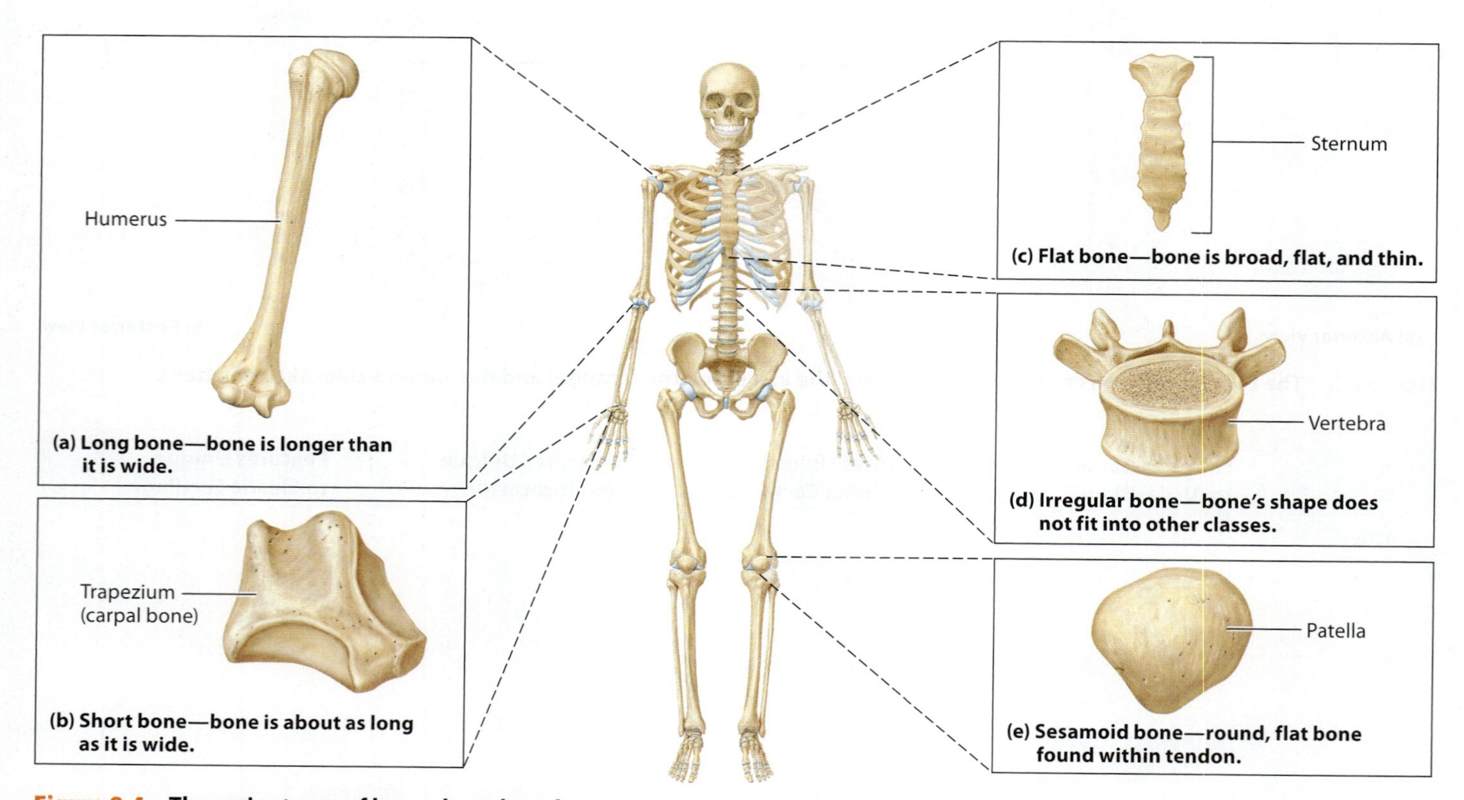

Figure 8-4 The major types of bones based on shape.

Table 8-1 Bone Markings

Bone Marking	Description	Example
DEPRESSIONS: Clefts of varying depth in a bone; located where a bone meets another structure, such as another bone or a blood vessel.		
Facet	Shallow convex or concave surface where two bones articulate	**Rib:** Articular facet for articulation with a transverse process 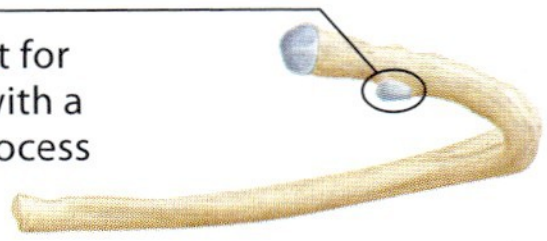
***Fossa* (plural, *fossae*)**	Indentation in a bone into which another structure fits	**Humerus:** Distal portion with olecranon fossa 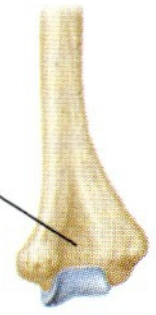
Fovea	Shallow pit	**Femur:** Fovea capitis 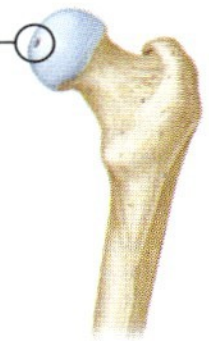
Groove (or sulcus)	Long indentation along which a narrow structure travels	**Rib:** Costal groove 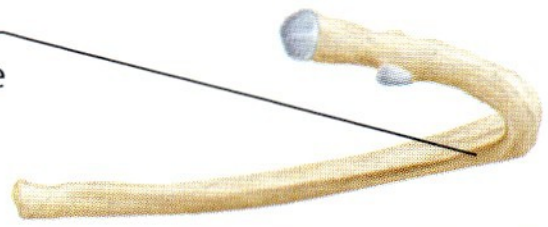
OPENINGS: Holes that allow blood vessels and nerves to travel through a bone; permit access to the middle and inner ear; encase delicate structures and protect them from trauma.		
Canal (or meatus)	Tunnel through a bone	**Temporal bone:** External acoustic meatus 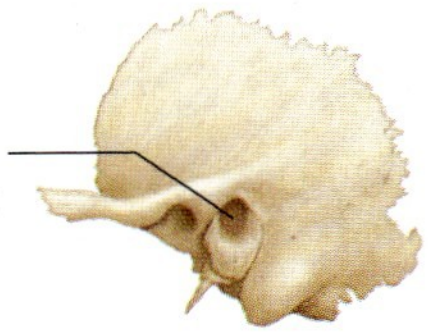
Fissure	Narrow slit in a bone or between adjacent parts of bones	**Sphenoid bone:** Superior orbital fissure 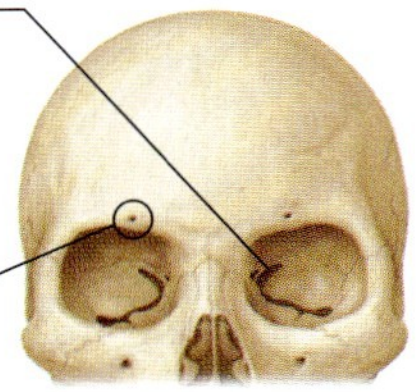
***Foramen* (plural, *foramina*)**	Hole in a bone	**Frontal bone:** Supraorbital foramen
PROJECTIONS: Bony extensions of varying shapes and sizes; some provide locations for attachment of muscles, tendons, and ligaments; some fit into depressions of other bones to stabilize joints.		
Condyle	Rounded end of a bone that articulates with another bone	**Mandible:** Mandibular condyle 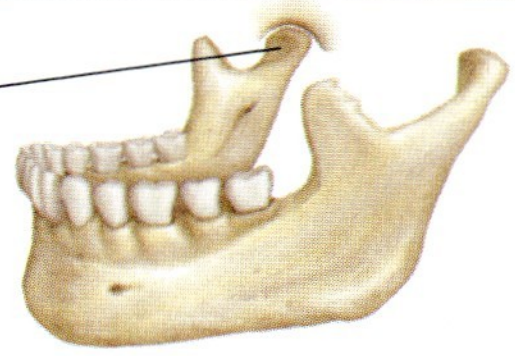

(Continued)

Table 8-1 Bone Markings (*Continued*)

Bone Marking	Description	Example
Crest	Ridge or projection	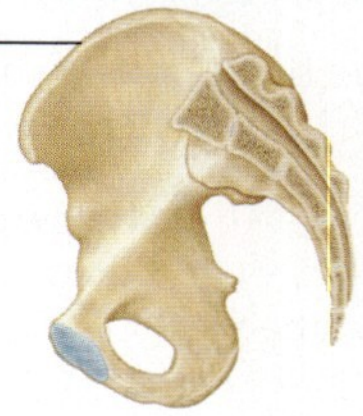 **Ilium:** Iliac crest
Head	Round projection from a bone's epiphysis	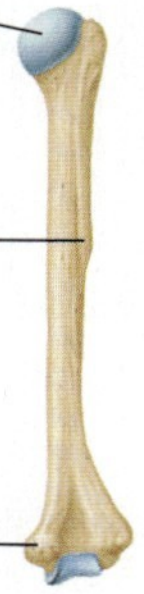 **Humerus:** Head **Humerus:** Deltoid tuberosity **Humerus:** Medial epicondyle
Tubercle and tuberosity	Small, rounded bony projection; a tuberosity is a large tubercle	
Epicondyle	Small projection usually proximal to a condyle	
Process	Prominent bony projection	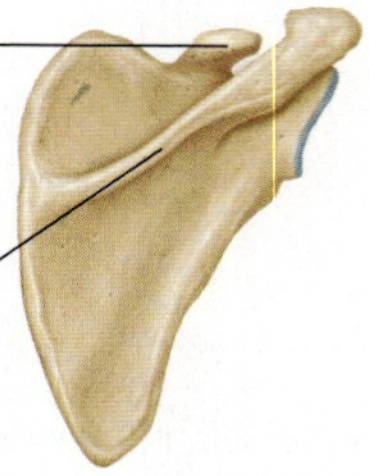 **Scapula:** Coracoid process **Scapula:** Spine
Spine	Sharp process	
Protuberance	Outgrowth from a bone	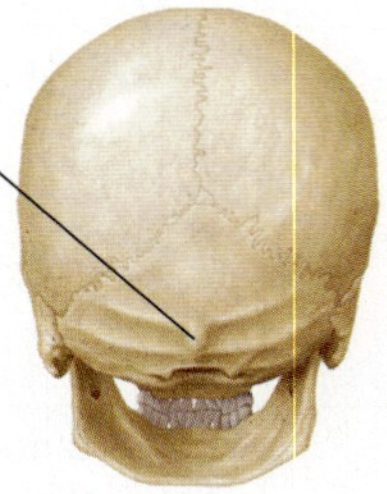 **Occipital bone:** External occipital protuberance
Trochanter	Large projection found only on the femur	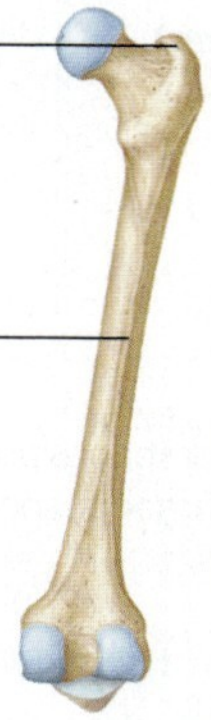 **Femur:** Greater trochanter **Femur:** Linea aspera
Line	Long, narrow ridge	

ACTIVITY 2

Classifying and Identifying the Bones of the Skeleton

Learning Outcomes

1. Identify the major bones of the skeleton and classify each bone according to shape.
2. Recognize and define the various bone markings.
3. Identify disarticulated bones by touch only, and justify your identification based on bone size, shape, and markings.

Materials Needed

- ☐ Articulated skeleton
- ☐ Labeling tape
- ☐ Markers
- ☐ Bag of disarticulated bones

Instructions

A. Classifying Bones by Shape

1. CHART Identify on the articulated skeleton each of the bones in the chart at right. In the space provided, write (1) its name, (2) whether it belongs to the axial skeleton (AX) or the appendicular skeleton (AP), and (3) its shape (either long, short, flat, irregular, or sesamoid). The first bone has been completed as an example.
2. Next, copy your three-part identification of the bone onto a piece of labeling tape, and then stick the tape onto the appropriate bone.

B. Bones-in-a-Bag

In this activity you and your group members will take turns identifying each of the five disarticulated bones in a bag provided by your instructor.

1. Reach into the bag with one hand (without looking at the bones—no peeking!) and pick up a bone, but do not take it out of the bag.
2. Based on feel alone, describe the bone's size, shape, and bone markings to the other members of your lab group. Then record your description in **Table 8-2** on the next page.
3. Name the bone you are holding in the bag and then pull it out of the bag to see if your identification was correct.
4. Record in Table 8-2 whether the bone belongs to the axial (AX) or appendicular (AP) skeleton.
5. Pass the bag to another group member, who will repeat steps 1–4.
6. Repeat steps 1–5 until all of the bones in the bag have been identified.

Bone	Label
Carpal bone	*carpal bone / AP / short*
Clavicle	
Coxal bone	
Cranial bones	
Femur	
Fibula	
Humerus	
Metacarpals	
Metatarsals	
Patella	
Phalanges	
Radius	
Ribs	
Sacrum	
Scapula	
Sternum	
Tarsal	
Tibia	
Ulna	
Vertebra	

Table 8-2 Bones-in-a-Bag Identification and Classification

Description of Bone Based on Touch Alone	Name of Bone	Shape of Bone	AX or AP?

Gross Anatomy of Long Bones

Figure 8-5 illustrates the gross anatomical features of a long bone. The enlarged ends of long bones are called **epiphyses**; the shaft of the bone is called the **diaphysis**. Epiphyses contain a thin, outer layer of compact bone surrounding a spongy bone interior often filled with blood cell–producing **red bone marrow**. The diaphysis contains a thick collar of compact bone surrounding a central **medullary cavity** filled with fat (**yellow bone marrow**).

Located between the epiphysis and the diaphysis in mature long bones are **epiphyseal lines**, which mark the sites where bands of actively dividing hyaline cartilage—called **epiphyseal plates**—were once located in growing bones. Eventually, the epiphyseal lines are no longer visible due to the remodeling of bone throughout adulthood. Hyaline cartilage also forms the **articular cartilage** that covers the epiphyses. The remainder of the bone is surrounded by a fibrous membrane called the periosteum.

The **periosteum** is composed of two layers: an outer fibrous layer that serves as an attachment site for tendons and ligaments, and an inner layer that contains specialized cells involved in bone growth, repair, and remodeling. The periosteum is anchored to the bone surface via perforating (Sharpey's) fibers. Through the periosteum runs a rich supply of nerves, blood vessels, and lymphatic vessels that enter the diaphysis via an opening called the **nutrient foramen**. The **endosteum** forms the inner lining of the diaphysis; it also covers the trabeculae of spongy bone and lines the central (Haversian) canals of compact bone. Note that bones of other shapes also have articular cartilages and are covered by periosteum and lined by endosteum.

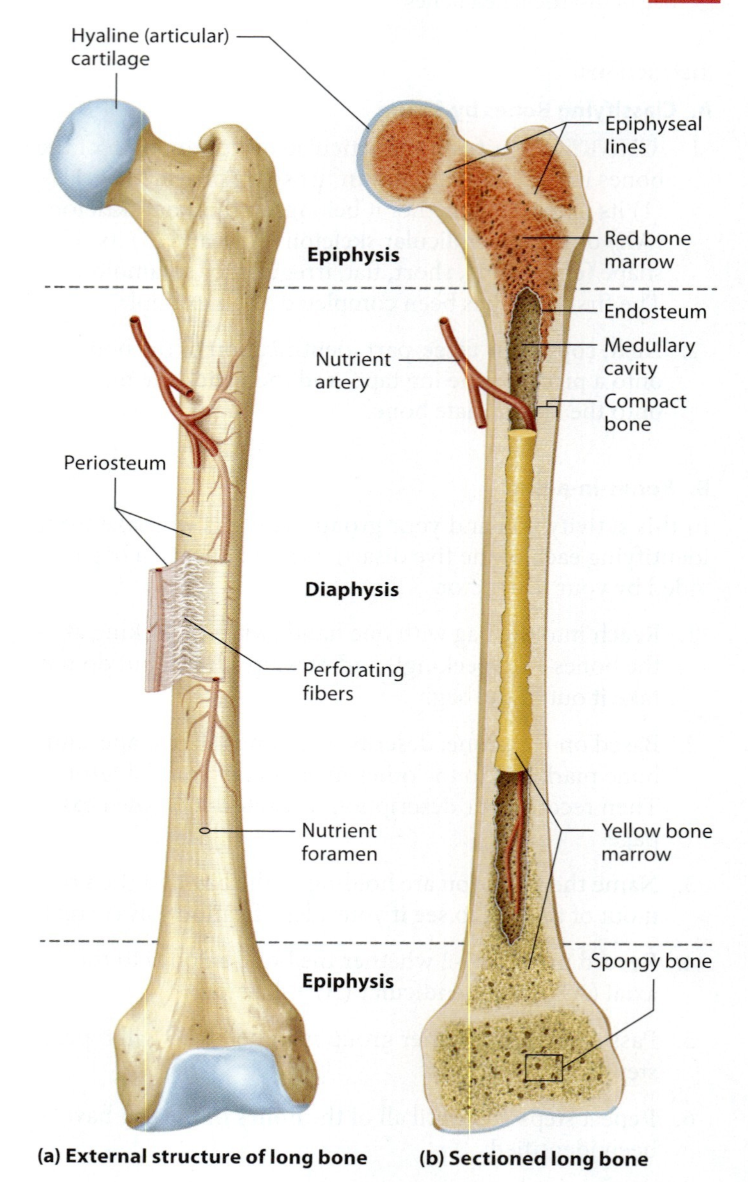

Figure 8-5 **The gross anatomy of a long bone.**

ACTIVITY 3

Examining the Gross Anatomy of a Long Bone

Learning Outcome

1. Sketch a typical long bone, label its major components, and describe the function of each component.

Materials Needed

- ☐ Long bone (or fresh beef bone) sectioned longitudinally
- ☐ Disposable gloves and safety glasses

Instructions

1. If you are using a fresh beef bone, put on disposable gloves and safety glasses.
2. Using Figure 8-5 as an aid, sketch and label each of the components of typical long bones included in the following list.
   - ☐ diaphysis
   - ☐ epiphysis
   - ☐ articular cartilage
   - ☐ epiphyseal line
   - ☐ medullary cavity
   - ☐ periosteum
   - ☐ endosteum
3. Write down two or three words or phrases that define or describe the structure and/or function of each component. The first item has been completed as an example.

   Diaphysis shaft; consists of compact bone surrounding medullary cavity (containing yellow bone marrow)

   Epiphysis ______________________________

   ____

   ____

   Articular cartilage ______________________________

   ____

   ____

   Epiphyseal line ______________________________

   ____

   ____

   Medullary cavity ______________________________

   ____

   ____

   Periosteum* ______________________________

   ____

   ____

   Endosteum* ______________________________

   ____

   ____

   * Visible for fresh bones, but not dry ones. If you do not have a fresh bone, identify this structure on Figure 8-5 or on diagrams provided by your instructor.

4. Answer the following questions concerning long bones.
   a. Describe the similarities and differences in the locations and functions of red bone marrow and yellow bone marrow.

   ____

   ____

   ____

   b. Describe the similarities and differences between an epiphyseal plate and an epiphyseal line.

   ____

   ____

   ____

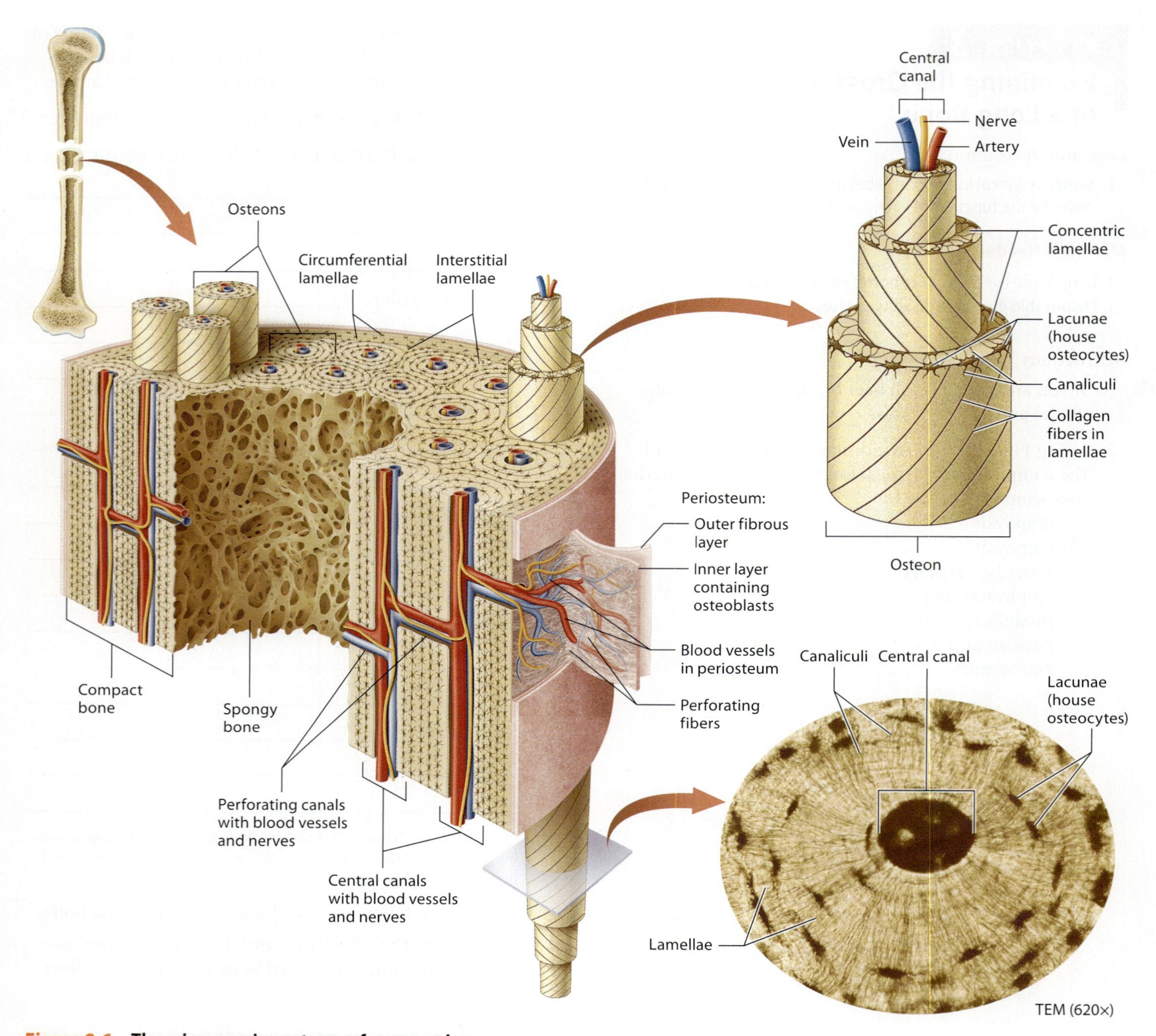

Figure 8-6 The microscopic anatomy of compact bone.

Microscopic Anatomy of Compact Bone

The functional unit of compact bone is a microscopic structure called the **osteon**, or Haversian system (**Figure 8-6**). Each osteon consists of a **central canal**, which conducts blood vessels, nerves, and lymphatic vessels, surrounded by concentric layers of mineralized extracellular matrix called **concentric lamellae**. In the spaces between the osteons are **interstitial lamellae**. Just inside the periosteum and outside the spongy bone are outer and inner rings of **circumferential lamellae**. Embedded within the lamellae are cavities called **lacunae**, which house **osteocytes**, mature bone cells that maintain the bone matrix. Cytoplasmic extensions of osteocytes project into tiny canals called **canaliculi**, which radiate from each lacuna and connect with the central canal. Note that osteocytes and their cytoplasmic extensions are found only in living tissue, not in prepared slides. The canaliculi and lacunae provide pathways for oxygen, nutrients, and wastes to move through the osseous tissue. **Perforating canals** lie perpendicular to the shaft of the bone; they carry blood vessels into the bone from the periosteum and connect the central canals of adjacent osteons.

LabBOOST >>>

Osteon Model Compact bone is made up of *many* osteons connected to each other. Not only are osteons very long; they are quite narrow—each only about 0.2 mm in diameter, a small fraction of the width of compact bone in the longitudinally sectioned bone you examined in Activity 3. When you look at an osteon model, remember that osteons are microscopic structures found within compact bone, and that the model has been greatly enlarged to make observing the structures in each osteon easier. And remember too that because perforating canals run perpendicular to the long axis of the bone and connect one central canal to another, they may not be visible on your model. Note the line that has been added to the photo below to represent a perforating canal.

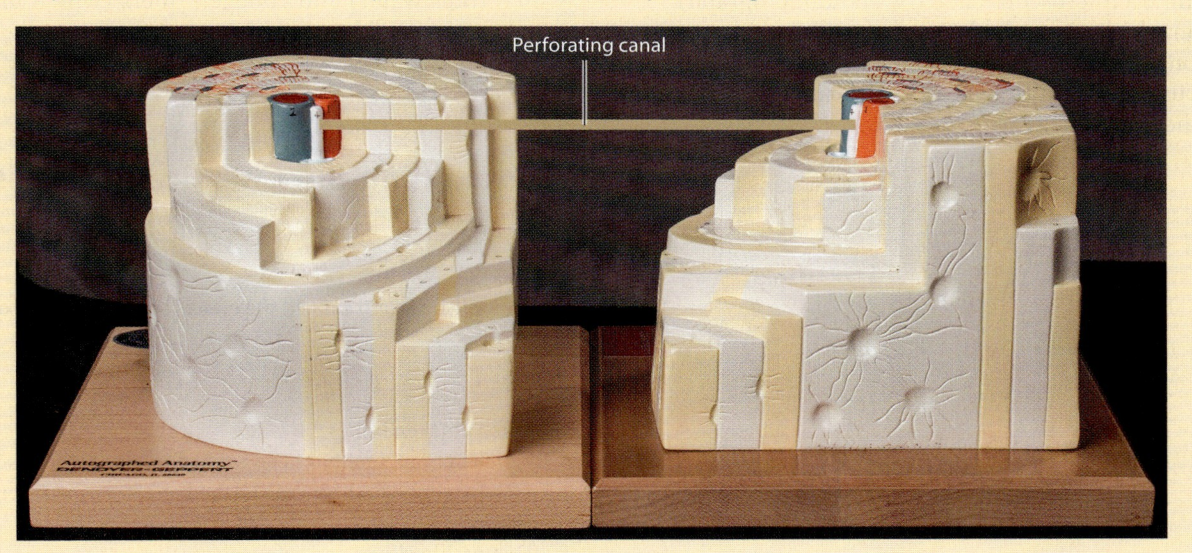

ACTIVITY 4

Exploring the Microscopic Anatomy of Compact Bone—The Osteon

Learning Outcome

1. Sketch an osteon, label its major components, and describe the function of each component.

Materials Needed

- ☐ Osteon model or diagram
- ☐ Microscope and microscope slide (or photomicrograph) of compact bone

Instructions

1. Identify each of the following structures on an osteon model or diagram, and then write two or three words or phrases that define or describe its structure and/or function. The first item has been completed as an example.

   Osteon microscopic functional unit; a component of compact bone; consists of a central canal surrounded by lamellae

   Central canal ______________________________

   Osteocyte ______________________________

   Lacuna ______________________________

   Canaliculus ______________________________

   Interstitial lamella ______________________________

Concentric lamella ______________________________

____

____

Perforating canal ______________________________

____

____

2. Examine a microscope slide (or photomicrograph) of compact bone under low power and then draw an osteon within the circle provided. Label the following structures on your drawing: central canal, lacuna, canaliculus, lamella.

Total magnification: __________ ×

Which cells are housed in the lacunae?

____

Which structures are housed in the canaliculi?

____

Given that oxygen and nutrients are unable to diffuse through the hardened, mineralized extracellular matrix of compact bone, how do osteocytes survive?

____

____

3. **Optional Activity**

View and label histology slides of compact bone at MasteringA&P® > Study Area > Practice Anatomy Lab > Histology > Connective Tissue > Images 27–29

Chemical Composition of Bone

Bone contains both inorganic and organic components. Sixty-five percent of bone tissue consists of an *inorganic* extracellular matrix made of mineral salts, or hydroxyapatites, predominantly calcium phosphate and calcium carbonate. These mineral salts give bone its exceptional hardness and enable it to resist compression. The remaining 35% of bone tissue consists of an *organic* matrix, known as **osteoid**, plus bone cells: **osteoblasts** (bone-building cells), **osteoclasts** (bone-destroying cells), and **osteocytes** (mature bone cells). Osteoid includes ground substance (such as the proteoglycan hyaluronic acid, which traps water) and collagen fibers (which provide the bone with a flexible strength).

ACTIVITY 5

Examining the Chemical Composition of Bone

Learning Outcome

1. Demonstrate and explain the effects of heat and acid on bone tissue.

Materials Needed

- ☐ Fresh chicken bones
- ☐ Chicken bone that has been soaked in vinegar for 5–7 days
- ☐ Chicken bone that has been baked at 250 °F for 2 hours
- ☐ Mallet
- ☐ Disposable gloves and safety glasses

Instructions

1. Put on disposable gloves and safety glasses.
2. Examine a fresh chicken bone and one that has been soaked in vinegar for several days. Note the texture and flexibility of each bone. The acidity of vinegar removes the inorganic salts from the extracellular matrix of the bone.

   Which component of bone is damaged or changed as a result of exposure to acid?

   ____

3. Now examine a chicken bone that has been baked at 250 °F for 2 hours. Exposing the bone to high temperature denatures the protein and other organic substances from the extracellular matrix of the bone.

   Which component of bone is damaged or changed as a result of exposure to high temperature?

   ____

4. Strike each bone with a mallet and describe what happens in each case.

   Vinegar-soaked bone: ______________________________

   Baked bone: ______________________________

5. Follow your instructor's directions for the cleanup and disposal of the materials used in this activity.

POST-LAB ASSIGNMENTS

Name: ______________________ Date: ____________ Lab Section: ________

PART I. Check Your Understanding

Activity 1: Reviewing Skeletal Cartilages

1. Which specific cartilage type:

   ______________ a. covers the ends of moveable bones?

   ______________ b. is found in the epiphyseal plate?

   ______________ c. is predominant in the pubic symphysis?

   ______________ d. connects the vertebrae to one another?

   ______________ e. forms the meniscus?

   ______________ f. connects the ribs to the sternum?

2. State one structural difference between fibrocartilage and elastic cartilage.

   ____

   ____

Activity 2: Classifying and Identifying the Bones of the Skeleton

1. For each of the five bones indicated in the figure at the right, complete the following chart:

	Name of Bone	Shape of Bone	Axial or Appendicular?
a.			
b.			
c.			
d.			
e.			

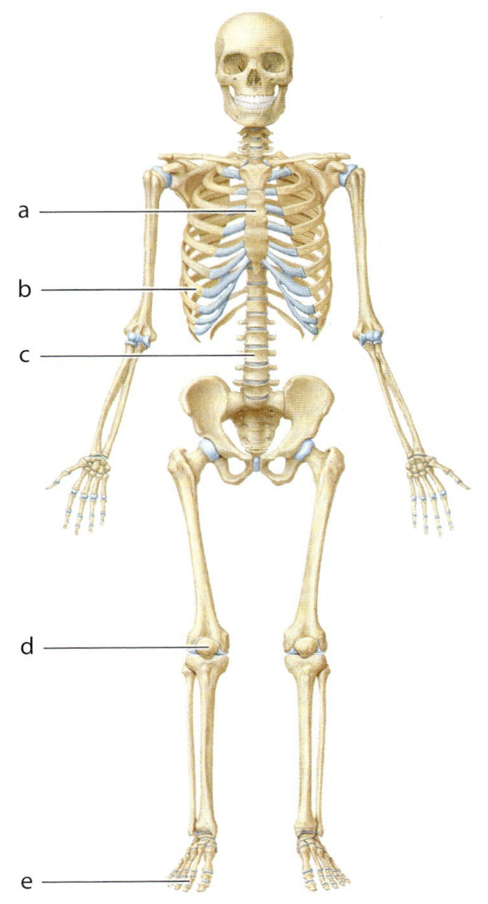

2. A smooth, nearly flat articular surface is called a:
   a. ramus.
   b. facet.
   c. condyle.
   d. notch.
   e. fossa.

3. A shallow depression on a bone surface is called a:
   a. foramen.
   b. fissure.
   c. fossa.
   d. facet.
   e. fontanel.

Activity 3: Examining the Gross Anatomy of a Long Bone

1. What is the function of the periosteum? ______________________________

____

2. The endosteum lines the:
   a. outside of the diaphysis.
   b. outside of the epiphysis.
   c. medullary cavity.
   d. trabeculae.
   e. medullary cavity and the trabeculae.

3. Yellow marrow is found in the:
   a. spongy bone of the epiphysis.
   b. epiphyseal plate.
   c. medullary cavity of the epiphysis.
   d. medullary cavity of the diaphysis.

Activity 4: Exploring the Microscopic Anatomy of Compact Bone—The Osteon

1. Identify the parts of an osteon in the accompanying photomicrograph:

   a. ______________________________

   b. ______________________________

   c. ______________________________

   d. ______________________________

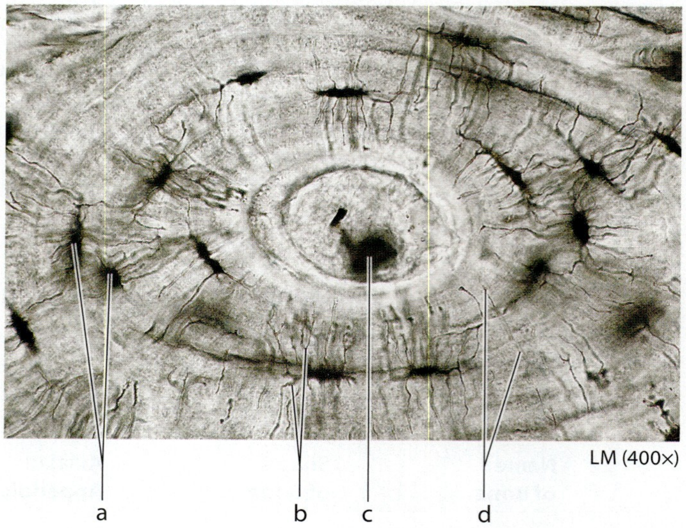

a b c d

LM (400×)

2. Which of the following structures routes blood from the periosteum to the central canal of each osteon?
   a. canaliculi
   b. endosteum
   c. osteocytes
   d. perforating canals

3. How do osteocytes obtain oxygen through the hardened minerals of osteon?
   a. They don't; oxygen diffuses through the canaliculi.
   b. They don't because mature osteocytes don't need oxygen.
   c. Perforating canals that connect to each osteocyte bring nutrients and gases to the osteocytes.
   d. The minerals are spaced such that gases can diffuse through the bone.

4. Differentiate between a lacuna and a canaliculus. ______________________________

____

5. Distinguish between interstitial lamellae and concentric lamellae. ______________________________

____

Activity 5: Examining the Chemical Composition of Bone

1. Match each of the following functions/descriptions with its correct term. More than one correct answer may be possible for each numbered item; include all correct answers.

________	a. component of bone dissolved by acid	1. osteoblasts
________	b. component of bone broken down by heat	2. osteoclasts
________	c. breaks down matrix	3. collagen
________	d. produces matrix	4. inorganic matrix
________	e. provides flexible strength	5. organic matrix
________	f. hyaluronic acid is part of this component	6. calcium minerals

2. Briefly describe the effect of heat on bone and explain the chemical basis of that effect.

____

____

3. Briefly describe the effect of acid on bone and explain the chemical basis of that effect.

____

____

PART II. Putting It All Together

A. Review Questions

Answer the following questions using your lecture notes, your textbook, and your lab notes:

1. Indicate whether each of the following descriptions applies to bone (B) or cartilage (C) or to both tissue types (B and C):

______________ a. Contains lacunae.

______________ b. Contains collagen fibers.

______________ c. Lacks blood supply.

______________ d. Has mineralized matrix.

______________ e. Produces blood cells.

2. Indicate whether each of the following descriptions applies to the organic component of bone (O) or the inorganic component of bone (I) or to both components of bone (O and I):

______________ a. Contains osteoid.

______________ b. Contains mineral salts.

______________ c. Contains osteoblasts.

______________ d. Contains collagen fibers.

______________ e. Comprises 65% of bone tissue.

3. Most of the human skeleton begins as a hyaline cartilage model, which is later replaced by bone. When chondrocytes become surrounded by mineralized matrix, they die, but when osteocytes become surrounded by mineralized matrix, they do not die. Based on your understanding of the microscopic anatomy of cartilage and bone, explain why this is the case.

____

____

B. Concept Mapping

1. Fill in the blanks to complete this concept map outlining the structure of compact bone.

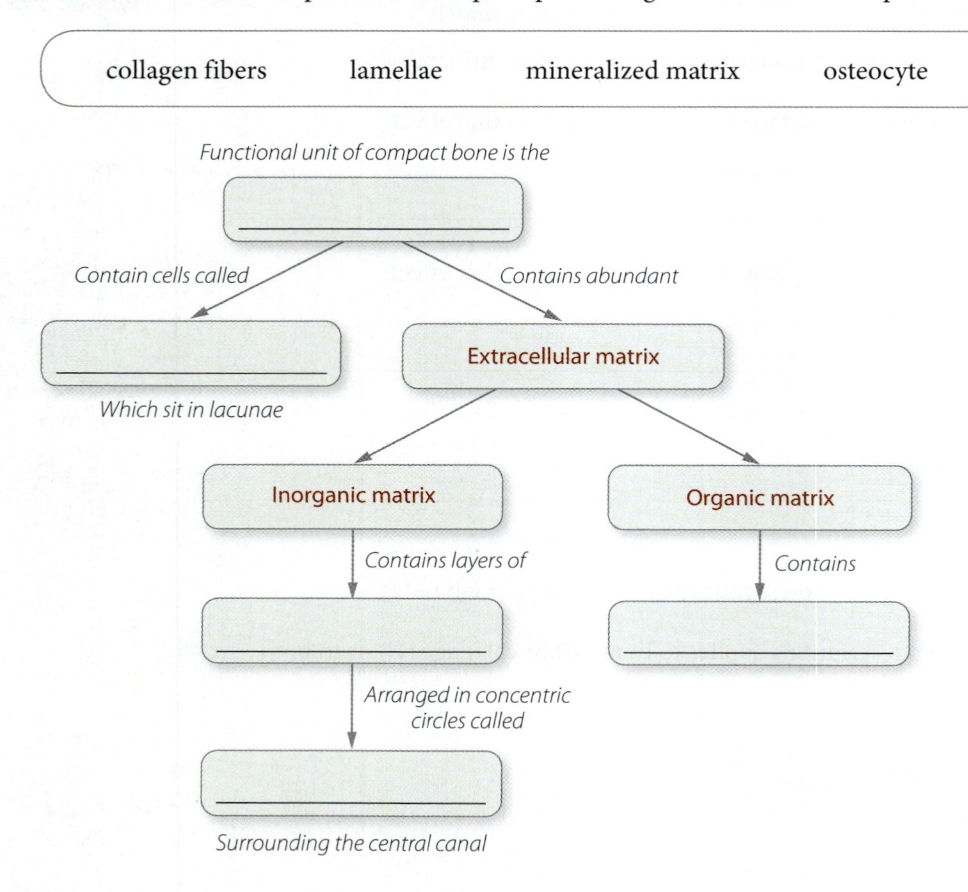

2. Construct a unit concept map to show the relationships among the following set of terms. Include all of the terms in your diagram. Your instructor may choose to assign additional terms.

collagen fibers	compact bone	diaphysis	femur	lamellae
mineralized	osteoblast	osteoclast	osteocyte	osteon
periosteum	red bone marrow	rib	sternum	yellow bone marrow

9

The Axial Skeleton

UNIT OUTLINE

In this unit we will continue our study of the skeleton by looking at the axial skeleton. In Unit 10 we will study the appendicular skeleton; in Unit 11 we will study joints (articulations). The **axial skeleton** forms the longitudinal axis of the body and is composed of 80 bones, including the skull, vertebral column, and thoracic cage. Its central location, combined with limitations on movement between its joints, provides a strong structural support column for the body. Additionally, bones of the axial skeleton encase major organs, including the brain, spinal cord, lungs, and heart and provide protection for these soft tissues.

THINK ABOUT IT *Why does cardiopulmonary resuscitation (CPR) include compressions on the sternum (part of the thoracic cage)?*

____

____

PRE-LAB ASSIGNMENTS

Pre-lab quizzes are also assignable in MasteringA&P®

To maximize learning, BEFORE your lab period carefully read this entire lab unit and complete these pre-lab assignments using your textbook, lecture notes, and prior knowledge.

PRE-LAB Activity 1: Studying the Bones of the Skull

1. Use the list of bones provided to label the accompanying illustration of the skull. Check off each bone as you label it.

☐ parietal bone	☐ sphenoid bone	☐ zygomatic bone
☐ frontal bone	☐ temporal bone	☐ lacrimal bone
☐ occipital bone	☐ mandible bone	☐ ethmoid bone
☐ nasal bone	☐ maxilla bone	

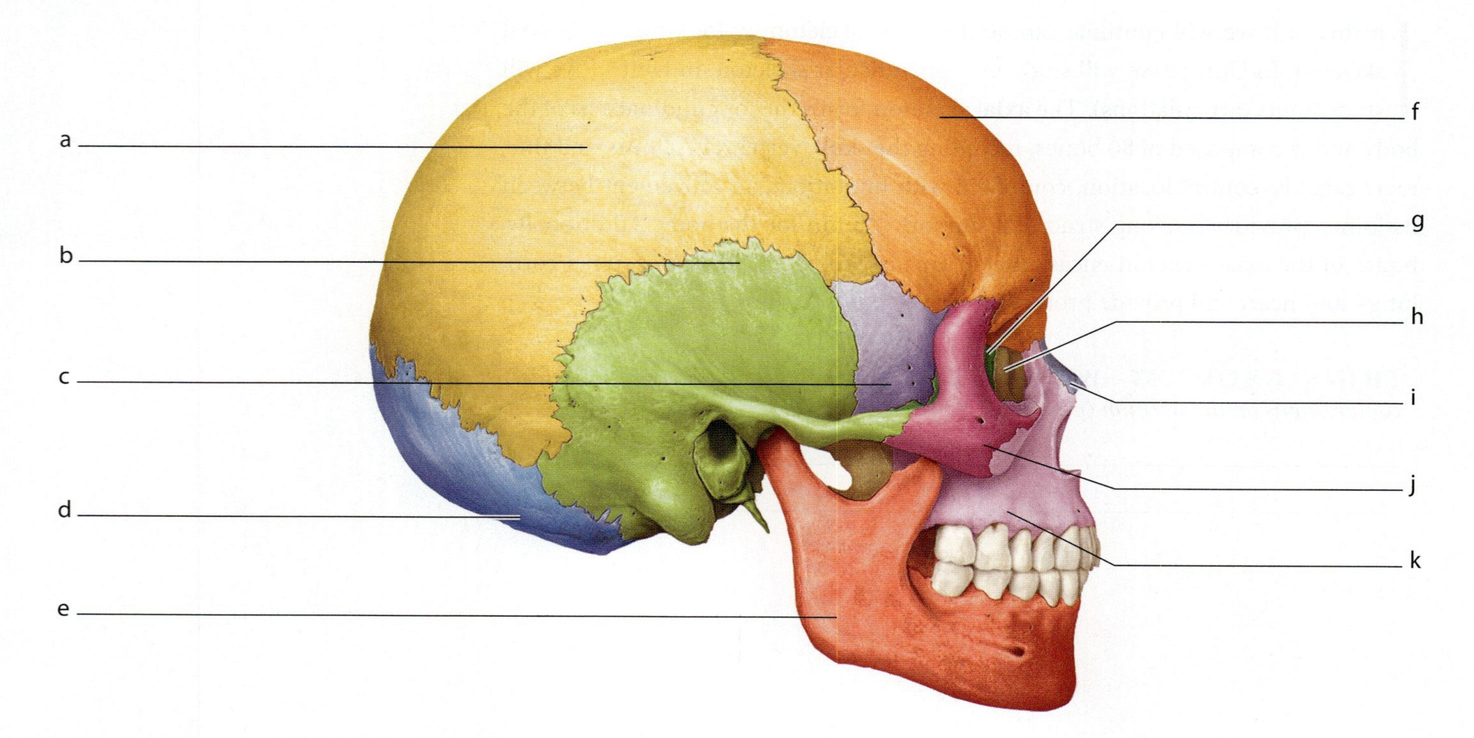

2. Match each of the following skull structures with its correct bone:

_____	a. styloid process	1. ethmoid bone
_____	b. infraorbital foramen	2. sphenoid bone
_____	c. sella turcica	3. temporal bone
_____	d. foramen magnum	4. mandible
_____	e. coronoid process	5. maxilla
_____	f. crista galli	6. occipital bone

PRE-LAB Activity 2: Examining the Fetal Skull

1. The anterior fontanel is located:
   a. between the occipital and parietal bones.
   b. where the coronal and frontal sutures meet.
   c. where the lambdoid and sagittal sutures meet.
   d. between the sphenoid and temporal bones.
2. Of what tissue are the "soft spots" composed?
   a. hyaline cartilage
   b. osseous tissue
   c. epithelium
   d. fibrous connective tissue

PRE-LAB Activity 3: Studying the Bones of the Vertebral Column and Thoracic Cage

1. Each human has ____________ cervical vertebrae.
2. Which of the following bony features is found on thoracic vertebrae *only?*
   a. costal facets
   b. transverse foramina
   c. vertebral foramen
   d. lamina
3. Each human has ____________ pairs of ribs.

PRE-LAB Activity 4: Identifying Bones-in-a-Bag

1. If you feel a long, thin, curved bone, it is most likely:
   a. a skull bone.
   b. a vertebra.
   c. the sternum.
   d. a rib.
2. If a bone feels like it has posts on either side and teeth in the middle, it is most likely:
   a. a temporal bone.
   b. the zygomatic bone.
   c. the mandible.
   d. the palatine bone.

The Skull

The **skull** protects the delicate brain, provides attachment sites for head and neck muscles, and houses the major sensory organs for vision, hearing, balance, taste, and smell. The skull and associated bones total 29 bones: 22 skull bones (8 cranial bones and 14 facial bones), 6 auditory ossicles (discussed in Unit 18), and the hyoid bone (discussed with the skull because of its proximity). The **vertebral column** supports the trunk, protects the spinal cord, and provides attachment sites for the ribs and for the muscles of the neck and back. In fetuses and infants, the vertebral column consists of 33 separate vertebrae. However, adults have only 26 vertebral bones because nine of these bones eventually fuse to form two composite bones—the sacrum and coccyx. The **thoracic cage** (or rib cage) forms a protective structure around the organs of the thoracic cavity. It consists of the sternum and the paired ribs (a total of 25 bones), the costal cartilages, and the thoracic vertebrae of the vertebral column.

The eight bones that form the cranium—the **cranial bones**—are connected by immovable articulations called **sutures**. The cranium can be divided into two major areas: the **cranial vault** (or calvarium), which forms the superior, lateral, and posterior walls of the skull, and the **cranial base** (or cranial floor), which forms the bottom of the skull (**Figure 9-1**).

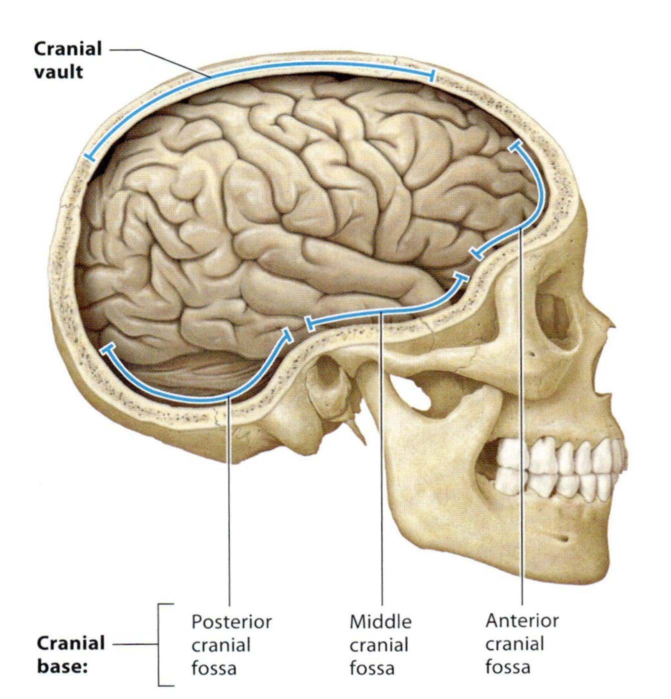

Figure 9-1 The cranial vault and the cranial base, with the locations of the three cranial fossae indicated.

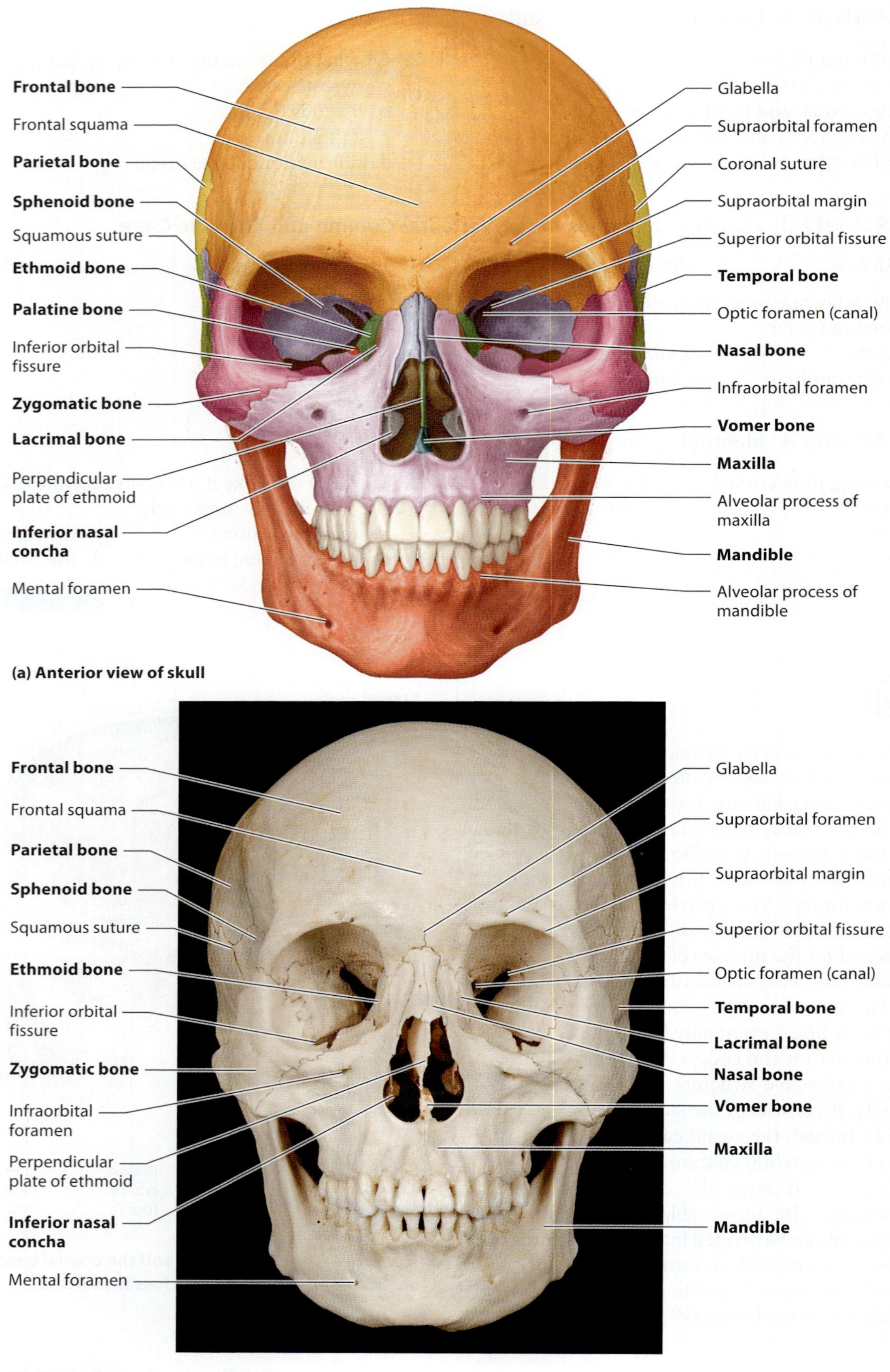

Figure 9-2 **Anterior view of the skull.**

The cranial base has three distinct depressions: the anterior, middle, and posterior cranial fossae. The brain fits securely in these fossae and is completely surrounded by the cranial vault. The 14 **facial bones** form the framework for the face, anchor the facial muscles, secure the teeth, and house special sense organs for vision, taste, and smell. All of the facial bones are joined by sutures (except for the mandible, which is joined to the cranium by a freely movable joint).

The Cranial Bones

The **frontal bone** (**Figure 9-2**) forms the anterior portion of the cranium. Features of the frontal bone include the **frontal squama** or forehead; the **supraorbital foramen** (notch), an opening above each orbit that serves as a passageway for blood vessels and nerves; and the **glabella**, the smooth area between the eyes.

The **parietal bones** form the superior portion and part of the lateral walls of the cranium. The **sagittal suture** (**Figure 9-3**) connects the two parietal bones; the **coronal suture**) connects the parietal bones to the frontal bone; the **squamous suture** connects the temporal bone with the parietal bone; the **lambdoid suture** connects the occipital bone to the parietal bones; and the **occipitomastoid suture** (see Figure 9-4) connects the temporal bone (near the mastoid process) to the occipital bone.

The **temporal bones**, which form part of the lateral walls of the cranium inferior to the parietal bones (**see Figures 9-4 through 9-6**), are divided into the squamous, tympanic, mastoid, and petrous regions. The temporal bones also house a total of six tiny bones called auditory ossicles, which function in the special sense of hearing.

One important bone marking of the squamous region of the temporal bone is the **zygomatic process** (see Figure 9-4), a bar-like structure that projects anteriorly to articulate with the zygomatic bone; along with the temporal process of the zygomatic bone, it forms the zygomatic arch (cheekbone). Another marking is the **mandibular fossa** (see Figure 9-5), a depression where the mandibular condyle of the mandible articulates with the temporal bone.

Important bone markings of the tympanic region of the temporal bone include the **external acoustic meatus** (see Figure 9-4), which conducts sound waves toward the eardrum; and the **styloid process**, a sharp projection that serves as the attachment site for some muscles of the tongue and pharynx and is also the attachment site for the ligament that anchors the hyoid bone to the skull. The mastoid region contains the **mastoid process** (see Figures 9-3 and 9-4), a prominent projection that serves as an attachment site for some neck muscles. Important bone markings of the petrous region of the temporal bone include the **jugular foramen** (see Figure 9-5), a passageway for three cranial nerves and for the internal jugular vein, which drains blood from the brain; the **carotid canal** (see Figure 9-5), a passageway for the internal carotid artery, which delivers blood to the brain; the **foramen lacerum** (see Figure 9-6), a jagged opening that serves as a passageway for small arteries supplying blood to the inner surface of the cranium; and the **internal acoustic meatus** (see **Figure 9-7**), a passageway for two cranial nerves. In a living person, the foramen lacerum is mostly filled with hyaline cartilage.

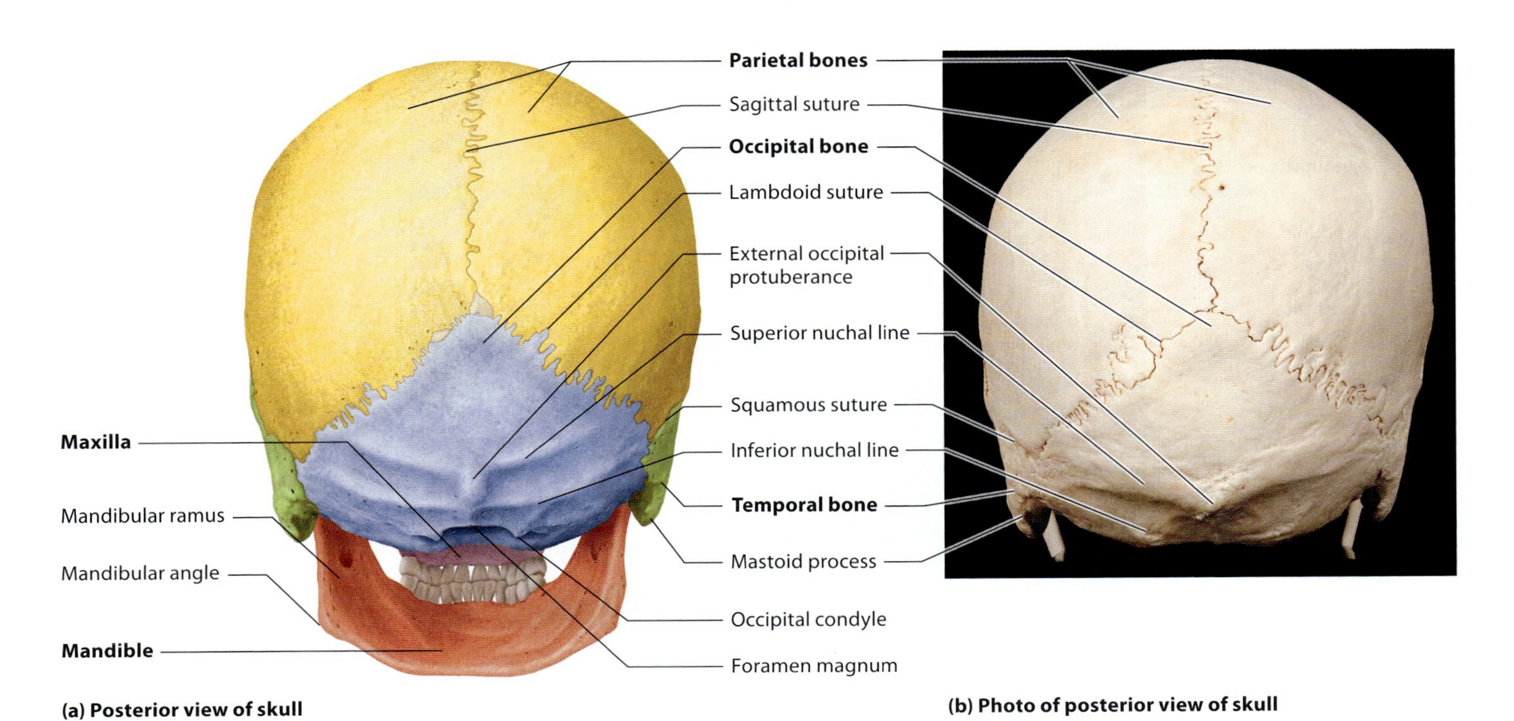

Figure 9-3 Posterior view of the skull.

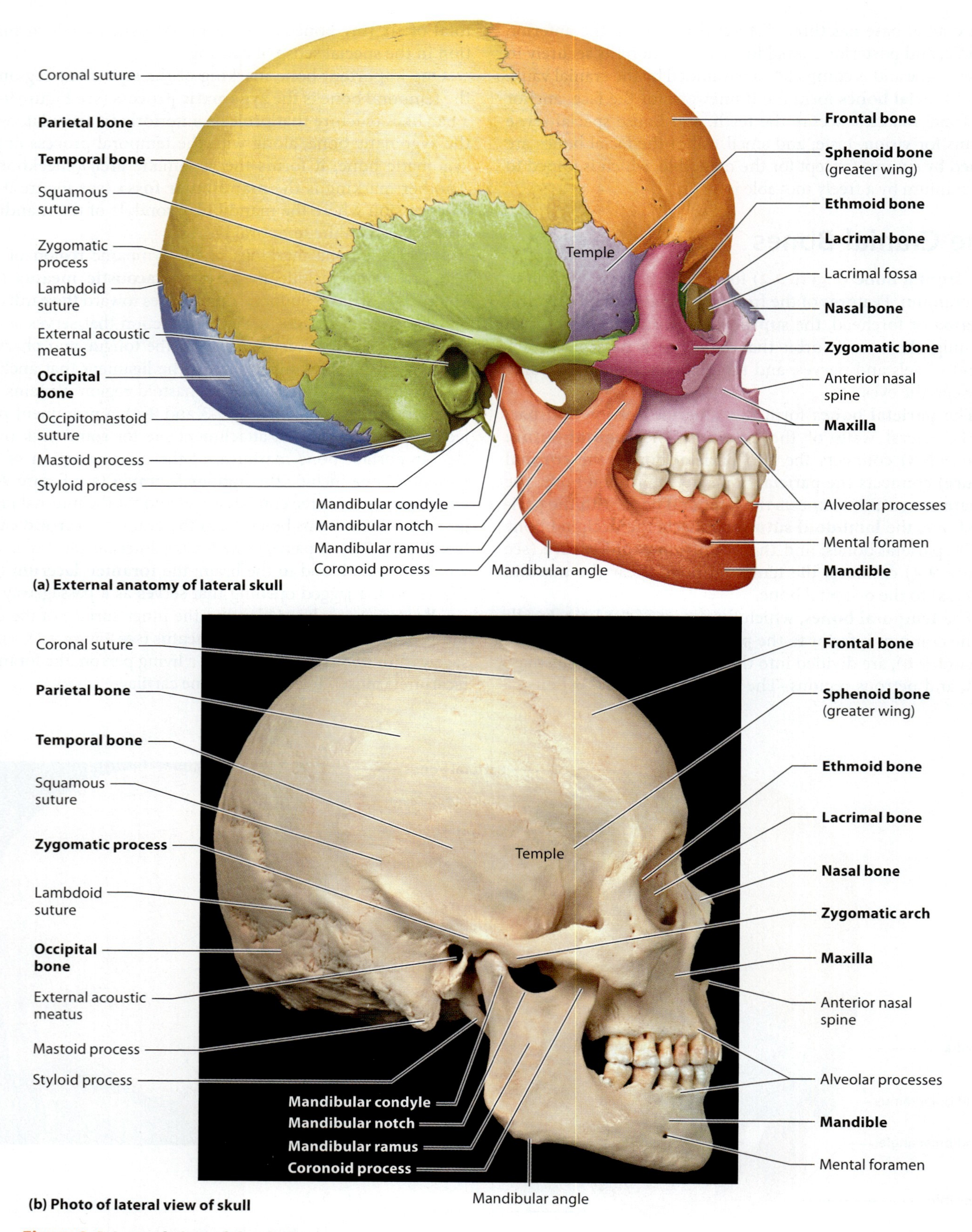

Figure 9-4 **Lateral view of the skull.**

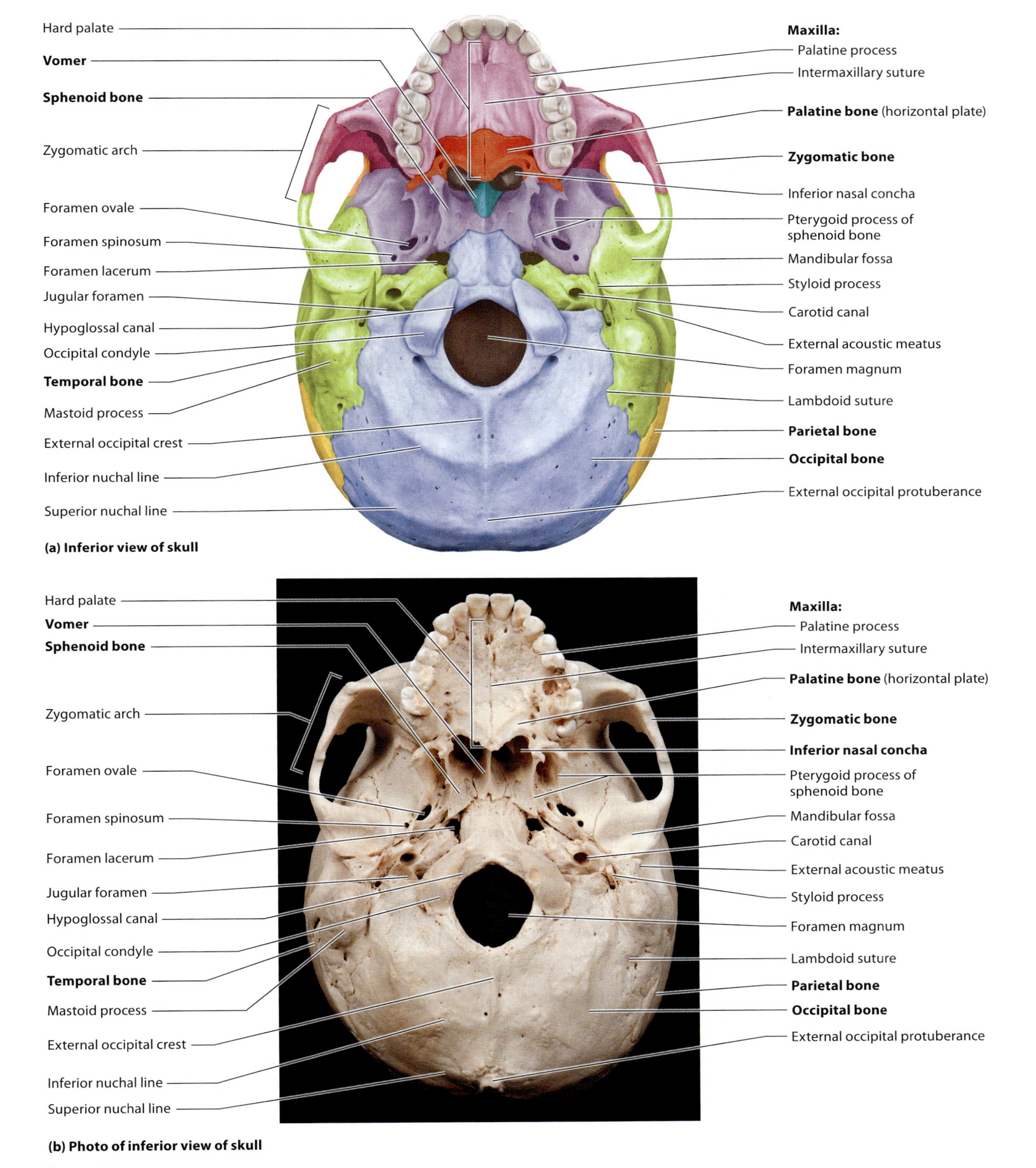

Figure 9-5 Inferior view of the skull.

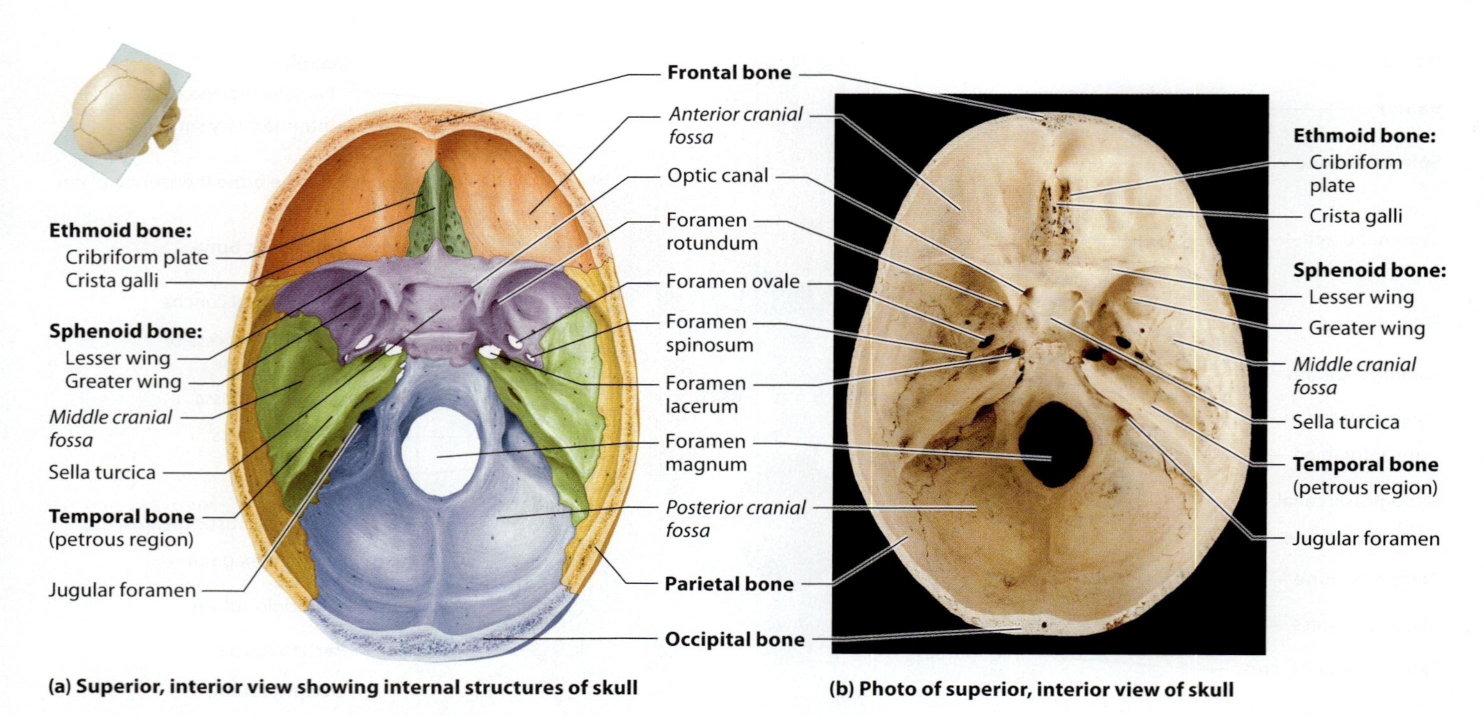

(a) Superior, interior view showing internal structures of skull

(b) Photo of superior, interior view of skull

Figure 9-6 Superior, interior view of the skull.

The **occipital bone** forms the posterior part and most of the base of the cranium. The **foramen magnum** (see Figure 9-5) is a large opening on the inferior surface of the skull where the brain and spinal cord meet. The **occipital condyles** are rounded projections that articulate with the first cervical vertebra (atlas) to form the atlanto-occipital joint, which allows you to nod "yes." The **hypoglossal canals** are openings through which a cranial nerve passes. The **external occipital crest** is a ridge of bone that extends posteriorly from the foramen magnum and ends at the external occipital protuberance, whereas the **external occipital protuberance** (see Figures 9-3 and 9-5) is a small, mid-line bump at the end of the external occipital crest; it occurs at the junction between the base and the posterior wall of the skull. Finally, the **superior** and **inferior nuchal lines** (see Figure 9-5) are small transverse ridges on either side of the external occipital protuberance; they are created by neck muscle attachments.

The **sphenoid bone** (see Figures 9-4, 9-5, and 9-6), a bat-shaped bone located posterior to the frontal bone, articulates with every other cranial bone; thus it is considered the "keystone" bone of the cranium. The **greater wings** (see Figures 9-6 and 9-7) project laterally from the central body of the sphenoid bone and form part of the floor of the middle cranial fossa; the **lesser wings** (see Figures 9-6 and 9-7) are horn-shaped projections that form part of the floor of the anterior cranial fossa. The **pterygoid processes** (see Figure 9-5) project inferiorly from the greater wings and serve as attachment sites for muscles. The **superior orbital fissure** (see Figure 9-2), a long, slit-like opening between the greater and lesser wings, is a passageway for three cranial nerves. The **sella turcica** (see Figures 9-6 and 9-7) is a saddle-shaped area in the midline of the sphenoid bone containing the hypophyseal fossa (the seat of the saddle), which houses the pituitary gland (also called the hypophysis). The **optic canals** (see Figure 9-6) are openings at the base of the lesser wings that serve as passageways for the optic nerves. Finally, six openings—the oval-shaped **foramina ovale** posterior to the sella turcica, the round **foramina rotundum** lateral to the sella turcica, and the small foramina spinosum lateral to the foramina ovale (see Figure 9-6)—serve as passageways for another cranial nerve and an artery.

The **ethmoid bone** (see Figures 9-6 and 9-7) is the most deeply situated bone of the skull. The **crista galli** (see Figures 9-6 and 9-7), a superior projection in the midline of the ethmoid bone, is attached to the brain by connective tissue wrappings and helps secure the brain within the cranial cavity. Located on either side of the crista galli are the **cribriform plates** (see Figure 9-6), bony plates studded with olfactory foramina that serve as passageways for fibers of the olfactory cranial nerves. The **perpendicular plate** (see Figure 9-7) forms the superior portion of the nasal septum. Finally, the **superior** and **middle nasal conchae**

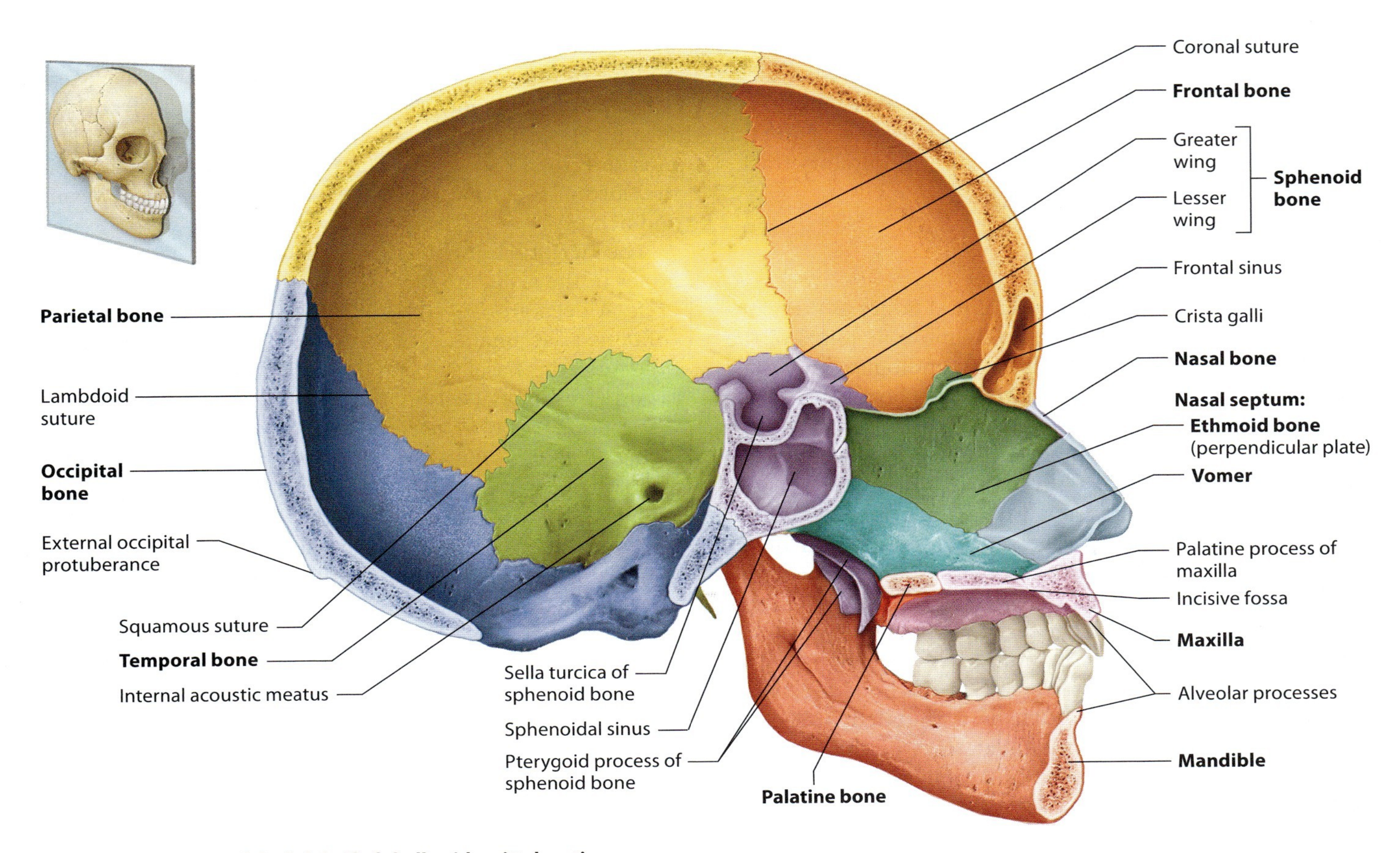

(a) Internal anatomy of the left half of skull, midsagittal section

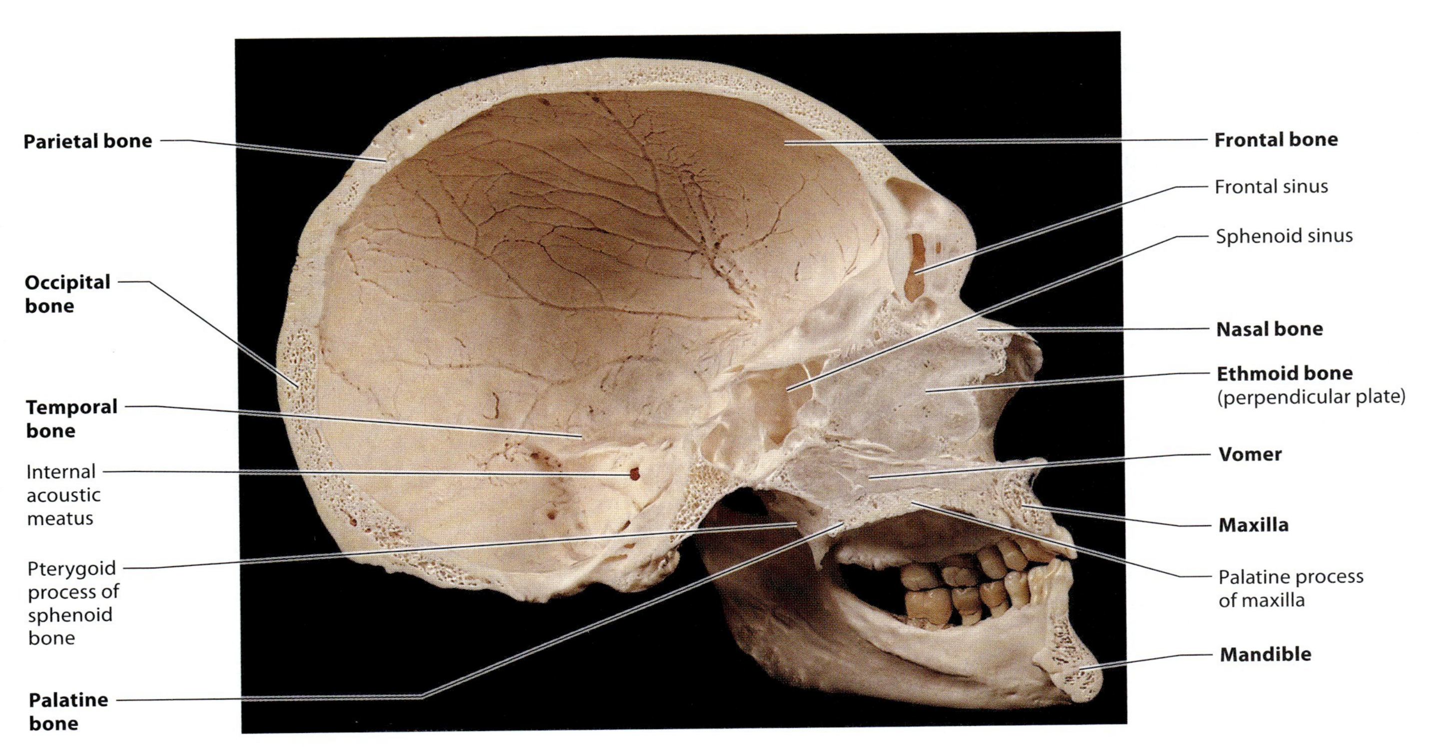

(b) Photo of the skull, midsagittal section

Figure 9-7 Midsagittal section of the skull.

(or turbinates; not illustrated) are thin scrolls of bone that project into the nasal cavity on either side of the perpendicular plate; they create air turbulence that slows air movement, warms and moistens inhaled air, and removes dust before the air reaches the delicate tissues of the lower respiratory tract.

The Facial Bones

The **mandible** or lower jawbone (see Figures 9-2, 9-4, and **Figure 9-8**), articulates with the temporal bones at the mandibular fossae to form the only freely movable joints of the skull. The **body of the mandible** (mandibular body; see Figure 9-8) is the horizontal part of the bone that forms the inferior jawline (chin) and anchors the lower teeth. The **alveolar process** (see Figures 9-4 and 9-8), the superior border of the mandibular body, contains the tooth sockets. The **mental foramina** (see Figure 9-8) are openings in the mandible that serve as passageways for nerves and blood vessels supplying the lips and chin.

The **mandibular ramus** (see Figure 9-8) is an upright "branch" of the mandible that serves as the attachment site for a muscle that assists in elevating (closing) the jaw. The **mandibular angle** is the area at which the body meets the ramus; the **mandibular condyle** is a rounded projection that articulates with the mandibular fossa of the temporal bone to form the temporomandibular joint (TMJ). The **coronoid process** is the insertion site for a muscle that closes the mouth. Finally, the **mandibular notch** is the indentation between the coronoid process and the mandibular condyle.

The **maxillae** (see Figures 9-2 and 9-5), which form the upper jawbone, are considered the "keystone bones" of the face because they articulate with all other facial bones except the mandible. Note that the two maxillae fuse medially so that a maxilla is one bone of the fused pair. The inferior border of the bones—the **alveolar process** (see Figure 9-4)—contains the tooth sockets. The **palatine processes** (see Figure 9-5) project medially from the alveolar margin to form the anterior portion of the hard palate (the bony roof of the mouth).

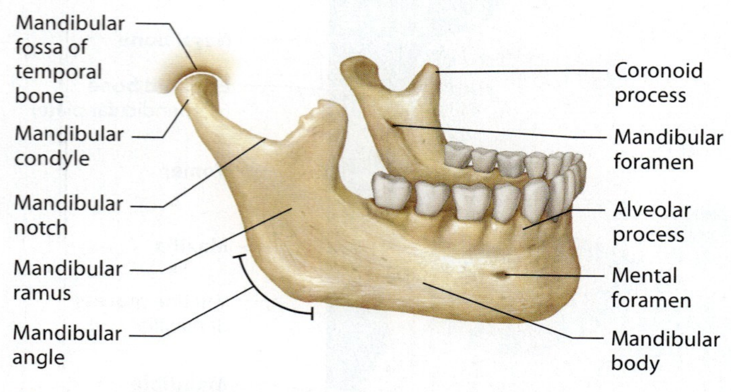

Figure 9-8 Mandible.

The **incisive fossa** (see Figure 9-7), an opening on the inferior midline of the palatine process, leads into the incisive canal, which contains blood vessels and nerves. The **infraorbital foramen** (see Figure 9-2), an opening below the orbit, and the inferior orbital fissure (see Figure 9-2), a slit-like opening in the floor of the orbit, provide passageways for blood vessels and nerves.

The **lacrimal bones** (see Figures 9-2 and 9-4), the smallest bones in the skull, are located in the medial portion of each orbit. A deep groove—the **lacrimal fossa** (see Figure 9-4)—contains a lacrimal sac, which collects tears and drains the fluid into the nasal cavity. The **temporal processes** of the **zygomatic bones** (see Figures 9-4 and 9-5) plus the zygomatic processes of the temporal bones form the zygomatic arches (see Figure 9-4), which form the lateral rims of the orbits. The paired **nasal bones** (see Figure 9-2) form the bridge of the nose. The **vomer bone** (see Figures 9-5 and 9-7) forms the inferior portion of the nasal septum. Posterior to the palatine processes of the maxillae are the paired **palatine bones** (see Figures 9-5 and 9-7), which, with the palatine processes of the maxillae, form the hard palate (the bony roof of the mouth). The **inferior nasal conchae** (turbinates) (see Figure 9-2) are paired bones that project medially to form the lower, lateral walls of the nasal cavity.

The **paranasal sinuses** (**Figure 9-9**), located in the frontal, ethmoid, sphenoid, and maxillary bones, are air-filled, mucus-lined cavities that connect with the nasal cavity. They lighten the skull, add resonance to the voice, and warm, moisten, and clean inhaled air. The frontal and sphenoidal sinuses are also visible in Figure 9-7.

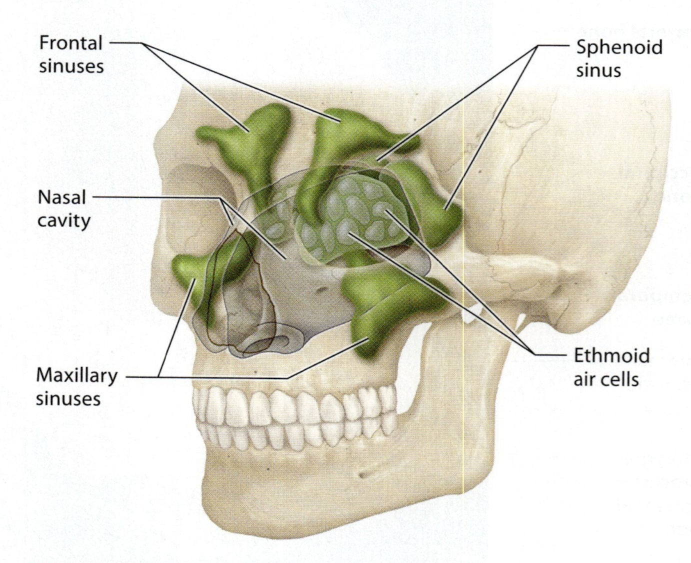

Figure 9-9 Paranasal sinuses.

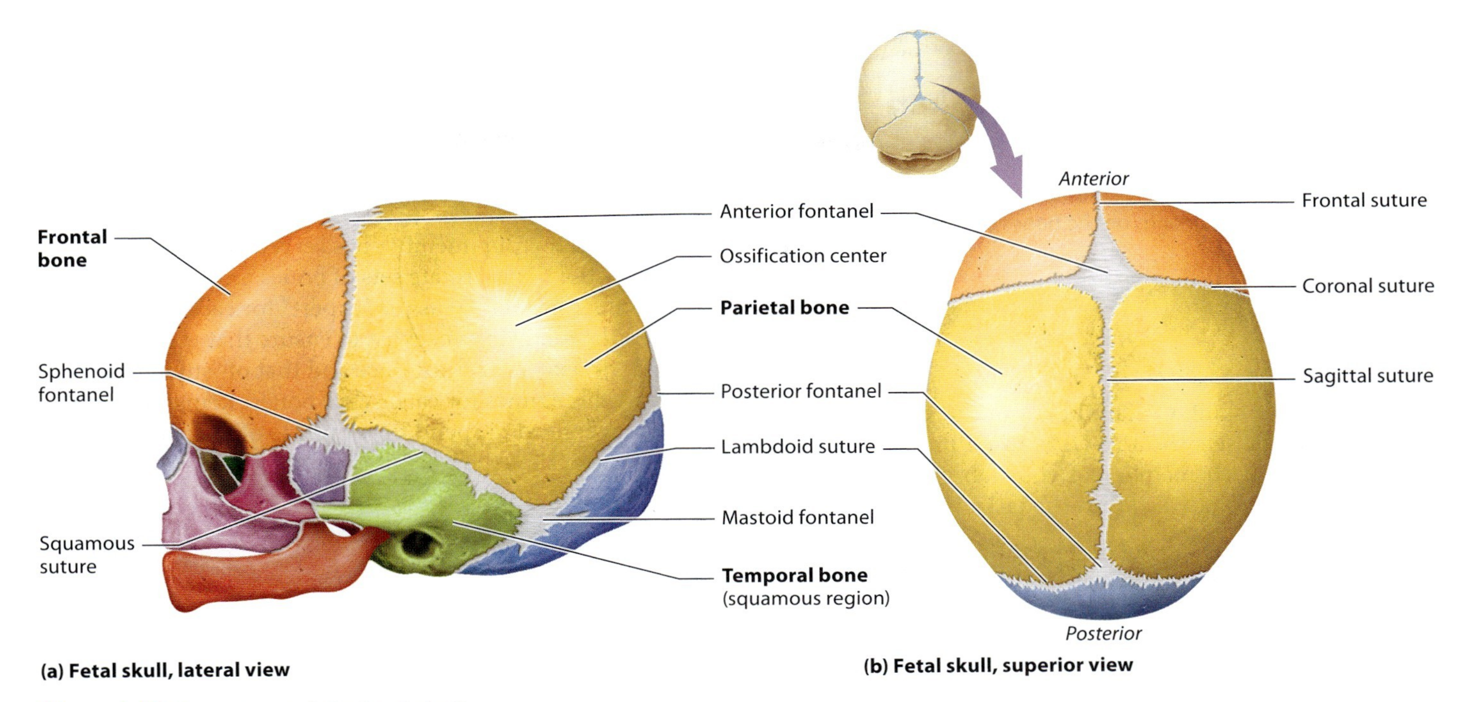

(a) Fetal skull, lateral view **(b) Fetal skull, superior view**

Figure 9-10 **Two views of the fetal skull.**

The Fetal Skull

The bones of the skull develop from fibrous membranes in the developing embryo. At birth, fontanels—commonly called "soft spots"—remain between the cranial bones. These unossified fibrous membranes allow the fetal skull (**Figure 9-10**) to change shape as it passes through the birth canal and then permit brain growth during infancy.

The **anterior fontanel**—the largest fontanel—is diamond shaped and is located along the midline of the skull between the frontal bone and the two parietal bones; it usually closes 18–24 months after birth. The **posterior fontanel**, located along the midline between the two parietal bones and the occipital bone, usually closes around 2 months after birth. The small, irregularly shaped **sphenoid fontanels** are located between the frontal, parietal, temporal, and sphenoid bones; they usually close around 3 months after birth. The small, irregularly shaped **mastoid fontanels** are located between the parietal, occipital, and temporal bones; their complete closure occurs around 12 months after birth. Note as well the **frontal suture**, which divides the left and right frontal bones and closes completely at 3–9 month of age.

Hyoid Bone

Just inferior to the mandible is the C-shaped **hyoid bone** (**Figure 9-11**), unique because it does not articulate with any other bone but instead is suspended from the styloid processes of the temporal bone by ligaments and muscles. The hyoid bone serves as a moveable base for the tongue. The horizontal part of the bone—the **body**—is an attachment site for neck muscles that raise and lower the larynx during swallowing and speech. The paired **greater** and **lesser horns** are attachment sites for other muscles and ligaments.

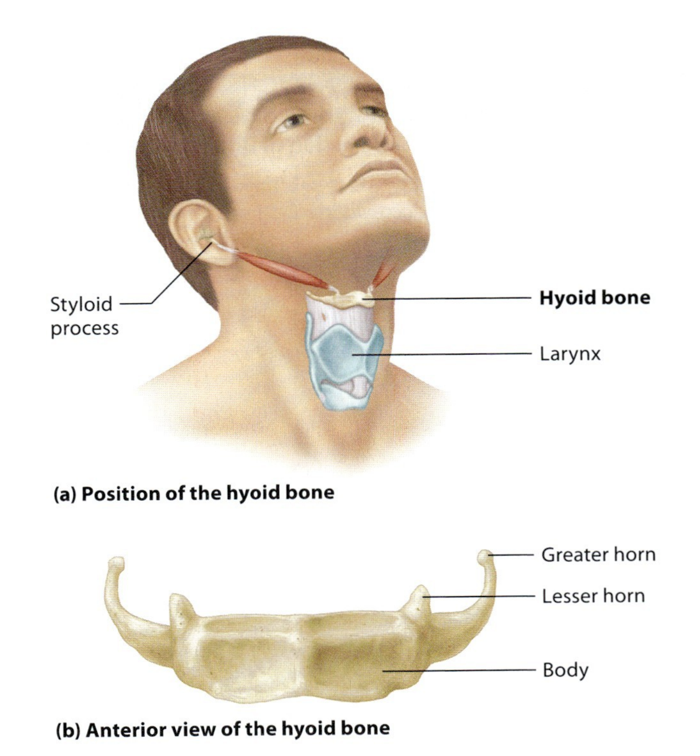

(a) Position of the hyoid bone

(b) Anterior view of the hyoid bone

Figure 9-11 **Hyoid bone.**

ACTIVITY 1

Studying the Bones of the Skull

Learning Outcomes

1. Identify the bones and major sutures of the skull.
2. Identify the bone markings on the bones of the skull.

Materials Needed

☐ Intact skull

Instructions

Obtain an intact skull. Certain bones of the skull have been divided into four sets (Sets A–D, as listed in the charts that follow). Assign each member of your lab group one set of bones. Then, complete the following tasks:

1. CHART Study your set of bones, filling out the description column in the appropriate Making Connections chart.
2. After you have studied your bones, you will be the "expert" as you teach the bones and bone markings that you have learned to the other members of your lab group, the "students."
3. As you are learning the bones and their markings, brainstorm with your group members to make connections between what you are learning and what you have learned in lectures, labs, and from your textbook. Fill out the connections column in the Making Connections chart.
4. After each group member has served as the "expert" for one set of bones, rotate sets of bones until each group member has served as the "expert" for each set of bones. Use the following checklist to ensure that you play to role of "expert" at least once and that of "student" at least three times for each set of bones. Remember that you will need to spend time outside of normal lab time to master these bones.

Set A	Set B	Set C	Set D
☐ "Expert"	☐ "Expert"	☐ "Expert"	☐ "Expert"
☐ "Student"	☐ "Student"	☐ "Student"	☐ "Student"
☐ "Student"	☐ "Student"	☐ "Student"	☐ "Student"
☐ "Student"	☐ "Student"	☐ "Student"	☐ "Student"

Making Connections: The Skull, Set A

Bone	Description of Bone and/or Bone Marking(s)	Connections to Things I Have Already Learned
Cranial vault of skull	*Superior half of the skull; composed of the superior portions of occipital, parietals, and frontals*	*Sutures join together the flat bones; the vault houses the brain.*
Cranial base		
• Anterior cranial fossa		
• Middle cranial fossa		
• Posterior cranial fossa		
Frontal bone		
• Squama		
• Glabella		
• Supraorbital foramen		

Bone	Description of Bone and/or Bone Marking(s)	Connections to Things I Have Already Learned
Parietal bone		
• Coronal suture		
• Sagittal suture		
• Squamous suture		
• Lambdoid suture		
• Occipitomastoid suture		

Making Connections: The Skull, Set B

Bone	Description of Bone and/or Bone Marking(s)	Connections to Things I Have Already Learned
Temporal bone		
• Zygomatic process		
• Mandibular fossa	*Mandibular fossa is an articular depression.*	*Articulates with mandibular condyles; only moveable joint of skull.*
• External acoustic meatus		
• Styloid process		
• Mastoid process		
• Jugular foramen		
• Carotid canal		
• Internal acoustic meatus		
• Foramen lacerum		

(Continued)

Making Connections: The Skull, Set B (*Continued*)		
Bone	**Description of Bone and/or Bone Marking(s)**	**Connections to Things I Have Already Learned**
Occipital bone		
• Foramen magnum	*Literally "hole, large"; located in cranial base*	
• Occipital condyle		
• Hypoglossal canal		
• External occipital crest		
• External occipital protuberance		
• Superior nuchal line		
• Inferior nuchal line		*Created by neck muscles.*

Making Connections: The Skull, Set C		
Bone	**Description of Bone and/or Bone Marking(s)**	**Connections to Things I Have Already Learned**
Ethmoid bone		
• Crista galli	*Projection of bone*	*Literally means "comb, rooster."*
• Cribriform plate		
• Perpendicular plate		
• Superior and middle nasal conchae		
Sphenoid bone		
• Greater wings		
• Lesser wings		

Bone	Description of Bone and/or Bone Marking(s)	Connections to Things I Have Already Learned
• Pterygoid process		
• Superior orbital fissure	*Slit or crack in a greater wing; passage for cranial nerves*	
• Sella turcica		
• Hypophyseal fossa		
• Optic canal		
• Foramen ovale		
• Foramen rotundum		
• Foramen spinosum		

Making Connections: The Skull, Set D

Bone	Description of Bone and/or Bone Marking(s)	Connections to Things I Have Already Learned
Mandible		
• Body of mandible		
• Mandibular ramus		
• Mandibular angle		
• Mandibular condyle		*Part of the temporomandibular joint*
• Coronoid process		
• Mental foramen		
• Mandibular notch		

(Continued)

Making Connections: The Skull, Set D (*Continued*)		
Bone	**Description of Bone and/or Bone Marking(s)**	**Connections to Things I Have Already Learned**
• Alveolar process		
• Mandibular symphysis		
Maxillae		
• Alveolar process		
• Palatine process		
• Infraorbital foramen		
• Incisive fossa		
Palatine bone		
Zygomatic bone		
Lacrimal bone		
• Lacrimal fossa		
Vomer bone		*Inferior part of the nasal septum*
Inferior nasal concha		
Nasal bone		

5. **Optional Activity**

Practice labeling skull bones and their structural features at MasteringA&P® > Study Area > Practice Anatomy Lab > Human Cadaver: Axial Skeleton > Skull

ACTIVITY 2

Examining the Fetal Skull

Learning Outcome

1. Describe the differences between the adult and fetal skulls.

Materials Needed

- ☐ Intact adult skull
- ☐ Intact fetal or neonate skull
- ☐ Labeling tape
- ☐ Marker

Instructions

1. Write the names of each of the following structures of the fetal skull on small pieces of tape: *frontal bones, frontal suture, parietal bones, sagittal suture, temporal bones, squamosal suture, occipital bone, lambdoid suture, mastoid process, occipital condyles, external occipital protuberance, maxilla, mandible, zygomatic bone, zygomatic arch, nasal bones, anterior fontanel,* and *posterior fontanel.*
2. Place each piece of tape on the proper feature of the fetal skull. At the same time, find each structure on the adult skull.
3. CHART Write the differences you observe between the fetal and adult skulls in the following chart. Some examples of entries have been provided.
4. **Optional Activity**

Practice labeling fetal skull features at MasteringA&P® > Study Area > Practice Anatomy Lab > Human Cadaver: Axial Skeleton > Skull > Images 13–15

Bone or Feature	Fetus/Neonate	Adult
Frontal bones		
Frontal suture	*Divides the left and right frontal bones.*	*Typically fuses by age 2 and is thus absent in most adults.*
Occipital bone		
Mastoid process		
Maxilla		
Mandible		
Anterior fontanel		
Posterior fontanel		
Sutures	*In a living fetus, these are gaps filled with fibrous connective tissue.*	*In adults, these are nonmoveable joints with no joint cavity.*

The Vertebral Column

The adult vertebral column, or spine, is a flexible chain of 26 bones called **vertebrae** (singular = vertebra). Fibrocartilage pads called **intervertebral discs** cushion adjacent vertebrae (except the first two vertebrae, the atlas and the axis). The vertebral column provides support for the head, neck, and trunk, and protection for the spinal cord.

Based on their location and structural features, the bones of the vertebral column are grouped into five regions: the cervical vertebrae (7), the thoracic vertebrae (12), the lumbar vertebrae (5), plus the sacrum (1) and the coccyx (1) (**Figure 9-12**). When viewed from the side, the adult vertebral column exhibits four normal curves. The cervical and lumbar curvatures are concave posteriorly, whereas the thoracic and sacral curvatures are convex posteriorly. These curves increase the strength, resilience, and flexibility of the vertebral column.

(a) Vertebral column, anterior view

(b) Normal spinal curvatures, lateral view

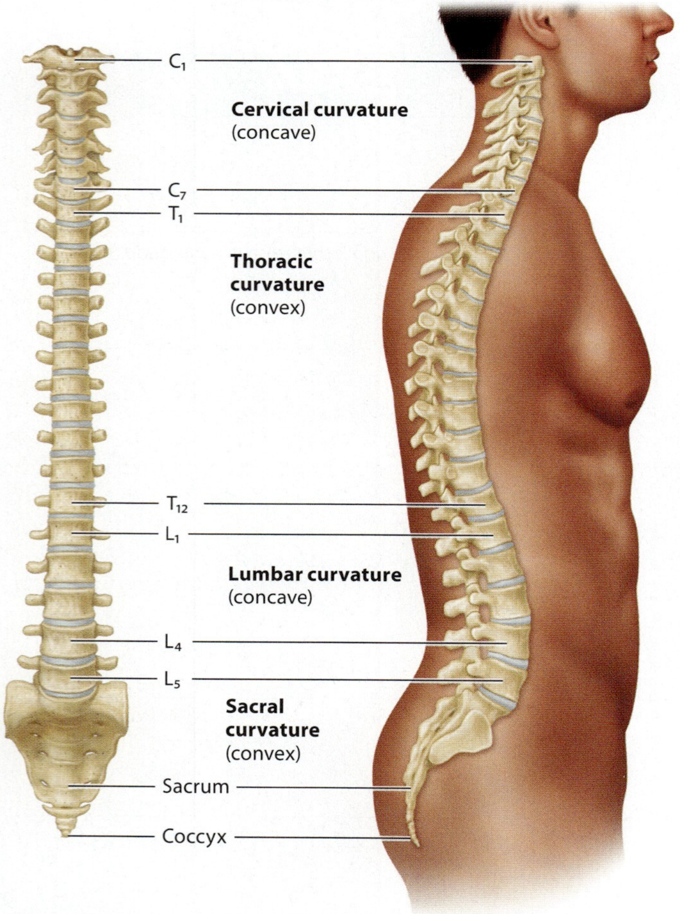

Figure 9-12 The vertebral column showing its five major regions (at left) and its four curvatures (at right).

General Structure of Vertebrae

A typical vertebra (**Figure 9-13**) has a rounded **body** (or centrum) that forms the vertebra's anterior portion and bears the weight of superior structures. A vertebra's posterior portion is the **vertebral arch** (see Figure 9-13a), a composite structure composed of two **pedicles**, which connect the arch to the body, and two **laminae**, which form the rest of the arch. The inferior sides of the pedicles curve superiorly to create an **inferior vertebral notch** (see Figure 9-13b), and the superior sides of the pedicles curve inferiorly to create a **superior vertebral notch**.

Each vertebral arch also has the following seven features: a **spinous process**, a medial posterior projection that arises from the junction of the two laminae; two **transverse processes**, which project laterally from either side of the vertebral arch; two **superior articular processes**, which project superiorly from the pedicle-lamina junction and articulate with the inferior articular processes of the vertebra immediately superior to it at facets; and two **inferior articular processes** (see Figure 9-13b), which project inferiorly from the pedicle-lamina junction and

CLINICAL CONNECTION

Ruptured Disc

An intervertebral disc, which lies between the bodies of adjacent vertebrae (see Figure 9-14b), consists of a gel-like inner nucleus pulposus surrounded by a tough, primarily fibrocartilage collar called the annulus fibrosus. If you liken an intervertebral disc to a jelly doughnut, the jelly inside the doughnut is analogous to the nucleus pulposus, and the doughnut itself is analogous to the annulus fibrosus. When a disc is compressed under normal conditions, it flattens, and the annulus fibrosus limits the expansion of the nucleus pulposus within it. However, severe forces to the back area can cause the fibrocartilage collar to rupture and the nucleus pulposus to bulge out through the fibrocartilage (just as compressing a jelly doughnut too much would rupture it and cause the jelly to squirt out). Such a condition is called a herniated (or slipped) disc and can cause compression on the spinal nerve, leading to extreme pain.

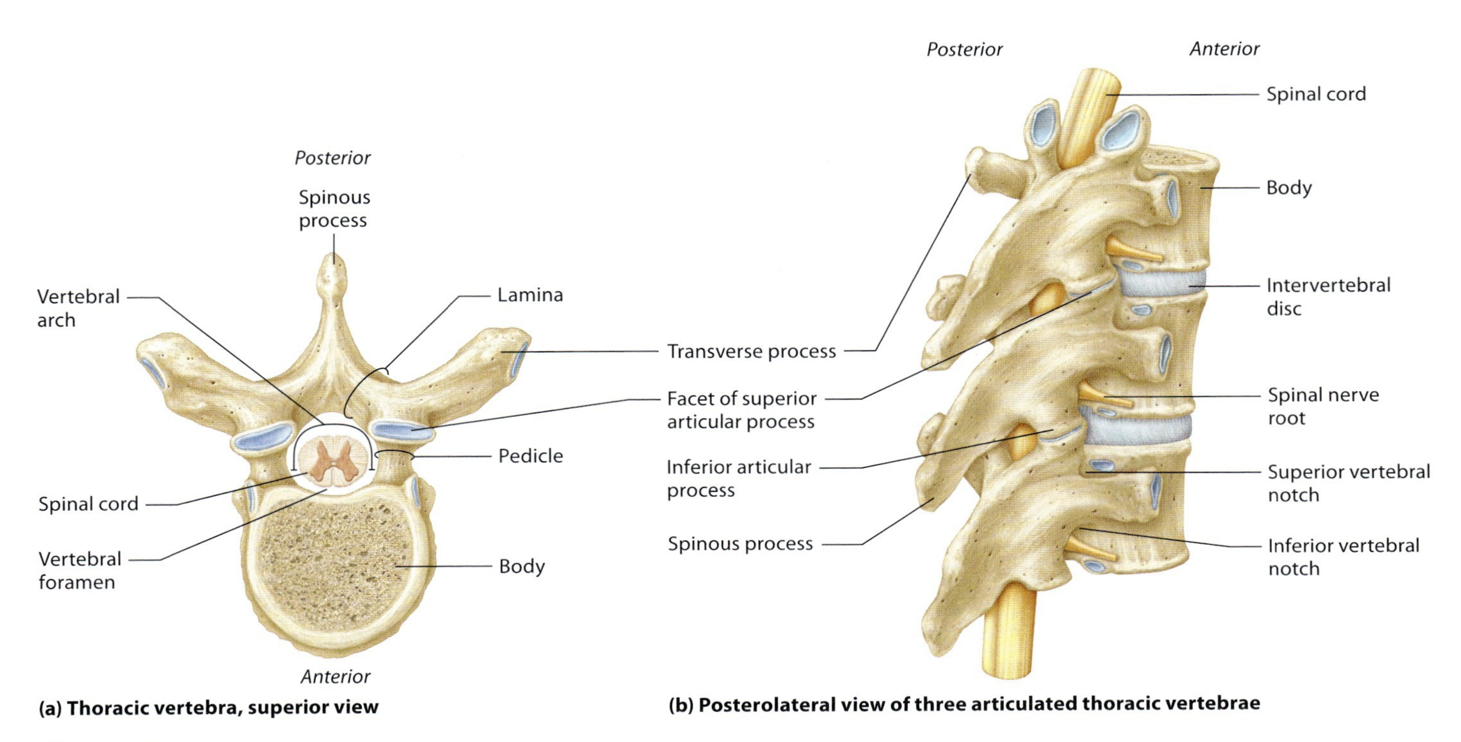

(a) Thoracic vertebra, superior view

(b) Posterolateral view of three articulated thoracic vertebrae

Figure 9-13 The general structure of a vertebra.

articulate with the superior articular processes of the vertebra immediately inferior to it at facets.

Each vertebra also has a **vertebral foramen** (see Figure 9-13a), an opening that lies between the body and the vertebral arch; adjacent vertebral foramina form the vertebral canal, which houses the spinal cord.

Characteristics of Cervical, Thoracic, and Lumbar Vertebrae

A typical **cervical vertebra** (**Figure 9-14a**) is characterized by (1) an oval-shaped body; (2) transverse foramina for the passage of the vertebral arteries, which supply blood to the brain, and of the vertebral veins, which drain blood from the brain; (3) a bifid (forked) spinous process; and (4) a triangular vertebral foramen.

The first two cervical vertebrae, the **atlas** (C_1) and the **axis** (C_2), are highly modified to enable special movements. The atlas is named after Atlas, the mythical Greek god who carried the world on his shoulders. It is the only vertebra that articulates with the skull. The atlas (**Figure 9-14b**) lacks a vertebral body and has no spinous process. It is essentially a ring of bone containing **anterior arches** that meet at the **anterior tubercle** and **posterior arches** that meet at the **posterior tubercle**, with **lateral masses** on either side. Each lateral mass has articular facets on both its superior and inferior surfaces. The **superior articular facets** receive the occipital condyles of the occipital bone of the skull, and the inferior articular facets articulate with the axis (**Figure 9-14c**), which is inferior to it.

The atlantoaxial joint—the joint between C_1 and C_2 (**Figure 9-14d**)—is unique among vertebral articulations. There is no intervertebral disc between the atlas and the axis. Instead, the axis is modified to articulate with the atlas, which is superior to it. One of the axis's unique features is a peg-like **dens** (odontoid process) that projects superiorly from the body. The dens articulates with the enlarged vertebral foramen of the atlas.

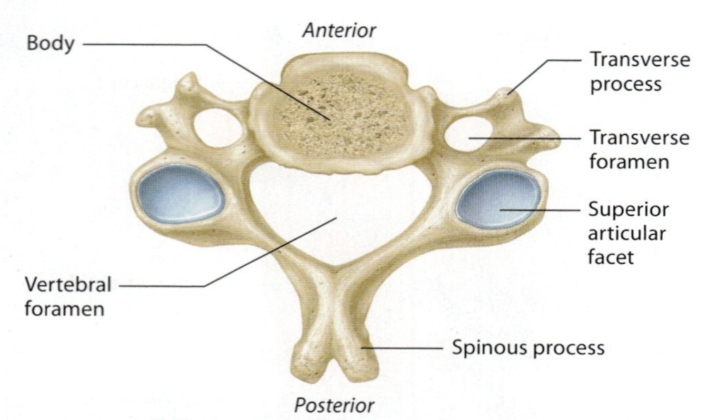

(a) Typical cervical vertebra (C_5)

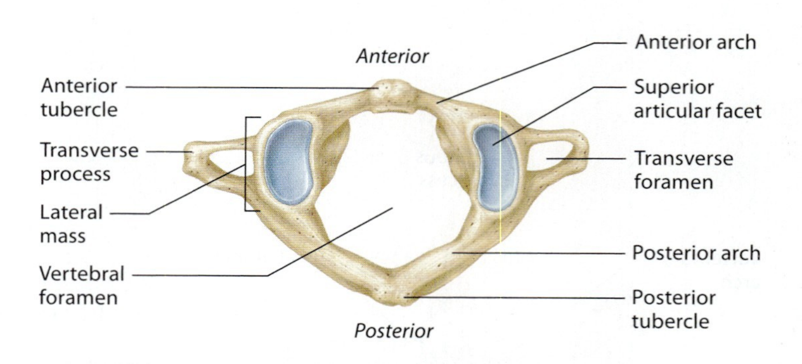

(b) Atlas (C_1), superior view

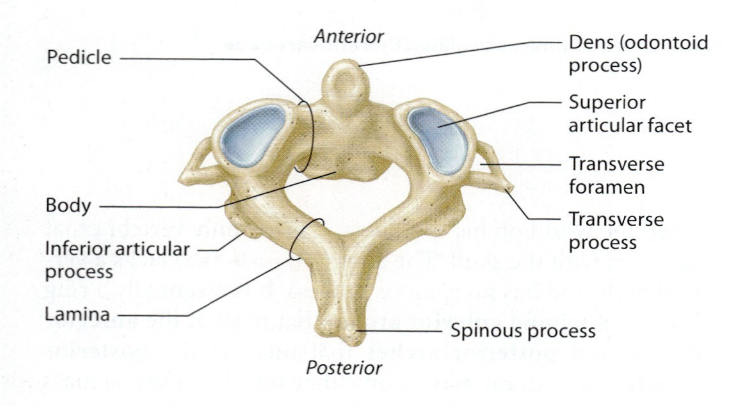

(c) Axis (C_2), superior view

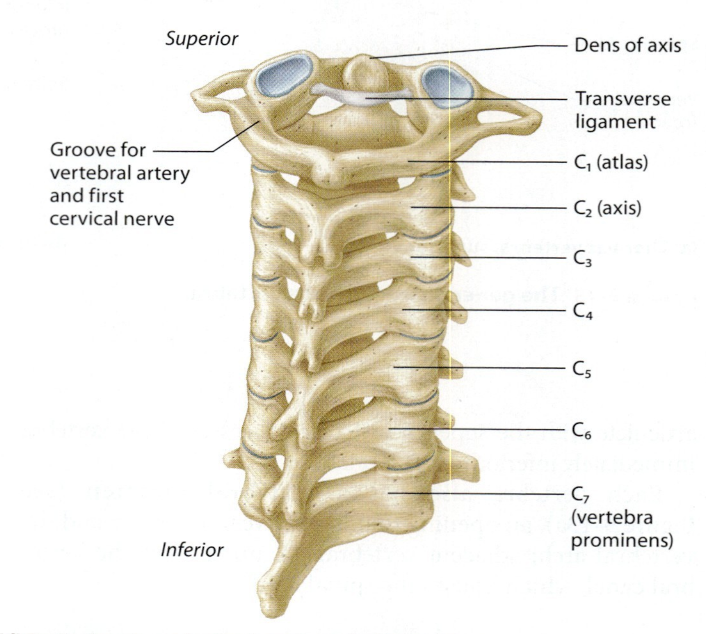

(d) Posterior view of articulated cervical vertebrae

Figure 9-14 Cervical vertebrae.

The most inferior cervical vertebra (C_7) has a spinous process that is larger than those of other cervical vertebra, and it is not bifid. Because it can be felt (palpated) through the skin, C_7 is called the **vertebra prominens** and is used as an anatomical landmark for counting the vertebrae.

Each of the 12 **thoracic vertebrae** is characterized by (1) a heart-shaped body bearing four (two inferior and two superior) small facets (costal facets), where a rib articulates with the vertebra; (2) a spinous process that points sharply inferiorly; and (3) a circular vertebral foramen. The five **lumbar vertebrae** are characterized by having (1) large sizes and weights; (2) wide, kidney-shaped bodies; (3) blunt, hatchet-shaped spinous processes; and (4) triangular vertebral foramina.

Table 9-1 summarizes the characteristics of the cervical, thoracic, and lumbar vertebrae in humans.

Table 9-1 Characteristics of Cervical, Thoracic, and Lumbar Vertebrae

	Cervical	Thoracic	Lumbar
Body	Small, oval-shaped	Medium-sized, heart-shaped; have facets for ribs	Large, kidney-shaped
Vertebral foramen	Triangular	Round	Triangular
Spinous process	Short, bifid (forked); projects directly posteriorly	Long, sharp; projects sharply inferiorly	Short, blunt, hatchet-shaped; projects directly posteriorly
Transverse processes	Contain foramina	Have facets for ribs (except T_{11} and T_{12})	Thin and tapered
Lateral view			
Superior view			

Characteristics of the Sacrum

The **sacrum** is an irregular, wedge-shaped bone composed of five fused vertebrae that articulates with the fifth lumbar vertebra at the flattened **base** and **superior articular processes**, and with the coccyx at the **apex** (**Figure 9-15**). The **auricular surfaces** (see Figure 9-15b) articulate with the ilium of each coxal bone to form the posterior wall of the bony pelvis. The fusion of the sacral bones before birth consolidates the vertebral canal into the **sacral canal**; at the fifth sacral vertebra the sacral canal opens as the **sacral hiatus**.

Among the sacrum's important bone markings are the **sacral foramina**, through which sacral nerves pass. The **median sacral crest** (see Figure 9-15b), formed by the spinous processes of the five fused vertebrae, provides attachment sites for muscles of the lower back and hip. The **lateral sacral crest**, formed by the transverse processes of the five fused vertebrae, provides attachment sites for muscles of the lower back and hip. A prominent bulge on the anterosuperior margin—the **sacral promontory** (see Figure 9-15a)—is an important anatomical landmark in females during pelvic exams and during labor and delivery. On either side of the sacral promontory are **alae** (singular = ala). Finally, **transverse lines** mark the sites of vertebral fusion.

Characteristics of the Coccyx

The **coccyx** (see Figure 9-15) is composed of three to five fused vertebrae with hornlike projections, each called a **coccygeal cornu**. The coccyx offers slight support to the pelvic organs and serves as a point of attachment for the muscles of the pelvic floor.

The Thoracic Cage

The **thoracic cage** (rib cage) consists of the sternum, ribs, costal cartilages, and thoracic vertebrae (**Figure 9-16**). It serves three major functions: It (1) forms a protective cage around the organs of the thoracic cavity, (2) provides support for the pectoral (shoulder) girdle and the upper appendages, and (3) provides attachment sites for skeletal muscles of the neck, back, chest, and shoulders.

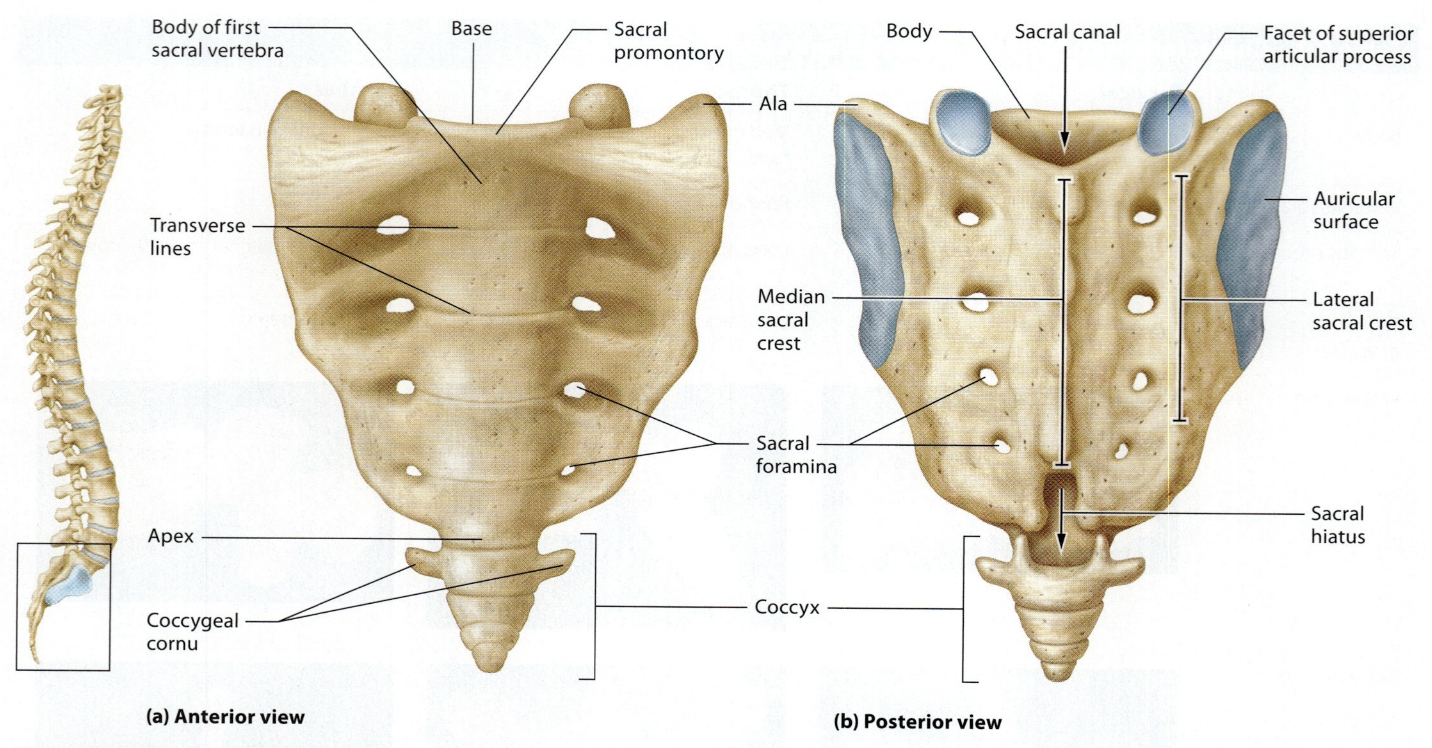

Figure 9-15 **Sacrum and coccyx.**

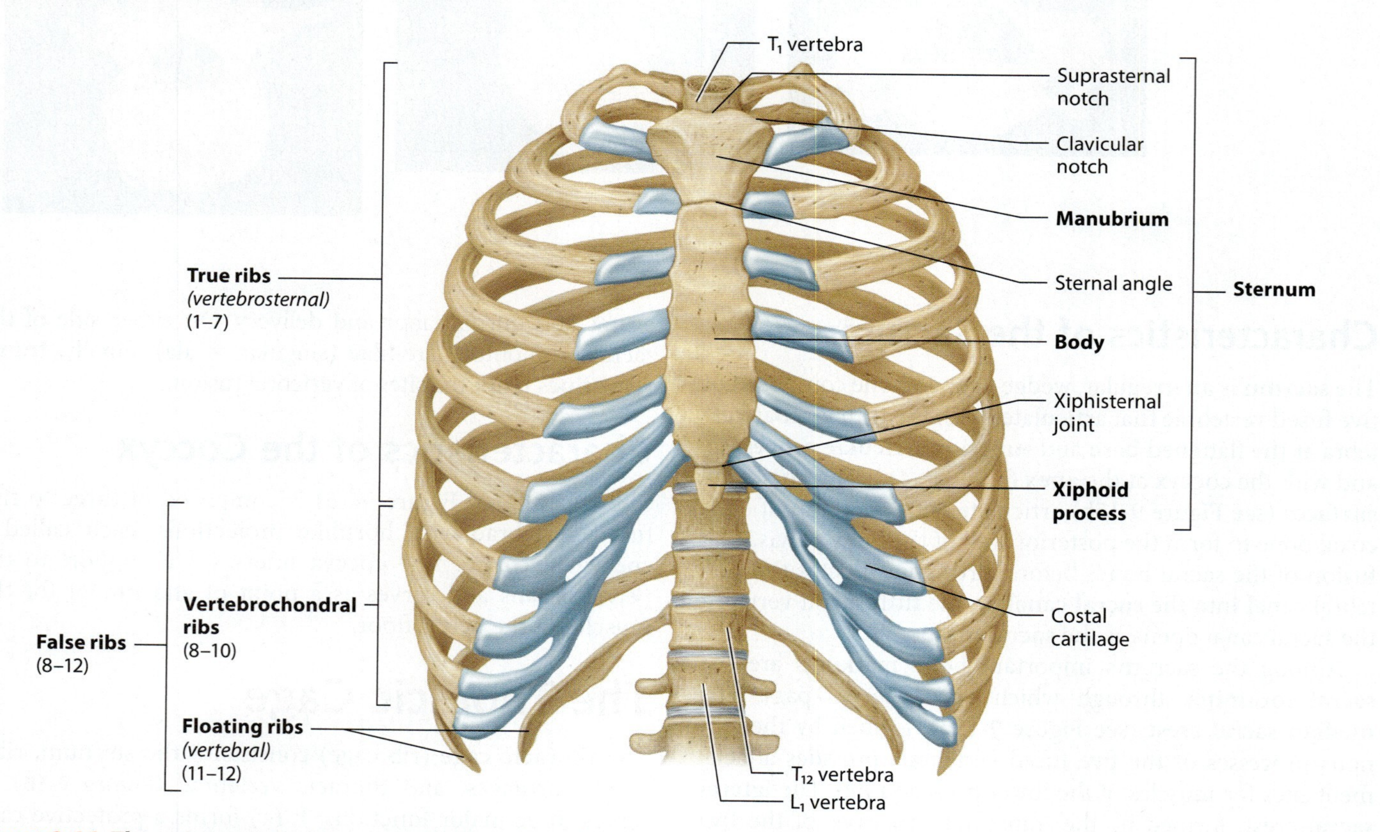

Figure 9-16 **Thoracic cage: the sternum, ribs, costal cartilages, and thoracic vertebrae.**

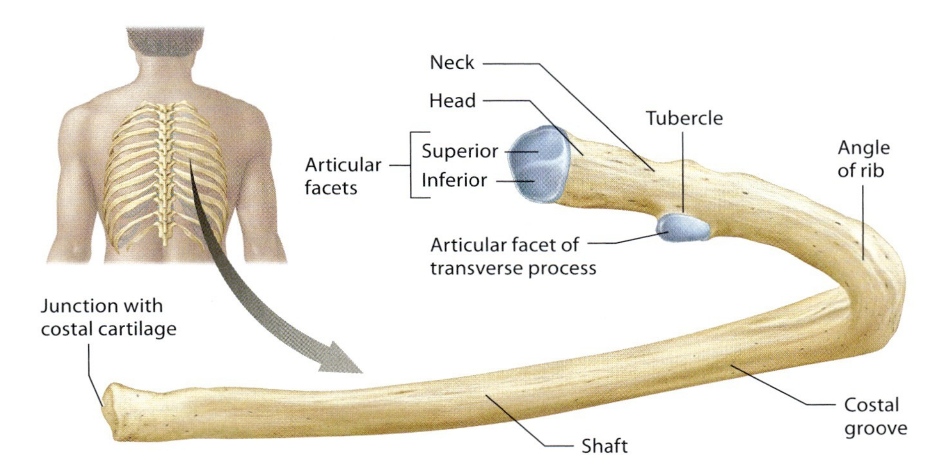

Figure 9-17 Structure of a typical rib, superior view.

Sternum

The **sternum**, a "dagger-shaped" bone located in the anterior midline of the thorax, is a composite bone resulting from the fusion of three bones: the superior **manubrium**, the middle **body**, and the inferior **xiphoid process** (see Figure 9-16). Features of the sternum include the **suprasternal notch** (or jugular notch), which is located medially at the level of the intervertebral disc between T_2 and T_3; the **clavicular notch**, where the clavicle forms a joint with the manubrium; the **sternal angle**, a landmark for finding rib 2 when using a stethoscope to listen to heart sounds; and the **xiphisternal joint**, behind which the heart sits on the diaphragm.

The Ribs

Ribs (Figure 9-17) have several general features. The wedge-shaped vertebral (medial) end of a rib bone is the **head**; the bulk of each rib is the **shaft**. The superior border of the shaft is smooth, whereas its inferior border is sharp and thin and contains a **costal groove**, which houses intercostal nerves and blood vessels. The constricted area between the head and the shaft is the **neck**. Each rib has two **articular facets**: one that articulates with a vertebral body and one that articulates with the vertebral transverse process. The **tubercle**, the portion of a rib that is immediately lateral to the neck, contains a facet that articulates with a transverse process of a thoracic vertebra. Finally, the **angle** is a bend where the shaft begins to curve toward the sternum.

Ribs are also categorized according to location and type of attachment points. Rib pairs 1–7 are called **true ribs**, or vertebrosternal ribs (see Figure 9-16); they attach to the sternum by their own cartilage. Rib pairs 8–12 are called **false ribs** because they either attach indirectly to the sternum (by way of cartilage immediately above each rib) or entirely lack a sternal attachment. Rib pairs 8 through 10 are also called vertebrochondral ribs and rib pairs 11 and 12 are also called **floating** (or vertebral) **ribs**.

ACTIVITY 3
Studying the Bones of the Vertebral Column and Thoracic Cage

Learning Outcomes

1. Identify the five regions of the vertebral column.
2. Describe the normal curvatures of the spine.
3. Identify the features of a typical vertebra.
4. Compare and contrast the features used to identify vertebrae as cervical, thoracic, or lumbar.
5. Identify the features of the sacrum and the coccyx.
6. Compare and contrast the atlas and the axis with respect to structure and function.
7. Identify the three regions of the sternum.
8. Identify the general features of a rib.
9. Distinguish among true ribs, false ribs, and floating ribs.

Materials Needed

☐ Disarticulated bones of the vertebral column and the thoracic cage

Instructions

CHART Obtain disarticulated bones of the vertebral column and the thoracic cage. These bones have been divided into four sets (Sets A–D, as listed in the charts that follow). Your instructor will assign each member of your lab group one of the four sets of bones. Then, complete the tasks outlined in Activity 1 for these sets of bones.

Set A	Set B	Set C	Set D
☐ "Expert"	☐ "Expert"	☐ "Expert"	☐ "Expert"
☐ "Student"	☐ "Student"	☐ "Student"	☐ "Student"
☐ "Student"	☐ "Student"	☐ "Student"	☐ "Student"
☐ "Student"	☐ "Student"	☐ "Student"	☐ "Student"

Making Connections: The Vertebral Column and Thoracic Cage, Set A		
Bone	**Description of Bone and/or Bone Marking(s)**	**Connections to Things I Have Already Learned**
General vertebral structure		
• Body	*Disc-shaped with usually large, flat anterior and posterior surfaces*	
• Spinous process		
• Transverse process		
• Vertebral foramen		
• Vertebral arch	*Forms the posterior and lateral walls of the vertebral foramen*	
• Superior articular process (and facet)		
• Inferior articular process (and facet)		*Facets are shallow convex or concave surfaces where two bones articulate*
• Pedicle		
• Lamina		
Cervical vertebra and its features		
• Transverse foramen		*Vertebral arteries pass through the transverse foramen*
• C_2–C_6 spinous process direction/shape	*Direction is posterior, but not inferior and shape is bifid*	
• Triangular vertebral foramen		
• Oval-shaped body		
Atlas		
• Anterior arch		
• Posterior arch		
Axis		
• Dens (odontoid process)		*Odontoid means tooth-shaped*

Making Connections: The Vertebral Column and Thoracic Cage, Set B		
Bone	**Description of Bone and/or Bone Marking(s)**	**Connections to Things I Have Already Learned**
Thoracic vertebra structure		
• Body	*Include articular facets for ribs*	
• Spinous process		
• Transverse process		*Articular facets on transverse processes articulate with the tubercles of the ribs*
• Vertebral foramen		
• Vertebral arch		
• Superior articular process (and facet)		
• Inferior articular process (and facet)		
• Pedicle		
• Lamina		
Thoracic special features		
• Heart-shaped body		
• Costal facet (head and tubercle)		
• Circular vertebral foramen		
• Spinous process direction		

Making Connections: The Vertebral Column and Thoracic Cage, Set C		
Bone	**Description of Bone and/or Bone Marking(s)**	**Connections to Things I Have Already Learned**
Lumbar vertebra structure		
• Body	*Large, thick bodies*	
• Spinous process		
• Transverse process		
• Vertebral foramen		
• Vertebral arch		
• Superior articular process (and facet)		
• Inferior articular process (and facet)		
• Pedicle		
• Lamina		
Lumbar special features		
• Size, weight		
• Triangular vertebral foramen		
• Spinous process direction/shape		
Sacrum		
• Sacral promontory		
• Median sacral crest		
• Lateral sacral crest		
• Sacral foramen		

Bone	Description of Bone and/or Bone Marking(s)	Connections to Things I Have Already Learned
• Superior articular process		
Coccyx		
• Coccygeal cornu		

Making Connections: The Vertebral Column and Thoracic Cage, Set D

Bone	Description of Bone and/or Bone Marking(s)	Connections to Things I Have Already Learned
Sternum		
• Manubrium		
• Body		
• Xiphoid process		
• Suprasternal notch		
• Xiphisternal joint		
• Sternal angle		
Rib		
• Head		
• Neck		
• Shaft		
• Tubercle		
Types of ribs:		
• True (vertebrosternal) ribs		
• False ribs		
• Floating ribs		

Optional Activity

Practice labeling bones of the vertebral column, bony thorax, and pelvic girdle at MasteringA&P® > Study Area > Practice Anatomy Lab > Human Cadaver: Axial Skeleton

ACTIVITY 4 Identifying Bones-in-a-Bag

Learning Outcome

1. Identify disarticulated bones of the vertebral column and the bony thorax by touch only, and justify your identifications based on bone size, shape, and markings.

Materials Needed

☐ Disarticulated bones of the axial skeleton

Instructions

In this activity you and your group members will take turns identifying each of the eight disarticulated bones in a bag your instructor will give you.

1. Reach into the bag with one hand (without looking at the bones—no peeking!) and pick up a bone, but do not take it out of the bag.
2. Based on feel alone, describe the bone's size, shape, and special features to the other members of your lab group. Then record your description in **Table 9-2**.
3. Write in Table 9-2 the name of the bone you are holding, and then pull it out of the bag to see if your identification was correct.
4. List any features of the bone that you missed while feeling it inside the bag. Record this information in Table 9-2 as well.
5. Pass the bag to another group member, who will repeat steps 1–4.
6. Repeat steps 1–5 until all of the bones in the bag have been identified.

Table 9-2 Bones-in-a-Bag Identification

	Description of Bone Based on Touch Alone	Name of Bone	Additional Bone Features
1.			
2.			
3.			
4.			
5.			
6.			
7.			
8.			

Post-lab quizzes are also assignable in MasteringA&P®

POST-LAB ASSIGNMENTS

Name: ______________________ Date: ______________ Lab Section: ________

PART I. Check Your Understanding

Activity 1: Studying the Bones of the Skull

1. Identify the bones and bone markings in the accompanying lateral view of the skull.

a. ______________________
b. ______________________
c. ______________________
d. ______________________
e. ______________________
f. ______________________
g. ______________________
h. ______________________
i. ______________________
j. ______________________
k. ______________________
l. ______________________
m. ______________________
n. ______________________
o. ______________________

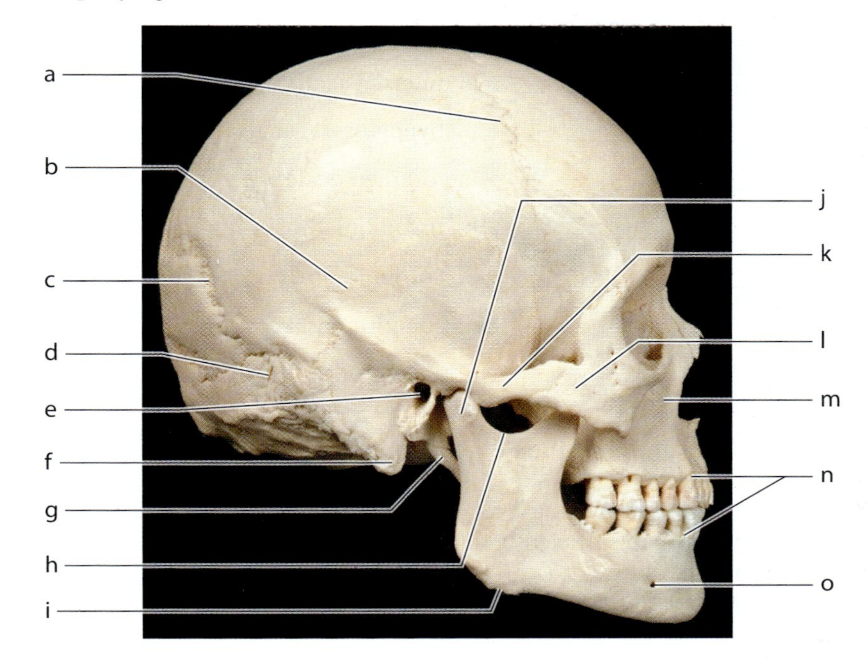

2. Identify the bones and bone markings in the accompanying superior, internal view of the skull.

a. ______________________
b. ______________________
c. ______________________
d. ______________________
e. ______________________
f. ______________________
g. ______________________
h. ______________________
i. ______________________
j. ______________________
k. ______________________
l. ______________________
m. ______________________
n. ______________________
o. ______________________

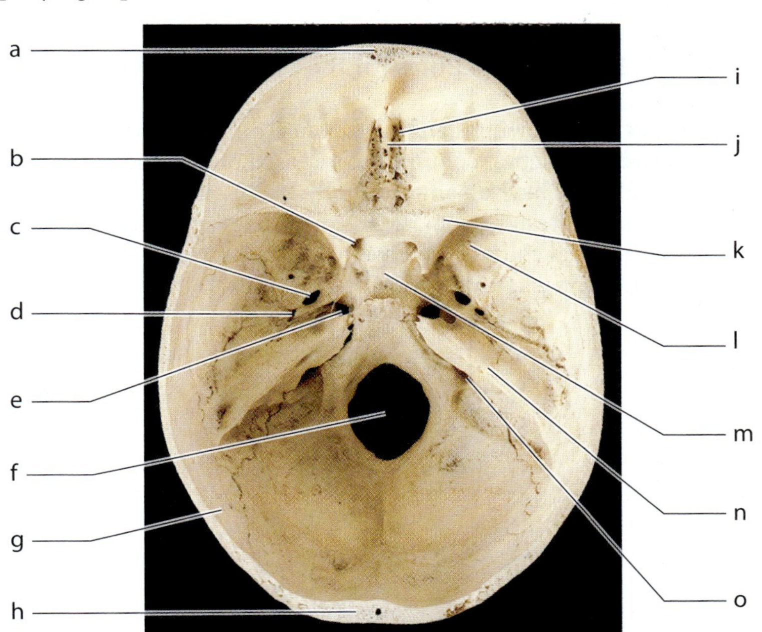

3. On which skull bone is each of the following features found?

_________________________ a. hypophyseal fossa

_________________________ b. carotid canal

_________________________ c. styloid process

_________________________ d. palatine process

_________________________ e. infraorbital foramina

_________________________ f. zygomatic process

_________________________ g. olfactory foramina

_________________________ h. internal acoustic meatus

_________________________ i. hypoglossal canal

_________________________ j. foramen ovale

_________________________ k. mandibular fossa

_________________________ l. coronoid process

_________________________ m. superior nasal conchae

_________________________ n. styloid process

Activity 2: Examining the Fetal Skull

1. Which bony feature on the fetal/neonate skull is undeveloped (is smooth or flat) until the neck muscles develop sufficiently to create it?
   a. lambdoid suture
   b. mandibular fossa
   c. mastoid process
   d. nasal septum

2. Which suture is typically found only in newborns and very young children?
   a. lambdoid suture
   b. squamosal suture
   c. frontal suture
   d. sagittal suture

Activity 3: Studying the Bones of the Vertebral Column and Thoracic Cage

1. Which of the following statements concerning the vertebral column is true?
   a. The adult vertebral column is composed of 24 bones.
   b. All vertebrae have an anterior body.
   c. The sacrum results from the fusion of five vertebrae.
   d. The vertebral column has five major curvatures.
   e. The thoracic vertebrae are located inferior to the lumbar vertebrae.

2. Identify the bones and bone markings in the accompanying photo of a vertebral column:

   a. ______________________________

   b. ______________________________

   c. ______________________________

   d. ______________________________

   e. ______________________________

   f. ______________________________

   g. ______________________________

   h. ______________________________

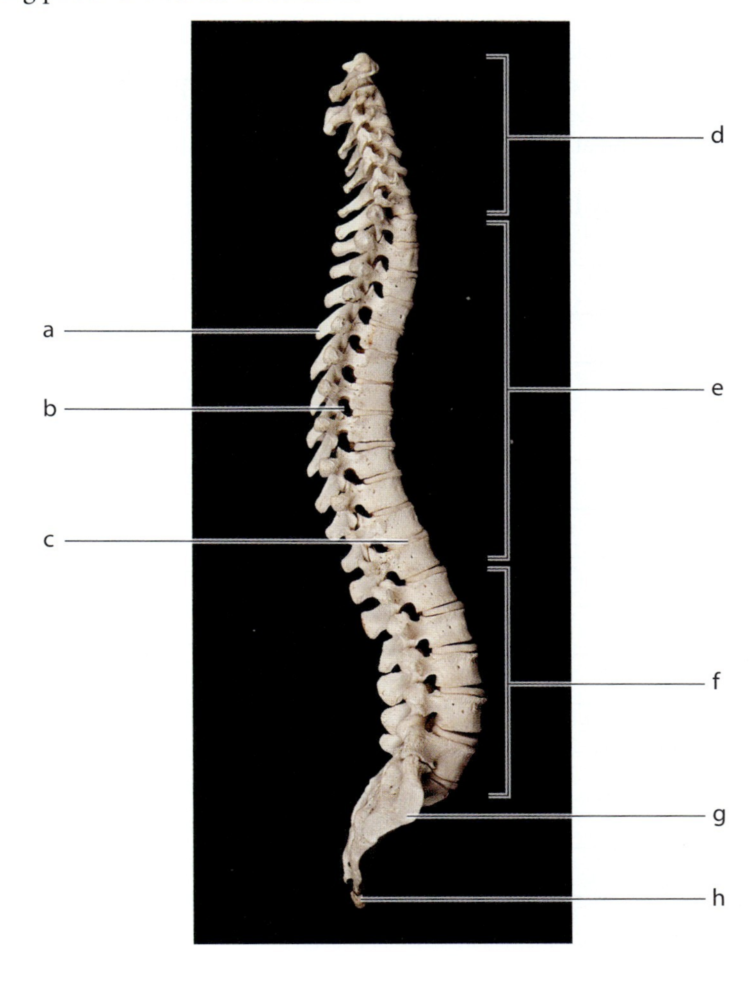

3. Identify the bone markings in the accompanying photo of a typical vertebra:

   a. ______________________________

   b. ______________________________

   c. ______________________________

   d. ______________________________

   e. ______________________________

   f. ______________________________

   g. ______________________________

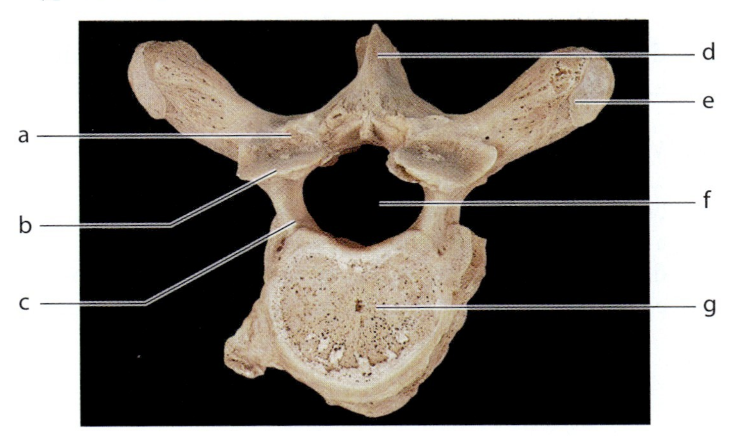

4. Which of the following statements is true?
   a. There is an intervertebral disc between the atlas and the axis.
   b. The atlas lacks transverse foramina.
   c. The axis articulates directly with the occipital condyles.
   d. The dens articulates where the atlas's absent body would be.
   e. The atlas has a longer spinous process than does the axis.

5. Identify the bones and bone markings in the accompanying illustration of a bony thorax:

   a. ______________________________
   b. ______________________________
   c. ______________________________
   d. ______________________________
   e. ______________________________
   f. ______________________________
   g. ______________________________
   h. ______________________________
   i. ______________________________
   j. ______________________________
   k. ______________________________
   l. ______________________________
   m. ______________________________
   n. ______________________________
   o. ______________________________

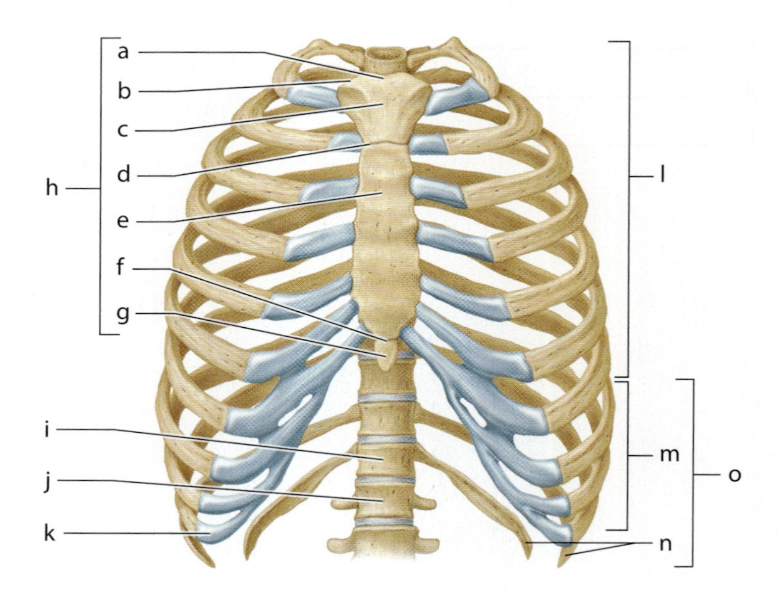

6. Which of the following statements is true? Choose all answers that are correct.
   a. The manubrium is inferior to the xiphoid process.
   b. Floating ribs are not attached to the vertebral column.
   c. Vertebrochondral ribs are classified as false ribs.
   d. The sternal angle is a result of the fusion of the body and the xiphoid process.
   e. All of the ribs articulate posteriorly with the vertebral column.

7. Which of the following statements is true?
   a. Each rib articulates with a lumbar vertebra.
   b. Transverse foramina are found on cervical vertebrae only.
   c. Costal facets are found on cervical vertebrae only.
   d. Thoracic vertebrae are larger and more block-shaped than lumbar vertebrae.
   e. Lumbar vertebrae have longer, sharper spinous processes.

Activity 4: Identifying Bones-in-a-Bag

1. The bone that has five bulges and a midline crest over a tube that opens inferiorly is the:
   a. coccyx.
   b. sacrum.
   c. atlas.
   d. occipital bone.
   e. frontal bone.

2. The bone with a relatively flat surface, a sharp arch, a hole under the arch just behind a smooth surface, and a large projection or bump on the inferoposterior edge is the:
   a. coccyx.
   b. ethmoid bone.
   c. vomer.
   d. parietal bone.
   e. temporal bone.

PART II. Putting It All Together

A. Review Questions

Answer the following questions using your lecture notes, your textbook, and your lab notes.

1. For each of the following characteristics, indicate whether it pertains to a cervical, thoracic, or lumbar vertebra.

   ____________ a. Bifid spinous process

   ____________ b. Facets for ribs

   ____________ c. Spinous process that projects sharply inferiorly

   ____________ d. Large, kidney-shaped body

   ____________ e. Transverse foramina

2. Why is the sphenoid bone considered to be the "keystone" bone of the cranium?

   ____

   ____

3. Explain how the atlas and axis differ from a typical cervical vertebra, and how each difference reflects the bone's function. ______________________________

   ____

   ____

   ____

4. Are all false ribs classified as vertebrochondral ribs? Explain. ______________________

   ____

   ____

5. Which two bones form the nasal septum? ______________________________

6. Which two bones form the hard palate? ______________________________

B. Concept Mapping

1. Fill in the blanks to complete this concept map outlining structures that are found in the midline.

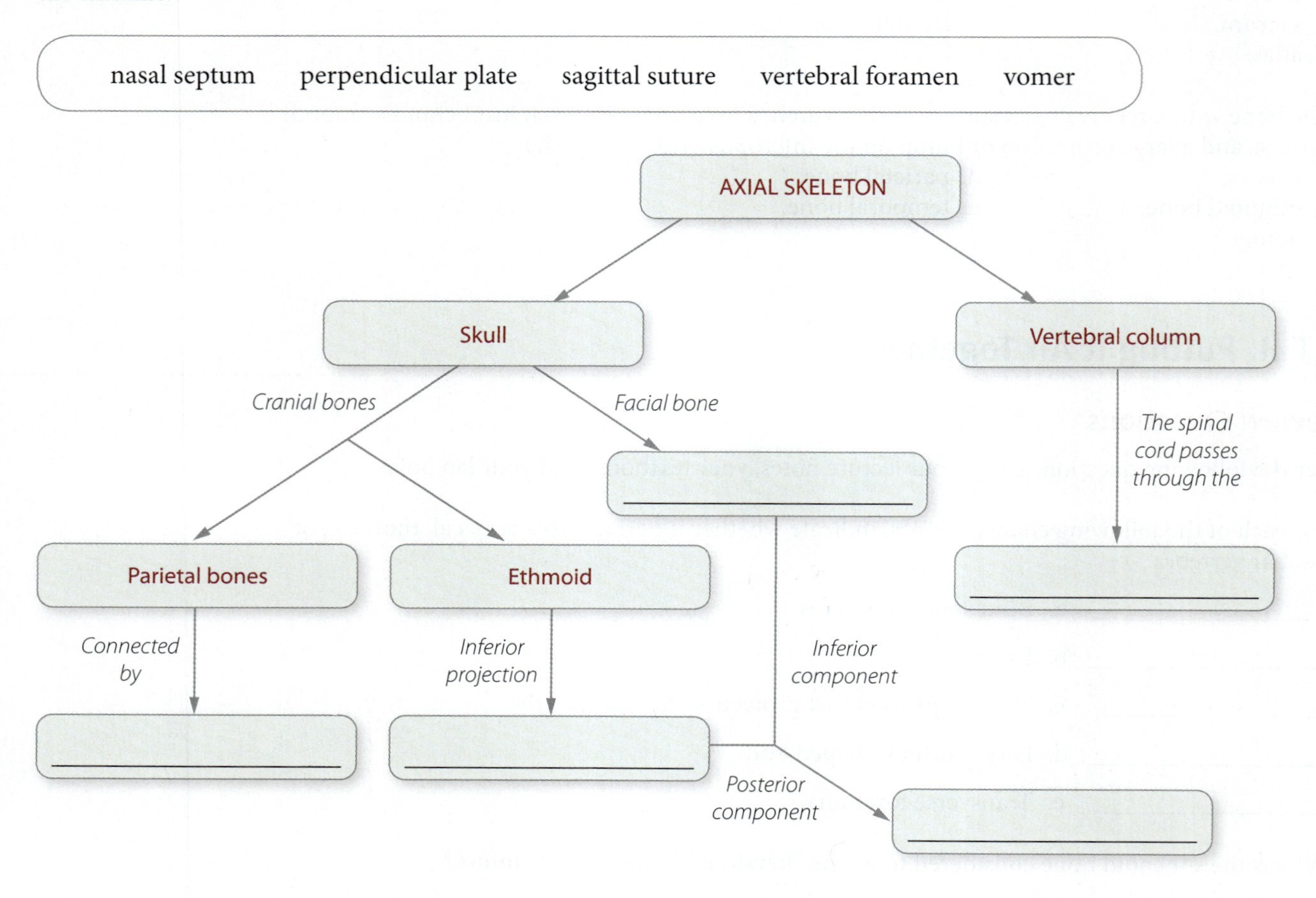

2. Construct a unit concept map to show the relationships among the following set of terms. Include all of the terms in your diagram. Your instructor may choose to assign additional terms.

coronal	costal cartilage	foramen ovale	foramen rotundum
inferior nasal conchae	nasal septum	optic canal	perpendicular plate
rib	sagittal	spinous process	squamosal
thoracic vertebra	transverse process	vertebral foramen	

10 The Appendicular Skeleton

UNIT OUTLINE

The **appendicular skeleton** is composed of two **pectoral girdles** (each consisting of a clavicle and a scapula) with the attached **upper limbs**, plus the **pelvic girdle** with the attached **lower limbs**. The 126 bones in the appendicular skeleton enable us to move and manipulate our surroundings.

Due to their structures, the pectoral and pelvic girdles differ in both mobility and strength/stability. Because the head of the humerus is received by the shallow glenoid cavity of the scapula, and these structures do not directly articulate with the vertebral column, the pectoral girdle provides relatively greater mobility and less strength and stability. By contrast, because the deep acetabulum cups the head of the femur, and the pelvic girdle articulates with the sacrum to form the sacroiliac joint, the pelvic girdle provides less mobility and greater strength and stability.

THINK ABOUT IT *Would you expect animals that walk on four legs to have more mobility or less mobility in their pectoral girdles than humans? Explain your answer.*

____

____

PRE-LAB ASSIGNMENTS

Pre-lab quizzes are also assignable in MasteringA&P®

To maximize learning, BEFORE your lab period carefully read this entire lab unit and complete these pre-lab assignments using your textbook, lecture notes, and prior knowledge.

PRE-LAB Activity 1: Studying the Bones of the Appendicular Skeleton

1. Use the list of bones provided to label the accompanying illustration of the posterior view of the skeleton; check off each bone as you label it.

   - ☐ humerus
   - ☐ tibia
   - ☐ ulna
   - ☐ scapula
   - ☐ tarsal
   - ☐ femur
   - ☐ metacarpals
   - ☐ pelvic bone

   a ______
   b ______
   c ______
   d ______
   e ______
   f ______
   g ______
   h ______

2. Match each of the following structures with the appropriate bone.

_____	a. infraspinous fossa	1. humerus
_____	b. trochlea	2. pelvic bone
_____	c. radial notch	3. tibia
_____	d. gluteal tuberosity	4. femur
_____	e. medial malleolus	5. ulna
_____	f. acromion end	6. scapula
_____	g. lateral malleolus	7. fibula
_____	h. acetabulum	8. clavicle

PRE-LAB Activity 2: Identifying Bones-in-a-Bag

1. If you feel a relatively long bone with a short, sharp process at its narrow end and a large hook around a groove at its larger end, it is mostly likely the:
   a. humerus.
   b. radius.
   c. ulna.
   d. clavicle.

2. If you feel a large, very long bone with a ball angled off to one side on one end and two rounded edges on the other end, it is most likely the:
   a. humerus.
   b. femur.
   c. pelvic bone.
   d. scapula.

The Pectoral Girdle and Upper Appendages

The paired **pectoral girdles**, each consisting of an anterior **clavicle** and a posterior **scapula**, attach the upper limbs to the axial skeleton and provide attachment sites for many muscles of the trunk and neck. **Figure 10-1** shows the pectoral girdle and its anatomical relationship to the thoracic cage and the humerus.

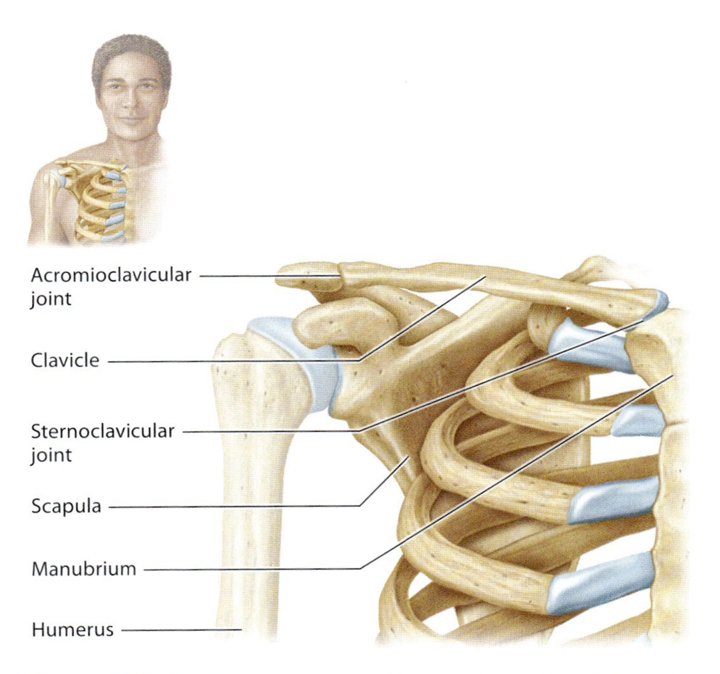

Figure 10-1 The right pectoral girdle and its relationship to the humerus and the thoracic cage.

Clavicle

The **clavicle** (**Figure 10-2**), commonly called the collarbone, is an S-shaped flat bone and the only bony connection between the pectoral girdle and the axial skeleton. It anchors the arm to the body (such as when you hang from a tree branch). The major features of the clavicle include the

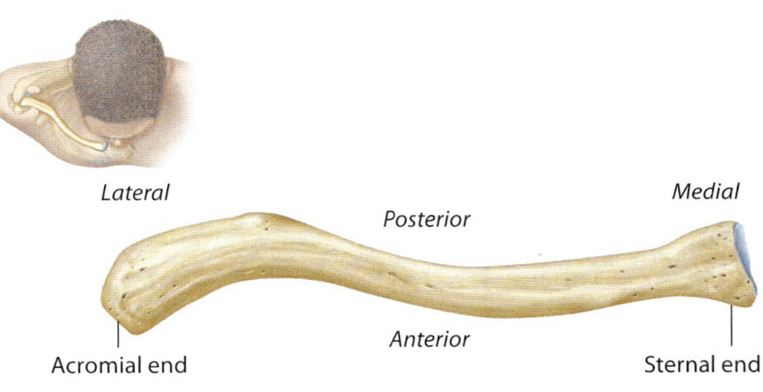

(a) Right clavicle, superior view

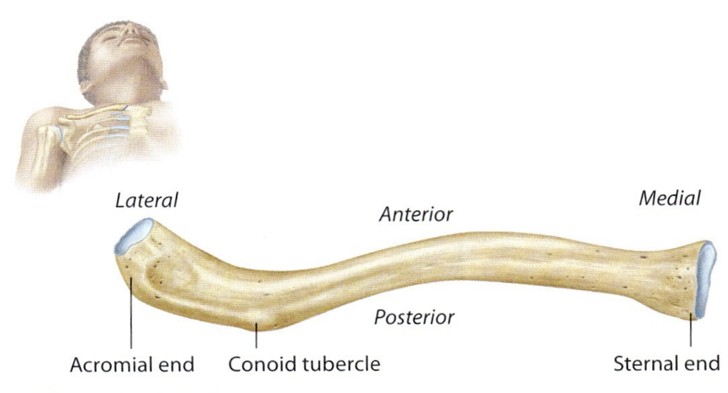

(b) Right clavicle, inferior view

Figure 10-2 The clavicle.

sternal (medial) **end**, which articulates with the manubrium of the sternum, and the **acromial** (lateral) **end**, the broader, flattened, roughened end that articulates with the acromion of the scapula (see Figure 10-2). The **conoid tubercle** (see Figure 10-2b), a point on the inferior surface of the lateral end of the clavicle, attaches the clavicle to the coracoid process of the scapula.

Scapula

The **scapula** (**Figure 10-3**), commonly called the shoulder blade, is a large, triangular flat bone with **superior**, **medial**, and **lateral borders**. The corners where the borders meet are the **superior**, **lateral**, and **inferior angles**. The **suprascapular notch** on the superior border serves as a passageway for nerves. The **acromion**, a flattened, expanded process that projects from the lateral end of the spine and articulates with the clavicle, is easily felt as the high point of the shoulder. The **glenoid cavity**, a shallow depression inferior to the acromion, receives the head of the humerus to form the glenohumeral (shoulder) joint. The **coracoid process** extends anteriorly from the scapula and is an attachment site for several tendons and ligaments. The **spine** is a prominent ridge that divides the posterior surface of the scapula into the **supraspinous fossa** and the **infraspinous fossa**; they and the **subscapular fossa** on the anterior surface are sites of muscle attachment.

Humerus

The **humerus** (**Figure 10-4**), the bone of the arm, has several important features. The **head** articulates with the shallow glenoid cavity to form the glenohumeral (shoulder) joint. The **anatomical neck** separates the **diaphysis** of the bone from the head. The more distal **surgical neck** is so named because it is a common site of fractures that require surgical repair. A relatively large proximal process called the **greater tubercle**, the smaller **lesser tubercle**, and a roughened projection on the lateral surface of the diaphysis called the **deltoid tuberosity** are attachment sites for shoulder muscles. The **radial groove**, a depression that runs along the posterior margin of the deltoid tuberosity, marks the course of a nerve that extends to the forearm.

The **trochlea**, the medial condyle on the distal end of the humerus, articulates with the ulna, whereas the **capitulum**, the lateral condyle on the distal end of the humerus, articulates with the radius. The **coronoid fossa**, a depression on the anterodistal surface of the humerus, receives the coronoid process of the ulna, whereas the **olecranon fossa**,

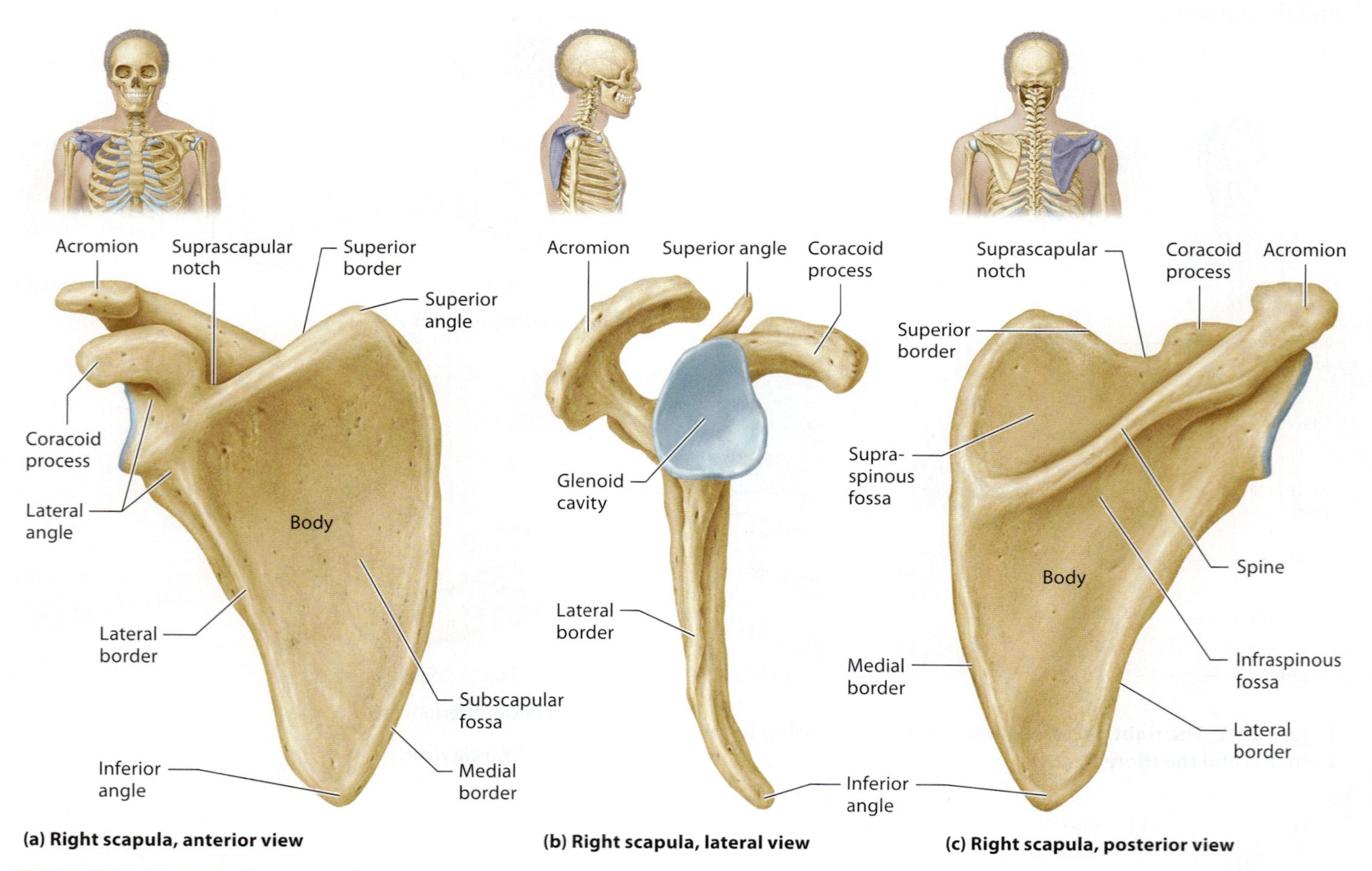

(a) Right scapula, anterior view **(b) Right scapula, lateral view** **(c) Right scapula, posterior view**

Figure 10-3 The scapula.

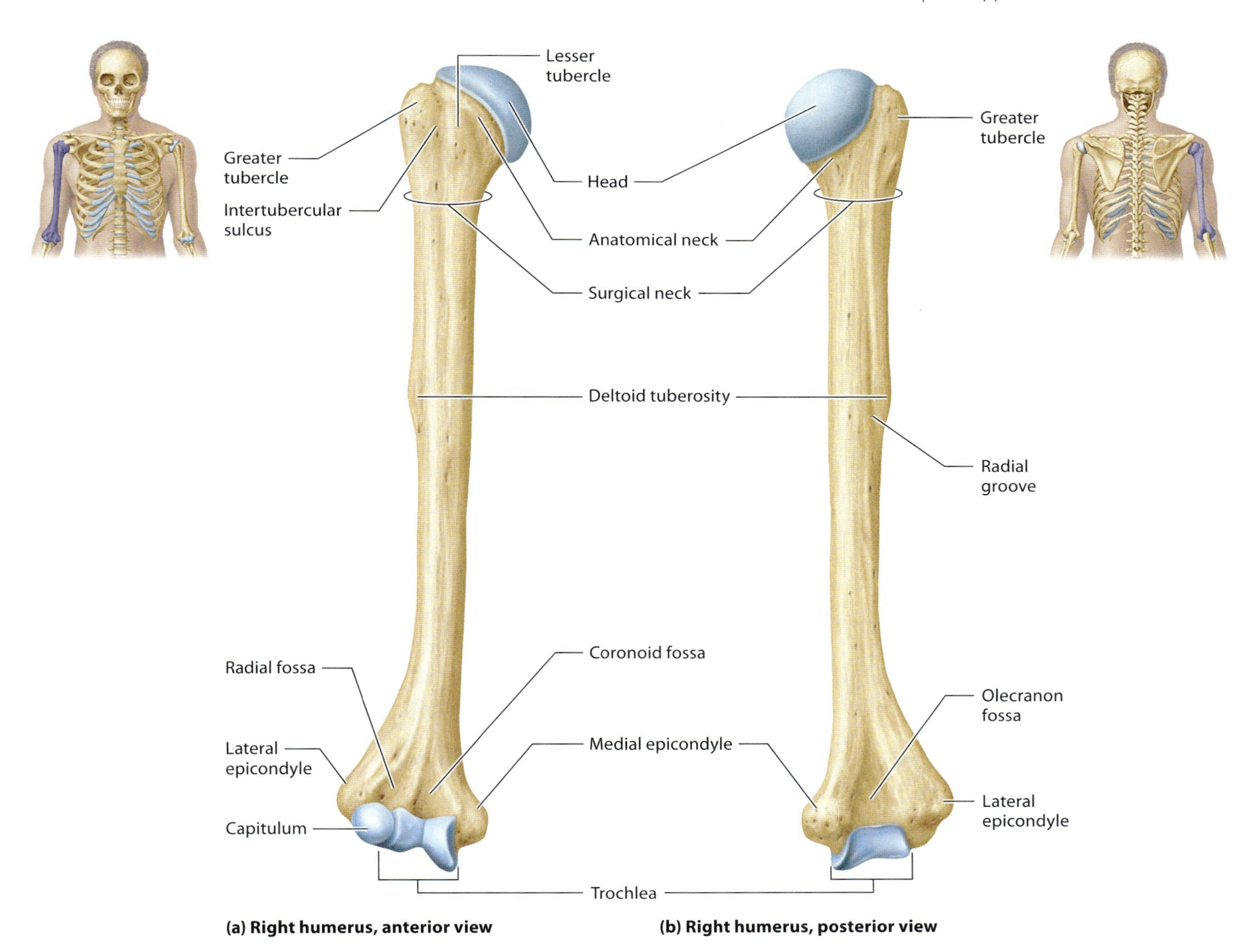

Figure 10-4 The humerus.

a depression on the posterodistal surface of the humerus, receives the olecranon process of the ulna. The **lateral epicondyle**, a small process proximal to the lateral condyle, and the **medial epicondyle**, a small process proximal to the medial condyle, are attachment sites for forearm muscles.

Radius

The **radius** (**Figure 10-5**) is located on the lateral aspect (thumb side) of the forearm. The proximal, disc-shaped **radial head** articulates with the capitulum of the humerus. The **radial neck** is a constricted region distal to the head. The **radial tuberosity** a projection just distal to the neck, is an attachment site for the biceps brachii, the major anterior arm muscle. The **ulnar notch**, located at the distal end of the radius, receives the head of the ulna. The **styloid process** forms the lateral boundary of the wrist.

Ulna

The **ulna** (see Figure 10-5), the medial bone of the forearm, is longer than the radius and connects to the radius through an **interosseous membrane**. The **olecranon**, a process on the posterior surface of the proximal end, articulates with the humerus at the olecranon fossa. The **coronoid process** on the anterior surface of the proximal end extends into the coronoid fossa of the humerus during flexion. The **trochlear notch**, a large, curved area between the olecranon and coronoid process, articulates with the humerus at the trochlea. The **radial notch** at the proximal end of the ulna receives the head of the radius. The **ulnar head**, the distal end of the ulna, articulates with the lunate of the carpals. Finally, the **styloid process** is a short projection of the ulnar head that creates the medial boundary of the wrist.

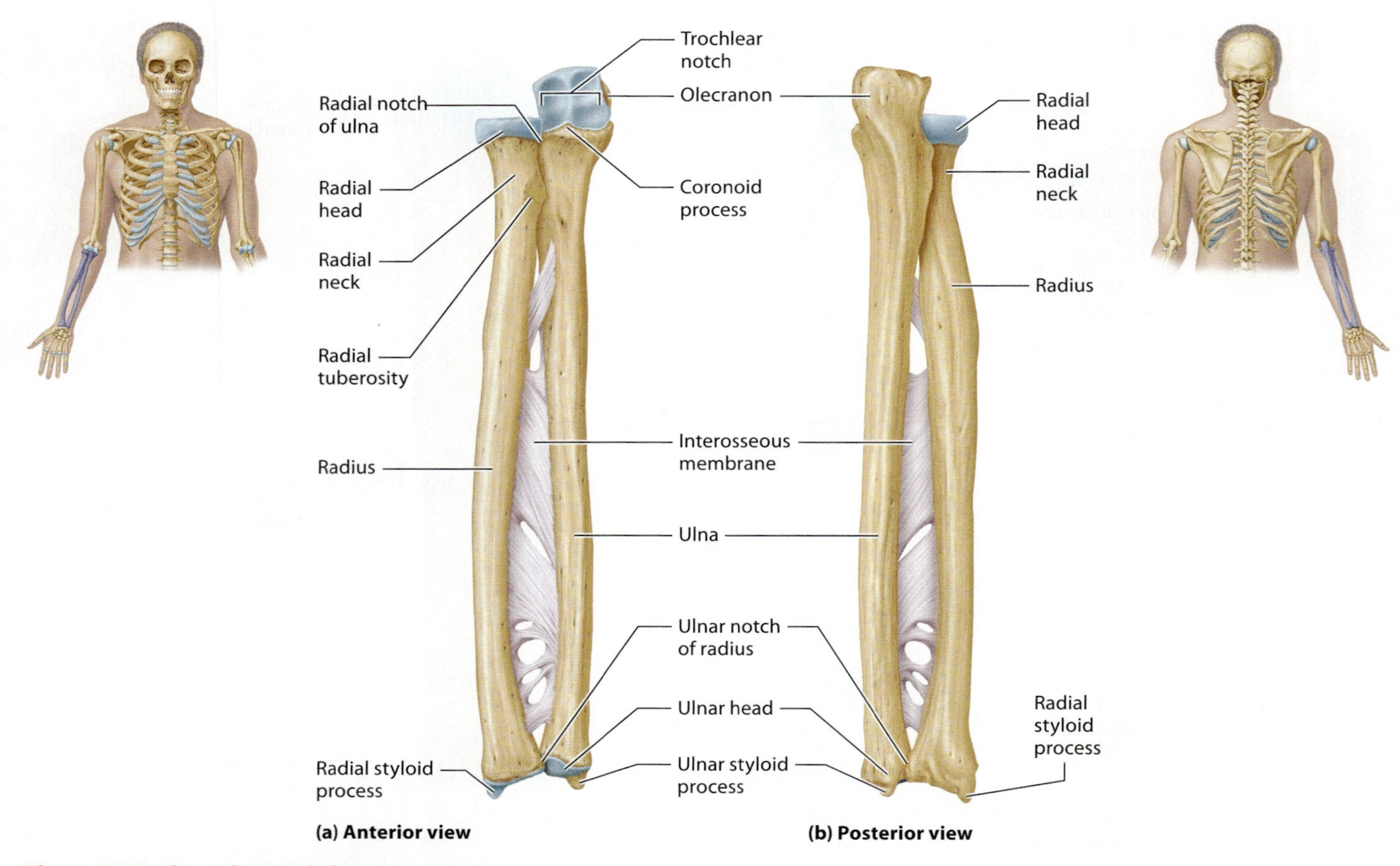

Figure 10-5 The radius and ulna.

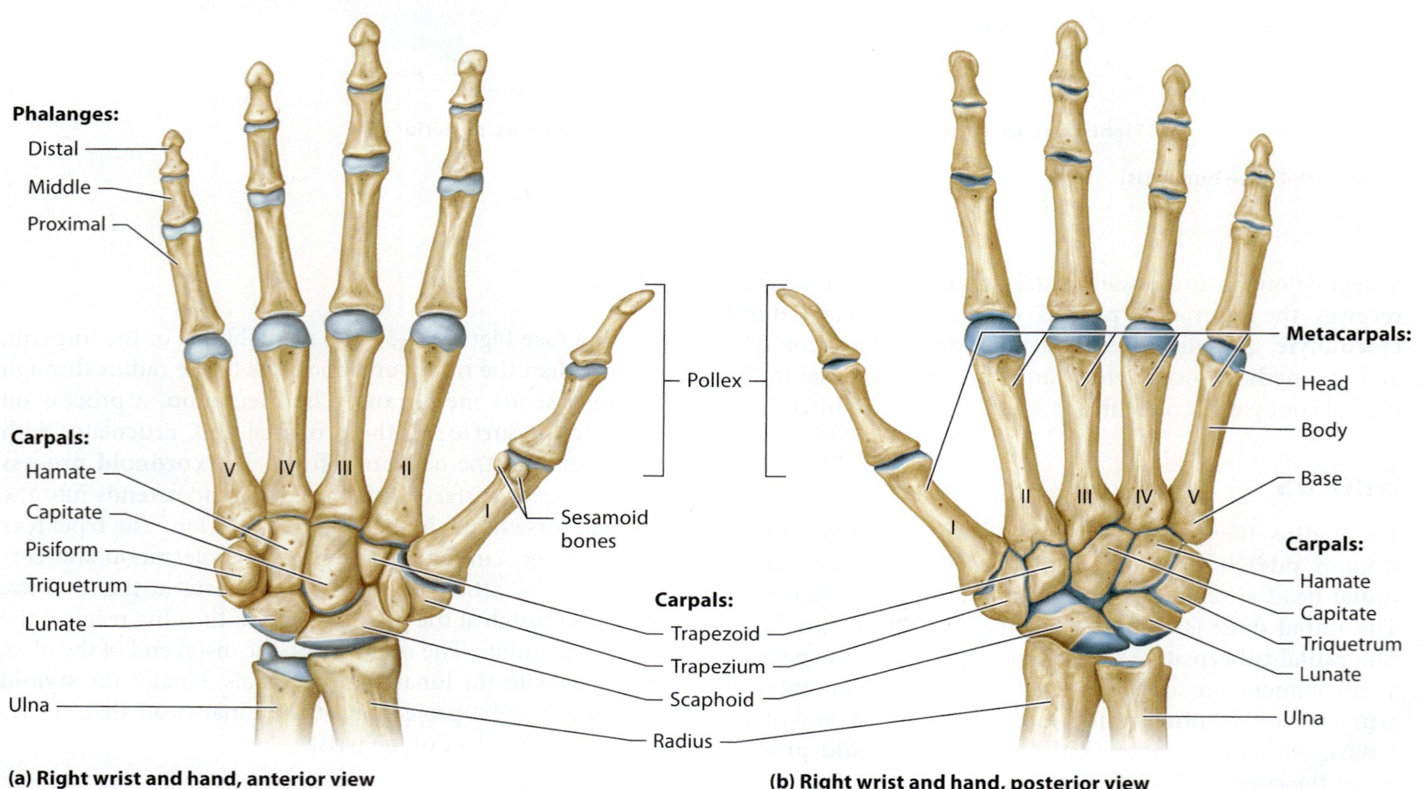

Figure 10-6 Bones of the wrist and hand.

Bones of the Wrist, Hand, and Fingers

The wrist, hand, and fingers consist of three sets of bones: the carpals, metacarpals, and phalanges (Figure 10-6).

The wrist (carpus) consists of eight short **carpal bones** arranged into two transverse rows of four bones each. In the proximal row (lateral to medial) are the **scaphoid**, **lunate**, **triquetrum**, and **pisiform**. In the distal row (lateral to medial) are the **trapezium**, **trapezoid**, **capitate**, and **hamate**.

The hand (manus) consists of five **metacarpal bones** numbered I to V starting with the pollex (or thumb). Each metacarpal has a proximal epiphysis called the base, a diaphysis called the body, and a distal epiphysis called the head or knuckle.

The fingers consists of 14 **phalanges** (singular = phalanx), or bones of the digits. Like the metacarpals, the digits are numbered I to V starting with the thumb, which consists of a proximal and a distal phalanx. The other four digits (fingers) each contain a proximal, a middle, and a distal phalanx.

The Pelvic Girdle and Lower Appendages

The **pelvic girdle** (Figure 10-7) consists of two large, irregular **pelvic** (hip) bones (also called coxal bones). The pelvic girdle articulates with the sacrum of the vertebral column and attaches the lower limbs to the axial skeleton. Together, the pelvic bones and the sacrum and coccyx of the axial skeleton form the **bony pelvis**. The skeletons of females and males are very similar—with the exception of the pelvis. Because the female pelvis is structurally adapted for childbearing, it is lighter, wider, shallower, and rounder than the male pelvis. The major differences between the female and male pelves are presented in Table 10-1.

Pelvic Bone

Each pelvic bone is actually three separate bones—the ilium, ischium, and pubis (pubic bone)—that fuse together at the acetabulum during childhood to form one composite bone (see Figure 10-7). The **acetabulum**, a deep socket at the point of fusion of the ilium, ischium, and pubis, receives the head of the femur to form the hip joint. The **obturator foramen**, the largest foramen in the skeleton, is formed by the fusion of the ischium and pubis; blood vessels and nerves travel through it. Finally, the joint where the two pubic bones meet is the **pubic symphysis**.

Ilium

The **ilium** is the largest and most superior bone of the pelvic bone. The **iliac crest** is the long, superior ridge of the ilium you feel when you place your hands on your hips. The iliac crest ends anteriorly as the **anterior superior iliac spine** and posteriorly as the **posterior superior iliac spine**. Inferior to them are the **anterior inferior iliac spine** and the **posterior inferior iliac spine**, respectively; these spines are attachment sites for tendons of the muscles of the trunk, hip, and thighs. The **greater sciatic notch** is an indention between the ilium and ischium through which the sciatic nerve travels. The **iliac fossa** is a depression on the medial surface of the ilium marked by the **posterior**, **anterior**, and **inferior gluteal lines** which are sites of muscle attachment. Finally, the **auricular surface** is a roughened surface where the ilium articulates with the sacrum to form the **sacroiliac joint**.

Ischium

The **ischium** is the posterior, inferior portion of the pelvic bone made of an **ischial body** and an **ischial ramus** and is the part of the hip bone on which you sit. The **ischial tuberosity** is a roughened projection that is an attachment site for three of the four hamstring muscles. The **ischial spine**, a sharp projection, is an important anatomical landmark for clinicians such as physical therapists.

Pubis

The **pubis** (or pubic bone) is the anterior, inferior portion of the pelvic bone. Both the **superior pubic ramus**, the superior branch of the pubis extending from the pubic body, and the **inferior pubic ramus**, the inferior branch of the pubis

LabBOOST >>>

The Pelvic Bones In identifying the structures of the pelvic bones, the first step is to determine whether the bone is the right pelvic bone or the left pelvic bone. First, hold the bone in front of you so that the acetabulum, which articulates with the femur (thigh bone), angles inferiorly. Second, position the large, curved iliac crest (what you touch when you put your hands on your hips) above the acetabulum, and position the rounded ischium so that it faces inferiorly and posteriorly. Now, if the acetabulum is on the left, you have the left pelvic bone; if it is on the right, you have the right pelvic bone. Knowing which side of the pelvis you are viewing makes identifying the structures considerably easier.

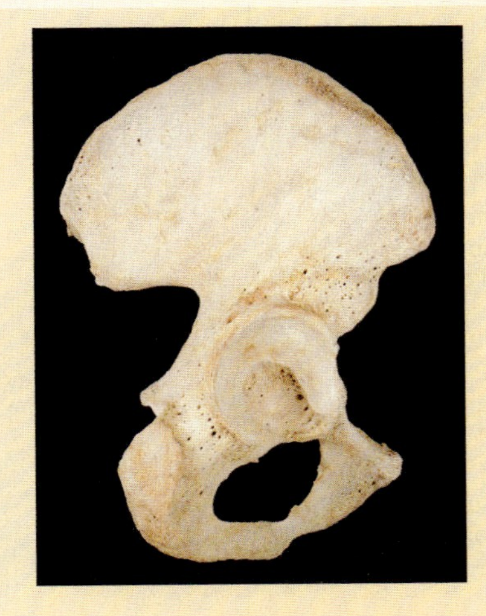

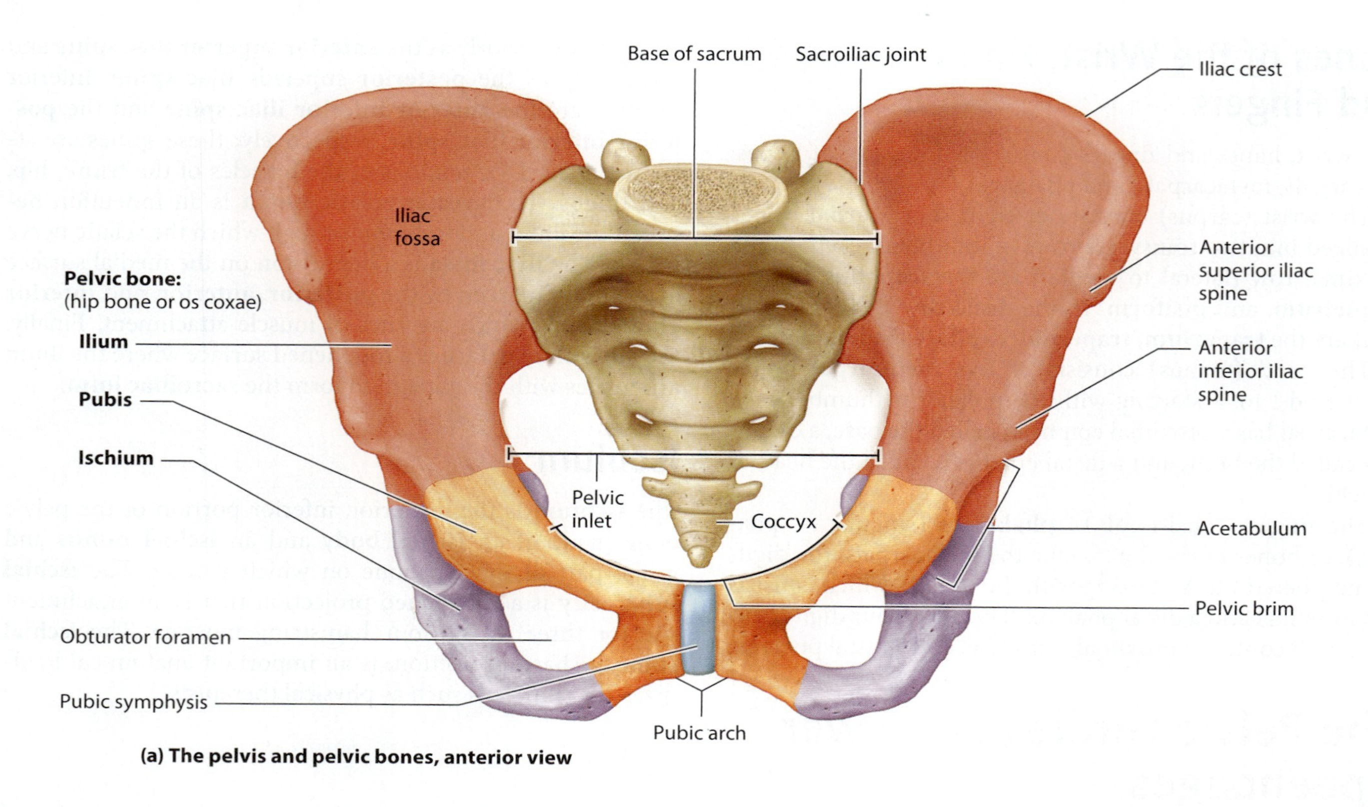

(a) The pelvis and pelvic bones, anterior view

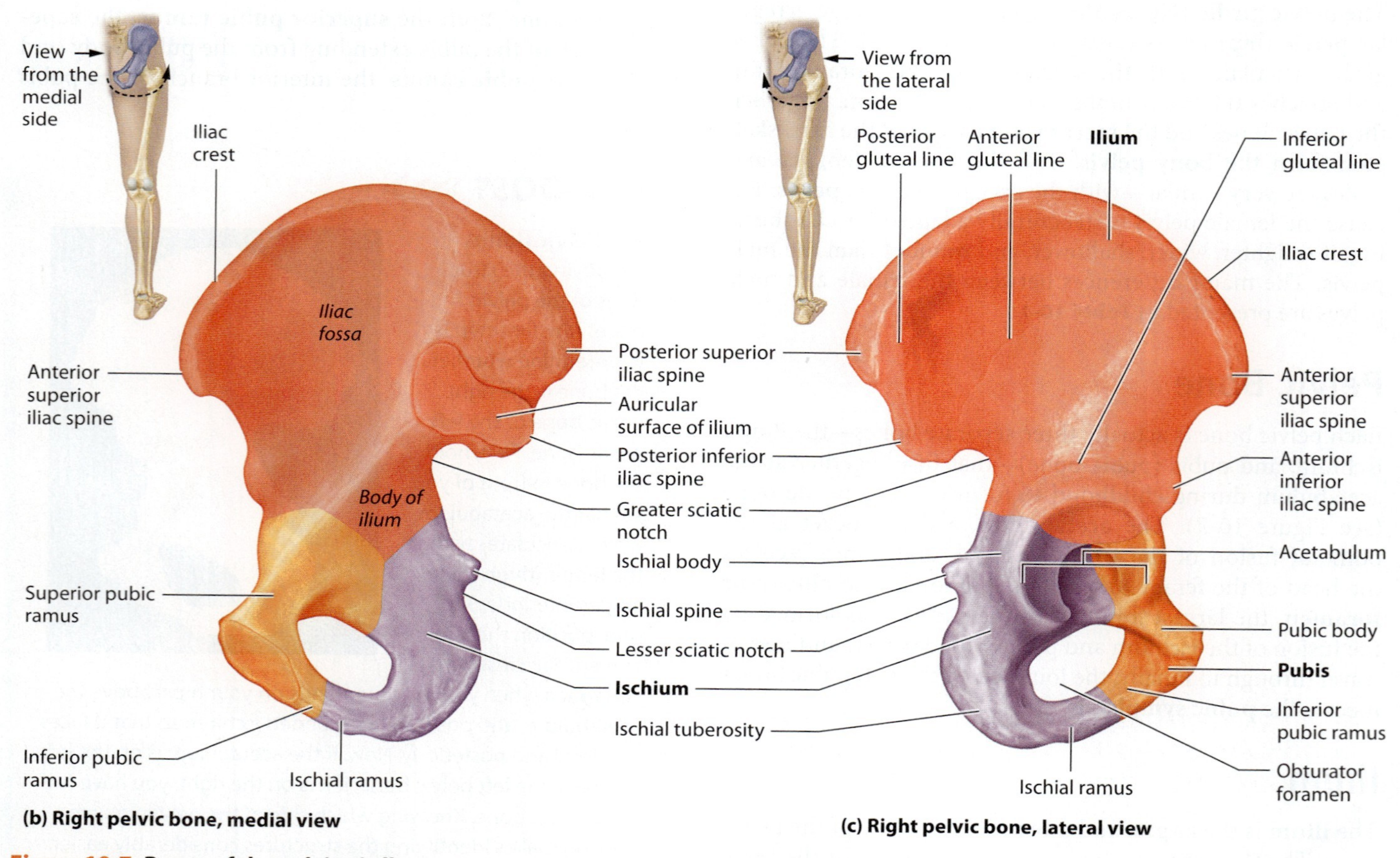

(b) Right pelvic bone, medial view

(c) Right pelvic bone, lateral view

Figure 10-7 Bones of the pelvic girdle.

Table 10-1 Comparison of Female and Male Pelves

Characteristic	Female Pelvis	Male Pelvis
General structure	Lighter and thinner	Heavier and thicker
Greater sciatic notch	Very wide	Narrow
Pubic arch (subpubic angle)	Broader (90–100°)	More acute (60–70°)
Sacrum	Wider and shorter	Narrower and longer
Coccyx	More moveable, straighter	Less moveable, curves anteriorly
Pelvic inlet	Wider and more oval	Narrower and heart shaped
Anterior view	Wider greater pelvis; Wider, shorter sacrum; Farther apart acetabula; Oval shaped pelvic inlet; 90°–100° angle on pubic arch	Narrower greater pelvis; Narrower, longer sacrum; Closer together acetabula; Heart shaped pelvic inlet; 60°–70° angle on pubic arch
Right lateral view		

extending from the pubic body, are important muscle attachment sites. The **pubic symphysis** is the joint between the pubis of each pelvic bone. Just inferior to it is the **pubic arch**, formed where the inferior ramus of each pubic bone unites.

Femur

The **femur** (**Figure 10-8**), or thigh bone, is the heaviest, longest, and strongest bone in the body. Each femur articulates proximally with a pelvic bone to form a hip joint, and distally with the tibia and patella to form the knee joint. The **head** is a ball-shaped projection at the proximal end of the femur that contains a small, central depression called the **fovea capitis**, where a ligament extending from the acetabulum attaches to secure the femur in the hip joint. The **neck**, a constricted region of the bone distolateral to the head, is the weakest part of the femur; its fracture results in a "broken hip." Two projections—the **greater trochanter** and the **lesser trochanter**—are attachment sites for hip muscles. The **gluteal tuberosity** is a roughened projection on the posteroproximal surface that blends into a roughened line called the **linea aspera**; both structures serve as attachment sites for hip muscles.

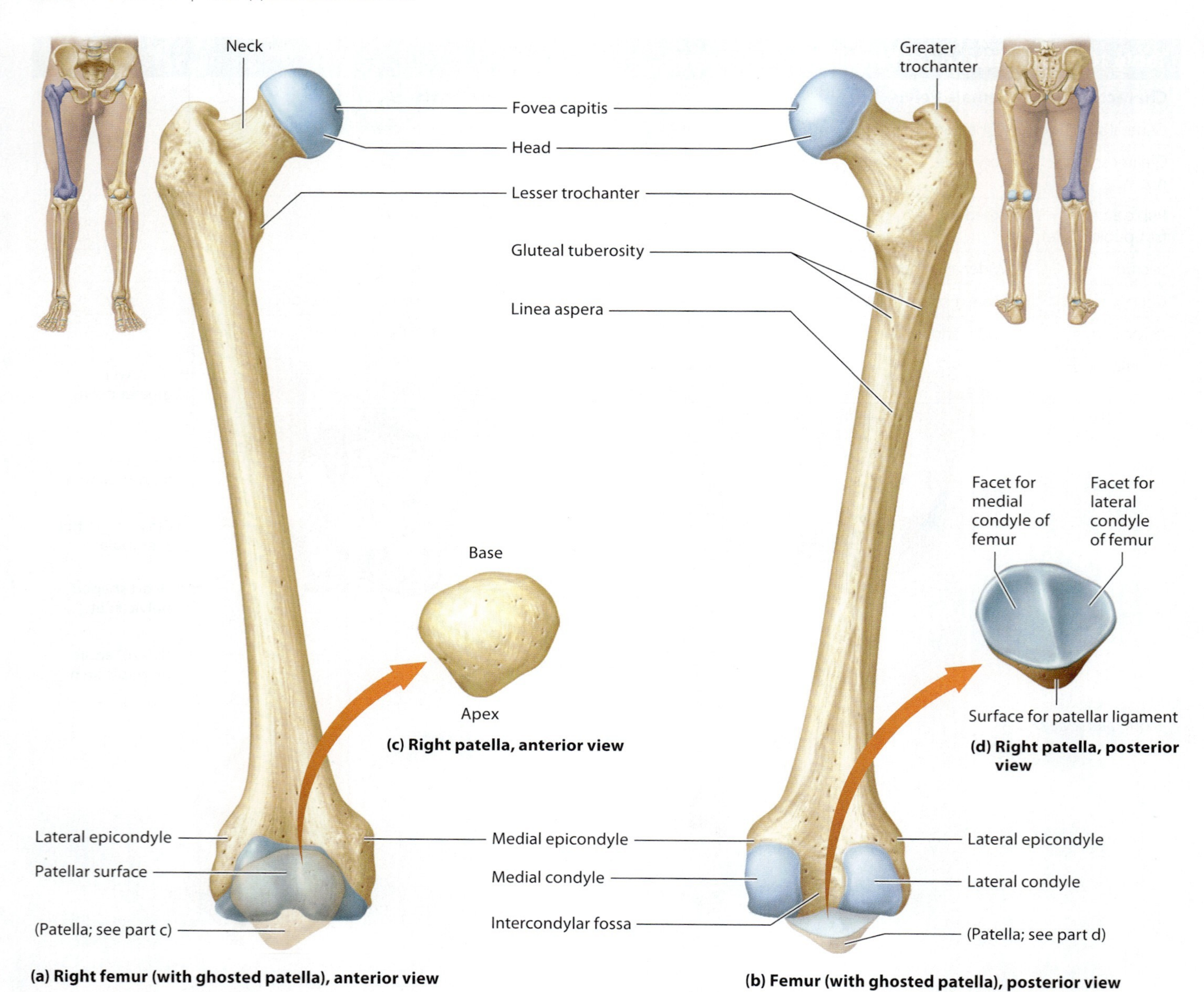

Figure 10-8 **The femur and patella.**

The **lateral condyle** and the **medial condyle**—large rounded projections on the distal femur—articulate with the lateral condyle of the tibia and the medial condyle of the tibia, respectively. The **lateral** and **medial epicondyles**, projections located just superior to the condyles, serve as attachment sites for knee and hip muscles. Finally, the **patellar surface**, the area between the condyles on the anterior surface, articulates with the patella.

Patella

The **patella**, or kneecap, is a small, triangular, sesamoid bone (see Figure 10-8c and d). The broad proximal end of this bone is called the **base**, and the pointed distal end is called the **apex**. The posterior surface contains two articular **facets**, one for the medial condyle of the femur and one for the lateral condyle of the femur. The patellar ligament attaches the patella to the tibia.

Tibia and Fibula

The leg contains two bones, the tibia and the fibula, connected by an **interosseous membrane** (**Figure 10-9**). The **tibia**, or shin bone, is second only to the femur in size and strength. It articulates proximally with the fibula and the femur, and distally with the fibula and the talus bone of the ankle. The **lateral** and **medial condyles** are concave projections on the proximal end of the tibia that articulate with the corresponding condyles on the femur to form the knee joint. The **intercondylar eminence**, an irregular projection

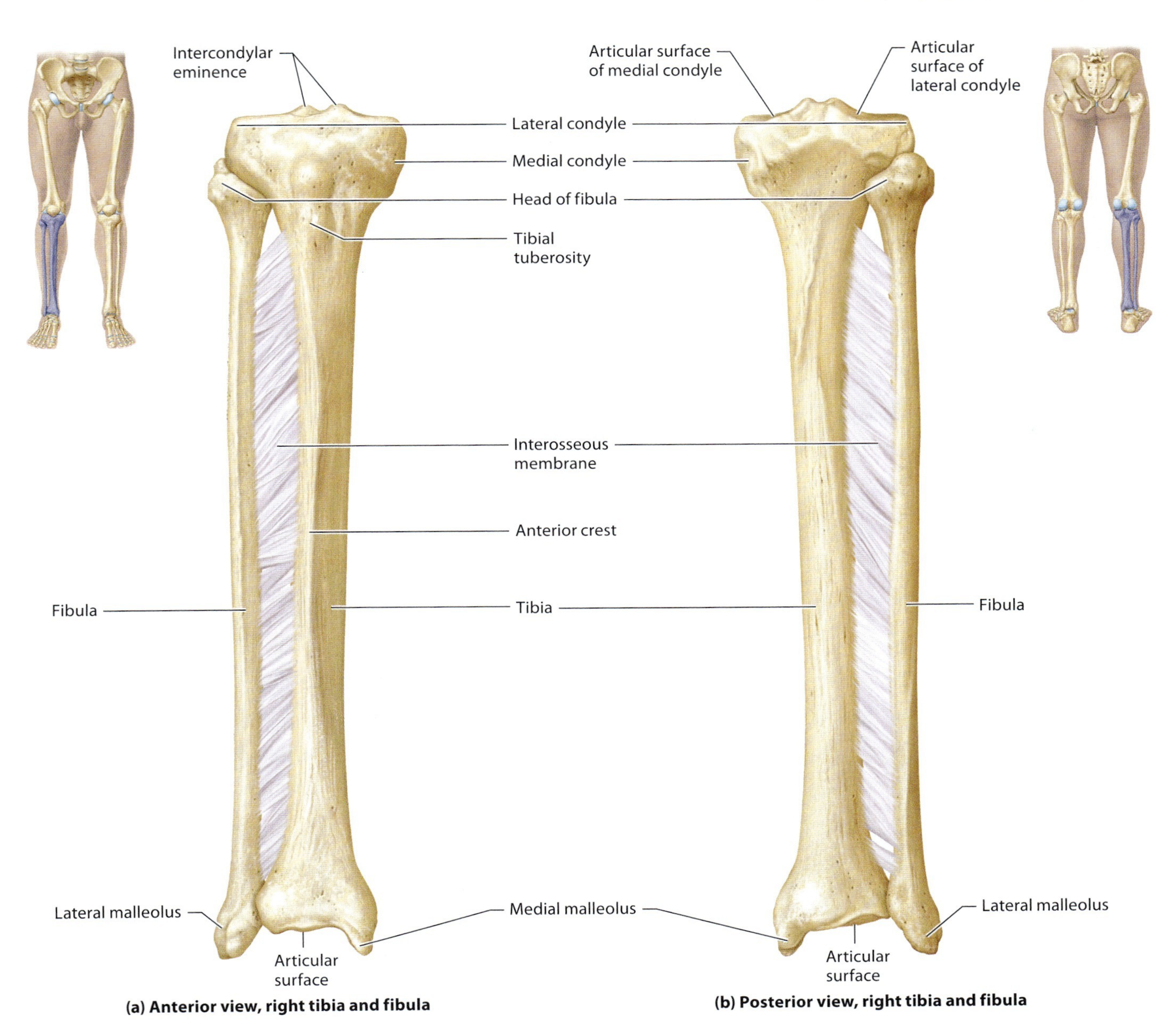

Figure 10-9 The tibia and fibula.

created by the ligaments of the knee joint, separates the lateral and medial condyles. The **tibial tuberosity**, a projection on the anterior proximal surface of the tibia, is an attachment site for the patellar ligament. The **anterior crest** (or anterior border) is a sharp ridge inferior to and continuous with the tibial tuberosity. Finally, the **medial malleolus**, an expansion of the distal end of the tibia, articulates with the talus bone of the ankle and forms the prominence that can be felt on the medial surface of the ankle.

The **fibula**, a thin, stick-like bone that lies parallel and lateral to the tibia, articulates both proximally and distally with the tibia and is part of the knee joint. The proximal end is the **head**; the distal end, the **lateral malleolus**, can be felt as the lateral prominence of the ankle.

Bones of the Ankle, Foot, and Toes

The ankle, foot, and toes consist of three sets of bones: the tarsals, metatarsals, and phalanges (Figure 10-10). The ankle (tarsus) consists of seven short **tarsal bones**: the **talus**, the **calcaneus**, the **navicular**, the **cuboid**, and three cuneiform bones called the **medial**, **intermediate**, and **lateral cuneiforms**. The talus is the only ankle bone that articulates with the tibia and the fibula. The calcaneus, or heel bone, is the attachment site for the Achilles (calcaneal) tendon.

The foot (metatarsus) consists of five **metatarsal bones** numbered I to V starting with the hallux (great toe). The toes consist of 14 **phalanges**, or bones of the digits. Like the metatarsals, the digits are numbered I to V starting with the

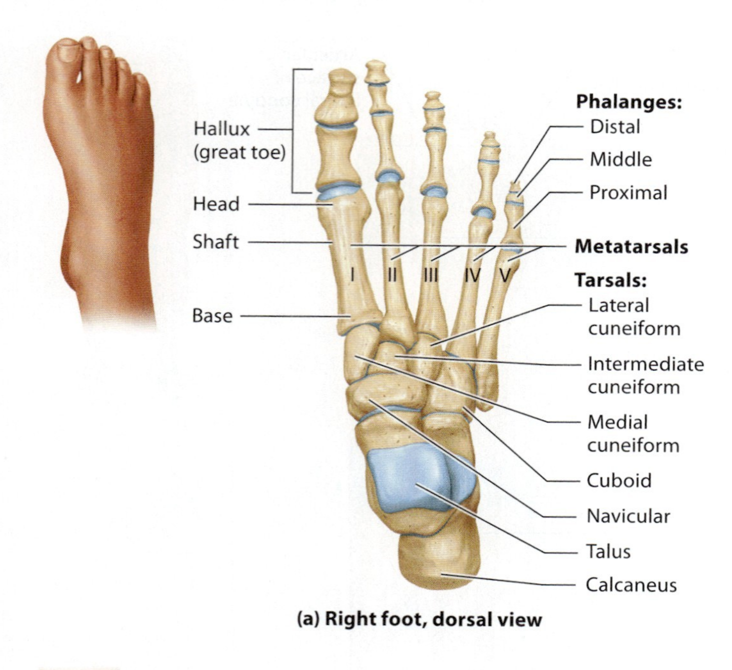

(a) Right foot, dorsal view

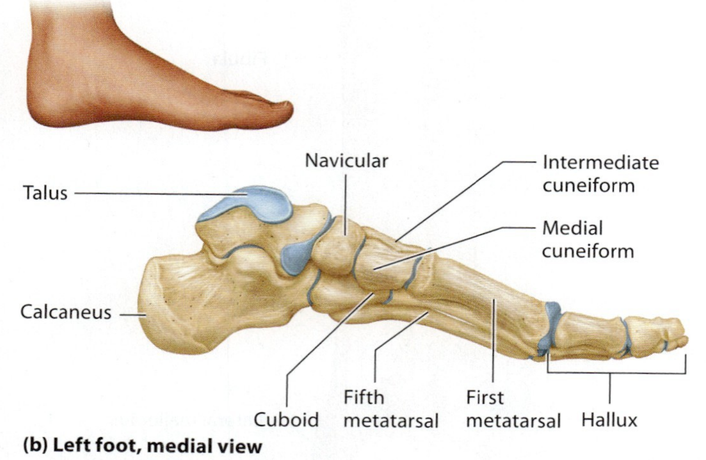

(b) Left foot, medial view

Figure 10-10 **Bones of the ankle and foot.**

great toe, which consists of a proximal and a distal phalanx; the other four digits each contain a proximal, a middle, and a distal phalanx.

ACTIVITY 1

Studying the Bones of the Appendicular Skeleton

Learning Outcomes

1. Identify the components of the pectoral girdle, and describe its functions.
2. Identify the major bone features of the scapula and the clavicle, and describe the functions of each.
3. Identify the bones and bone features of the upper limbs, and describe the functions of each bone feature.
4. Identify the components and describe the functions of the pelvic girdle.
5. Identify the major bone features of the pelvic bones, and describe the functions of each.
6. Identify the bones and bone features of the lower limbs, and describe the functions of each bone feature.
7. Compare and contrast the female and male pelves.

Materials Needed

- ☐ Disarticulated bones of the appendicular skeleton
- ☐ Male pelvis
- ☐ Female pelvis

Instructions

1. CHART Obtain disarticulated appendicular bones. These bones have been divided into four sets (Sets A-D, as listed in the charts that follow). Assign each member of your lab group one set of bones. Then, complete the following tasks:
   a. Study your set of bones, filling out the description column in the appropriate Making Connections chart as you go.
   b. After you have studied your bones, you will be the "expert" as you teach the bones and bone markings that you have learned to the other members of your lab group, the "students."
   c. As you are learning the bones and their markings, brainstorm with your group members to make connections between what you are learning and what you have learned in lectures, labs, and from your textbook. Fill out the connections column in the Making Connections chart.
   d. After each group member has served as the "expert" for one set of bones, rotate sets of bones until each group member has served as the "expert" for each set of bones. Use the following checklist to ensure that you play the role of "expert" at least once and that of "student" at least three times for each set of bones. Remember that you will need to spend time outside of normal lab time to master these bones.

Set A	Set B	Set C	Set D
☐ "Expert"	☐ "Expert"	☐ "Expert"	☐ "Expert"
☐ "Student"	☐ "Student"	☐ "Student"	☐ "Student"
☐ "Student"	☐ "Student"	☐ "Student"	☐ "Student"
☐ "Student"	☐ "Student"	☐ "Student"	☐ "Student"

Making Connections: Bones of the Appendicular Skeleton, Set A		
Bone	**Description of Bone and/or Bone Marking(s)**	**Connections to Things I Have Already Learned**
Clavicle	*S-shaped, fairly flat bone*	*Also called the collar bone; is part of the pectoral girdle; helps anchor the arm to the axial skeleton*
• Medial (sternal) end		
• Acromial (lateral) end		
• Conoid tubercle		
Scapula		
• Borders (superior, middle, lateral)		
• Angles (superior, inferior, lateral)		
• Subscapular fossa	*Depression on anterior surface*	*Origin of the subscapularis muscle; posterior ridge of ribs is just anterior to this muscle*
• Spine		
• Acromion		
• Infraspinous fossa		
• Supraspinous fossa		
• Coracoid process		
• Suprascapular notch		
Humerus		
• Head		
• Neck (anatomical, surgical)		

(*Continued*)

Making Connections: Bones of the Appendicular Skeleton, Set A (*Continued*)

Bone	Description of Bone and/or Bone Marking(s)	Connections to Things I Have Already Learned
• Greater tubercle		
• Lesser tubercle		
• Deltoid tuberosity		
• Radial groove		
• Trochlea		
• Capitulum		
• Medial epicondyle		
• Lateral epicondyle		
• Coronoid fossa		
• Olecranon fossa		

Making Connections: Bones of the Appendicular Skeleton, Set B

Bone	Description of Bone and/or Bone Marking(s)	Connections to Things I Have Already Learned
Radius		
• Head		
• Neck		
• Ulnar notch		
• Radial tuberosity		*Attachment site for the biceps brachii muscle*
• Styloid process	*Short projection on the anterior, distal surface*	

Bone	Description of Bone and/or Bone Marking(s)	Connections to Things I Have Already Learned
Ulna		
• Head		
• Olecranon process		
• Coronoid process	*Bottom ridge or lip of the trochlear notch*	*Articulates with the coronoid fossa of the humerus*
• Trochlear notch		
• Radial notch		
• Styloid process		
Carpals		
• Scaphoid		
• Lunate		
• Triquetrum		
• Pisiform		
• Trapezium		
• Trapezoid		
• Capitate		
• Hamate		
Metacarpals		
Phalanges		

Making Connections: Bones of the Appendicular Skeleton, Set C		
Bone	**Description of Bone and/or Bone Marking(s)**	**Connections to Things I Have Already Learned**
Femur		
• Head		
• Fovea capitis	*Shallow depression in the center of the femoral head*	*Created by ligament attaching to the acetabulum; helps anchor femur to hip; pathway for blood supply to the femoral head*
• Neck		
• Greater trochanter		
• Lesser trochanter		
• Gluteal tuberosity		
• Linea aspera		
• Lateral condyle		
• Medial condyle		
• Medial epicondyle		
• Lateral epicondyle		
• Patellar surface		
Patella		
Tibia		
• Medial condyle		
• Lateral condyle		
• Intercondylar eminence	*Short, sharp projections located between the condyles*	*Attachment sites for the ACL and PCL*

Bone	Description of Bone and/or Bone Marking(s)	Connections to Things I Have Already Learned
• Tibial tuberosity		
• Anterior border		
• Medial malleolus		
Fibula		
• Head		
• Lateral malleolus		
Tarsals		
• Talus		
• Calcaneus		
• Cuboid		
• Navicular		
• Medial cuneiform		
• Intermediate cuneiform		
• Lateral cuneiform		
• Medial		
• Intermediate		
• Lateral		
Metatarsals		
Phalanges		

Making Connections: Bones of the Appendicular Skeleton, Set D		
Bone	**Description of Bone and/or Bone Marking(s)**	**Connections to Things I Have Already Learned**
Pelvic bone		
• Ilium		
• Iliac crest		
• Anterior superior iliac spine	*Point on anterior edge of ilium*	*Origin for several hip muscles; attachment site for inguinal ligament; a body landmark visible above the bikini line*
• Posterior superior iliac spine		
• Greater sciatic notch		
• Iliac fossa		
• Auricular surface		
• Acetabulum		
• Ischium		
• Ischial tuberosity		
• Ischial spine		
• Lesser sciatic notch		
• Pubis		*Origin for three of four hamstring muscles; when you sit up straight, this is what you are sitting on*
• Superior ramus		
• Inferior ramus		
• Pubic crest	*Roughened surface on most inferior surface of the pelvic bone*	
• Pubic symphysis		
• Pubic arch		

2. CHART Obtain a male pelvis and a female pelvis. Then, complete the following chart as you compare and contrast the two pelves:

Comparing the Male and Female Pelves

Bone Feature	Male Pelvis	Female Pelvis
Sacrum width		
Subpubic angle		
Greater sciatic notch		
Coccyx shape		
Pelvic inlet shape		

3. **Optional Activity**

Practice labeling bones of the appendicular skeleton at > MasteringA&P® > Study Area > Practice Anatomy Lab > Human Cadaver > Appendicular Skeleton

ACTIVITY 2
Identifying Bones-in-a-Bag

Learning Outcome

1. Identify disarticulated bones of the appendicular skeleton by touch only, and justify your identifications based on bone size, shape, and markings.

Materials Needed

☐ Bag containing disarticulated bones of the appendicular skeleton

Instructions

In this activity you and your group members will take turns identifying each of the eight disarticulated bones in a bag your instructor will give you.

1. Reach into the bag with one hand (without looking at the bones—no peeking!) and pick up a bone, but do not take it out of the bag.
2. Based on feel alone, describe the bone's size, shape, and special features to the other members of your lab group. Then record your description in **Table 10-2**.
3. Write in Table 10-2 the name of the bone you are holding, then pull it out of the bag to see if your identification was correct.
4. List any features of the bone that you missed while feeling it inside the bag. Record this information in Table 10-2 as well.
5. Pass the bag to another group member, who will repeat steps 1–4.
6. Repeat steps 1–5 until all of the bones in the bag have been identified.

Table 10-2 Bones-in-a-Bag Identification

	Description of Bone Based on Touch Alone	Name of Bone	Additional Bone Features
1.			
2.			
3.			
4.			
5.			
6.			
7.			
8.			

Post-lab quizzes are also assignable in MasteringA&P®

POST-LAB ASSIGNMENTS

Name: ______________________ Date: ______________ Lab Section: __________

PART I. Check Your Understanding

Activity 1: Studying the Bones of the Appendicular Skeleton

1. Which of the following bone features is/are correctly matched with the bone on which it is found?
   a. greater trochanter – femur
   b. obturator foramen – sacrum
   c. medial malleolus – fibula
   d. conoid tubercle – scapula
   e. radial notch – radius

2. Which of the following bone features are articular surfaces? (Choose all that are correct.)
   a. acetabulum
   b. lateral epicondyle
   c. glenoid cavity
   d. tibial tuberosity
   e. iliac fossa

3. Which of the following bones have a styloid process? (Choose all that are correct.)
   a. radius
   b. tibia
   c. humerus
   d. clavicle
   e. ulna

4. Indicate whether each of the following statements is true for the male pelvis or the female pelvis:
   a. The greater sciatic notch is wider. ______________________
   b. The coccyx curves anteriorly. ______________________
   c. The angle of the pubic arch is more acute. ______________________
   d. The bones are lighter and thinner. ______________________
   e. The sacrum is wider and shorter. ______________________

5. Identify the bone in the accompanying figure: ______________

   Identify each of the labeled bone features:
   a. ______________________
   b. ______________________
   c. ______________________
   d. ______________________
   e. ______________________
   f. ______________________
   g. ______________________
   h. ______________________
   i. ______________________

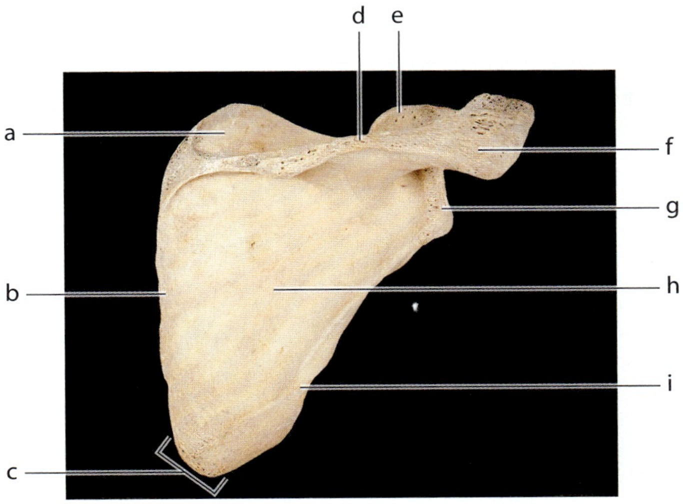

6. Identify the bone in the accompanying figure: ____________________

Identify each of the labeled bone features:

a. ____
b. ____
c. ____
d. ____
e. ____
f. ____
g. ____
h. ____
i. ____

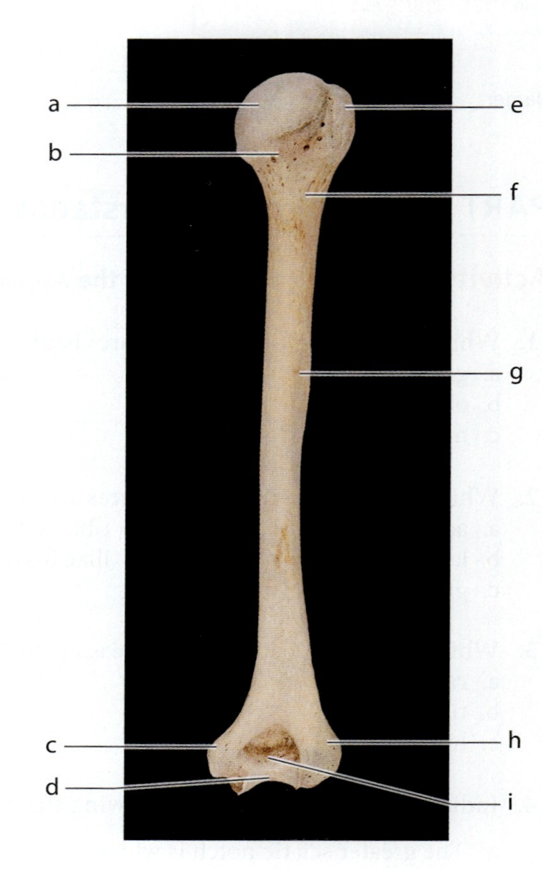

7. Identify the bone in the accompanying figure: ____________________

Identify each of the labeled bone features:

a. ____
b. ____
c. ____
d. ____
e. ____
f. ____

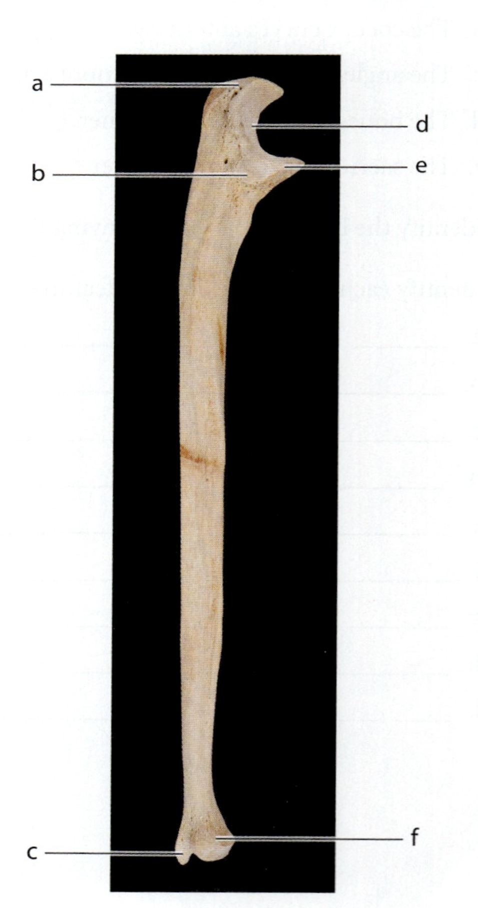

8. Identify each of the bones in the accompanying figure:

a. ______________________________

b. ______________________________

c. ______________________________

d. ______________________________

e. ______________________________

f. ______________________________

g. ______________________________

h. ______________________________

i. ______________________________

j. ______________________________

k. ______________________________

l. ______________________________

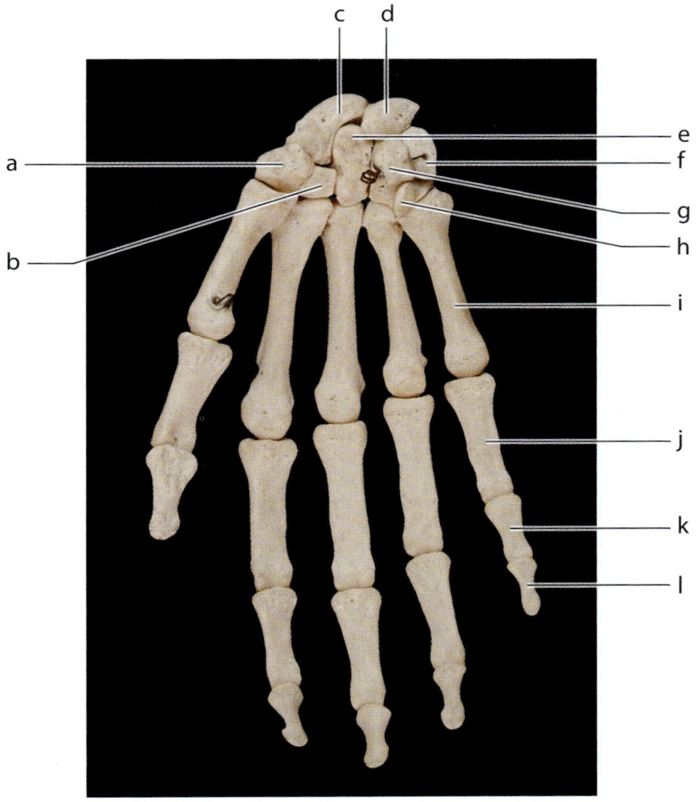

9. Identify the bone in the accompanying figure: ______________

Identify each of the labeled bone features:

a. ______________________________

b. ______________________________

c. ______________________________

d. ______________________________

e. ______________________________

f. ______________________________

g. ______________________________

h. ______________________________

i. ______________________________

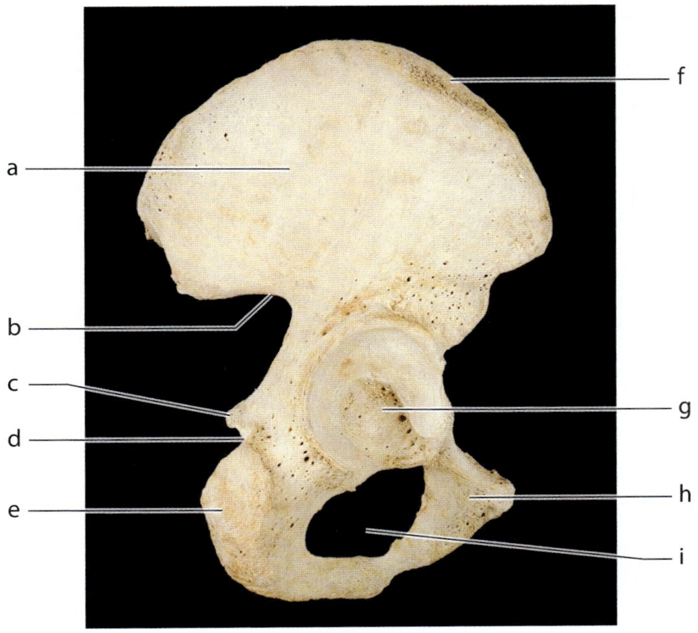

10. Identify the bone in the accompanying figure: ____________________

    Identify each of the labeled bone features:

    a. ______________________________
    b. ______________________________
    c. ______________________________
    d. ______________________________
    e. ______________________________
    f. ______________________________
    g. ______________________________
    h. ______________________________
    i. ______________________________
    j. ______________________________
    k. ______________________________

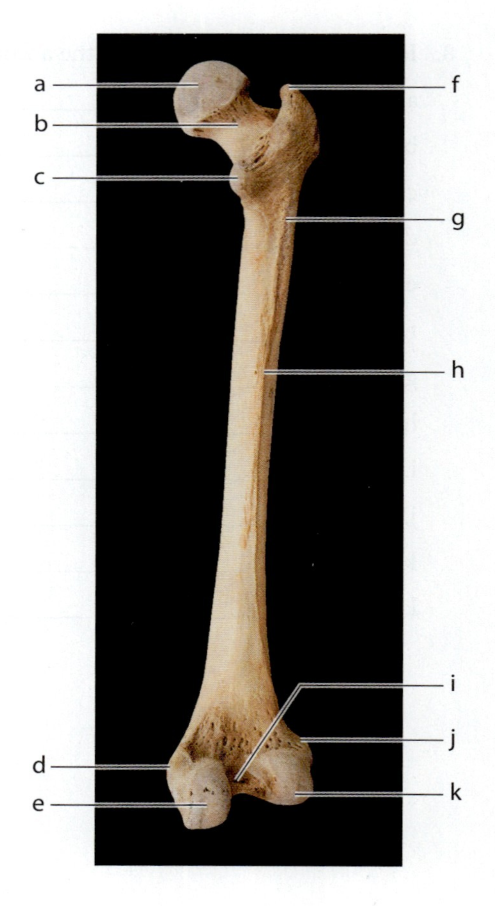

11. Identify the bone in the accompanying figure: ____________________

    Identify each of the labeled bone features:

    a. ______________________________
    b. ______________________________
    c. ______________________________
    d. ______________________________

12. Identify the bone in the accompanying figure: ____________________

Identify each of the labeled bone features:

a. ______________________________

b. ______________________________

c. ______________________________

d. ______________________________

e. ______________________________

f. ______________________________

g. ______________________________

h. ______________________________

i. ______________________________

j. ______________________________

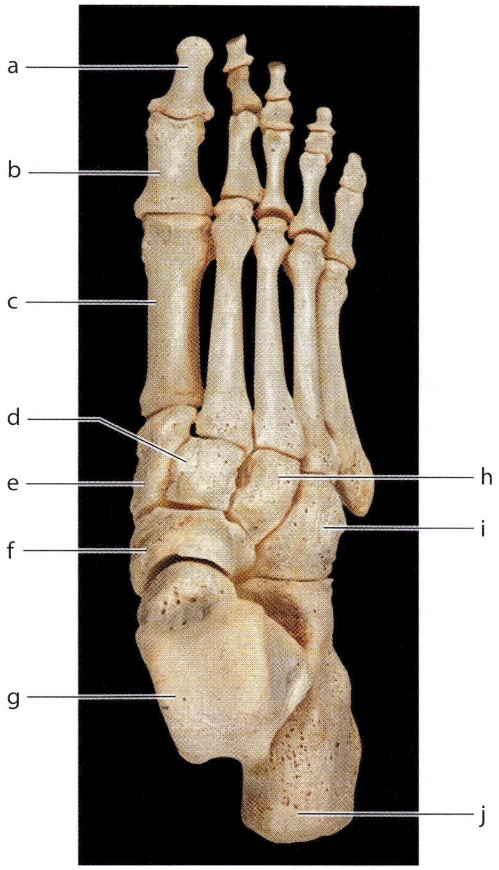

Activity 2: Identifying Bones-in-a-Bag

1. An oddly shaped bone with blades going in three different directions, a large hole, and a very deep cup is the:
   a. humerus.
   b. tibia.
   c. pelvic bone.
   d. fibula.
   e. calcaneus.

2. A bone with a relatively thin shaft and tiny head, a large "mouth" with a smooth inner surface at one end, and processes on either edge of this surface is the:
   a. radius.
   b. ulna.
   c. talus.
   d. ilium.
   e. fibula.

PART II. Putting It All Together

A. Review Questions

Answer the following questions using your lecture notes, your textbook, and your lab notes.

1. Explain how you can determine if a femur is the left femur or the right femur. Include specific bone markings in your answer. ____________________

2. Explain how the radius rotates around the ulna in the movements called "pronation" and "supination." Include specific bone markings in your answer. ____________________

3. For each of the following joints, write the two bones that form it. Include specific bone markings or features in your answer. The first row has been filled in for you.

Sacroiliac joint	*Auricular surface of sacrum and auricular surface of ilium*
Pubic symphysis	
Glenohumeral joint	
Distal tibiofibular joint	
Knee joint	
Ankle joint	
Elbow joint	

4. Identify each bone or bone feature on the accompanying x-ray.

   a. ______________________________

   b. ______________________________

   c. ______________________________

   d. ______________________________

   e. ______________________________

   f. ______________________________

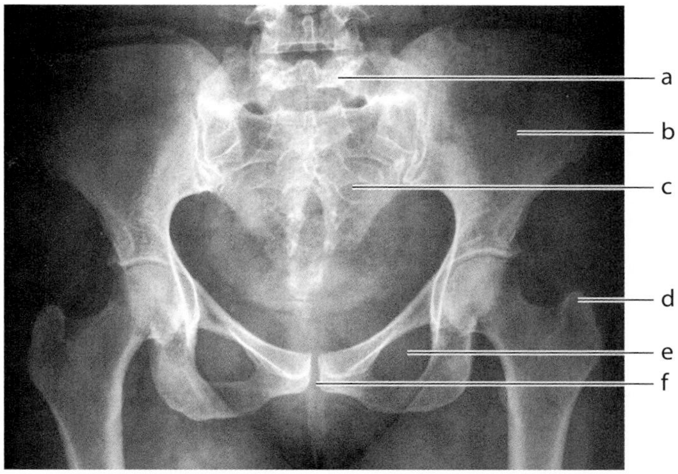

5. Identify each bone or bone feature on the accompanying x-ray.

   a. ______________________________

   b. ______________________________

   c. ______________________________

   d. ______________________________

   e. ______________________________

   f. ______________________________

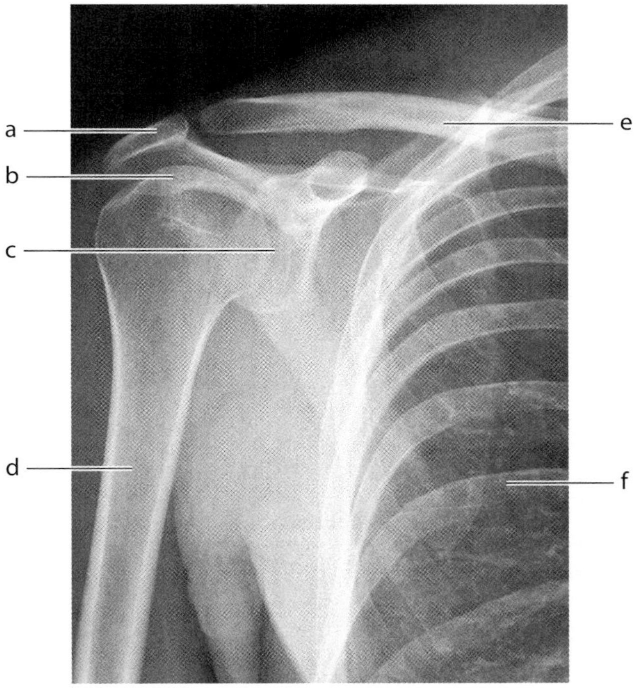

B. Concept Mapping

1. Fill in the blanks to complete this concept map outlining features of the humerus and ulna.

coronoid process	deltoid tuberosity	greater tubercle	olecranon process	trochlea

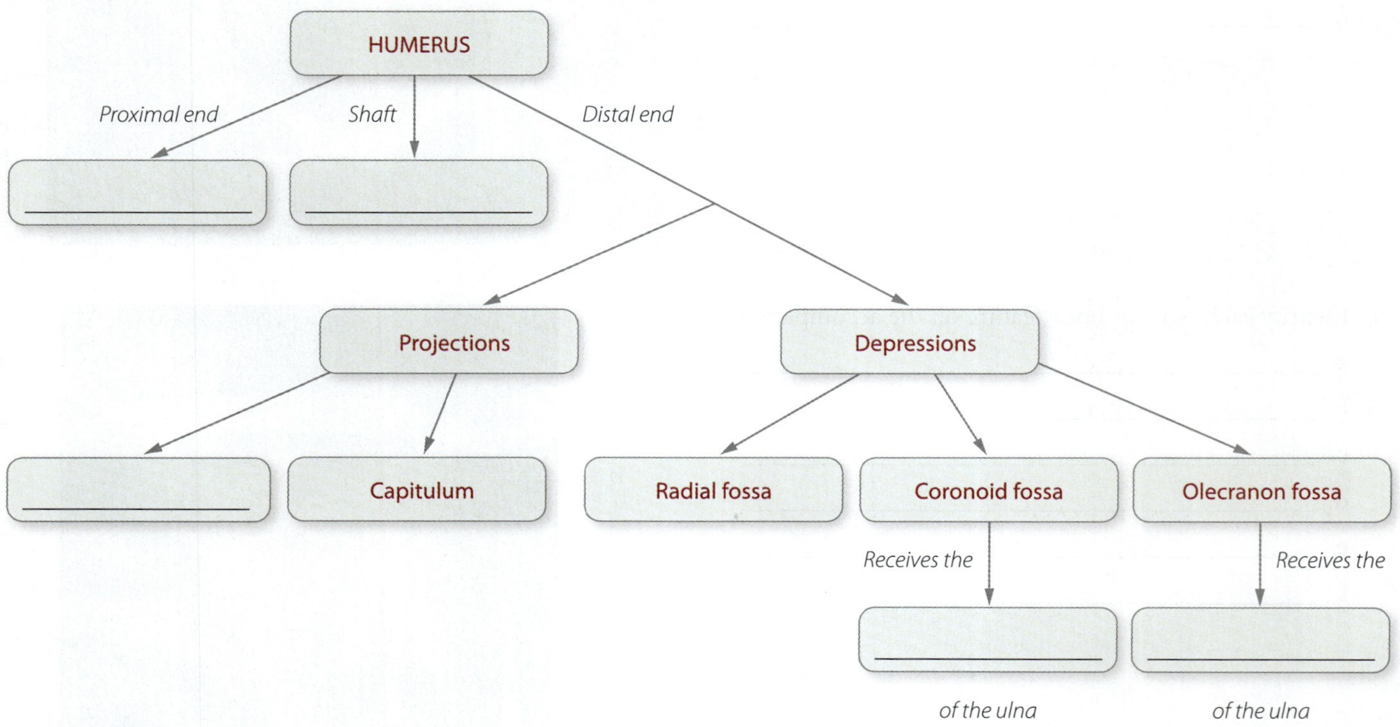

2. Construct a unit concept map to show the relationships among the following set of terms. Include all of the terms in your diagram. Your instructor may choose to assign additional terms.

acetabulum	coronoid process	deltoid tuberosity	femur	greater tubercle
head of femur	intertubercular groove	ligament	metacarpal	olecranon process
patella	phalanx	tibia	trapezium	trochlea

11 Joints

UNIT OUTLINE

Articulations, or **joints**, are points of contact between two bones, bone and cartilage, or bone and teeth. Joints perform a wide variety of functions: They bind bones together, allow movement of the skeleton in response to skeletal muscle contraction, participate in the linear growth of bones, and permit parts of the skeleton to change shape during childbirth.

THINK ABOUT IT *Using your knowledge of the structure of the pelvic girdle, identify the joint that changes shape during childbirth.*

____

PRE-LAB ASSIGNMENTS

Pre-lab quizzes are also assignable in MasteringA&P®

To maximize learning, BEFORE your lab period carefully read this entire lab unit and complete these pre-lab assignments using your textbook, lecture notes, and prior knowledge.

PRE-LAB Activity 1: Identifying and Classifying Joints

1. Use the list of structures provided to label the accompanying illustration of a synovial joint. Check off each structure as you label it.

   - ☐ synovial cavity
   - ☐ articular cartilage
   - ☐ periosteum
   - ☐ fibrous outer layer

a ____________

b ____________

c ____________

d ____________

2. Match each of the following joint structures with its description/function:

_____	a. syndesmosis	1. contains periodontal ligament
_____	b. suture	2. organized into a band or sheet
_____	c. synovial joint	3. contains joint cavity
_____	d. gomphosis	4. is found only in the skull
_____	e. synchondrosis	5. epiphyseal plate

PRE-LAB Activity 2: Demonstrating Movements Allowed by Joints

1. If you "open up" your elbow, making your arm straight, then your elbow is exhibiting:
   a. flexion. c. abduction.
   b. extension. d. adduction.

2. If you stand on your tippy-toes like a ballet dancer, then your ankle is exhibiting:
   a. dorsiflexion. c. pronation.
   b. inversion. d. plantarflexion.

PRE-LAB Activity 3: Comparing and Contrasting the Structure and Function of Selected Synovial Joints

1. Which of the following is an amphiarthrosis?
   a. humeroulnar joint c. femoroacetabular joint
   b. femorotibial joint d. pubic symphysis

2. Which of the following joints includes a semicircular pad of fibrocartilage?
   a. femoroacetabular joint c. femorotibial joint
   b. tibiofibular joint d. talocrural joint

Classification of Joints

Joints can be classified according to structure or function. Structurally, joints are classified based either on the type of connective tissue that binds the bones or on the presence of a fluid-filled space (synovial cavity) between the bones. The body has three structural types of joints: fibrous joints, cartilaginous joints, and synovial joints. Additionally, joints can also be classified into three functional types according to the amount of movement they allow: synarthroses, which are immoveable joints; amphiarthroses, which are slightly moveable joints; and diarthroses, which are freely moveable joints.

Fibrous Joints

In **fibrous joints** (**Figure 11-1**), there is no synovial cavity, and the bones are held together by dense regular collagenous connective tissue. There are three types of fibrous joints: sutures, gomphoses, and syndesmoses.

In a **suture** (see Figure 11-1a), irregular, interlocking edges of bone are joined by a thin layer of dense regular collagenous connective tissue. This type of fibrous joint, which is immoveable (a synarthrosis), is found only in the skull. In a **gomphosis** (or dentoalveolar joint; see Figure 11-1b), a cone-shaped tooth is held in a bony socket (alveolus) of the mandible or the maxillae by a fibrous periodontal ligament. Gomphoses are synarthroses. In a **syndesmosis**, bones are held together by either a band (interosseous ligament) or a sheet (interosseous membrane) of dense regular collagenous connective tissue; an example is the interosseous membrane between the radius and ulna in the forearm (see Figure 11-1c). Syndesmoses are classified functionally as either synarthroses or amphiarthroses depending on the length of the connecting fibers.

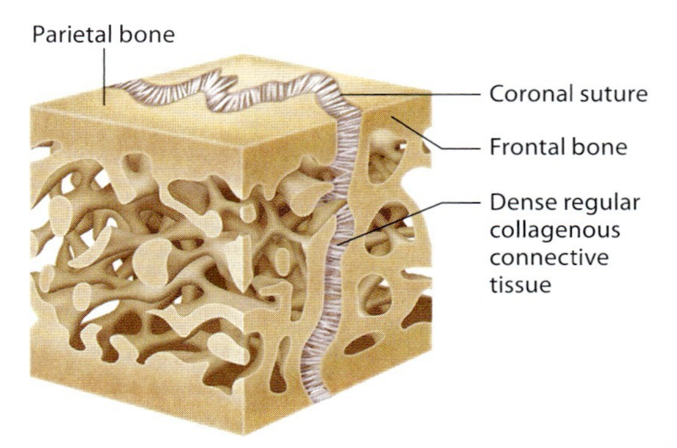

(a) Suture between parietal and frontal bones

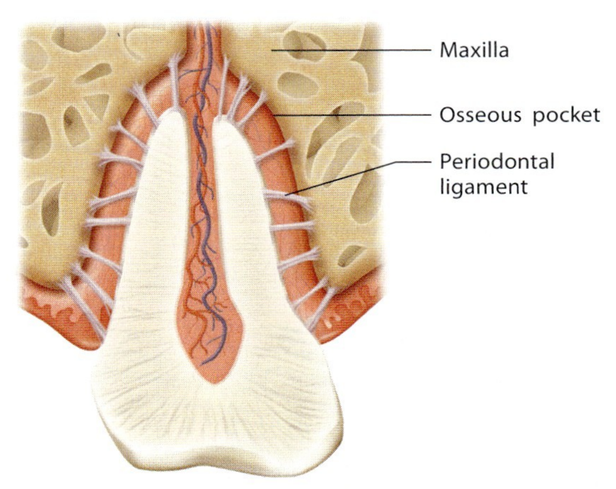

(b) Gomphosis in an upper tooth

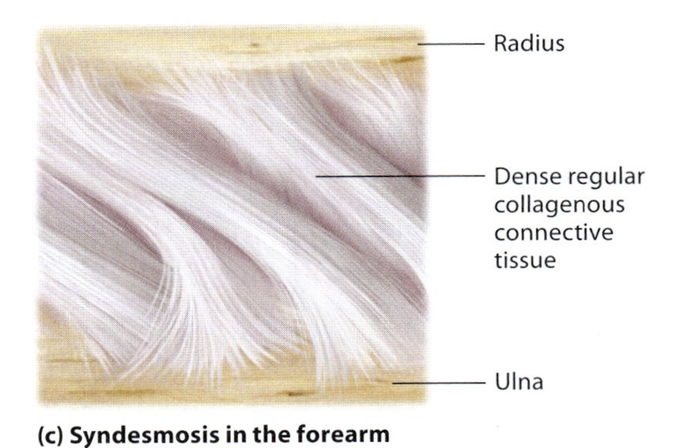

(c) Syndesmosis in the forearm

Figure 11-1 Fibrous joints.

Cartilaginous Joints

In cartilaginous joints, which lack a synovial cavity and allow little or no movement, the bones are held together by either hyaline cartilage or fibrocartilage. There are two types of cartilaginous joints: synchondroses and symphyses (**Figure 11-2**).

In a **synchondrosis**, bones are held together by a band of hyaline cartilage; examples include the epiphyseal plate of a growing long bone (see Figure 11-2a), the joint between the first rib and the manubrium of the sternum (the first sternocostal joint), and the joints between ribs and costal cartilages (costochondral joints; see Figure 11-2b). Synchondroses are synarthroses. In a **symphysis**, bones are held together by a broad, flat disc of fibrocartilage. Examples of symphyses, all of which are found in the midline of the body, include the intervertebral joints (see Figure 11-2c) and the pubic symphysis (see Figure 11-2d). Symphyses are amphiarthroses.

Synovial Joints

Synovial joints have a fluid-filled synovial cavity between the articulating bones. Most of the body's joints are synovial joints, and all synovial joints are diarthroses.

Synovial joints have several distinguishing structural features (**Figure 11-3**). A thin layer of hyaline cartilage—the **articular cartilage**—covers the articulating surfaces of the

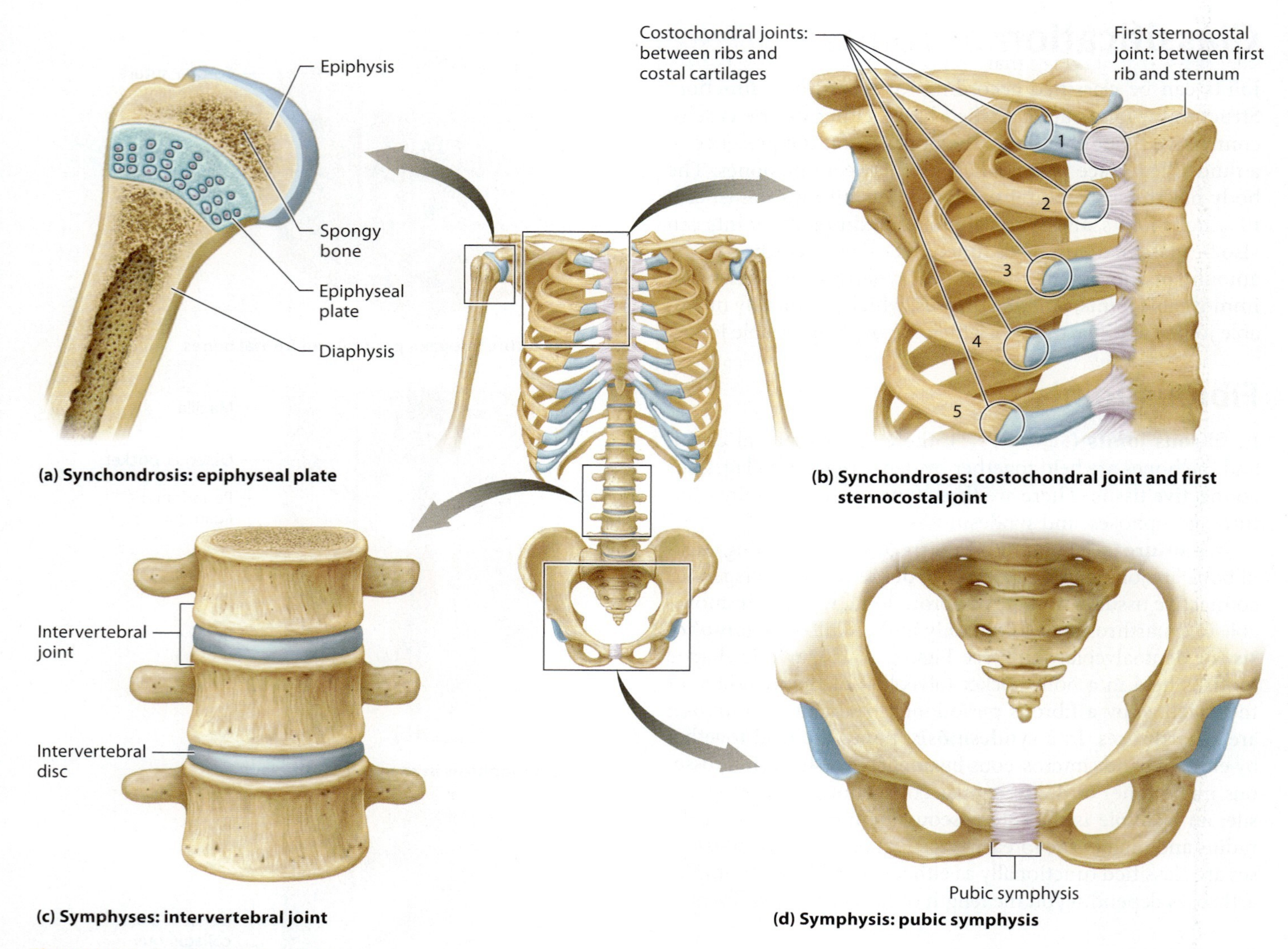

(a) Synchondrosis: epiphyseal plate

(b) Synchondroses: costochondral joint and first sternocostal joint

(c) Symphyses: intervertebral joint

(d) Symphysis: pubic symphysis

Figure 11-2 Cartilaginous joints.

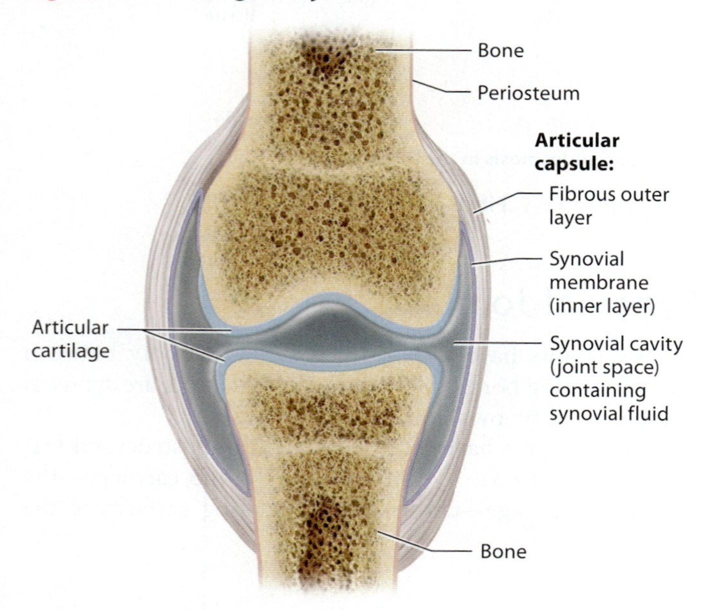

Figure 11-3 Structure of a synovial joint.

bones, reducing friction between them and absorbing shock. Between the articulating bones is the **synovial cavity**, a small space filled with **synovial fluid**, a thick, colorless liquid that lubricates the joint surfaces and absorbs shock. Additionally, it supplies oxygen and nutrients to and removes carbon dioxide and metabolic wastes from the chondrocytes of the articular cartilage. Surrounding each synovial joint is an **articular (joint) capsule** consisting of two layers. The outer layer is composed of dense irregular connective tissue with abundant collagen fibers and is continuous with the periosteum of the articulating bones; the inner layer—the synovial membrane—is composed of areolar connective tissue with abundant elastic fibers.

Another component of many synovial joints is **ligaments**, including extrinsic ligaments, which lie outside the joint capsule, and intrinsic ligaments, which are located within the joint capsule. Other accessory structures present within certain synovial joints include **menisci** (singular = meniscus; also called **articular discs**), which are fibrocartilage pads that act as shock absorbers and provide a better "fit" between articulating surfaces; **bursae** (singular = bursa), which are flattened sacs filled with synovial fluid that provide extra cushioning and reduce friction between structures that rub against each other during movement; **fat pads**, which are shock-absorbing adipose-filled structures that fill the space between the articular capsule and the synovial membrane; and **tendons**, which usually connect muscle to bone and stabilize joints.

Types of Synovial Joints Based on Shape of Articular Surfaces

Even though all synovial joints are diarthroses, the direction and range of motion allowed by specific joints vary. One way to describe the type of motion allowed by a synovial joint is in terms of the movement of bones with respect to one or more invisible axes. **Nonaxial** joints allow slipping or gliding movements only. As the name implies, there is no axis around which movement can occur. **Uniaxial** joints allow movements around only one axis, **biaxial** joints allow movement around two axes, and **multiaxial** joints allow movements around all three axes. The structure of a synovial joint's articular surface helps determine the type of movement permitted at that joint. Figure 11-4 illustrates, from least moveable to most moveable, the six types of synovial joints based on shape of articular surface.

Plane (gliding) joints consist of flat (or nearly flat) articular surfaces that slide over one another, allowing nonaxial movement; examples include the intercarpal joints, the intertarsal joints, and the sacroiliac joints. In **hinge joints**,

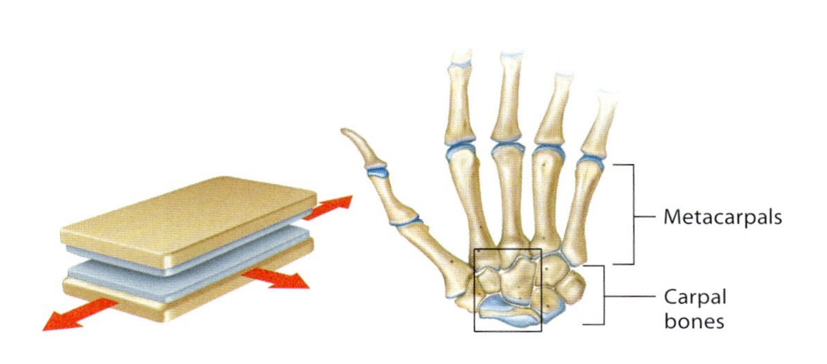

(a) Plane joint, nonaxial: intercarpal joint

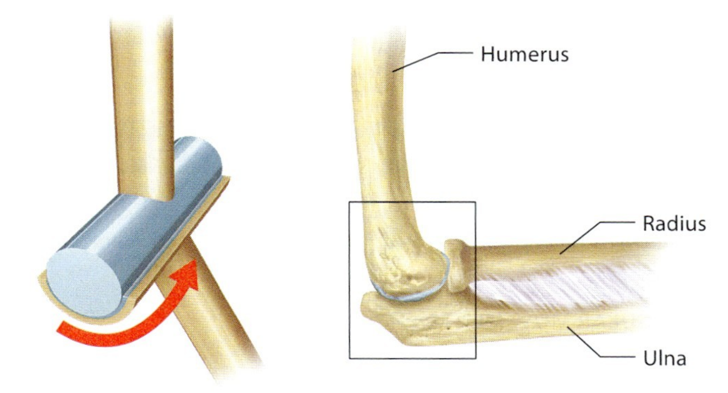

(b) Hinge joint, uniaxial: elbow joint

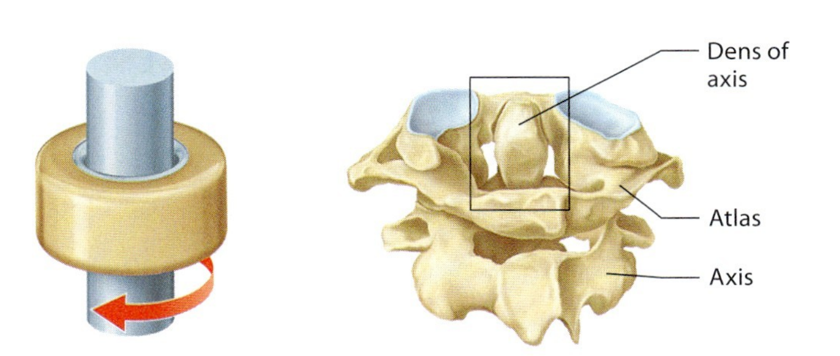

(c) Pivot joint, uniaxial: atlantoaxial joint

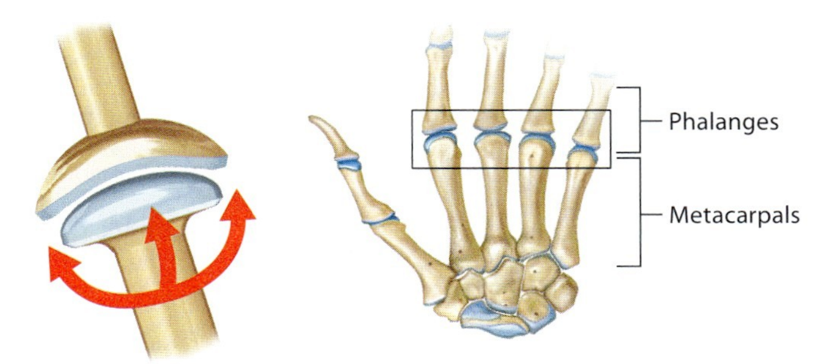

(d) Condylar joint, biaxial: metacarpophalangeal joint

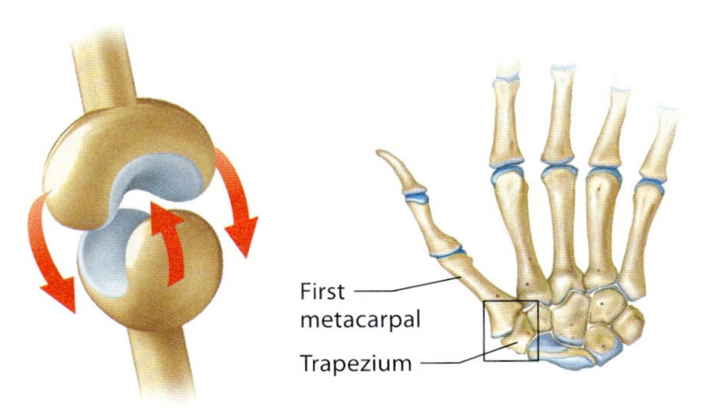

(e) Saddle joint, biaxial: carpometacarpal joint of thumb

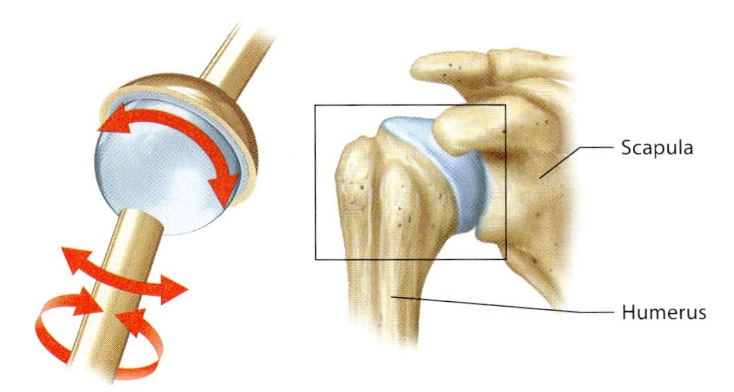

(f) Ball-and-socket joint, multiaxial: shoulder joint

Figure 11-4 The six types of synovial joints.

one concave articular surface and one convex articular surface articulate to allow movement around one axis (uniaxial). Examples of hinge joints include the elbow, the knee, the interphalangeal joints, and the temporomandibular joint. In **pivot joints**, the rounded or conical articular surface of one bone articulates with a shallow depression on another bone, allowing uniaxial rotation; examples include the atlantoaxial joint and the proximal radioulnar joint. **Condylar (ellipsoid) joints**, in which an oval-shaped condyle articulates with an elliptical cavity, allow biaxial movement; examples include the metacarpophalangeal joints (knuckles) and the atlanto-occipital joint. **Saddle joints**, in which two saddle-shaped articulating surfaces articulate with each other, allow biaxial movement; an example is the carpometacarpal joint between the metacarpal of the thumb and the trapezium of the wrist. In **ball-and-socket joints**, the ball-shaped head of one bone articulates with a cuplike depression in another bone, allowing multiaxial movement; the only examples in the human body are the shoulder (glenohumeral) joints and the hip (acetabulofemoral) joints.

Types of Movements Allowed by Synovial Joints

Movements of the body are possible because skeletal muscles are typically attached to bones at two sites and move the bones when they contract. The more stationary (or less moveable) site of muscle attachment is called the **origin**, and the more moveable site of muscle attachment is called the **insertion**. When skeletal muscles contract across a diarthrosis, each contracted muscle moves the insertion toward the origin, a result termed the **action**. The most common types of movements that synovial joints allow are described in **Table 11-1** and illustrated in **Figure 11-5**.

Table 11-1 Types of Movements Allowed by Synovial Joints

Type of Movement	Description of Movement
GLIDING MOVEMENT	Sliding of a flat (or nearly flat) bone surface over another bone in a back-and-forth or side-to-side direction
ANGULAR MOVEMENTS	
• Flexion (see Figure 11-5a)	Movement that decreases the angle between articulating bones
• Extension (see Figure 11-5a)	Movement that increases the angle between articulating bones
• Hyperextension (see Figure 11-5a)	Continuation of extension beyond anatomical position
• Abduction (see Figure 11-5b)	Movement of a bone away from the midline of the body
• Adduction (see Figure 11-5c)	Movement of a bone toward the midline of the body
• Circumduction (see Figure 11-5d)	Circular movement of the distal end of a body part
• Rotation (see Figure 11-5e)	The turning of a bone around its own longitudinal axis
SPECIAL MOVEMENTS	
• Supination (see Figure 11-5f)	Rotation of the forearm and hand laterally so that the palm faces anteriorly
• Pronation (see Figure 11-5g)	Rotation of the forearm and hand medially so that the palm faces posteriorly
• Dorsiflexion (see Figure 11-5h)	Lifting of the foot so that the toes are pulled up toward the head
• Plantarflexion (see Figure 11-5i)	Depression of the foot by pointing the toes toward the ground
• Inversion (see Figure 11-5j)	Movement of the sole of the foot medially
• Eversion (see Figure 11-5k)	Movement of the sole of the foot laterally
• Protraction	Anterior movement of a body part
• Retraction	Posterior movement of a body part
• Elevation	Superior movement of a body part
• Depression	Inferior movement of a body part

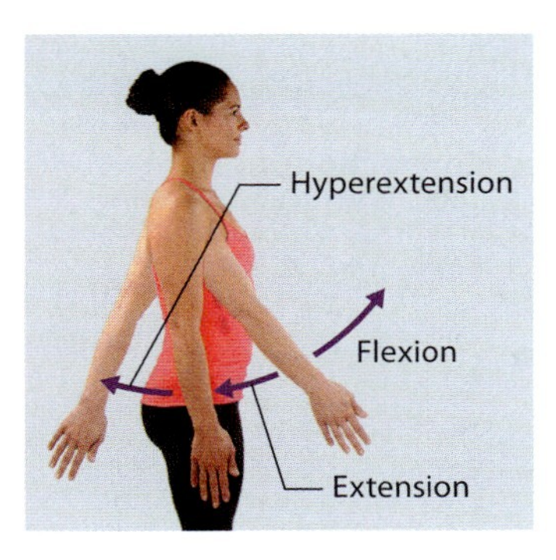

(a) **Shoulder flexion, extension, and hyperextension**

(b) **Abduction of shoulders and hips**

(c) **Adduction of shoulders and hips**

(d) **Circumduction of shoulder**

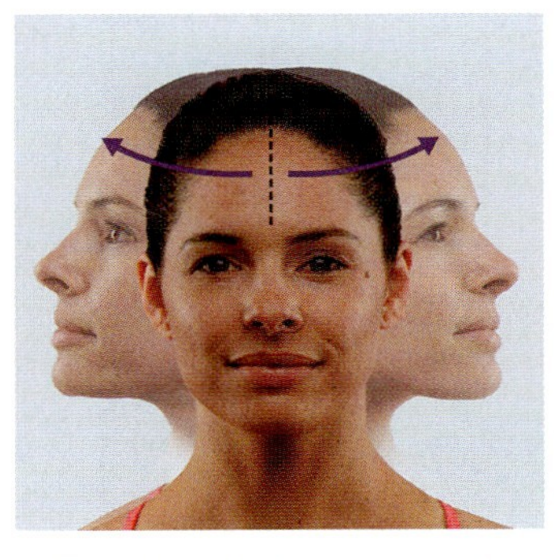

(e) **Rotation of atlantoaxial joint in neck**

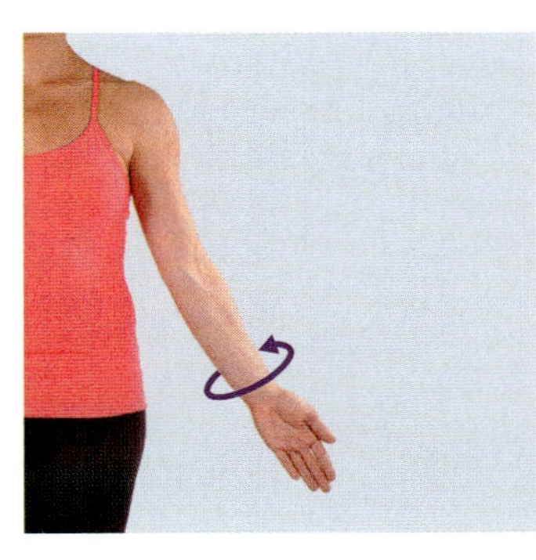

(f) **Supination of forearm**

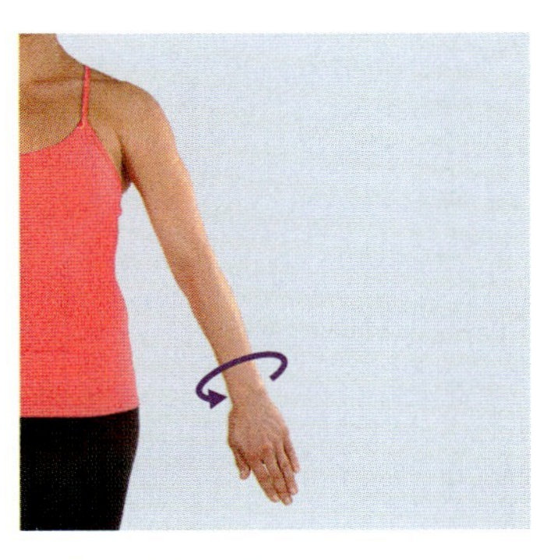

(g) **Pronation of forearm**

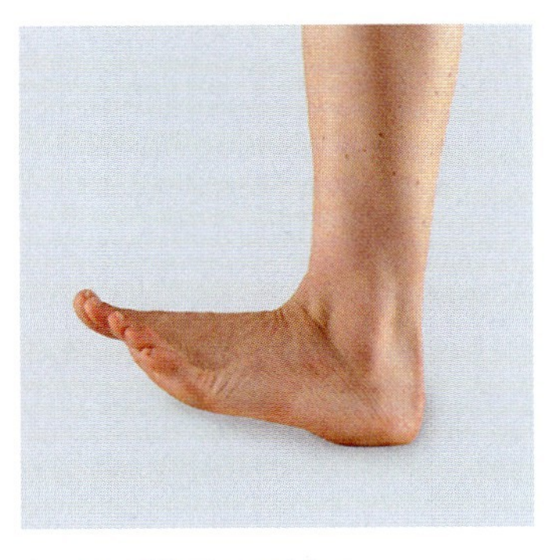

(h) **Dorsiflexion of foot**

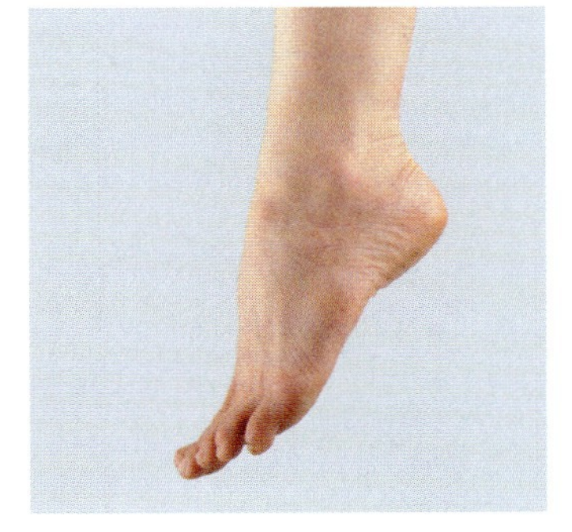

(i) **Plantarflexion of foot**

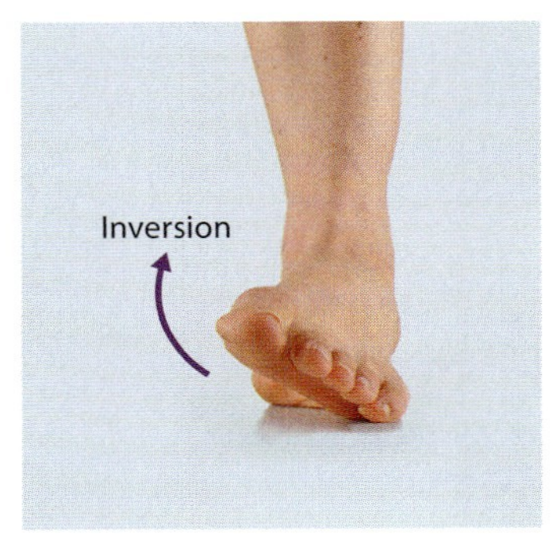

(j) **Inversion of foot**

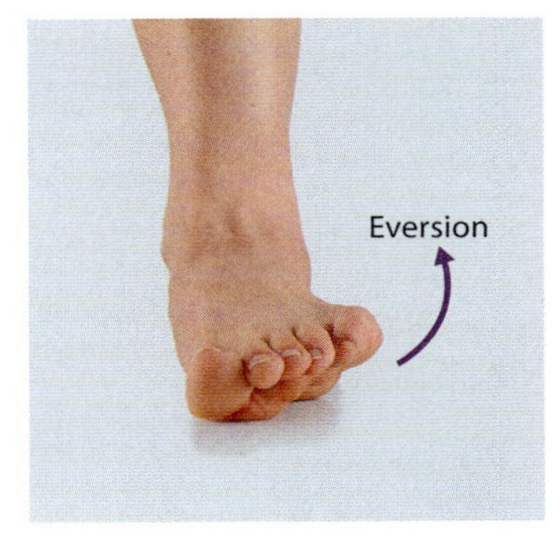

(k) **Eversion of foot**

Figure 11-5 Movements allowed by synovial joints.

ACTIVITY 1

Identifying and Classifying Joints

Learning Outcomes

1. Classify joints based on structure and function.
2. Identify joints on a skeleton and classify them according to structure and function.
3. Name the major structural features of a synovial joint, and describe the function of each.

Materials Needed

☐ Skeleton
☐ Synovial joint model or diagram

Instructions

1. CHART Locate on a skeleton each of the joints in the Making Connections chart on the next page and then fill in the chart. As you write the terms for the movements allowed by each joint, demonstrate the movements on the skeleton and your own body. Refer to Figures 11-4 and 11-5 as needed. Finally, write any "connections" to things you have already learned in the final column. After you have completed the other activities in this unit, you should return to this chart to add further details you have learned.

Making Connections: Joints

Joint	Articulating Bones	Classifications		Movements Allowed; Connections to Things I Have Already Learned
		Structural	Functional	
Talocrural (ankle)	*Tibia and talus*	*Synovial, hinge*	*Diarthrosis*	*Dorsiflexion and plantarflexion of the foot; talus = tarsal*
Atlantoaxial		*Synovial, pivot*	*Diarthrosis*	*Rotation of head (shaking head "no"); atlas = C_1; lacks a body, lacks a spinous process; axis = C_2, dens*
Atlanto-occipital				
Trapeziometacarpal				
Coxal (hip)				
Distal tibiofibular	*Tibia and fibula*	*Fibrous; syndesmosis*	*Synarthrosis*	*Lateral malleolus of fibula creates "bump" on outside of "ankle"*
Proximal tibiofibular	*Tibia and fibula*	*Synovial*		
Tibiofemoral (knee)				
Elbow				
Glenohumeral (shoulder)				
Dentoalveolar				
Sagittal suture (skull)				
Intercarpal	*Between two carpals*	*Synovial, plane*	*Diarthrosis*	
Intervertebral				
Metacarpophalangeal				
Pubic symphysis				
Sternocostal (first)				
Temporomandibular				

2. Locate each of the following features of synovial joints on a model or diagram, and then briefly describe (in your own words) the feature's structure and function in the blanks provided.
   a. Articular (joint) capsule ______________________
   ____
   b. Outer fibrous capsule ______________________
   ____
   c. Inner synovial membrane ______________________
   ____
   d. Synovial cavity ______________________
   ____
   e. Synovial fluid ______________________
   ____
   f. Articular cartilage ______________________
   ____
   g. Ligaments ______________________
   ____

ACTIVITY 2
Demonstrating Movements Allowed by Joints

Learning Outcomes

1. Demonstrate movements allowed by synovial joints.

Materials Needed

☐ Muscle models or diagrams

Instructions

CHART Locate each of the skeletal muscles in the following Making Connections chart on a muscle model or diagram. Then locate that muscle on your own body, demonstrate its major action, and describe that action as you perform it. Next, write descriptions of the muscle's action in the chart. Finally, write under "Connections to Things I Have Already Learned" any helpful information that you have previously learned about specific muscles. The first muscle has been completed as an example.

Optional Activity

PAL™ Play animations of joint movements and actions at > MasteringA&P® > Study Area > Practice Anatomy Lab > Animations

Making Connections: Movements Allowed by Joints

Skeletal Muscle	Major Action of the Muscle	Description of the Action	Connections to Things I Have Already Learned
Biceps brachii	*Flexes the forearm*	*Decreases the angle of the elbow joint*	*Located in brachial region; triceps brachii extends the forearm*
Deltoid		*Moves the arm superiorly and away from the midline of the body*	
Gastrocnemius			
Gluteus maximus			
Rectus femoris	*Flexes the hip, extends the knee*		

Selected Synovial Joints

The Hip

The hip, or coxal joint (**Figure 11-6**), is a multiaxial ball-and-socket joint that allows movements around all possible axes. The rounded head of the femur fits into the deeply cupped **acetabulum** of the pelvic bone (see Figure 11-6c). The depth of the acetabulum is increased by the presence of a fibrocartilaginous ring called the **acetabular labrum** (see Figures 11-6a and c).

The exceptional stability of the coxal joint is a result of both this deep socket and the presence of several strong, reinforcing ligaments. The **ligament of the head of the femur** (see Figure 11-6c) is a short ligament that runs from the pitlike fovea capitis on the head of the femur to the acetabulum; it secures the femur to the pelvic bone. The Y-shaped **iliofemoral ligament** (see Figure 11-6b) is an anterior

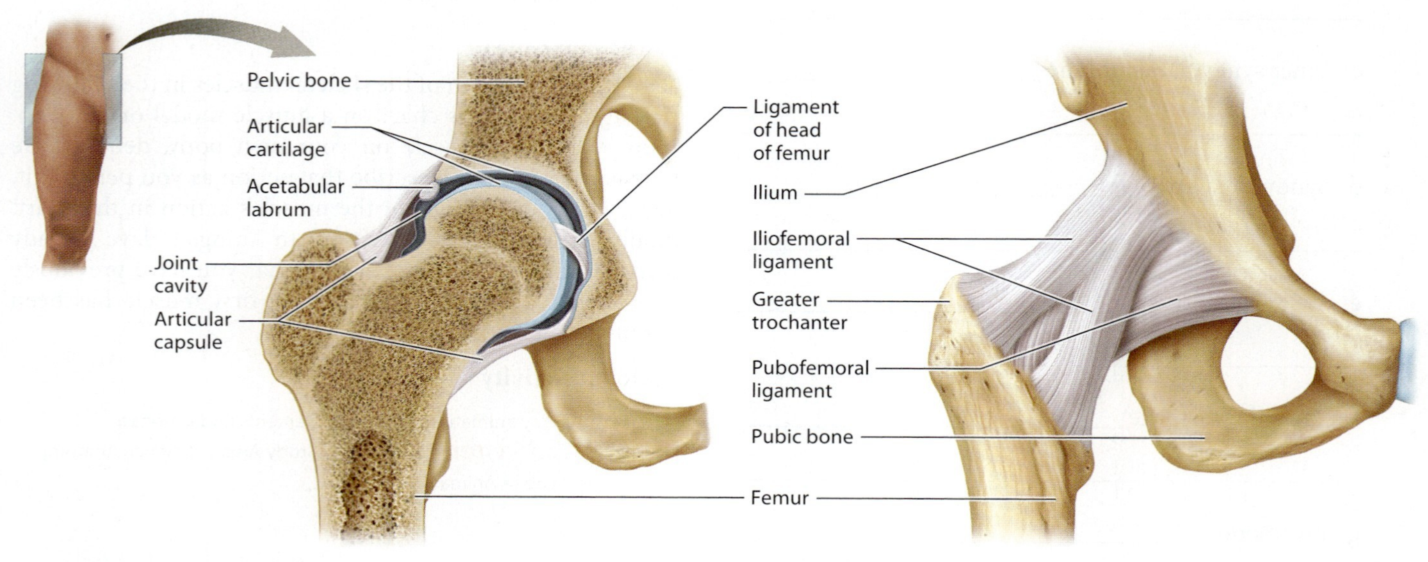

(a) Frontal section, anterior view

(b) Anterior view

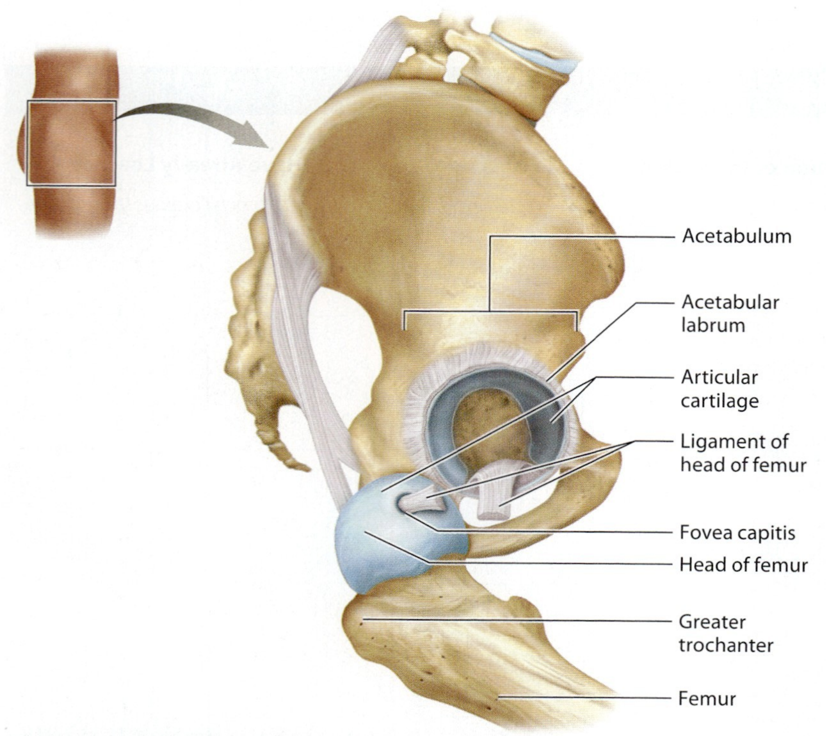

(c) Lateral view with head of femur removed from acetabulum

Figure 11-6 Three views of the right hip (coxal joint).

ligament that attaches proximally to the ilium at the anterior inferior iliac spine and the rim of the acetabulum and extends distally to attach to the greater trochanter of the femur. The **pubofemoral ligament** (see Figure 11-6b) is another anterior ligament that attaches proximally to the pubic portion of the acetabular rim and extends distally, where it blends in with the iliofemoral ligament. Finally, the **ischiofemoral ligament** (not shown) is a posterior ligament that attaches proximally to the ischial portion of the acetabular rim and extends to the neck of the femur, medial to the base of the greater trochanter.

The Shoulder

The shoulder (or glenohumeral) joint is a ball-and-socket joint (**Figure 11-7**) and is the most freely moveable joint in the body. The shoulder joint is formed by the articulation between the glenoid cavity of the scapula and the rounded articular surface of the head of the humerus (see Figure 11-7c). The glenoid cavity is deepened by the presence of a rim of fibrocartilage called the **glenoid labrum**. The shoulder joint is stabilized predominantly by the rotator cuff, composed of the scapular muscles and their tendons (see Figure 11-7b).

Several ligaments (not shown in Figure 11-7) also reinforce the shoulder. The **coracohumeral ligament** attaches the coracoid process of the scapula to the anatomical neck of the humerus. The **glenohumeral ligaments** extend from the superior margin of the glenoid cavity to the lesser tubercle and the anatomical neck of the humerus. The **coracoacromial ligament** connects the coracoid process and the acromion of the scapula, and the **transverse humeral ligament** forms a tunnel that covers the intertubercular groove.

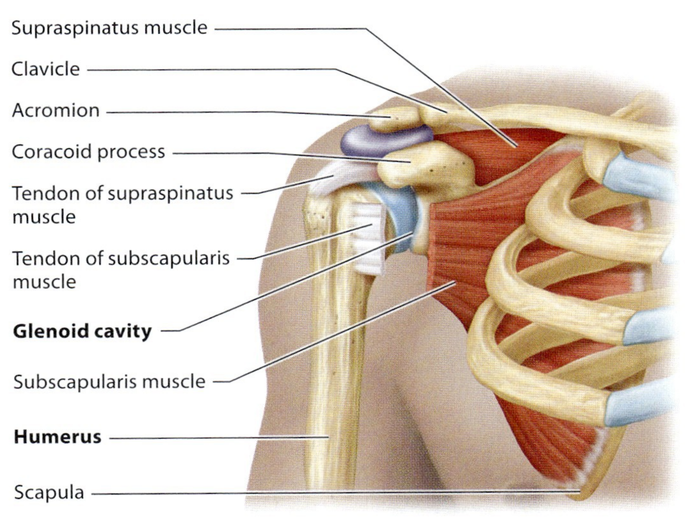

(a) Anterior view

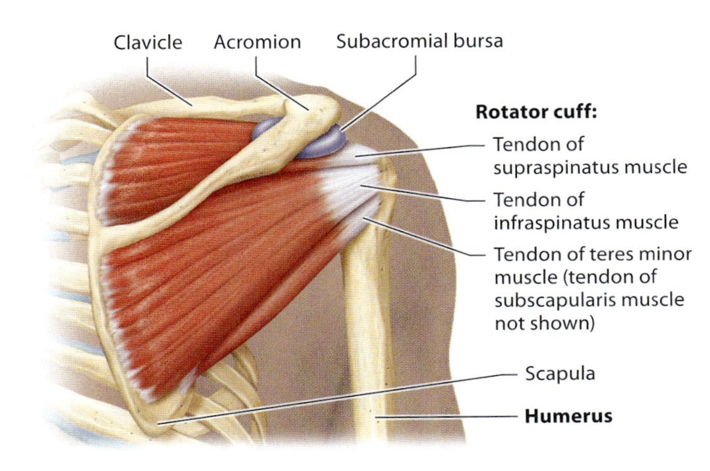

(b) Posterior view

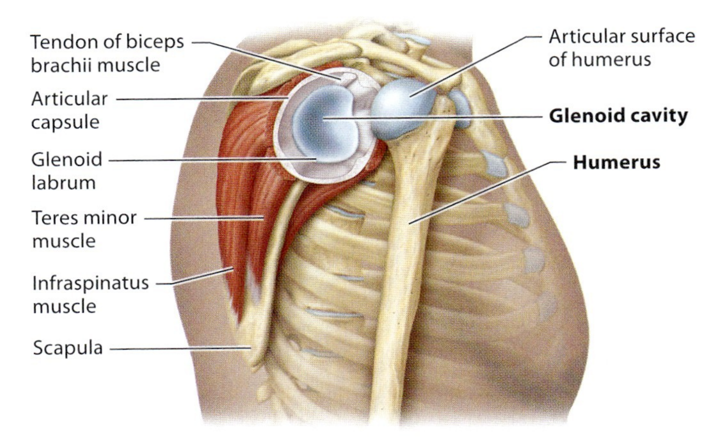

(c) Lateral view with head of humerus removed from glenoid cavity

Figure 11-7 Three views of the right shoulder.

The Knee

The knee is the most complex joint in the body, and as a result it is more likely to be injured than any other joint. The knee (**Figure 11-8**) is composed of articulations involving four bones: the femur, tibia, fibula, and patella. The knee joint is classified as a modified hinge joint, and its main movements are flexion and extension, but limited medial/lateral rotation and limited lateral gliding can also occur.

The knee actually consists of two separate articulations: the **tibiofemoral joint** between the tibia and the femur, and the **patellofemoral joint** between the patella and the femur (see Figure 11-8c). The joint capsule of the tibiofemoral joint is unique in that it encloses only the lateral and posterior aspects of the knee joint. The knee is stabilized anteriorly by the tendon of the anterior thigh muscles, collectively known as the quadriceps femoris, which attaches the muscle to the patella. Note that the patellar ligament is a continuation of the tendon of the quadriceps femoris, and that it attaches the patella to the tibial tuberosity.

Several ligaments also stabilize the knee joint (see Figure 11-8a and b). The **fibular (lateral) collateral ligament** is an extrinsic ligament that connects the femur to

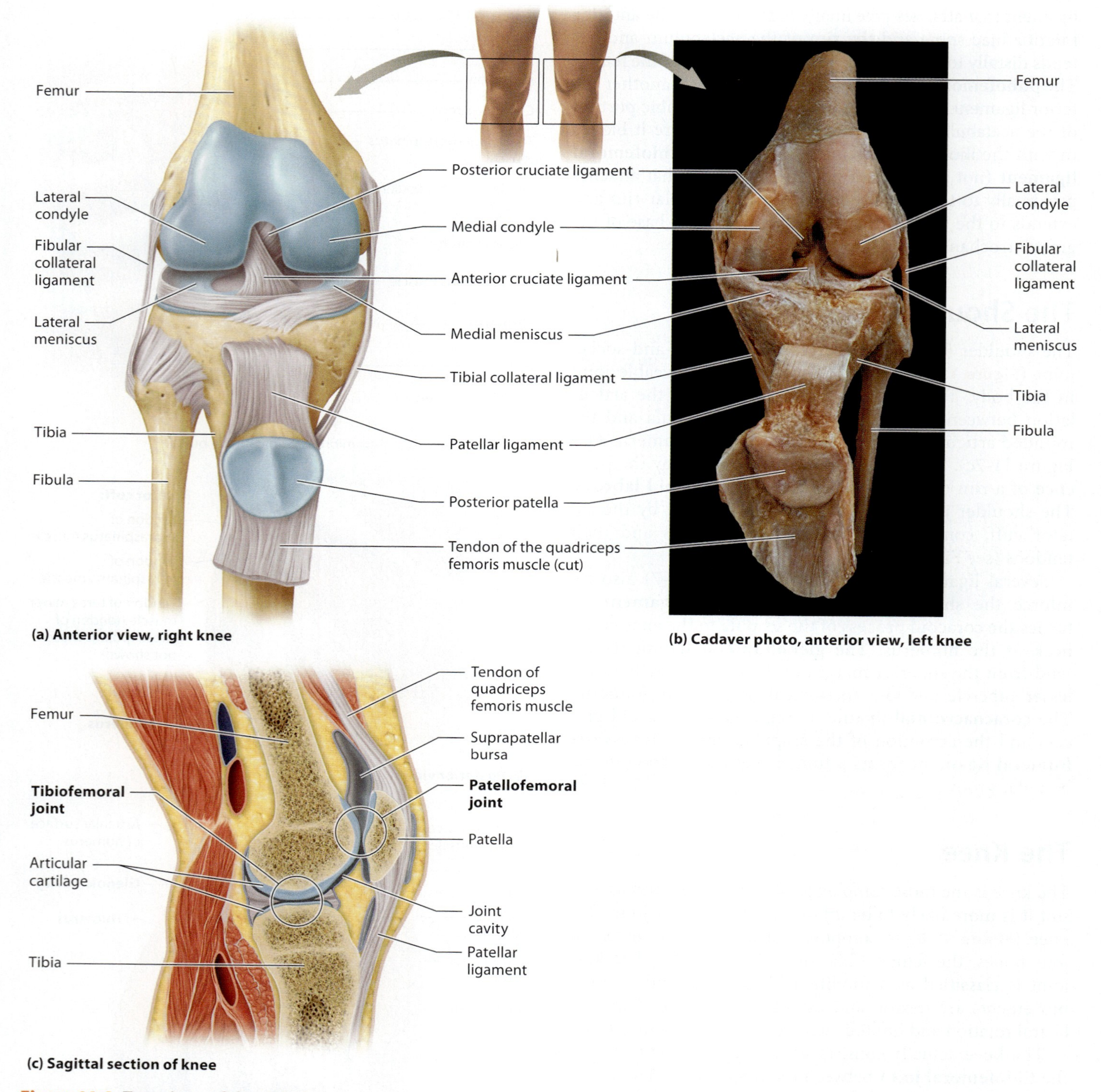

Figure 11-8 **Two views of the right knee.**

the fibula, whereas the **tibial (medial) collateral ligament** is an extrinsic ligament that connects the femur to the tibia. The **anterior cruciate ligament**, an intrinsic ligament that attaches to the anterior intercondylar area of the tibia and passes posteriorly to attach to the femur, prevents anterior displacement of the tibia. The **posterior cruciate ligament**, an intrinsic ligament that attaches to the posterior intercondylar area of the tibia and passes anteriorly to attach to the femur, prevents posterior displacement of the tibia. Finally, two **popliteal ligaments** (not shown) extend from the head of the femur to the fibula and tibia and support the posterior portion of the knee.

Also associated with the knee are the **lateral meniscus** and **medial meniscus** (see Figure 11-8a and b), which are C-shaped fibrocartilage pads on the articular surface of the tibia; they receive the condyles of the femur, increase the stability of the joint, and function as shock absorbers. The knee also contains several **bursae**—synovial-fluid–filled sacs that cushion the joint—including the suprapatellar bursa (see Figure 11-8c).

The Elbow

The elbow joint (**Figure 11-9**) is a hinge joint composed of two articulations: the **humeroradial joint** and the **humeroulnar joint**. The combination of the structure of the articulating bones and the presence of strong capsular ligaments (the **radial collateral** and **ulnar collateral ligaments**) results in a stable joint that allows substantial freedom of movement during elbow flexion and extension.

The proximal radioulnar articulation—in which the head of the radius articulates with the radial notch of the ulna to form a pivot joint (not shown)—is not a part of the elbow joint proper, but it is associated with the elbow complex. The **anular ligament** attaches to the radial notch and wraps around the head of the radius, holding the radial head in its correct position. The radial head rotates within the anular ligament during pronation and supination of the forearm.

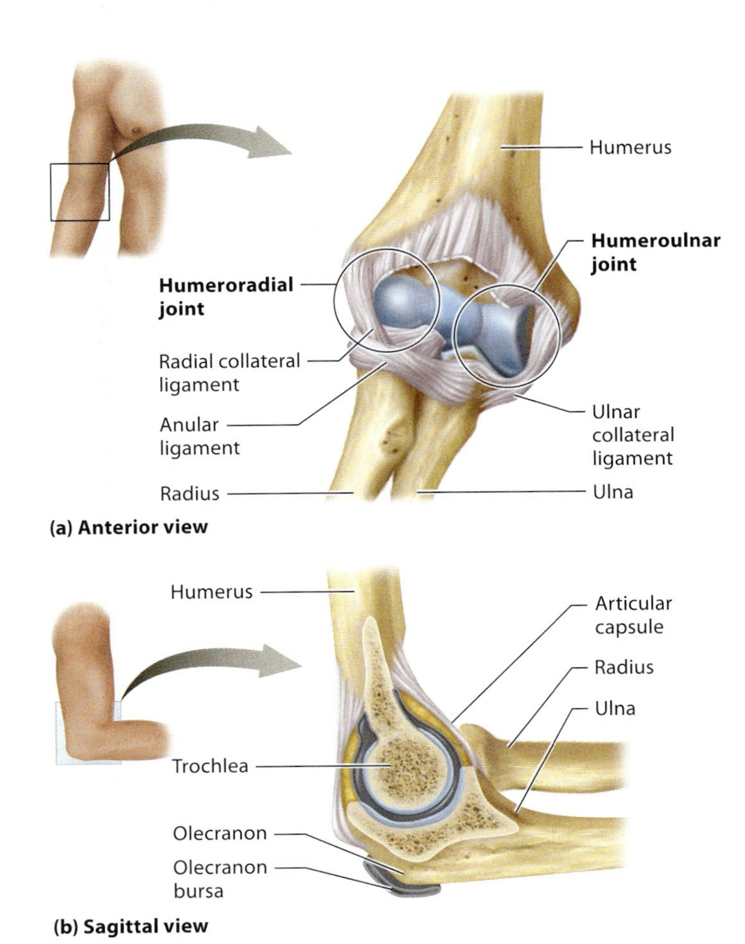

Figure 11-9 **Two views of the right elbow.**

ACTIVITY 3

Comparing and Contrasting the Structure and Function of Selected Synovial Joints

Learning Outcome

1. Compare and contrast two synovial joints with respect to both structure and function.

Materials Needed

- ☐ Hip joint model or diagram and disarticulated bones (pelvic bone and femur)
- ☐ Shoulder joint model or diagram and disarticulated bones (scapula and humerus)
- ☐ Knee joint model or diagram and disarticulated bones (femur, tibia, fibula, and patella)
- ☐ Elbow joint model or diagram and disarticulated bones (humerus, radius, and ulna)

Instructions

CHART Your instructor will assign two of the joint models or diagrams for your group to study. For the joints assigned to your group, locate each of the structures in the following charts on the model or diagram and on the disarticulated bones. Then, write each structure's description and location in the chart. The first structure for each joint has been completed as examples.

A. Hip Joint

Structure	Description and Location
Acetabulum	*Cup-like depression in pelvic bone that receives head of femur*
Femoral head	
Acetabular labrum	
Fovea capitis	
Ligament of the head of the femur	
Iliofemoral ligament	
Pubofemoral ligament	
Ischiofemoral ligament	

B. Shoulder Joint

Structure	Description and Location
Glenoid cavity	*Shallow depression on superomedial scapula that receives head of humerus*
Humeral head	
Glenoid labrum	
Coracohumeral ligament	
Glenohumeral ligament	
Coracoacromial ligament	
Transverse humeral ligament	

C. Knee Joint

Structure	Description and Location
Lateral condyles of femur	*Articular surface of distal, lateral epiphysis of femur; curved shape fits into lateral meniscus of tibia*
Medial condyles of femur	
Patellar surface	
Tibial tuberosity	
Fibular collateral ligament	
Tibial collateral ligament	
Anterior cruciate ligament	
Posterior cruciate ligament	
Lateral and medial menisci	

D. Elbow Joint

Structure	Description and Location
Olecranon process	*Projection on proximal end of ulna that articulates with olecranon fossa of humerus*
Olecranon fossa	
Coronoid process	
Coronoid fossa	
Trochlea	
Trochlear notch	
Radial head	
Radial (lateral) collateral ligament	
Ulnar (medial) collateral ligament	
Anular ligament	

Post-lab quizzes are also assignable in MasteringA&P®

POST-LAB ASSIGNMENTS

Name: ______________________________ Date: ______________ Lab Section: ________

PART I. Check Your Understanding

Activity 1: Identifying and Classifying Joints

1. Match each of the following joints with its structural classification(s).

_____	a. atlantoaxial joint	1. suture
_____	b. intervertebral joint	2. pivot joint
_____	c. glenohumeral joint	3. ball-and-socket joint
_____	d. interphalangeal joint	4. saddle joint
_____	e. dentoalveolar joint	5. syndesmosis
_____	f. sacroiliac joint	6. hinge joint
_____	g. frontonasal joint	7. synchondrosis
_____	h. intertarsal joint	8. gomphosis
_____	i. metacarpophalangeal joint	9. symphysis
		10. condyloid joint
		11. plane joint

2. Which of the following statements is true?
   a. An articular capsule covers the articular surface of each bone in a synovial joint and reduces friction.
   b. A labrum is the innermost layer of an articular capsule.
   c. The innermost layer of the articular capsule secretes synovial fluid.
   d. Articular capsules are only found in synchondroses and symphyses.
   e. An articular capsule is composed primarily of hyaline cartilage.

Activity 2: Demonstrating Movements Allowed by Joints

1. Fill in the blank with the appropriate term:

_______________________ a. Movement of a body part away from the midline

_______________________ b. Movement of the sole of the foot laterally

_______________________ c. Movement of a bone around its own longitudinal axis

_______________________ d. Movement of a bone toward the midline of the body

_______________________ e. A decrease in the angle between articulating bones

_______________________ f. Continuation of extension beyond anatomical position

_______________________ g. Rotation of the forearm and hand medially so that the palm faces posteriorly

_______________________ h. Movement of the foot so that the toes are pulled up toward the head

_______________________ i. Anterior movement of a body part

_______________________ j. Movement of the distal end of a body part in a circle

Activity 3: Comparing and Contrasting the Structure and Function of Selected Synovial Joints

1. Identify the structures of the hip joint.

 a. ______________________________

 b. ______________________________

 c. ______________________________

 d. ______________________________

 e. ______________________________

 f. ______________________________

a

b

c

d

e

f

2. Which of the following joints has the most bones?
 a. hip
 b. knee
 c. shoulder
 d. elbow

3. Based on the structure of its joint surfaces, which of the following synovial joints has the greatest mobility?
 a. knee (femorotibial) joint
 b. elbow (humeroulnar) joint
 c. hip (femoroacetabular) joint
 d. shoulder (glenohumeral) joint

PART II. Putting It All Together

A. Review Questions

Answer the following questions using your lecture notes, your textbook, and your lab notes.

1. What are the two general ways in which joints can be classified? ______________________________

____

2. Why is the coxal joint more stable than the shoulder joint? ______________________________

____

3. For each of the following exercises, name two joints involved and one type of movement that occurs at each joint:
   a. Chin-ups: ______________________________
   ____
   b. Squats: ______________________________
   ____

4. Carol is a 22-year-old professional downhill skier who recently had knee surgery to repair a torn ligament; the injury was allowing the tibia to move anteriorly. Which ligament was repaired during surgery?
   a. posterior cruciate
   b. fibular collateral
   c. anterior cruciate
   d. tibial collateral

B. Concept Mapping

1. Fill in the blanks to complete this concept map outlining the relationship of articular surface shape to the action that occurs at a joint.

flexion hinge hyperextension plane saddle

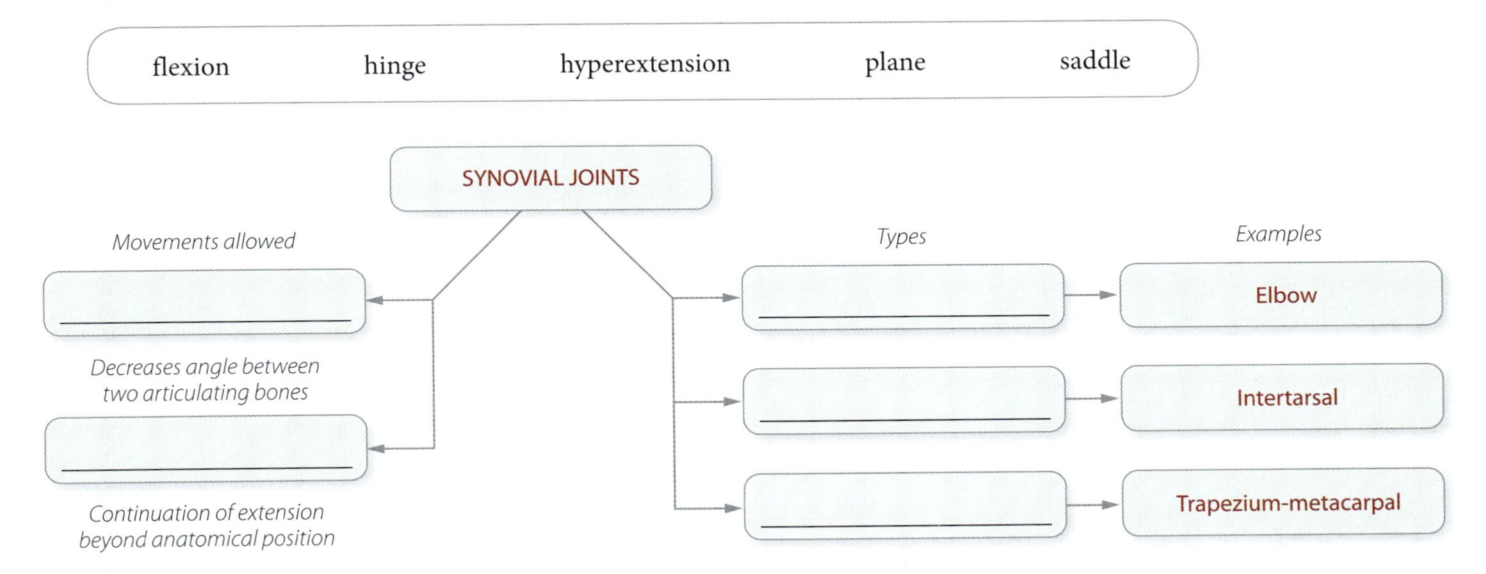

2. Construct a unit concept map to show the relationships among the following set of terms. Include all of the terms in your diagram. Your instructor may choose to assign additional terms.

acetabular labrum anterior cruciate ligament articular cartilage
extension flexion fovea capitis hinge hyperextension
ligamentum teres meniscus periosteum plane
saddle synovial fluid tibial collateral ligament

12

Introduction to the Muscular System: Muscle Tissue

UNIT OUTLINE

In Unit 6 we examined the three different types of muscle tissue: cardiac, smooth, and skeletal. Cardiac muscle tissue is confined to the heart, where its contractions pump blood. Smooth muscle is found in hollow organs, where it propels substances such as food or urine along internal passageways. Skeletal muscle tissue, by contrast, is the major tissue type in the skeletal muscles that make up the muscular system. Skeletal muscle functions in movement, maintenance of posture, stabilization of joints, and generation of heat. In this unit you will be studying the anatomy of a skeletal muscle and the basic physiology of skeletal muscle contraction. In Unit 13 we will first consider how muscles are named and then will survey the major muscles of the body.

THINK ABOUT IT *Based on what you have learned about the three types of muscle tissue (Unit 6), make a sketch of smooth muscle, cardiac muscle, and skeletal muscle, label the structures listed for each tissue type, and describe the shape of the cells in each tissue:*

smooth muscle (nucleus)

cardiac muscle (nucleus, striations, intercalated disc)

skeletal muscle (nuclei, striations)

PRE-LAB ASSIGNMENTS

Pre-lab quizzes are also assignable in MasteringA&P®

To maximize learning, BEFORE your lab period carefully read this entire lab unit and complete these pre-lab assignments using your textbook, lecture notes, and prior knowledge.

PRE-LAB Activity 1: Identifying the Structural Components of a Skeletal Muscle

1. Use the list of terms provided to label the accompanying illustration of a skeletal muscle; check off each term as you label it.

- ☐ endomysium
- ☐ epimysium
- ☐ fascicle
- ☐ perimysium
- ☐ muscle
- ☐ muscle fiber

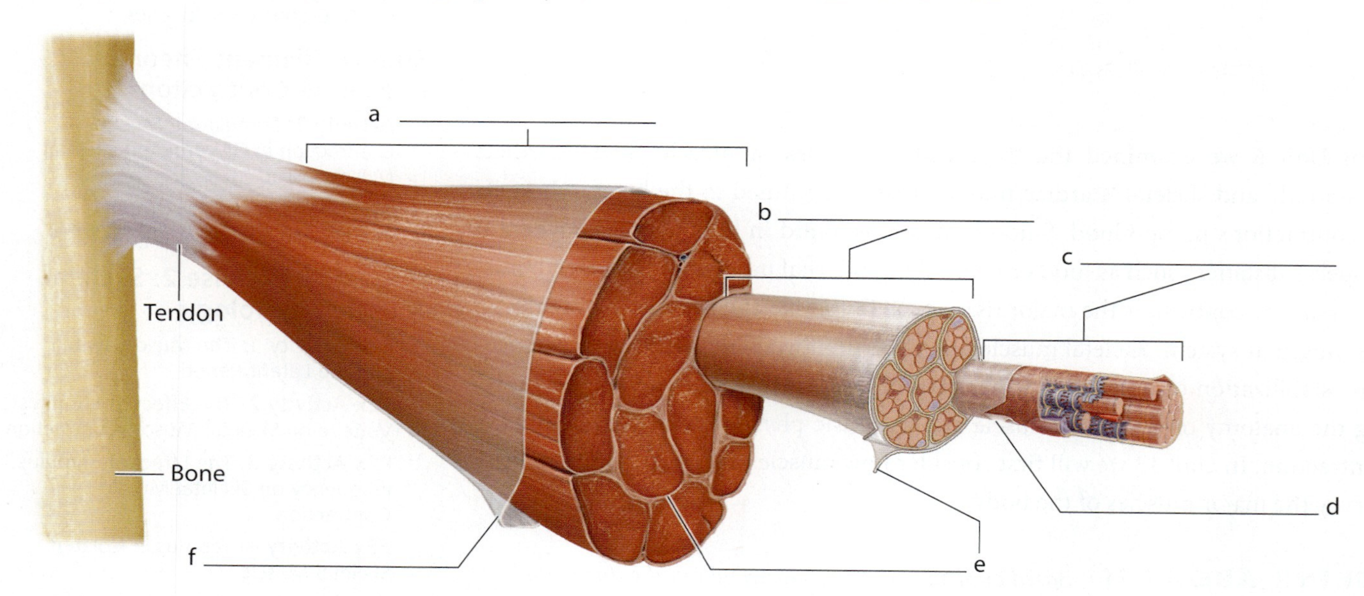

2. Use the list of terms provided to label the accompanying illustration of a skeletal muscle fiber; check off each term as you label it.

- ☐ mitochondrion
- ☐ myofibrils
- ☐ nucleus
- ☐ sarcolemma
- ☐ sarcoplasmic reticulum
- ☐ T-tubule
- ☐ terminal cisternae
- ☐ triad

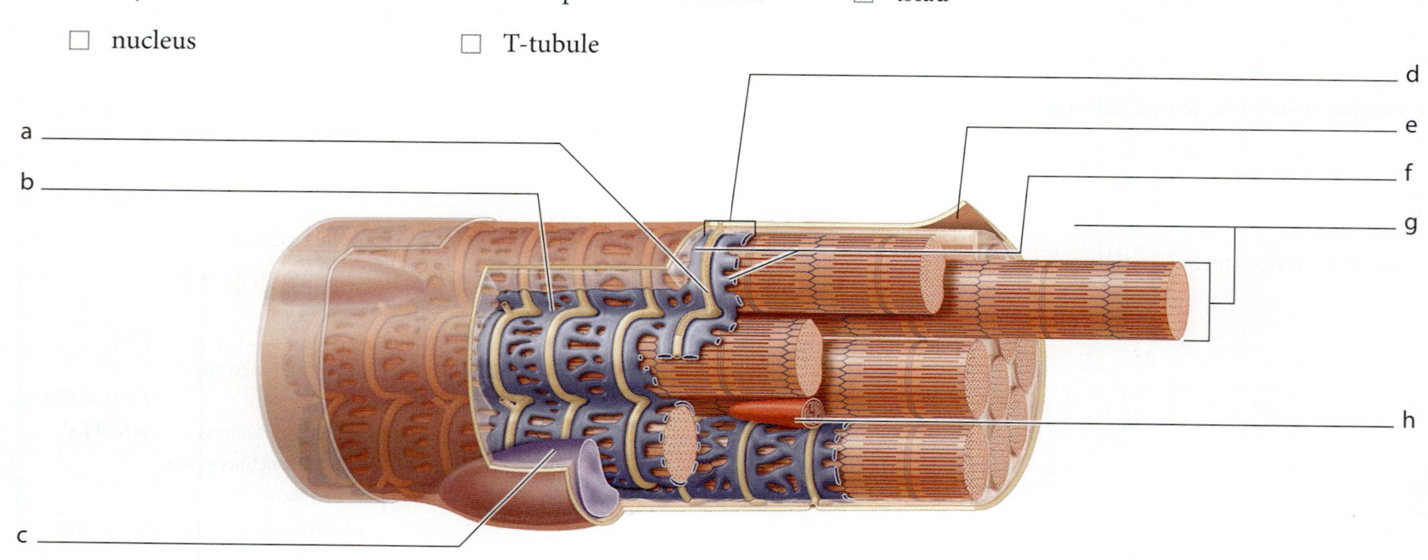

PRE-LAB Activity 2: Examining the Microscopic Anatomy of Skeletal Muscle Tissue and the Neuromuscular Junction

1. Which of the following descriptions characterizes skeletal muscle tissue?
   a. It contains striations.
   b. It contains uninucleate cells.
   c. It contains spindle-shaped cells.
   d. It contains intercalated discs.
   e. It contains branched cells.

2. A muscle action potential travels deep into the interior of the cell via:
   a. myofibrils.
   b. the sarcoplasm.
   c. T-tubules.
   d. terminal cisternae.
   e. motor axons.

3. A neuromuscular junction is a junction between a ______________________________ and a ______________________________.

4. Acetylcholine is a ______________________________ released from the axon terminal of a motor neuron.

PRE-LAB Activity 3: Stimulating Muscle Contraction in Glycerinated Skeletal Muscle Tissue

1. The functional unit of muscle contraction is the:
   a. sarcomere.
   b. myofibril.
   c. myofilament.
   d. muscle fiber.
   e. T-tubule.

2. Which ion is responsible for exposing the active sites on actin?
   a. Sodium
   b. Potassium
   c. Calcium
   d. Hydrogen
   e. Chloride

3. Exposing skeletal muscle fibers to __________________ removes ions and ATP from the cells and disrupts the troponin–tropomyosin complex such that it can no longer block the active sites on the actin molecules.

4. The boundaries of a sarcomere are formed by ____.

5. When muscle contraction occurs, what happens to the thick filaments and the thin filaments?

   ____

PRE-LAB Activity 4: Electromyography in a Human Subject Using BIOPAC®

1. How many electrodes are attached to the forearm to record the EMG?
   a. 2
   b. 3
   c. 4
   d. 5

2. The recording of muscle activity obtained by using electrodes on the skin is called
   a. electroencephalagraphy.
   b. an electrocardiogram.
   c. electromyography.
   d. None of these is correct.

Skeletal Muscle Anatomy

A **skeletal muscle** is an organ (which, by definition, is composed of two or more tissue types that work together to perform a common function). Skeletal muscles are mostly skeletal muscle tissue, but they also contain many connective tissue wrappings that separate and electrically insulate portions of the muscle so that electrical impulses that stimulate muscle contraction in one portion do not spread to neighboring portions.

Figure 12-1 depicts a skeletal muscle with its connective tissue wrappings. The outermost connective tissue layer, which surrounds the entire muscle, is the **epimysium**. Groups of muscles fibers are surrounded by a **perimysium**, producing bundles called fascicles. Both the epimysium and the perimysium are composed of dense irregular connective tissue. Individual muscle fibers (individual muscle cells) are surrounded by **endomysium**, which is composed of areolar connective tissue. Collagen fibers within all three connective tissue wrappings fuse to form strong cordlike **tendons** or sheet-like **aponeuroses**, which attach muscles to each other or to bones.

The Skeletal Muscle Fiber

Each muscle fiber (**Figure 12-2**) within its sheath of endomysium is an individual multinucleate cell surrounded by a plasma membrane called a **sarcolemma**. The cytoplasm of the muscle fiber, called sarcoplasm, contains abundant stores of glycogen (a storage form of carbohydrate), as well as a red pigment called myoglobin, which binds oxygen molecules that diffuse into the muscle fiber from the interstitial fluid. The sarcolemma of a muscle fiber invaginates in many places to form transverse tubules (**T-tubules**) that extend deep into the interior of the cell. Each muscle fiber is packed with cylindrical organelles called **myofibrils**, which extend the entire length of the fiber. Myofibrils contain thin filaments (composed of the contractile protein actin and other proteins) and thick filaments (composed of the contractile protein myosin). Myofibrils are surrounded by fluid-filled membranous sacs called the **sarcoplasmic reticulum (SR)**, which store calcium ions. Expanded end sacs of the SR called **terminal cisternae** flank a T-tubule on either side. A T-tubule with its two adjacent terminal cisternae is called a **triad**.

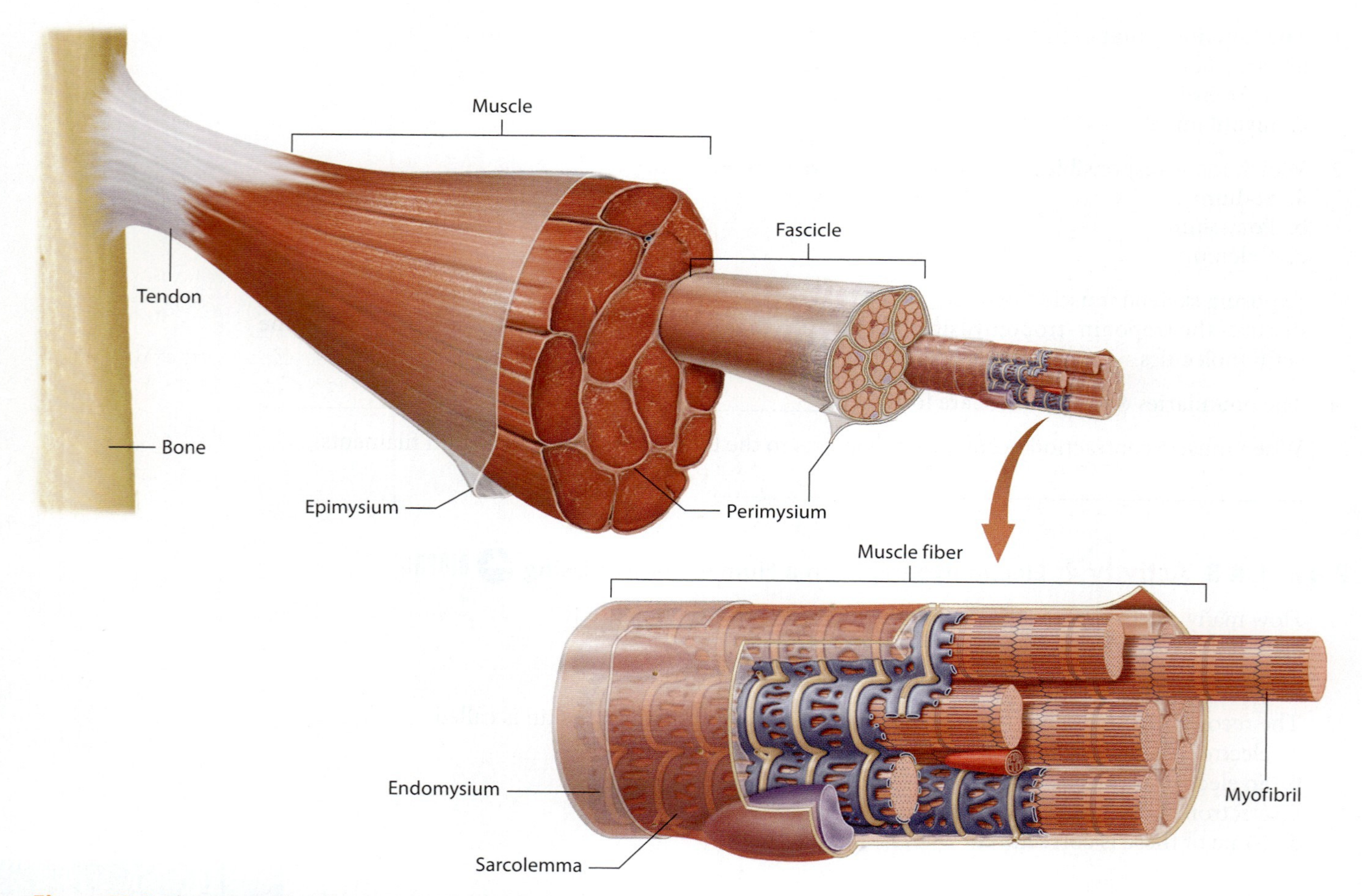

Figure 12-1 The connective tissue wrappings of a skeletal muscle.

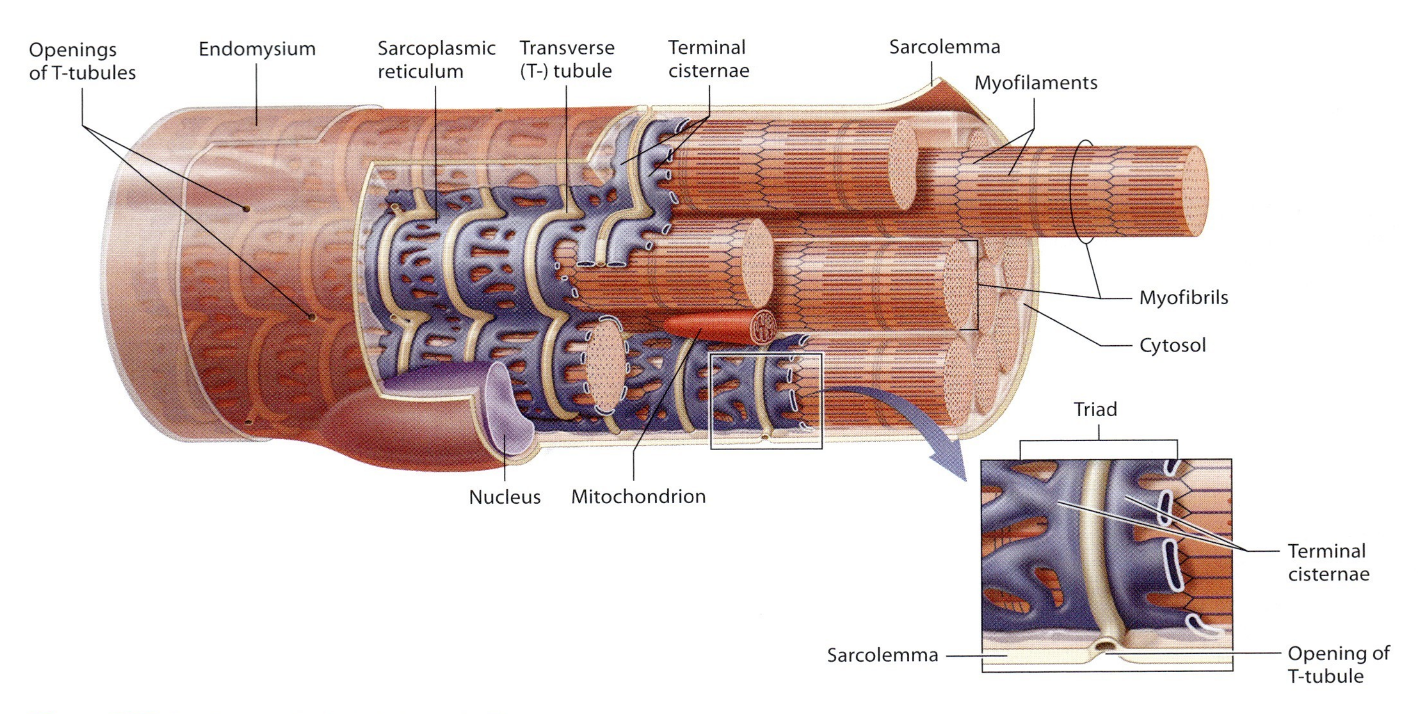

Figure 12-2 **Anatomy of a skeletal muscle fiber.**

CLINICAL CONNECTION

Muscle Growth and Repair

Skeletal muscle fibers are multinucleate because during embryonic development 100 or more nucleated stem cells, called myoblasts, fuse to produce each muscle fiber. Once fusion has occurred, a muscle fiber loses the ability to undergo mitosis, so the total number of muscle fibers an individual has is determined before birth. Throughout life some myoblasts remain between the muscle fibers and the endomysium. When a muscle is injured, these myoblasts can undergo mitosis to a certain extent and produce functional muscle fibers. However, the number of new skeletal muscle fibers that can be produced cannot compensate for significant muscle damage. In that case, instead of regeneration of functional muscle tissue, most muscle repair takes the form of fibrosis, the formation of scar tissue. The dramatic muscle growth that occurs during childhood and adolescence is mainly due to hypertrophy (enlargement) of existing muscle fibers. Hypertrophy results from increased production of myofibrils and other organelles within muscle fibers. Forceful, repetitive muscular activity such as strength training stimulates hypertrophy of muscle fibers in adults.

ACTIVITY 1

Identifying the Structural Components of a Skeletal Muscle

Learning Outcomes

1. Diagram, label, and explain the functions of the connective tissue wrappings of a skeletal muscle.
2. Identify the structural features of a skeletal muscle fiber on an anatomical model or chart and describe the function of each.

Materials Needed

☐ Skeletal muscle fiber model and/or anatomical charts

Instructions

1. Sketch a cross section of a muscle and label the following structures: *epimysium, perimysium, endomysium, fascicle, muscle fiber.*

2. CHART For each of the structural components of a skeletal muscle fiber listed in the Making Connections chart, locate the structure (on an anatomical model or an anatomical chart), write its function in the chart, and then make "connections" to what you have already learned in your lectures, your assigned reading, and your textbook.

Making Connections: Structural Components of a Skeletal Muscle Fiber

Muscle Fiber Structure	Function	Connections to Things I Have Already Learned
Sarcolemma		*Another name for the plasma membrane is the plasmalemma; "sarco" means flesh and refers to skeletal muscle.*
Sarcomere		
Sarcoplasm	*The cytoplasm within a skeletal muscle fiber*	
Sarcoplasmic reticulum		*Modified endoplasmic reticulum (synthesis and transport of proteins)*
Terminal cisternae		
T-tubule	*Invagination of the plasma membrane that carries muscle action potential deep into the muscle fiber's interior*	
Mitochondrion		
Nucleus		*Hundreds of embryonic cells fuse to produce muscle fibers; therefore, muscle fibers are multinucleate.*
Myofibril	*Specialized organelle in skeletal muscle fibers that contains thick filaments and thin filaments*	
Triad		
Thin filament		*Slides past the thick filament during muscle contraction.*
Thick filament		

The Neuromuscular Junction

Skeletal muscles are innervated by somatic motor neurons. Each somatic motor neuron has a cell body located in the brain or spinal cord, plus an axon that extends to a group of skeletal muscle fibers. The region at which the bulbous ending of an axon (the axon terminal) abuts a skeletal muscle fiber is the **neuromuscular junction** (**Figure 12-3**).

The events that occur at the neuromuscular junction leading to muscle contraction are outlined in **Figure 12-4**. Synaptic vesicles within the cytosol of the axon terminal contain the neurotransmitter acetylcholine (ACh). When an action potential (an electrical impulse) traveling along the axon reaches the axon terminal (step 1), it triggers an influx of calcium ions into the axon terminal. This calcium influx then triggers exocytosis of the synaptic vesicles, causing the release of ACh into the synaptic cleft (step 2). ACh then diffuses across the synaptic cleft. The region of the sarcolemma across from the axon terminal, called the motor end plate, contains millions of transmembrane proteins that serve as both acetylcholine receptors (ACh-Rs) and sodium ion (Na^+) channels. When ACh binds to an ACh-R on the motor end plate (step 3), the sodium channel opens, and Na^+ rushes into the skeletal muscle fiber (step 4). This influx of Na^+ results in a localized depolarization called an end-plate potential, or a muscle action potential (step 5). This muscle action potential subsequently travels along the sarcolemma of the muscle fiber and down into the T-tubules. The resulting release of calcium into the sarcoplasm leads to muscle contraction. The physiological effects of ACh binding with ACh-R are brief, because ACh in the synaptic cleft is rapidly degraded by an enzyme called acetylcholinesterase.

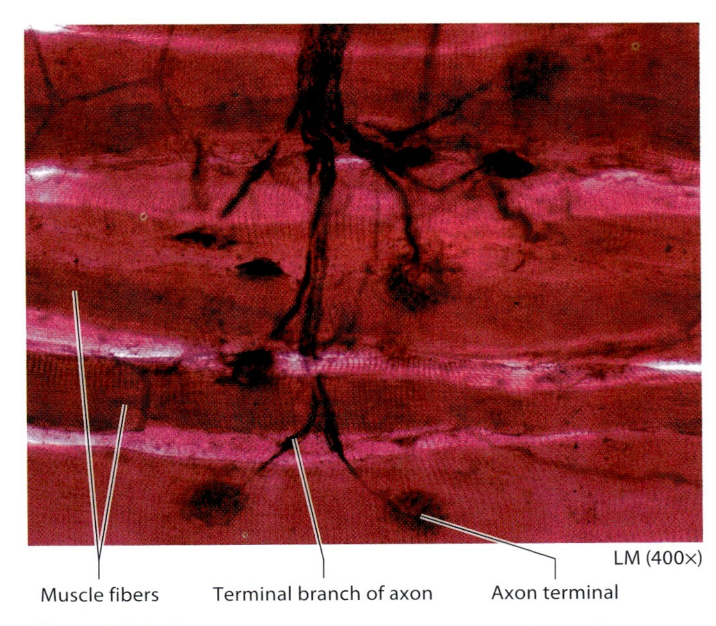

Figure 12-3 Photomicrograph of neuromuscular junctions.

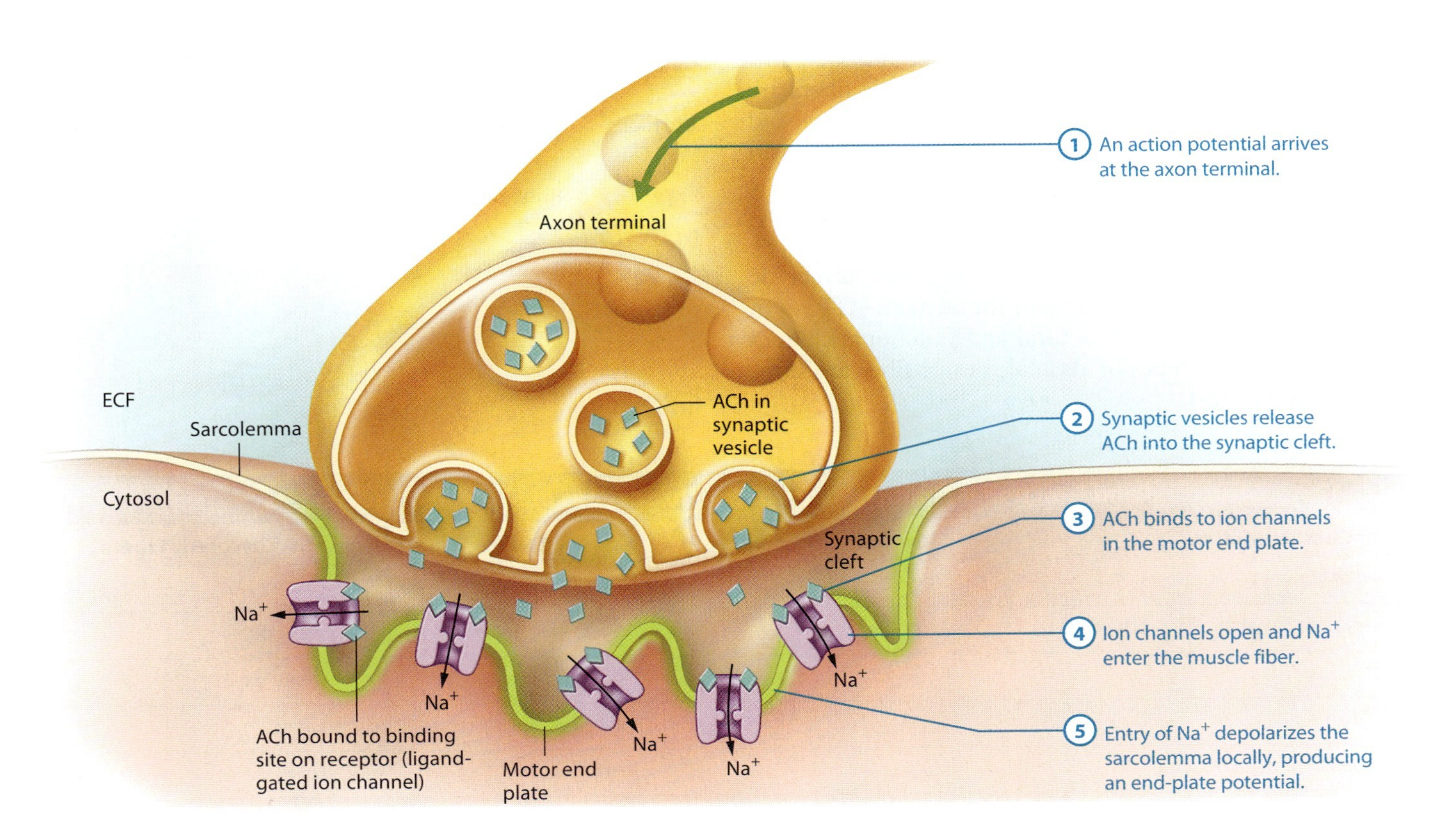

Figure 12-4 The structure and function of the neuromuscular junction.

THINK ABOUT IT Unbeknownst to him, Chris Canner improperly canned some homegrown fruit. As a result, certain bacteria survived and produced a toxin (called botulinum toxin) in the canned fruit, and Chris developed botulism after eating the fruit. This toxin somehow interferes with muscle contraction, producing difficulty in swallowing and breathing, among other signs and symptoms. Eventually, Chris died of respiratory failure when his respiratory muscles could no longer contract.

Assuming that botulinum toxin affects the neuromuscular junction, which of the following statements might explain the physiological basis for his death? Check all that apply.

- ☐ The toxin blocked ACh receptors on the motor end plate.
- ☐ The toxin stimulated the release of calcium from the sarcoplasmic reticulum.
- ☐ The toxin blocked the activity of acetylcholinesterase.
- ☐ The toxin stimulated the influx of calcium into the axon terminal. ■

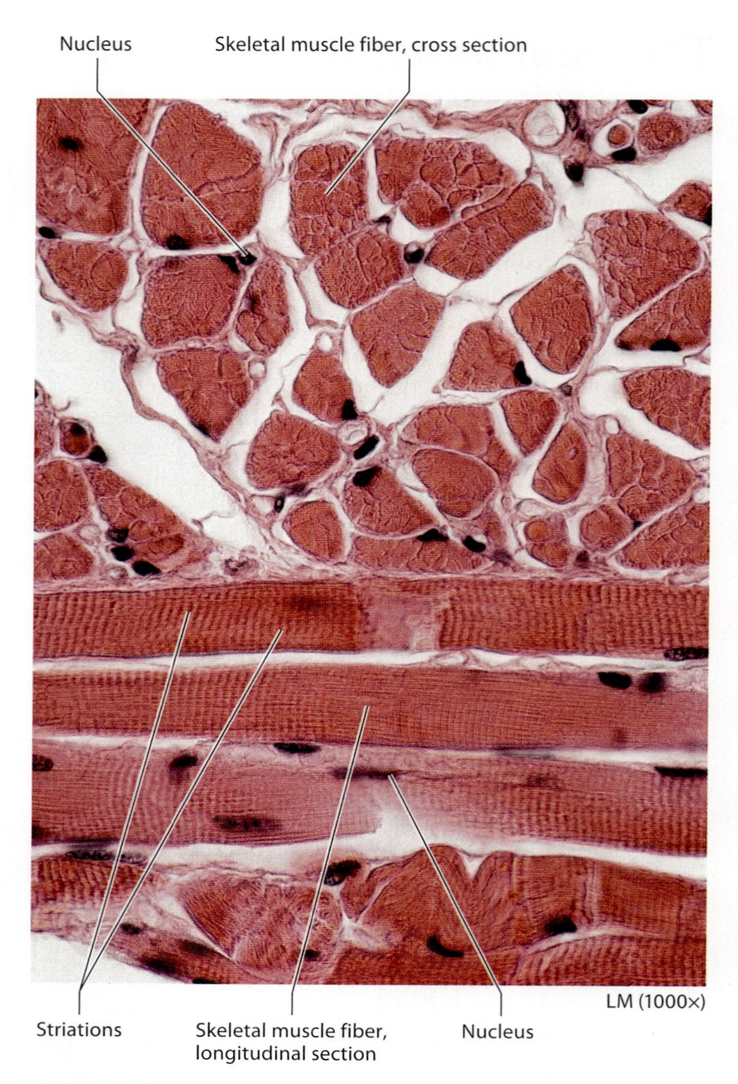

Figure 12-5 Photomicrograph of skeletal muscle fibers.

Types of Skeletal Muscle Fibers

Recall from your study of muscle tissue types (Unit 6) that skeletal muscle fibers are elongated, cylinder-shaped, striated, multinucleate cells. A photomicrograph of skeletal muscle fibers is shown in **Figure 12-5**.

Based on the speed at which they contract and the pathways by which they produce ATP, skeletal muscle fibers are of three types: slow oxidative fibers, fast oxidative fibers, and fast glycolytic fibers. Skeletal muscle fibers contain myosin ATPase, an enzyme that breaks down ATP to power muscle contraction. A muscle fiber's speed of contraction is determined by how fast this enzyme can break down ATP. The myosin ATPase in fast oxidative fibers and fast glycolytic fibers breaks down ATP faster than the myosin ATPase in slow oxidative fibers. The structural and functional characteristics of the three types of skeletal muscle fiber are summarized in **Table 12-1**.

Table 12-1 Structural and Functional Characteristics of the Three Types of Skeletal Muscle Fibers

Fiber Characteristics	Slow Oxidative Fibers	Fast Oxidative Fibers	Fast Glycolytic Fibers
Speed of contraction	Slow	Fast	Fast
Primary pathway(s) for ATP production	Aerobic	Aerobic and anaerobic	Anaerobic
Color	Red	Pink	White (pale)
Fiber diameter	Small	Intermediate	Large
Abundance of mitochondria	Many	Many	Few
Abundance of capillaries	Many	Many	Few

ACTIVITY 2

Examining the Microscopic Anatomy of Skeletal Muscle Tissue and the Neuromuscular Junction

Learning Outcomes

1. Sketch two microscopic views of skeletal muscle tissue and label the following structures: part of skeletal muscle fiber, striations, nucleus (longitudinal view) and skeletal muscle fiber, fascicle, epimysium, periomysium, endomysium (cross-sectional view).
2. Sketch and label a microscopic view of a neuromuscular junction, label the following structures—axon, axon terminal, and skeletal muscle fiber—and briefly explain the events that occur there.
3. Distinguish among different types of skeletal muscle fiber types and identify each microscopically.

Materials Needed

☐ Photomicrographs of skeletal muscle tissue (in longitudinal section and in cross-section), a neuromuscular junction, and a skeletal muscle showing all three fiber types in cross-section

OR

☐ Microscope and prepared slides of skeletal muscle tissue (in longitudinal section and in cross-section), a neuromuscular junction, and a skeletal muscle showing all three fiber types in cross-section

Instructions

1. Examine a slide or photomicrograph of skeletal muscle tissue in longitudinal section. Make a sketch of your observations and label a skeletal muscle fiber, a nucleus, and striations.

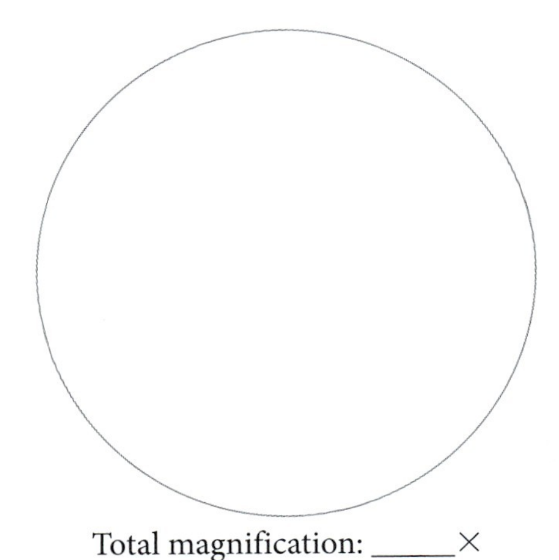

Total magnification: _____×

What causes the striations in this tissue?

____

Are these cells uninucleate or multinucleate?

____

2. Examine a slide or photomicrograph of skeletal muscle tissue in cross-section. Make a sketch of your observations and label a skeletal muscle fiber, epimysium, perimysium, endomysium, and fascicle.

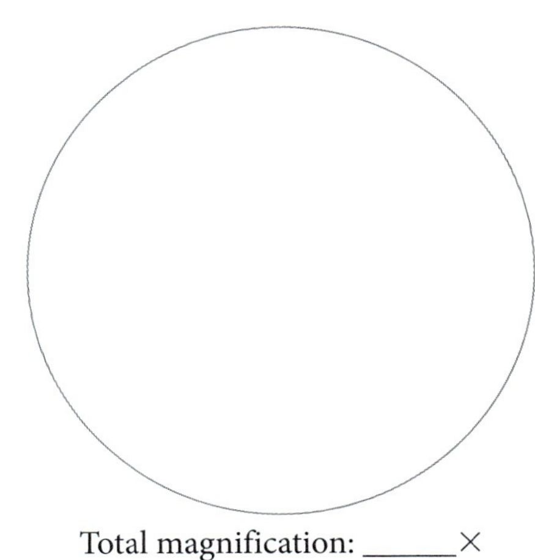

Total magnification: _____×

What are two functions of the connective tissue wrappings of muscles?

____

____

What is a fascicle?

____

3. Examine a slide or photomicrograph of a neuromuscular junction. Make a sketch of your observations and label an axon, an axon terminal, and a skeletal muscle fiber.

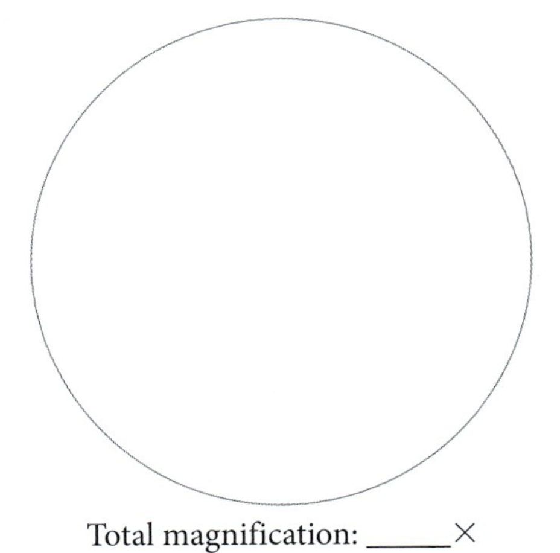

Total magnification: _____×

What is released by the axon terminal?

____

Name the two specific cell types visible on this slide.

____

4. Examine a slide or photomicrograph of a skeletal muscle in cross-section. Make a sketch of your observations and label a slow oxidative fiber, a fast oxidative fiber, and a fast glycolytic fiber.

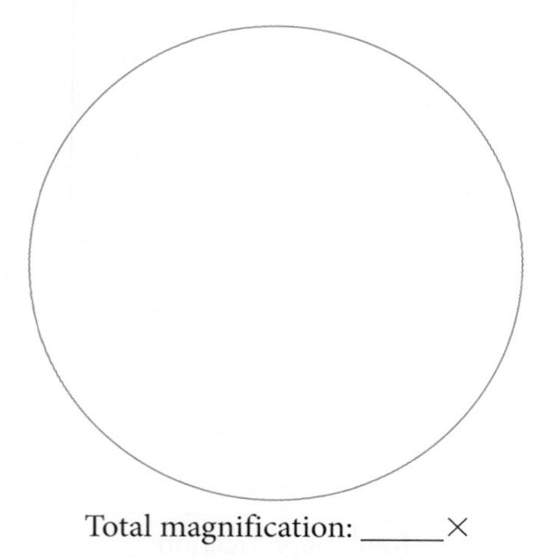

Total magnification: _____×

Which fiber type has more mitochondria: slow oxidative fibers or fast glycolytic fibers? Why?

____

____

*Why are slow oxidative fibers red?*______________________

____

5. **Optional Activity**

View and label histology slides of skeletal muscle tissue at MasteringA&P® > Study Area > Practice Anatomy Lab > Histology > Muscular System

Sliding Filament Theory of Muscle Contraction

The functional unit of a myofibril is the **sarcomere** (**Figure 12-6**). The striated appearance of skeletal muscle results from the

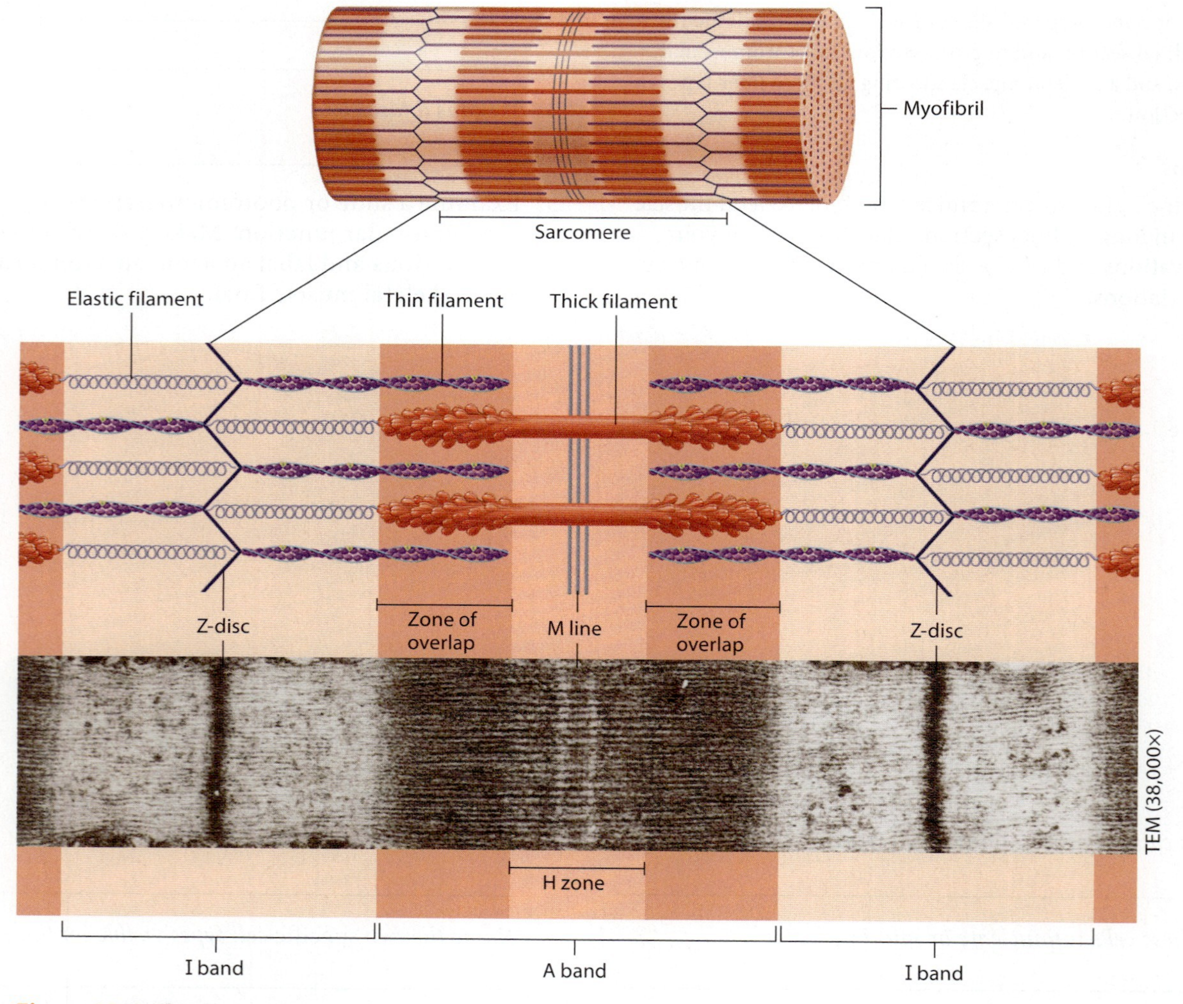

Figure 12-6 The sarcomere.

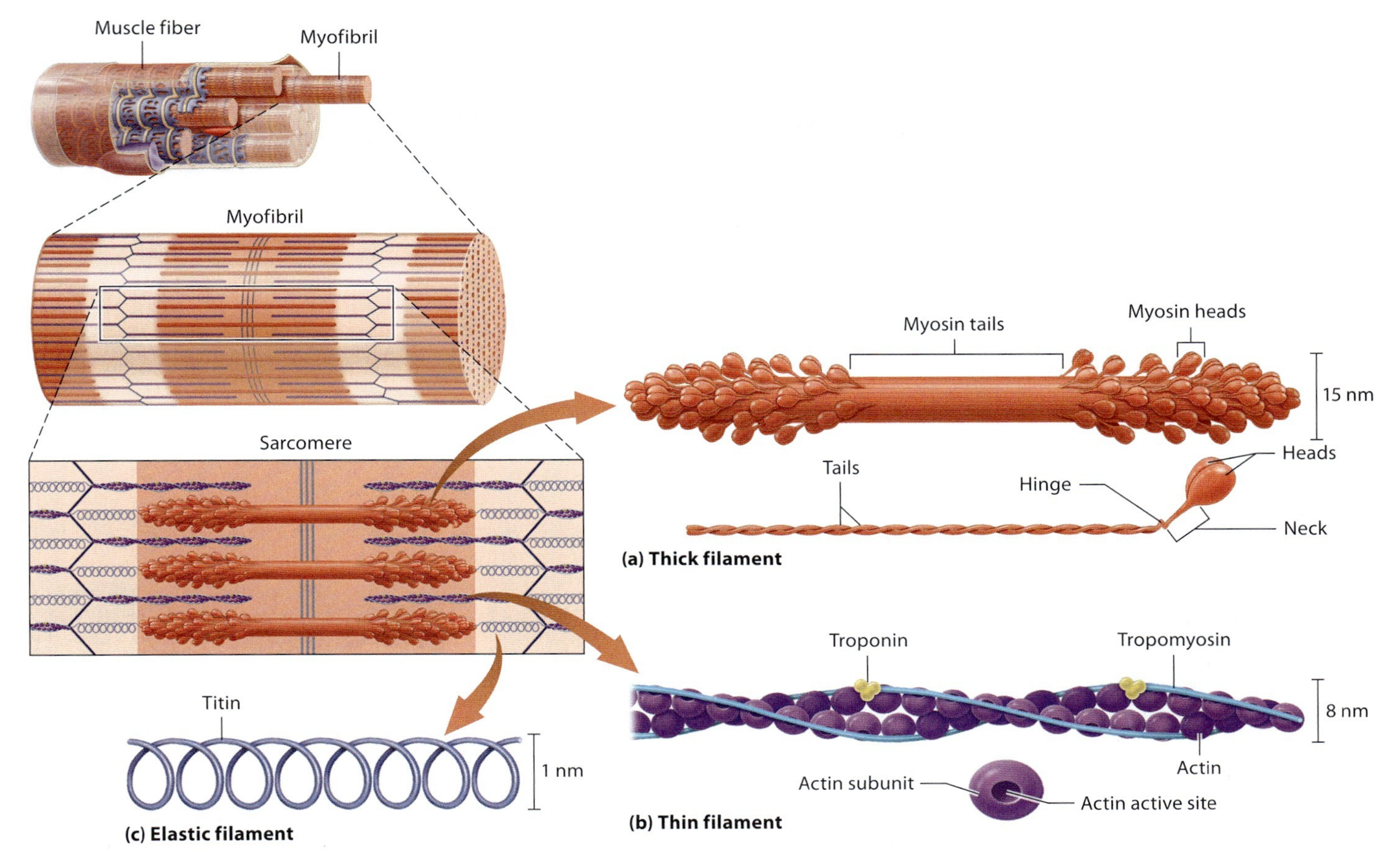

Figure 12-7 **The structure of the three types of myofilaments.**

arrangement of structures in the myofibrils called **myofilaments**, which are of three types: thick filaments, thin filaments, and elastic filaments (**Figure 12-7**). **Thick filaments** are composed of several hundred molecules of the contractile protein **myosin**, each shaped like a two-headed golf club. **Thin filaments** are composed of two intertwined strands of the contractile protein **actin** and a complex of two regulatory proteins called **troponin** and **tropomyosin**. Each **elastic filament** is composed of the giant protein **titin**. The elastic filament attaches to the Z-disc, runs through the core of the myosin molecule, and then attaches to the M line (see Figure 12-6). Elastic filaments hold the thick filaments in place and maintain the integrity of the A band.

The boundaries of a sarcomere are formed by **Z-discs** (see Figure 12-6), which function as anchors for the thin filaments. The darker middle portion of the sarcomere is the **A band**, which extends the entire length of the thick filaments. Within the A band is the narrow **H zone**, which contains thick filaments only. The **M line** (so named because it is at the midpoint of the sarcomere) contains supporting proteins that hold the thick filaments together. The **I bands**, which straddle adjacent sarcomeres, are lighter-colored bands that contain thin filaments and elastic filaments only. As we will see

LabBOOST »»

Visualizing Sliding Filaments To visualize the sliding of filaments during muscle contraction, place your hands in front of you with palms facing you and fingers loosely interlaced with minimal overlap. Your wrists represent Z-discs and your fingers represent the contractile proteins actin and myosin. Now, slide your fingers past one another as your wrists move closer together to represent muscle contraction. Note that your fingers do not get shorter, but the sarcomere (Z-disc to Z-disc) does get shorter during muscle contraction.

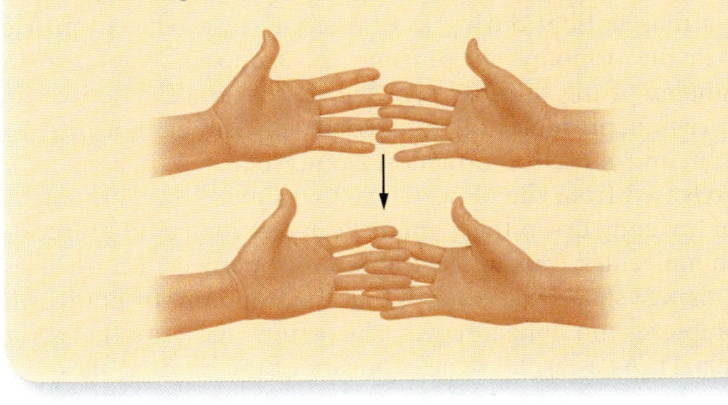

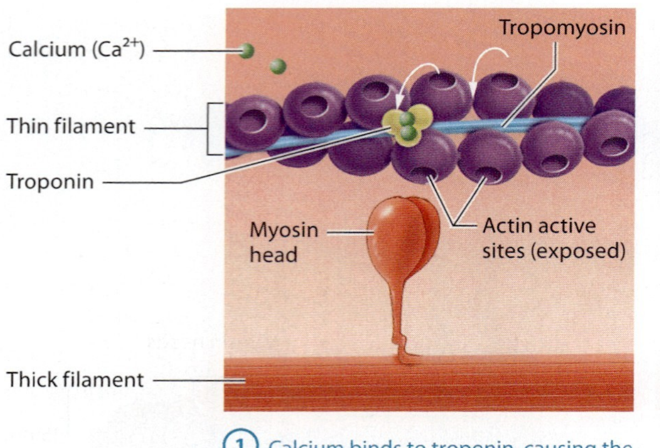

(1) Calcium binds to troponin, causing the troponin–tropomyosin complex to change shape, thereby exposing the active sites on actin.

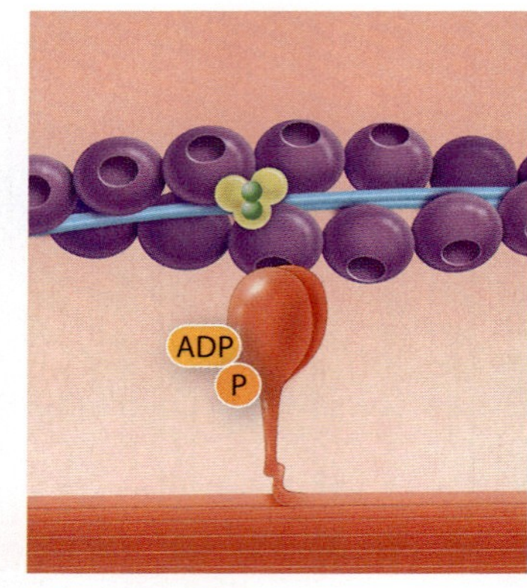

(2) Myosin head binds to actin, forming a crossbridge.

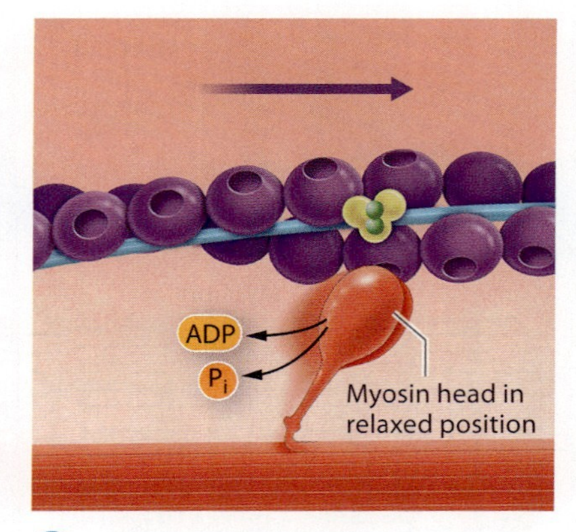

(3) The power stroke occurs as myosin pulls actin toward the center of the sarcomere.

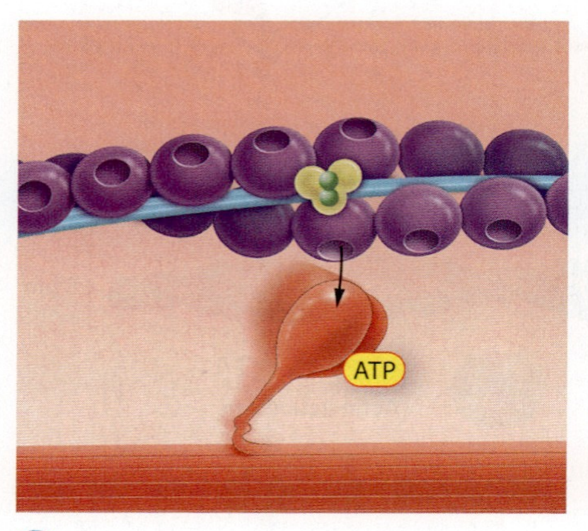

(4) Binding of ATP to the crossbridge causes detachment of myosin from actin.

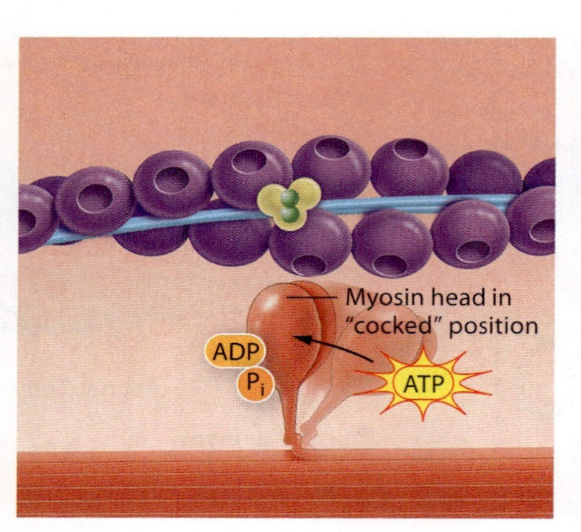

(5) Myosin head is reenergized and repositioned.

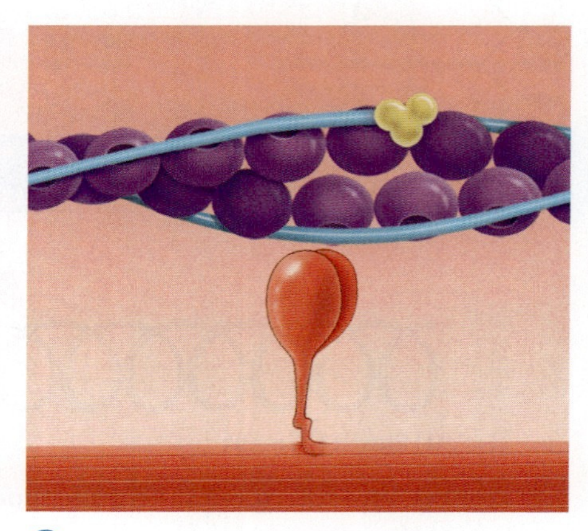

(6) Calcium ions are pumped back into SR. The troponin–tropomyosin complex once again covers the myosin-binding sites on actin.

Figure 12-8 The sliding filament theory of muscle contraction.

next, an understanding of the relationships between the thick and thin filaments—more precisely, between their protein components—is crucial to understanding the **sliding filament theory** of muscle contraction, which describes how the binding of actin and myosin results in thin filaments sliding past the thick filaments, enabling a muscle fiber to contract.

Figure 12-8 summarizes the sliding filament theory of muscle contraction, which explains the events that occur in a muscle fiber during a contraction. In a relaxed muscle fiber, the tropomyosin portion of a troponin–tropomyosin complex of the thin filament blocks (covers) the active site on each actin molecule. As a result, myosin cannot bind to actin, and muscle contraction cannot occur. When calcium is released from the SR as a result of a muscle action potential traveling down the T-tubule, calcium ions bind to troponin molecules, causing the troponin–tropomyosin complexes to change shape; the result is that the troponin–tropomyosin complexes no longer cover the active sites on the actin (step 1). As a result of the exposure of the binding sites, the already-energized myosin heads (energized by the splitting of ATP molecules in the previous contraction cycle) can achieve binding to the active sites on the actin, forming what are known as cross-bridges (step 2). During the next step, called the power stroke (step 3), the energized myosin heads swivel, pulling the thin filaments past the thick filaments and toward the center of the sarcomere, causing the H zone to narrow. Next, the binding of an ATP molecule to myosin causes the myosin head to detach from actin, disconnecting the cross-bridge (step 4). When the ATPase on the myosin head splits the bound ATP molecule into ADP and phosphate (P_i), the myosin head is reenergized and repositioned ("cocked") for the next round of the contraction cycle (step 5). Finally, when calcium ions are pumped back into the sarcoplasmic reticulum, the troponin–tropomyosin complex slides back into its original position, covering the active sites on actin (step 6), and the muscle fiber relaxes until a new electrical impulse stimulates a new contraction. It is important to note that the myofilaments themselves do not become shorter during contraction; instead, the thin filaments slide past the thick filaments, causing the sarcomere as a whole to shorten.

ACTIVITY 3

Stimulating Muscle Contraction in Glycerinated Skeletal Muscle Tissue

Learning Outcomes

1. Sketch a sarcomere, label its parts, and describe what occurs to the Z-discs, the H zone, the A band, and the I bands during muscle contraction.
2. Stimulate glycerinated skeletal muscle tissue, observe the resulting contractions, and discuss the importance of calcium ions and ATP to contraction in living muscle tissue.

Materials Needed

- ☐ ATP-glycerinated muscle kit
- ☐ Petri dish
- ☐ Teasing needle
- ☐ 4 Microscope slides
- ☐ 4 Coverslips
- ☐ Marker
- ☐ Ruler (marked in millimeters)
- ☐ Scissors
- ☐ Dissecting microscope
- ☐ Compound microscope

Instructions

Glycerinated muscle behaves differently than muscle in living tissue. Exposing skeletal muscle fibers to glycerin removes ions and ATP from the cells and disrupts the troponin–tropomyosin complex such that it can no longer block the active sites on the actin molecules. As a result, the presence of calcium ions is not needed for contraction to occur. However, ATP and other ions are required for the proper functioning of various enzymes involved in contraction.

For this activity, complete the following steps to test the effects of three different solutions—ATP only, ATP and ions, and ions only—on glycerinated muscle fibers.

1. Predict the effects of each solution on a glycerinated muscle fiber, and explain the basis of (rationale for) each prediction.

	Will the Muscle Fiber Contract?	Rationale
ATP Only		
ATP and Ions		
Ions Only		

2. Use scissors to cut a segment (1–2 cm in length) of skeletal muscle tissue. Place the segment of tissue in a petri dish and place the dish on the stage of a dissecting microscope. Use a teasing needle to tease the segments into 10 very thin strands of two to four fibers each. Strands of muscle exceeding 0.2 mm in cross-sectional diameter are too thick to use.
3. Place a thin strand of muscle tissue on a microscope slide, cover the tissue with a coverslip, and examine the strand using a compound microscope at low magnification. Note the striations in the muscle fibers.
4. Use a marker to label three other microscope slides "1," "2," and "3."
5. Now, transfer three of the thin strands to a tiny drop of glycerol on microscope slide 1. Note that the less glycerol used, the easier the fibers are to measure. Arrange the strands so that they are straight and parallel to one another. Do not put a coverslip on this slide.
6. Using a ruler, measure the length of the strands in millimeters using a dissecting microscope, and record your results in the following table under "Resting length."
7. Now, flood the strands on microscope slide 1 with several drops of 0.25% ATP in distilled water, and observe the reaction of the fibers. After 45 seconds, measure each strand again using a dissecting microscope, and record your results in the table under "Contracting length."
8. Put a coverslip over slide 1 and examine it under low power using a compound microscope. Compare these strands treated with the ATP solution to the untreated strand.

Effects of ATP and Ions on Glycerinated Skeletal Muscle Fibers

Muscle Strand		ATP Only	ATP and Ions	Ions Only
1	Resting length			
	Contracting length			
	% Contraction			
2	Resting length			
	Contracting length			
	% Contraction			
3	Resting length			
	Contracting length			
	% Contraction			
Average % Contraction				

9. Repeat steps 5 through 8 using a clean slide, three new muscle fiber strands, and the solution containing both ATP and ions.
10. Repeat steps 5 through 8 using a clean slide, three new muscle fiber strands, and the solution containing ions only.
11. Follow directions given by your instructor for disposal of materials. Wash all instruments (other than the microscopes!) with warm, soapy water and dry them thoroughly. Clean your lab space and wipe the lab bench with disinfectant.
12. Analyze the data:
    a. Calculate the % original length of each muscle strand, as follows:

$$\% \text{ Original length} = \frac{\text{contracting length}}{\text{resting length}} \times 100\%$$

    b. Then subtract % original length from 100% to calculate the % contraction for each muscle strand.
    c. Average the % contraction for each treatment and record in the following table.
13. Answer each of the following questions based on your results:
    a. Which solution caused the greatest contraction?

    ____

    Why? ______________________________

    ____

    ____

    b. Why is calcium not needed in order to stimulate contraction in glycerinated skeletal muscle tissue?

    ____

    ____

    ____

    c. Sketch a sarcomere "at rest" and label its parts.

    d. Sketch a sarcomere "during contraction" and label its parts.

    e. What happened to sarcomere length during contraction?

    ____

    ____

    f. Describe what happened to the lengths of the thick and thin filaments during contraction.

    ____

    ____

ACTIVITY 4

Electromyography in a Human Subject Using BIOPAC®

Learning Outcomes

1. To record a skeletal muscle tonus measured against baseline activity level associated with the resting state.
2. To record how motor unit recruitment changes as the power of a skeletal muscle contraction increases.
3. To correlate an EMG "sound" intensity with the level of motor unit recruitment.

Materials Needed

- ☐ Computer system (Windows 7, Vista, XP, Mac OS X 10.5–10.7)
- ☐ Biopac Student Lab system (MP35 unit and software)
- ☐ Electrode lead set (SS2L lead set)
- ☐ Disposable vinyl electrodes (EL503), six electrodes per subject
- ☐ Biopac Electrode Gel (GEL1) and Abrasive Pad (ELPAD) or alcohol prep
- ☐ Biopac wall transformer (AC100A)
- ☐ Biopac serial cable (CBLSERA)
- ☐ Biopac USB adapter (USB1W)

Instructions

When a skeletal muscle is stimulated by the nervous system, action potentials are generated along the cell membranes of the muscle fibers leading to muscle contraction. These electrical changes are conducted to the skin surface and can be detected, amplified, and recorded by a technique called **electromyography**. An **electromyogram (EMG)** is the graph produced by such a recording.

A **motor unit** is a single motor neuron and all of the skeletal muscle fibers within a skeletal muscle that it activates. A single skeletal muscle contains many motor units, and force of muscle contraction in a muscle is directly related (1) to the number of motor units activated and (2) to the frequency of nerve impulses delivered to the fibers of a motor unit. The gradual recruitment of more and more motor units is called **multiple motor unit summation**, and an increase in the frequency of nerve impulses delivered to a motor unit is called **temporal (wave) summation**. In this activity, you will use the Biopac Student Lab system to record an EMG and to correlate motor unit recruitment with force of muscle contraction.

A. Setup

1. Turn the computer ON.
2. Turn the MP35 unit OFF.
3. Plug the electrode leads (SS2L) into the CH 3 channel and the headphones jack (OUT1) into the back of the unit.
4. Turn the MP35 unit ON.
5. Attach three electrodes either to your own dominant forearm or, if you are working with a partner, to your partner's dominant forearm (**Figure 12-9**). This will be forearm 1, and the nondominant forearm will be forearm 2. For optimal signal quality, place the electrodes on the skin at least 5 minutes before starting the calibration section.
6. Attach the electrode lead set (SS2L) to the electrodes on forearm 1. Make sure the electrode lead colors match.
7. Start the Biopac Student Lab program on the computer.
8. Choose lesson "L01-EMG-1" and click OK.
9. Type in a unique file name and click OK to end the Setup section.

B. Calibration

The calibration procedure establishes the hardware's internal parameters (such as gain, offset, and scaling) and is critical for optimal performance.

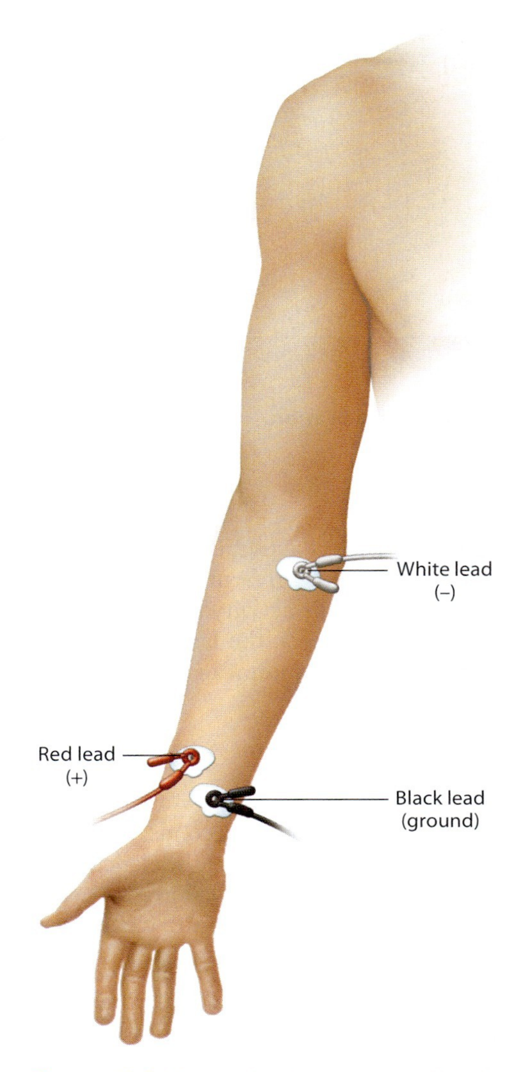

Figure 12-9 Electrode placement and lead connections.

1. Click Calibrate.
2. Read the dialog box, and click OK when ready.
3. Wait 2 seconds. Then clench your forearm 1 fist as hard as possible for 5–6 seconds and release. The calibration will last 8 seconds and then stop automatically.
4. Your computer screen should resemble **Figure 12-10**. If your screen is different, repeat calibration steps 1 through 3. If your calibration recording did not begin with a zero baseline, this means that you clenched your fist before waiting 2 seconds; repeat the calibration.

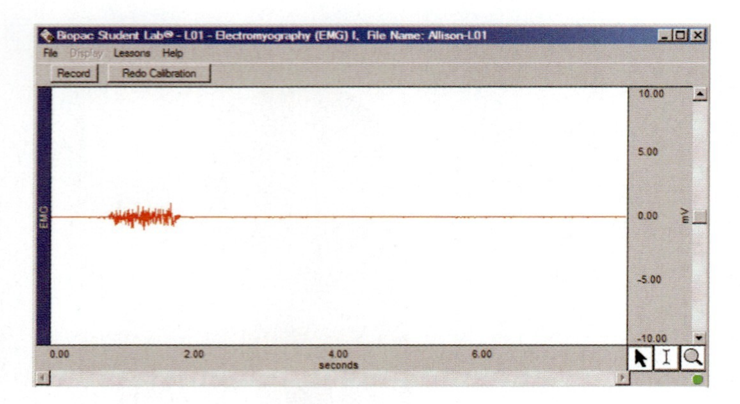

Figure 12-10 **Sample calibration.**

C. Data Recording

You will perform two EMG activities: segment 1 from forearm 1 and segment 2 from forearm 2. To work efficiently, read through the rest of this activity so you will know what to do before recording. Check the last line of the software journal, and note the total amount of time available for the recording.

Segment 1: Forearm 1, Dominant Arm

1. On the computer, click Record.
2. Clench your fist and hold for 2 seconds. Release and clench and wait 2 seconds. Repeat the clench–release–wait sequence while increasing the force in each sequence by equal increments so that the fourth clench involves maximum force.
3. On the computer, click Suspend, and then review the recording on the screen. If it looks similar to **Figure 12-11**, go to step 5. If your recording looks different, go to step 4.
4. Click Redo, and repeat steps 1 through 3.
5. Remove the electrode cable pinch connectors. Peel the electrodes off your arm and dispose of them. Use soap and water to wash the electrode gel residue from your skin.

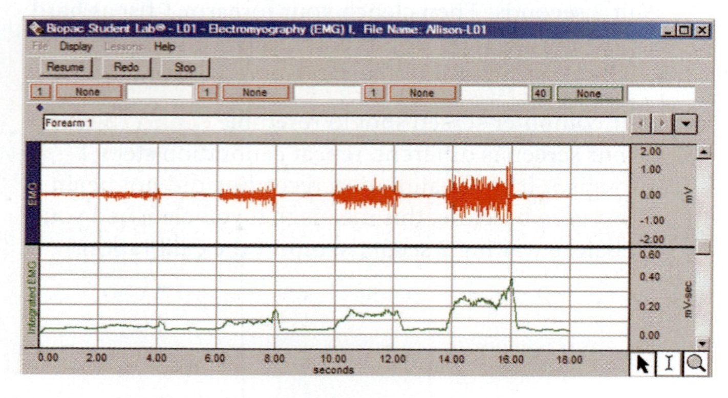

Figure 12-11 **Clench–wait–release sequence.**

Segment 2: Forearm 2, Nondominant Arm

1. Attach the remaining three electrodes to your nondominant arm, again placing them as shown in Figure 12-9.
2. On the computer, click Resume. A marker labeled "Forearm 2" will automatically be inserted when you do this.
3. Repeat a cycle of clench–release–wait, holding each clench for 2 seconds and waiting 2 seconds after release before beginning the next cycle. Increase the strength of your clench by the same amount for each cycle, with the fourth clench having maximum force.
4. On the computer, click Suspend, and then review the recording on the screen. If it looks similar to **Figure 12-12**, go to step 6. If your recording looks different from Figure 12-12, go to step 5.

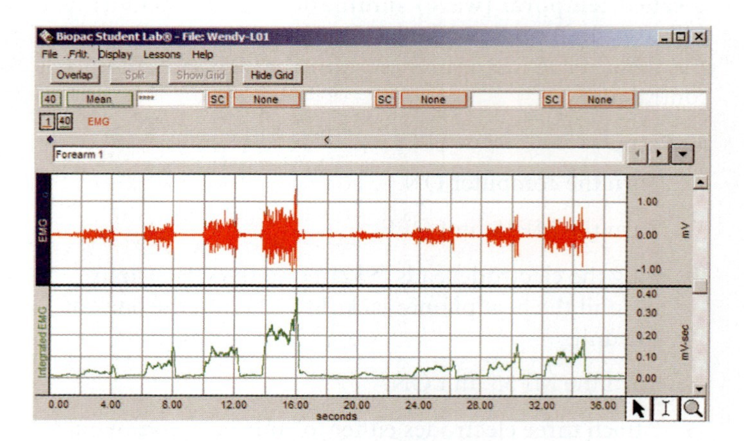

Figure 12-12 **Sample data from nondominant arm.**

5. Click Redo, and repeat steps 3 and 4.
6. Click Stop. Clicking "yes" in the dialog box will end the data-recording section and will automatically save the data. Clicking "no" will bring you back to either Resume or Stop. Listening to the EMG can be a valuable tool in detecting muscle abnormalities and is performed here for the general interest. If you want to listen to the EMG signal, go to step 7. If you want to end the recording, go to step 11.
7. Put on headphones, and click Listen. The volume through the headphones may be very loud because of system feedback. Because the volume cannot be adjusted, you may have to position the headphones slightly off your ears for comfort.
8. Experiment by changing the clench force during the clench–release–wait cycles as you watch the screen and listen. You will hear the EMG signal through the headphones as it is being displayed on the screen. The screen will display two channels: CH 3 EMG and CH 40 integrated EMG. The data on the screen will not be saved. The signal will run until you click Stop.

Cluster Number	Forearm 1 (Dominant)				Forearm 2 (Nondominant)			
	Min (CH 3 min)	Max (CH 3 max)	p-p (CH 3 p-p)	Mean (CH 40 mean)	Min (CH 3 min)	Max (CH 3 max)	p-p (CH 3 p-p)	Mean (CH 40 mean)
1								
2								
3								
4								

9. Click Stop to end the listening-to-EMG portion.
10. To listen again, click Redo.
11. Click Done. If choosing the "Record from another subject" option, you must attach electrodes per the Setup section and continue the activity from the beginning.

D. Data Analysis

1. Enter the Review Saved Data option, and choose the correct file. Note the channel settings in the small menu boxes as in the string of small boxes across the top containing numbers.

   Note the channel (CH) designations:

CH 3	Raw EMG
CH 40	Integrated EMG

2. Set up your display window for optimal viewing of the first data segment (forearm 1, dominant). The following tools help you adjust the data window:

Autoscale	Zoom previous
Horizontal	Horizontal (Time) scroll bar
Autoscale	Vertical (Adjustment) scroll bar
Waveforms	Overlap button
Zoom tool	Split button

3. Set up the measurements boxes as follows:

CH 3	Min
CH 3	Max
CH 3	p-p
CH 40	Mean

   The measurement boxes are above the marker region in the data window. Each measurement has three sections: channel number, measurement type, and result. The first two sections are pull-down menus that are activated when you click on them. The following is a brief description of these measurements; "selected area" is the area selected by the I-beam cursor tool.

Min	displays the minimum value in the selected area
Max	displays the maximum value in the selected area
p-p	finds the maximum value in the selected area and subtracts the minimum value found in the selected area
Mean	displays the average value in the selected area.

4. Using the I-beam cursor, select an area enclosing the first EMG cluster. Record your data in the following table.
5. Repeat step 4 on each successive EMG cluster.
6. Scroll to the sound recording segment, which is for forearm 2 (nondominant) and begins after the first marker.
7. Repeat steps 4 and 5 for the forearm 2 data.
8. Save or print the data file. You may save data, save notes that are in the software journal, or print the data file.
9. Quit the program.

EMG Measurements

1. Complete the table using the data obtained during the EMG 1 experiment.
2. Use the mean measurement from the table to compute the percentage increase in the EMG activity recorded between the weakest clench and the strongest clench of forearm 1.

   Calculation: Answer: ______ %

PhysioEx EXERCISE 2
Skeletal Muscle Physiology

The PhysioEx 9.1 Laboratory Simulations in Physiology are easy-to-use laboratory simulations that can be used as an alternative to or as a supplement to wet lab activities in this unit. Each simulation allows you to investigate important physiological concepts, repeat labs as often as you like, and conduct experiments that are difficult to perform in a wet lab environment because of time, cost, or safety concerns.

Access the simulations in these activities at MasteringA&P® > Study Area > PhysioEx 9.1

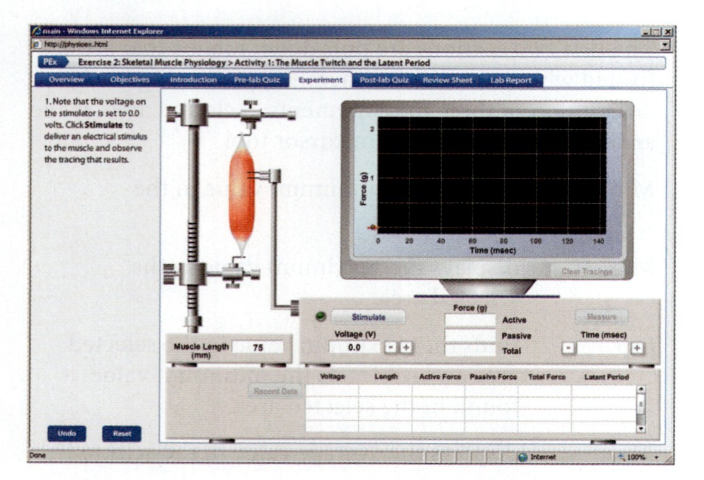

There you will find the following activities:

- **PEx Activity 1:** The Muscle Twitch and the Latent Period
- **PEx Activity 2:** The Effect of Stimulus Voltage on Skeletal Muscle Contraction
- **PEx Activity 3:** The Effect of Stimulus Frequency on Skeletal Muscle Contraction
- **PEx Activity 4:** Tetanus in Isolated Skeletal Muscle
- **PEx Activity 5:** Fatigue in Isolated Skeletal Muscle
- **PEx Activity 6:** The Skeletal Muscle Length–Tension Relationship
- **PEx Activity 7:** Isotonic Contractions and the Load–Velocity Relationship

Post-lab quizzes are also assignable in MasteringA&P®

POST-LAB ASSIGNMENTS

Name: ______________________ Date: ______________ Lab Section: ________

PART I. Check Your Understanding

Activity 1: Identifying the Structural Components of a Skeletal Muscle

1. Which connective tissue wrapping surrounds:
   a. an individual skeletal muscle fiber? ______________________
   b. a bundle of skeletal muscle fibers? ______________________
   c. a skeletal muscle? ______________________

2. Identify the components of a skeletal muscle fiber.

a. ______________________
b. ______________________
c. ______________________
d. ______________________
e. ______________________
f. ______________________
g. ______________________
h. ______________________
i. ______________________
j. ______________________
k. ______________________
l. ______________________

2. Match each of the following terms with its description/function:

a. ______ transverse tubule
b. ______ sarcoplasmic reticulum
c. ______ epimysium
d. ______ thin filament
e. ______ myofibril
f. ______ triad
g. ______ endomysium
h. ______ motor end plate
i. ______ sarcomere
j. ______ perimysium
k. ______ thick filament
l. ______ fascicle

1. connective tissue wrapping around a muscle fiber
2. organelle packed with myofilaments
3. transverse tubule and its flanking terminal cisternae
4. invagination of the sarcolemma
5. connective tissue wrapping around fascicle
6. the region of sarcolemma across from the axon terminal at the neuromuscular junction
7. consists of actin, troponin, and tropomyosin
8. stores calcium
9. consists of myosin
10. a bundle of muscle fibers
11. connective tissue wrapping around a muscle
12. functional unit of muscle contraction

Activity 2: Examining the Microscopic Anatomy of Skeletal Muscle Tissue and the Neuromuscular Junction

1. Briefly describe the microscopic appearance of skeletal muscle tissue.

____

____

2. Identify the structures of the neuromuscular junction in the following photomicrograph.

a. ______________________

b. ______________________

c. ______________________

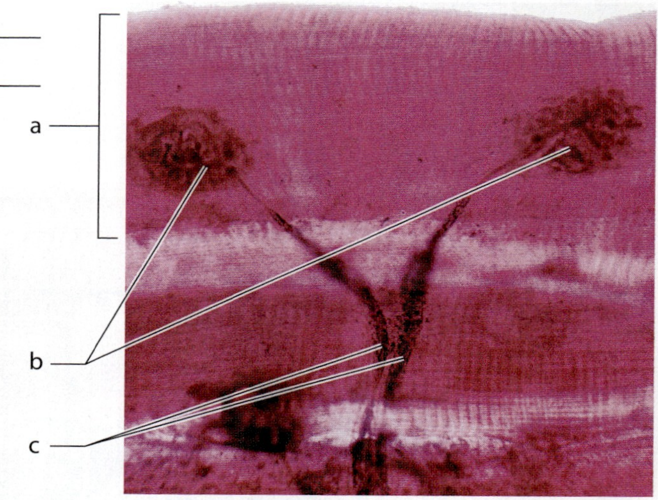

LM (1000×)

3. Which skeletal muscle fiber type shown in the following photomicrograph:

    ______ a. contains abundant myoglobin?

    ______ b. relies on aerobic respiration?

    ______ c. contains abundant glycogen stores?

    ______ d. has slow myosin ATPase activity?

    ______ e. has few capillaries

1

2

3

TEM (850×)

Activity 3: Stimulating Muscle Contraction in Glycerinated Skeletal Muscle Tissue

1. Put the following events of the sliding filament theory into their proper sequence:
    1. power stroke of the cross bridge, causing sliding of the thin filaments
    2. influx of calcium ions into the sarcoplasm
    3. hydrolysis of ATP, leading to reenergizing and repositioning of cross bridges
    4. binding of myosin to actin
    5. transport of calcium ions back into the sarcoplasmic reticulum
    6. binding of ATP to a cross bridge, resulting in the cross bridge disconnecting from actin

    a. 2, 4, 1, 6, 3, 5
    b. 2, 6, 1, 4, 3, 5
    c. 2, 4, 1, 3, 6, 5
    d. 2, 1, 4, 6, 3, 5
    e. 2, 4, 3, 1, 6, 5

2. What happens to each of the following components of a sarcomere during muscle contraction?

    a. Z-discs ______________________________

    b. H zone ______________________________

    c. A band ______________________________

    d. I bands ______________________________

3. Describe the effects of glycerin on skeletal muscle fibers. ______________________________

    ____

Activity 4: Electromyography in a Human Subject Using BIOPAC®

1. Explain any differences in tonus between the two forearm clench muscles. ______________________

____

____

2. Are the mean measurements for the right and left maximum clench EMG cluster the same or different? ______________ Which one suggests the greater clench strength? ______________

3. What other factors besides gender may contribute to observed differences in clench strength?

____

____

4. Define motor recruitment. ______________________________________

____

____

5. Define electromyography. ______________________________________

____

____

PART II. Putting It All Together

A. Review Questions

Answer the following questions using your lecture notes, your textbook, and your lab notes.

1. Distinguish between a skeletal muscle and a skeletal muscle fiber. ______________________

____

____

2. State two ways in which a skeletal muscle fiber is structurally adapted to perform its function.

____

____

____

3. Explain the physiological significance of each of the following characteristics of fast glycolytic skeletal muscle fibers:

   a. pale appearance ____

   ____

   ____

   b. abundant glycogen stores ____

   ____

   ____

   c. few mitochondria ____

   ____

   ____

   d. little myoglobin ____

   ____

   ____

4. Predict the effect on muscle contraction of each of the following situations, and then explain the basis for each prediction:

   a. a poison that blocks ACh receptors ____

   ____

   ____

   b. a drug that blocks calcium pumps on the terminal cisternae ______________________________

   ____

   ____

5. Explain a possible molecular basis for the onset of fatigue during strenuous exercise. ____________

   ____

   ____

   ____

6. Based on your knowledge of the sliding filament theory, why does rigor mortis occur following death?

   ____

   ____

   ____

   ____

B. Concept Mapping

1. Fill in the blanks to complete this concept map outlining the anatomy of a skeletal muscle fiber.

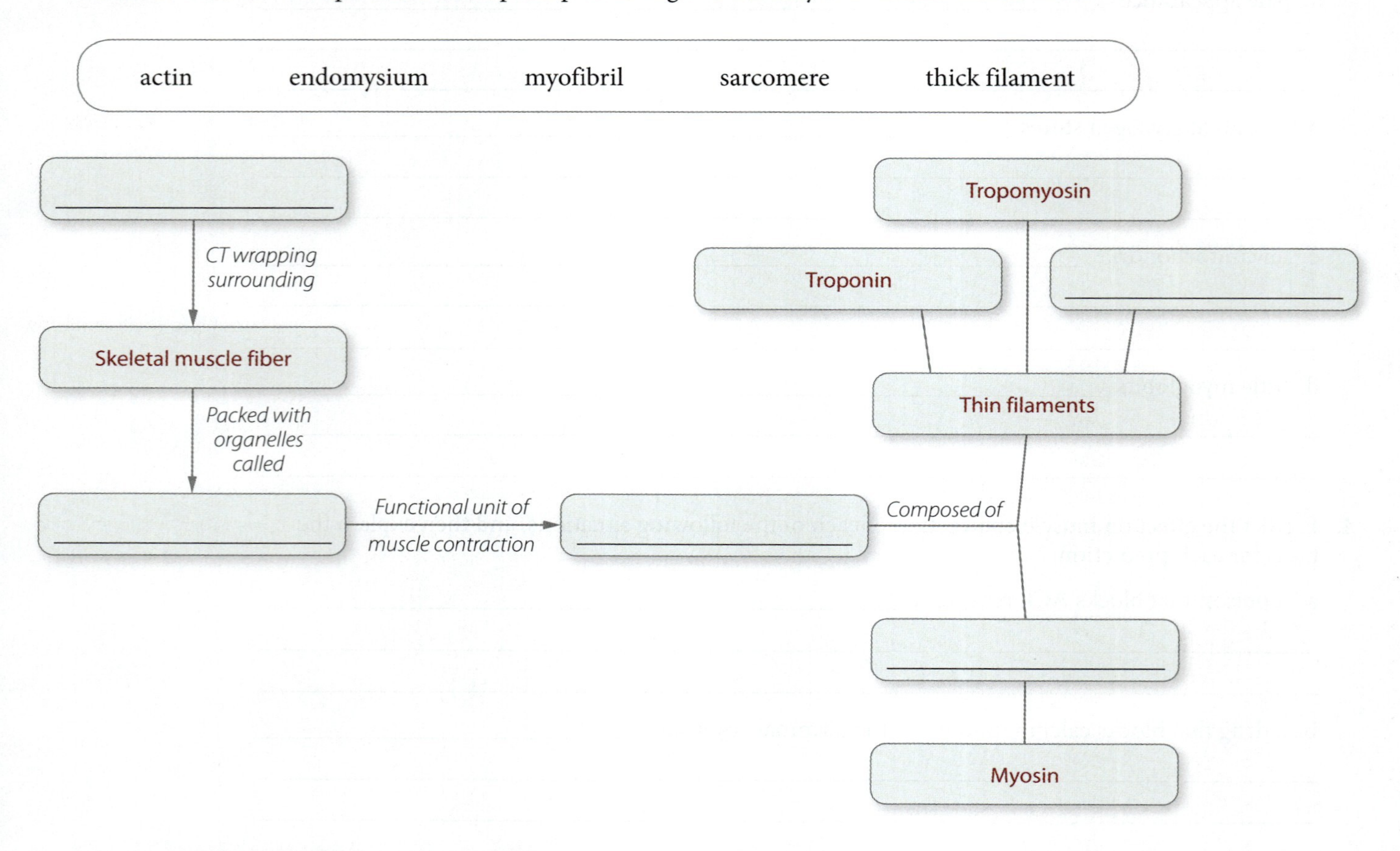

2. Construct a unit concept map to show the relationships among the following set of terms. Include all of the terms in your diagram. Your instructor may choose to assign additional terms.

actin	calcium	endomysium	epimysium	muscle action potential
myofibril	myosin head	perimysium	power stroke	sarcomere
sarcoplasmic reticulum	T-tubule	thick filament	tropomyosin	troponin

13

Gross Anatomy of the Muscular System

UNIT OUTLINE

The human body contains more than 600 skeletal muscles, the major organs of the muscular system. In this unit you will examine some of the body's major muscles, which we will divide into four groups: muscles of the head and neck, muscles of the trunk, muscles of the upper limb, and muscles of the lower limb.

THINK ABOUT IT *Many skeletal muscles are named based on their location. Describe the general location of the following muscles based on their name:*

Epicranius ______________________________

Biceps femoris ______________________________

Flexor carpi radialis ______________________________

Zygomaticus ______________________________

Infraspinatus ______________________________

Transversus abdominis ______________________________ ■

PRE-LAB ASSIGNMENTS

Pre-lab quizzes are also assignable in MasteringA&P®

To maximize learning, BEFORE your lab period carefully read this entire lab unit and complete these pre-lab assignments using your textbook, lecture notes, and prior knowledge.

PRE-LAB Activity 1: Determining How Skeletal Muscles Are Named

1. Use the list of terms provided to label the accompanying diagram of skeletal muscles. Check off each term as you label it.

- ☐ biceps brachii
- ☐ biceps femoris
- ☐ deltoid
- ☐ epicranius, occipital belly
- ☐ extensor digitorum
- ☐ flexor carpi radialis
- ☐ gastrocnemius
- ☐ gluteus maximus
- ☐ latissimus dorsi
- ☐ infraspinatus
- ☐ orbicularis oculi
- ☐ pectoralis major
- ☐ rectus abdominis
- ☐ rectus femoris
- ☐ sternocleidomastoid
- ☐ tibialis anterior
- ☐ trapezius
- ☐ triceps brachii

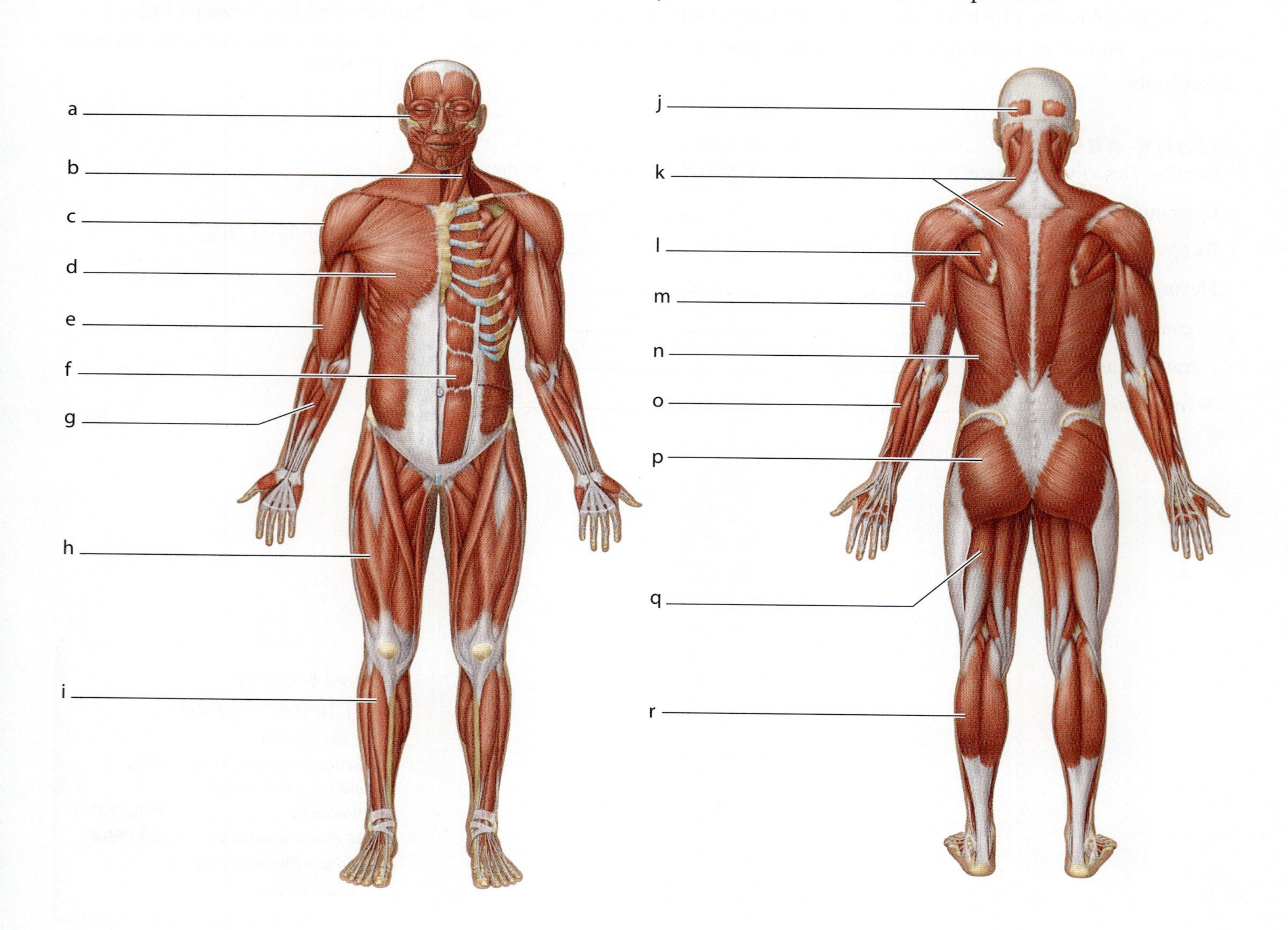

PRE-LAB Activity 2: Mastering the Muscles of the Head and Neck

1. Match each of the following muscles with its correct description:

	Muscle		Description
_____	a. sternohyoid	1.	originates on the sternum and clavicle
_____	b. hyoglossus	2.	classified as an infrahyoid muscle
_____	c. sternocleidomastoid	3.	consists of two bellies
_____	d. orbicularis oris	4.	elevates and protrudes lower lip
_____	e. splenius capitis	5.	is a major chewing muscle
_____	f. mentalis	6.	classified as a suprahyoid muscle
_____	g. stylohyoid	7.	depresses the tongue
_____	h. zygomaticus major	8.	closes the lips
_____	i. digastric	9.	extends the head
_____	j. masseter	10.	one of the smiling muscles

PRE-LAB Activity 3: Mastering the Muscles of the Trunk

1. Match each of the following muscles with its correct description:

	Muscle		Description
_____	a. internal intercostals	1.	dome-shaped muscle that functions in inspiration
_____	b. rectus abdominis	2.	eleven pairs of muscles that elevate the rib cage
_____	c. quadratus lumborum	3.	tripartite muscle that runs from the sacrum to the skull
_____	d. semispinalis	4.	flexes the vertebral column
_____	e. trapezius	5.	large, triangular muscle that extends head
_____	f. subclavius	6.	eleven pairs of muscles that depress the rib cage
_____	g. erector spinae	7.	square-shaped muscle in the lumbar region
_____	h. external intercostals	8.	tripartite muscle that runs from the cervical or thoracic vertebrae to the skull
_____	i. diaphragm	9.	located below the clavicle

PRE-LAB Activity 4: Mastering the Muscles of the Upper Limb

1. Which of the following muscles is one of the rotator cuff muscles?
   a. trapezius
   b. pectoralis major
   c. supraspinatus
   d. latissimus dorsi
   e. triceps brachii

2. Distinguish between the triceps brachii and the biceps brachii in terms of location and function.

   ____

   ____

3. Indicate whether each of the following muscles is located on the anterior (A) or the posterior (P) surface of the body, and then write the action of the muscle:

   _____ a. palmaris longus: ______________________________

   _____ b. extensor carpi ulnaris: ______________________________

   _____ c. extensor digitorum: ______________________________

   _____ d. flexor carpi radialis: ______________________________

   _____ e. pronator teres: ______________________________

PRE-LAB Activity 5: Mastering the Muscles of the Lower Limb

1. Indicate whether each of the following muscles is located on the anterior (A) or the posterior (P) surface of the thigh, and then write its action (either flexes or extends):

_____ a. vastus lateralis: _______________ the leg.

_____ b. biceps femoris: _______________ the leg.

_____ c. rectus femoris: _______________ the leg.

_____ d. semitendinosus: _______________ the leg.

_____ e. vastus medialis: _______________ the leg.

2. Match each of the following muscles with its correct description:

_____	a. semimembranosus	1. extends the toes
_____	b. sartorius	2. one of the quadricep muscles
_____	c. extensor digitorum longus	3. located deep to the gastrocnemius
_____	d. soleus	4. one of the hamstring muscles
_____	e. rectus femoris	5. flexes the toes
_____	f. flexor digitorum longus	6. long, narrow, strap-like muscle located in the thigh region

Classification and Naming of Skeletal Muscles

Studying the body's individual skeletal muscles requires some basic terminology. Recall from Unit 11 that most skeletal muscles cross at least one joint and are usually attached to the bones that form that joint. During movements, one of the bones remains relatively stationary while the other bone moves. The site of attachment of a muscle's tendon to the stationary bone is called the muscle's **origin**, and the site of attachment to the moveable bone is called the muscle's **insertion**. During muscle contraction, the insertion moves toward the origin. The thicker middle region of the muscle between the origin and the insertion is called the **belly** of the muscle. A muscle's **innervation** refers to the identity of the nerve that stimulates it to contract, producing that muscle's **action**. **Figure 13-1** illustrates the origins of the biceps brachii on the scapula and its insertion onto the radius, as well as the action that results from its contraction.

Understanding innervation requires some information about nerves. Skeletal muscles are innervated by two groups of nerves: cranial nerves and spinal nerves.

Cranial nerves arise from the base of the brain and emerge through the foramina of the skull; these nerves innervate the skeletal muscles of the head and neck. The 12 cranial nerves are identified in two ways: by the abbreviation CN followed by Roman numerals (CN I–CN XII) and by names. Note that not all cranial nerves innervate skeletal muscles; some cranial nerves are sensory nerves that transmit nerve impulses from the sensory receptors of the eye, ear, and nose to the brain.

Spinal nerves arise from the spinal cord and emerge through the intervertebral foramina; these nerves innervate skeletal muscles below the neck. The 31 pairs of spinal nerves are identified by combinations of letters (C = cervical, T = thoracic, L = lumbar, S = sacral, C_o = coccygeal) and numbers that correspond to the spinal cord segment from which they arise. Thus, C_1 refers to the first cervical spinal nerve, T_8 to the eighth thoracic spinal nerve, and so forth.

Movements are usually the result of the actions of several skeletal muscles acting as a group. Muscles can be classified into four functional groups: prime movers, antagonists, synergists, and fixators. A **prime mover**, or agonist, is the muscle that is most responsible for producing a particular movement, termed that muscle's action. The term agonist can be applied to a prime mover or to any other muscle that performs the same action as the prime mover. An **antagonist** is a muscle that opposes the action of a prime mover. For example, during flexion of the elbow, the biceps brachii serves as the prime mover, and the triceps brachii serves as the antagonist. These two muscles constitute an antagonistic pair that acts on the opposite sides of a bone or joint.

A **synergist** is a muscle that aids a prime mover and is usually located close to it. A synergist can stabilize a joint, prevent unwanted movements, or add extra force so that the action of the prime mover is more coordinated and specific. Consider the role of the wrist extensor muscles as synergists in flexion of the fingers. The prime movers of finger flexion cross the intercarpal and the radiocarpal joints. You are able to flex your fingers without flexing your wrist because the

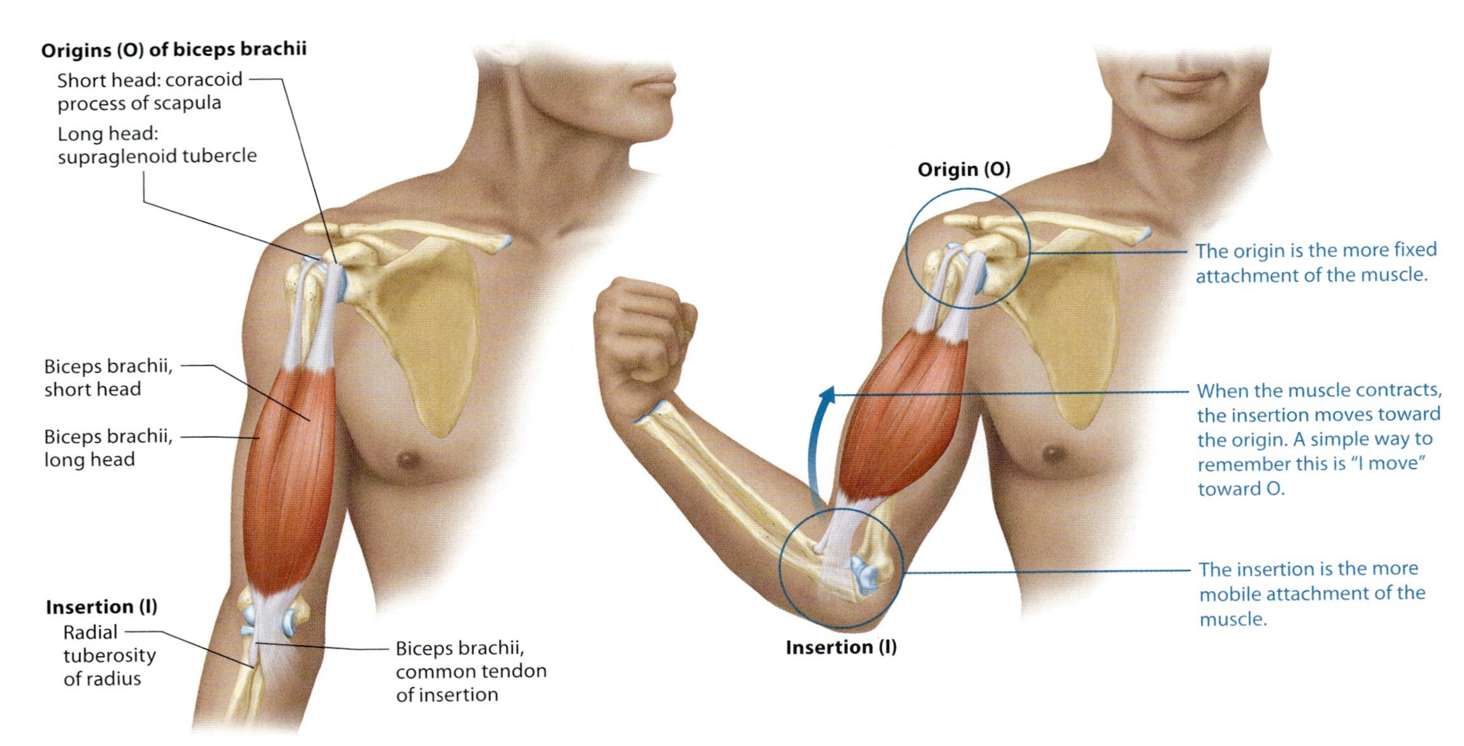

Figure 13-1 Origins and insertion of the biceps brachii muscle.

synergistic contraction of the wrist extensor muscles stabilizes the wrist joint and prevents it from flexing.

A **fixator** is a muscle that stabilizes the origin of the prime mover so that the prime mover can act more efficiently. A fixator "fixes" a bone—that is, it holds it steady.

Skeletal muscles are named based on one or more of the following seven criteria: location, shape, principal action, relative size, number of origins, locations of origins and insertions, and direction of muscle fibers. In the following activity you will identify skeletal muscles that are named based on each criterion.

ACTIVITY 1

Determining How Skeletal Muscles Are Named

Learning Outcomes

1. Define origin, insertion, action, and innervation.
2. Distinguish among prime movers (agonists), antagonists, synergists, and fixators.
3. Describe the criteria by which skeletal muscles are named.

Materials Needed

☐ Anatomical models and/or anatomical charts

Instructions

Refer to Figures 13-2 and 13-3 as you learn about the criteria by which skeletal muscles are named and identify examples of muscles named based on each criterion.

1. **Location of the muscle:** The names of some muscles refer to the bone or body region in which the muscle is found. For example, the brachialis overlies the brachium (arm).

List three additional muscles in Figures 13-2 and 13-3 that are named based on their locations. Then, locate each of these muscles on an anatomical model. ______________________

____

____

2. **Shape of the muscle:** The names of some muscles refer to their shapes. For example, the deltoid is roughly triangular (*delta* = triangle).

List three additional muscles in Figures 13-2 and 13-3 that are named according to their shapes. Then, locate each of these muscles on an anatomical model. ______________________

____

____

3. **The muscle's principal action:** The names of some muscles reflect the actions they perform. For example, the extensor digitorum extends the digits (fingers).

List three additional muscles in Figures 13-2 and 13-3 that are named according to their actions. Then, locate each of these muscles on an anatomical model. ______________________

____

____

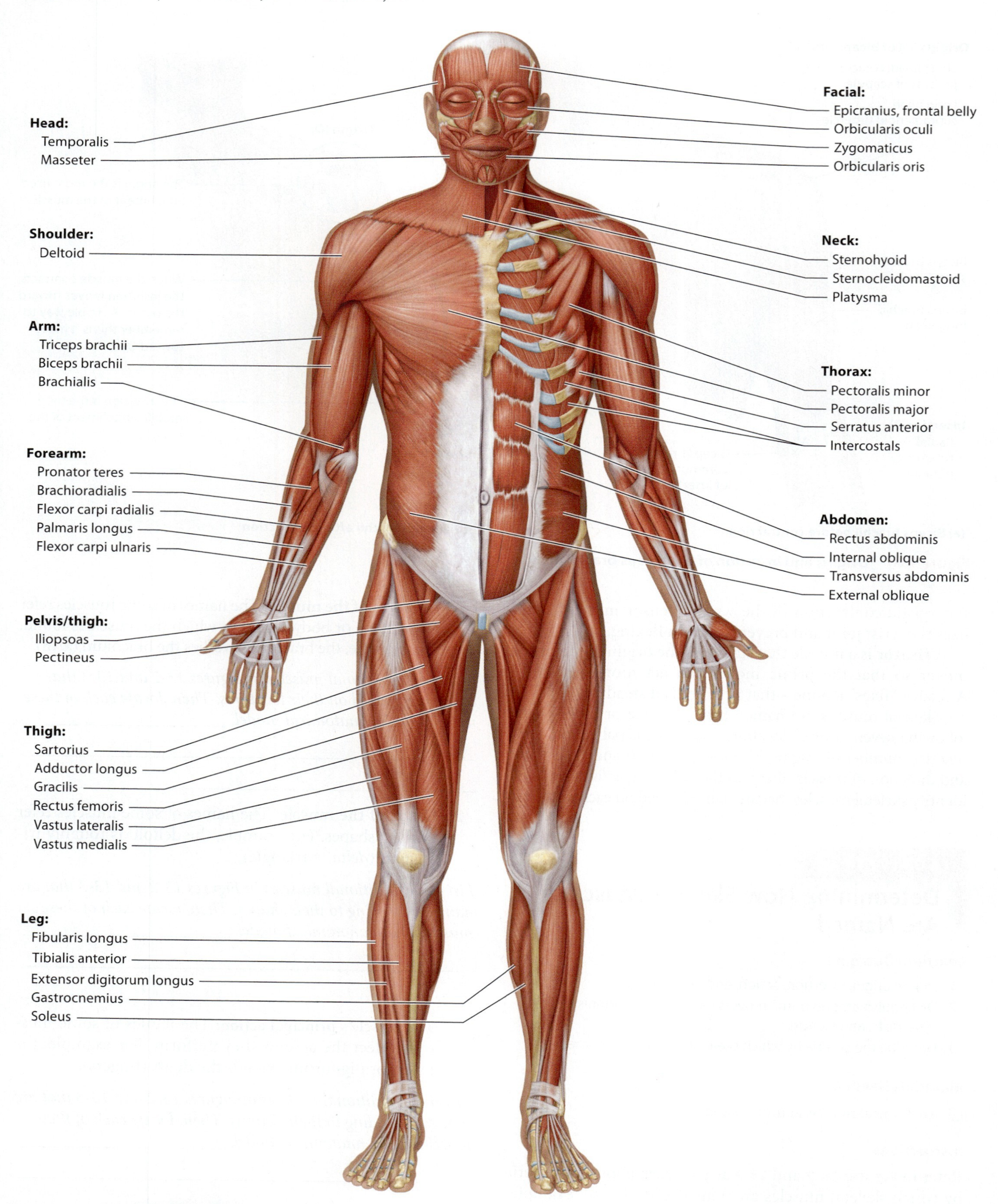

Figure 13-2 Major skeletal muscles, anterior view.

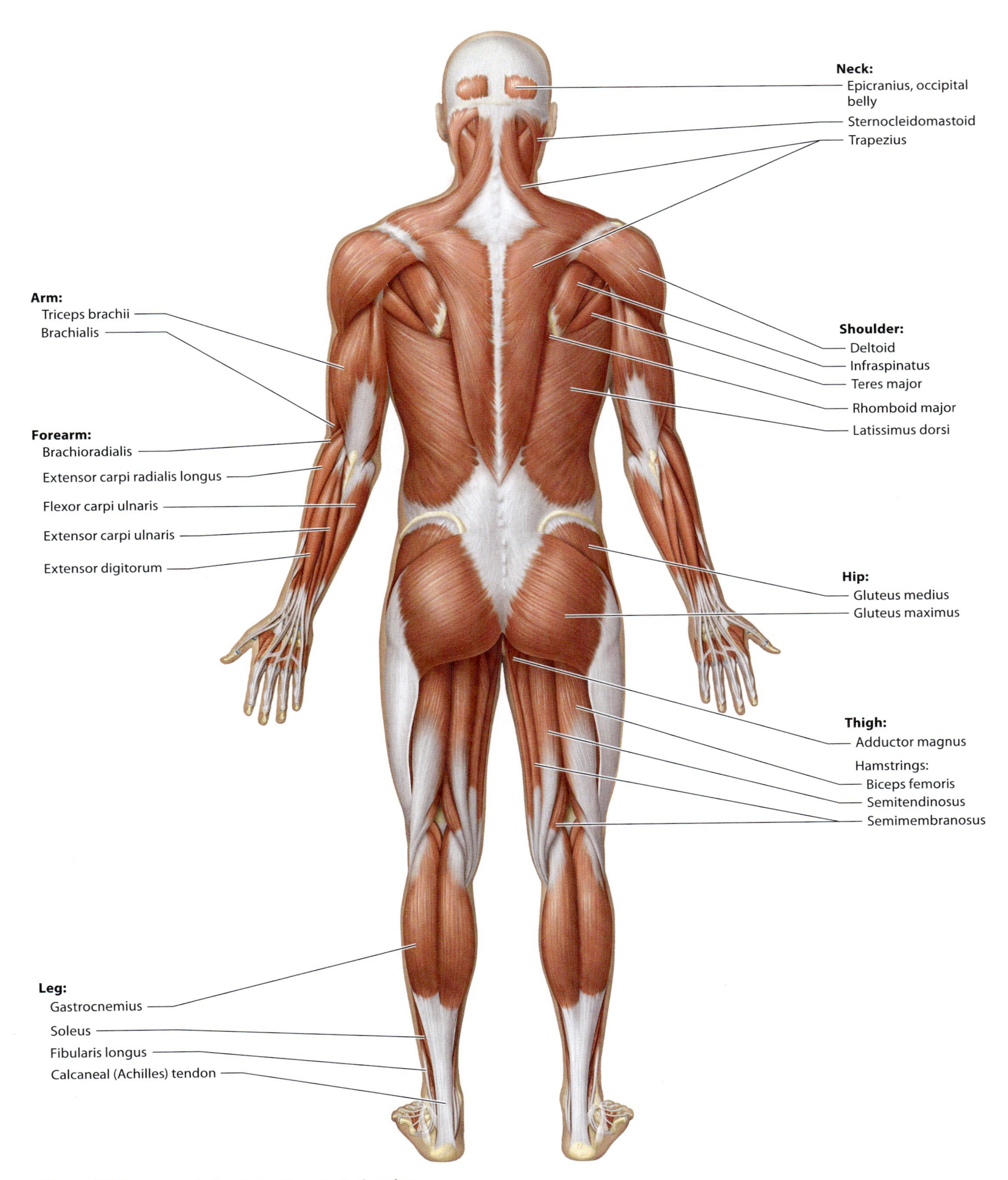

Figure 13-3 **Major skeletal muscles, posterior view.**

a. Review the types of movements allowed at synovial joints in Table 11-1 and Figure 11-6.

Describe and demonstrate the following movements:

flexion and extension ______________________

____

adduction and abduction ______________________

____

pronation and supination ______________________

____

4. **Relative size of the muscle:** The names of some muscles reflect the relative size of the muscle. For example, the gluteus maximus is the largest of the buttocks muscles, whereas the gluteus minimus is the smallest of the buttocks muscles.

List three additional muscles in Figures 13-2 and 13-3 that are named according to their relative size. Then, locate each of these muscles on an anatomical model. ______________________

____

____

5. **Number of origins:** Some muscle names refer to the number of origins the muscle has. For example, the biceps brachii has two origins.

List two additional muscles in Figures 13-2 and 13-3 that are named according to the number of origins they have. Then, locate each of these muscles on an anatomical model.

____

____

6. **Locations of the muscle's origins and insertions:** Some muscles are named according to their points of attachment. For example, the sternocleidomastoid originates on the sternum and the clavicle, and it inserts onto the mastoid process of the temporal bone.

List one additional muscle in Figures 13-2 and 13-3 that is named according to its points of attachment. Then, locate this muscle on an anatomical model. ______________________

____

7. **Direction of the muscle's fibers (and fascicles):** Some muscle names refer to the orientation of the muscle's fascicles relative to the body's midline. For example, in muscles whose names include the term rectus, meaning "straight," the fascicles run parallel to the body's midline. In muscles whose names include the term oblique, the fascicles run obliquely with respect to the body's midline.

List three additional muscles in Figures 13-2 and 13-3 that are named according to the direction of their fascicles. Then, locate each of these muscles on an anatomical model.

____

____

Muscles of the Head and Neck

The muscles of the head and neck can be divided into four functional groups: the muscles of facial expression, the muscles of mastication (chewing) and tongue movements, the muscles that move the hyoid bone and larynx, and the muscles that move the head and neck.

Muscles of Facial Expression

The muscles that enable us to communicate emotions through facial expression are located just deep to the skin in the scalp, face, and neck (**Figure 13-4**). These muscles are unusual in that they originate on the bones of the skull and then insert into the skin. Thus, when they contract, the skin moves. All facial muscles are innervated by the facial nerve (CN VII).

The **epicranius** is a bipartite (two-part) muscle with two bellies (an occipital belly and a frontal belly) connected by the epicranial aponeurosis (see Figure 13-4a and b). The frontal belly (also called the **frontalis**) overlies the frontal bone and raises the eyebrows (as when you are surprised), whereas the occipital belly (also called the **occipitalis**) draws the scalp back toward the posterior neck. The **orbicularis oculi** (see Figure 13-4a and b) encircles the eye and closes it during blinking. Both the **zygomaticus major** and **zygomaticus minor** (see Figure 13-4a and b) originate on the zygomatic bone and draw the lateral corners of the mouth upward (as in smiling). The **risorius** (see Figure 13-4a and b) is a narrow muscle located inferior and lateral to the zygomaticus major. It pulls the lips laterally and is considered to be a synergist of the zygomaticus muscles. The **buccinator** (see Figure 13-4a and b) is a thin, horizontal cheek muscle that lies deep to the masseter muscle; it compresses the cheeks during whistling, blowing, or sucking on a straw. This muscle is well developed in nursing infants. The **orbicularis oris** (see Figure 13-4a and b) is a circular muscle that closes and protrudes the lips as in kissing. The **depressor labii inferioris** (see Figure 13-4a and b) depresses the lower lip; the **levator labii superioris** (see Figure 13-4a and b) raises the upper lip. The **mentalis** (see Figure 13-4a and b) elevates and protrudes the lower lip and pulls the skin of the chin up as in pouting. Finally, the **platysma** (see Figure 13-4a and b) is an unpaired, thin sheetlike superficial neck muscle that depresses the mandible and draws the outer part of the lower lip inferiorly and posteriorly, as in pouting. **Table 13-1** summarizes the muscles of facial expression.

Table 13-1 Muscles of Facial Expression

Muscle	Action(s)	Origin/Insertion/Nerve(s)
Frontalis	**Raises** eyebrows; wrinkles skin of forehead horizontally	O: Epicranial aponeurosis I: Skin of eyebrows N: Facial nerve
Occipitalis	Pulls scalp **posteriorly**	O: Occipital bone I: Epicranial aponeurosis N: Facial nerve
Corrugator supercilii	Pulls eyebrows **inferiorly** and **medially** (as in squinting)	O: Medial supraorbital margin of frontal bone I: Skin of medial eyebrows N: Facial nerve
Orbicularis oculi	**Closes** eye; pulls skin around the eyes, as in blinking and winking	O: Orbital portions of the frontal bone and maxilla I: Skin of the orbital area and eyelids N: Facial nerve
Levator labii superioris	**Elevates** the upper lip; everts and furrows upper lip (as in sneering)	O: Zygomatic and upper maxilla near orbit I: Skin and muscle of the upper lip N: Facial nerve
Zygomaticus minor	**Raises** lateral portion of the upper lip to expose upper teeth (as in smiling)	O: Zygomatic I: Skin and muscle of the lateral upper lip N: Facial nerve
Zygomaticus major	Pulls the angle of the mouth **superiorly** and **laterally** (as in smiling or laughing)	O: Zygomatic I: Lateral muscle fibers of corner/angle of mouth N: Facial nerve
Risorius	Pulls the angle of the mouth **laterally** to make a closed-mouth smile	O: Connective tissue anterior to the ear I: Modiolus* N: Facial nerve
Orbicularis oris	**Closes** and **protrudes** lips (as in puckering the lips for a kiss)	O: Maxilla and mandible I: Skin and connective tissue of the lips N: Facial nerve
Depressor anguli oris	Draws corners of the mouth **inferiorly** (unhappy face)	O: Lower body of mandible I: Modiolus* N: Facial nerve
Depressor labii inferioris	**Protrudes** lower lip (sad or pouting expressions)	O: Medial mandible near mental foramen I: Skin and connective tissue of lower lip N: Facial nerve
Mentalis	**Protrudes** the lower lip and chin for drinking and "doubtful" expression	O: Anterior mandible I: Skin of the chin near lower lip N: Facial nerve
Buccinator	Helps manipulate food during chewing and **expels** air through pursed lips (as in blowing a trumpet)	O: Molar regions of maxilla and mandible I: Orbicularis oris and connective tissue of cheek/lips N: Facial nerve
Platysma	Lowers lower lip and opens mouth by **depressing** the mandible	O: Connective tissue of deltoid and pectoralis major I: Mandible; skin and connective tissue below mouth N: Facial nerve

Concept Figures

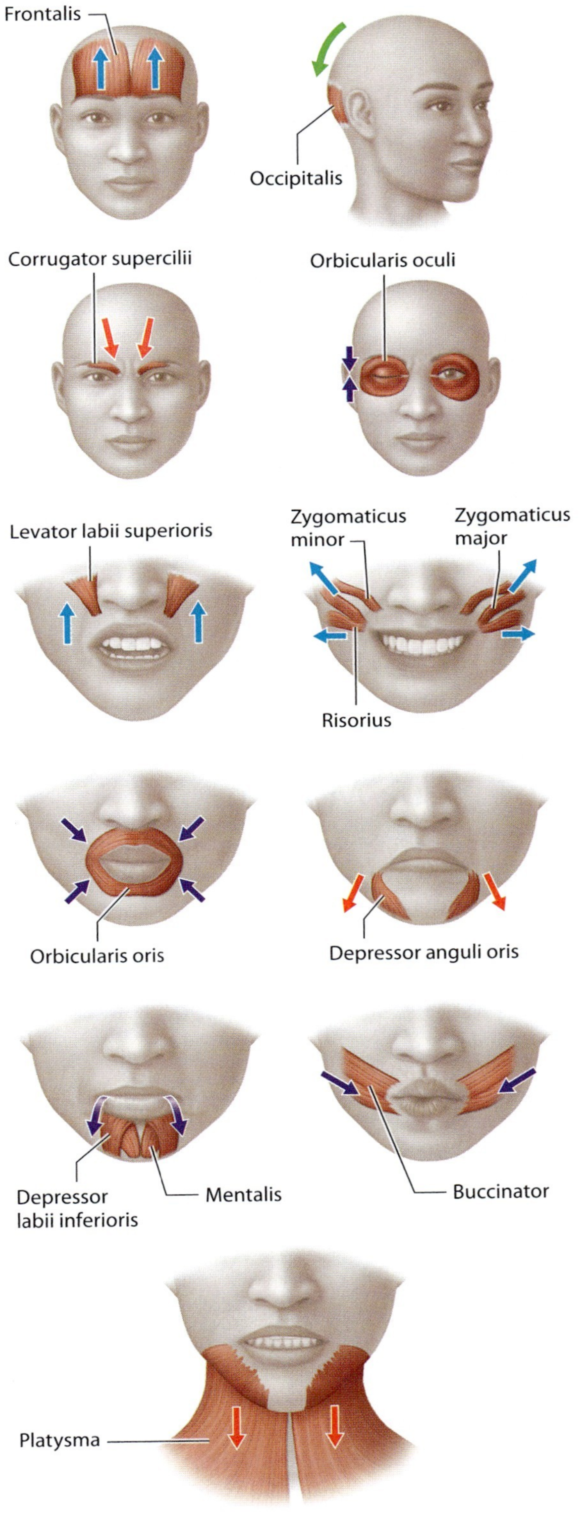

Note: Colors of actions and/or directions of action in Action(s) column match colors of directional arrow(s) in Concept Figures.

*Mix of muscle and connective tissue at the corners of the mouth.

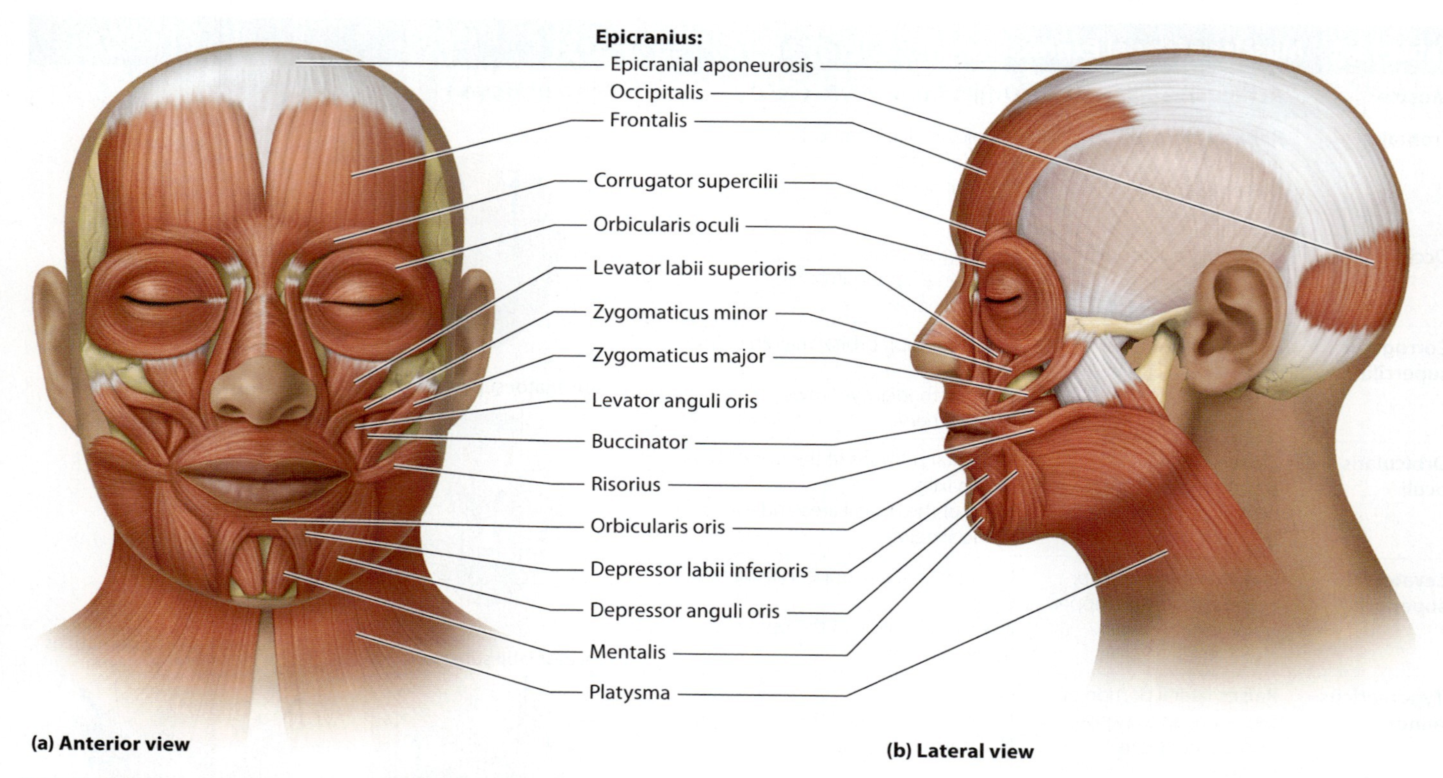

Figure 13-4 **Muscles of the head and neck.**

Muscles of Mastication (Chewing) and Tongue Movements

The four muscles of mastication (**Figure 13-5**) produce chewing movements by depressing and elevating the mandible, and all are innervated by the trigeminal nerve (CN V). The short and thick **masseter** (see Figure 13-5a) is a prime mover of jaw closure and biting. The fan-shaped **temporalis** (see Figure 13-5a) elevates the mandible and functions as an agonist to the masseter. The **lateral pterygoid** (see Figure 13-5b) depresses and protracts the mandible and moves it from side to side. Finally, the **medial pterygoid** (see Figure 13-5b) elevates the mandible and moves it from side to side. **Table 13-2** summarizes the muscles of chewing.

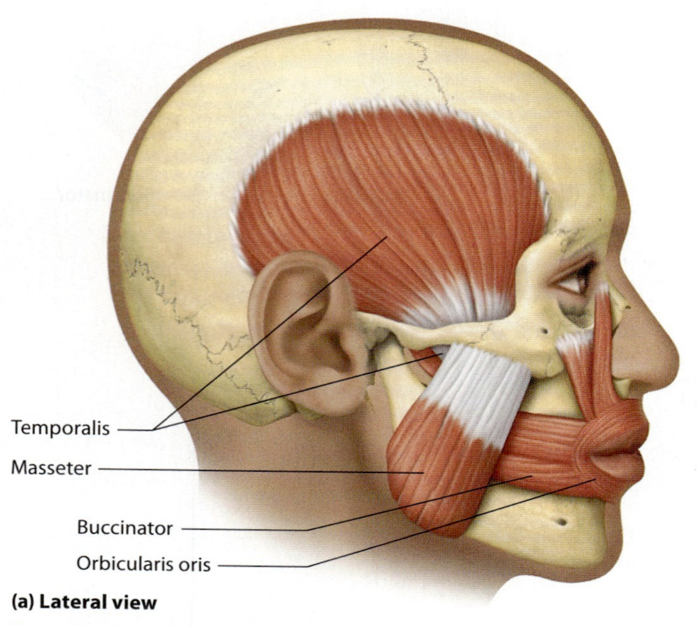

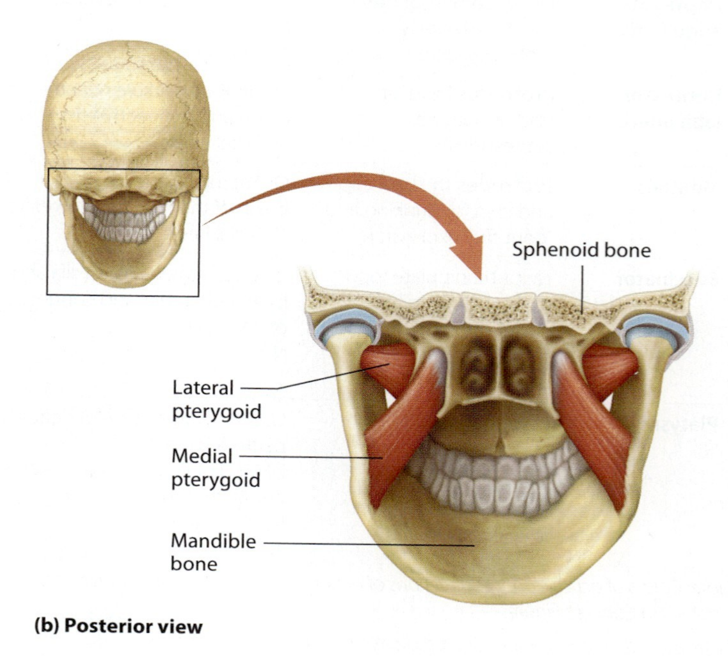

Figure 13-5 **Muscles of mastication.**

Table 13-2 Muscles of Chewing

Muscle	Action(s)	Origin/Insertion/Nerve(s)	Concept Figures
Masseter	Elevates the mandible	O: Zygomatic arch I: Angle and lateral surface of ramus of mandible N: Mandibular nerve (branch of the trigeminal nerve)	Temporalis Masseter
Temporalis	Elevates and retracts the mandible	O: Lateral surface of skull I: Coronoid process of the mandible N: Mandibular nerve	
Medial pterygoid	Elevates and protracts the mandible; assists in lateral movements to grind food	O: Pterygoid plate of the sphenoid and a small portion of the maxilla I: Medial angle and ramus of the mandible N: Mandibular nerve	Lateral pterygoid Medial pterygoid
Lateral pterygoid	Protracts and depresses the mandible; lateral movements to grind food	O: Pterygoid plate and greater wing of the sphenoid I: Condyle and neck of mandible N: Mandibular nerve	

The tongue is involved in mastication, detection of taste, and deglutition (swallowing). Muscles that control the tongue are classified as either intrinsic (originate and insert within the tongue and change the shape of the tongue) or extrinsic (originate outside the tongue and insert onto it). The three extrinsic tongue muscles (Figure 13-6) move the tongue in various directions and are innervated by the hypoglossal nerve (CN XII). The **genioglossus** depresses the tongue and thrusts it anteriorly (protraction); the **hyoglossus** also depresses the tongue; and the **styloglossus** elevates the tongue and draws it posteriorly (retraction). Note that the names of all these muscles end in –glossus, which means "tongue." The muscles of the tongue are summarized in **Table 13-3**.

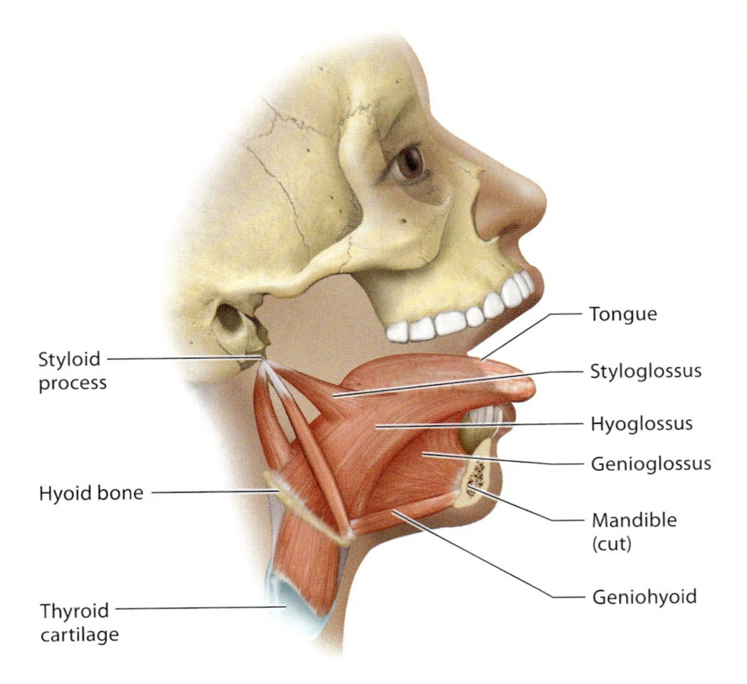

Figure 13-6 The extrinsic muscles of the tongue.

Muscles That Move the Hyoid Bone and Larynx

The muscles that move the hyoid bone and the larynx (Figure 13-7) consist of two groups called the suprahyoid and infrahyoid muscles.

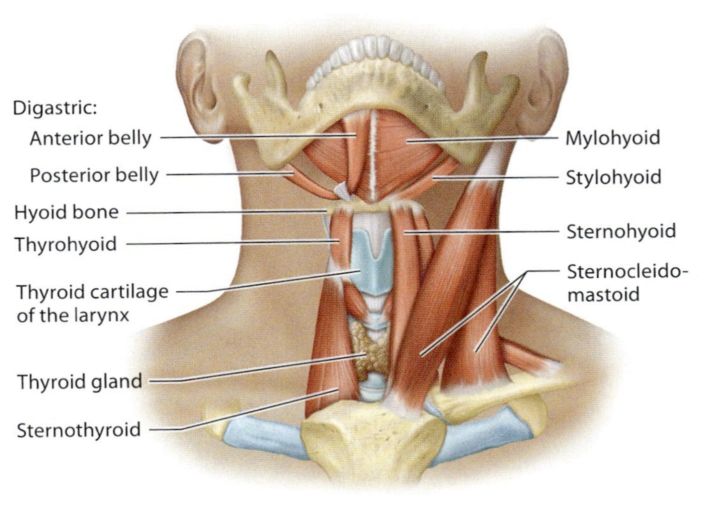

Figure 13-7 Muscles that move the hyoid bone and larynx, anterior view.

Table 13-3 Muscles of the Tongue

Muscle	Action(s)	Origin/Insertion/Nerve(s)	Concept Figure
Genioglossus	**Protrudes** tongue	O: Internal/anterior mandible (near symphysis) I: Majority of muscle blends into the tongue; body of hyoid bone N: Hypoglossal nerve	Styloglossus; Hyoglossus; Genioglossus
Hyoglossus	**Depresses** tongue	O: Hyoid bone (greater horn) I: Posterior lateral surface of the tongue N: Hypoglossal nerve	
Styloglossus	**Retracts** and **elevates** tongue	O: Styloid process of temporal bone I: Lateral/posterior border of tongue N: Hypoglossal nerve	

The suprahyoid muscles originate superior to the hyoid bone and elevate it, the floor of the oral cavity, and the tongue during swallowing. Among the suprahyoid muscles are the **digastric**, which consists of an anterior belly and a posterior belly united by a central tendon and which depresses the mandible and elevates the hyoid bone during swallowing and talking; the **mylohyoid**, which elevates the hyoid bone and the base of the tongue during swallowing; and the **stylohyoid**, which elevates the hyoid bone.

The infrahyoid muscles originate inferior to the hyoid bone and depress it and the larynx during swallowing and talking. Among the infrahyoid muscles are the **sternohyoid**, which depresses the hyoid bone and larynx during swallowing; the **sternothyroid**, which depresses the larynx; and the **thyrohyoid**, which depresses the hyoid bone. The muscles that move the hyoid bone and larynx are summarized in **Table 13-4**.

Table 13-4 Muscles That Move the Hyoid Bone and the Larynx

Muscle(s)	Action(s)	Origin/Insertion/Nerve(s)	Concept Figures
Stylohyoid	**Elevates** and **retracts** the hyoid and floor of the mouth during swallowing	O: Styloid process of temporal bone I: Hyoid bone N: Facial nerve	Oral cavity; Food; Pharynx; Hyoid
Mylohyoid	**Elevates** hyoid and floor of mouth, assisting tongue to push food toward the pharynx	O: Medial portion of the mandible I: Hyoid bone and midline connective tissue raphe N: Mylohyoid nerve	
Geniohyoid	**Elevates** and **protracts** hyoid	O: Inner surface of mandibular symphysis I: Hyoid bone N: Branch of the cervical plexus*	
Sternohyoid	**Depresses** hyoid bone and larynx	O: Posterior manubrium of sternum I: Lower portion of hyoid bone N: Branch of the cervical plexus	Oral cavity; Food; Hyoid; Larynx
Sternothyroid	**Depresses** larynx and hyoid bone	O: Posterior manubrium of sternum I: Thyroid cartilage of the larynx N: Branch of the cervical plexus	
Omohyoid	**Depresses** and **retracts** hyoid bone	O: Superior border of scapula I: Lower portion of hyoid bone N: Branch of the cervical plexus	
Thyrohyoid	**Depresses** hyoid bone; may elevate larynx	O: Thyroid cartilage of the larynx I: Hyoid bone N: Branch of the cervical plexus*	Geniohyoid; Digastric; Mylohyoid; Stylohyoid; Thyrohyoid; Omohyoid; Sternothyroid; Sternohyoid
Digastric	**Depresses** the mandible; fixator of the hyoid during swallowing	O: Anterior belly: anterior inner margin of the mandible; posterior belly: into mastoid notch, medial to the mastoid process I: Connective tissue of the hyoid bone N: Mylohyoid nerve (anterior belly) and facial nerve (posterior belly)	

*Cervical nerve fibers are distributed to this muscle via the hypoglossal nerve.

Muscles That Move the Head and Neck

The muscles that flex the head are arranged in pairs on the anterior and lateral sides of the neck; the muscles that extend the head are located on the posterior side of the neck (**Figure 13-8**). The pair of **sternocleidomastoid** muscles constitutes the main muscles of the anterior neck (see Figures 13-7 and 13-8a). Each sternocleidomastoid originates on the sternum and clavicle and inserts onto the mastoid process of the temporal bone. When acting together (bilaterally), these muscles flex the neck; when acting singularly (unilaterally), the sternocleidomastoid rotates the head toward the side opposite the contracting muscle. The **splenius** (see Figure 13-8b) is a bipartite muscle that extends the head; it consists of the splenius capitis and the splenius cervicis (deep to the trapezius). The **scalenes** (anterior, middle, and posterior) are located on the lateral aspect of the neck deep to both the platysma and the sternocleidomastoid; these muscles flex the neck and also elevate ribs 1 and 2 to aid in respiration.

Note that the levator scapulae, which are located on the posterolateral side of the neck (not shown), do not move the head; instead, they elevate the scapula (as their name suggests). **Table 13-5** summarizes the muscles that move the head and neck.

Table 13-5 Muscles That Move the Head and Neck

Muscle(s)	Action(s)	Origin/Insertion/Nerve(s)	Concept Figures
Sternocleidomastoid	Together: **flex** head (chin moves toward the chest); individually: flex and **rotate** the head toward opposite side; accessory muscles of respiration	O: Manubrium of sternum, medial portion of clavicle I: Mastoid process of temporal bone N: Accessory nerve	
Scalenes	A mix of actions, depending on the fixators in play; move head **laterally** when contracted individually; **elevate** rib cage with vertebral column fixed; accessory muscles of respiration	O: Tubercles of transverse processes of C_2–C_7* I: Laterally on the first two ribs N: Spinal nerve branches of C_3–C_8	
Trapezius (superior section)†	**Extends** the head (raises head from "bowed" position); other actions covered in the section on muscles of the arm	O: External occipital protuberance and cervical vertebrae I: Lateral portion of clavicle N: Accessory nerve	
Splenius capitis	**Extends** the head; with other muscles, **rotates** the head to the same side	O: Spinous processes of cervical vertebrae and connective tissue of the posterior neck (the ligamentum nuchae) I: Mastoid process and occipital bone N: Branches of C_2–C_3 dorsal rami (posterior branches of spinal nerves)	
Splenius cervicis	**Extends** the head and neck; **rotates** the upper cervical vertebrae to the same side as the muscle that is contracting	O: Spinous processes of T_3–T_6 I: Transverse processes of C_1–C_4 N: Branches of the lower cervical dorsal rami	
Semispinalis capitis	**Extends** the head; with other muscles, **rotates** the head to the opposite side from the muscle that is contracting	O: Transverse and articular processes of C_4–T_6 I: Occipital bone N: Variable dorsal rami of cervical and thoracic nerves	

*Some sources disagree about which is the origin or insertion, since both ends can move, depending on the fixators in play.

†The trapezius is covered in more detail with the muscles of the shoulder and arm; however, it is important to note the substantial role of the superior portion of the trapezius in moving the head here.

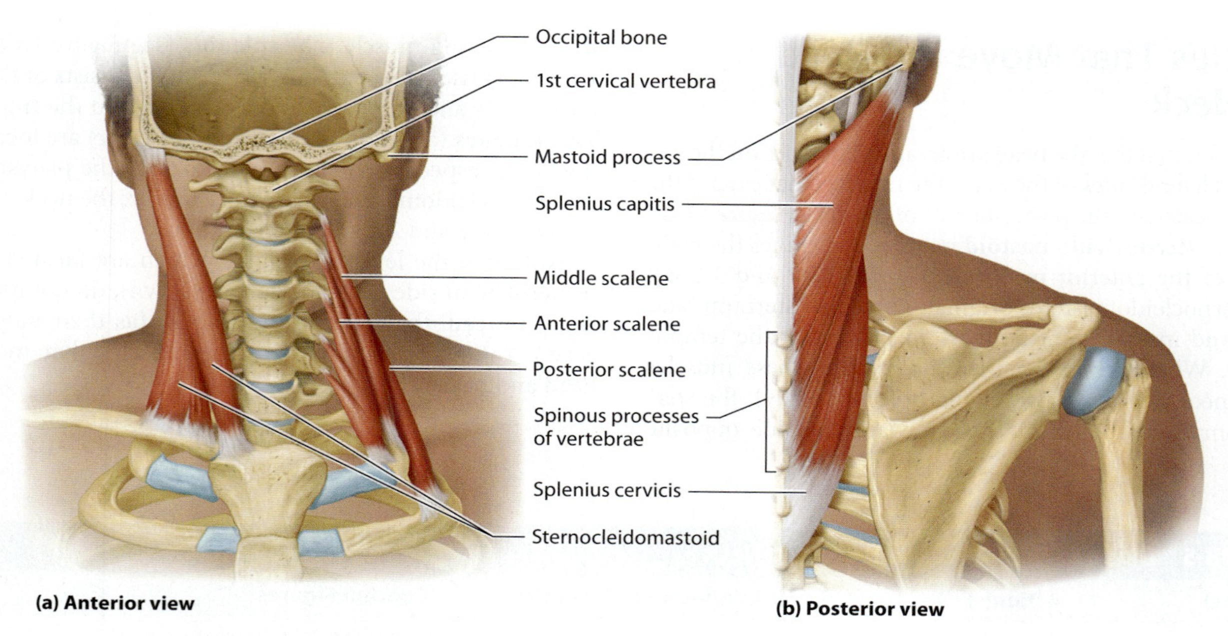

Figure 13-8 **Muscles that move the head and neck.**

ACTIVITY 2

Mastering the Muscles of the Head and Neck

Learning Outcomes

1. Locate assigned skeletal muscles of the head and neck on anatomical models, and describe and demonstrate the major action(s) of each.
2. Make connections between each assigned muscle and things you are learning in this unit. If required by your instructor, make connections between each assigned muscle and its origins and insertions.

Materials Needed

☐ Muscle models or anatomical charts
☐ Articulated skeleton

Instructions

CHART For each muscle assigned by your instructor, (1) locate the muscle on a muscle model or anatomical chart, (2) describe the action(s) of the muscle, and (3) perform the action(s) of the muscle using both your body and an articulated skeleton. (Refer to Unit 11 if you need a review of movements allowed by synovial joints.) Then design a chart similar to the one below so that you can organize the important information about each muscle assigned. Include in each "Connections to Things I'm Learning about Muscles" box relevant information from other sources, including lecture. Examples of "connections" include information conveyed by the name of the muscle, its agonists and antagonists, and perhaps its origin, insertion, and innervation. If knowing origins, insertions, and innervations is not required, the first example may be more useful. If that information is required, the second

Making Connections: Muscles of the Head and Neck

Muscle	Action(s)	Connections to Things I'm Learning about Muscles
Epicranius	*Frontal belly raises eyebrows; occipital belly pulls scalp posteriorly*	*Epi = above, so this muscle is located above the cranium; bipartite muscle; two bellies connected by epicranial aponeurosis*
Orbicularis oculi	*Closes the eyelids*	*Circular muscle; originates on medial aspect of orbit and inserts on skin of eyelid; innervated by the facial nerve*

example may be more useful. In that case, locate the origin and insertion of each assigned muscle on the articulated skeleton.

Optional Activity

Practice labeling muscles on human cadavers at MasteringA&P® > Study Area > Practice Anatomy Lab > Human Cadaver > Muscular System: Head and Neck

Play animations of muscle origins, insertions, and actions at MasteringA&P® > Study Area > Practice Anatomy Lab > Animations > Head and Neck

Muscles of the Trunk

The muscles of the trunk include muscles of the vertebral column, the muscles of respiration, muscles of the abdominal wall, and muscles that act on the pectoral girdle. The superficial back muscles play major roles in moving the limbs, and we consider these muscles in greater depth later in this unit. First, however, we focus on the deep muscles of the back, those that are associated with the vertebral column.

Muscles of the Vertebral Column

The muscles associated with the vertebral column (**Figure 13-9**) include the erector spinae group, the transversospinal group, and the quadratus lumborum. The **erector spinae group**, which extends and flexes the vertebral column laterally, consists of three separate muscle groups that run between the sacrum and the posterior surface of the skull on either side of the vertebral column (see Figure 13-9a). These three muscle groups are (from medial to lateral) the **spinalis, longissimus**, and **iliocostalis**. Each of these muscle groups in turn contains three muscles. The spinalis and longissimus groups each consist of a capitis muscle, a cervicis muscle, and a thoracis muscle. The iliocostalis group consists of a cervicis muscle, a thoracis muscle, and a lumborum muscle. The **transversospinal group**, which also extends the vertebral column, is a similar but deeper group of muscles (see Figure 13-9b). It includes the **semispinalis** group consisting of capitis, cervicis, and thoracis portions; the **multifidus** muscle; and the **rotatores** muscle (not shown). Finally, the square-shaped **quadratus lumborum** (see Figure 13-9b) flexes the vertebral column laterally. **Table 13-6** summarizes the muscles of the vertebral column.

Muscles of Respiration

A major function of the muscles of the trunk is respiration. Together the diaphragm, which separates the thoracic and abdominopelvic cavities, and the intercostal muscles, located between the ribs, move air in and out of the lungs during ventilation.

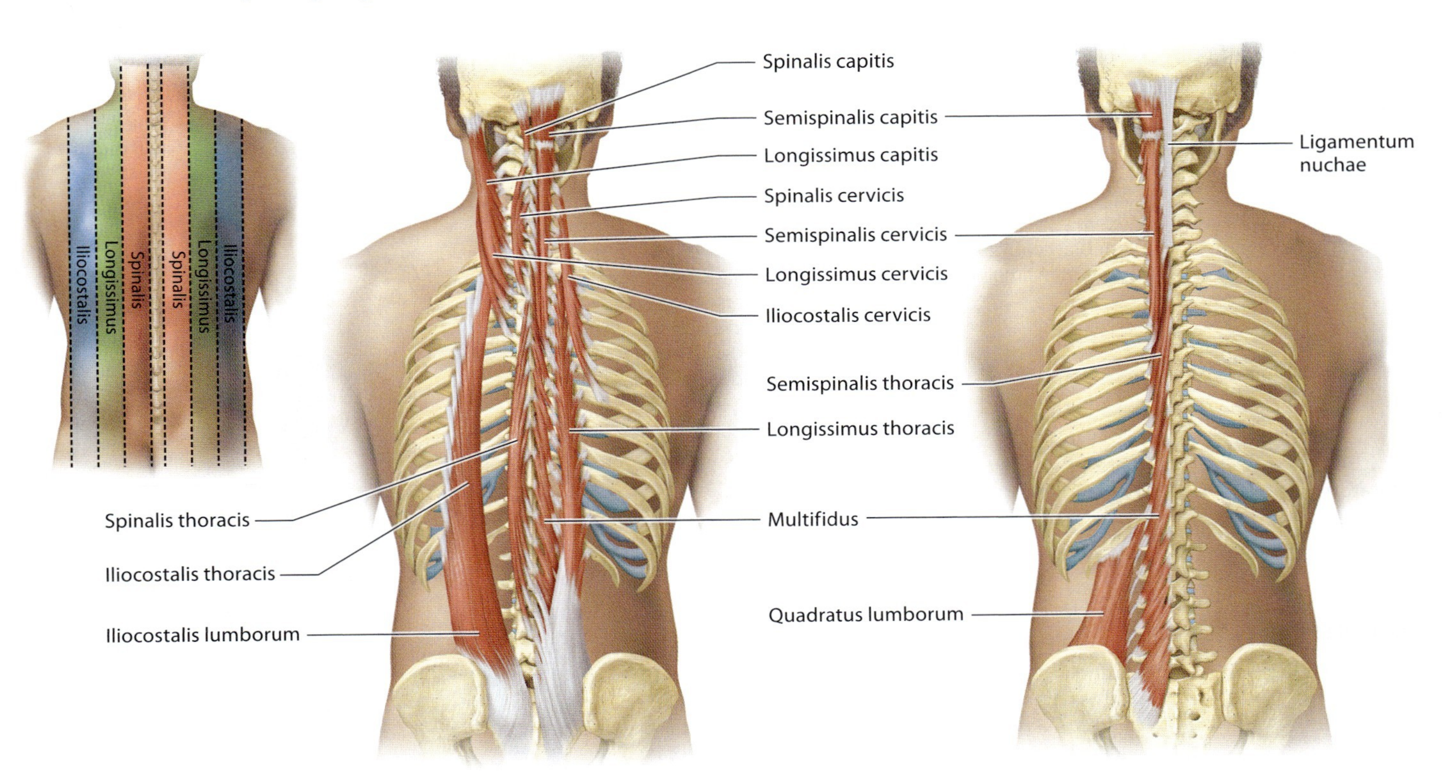

Figure 13-9 Muscles associated with the vertebral column.

Table 13-6 Muscles of the Vertebral Column

Muscle(s)	Action(s)	Origin/Insertion/Nerve(s)	Concept Figures
ERECTOR SPINAE GROUP			
Spinalis column Capitis Cervicis Thoracis	**Extend** the vertebral column (particularly the thoracis); capitis and cervicis may play a minor role in **extending** the head	O: Spinous processes of upper lumbar, thoracic, and lower cervical vertebrae I: Spinous processes of thoracic and cervical vertebrae and skull N: Dorsal rami of cervical, thoracic, and lumbar spinal nerves	
Longissimus column Capitis Cervicis Thoracis (largest and most powerful component of the erector spinae)	**Extend** the vertebral column (particularly the thoracis) and maintain posture; capitis and cervicis may play a minor role in **extending** the head and rotating it to the same side; **laterally bend** the vertebral column when contracted on one side only	O: Transverse processes of vertebrae—usually several vertebrae lower than insertions I: Mastoid process (capitis); transverse processes of vertebrae; medial posterior portions of ribs in the thoracic area N: Dorsal rami of cervical, thoracic, and lumbar spinal nerves	
Iliocostalis column Cervicis Thoracis Lumborum	**Extend** the vertebral column (particularly the thoracis and lumborum); maintain **posture**; **laterally bend** the vertebral column when contracted on one side only	O: Posterior surfaces of ribs; portions of sacrum and ilium I: Transverse processes of lower cervical vertebrae; posterior surfaces of ribs N: Dorsal rami of cervical, thoracic, and lumbar spinal nerves	
TRANSVERSOSPINAL GROUP			
Semispinalis capitis (see Table 13-5), cervicis, and thoracis	**Extend** the vertebral column; capitis and cervicis **extend** the head	O: Transverse and articular processes of C_4–T_6 I: Spinous processes (several vertebrae higher than origins for cervicis and thoracis) and skull N: Dorsal rami of cervical and thoracic nerves	
Multifidus	Synergist of other muscles of vertebral column movement and assists in maintaining **posture**	O: In its inferior half, sacrum and transverse processes of lumbar vertebrae; in its superior half, transverse processes of thoracic vertebrae and articular processes of lower cervical vertebrae I: Spinous processes of vertebrae from L_5 to C_2 N: Dorsal rami of cervical and thoracic nerves	
Rotatores	**Extend** the vertebral column	O: Transverse process of one vertebra I: Spinous process of next one or two vertebrae above N: Dorsal rami of cervical, thoracic, and lumbar nerves	
OTHER SPINAL EXTENSORS			
Quadratus lumborum	**Extend** the vertebral column; maintain **posture**; **laterally bend** the vertebral column when contracted on one side only	O: Iliac crest and connective tissue of lumbar region I: Rib 12 and transverse processes of lumbar vertebrae N: Ventral rami of T_{12} and lumbar spinal nerves	

The most important muscle of inspiration (**Figure 13-10a**) is the dome-shaped **diaphragm**, innervated by the phrenic nerve. The fibers of the 11 pairs of **external intercostals** run obliquely (inferiorly and anteriorly) between the ribs (**Figure 13-10b**). The external intercostals, which pull the ribs toward one another and elevate the rib cage, are innervated by the intercostal nerves. The fibers of the 11 pairs of the **internal intercostals** run deep to and at right angles to those of the external intercostal muscles and serve as antagonists to the external intercostals. The internal intercostals depress the rib cage and draw the ribs together, and they are innervated by the intercostal nerves. **Table 13-7** summarizes the muscles of respiration.

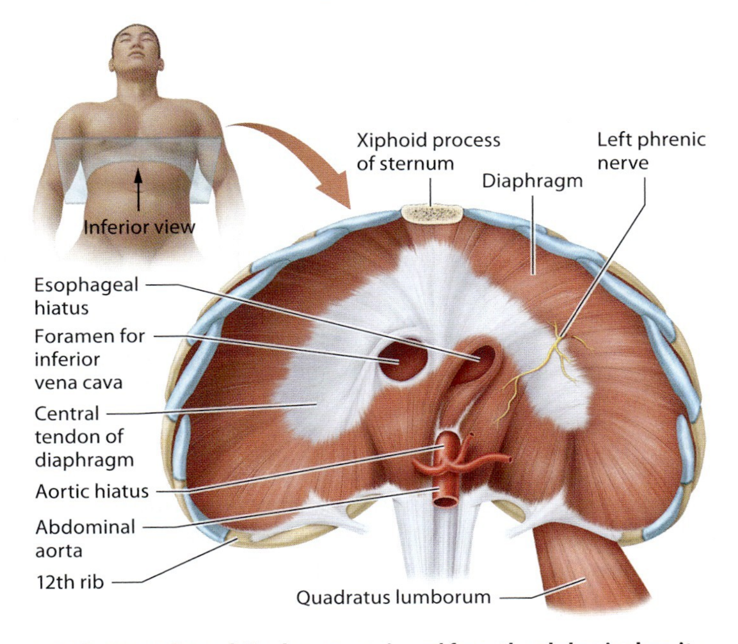

(a) Inferior surface of diaphragm as viewed from the abdominal cavity

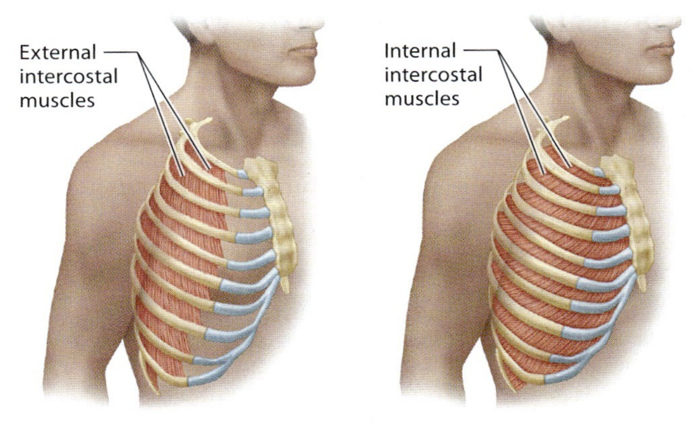

(b) The external and internal intercostal muscles

Figure 13-10 Muscles of respiration.

Table 13-7 Muscles of Respiration (Breathing)*

Muscle	Action	Origin/Insertion/Nerve(s)	Concept Figure
Diaphragm	**Flattens** to lower the floor of the thoracic cavity to increase volume and decrease pressure, thereby causing inspiration (inhalation)	O: Xiphoid process of sternum, lower ribs and costal cartilages, lumbar vertebrae I: Central tendon of diaphragm itself N: Phrenic nerve	Diaphragm
External intercostal muscles	**Elevates** the rib cage, spreading the ribs, assisting inspiration	O: Lower edge of rib superior to its insertion I: Upper edge of rib inferior to its origin N: Intercostal nerves	External intercostal muscles
Internal intercostal muscles	**Depresses** the rib cage, pulling ribs closer together, assisting expiration	O: Upper edge of rib inferior to its insertion I: Lower edge of rib superior to its origin, deep to the insertion of the external intercostals N: Intercostal nerves	Internal intercostal muscles

*See Table 13-5 for scalenes and sternocleidomastoid, which also assist with forced inspiration.

Muscles of the Abdominal Wall

The abdominal muscles (**Figure 13-11**) support and protect the organs within the abdominal cavity; they also flex the vertebral column and rotate and laterally flex the trunk. The medial superficial **rectus abdominis** flexes the vertebral column. The **external oblique** and **internal oblique** also flex the vertebral column, and they also compress the abdominal wall. Finally, the **transversus abdominis** compresses the abdominal contents. All four abdominal muscles are innervated by the intercostal nerves. **Table 13-8** summarizes the muscles of the abdominal wall.

Muscles That Act on the Pectoral Girdle

The muscles that act on the pectoral girdle (**Figure 13-12**) support and position the scapula and the clavicle. They stabilize the shoulder joint by reinforcing the articulation between the humerus and the scapula.

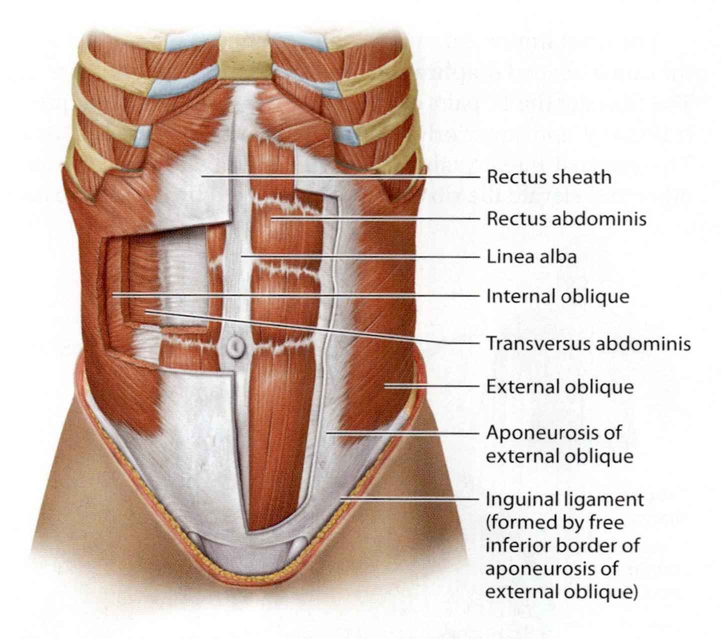

Figure 13-11 Muscles of the anterior trunk.

Table 13-8 Abdominal Muscles

Muscle	Action	Origin/Insertion/Nerve(s)	Concept Figures
Rectus abdominis	**Flexes** the trunk; with the vertebral column fixed, **compresses** abdominal cavity	O: Superior aspect of pubic bones I: Costal cartilages of lower ribs N: Branches of ventral rami of the lower six to seven thoracic nerves	Rectus abdominis External and internal obliques
External oblique	**Flexes** and **laterally bends** the trunk; with the vertebral column fixed, **compresses** abdominal cavity	O: Lower eight ribs I: Iliac crest, pubic tubercle, and linea alba N: Branches of the lower six thoracic nerves	External oblique
Internal oblique	**Flexes** and **laterally bends** the trunk; **compresses** abdominal cavity	O: Iliac crest and inguinal ligament I: Lower three to four ribs, linea alba, and pubic bone N: Branches of the lower six thoracic nerves and first lumbar nerve	
Transversus abdominis	**Compresses** abdominal cavity	O: Lower six costal cartilages, thoracolumbar fascia, crest and inguinal ligament I: Linea alba and pubic bone N: Branches of the lower six thoracic nerves and first lumbar nerve	Transversus abdominis Internal oblique

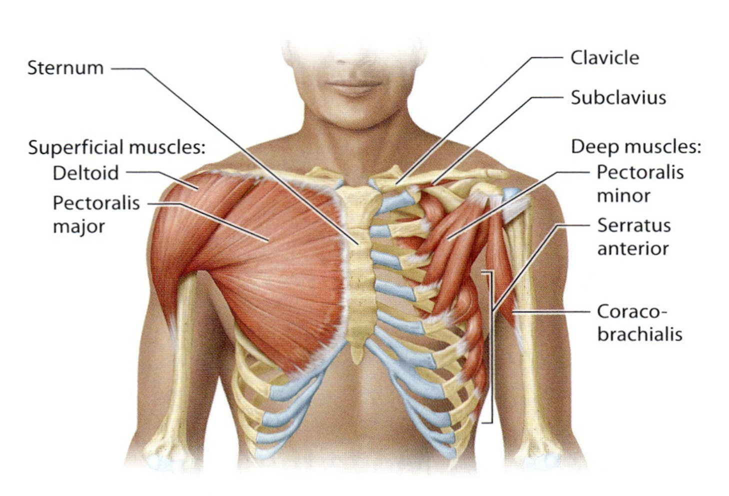

(a) Anterior view: superficial muscles shown on left; deep muscles shown on right, with superficial removed

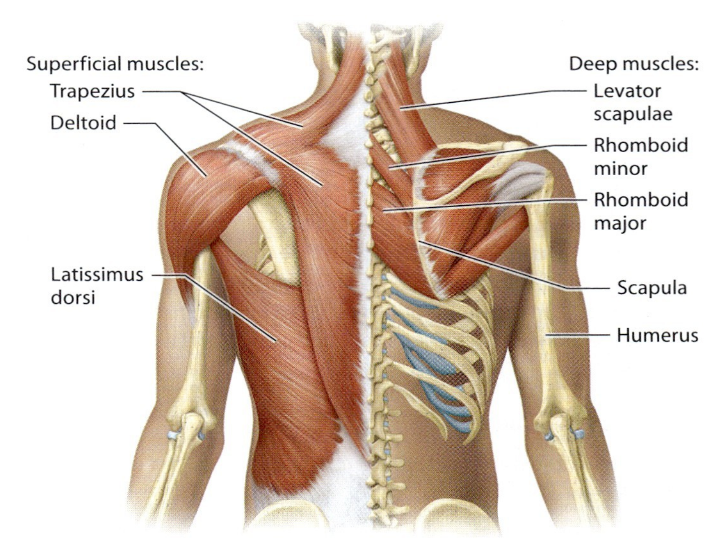

(b) Posterior view: superficial muscles shown on left; deep muscles shown on right, with superficial removed

Figure 13-12 Muscles that act on the pectoral girdle.

The thin, flat **pectoralis minor** (see Figure 13-12a) is deep to the **pectoralis major**; it pulls the pectoral girdle inferiorly and anteriorly and helps raise the ribs during forced inspiration. The large, flat, fan-shaped **serratus anterior** (see Figure 13-12a), located between the ribs and the scapula, is so named because of the serrated appearance of its origins on the ribs. It rotates the scapula such that the inferior angle moves laterally and superiorly. The **trapezius** (see Figure 13-12b) is one of a pair of large triangular muscles that span the gap between the two scapulae and extend from the posterior of the head to the inferior thoracic vertebrae. It extends the head and elevates, superiorly rotates, adducts, and stabilizes the scapula. The thick, strap-like **levator scapulae** (see Figure 13-12b) elevates and adducts the scapula. The diamond-shaped **rhomboid major** and **rhomboid minor** (see Figure 13-12b) lie inferior to the levator scapulae and deep to the trapezius. They adduct, elevate, and rotate the scapula inferiorly and stabilize its position. Finally, the small cylindrical **subclavius** (not shown) lies inferior to the clavicle and stabilizes and depresses the pectoral girdle. **Table 13-9** summarizes the muscles that act on the pectoral girdle.

Table 13-9 Muscles That Move the Scapula

Muscle(s)	Action(s)	Origin/Insertion/Nerve(s)	Concept Figures
Trapezius (superior section)	**Elevates** the scapula; **rotates** scapula upward in collaboration with inferior portion	O: External occipital protuberance and cervical vertebrae I: Lateral clavicle N: Spinal accessory nerve	*Posterior* Trapezius (superior section) Levator scapulae Rhomboid minor Rhomboid major
Levator scapulae	**Elevates** scapula	O: Transverse processes of C_1–C_4 I: Superior angle of scapula N: Dorsal scapular nerve	
Rhomboid major and rhomboid minor	**Retracts** scapula	O: Spinous processes of: T_2–T_5 (major); C_7–T_1 (minor) I: Medial border of scapula N: Dorsal scapular nerve	
Trapezius (middle and inferior section)	**Retracts** (via middle portion) and **depresses** (via inferior section) the scapula; **rotates** scapula upward in collaboration with upper fibers	O: Spinous processes of thoracic vertebrae I: Spine and acromion of scapula N: Spinal accessory nerve	*Posterior* *Anterior* Trapezius (mid/inferior section) Pectoralis minor Serratus anterior
Pectoralis minor	**Protracts** and **depresses** scapula	O: Anterior part of ribs 3–5 I: Coracoid process N: Medial pectoral nerve	
Serratus anterior	Prime mover of scapula in **protraction**; assists in upward rotation of scapula	O: First nine ribs I: Medial border of scapula N: Long thoracic nerve	

Note: Some of these actions vary depending on which fibers are stimulated.

ACTIVITY 3

Mastering the Muscles of the Trunk

Learning Outcomes

1. Locate assigned skeletal muscles of the trunk on anatomical models, and describe and demonstrate the major action(s) of each.
2. Make connections between each assigned muscle and things you are learning in this unit. If required by your instructor, make connections between each assigned muscle and its origins and insertions.

Materials Needed

- ☐ Muscle models or anatomical charts
- ☐ Articulated skeleton

Instructions

CHART For each muscle assigned by your instructor, (1) locate the muscle on a muscle model or anatomical chart, (2) describe the action(s) of the muscle, and (3) perform the action(s) of the muscle using both your body and an articulated skeleton. (Refer to Unit 11 if you need a review of movements allowed by synovial joints.) Then design a chart similar to the following chart so that you can organize the important information about each muscle assigned. Include in each "Connections to Things I'm Learning about Muscles" box relevant information from other sources, including lecture. Examples of "connections" include information conveyed by the name of the muscle, its agonists and antagonists, and perhaps its origin, insertion, and innervation. If knowing origins, insertions, and innervations is not required, the first example may be more useful. If that information is required, the second example may be more useful. In that case, locate the origin and insertion of each assigned muscle on the articulated skeleton.

Optional Activity

Practice labeling muscles on human cadavers at MasteringA&P® > Study Area > Practice Anatomy Lab > Human Cadaver > Muscular System: Trunk

Play animations of muscle origins, insertions, and actions at MasteringA&P® > Study Area > Practice Anatomy Lab > Animations > Trunk

Muscles of the Upper Limb

The muscles of the upper limb include muscles that move the arm, muscles that move the forearm, and muscles that move the wrist, hand, and digits.

Muscles That Move the Arm

Muscles that move the arm originate on the scapula or on the vertebral column, cross the shoulder joint, and insert onto the humerus (**Figure 13-13**). The large, superficial, fan-shaped **pectoralis major** (see Figure 13-13a) covers the superior portion of the chest and is a prime mover of arm flexion; it also adducts and medially rotates the arm. The **deltoid** (see Figure 13-13a and b) is a superficial triangular shoulder muscle and a prime mover of arm abduction. The broad flat **latissimus dorsi** (see Figure 13-13b), located in the lumbar region, is a prime mover of arm extension. It also adducts and medially rotates the arm. The thick rounded **teres major** (see Figure 13-13b) lies inferior to the teres minor (discussed shortly) and it extends, medially rotates, and adducts the humerus. The small cylindrical **coracobrachialis** (see Figure 13-13a) flexes and adducts the humerus.

The **supraspinatus** (see Figure 13-13b) lies deep to the trapezius and is one of the four rotator cuff muscles. The **infraspinatus** (see Figure 13-13b) is a rotator cuff muscle

Making Connections: Muscles of the Trunk

Muscle	Action(s)	Connections to Things I'm Learning about Muscles
Erector spinae	*Extends the vertebral column and bends it laterally*	*Tripartite muscle consisting of (from lateral to medial) the iliocostalis, longissimus, and spinalis muscles*
Rectus abdominis	*Flexes the vertebral column*	*Fibers run parallel to the midline of the body; most medial superficial abdominal muscle; originates on the pubic crest and pubic symphysis and inserts on the xiphoid process and costal cartilages of ribs 5–7; innervated by the intercostal nerves; antagonist to the erector spinae group*

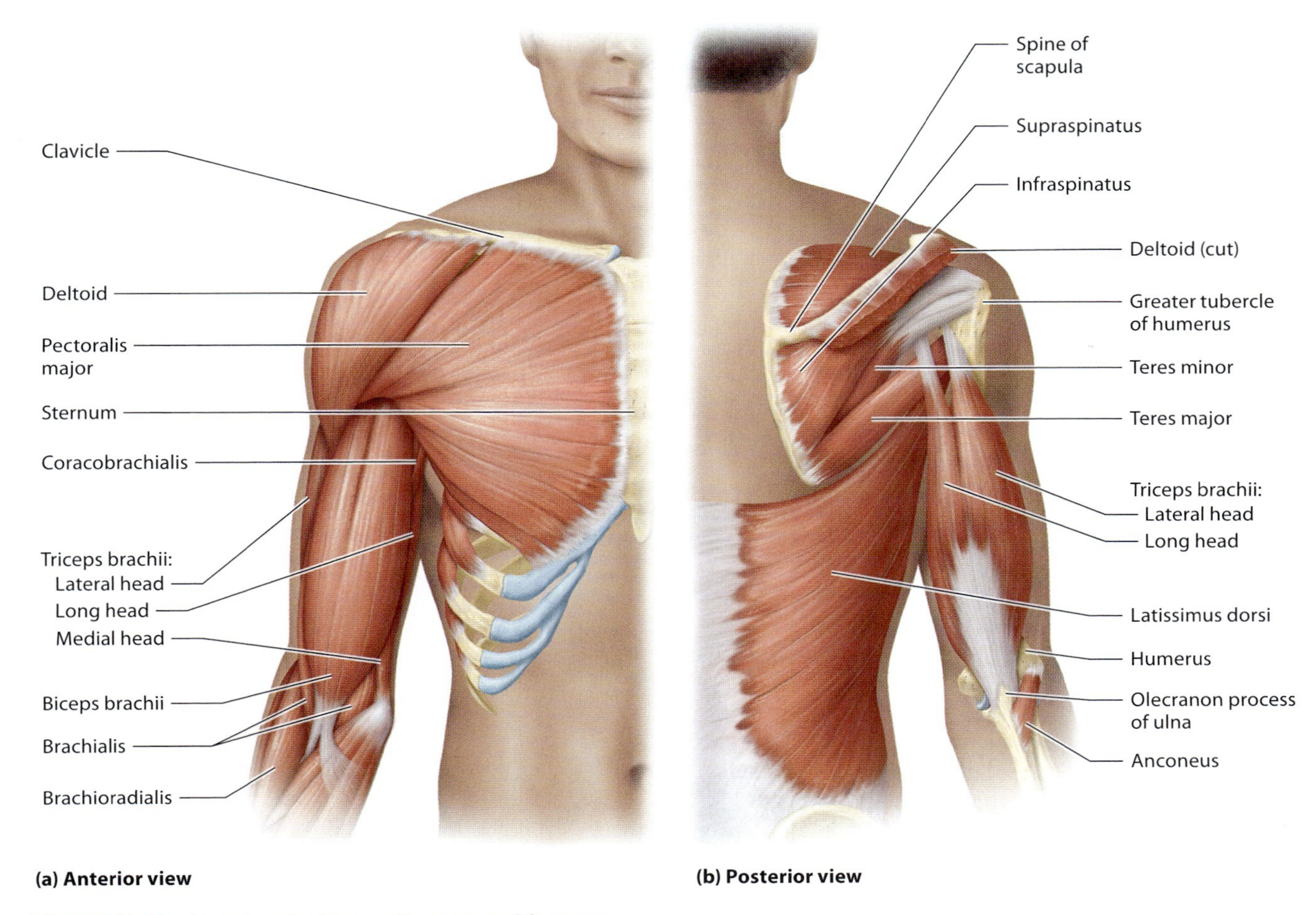

Figure 13-13 **Muscles that move the arm and forearm.**

that rotates the humerus laterally and stabilizes the shoulder. The **teres minor** (see Figure 13-13b) is a rotator cuff muscle that laterally rotates the humerus and stabilizes the shoulder. The **subscapularis** (not shown) is a rotator cuff muscle that medially rotates the humerus and stabilizes the shoulder. **Table 13-10** summarizes the muscles that move the arm.

Muscles That Move the Forearm

The muscles that move the forearm originate on the humerus, cross over the elbow joint, and insert onto the ulna and/or radius (see Figure 13-13). The large, three-headed **triceps brachii** (see Figure 13-13a and b) is the only muscle on the posterior side of the arm; it is a powerful forearm extensor. The two-headed **biceps brachii** (see Figure 13-13a) flexes and supinates the forearm. The superficial **brachioradialis** (see Figure 13-13a) of the lateral forearm acts as a synergist in forearm flexion. Finally, the **brachialis** (see Figure 13-13a) lies deep to the biceps brachii muscle and flexes the forearm. **Table 13-11** summarizes the muscles that move the forearm.

Muscles That Move the Wrist, Hand, and Digits

Muscles of the forearm that move the wrist, hand, and digits are illustrated in **Figure 13-14**. The superficial **pronator teres** (see Figure 13-14a) pronates the forearm and is also a weak forearm flexor. The **flexor carpi radialis** (see Figure 13-14a) is a superficial forearm muscle named for both its proximity to the radius and its actions (wrist flexion and abduction). The **palmaris longus** (see Figure 13-14a) is a weak wrist flexor. The superficial **extensor carpi ulnaris** (see Figure 13-14b) is named for both its proximity to the ulna and its actions (wrist extension and adduction). Both the **extensor carpi radialis longus** and the **extensor carpi radialis brevis** (see Figure 13-14b) extend and abduct the wrist. The superficial **extensor digitorum** (see Figure 13-14b) on the posterior surface of the forearm extends the digits (fingers). Finally, the **flexor carpi ulnaris** (see Figure 13-14b) flexes and adducts the wrist. **Table 13-12** summarizes the muscles that move the wrist, hand, and digits.

Table 13-10 Muscles That Move the Arm

Muscle	Action(s)	Origin/Insertion/Nerve(s)
Pectoralis major	**Flexes** and **adducts** the arm; **rotates** arm medially	O: Medial clavicle, sternum, costal cartilages 1–7 I: Greater tubercle and lateral lip of intertubercular sulcus of humerus N: Medial and lateral pectoral nerves
Latissimus dorsi	**Adducts** and **extends** the arm; **rotates** arm medially	O: Iliac crest, spinous processes of lower thoracic and all lumbar vertebrae I: Floor of intertubercular sulcus of the humerus N: Thoracodorsal nerve
Deltoid	**Abducts** the arm; secondarily **flexes** and **extends** arm	O: Acromion and spine of scapula; lateral clavicle I: Deltoid tuberosity of humerus N: Axillary nerve
Teres major	**Adducts**, **extends**, and **rotates** arm laterally	O: Posterior, inferior portion of scapula I: Medial lip of intertubercular sulcus of the humerus N: Lower subscapular nerve
Coracobrachialis	**Flexes** and **adducts** the arm	O: Coracoid process I: Medial diaphysis of humerus N: Musculocutaneous nerve
Supraspinatus*	Assists **abduction**; holds the humerus and stabilizes shoulder joint	O: Supraspinous fossa of scapula I: Greater tubercle of humerus N: Suprascapular nerve
Infraspinatus*	Laterally **rotates** the humerus; stabilizes shoulder joint	O: Infraspinous fossa of scapula I: Greater tubercle of humerus N: Suprascapular nerve
Teres minor*	Laterally **rotates** the humerus; stabilizes shoulder joint	O: Posterior, lateral border of scapula I: Greater tubercle of humerus N: Axillary nerve
Subscapularis*	**Adducts** and **rotates** the humerus medially; stabilizes shoulder joint	O: Subscapular fossa of scapula I: Lesser tubercle of humerus N: Subscapular nerves

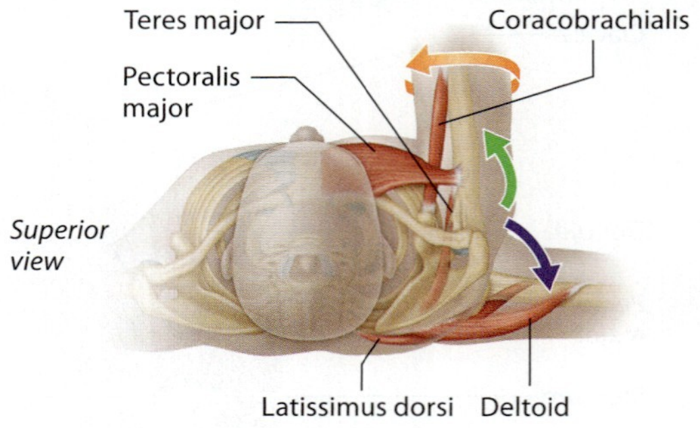

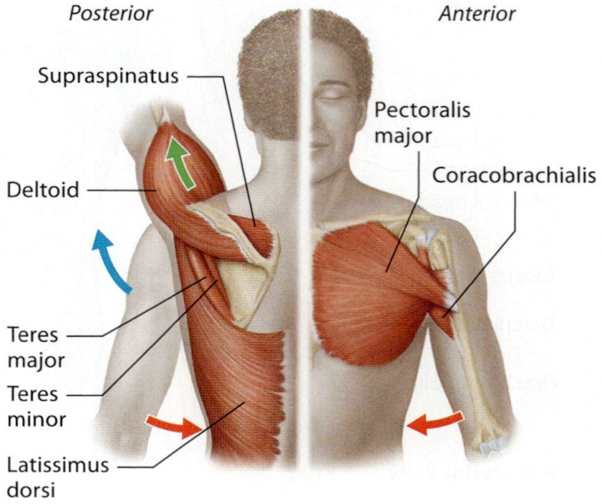

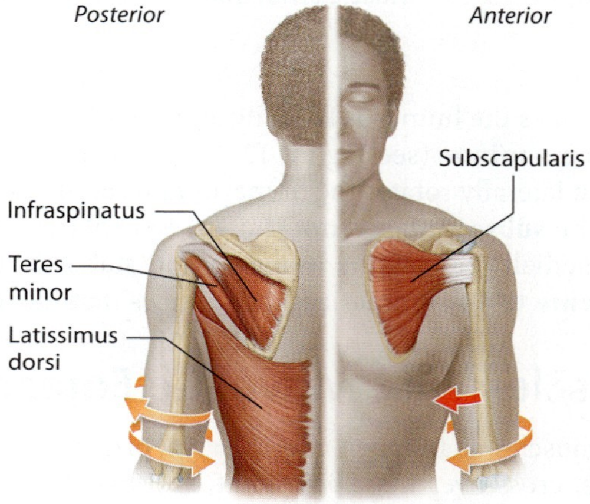

Concept Figures

*Rotator cuff muscles.

Table 13-11 Muscles That Move the Forearm at the Elbow Joint

Muscle	Action(s)	Origin/Insertion/Nerve(s)	Concept Figure
Biceps brachii	**Flexes** the elbow; **supinates** the forearm	O: Supraglenoid tubercle (long head) and coracoid process (short head) I: Radial tuberosity N: Musculocutaneous nerve	
Brachialis	**Flexes** the elbow (prime mover)	O: Distal half of the diaphysis of the anterior humerus I: Coronoid process of ulna N: Musculocutaneous nerve	
Brachioradialis	**Flexes** the elbow (synergist)	O: Ridge above the lateral condyle of the distal humerus I: Distal radius above styloid process N: Radial nerve	
Triceps brachii	**Extends** the elbow	O: Infraglenoid tubercle (long head); posterior, proximal diaphysis (lateral head); most of posterior diaphysis (medial head) I: Olecranon process N: Radial nerve	
Anconeus	Assists **extension** of the elbow	O: Lateral epicondyle of humerus I: Olecranon process N: Radial nerve	

CLINICAL CONNECTION

Intramuscular Injections

Intramuscular (IM) injections are injections given directly into the center of a specific muscle. As a result, the injected substance is absorbed into the blood vessels supplying the muscle. Two of the most commonly used muscles for IM injections are the deltoid and the gluteus medius. If the volume to be injected is 2 ml or less, the deltoid muscle is often the preferred injection site. In this case, the needle is inserted approximately two finger widths below the acromion process of the scapula and lateral to the tip of the acromion (**Figure A**). If the volume to be injected is greater than 2 ml, then the gluteus medius is often the preferred injection site. Locating the proper injection site involves visualizing an imaginary line extending from the posterior superior iliac spine to the greater trochanter of the femur. The injection is given about three finger widths superior to and one-third of the way down the imaginary line (**Figure B**). Clearly, knowledge of musculoskeletal anatomy is crucial for administering IM injections properly!

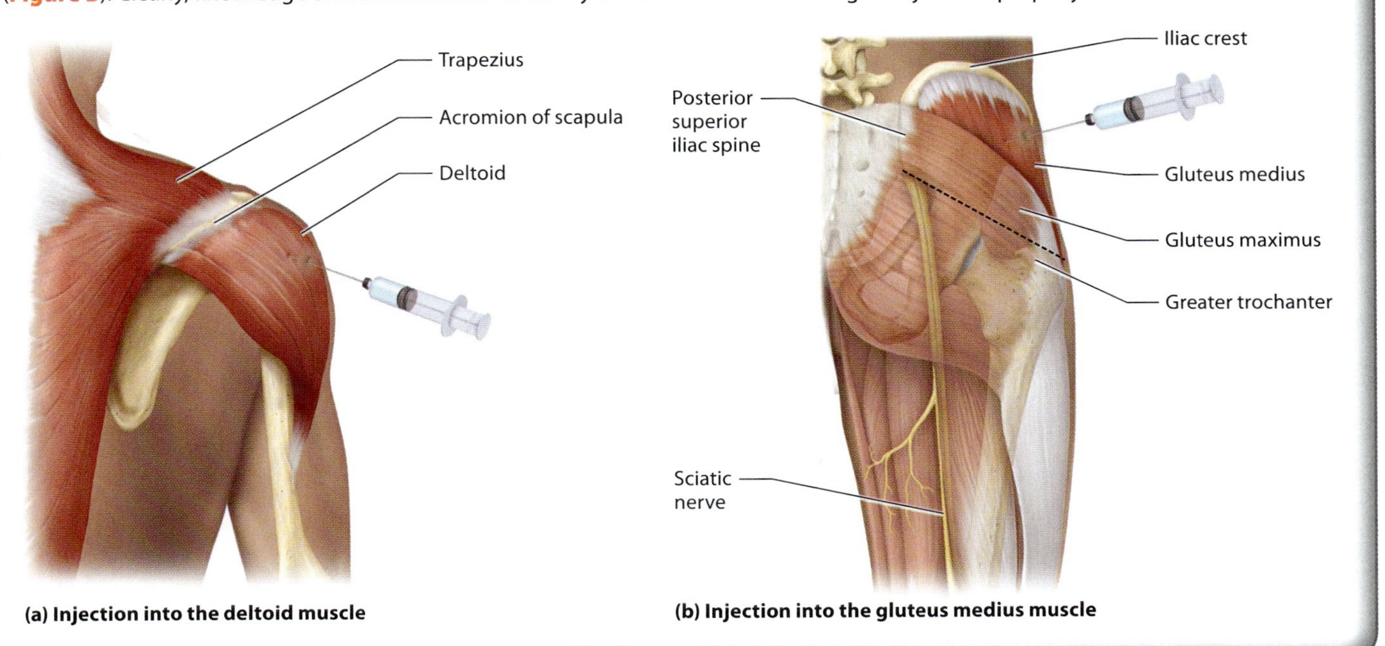

(a) Injection into the deltoid muscle

(b) Injection into the gluteus medius muscle

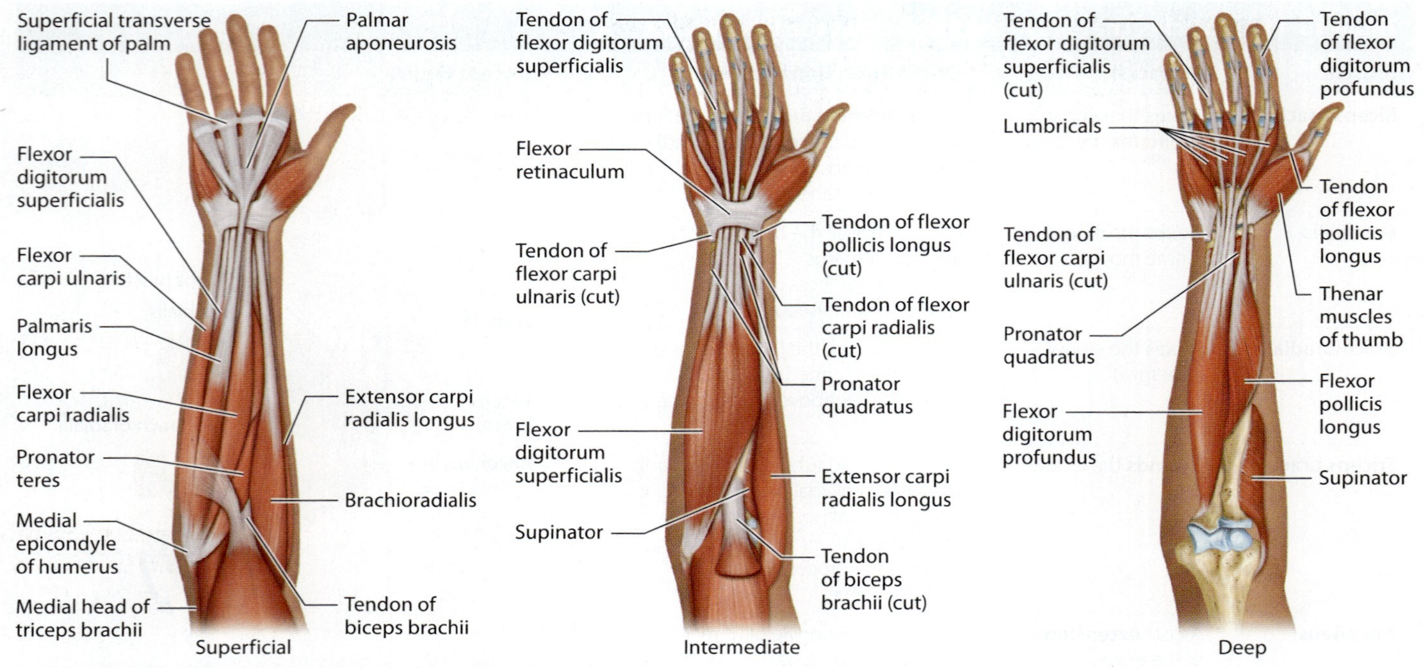

(a) Anterior view with muscles removed as necessary to reveal deeper muscles

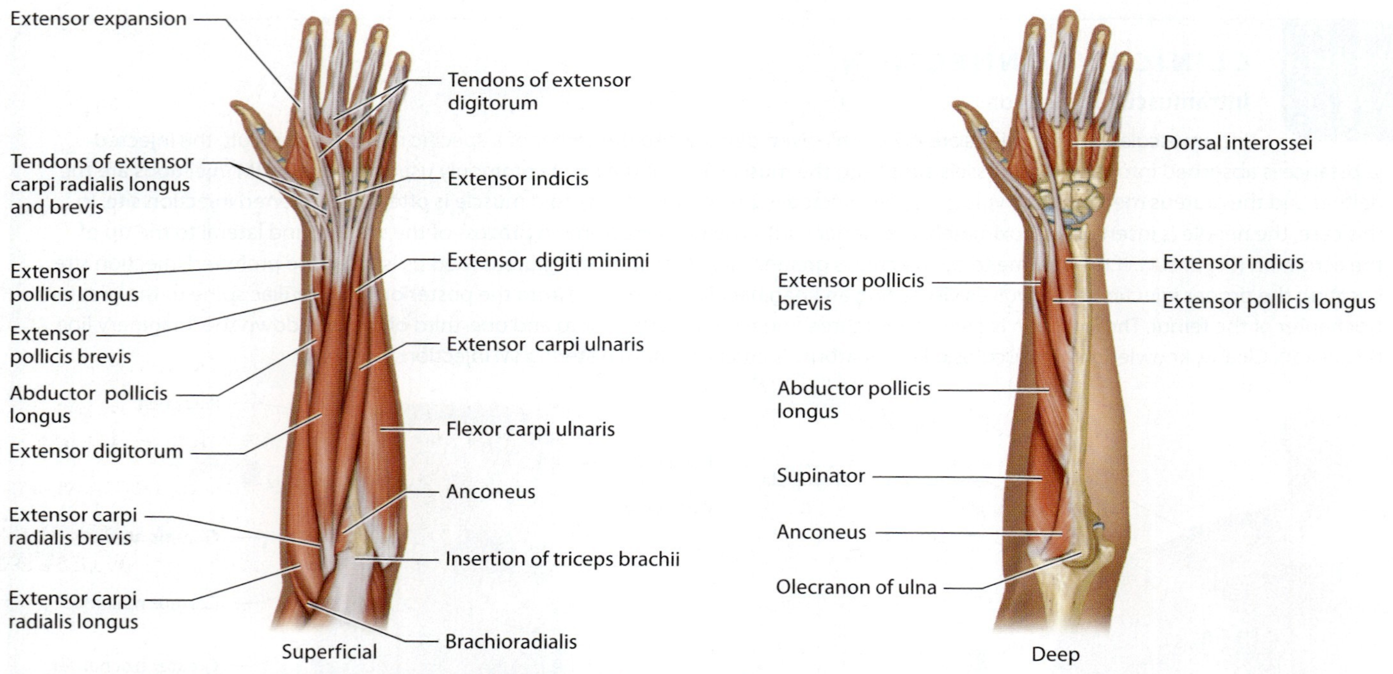

(b) Posterior view with muscles removed as necessary to reveal deeper muscles

Figure 13-14 Muscles that move the wrist, hand, and digits.

Table 13-12 Muscles That Move the Wrist and Hand

Muscle	Action(s)	Origin/Insertion/Nerve(s)
Pronator teres	Pronates forearm (palm posterior)	O: Medial epicondyle of humerus, coronoid process of ulna I: Lateral surface of radius N: Median nerve
Flexor carpi radialis	Flexes wrist (bends wrist toward anterior forearm); abducts hand	O: Medial epicondyle of humerus I: Second and third metacarpals N: Median nerve
Palmaris longus	Flexes wrist; tenses dense connective tissue in palm of hand	O: Medial epicondyle of humerus I: Flexor retinaculum and surrounding connective tissue in palm of hand N: Median nerve
Flexor carpi ulnaris	Flexes wrist; adducts hand	O: Medial epicondyle of humerus, medial olecranon process of ulna I: Pisiform, hamate, fifth metacarpal N: Ulnar nerve
Flexor digitorum superficialis	Flexes fingers; flexes wrist	O: Medial epicondyle of humerus, coronoid process of ulna, anterior proximal radius I: Middle phalanges of fingers 2–5 N: Median nerve
Flexor pollicis longus	Flexes thumb	O: Anterior diaphysis of radius I: Distal phalanx of thumb N: Median nerve
Flexor digitorum profundus	Flexes fingers; flexes wrist	O: Medial/anterior ulna and coronoid process I: Distal phalanges of fingers 2–5 N: Median and ulnar nerve branches
Pronator quadratus	Pronates forearm	O: Anterior distal ulna I: Anterior distal radius N: Median nerve
Extensor carpi radialis longus	Extends wrist; abducts hand	O: Lateral lower diaphysis of humerus I: Second metacarpal N: Radial nerve
Extensor carpi radialis brevis	Extends wrist; abducts hand	O: Lateral epicondyle of humerus I: Third metacarpal N: Deep radial nerve
Extensor digitorum	Extends fingers; extends wrist	O: Lateral epicondyle of humerus I: Posterior phalanges of fingers 2–5 N: Posterior interosseous nerve
Extensor digiti minimi	Extends little finger	O: Lateral epicondyle of humerus I: Phalanges of finger 5 N: Posterior interosseous nerve
Extensor carpi ulnaris	Extends wrist; adducts hand	O: Lateral epicondyle of humerus, lateral proximal ulna I: Fifth metacarpal N: Posterior interosseous nerve
Supinator	Supinates forearm (palm anterior)	O: Lateral epicondyle of humerus, area of the radial notch of ulna I: Anterior/lateral area of radius below tuberosity N: Deep radial nerve
Abductor pollicis longus	Abducts and extends thumb	O: Posterior diaphysis of radius and ulna I: First metacarpal N: Posterior interosseous nerve
Extensor pollicis longus	Extends thumb ("thumbs up" motion)	O: Posterior diaphysis of ulna I: Distal phalanx of thumb N: Posterior interosseous nerve
Extensor pollicis brevis	Extends thumb	O: Lower posterior diaphysis of radius I: Proximal phalanx of thumb N: Posterior interosseous nerve
Extensor indicis	Extends index finger	O: Lower posterior diaphysis of ulna I: Phalanges of finger 2 N: Posterior interosseous nerve

ACTIVITY 4

Mastering the Muscles of the Upper Limb

Learning Outcomes

1. Locate assigned skeletal muscles of the upper limb on anatomical models, and describe and demonstrate the major action(s) of each.
2. Make connections between each assigned muscle and things you are learning in this unit. If required by your instructor, make connections between each assigned muscle and its origins and insertions.

Materials Needed

☐ Muscle models or anatomical charts
☐ Articulated skeleton

Instructions

CHART For each muscle assigned by your instructor, (1) locate the muscle on a muscle model or anatomical chart, (2) describe the action(s) of the muscle, and (3) perform the action(s) of the muscle using both your body and an articulated skeleton. (Refer to Unit 11 if you need a review of movements allowed by synovial joints.) Then design a chart similar to the following chart so that you can organize the important information about each muscle assigned. Include in each "Connections to Things I'm Learning about Muscles" box relevant information from other sources, including lecture. Examples of "connections" include information conveyed by the name of the muscle, its agonists and antagonists, and perhaps its origin, insertion, and innervation. If knowing origins, insertions, and innervations is not required, the first example may be more useful. If that information is required, the second example may be more useful. In that case, locate the origin and insertion of each assigned muscle on the articulated skeleton.

Optional Activity

Practice labeling muscles on human cadavers at MasteringA&P® > Study Area > Practice Anatomy Lab > Human Cadaver > Muscular System: Upper Limb

Play animations of muscle origins, insertions, and actions at MasteringA&P® > Study Area > Practice Anatomy Lab > Animations > Upper Limb

Making Connections: Muscles of the Upper Limb

Muscle	Action(s)	Connections to Things I'm Learning about Muscles
Deltoid	*Prime mover of arm abduction*	*Triangular (delta = Δ = triangle) muscle located on shoulder; common site of intramuscular injections*
Biceps brachii	*Flexes forearm*	*Two-headed muscle located in the brachial region; short head originates on the coracoid process of the scapula and long head originates on the supraglenoid tubercle; inserts on the radial tuberosity; innervated by the musculocutaneous nerve*

Muscles of the Lower Limb

The muscles of the lower limb include muscles that move the thigh, muscles that move the leg, and muscles that move the ankle, foot, and digits.

Muscles That Move the Thigh

Among the muscles that move the thigh (**Figure 13-15**) is the **iliopsoas** (see Figure 13-15a), a composite muscle consisting of two muscles, the **psoas major** and the **iliacus**, that flex the trunk on the thigh (as in taking a bow). Butchers refer to this muscle as the tenderloin. The **gluteus maximus** (see Figure 13-15b) is the most superficial and the most prominent of the three gluteal muscles that form the buttocks. It is a powerful extensor of the thigh. The **gluteus medius** (see Figure 13-15b) lies deep to the gluteus maximus and abducts and medially rotates the thigh. It is a common site for intramuscular injections. The **gluteus minimus** (not shown), the smallest of the three gluteal muscles, lies deep to the gluteus medius; it acts as an agonist to the gluteus medius in abducting and medially rotating the thigh.

The **tensor fasciae latae** (see Figure 13-15a), located on the lateral surface of the thigh, flexes and abducts the thigh.

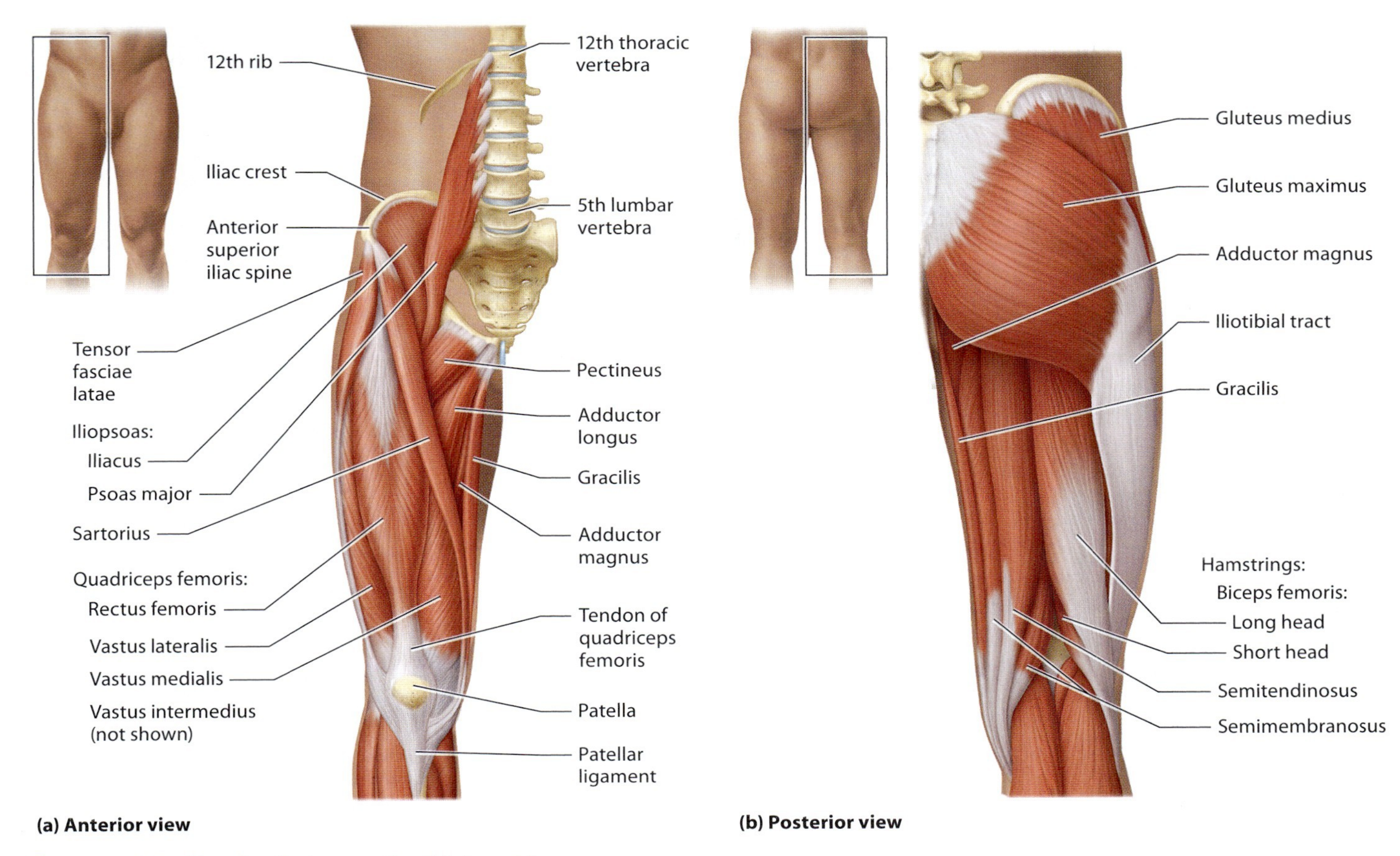

Figure 13-15 **Muscles that move the thigh and leg.**

The pear-shaped **piriformis** (not shown) lies inferior to the gluteus minimus and laterally rotates and abducts the thigh. The **adductor longus** (see Figure 13-15a), a medial thigh muscle, belongs to the adductor group, which adducts, flexes, and medially rotates the thigh. The anterior part of the **adductor magnus** (see Figure 13-15a and b), another medial thigh muscle, adducts, medially rotates, and flexes the thigh, whereas the posterior part is a thigh extensor. The **adductor brevis** (not shown) is a medial thigh muscle that adducts and medially rotates the thigh. The small flat **pectineus** (see Figure 13-15a) overlies the adductor brevis and adducts, flexes, and medially rotates the thigh. Finally, the long, thin, superficial **gracilis** (see Figure 13-15a and b) is a medial thigh muscle that adducts the thigh and flexes and medially rotates the leg. **Tables 13-13** and **13-14** summarize the muscles that move the thigh.

Muscles That Move the Leg

Among the thigh muscles that move the leg is the **rectus femoris** (see Figure 13-15a), which is on the anterior aspect of the thigh and is the most superficial of the four muscles that belong to the group known as the *quadriceps femoris*, which extends the leg. Because the rectus femoris crosses the hip joint, it also flexes the thigh. The **vastus lateralis** (see Figure 13-15a), on the lateral aspect of the thigh, is a member of the quadriceps femoris that extends the knee. The **vastus medialis** (see Figure 13-15a), a member of the quadriceps femoris on the medial aspect of the thigh, extends the knee. The **vastus intermedius** (not shown), a member of the quadriceps femoris deep to the rectus femoris, extends the knee.

The long, narrow, strap-like **sartorius** (see Figure 13-15a) forms a band across the thigh from the ilium to the medial

Table 13-13 Muscles That Move the Thigh and Knee: Anterior and Medial Muscles

Muscle	Action(s)	Origin/Insertion/Nerve(s)	Concept Figures
Iliacus*	**Flexes** the thigh	O: Iliac fossa I: Lesser trochanter of the femur N: Femoral nerve	
Psoas major*	**Flexes** the thigh; **laterally bends** the vertebral column	O: Lateral sides and transverse processes of T_{12} and lumbar vertebrae I: Lesser trochanter of the femur N: L_1–L_3	
Sartorius	**Flexes** the thigh and **flexes** the leg at the knee; **abducts** and laterally **rotates** the thigh; actions produce a "tailor's" seated posture	O: Anterior superior iliac spine I: Proximal portion of the medial condyle of the tibia N: Femoral nerve	
Adductor magnus	**Adducts** the thigh; **rotates** the thigh medially *(Lower part is a synergist of hamstring action, see Table 13-14.)*	O: Ischial tuberosity, ischial ramus, inferior ramus of pubis I: Linea aspera and adductor tubercle of femur N: Obturator and sciatic nerves	
Adductor longus	**Adducts** the thigh; **flexes** the thigh; **rotates** the thigh medially	O: Body of pubis I: Linea aspera N: Obturator nerve	
Adductor brevis	**Adducts** the thigh; **flexes** the thigh; **rotates** the thigh medially	O: Inferior ramus of the pubis I: Superior linea aspera N: Obturator nerve	
Pectineus	**Adducts** the thigh; **flexes** the thigh; **rotates** the thigh medially	O: Superior ramus of pubis I: Posterior femur between the lesser trochanter and linea aspera N: Femoral nerve	
Gracilis	**Adducts** the thigh; **flexes** the leg; **rotates** the thigh medially	O: Body and inferior ramus of pubis; ischial ramus I: Medial surface of upper tibia N: Obturator nerve	
Rectus femoris†	**Extends** the knee; **flexes** the thigh	O: Anterior inferior iliac spine, superior margin of acetabulum I: Patella and tibial tuberosity N: Femoral nerve	
Vastus lateralis†	**Extends** the knee; stabilizes patella	O: Greater trochanter, intertrochanteric line, proximal half of the linea aspera I: Patella and tibial tuberosity N: Femoral nerve	
Vastus medialis†	**Extends** the knee; stabilizes patella	O: Linea aspera, intertrochanteric line I: Patella and tibial tuberosity N: Femoral nerve	
Vastus intermedius†	**Extends** the knee	O: Anterior and lateral portions of proximal two-thirds of the diaphysis of femur I: Patella and tibial tuberosity N: Femoral nerve	

*Part of the group of iliopsoas muscles that share a common insertion tendon.

†Part of the group of quadriceps femoris muscles that share a common insertion tendon.

Table 13-14 Muscles That Move the Thigh and Knee: Posterior Muscles

Muscle(s)	Action(s)	Origin/Insertion/Nerve(s)	Concept Figures
Gluteus maximus	**Extends** the thigh (especially when the thigh is in flexed position); laterally **rotates** the thigh; **abducts** the thigh	O: Posterior and lateral portions of the ilium, sacrum, and coccyx I: Gluteal tuberosity of femur N: Inferior gluteal nerve	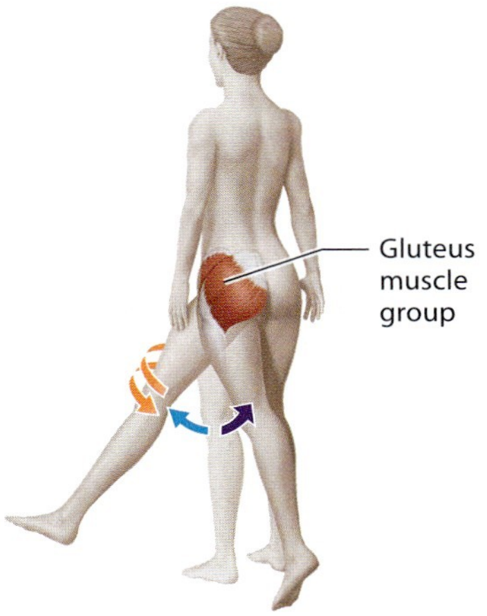 Gluteus muscle group
Gluteus medius	**Abducts** the thigh; medially **rotates** the thigh; stabilizes pelvis while walking	O: Between posterior and anterior gluteal lines on the outer surface of ilium I: Greater trochanter of the femur N: Superior gluteal nerve	
Gluteus minimus	**Abducts** the thigh; medially **rotates** the thigh; stabilizes pelvis while walking	O: Between the anterior and inferior gluteal lines on the outer surface of ilium I: Greater trochanter of the femur N: Superior gluteal nerve	
Piriformis	**Abducts** the thigh; laterally **rotates** the thigh	O: Anterior/lateral sacrum I: Greater trochanter of femur N: L_5, S_1–S_2	Piriformis
Obturator externus	Laterally **rotates** the thigh	O: Edges of and outer surface of obturator foramen and membrane I: Greater trochanter of femur N: Obturator nerve	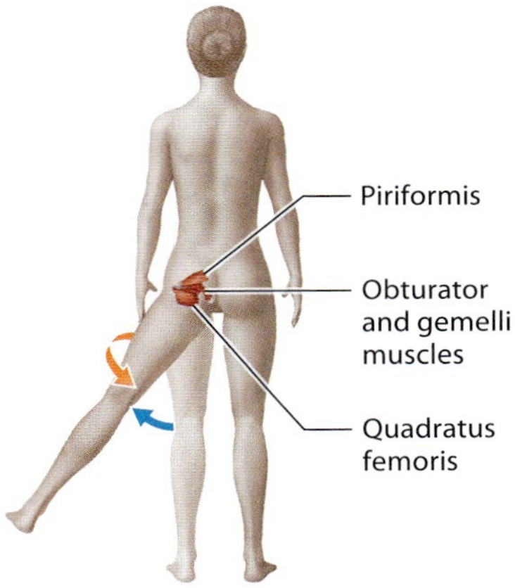 Obturator and gemelli muscles
Obturator internus	Laterally **rotates** the thigh	O: Edges of and inner surface of obturator foramen and membrane I: Greater trochanter of femur N: L_5, S_1	
Gemelli (superior gemellus and inferior gemellus)	Laterally **rotate** the thigh	O: Ischial spine and tuberosity I: Greater trochanter of femur N: L_5, S_1	
Quadratus femoris	Laterally **rotates** the thigh	O: Ischial tuberosity I: Intertrochanteric crest of femur N: L_5, S_1	Quadratus femoris
Biceps femoris*	**Extends** the thigh; **flexes** the knee	O: Ischial tuberosity and lower half of posterior femur I: Head of fibula; lateral condyle of tibia N: Sciatic nerve	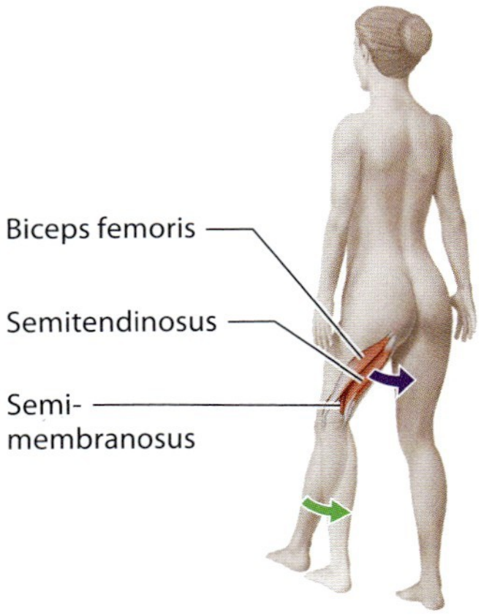 Biceps femoris
Semitendinosus*	**Extends** the thigh; **flexes** the knee	O: Ischial tuberosity I: Upper medial surface of tibia N: Sciatic nerve	Semitendinosus
Semimembranosus*	**Extends** the thigh; **flexes** the knee	O: Ischial tuberosity I: Posterior surface of medial condyle of tibia N: Sciatic nerve	Semi-membranosus

*Part of the hamstring muscle group.

side of the tibia; it flexes, abducts, and laterally rotates the thigh. This is the muscle you use to sit cross-legged. The two-headed **biceps femoris** (see Figure 13-15b) is one of three muscles on the posterior aspect of the thigh that collectively are commonly called the hamstrings, a name derived from an old butchers' practice of curing pork thighs in a smokehouse by hanging them from their tendons (their hamstrings). Because this muscle crosses both the hip and knee joints, it can both extend the thigh and flex the leg. The **semitendinosus** (see Figure 13-15b), a hamstring muscle medial to the biceps femoris, flexes the leg at the knee joint. The **semimembranosus** (see Figure 13-15b) is a hamstring muscle that lies deep to the semitendinosus and flexes the leg at the knee joint. Tables 13-13 and 13-14 summarize the muscles that move the leg.

Muscles That Move the Ankle, Foot, and Digits

Among the leg muscles that move the ankle, foot, and digits (**Figure 13-16**) is the **tibialis anterior** (see Figure 13-16a and b), a superficial anterior leg muscle just lateral to the anterior crest of the tibia. It inverts the foot and is a prime mover of dorsiflexion. The **extensor digitorum longus** (see Figure 13-16a and b), which is lateral to the tibialis anterior, is a prime mover of toe extension. The **fibularis longus** (see Figure 13-16a and b), a superficial lateral muscle that overlies the fibula, produces plantar flexion and everts the foot. The **gastrocnemius** (see Figure 13-16a and b), a superficial muscle that forms the bulk of what is commonly called the calf muscle, produces plantar flexion; because it crosses the

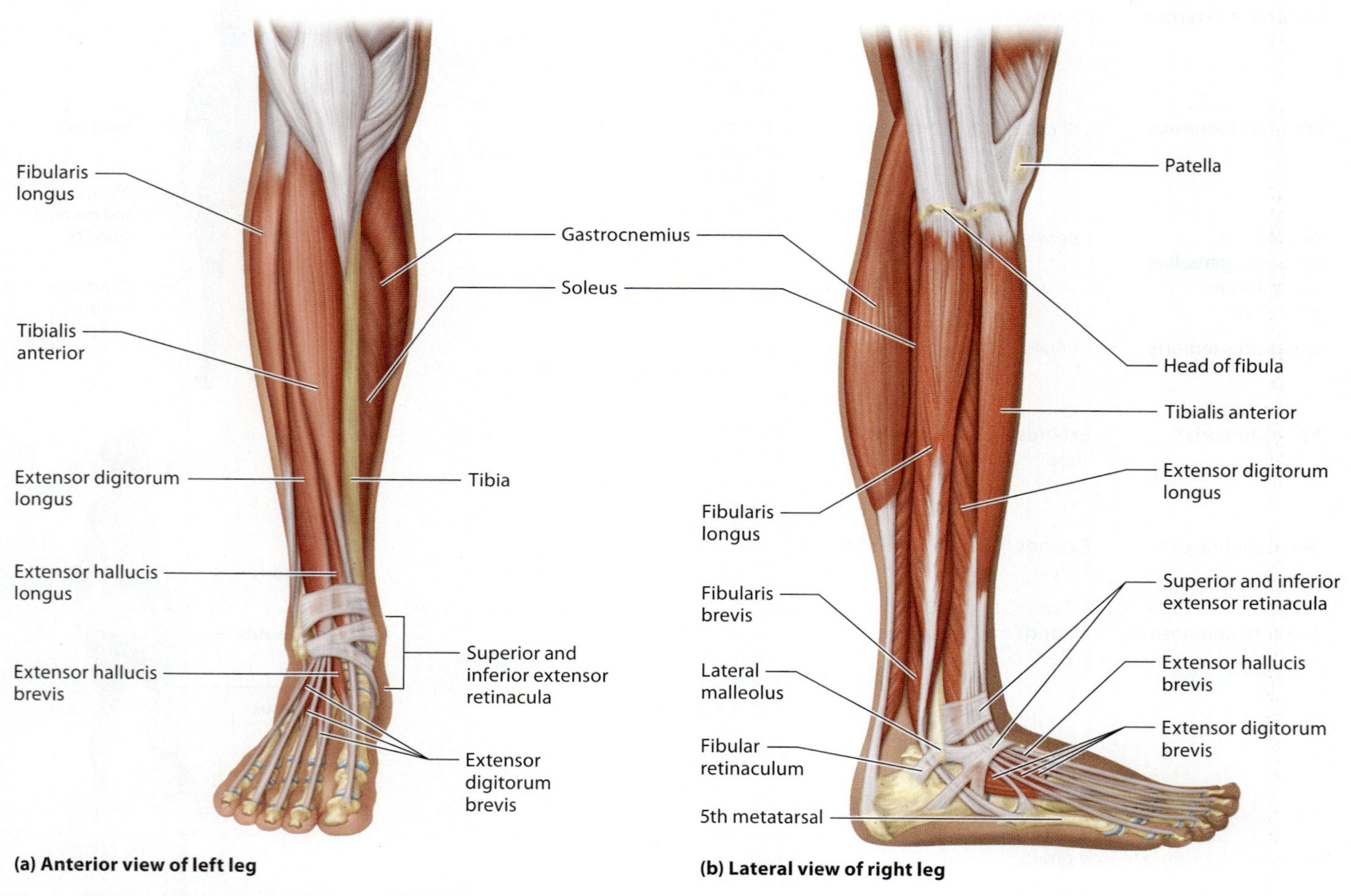

Figure 13-16 **Muscles that move the ankle, foot, and digits.**

knee joint, it can flex the knee when the foot is dorsiflexed. The broad, flat **soleus** (see Figure 13-16a and b), named for its resemblance to a sole (a type of flat fish), is located deep to the gastrocnemius and produces plantar flexion of the foot at the ankle joint. The thick, flat **tibialis posterior** (not shown), which is deep to the soleus, is a prime mover of foot inversion and produces plantar flexion of the foot at the ankle joint. Finally, the long narrow **flexor digitorum longus** (not shown) runs medial to the tibialis anterior muscle; it produces plantar flexion, inverts the foot, and flexes the toes. **Table 13-15** summarizes the muscles that move the ankle, foot, and digits.

Table 13-15 Muscles That Move the Foot and Toes

Muscle	Action(s)	Origin/Insertion/Nerve(s)
Tibialis anterior	**Dorsiflexes** the foot; **inverts** the foot	O: Lateral condyle and upper diaphysis of the tibia I: Cuneiform and first metatarsal N: Deep fibular nerve
Extensor digitorum longus	**Extends** the toes; **dorsiflexes** the foot	O: Lateral condyle of the tibia and upper portion of the fibula I: Phalanges and connective tissues of toes 2–5 N: Deep fibular nerve
Extensor hallucis longus	**Extends** the great toe; **dorsiflexes** the foot	O: Diaphysis of the fibula and interosseous membrane I: Distal phalanx of great toe N: Deep fibular nerve
Fibularis (peroneus) longus	**Everts** the foot; **plantar flexes** the foot	O: Upper lateral fibula I: Cuneiform and first metatarsal N: Superficial fibular nerve
Fibularis brevis	**Everts** the foot; **plantar flexes** the foot	O: Lower fibula I: Fifth metatarsal N: Superficial fibular nerve
Gastrocnemius*	**Plantar flexes** the foot; assists with flexing knee	O: Medial and lateral condyles of the femur I: Posterior surface of calcaneus N: Tibial nerve
Soleus*	**Plantar flexes** the foot	O: Head of the fibula, proximal tibia, and interosseous membrane I: Posterior surface of calcaneus N: Tibial nerve
Plantaris	**Plantar flexes** the foot; assists with flexing knee (very weak synergist)	O: Posterior femur near lateral condyle I: Calcaneus and calcaneal tendon N: Tibial nerve
Tibialis posterior	**Inverts** the foot; **plantar flexes** the foot; stabilizes foot	O: Upper tibia, fibula, and interosseous membrane I: Metatarsals 2–4 N: Tibial nerve
Flexor digitorum longus	**Flexes** toes; stabilizes foot	O: Posterior tibia I: Distal phalanges of toes 2–5 N: Tibial nerve
Flexor hallucis longus	**Flexes** the great toe	O: Posterior fibula and interosseous membrane I: Distal phalanx of great toe N: Tibial nerve
Popliteus	**Flexes** leg at the knee; rotates leg and other minor movements, depending on which fixators are in use	O: Lateral condyle of the femur I: Posterior surface of upper tibia N: Tibial nerve

Concept Figures

*Part of the group of triceps surae muscles that share a common insertion tendon, the calcaneal (Achilles) tendon.

ACTIVITY 5

Mastering the Muscles of the Lower Limb

Learning Outcomes

1. Locate assigned skeletal muscles of the lower limb on anatomical models, and describe and demonstrate the major action(s) of each.
2. Make connections between each assigned muscle and things you are learning in this unit. If required by your instructor, make connections between each assigned muscle and its origins and insertions.

Materials Needed

- ☐ Muscle models or anatomical charts
- ☐ Articulated skeleton

Instructions

CHART For each muscle assigned by your instructor, (1) locate the muscle on a muscle model or anatomical chart, (2) describe the action(s) of the muscle, and (3) perform the action(s) of the muscle using both your body and an articulated skeleton. (Refer to Unit 11 if you need a review of movements allowed by synovial joints.) Then design a chart similar to the following chart so that you can organize the important information about each muscle assigned. Include in each "Connections to Things I'm Learning about Muscles" box relevant information from other sources, including lecture. Examples of "connections" include information conveyed by the name of the muscle, its agonists and antagonists, and perhaps its origin, insertion, and innervation. If knowing origins, insertions, and innervations is not required, the first example may be more useful. If that information is required, the second example may be more useful. In that case, locate the origin and insertion of each assigned muscle on the articulated skeleton.

Optional Activity

Practice labeling muscles on human cadavers at MasteringA&P® > Study Area > Practice Anatomy Lab > Human Cadaver > Muscular System: Lower Limb

Play animations of muscle origins, insertions, and actions at MasteringA&P® > Study Area > Practice Anatomy Lab > Animations > Lower Limb

Making Connections: Muscles of the Lower Limb

Muscle	Action(s)	Connections to Things I'm Learning about Muscles
Gluteus maximus	*Powerful thigh extensor*	*Most superficial and largest of three gluteal muscles*
Gastrocnemius	*Plantar flexes foot*	*Large two-headed calf muscle; originates on lateral and medial condyles of the femur; inserts on calcaneus via calcaneal (Achilles) tendon; innervated by the tibial nerve*

Post-lab quizzes are also assignable in MasteringA&P®

POST-LAB ASSIGNMENTS

Name: ______________________ Date: ____________ Lab Section: ________

PART I. Check Your Understanding

Activity 1: Determining How Skeletal Muscles Are Named

1. Match the following terms with their descriptions:

_____	a. origin	1. nerve supply
_____	b. prime mover	2. stabilizes the origin
_____	c. innervation	3. aids a prime mover
_____	d. antagonist	4. site of muscle attachment to a moveable bone
_____	e. synergist	5. opposes a prime mover
_____	f. fixator	6. agonist
_____	g. insertion	7. site of muscle attachment to a stationary bone

2. Name seven criteria by which muscles are named:

____

____

Activity 2: Mastering the Muscles of the Head and Neck

1. Identify the muscles of the head and neck.

a. ______________________

b. ______________________

c. ______________________

d. ______________________

e. ______________________

f. ______________________

g. ______________________

h. ______________________

i. ______________________

j. ______________________

k. ______________________

l. ______________________

m. ______________________

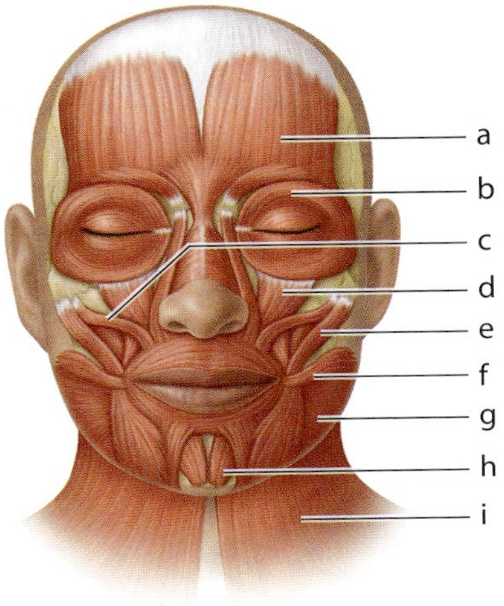

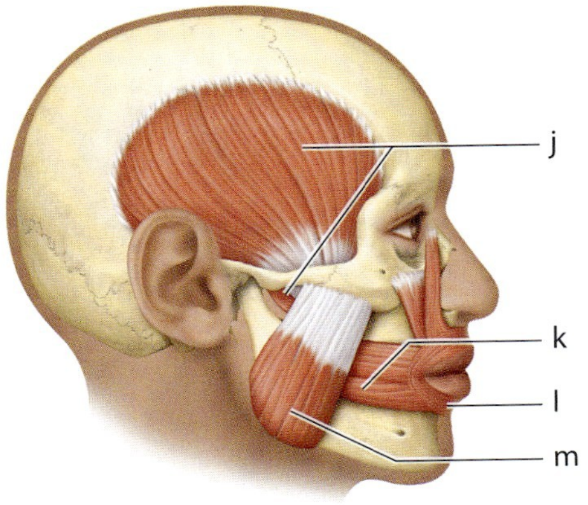

2. Name a muscle of the head and neck that fits each of the following descriptions:

_______________ a. Is a synergist of the zygomaticus muscles.

_______________ b. Raises the upper lip.

_______________ c. Is a bipartite muscle with two bellies.

_______________ d. Elevates and protrudes the lower lip.

_______________ e. Closes the eye.

_______________ f. Elevates the mandible.

_______________ g. Compresses the cheeks when whistling.

_______________ h. Depresses mandible and moves it side to side.

_______________ i. Extends the head.

_______________ j. Originates on the styloid process of the temporal bone.

Activity 3: Mastering the Muscles of the Trunk

1. Identify the muscles of the trunk.

a. _______________

b. _______________

c. _______________

d. _______________

e. _______________

f. _______________

g. _______________

h. _______________

i. _______________

j. _______________

k. _______________

l. _______________

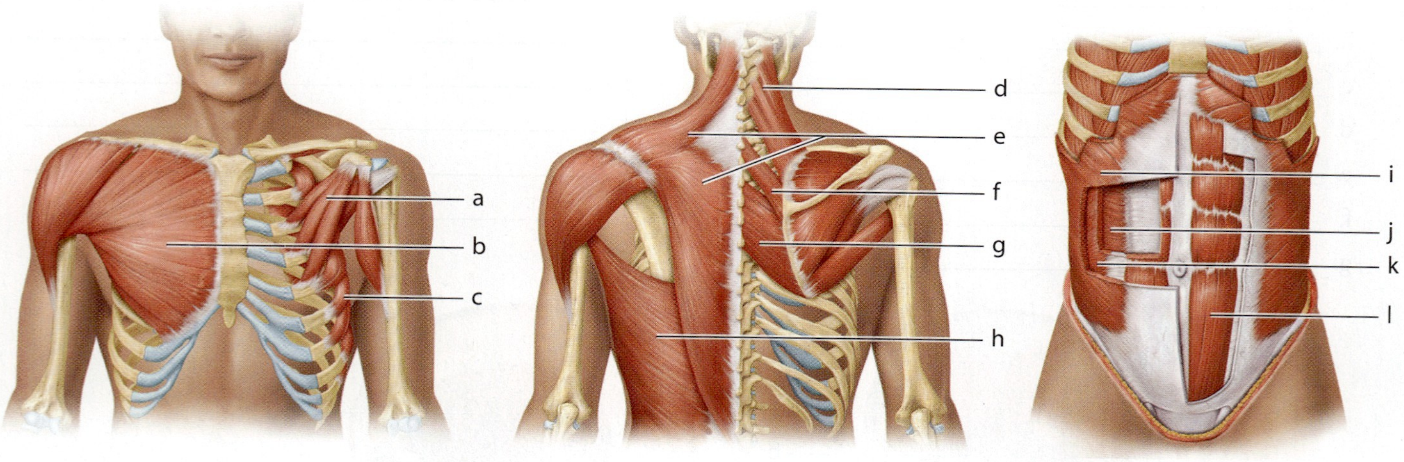

2. Name a trunk muscle that fits each of the following descriptions:

______________________ a. Extends the head.

______________________ b. Acts as an antagonist to the diaphragm.

______________________ c. Is the deepest muscle of the abdominal wall.

______________________ d. Is a tripartite muscle that extends from the sacrum to the skull.

______________________ e. Is a square-shaped muscle of the lower back.

______________________ f. Elevates the scapula.

______________________ g. Is located inferior to the clavicle.

______________________ h. Is the abdominal muscle that originates on pubic crest and pubic tubercle; inserts onto xiphoid process and the costal cartilages of ribs 5–7.

______________________ i. Is innervated by the phrenic nerve.

______________________ j. Is a large, flat, fan-shaped muscle located between ribs and scapula.

Activity 4: Mastering the Muscles of the Upper Limb

1. Identify the muscles of the upper limb.

a. ______________________

b. ______________________

c. ______________________

d. ______________________

e. ______________________

f. ______________________

g. ______________________

h. ______________________

i. ______________________

j. ______________________

k. ______________________

l. ______________________

m. ______________________

n. ______________________

o. ______________________

p. ______________________

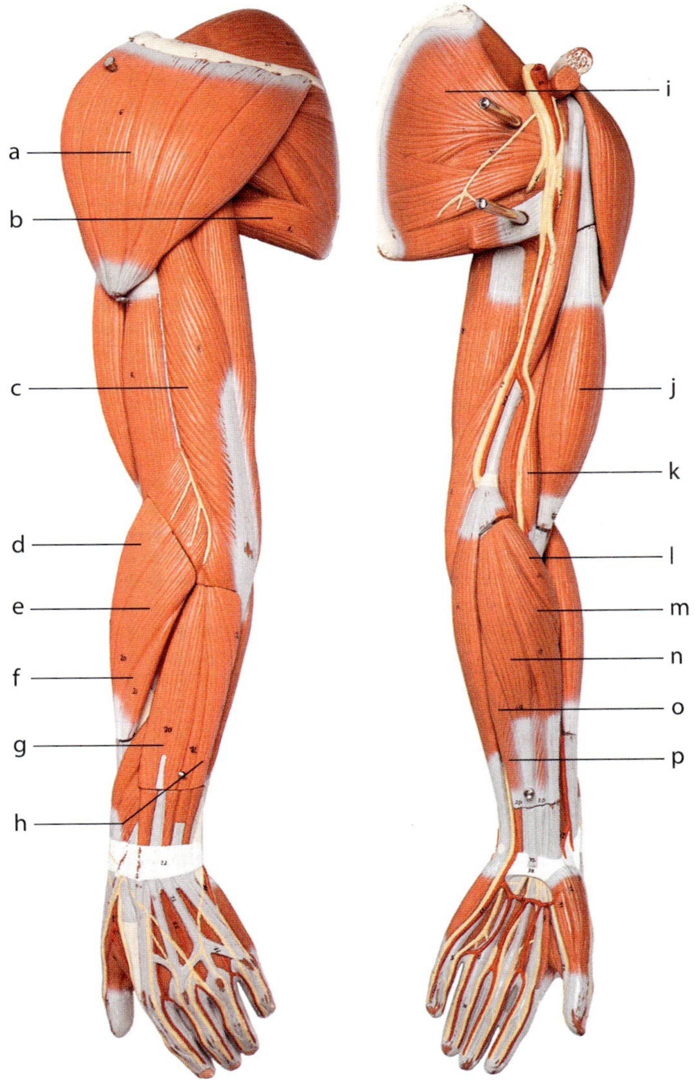

2. Name a muscles that fits each of the following descriptions.

_______________ a. Acts as an antagonist to the flexor carpi ulnaris.

_______________ b. Is a rotator cuff muscle that originates on the lateral border of the scapula.

_______________ c. Is a large, superficial, fan-shaped muscle.

_______________ d. Is a forearm extensor innervated by the radial nerve.

_______________ e. Is a prime mover of arm flexion.

_______________ f. Is located deep to the biceps brachii.

_______________ g. Is a powerful forearm extensor.

_______________ h. Extends and adducts the wrist.

_______________ i. Pronates the forearm.

_______________ j. Is a prime mover of arm extension.

Activity 5: Mastering the Muscles of the Lower Limb

1. Identify the muscles of the lower limb.

a. _______________

b. _______________

c. _______________

d. _______________

e. _______________

f. _______________

g. _______________

h. _______________

i. _______________

j. _______________

k. _______________

l. _______________

m. _______________

n. _______________

o. _______________

p. _______________

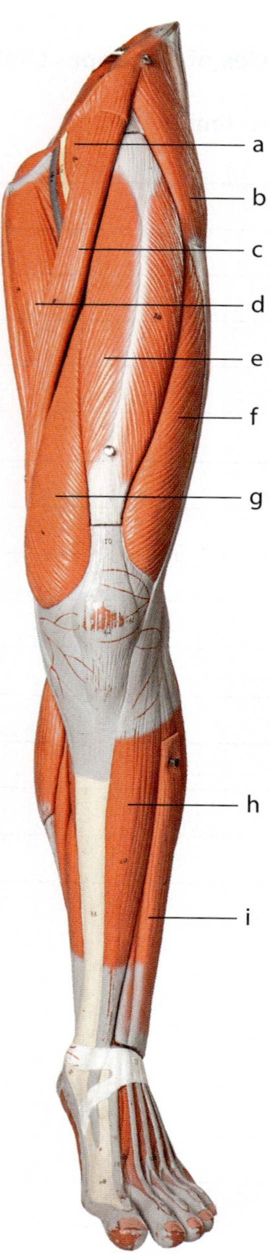

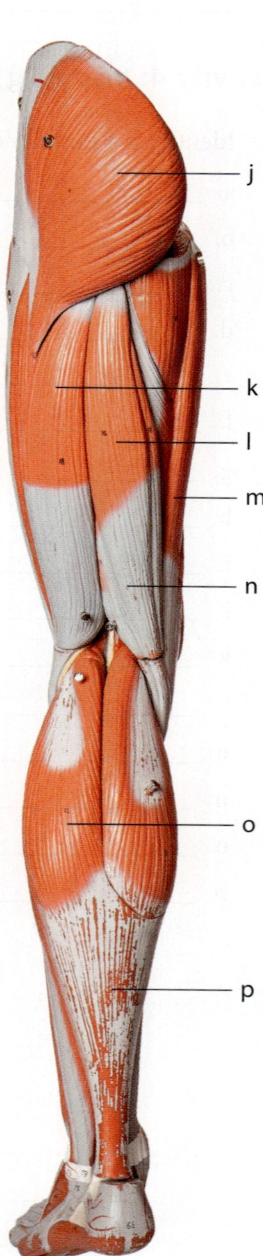

2. Name a muscle that fits each of the following descriptions:

_______________ a. Is the deepest quadriceps muscle.

_______________ b. Plantar flexes and everts the foot.

_______________ c. Flexes the thigh on the trunk.

_______________ d. Is a powerful thigh extensor.

_______________ e. Is a lateral hamstring muscle.

_______________ f. Laterally rotates and flexes the thigh.

_______________ g. Inserts onto the linea aspera and the adductor tubercle.

_______________ h. Extends the digits.

_______________ i. Is located deep to the gastrocnemius.

_______________ j. Is located just lateral to the anterior crest of the tibia.

PART II. Putting It All Together

A. Review Questions

Answer the following questions using your lecture notes, your textbook, and your lab notes.

1. Explain how it is possible to predict the action of a muscle if you know its origin and insertion.

____

____

2. A patient is rushed to the hospital, where an emergency appendectomy is performed. Indicate the abdominal quadrant and the skin layers that must be penetrated, and then list (in order from most superficial to deepest) each muscle that the surgeon must cut through to make the appendix accessible for removal.

   Abdominal quadrant: _______________

   Skin layers penetrated (superficial to deep): _______________

   Abdominal muscles (superficial to deep): _______________

   ____

3. Explain how the flexor carpi ulnaris can be classified as both an agonist and an antagonist.

____

____

4. Describe the major muscles and movements involved in each of the following actions:

   a. Doing jumping jacks _______________

   ____

   ____

   ____

   b. Drinking water through a straw and eating soup with a spoon _______________

   ____

   ____

   ____

5. Identify the following skeletal muscles (in anterior view):

a. ______
b. ______
c. ______
d. ______
e. ______
f. ______
g. ______
h. ______
i. ______
j. ______
k. ______
l. ______
m. ______
n. ______
o. ______
p. ______
q. ______
r. ______
s. ______
t. ______
u. ______
v. ______
w. ______
x. ______
y. ______
z. ______
aa. ______
bb. ______
cc. ______
dd. ______
ee. ______
ff. ______
gg. ______
hh. ______
ii. ______ 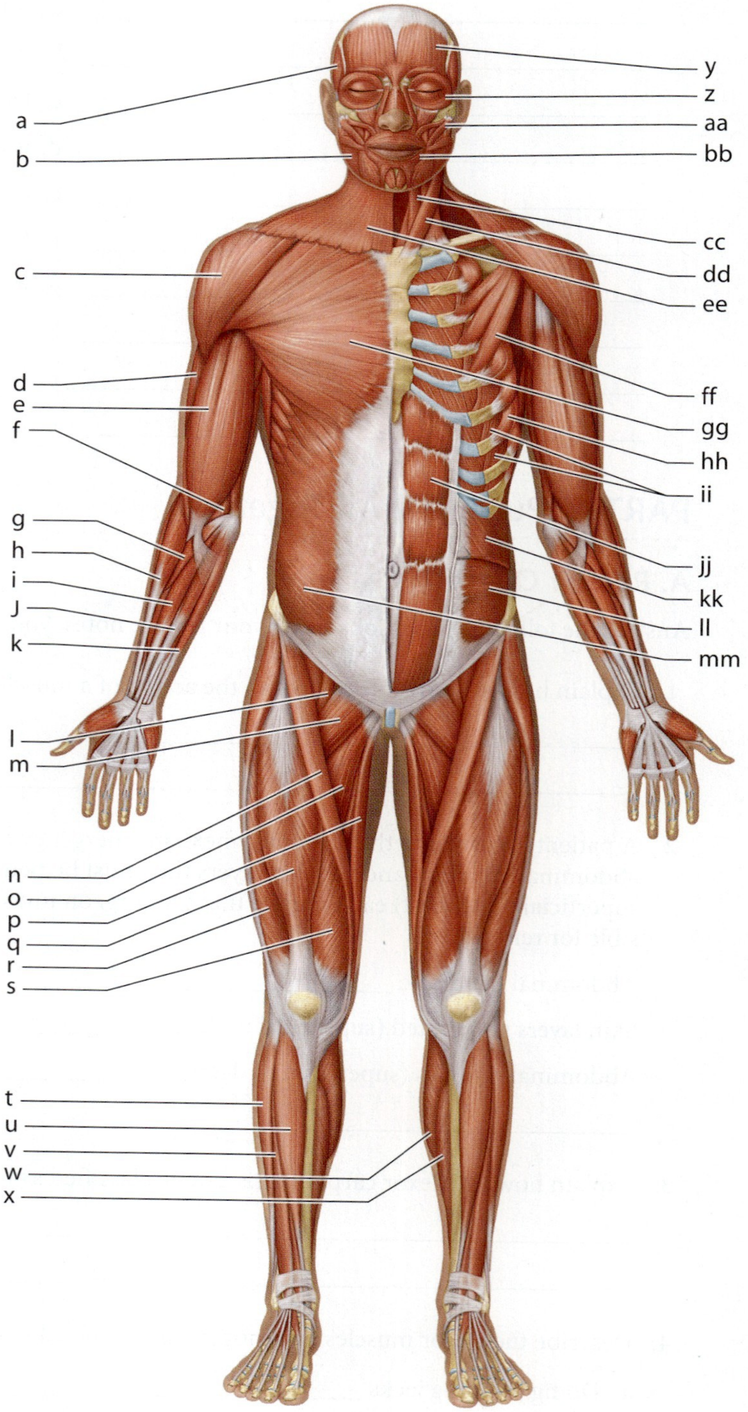
jj. ______
kk. ______
ll. ______
mm. ______

6. Identify the following skeletal muscles (in posterior view):

a. ______________________________
b. ______________________________
c. ______________________________
d. ______________________________
e. ______________________________
f. ______________________________
g. ______________________________
h. ______________________________
i. ______________________________
j. ______________________________
k. ______________________________
l. ______________________________
m. ______________________________
n. ______________________________
o. ______________________________
p. ______________________________
q. ______________________________
r. ______________________________
s. ______________________________
t. ______________________________
u. ______________________________
v. ______________________________
w. ______________________________

a
b
c
d
e
f
g
h
i
j
k
l
m
n
o
p
q
r
s
t
u
v
w

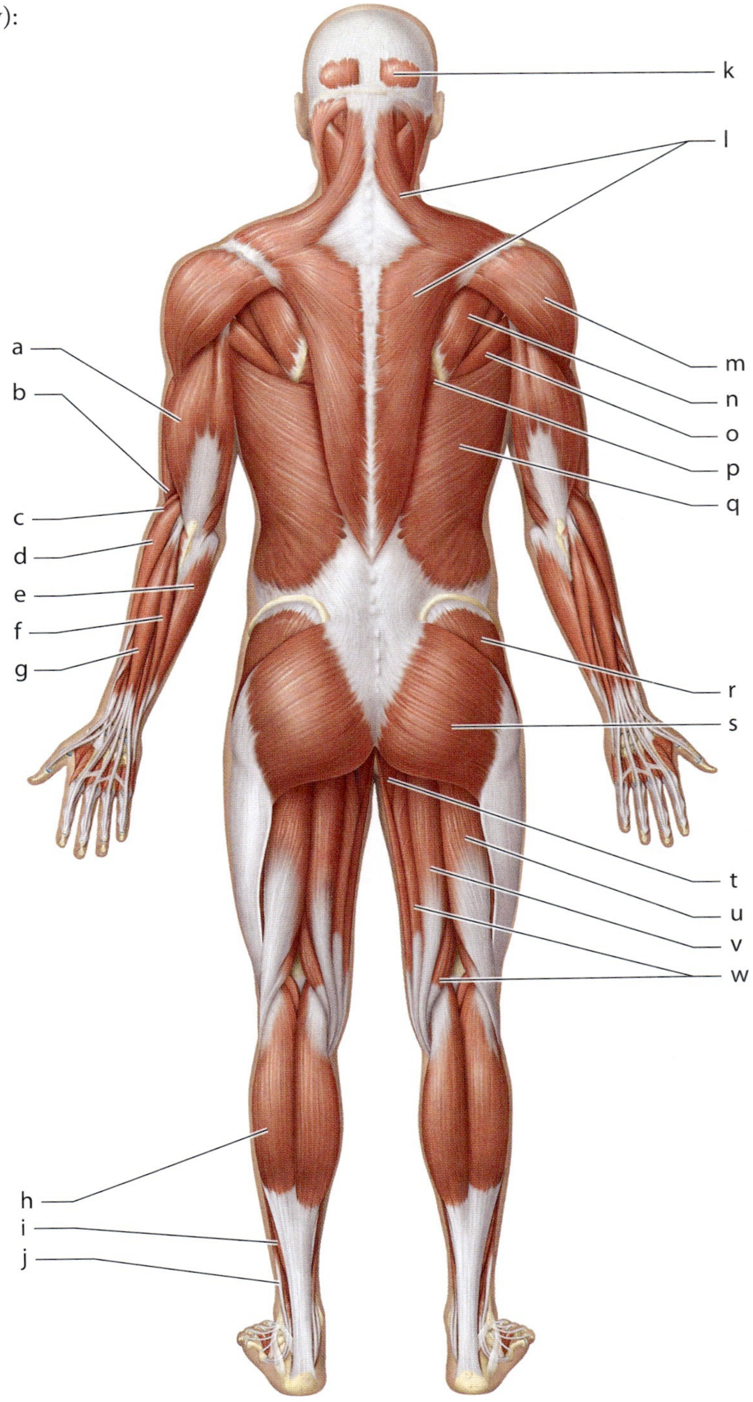

B. Concept Mapping

1. Fill in the blanks to complete this concept map outlining ways in which skeletal muscles are named.

gluteus maximus   infraspinatus   sternocleidomastoid   transversus abdominis   triceps brachii

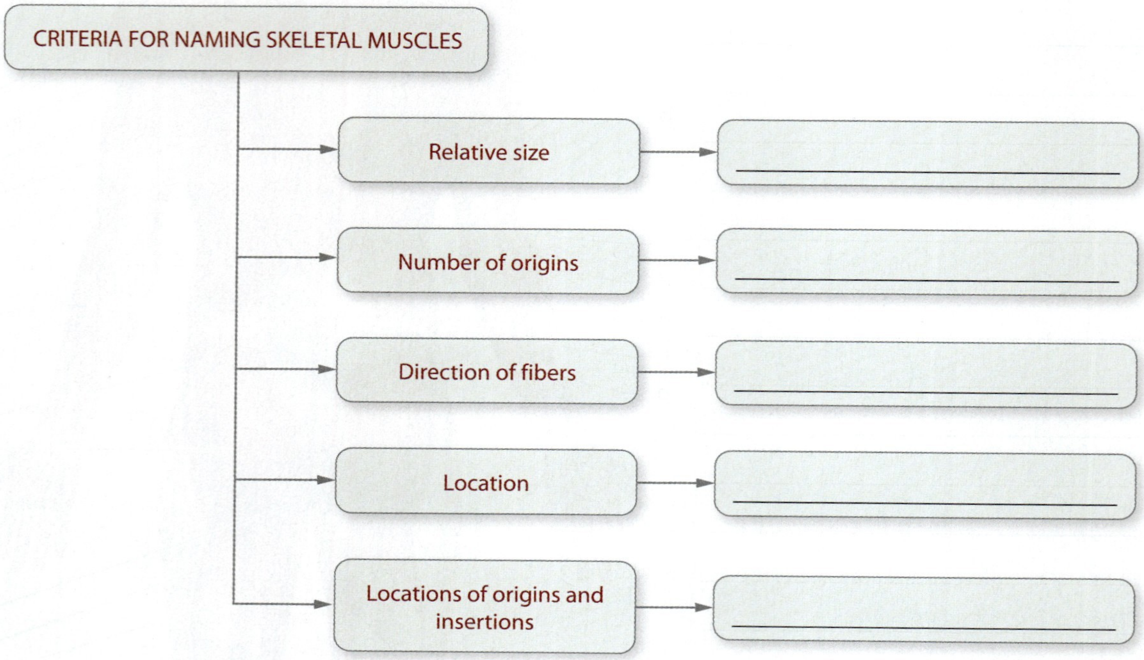

CRITERIA FOR NAMING SKELETAL MUSCLES

- Relative size → ______________________
- Number of origins → ______________________
- Direction of fibers → ______________________
- Location → ______________________
- Locations of origins and insertions → ______________________

2. Construct a unit concept map to show the relationships among the following set of terms. Include all of the terms in your diagram. Your instructor may choose to assign additional terms.

diaphragm   digastric   external intercostals   gluteus maximus   iliocostalis

infraspinatus   rectus abdominis   sartorius   semimembranosus   splenius capitis

sternocleidomastoid   teres minor   triceps brachii   transversus abdominis   vastus lateralis

14

Introduction to the Nervous System

UNIT OUTLINE

The nervous system consists of the brain, the spinal cord, and the nerves. The major functions of the nervous system involve receiving stimuli and relaying sensory information, integrating incoming information, and sending out efferent messages to muscles and glands that carry out responses. The nervous system is sometimes known as the body's fast-acting control system because it regulates body functions via rapid electrical impulses.

This unit focuses on the general organization and function of the nervous system and a histological study of the structure and function of nervous tissue. Unit 15 entails a detailed study of the structure and function of the central nervous system (the brain and spinal cord), and Unit 16 provides a detailed study of the structure and function of the peripheral nervous system (nerves and the autonomic nervous system).

THINK ABOUT IT *Which organ system is known as the slow-acting control system?*

____

How does this organ system differ from the nervous system?

____

PRE-LAB ASSIGNMENTS

Pre-lab quizzes are also assignable in MasteringA&P®

To maximize learning, BEFORE your lab period carefully read this entire lab unit and complete these pre-lab assignments using your textbook, lecture notes, and prior knowledge.

PRE-LAB Activity 1: Calculating Reaction Time

1. Which motor response will be used to calculate reaction time?

   ____

2. Which three types of stimuli will be tested during this activity?

   ____

3. Place the following components of a neural pathway in the correct order by assigning each a number from 1 (first) to 7 (last):

   _____ a. Transmission of electrical impulse to the precentral gyrus

   _____ b. Transmission of electrical impulse to the thalamus

   _____ c. Transmission of electrical impulse to the effector

   _____ d. Transduction of the stimulus into an electrical impulse

   _____ e. Motor response

   _____ f. Relay of electrical impulse to cerebral cortex

   _____ g. Interpretation of meaning of stimulus by association area

PRE-LAB Activity 2: Investigating the Motor Neuron

1. Use the list of terms provided to label the accompanying illustration of a motor neuron; check off each term as you use it.

☐ axolemma	☐ cell body	☐ neurofibrils	☐ nucleus
☐ axon hillock	☐ dendrites	☐ Nissl bodies	☐ telodendria
☐ axon terminals	☐ myelin sheath	☐ nodes of Ranvier	

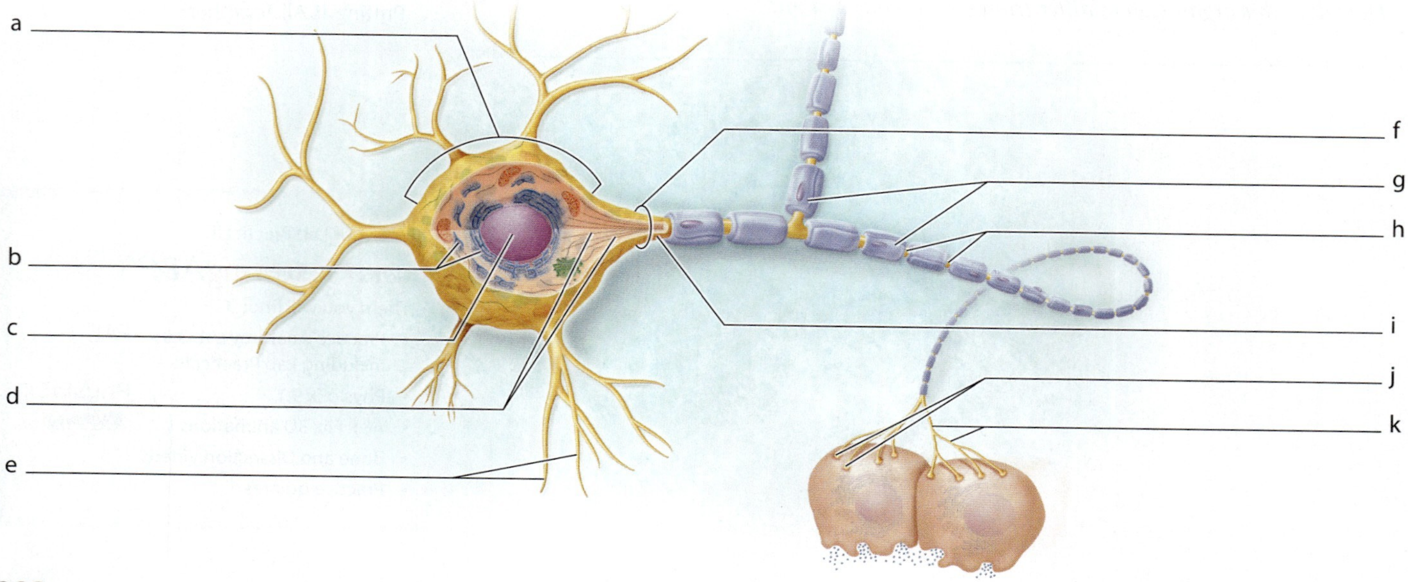

a ______________________________

b ______________________________

c ______________________________

d ______________________________

e ______________________________

f ______________________________

g ______________________________

h ______________________________

i ______________________________

j ______________________________

k ______________________________

2. What is the physiological significance of the presence of myelin on an axon? ____________________

____

3. How does a motor neuron differ from a sensory neuron? ____________________

____

PRE-LAB Activity 3: Investigating the Chemical Synapse

1. Match each of the following descriptions with the correct term:

_____	a. located on presynaptic membrane	1. synaptic cleft
_____	b. space between two neurons	2. neurotransmitter
_____	c. contains neurotransmitters	3. synaptic vesicle
_____	d. located on postsynaptic membrane	4. receptor
_____	e. bulbous ending of axon	5. axon terminal
_____	f. chemical messenger	6. calcium channel

2. Which neurotransmitter is released at the neuromuscular junction? ____________________

3. Put the following events that occur at the neuromuscular junction in their correct order by assigning each a number from 1 (first) to 5 (last):

_____ a. Sodium enters the muscle fiber, creating an end-plate potential.

_____ b. Neurotransmitter is released into the synaptic cleft.

_____ c. An action potential reaches the presynaptic membrane.

_____ d. Neurotransmitter binds to a receptor on the postsynaptic membrane.

_____ e. Sodium channels on the motor end plate open.

PRE-LAB Activity 4: Exploring the Histology of Nervous Tissue

1. Nervous tissue is the predominant tissue type of which three organs? ____________________

____

2. Nerve cells that conduct electrical impulses are called ____________________, whereas supporting cells are called ____________________ cells.

3. Cells in the peripheral nervous system that produce myelin are called ________________ cells.

4. Match each connective tissue wrapping of a nerve with its correct description:

_____	a. endoneurium	1. surrounds an entire nerve
_____	b. epineurium	2. surrounds a bundle of nerve fibers
_____	c. perineurium	3. surrounds an individual nerve fiber

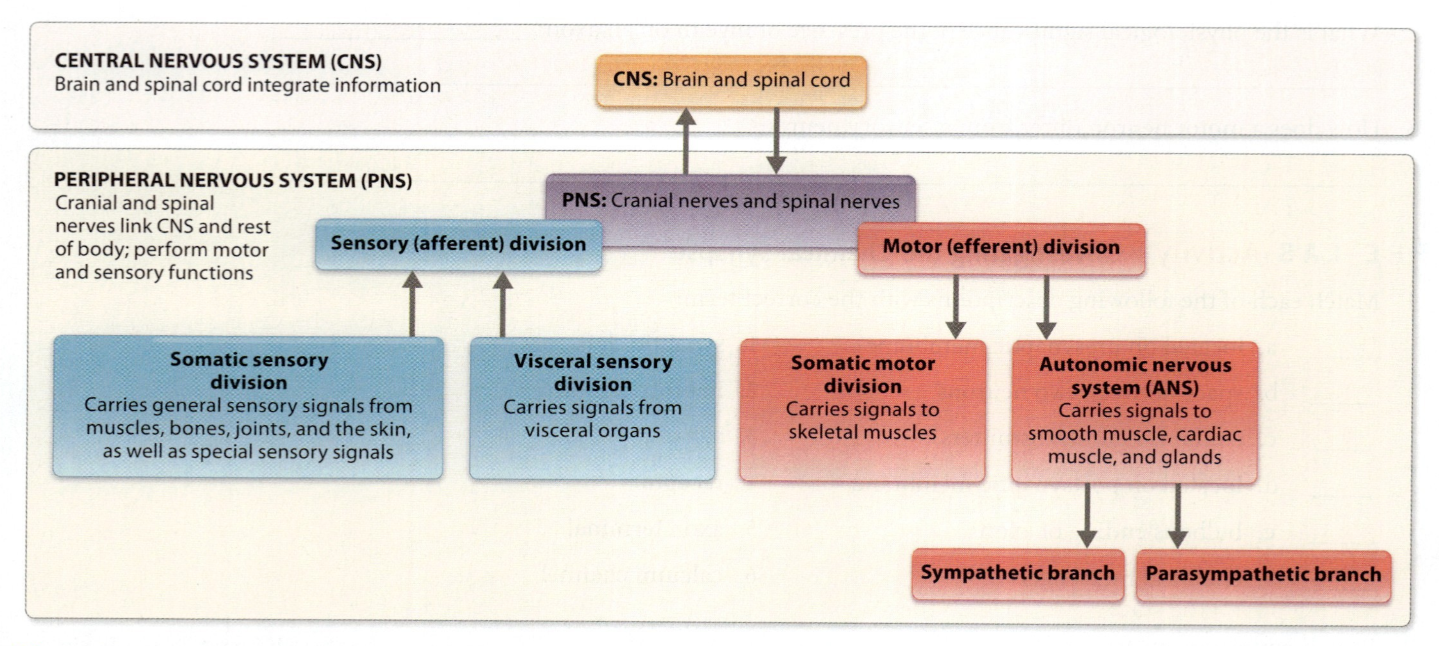

Figure 14-1 Organization of the nervous system.

Organization of the Nervous System

The nervous system (**Figure 14-1**) is divided into two major divisions: the **central nervous system (CNS)** and the **peripheral nervous system (PNS)**. The CNS consists of the brain and spinal cord; the PNS consists of cranial nerves and spinal nerves. The nerves of the PNS serve as communication lines between the CNS and the rest of the body. The PNS is divided into a **sensory (afferent) division**, which relays messages toward the CNS, and a **motor (efferent) division**, which relays messages from the CNS to the effectors (muscles and glands).

The sensory division is further subdivided into the somatic sensory division, which consists of nerves that carry signals from muscles, bones, joints, and the skin, and the visceral sensory division, which consists of nerves that carry signals from organs, including those of the special senses.

The motor division is further subdivided into the somatic motor division and the autonomic nervous system. The **somatic motor division** innervates skeletal muscles, whereas the **autonomic nervous system** innervates smooth muscle, cardiac muscle, and glands. The autonomic nervous system is further subdivided into the **sympathetic branch** and the **parasympathetic branch**. The sympathetic branch is often called the "E" division because it regulates physiological events associated with emergency, excitement, embarrassment, and exercise. The parasympathetic branch, by contrast, is often called the "D" division because it regulates physiological events associated with digestion, diuresis, and defecation.

Neural Pathways

We continue our examination of the structure of the nervous system by considering the pathways by which the nervous system operates. A **neural pathway** consists of a very rapid set of events that occur in a series of nervous system sites and produce a response. The following list outlines a typical neural pathway:

1. Transduction of a stimulus (whether visual, auditory, or tactile) into an electrical impulse
2. Transmission of the electrical impulse to the thalamus (an afferent message)
3. Relay of the electrical impulse to the cerebral cortex
4. Interpretation of the meaning of the electrical impulse by an association area of the cerebral cortex
5. Transmission of the electrical impulse from the association area to the precentral gyrus
6. Transmission of the electrical impulse from the precentral gyrus to muscles of the hand (an efferent message)
7. A motor response (typically, a set of muscle contractions)

The time it takes to complete these steps, from stimulus to response, is called the **reaction time**.

THINK ABOUT IT For the previous list of events involved in a typical neural pathway, underline in blue the components that involve the peripheral nervous system, and underline in red the components that involve the central nervous system. ■

In the activity that follows, you will investigate the neural pathways involved in catching a vertically hanging meter stick dropped by another student. You will compare the reaction times that occur with a visual, an auditory, or a tactile stimulus.

ACTIVITY 1

Calculating Reaction Time

Learning Outcomes

1. Outline the general neural pathway involved in a motor response to a visual, auditory, or tactile stimulus.
2. Calculate reaction times and determine which stimulus type (visual, auditory, or tactile) elicits the fastest motor response.

Materials Needed

- ☐ Meter stick
- ☐ Blindfold

Instructions

In this activity, you are going to perform an experiment to determine if your reaction time is faster when the initial stimulus is visual, auditory, or tactile.

1. Work in pairs. One student will be the subject, and one student will be the experimenter.
2. The experimenter holds a meter stick straight up and down (with the zero end down) so that the 50-cm mark is even with the top of the lab bench.
3. The subject places the thumb and index finger of his/her dominant hand on either side of the meter stick at the 50-cm mark, as close to the stick as possible without touching it.
4. The experimenter then releases the meter stick (without warning).
5. When the subject sees the meter stick dropping, he/she grasps it between thumb and index finger as quickly as possible, and keeps his/her grip on the stick.
6. The experimenter then determines the distance that the meter stick dropped by noting the position of the subject's fingers on the meter stick and subtracting 50 cm (the starting point).
7. Repeat steps 2–6 four additional times, and then calculate the average distance dropped when the stimulus is visual. Record data in the appropriate column in the following table.
8. Repeat steps 2–7 using an auditory stimulus rather than a visual stimulus. This time, blindfold the subject. At the same time that the experimenter releases the meter stick, he/she says "Now," and the subject grasps the meter stick as quickly as possible.
9. Repeat steps 2–7 using a tactile stimulus. Again, blindfold the subject. This time, at the same time that the experimenter releases the meter stick, he/she taps the subject on the shoulder, and the subject grasps the meter stick as quickly as possible.
10. For each of the three stimulus types, calculate the average distance dropped, and enter those numbers in the appropriate places in the table. The average distances dropped ("dd") will be used to calculate reaction times in the next step.
11. To calculate reaction time, we take the square root of twice the distance the meter stick dropped (dd) divided by the rate at which it fell (which is 980 cm/sec). Then convert from seconds to milliseconds (1 sec = 1000 msec). So, for each stimulus type, use the following equation to calculate reaction times.

$$\text{Reaction time (msec)} = \sqrt{2(\text{dd})\text{ cm} \times \frac{1\text{ sec}}{980\text{ cm}}} \times \frac{1000\text{ msec}}{1\text{ sec}}$$

Then enter the reaction times in the appropriate places in the table.

Trial	Distance Meter Stick Dropped (in cm)		
	Visual Stimulus	Auditory Stimulus	Tactile Stimulus
1			
2			
3			
4			
5			
Average distance			
Reaction time			

12. Answer each of the following questions based on your results:

a. Which stimulus type (visual, auditory, or tactile) resulted in the fastest reaction time?

____

b. Provide an explanation for your results.

____

____

____

13. Answer each of the following questions concerning the structures of a neural pathway:

    a. Describe the location and function of the thalamus.

    ____

    ____

    ____

    b. Describe the location and function of the visual association area. ______________________

    ____

    c. Describe the location and function of the precentral gyrus. ______________________

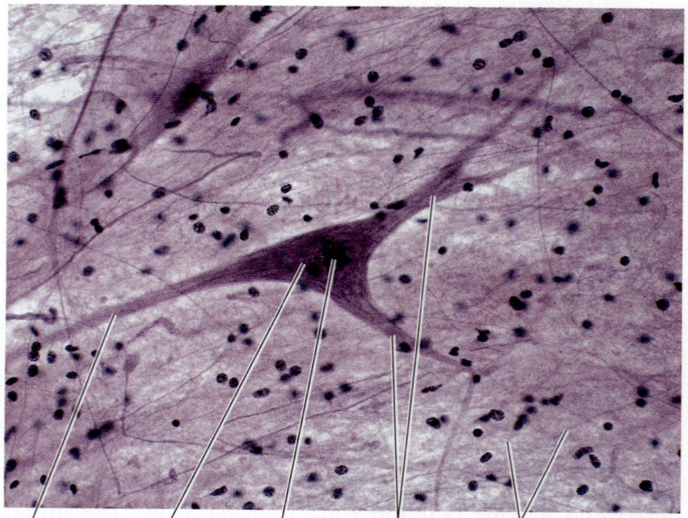

    ____

Nerves and Nervous Tissue

Recall from your study of muscle tissue in Unit 12 that a muscle is an organ composed primarily of skeletal muscle tissue and connective tissue wrappings. Similarly, a **nerve** (**Figure 14-2**) is an organ composed largely of nervous tissue and connective tissue wrappings. Nervous tissue contains two major populations of cells: **neurons** (impulse-conducting cells) and **neuroglial cells** (supporting cells). Three separate connective tissue wrappings surround and compartmentalize a nerve. The **epineurium**, a tough fibrous connective tissue, surrounds the entire nerve. The **perineurium** surrounds bundles of nerve fibers called fascicles, and the **endoneurium** is a delicate loose connective tissue that surrounds individual nerve fibers.

Axon Cell body Nucleus Dendrites Neuroglia LM (400×)

Figure 14-3 Neuron and neuroglial cells.

Neurons

The billions of neurons in your body transmit electrical messages. **Figure 14-3** is a photomicrograph showing a neuron surrounded by neuroglial cells. Although neurons vary in structure, they all have a cell body (soma) and one or more slender processes (dendrites and axons). The cell body consists of a nucleus and cytoplasm filled with a variety of organelles such as mitochondria, neurotubules (microtubules), and a specialized rough endoplasmic reticulum called either chromatophilic substance or Nissl bodies.

Clusters of neuron cell bodies in the CNS are called **nuclei** (singular = nucleus); clusters of neuron cell bodies in the PNS are called **ganglia** (singular = ganglion). The numerous dendrites of a neuron receive incoming stimuli and carry electrical information in the form of **graded potentials** toward the

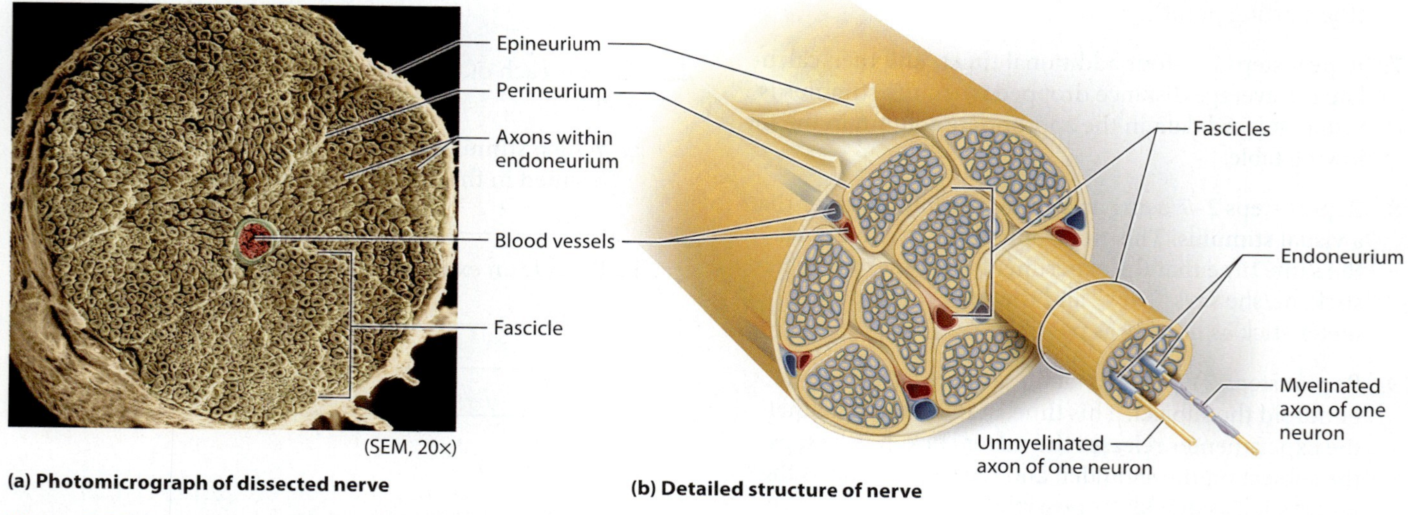

(a) Photomicrograph of dissected nerve **(b) Detailed structure of nerve**

Figure 14-2 A nerve and its connective tissue wrappings.

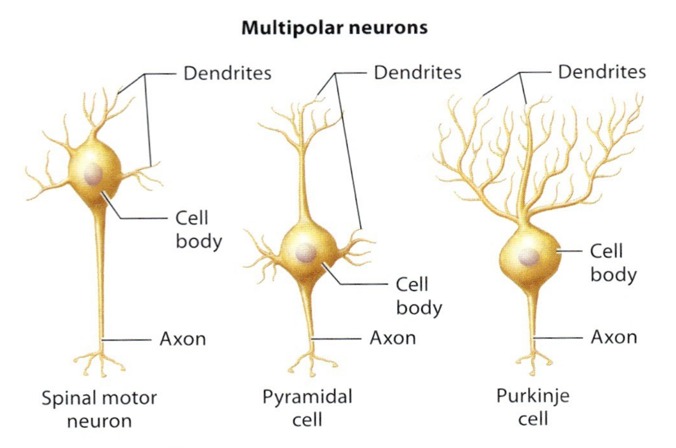

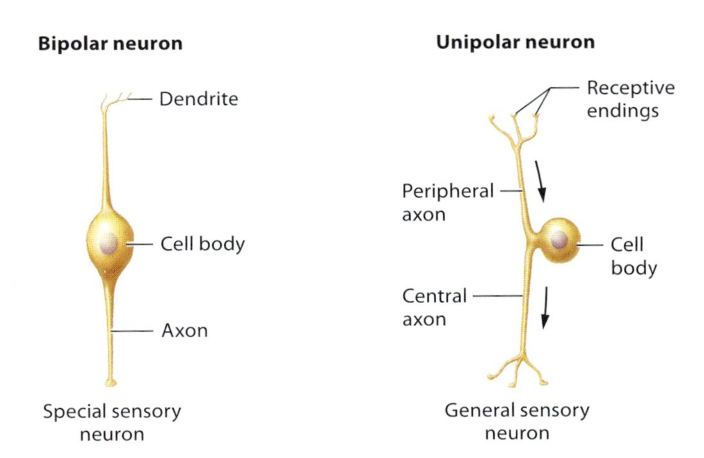

Figure 14-4 Classification of neurons based on structure.

cell body. Axons, by contrast, generate and then transmit electrical impulses (**action potentials**) away from the cell body.

Neurons can be classified based on both structure and function. Structurally, neurons are classified as either multipolar, bipolar, or unipolar (**Figure 14-4**). **Multipolar neurons** are the most common structural type of neuron and are the major neuron type of the CNS. These neurons have many dendrites and a single axon. **Bipolar neurons** are rare and are found only in specialized sensory structures such as the retina of the eye and the olfactory epithelium. These neurons have only two processes (an axon and a dendrite) that extend from the cell body. **Unipolar neurons** have a single short process that extends from the cell body and then divides into a peripheral axon that is associated with the cell's sensory receptors and a central axon that enters the CNS. Unipolar neurons are found primarily in the PNS and function as sensory neurons.

Functionally, neurons can be classified as **sensory neurons**, **motor neurons**, or **interneurons**. Sensory (or afferent) neurons carry electrical messages from sensory receptors toward the CNS. Motor (or efferent) neurons carry electrical messages away from the CNS to effector organs (muscles or glands). Interneurons, or association neurons, lie between sensory neurons and motor neurons and relay electrical signals between these two cells. Most interneurons are confined to the CNS.

Neuroglial Cells

Neuroglial cells, the second major category of cells found in nervous tissue, are the most abundant type of cells in the nervous system. Neuroglial cells perform a wide variety of functions. Collectively, they make up a network of cells called neuroglia ("nerve glue"). Four types of neuroglial cells are found in the CNS (**Figure 14-5**):

- *Astrocytes* are star shaped. They anchor neurons to capillaries and maintain the extracellular environment.
- *Oligodendrocytes* produce myelin sheaths that insulate certain axons and increase the speed of electrical impulses.

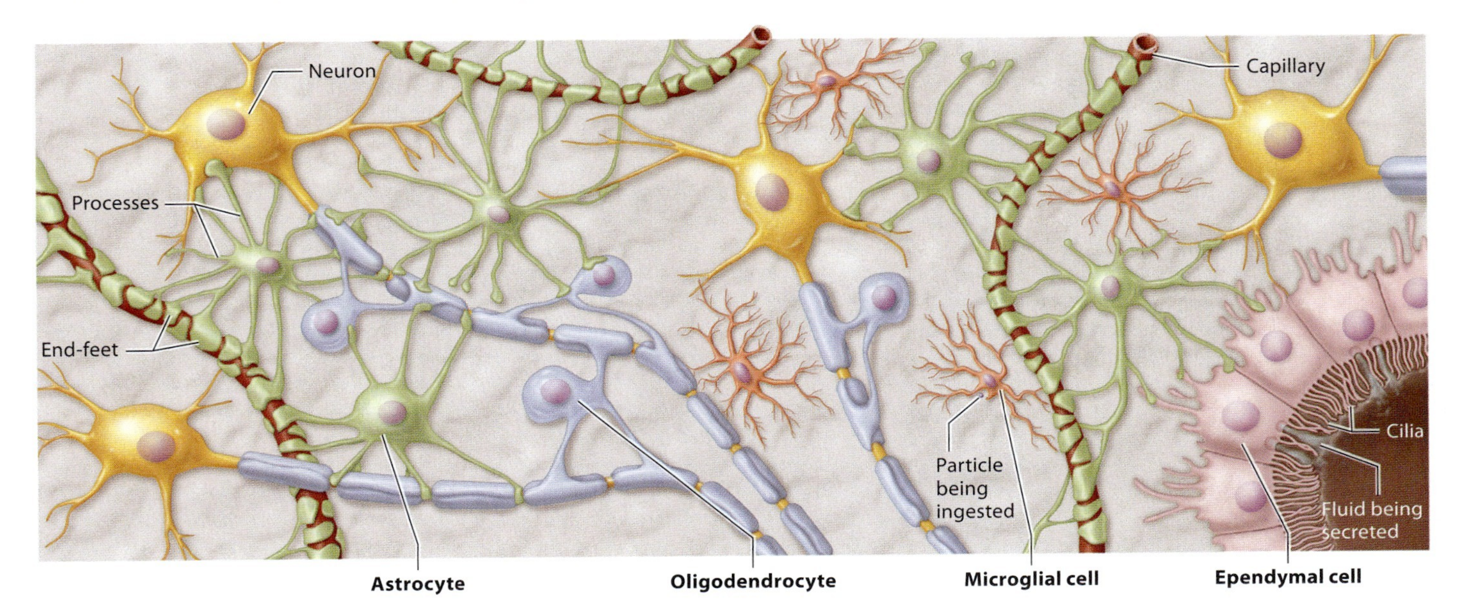

Figure 14-5 Neuroglial cells of the CNS.

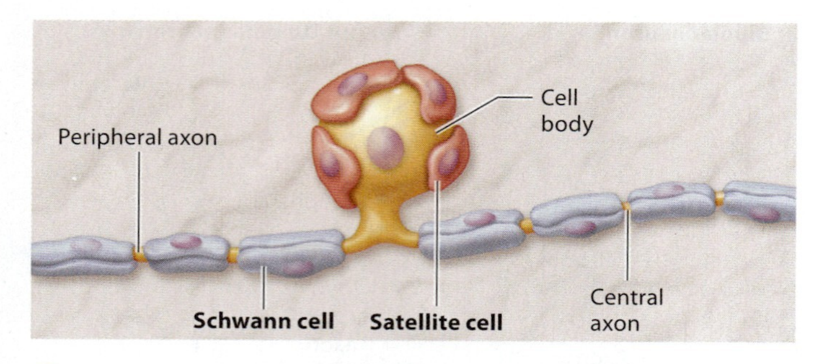

Figure 14-6 Neuroglial cells of the PNS.

- *Microglial cells* phagocytize foreign particles.
- *Ependymal cells* line hollow cavities and produce cerebrospinal fluid (a fluid that bathes neural tissue and cushions the brain and spinal cord).

Two types of neuroglial cells are found in the PNS (Figure 14-6):

- *Schwann cells* produce myelin sheaths around certain axons.
- *Satellite cells* maintain a proper ionic balance in the extracellular environment.

A Typical Motor Neuron

Figure 14-7 depicts a typical motor neuron. All neurons are surrounded by a plasma membrane called an **axolemma**, and most neurons consist of three distinct parts: dendrites, cell body, and axon. **Dendrites** (*dendros* = tree), branching structures that resemble a tree, receive incoming stimuli and carry electrical information toward the cell body. Dendrites are called receptive regions because they contain receptors for neurotransmitters and "receive" the chemical messengers released by other neurons.

The **cell body**, also called the perikaryon or soma, contains a nucleus surrounded by cytoplasm. The cytoplasm contains organelles (mitochondria, chromatophilic substance or Nissl bodies, and Golgi apparatus) that function as the metabolic machinery of the cell. The supporting cytoskeleton includes intermediate filaments (neurofilaments) and neurofibrils (microtubules). These cytoskeletal elements play important roles in maintaining cell shape and transporting materials.

The **axon** extends from a region of the cell body called the **axon hillock** and may divide into several collateral branches, each of which divides into terminal branches called **telodendria**. At the distal end of each telodendrion is an enlarged bulbous structure called the **axon terminal** (or synaptic knob), which contains synaptic vesicles filled with neurotransmitters (chemical messengers). Axons can be myelinated or unmyelinated. Myelin is a lipoprotein wrapped around an axon. Gaps in these myelin sheaths, called **nodes of Ranvier**, allow electrical impulses to "jump" along the axon from one node to the next, a process called saltatory conduction, thus increasing the speed of nerve impulse transmission.

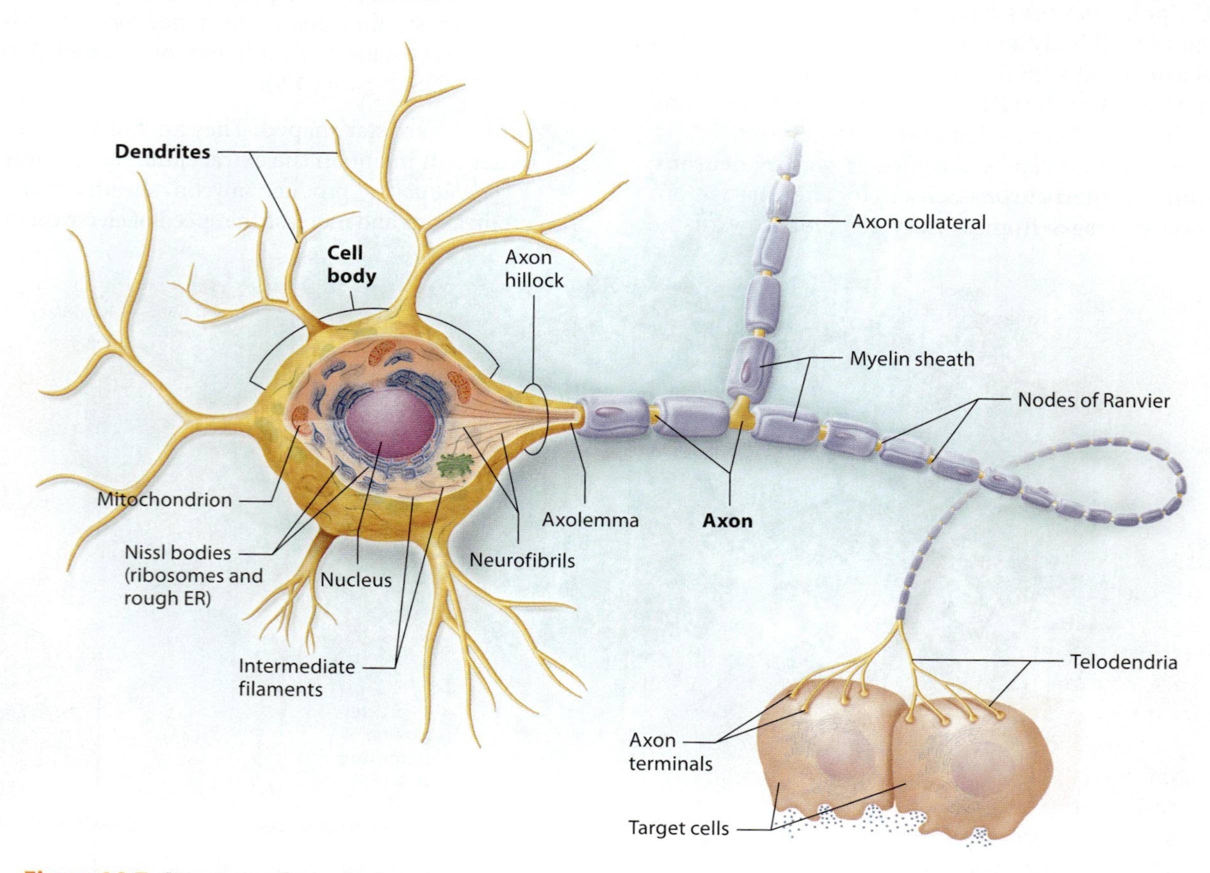

Figure 14-7 Structure of a typical motor neuron.

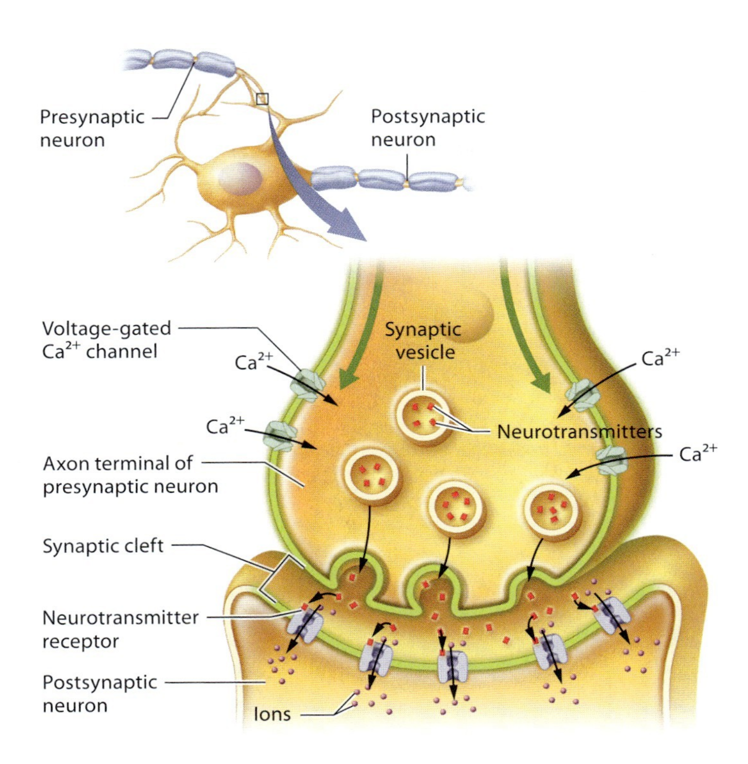

Figure 14-8 **A chemical synapse.**

The Chemical Synapse

The junction between two neurons is called a **synapse** (**Figure 14-8**). A synaptic cleft (a narrow extracellular space) typically separates the two neurons. The first neuron (the neurotransmitter-releasing neuron) is called the presynaptic neuron, and the second neuron (the neuron that responds to the neurotransmitter) is called the postsynaptic neuron. If the presynaptic neuron communicates with a muscle cell or a gland cell, then this cell is referred to as a postsynaptic cell. Recall from Unit 12 that a synapse between a motor neuron and a skeletal muscle fiber is called a neuromuscular junction. To review the events that occur at the neuromuscular junction, refer to Figure 12-4.

Transmission of an impulse from one neuron to another neuron occurs in a process that is similar to that at a neuromuscular junction. The arrival of an action potential at the axon terminal of the presynaptic neuron (see Figure 14-8) triggers an influx of calcium into the axon terminal. This calcium influx triggers the synaptic vesicles to fuse with the membrane of the presynaptic neuron and to release their contents (neurotransmitters) by exocytosis into the synaptic cleft. The neurotransmitters diffuse across the synapse and bind to receptors on the membrane of the postsynaptic neuron. Binding of neurotransmitter to receptors on the membrane of the postsynaptic neuron may lead to the propagation of an action potential throughout that neuron.

At a neuromuscular junction, the neurotransmitter is always acetylcholine, the receptor is always a sodium channel, and the response is always excitatory. In a synapse between two neurons, however, a variety of different neurotransmitters can be involved, the receptor can directly or indirectly open different types of ion channels, and the response can be either stimulatory or inhibitory.

ACTIVITY 2

Investigating the Motor Neuron

Learning Outcomes

1. Identify the components of a motor neuron on an anatomical model or chart and describe the function of each.
2. Demonstrate the role of myelin in the conduction of electrical impulses.

Materials Needed

☐ Motor neuron model
☐ Anatomical charts
☐ Two balls

Instructions

A. Components of the Motor Neuron

1. CHART Obtain a motor neuron model, and then identify the features of a motor neuron as you complete the Making Connections chart on the next page.

B. Demonstrating the Effect of the Myelin Sheath

Demonstrate the importance of the myelin sheath by completing the following interactive class activity. Read the instructions for the activity and form a hypothesis before you begin.

1. Your instructor will select 12 students to represent axon 1 and 6 students to represent axon 2.

2. Each group of students will arrange themselves into two parallel rows that each extends approximately 25 feet. The first student of each group, representing the initial segment of the axon, will stand side by side at the 0-foot mark. The last person in each group, representing the axon terminal, will stand side by side at the 25-foot mark. Students in each group will then position themselves in such a way that there is an equal distance between each student.

3. The instructor will give one ball to the first person in axon 1 and another ball to the first person in axon 2.

4. When the instructor shouts "Action!" the students representing axon 1 will *hand* the ball from person to person until the ball reaches the last person in line. The students representing axon 2 will *pass* (toss) the ball from person to person until the ball reaches the last person in line. This is a race: Note in which "axon" the ball first reaches the last person.

Making Connections: The Motor Neuron

Feature	Function	Connections to Things I Have Already Learned
Cell body (soma)		*Clusters of cell bodies in the CNS = nuclei; clusters of cell bodies in the PNS = ganglia*
Nissl bodies	*Synthesize proteins*	
Neurofibrils		*Form "railroad tracks" that transport materials within cell; for example, synaptic vesicles filled with neurotransmitter are transported from the cell body to the axon terminal via neurofibrils.*
Dendrites	*Short, branched projections from the cell body that typically transmit electrical messages toward the cell body*	
Axon		*Also referred to as a nerve fiber*
Axon hillock	*Cone-shaped region of the cell body from which the axon arises*	
Axolemma		*Sarcolemma = plasma membrane surrounding a skeletal muscle fiber*
Telodendria	*Terminal branches of axons*	
Axon terminals	*Bulbous endings of axon terminals containing neurotransmitter-filled synaptic vesicles*	
Myelin sheath		*Myelin is produced by Schwann cells in the PNS and by oligodendrocytes in the CNS.*
Nodes of Ranvier	*Gaps in the myelin sheaths surrounding an axon*	

5. You will repeat step 4 three more times, each time noting in which "axon" the ball reached the end of the line faster.
6. Before you do this demonstration, predict which group of students (the axon 1 or axon 2 group) will move the ball the entire length of the line (axon) more quickly, and provide an explanation for your prediction.

____

____

____

7. Do the demonstration and then record your data in the following table:

Trial	Which Axon Is Faster (Axon 1 or Axon 2)?
1	
2	
3	
4	

8. Answer the following questions based on your results:
   a. Which axon (axon 1 or axon 2) represents the "myelinated" axon? Why? _______________

   ____

   ____

   b. What does the movement of the ball in both axons represent? _______________

   ____

   c. What type of conduction is represented by axon 2?

   ____

9. **Optional Activities**

View and label histology slides of a nerve at MasteringA&P® > Study Area > Practice Anatomy Lab > Histology > Nervous Tissue > images 9 and 10

Play an animation of the propagation of an action potential at MasteringA&P® > Study Area > A&P Flix > A&P > Propagation of an Action Potential

ACTIVITY 3
Investigating the Chemical Synapse

Learning Outcomes

1. Label the components of a chemical synapse.
2. Outline the sequence of events that occur at a chemical synapse.
3. Predict the effects of certain drugs on a chemical synapse.

Materials Needed

None

Instructions

1. On the figure below, label the following parts of the chemical synapse and outline the events that occur at the synapse (steps 1–4):
   - ☐ neurotransmitter receptor
   - ☐ calcium channel
   - ☐ postsynaptic neuron
   - ☐ presynaptic neuron
   - ☐ synaptic cleft
   - ☐ axon terminal
   - ☐ neurotransmitter
   - ☐ synaptic vesicle

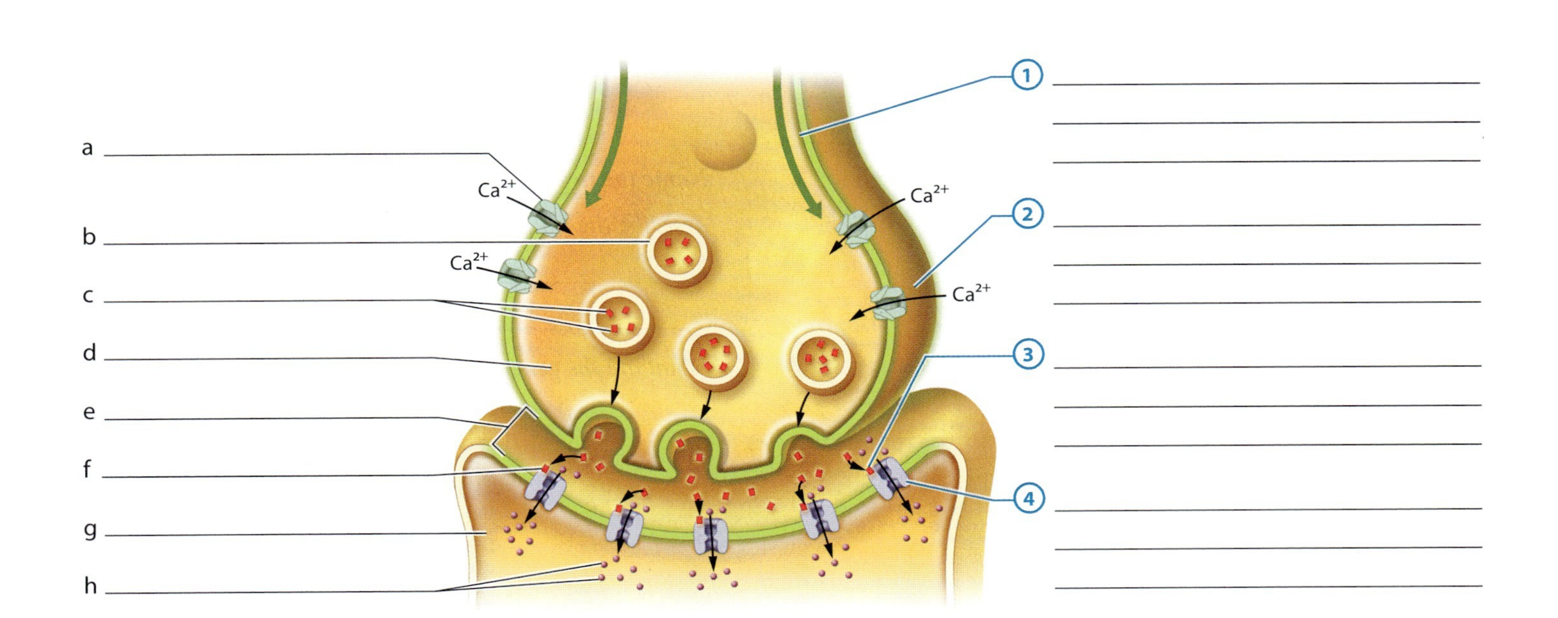

2. Answer the following question set:
   a. What causes the calcium influx into the axon terminal? ______________________
   ____
   b. What does the calcium influx trigger?
   ____
   c. What happens to the neurotransmitter after it is released from the axon terminal via exocytosis?
   ____
   ____
   d. What happens to the postsynaptic membrane if a chloride channel opens? ______________________
   ____
   e. What happens to the postsynaptic membrane if a potassium channel opens? ______________________
   ____
   f. What happens to the postsynaptic membrane if a sodium channel opens? ______________________
   ____
   g. When does an EPSP (excitatory postsynaptic potential) become an action potential? ______________________
   ____
   h. Drug X has a stimulatory effect on the postsynaptic neuron. Describe two possible mechanisms by which drug X might exert its effect.
   ____
   ____
   ____
   ____
   i. Drug Y has an inhibitory effect on the postsynaptic neuron. Describe two possible mechanisms by which drug Y might exert its effect.
   ____
   ____
   ____
   ____

ACTIVITY 4

Exploring the Histology of Nervous Tissue

Learning Outcomes

1. Identify and label the following structures on a motor neuron smear: motor neuron (cell body, nucleus, nucleolus, dendrites, axon) and nuclei of neuroglial cells.
2. Identify and label the following structures on a cross-section of a nerve: epineurium, perineurium, endoneurium, fascicle, myelin sheath, axon.
3. Identify and label the following structures on a longitudinal section of a nerve: axon, myelin, nucleus of Schwann cell, node of Ranvier.

Materials Needed

☐ Microscope and prepared microscope slides, or photomicrographs, of a motor neuron smear, a peripheral nerve in cross section, and a peripheral nerve in longitudinal section.

Instructions

1. Observe a motor neuron smear. Make a sketch of the slide and label the following structures: motor neuron (cell body, nucleus, nucleolus, cellular processes) and nuclei of neuroglial cells.

   Total magnification: _____ ×

   Name the two types of cellular processes of a neuron, and describe the function of each. ______________________
   ____
   ____

   What are the general functions of neuroglial cells?
   ____
   ____

Is the neuron in the smear a unipolar, bipolar, or multipolar neuron? Explain. ______________________________

____

____

2. Observe a cross-section of a peripheral nerve. Make a sketch of the slide and label the following structures: epineurium, perineurium, endoneurium, fascicle, myelin sheath, axon.

Total magnification: _____×

What are the functions of the connective tissue wrappings found in a nerve? ______________________________

____

3. Observe a longitudinal section of a peripheral nerve. Make a sketch of the slide and label the following structures: axon, myelin, nucleus of Schwann cell, node of Ranvier.

Total magnification: _____×

What is the function of the node of Ranvier?

____

Which neuroglial cell type produces myelin in the CNS?

____

4. **Optional Activity**

View and label histology slides of nervous tissue at MasteringA&P® > Study Area > Practice Anatomy Lab > Histology > Nervous Tissue

PhysioEx EXERCISE 3
Neurophysiology of Nerve Impulses

The PhysioEx 9.1 Laboratory Simulations in Physiology are easy-to-use laboratory simulations that can be used as an alternative to or as a supplement to wet lab activities in this unit. Each simulation allows you to investigate important physiological concepts, repeat labs as often as you like, and conduct experiments that are difficult to perform in a wet lab environment because of time, cost, or safety concerns.

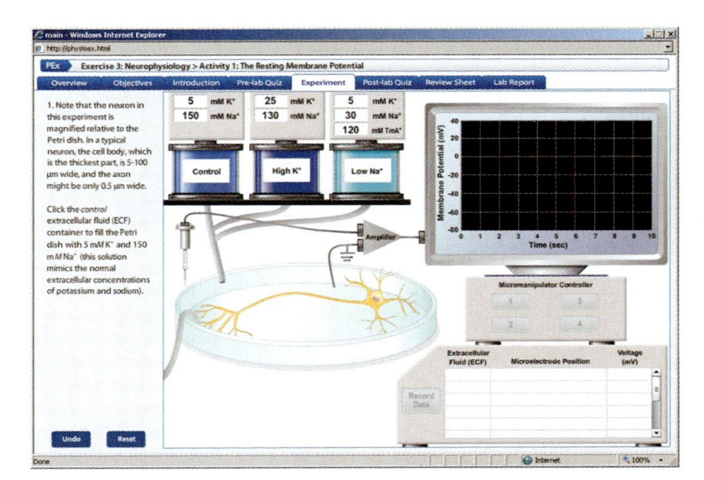

Access the simulations in these activities at MasteringA&P® > Study Area > PhysioEx 9.1

There you will find the following activities:

- **PEx Activity 1:** The Resting Membrane Potential
- **PEx Activity 2:** Receptor Potential
- **PEx Activity 3:** The Action Potential: Threshold
- **PEx Activity 4:** The Action Potential: Importance of Voltage-Gated Na^+ Channels
- **PEx Activity 5:** The Action Potential: Measuring Its Absolute and Relative Refractory Periods
- **PEx Activity 6:** The Action Potential: Coding for Stimulus Intensity
- **PEx Activity 7:** The Action Potential: Conduction Velocity
- **PEx Activity 8:** Chemical Synaptic Transmission and Neurotransmitter Release
- **PEx Activity 9:** The Action Potential: Putting It All Together

Post-lab quizzes are also assignable in MasteringA&P®

POST-LAB ASSIGNMENTS

Name: ______________________ Date: ______________ Lab Section: ________

PART I. Check Your Understanding

Activity 1: Calculating Reaction Time

1. Based on the neural pathway involved in the reaction time experiment, match each of the following structures with its function.

_____	a. sensory neuron	1. interprets meaning of stimulus
_____	b. effector	2. initiates efferent message to the muscle
_____	c. thalamus	3. transmits information from the receptor to the brain
_____	d. motor neuron	4. relays sensory information to the association area
_____	e. visual association area	5. carries out the motor response
_____	f. primary motor area	6. transmits electrical impulse to the muscle

2. Based on your data, which stimulus type led to the fastest reaction time? Provide a possible explanation for this result. ____

____

Activity 2: Investigating the Motor Neuron

1. Name two factors that increase the speed of nerve impulse conduction.

____

2. Match each of the following neuron structures with its correct description:

_____	a. neurofibrils	1. cytoskeletal elements
_____	b. Nissl bodies	2. gaps between Schwann cells
_____	c. telodendria	3. contain synaptic vesicles
_____	d. axon terminals	4. produce neurotransmitters
_____	e. nodes of Ranvier	5. terminal branches of axon

Activity 3: Investigating the Chemical Synapse

1. Which neuron structure stores neurotransmitters?
   a. neurofibrils
   b. synaptic vesicles
   c. axon hillock
   d. Nissl bodies
   e. telodendria

2. Put the following events that lead to the generation of an action potential in the postsynaptic neuron in the proper order in which they occur by assigning each a number from 1 (first event) to 6 (last event):

_____ a. The neurotransmitter binds to a receptor.

_____ b. Sodium enters the postsynaptic neuron, generating an EPSP.

_____ c. An action potential reaches the axon terminal.

_____ d. When the EPSP reaches threshold, an action potential fires.

_____ e. An influx of calcium triggers the release of neurotransmitter.

_____ f. The neurotransmitter diffuses across the synapse.

3. Drug X exerts a stimulatory effect at a synapse. Briefly describe two ways in which drug X might exert its stimulatory effect. ______________________________

____

4. Drug Y exerts an inhibitory effect at a synapse. Briefly describe two ways in which drug Y might exert its inhibitory effect. ______________________________

____

Activity 4: Exploring the Histology of Nervous Tissue

1. Identify each structure in this neuron smear:

a. ______________________________

b. ______________________________

c. ______________________________

d. ______________________________

e. ______________________________

f. ______________________________

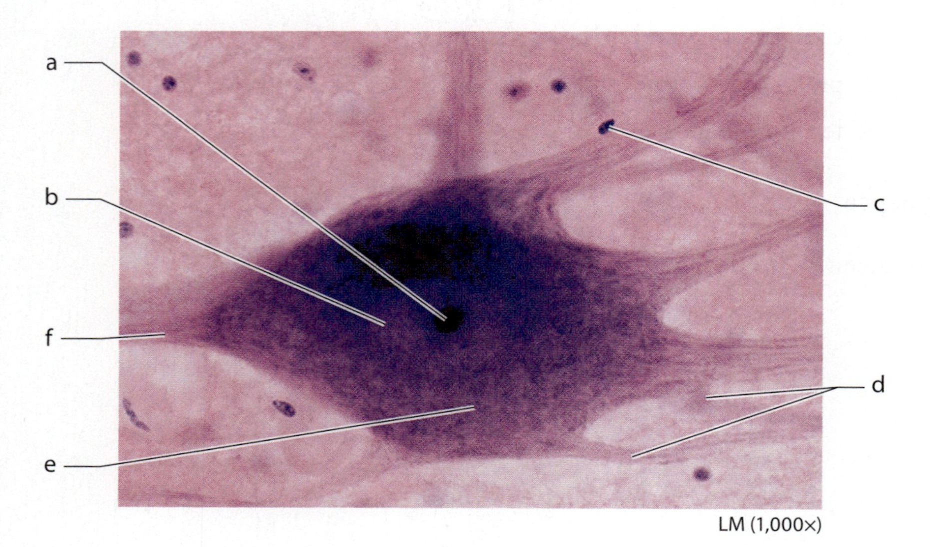

LM (1,000×)

2. Identify the components of nerve shown in the photomicrograph (cross section):

   a. ______________________

   b. ______________________

   c. ______________________

   d. ______________________

   e. ______________________

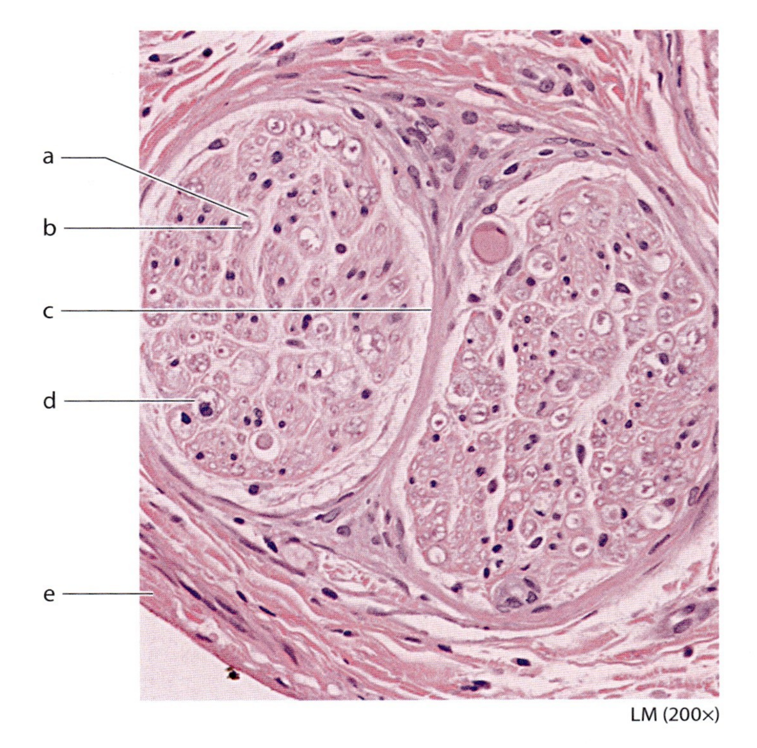

LM (200×)

3. Identify the components of nerve shown in the photomicrograph (longitudinal section):

   a. ______________________

   b. ______________________

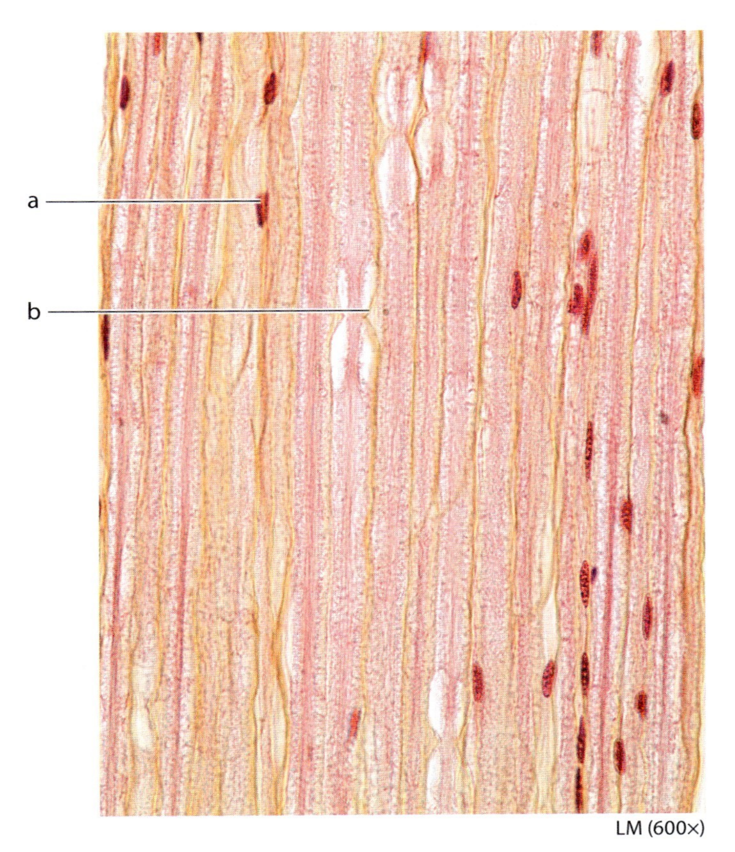

LM (600×)

PART II. Putting It All Together

A. Review Questions

Answer the following questions using your lecture notes, your textbook, and your lab notes.

1. Which of the following structures is a component of the peripheral nervous system?
   a. brain
   b. spinal cord
   c. cranial nerves
   d. spinal nerves
   e. More than one of these answers is correct.

2. Which of the following cell types produces myelin?
   a. oligodendrocyte
   b. astrocyte
   c. ependymal cell
   d. schwann cell
   e. More than one of these answers is correct.

3. The nervous system is divided into the CNS and the PNS. The PNS is divided into the ________________________ and the ________________________ divisions. The ________________________ division is further divided into somatic and autonomic portions. The autonomic nervous system innervates ________________________, ________________________, and ________________________, whereas the somatic nervous system innervates ________________________.

4. Is a nerve an organ or a tissue? Explain. ____

____

5. List the three categories of neurons based on structure: ________________________________

____

6. List the three categories of neurons based on function: ________________________________

____

7. Multiple sclerosis is an autoimmune disease that affects the brain and spinal cord. It causes damage to the myelin surrounding nerve fibers as well as to the nerve fibers themselves. The damaged myelin forms scar tissue (a process called sclerosis), which gives the disease its name. How does this disease affect nerve impulse conduction? Why? ____________________________________

____

____

8. Does each description apply to an astrocyte (A) or to a satellite cell (S) or to both cell types (A and S)?

   _____ a. Maintains proper ionic environment

   _____ b. Anchors capillaries to neurons

   _____ c. Found in central nervous system

9. Does each description apply to a hormone (H) or to a neurotransmitter (N) or to both a hormone and a neurotransmitter (H and N)?

   _____ a. Function as chemical messengers

   _____ b. Released into the bloodstream

   _____ c. Produced by endocrine glands

10. Does each description apply to a bipolar neuron (B) or to a unipolar neuron (U) or to both neuron types (B and U)?

   _____ a. Classified as a sensory neuron

   _____ b. Consists of one dendrite and one axon

   _____ c. Found in the retina and olfactory mucosa

11. Does each description apply to a graded potential (G) or to an action potential (A) or to both potential types (G and A)?

   _____ a. Classified as an electrical event

   _____ b. Can vary in intensity

   _____ c. Classified as an all-or-none event

B. Concept Mapping

1. Fill in the blanks to complete this concept map outlining the organization of the nervous system.

axon	cerebrum	nerve	neuron	Schwann cell

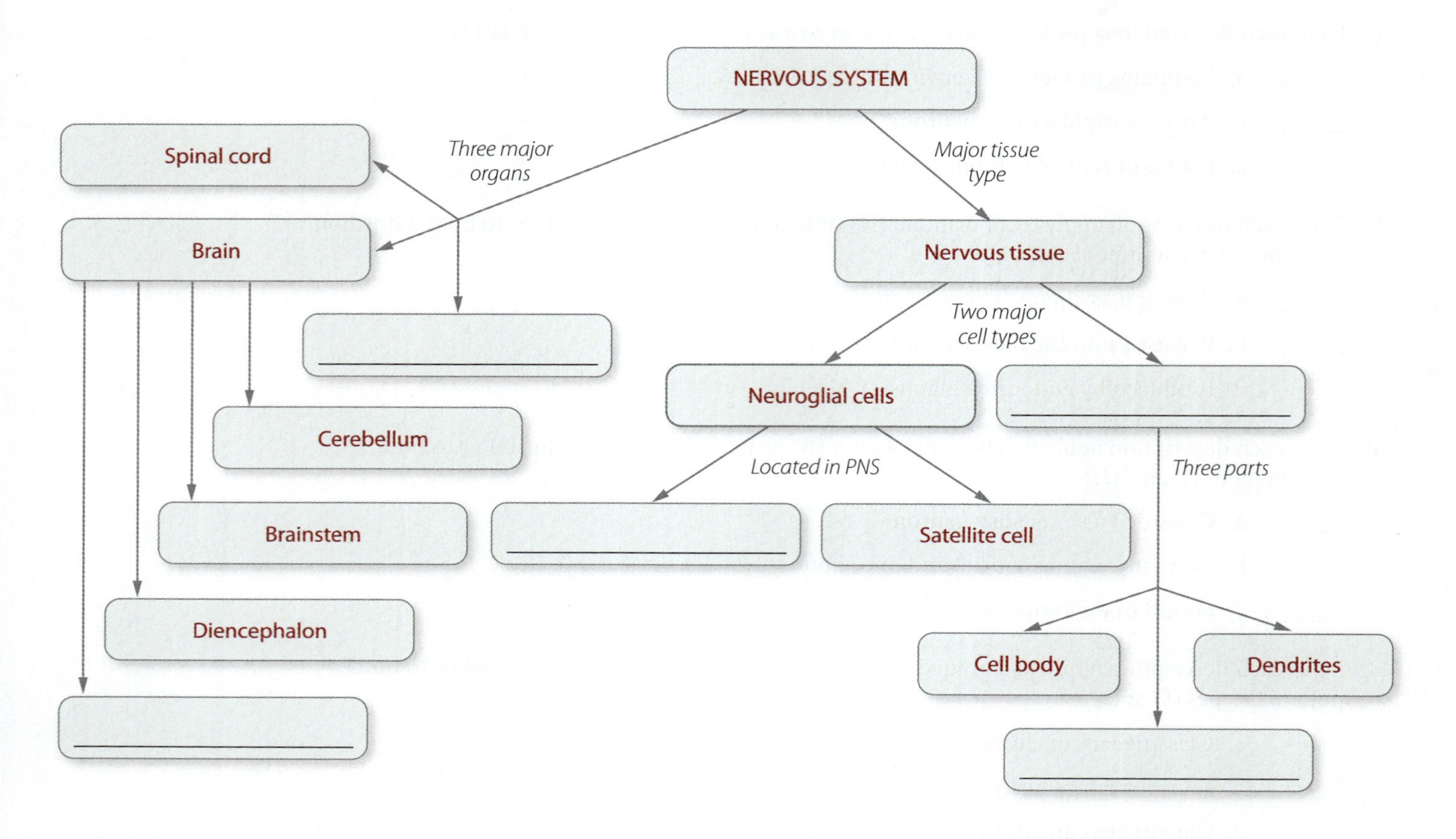

2. Construct a unit concept map to show the relationships among the following set of terms. Include all of the terms in your diagram. Your instructor may choose to assign additional terms.

action potential	arbor vitae	axon	brainstem	cerebrum
diencephalon	epineurium	gray matter	lateral horn	nerve
nerve fiber	neuron	node of Ranvier	Schwann cell	synaptic cleft

15

The Central Nervous System: Brain and Spinal Cord

UNIT OUTLINE

In Unit 14 we explored the basic organization of the nervous system, and you studied the structure and function of nervous tissue. In this unit we focus on the functional anatomy of the brain and spinal cord. In Unit 16 we will engage in a detailed study of the structure and function of the peripheral nervous system (nerves and the autonomic nervous system).

THINK ABOUT IT *In which body cavities are the brain and spinal cord located?*

brain: ______________________ spinal cord: ______________________

The skull protects the brain and the vertebral column protects the spinal cord. Describe the function of each of the following bone features:

foramen magnum ____

____

crista galli ____

____

vertebral foramen ____

____ ■

PRE-LAB ASSIGNMENTS

Pre-lab quizzes are also assignable in MasteringA&P®

To maximize learning, BEFORE your lab period carefully read this entire lab unit and complete these pre-lab assignments using your textbook, lecture notes, and prior knowledge.

PRE-LAB Activity 1: Exploring the Functional Anatomy of the Brain

1. Use the list of terms provided to label the accompanying figure; check off each term as you use it.

- ☐ central sulcus
- ☐ frontal lobe
- ☐ insula
- ☐ occipital lobe
- ☐ parietal lobe
- ☐ precentral gyrus
- ☐ postcentral gyrus
- ☐ temporal lobe

a ______________________

b ______________________

c ______________________

d ______________________

e ______________________

f ______________________

g ______________________

h ______________________

2. Match each of the following descriptions with the correct brain structure(s):

______	a. houses the corpora quadrigemina	1. pons
______ ______ ______	b. components of the diencephalon	2. thalamus
______	c. connects the third and fourth ventricles	3. cerebellum
______	d. continuous with the spinal cord	4. hypothalamus
______	e. produces cerebrospinal fluid	5. cerebral aqueduct
______ ______ ______	f. components of the brainstem	6. medulla oblongata
______	g. major homeostatic organ	7. midbrain
______	h. contains Broca's area	8. epithalamus
______	i. structure from which pineal gland extends	9. cerebrum
______	j. contains lateral ventricles	10. choroid plexus
______	k. contains arbor vitae	

PRE-LAB Activity 2: Electroencephalography in a Human Subject Using BIOPAC®

1. How many electrodes are attached to the scalp to record the EEG?
   - a. 2
   - b. 3
   - c. 4
   - d. 5

2. The recording of the brain's activity obtained by using electrodes is called
   - a. electroencephalography.
   - b. an electrocardiogram.
   - c. electromyography.
   - d. None of these is correct.

PRE-LAB Activity 3: Identifying the Meninges/Ventricles and Tracing the Flow of Cerebrospinal Fluid

1. Match each of the following brain structures with its description:

_____ a. dura mater
_____ b. pia mater
_____ c. arachnoid mater
_____ d. first and second ventricles
_____ e. apertures
_____ f. superior sagittal sinus
_____ g. fourth ventricle
_____ h. cerebral aqueduct
_____ i. third ventricle
_____ j. arachnoid villus
_____ k. subarachnoid space
_____ l. interventricular foramina

1. connects the lateral ventricles and the third ventricle
2. middle meninx
3. structure that projects through dura mater into arachnoid mater
4. passageway for cerebrospinal fluid between fourth ventricle and subarachnoid space
5. ventricle associated with the diencephalon
6. outermost meninx
7. lateral ventricles
8. meninx that follows every convolution of brain
9. connects third ventricle and fourth ventricle
10. ventricle located between brainstem and cerebellum
11. collects venous blood and cerebrospinal fluid
12. located between the arachnoid mater and the pia mater

PRE-LAB Activity 4: Examining the Functional Anatomy of the Spinal Cord

1. Match each of the following spinal cord structures with its description:

_____ a. filum terminale
_____ b. conus medullaris
_____ c. cauda equina
_____ d. denticulate ligament

1. tapered end of the spinal cord
2. lateral extension of pia mater that anchors the spinal cord
3. connective tissue extending from the tip of the spinal cord to the coccyx
4. name given to spinal nerves extending from the inferior end of the spinal cord as a result of their appearance

PRE-LAB Activity 5: Analyzing a Spinal Reflex

1. Match each of the following components of a reflex arc with its description:

_____ a. sensory neuron
_____ b. effector
_____ c. motor neuron
_____ d. receptor
_____ e. integration center

1. receives the stimulus
2. transmits message to the effector
3. carries out the motor response
4. transmits message to the CNS
5. nerve impulse is processed

PRE-LAB Activity 6: Dissecting a Sheep Brain and Spinal Cord

1. A midsagittal cut will divide the sheep brain into _______________________________.
2. A coronal (or frontal) section will divide the sheep brain into _______________________________.
3. The connective tissue wrappings surrounding a mammalian brain are collectively called _______________.

The Brain

The brain consists of four major components: the cerebrum, the diencephalon, the brainstem, and the cerebellum. Here we consider the functional anatomy of each of these four major components. Then, we explore the structure and function of the meninges and examine the role of cerebrospinal fluid.

Cerebrum

The **cerebrum**, the largest and most prominent part of the brain, is composed of two hemispheres, each of which is composed of five lobes: frontal, parietal, temporal, occipital, and insula (**Figure 15-1c**). The functions of the cerebral lobes are listed in **Table 15-1**.

The cerebrum consists of a thin layer of gray matter called the cerebral cortex surrounding an inner core of white matter. The surface of the cerebrum is thrown into folds called convolutions, each of which consists of an elevated ridge called a **gyrus** and a shallow groove called a **sulcus** (**Figure 15-1a**). In addition to these convolutions, the surface

Table 15-1 Functions of the Five Cerebral Lobes

Cerebral Lobe	Functions
Frontal	Controls voluntary skeletal muscles; intellect, decision making, personality
Parietal	General sensation
Temporal	Hearing and smell
Occipital	Vision
Insula	Memory, taste, and integration of the activities of the other cerebral lobes

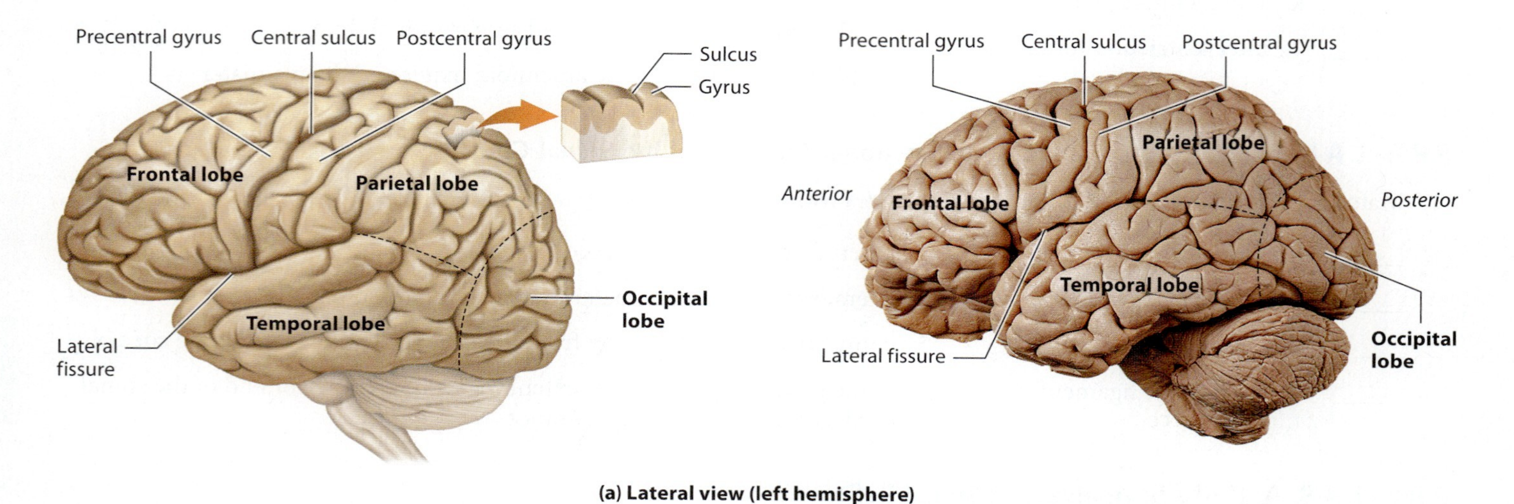

(a) Lateral view (left hemisphere)

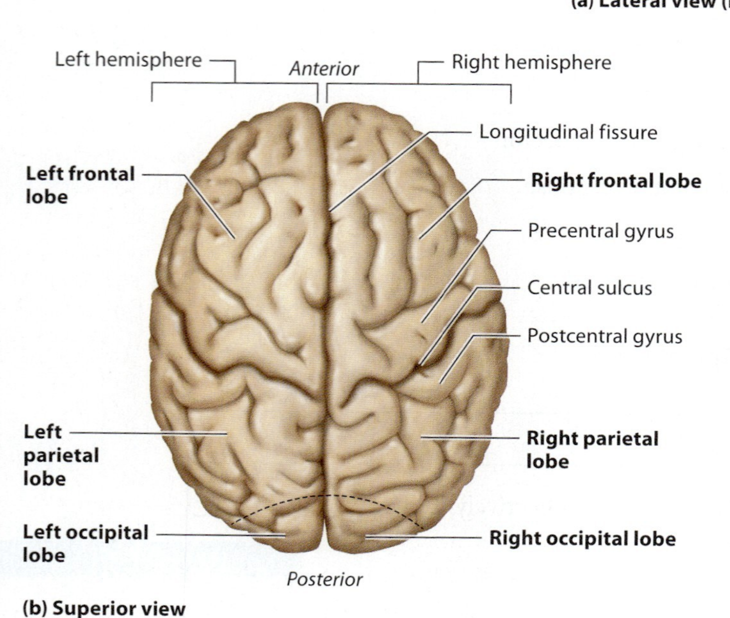

(b) Superior view

Central sulcus
Frontal lobe
Parietal lobe
Insula
Occipital lobe
Temporal lobe

(c) Lateral view (frontal, parietal, and temporal lobes pulled back)

Figure 15-1 The lobes and major anatomical landmarks of the cerebrum.

of the cerebrum also contains deep grooves called **fissures**. The longitudinal fissure travels along the midsagittal plane of the brain and separates the right and left cerebral hemispheres (**Figure 15-1b**).

QUICK TIP Here's a way to remember that the insula is located deep within the cerebrum: The insula is "insula-ted" by the other cerebral lobes.

The midsagittal view of the brain in **Figure 15-2** shows three additional cerebral structures. The large, comma-shaped **corpus callosum** is a tract that connects the two cerebral hemispheres. Recall that a tract is a bundle of nerve fibers in the CNS, and that a similar bundle of nerve fibers in the PNS is called a nerve. Inferior to the corpus callosum is the membranous **septum pellucidum**, which separates fluid-filled cavities called ventricles located within each cerebral hemisphere. You will learn about ventricles in the next section of this unit. Finally, inferior to the septum pellucidum is the **fornix**, which plays a role in olfaction.

The cerebral hemispheres can be divided into three basic regions: the cerebral cortex, the cerebral white matter, and the basal nuclei.

The **cerebral cortex** is a thin layer (2–4 mm thick) of gray matter containing predominantly neuronal cell bodies and dendrites. The cerebral cortex contains three types of functional areas: sensory areas, motor areas, and association areas. Sensory areas receive and interpret sensory impulses from sensory receptors; motor areas initiate motor impulses that travel from the brain to the skeletal muscles; and association areas receive impulses from and send impulses to different areas of the cerebral cortex. Some of the major functional areas of the cerebral cortex are illustrated in **Figure 15-3** and are described below:

Sensory Areas

- **Primary somatosensory cortex**—This functional area is located in the postcentral gyrus of the parietal lobes. Impulses that originate from a variety of sensory receptors (sensory receptors in the skin and proprioceptors in skeletal muscles, joints, and tendons) travel to this area of the brain.
- **Auditory areas**—These functional areas, located in the temporal lobes, receive and interpret sensory impulses that originate from the receptors of the inner ear.
- **Gustatory cortex** (not shown)—This functional area is located in the insulas, just deep to the temporal lobes. It receives sensory impulses that originate in the taste buds.
- **Visual areas**—These functional areas, located in the occipital lobes, receive and interpret the meaning of impulses that originate in the retina.
- **Olfactory cortex** (not shown)—This functional area, located in the medial temporal lobes, is involved in processing the sense of smell. It receives impulses from the olfactory receptors of the nasal epithelium.

Motor Areas

- **Primary motor cortex**—This functional area, located in the precentral gyrus of the frontal lobe, controls voluntary movement of skeletal muscles.
- **Frontal eye field**—This functional area is located superior to Broca's area and controls the voluntary movements of the eyes.
- **Premotor cortex**—This functional area, located just anterior to the precentral gyrus in the frontal lobe, plays a role in planning movements.

Association Areas

- **Broca's area**—This functional area is located at the base of the precentral gyrus just superior to the lateral sulcus. It controls the muscles involved in the production of speech and is also called the motor speech area. Broca's area is located in only one cerebral hemisphere, usually the left hemisphere.
- **Prefrontal cortex**—This area of the brain, located in the anterior portions of the frontal lobes, functions in decision making, personality, and higher thought processing.
- **Wernicke's area**—This area of the brain is located at the junction of the parietal and temporal lobes. It plays a role in sounding out unfamiliar words. Like Broca's area, Wernicke's area is located in only one cerebral hemisphere, typically the left.

Underlying the cerebral cortex is the second basic region of the cerebrum—the **cerebral white matter**—which consists of myelinated nerve fibers. These nerve fibers transmit impulses between areas of the cerebral cortex and lower brain areas.

The third basic region of the cerebrum, the **basal nuclei** (**Figure 15-4**), consists of clusters of gray matter embedded deep within the cerebral white matter. These subcortical nuclei (caudate nucleus, putamen, and globus pallidus) are involved in regulating voluntary motor impulses sent from the cerebral cortex to the brainstem and spinal cord. The caudate nucleus and the putamen are sometimes collectively called the striatum ("striped body") because of their striped appearance.

Diencephalon

The diencephalon (see Figure 15-2), which is located superior to the brainstem, deep within the brain, consists of several parts.

The **thalamus** is an egg-shaped region of the diencephalon that consists of two lobes connected by an intermediate

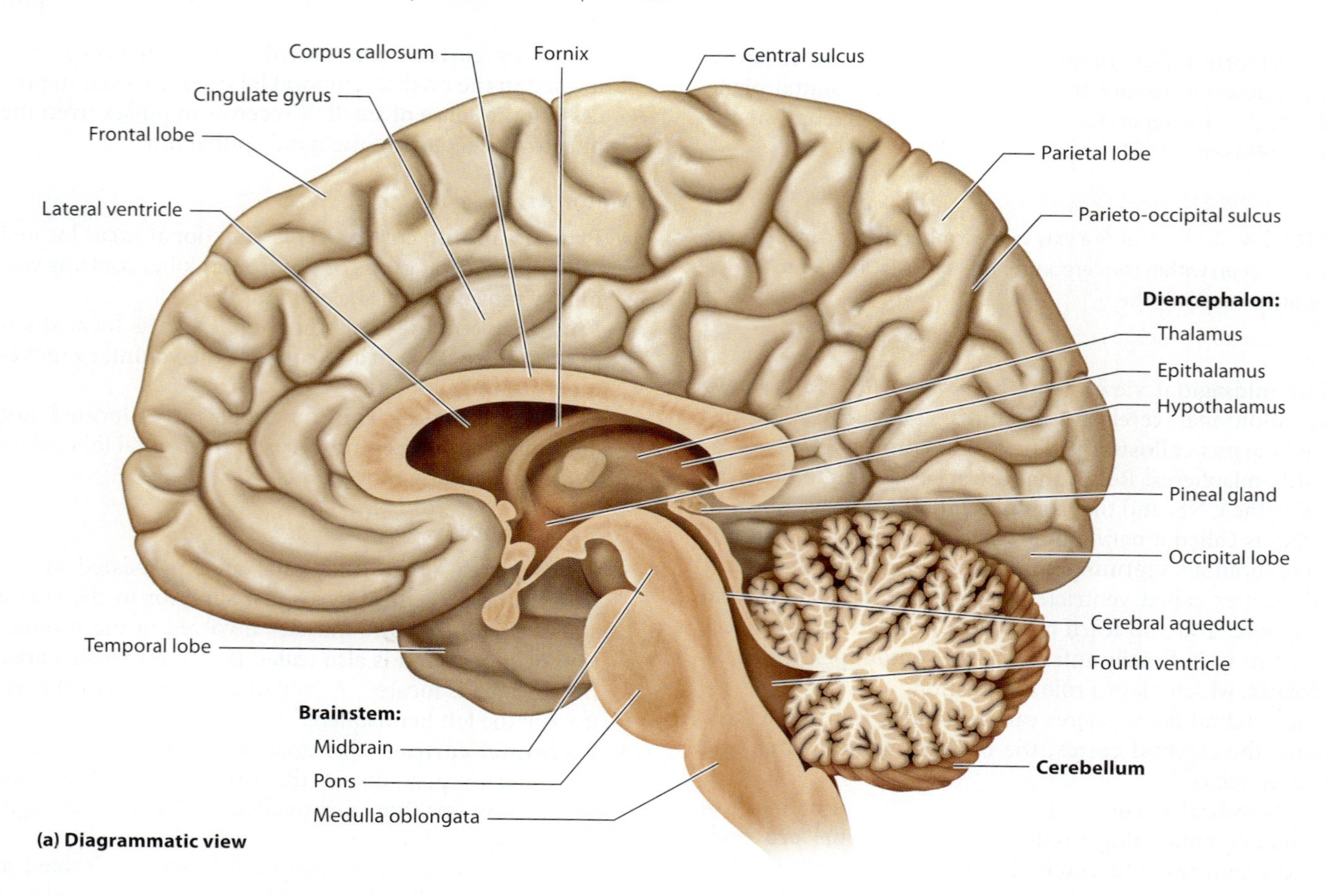

(a) Diagrammatic view

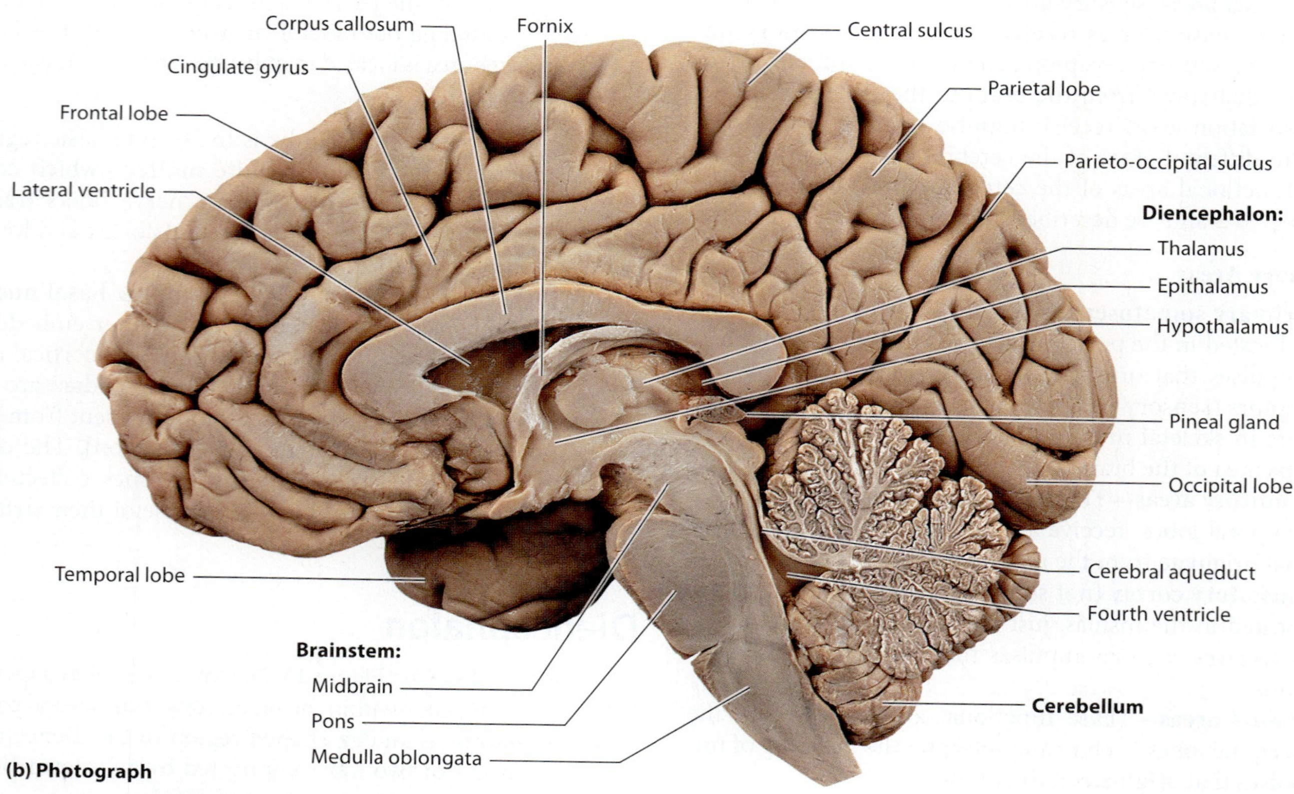

(b) Photograph

Figure 15-2 Midsagittal section of the brain.

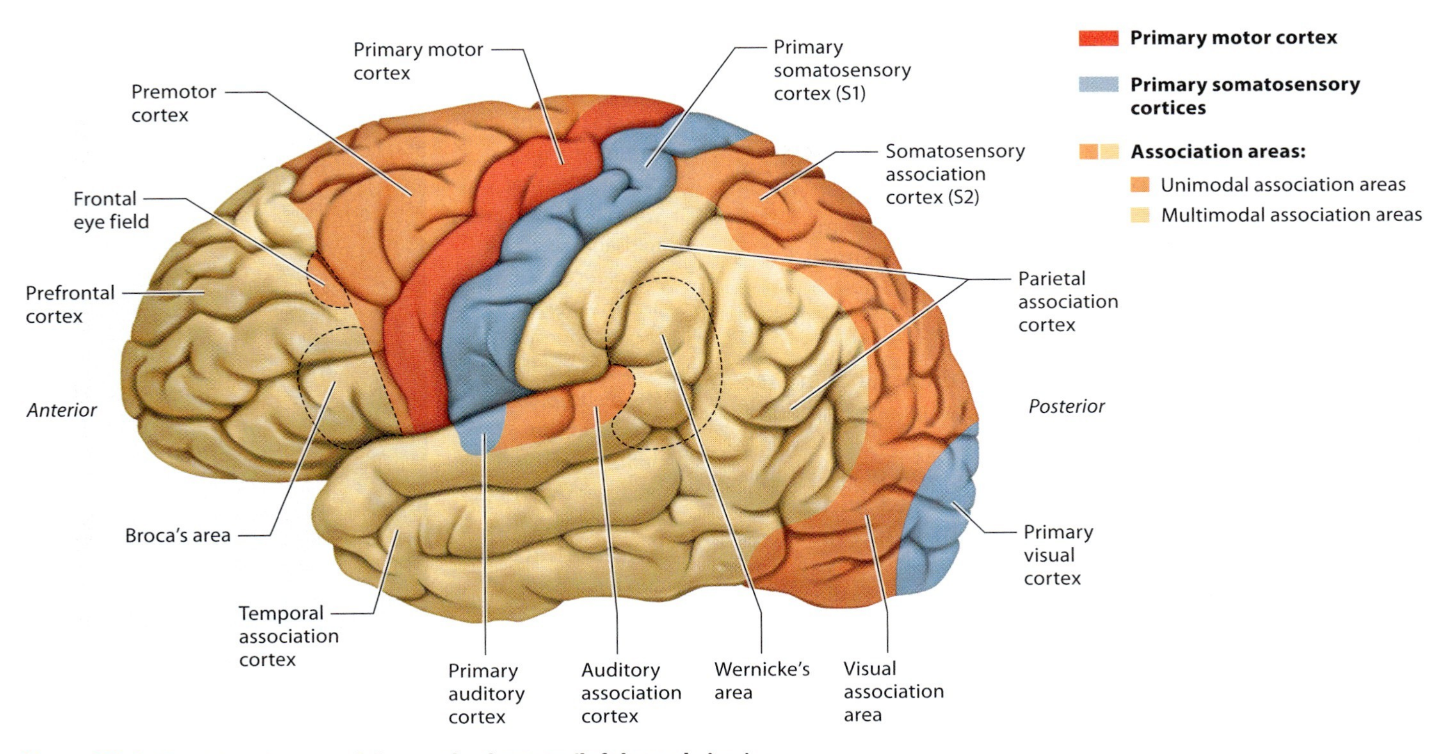

Figure 15-3 Functional areas of the cerebral cortex (left lateral view).

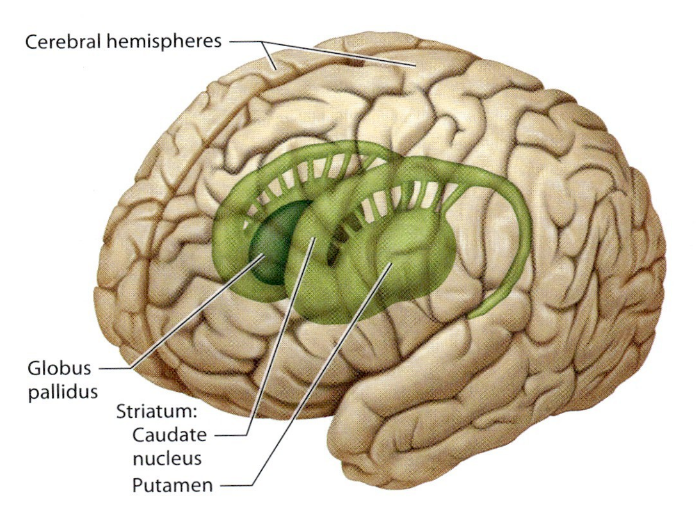

Figure 15-4 The basal nuclei (anterolateral view).

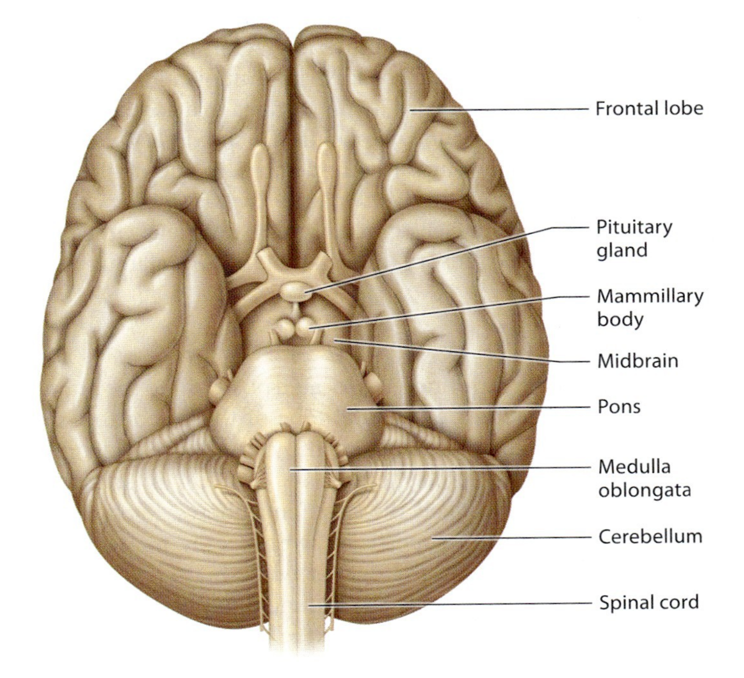

Figure 15-5 Inferior view of the brain and brainstem.

mass (see Figure 15-2). It functions as a sensory relay station for all sensory impulses (except smell) by receiving sensory messages and directing them to the appropriate areas of the cerebral cortex to be interpreted.

The **hypothalamus** (see Figure 15-2) is a major homeostatic organ that regulates many physiological events, including those that relate to hunger, thirst, body temperature, and blood pressure. The hypothalamus extends from the **optic chiasma** (the point at which the optic nerves cross over) to the pea-shaped **mammillary bodies**, nuclei that serve as relay stations for olfactory pathways. The **pituitary gland**, which hangs from the infundibulum of the hypothalamus, is an endocrine organ that secretes a variety of hormones. These three structures associated with the hypothalamus are shown in **Figure 15-5**.

The **epithalamus** (see Figure 15-2), located superior to the thalamus, contains the choroid plexus, a structure lined with ependymal cells, which produce cerebrospinal fluid (CSF). Also associated with the epithalamus is the pineal gland, a neuroendocrine organ that releases melatonin, a hormone that regulates sleep-waking cycles. Another structure, called the subthalamus, is inferior to the thalamus and works with the basal nuclei in the control of movement.

Brainstem

The brainstem, located inferior to the diencephalon (see Figure 15-2), consists of the midbrain, pons, and medulla oblongata (see Figures 15-2 and 15-5).

The **midbrain** is the most superior portion of the brainstem. On its dorsal surface are four pea-shaped nuclei. The two superior nuclei—the **superior colliculi**—function as visual reflex centers; the two inferior nuclei—the **inferior colliculi**—function as auditory reflex centers. Collectively these four nuclei are called the corpora quadrigemina. A slender passageway running through the middle of the midbrain, called the cerebral aqueduct, allows CSF to pass from the third ventricle to the fourth ventricle. On the ventral surface of the midbrain are the **cerebral peduncles**, the "little feet of the cerebrum," which contain the lateral corticospinal tracts.

The **pons** (which means "bridge") links the cerebellum with the brainstem, diencephalon, cerebrum, and spinal cord. The **medulla oblongata**, the most inferior region of the brain, is continuous with the spinal cord. All communication between the brain and the spinal cord involves tracts that ascend or descend through the medulla oblongata. The medulla oblongata contains many important autonomic nuclei, including the cardiac center, the respiratory center, and the vasomotor center. It also contains the nucleus gracilis and the nucleus cuneatus, which relay somatic sensory information to the thalamus.

Cerebellum

The **cerebellum** ("little brain") is located posterior to the brainstem and inferior to the occipital lobe of the cerebrum (see Figure 15-2). Like the cerebrum, the cerebellum is divided into left and right hemispheres (**Figure 15-6**). A structure called the **vermis** connects the two hemispheres. Each hemisphere contains a thin superficial layer of gray matter (cerebellar cortex) overlying the deeper white matter. A sagittal section of the cerebellum reveals a branching tree-like pattern of white matter called the **arbor vitae**. The cerebellar white matter converges into three large tracts called cerebellar peduncles that connect the cerebellum to the brainstem. The cerebellum coordinates skeletal muscle activity and regulates both balance and posture.

LabBOOST >>>

Visualizing the Brain Students often notice that, when examined in a midsagittal section, the diencephalon and the brainstem together form the shape of a seahorse. The diencephalon is its head, the midbrain of the brainstem is its neck, the pons of the brainstem is its belly, and the medulla oblongata of the brainstem is its tail.

Just liken the cerebrum to an oversized cap and the cerebellum to a backpack, and you have a great word picture to help you remember the basic organization of the brain. Make a sketch of the brain based on this word picture. Then label the four major brain regions (cerebrum, diencephalon, brainstem, and cerebellum), the three components of the diencephalon, and the three components of the brainstem.

ACTIVITY 1

Exploring the Functional Anatomy of the Brain

Learning Outcomes

1. Identify the following brain structures on a human brain model and/or anatomical charts, and describe the function(s) of each:
   a. Cerebrum—cerebral cortex, cerebral white matter, basal nuclei, corpus callosum, fornix, septum pellucidum, cerebral lobes (frontal, parietal, occipital, temporal, and insula), lateral ventricles.
   b. Diencephalon—epithalamus, choroid plexus, pineal gland, thalamus, intermediate mass, hypothalamus, optic chiasma, infundibulum, mammillary bodies, third ventricle.
   c. Brainstem—midbrain, cerebral peduncles, corpora quadrigemina (superior colliculi and inferior colliculi), cerebral aqueduct, pons, medulla oblongata, fourth ventricle.
   d. Cerebellum—cerebellar cortex, cerebellar hemispheres, vermis, arbor vitae.

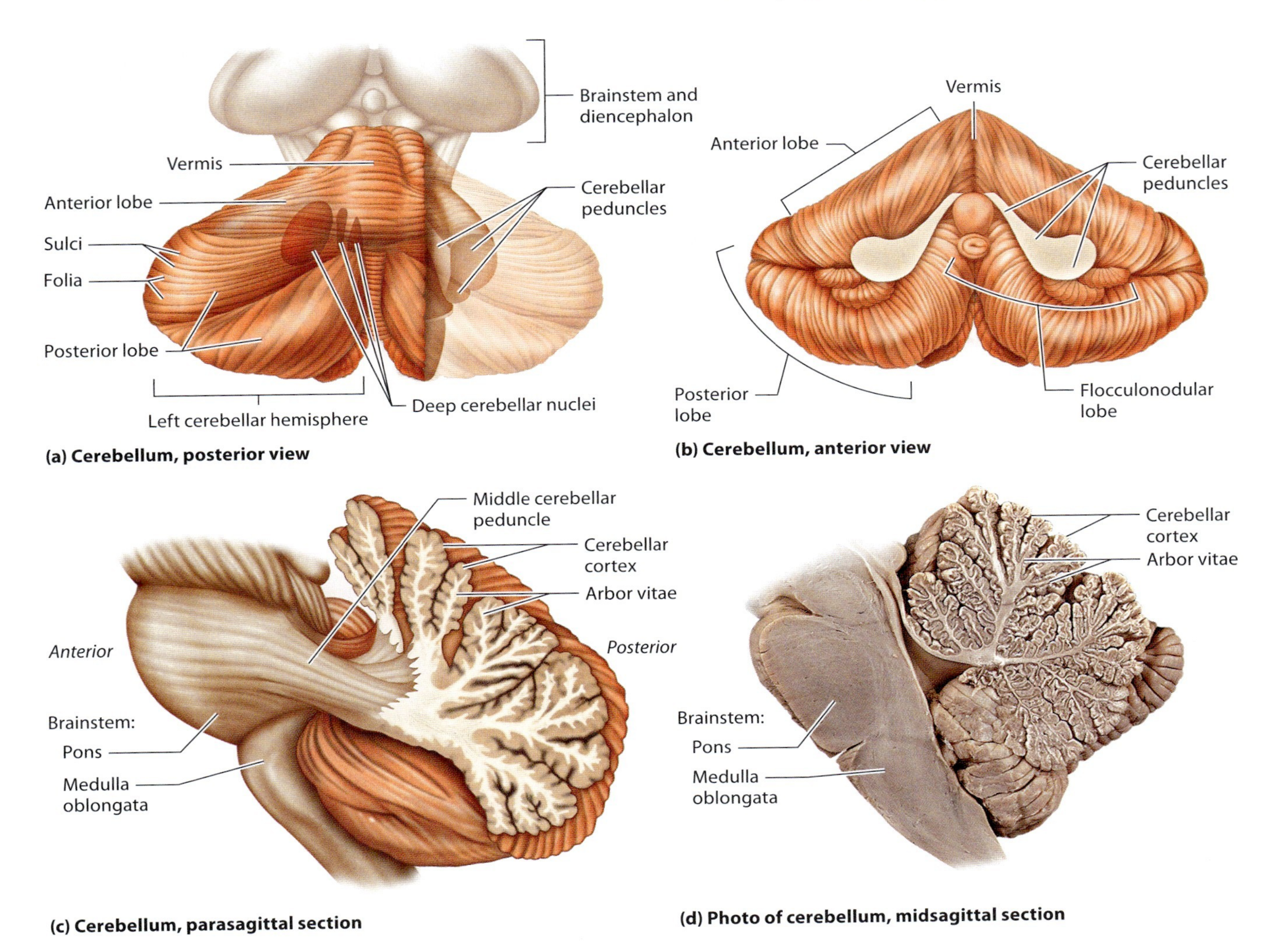

Figure 15-6 **The cerebellum.**

2. Locate the major functional areas of the cerebral cortex, and describe the function(s) of each: primary motor cortex, Broca's area, frontal eye field, premotor cortex, primary somatosensory cortex, visual areas, auditory areas, gustatory cortex, olfactory cortex, prefrontal cortex, and Wernicke's area.

Materials Needed

☐ Brain models and anatomical charts

Instructions

CHART In this activity you will focus on both the structure and function of the parts of the four major brain regions. First, locate on a brain model each of the structures listed in the Making Connections charts on the next two pages, and then discuss the function(s) of each structure with the other members of your lab group. Next, write a few words in the appropriate cells of the following charts to help you remember the descriptions and functions of each structure. Finally, brainstorm "connections to things I have already learned" with your lab group, and write the most helpful ones in the charts. Various cells of the charts have been completed for you as examples.

Optional Activity

Practice labeling brain structures on human brains at MasteringA&P® > Study Area > Practice Anatomy Lab > Human Cadaver: Nervous System > Central Nervous System > images 21 through 44

Making Connections: Cerebrum

Specific Structure	Description (Structure and/or Function)	Connections to Things I Have Already Learned
Cerebral cortex		
Cerebral white matter		*Contains bundles of myelinated nerve fibers called tracts*
Basal nuclei		
Corpus callosum	*The largest tract; myelinated nerve fibers that connect the right and left cerebral hemispheres*	
Fornix		*Relatively larger in sheep brain—sheep rely more heavily on olfaction than do humans*
Septum pellucidum		
Frontal lobe		
Parietal lobe		
Occipital lobe		*Occipital region of the body; occipitalis muscle, occipital bone*
Temporal lobe	*Lateral cerebral lobe*	
Insula		
Central sulcus		*Other sulci: lateral sulcus, parieto-occipital sulcus*
Primary motor cortex		
Broca's area		
Frontal eye field	*Controls voluntary muscles that move the eyes*	
Primary somatosensory cortex		*Receives sensory input from sensory receptors such as lamellated corpuscles and tactile corpuscles*
Visual areas		
Auditory areas	*Functional areas located in the temporal lobes that receive and interpret sensory impulses originating in the inner ear*	
Gustatory cortex		
Olfactory cortex		
Prefrontal cortex		
Wernicke's area		

Making Connections: Diencephalon

Specific Structure	Description (Structure and/or Function)	Connections to Things I Have Already Learned
Epithalamus • Choroid plexus • Pineal gland		*"Epi" means "above," for example, "epidermis"*
Thalamus • Intermediate mass		
Hypothalamus • Optic chiasma • Infundibulum • Mammillary bodies	*Located below the thalamus in the diencephalon* • *Optic nerves cross over* • *Stalk from which PG hangs* • *Pea-shaped nuclei posterior to infundibulum*	

Making Connections: Brainstem

Specific Structure	Description (Structure and/or Function)	Connections to Things I Have Already Learned

Midbrain
• Cerebral peduncles
• Cerebral aqueduct
• Corpora quadrigemina
  • Superior colliculi
  • Inferior colliculi
	Located inferior to the diencephalon
	• *On anterior surface*
	• *Centrally located canal*
	• *Four nuclei on posterior surface*	

Pons		
Medulla oblongata		*Contains important visceral control centers (nuclei): cardiac, respiratory, and vasomotor*

Making Connections: Cerebellum

Specific Structure	Description (Structure and/or Function)	Connections to Things I Have Already Learned
Cerebellar cortex		
Cerebellar hemispheres		
Vermis		*Vermis connects cerebellar hemispheres, whereas corpus callosum connects cerebral hemispheres*
Arbor vitae		

ACTIVITY 2

Electroencephalography in a Human Subject Using BIOPAC®

Learning Outcomes

1. To record an EEG from an awake, resting subject with eyes open and eyes closed.
2. To identify and examine alpha, beta, delta, and theta waves of the EEG recordings.
3. To compare differences between male and female students.

Materials Needed

- ☐ Computer system (Windows 7, Vista, XP, Mac OS X 10.5–10.7)
- ☐ Biopac Science Lab system (MP35 unit and software)
- ☐ Electrode lead set (SS2L lead set)
- ☐ Disposable vinyl electrodes (EL503), three electrodes per subject
- ☐ Biopac Electrode Gel (GEL1) and Abrasive Pad (ELPAD) or alcohol prep
- ☐ Optional: A supportive wrap to press the electrodes against the head for improved contact.

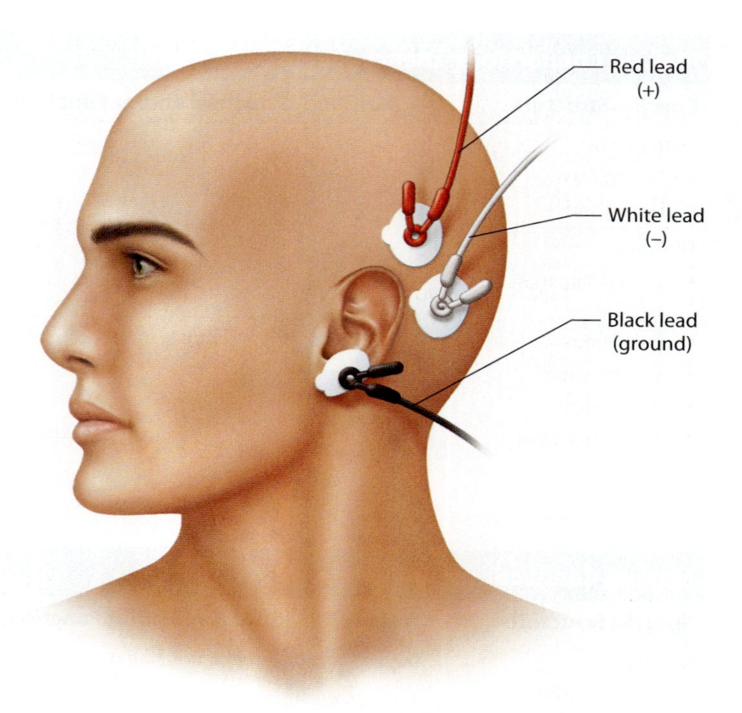

Figure 15-7 Electrode placement and lead connections.

Instructions

An **electroencephalogram (EEG)** is a graphic recording of the brain's electrical activity. Four basic wave patterns are produced by an EEG: alpha, beta, delta, and theta. **Alpha waves** (average frequency and high amplitude) are detected in healthy adults who are awake, but relaxed with their eyes closed. **Beta waves** (similar to alpha waves but with higher frequency and lower amplitude) are detected in individuals who are mentally alert such as during times of concentration or stress. **Delta waves** (low frequency and very high amplitude) are detected during deep sleep. **Theta waves** (low frequency, abnormally high amplitude waves) are abnormal in awake adults although they are detected in children. In this activity, you will use the BIOPAC student lab system to record your partner's EEG, examine the basic EEG waveforms, and explore differences between male and female students.

A. Setup

1. Turn the computer ON.
2. Turn the MP35 unit OFF.
3. Plug the electrode leads (SS2L) into the Electrode Check of the MP35.
4. Turn the MP35 unit ON.
5. Attach three electrodes to the subject as follows (see Figure 15-7):
   a. White: 4–5 cm behind the top of the ear
   b. Red: on the side of the head, 4–5 cm above the first electrode
   c. Black: on the earlobe, with adhesive folded under the ear.

   Hints for obtaining optimal data from the subject:
   - Part the hair so that the electrodes make contact with the scalp as much as possible.
   - Gently abrade the skin at the electrode sites with your finger.
   - Apply a drop of electrode gel.
   - Apply pressure to the electrode for approximately 1 minute after initial placement.
   - Have the subject remain still. Even blinking can interfere with the recordings.
   - Use a wrap to secure the electrodes.
6. Clip the electrode set following the color code.
7. Ask the subject to be seated and to relax. The room should be relatively quiet so that the subject can mentally relax. Wait 5 minutes before starting the recording to ensure that the subject is fully relaxed.
8. Start the Biopac Student Lab Program on the computer.
9. Choose lesson "L03-Electroencephalography I (EEGI)" and click OK.
10. Type in a unique file name and click OK to end the Setup section.

B. Calibration

The calibration procedure establishes the hardware's internal parameters (such as gain, offset, and scaling) and is critical for optimal performance.

1. Have the subject remain relaxed with eyes closed during the calibration process.
2. Click Calibrate. The subject must sit relaxed with eyes closed while waiting for the calibration process to finish.
3. Verify that the recording resembles the example data in **Figure 15-8**. If it does, click Continue and proceed to data recording. If the data does not look like the calibration data, click Redo Calibration.

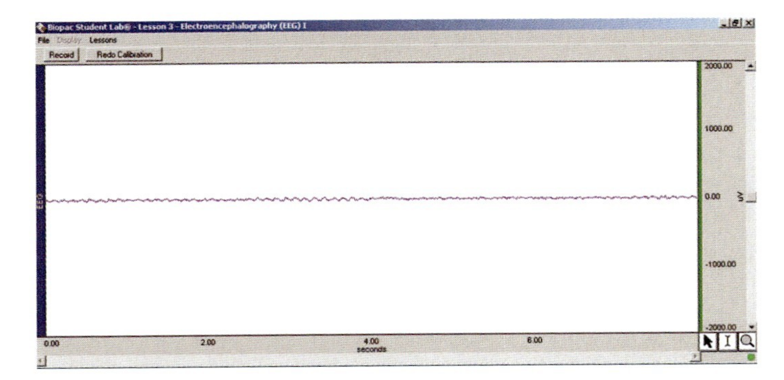

Figure 15-8 Example of calibration data.

C. Data Recording

1. Have the subject remain seated with eyes closed.
2. Review the recording steps.
3. While the subject remains seated, relaxed, and still with eyes closed, click Record.
4. Record for 20 seconds. Have a lab partner press F4 (this will insert a marker) and cue subject to open the eyes.
5. Record for an additional 20 seconds. Have the lab partner press F5 (this will insert a marker) and cue the subject to close the eyes.
6. Record for an additional 20 seconds.
7. Click Suspend.
8. Verify that the recording resembles the sample data in **Figure 15-9**. If it does, click Done to finish. If the data do not look similar, click Redo.

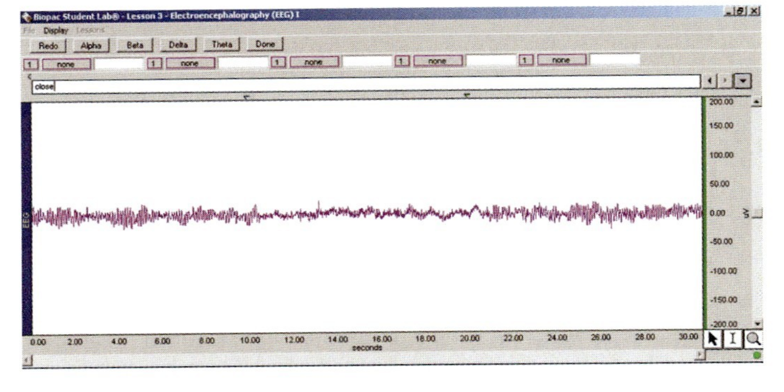

Figure 15-9 Example of sample data.

D. Data Analysis

1. Enter the Review Saved Data option.
2. Note the channel (CH) designations:

   CH 1 EEG (hidden)
   CH 40 Alpha
   CH 41 Beta
   CH 42 Delta
   CH 43 Theta

   Note as well the measurements for box settings:

   CH 40 Stddev
   CH 41 Stddev
   CH 42 Stddev
   CH 43 Stddev
   SC Freq

   Stddev = standard deviation, a measurement of the variability of the data points.

   Freq converts the time segment of the selected area to frequency in cycles/sec.
3. Set up your display window for optimal viewing of channels 40–43.
4. Use the I-beam cursor tool to select the first "Eyes Closed" data (see **Figure 15-10**), and enter the data in **Table 15-2**.

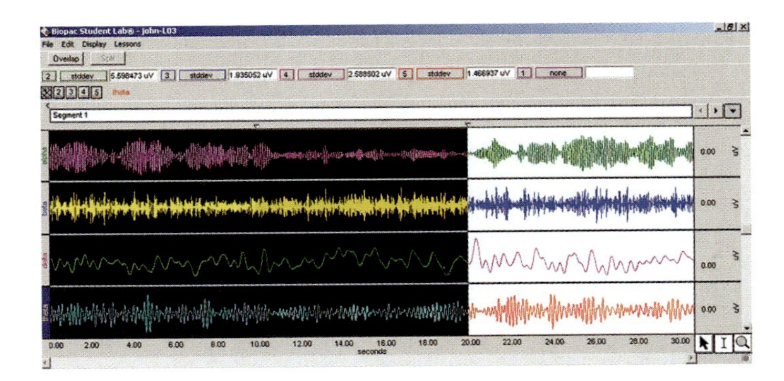

Figure 15-10 An example of completed data with first data segment highlighted.

Table 15-2 Standard Deviation [Stddev] of Signals in Each Segment

Rhythm	Channel	Eyes Closed	Eyes Open	Eyes Reclosed
Alpha	CH 40			
Beta	CH 41			
Delta	CH 42			
Theta	CH 43			

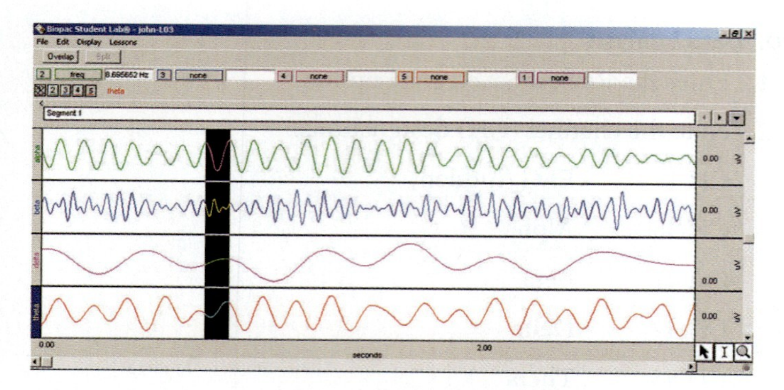

Figure 15-11 Selected area showing one cycle of the alpha wave.

5. Repeat step 4 for "Eyes Open" data.
6. Repeat step 4 for the second "Eyes Closed" data.
7. Zoom in on a 3–4 second section of the "Eyes Closed" data.
8. Use the I-beam cursor to select an area that represents one cycle in the alpha wave (see **Figure 15-11**). Enter the data in **Table 15-3**.

Table 15-3 Frequency (Hz) of Signals in Each Segment

Rhythm	Channel	Cycle 1	Cycle 2	Cycle 3	Mean
Alpha	CH 40				
Beta	CH 41				
Delta	CH 42				
Theta	CH 43				

9. Repeat step 8 for two other alpha-wave cycles.
10. Repeat steps 8 and 9 using the beta-wave data.
11. Repeat steps 8 and 9 using the delta-wave data.
12. Repeat steps 8 and 9 using the theta-wave data.
13. Save the data report.
14. Quit the program.

The Meninges and Cerebrospinal Fluid

Connective tissue wrappings surround the brain, and the spinal cord as well (discussed later in this unit). Collectively, these wrappings are called the **meninges** (**Figure 15-12**). The outermost, toughest meninx is the **dura mater** (see Figure 15-12a). The dura mater invaginates to form three major partitions that both subdivide the cranial cavity and limit excessive movement of the brain:

- **Falx cerebri** (see Figure 15-12a–c)—dips down into the longitudinal fissure between the cerebral hemispheres. This partition attaches to the crista galli of the ethmoid bone (see Figure 9-6).
- **Falx cerebelli** (not shown)—separates the cerebellar hemispheres.
- **Tentorium cerebelli** (see Figure 15-12b)—dips down into each transverse fissure and separates the occipital lobes of the cerebrum from the underlying cerebellar hemispheres.

The brain's dura mater consists of two layers of fibrous connective tissue: a superficial periosteal dura and a deep meningeal dura (see Figure 15-12a and c). In some areas, these two layers separate to form **dural sinuses** (see Figure 15-12a–c), which are vascular structures that collect venous blood and direct it toward the internal jugular vein and on to the heart.

The **arachnoid mater**—the middle meninx—forms a loose brain covering (see Figure 15-12a and c). Knoblike projections of the arachnoid mater, called arachnoid granulations (or arachnoid villi; see Figure 15-12c), protrude through the dura mater and project into the superior sagittal sinus, providing a pathway for CSF to return to the bloodstream. Deep to the arachnoid mater is the **subarachnoid space** (see Figure 15-12c), which is filled with cerebrospinal fluid (discussed next). The subarachnoid space separates the arachnoid mater from the innermost meninx, the **pia mater** (see Figure 15-12a and c). The pia mater is composed of a very delicate, richly vascular connective tissue that clings tightly to the brain surface, following its every convolution.

Cerebrospinal fluid (CSF) is located in and around the brain and spinal cord. In the brain it fills the ventricles

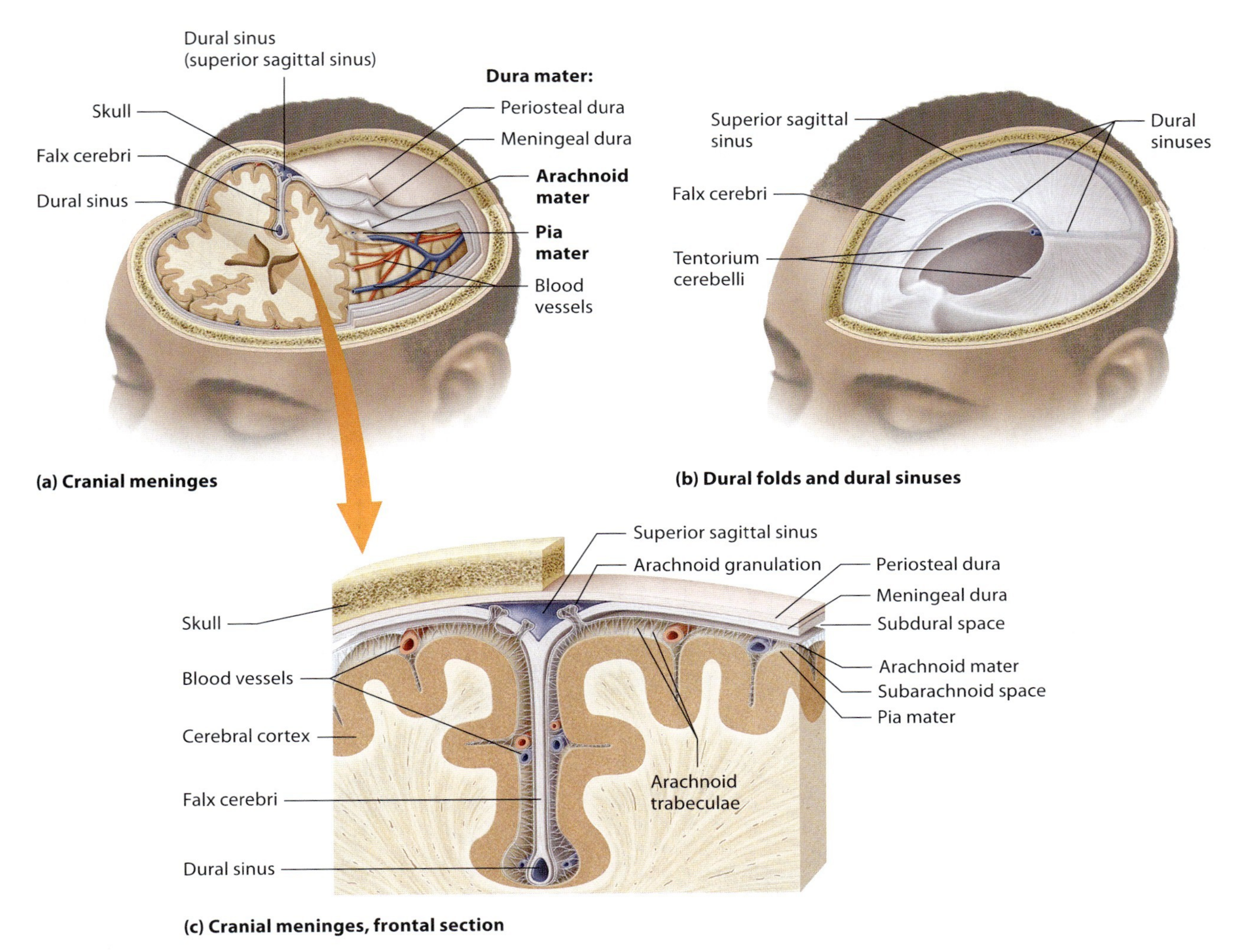

Figure 15-12 **The cranial meninges and dural folds and sinuses.**

(**Figure 15-13**), where it is produced continuously and is completely replaced approximately every 8 hours. The total volume of CSF is approximately 150 ml (about one-half cup). CSF provides buoyancy (thereby reducing the weight of the brain, which prevents the brain from crushing under its own weight), protects (cushions) the brain, and regulates the chemical environment of the CNS.

CSF is produced in each of the ventricles within structures called choroid plexuses, which hang from the roof of each ventricle. Each **choroid plexus** consists of ependymal cells embedded in pia mater and surrounded by capillaries. As CSF is produced by the ependymal cells of the choroid plexuses of the **lateral ventricles** (see Figure 15-13a and b), it flows through the **interventricular foramina** into the third ventricle (see Figure 15-13b). From the **third ventricle**, it flows through the **cerebral aqueduct** into the fourth ventricle. From the **fourth ventricle**, most of the CSF enters the subarachnoid space via two **lateral apertures** and one **median aperture** (not shown). Some CSF also enters the central canal of the spinal cord. CSF in the **subarachnoid space** eventually returns to the venous circulation via the **arachnoid granulations (villi)**, which project through the dura mater into the **superior sagittal sinus** (see Figure 15-12c). It then travels into the transverse sinuses, into the sigmoid sinuses, and finally into the internal jugular veins, the blood vessels that return venous blood to the heart.

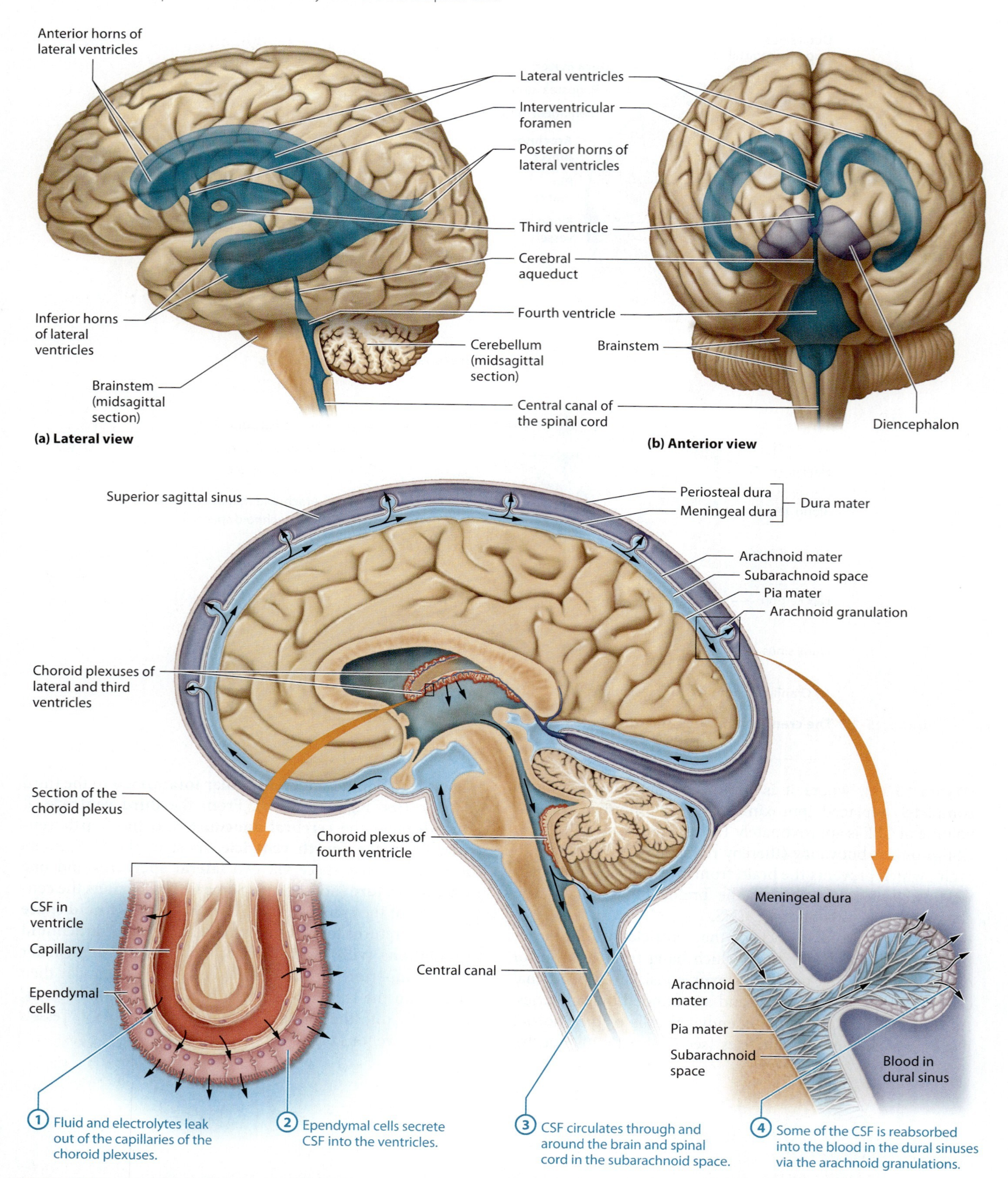

Figure 15-13 Ventricles of the brain and the flow (production, circulation, and reabsorption) of cerebrospinal fluid.

ACTIVITY 3

Identifying the Meninges/Ventricles and Tracing the Flow of Cerebrospinal Fluid

Learning Outcomes

1. Identify the three meningeal layers on anatomical models and/or charts.
2. Locate the dural partitions (falx cerebri, falx cerebelli, and tentorium cerebelli), and describe the functions of each.
3. Locate the four ventricles and the structures that connect the ventricles.
4. Trace the flow of cerebrospinal fluid from its production by the choroid plexus until it drains into the venous circulation.

Materials Needed

- ☐ Anatomical models and charts

Instructions

A. Meninges and Ventricles

Locate the following structures on anatomical models and/or charts. Check off each structure as you locate it, and then write a brief description of it.

- ☐ Dura mater ________________
- ☐ Arachnoid mater ________________
- ☐ Subarachnoid space ________________
- ☐ Pia mater ________________
- ☐ Lateral ventricles ________________
- ☐ Septum pellucidum ________________
- ☐ Third ventricle ________________
- ☐ Fourth ventricle ________________
- ☐ Choroid plexus ________________
- ☐ Interventricular foramina ________________
- ☐ Cerebral aqueduct ________________
- ☐ Median and lateral apertures ________________
- ☐ Central canal ________________
- ☐ Arachnoid granulation ________________
- ☐ Superior sagittal sinus ________________

B. The Flow of Cerebrospinal Fluid

1. Watch the "CSF dance" video at MasteringA&P® > Study Area > Unit 15 > CSF Dance Video, and then practice making the movements of the "dance" as you explain the pathway (production, circulation, and reabsorption) of CSF.
2. Answer the following questions related to what you have learned in this activity:
   a. The choroid plexus is lined with which type of glial cell?

   ____

   b. What is the function of these cells?

   ____

   c. The third ventricle is associated with which major brain region?

   ____

   d. Which structure connects the third ventricle and the fourth ventricle?

   ____

   e. How does CSF reach the subarachnoid space?

   ____

   f. Through which meningeal layer does the arachnoid villus protrude?

   ____

Optional Activity

Practice labeling brain ventricles on anatomical models at MasteringA&P® > Study Area > Practice Anatomy Lab > Anatomical Models: Nervous System > Central Nervous System > images 10 and 11

The Spinal Cord

The **spinal cord** (**Figure 15-14**) is continuous with the medulla oblongata at its superior end, and it terminates between lumbar vertebrae L1 and L2. The spinal cord is not uniform in diameter along its length. Instead, it has two enlarged regions: the **cervical enlargement**, which gives rise to nerves that supply the upper limbs, and the **lumbar enlargement**, which gives rise to nerves that innervate the pelvic region and the lower limbs. Along both lateral sides of the cord are **denticulate ligaments** (not shown), which are extensions of pia mater that provide lateral support to the spinal cord. Inferior to the lumbar enlargement, the spinal cord narrows and terminates as the cone-shaped **conus medullaris**. Spinal nerves extend from the conus medullaris in a parallel group called the **cauda equina**, so named because of their resemblance to a horse's tail. The conus medullaris is attached to the coccyx by the **filum terminale**, an extension of the conus covered by the pia mater. Because the spinal cord does not extend to the end of the vertebral column, the spinal nerves extending from the inferior end of the spinal cord actually travel through the vertebral canal before exiting through the intervertebral foramina.

Figure 15-15 depicts a transverse (cross) section of the spinal cord showing a core of butterfly-shaped gray matter

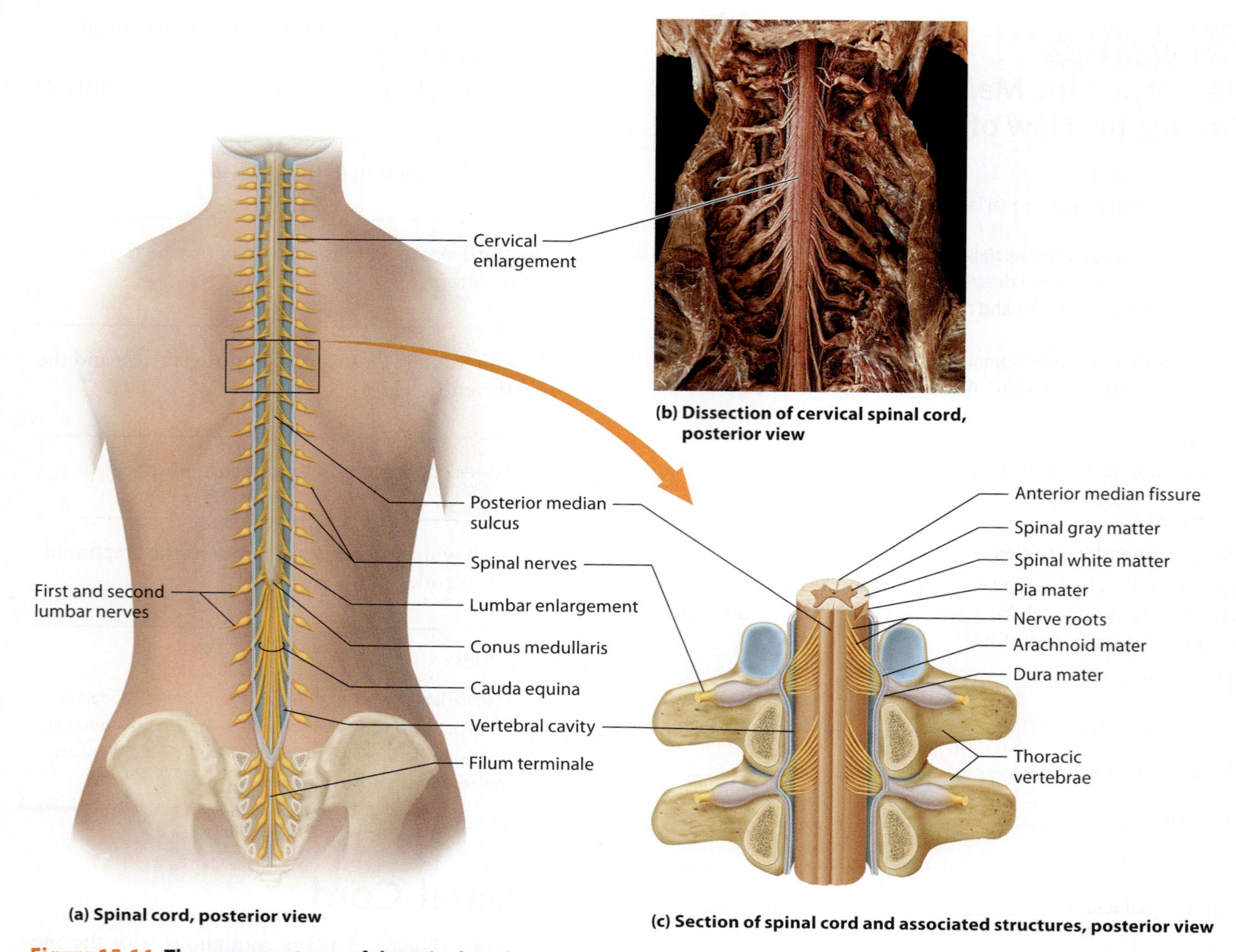

Figure 15-14 The gross anatomy of the spinal cord.

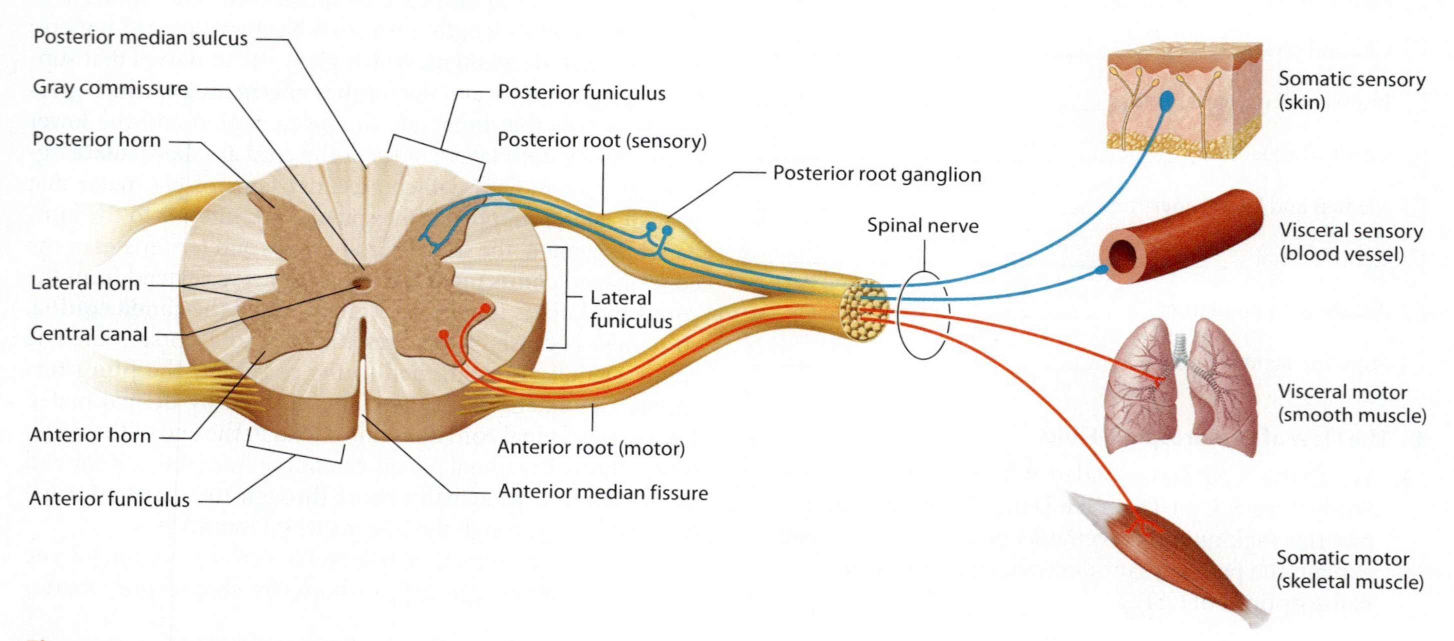

Figure 15-15 The spinal cord in cross-section.

surrounded by white matter. The cord is divided by a deep **anterior median fissure** and a shallow **posterior median sulcus**. The gray matter is divided into **anterior** (ventral), **lateral**, and **posterior** (dorsal) **horns** and contains nerve cell bodies and dendrites. The posterior horns contain interneurons and sensory neurons that enter the cord via the posterior root. The cell bodies of these sensory neurons synapse with interneurons and/or motor neuron cell bodies and are located in the **posterior root ganglion**. The anterior horns contain somatic motor neuron cell bodies. The lateral horns, which are present only in the thoracic and lumbar regions of the cord, contain cell bodies of autonomic motor neurons.

The **gray commissure** is a narrow band of gray matter in the center of the spinal cord. In the middle of the gray commissure is the **central canal**, which contains CSF. The white matter is divided into three columns called the anterior (ventral), lateral, and posterior (dorsal) **funiculi**. Each funiculus consists of tracts (bundles of nerve fibers in the CNS) containing neurons with the same origin, destination, and function. Ascending (sensory) tracts transmit sensory impulses to the brain. Descending (motor) tracts transmit motor impulses from the brain to skeletal muscles.

Extending laterally from the spinal cord are the anterior roots and the posterior roots. An **anterior root** and a **posterior root** fuse to form a **spinal nerve**. The anterior root contains motor neurons and the posterior root contains sensory neurons. A swelling in the posterior root, the posterior root ganglion, contains the nerve cell bodies of the sensory neurons. Because the nerve fibers of the motor neurons and the nerve fibers of the sensory neurons fuse to form the spinal nerves, all spinal nerves contain both sensory and motor neurons and are, therefore, classified as mixed nerves.

CLINICAL CONNECTION

During childbirth, a woman might receive an epidural block. During this procedure, an anesthetic drug is inserted into the epidural space between two lumbar vertebrae. The drug reduces the pain experienced during labor and childbirth by numbing the spinal nerves of the pelvis and lower limbs.

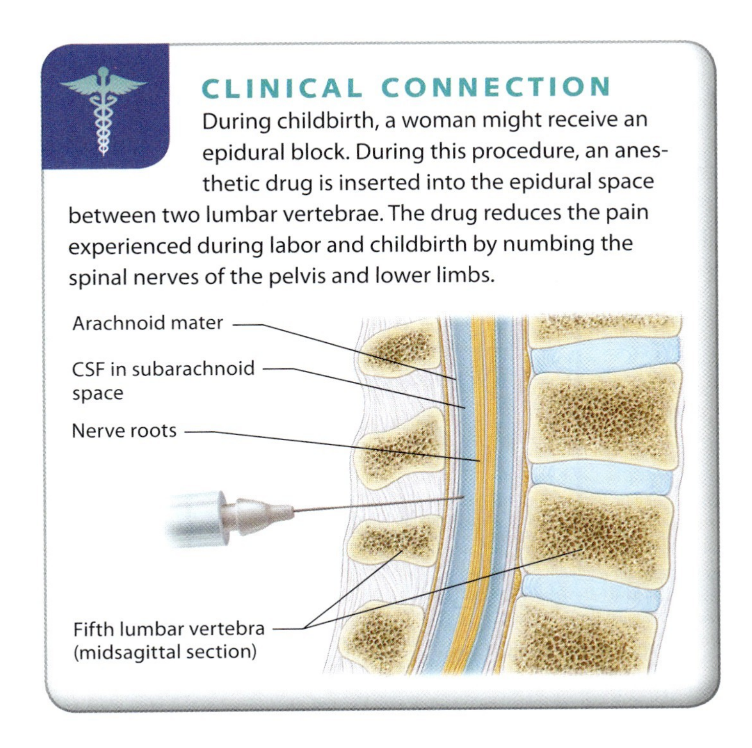

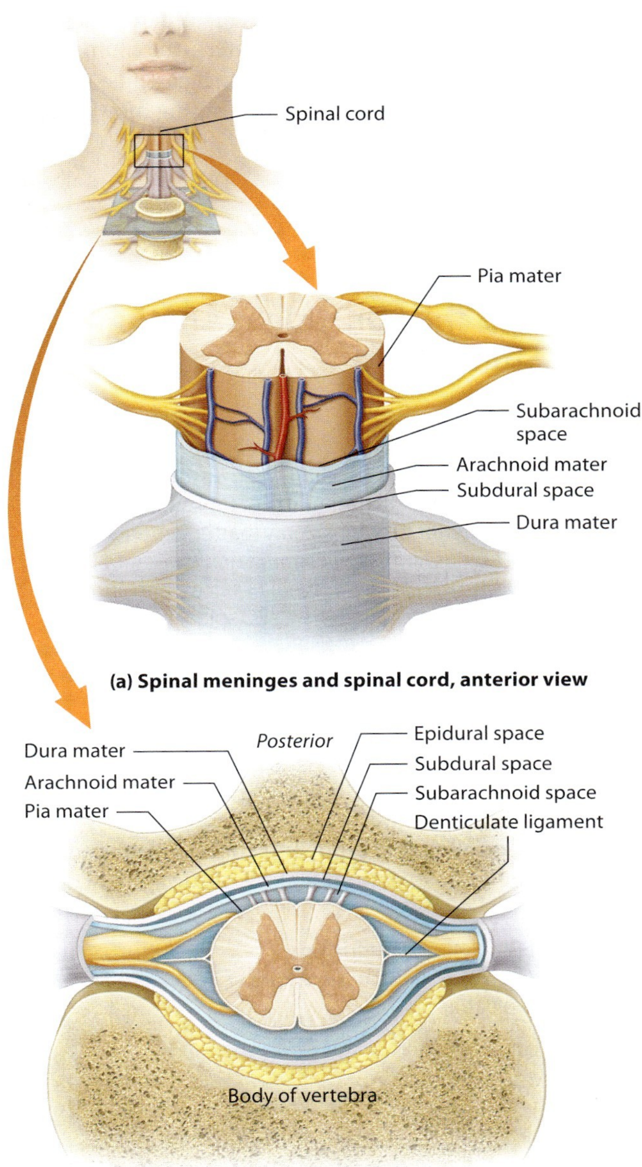

(a) Spinal meninges and spinal cord, anterior view

(b) Spinal meninges and spinal cord, transverse section

Figure 15-16 The spinal meninges.

The basic organization of the spinal meninges (**Figure 15-16**) is similar to that seen in the brain. However, several differences exist between the spinal meninges and the cranial meninges. First, the spinal dura mater consists of a single layer of connective tissue that is continuous with the meningeal dura of the brain. Second, the epidural space, located between the vertebra and the dura mater (see Figure 15-12b), is filled with adipose tissue that cushions the spinal cord. (Recall that in the brain, the outer layer of the dura mater—the periosteal dura—attaches to the periosteum of the skull bones.) Third, whereas the spinal cord ends between the first and second lumbar vertebrae, the dura mater and the arachnoid mater extend down to the sacral vertebrae. As a result, the subarachnoid space terminates as a blind sac filled with CSF. A fourth difference involves the pia mater. As previously mentioned, extensions of the pia mater—the filum terminale (see Figure 15-10a) and the denticulate ligaments—anchor the spinal cord to the dura mater, thereby limiting movement of the spinal cord.

ACTIVITY 4

Examining the Functional Anatomy of the Spinal Cord

Learning Outcome

1. Identify the following spinal cord structures on human spinal cord models and/or anatomical charts, and describe the function(s) of each: white matter (anterior funiculus, lateral funiculus, and posterior funiculus), gray matter (anterior horn, lateral horn, posterior horn, gray commissure, and central canal), anterior median fissure, posterior median sulcus, posterior root, anterior root, posterior root ganglion, and spinal nerve.

Materials Needed

☐ Spinal cord model or anatomical chart
☐ Colored pencils

Instructions

1. CHART Complete the following chart as you identify and describe each structural feature of the spinal cord:

Structural Feature	Description (Structure and/or Function)
White matter • Anterior funiculus • Lateral funiculus • Posterior funiculus	
Gray matter • Anterior horn • Lateral horn • Posterior horn • Gray commissure • Central canal	
Anterior median fissure	
Posterior median sulcus	
Posterior root	
Anterior root	
Posterior root ganglion	
Spinal nerve	

2. Sketch a cross-section of the spinal cord and label the structures listed in the previous chart.

3. **Optional Activity**

Practice labeling spinal cord structures on human spinal cords at MasteringA&P® > Study Area > Practice Anatomy Lab > Human Cadaver: Nervous System > Central Nervous System > images 1 through 20

Reflexes

The spinal cord serves as a center for spinal reflexes, which are rapid, predictable, and automatic responses to a stimulus. The neural pathway producing such a response is called the reflex arc (**Figure 15-17**).

A reflex arc consists of a series of three events:

1. **Detection and delivery of stimulus.** This event involves a *receptor*—a structure that is activated when it detects some stimulus—and a *sensory neuron*, which sends a nerve impulse to the central nervous system.
2. **Integration.** This event occurs at a site, called an *integration center*, at which the nerve impulse is processed.
3. **Delivery of motor response.** This event involves a *motor neuron*, which sends a nerve impulse away from the central nervous system, and an *effector*, a structure (a muscle or gland) that produces some response to the stimulus.

When the integration center is located in the spinal cord, the reflex is called a spinal reflex. In the following activity you will elicit a reflex that physicians commonly use to test spinal cord function.

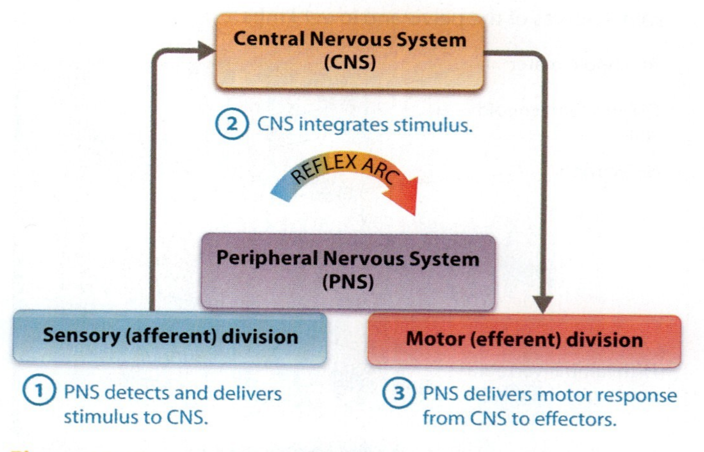

Figure 15-17 Events of a reflex arc.

ACTIVITY 5

Analyzing a Spinal Reflex

Learning Outcomes

1. Identify the components of a reflex arc on an anatomical model or chart.
2. Describe the function of each component of a reflex arc.
3. Demonstrate the patellar reflex.

Materials Needed

☐ Reflex hammer

Instructions

In this activity, you will elicit the patellar reflex by tapping on the patellar ligament, located just inferior to the patella.

1. Work in pairs. One student will be the "subject," and one student will be the "clinician."
2. The subject sits on the lab bench, with legs hanging free.
3. The clinician uses a reflex hammer to tap the subject's patellar tendon, located just inferior to the patella, as shown in **Figure 15-18**.

   Describe the subject's response. ______________________

   ____

4. The tapping of the tendon (the stimulus) causes the attached skeletal muscle to stretch slightly, which in turn stimulates specialized sensory receptors in the muscle, called muscle spindles. In response to this stimulus, the muscle spindles generate a nerve impulse, which then travels via a neuron to the spinal cord. Within the spinal cord the information in the nerve impulse is processed, and then another nerve impulse leaves the spinal cord via a different neuron. The outgoing nerve impulse then triggers the contraction of the slightly stretched muscle (the effector).

   For this reflex, in which skeletal muscle are the sensory receptors located?

   ____

   Which functional neuron type carries the nerve impulse to the spinal cord?

   ____

   Which functional neuron type carries the nerve impulse to the effector?

   ____

   What is the specific effector in this reflex?

   ____

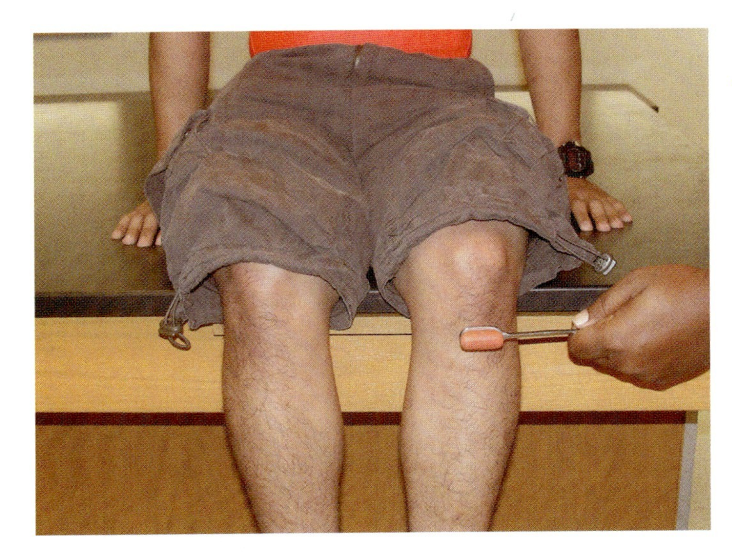

Figure 15-18 Patellar reflex testing.

Dissection Guidelines

As you prepare to dissect a sheep brain, it is important for you to consider the goal of dissecting a mammalian organ. The sheep brain is very similar in structure to the human brain, and dissecting it will give you the opportunity to gain an understanding of the three-dimensional structure of the brain. Although a sheep brain and a human brain do have differences, the similarities are striking. As you dissect the sheep brain and identify brain structures, discuss with the members of your lab group the functions of each structure as you begin to better understand the relationship between structure and function.

The following guidelines will help make your dissection experience a positive and educationally productive one:

- Read ALL instructions carefully before making any incisions.
- Use scissors and a dissecting probe whenever possible. Scalpels sometimes do more harm than good when you use them as your primary dissecting tool.
- A good strategy to use when you are working in pairs is to have one student read aloud the directions for the dissection step by step and to answer questions found throughout the activity. Another student can then do the dissection. These roles can be switched halfway through the activity so that each student can participate fully in the dissection.
- In addition to identifying each assigned structure, read the descriptions of each structure in your text and discuss the relationships that you see between structure and function. These discussions will enable you to answer the questions in the activity.

ACTIVITY 6

Dissecting a Sheep Brain and Spinal Cord

Learning Outcomes

1. Locate major brain structures on a dissected sheep brain, and describe the function of each structure.
2. Locate the major spinal cord structures on a dissected sheep spinal cord, and describe the function of each structure.

Materials Needed

- ☐ Preserved sheep brain
- ☐ Preserved sheep spinal cord
- ☐ Dissection supplies
- ☐ Latex gloves

Instructions

1. Put on latex gloves. Examine the preserved sheep brain and identify the cerebrum, the brainstem, and the cerebellum. (Refer to **Figure 15-19**.)

   How are these structures similar to those in the human brain?

   ____

   How are these structures different from those in the human brain?

   ____

2. Place your specimen dorsal side up on the dissecting tray. Carefully cut through the dura mater along the longitudinal fissure and remove it from the brain. Note that you will cut into the superior sagittal sinus as you cut the dura mater.

   What is contained within the superior sagittal sinus?

   ____

   a. Identify the arachnoid mater, which is a stringy, web-like layer beneath the dura mater.
   b. Remove a portion of the arachnoid mater to reveal the thin, shiny, inner connective tissue membrane called the pia mater. Note that the pia mater follows every convolution (gyrus + sulcus) of the brain surface. Collectively, these three connective tissue membranes (dura mater, arachnoid mater, and pia mater) form the meninges.

   Describe two functions of the meninges.

   ____

   ____

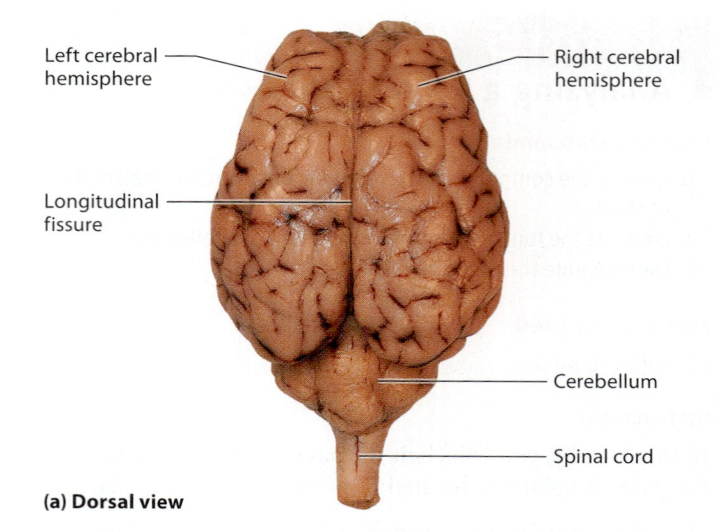

(a) Dorsal view

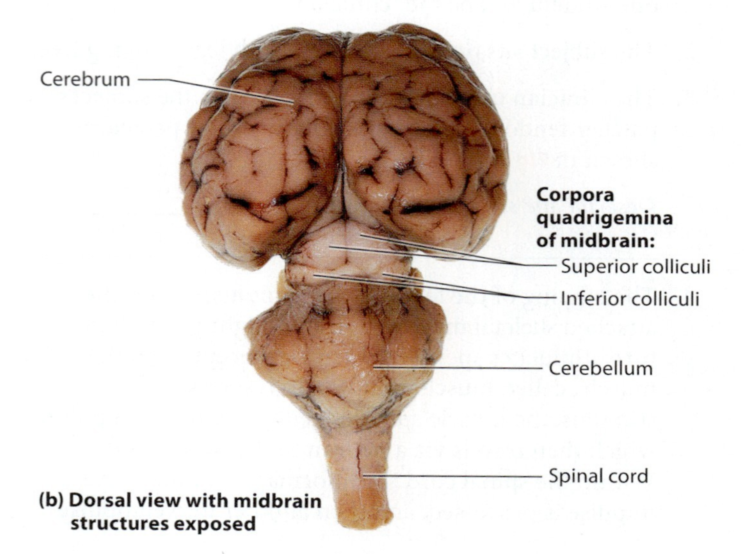

(b) Dorsal view with midbrain structures exposed

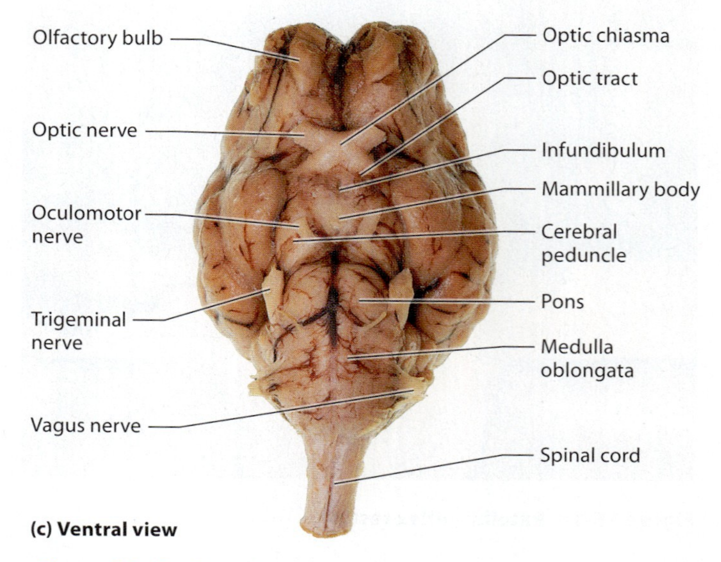

(c) Ventral view

Figure 15-19 **Sheep brain.**

3. While still examining the dorsal side of the brain, use your thumbs and fingers to push downward on the cerebrum and the cerebellum to expose the dorsal surface of the midbrain (see Figure 15-19b).
   a. Identify the superior colliculi and the inferior colliculi.

      Collectively, these four nuclei make up the

      ____.

   b. You can also see the pineal gland in this view. It is located just superior to the midline of the corpora quadrigemina.
4. Now, examine the ventral surface of the brain (see Figure 15-19c).
   a. Locate the olfactory bulbs. Olfactory neurons traveling from the nasal epithelium synapse with the olfactory bulbs.

      Through which skull bone and openings do these olfactory neurons pass?

      ____

   b. Locate the optic chiasma, the optic nerves, and the optic tracts.

      What is the function of the optic nerve?

      ____

      ____

      From which brain structure does the optic chiasma extend?

      ____
5. Separate the cerebral hemispheres enough so that you can identify the corpus callosum.

   What is the function of the corpus callosum?

   ____

   ____
6. Now, cut the brain in the midsagittal plane (refer to **Figure 15-20**) and separate it into two halves. As you identify each of the following structures, place a check mark in the appropriate box, and then write a brief description of that structure:

☐ cerebrum ________________________________

____

☐ corpus callosum __________________________

____

☐ fornix __________________________________

____

(a) Diagramatic view

(b) Photograph

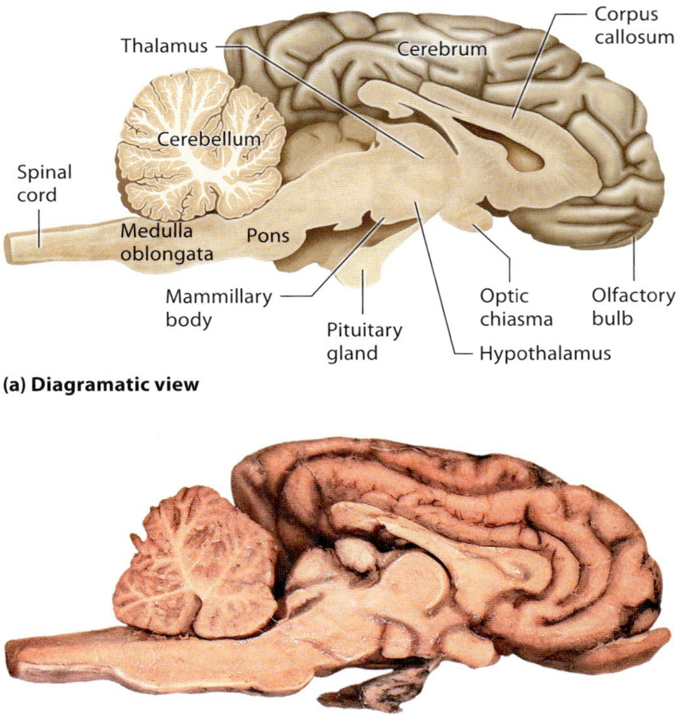

Figure 15-20 Sheep brain, midsagittal section.

☐ septum pellucidum ________________________

____

☐ hypothalamus ____________________________

____

☐ epithalamus _____________________________

____

☐ thalamus _______________________________

____

☐ optic chiasma ____________________________

____

☐ mammillary body _________________________

____

☐ pineal body ______________________________

____

☐ superior colliculi _________________________

____

☐ inferior colliculi __________________________

____

- ☐ transverse fissure ____________________
- ☐ cerebellum ____________________
- ☐ fourth ventricle ____________________
- ☐ medulla oblongata ____________________
- ☐ midbrain ____________________
- ☐ diencephalon ____________________
- ☐ brainstem ____________________
- ☐ pons ____________________

7. Obtain a preserved spinal cord and identify the dura mater.

   Describe the appearance of the dura mater.

   ____

8. Cut a transverse section of the spinal cord (refer to **Figure 15-21**).

   a. Identify the arachnoid mater.

   Describe the appearance of the arachnoid mater.

   ____

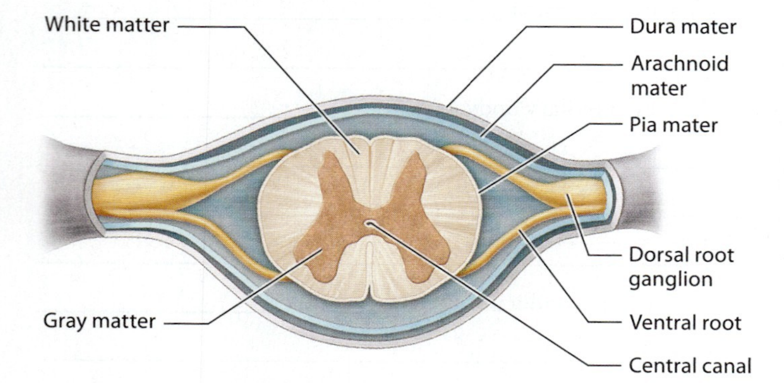

Figure 15-21 Sheep spinal cord, transverse section.

   b. Identify the pia mater.

   Describe the appearance of the pia mater.

   ____

   c. Identify the denticulate ligaments.

   What is the function of these ligaments?

   ____

   d. Peel back the spinal meningeal layers to expose the posterior roots and the anterior roots.

   A posterior root and an anterior root fuse together to form a

   ____________________.

   e. As you identify each of the following structures, place a check mark in the appropriate box, and review the functions of each structure with the members of your lab group:

   - ☐ White matter
     - ☐ Anterior funiculus
     - ☐ Lateral funiculus
     - ☐ Posterior funiculus
   - ☐ Gray matter
     - ☐ Anterior horn
     - ☐ Lateral horn
     - ☐ Posterior horn
     - ☐ Gray commissure
     - ☐ Central canal
   - ☐ Anterior median fissure
   - ☐ Posterior median sulcus

9. Follow the instructions given by your instructor for disposing of the specimen and cleaning up.

10. **Optional Activity**

Practice labeling brain structures on sheep brains at MasteringA&P® > Study Area > Practice Anatomy Lab > Cat > Nervous System > images 1 thru 10

Post-lab quizzes are also assignable in MasteringA&P®

POST-LAB ASSIGNMENTS

Name: ____________________ Date: ____________ Lab Section: ________

PART I. Check Your Understanding

Activity 1: Exploring the Functional Anatomy of the Brain

1. Identify the components of the brain (midsagittal section):

   a. ____________________

   b. ____________________

   c. ____________________

   d. ____________________

   e. ____________________

   f. ____________________

   g. ____________________

   h. ____________________

   i. ____________________

   j. ____________________

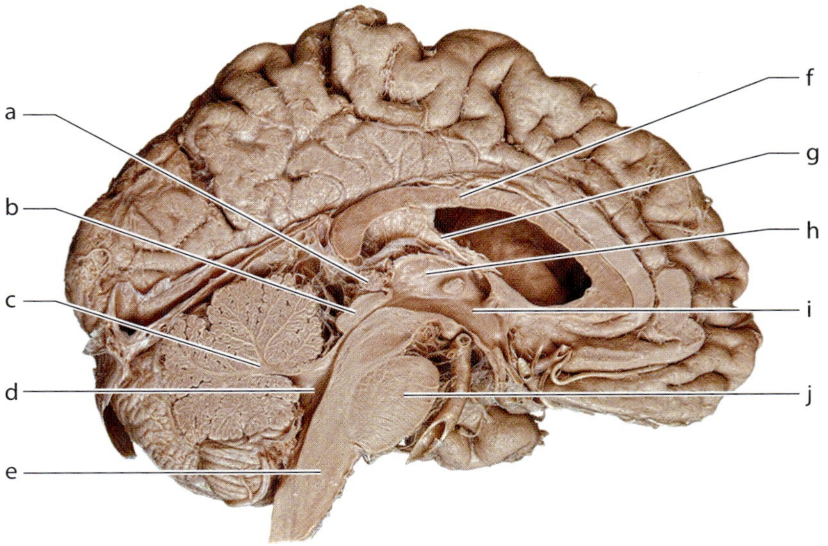

a
b
c
d
e
f
g
h
i
j

2. Match each of the following brain structures with its correct description:

_____	a. corpus callosum	1. separates lateral ventricles
_____	b. septum pellucidum	2. continuous with the spinal cord
_____	c. thalamus	3. a "bridge" between the midbrain and medulla oblongata
_____	d. midbrain	4. contains two hemispheres connected by the vermis
_____	e. medulla oblongata	5. large tract
_____	f. cerebellum	6. masses of gray matter within cerebral white matter
_____	g. pons	7. contains corpora quadrigemina
_____	h. hypothalamus	8. gateway for most sensory impulses en route to the cerebral cortex
_____	i. basal nuclei	9. secretes hormones that control pituitary function

3. Visual areas are located in which cerebral lobe? ____________________

4. Which association area controls the muscles used during speech? ____________________

5. The gustatory cortex is located in which cerebral lobe? ____________________

Activity 2: Electroencephalography in a Human Subject Using BIOPAC®

1. List and define two characteristics of regular, periodic waveforms.

____

____

2. Examine the recordings of the alpha and beta waves for changes between "Eyes Open" and "Eyes Closed." Does the desynchronization of the alpha rhythm occur when the eyes are open? Explain.

____

____

Does the beta rhythm become more pronounced in the "Eyes Open" state? Explain.

____

3. Examine the wavelengths for the delta and theta rhythms. Is there an increase in delta and theta activity when the eyes are open? Explain.

____

____

4. Define the following:

a. Alpha rhythm _______________

b. Beta rhythm _______________

c. Delta rhythm _______________

d. Theta rhythm _______________

Activity 3: Identifying the Meninges/Ventricles and Tracing the Flow of Cerebrospinal Fluid

1. Name the three meningeal layers from most superficial layer to deepest layer.

____

2. State two functions of cerebrospinal fluid. _______________

____

3. Identify ventricles and associated CSF passageways in the accompanying figure:

a. _______________

b. _______________

c. _______________

d. _______________

e. _______________

f. _______________

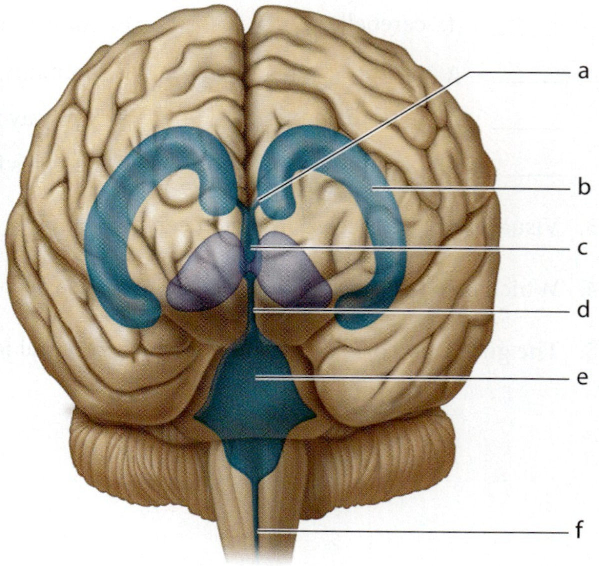

4. Which ventricle(s) is(are):

    ____________________ a. located between the brainstem and the cerebellum?

    ____________________ b. located in the cerebrum?

    ____________________ c. located in the midbrain?

Activity 4: Examining the Functional Anatomy of the Spinal Cord

1. Identify the structures of the spinal cord (transverse section):

    a. ____________________________________

    b. ____________________________________

    c. ____________________________________

    d. ____________________________________

    e. ____________________________________

a
b
c
d
e

2. The spinal cord tapers to a tip called the ____________________________________.

3. The ____________________ is an extension of the pia mater that anchors the spinal cord to the coccyx.

4. Name the two enlargements of the spinal cord: ____________________________________

    ____

5. The space between a vertebra and the dura mater is called the ____________________.

Activity 5: Analyzing a Spinal Reflex

1. Name the five components that participate in the three events of a typical reflex arc.

    ____

    ____

2. In the patellar reflex,

    a. which structure serves as the receptor? ____________________________________

    b. which structure serves as the effector? ____________________________________

Activity 6: Dissecting a Sheep Brain and Spinal Cord

1. Identify the structures of a sheep brain (midsagittal section), and write the brain region with which each is associated:

Structure	Brain Region
a. ______________	______________
b. ______________	______________
c. ______________	______________
d. ______________	______________
e. ______________	______________

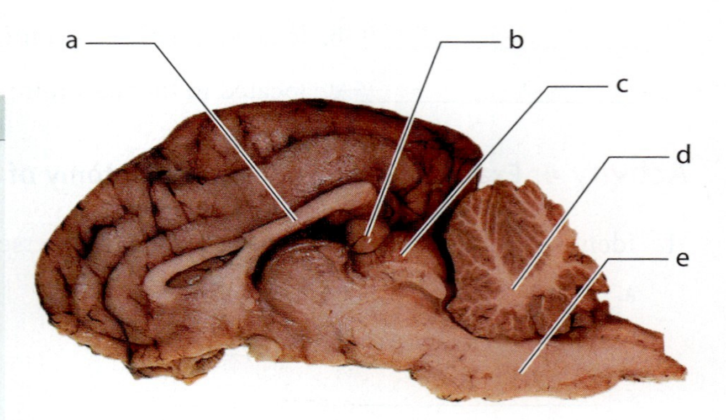

2. State two differences between a sheep brain and a human brain. ______________________________
____
____

PART II. Putting It All Together

A. Review Questions

Answer the following questions using your lecture notes, your textbook, and your lab notes.

1. List the four major regions of the brain. ______________________________

2. Name the brain structure that:
   a. relays messages between the right and left cerebral hemispheres. ______________
   b. serves as a visual reflex center. ______________________________
   c. interprets sensory impulses for smell. ______________________________
   d. contains neurosecretory cells that extend to the posterior pituitary gland. ______________

3. Do each of the following descriptions characterize the cranial meninges (C), the spinal meninges (S), or both (C and S)?

_____ a. Consists of dura, arachnoid, and pia maters.

_____ b. Dura mater is composed of two layers.

_____ c. Contains epidural space.

_____ d. Is anchored by denticulate ligaments.

4. List one way in which the conus medullaris and filum terminale are similar: ____________________

   ____

   List one way in which they are different: ____

   ____

5. List one way in which the frontal eye field and Broca's area are similar: ____________________

   ____

   List one way in which they are different: ____

   ____

6. When a spinal tap is performed on a patient, a needle is inserted below the second lumbar vertebra to remove a sample of spinal fluid. Why is the needle not inserted above this level?

   ____

   ____

7. Which type of nerve fiber (sensory or motor) travels through each of the following structures?

   ____________________ a. anterior gray horn

   ____________________ b. posterior gray horn

   ____________________ c. lateral gray horn

   ____________________ d. lateral corticospinal tract

   ____________________ e. posterior spinocerebellar tract

8. Put the following events describing the flow of cerebrospinal fluid in order by writing the correct numeral (1–8) in the blanks.

   __________ CSF enters the subarachnoid space.

   __________ CSF travels down the cerebral aqueduct.

   __________ CSF passes through the interventricular foramina.

   __________ CSF enters the fourth ventricle.

   __________ CSF passes through the two lateral and one median apertures.

   __________ CSF is released by ependymal cells.

   __________ CSF enters the third ventricle.

   __________ CSF enters the superior sagittal sinus.

B. Concept Mapping

1. Fill in the blanks to complete this concept map outlining the production and flow of cerebrospinal fluid from the third ventricle to the fourth ventricle.

cerebellum | cerebral aqueduct | cerebrospinal fluid | ependymal cells | epithalamus

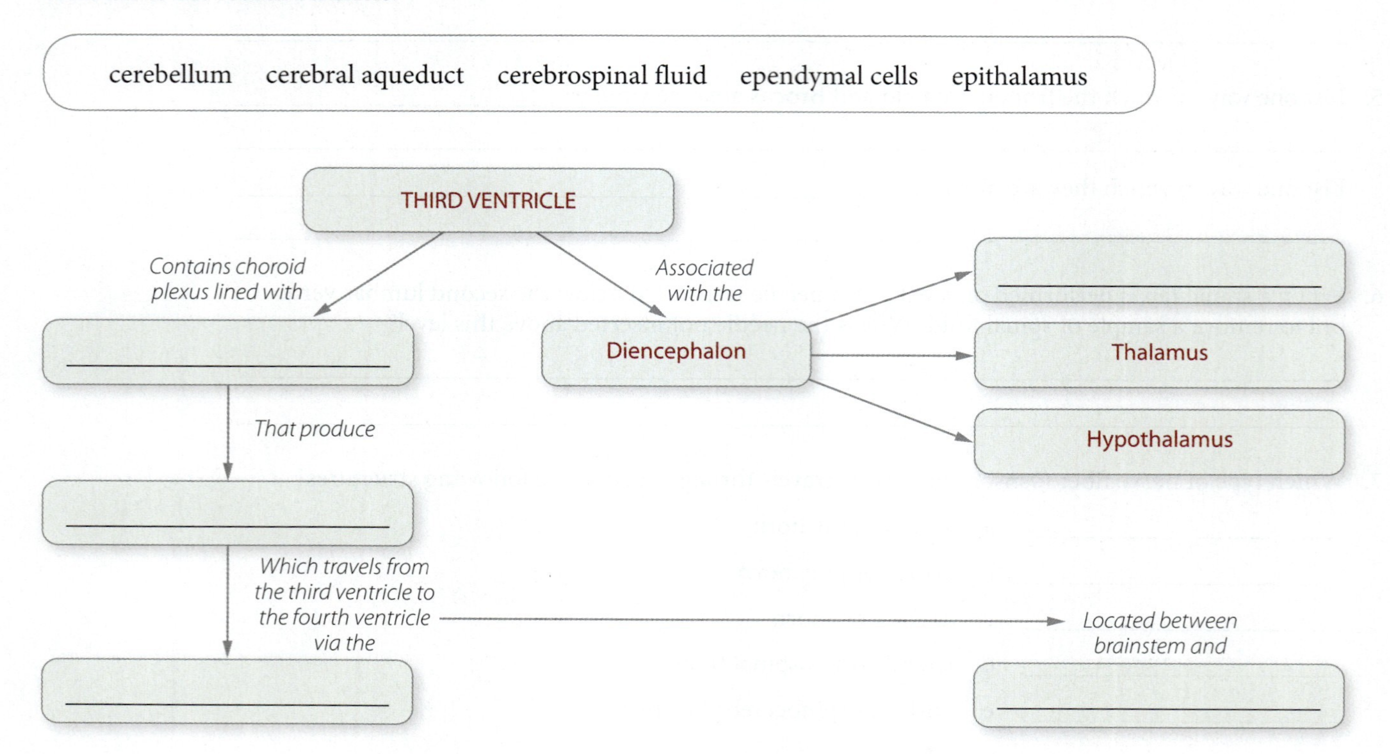

2. Construct a unit concept map to show the relationships among the following set of terms. Include all of the terms in your diagram. Your instructor may choose to assign additional terms.

cardiac center	cerebellum	cerebral aqueduct	cerebrospinal fluid
corpora quadrigemina	corpus callosum	ependymal cells	epithalamus
funiculus	intermediate mass	longitudinal fissure	
mammillary bodies	meninges	myelin	vermis

16

The Peripheral Nervous System: Nerves and Autonomic Nervous System

UNIT OUTLINE

In Unit 15 we studied the structure and function of the central nervous system (brain and spinal cord). This unit focuses on the structure and function of the peripheral nervous system (PNS): nerves and the autonomic nervous system. Recall from Unit 14 that a typical nerve contains thousands of nerve fibers (some myelinated and others unmyelinated) organized into bundles called fascicles and surrounded by connective tissue wrappings (epineurium, perineurium, and endoneurium). Recall as well that the PNS is subdivided into the sensory (afferent) and motor (efferent) divisions, and that the motor division is further divided into the somatic nervous system and the autonomic nervous system.

THINK ABOUT IT

What is a nerve? ____________________

How is a nerve similar to a tract? ____________________

How is a nerve different from a tract? ____________________

What is a ganglion? ____________________

How is a ganglion similar to a nucleus? ____________________

How is a ganglion different from a nucleus? ____________________ ■

PRE-LAB ASSIGNMENTS

To maximize learning, BEFORE your lab period carefully read this entire lab unit and complete these pre-lab assignments using your textbook, lecture notes, and prior knowledge.

PRE-LAB Activity 1: Learning the Cranial Nerves

1. Use the list of terms provided to label the accompanying illustration of the cranial nerves. Check off each term as you label it.

- ☐ accessory nerve
- ☐ optic nerve
- ☐ vestibulocochlear nerve
- ☐ glossopharyngeal nerve
- ☐ trigeminal nerve
- ☐ hypoglossal nerve
- ☐ olfactory nerve
- ☐ vagus nerve
- ☐ trochlear nerve
- ☐ abducens nerve
- ☐ facial nerve
- ☐ oculomotor nerve

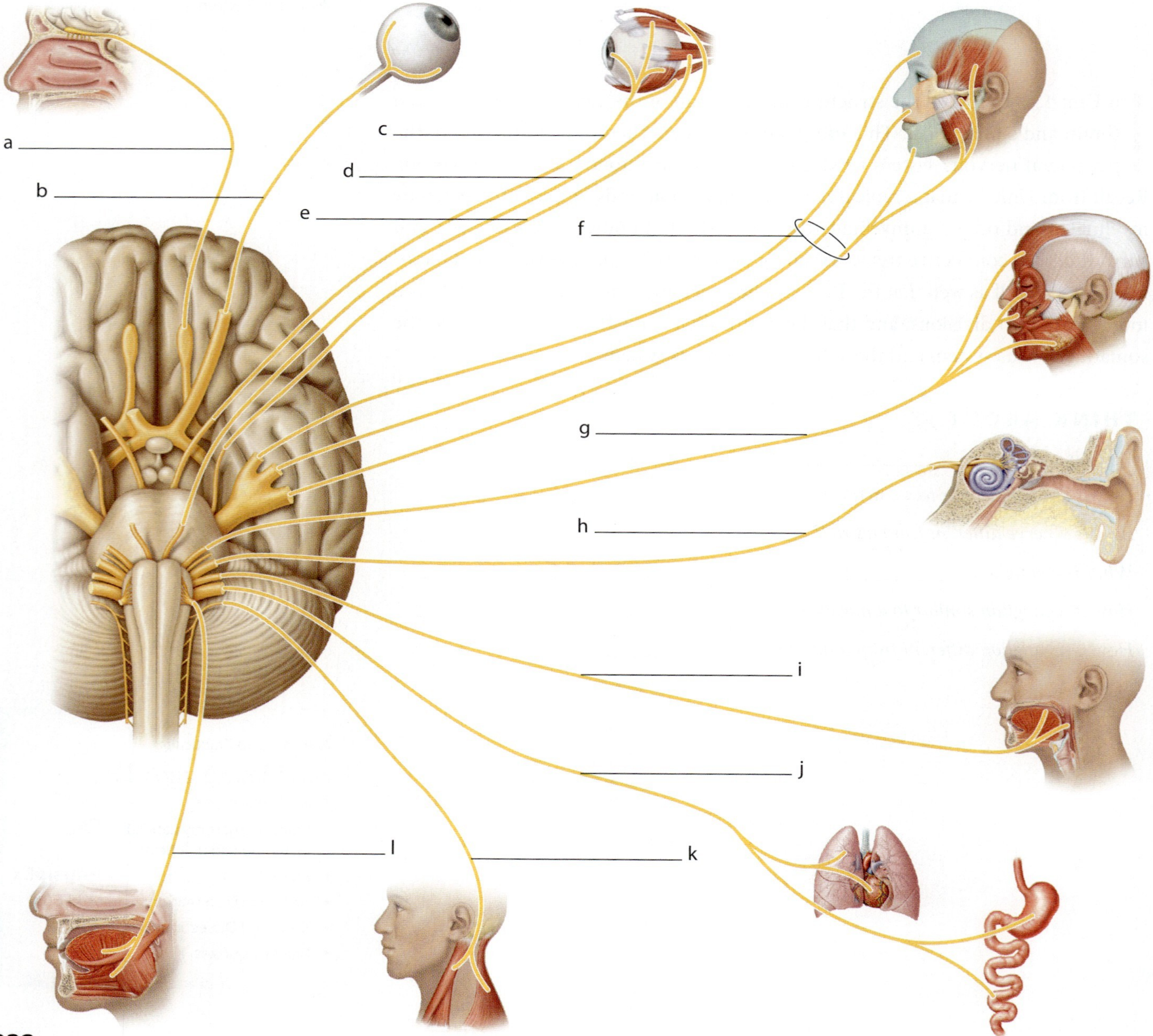

a ______

b ______

c ______

d ______

e ______

f ______

g ______

h ______

______ i

______ j

______ l

______ k

2. Which cranial nerve passes through the:
   a. optic canal? ____________________
   b. hypoglossal canal? ____________________
   c. foramina of the cribriform plate? ____________________
   d. internal acoustic meatus? ____________________

PRE-LAB Activity 2: Evaluating the Function of the Cranial Nerves

1. Match each of the following cranial nerves with its description:

_____ a. olfactory nerve (CN I)
_____ b. optic nerve (CN II)
_____ c. oculomotor nerve (CN III)
_____ d. trochlear nerve (CN IV)
_____ e. trigeminal nerve (CN V)
_____ f. abducens nerve (CN VI)
_____ g. facial nerve (CN VII)
_____ h. vestibulocochlear nerve (CN VIII)
_____ i. glossopharyngeal nerve (CN IX)
_____ j. vagus nerve (CN X)
_____ k. accessory nerve (CN XI)
_____ l. hypoglossal nerve (CN XII)

1. transmits impulses from the retina to the occipital lobe of the cerebrum
2. innervates extrinsic eye muscles
3. innervates muscles of facial expression
4. wanders into the thorax
5. innervates the tongue muscles
6. innervates the chewing muscles
7. innervates muscles involved in swallowing
8. fibers travel through tiny openings in the cribriform plate of the ethmoid bone
9. functions in hearing and equilibrium
10. innervates the sternocleidomastoid muscle

PRE-LAB Activity 3: Identifying the Spinal Nerves and Nerve Plexuses

1. Match each of the following spinal nerves with the plexus from which it arises.

_____ a. saphenous nerve
_____ b. radial nerve
_____ c. phrenic nerve
_____ d. sciatic nerve
_____ e. median nerve
_____ f. obturator nerve
_____ g. ulnar nerve
_____ h. tibial nerve
_____ i. musculocutaneous nerve
_____ j. axillary nerve
_____ k. common fibular nerve
_____ l. femoral nerve

1. cervical plexus
2. brachial plexus
3. lumbar plexus
4. sacral plexus

PRE-LAB Activity 4: Exploring the Autonomic Nervous System

1. The nervous system is divided into the central nervous system (CNS) and the peripheral nervous system (PNS). The PNS is divided into the ________________ and the ________________ divisions. The ________________ division is further divided into the somatic and autonomic divisions. The ________________ division innervates cardiac muscle, smooth muscle, and glands, whereas the ________________ division innervates skeletal muscle.

2. How does the efferent pathway in an autonomic reflex differ from that of a somatic reflex? ________________
____

3. An example of an autonomic reflex is the ________________
________________.

Cranial Nerves

With the exception of the vagus nerves, the cranial nerves serve structures of the head and neck only. The 12 pairs of cranial nerves pass through the various foramina of the skull. Three of the cranial nerves (I, II, and VIII) are classified as entirely sensory (S) (**Figure 16-1**), five (III, IV, VI, XI, and XII) as motor nerves (M) (**Figure 16-2**), and four (V, VII, IX, and X) as mixed nerves (**Figures 16-3** and **16-4**). Note that the cranial nerves classified as motor nerves do contain a few sensory fibers that inform the brain about proprioception, but these cranial nerves are primarily motor in function.

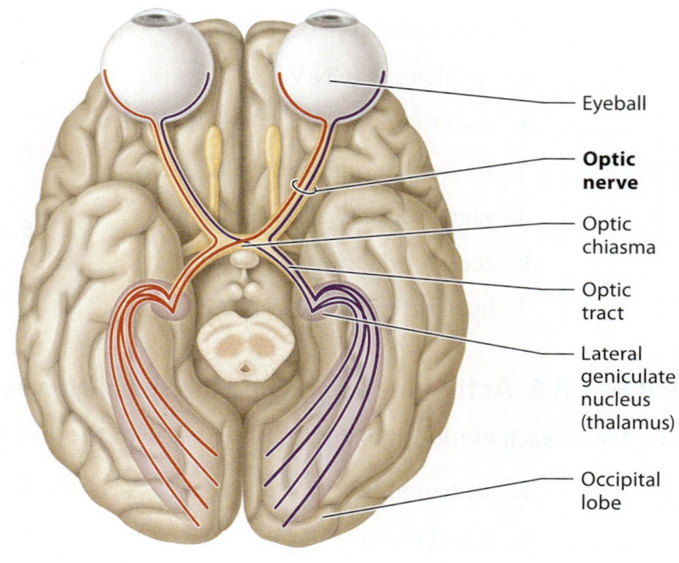

(b) Optic nerve (II)

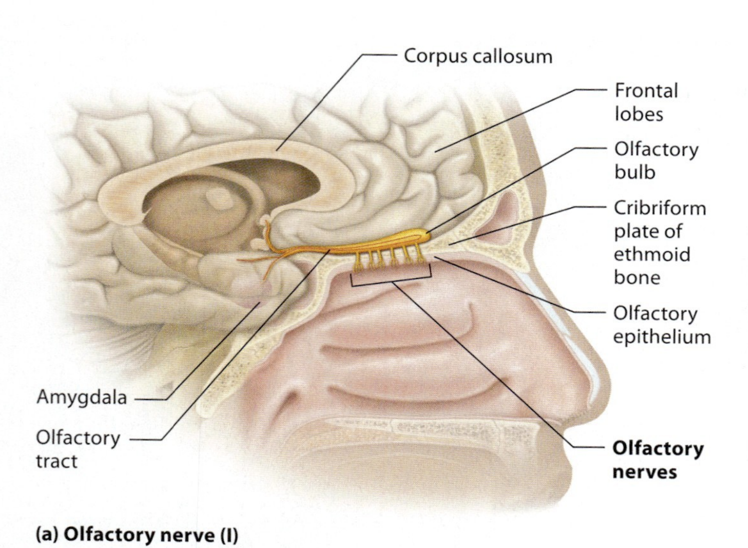

(a) Olfactory nerve (I)

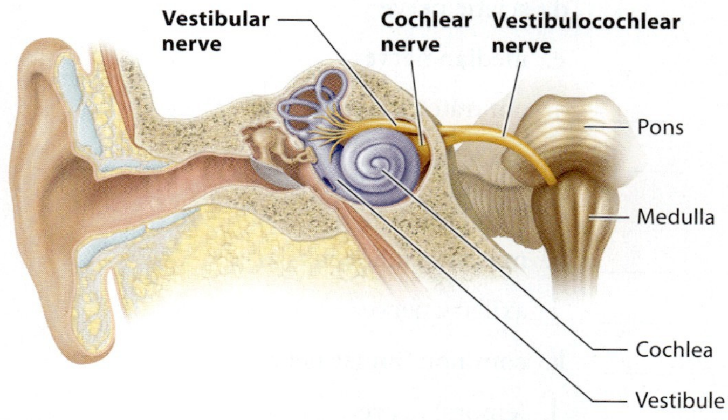

(c) Vestibulocochlear nerve (VIII)

Figure 16-1 The sensory cranial nerves.

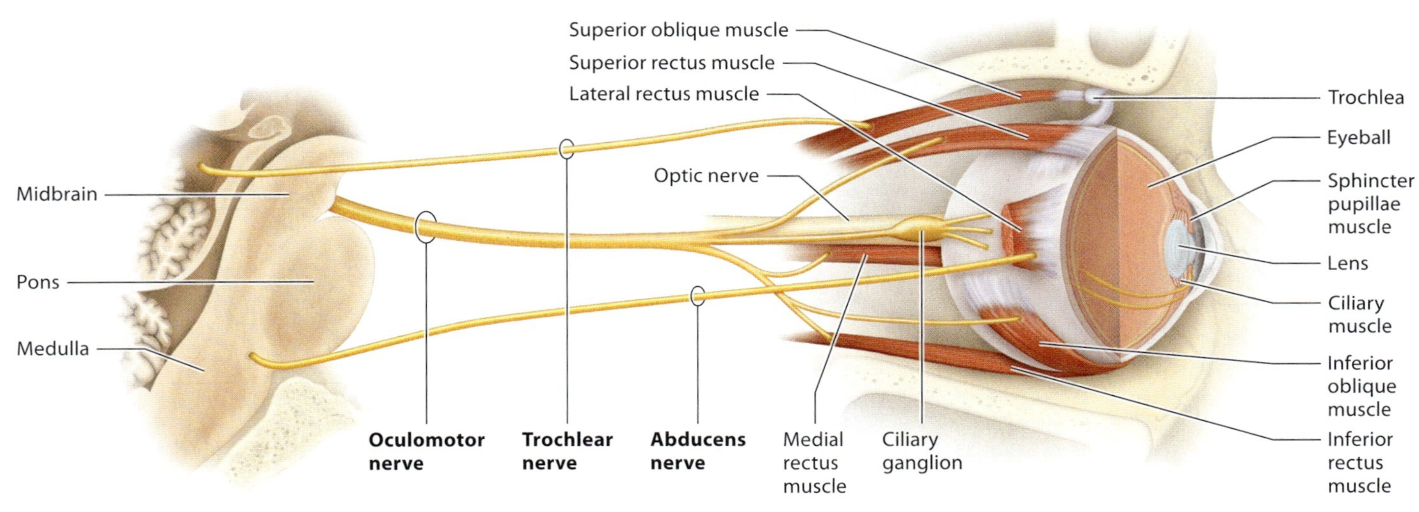

(a) Oculomotor (III), trochlear (IV), and abducens (VI) nerves

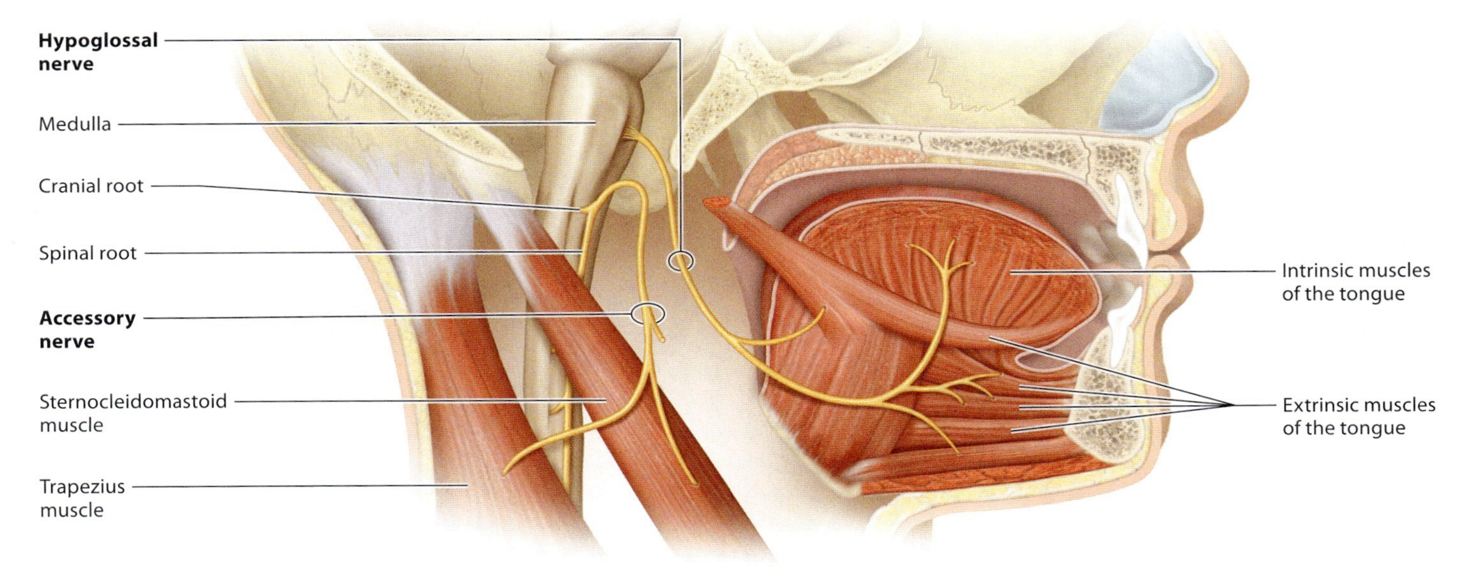

(b) Accessory (XI) and hypoglossal (XII) nerves

Figure 16-2 The motor cranial nerves.

Next we consider each of the 12 pairs of cranial nerves in order by their number.

Cranial Nerve I: Olfactory Nerve

(Sensory Only)

Sensory fibers for the sense of smell (olfaction) originate in axons in the olfactory epithelium of the superior nasal cavity (see Figure 16-1a). Nerve fibers pass through the olfactory foramina of the cribriform plate of the ethmoid bone and synapse with neurons in the olfactory bulbs. From there, nerve fibers travel along the olfactory tracts to the structures of the limbic system in temporal lobes of the cerebrum.

Cranial Nerve II: Optic Nerve

(Sensory Only)

Optic nerve fibers that arise from the retina of each eye form the optic nerves (see Figure 16-1b), which pass through the optic canals of the sphenoid bone and fuse to form the X-shaped optic chiasma, in which some fibers cross over to the contralateral side, and then continue as the optic tracts to the thalamus where they synapse with neurons there. Fibers then travel to the thalamus, midbrain, and eventually the primary visual cortex in the occipital lobe of the cerebrum.

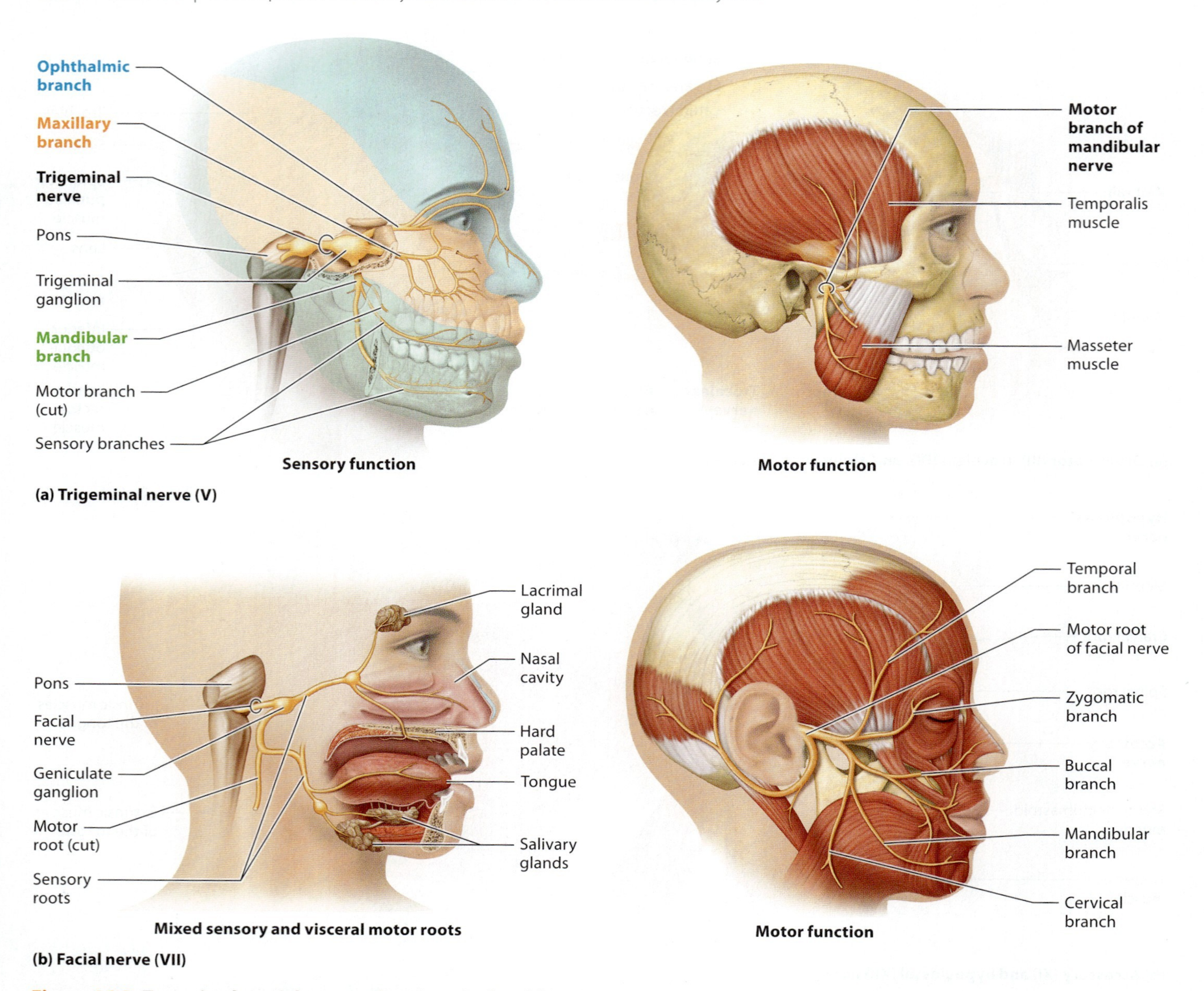

Figure 16-3 **Two mixed cranial nerves: the trigeminal and facial nerves.**

Cranial Nerve III: Oculomotor Nerve

(Motor Only)

Fibers of CN III (see Figure 16-2a) travel from the ventral midbrain and pass through the bony orbit of the eye via the superior orbital fissure to innervate four of the six extrinsic eye muscles (somatic motor fibers) and the smooth muscles of the iris that constrict the pupil (autonomic motor fibers). These nerves also control the shape of the lens (lens accommodation).

Cranial Nerve IV: Trochlear Nerve

(Motor Only)

Fibers of CN IV (see Figure 16-2a), the trochlear nerves—so named because each passes through a pulley-shaped structure called the trochlea—emerge from the dorsal midbrain, enter the orbit via the superior orbital fissure, and innervate the superior oblique muscle.

Cranial Nerve V: Trigeminal Nerve

(Mixed)

The trigeminal nerves (see Figure 16-3a), the largest cranial nerves, are mixed nerves originating from the midbrain and medulla oblongata that branch into three parts: the ophthalmic, maxillary, and mandibular branches. Both the ophthalmic branch, which passes through the superior orbital fissure of the eye socket, and the maxillary branch, which passes through the foramen rotundum of the sphenoid bone, contain sensory nerve fibers serving the orbital structures, cornea, nasal cavity, skin of the face and anterior scalp, eyelids, and mucosae of the mouth. The mandibular branch, which

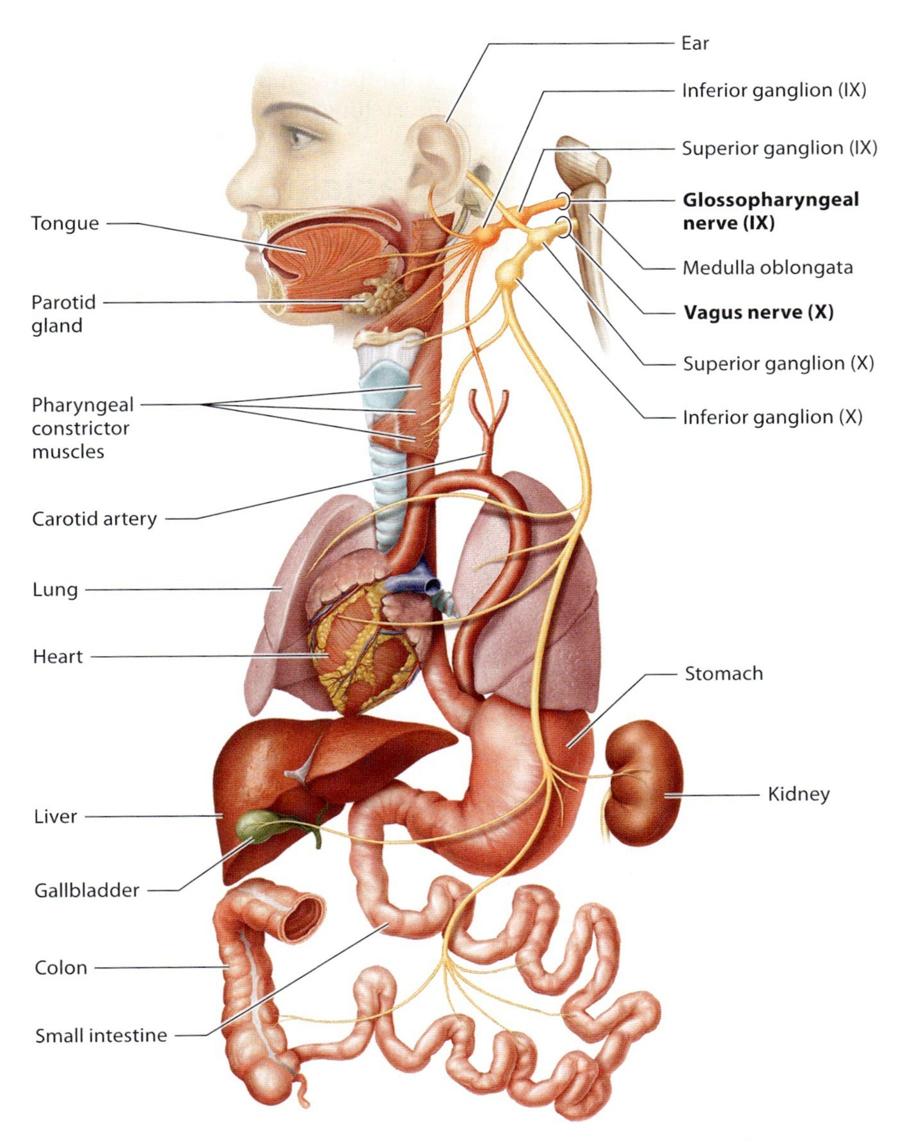

Figure 16-4 Two mixed cranial nerves: the glossopharyngeal and vagus nerves (not to scale).

passes through the foramen ovale of the sphenoid bone, contains sensory fibers that service the lower gums, teeth, lips, palate, and some areas of the tongue, and motor fibers that innervate the muscles of mastication.

Cranial Nerve VI: Abducens Nerve (Motor Only)

Each abducens nerve (CN VI) emerges from the pons, enters the orbit via the superior orbital fissure, and controls the lateral rectus muscle, which moves the eye laterally (see Figure 16-2a).

Cranial Nerve VII: Facial Nerve (Mixed)

The facial nerves (see Figure 16-3b) are mixed cranial nerves that arise from the pons, enter the temporal bones via the internal acoustic meatus, exit the skull via the stylomastoid foramen, and service the face. Sensory fibers of CN VII carry information from taste receptors on the anterior two-thirds of the tongue to the brain. Five motor branches—the temporal, zygomatic, buccal, mandibular, and cervical branches—innervate the muscles of facial expression, the lacrimal glands, and two pairs of salivary glands (submandibular and sublingual).

Cranial Nerve VIII: Vestibulocochlear Nerve (Sensory Only)

The vestibulocochlear nerves (see Figure 16-1c), which function in hearing and equilibrium, receive sensory input from the cochlea and the vestibule of the inner ear and travel through the internal acoustic meatus to nuclei at the junction of the pons and medulla oblongata and to various auditory areas.

Cranial Nerve IX: Glossopharyngeal Nerve (Mixed)

The glossopharyngeal nerves (see Figure 16-4) are small mixed nerves. They contain motor fibers that arise from the medulla oblongata, exit the skull via the jugular foramen, and innervate the pharyngeal muscles involved in swallowing. Sensory fibers of CN IX receive input from the pharynx, the taste buds of the posterior one-third of the tongue, and the carotid arteries.

Cranial Nerve X: Vagus Nerve (Mixed)

The vagus nerves are unique cranial nerves in that they innervate many structures located in the thoracic and abdominal cavities (see Figure 16-4). Motor fibers of CN X arise from the medulla oblongata, exit the skull via the jugular foramen, and descend into the thorax and abdomen. Somatic fibers of the vagus nerves innervate the pharyngeal and laryngeal skeletal muscles involved in swallowing. Sensory fibers of CN X transmit information from the thoracic and abdominal viscera and taste sensation from the posterior tongue and pharynx. The vagus nerves are the predominant pathway by which parasympathetic impulses travel from the brain to the heart, lungs, and abdominal viscera.

Cranial Nerve XI: Accessory Nerve (Motor Primarily)

The accessory nerves (see Figure 16-2b) are unique among the cranial nerves in that they arise from the spinal cord (instead of the brainstem). From there they enter the skull via the foramen magnum and exit the skull (along with the vagus nerves) via the jugular foramina. They innervate trapezius and sternocleidomastoid muscles and certain skeletal muscles of the palate, pharynx, and larynx involved in speech.

Cranial Nerve XII: Hypoglossal Nerve (Motor Primarily)

The hypoglossal nerves (see Figure 16-2b) originate in the medulla and supply the muscles of the tongue.

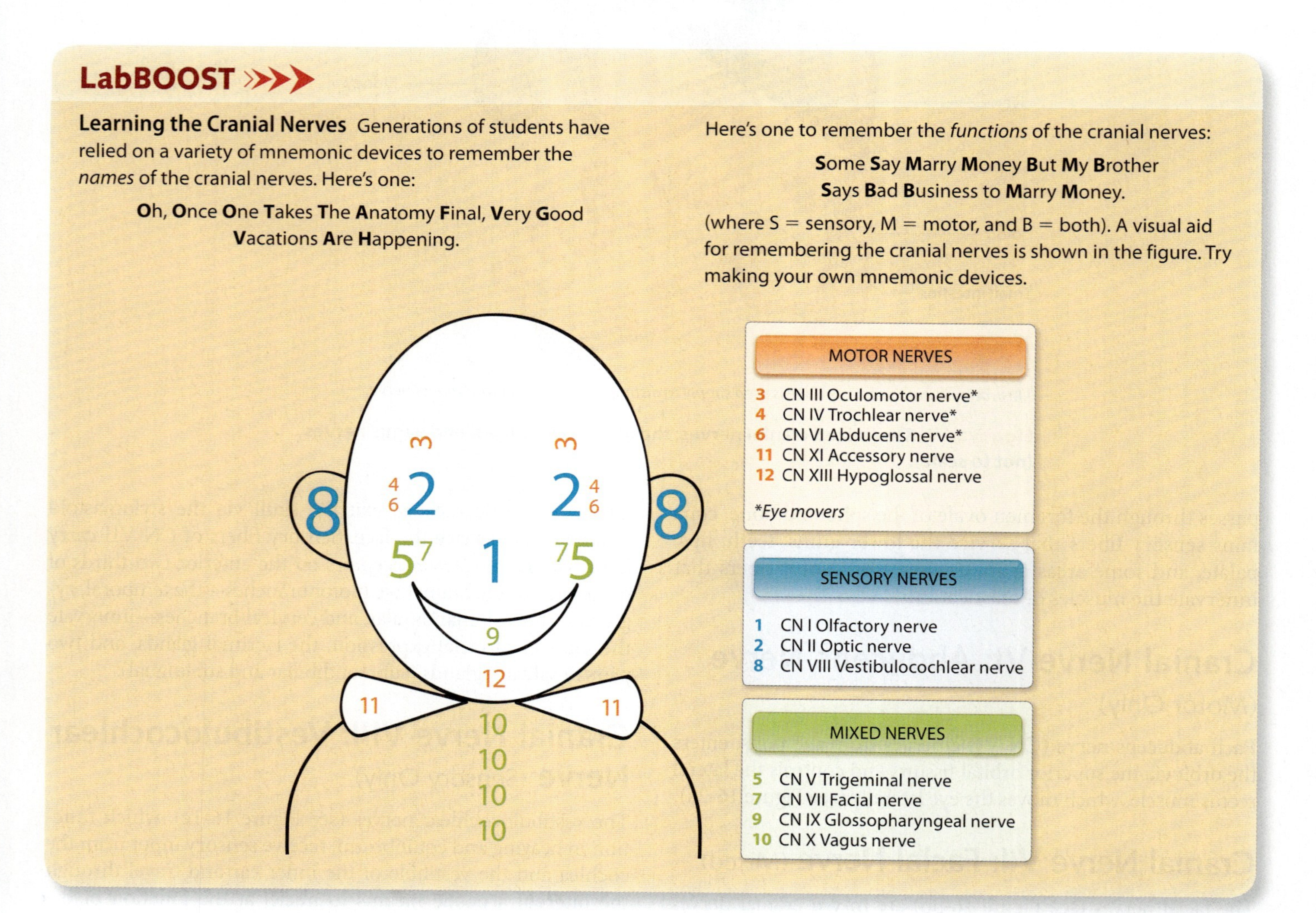

LabBOOST
Learning the Cranial Nerves Generations of students have relied on a variety of mnemonic devices to remember the names of the cranial nerves. Here's one:
Oh, Once One Takes The Anatomy Final, Very Good Vacations Are Happening.
Here's one to remember the functions of the cranial nerves:
Some Say Marry Money But My Brother Says Bad Business to Marry Money.
(where S = sensory, M = motor, and B = both). A visual aid for remembering the cranial nerves is shown in the figure. Try making your own mnemonic devices.
MOTOR NERVES
3 CN III Oculomotor nerve*
4 CN IV Trochlear nerve*
6 CN VI Abducens nerve*
11 CN XI Accessory nerve
12 CN XIII Hypoglossal nerve
*Eye movers
SENSORY NERVES
1 CN I Olfactory nerve
2 CN II Optic nerve
8 CN VIII Vestibulocochlear nerve
MIXED NERVES
5 CN V Trigeminal nerve
7 CN VII Facial nerve
9 CN IX Glossopharyngeal nerve
10 CN X Vagus nerve

ACTIVITY 1

Learning the Cranial Nerves

Learning Outcome

1. Identify the 12 cranial nerves, and describe the action of each.

Materials Needed

☐ Brain models and/or anatomical charts

Instructions

CHART Identify each of the cranial nerves on a brain model (or on anatomical charts) and then complete the following Making Connections chart. For each cranial nerve, use information you have learned in Units 14, 15, and 16 to make "connections" to things you have already learned.

Making Connections: The Cranial Nerves

Cranial Nerve (Numeral)	Cranial Nerve (Name)	Sensory, Motor, or Mixed?	Connections to Things I Have Already Learned
I	*Olfactory nerve*	*Sensory*	*Passes through olfactory foramina of ethmoid bone; transmits information concerning smell to the gustatory cortex of the insula.*
II			
III			
IV			
V			
VI			
VII			
VIII			
IX			
X			
XI			
XII			

Optional Activity

Practice labeling cranial nerves on human brains at > MasteringA&P® > Study Area > Practice Anatomy Lab > Human Cadaver: Nervous System > Peripheral Nervous System > images 1 thru 5

ACTIVITY 2

Evaluating the Function of the Cranial Nerves

Learning Outcomes

1. Evaluate cranial nerve function by performing some common cranial nerve tests.

Materials Needed

- ☐ Snellen eye examination chart
- ☐ Small vial containing vanilla extract
- ☐ Small vial containing cinnamon
- ☐ Small vial containing coffee
- ☐ Blindfold
- ☐ Tuning fork

Instructions

In this activity, you and a lab partner will evaluate the function of the cranial nerves by performing eight tests. For some of the tests you will be the "clinician," and for some you will be the "subject." For each test, record both your observations and the cranial nerve(s) that were evaluated. Then, use your textbook to describe the abnormal response you would expect to observe if the evaluated nerve(s) was (were) damaged.

1. The clinician asks the subject to protrude and retract the tongue.

   Observations: ______________________

   Cranial nerve(s) evaluated: ______________________

   Abnormal response: ______________________

2. The clinician asks the subject to stand 20 feet away from a Snellen eye examination chart and then cover one eye at a time with a hand. The subject then reads the chart, starting with the big letter "E" and continuing to the smallest line that can be read correctly. Record the number of the last line that was identified correctly for each eye.

   Observations: Right eye: __________ Left eye: __________

   Cranial nerve(s) evaluated: ______________________

   Abnormal response: ______________________

3. The clinician stands behind the subject and firmly places her or his hands on the tops of the subject's shoulders. The clinician then instructs the subject to move the shoulders upward while the clinician applies resistance with the hands.

   Observations: ______________________

   Cranial nerve(s) evaluated: ______________________

   Abnormal response: ______________________

4. The clinician blindfolds the subject, places one of the three vials (containing "unknown substance A") under the subject's nose, and asks the subject to identify the substance. Repeat this procedure with the other vials (containing "unknown substance B" and "unknown substance C").

   Observations: "Unknown substance A" ______________________

   "Unknown substance B" ______________________

   "Unknown substance C" ______________________

   Cranial nerve(s) evaluated: ______________________

   Abnormal response: ______________________

5. The clinician instructs the subject to follow the clinician's finger with the eyes while keeping the head still. The clinician then slowly draws a large "Z" in the air in front of the subject.

   Observations: ______________________

   Cranial nerve(s) evaluated: ______________________

   Abnormal response: ______________________

6. The clinician palpates the subject's masseter muscle and then instructs the subject to clench his or her teeth.

   Observations: ______________________

   Cranial nerve(s) evaluated: ______________________

   Abnormal response: ______________________

7. The clinician instructs the subject to perform each of the following actions one at a time: smiling, frowning, whistling, and raising the eyebrows.

   Observations: ______________________

   Cranial nerve(s) evaluated: ______________________

   Abnormal response: ______________________

8. After blindfolding the subject, the clinician gently strikes a tuning fork on the lab bench and holds it near first one of the subject's ears, and then the other ear. In each case the clinician asks if the subject hears anything.

   Observations: ______________________

   Cranial nerve(s) evaluated: ______________________

   Abnormal response: ______________________

Spinal Nerves

In the human body, 31 pairs of **spinal nerves** emerge from the spinal cord (**Figure 16-5**): 8 pairs of **cervical nerves**, 12 pairs of **thoracic nerves**, 5 pairs of **lumbar nerves**, 5 pairs of **sacral nerves**, and 1 pair of **coccygeal nerves**. The name of each spinal nerve is based on its location of origin. Note, however, that there are eight cervical nerves but only seven cervical vertebrae. The first seven pairs of cervical nerves emerge superior to the vertebra for which they are named, but the eighth pair emerges inferior to the seventh cervical vertebra. The remaining pairs of spinal nerves emerge inferior to the vertebra for which they are named.

Spinal nerves are formed from the uniting of the posterior and anterior roots of the spinal cord (**Figure 16-6**). Spinal nerves are mixed nerves because the posterior roots contain sensory neurons entering the spinal cord, and the anterior roots contain the myelinated axons of motor (efferent) neurons whose cell bodies are located in the spinal cord. With the exception of the first pair, all spinal nerves leave the vertebral canal via the intervertebral foramina.

Soon after passing through the intervertebral foramina, each spinal nerve divides into several branches, or rami (singular = ramus). The **posterior rami** (see Figure 16-6) innervate the skin and muscles of the back. The **anterior rami** of all spinal nerves except T_2–T_{12} form complicated branching networks called nerve plexuses (discussed shortly), which supply both sensory and motor fibers to the skin and muscles of the upper and lower appendages. Spinal nerves

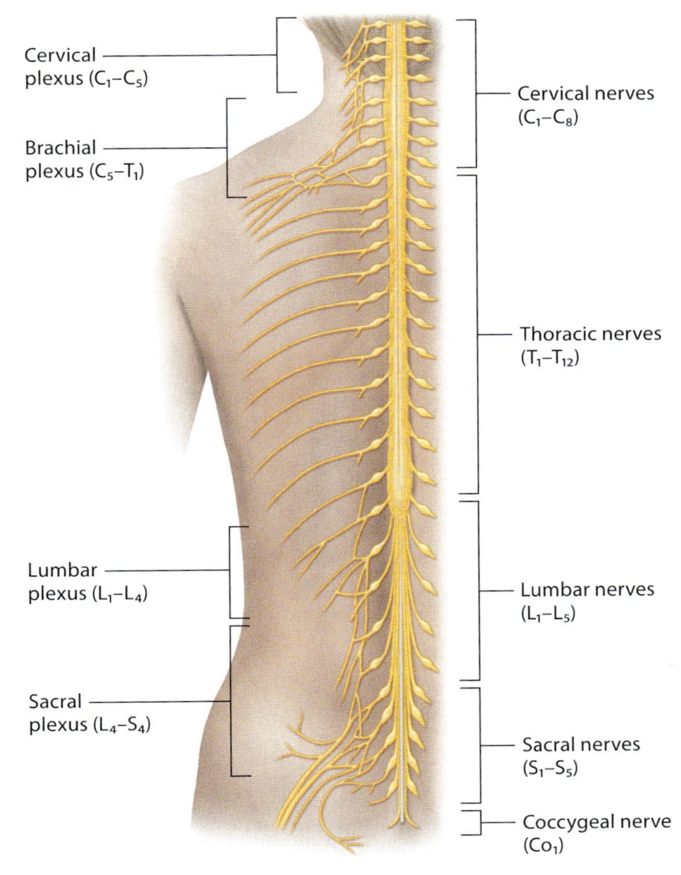

Figure 16-5 **The spinal nerves (at right) and nerve plexuses (at left).**

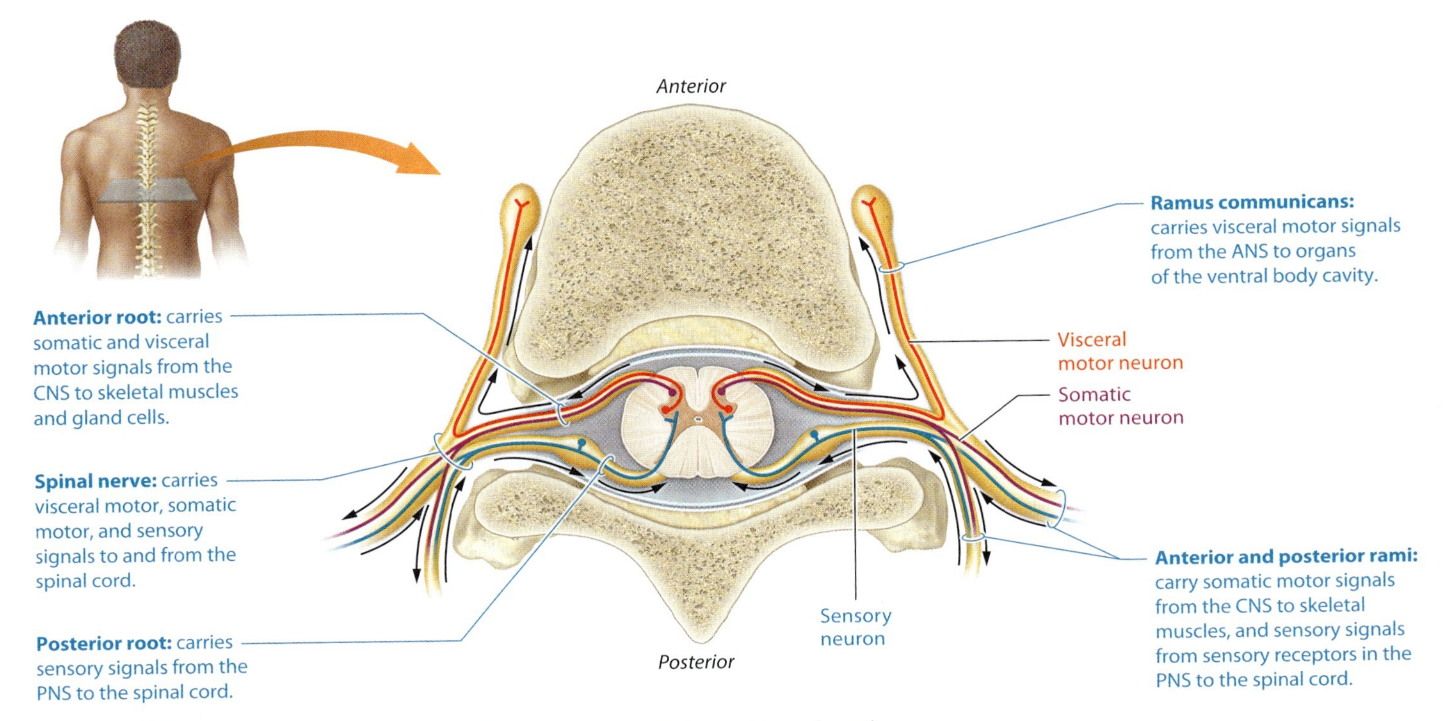

Figure 16-6 **Structure and function of spinal cord roots, spinal nerves, and rami.**

T_2–T_{12} travel anteriorly as the intercostal nerves, which innervate both the intercostal and abdominal muscles and relay sensory information from the lateral and anterior trunk. **Rami communicantes** (see Figure 16-6) are branches in some parts of the spinal cord that innervate the sympathetic ganglia of the autonomic nervous system. Finally, the meningeal rami pass back into the vertebral canal to innervate the vertebrae and meninges.

Nerve Plexuses

Nerve plexuses are interlacing nerve networks, located lateral to the vertebral column, that are formed from the joining of anterior rami from all vertebrae except T_2–T_{12}. The four major nerve plexuses are the cervical plexus, the brachial plexus, the lumbar plexus, and the sacral plexus (see Figure 16-5).

The **cervical plexus** (**Figure 16-7**) arises from the anterior rami of C_1–C_4 and parts of C_5, and it is located deep to the sternocleidomastoid muscle. The major motor branch of this plexus is the **phrenic nerve**, which arises from C_3, C_4, and parts of C_5. The phrenic nerve supplies sensory and motor nerve fibers to the diaphragm. Other nerves arising from this plexus serve the scalp, neck, shoulders, and chest.

QUICK TIP The saying "cervical nerves 3, 4, and 5 keep the diaphragm alive" can help you remember which rami form the phrenic nerve.

The large and complicated **brachial plexus** (**Figure 16-8**) arises from the ventral rami of C_5–C_8 and T_1. It successively splits into trunks, then into cords, and finally into five major nerves that serve the shoulders and upper limbs. The lateral cord gives rise to the **musculocutaneous nerve**, which supplies the forearm flexors and the skin of the lateral aspect of the forearm. The **median nerve**—formed from a branch of the lateral cord and a branch of the medial cord—supplies the forearm flexor muscles, several hand muscles, and the skin over the anterior and lateral hand. The medial cord also gives rise to the **ulnar nerve**, which supplies the flexor muscles in the anterior forearm, most intrinsic muscles of the hand, and the skin over the medial hand. The posterior cord gives rise to the **axillary nerve**, which innervates the shoulder, and the **radial nerve**, which innervates the triceps brachii muscle, the extensor muscles of the forearm and hand, and the skin in these regions.

The **lumbar plexus** (**Figure 16-9**), which arises from L_1–L_4, serves the lower abdominopelvic region and the anterior thigh. The largest nerves that arise from the lumbar plexus are the **femoral nerve**, which innervates the muscles of the anterior thigh, and the **obturator nerve**, which innervates the adductor muscles of the medial thigh. Additionally, a cutaneous branch of the femoral nerve, the saphenous nerve (not shown), innervates the skin of the anteromedial surface of the thigh and leg.

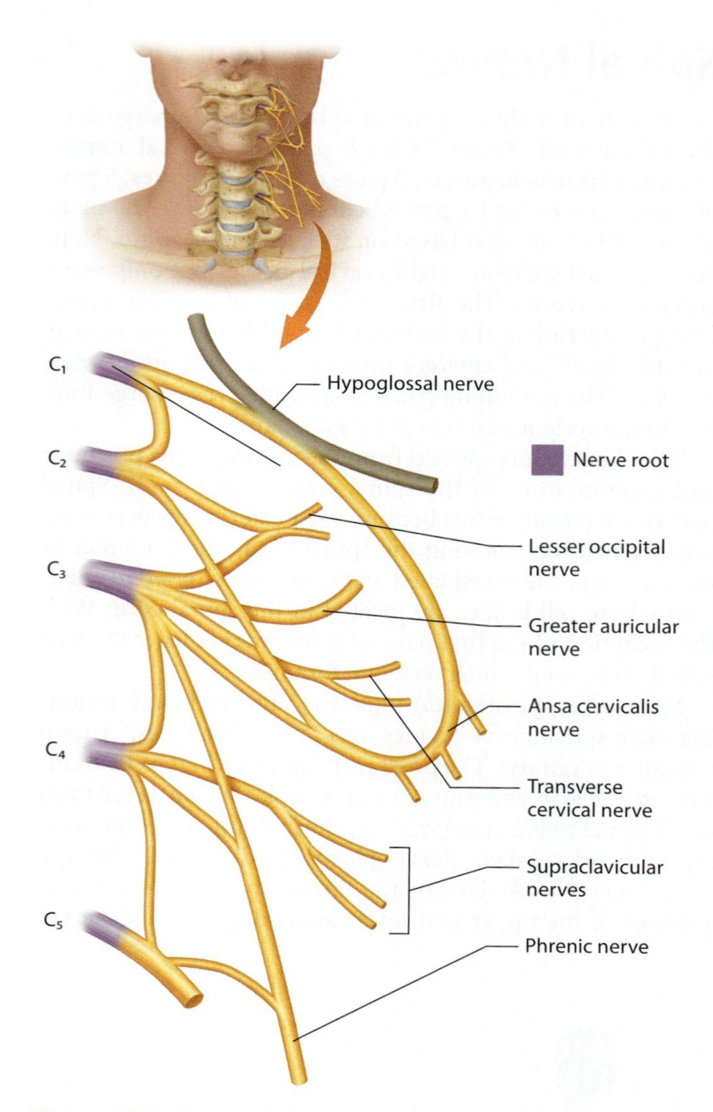

Figure 16-7 The cervical plexus.

The **sacral plexus** (**Figure 16-10**) arises from L_4–S_4. It serves the buttocks and posterior thigh, and gives rise to most of the sensory and motor fibers of the leg and foot. The largest nerve arising from the sacral plexus—the **sciatic nerve**—is the largest nerve in the body. It travels along the posterior thigh and divides in the popliteal region to form the **common fibular nerve** and the **tibial nerve**.

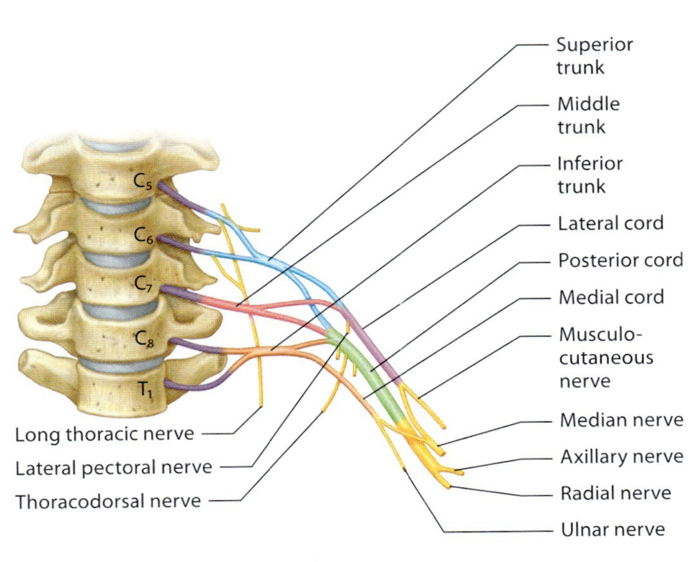

(a) Brachial plexus, anterior view

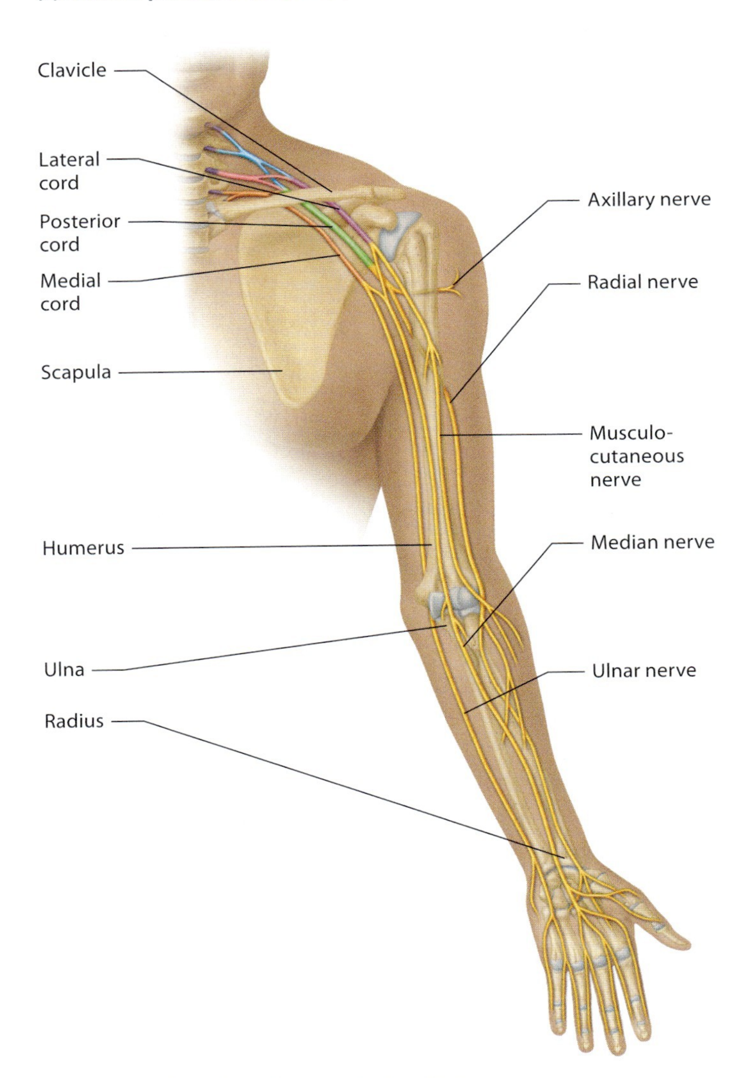

(b) Nerves of brachial plexus, anterior view

Figure 16-8 The brachial plexus.

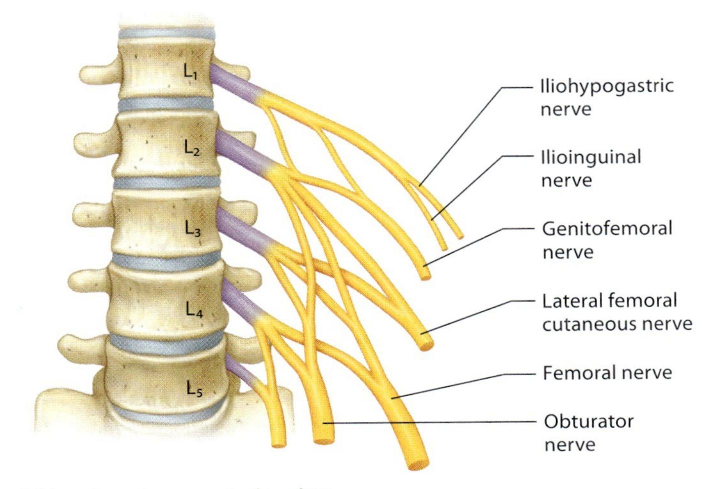

(a) Lumbar plexus, anterior view

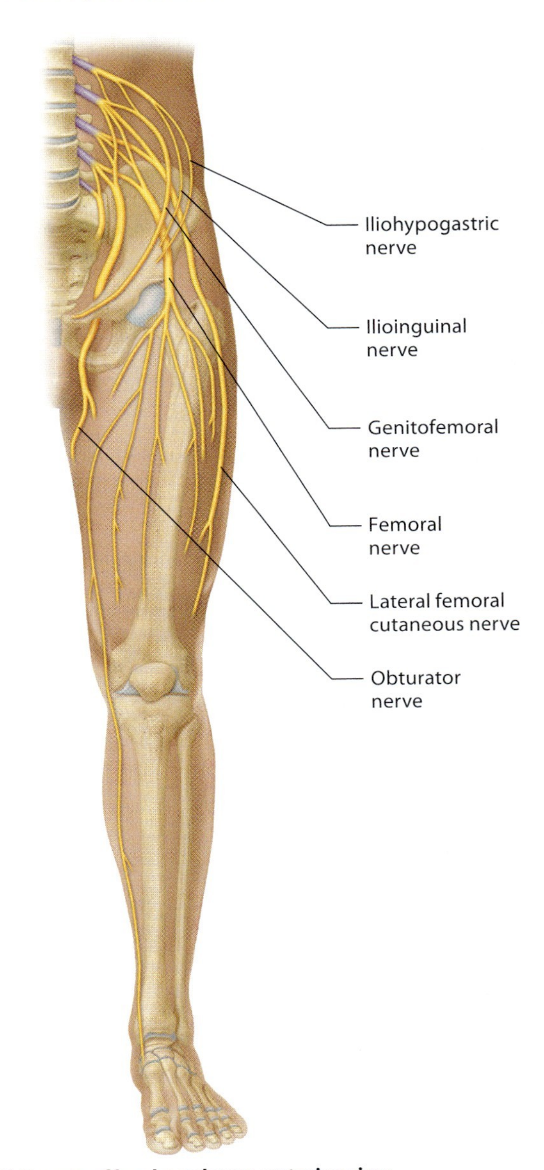

(b) Nerves of lumbar plexus, anterior view

Figure 16-9 The lumbar plexus.

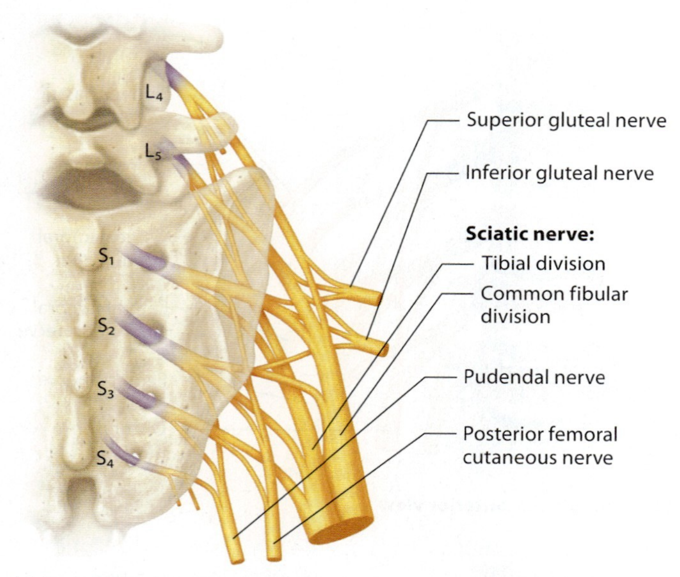

(a) Sacral plexus, posterior view

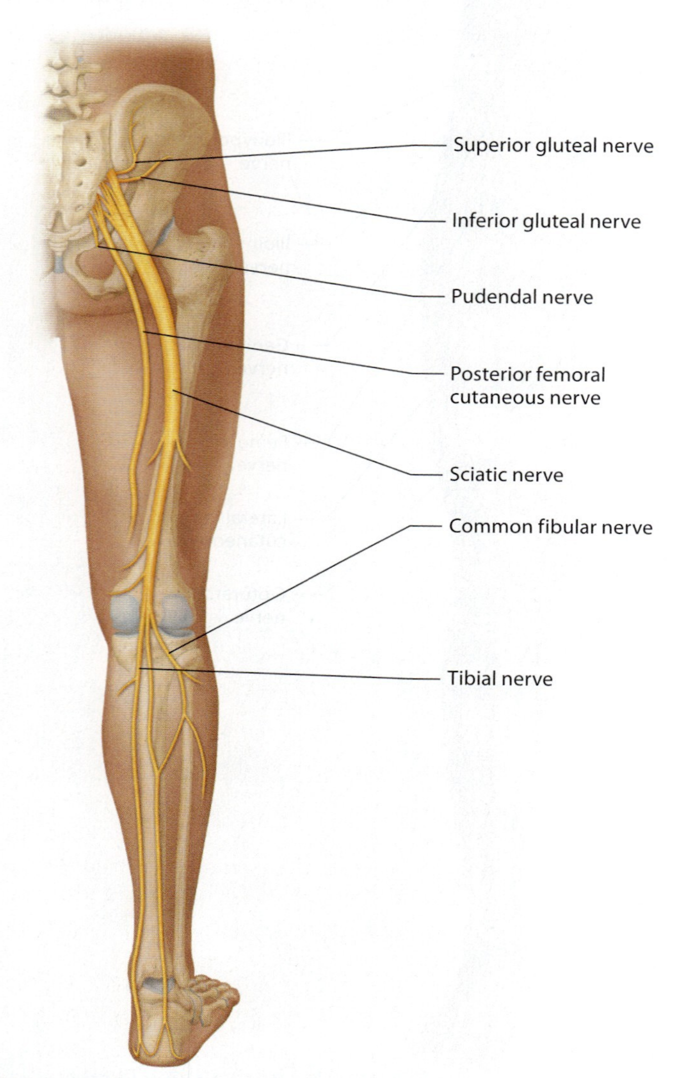

(b) Nerves of sacral plexus, posterior view

Figure 16-10 The sacral plexus.

ACTIVITY 3

Identifying the Spinal Nerves and Nerve Plexuses

Learning Outcomes

1. Identify the four plexuses on an anatomical chart.
2. Locate the major nerves arising from each plexus on an anatomical model and describe the function of each nerve.

Materials Needed

- ☐ Anatomical models
- ☐ Anatomical charts

Instructions

CHART Locate each of the following nerves on an anatomical model. Then, using your textbook, write in the following chart each nerve's plexus of origin and function.

Nerve	Plexus from Which It Arises	Function
Phrenic	*cervical*	*innervates the diaphragm*
Axillary		
Musculocutaneous		
Median		
Radial		
Ulnar		

Nerve	Plexus from Which It Arises	Function
Femoral		
Saphenous		
Obturator		
Sciatic		
Tibial		
Common fibular		

Optional Activity

Practice labeling nerves and plexuses on human cadavers at > MasteringA&P® > Study Area > Practice Anatomy Lab > Human Cadaver: Nervous System > Peripheral Nervous System > images 6 thru 13

The Autonomic Nervous System

Recall that the **peripheral nervous system** consists of the **sensory (afferent) division** and the **motor (efferent) division**, and that the motor division is further divided into the **somatic nervous system (SNS)** and the **autonomic nervous system (ANS)** (see Figure 14-1). Whereas the somatic nervous system is the voluntary nervous system that innervates skeletal muscles, the autonomic nervous system is involuntary and innervates cardiac muscle, smooth muscle, and glands.

The autonomic nervous system is further subdivided into the **parasympathetic ("resting and digesting") branch** and the **sympathetic ("fight or flight") branch**. In general, the sympathetic branch prepares the body for increased activity.

Most visceral organs have dual innervation—that is, they are innervated by both sympathetic and parasympathetic nerve fibers. Both branches work together to maintain homeostasis. Typically one branch stimulates an effector while the other branch inhibits the same effector.

QUICK TIP The sympathetic branch is sometimes known as the "E" division because it is activated during emergency, exercise, excitement, and embarrassment. The parasympathetic branch is sometimes known as the "D" division because it regulates such housekeeping activities as digestion, defecation, and diuresis (urination).

An autonomic motor pathway consists of two neurons—a preganglionic fiber and a postganglionic fiber—that synapse in a ganglion (a collection of nerve cell bodies in the PNS). Sympathetic preganglionic fibers arise from cell bodies of preganglionic neurons in spinal cord segments T_1–L_2 and are therefore considered part of the **thoracolumbar division**. These preganglionic fibers exit the spinal cord via the anterior root, pass into a spinal nerve, and then synapse with a postganglionic fiber in a sympathetic chain ganglion, which is part of the sympathetic chain flanking either side of the vertebral column. The axon of the postganglionic neuron then travels to its target organ. Parasympathetic preganglionic neurons originate in the brainstem and sacral spinal cord, and thus the parasympathetic branch is also called the **craniosacral division**. These preganglionic neurons extend from the CNS almost all the way to the structures they innervate. They synapse with postganglionic fibers in ganglia that are very close to or within their target organs.

Autonomic reflexes are mediated by the ANS. An autonomic reflex arc contains the same five components as a somatic reflex arc: a receptor, a sensory neuron, an integration center, a motor neuron, and an effector. The integration centers for most autonomic reflexes are located in the hypothalamus and brainstem (these are cranial reflexes), but the reflex arcs that control urination, defecation, erection, and ejaculation have integration centers located in the spinal cord (these are spinal reflexes).

Unlike somatic reflex arcs, autonomic reflex arcs contain two motor neurons—a preganglionic fiber and a postganglionic fiber—that synapse in a ganglion. An example of an autonomic reflex is the pupillary light reflex. Light hitting a photoreceptor (*receptor*) in the retina of the eye stimulates the receptor, leading to the transmission of electrical impulses via the optic nerve (*sensory neuron*) to the brain. There an interneuron (*integration center*) transmits the impulses to the preganglionic fiber (*motor neuron #1*) in the midbrain. The axon of the preganglionic fiber travels along the oculomotor nerve (CN III) to the ciliary ganglion, where it synapses with the postganglionic fiber (*motor neuron #2*), which then targets the smooth muscle (*effector*) in the iris of the eye, causing the muscle to contract and the pupil to constrict.

ACTIVITY 4
Exploring the Autonomic Nervous System

Learning Outcomes

1. Describe a typical autonomic pathway.
2. Describe the general effects of the parasympathetic and sympathetic branches of the ANS.
3. Distinguish between a somatic reflex arc and an autonomic reflex arc.
4. Demonstrate and explain the pupillary light reflex.
5. Demonstrate the effects of exercise on sympathetic nervous system activity.

Materials Needed

- ☐ Penlight
- ☐ Millimeter ruler
- ☐ Stethoscope
- ☐ Sphygmomanometer (blood pressure cuff)

Instructions

A. Comparing Sympathetic and Parasympathetic Effects

CHART Using your textbook as a reference, compare sympathetic and parasympathetic effects by completing the following chart.

	Sympathetic	Parasympathetic
Function		
Division		
Neurotransmitter(s)		
Effect on: • Iris of the eye		
Effect on: • Heart		
Effect on: • Bronchioles of the lungs		
Effect on: • Digestive tract activity		
Effect on: • Metabolic rate		
Effect on: • Adrenal medulla		
Effect on: • Sweat glands		

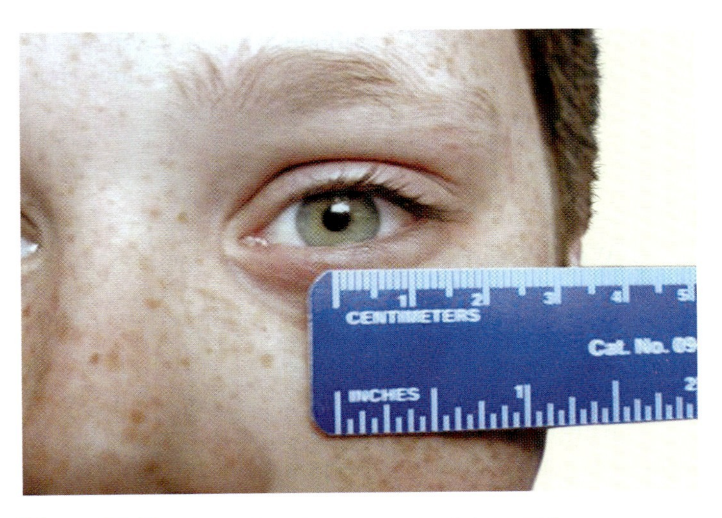

Figure 16-11 Measuring the diameter of the pupil.

B. Testing the Pupillary Light Reflex

1. Obtain a penlight and a millimeter ruler, and designate one of your lab partners to be the test subject.
2. Measure the diameter of the subject's pupils using the ruler (see **Figure 16-11**), and record the measurements in the first row in the following table.
3. Instruct the subject to cover his or her right eye with a hand.
4. Using the penlight, shine light into the subject's left eye.

   What happens to the pupil? ____________________
5. While the light is still shining into the subject's eye, measure the left pupil and record the measurement in the table.
6. Now, uncover the right eye and measure the right pupil. Record the measurement in the table.
7. Repeat steps 3–6, but this time have the subject cover the left eye and then shine the light into the right eye.

	Right Pupil (diameter in mm)	Left Pupil (diameter in mm)
Initial measurement		
Light in left eye		
Light in right eye		

8. Answer the following questions based on your results:
   a. Name each component of the reflex arc in the pupillary reflex.

      Receptor: ____________________

      Sensory neuron: ____________________

      Integration center: ____________________

      Motor neuron: ____________________

      Effector: ____________________

   b. State two differences between a somatic reflex and an autonomic reflex: ____________________

      ____

      ____

      ____

      ____

   c. What conclusions can you draw from your data for this activity? ____________________

      ____

      ____

      ____

      ____

C. Observing the Effects of Exercise on the Sympathetic Nervous System

1. Obtain a stethoscope and a sphygmomanometer, and designate someone in your lab group to be the test subject.
2. Instruct the subject to sit quietly for 3 minutes and then measure the subject's resting heart rate. Place your index and middle fingers on the subject's pulse point over the radial artery, count the number of beats for 30 seconds, and then multiply that number by 2 to determine the number of beats per minute (bpm).

   Resting heart rate: ____________ *bpm*
3. Obtain a sphygmomanometer and a stethoscope, and clean the earpiece and diaphragm of the stethoscope. Have your lab partner sit quietly in a chair for five minutes and then perform the following procedure to measure his/her resting blood pressure (BP):
   a. Wrap the cuff around your lab partner's right arm as shown in **Figure 16-12** so that its lower edge is just superior to the antecubital region of the arm.

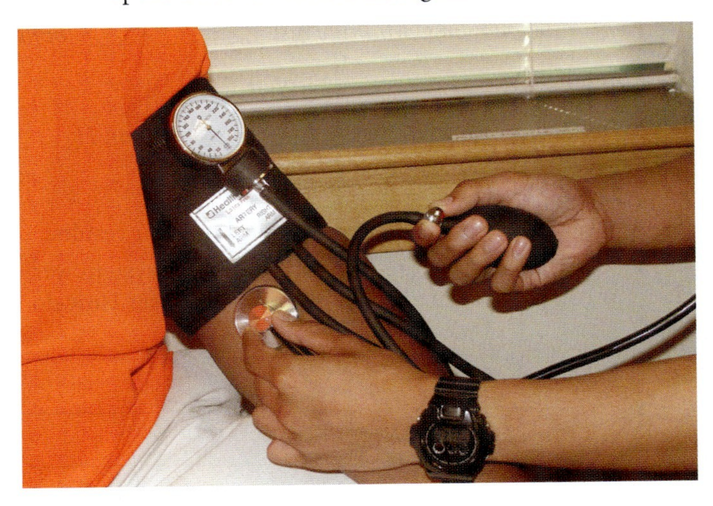

Figure 16-12 Placement of sphygmomanometer and stethoscope for measuring blood pressure.

b. Position the diaphragm of the stethoscope over the brachial artery distal to the cuff.

c. Close the valve on the cuff and squeeze the rubber bulb to inflate the cuff to a pressure of approximately 160 mm Hg.

⚠ Do not overinflate the cuff or leave the cuff inflated for more than 1 minute because it interrupts blood flow to the forearm and could cause fainting.

d. Turn the knob at the base of the bulb to slowly reduce the pressure in the cuff, deflating it at a rate of about 2 or 3 mm Hg per second. During this time listen to the brachial artery through the stethoscope.

e. When the cuff pressure drops below arterial pressure, blood resumes flowing through the artery. This turbulent blood flow makes characteristic sounds called Korotkoff's sounds, which can be heard through the stethoscope. Make a mental note of the pressure at which you begin to hear Korotkoff's sounds. That pressure corresponds to the systolic pressure.

f. Continue deflating the cuff. The thumping Korotkoff's sounds will become faint and will eventually disappear. Note the pressure at which the last sound is heard; this is the diastolic pressure.

g. Release all of the pressure from the cuff, and record the results.

Resting BP: ______________ *mm Hg*

4. Instruct the subject to engage in vigorous exercise for 5 minutes and then immediately measure the subject's heart rate and BP again.

Heart rate immediately after exercising:

______________ *bpm*

BP immediately after exercising: ______________ *mm Hg*

5. Instruct the subject to rest. Repeat the heart rate and BP measurements after 3 minutes and 6 minutes of rest.

Heart rate following: 3-minute rest ______________ *bpm*

6-minute rest ______________ *bpm*

BP following: 3-minute rest ______________ *mm Hg*

6-minute rest ______________ *mm Hg*

6. Answer the following questions based on your results:

a. Which branch of the autonomic nervous system was most active immediately following exercise? How do you know? ______________________________

____

____

b. Which branch of the autonomic nervous system was most active following the 6-minute rest period? How do you know? ______________________________

____

____

c. What effect does the sympathetic nervous system have on the digestive system? ______________________________

____

d. Sketch the motor pathway that functions in the sympathetic nervous system, and label the preganglionic fiber, ganglion, postganglionic fiber, and effector.

e. Which receptor types are found on sympathetic target organs? ______________________________

____

D. Optional Activity

Practice labeling structures of the autonomic nervous system on human cadavers at MasteringA&P® > Study Area > Practice Anatomy Lab > Human Cadaver: Nervous System > Autonomic Nervous System

Post-lab quizzes are also assignable in MasteringA&P®

POST-LAB ASSIGNMENTS

Name: ______________________ Date: ______________ Lab Section: __________

PART I. Check Your Understanding

Activity 1: Learning the Cranial Nerves

1. Identify the cranial nerves on the accompanying photograph of a human brain.

   a. ______________________
   b. ______________________
   c. ______________________
   d. ______________________
   e. ______________________
   f. ______________________
   g. ______________________
   h. ______________________
   i. ______________________
   j. ______________________
   k. ______________________

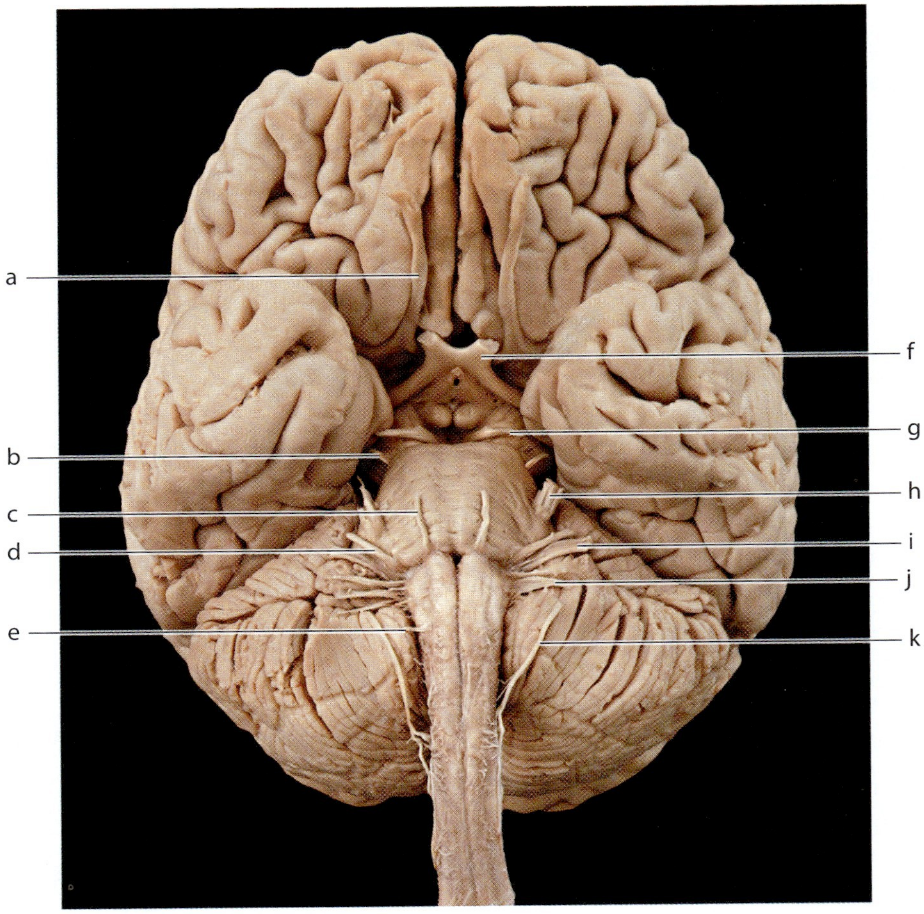

Activity 2: Evaluating the Function of the Cranial Nerves

1. Which of the following cranial nerves is classified as a sensory nerve?
   a. hypoglossal nerve
   b. vagus nerve
   c. oculomotor nerve
   d. olfactory nerve
   e. More than one of these answers is correct.

2. Which of the following cranial nerves is classified as a mixed nerve?
   a. optic nerve
   b. trigeminal nerve
   c. vestibulocochlear nerve
   d. accessory nerve
   e. trochlear nerve

3. In an automobile accident, a woman suffers a crushing hip injury. Which nerve might be damaged if she is unable to flex her leg?
   a. saphenous nerve
   b. sciatic nerve
   c. femoral nerve
   d. obturator nerve
   e. median nerve

4. Which cranial nerve is the only cranial nerve to extend beyond the head and neck to the thorax and abdomen?
   a. trigeminal nerve
   b. glossopharyngeal nerve
   c. vagus nerve
   d. accessory nerve
   e. abducens nerve

5. Predict the cranial nerve(s) that might be damaged if someone could not:
   ______________________ a. shrug the shoulders
   ______________________ b. see
   ______________________ c. move the zygomaticus muscle
   ______________________ d. maintain equilibrium
   ______________________ e. raise the eyebrows
   ______________________ f. track objects with the eyes

Activity 3: Identifying the Spinal Nerves and Nerve Plexuses

1. Identify the spinal nerves and nerve plexuses on the accompanying illustration.
   a. ______________________
   b. ______________________
   c. ______________________
   d. ______________________
   e. ______________________
   f. ______________________
   g. ______________________
   h. ______________________
   i. ______________________

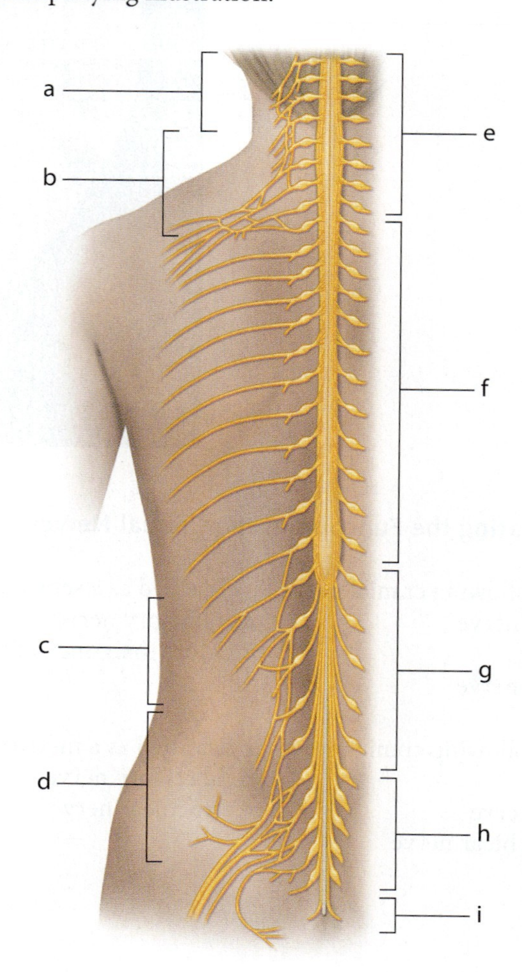

2. Which of the following statements is *false?*
   a. There are eight pairs of cervical nerves.
   b. The lumbar plexus gives rise to the obturator nerve.
   c. The brachial plexus is deep to the sternocleidomastoid muscle.
   d. The sacral plexus serves the posterior surface of the thigh.
   e. The cervical plexus gives rise to the phrenic nerve.

3. In the popliteal region, the sciatic nerve splits to form the ____________________ nerve and the ____________________ nerve.

4. Name the five major nerves that arise from the brachial plexus. ____
____

Activity 4: Exploring the Autonomic Nervous System

1. Using the following list of terms, fill in the blanks to complete an outline of the organization of the nervous system:

☐ sympathetic branch ☐ parasympathetic branch
☐ somatic nervous system ☐ CNS
☐ PNS ☐ ANS
☐ motor (efferent) division ☐ sensory (afferent) division

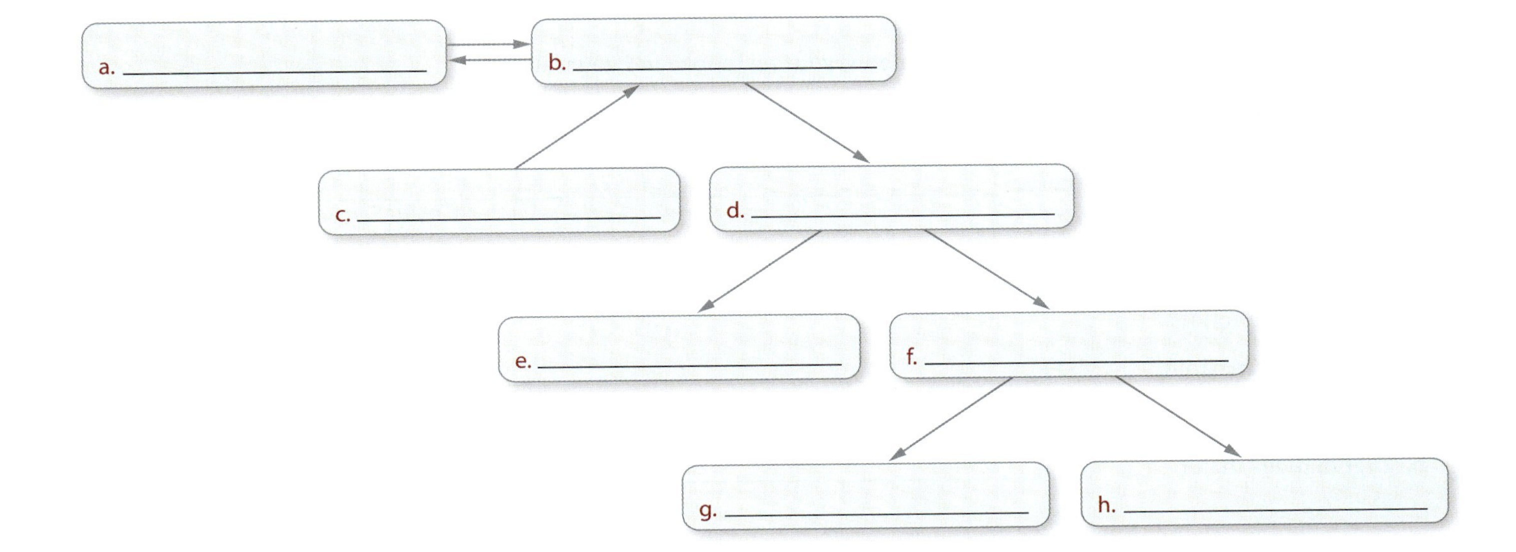

2. Which of the following reflexes is mediated by the spinal cord?
   a. patellar reflex
   b. micturition (urination) reflex
   c. ejaculation reflex
   d. erection reflex
   e. All of these reflexes are mediated by the spinal cord.

3. In the light pupillary reflex,
   a. which type of receptor is stimulated? ____________________
   b. which nerve transmits sensory information to the brain? ____________________
   c. which brain region serves as the integration center? ____________________
   d. which nerve transmits motor information to the effector? ____________________
   e. preganglionic and postganglionic fibers synapse in which ganglion? ____________________
   f. what is the motor response? ____________________

   ____

4. Are sympathetic effects on the heart similar to or different from sympathetic effects on the activity of the digestive tract? ____________________ Explain the physiological significance of this relationship.

   ____

   ____

PART II. Putting It All Together

A. Review Questions

Answer the following questions using your lecture notes, your textbook, and your lab notes.

1. A dentist desensitizes the teeth by injecting a local anesthetic into the alveolar branches of which two divisions of the trigeminal nerve? ____________________

   ____

2. For each of the following skeletal muscles, (1) name the cranial nerve that innervates it and (2) state one of its major actions:
   a. Sternocleidomastoid (1) ____________________
      (2) ____________________
   b. Zygomaticus major (1) ____________________
      (2) ____________________
   c. Lateral rectus (1) ____________________
      (2) ____________________
   d. Masseter (1) ____________________
      (2) ____________________

e. Hyoglossus (1) ____________________

(2) ____________________

3. Which cerebral lobe receives impulses from the vestibular branch of CN VIII? ____________

4. Which cerebral lobe receives impulses from CN I? ____________________

5. From which two brain regions do the cranial nerves classified as "eye-movers" emerge? ________

____

B. Concept Mapping

1. Fill in the blanks to complete this concept map outlining the anatomy of a spinal nerve.

anterior rami	anterior roots	plexuses	posterior roots	spinal nerves

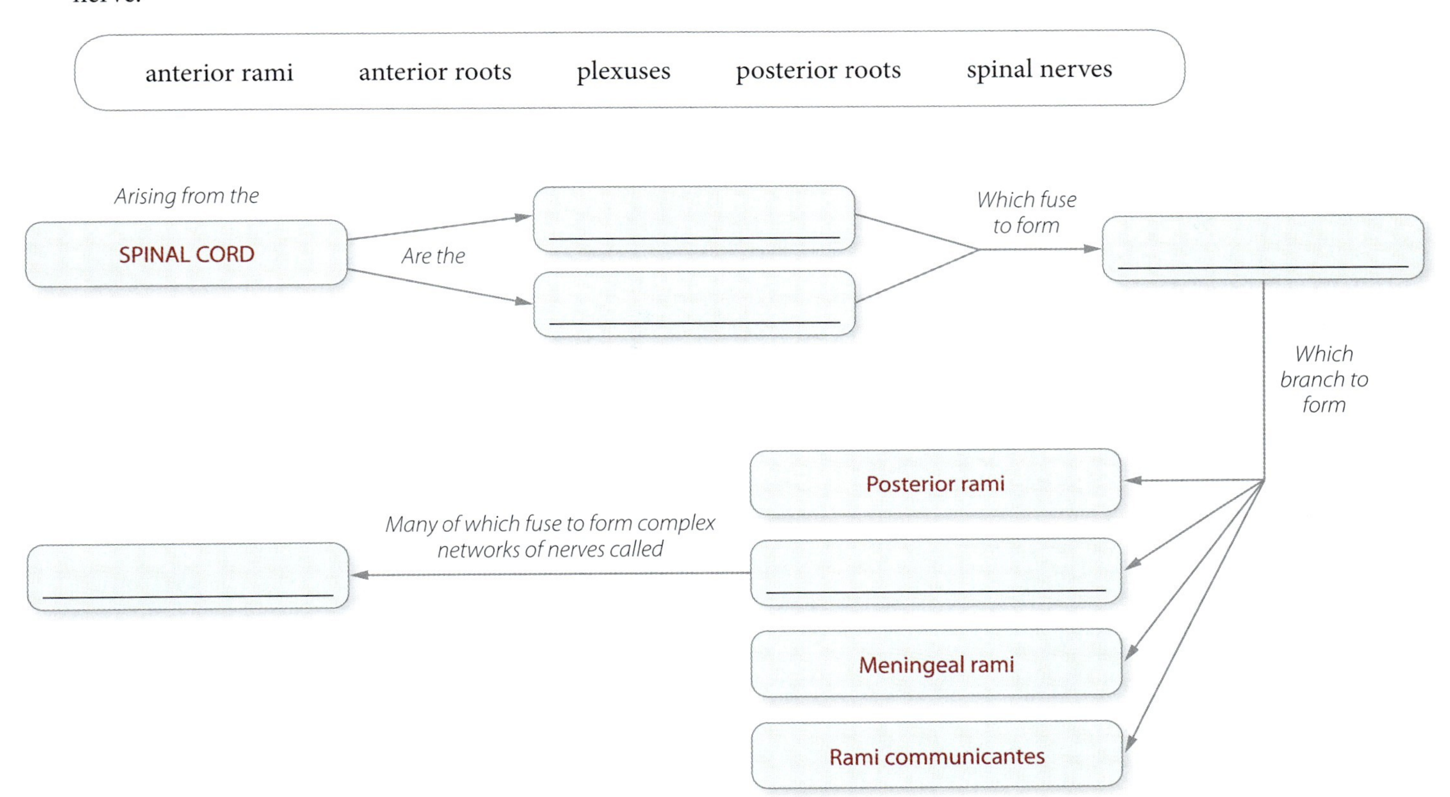

2. Construct a unit concept map to show the relationships among the following set of terms. Include all of the terms in your diagram. Your instructor may choose to assign additional terms.

accessory nerve	anterior rami	anterior roots	axillary nerve	brainstem
foramina	hypoglossal nerve	median nerve	plexuses	posterior rami
posterior roots	radial nerve	skeletal muscle	spinal nerves	trochlear nerve

17 General Senses

UNIT OUTLINE

In this unit we investigate **sensory receptors**, structures that respond to changes, or stimuli, in the external environment and within the body. Most sensory receptors—those that respond to touch, pressure, cold, heat, pain, stretch, vibration, and changes in body position—are widely distributed throughout the body and are classified as general sensory receptors. **Somatic sensory receptors** are found in the skin and in musculoskeletal organs, whereas **visceral sensory receptors** are located in visceral organs. Sensory receptors for the special senses, by contrast, are housed in specialized sense organs in the head. The special senses, which include gustation (taste), olfaction (smell), vision, hearing, and equilibrium, will be our focus in Unit 18.

THINK ABOUT IT

Which brain structure serves as a relay station for most sensory impulses?

____

Sensory impulses originating from general sensory receptors in the skin are relayed to which functional area of the cerebral cortex? ____________________

What is the difference between sensation and perception and where does each occur?

____

____

PRE-LAB ASSIGNMENTS

Pre-lab quizzes are also assignable in MasteringA&P®

To maximize learning, BEFORE your lab period carefully read this entire lab unit and complete these pre-lab assignments using your textbook, lecture notes, and prior knowledge.

PRE-LAB Activity 1: Identifying General Sensory Receptors

1. Use the list of terms provided to label the accompanying illustration of cutaneous receptors. Check off each term as you label it.

   ☐ lamellated corpuscle

   ☐ tactile corpuscle

   ☐ free nerve endings

   ☐ Ruffini ending

   a ____________________

   b ____________________

   c ____________________

   d ____________________

2. Fill in the blanks with the appropriate term(s):

   ____________________ a. light touch receptor located in the papillary layer of the dermis

   ____________________ b. deep pressure receptor

   ____________________ c. light touch receptor located in the epidermis

   ____________________ d. receptors that detect pain and temperature

PRE-LAB Activity 2: Examining the Microscopic Structure of General Sensory Receptors

1. Where specifically in the skin are tactile (Meissner) corpuscles located? ____________________

   ____

2. Where specifically in the skin are lamellated (Pacinian) corpuscles located? ____________________

   ____

PRE-LAB Activity 3: Performing a Two-Point Discrimination Test

1. Which instrument will be used to determine the ability of a subject to discriminate between two points?

   ____

2. What is the relationship between two-point receptor density and tactile sensitivity? ____________________

   ____

General Sensory Receptors

Sensory receptors respond to various stimuli, both external and internal, and they are often classified according to the source of the stimulus. These receptors act as transducers—that is, they convert stimuli into afferent nerve impulses. Because action potentials, once generated, are identical, the body distinguishes among different stimuli entirely by the region of the cerebral cortex that receives the sensory input originating from the receptor.

Sensory receptors can be classified in various ways. One classification method groups sensory receptors according to the type of stimulus to which they respond. Based on this criterion, sensory receptors can be divided into five groups: **mechanoreceptors**, which respond to touch or pressure; **chemoreceptors**, which respond to chemicals; **photoreceptors**, which respond to light; **nociceptors**, which respond to intense thermal, chemical, or mechanical stimuli that result in pain; and **thermoreceptors**, which respond to temperature.

Sensory receptors can also be classified based on the source of the stimulus. **Exteroceptors**, which respond to stimuli from the external environment, are usually located near the body surface. Examples are the sensory receptors found in the skin and the specialized sensory receptors for vision and hearing located in the eye and ear, respectively. **Interoceptors** (also called visceroceptors) respond to stimuli within the body. Examples include chemoreceptors (such as those located in the medulla oblongata that respond to blood carbon dioxide levels), stretch receptors in the walls of hollow organs such as the urinary bladder, and baroreceptors in arterial walls that respond to changes in blood pressure.

A third type of receptor based on stimulus type is proprioceptors. Like interoceptors, **proprioceptors** respond to internal stimuli, but these receptors are located only in musculoskeletal organs and their associated connective tissue coverings. Two types of proprioceptors—muscle spindles and Golgi tendon organs—respond to changes in muscle length and in joint position and movement, respectively.

Receptors in the skin, or cutaneous receptors (**Figure 17-1**), are of two general types: free nerve endings and encapsulated receptors. **Free nerve endings**, which are the dendrites of sensory neurons, are the least specialized receptors; they respond primarily to temperature and pain. Some free nerve endings are associated with epidermal cells and are called **Merkel cell fibers** (or tactile discs); others are wrapped around a hair follicle (found only in thin skin) and are called **hair follicle receptors** (or root hair plexuses, not shown). Both Merkel cell fibers and hair follicle receptors respond to light touch. Encapsulated receptors are nerve endings that are encapsulated by connective tissue cells. **Tactile corpuscles** (also called Meissner corpuscles) are located in the dermal papillae and function as light touch receptors. **Lamellated corpuscles** (also called Pacinian corpuscles) are located deeper in the reticular layer of the dermis and respond to deep pressure. **Ruffini endings** (also called bulbous corpuscles) lie in the dermis, hypodermis, and joint capsule and respond to deep and continuous pressure.

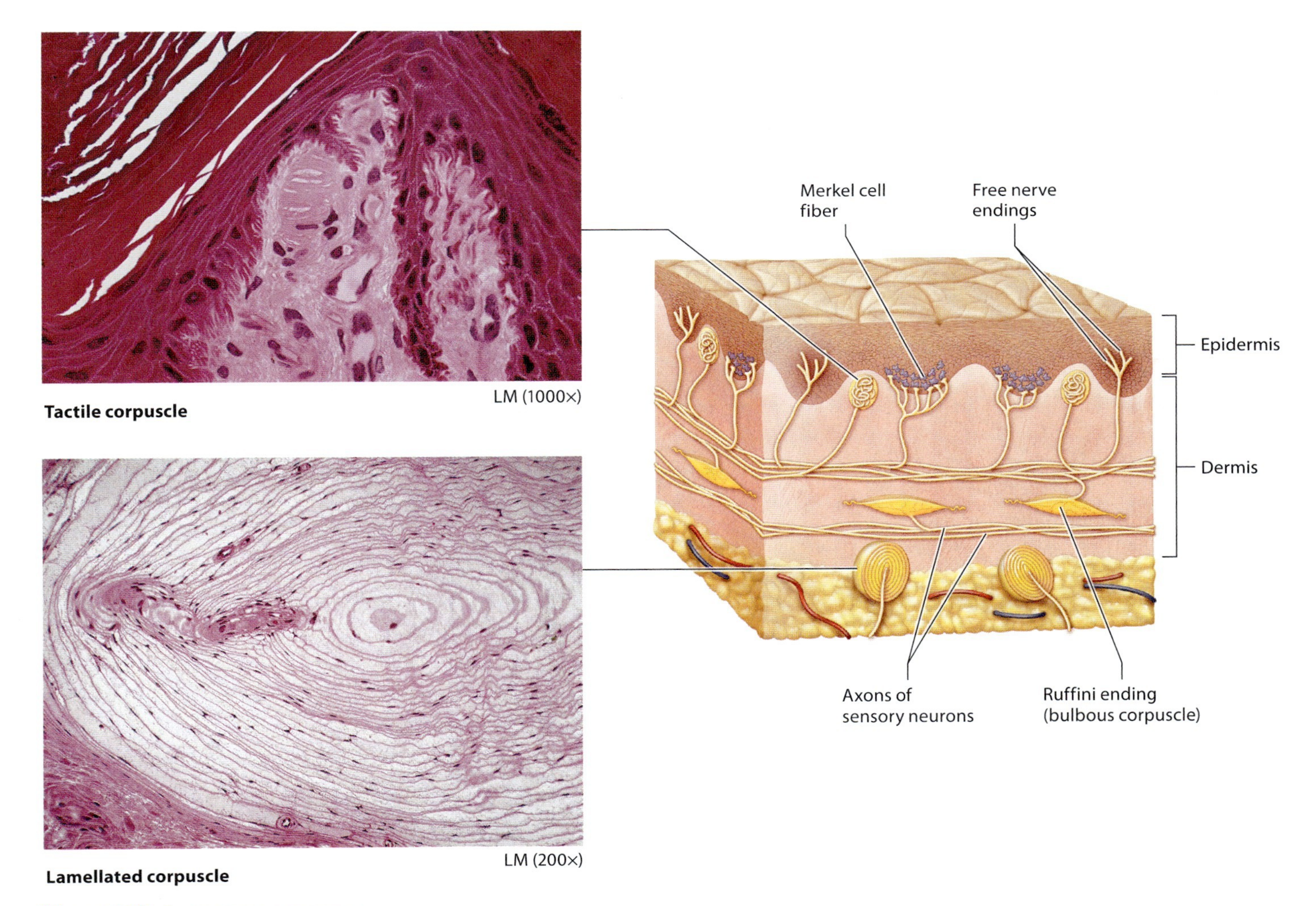

Figure 17-1 Cutaneous receptors.

ACTIVITY 1

Identifying General Sensory Receptors

Learning Outcomes

1. Identify cutaneous receptors—free nerve ending, lamellated corpuscle, hair follicle receptor, Ruffini ending, tactile corpuscle, and Merkel cell fiber—on a skin model or anatomical chart, and then describe the location and function of each.
2. Identify proprioceptors (Golgi tendon organ and muscle spindle) on an anatomical model or chart, and then describe the location and function of each.

Materials Needed

- ☐ Skin model and anatomical charts
- ☐ Muscle model or anatomical charts
- ☐ Colored pencils

Instructions

1. CHART Complete the following chart as you identify each of the types of sensory receptor on an anatomical model (or on an anatomical chart). Information for the first structure has been filled in for you.

Sensory Receptor	Location	Stimulus Type Received	In What Way(s) Can This Receptor Be Classified?
Free nerve ending	*Skin (epidermis and dermis)*	*Temperature; pain; light touch*	*Exteroceptor: responds to external stimulus* *Thermoreceptor: responds to temperature* *Nociceptor: responds to pain* *Mechanoreceptor: responds to touch*
Golgi tendon organ			
Lamellated corpuscle			
Muscle spindle			
Hair follicle receptor			
Ruffini ending			
Tactile corpuscle			
Merkel cell fiber			

2. Sketch a longitudinal section of the skin.
   a. Label and then color the following structures: epidermis, blue; dermis, green; free nerve endings, yellow; tactile corpuscle, brown; lamellated corpuscle, purple; Ruffini ending, red; hair follicle receptor, orange.
   b. Draw a black box around the labels for the receptors classified as free nerve endings.
   c. Draw a black circle around the labels for the receptors classified as encapsulated receptors.

ACTIVITY 2

Examining the Microscopic Structure of General Sensory Receptors

Learning Outcomes

1. Distinguish between a tactile corpuscle and a lamellated corpuscle in a longitudinal section of the skin.
2. Compare and contrast a tactile corpuscle and a lamellated corpuscle with respect to structure and function.

Materials Needed

- ☐ Anatomical models
- ☐ Microscope and skin sections (or photomicrographs) showing a tactile corpuscle and a lamellated corpuscle

Instructions

1. Observe a longitudinal section (or a photomicrograph) of the skin showing a tactile corpuscle under low power. Sketch and label the following structures: epidermis, dermal papilla, dermis, tactile corpuscle.

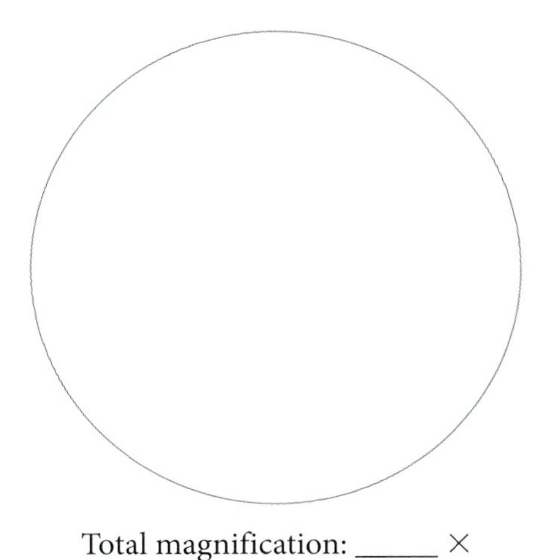

Total magnification: _____ ×

Describe the location of the tactile corpuscle.

____

Switch to the high-power objective (or view a more-magnified photomicrograph), observe the cells of the tactile corpuscle, and describe their appearance.

____

____

2. Observe a histological section (or a photomicrograph) of the skin showing a lamellated corpuscle under low power. Sketch and label the following structures: dermis, lamellated corpuscle.

Total magnification: _____ ×

Describe the location of the lamellated corpuscle.

____

Switch to the high-power objective (or view a more-magnified photomicrograph), observe the cells of the lamellated corpuscle, and describe their appearance.

____

How are the locations of the tactile corpuscle and the lamellated corpuscle related to their functions?

____

____

____

3. CHART Complete the following chart to note the similarities and differences between these two types of cutaneous receptors:

	Tactile Corpuscle	Lamellated Corpuscle
In papillary layer or reticular layer of the dermis?		
Detects light pressure or deep pressure?		
Exteroceptor or interoceptor?		
Free nerve ending or encapsulated ending?		

ACTIVITY 3 Performing a Two-Point Discrimination Test

Learning Outcomes

1. Perform a two-point discrimination test on various areas of the body.
2. Relate the results of a two-point discrimination test to receptor density.

Materials Needed

☐ Blunt-tip drawing compass with millimeter scale

Instructions

In this activity, you will use the two-point discrimination test to determine the density of tactile receptors in five different areas of the body. The greater the touch receptor density in a given area, the greater the tactile sensitivity in that area.

Before you begin the test, predict the relative density of tactile receptors by ranking the five locations in the following table from 1 to 5 (1 has the greatest receptor density, and 5 has the lowest receptor density). Then write a brief rationale for your prediction.

Location	Prediction (Ranking)
Cheek	
Palm	
Fingertip	
Posterior neck	
Anterior thigh	

Rationale for prediction: ______________________________

____

____

Now perform the following steps:

1. Designate one student in your lab group as the test subject.
2. Obtain a compass and instruct the subject to close both eyes. Starting with the compass closed all the way, gently touch the subject's cheek. Ask the subject how many points he or she can discriminate: one or two.
3. If the subject reports feeling only one point, remove the compass and open it one millimeter. Now, touch the subject's cheek again and ask how many points the subject can discriminate: one or two.
4. Continue this process until the subject reports feeling two points. Then carefully remove the compass, measure the distance between the two points in millimeters, and record the distance in the following table.
5. Repeat steps 1–4 for each of the four remaining locations.

Location	Two-Point Discrimination Distance (mm)
Cheek	
Palm	
Fingertip	
Posterior neck	
Anterior thigh	

6. Work with your lab group members to answer the following questions based on the data you collected:
   a. State which body area has the greatest density of receptors and which body area the least?

   ____

   b. Draw a conclusion with respect to two-point discrimination distance and receptor density.

   ____

   ____

POST-LAB ASSIGNMENTS

Name: ______________________ Date: ______________ Lab Section: __________

PART I. Check Your Understanding

Activity 1: Identifying General Sensory Receptors

1. Match each of the following descriptions with its receptor. More than one correct answer is possible for each question; include all correct answers.

Answers	Description	Receptor
_____ _____ _____	a. function as light touch receptors	1. hair follicle receptors
_____ _____ _____	b. are located in the dermis	2. muscle spindles
_____ _____	c. respond to stretch	3. free nerve endings
_____	d. respond to pain and temperature	4. lamellated corpuscles
_____ _____	e. function as deep pressure receptors	5. tactile corpuscles
_____	f. are also called "naked" receptors	6. Golgi tendon organs
_____	g. lie deepest in the dermis	7 Ruffini endings
_____ _____	h. are located in the epidermis	8. Merkel cell fibers
_____ _____	i. are classified as proprioceptors	
_____ _____ _____ _____ _____	j. are classified as encapsulated receptors	

Activity 2: Examining the Microscopic Structure of General Sensory Receptors

1. Identify the following cutaneous receptor, describe its specific location in the skin, and state its function.

____

____

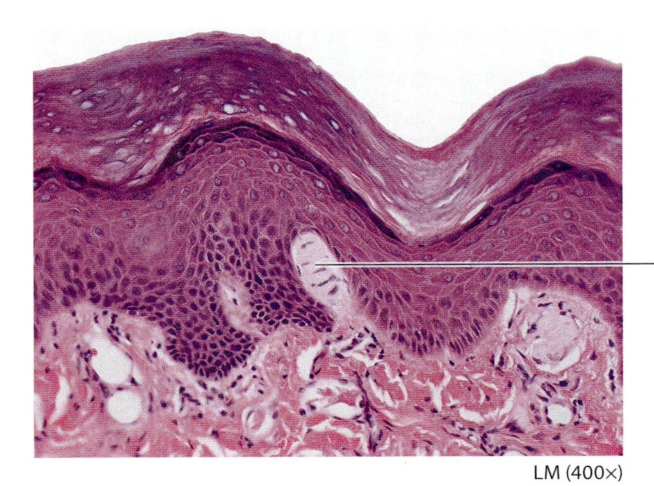

LM (400×)

2. Identify the following cutaneous receptor, describe its specific location in the skin, and state its function.

____

____

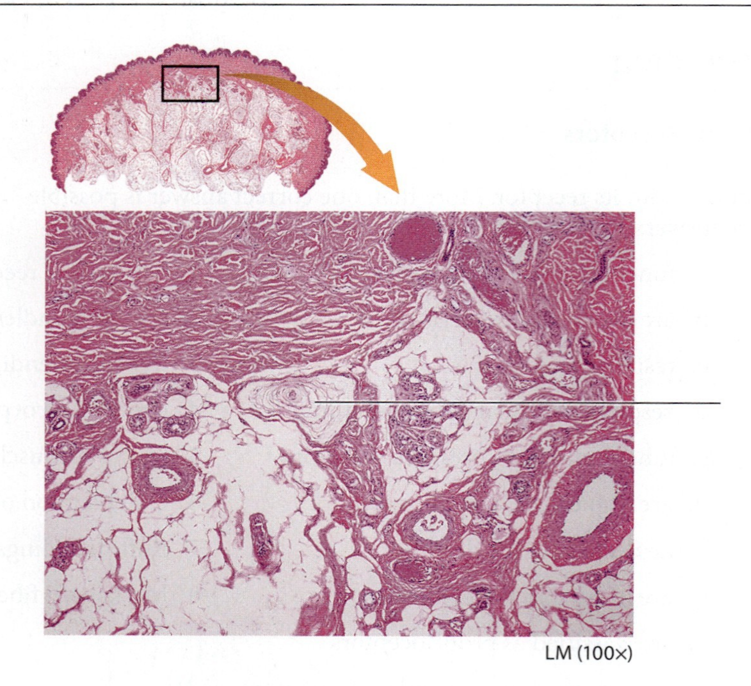

LM (100×)

Activity 3: Performing a Two-Point Discrimination Test

1. A two-point discrimination test is used to map the distribution of _______________ receptors in the skin. Which area of the skin has the highest concentration of receptors?

   ____

   What is the physiological significance of these data? _______________________

   ____

PART II. Putting It All Together

A. Review Questions

Answer the following questions using your lecture notes, your textbook, and your lab notes.

1. Explain what it means to say that sensory receptors act as transducers. _______________

   ____

2. Why is the thalamus called a "gateway" for sensory impulses? _______________

   ____

3. How does the size of the somatosensory cortex area that receives sensory input from a particular area of the body relate to cutaneous receptor density? ______________________________

____

4. Why is it important for tactile receptors to exhibit adaptation? ______________________________

____

5. Why are you not aware of sensory signals originating in proprioceptors? ______________________________

____

B. Concept Mapping

1. Fill in the blanks to complete this concept map outlining the classification of sensory receptors.

exteroceptor mechanoreceptor muscle spindle photoreceptor proprioceptor

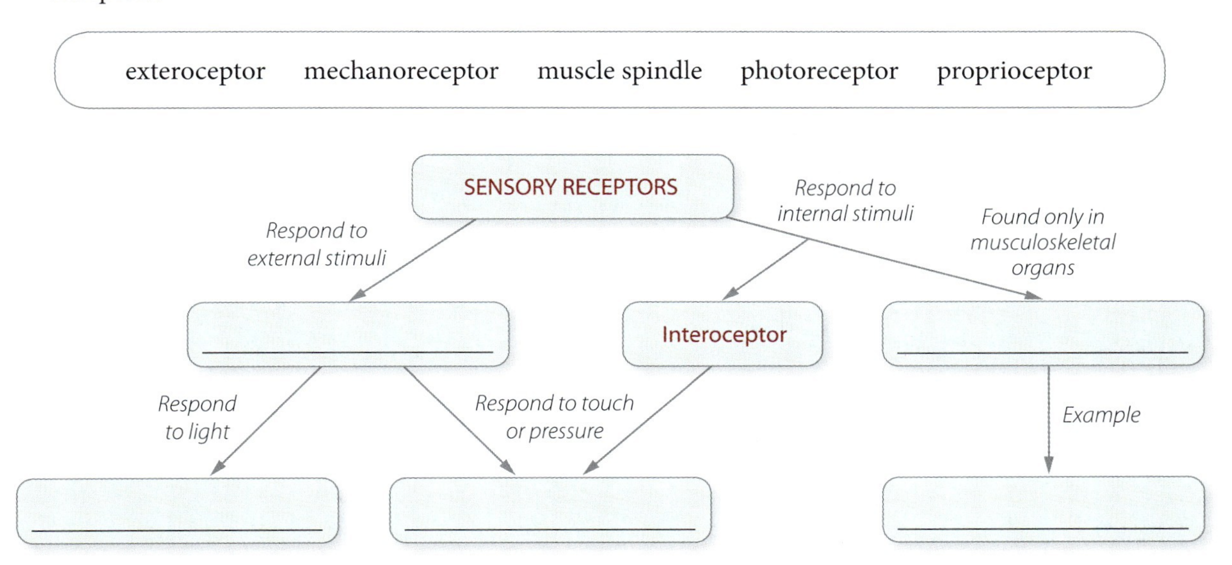

2. Construct a unit concept map to show the relationships among the following set of terms. Include all of the terms in your diagram. Your instructor may choose to assign additional terms.

cerebellum	cerebrum	encapsulated ending	epidermis	exteroceptor
free nerve ending	lamellated corpuscle	mechanoreceptor	muscle spindle	
photoreceptor	primary somatosensory cortex	proprioceptor		
tactile corpuscle	thalamus	thermoreceptor		

18 Special Senses

UNIT OUTLINE

This unit entails a detailed study of the structure and function of the body's complex **special senses**: olfaction, gustation, vision, hearing, and equilibrium. Whereas most general sensory receptors are modified nerve endings of sensory neurons, special sense receptors are highly specialized receptor cells. Sensory information from these specialized receptor cells is transmitted via complex neural pathways to specific areas of the cerebral cortex for integration and interpretation.

THINK ABOUT IT *Which cerebral lobe receives and interprets incoming sensory impulses for each of the following?*

smell ______________________________

taste ______________________________

hearing ______________________________

equilibrium ______________________________

vision ______________________________

PRE-LAB ASSIGNMENTS

Pre-lab quizzes are also assignable in MasteringA&P®

To maximize learning, BEFORE your lab period carefully read this entire lab unit and complete these pre-lab assignments using your textbook, lecture notes, and prior knowledge.

PRE-LAB Activity 1: Exploring the Gross Anatomy of Olfactory and Gustatory Structures and Demonstrating the Effect of Olfaction on Gustation

1. Olfactory and gustatory receptors are classified as which receptor type based on stimulus type received? _______________
2. The nerve fibers of the olfactory nerves pass through the _______________ foramina of the _______________ plate of the _______________ bone en route to the cerebral cortex.
3. When gustatory cells in taste buds are stimulated by dissolved chemicals, sensory messages are relayed to the medulla oblongata via which cranial nerves? _______________

PRE-LAB Activity 2: Examining the Gross Anatomy of the Eye

1. Use the list of terms provided to label the accompanying illustration of the eye. Check off each term as you label it.

☐ aqueous humor	☐ iris	☐ retina
☐ ciliary body	☐ lens	☐ sclera
☐ choroid	☐ optic nerve	☐ vitreous humor
☐ cornea	☐ pupil	

a _______________

b _______________

c _______________ (opening)

d _______________ (watery fluid)

e _______________

f _______________

g _______________

h _______________

i _______________

j _______________

k _______________ (gel-like substance)

2. Match each eye structure with its correct description:

_____ a. structure filled with aqueous humor
_____ b. structure continuous with the optic nerve
_____ c. attachment site for extrinsic eye muscles
_____ d. retina
_____ e. iris, ciliary body, and choroid
_____ f. structure that houses the photoreceptors
_____ g. structure filled with vitreous humor
_____ h. sclera and cornea

1. outer fibrous layer
2. middle vascular layer
3. inner neural layer
4. anterior cavity
5. posterior cavity

PRE-LAB Activity 3: Dissecting a Mammalian Eye

1. How does the cow eye differ from the human eye? ____

____

2. Which tissue type surrounding the eyeball must be cut away before dissecting it? ______________

PRE-LAB Activity 4: Performing Visual Tests

1. Color-blind individuals are missing one or more of which type of photoreceptor? ____________
2. ______________________ is an eye disorder caused by an irregular curvature of the cornea or lens.
3. As people age, why does their near point of accommodation increase? ____________________

____

PRE-LAB Activity 5: Examining the Gross Anatomy of the Ear

1. Use the list of terms provided to label the accompanying illustration of the ear. Check off each term as you label it.

☐ auricle
☐ cochlea
☐ external auditory canal
☐ tympanic membrane
☐ incus
☐ malleus
☐ semicircular canals
☐ stapes
☐ vestibule

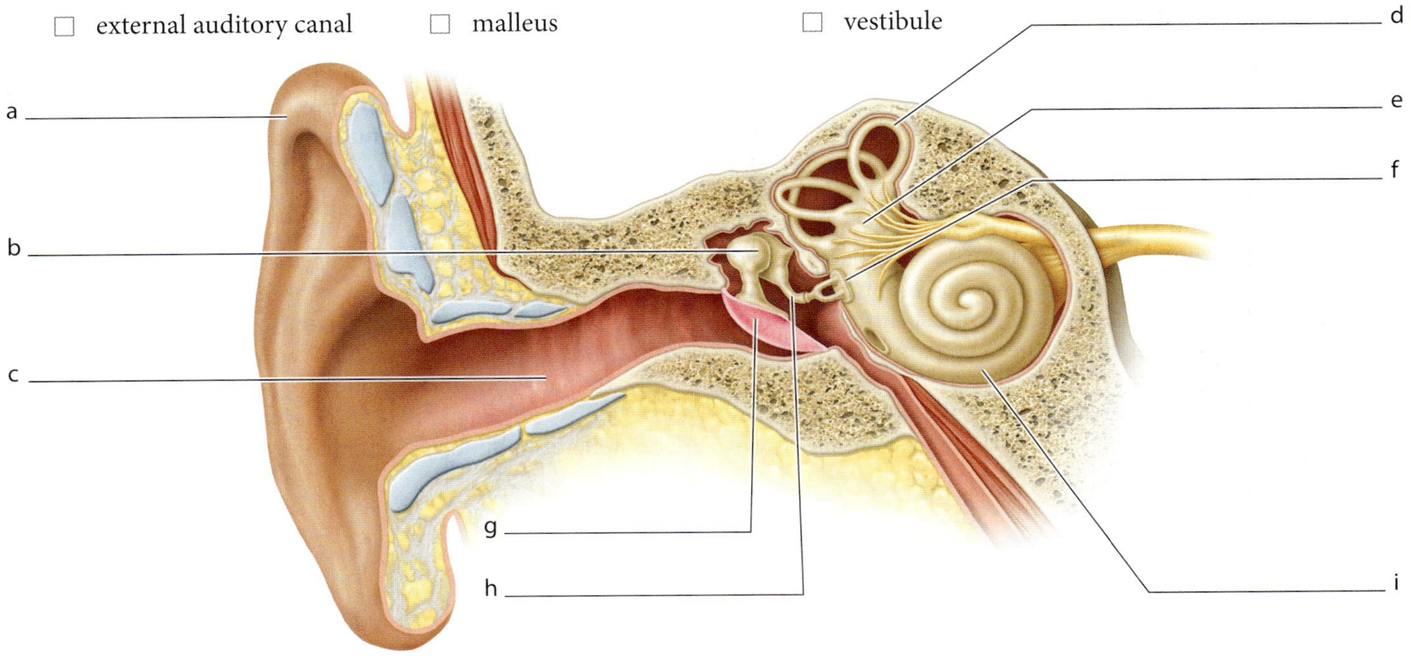

a ____________
b ____________
c ____________
d ____________
e ____________
f ____________
g ____________
h ____________
i ____________

2. Match each of the following ear structures with its description:

_____	a. malleus, incus, and stapes	1. tympanic membrane
_____	b. houses receptor organ for hearing	2. auditory ossicles
_____ _____	c. function in equilibrium	3. cochlea
_____	d. connects middle ear to nasopharynx	4. semicircular canal
_____	e. is found in middle ear	5. vestibule
_____	f. eardrum	6. auditory tube
_____ _____ _____	g. is found in the inner ear	
_____	h. snail-like structure	

3. Which of the following structures vibrates in response to sound waves?
   a. vestibulocochlear nerve
   b. tympanic membrane
   c. vestibule
   d. semicircular canal
   e. auditory tube

PRE-LAB Activity 6: Performing Hearing and Equilibrium Tests

1. The _______________ and _______________ tests are tests health care providers use to determine if hearing loss is conductive or sensorineural.
2. The _______________ test evaluates the ability of the vestibular apparatus to maintain equilibrium in the absence of visual stimuli.

Olfaction and Gustation

The **olfactory epithelium**, a specialized area located in the superior portion of the nasal cavity (**Figure 18-1a**), consists of three types of cells: olfactory neurons, basal cells, and supporting cells (see Figure 18-1b). **Olfactory neurons**, which are bipolar neurons, possess cilia that project into the mucous layer overlying the surface of the olfactory epithelium. The cilia (see Figure 18-1c) contain olfactory receptors to which chemical odorants bind as they enter the nasal cavity. **Basal cells** are stem cells that divide to produce new olfactory neurons, and **supporting cells** support and nourish the olfactory neurons (see Figure 18-1b). The olfactory epithelium contains olfactory glands that secrete mucus. Chemical odorants entering the nasal cavity must diffuse through this watery mucous layer before they can stimulate receptors on the olfactory cilia.

The axons of olfactory neurons form the **olfactory nerves** (CN I), which pass through the olfactory foramina of the cribriform plate of the ethmoid bone and synapse in the **olfactory bulbs** (see Figure 18-1b). There olfactory nerve fibers synapse with mitral cells in complex structures called glomeruli ("little balls"). When stimulated, the mitral cells trigger impulses that travel via the **olfactory tracts** to the primary olfactory cortex in the temporal lobe, where awareness and identification of odors occur.

The receptors for gustation, called **gustatory cells** (**Figure 18-2c**), are in **taste buds** located primarily on the superior surface of the tongue (see Figure 18-2a), but a few are also located on the soft palate, pharynx, epiglottis, and inner surface of the cheeks. A taste bud contains up to 100 gustatory cells, as well as basal cells and supporting cells. Each gustatory cell has a lifespan of about 10 days; basal cells divide regularly to form supporting cells, which then differentiate into new gustatory cells.

Taste buds on the tongue are located within surface projections called **papillae**, which are of four types: vallate, foliate, fungiform, and filiform. The largest, called

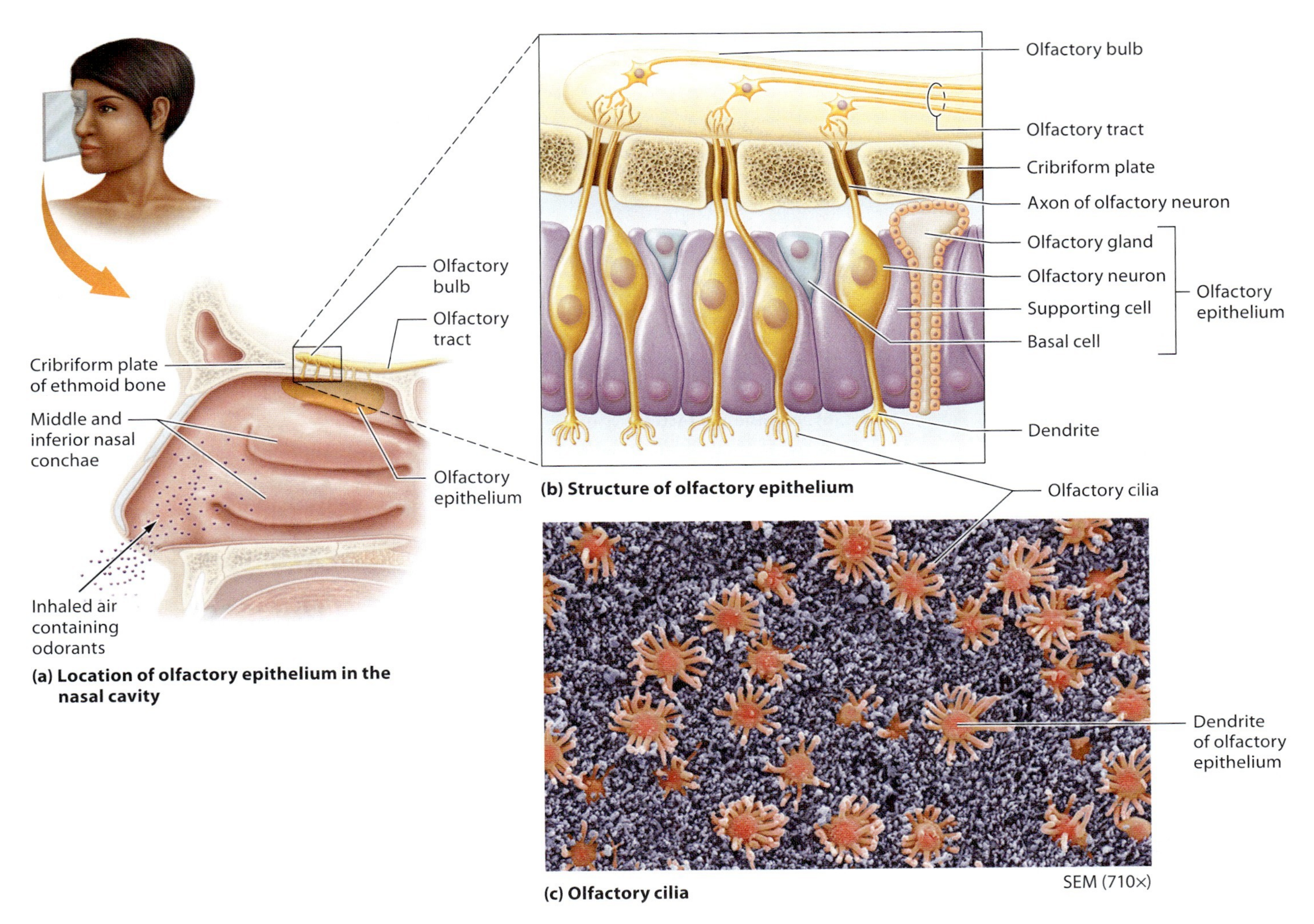

Figure 18-1 **The olfactory epithelium.**

vallate papillae, are arranged in an inverted V shape on the posterior surface of the tongue (see Figure 18-2a). **Foliate papillae**, located on the lateral surfaces of the tongue, are present mostly in children. **Fungiform papillae** (see Figure 18-2a) and **filiform papillae** (not shown) are scattered across the anterior two-thirds of the tongue's surface. Although filiform papillae are the most abundant, they contain tactile receptors instead of taste buds; these papillae give the tongue a rough texture that aids in manipulating food. When gustatory cells in taste buds are stimulated by dissolved chemicals, sensory messages are relayed to the medulla oblongata via the facial (CN VII), glossopharyngeal (CN IX), and vagus (CN X) nerves. Impulses are then transmitted to the thalamus and ultimately to the gustatory cortex located in the insula.

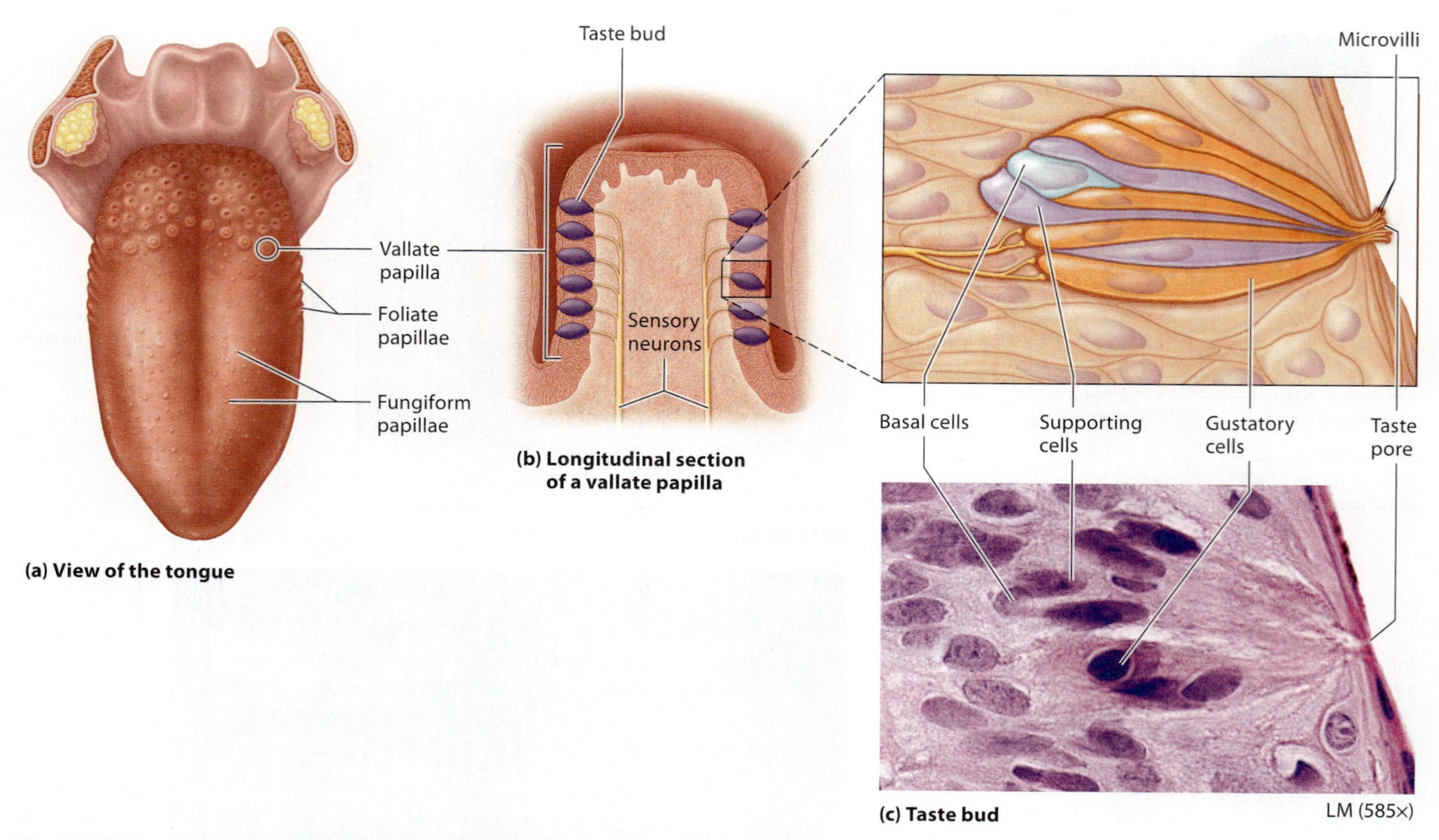

Figure 18-2 The location and structure of gustatory cells in taste buds.

ACTIVITY 1

Exploring the Gross Anatomy of Olfactory and Gustatory Structures and Demonstrating the Effect of Olfaction on Gustation

Learning Outcomes

1. Identify the olfactory structures associated with the nasal cavity, skull, and brain.
2. Identify the locations of the four types of papillae on the tongue.
3. Identify the cranial nerves that transmit sensory information from the taste buds to the cerebral cortex.
4. Demonstrate the effect of olfaction on taste by identifying unknown food samples with nostrils closed and with nostrils open.

Materials Needed

- ☐ Mirror
- ☐ Midsagittal head model
- ☐ Skull
- ☐ Brain model
- ☐ Paper towels
- ☐ Plastic container containing cubes of four different foods (provided by instructor and unknown to test subject)
- ☐ Toothpicks

Instructions

A. Olfactory and Gustatory Structures

1. Look into a mirror and identify vallate and fungiform papillae on your tongue. Sketch your tongue and indicate the locations of these two types of papillae.

2. On a brain model, identify the three cranial nerves that carry sensory information from the taste buds to the brain. List the three nerves here: ______________________

____

3. Identify each of the following olfactory structures on an appropriate anatomical model or chart, and then indicate the function of each:
   a. Olfactory epithelium ______________________
   ____
   b. Fibers of olfactory nerve ______________________
   ____
   c. Olfactory foramina ______________________
   ____
   d. Olfactory bulb ______________________
   ____
   e. Olfactory tract ______________________
   ____

4. **Optional Activity**

PAL™ Practice labeling gustatory and olfactory structures on human cadavers at > MasteringA&P® > Study Area > Practice Anatomy Lab > Human Cadaver: Nervous System > Special Senses > Images 1–5

B. Demonstrating the Effect of Olfaction on Gustation

Demonstrate the relationship between olfaction and taste by completing the following activity:

1. Before performing the experiment, make the following predictions:
   a. What will be the effect of an inability to smell on your ability to taste and identify unknown food samples? ______________________
   ____
   b. What is the basis of this prediction? ______________________
   ____
2. Work in pairs and choose one student to be the test subject.
3. Have the subject dry her or his tongue with a clean paper towel, close both eyes, and pinch both nostrils shut.
4. Using a clean toothpick, place a bit of the first food sample (sample #1) on the subject's tongue. Then ask the subject to identify the food (1) immediately—no chewing allowed!—with eyes and nostrils closed; (2) after chewing, with eyes and nostrils closed; and (3) after chewing with eyes closed and nostrils open. In each case where the subject is able to correctly identify the food, write "Yes" in the following table; if the subject could not correctly identify the food, write "No."
5. Repeat steps 3 and 4 for the remaining three food samples.

Food Sample	(1) No Chewing, Eyes and Nostrils Closed	(2) Chewing, Eyes and Nostrils Closed	(3) Chewing, Eyes Closed and Nostrils Open
#1			
#2			
#3			
#4			

6. Answer the following questions based on your results:
   a. Did your experimental results match your prediction? ______________________
   Explain: ______________________
   ____
   ____
   b. Compare and explain the effects of no chewing and chewing on the ability to taste an unknown food.
   ____
   ____
   ____

Vision

Accessory Structures of the Eye

The accessory structures of the eye (**Figure 18-3**) include the eyelids, conjunctiva, lacrimal apparatus, and the extrinsic eye muscles. The eyes are covered anteriorly by the superior and inferior eyelids (see Figure 18-3a), which meet medially and laterally at the **medial commissure** and **lateral commissure**, respectively. The medial commissure contains a small fleshy mass of tissue called the **lacrimal caruncle**, which contains modified sweat glands that release a lubricating secretion. Each eyelid contains tarsal glands (see Figure 18-3b), modified sebaceous glands that produce an oily secretion that keeps the eye's surfaces lubricated. The conjunctiva, a thin mucous membrane, lines the internal surface of the eyelids (the **palpebral conjunctiva**) and then folds to cover much of the anterior surface of the eyeball (the **bulbar conjunctiva**). The conjunctiva secretes mucus that lubricates the surface of the eyeball.

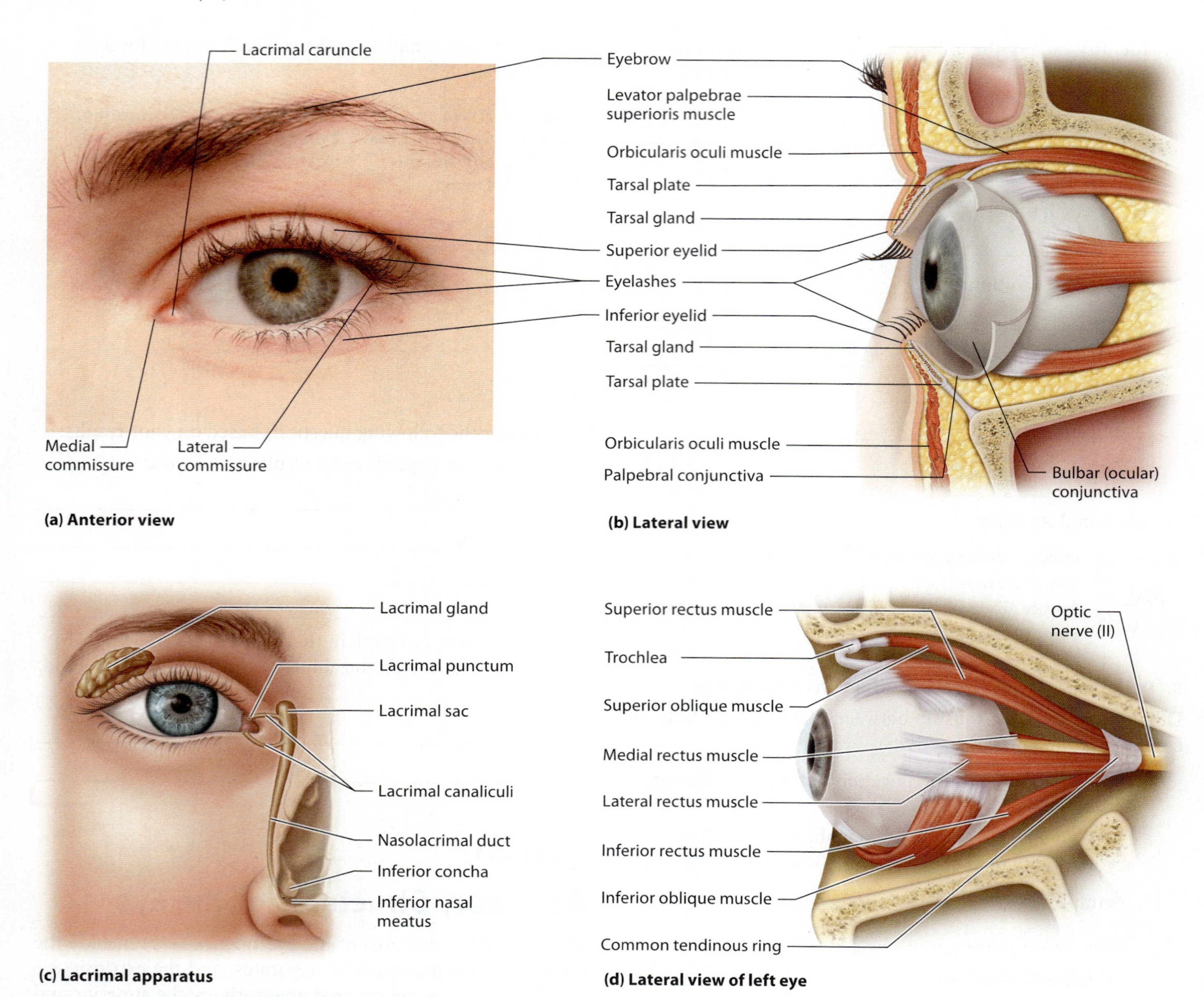

Figure 18-3 Accessory structures of the eye.

The lacrimal apparatus (see Figure 18-3c) consists of the lacrimal gland and its accessory structures. The **lacrimal gland** secretes watery, alkaline tears containing an antibacterial enzyme called lysozyme. The tears clean, lubricate, and protect the eye. Tears travel across the surface of the eye and drain through small pores called **lacrimal puncta** (singular = punctum) into the **lacrimal canaliculi** (singular = canaliculus) and then into the **lacrimal sac**. The **nasolacrimal duct** receives tears from the lacrimal sac and transports them into the nasal cavity.

The actions and innervation of each of the six extrinsic eye muscles (see Figure 18-3d)—skeletal muscles that move the eyeball—are summarized in **Table 18-1**.

Table 18-1 Extrinsic Eye Muscles

Extrinsic Eye Muscle	Action	Innervation
Inferior oblique	Elevates the eye and rotates it laterally.	CN III
Inferior rectus	Depresses the eye.	CN III
Medial rectus	Moves the eye medially.	CN III
Lateral rectus	Moves the eye laterally.	CN VI
Superior oblique	Depresses the eye and rotates it laterally.	CN IV
Superior rectus	Elevates the eye.	CN III

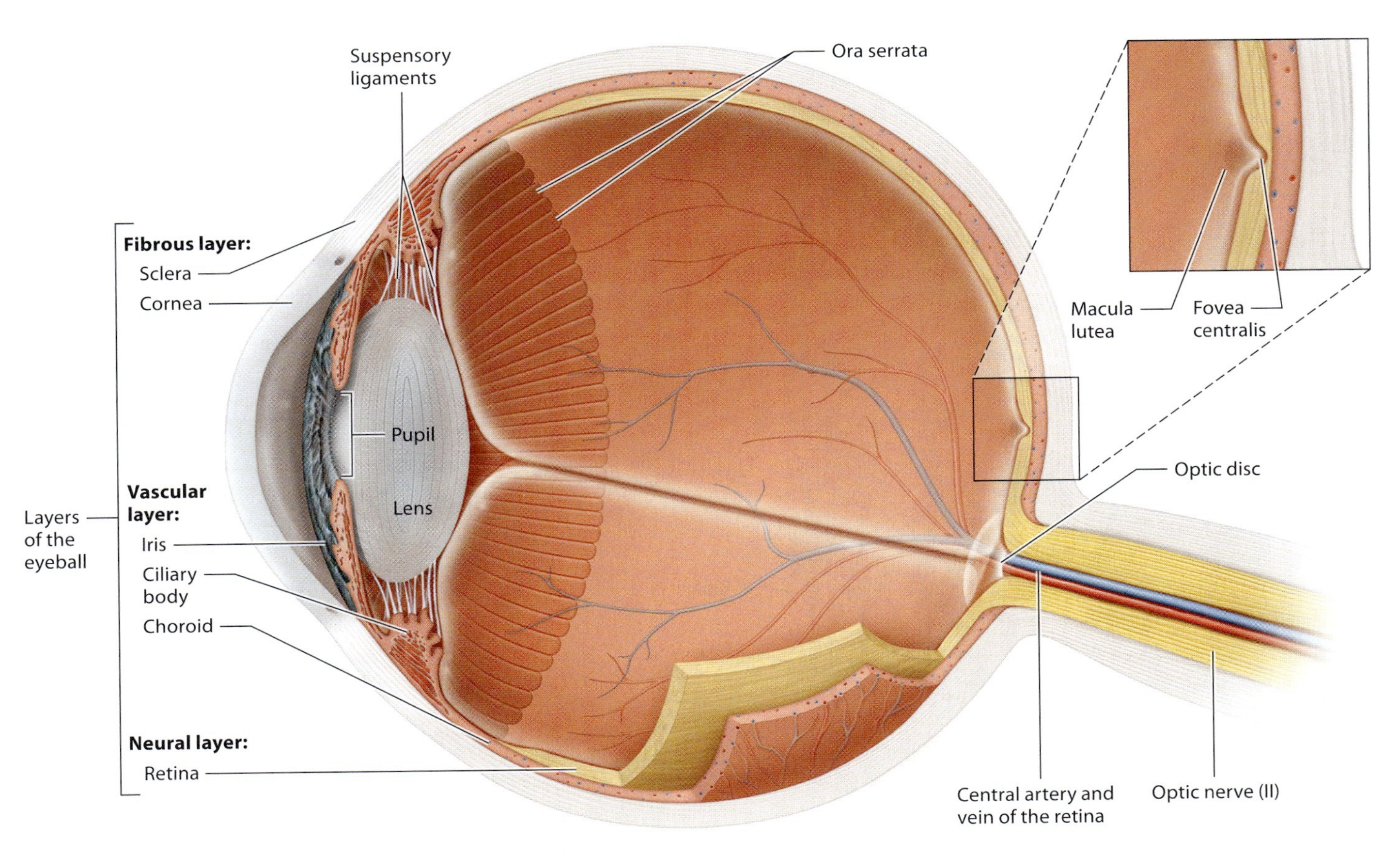

Figure 18-4 **The internal structure of the eye (sagittal section).**

Structure of the Eyeball

The eye (**Figure 18-4**) is a complex sensory organ composed of three layers: the outer fibrous layer, the middle vascular layer, and the inner neural layer.

The **fibrous layer** consists of the cornea and the sclera. The transparent **cornea**, the anterior one-sixth of the fibrous layer, bends light as it enters the eye. The **sclera**—the white of the eyeball—makes up the posterior five-sixths of the fibrous layer; it protects the eye and serves as an attachment site for the extrinsic eye muscles.

The **vascular layer** consists of the choroid, ciliary body, and iris. The **choroid** is a highly vascular structure that supplies oxygen and nutrients to the cells of the eye. It contains an abundance of melanocytes, cells that produce the pigment melanin, which absorbs excess light. Anteriorly, the choroid coat thickens to form the **ciliary body**, a structure that regulates the shape of the lens to enable light to be focused on the retina. The ciliary body also secretes aqueous humor, the clear watery fluid that fills the anterior cavity of the eye. The most anterior portion of the vascular layer is the **iris**, which contains pigmented cells that are responsible for eye color and smooth muscle fibers that regulate the size of the pupil—and thus the amount of light that enters the eye.

The **neural layer** consists of the retina, which begins at the ora serrata, the serrated boundary between the retina and the ciliary body. The **retina** is composed of two layers: an outer pigmented epithelium that absorbs light and prevents it from scattering, and an inner neural layer that contains the photoreceptors (rods and cones). When stimulated by light, the photoreceptors trigger stimulation of the optic nerve and the transmission of impulses to the thalamus and eventually to the occipital lobe of the cerebrum. Photoreceptors are located throughout the retina, except at the **optic disc** (also called the blind spot), the point at which the optic nerve leaves the eyeball. Just lateral to the optic disc is the **macula lutea**; in the center of it is the **fovea centralis**, a small area where the concentration of cones is greatest.

The retina (**Figure 18-5**) contains three major populations of cells: photoreceptors (rods and cones), bipolar cells, and retinal ganglion cells. **Rods** enable vision in low-light conditions and are important to peripheral vision; they are highly concentrated along the periphery of the retina. By contrast, **cones** are important for both visual acuity (sharpness) and color vision; their concentration increases as you move from the periphery of the retina to its center, where the macula lutea is located. The fovea centralis contains cones only. Rods and cones transmit electrical signals to **bipolar cells**, which

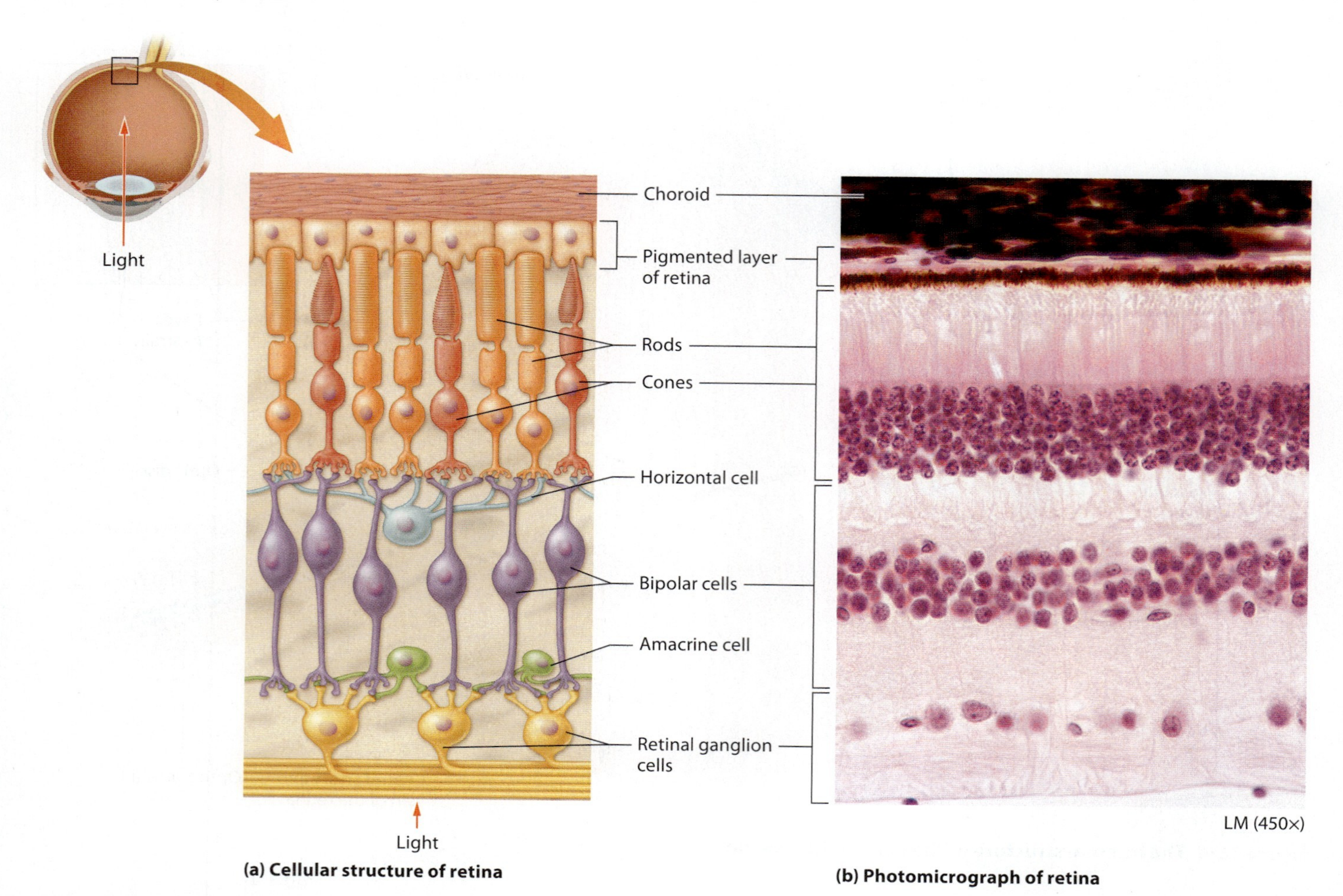

(a) Cellular structure of retina

(b) Photomicrograph of retina

Figure 18-5 The retina.

synapse with the **retinal ganglion cells**. The axons of the ganglion cells fuse to form the optic nerve.

Figure 18-6 shows the structures within the internal cavities and chambers of the eye. The **lens**, a biconcave structure that bends and focuses incoming light on the retina, is composed of highly specialized lens fiber cells packed with transparent proteins called crystallins. The lens is attached to the ciliary body via the suspensory ligament.

The lens separates the eye into anterior and posterior cavities. The **anterior cavity** is subdivided by the iris into anterior and posterior chambers. The ciliary body continuously secretes aqueous humor into the posterior chamber. This watery fluid, which is formed as a filtrate of the blood from capillaries in the ciliary body, flows through the pupil into the anterior chamber and then drains into the scleral venous sinus, an enlarged vessel that returns the aqueous humor to the blood. **Aqueous humor** contributes to intraocular pressure and provides oxygen and nutrients to the avascular cornea and lens. The **posterior cavity** is filled with vitreous humor, a gel-like substance formed prior to birth. The **vitreous humor** provides internal support for the posterior part of the eyeball and helps keep the retina pushed firmly against the vascular choroid.

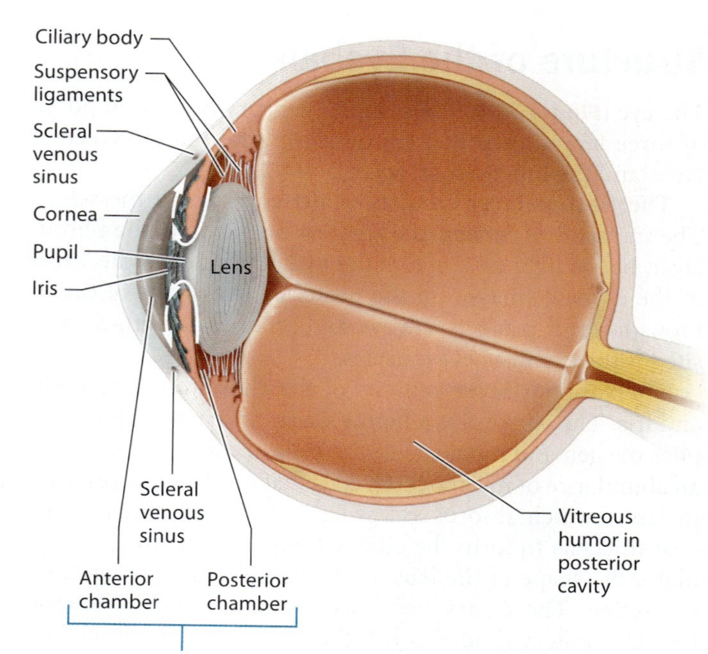

Anterior cavity: Anterior and posterior chambers form the anterior cavity and are filled with aqueous humor. The arrows show the flow of aqueous humor before it drains into the scleral venous sinus.

Figure 18-6 Cavities and chambers of the eye.

ACTIVITY 2

Examining the Gross Anatomy of the Eye

Learning Outcomes

1. Identify the accessory structures of the eye on an eye model, and describe the function of each.
2. Locate each extrinsic eye muscle on an eye model; then describe each muscle's action and the cranial nerve that innervates it.
3. Identify the lens and the components of each of the layers of the eye on a model, and describe the function of each structure.
4. Distinguish between the cavities and chambers of the eye on a model, and describe what is found within each.

Materials Needed

- ☐ Eye model (or anatomical chart)

Instructions

CHART Locate each of the structures in the following Making Connections charts on an eye model or anatomical chart. Then review its description and/or function(s) and make "Connections to Things I Have Already Learned" based on your lectures, assigned reading, and lab. Parts of each chart have been filled out for you.

Making Connections: Accessory Structures of the Eye

Structure	Description (Structure and/or Function)	Connections to Things I Have Already Learned
Lacrimal gland		
Lacrimal canaliculi		*Canaliculi = "little canals"; are similar to canaliculi in compact bone, which house the cytoplasmic extensions of osteocytes.*
Lacrimal sac		*Sits in the lacrimal fossa of the skull.*
Nasolacrimal duct	*Receives tears from the lacrimal sac and drains them into the nasal cavity.*	
Palpebrae (eyelids)		
Medial and lateral commissures		*Commissural tracts are bundles of nerve fibers in the CNS that connect the right and left cerebral hemispheres; the medial and lateral commissures "connect" the upper and lower eyelids.*
Lacrimal caruncle		
Conjunctiva	*Thin mucous membrane lining the internal surface of the eyelids and the anterior surface of the eyeball*	
Conjunctival sac		

Making Connections: Extrinsic Eye Muscles

Muscle	Action	Connections to Things I Have Already Learned
Inferior oblique		*Oblique = muscle fibers run diagonally; innervated by CN III (oculomotor nerve).*
Inferior rectus		

(continued)

Making Connections: Extrinsic Eye Muscles (*Continued*)

Muscle	Action	Connections to Things I Have Already Learned
Medial rectus		
Lateral rectus	*Moves eye laterally.*	
Superior oblique		*Passes through the trochlea, a loop suspended from the frontal bone; innervated by CN IV (trochlear nerve).*
Superior rectus	*Elevates eye and moves it medially.*	

Making Connections: The Structure of the Eyeball

Structure	Description (Structure and/or Function)	Connections to Things I Have Already Learned
Cornea		
Sclera		
Choroid		*Contains abundant melanocytes; melanin in melanocytes absorbs stray light; melanin is also produced by melanocytes in the epidermis.*
Ciliary body	*Produces aqueous humor; controls shape of lens.*	
Iris		*Pupillary reflex is controlled by the ANS, which acts on smooth muscle of the iris to control the size of the pupil.*
Retina	*Inner neural layer containing rods and cones (photoreceptors)*	
Lens		*Composed of anucleate cells packed with transparent proteins called crystallins; RBCs are also anucleate and are packed with hemoglobin for the transport of oxygen.*
Anterior cavity	*Located between the cornea and lens; divided into anterior and posterior chambers; filled with aqueous humor*	
Posterior cavity		
Optic nerve		*Continuous with retina; leaves eye at the optic disc (or "blind spot" because it lacks photoreceptors); optic nerves cross over to form the optic chiasma.*

Optional Activity

Practice labeling eye structures on human cadavers at > MasteringA&P® > Study Area > Practice Anatomy Lab > Human Cadaver: Nervous System > Special Senses > Images 8–11

ACTIVITY 3

Dissecting a Mammalian Eye

Learning Outcomes

1. Dissect a cow eye.
2. Locate on a dissected cow eye and describe the function of each of the following structures: extrinsic eye muscle, optic nerve, sclera, cornea, pupil, lens, iris, ciliary body, choroid, retina, optic disc, vitreous humor, and tapetum lucidum.

Materials Needed

☐ Fresh or preserved cow eye
☐ Dissection tray
☐ Dissection instruments (scissors, scalpel)
☐ Disposable gloves
☐ Safety goggles

Instructions

1. Put on disposable gloves and safety goggles.
2. Obtain a dissecting tray, dissecting instruments, and a cow eye. Refer to **Figure 18-7** while dissecting the eye and identifying eye structures.
3. Use scissors to carefully remove excess adipose tissue surrounding the eyeball, and identify the optic nerve.

   The optic nerve is continuous with which layer of the eyeball? ______________________

4. Identify the sclera and also the cornea, which is normally transparent but it will appear cloudy in a preserved specimen.

   List two functions of the cornea. ______________________

   ____

   List two functions of the sclera. ______________________

   ____

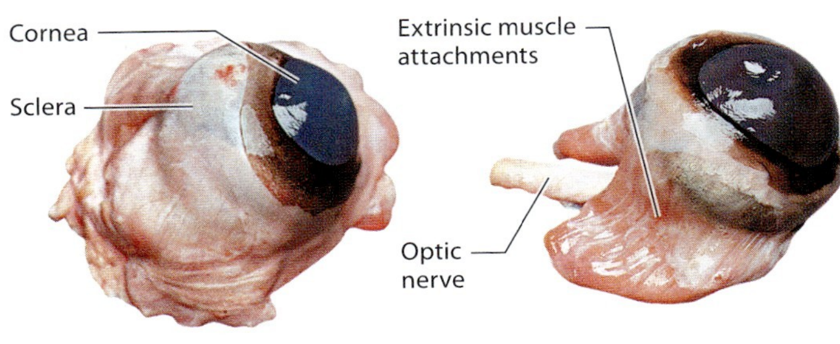

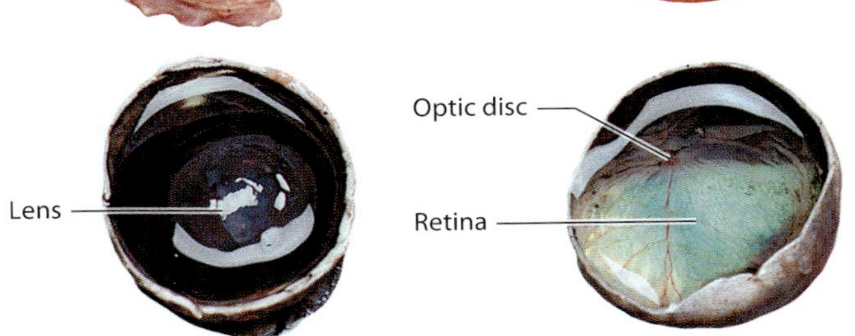

Figure 18-7 **A cow eye.**

5. Now, use a sharp scalpel to carefully pierce the eyeball approximately one-quarter inch posterior to the cornea.

   Safety Note: Wear safety goggles! Because substantial pressure must be applied to the sclera to penetrate it, fluid is likely to squirt out when you cut into the eye.

6. Insert scissors into the opening you made and cut all the way around the eye in the coronal plane, separating it into anterior and posterior parts (see Figure 18-7). As you identify each of the following structures on the anterior part of the eyeball, check the adjacent box and then write a brief description of it.

   ☐ Lens ______________________

   ☐ Ciliary body ______________________

   ☐ Iris ______________________

   ☐ Pupil ______________________

7. As you identify each of the following structures on the posterior part of the eyeball, check the adjacent box and then write a brief description of it. The tapetum lucidum, which is absent in the human eye, is an iridescent blue reflective surface in the eyes of animals that is active in low-light conditions.

   ☐ Vitreous humor ______________________

   ☐ Retina ______________________

   ☐ Optic disc ______________________

   ☐ Tapetum lucidum ______________________

   ☐ Choroid ______________________

   How does the choroid in the cow eye differ from the choroid in the human eye? ______________________

   ____

8. **Optional Activity**

Practice labeling structures of a cow eye at > MasteringA&P® > Study Area > Practice Anatomy Lab > Cat > Nervous System

ACTIVITY 4

Performing Visual Tests

Learning Outcomes

1. Demonstrate the blind spot, and explain the anatomical basis for it.
2. Test your visual acuity.
3. Define astigmatism, and perform an astigmatism test.
4. Perform a test for color blindness.
5. Determine your near point of accommodation, and compare your results to the results expected for your age group.

Materials Needed

- ☐ Snellen eye chart
- ☐ Ishihara color plates
- ☐ Dissecting pin
- ☐ Meter stick

Instructions

A. Demonstrating the Blind Spot

1. Hold **Figure 18-8** about 45 cm (18 inches) from your eyes.
2. Follow the directions in the figure to demonstrate the blind spot.

   What causes the X to disappear? ______________________

   ____

B. Testing Visual Acuity

1. Designate one of the members of your lab group to be the subject. If the subject normally wears eyeglasses, perform the test twice: without glasses and then again with glasses.
2. Instruct the subject to stand 6 m (20 feet) from a posted Snellen eye chart and to cover the left eye with the left hand.
3. Instruct the subject to read each consecutive line of the eye chart aloud.
4. Note the line with the smallest print for which the subject made no errors. Record in the blanks in step 6 the numbers associated with this line, which represent visual acuity.
5. Repeat steps 2–4 with glasses.
6. Repeat steps 2–4 covering the right eye, with and without glasses. Note the visual acuity of:

   right eye without glasses ______________________

   right eye with glasses ______________________

   left eye without glasses ______________________

   left eye with glasses ______________________

7. Vision designated 20/20 is considered normal. Individuals with 20/20 vision see the same details at 20 feet that most other people can see at the same distance.

   What would be the definition of 20/200 vision? ________

   ____

C. Testing for Astigmatism

Astigmatism is an eye disorder in which an irregular curvature of the cornea or lens prevents light rays entering the eye from being focused properly on the retina, resulting in blurred vision. In this activity you will perform a simple test to determine if you have astigmatism.

1. Close or cover one eye and focus on the center of the astigmatism test chart in **Figure 18-9**.

(a) Find your "blind spot."

1. To test your left eye, cover your right eye and focus on the X. Slowly move your book toward (or away from) you until the dot disappears.
2. To test your right eye, cover your left eye and focus on the dot. Move the book until the X disappears.

(b) Fill in your "blind spot."

1. To test your left eye, cover your right eye and focus on the X. Slowly move your book toward (or away from) you until the pencil appears intact.
2. When the gap in the pencil hits your blind spot, areas in your brain fill in the missing information so the pencil appears intact.
3. To test your right eye, turn your book upside down, cover your left eye, and repeat the process.

Figure 18-8 **An image for demonstrating the blind spot.**

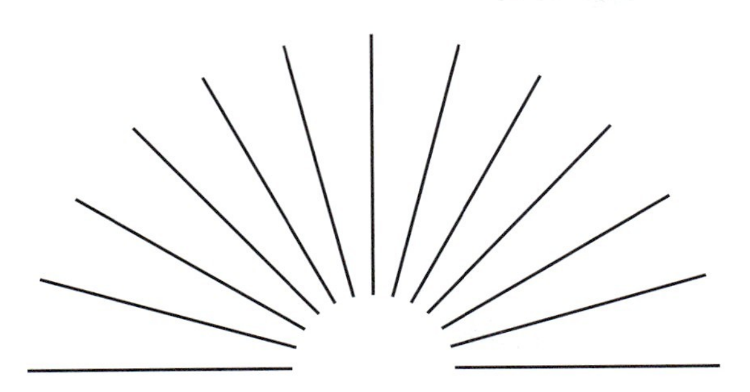

Figure 18-9 Astigmatism test chart.

2. If all the parallel lines radiating from the center appear equally distinct and dark, then you do not have astigmatism. If, however, some of the lines appear blurred or lighter than others, then some astigmatism is present.
3. Repeat steps 1 and 2 with the other eye.

Based on this test, do you have astigmatism? ___________

How is astigmatism corrected? ___________

____

D. Testing for Color Blindness

Three types of cones in the retina are responsible for color vision. Each cone type is able to absorb light at specific wavelengths. Color blindness is a sex-linked genetic trait resulting in the absence of one or more cone types. Test each student in your lab group for color blindness according to the following instructions:

1. View Ishihara color plates from a distance of 80 cm (31.5 inches) in a brightly lit room or in sunlight. Within 3 seconds of observing the plate, report to a lab partner what you see in each plate.
2. Have the lab partner compare your response to the answers provided with the color plates. The inability to correctly identify the figures in the plates indicates some degree of color blindness.
3. Report your group's results to your instructor, and then record the class-wide data in the following blanks.

Number of females with normal color vision: ___________

Number of females who are color blind: ___________

Number of males with normal color vision: ___________

Number of males who are color blind: ___________

4. Explain why color blindness is more common in males than in females. ___________

____

E. Determining Near Point of Accommodation

As people age, the elasticity of the lens in the eye decreases dramatically, making it difficult to focus for near vision. This condition, called presbyopia (literally, "old vision"), is why people need reading glasses as they age. Determine your near point of accommodation—the minimum distance at which an object can remain in focus—by performing the following steps:

1. Hold a dissecting pin vertically at arm's length in front of one eye.
2. Slowly move the pin toward the eye until it just starts to become blurry.
3. Have a lab partner measure the distance (in cm) between your eye and the pin.
4. Repeat the procedure with your other eye and record your results:

Near point of accommodation of right eye: ___________ *cm*

Near point of accommodation of left eye: ___________ *cm*

5. Your instructor will provide you with a graph so you can compare your results with those expected for someone your age.

Hearing and Equilibrium

The ear (**Figure 18-10**) is divided into three regions: the outer ear, middle ear, and inner ear. The **outer ear** consists of the **auricle**, which directs sound waves to the **external auditory canal**, which then conducts them to the **tympanic membrane**, or eardrum. This partition between the outer ear and the middle ear vibrates in response to sound waves.

The **middle ear** is an air-filled cavity within the temporal bone that extends from the tympanic membrane to the oval window of the inner ear. It contains three small bones—the auditory ossicles: the **malleus**, **incus**, and **stapes**—which transmit and amplify sound waves. The **pharyngotympanic tube** (also called the auditory tube or the Eustachian tube) connects the middle ear to the nasopharynx and equalizes pressure between the middle ear and the external environment.

The **inner ear** houses the sensory organs for both hearing and equilibrium within the labyrinth, which is composed of three regions: three semicircular canals, the snail-shaped cochlea, and the vestibule (see Figure 18-10). The outer bony labyrinth is filled with **perilymph**, whereas the inner membranous labyrinth (**Figure 18-11**) is filled with **endolymph**.

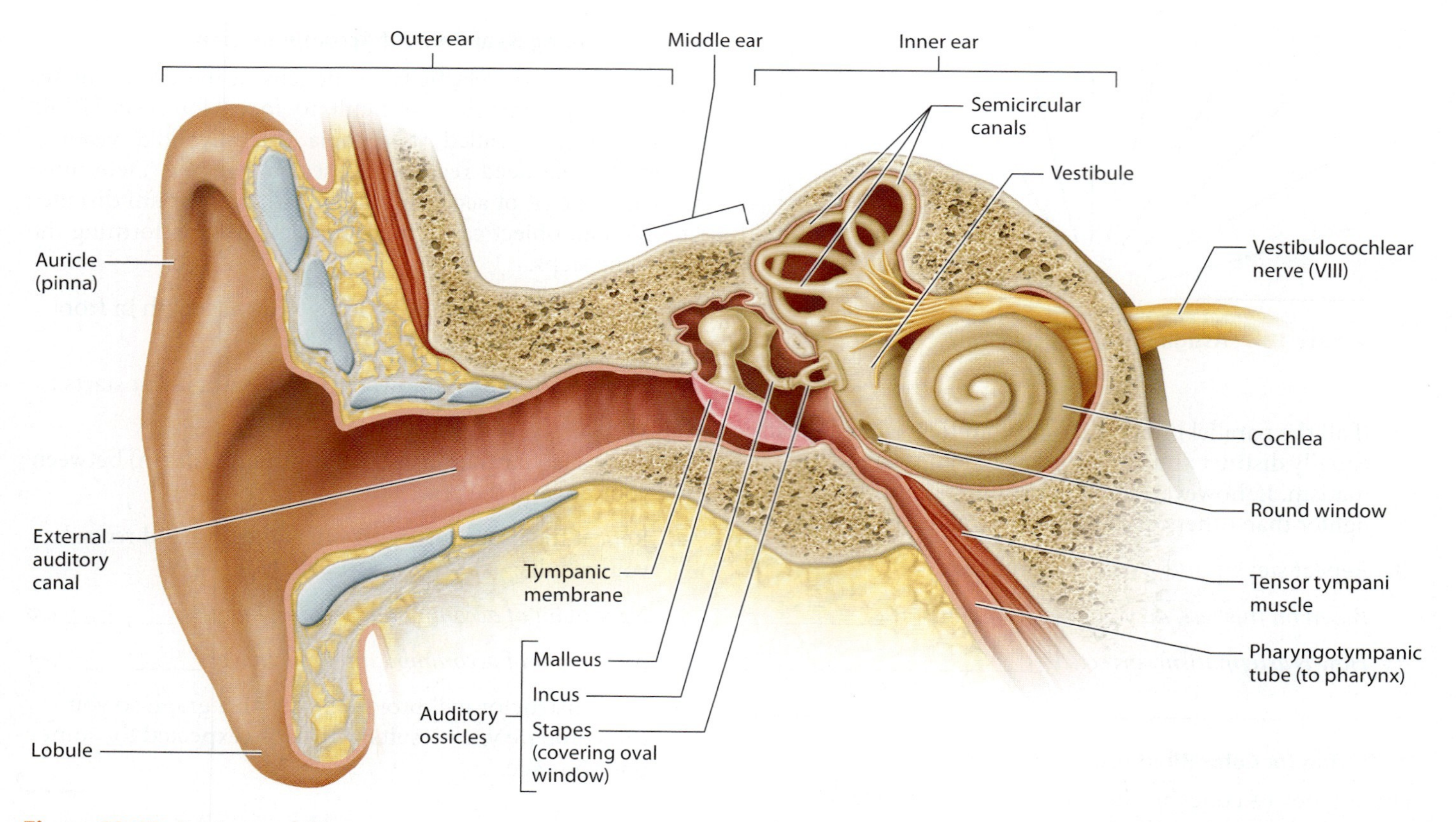

Figure 18-10 Anatomy of the ear.

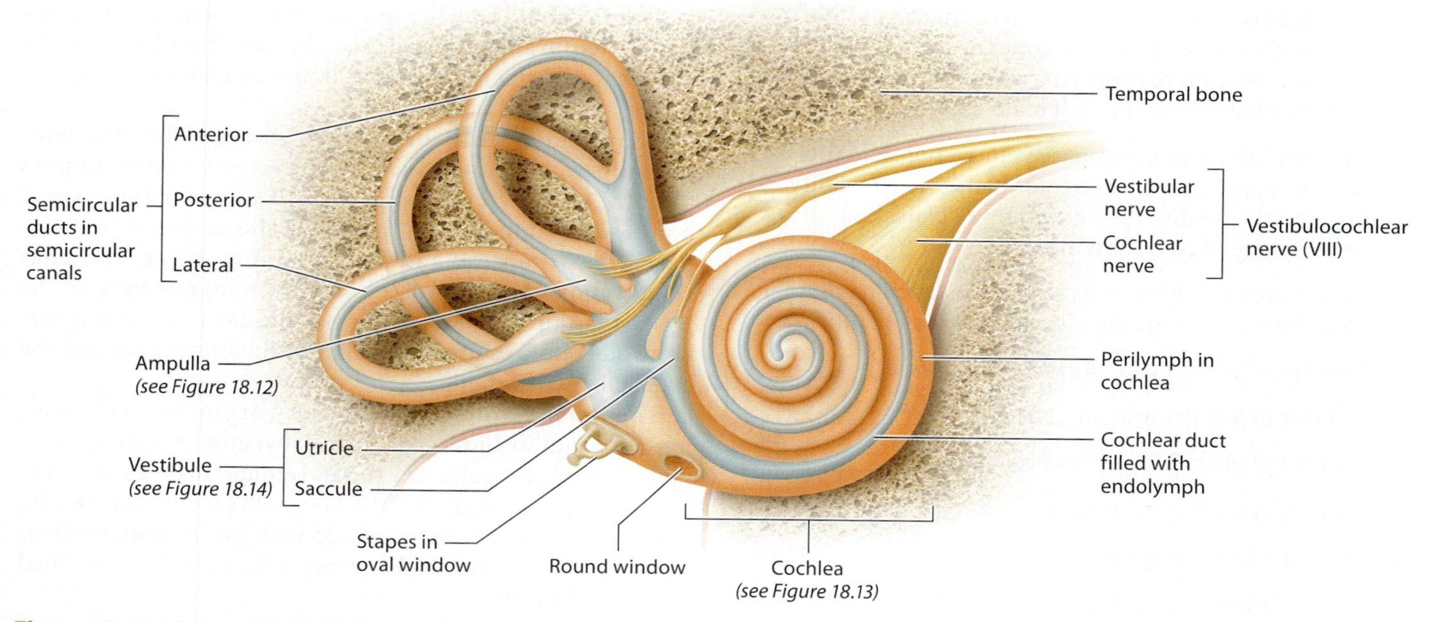

Figure 18-11 The membranous labyrinth.

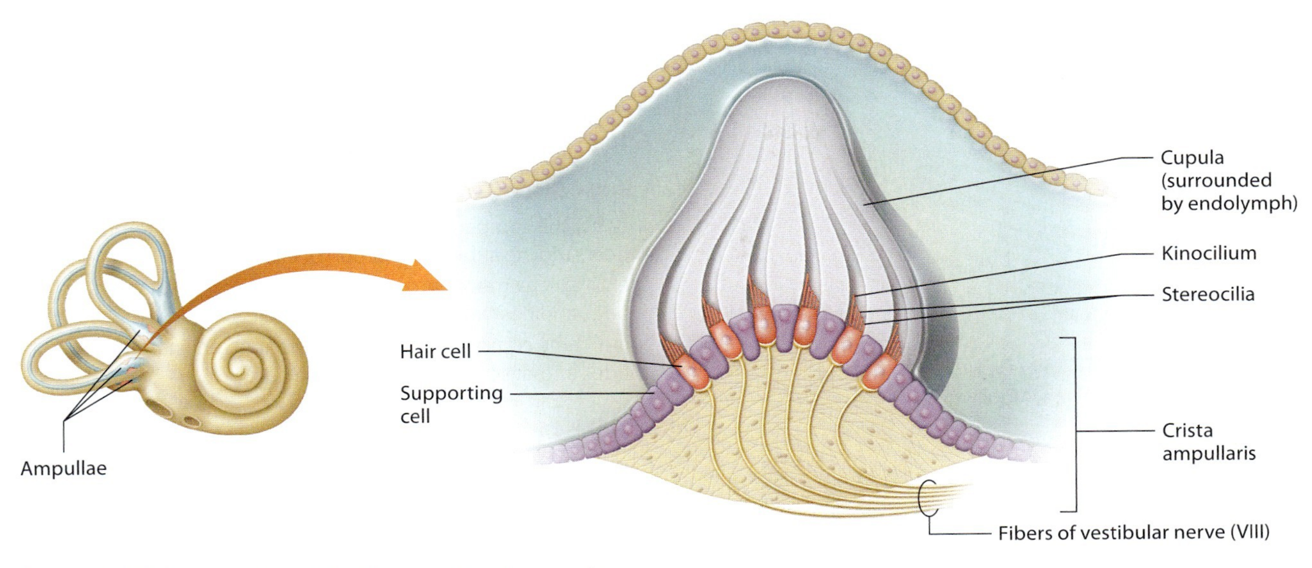

Figure 18-12 An ampulla within the semicircular canals.

Within each ear are three **semicircular canals**, each of which is expanded at its base to form an **ampulla** (**Figure 18-12**). Each ampulla houses a receptor organ called the **crista ampullaris**, which contains both hair cells and supporting cells. The stereocilia of the hair cells project upward into a gelatinous membrane called the cupula. The crista ampullaris functions in dynamic equilibrium (angular acceleration or deceleration). When the stereocilia of a hair cell rub up against the cupula, as when the head rotates, a graded potential occurs in the hair cell. If the graded potential reaches threshold, then an action potential will transmit sensory messages concerning dynamic equilibrium to the cerebral cortex via the vestibular nerve, a branch of cranial nerve VIII.

The **cochlea** (see Figure 18-11) contains the receptor organ for hearing: the spiral organ (organ of Corti). The **spiral organ** (**Figure 18-13**) sits on the basilar membrane and contains specialized hair cells with stereocilia (hairs) projecting from their apical surface into the endolymph of the cochlear duct. The tectorial membrane overlies these hair cells and comes into contact with the stereocilia of the outer hair cells. When the basilar membrane vibrates, as in response to sound waves, stereocilia of a hair cell bend in such a way that potassium channels on the hair cell open, potassium ions flow into the cell, and the hair cell is depolarized. This depolarization leads to the release of neurotransmitters that trigger an action potential in a neuron of the cochlear nerve, which joins the vestibular nerve to form the vestibulocochlear nerve (CN VIII), and a sensory message concerning sound will be sent to the brain.

The **vestibule**, which is subdivided into the **utricle** and **saccule** (see Figure 18-11), contains receptor organs called maculae. Each **macula** (**Figure 18-14**) contains hair cells and supporting cells and provides sensory information concerning static equilibrium (head position and linear acceleration). The hair cells contain "hairs" (stereocilia and kinocilia) that project upward into a gelatinous otolithic membrane

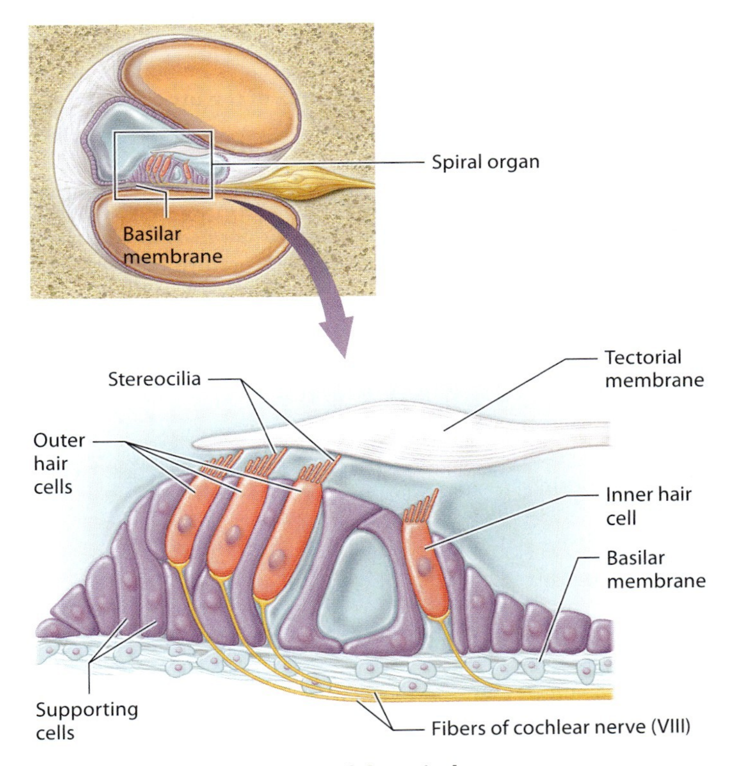

Figure 18-13 The structure of the spiral organ.

containing calcium carbonate "rocks" (called otoliths). Tilting of the head causes the hairs of the hair cells to rub against the otolithic membrane, generating a graded potential in the hair cells. If these graded potentials reach threshold, they will trigger the firing of an action potential along the fibers of the vestibular nerve, which joins the cochlear nerve to form the vestibulocochlear nerve (CN VIII), and transmits electrical impulses concerning static equilibrium to the brain.

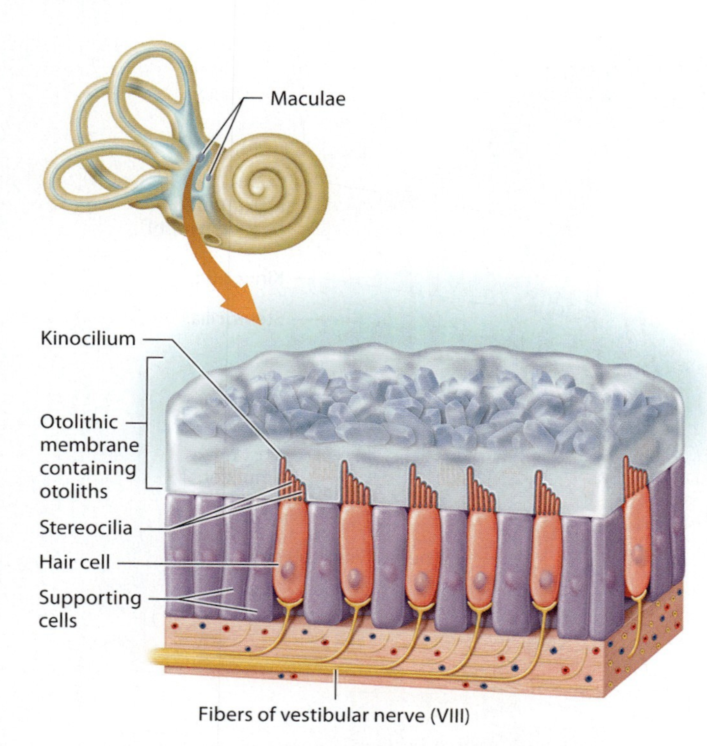

Figure 18-14 The structure of a macula of the utricle and saccule.

ACTIVITY 5

Examining the Gross Anatomy of the Ear

Learning Outcomes

1. Identify the components of the outer ear on an anatomical model, and describe the function of each.
2. Identify the components of the middle ear on an anatomical model, and describe the function of each.
3. Identify the components of the inner ear on an anatomical model, and describe the function of each.

Materials Needed

☐ Ear model

Instructions

1. CHART Locate each of the structures in the following Making Connections chart on an ear model or anatomical chart. Then review its description and/or function(s) and make "Connections to Things I Have Already Learned" based on your lectures, assigned reading, and lab. Parts of the chart have been filled out for you.

Making Connections: The Structure of the Ear

Ear Structure	Description (Structure and/or Function)	Connections to Things I Have Already Learned
Auricle (pinna)		*Is largely composed of elastic cartilage; elastic fibers give it flexibility.*
External auditory canal	*Component of external ear; located within the temporal bone; is a passageway for sound waves.*	
Tympanic membrane		
Auditory ossicles		*3 smallest bones in the body: malleus (hammer), incus (anvil), stapes (stirrup); names are based on the shapes of the bones*
Oval window		
Pharyngotympanic tube		*Old names = Eustachian tube, auditory tube*
Bony labyrinth	*Outer, bony part of the inner ear; is filled with perilymph.*	
Membranous labyrinth		

Ear Structure	Description (Structure and/or Function)	Connections to Things I Have Already Learned
Cochlea	*Snail-shaped structure that contains the receptor organ for hearing—the spiral organ*	
Vestibule		
Semicircular canals		
Vestibulocochlear nerve		*CN VIII; passes through the internal acoustic meatus (passageway) of the temporal bone on its way to the thalamus.*

2. Name the receptor organ in the cochlea, and describe its function. ______________________________

____

3. Name the receptor organs in the vestibule, and describe their function. ______________________________

____

4. Name the receptor organs in the semicircular canals, and describe their function. ______________________________

____

5. **Optional Activity**

Practice labeling ear structures on human cadavers at > MasteringA&P® > Study Area > Practice Anatomy Lab > Human Cadaver: Nervous System > Special Senses > Images 12–17

ACTIVITY 6
Performing Hearing and Equilibrium Tests

Learning Outcomes

1. Distinguish between conductive deafness and sensorineural deafness.
2. Perform a Weber test and a Rinne test, and explain the physiological basis for each.
3. Perform a Romberg test, and explain how it is used to detect impairment of the vestibular apparatus.

Materials Needed

- ☐ Tuning fork
- ☐ Whiteboard or chalkboard
- ☐ Whiteboard markers or chalk

Instructions

Hearing loss is classified as either conduction deafness or sensorineural deafness. Conduction deafness results from damage to the tympanic membrane or auditory ossicles; sensorineural deafness results from damage to the hair cells of the cochlea, to the cochlear nerve, or to the neural pathway that transmits neural impulses from the vestibulocochlear nerve to the primary auditory cortex in the temporal lobe. To determine if hearing loss is conductive or sensorineural, health care providers use two simple tests: the Weber test and the Rinne test. These tests involve placing a tuning fork that vibrates at a particular frequency either on the skull (to evaluate the ability to hear the vibrations transmitted through the bone, or bone conduction) or near the opening of the external auditory canal (to evaluate the conduction of sound through air).

A. Performing the Weber Test

1. Assign one person in your lab group to be the "clinician," and another to be the subject. The clinician strikes a tuning fork, gently places the end of the handle on the top center of the subject's head (**Figure 18-15a**), and asks the subject if the sound is equally loud in both ears. If the sound is equally loud in both ears, then the subject has normal hearing—or possibly some degree of bilateral deafness (equal hearing loss in both ears). If the sound is louder in one ear, then the subject may have some degree of unilateral deafness.

   Can you tell by performing this test whether the unilateral deafness is conductive hearing loss or sensorineural hearing loss? ______________________________

   Explain. ______________________________

   ____

   ____

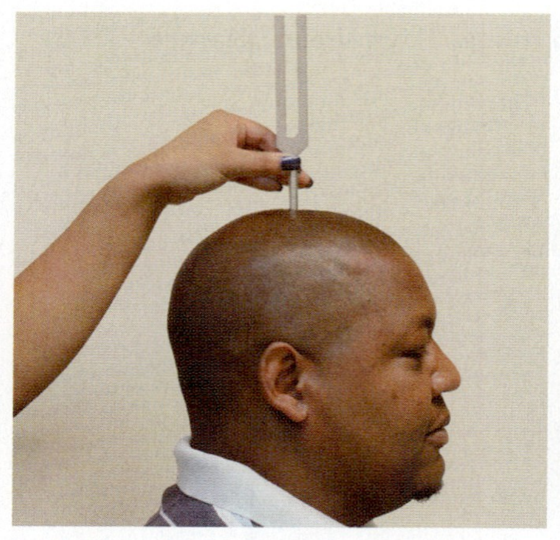
(a) Weber test: Place handle of vibrating tuning fork on top of head

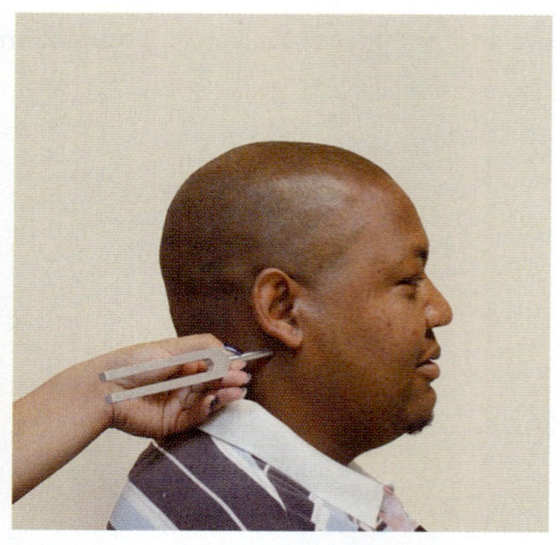
(b) Rinne test: Place handle of vibrating tuning fork on mastoid process

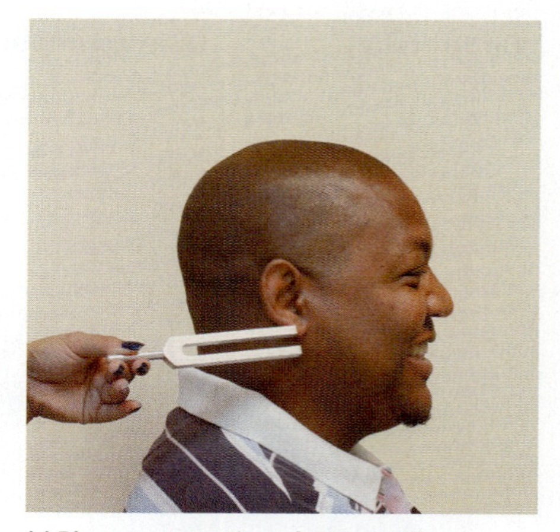
(c) Rinne test, continued: Move still vibrating tuning fork next to external auditory canal

Figure 18-15 **The Weber and Rinne tests.**

2. Mimic unilateral conduction deafness by placing a cotton ball in one external auditory canal and repeating the Weber test.

   If an individual has unilateral conduction deafness, in which ear do you think the sound will be loudest: the normal ear or the deaf ear? ______________________

   Explain. ______________________

   ____

   If an individual has unilateral sensorineural deafness, in which ear do you think the sound will be loudest: the normal ear or the deaf ear? ______________________

   Explain. ______________________

   ____

   ____

B. Performing the Rinne Test

1. Choose a new clinician and subject. The clinician strikes a tuning fork and places the handle on the subject's mastoid process, as shown in **Figure 18-15b**. When the subject reports no longer hearing the sound, the clinician moves the still-vibrating tuning fork close to the subject's external auditory canal, as shown in **Figure 18-15c**, to test hearing by air conduction. If the subject hears the sound once again, the subject's hearing is not impaired. If the subject cannot hear the sound again, the subject may have conduction deafness.
2. Once again mimic conduction deafness by placing a cotton ball in one of the subject's external auditory canals, and test this ear again. This time, however, the clinician will test hearing by air conduction first by striking the tuning fork and placing it next to (but not touching) the subject's ear. The clinician asks the subject to indicate when the sound can no longer be heard and then places the tuning fork on the subject's mastoid process. If the subject hears the sound again, then conduction deafness is present in that ear.

C. Performing the Romberg Test

The Romberg test evaluates the ability of the vestibular apparatus to maintain equilibrium in the absence of visual stimuli.

1. Choose a new clinician and subject. The clinician instructs the subject to stand erect, the back next to a chalkboard or whiteboard.
2. The clinician draws a vertical line on the board, immediately next to each side of the subject's torso.
3. The clinician instructs the subject to stand erect, with eyes open and staring straight ahead, for 1 minute.

   Did the subject sway slightly or significantly? ______________________
4. Repeat the procedure, but this time the clinician instructs the subject to keep both eyes closed for 1 minute.

   Did the subject sway slightly more or considerably more than with eyes open? ______________________
5. Based on your observations, does vision play a role in maintaining equilibrium? Explain. ______________________

   ____
6. Predict how the results would have been affected if the subject has a damaged vestibular apparatus. ______________________

   ____

Name: ______________________ Date: ______________ Lab Section: ________

PART I. Check Your Understanding

Activity 1: Exploring the Gross Anatomy of Olfactory and Gustatory Structures and Demonstrating the Effect of Olfaction on Gustation

1. The receptors for olfaction and gustation are classified as ______________ because they respond to chemicals in an aqueous solution.

2. Name the cranial nerves that transmit sensory information from the taste buds to the gustatory cortex.
   a. ______________
   b. ______________
   c. ______________

3. What effects does olfaction have on a person's ability to taste food? ______________

____

____

Activity 2: Examining the Gross Anatomy of the Eye

1. Which three cranial nerves innervate the extrinsic eye muscles?

____

2. Identify each labeled eye structure in the accompanying illustration. Then circle the components of the outer fibrous layer in green, the components of the middle vascular layer in red, and the structure that comprises the inner neural layer in yellow.
   a. ______________
   b. ______________
   c. ______________
   d. ______________
   e. ______________
   f. ______________
   g. ______________
   h. ______________
   i. ______________
   j. ______________
   k. ______________

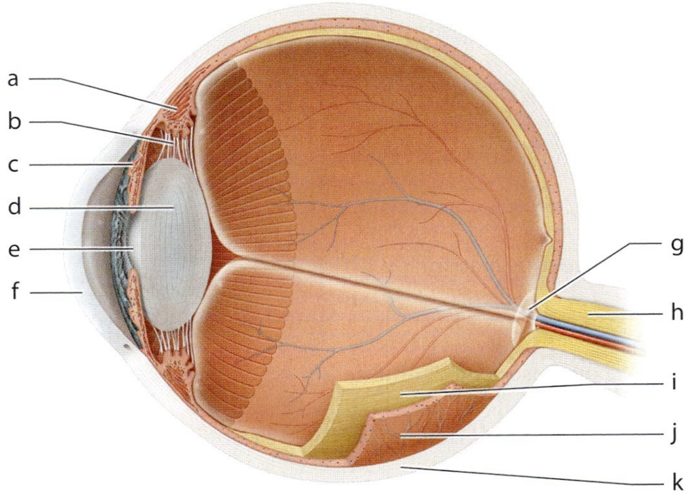

3. On the accompanying illustration:

   a. Color the anterior cavity blue. Which fluid fills it?

   ____

   b. Which structure produces this fluid?

   ____

   c. What is the function of this fluid?

   ____

   d. Which structure divides the anterior cavity into an anterior chamber and a posterior chamber? ______________

   e. Color the posterior cavity orange. What substance fills it?

   ____

   f. What are two functions of this substance? ______________________________

   ____

Activity 3: Dissecting the Mammalian Eye

1. Identify the labeled structures in the accompanying photographs of a dissected cow eye:

   a. ______________________

   b. ______________________

   c. ______________________

   d. ______________________

   e. ______________________

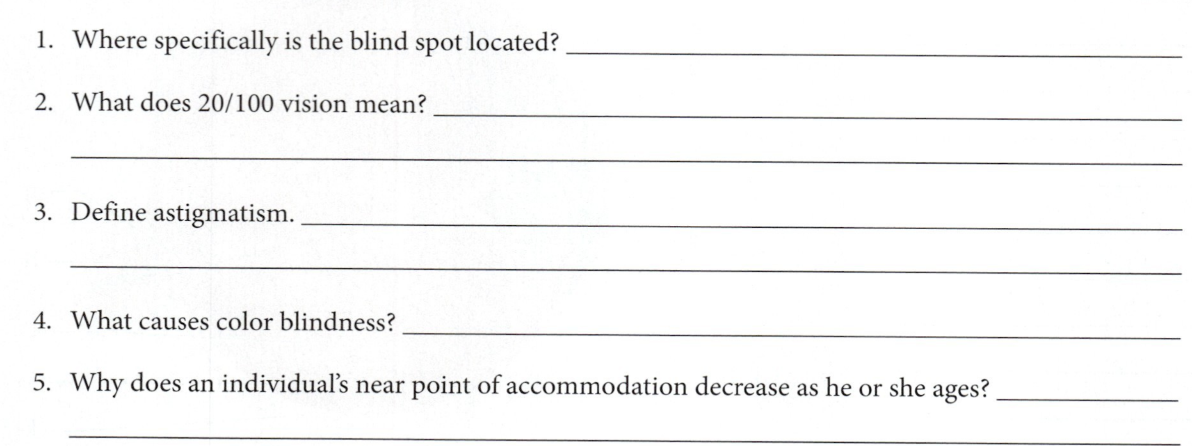

Activity 4: Performing Visual Tests

1. Where specifically is the blind spot located? ______________________________

2. What does 20/100 vision mean? ______________________________

   ____

3. Define astigmatism. ______________________________

   ____

4. What causes color blindness? ______________________________

5. Why does an individual's near point of accommodation decrease as he or she ages? ______________

   ____

Activity 5: Examining the Gross Anatomy of the Ear

1. Identify each of the structures of the ear shown in the accompanying illustration.

   a. ______________________

   b. ______________________

   c. ______________________

   d. ______________________

   e. ______________________

   f. ______________________

   g. ______________________

   h. ______________________

   i. ______________________

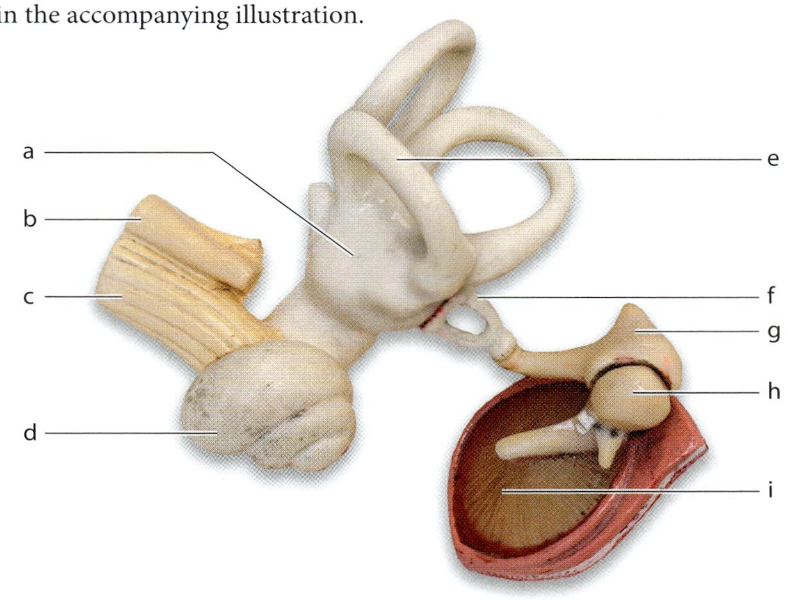

LT-E10: Classic Giant Ear, 4-part, 3B Scientific®

Activity 6: Performing Hearing and Equilibrium Tests

1. List two possible causes of conduction deafness. ______________________

2. List two possible causes of sensorineural deafness. ______________________

3. What was the purpose of placing a cotton ball in the subject's external auditory canal during the Weber test and the Rinne test? ______________________

PART II. Putting It All Together

A. Review Questions

Answer the following questions using your lecture notes, your textbook, and your lab notes.

1. What are the functions of the gustatory hairs that project from a taste pore? ______________________

2. Why is it impossible for a contact lens to get "lost" behind the eyeball? ______________________

3. List (in order) the eye structures (internal structures, fluids, and retinal layers) through which light travels from the point at which it enters the eye until it reaches the photoreceptors in the retina.

____

4. A retinal detachment occurs when the retina separates from the choroid. Why is this condition considered to be a medical emergency? ______________________________

____

5. Why can the inner ear be described as a "tube within a tube"? ______________________________

____

6. Melanocytes in both the choroid and the cells of the pigmented epithelium contain melanin. How is this pigment important to eye function? ______________________________

____

7. Name another location of melanocytes, and describe their function at that site. ______________________________

____

B. Concept Mapping

1. Fill in the blanks to complete this concept map outlining the anatomy of the eye.

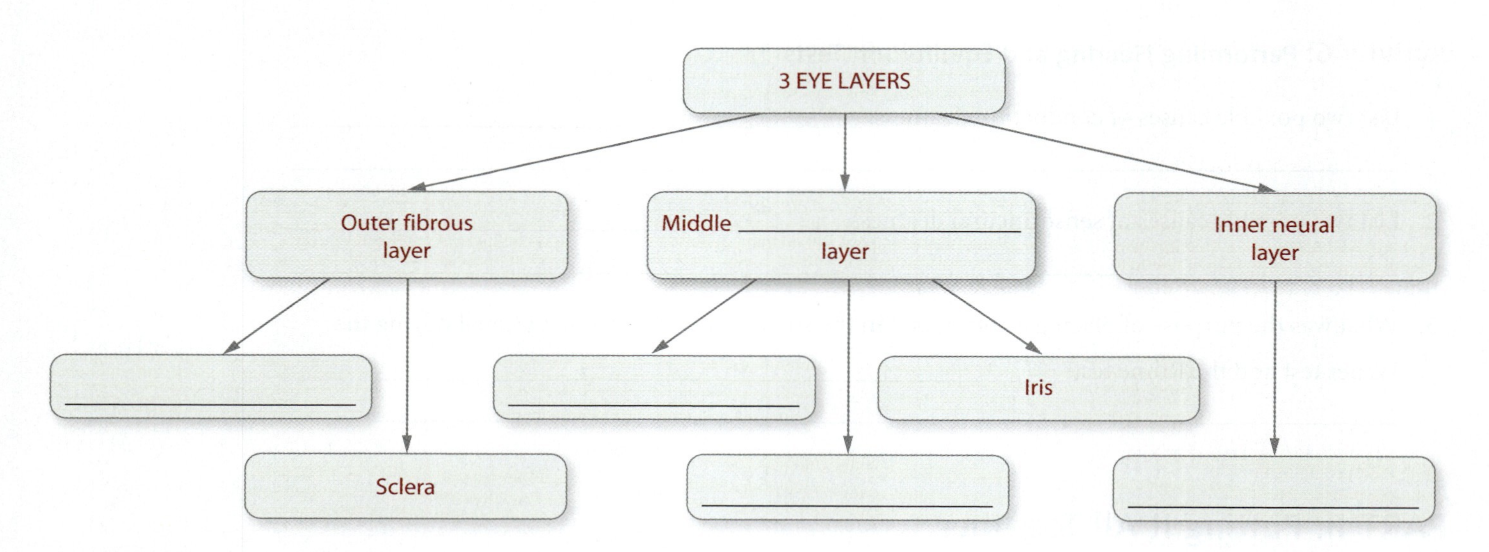

2. Construct a unit concept map to show the relationships among the following set of terms. Include all of the terms in your diagram. Your instructor may choose to assign additional terms.

aqueous humor	cerebral cortex	foramina	incus	malleus
mechanoreceptor	neural	optic nerve	photoreceptor	rod
taste bud	thalamus	utricle	vascular	vestibulocochlear nerve

19

The Endocrine System

UNIT OUTLINE

The endocrine system consists of ductless endocrine glands that secrete hormones directly into the bloodstream. **Hormones** are chemical signals that regulate the functions of other cells as long as those cells have receptors for the specific hormone. Some endocrine structures are composed of nervous tissue and are called neuroendocrine organs; their major function is the release of **neurohormones**, chemicals produced by nervous tissue that act as hormones. Together, the endocrine and nervous systems maintain homeostasis by regulating physiological events throughout the body.

In this unit you will be studying the gross and microscopic anatomy of endocrine and neuroendocrine organs. As you explore the anatomy of each organ, you will identify the major hormones produced by each. Additionally, you will describe the specific source, target, and biological action of each hormone and then use this knowledge to investigate some endocrine case studies.

THINK ABOUT IT *Compare and contrast endocrine regulation and nervous regulation of physiological events.*

____

____

____

____

PRE-LAB ASSIGNMENTS

To maximize learning, BEFORE your lab period carefully read this entire lab unit and complete these pre-lab assignments using your textbook, lecture notes, and prior knowledge.

PRE-LAB Activity 1: Exploring the Organs of the Endocrine System

1. Use the list of terms provided to label the accompanying illustration of the organs of the endocrine system. Check off each term as you label it.

- ☐ testes
- ☐ adrenal cortex
- ☐ adrenal medulla
- ☐ anterior pituitary gland
- ☐ posterior pituitary gland
- ☐ hypothalamus
- ☐ pancreas
- ☐ thyroid gland
- ☐ thymus gland
- ☐ pineal gland
- ☐ ovaries
- ☐ parathyroid glands

Neuroendocrine organs

a ______

b ______

c ______

Endocrine organs

d ______

e ______

f ______

g ______

h ______

i ______

j ______

k ______

l ______

Anterior view

Posterior view

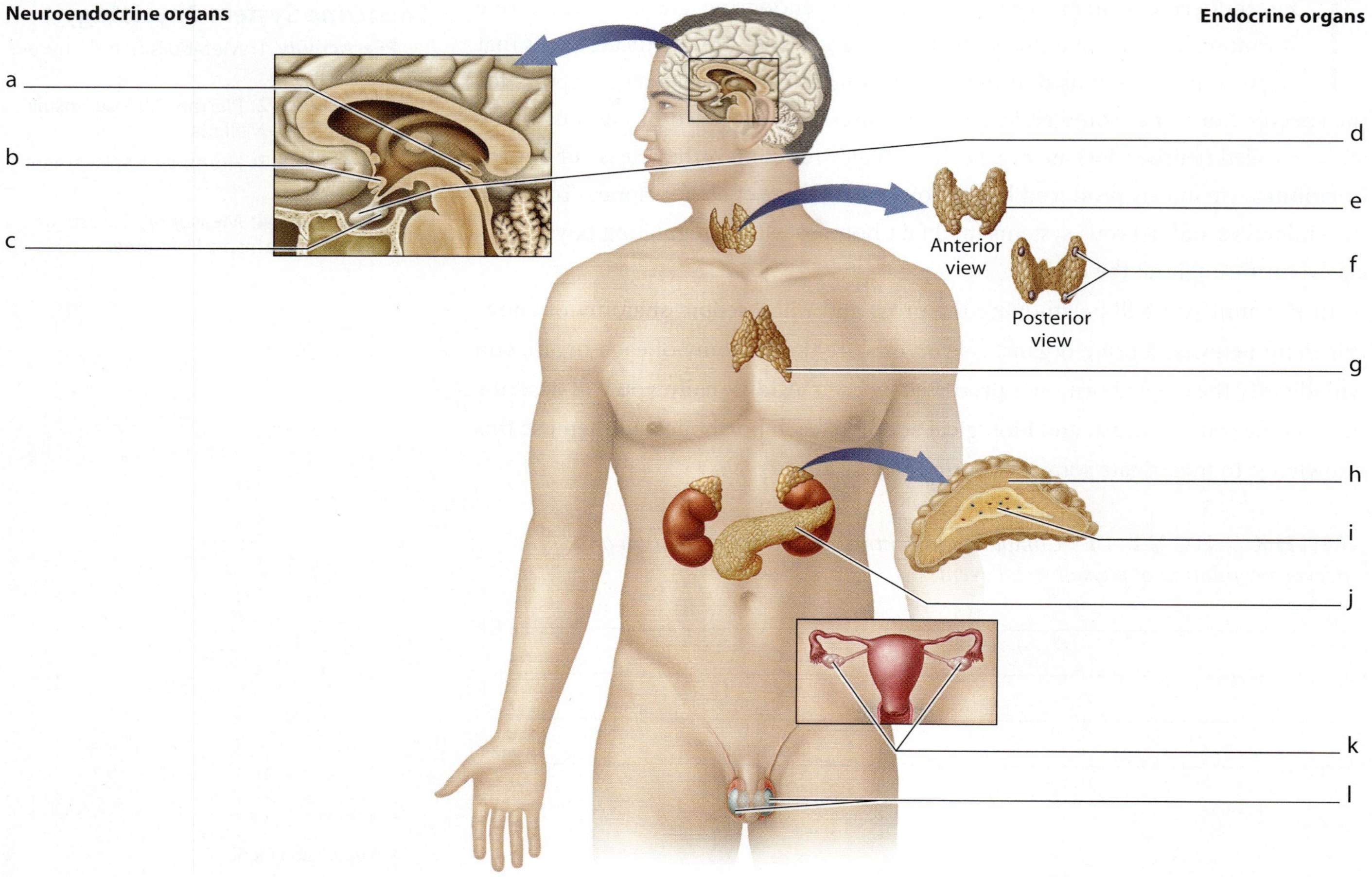

2. Endocrine glands:
   a. release substances onto internal and external body surfaces.
   b. are also classified as exocrine glands.
   c. contain ducts.
   d. synthesize and secrete hormones.
   e. produce enzymes.

3. Which endocrine organ is regulated *both* hormonally and neurally?
   a. pituitary gland
   b. pancreas
   c. thyroid gland
   d. testis
   e. ovary

4. Which hormone is correctly matched with its source?
   a. luteinizing hormone – ovary
   b. epinephrine – adrenal medulla
   c. glucagon – liver
   d. thyroid-stimulating hormone – thyroid gland
   e. cortisol – pineal gland

PRE-LAB Activity 2: Examining the Microscopic Anatomy of the Pituitary Gland, Thyroid Gland, Parathyroid Gland, Adrenal Gland, and Pancreas

1. Which zone of the adrenal cortex is closest to the adrenal medulla?
   a. zona glomerulosa
   b. zona reticularis
   c. zona fasciculata

2. Match each of the following endocrine glands with the correct hormone-secreting cell.

_____	a. thyroid gland	1. chromaffin cells
_____	b. parathyroid gland	2. follicular cells
_____	c. pancreas	3. islet cells
_____	d. adrenal gland	4. chief cells

PRE-LAB Activity 3: Investigating Endocrine Case Studies: Clinician's Corner

1. Endocrine disorders result when a gland produces ______________________________ of a specific hormone—a situation called a *hormone imbalance*.

2. *True or false?:* Normally, if there is too much or too little of a given hormone, feedback mechanisms within the endocrine system have difficulty correcting hormone levels.

Functional Anatomy of the Endocrine System

The major endocrine and neuroendocrine organs are illustrated in **Figure 19-1**; we consider these structures in the sections that follow.

Hypothalamus

The **hypothalamus** (**Figure 19-2**), a component of the diencephalon of the brain, is a major homeostatic organ that regulates blood pressure, hunger, thirst, body temperature, and some reproductive functions. Additionally, the hypothalamus produces a variety of releasing and inhibiting hormones that regulate the release of hormones from the anterior pituitary gland (**Table 19-1**). For example, the hypothalamus synthesizes and releases gonadotropin-releasing hormone (GnRH), which stimulates the release of gonadotropins called follicle-stimulating hormone (FSH) and luteinizing hormone (LH) from the anterior pituitary gland. FSH and LH are **gonadotropins** because they target the gonads (ovaries and testes). In addition to secreting releasing and inhibiting factors, the hypothalamus synthesizes two other neurohormones, antidiuretic hormone (ADH) and oxytocin (OXY), which are stored in the posterior pituitary gland.

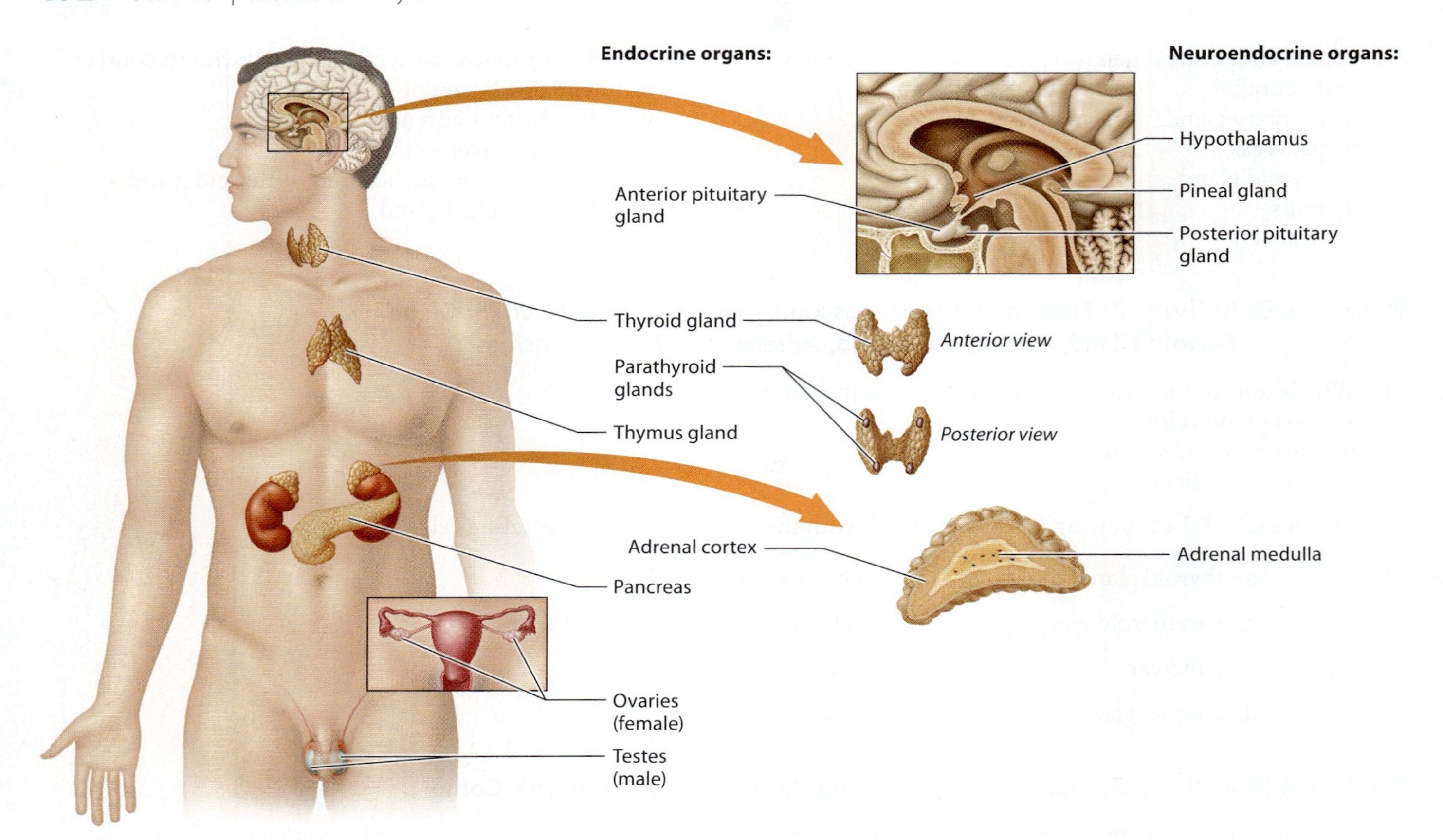

Figure 19-1 The major endocrine and neuroendocrine organs.

Table 19-1 Hypothalamic Hormones

Hormone	Target	Biological Action
Gonadotropin-releasing hormone (GnRH)	Anterior pituitary	Stimulates release of follicle-stimulating hormone (FSH) and luteinizing hormone (LH).
Corticotropin-releasing hormone (CRH)	Anterior pituitary	Stimulates release of adrenocorticotropic hormone (ACTH).
Thyrotropin-releasing hormone (TRH)	Anterior pituitary	Stimulates release of thyroid-stimulating hormone (TSH).
Prolactin-inhibiting hormone (dopamine)	Anterior pituitary	Inhibits release of prolactin (PRL).
Somatostatin	Anterior pituitary	Inhibits release of growth hormone (GH).
Growth hormone–releasing hormone (GHRH)	Anterior pituitary	Stimulates release of growth hormone (GH).
Antidiuretic hormone (ADH)	Kidney	Stimulates water retention (or decreased urine output).
Oxytocin (OXY)	Uterus and mammary glands	Stimulates uterine smooth muscle contraction; stimulates myoepithelial cells in mammary glands.

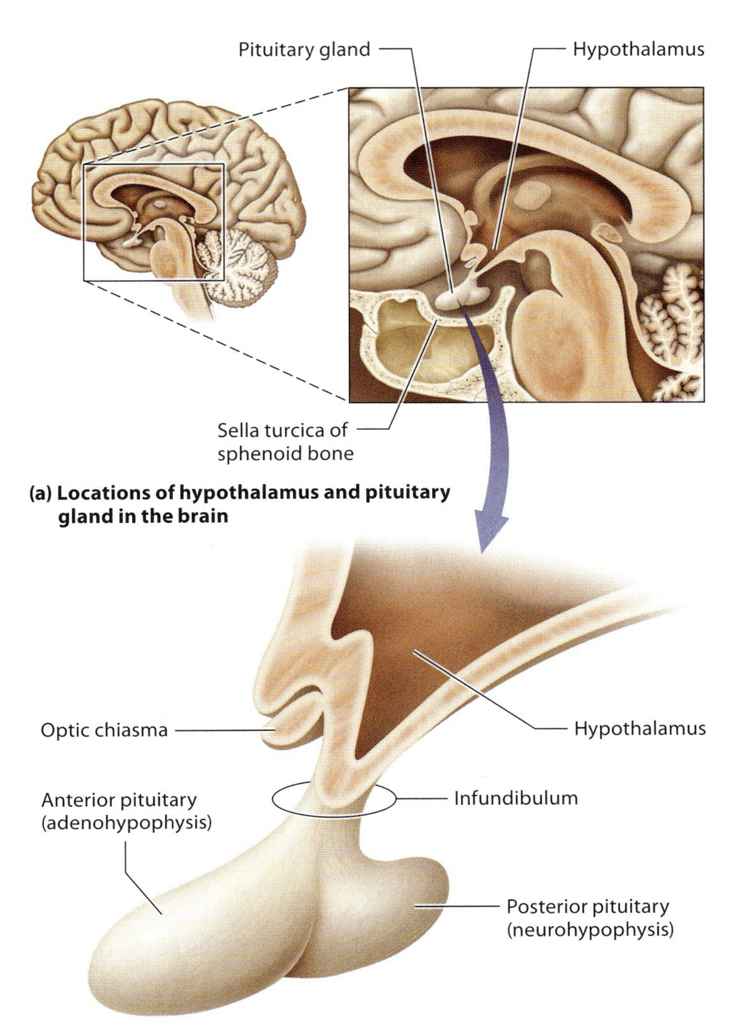

(a) Locations of hypothalamus and pituitary gland in the brain

(b) Structure of hypothalamus, and anterior and posterior pituitary glands

Figure 19-2 Structures of the hypothalamus and pituitary gland.

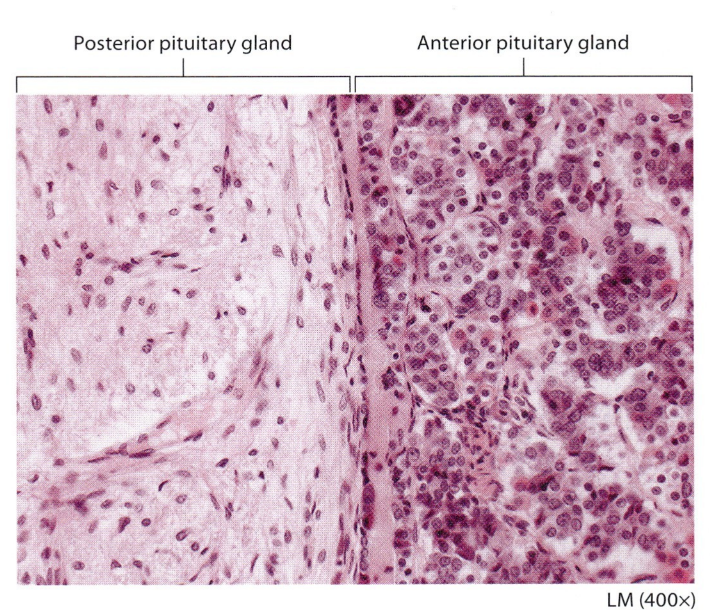

Figure 19-3 Microscopic anatomy of the anterior and posterior pituitary glands.

Pituitary Gland

The **pituitary gland** (or hypophysis) extends from the hypothalamus via the stalk-like infundibulum (see Figure 19-2b). It sits in the hypophyseal fossa of the sella turcica (sphenoid bone) and consists of two lobes: the **anterior pituitary gland** or adenohypophysis (*adeno* = gland) and the **posterior pituitary gland** or neurohypophysis (*neuro* = relating to nervous tissue). During development, the anterior pituitary gland arises as an outpocketing of the roof of the mouth and is composed of epithelial tissue; as a result, its cells resemble glandular epithelial tissue (**Figure 19-3**). The posterior pituitary gland, by contrast, is derived embryologically from an outgrowth of the diencephalon and consists of nervous tissue. **Table 19-2** summarizes the hormones produced by the pituitary gland.

Table 19-2 Pituitary Hormones

Hormone/Specific Source	Target	Biological Action
Follicle-stimulating hormone (FSH)	Ovaries, testes	Stimulates ovary to produce oocytes (eggs) and release estrogen; indirectly stimulates sperm production in testis.
Luteinizing hormone (LH)	Ovaries, testes	Stimulates ovary to produce estrogen and progesterone; stimulates testis to produce testosterone; stimulates ovulation.
Thyroid-stimulating hormone (TSH)	Thyroid gland	Stimulates the thyroid gland to secrete thyroid hormones (T_3 and T_4).
Adrenocorticotropic hormone (ACTH)	Adrenal gland (adrenal cortex)	Stimulates release of hormones such as cortisol and aldosterone from the adrenal cortex.
Growth hormone (GH)	Bone, muscle, adipose tissue, liver, cartilage	Stimulates growth.
Prolactin (PRL)	Mammary glands	Stimulates milk production.

Pineal Gland

The **pineal gland** (see Figure 19-1) is a small cone-shaped organ located in the epithalamus along the roof of the third ventricle in the brain. It secretes **melatonin**, a neurohormone thought to control the daily sleep/wake cycle, although its exact role in humans is still controversial.

Thymus

The **thymus gland** (see Figure 19-1) is an irregularly shaped gland in the mediastinum just posterior to the sternum. It produces hormones called **thymosin** and **thymopoietin**, which stimulate the maturation of T lymphocytes, a type of leukocyte involved in the immune response. The thymus is relatively large in children but begins to atrophy at puberty; eventually it is mostly replaced by adipose and fibrous connective tissues.

Thyroid Gland

The **thyroid gland** is a bi-lobed organ located anterior to the trachea and inferior to the thyroid cartilage (Adam's apple) of the larynx (**Figure 19-4a**). The two lobes are connected by a narrow isthmus. Histologically, the thyroid gland consists of abundant thyroid follicles that are filled with an iodine-rich gelatinous material called **colloid** and are lined with simple cuboidal epithelial cells called **follicle cells** (see Figure 19-4b). Thyroid-stimulating hormone (TSH) from the anterior pituitary gland stimulates the follicle cells to secrete thyroglobulin, a precursor molecule that is then combined with iodine molecules to produce the thyroid hormones **triiodothyronine (T_3)** and **thyroxine (T_4)**. The thyroid hormones regulate metabolic rate and thermoregulation and promote growth and development. Between the follicles are **parafollicular cells** that secrete **calcitonin**, a hormone that decreases blood calcium levels.

Parathyroid Glands

Embedded on the posterior surface of the thyroid gland are three to five pea-shaped structures called **parathyroid glands** (**Figure 19-5a**). These glands contain **chief cells** (see Figure 19-5b), which secrete **parathyroid hormone (PTH)**. PTH elevates blood calcium levels by targeting bones, the kidneys, and the small intestine. In bones, PTH stimulates osteoclasts to break down bone and release calcium into the blood; in the kidneys, PTH stimulates calcium reabsorption from the blood; and in the small intestine, PTH stimulates the absorption of calcium from food.

Adrenal Glands

The roughly pyramid-shaped **adrenal glands** (**Figure 19-6**) lie superior to the kidneys. Each adrenal gland is surrounded by a dense irregular connective tissue capsule that protects it and anchors it to the kidney. An adrenal gland is actually composed of two separate regions: an outer **adrenal cortex** and an inner **adrenal medulla** (Figure 19-6a). The outer adrenal cortex is composed of typical glandular epithelium arranged into three distinct zones—the zona glomerulosa, the zona fasciculata, and the zona reticularis (Figure 19-6b)—each of which produces steroid hormones derived from cholesterol.

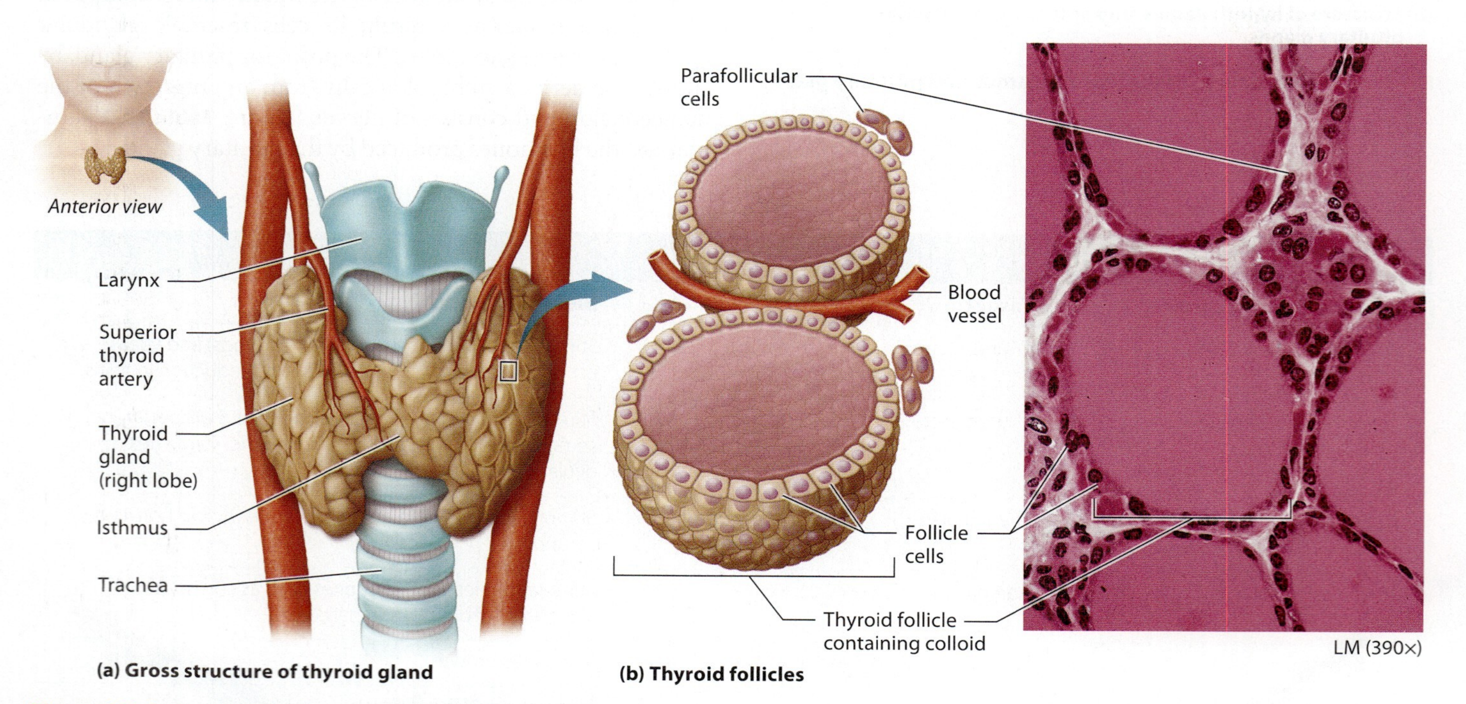

Figure 19-4 **Gross and microscopic anatomy of the thyroid gland.**

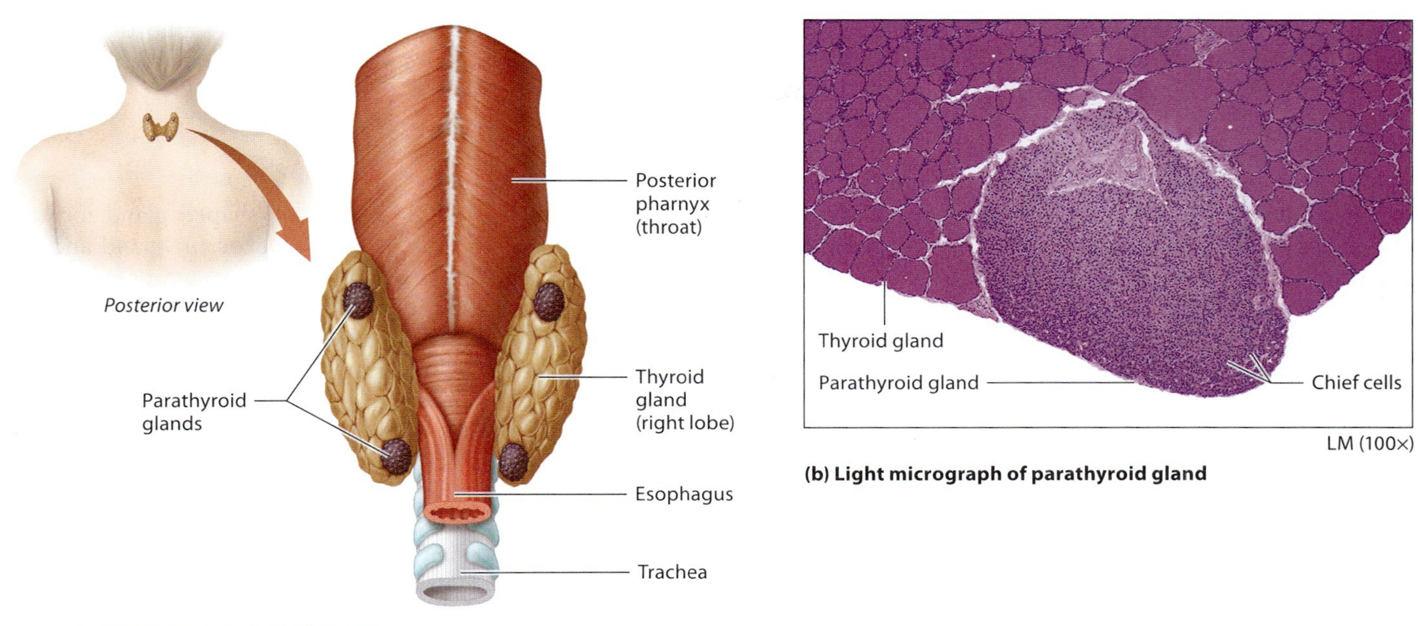

(a) **Anatomy of the parathyroid glands**

(b) **Light micrograph of parathyroid gland**

Figure 19-5 **Gross and microscopic anatomy of the parathyroid glands.**

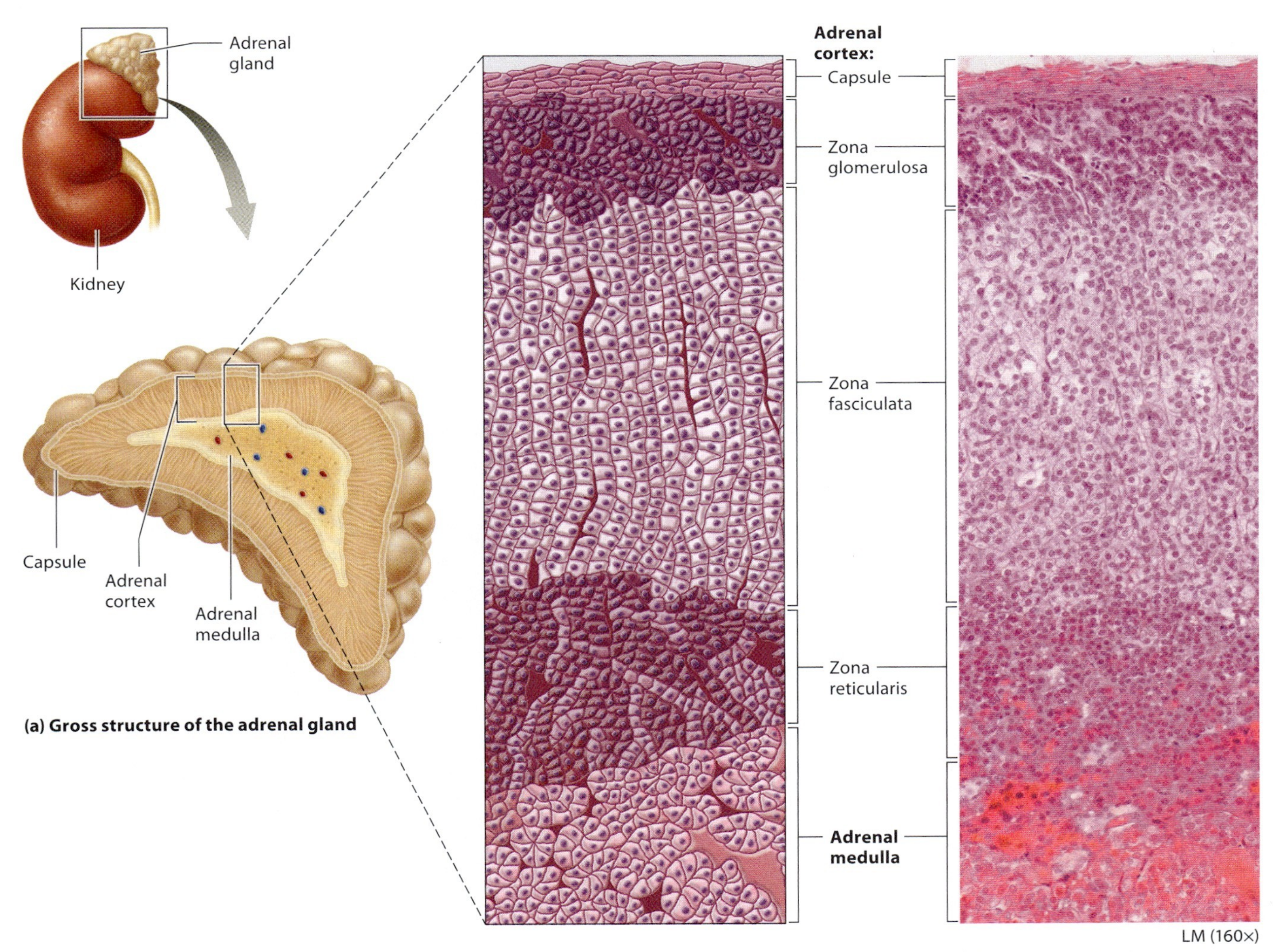

(a) Gross structure of the adrenal gland

(b) Histology of the adrenal gland: illustration (left) and light micrograph (right)

Figure 19-6 **Gross and microscopic anatomy of the adrenal gland.**

The outer **zona glomerulosa** consists of balls of cells that produce a class of hormones known as **mineralocorticoids**, which regulate mineral balance. The major mineralocorticoid—**aldosterone**—stimulates cells of the kidneys to reabsorb sodium ions; water follows the resorbed sodium ions via osmosis. As a result, both blood volume and blood pressure increase as urine volume decreases. The middle **zona fasciculata** consists of bundles of cells that produce a class of hormones known as **glucocorticoids**, which regulate glucose metabolism. The main glucocorticoid—**cortisol**—targets liver, muscle, and adipose cells. Cortisol's actions include gluconeogenesis (production of glucose from amino acids and fatty acids) in the liver, the breakdown of muscle proteins to release amino acids into the bloodstream, and the breakdown of lipids to release fatty acids into the bloodstream. The deep **zona reticularis** consist of a network of cells that predominantly produce **androgens** (male sex hormones).

The **chromaffin cells** of the adrenal medulla produce two catecholamines—**epinephrine** and **norepinephrine**—which are neurohormones that contribute to the fight-or-flight response to stress. These chemical messengers, whether they act as neurohormones released into the bloodstream or as neurotransmitters released into a synapse, increase heart rate, blood pressure, and respiratory rate, and decrease digestive activity.

Pancreas

The **pancreas** (**Figure 19-7a**), a club-shaped gland located posterior to the stomach, is both an exocrine gland and an endocrine gland. The exocrine cells are called **acinar cells**; we will study the exocrine pancreas, which produces digestive enzymes that reach the intestines via ducts, in Units 28 and 29. We focus here on the endocrine pancreas.

The hormone-producing portion of the pancreas, the **pancreatic islets** (see Figure 19-7b), contains several cell types; the most abundant are alpha cells and beta cells, which produce hormones that regulate blood glucose levels. The **alpha (α) cells** secrete **glucagon**, which raises blood glucose levels by triggering the release of glucose from glycogen stored in the liver. The **beta (β) cells** secrete **insulin**, which lowers blood glucose levels by stimulating body cells to take up glucose and liver cells to take up and convert glucose to glycogen for storage. The **delta (δ) cells** secrete somatostatin.

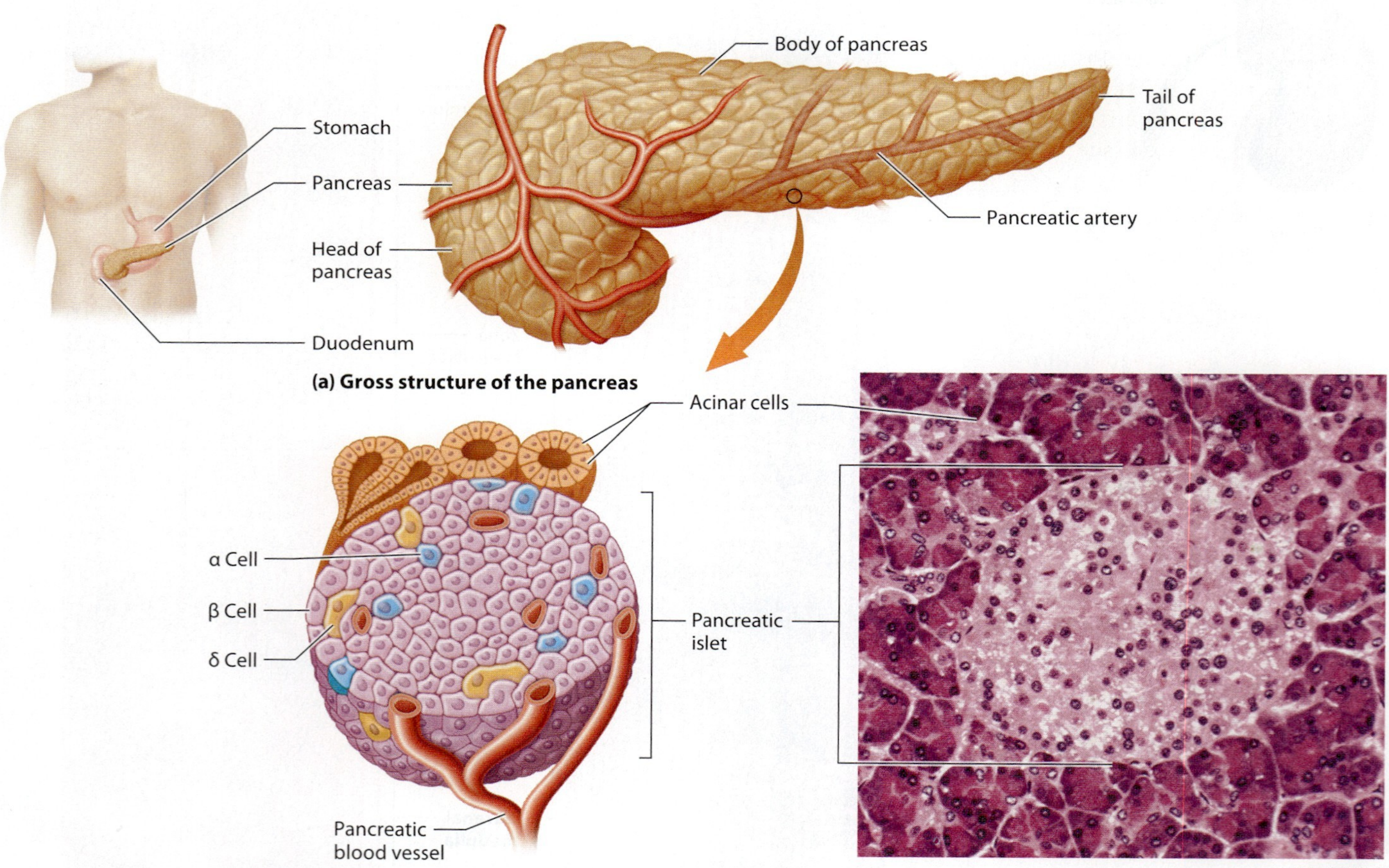

Figure 19-7 **Gross and microscopic anatomy of the pancreas.**

Testes

The two major functions of the **testes** (male gonads) are gamete (sperm) production and testosterone production. The microscopic anatomy of the testis is presented in Unit 32; here we focus on the pathways involving the testis and the endocrine system—designated the hypothalamic–hypophyseal–testicular axis. Recall that the hypothalamus produces releasing factors that control the release of hormones from the anterior pituitary gland. In this case, GnRH released into the bloodstream reaches the anterior pituitary gland, where it stimulates the secretion of the gonadotropins FSH and LH, which target the testis. FSH aids in sperm development, whereas LH stimulates testosterone secretion. Testosterone stimulates spermatogenesis (sperm production) and is responsible for such secondary sex characteristics as enlargement of the larynx, increased muscle mass, and facial hair.

Ovaries

The two major functions of the **ovaries** (female gonads) are production of gametes (oocytes) and production of estrogen and progesterone. As in males, GnRH stimulates the anterior pituitary gland to produce FSH and LH. In females, FSH stimulates gamete formation and the secretion of estrogens from developing follicles; LH stimulates the secretion of progesterone and estrogens. Progesterone and estrogens regulate the female reproductive cycle and are responsible for such secondary sex characteristics as breast development and the deposition of adipose tissue around the hips and thighs.

ACTIVITY 1

Exploring the Organs of the Endocrine System

Learning Outcomes

1. Identify the major endocrine organs on anatomical models.
2. Name selected hormones produced by endocrine organs; for each hormone, locate its target organ(s) and describe its biological action(s).
3. Compare and contrast the mechanisms by which the release of selected hormones is controlled.

Materials Needed

- ☐ Torso model or other anatomical models
- ☐ Anatomical charts

Instructions

A. Hypothalamus

1. Identify the hypothalamus on an anatomical model. Describe its location using at least three directional terms. ______________________________

____

____

2. Explain why the hypothalamus is a major homeostatic organ. ______________________________

____

____

____

3. Why is the hypothalamus classified as a neuroendocrine organ? ______________________________

____

____

B. Pituitary Gland

1. Identify the pituitary gland on an anatomical model. Describe its location using at least three directional terms. ______________________________

____

____

2. From which brain region does the pituitary gland extend? ______________________________

3. Do you agree that the pituitary gland could be considered to be two separate endocrine glands? Why or why not? ______________________________

____

____

____

C. Pineal Gland

1. Identify the pineal gland on an anatomical model. Describe its location using at least three directional terms. ______________________________

____

____

2. In which specific brain region is the pineal gland found?

____

3. Which hormone does the pineal gland produce?

____

D. Thymus

1. Identify the thymus on an anatomical model. Describe its location using at least three directional terms.

____

____

____

2. Name two hormones produced by the thymus.

3. What do these hormones regulate?

E. Thyroid Gland

1. Identify the thyroid gland on an anatomical model. Describe its location using at least three directional terms.

2. What is the isthmus of the thyroid gland?

3. Which hormones are secreted by the thyroid gland?

F. Parathyroid Glands

1. Identify the parathyroid glands on an anatomical model. Describe their location using at least three directional terms.

2. Which hormone is secreted by the parathyroid glands?

3. Name three target organs of this hormone.

G. Adrenal Glands

1. Identify the adrenal glands on an anatomical model. Describe their location using at least three directional terms.

2. The adrenal ________________ makes up the external portion of the adrenal gland; the adrenal ________________ makes up the internal portion of the adrenal gland.

3. How is hormone release from the adrenal cortex controlled?

4. How is hormone release from the adrenal medulla controlled?

H. Pancreas

1. Identify the pancreas on an anatomical model. Describe its location using at least three directional terms.

2. Why is the pancreas both an endocrine gland and an exocrine gland?

3. Name the pancreatic hormones that regulate blood glucose levels.

   Why are these hormones considered an antagonistic pair?

4. How is the release of these two hormones regulated?

I. Testes

1. Identify the testes on an anatomical model. Describe their location using at least three directional terms.

2. To which two organ systems do the testes belong?

3. Name the hormone produced by the testes.

J. Ovaries

1. Identify the ovaries on an anatomical model. Describe their location using at least three directional terms.

2. To which two organ systems do the ovaries belong?

3. Name two hormones produced by the ovaries.

K. Making Connections

CHART For each of the hormones in the following Making Connections chart, write its source, target organ(s), and biological action(s); then write some "connections" to things you have already learned in lecture, assigned reading, and lab. Some entries have been provided for you.

Making Connections: Hormones

Hormone	Source	Target(s)	Biological Action(s)	Connections to Things I Have Already Learned
Gonadotropin-releasing hormone (GnRH)	*Hypothalamus*	*Anterior pituitary*	*Stimulates release of FSH and LH from the anterior pituitary.*	*FSH and LH are gonadotropins; they target the gonads (testes and ovaries).*
Oxytocin		*Uterus and mammary glands*		
Luteinizing hormone (LH)			*Stimulates ovary to produce estrogen and progesterone; stimulates testis to produce testosterone.*	
Growth hormone (GH)		*Bone, muscle, adipose tissue, liver, cartilage*		
Triiodothyronine (T_3) and thyroxine (T_4)				
Calcitonin	*Thyroid gland*			
Parathyroid hormone (PTH)				
Aldosterone			*Stimulates cells of the kidneys to reabsorb sodium ions.*	
Cortisol				

(Continued)

Making Connections: Hormones (*Continued*)

Hormone	Source	Target(s)	Biological Action(s)	Connections to Things I Have Already Learned
Epinephrine and norepinephrine				
Insulin				
Glucagon				
Testosterone			*Spermatogenesis and development of male sex characteristics*	
Estrogen	*Ovary*			
Progesterone				

Optional Activity

Practice labeling endocrine structures on human cadavers at > MasteringA&P® > Study Area > Practice Anatomy Lab > Human Cadaver > Endocrine System

ACTIVITY 2

Examining the Microscopic Anatomy of the Pituitary Gland, Thyroid Gland, Parathyroid Gland, Adrenal Gland, and Pancreas

Learning Outcomes

1. Differentiate between the anterior pituitary and posterior pituitary in a histological section.
2. Differentiate among follicle cells, colloid, and parafollicular cells in a histological section of the thyroid gland.
3. Observe chief cells in a histological section of the parathyroid gland.
4. Differentiate among the three layers of the adrenal cortex and the adrenal medulla in a histological section.
5. Differentiate between the exocrine pancreas (acinar cells) and the endocrine pancreas (pancreatic islets) in a histological section.

Materials Needed

- ☐ Microscope and slides (or photomicrographs) of pituitary gland, thyroid gland, parathyroid gland, adrenal gland, and pancreas

Instructions

1. Observe a histological slide of the pituitary gland (or a photomicrograph) under low power.
   a. Sketch and label the anterior pituitary gland and the posterior pituitary gland.

Total magnification: _____ ×

   b. Switch to the high-power objective (or view a photomicrograph).

   How do the cells of the anterior pituitary and the posterior pituitary differ in appearance?

   ____

   ____

   ____

   Suggest a reason for this difference. ____________________

   ____

   ____

2. Observe a histological slide (or a photomicrograph) of the thyroid gland under low power. Sketch a histological view of the thyroid gland and label the follicle, follicular cells, colloid, and parafollicular cells.

Total magnification: _____ ×

What is the function of follicular cells? ____________________

____

What is the function of colloid? ____________________

____

What are the cells surrounding the follicles called?

____

Which hormone do these cells secrete? ____________________

3. Observe a histological section (or a photomicrograph) of the parathyroid gland under low power. Sketch a histological view of the parathyroid gland and label the chief cells.

Total magnification: _____ ×

Describe the appearance of the chief cells. ____________________

____

Which hormone is produced by the chief cells?

____

4. Observe a histological slide (or a photomicrograph) of the adrenal gland under low power.
   a. Sketch the adrenal gland and label the adrenal capsule, adrenal cortex, and adrenal medulla.

Total magnification: _____ ×

b. Sketch the adrenal gland under high power and label the adrenal capsule, adrenal cortex, zona glomerulosa, zona fasciculata, zona reticularis, and adrenal medulla.

Total magnification: _____ ×

Describe the appearance of the cells in each layer and name the major hormone secreted by each layer.

Zona glomerulosa ______________________________

____

Zona fasciculata ______________________________

____

Zona reticularis ______________________________

____

Describe the appearance of the cells in the adrenal medulla. ______________________________

____

Name the hormones produced in the adrenal medulla. ______________________________

____

LabBOOST >>>

Microscopic Anatomy of the Adrenal Cortex The microscopic anatomy of the adrenal cortex contains terms that you have encountered in previous units and one that you will encounter in Unit 30 when you study the urinary system. The zona *fasciculata* contains cord-like rows (bundles) of cells. The term *fascicle* means "bundle" and you learned about bundles of muscle fibers (Unit 11) and bundles of nerve fibers (Unit 14), both of which are called *fascicles*. The zona *reticularis* contains a net-like arrangement of cells. You learned about *reticular* connective tissue in Unit 6 when you studied different types of connective tissue. *Reticular* fibers form a support network on which cells of connective tissue sit. In Unit 30, you will learn that the kidney contains specialized "balls" of capillaries called *glomeruli*. These *glomeruli* appear similar in shape to the clusters of cells in the zona *glomerulosa* of the adrenal cortex.

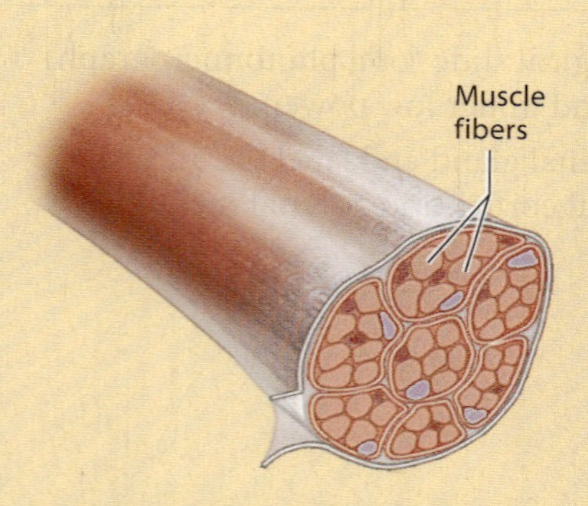

(a) Fascicle = bundle of muscle fibers

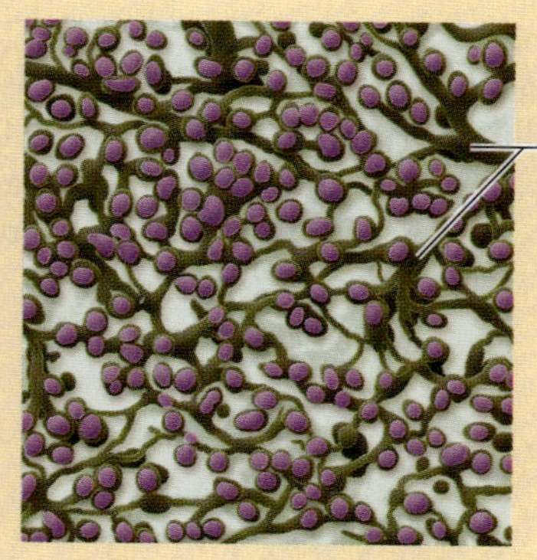

(b) Reticular tissue = net-like appearance of fibers and cells

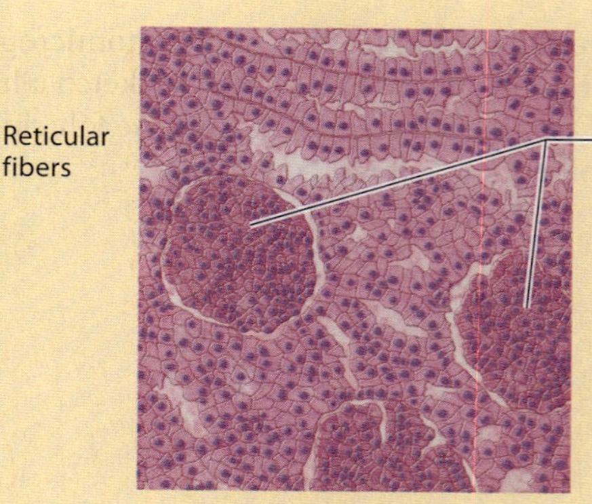

(c) Glomeruli = balls of capillaries

5. Observe a histological section (or a photomicrograph) of the pancreas under low power and distinguish between the acinar cells of the exocrine pancreas and the pancreatic islets of the endocrine pancreas.

   a. Sketch the pancreas under low power and label the acinar cells and pancreatic islets.

   Total magnification: _____ ×

   b. Now, increase the magnification to high power.

   Describe the appearance of the acinar cells.

   ____

   Describe the appearance of the pancreatic islets.

   ____

   Which hormone is produced by alpha cells? __________

   Which hormone is produced by beta cells? __________

6. **Optional Activity**

View and label histology slides of the endocrine system at > MasteringA&P® > Study Area > Practice Anatomy Lab > Histology > Endocrine System

Endocrine System Physiology

As you have seen, endocrine glands secrete a wide variety of hormones that function with the nervous system to regulate numerous aspects of homeostasis in the body. However, even slight changes in the balance of hormones can lead to endocrine disorders, which we explore next.

ACTIVITY 3

Investigating Endocrine Case Studies: Clinician's Corner

Learning Outcomes

1. Evaluate clinical scenarios and interpret the data presented.
2. Explain the role of negative feedback mechanisms in endocrine control.

Materials Needed

☐ Your class notes, lab manual, and textbook

Instructions

Endocrine disorders often result when a gland produces too much or too little of a specific hormone, producing a **hormone imbalance**. Sometimes such imbalances can result from a medication or synthetic hormone. Normally, if there is too much or too little of a given hormone in the body, feedback mechanisms within the endocrine system correct hormone levels. Hormone imbalances can also occur if the feedback system cannot correctly regulate hormone levels in the blood or if the body cannot properly clear the blood of hormones.

Read each of the following clinical scenarios and answer the accompanying question set:

1. Chad is an active, muscular 28-year-old man with a 3-year history of infertility. Upon physical examination, his body temperature is 98.8°F, blood pressure is 122/82, and pulse is 58 beats/min. Chad reports no family history of any endocrine disorder and no medical problems, although he has extensive acne. He is physically fit and exhibits a strong libido. Blood analysis indicates abnormally low levels of FSH and LH; a semen sample reveals a low sperm count. Upon further questioning, Chad admits to long-term use of anabolic steroids.

   a. Describe the normal hypothalamic–hypophyseal–testicular axis. ______________________________

   ____

   ____

   ____

   b. How do anabolic steroids disrupt this normal axis?

   ____

   ____

c. How can anabolic steroid abuse lead to infertility?

____

____

____

d. Can Chad's infertility be reversed? If so, how?

____

____

2. Molly, a 31-year-old mother of three children, has been working out with a trainer and following a moderate, balanced diet for approximately a year. Despite her disciplined approach to diet and exercise, Molly has experienced significant weight gain during the past 4 months. Especially noticeable are the fat that has accumulated between her shoulder blades and excess hair growth on her face, neck, and chest. Cuts take a long time to heal and she bruises easily. Her menstrual cycles have become very irregular and she feels anxious and stressed most of the time.

   a. What is your diagnosis? ______________________________

   b. What causes this endocrine disorder? ______________________________

   ____

   c. How can this endocrine disorder be treated?

   ____

   ____

3. Ellen is a 45-year-old woman complaining of severe fatigue and weight gain. She reports that her skin is very dry and that her hair has been falling out. Additionally, she has been experiencing bouts of irritability and depression. Blood analysis reveals the following:

Hormone	Results	Normal Values
TSH	8.0 milliunit/L	0.3–3.0 milliunit/L
T_3	50 mg/dL	100–200 ng/dL
T_4	3.88 mcg/dL	4.5–11.2 mcg/dL

   a. What is your diagnosis? ______________________________

   b. Explain why TSH levels are elevated and T_3/T_4 levels are low. ______________________________

   ____

   ____

   c. Identify different causes of this disorder.

   ____

   ____

   d. How is this disorder treated? ______________________________

   ____

PhysioEx EXERCISE 4

Endocrine System Physiology

The PhysioEx 9.1 Laboratory Simulations in Physiology are easy-to-use laboratory simulations that can be used as an alternative to or as a supplement to wet lab activities in this unit. Each simulation allows you to investigate important physiological concepts, repeat labs as often as you like, and conduct experiments that are difficult to perform in a wet lab environment because of time, cost, or safety concerns.

Access the simulations in these activities at MasteringA&P® > Study Area > PhysioEx 9.1

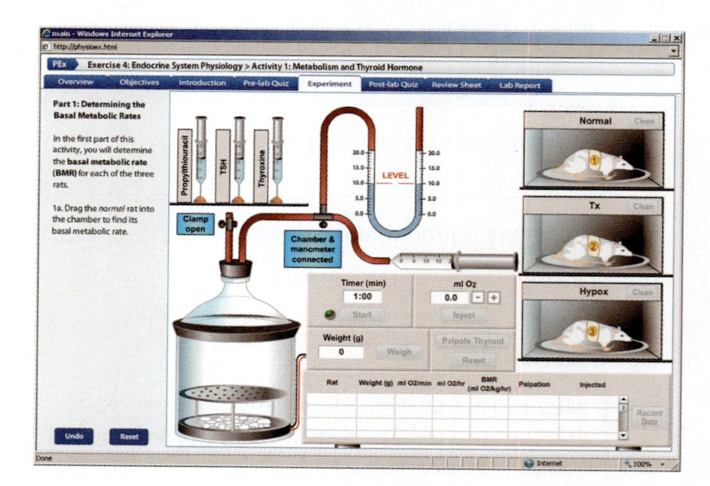

There you will find the following activities:

PEx Activity 1: Metabolism and Thyroid Hormone

PEx Activity 2: Plasma Glucose, Insulin, and Diabetes Mellitus

PEx Activity 3: Hormone Replacement Therapy

PEx Activity 4: Measuring Cortisol and Adrenocorticotropic Hormone

Post-lab quizzes are also assignable in MasteringA&P®

POST-LAB ASSIGNMENTS

Name: ______________________ Date: ______________ Lab Section: __________

PART I. Check Your Understanding

Activity 1: Exploring the Organs of the Endocrine System

1. Identify the structures of the endocrine organs in the accompanying diagrams:

   a. ______________________
   b. ______________________
   c. ______________________
   d. ______________________
   e. ______________________
   f. ______________________
   g. ______________________

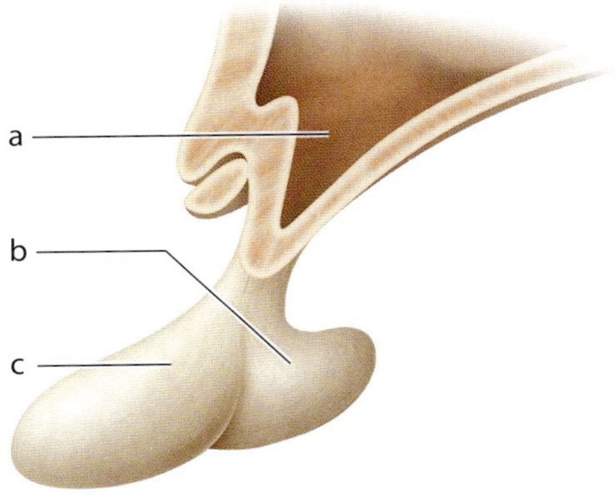

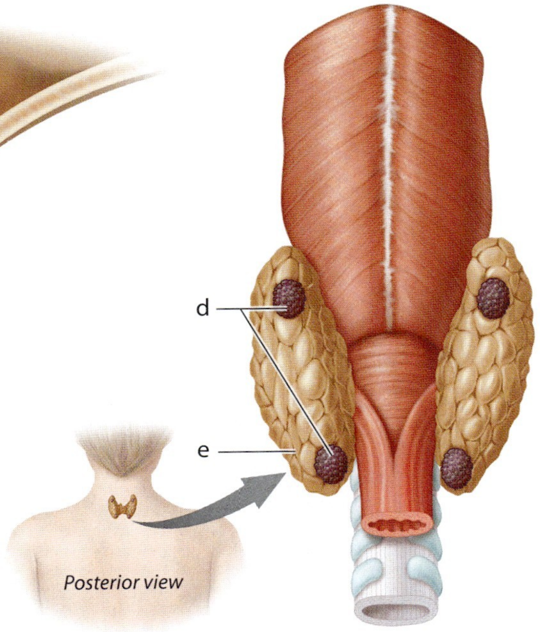

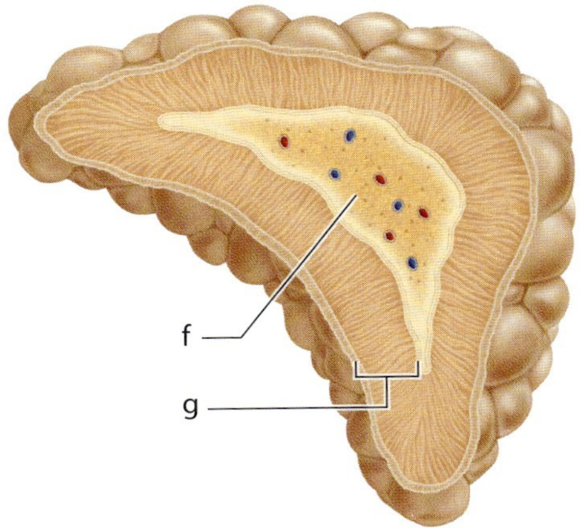

a
b
c
d
e
Posterior view
f
g

2. Match each of the following endocrine organs to the hormone(s) that it produces:

	Organ	Hormone	Hormone
________	a. thyroid gland	1. estrogen	10. T_3
________	b. pituitary gland	2. oxytocin	11. epinephrine
________	c. hypothalamus	3. glucagon	12. prolactin
________	d. pineal gland	4. parathyroid hormone	13. T_4
________	e. pancreas	5. calcitonin	14. cortisol
________	f. parathyroid gland	6. thyroid-releasing hormone	15. growth hormone
________	g. adrenal gland	7. testosterone	
________	h. ovary	8. aldosterone	
________	i. testis	9. melatonin	

3. Indicate whether the release of each of the following hormones is controlled neurally or hormonally.

_________________ a. Epinephrine

_________________ b. Thyroid-stimulating hormone

_________________ c. Insulin

_________________ d. Antidiuretic hormone

_________________ e. Parathyroid hormone

Activity 2: Examining the Microscopic Anatomy of the Pituitary Gland, Thyroid Gland, Parathyroid Gland, Adrenal Gland, and Pancreas

1. Identify the structures in the accompanying micrographs of endocrine tissue:

a. ______________________________

b. ______________________________

c. ______________________________

d. ______________________________

e. ______________________________

f. ______________________________

g. ______________________________

h. ______________________________

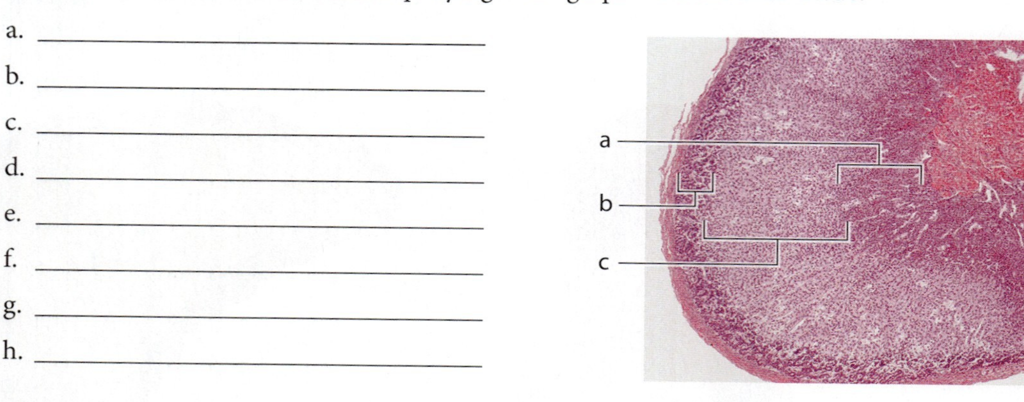

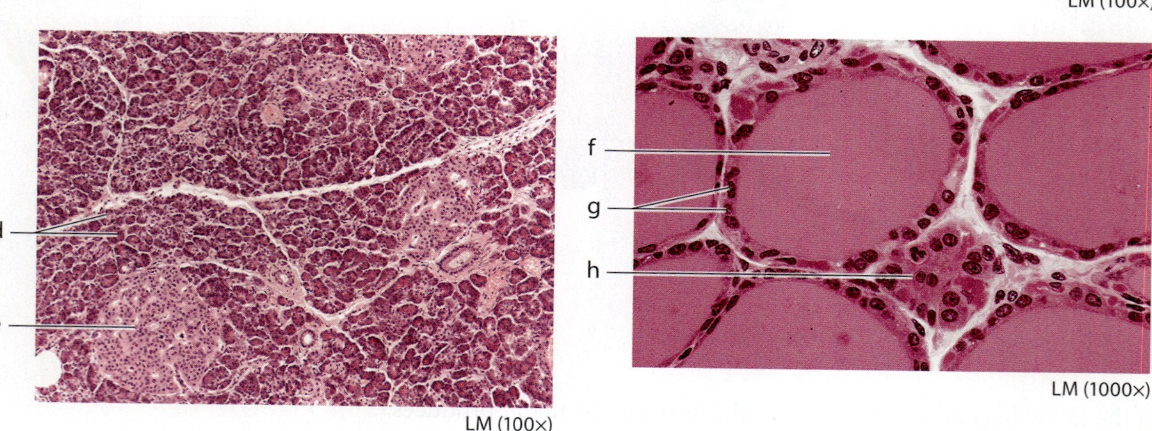

2. Match each of the following cell types to the hormone(s) that it produces.

_____ a. beta cells	1. parathyroid hormone
_____ b. follicle cells	2. epinephrine
_____ c. parafollicular cells	3. insulin
_____ d. chief cells	4. calcitonin
_____ e. neuroendocrine cells	5. oxytocin
_____ f. chromaffin cells	6. thyroid hormones

Activity 3: Investigating Endocrine Case Studies: Clinician's Corner

1. For each of the following pairs of endocrine disorders, list the gland(s) and hormone(s) involved, and then indicate one difference between the disorders.

   a. Diabetes mellitus and diabetes insipidus ______

   b. Cushing syndrome and Addison disease ______

   c. Hypothyroidism and hyperthyroidism ______

PART II. Putting It All Together

A. Review Questions

Answer the following questions using your lecture notes, your textbook, and your lab notes.

1. How are the adrenal gland and the pituitary gland similar? ______

2. Complete the following chart:

Hormone	Specific Source	Control of Release
Oxytocin	*Neurosecretory cells of the posterior pituitary*	*Nerve impulses traveling from hypothalamus to posterior pituitary*
Insulin		
T_3 and T_4		
Epinephrine		
Prolactin		
Aldosterone		

3. Explain how anabolic steroid abuse can lead to infertility. ______________________________

____

____

____

4. Explain why hypothyroidism is often associated with elevated TSH levels. ______________________________

____

____

B. Concept Mapping

1. Fill in the blanks to complete this concept map outlining the regulation of reproductive hormone function.

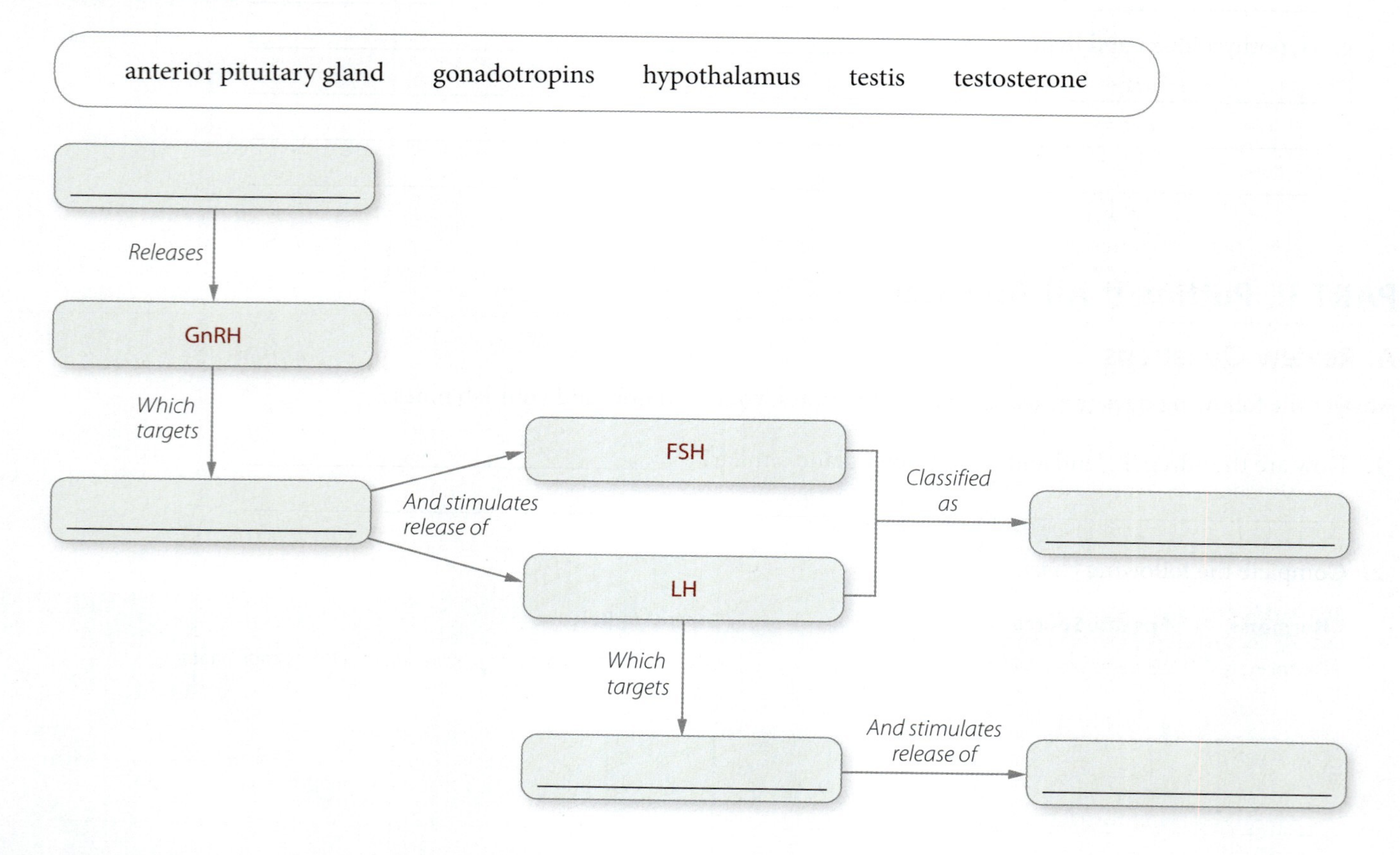

2. Construct a unit concept map to show the relationships among the following set of terms. Include all of the terms in your diagram. Your instructor may choose to assign additional terms.

adrenal cortex	chromaffin cells	endocrine	exocrine	gonadotropins
hypothalamus	melatonin	neurosecretory cells	ovary	pancreas
parafollicular cells	parathyroid gland	thymus gland	thyroid hormones	zona glomerulosa

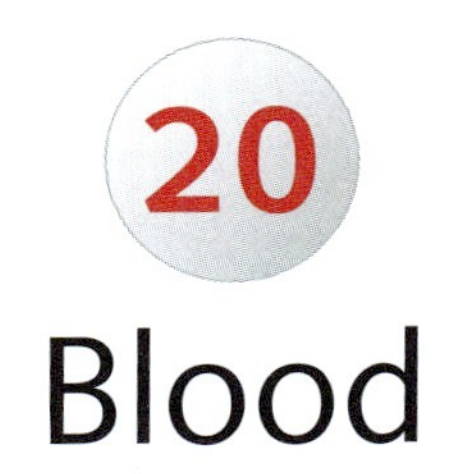

Blood

UNIT OUTLINE

Blood is a connective tissue with a fluid matrix, and it consists of two components: **formed elements**, which include blood cells and cell fragments called platelets, and **plasma**, which is over 90% water.

Blood's many important functions can be grouped into three major types: delivery, regulatory, and protective functions. Blood delivers oxygen and nutrients to body cells, carbon dioxide and waste products to their elimination sites (the lungs and kidneys), and hormones to their target organs. Blood also plays an important role in various aspects of homeostasis, such as the regulation of body temperature, pH, and blood pressure. Finally, blood's protective functions include immunity and blood clotting.

Blood tests are routinely performed in clinical settings as either general health assessments or as tools for diagnosing pathological conditions. In this lab, you have the opportunity to perform some routine hematological tests—a hematocrit, a differential white blood cell count, and a coagulation time test—and to learn about their clinical applications.

THINK ABOUT IT *How do each of the following organ systems interact with the blood to maintain homeostasis in the body?*

Integumentary system: ______________________________

____

Skeletal system: ______________________________

____

Nervous system: ______________________________

____

Endocrine system: ______________________________

____

PRE-LAB ASSIGNMENTS

Pre-lab quizzes are also assignable in MasteringA&P®

To maximize learning, BEFORE your lab period carefully read this entire lab unit and complete these pre-lab assignments using your textbook, lecture notes, and prior knowledge.

PRE-LAB Activity 1: Exploring the Formed Elements of Blood

1. Use the list of terms provided to label the accompanying illustration. Check off each term as you label it.

☐ erythrocytes

☐ platelets

☐ leukocytes

a ____

b ____

c ____

2. Which of the following types of formed element is produced when a megakaryocyte fragments?
   a. erythrocyte
   b. platelet
   c. neutrophil
   d. basophil
   e. All of these answers are correct.

3. Which of the following statements concerning blood is true?
   a. White blood cells outnumber red blood cells.
   b. Red blood cells function in immunity.
   c. All formed elements are produced in red bone marrow.
   d. Lymphocytes are the most numerous type of white blood cell.
   e. Platelets transport oxygen.

4. Which of the following substances is found in plasma?
   a. insulin
   b. oxygen
   c. glucose
   d. sodium ions
   e. All of these substances are found in plasma.

PRE-LAB Activity 2: Performing a Hematocrit

1. The hematocrit, or packed cell volume, is the percentage of ____________ in a volume of a blood sample.

2. In this activity you will be using a heparinized capillary tube. What is the function of heparin? ____________

____

3. In this activity the formed elements will be separated using what type of lab instrument? ____________

____

PRE-LAB Activity 3: Performing a Differential White Blood Cell Count

1. Which three leukocyte types are classified as granulocytes?

____

2. Which two leukocyte types are classified as agranulocytes?

____

3. How does a granulocyte differ from an agranulocyte?

____

PRE-LAB Activity 4: Determining Coagulation Time

1. What is coagulation? ______________________________

2. Does coagulation involve a positive feedback mechanism or a negative feedback mechanism?

____

PRE-LAB Activity 5: Determining Blood Types

1. Proteins located on the plasma membrane of red blood cells are called ____________.

2. Antibodies react with foreign proteins to cause ____________, or the clumping of cells.

3. An individual with type B blood has ____________ on the plasma membranes of the red blood cells and ____________ in the plasma.

4. Which blood type is known as the universal donor?

____

5. Which blood type is known as the universal recipient?

____

The Composition of Blood

Blood is a fluid connective tissue that circulates through blood vessels (arteries, capillaries, and veins). It consists of cells and cell fragments (collectively called formed elements) suspended in a nonliving extracellular matrix called plasma (**Figure 20-1a**). The formed elements of blood are erythrocytes (red blood cells), leukocytes (white blood cells), and thrombocytes (platelets).

Blood is a sticky, opaque fluid that varies in color depending on the amount of oxygen it contains. Oxygen-rich blood is a bright, scarlet red, whereas oxygen-poor blood is dark-red. Blood is a relatively viscous fluid (five times more viscous than water), and it has a slightly alkaline pH ranging from 7.35 to 7.45.

When centrifuged, blood separates into three components (**Figure 20-1b**). The largest, lightest component (about 55% of blood volume) is plasma. Among the more than 100 different substances found in the plasma are proteins, nutrients, electrolytes, hormones, wastes, and respiratory gases. Beneath the plasma is a thin layer, called the buffy coat, which contains leukocytes and platelets. The final

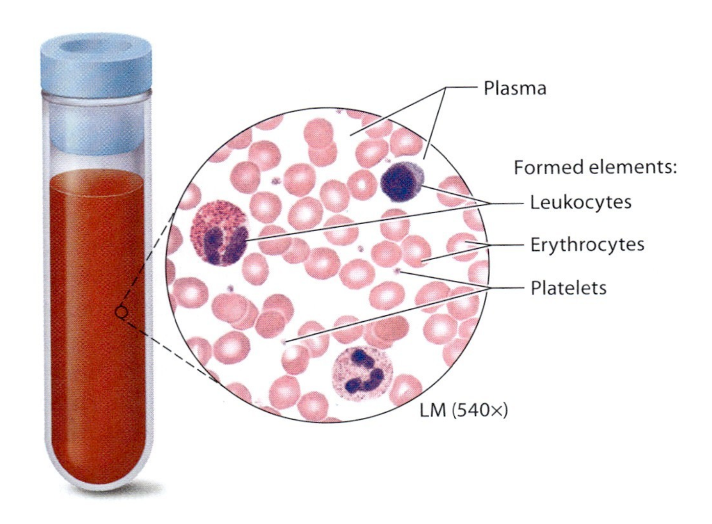

(a) Blood sample (non-centrifuged)

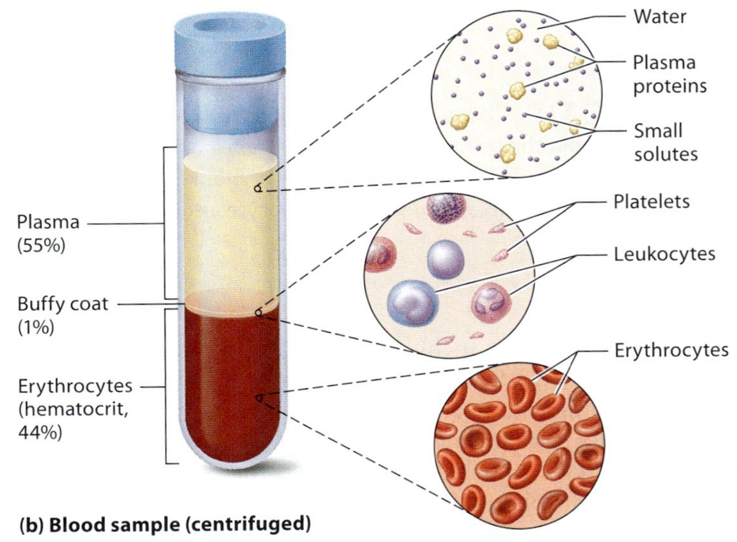

(b) Blood sample (centrifuged)

Figure 20-1 Components of blood.

CLINICAL CONNECTION

Blood Doping

When hypoxia (a low level of oxygen in the blood) occurs, the kidneys release erythropoietin (EPO), a hormone that induces the bone marrow to produce more oxygen-transporting RBCs (a process known as erythropoiesis). Some athletes try to take advantage of this physiological process by using a procedure called "blood doping." In blood doping, blood is removed from the body, reducing its oxygen-carrying capacity. The resulting hypoxia stimulates the kidneys to release EPO, which then induces the red bone marrow to increase RBC production. When the blood that was removed is injected back into the individual's bloodstream, the individual has many more RBCs than normal, thereby increasing the blood's oxygen-carrying capacity and potentially enhancing athletic performance.

component (about 44% of blood volume) is erythrocytes, which are propelled to the bottom when blood is centrifuged.

Mature **erythrocytes**, with shapes characterized as biconcave discs, are anucleate cells packed with hemoglobin, a pigment that transports both oxygen and a small percentage of carbon dioxide. The cell membrane of red blood cells (RBCs) is associated with a flexible protein called spectrin, which enables RBCs to squeeze through the smallest of capillaries. Like all formed elements, RBCs are produced in the red bone marrow.

The five different types of **leukocytes** (listed in order of abundance) are neutrophils, lymphocytes, monocytes, eosinophils, and basophils (**Figure 20-2**). Based on the presence or absence of cytoplasmic granules, these cells are divided into two major groups: granulocytes and agranulocytes.

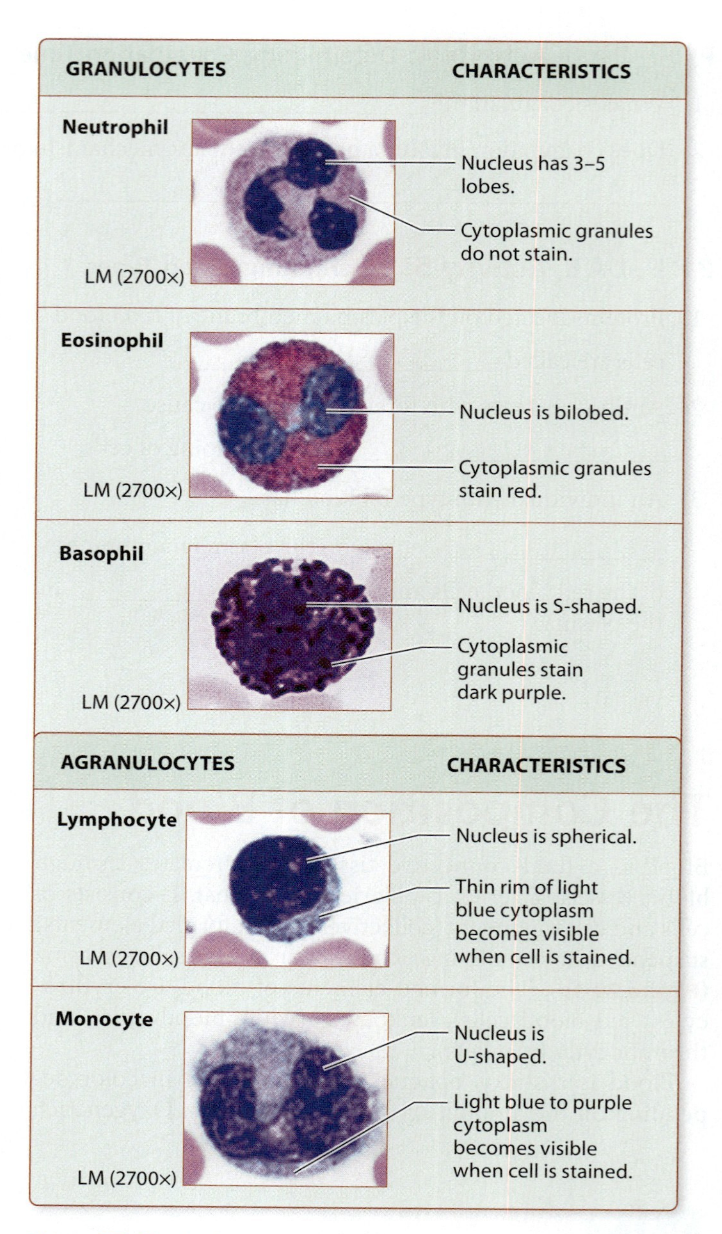

Figure 20-2 **Leukocyte types.**

QUICK TIP The following mnemonic device can help you remember the relative abundance of each leukocyte type, from most abundant to least: ***n****ever* ***l****et* ***m****onkeys* ***e****at* ***b****ananas* (***n****eutrophils,* ***l****ymphocytes,* ***m****onocytes,* ***e****osinophils,* ***b****asophils*).

The **granulocytes** include neutrophils, eosinophils, and basophils. The name of each of these granulocytes is based on the staining characteristics of granules in the cytoplasm. **Neutrophils** contain non-staining cytoplasmic granules and a multilobed nucleus. As a result, these cells are also called polymorphonuclear leukocytes. Neutrophils function primarily in phagocytosis, and their numbers rise during acute infections. **Eosinophils** have cytoplasmic granules that stain red with acidic dyes. They have a bilobed nucleus and are similar in size to neutrophils. Eosinophils increase in number during parasitic infections and allergic reactions. **Basophils**, which are the least numerous of the white blood cells, contain cytoplasmic granules that stain dark purple with basic dyes and a large S-shaped nucleus. Basophils sometimes pass through intact blood vessel walls into the surrounding tissues, a process termed diapedesis. When basophils migrate into the tissues, they become mast cells. Mast cells produce a variety of substances, including histamine, a chemical that enhances the inflammatory response.

The **agranulocytes**, which lack brightly staining granules, include lymphocytes and monocytes. **Lymphocytes**, which

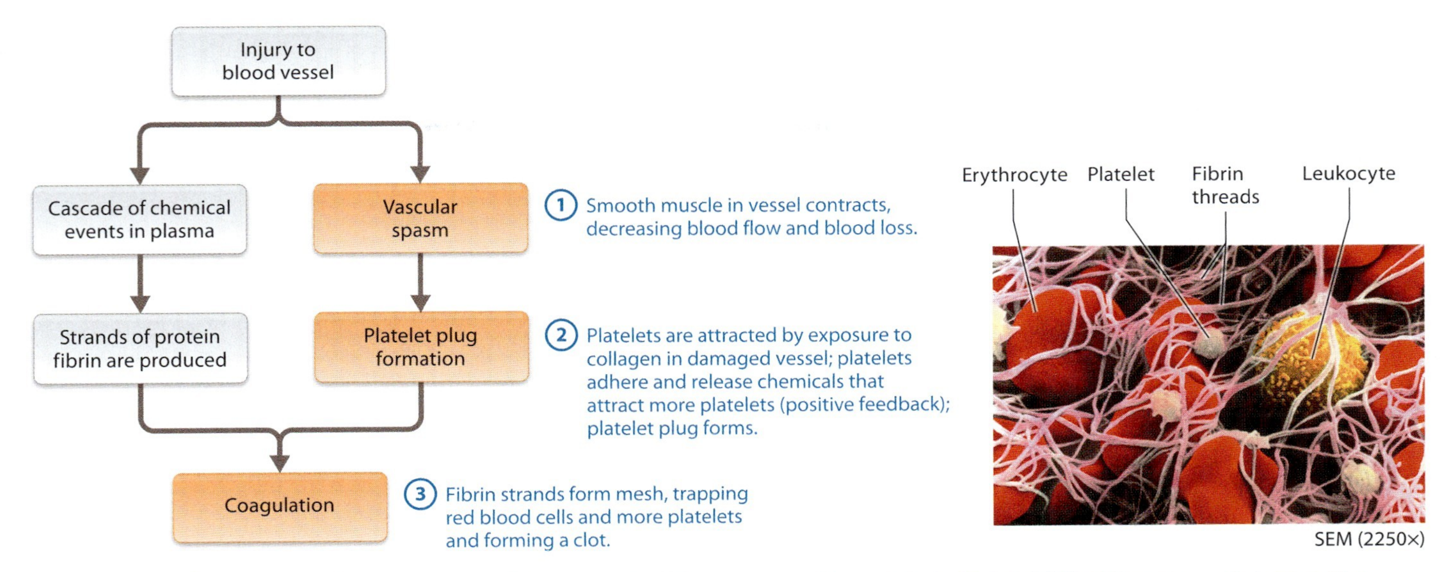

Figure 20-3 The events leading up to coagulation, a part of the process of hemostasis. Two additional events—clot retraction and thrombolysis—follow coagulation.

play active roles in the immune response, are small cells in which the nucleus takes up most of the cell's volume. There are two types of lymphocytes: B lymphocytes, which mediate the immune response via the production of antibodies, and T lymphocytes, which act via direct cellular attack. **Monocytes**, the largest of the leukocytes, have a large U-shaped nucleus. Monocytes also migrate via diapedesis into the tissues, where they become macrophages that attack and destroy bacteria and viruses via phagocytosis.

The formed elements called **platelets**, or thrombocytes, are actually fragments of large multinucleate cells called megakaryocytes, located in the red bone marrow. When blood vessels are damaged, platelets play an important role by mediating the process of blood clotting, or coagulation, to prevent excessive blood loss. Coagulation is but one part of hemostasis, a multistep process that results in the stoppage of bleeding. Hemostasis involves a positive feedback mechanism based on the interaction of a variety of substances, including plasma proteins and chemicals released by platelets and damaged body cells. **Figure 20-3** outlines the basic events that lead to coagulation.

ACTIVITY 1
Exploring the Formed Elements of Blood

Learning Outcomes

1. Identify erythrocytes, leukocytes, and platelets in a prepared blood smear slide.
2. Distinguish among the five types of leukocytes and identify each in a prepared blood smear slide.
3. Sketch each type of formed element and relate the structure of each to its function.

Materials Needed

- ☐ Microscope and blood smear slide (or photomicrographs)

Instructions

1. View a blood smear under high power (or view a photomicrograph). Then make a sketch of your observations and label the following: erythrocytes, leukocytes, and platelets.

Total magnification: ______×

State the function of each formed element:

Erythrocyte: ________________________________

____

Leukocyte: ________________________________

____

Platelet: ________________________________

____

2. Now, switch to the oil-immersion lens (or view a photomicrograph) and locate each type of leukocyte. Note that you will probably need to view several different areas of the slide to locate all five leukocyte types: neutrophil, basophil, eosinophil, monocyte, and lymphocyte. Sketch each cell type below and write a brief description of each in the blanks that follow.

Total magnification: _____×

Neutrophil: ________________________________

____

Total magnification: _____×

Basophil: ________________________________

____

Total magnification: _____×

Eosinophil: ________________________________

____

Total magnification: _____×

Monocyte: ________________________________

____

Total magnification: _____×

Lymphocyte: ________________________________

____

3. Based on your histological observations of blood:

   a. Distinguish between granulocytes and agranulocytes.

   ____

   ____

   b. Distinguish between monocytes and neutrophils.

   ____

   ____

   c. Distinguish between lymphocytes and erythrocytes.

   ____

   ____

   d. Distinguish between eosinophils and basophils.

   ____

   ____

   e. Distinguish between monocytes and lymphocytes.

   ____

   ____

4. **Optional Activity**

PAL™ View and label histology slides of blood at MasteringA&P® > Study Area > Practice Anatomy Lab > Histology > Cardiovascular System > Images 1–7

ACTIVITY 2
Performing a Hematocrit

Learning Outcomes

1. Perform a hematocrit.
2. Differentiate between normal and abnormal hematocrit results.
3. Explain possible causes of abnormal hematocrit values.

Materials Needed

- ☐ Biohazard waste container
- ☐ Sharps biohazard disposal container
- ☐ Disposable gloves
- ☐ Safety goggles
- ☐ 4 sterile, disposable lancets
- ☐ Alcohol wipes
- ☐ 10% bleach solution in spray bottle
- ☐ Sterile gauze pads
- ☐ 4 small circular adhesive bandages
- ☐ Human or animal blood
- ☐ Sterile, heparinized capillary tubes
- ☐ Sealing clay
- ☐ Microcentrifuge
- ☐ Microhematocrit reader (or millimeter ruler)

Instructions

The hematocrit, or packed cell volume, is the percentage of erythrocytes in a volume of blood sample. The procedure for determining hematocrit requires the use of centrifugation to separate erythrocytes from the leukocytes, platelets, and plasma in a sample of whole blood. Determine the hematocrit of a blood sample by performing the following steps:

1. ⚠ Put on safety goggles and disposable gloves.
2. Obtain a blood sample. If you are using your own blood, remove one glove, clean the fingertip of your third or fourth finger with an alcohol wipe, and prick your finger with a sterile lancet. Dispose of the lancet in the sharps biohazard disposal container. Use a small square of sterile gauze to wipe away the first drop of blood. Then place one end of a sterile, heparinized capillary tube (contains heparin, which prevents the blood from coagulating) just into the drop of blood, holding the tube at an angle so that blood will enter the tube by capillary action (see **Figure 20-4a**). Fill the tube at least two-thirds full. (If using a blood sample provided by your instructor, simply immerse the capillary tube in the blood sample and fill it two-thirds full.) Stop the flow of blood by holding a folded square of sterile gauze over the prick site for a minute or so, and then place an adhesive bandage over the site. Replace the glove that was previously removed. Throw alcohol wipes and gauze in the biohazard waste container.
3. While holding your index finger over the dry end of the capillary tube, seal the blood-containing end of the tube by gently pushing it into the sealing clay (see **Figure 20-4b**). Be very careful because capillary tubes are extremely fragile.
4. Place the tube into one of the numbered grooves of the microcentrifuge, with the sealed end of the tube against the outer rubber lining on the rim (see **Figure 20-4c**). Note that the centrifuge must be balanced by having another student place a capillary tube directly opposite yours. An empty capillary tube sealed with clay can be used to balance the centrifuge if another student's sample is not available. Be sure to note the numbered groove in which you place your tube so that you will be sure to remove the correct tube.
5. Place the metal top on the centrifuge and screw it into place. Close the cover of the centrifuge and spin the capillary tubes for 4 minutes.
6. When the centrifuge stops, carefully remove your capillary tube. The erythrocytes are now packed into the bottom of the tube, and the clear liquid on top is the

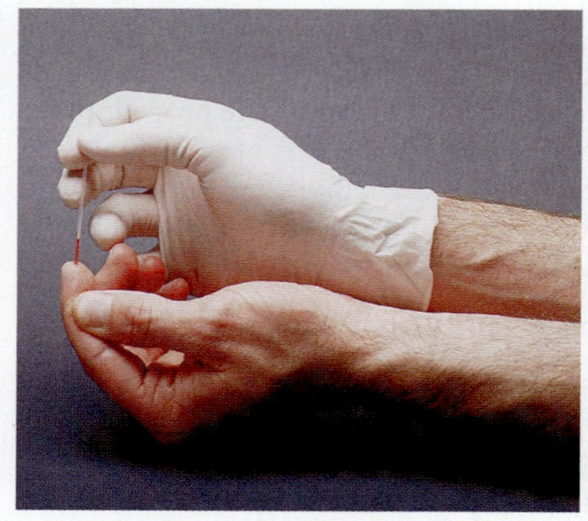

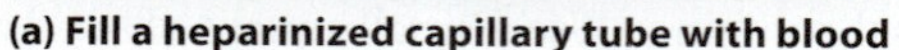

(a) Fill a heparinized capillary tube with blood

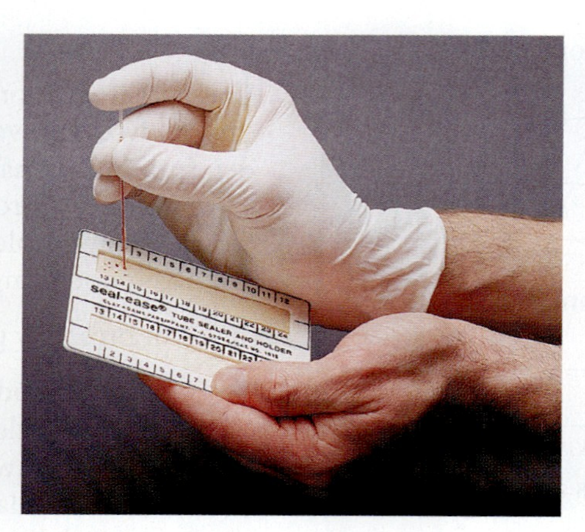

(b) Seal the end of the tube with clay

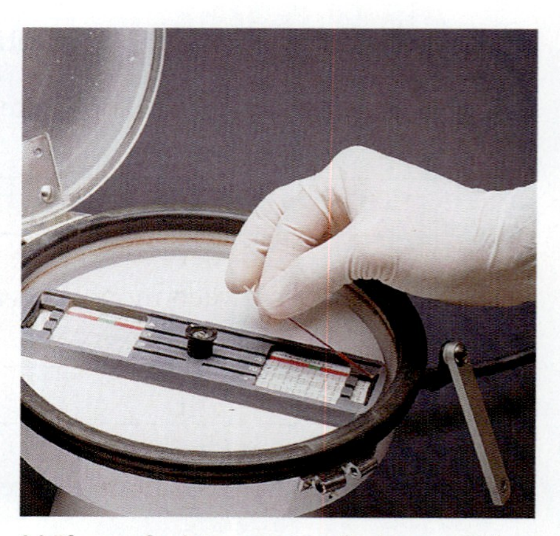

(c) Place tube into groove of microcentrifuge

Figure 20-4 Steps in preparing a blood sample for hematocrit determination.

plasma. Between the erythrocytes and the plasma is the buffy coat, which contains leukocytes and platelets.

7. Determine the hematocrit using a microhematocrit reader. Several types of readers are available; your instructor will show you how to use the reader in your lab. If a reader is not available, then you can still calculate the hematocrit for your blood sample by completing the following steps:
   a. Use a millimeter ruler to measure the length of the whole column of blood in the tube (including erythrocytes, buffy coat, and plasma).

      Length of whole column = ______________

   b. Now, measure the length of the packed erythrocytes in millimeters, and record your measurement:

      Length of packed erythrocytes = ______________

   c. Use the following formula to calculate the percentage of erythrocytes in whole blood (hematocrit):

$$\text{Hematocrit} = \frac{\text{Length of erythrocytes (in mm)}}{\text{Length of whole column (in mm)}} \times 100\% = \_\_\_\_\_\_\_\_$$

8. Dispose of your capillary tube in the sharps biohazard disposal container and clean your workspace with some 10% bleach solution.
9. Remove your gloves and put them in the biohazard waste container. Wash your hands thoroughly with soap and water.
10. Work with your lab group to answer the following question set:
    a. What is the normal hematocrit value for a healthy adult male? ______________
    b. What is the normal hematocrit value for a healthy adult female? ______________
    c. What might cause a low hematocrit value? ______________

       ____

ACTIVITY 3
Performing a Differential White Blood Cell Count

Learning Outcomes

1. Perform a differential white blood cell count.
2. Differentiate between normal and abnormal results for a differential white blood cell count.
3. Describe the significance of elevated and depressed white blood cell counts.

Materials Needed

- ☐ Prepared slide of normal blood
- ☐ Prepared slide of abnormal blood
- ☐ Compound microscope

Instructions

A differential white blood cell count is performed to determine the percentages of each white blood cell type in a blood sample. In a clinical laboratory, this procedure can be done rapidly with computers. In this lab activity, you will perform a differential white blood cell count manually using prepared microscope slides and a microscope. Complete the following steps:

1. Position your slide of a normal human blood smear in the microscope's field of view so that you are viewing the blood smear near the lower left corner of the slide.
2. Use the low-power objective to bring the blood smear into focus. Then switch to the high-power objective.

Although this activity can be completed with the high-power objective, the best results are obtained using the oil-immersion objective. If you are told by your instructor to use the oil-immersion objective, review the proper procedure for its use in Unit 4.

3. You are going to identify each leukocyte of a total of 100 white blood cells. Begin by identifying the white blood cells in the first field of view, and place tally marks in **Table 20-1** to keep track of the numbers of each white blood cell type. Each time you identify a white blood cell, place one tally mark in the "Number of Cells Observed" column and one tally mark in the "Running Tally" column. The "Running Tally" column will help you to know when you have identified 100 white blood cells. To avoid counting any white blood cells more than once, it is important to use a systematic approach to moving the slide, as shown in **Figure 20-5**. Begin moving the slide up, identifying the WBCs for each new field of view, until you reach the top of the smear; then move the slide to the right for several fields of view, then down to near the bottom of the slide, then right for several fields, and so on, until you have counted a total of 100 white blood cells.
4. Repeat steps 1–3 using an abnormal human blood smear this time. Once again, record your results in Table 20-1.

Table 20-1 Tally Chart for Differential WBC Count

WBC Type	Number of Cells Observed	Running Tally	Percent of Total Cells Observed
NORMAL BLOOD SMEAR			
Neutrophil			
Basophil			
Eosinophil			
Monocyte			
Lymphocyte			
TOTAL	100	100	100%
ABNORMAL BLOOD SMEAR			
Neutrophil			
Basophil			
Eosinophil			
Monocyte			
Lymphocyte			
TOTAL	100	100	100%

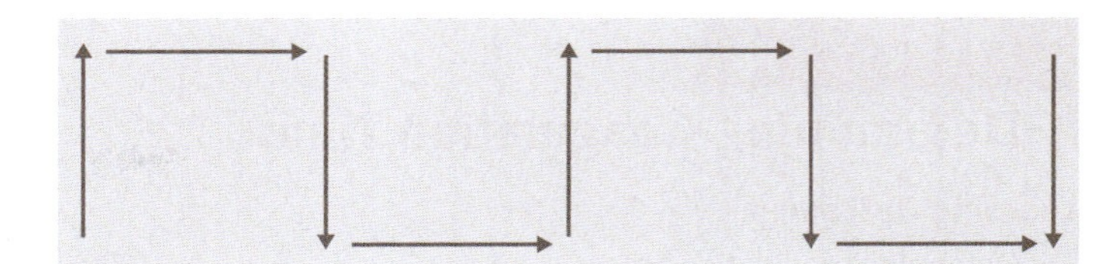

Figure 20-5 Method of moving the microscope slide when performing a differential white blood cell count.

5. Work with your lab group to answer the following question set:
   a. Based on your data for a normal blood smear, list the white blood cell types from most numerous to least numerous. Then, circle the agranulocytes.

   ____

   ____

   b. What might elevated numbers of white blood cells indicate? ______________________________

   ____

   c. What might decreased numbers of white blood cells indicate? ______________________________

   ____

   d. Did your results match the expected results for a normal blood smear? Explain. ______________________________

   e. In what ways did the white blood cell percentages in the table differ between the normal blood smear and the abnormal blood smear? ______________________________

   ____

   f. What might the white blood cell percentages of the abnormal blood smear indicate? Defend your answer.

   ____

   ____

   ____

ACTIVITY 4

Determining Coagulation Time

Learning Outcomes

1. Determine coagulation time.
2. Explain the importance of coagulation.
3. Outline the major events that occur during coagulation.

Materials Needed

- ☐ Biohazard waste container
- ☐ Sharps disposal container
- ☐ Disposable gloves
- ☐ Safety goggles
- ☐ Sterile, disposable lancet
- ☐ Alcohol wipes
- ☐ 10% bleach solution in spray bottle
- ☐ Sterile gauze pads
- ☐ 4 small circular adhesive bandages
- ☐ Human or animal blood
- ☐ Sterile, nonheparinized capillary tubes
- ☐ Paper towels
- ☐ Small triangular metal file
- ☐ Timer

Instructions

We saw in Figure 20-3 that coagulation, the formation of a blood clot, is an important part of the process of stopping bleeding. But when blood is removed from the body, it will also coagulate, typically within about 2–6 minutes. This process is prolonged in individuals with bleeding disorders such as hemophilia and in individuals being treated with anticoagulants such as heparin or aspirin. To determine coagulation time, complete the following steps:

1. Put on safety goggles and disposable gloves.
2. Obtain a blood sample. If you are using your own blood, remove one glove, clean the fingertip of your third or fourth finger with an alcohol wipe, and prick your finger with a sterile lancet. Dispose of the lancet in the sharps disposal container. Use a small square of sterile gauze to wipe away the first drop of blood. Then place one end of a sterile, nonheparinized capillary tube just into the drop of blood, holding the tube at an angle so that blood will enter the tube by capillary action. Fill the tube with blood. (If using a blood sample provided by your instructor, simply immerse the capillary tube in the blood sample and fill it with blood.) Stop the flow of blood by holding a folded square of sterile gauze over the prick site for a minute or so, and then place an adhesive bandage over the site. Replace the glove that was previously removed. Throw alcohol wipes and gauze in the biohazard waste container.
3. Place the capillary tube filled with blood on a paper towel, and start the timer.
4. After 30 seconds, scratch the tube with a metal file approximately 1 cm from one end, and then gently break the tube where you scratched it.
5. Gently pull the two ends of the tube apart and look for a thin fibrin thread.
6. If no fibrin is present, wait 30 seconds and repeat steps 4 and 5.
7. Repeat this process (wait 30 seconds, scratch and break the tube, look for a fibrin thread) until you see the fibrin thread, and record the elapsed time, which is the coagulation time for your blood sample: ____________
8. Clean up by placing all lancets and capillary tubes in the sharps biohazard disposal container and then wash your lab bench thoroughly with 10% bleach solution.
9. Remove your gloves and put them in the biohazard waste container. Wash your hands thoroughly with soap and water.
10. Work with your lab group to answer the following question set:
    a. What is the range for normal coagulation times?

    ____

    b. What might a prolonged coagulation time indicate?

    ____

    ____

    ____

    c. Why did you use a nonheparinized capillary tube?

    ____

    d. Outline the sequence of events that takes place to prevent blood loss. Begin with injury to a blood vessel and end with fibrin clot formation.

    ____

    ____

    ____

Blood Typing

Blood types are genetically determined traits based on the presence or absence of specific glycoproteins (antigens) embedded in the cell membranes of RBCs. Although human RBCs display many different antigens, only those in two groups—the ABO system and the Rh system—are routinely typed. There are four variants in the ABO system, and two variants in the Rh system; therefore, the combined use of these two blood typing systems produces eight major blood types, as shown in **Table 20-2**.

The four variants in the ABO system are A, B, AB, and O. Type A blood contains RBCs displaying A antigens. Type B blood contains RBCs displaying B antigens. Type AB blood contains RBCs displaying both A antigens and B antigens. Type O blood contains RBCs that display neither A antigens nor B antigens.

Table 20-2 The Eight Major Blood Types

Blood Type	Prevalence in the U.S. Population	Antigens Present on Erythrocyte Surface	Antibodies Present in Plasma	May Receive from	May Donate to
AB+	3%	A, B, Rh 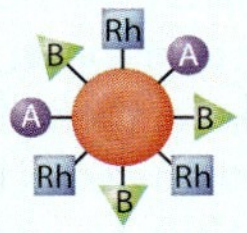	None	Universal recipient	AB+
AB−	1%	A, B 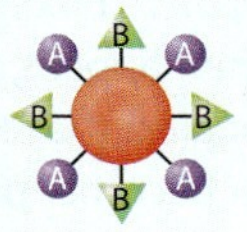	Anti-Rh 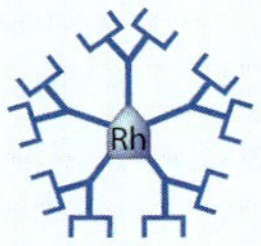	AB−, A−, B−, O−	AB+, AB−
A+	34%	A, Rh 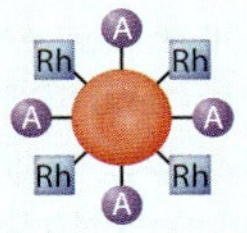	Anti-B 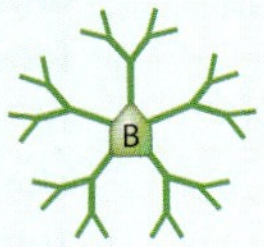	A+, A−, O+, O−	AB+, A+
A−	6%	A 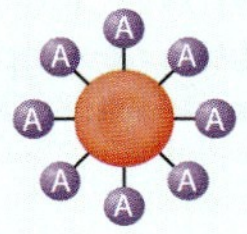	Anti-B, anti-Rh 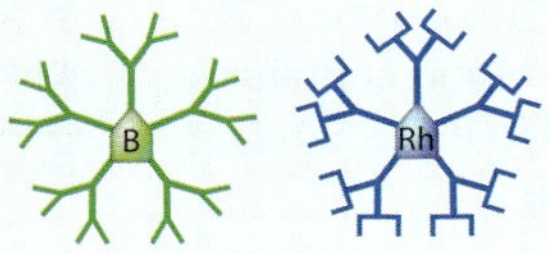	A−, O−	AB+, AB−, A+, A−
B+	9%	B, Rh 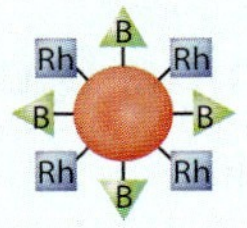	Anti-A 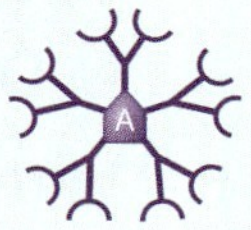	B+, B−, O+, O−	AB+, B+
B−	2%	B 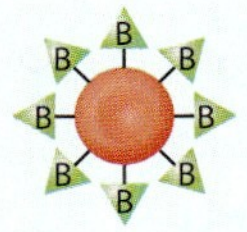	Anti-A, anti-Rh 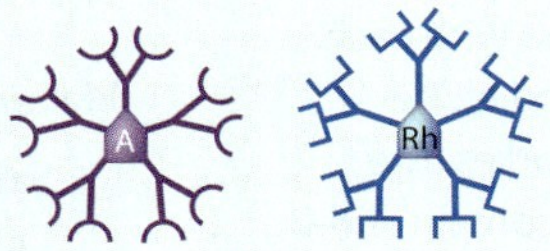	B−, O−	AB+, AB−, B+, B−
O+	38%	Rh 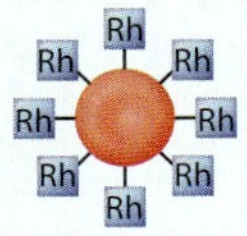	Anti-A, anti-B 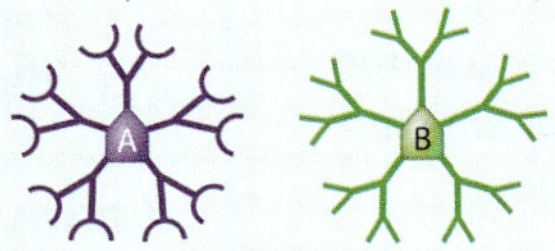	O+, O−	AB+, A+, B+, O+
O−	7%	None 	Anti-A, anti-B, anti-Rh 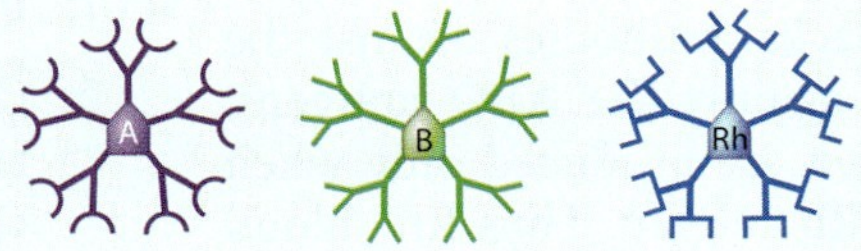	O−	Universal donor

QUICK TIP The O in type O blood looks a lot like a zero, and zero is how many antigens a type O RBC has.

The "Rh" in the Rh system stands for rhesus monkey because the Rh antigen (also called the Rh factor) was first discovered in this primate. There are only two blood types in the Rh system. Blood containing RBCs with Rh antigens is considered Rh positive (Rh^+), whereas blood with RBCs that lack Rh antigens is considered Rh negative (Rh^-).

The plasma portion of blood also contains specialized proteins called antibodies, which chemically react with (bind to) specific types of antigens. When an antibody encounters its specific antigen, the two will bind. In the case of blood, should an antibody in plasma meet its specific antigen on an RBC, the resulting binding produces an agglutination reaction (the clumping of RBCs). To cite one example, the type of antibody that binds with an A antigen is called an anti-A antibody.

The plasma of type A blood contains anti-B antibodies, that of type B blood contains anti-A antibodies, that of type AB blood contains no antibodies, and that of type O blood contains both anti-A antibodies and anti-B antibodies (see Table 20-2). These antibodies begin forming shortly after birth.

QUICK TIP Here's another way to remember which antibodies are in the plasma for each ABO blood type.

What would happen if type A blood, which has RBCs displaying A antigens, had plasma that contained anti-A antibodies? The blood would agglutinate (clump), because the A antigens and the anti-A antibodies would bind with each other! So type A blood *cannot* contain anti-A antibodies, because then blood could not flow throughout our bodies. The same reasoning applies for type B blood; it cannot contain anti-B antibodies, because if it did it would clump. Type AB blood has RBCs with both A and B antigens, so it cannot have either anti-A or anti-B antibodies in the plasma. Finally, because the RBCs of type O blood have no A or B antigens, its plasma can (and does) contain both anti-A and anti-B antibodies.

The plasma of Rh^+ blood lacks antibodies against the Rh antigen. Unlike antibodies against the ABO antigens, antibodies against the Rh antigen are produced only when an individual is exposed to the Rh antigen. Thus Rh^+ blood lacks antibodies against the Rh factor, and individuals with Rh^- blood do not produce antibodies against the Rh factor unless they are exposed to it. Such exposure can occur during a blood transfusion or during childbirth.

CLINICAL CONNECTION

Erythroblastosis Fetalis

Recall that individuals do not produce antibodies against Rh antigens unless exposed to the antigen. If a pregnant Rh^- woman's first child is Rh positive, it is likely that no problems will develop for the child, because normally, fetal blood does not mix with maternal blood. The fetus's Rh^+ RBCs do not pass through the placenta into the mother's circulation, so she is not exposed to Rh antigens. During the process of birth, however, the mother's blood may come into contact with fetal Rh antigens. If this happens, the Rh^- mother will begin producing antibodies against the Rh antigens in her child's blood. If this woman later becomes pregnant with a second Rh^+ child, a potentially life-threatening situation can develop. The mother's anti-Rh antibodies can pass through the placenta and enter the fetal circulation. In this case, the anti-Rh antibodies bind to the Rh antigens on the fetal RBCs, and agglutination occurs. This clumping of RBCs leads to erythroblastosis fetalis, a sometimes serious condition in which the destruction of RBCs can lead to anemia. Rh^- women are treated with a drug called RhoGAM, which destroys any Rh^+ fetal cells that enter her blood, thereby preventing her from developing anti-Rh antibodies.

ACTIVITY 5

Determining Blood Types

Learning Outcomes

1. Explain the physiological basis of the ABO and Rh blood groups.
2. Type unknown blood samples and interpret results.

Materials Needed

- ☐ Safety goggles
- ☐ Disposable gloves
- ☐ Four simulated blood samples (samples #1 through #4)
- ☐ Microscope slides
- ☐ Wax pencil
- ☐ Solutions containing anti-A, anti-B, or anti-Rh antibodies
- ☐ 12 toothpicks

Instructions

Procedure A:

1. Put on safety goggles and disposable gloves.
2. Obtain four clean microscope slides. Using a wax pencil, label the slides "1," "2," "3," and "4." Then, draw three circles on each slide and label the circles by writing "A," "B," and "Rh" next to them.
3. Place a drop of simulated blood (sample #1) in each of the three circles on slide 1.

BLOOD TYPE	**Anti-A** antibodies agglutinate A antigens.	**Anti-B** antibodies agglutinate B antigens.
A		
B		
AB		
O		

Figure 20-6 Examples of ABO blood type testing.

4. Add a drop of anti-A antibodies to the blood sample in circle "A." Add a drop of anti-B antibodies to the blood sample in circle "B," and add a drop of anti-Rh antibodies to the blood sample in circle "Rh."
5. Using a different toothpick for each blood sample, gently mix the drop containing antibodies and blood in each circle.
6. Examine each mixture to determine whether agglutination has occurred. See **Figure 20-6** for examples.
7. Record the data in the first row of **Table 20-3**, and identify the blood type of sample #1.

Table 20-3 Determining Blood Types

Blood Sample	Agglutination with Anti-A Antibodies? (yes / no)	Agglutination with Anti-B Antibodies? (yes / no)	Agglutination with Anti-Rh Antibodies? (yes / no)	Blood Type
#1				
#2				
#3				
#4				

8. Repeat steps 3–7 using blood sample #2, sample #3, and sample #4.
9. Follow your instructor's directions for cleanup.

Optional Procedure B:

Blood typing can be a confusing topic. Students often have a difficult time differentiating between antigens and antibodies, as well as interpreting blood typing results. If you are struggling with these concepts, the following hands-on activity might help you better understand the biology of blood types.

1. Copy (or trace) and cut out the "RBCs" and the "antibodies" in **Figure 20-7**. Place the two different kinds of cutouts representing blood types A, B, AB, and O into four separate plastic bags labeled "type A blood," "type B blood," "type AB blood," and "type O blood." Remember that type A blood contains anti-B antibodies, type B blood contains anti-A antibodies, type AB blood contains no antibodies, and type O blood contains both anti-A and anti-B antibodies.
2. Take the cutouts that represent **type A blood**—both cells and antibodies—out of their bag and place them on the lab table.

   Which antigens are located on these red blood cells?

   ____

   Which antibodies are in the plasma of this blood?

   ____

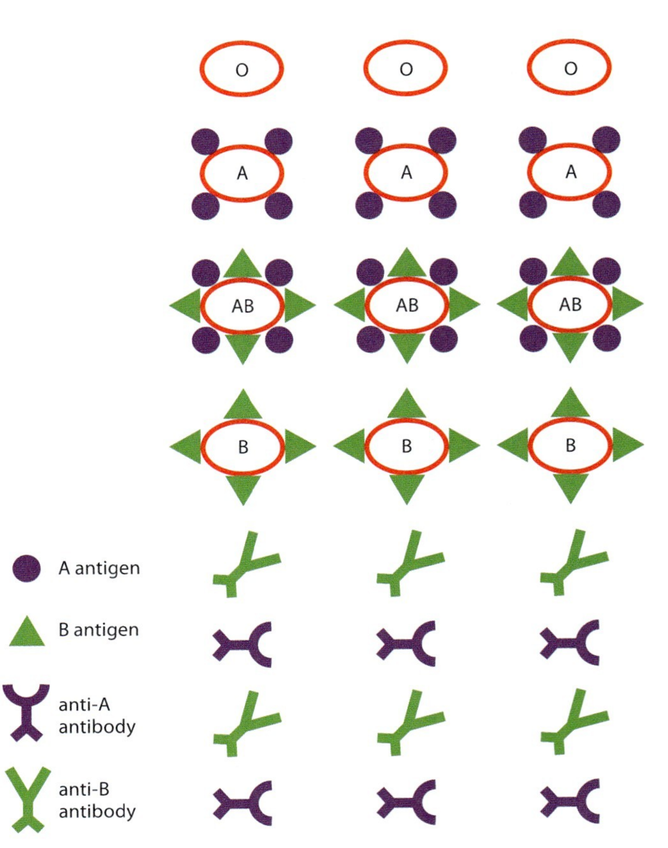

Figure 20-7 Worksheet for blood typing activity.

3. Now take only the cutouts representing **type B red blood cells** out of their bag, and place them on the table near the cutouts for type A blood.
4. Demonstrate "agglutination" by matching the anti-B antibodies from the type A blood with the RBCs bearing B antigens from the type B blood.
5. Return all of the cutouts for type A blood and type B blood to their respective bags.
6. Now place both kinds of cutouts representing type AB blood on the lab table. Then add only the cutouts representing **type O red blood cells** to the type AB blood.

   Does agglutination occur? ______________________

   Why or why not? ______________________

   ____

7. Return all of the cut-outs for type AB blood and type O blood to their respective bags.
8. Now, place the cutouts representing type A blood on the lab table, and add only the cutouts representing **type O red blood cells** to the type A blood.

   Does agglutination occur? ______________________

   Why or why not? ______________________

   ____

9. Answer the following questions:

   a. Why did you add only "RBCs" to "blood" instead of both "RBCs" and "antibodies"? ______________________

   ____

   ____

   b. Which blood type is the universal recipient? __________

   Why? ______________________

   ____

   c. Which blood type is the universal donor? __________

   Why? ______________________

   ____

PhysioEx EXERCISE 11

Blood Analysis

The PhysioEx 9.1 Laboratory Simulations in Physiology are easy-to-use laboratory simulations that can be used as an alternative to or as a supplement to wet lab activities in this unit. Each simulation allows you to investigate important physiological concepts, repeat labs as often as you like, and conduct experiments that are difficult to perform in a wet lab environment because of time, cost, or safety concerns.

Access the simulations in these activities at MasteringA&P® > Study Area > PhysioEx 9.1

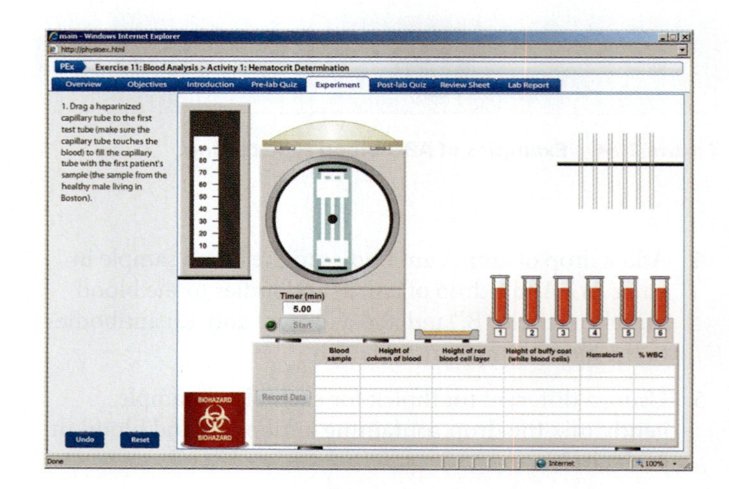

There you will find the following activities:

PEx Activity 1: Hematocrit Determination
PEx Activity 2: Erythrocyte Sedimentation Rate
PEx Activity 3: Hemoglobin Determination
PEx Activity 4: Blood Typing
PEx Activity 5: Blood Cholesterol

Post-lab quizzes are also assignable in MasteringA&P®

POST-LAB ASSIGNMENTS

Name: ______________________ Date: ______________ Lab Section: ________

PART I. Check Your Understanding

Activity 1: Exploring the Formed Elements of Blood

1. Identify the formed elements shown in the accompanying photomicrographs.

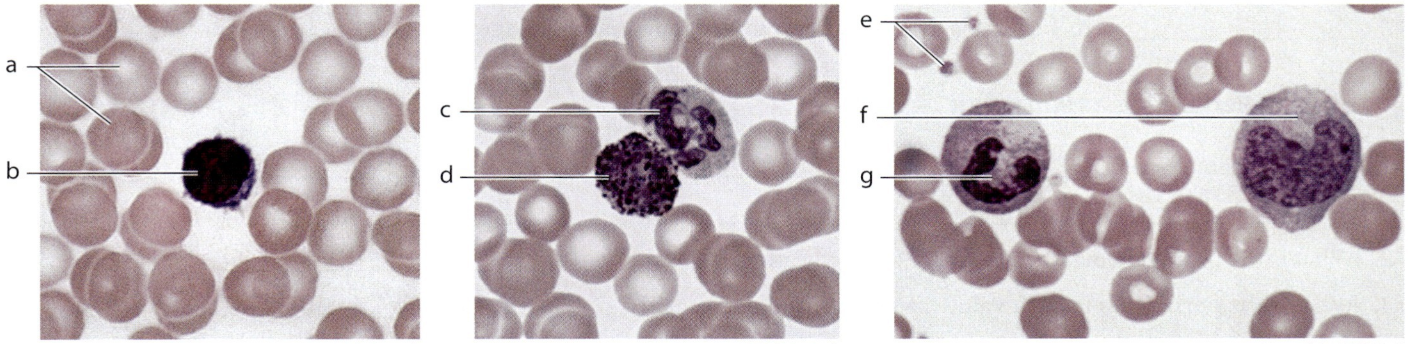

a
b

LM (1200×)

c
d

LM (1200×)

e
f
g

LM (1200×)

a. ______________________
b. ______________________
c. ______________________
d. ______________________
e. ______________________
f. ______________________
g. ______________________

2. Which formed element is not shown in the photomicrographs in Question 1? ______________________

3. What do formed elements "a" and "e" in the photomicrographs of Question 1 have in common?

____

4. What do formed elements "c" and "d" in the photomicrographs of Question 1 have in common?

____

5. Which formed elements (erythrocytes, leukocytes, or platelets):

______________ a. contain a nucleus?

______________ b. are packed with hemoglobin?

______________ c. are cellular fragments?

______________ d. function in immunity?

______________ e. function in blood clotting?

______________ f. contain spectrin?

6. Which leukocyte(s):

_______________ a. contain a bilobed nucleus?

_______________ b. are classified as agranulocytes?

_______________ c. function in the production of antibodies?

_______________ d. increase in number during parasitic infections?

_______________ e. are known as polymorphonuclear leukocytes?

_______________ f. become macrophages when they migrate into the tissues?

_______________ g. contain a large, U-shaped nucleus?

_______________ h. become mast cells when they migrate into the tissues?

Activity 2: Performing a Hematocrit

1. What is a hematocrit? _______________

____

2. What is the normal hematocrit value for a healthy adult male? _______________
3. What is the normal hematocrit value for a healthy adult female? _______________
4. What might a low hematocrit indicate? _______________

____

Activity 3: Performing a Differential White Blood Cell Count

1. What is a differential white blood cell count? _______________

____

2. List the leukocytes in order, from most abundant to least abundant, as seen in a normal blood smear.

a. _______________ b. _______________ c. _______________

d. _______________ e. _______________

3. How did the white blood cell count of the abnormal blood smear differ from that of the normal blood smear? _______________

____

4. Name two possible reasons for elevated white blood cell numbers. _______________

____

Activity 4: Determining Coagulation Time

1. What was the coagulation time determined for the blood sample? _______________
2. What role do platelets play in coagulation? _______________

Activity 5: Determining Blood Types

1. Distinguish between an antigen and an antibody. _______________

____

2. Complete the following table:

Blood Type	Antibodies Present on Surface of RBCs	Antibodies Present in Plasma	Can Receive Blood from Blood Type(s):	Can Donate Blood to Blood Type(s):
A−				
O+				
AB−				
B+				

3. Based on the following blood typing results, identify the blood type of each blood sample:

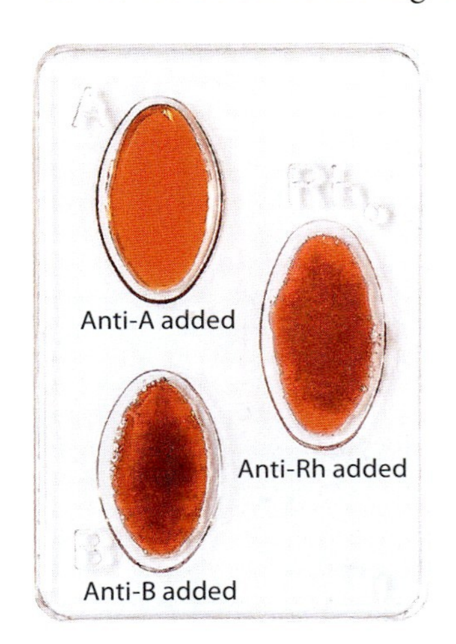

Blood type: ______________

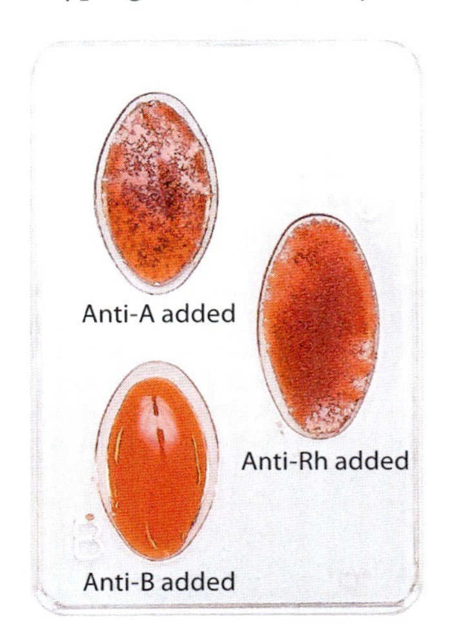

Blood type: ______________

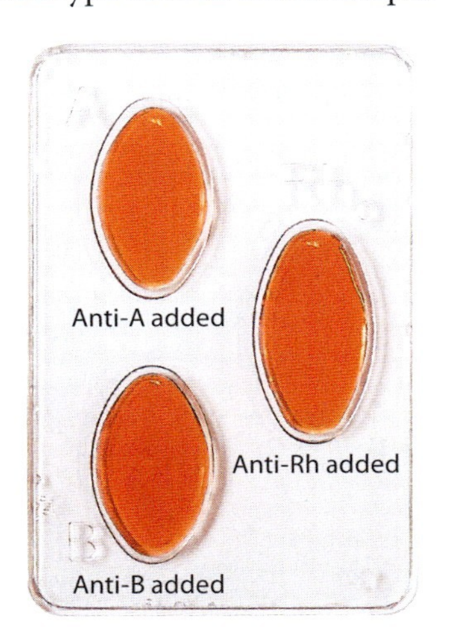

Blood type: ______________

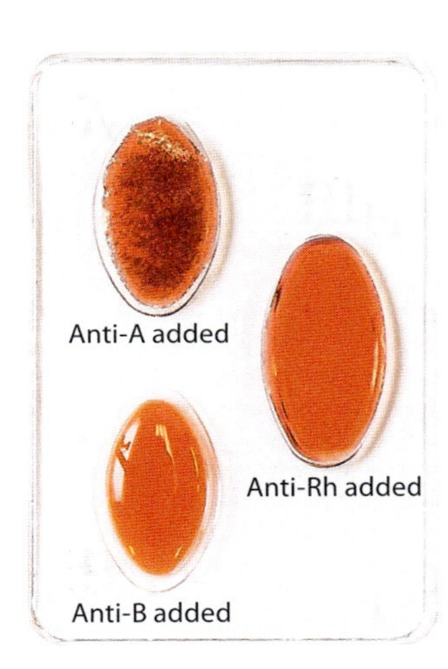

Blood type: ______________

PART II. Putting It All Together

A. Review Questions

Answer the following questions using your lecture notes, your textbook, and your lab notes:

1. Explain how each of the following formed elements is structurally adapted to its function:
   a. basophil ____
   ____
   b. monocyte ____
   ____
   c. erythrocyte ____
   ____

2. Provide an explanation for why normal hematocrit values are higher in males than in females.
   ____
   ____

3. An A− woman's first child is O+. Explain why this is usually not a problem for the developing baby.

____

____

____

4. If this woman has a second child (who is A+), the developing baby can develop erythroblastosis fetalis. Explain why the second child is in danger. _______________________

____

____

B. Concept Mapping

1. Fill in the blanks to complete this concept map outlining the white blood cell types.

eosinophil	lymphocyte	macrophage	mast cell	neutrophil

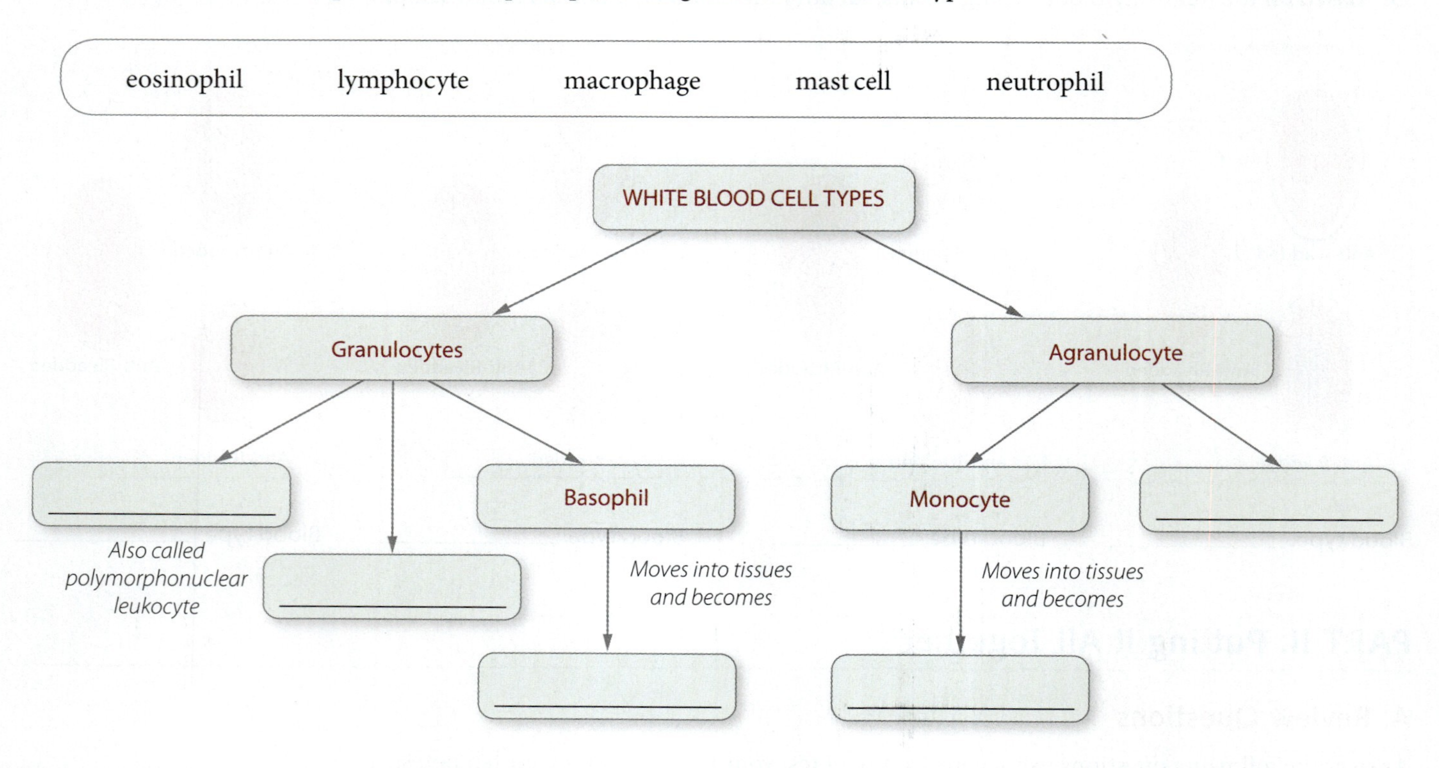

2. Construct a unit concept map to show the relationships among the following set of terms. Include all of the terms in your diagram. Your instructor may choose to assign additional terms.

agglutination	antibody	antigen	anucleate	diapedesis
eosinophil	hemoglobin	hormone	lymphocyte	macrophage
mast cell	neutrophil	plasma	plasma membrane	spectrin

21

Anatomy of the Heart

UNIT OUTLINE

The cardiovascular system—the blood, heart, and blood vessels—delivers oxygen, nutrients, hormones, and other vital substances to the tissues of the body and then transports waste materials from the body cells to their elimination sites. In the previous unit, you studied the components and functions of blood. In this unit you will examine the gross and microscopic features of the heart that enable it to beat 100,000 times each day and to pump over 5 liters of blood every minute—without ever resting—for your entire lifetime. Then, in the next unit, you will explore the physiology of the heart.

THINK ABOUT IT *Name three specific tissue types predominant in the heart, and describe the function of each.*

1. ______________________________

____

2. ______________________________

____

3. ______________________________

____

PRE-LAB ASSIGNMENTS

Pre-lab quizzes are also assignable in MasteringA&P®

To maximize learning, BEFORE your lab period carefully read this entire lab unit and complete these pre-lab assignments using your textbook, lecture notes, and prior knowledge.

PRE-LAB Activity 1: Examining the Functional Anatomy of the Heart

1. Use the list of terms to label the accompanying illustration of the heart. Check off each term as you label it.

☐ aorta	☐ left ventricle	☐ right ventricle
☐ mitral (bicuspid) valve	☐ pulmonary valve	☐ superior vena cava
☐ aortic valve	☐ pulmonary trunk	☐ tricuspid valve
☐ left atrium	☐ right atrium	

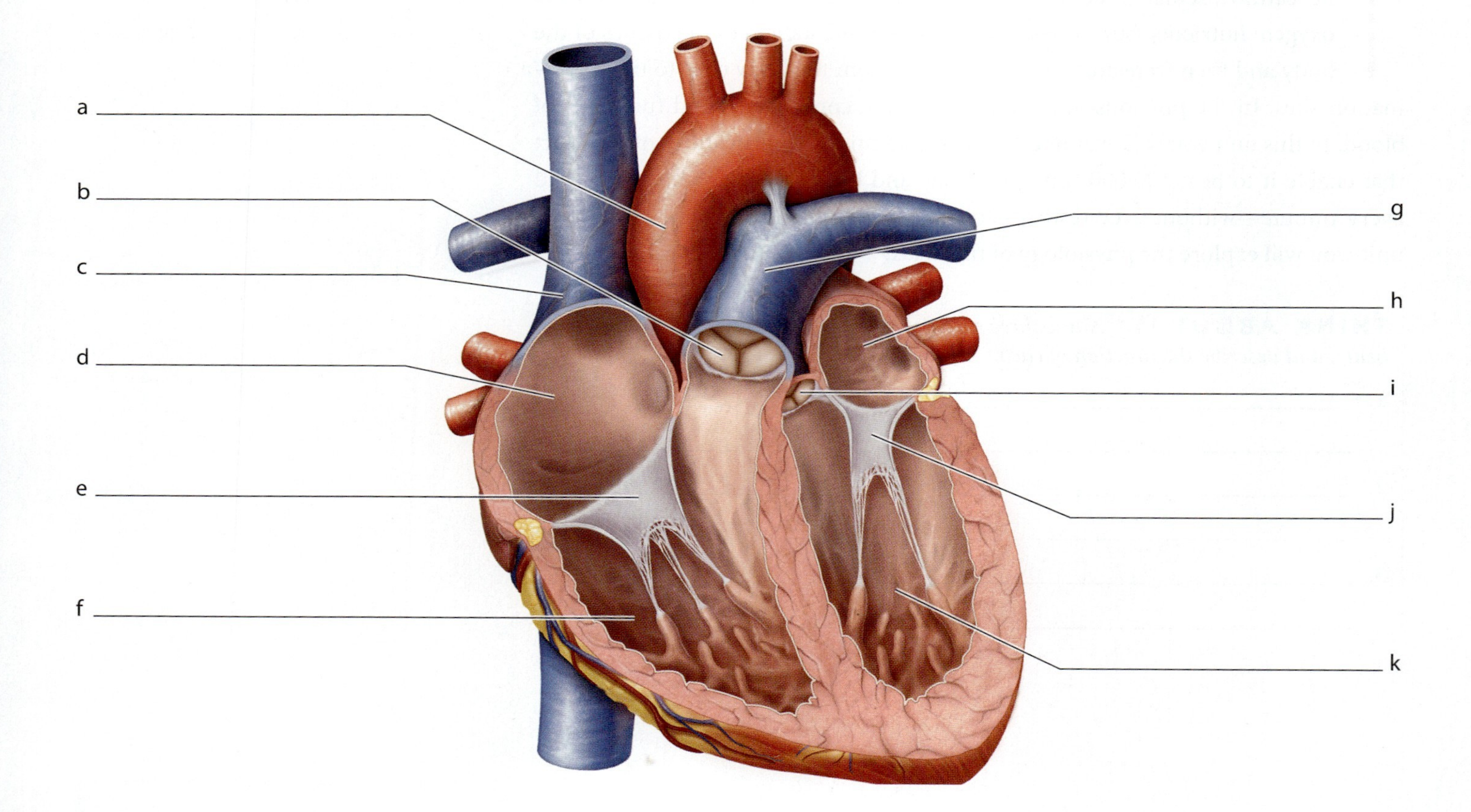

2. Match each of the following structures with its correct description.

______	a. right atrium	1. "heart strings"
______	b. chordae tendineae	2. receives blood from the pulmonary veins
______	c. bicuspid valve	3. pumps blood to the pulmonary trunk
______	d. aorta	4. arises from the right ventricle
______	e. left ventricle	5. left atrioventricular valve
______	f. left atrium	6. receives blood from the venae cavae
______	g. pulmonary trunk	7. largest artery in the body
______	h. tricuspid valve	8. right atrioventricular valve
______	i. right ventricle	9. pumps blood to the aorta

PRE-LAB Activity 2: Dissecting a Mammalian Heart

1. Distinguish between the base of the heart and the apex of the heart. ______________________________

____

2. Name the three layers of the heart wall and circle the layer that is also called the visceral pericardium. ______

____

3. To examine the internal anatomy of the heart, you will make which type of cut? ______________________________

PRE-LAB Activity 3: Reviewing the Microscopic Structure of Cardiac Muscle Tissue

1. Use the list of terms to label the accompanying photomicrograph of cardiac muscle tissue. Check off each term as you label it.

☐ cardiac muscle fibers
☐ nucleus
☐ intercalated disc
☐ striations

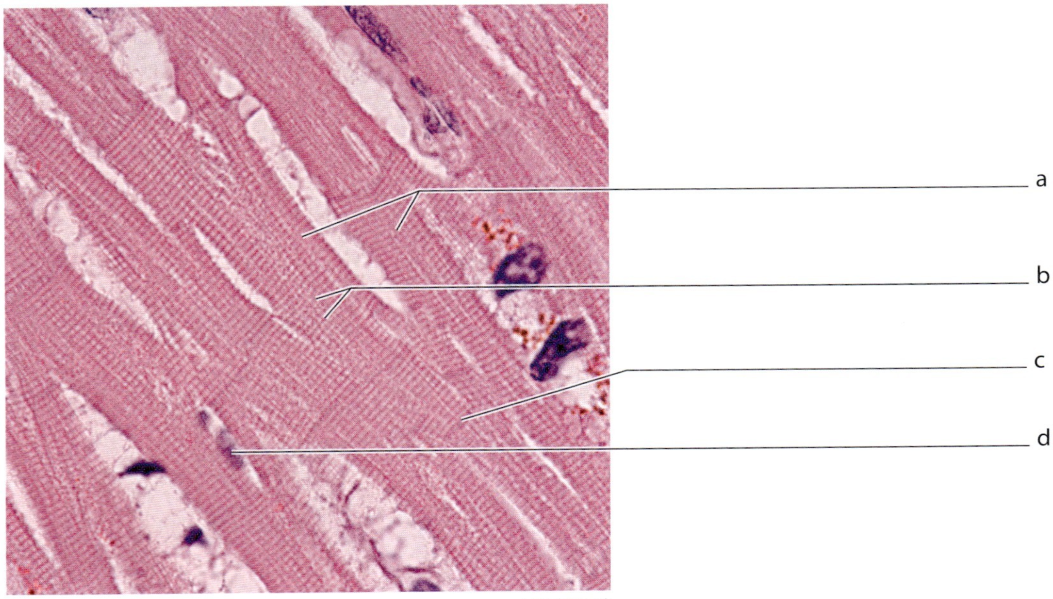

LM (775×)

2. Which layer of the heart wall is composed predominantly of cardiac muscle tissue?
   a. epicardium
   b. myocardium
   c. endocardium

PRE-LAB Activity 4: Tracing Circulatory Pathways

1. Which blood vessel returns deoxygenated blood to the right atrium?
   a. pulmonary vein
   b. aorta
   c. pulmonary artery
   d. inferior vena cava
   e. cardiac vein
2. Which structure prevents the backflow of blood from the right ventricle into the right atrium?
   a. pulmonary semilunar valve
   b. tricuspid valve
   c. aortic semilunar valve
   d. bicuspid valve
   e. mitral valve
3. Which of the following structures branches directly from the ascending aorta?
   a. marginal artery
   b. coronary sinus
   c. left coronary artery
   d. pulmonary trunk
   e. circumflex artery

Heart Anatomy

The heart is a fist-sized, four-chambered muscular pump lying within the mediastinum of the thoracic cavity (**Figure 21-1a**) and surrounded by a double-layered membrane called the **pericardium**. Blood vessels enter and leave the heart at its broad **base**, and the inferior tip of the heart is called the **apex** (Figure 21-2)

The pericardium (Figure 21-1b) consists of an outer, dense irregular connective tissue layer—the **fibrous pericardium**—that anchors the heart to surrounding structures, and an inner **serous pericardium** composed of two layers separated

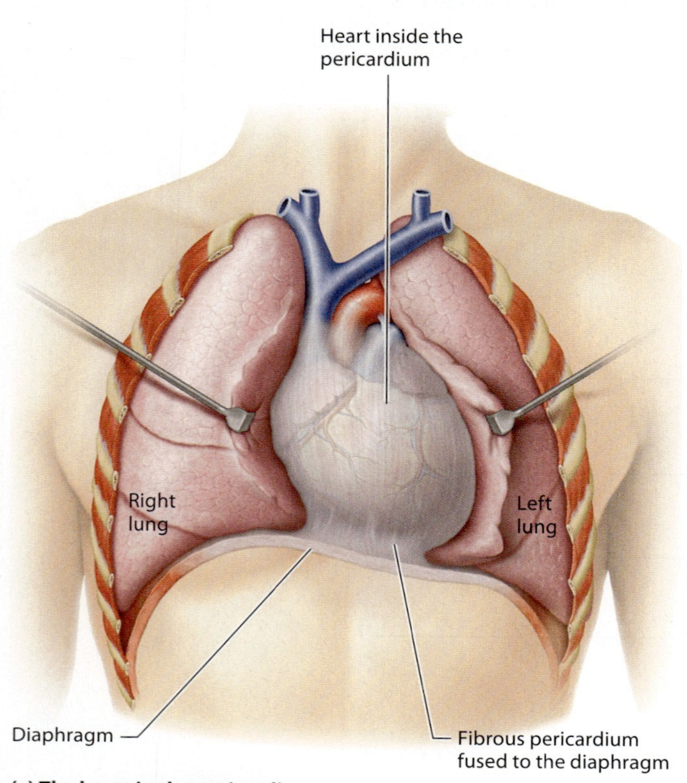

(a) The heart in the pericardium

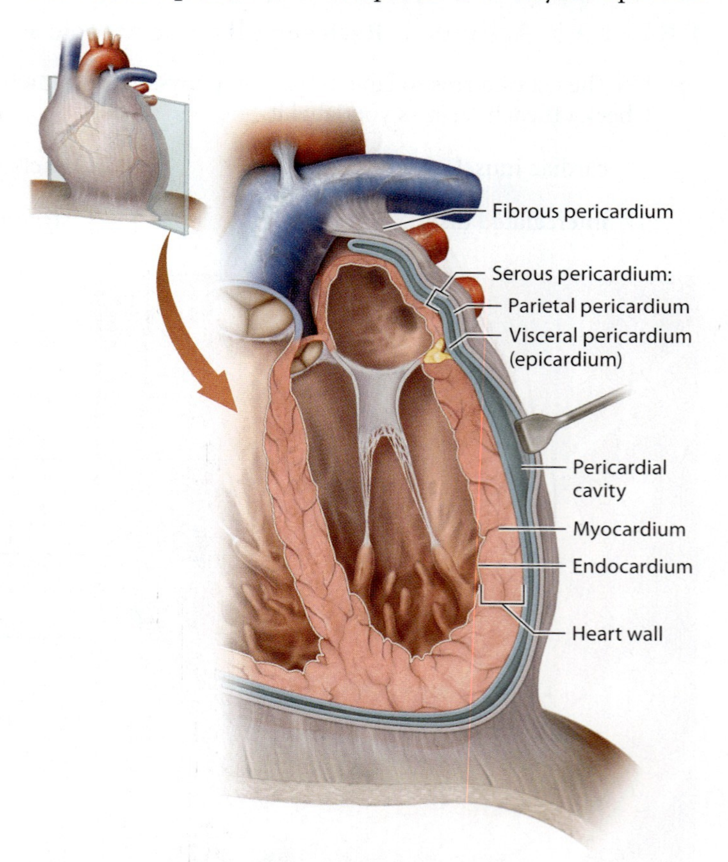

(b) The layers of the percardium and the heart wall

Figure 21-1 The pericardium and layers of the heart wall.

by a narrow serous fluid-filled cavity called the **pericardial cavity**. Pericardial fluid reduces friction during muscular contraction. The outer layer of the serous pericardium, the **parietal pericardium**, is fused to the fibrous pericardium, whereas the inner layer of the serous pericardium, the **visceral pericardium**, is attached to the heart muscle and forms the **epicardium**, the outer layer of the heart wall. The heart wall also contains a middle muscular layer called the **myocardium** and an inner layer called the **endocardium**. The endocardium is composed of a specialized simple squamous epithelium that rests on a thin layer of areolar connective tissue, and it is continuous with the innermost lining of blood vessels throughout the body.

Figure 21-2 illustrates the major structural features of the heart. Two superior thin-walled chambers, called **atria**, receive and direct blood coming into the heart. On the outer

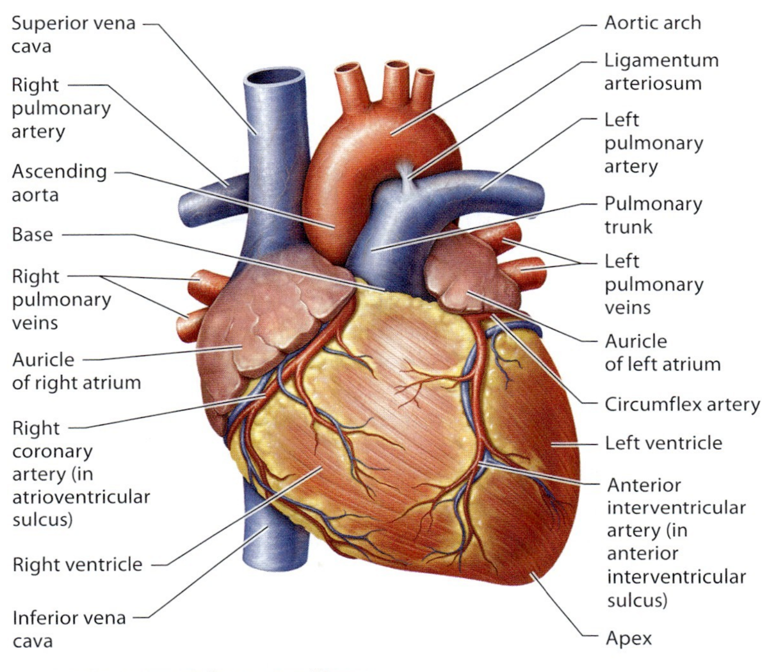

(a) Anterior view of the external heart

Aortic arch
Left pulmonary artery
Left pulmonary veins
Left atrium
Coronary sinus
Left ventricle
Ascending aorta
Superior vena cava
Right pulmonary artery
Right pulmonary veins
Right atrium
Inferior vena cava
Right ventricle

(b) Posterior view of the external heart

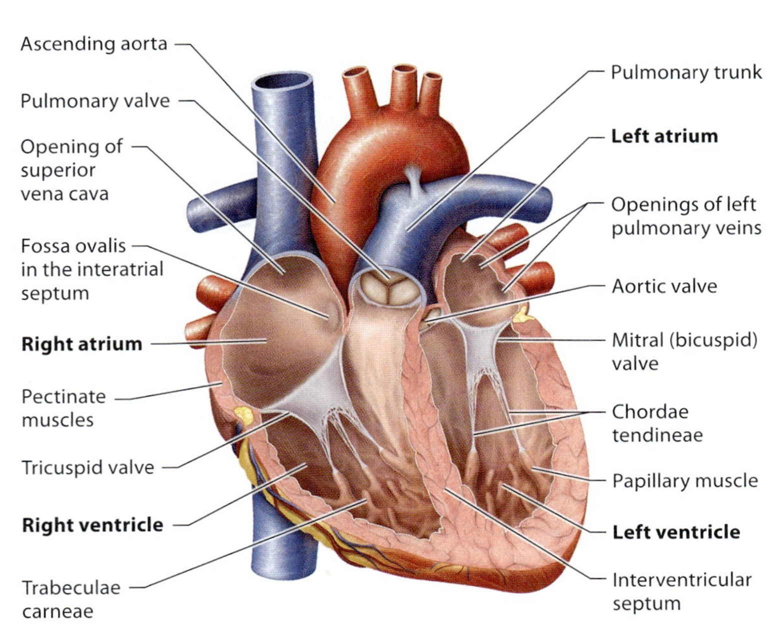

(c) Internal anatomy of the heart, frontal section

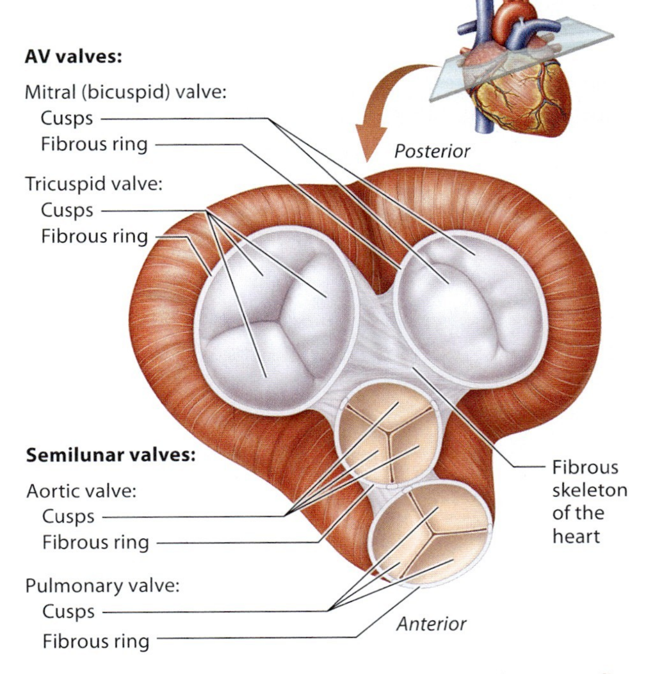

(d) The four valves of the heart (superior view, atria removed)

Figure 21-2 Heart anatomy.

surface of each atrium is a flap-like structure called an auricle. The **right atrium** receives deoxygenated blood from the superior and inferior venae cavae of the systemic circuit and the coronary sinus (Figure 21-2b) of the coronary circulation; the **left atrium** receives oxygenated blood from the four pulmonary veins of the pulmonary circuit. (*Note:* You will study the circulatory pathways—systemic circuit, pulmonary circuit, and coronary circulation—later in this unit.) Lining the right atrium are comb-like muscular ridges called **pectinate muscles**, which provide additional strength to the atrial wall. The atria are separated by a thin wall called the **interatrial septum**. A depression in the interatrial septum called the **fossa ovalis** (Figure 21-2c) is the remnant of the foramen ovale, an opening between the right and left atria by which the pulmonary circuit is bypassed in the fetus.

Two inferior thick-walled chambers, called **ventricles**, pump blood. The **right ventricle** receives deoxygenated blood from the right atrium and pumps it into the pulmonary circuit via the pulmonary trunk (see Figure 21-2a), which branches into the left and right pulmonary arteries. Also note a band of connective tissue at the point where the pulmonary trunk branches into the right artery and left pulmonary arteries. This structure, called the **ligamentum arteriosum**, is a remnant of the ductus arteriosus, a fetal blood vessel that shunts blood from the pulmonary trunk to the aorta, thereby bypassing the lungs. The **left ventricle** receives oxygenated blood from the left atrium and pumps it into the systemic circuit via the ascending aorta. Both ventricular walls are characterized by the presence of a distinct network of muscular ridges called **trabeculae carneae** (see Figure 21-2c).

Four valves within the heart ensure that blood continues flowing in one direction only (see Figure 21-2c and d). The **tricuspid valve** (or right atrioventricular valve) prevents backflow of blood from the right ventricle to the right atrium. The **mitral (bicuspid) valve** (or left atrioventricular valve) prevents backflow of blood from the left ventricle to the left atrium. In addition, two other valves—the **pulmonary valve** (or pulmonary semilunar valve) within the pulmonary trunk, and the **aortic valve** (or aortic semilunar valve) within the aorta—open when the contracting ventricles pump blood into the arteries and then close to prevent blood from flowing back into the ventricles. The tricuspid and mitral (bicuspid) valves are anchored to the **papillary muscles** of the ventricular wall by structures called **chordae tendineae** (see Figure 21-2c).

ACTIVITY 1
Examining the Functional Anatomy of the Heart

Learning Outcomes

1. Locate and identify the major anatomical structures of the heart on a heart model and describe the function of each.
2. Locate and identify the major blood vessels associated with the heart on a heart model.
3. Explain the changes that occur in the fetal heart after birth.

Materials Needed

☐ Heart models
☐ Anatomical charts

Instructions

CHART Locate on a heart model each of the structures in the following Making Connections chart. Then write brief descriptions of its structure and function. Finally, make "connections" to things you have already learned in lecture, assigned readings, and lab.

Making Connections: Heart Anatomy

Structure	Description (Structure and/or Function)	Connections to Things I Have Already Learned
Epicardium		
Myocardium		*"myo" = muscle myoglobin, myofilament*
Endocardium		
Right atrium	*Thin-walled receiving chamber that receives blood from the superior vena cava, inferior vena cava, and coronary sinus*	
Right ventricle		
Left atrium		*An atrium is an entry chamber*

Structure	Description (Structure and/or Function)	Connections to Things I Have Already Learned
Left ventricle		
Tricuspid valve	*Prevents backflow of blood from the right ventricle to the right atrium*	
Mitral (bicuspid) valve		
Pulmonary trunk		*A trunk is a blood vessel that gives rise to other blood vessels*
Pulmonary valve		
Aorta • Ascending aorta • Aortic arch • Descending aorta		
Aortic valve	*Prevents backflow of blood from the aorta to left ventricle*	
Chordae tendineae		
Papillary muscle		
Coronary sinus		*A sinus is an enlarged vein. Superior sagittal sinus collects venous blood in brain*
Inferior vena cava		
Superior vena cava	*Large vein that receives deoxygenated blood from the upper body and returns it to the right atrium*	
Pulmonary veins		
Fossa ovalis	*Depression in interatrial septum; remnant of the foramen ovale*	
Ligamentum arteriosum		

Optional Activity

Practice labeling heart structures on human hearts at MasteringA&P® > Study Area > Practice Anatomy Lab > Human Cadaver: Cardiovascular System > Heart

ACTIVITY 2

Dissecting a Mammalian Heart

Learning Outcomes

1. Dissect a mammalian heart and locate selected anatomical structures.
2. Describe the relationship between the fibrous pericardium and the serous pericardium.
3. Identify the epicardium, myocardium, and endocardium.

Materials Needed

- ☐ Dissecting tray
- ☐ Dissecting instruments
- ☐ Disposable gloves
- ☐ Safety goggles
- ☐ Preserved mammalian heart

Instructions

This activity involves the dissection of a preserved mammalian heart. Although the instructions and figures relate to the sheep heart—which is very similar to the human heart in both structure and size—they can be used for other mammalian hearts.

1. Put on disposable gloves and safety goggles.
2. Obtain a dissecting tray, dissecting instruments, and a preserved heart.
3. Rinse the heart thoroughly with cold water to remove excess preservative. Run water through the large blood vessels to flush blood clots out of the heart.

A. External Anatomy

1. Identify the pericardial sac (if still present). Using a pair of scissors, carefully cut through the pericardial sac and remove the fibrous pericardium, a thick layer of fibrous connective and adipose tissue that is the outer layer of the pericardial sac. Note that the parietal layer of the serous pericardium lines the fibrous pericardium and will be removed with the fibrous pericardium. The visceral layer of the serous pericardium—also known as the epicardium—is a thin, transparent layer on the surface of the heart.
2. Distinguish between the heart's broad superior aspect (base) and its pointed inferior aspect (apex) (**Figure 21-3a**). Now, orient the heart so that the auricles face you, and distinguish between the anterior and posterior surfaces of the heart. (The atria are beneath the auricles.)
3. Identify each of the following structures. Discuss the function of each structure with your lab group and then write a brief description of that function:

   a. Aorta ______________________________

   ____

   b. Pulmonary trunk ______________________________

   ____

   c. Superior vena cava ______________________________

   ____

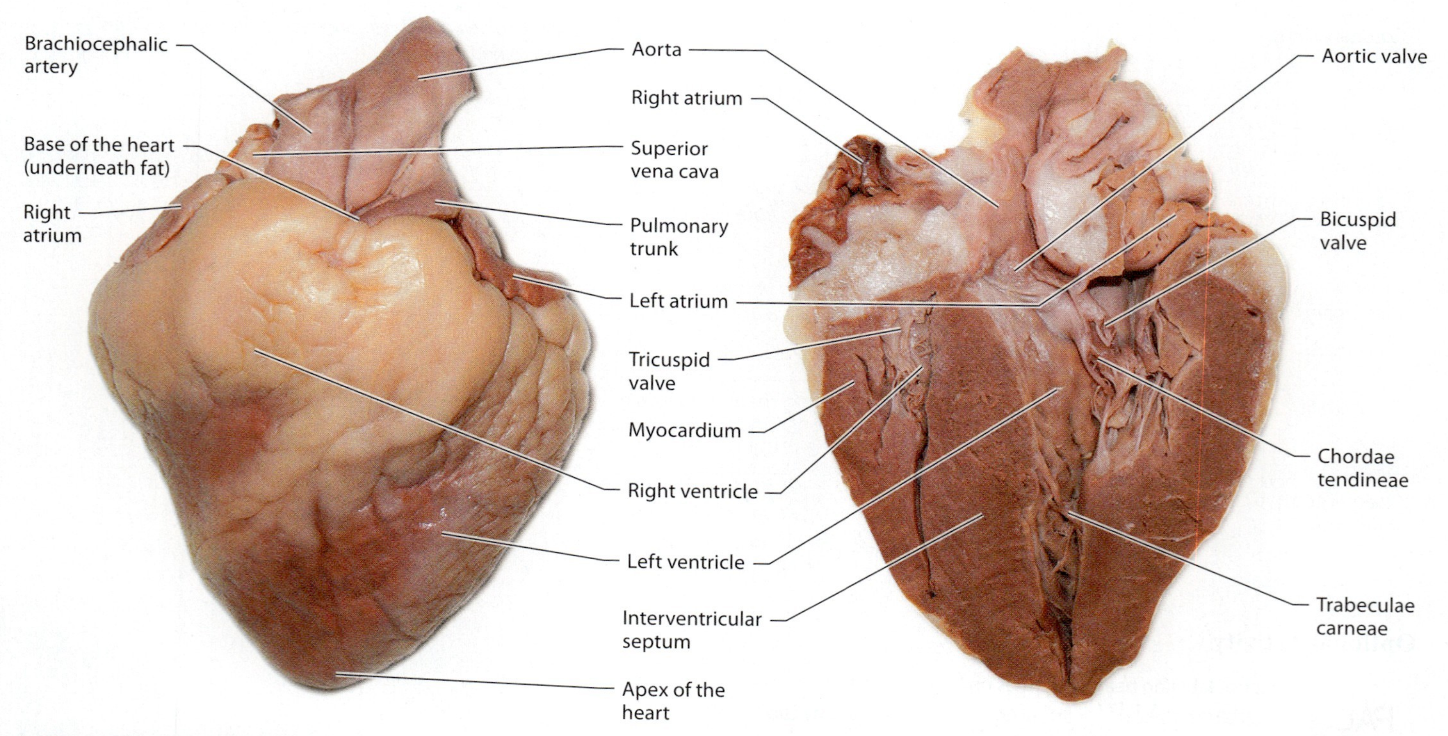

Figure 21-3 A sheep heart.

d. Inferior vena cava ______________________________

____

e. Pulmonary veins ______________________________

____

f. Ventricles ______________________________

____

g. Atria ______________________________

____

How do the locations of the superior and inferior venae cavae in a sheep heart differ from those of the human heart? ______________________________

____

4. Locate the adipose-filled anterior interventricular sulcus between the ventricles. Carefully remove some of the adipose tissue to expose the anterior interventricular artery and the cardiac vein.

   Predict the function of the anterior interventricular artery. ______________________________

   ____

   Predict the function of the cardiac vein. ______________________________

   ____

5. Locate the atrioventricular sulcus separating the atria from the ventricles. Carefully remove some of the adipose tissue to expose the thin-walled coronary sinus, which empties into the right atrium.

B. Internal Anatomy

1. Using a knife with a 5-inch blade, make a frontal (coronal) section (see Figure 21-3b) by cutting at the apex and continuing toward the base until the section is complete.
2. Identify each of the following structures. Discuss the function of each structure with your lab group and then write a brief description of that function:

   a. Epicardium ______________________________

   ____

   b. Myocardium ______________________________

   ____

   c. Endocardium ______________________________

   ____

   d. Interventricular septum ______________________________

   ____

   e. Tricuspid valve ______________________________

   ____

   f. Bicuspid valve ______________________________

   ____

   g. Pulmonary valve ______________________________

   ____

   h. Aortic valve ______________________________

   ____

   i. Chordae tendineae ______________________________

   ____

   j. Papillary muscles ______________________________

   ____

   Compare and contrast the structure of the tricuspid and bicuspid valves. ______________________________

   ____

   Compare and contrast the structure of the right and left ventricles. ______________________________

   ____

3. Dispose of the sheep heart, wash the dissecting tray and instruments, and disinfect the lab table as directed by your lab instructor.
4. **Optional Activity**

PAL™ Practice labeling heart structures on cat and sheep hearts at MasteringA&P® > Study Area > Practice Anatomy Lab > Cat > Cardiovascular System > Images 14–17

Cardiac Muscle Tissue

Cardiac muscle tissue (**Figure 21-4**) is the predominant tissue type of the heart. Cardiac muscle fibers are short, branched, uninucleate cells filled with the contractile proteins action and myosin. The regular arrangement of the actin and myosin produces **striations** giving the cells a striped appearance. Individual cardiac muscle fibers are connected by structures called **intercalated discs**. Intercalated discs contain both desmosomes (anchoring junctions that prevent cardiac muscle fibers from pulling apart during contraction) and gap junctions (junctions that allow the movement of materials from one cell to the next).

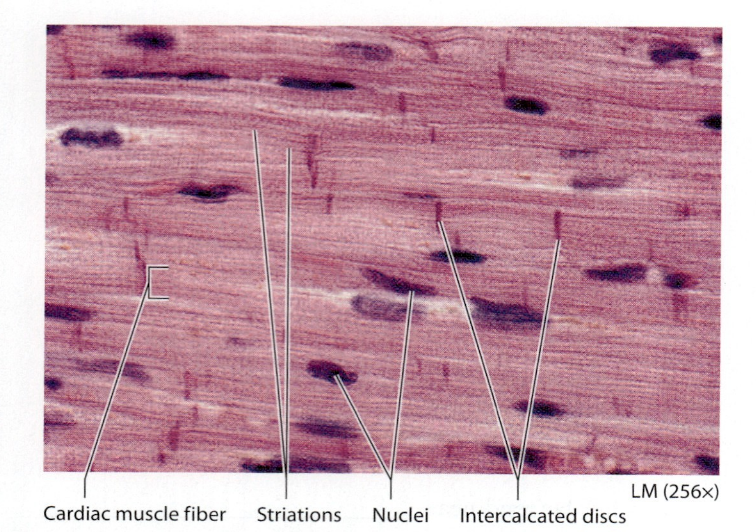

Figure 21-4 Cardiac muscle anatomy.

ACTIVITY 3

Reviewing the Microscopic Structure of Cardiac Muscle Tissue

Learning Outcomes

1. Review the histological characteristics of cardiac muscle tissue.
2. Explain the physiological significance of intercalated discs.

Materials Needed

☐ Microscope and slide (or photomicrograph) of cardiac muscle tissue

Instructions

1. View cardiac muscle tissue under high power (or in a photomicrograph). Sketch what you observe and label cardiac muscle fiber, intercalated disc, striations, nucleus, and endomysium.

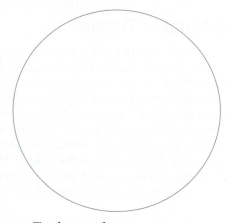

Total magnification: _____ ×

Cardiac muscle tissue makes up the bulk of which heart wall layer? ______________________

Describe the appearance of cardiac muscle fibers. ______________________

____

2. Name the two types of membrane junctions in intercalated discs. Then explain the physiological significance of each.

   a. ______________________

   ____

   b. ______________________

   ____

3. How are cardiac muscle and skeletal muscle similar?

   ____

   ____

4. How are cardiac muscle and skeletal muscle different?

   ____

   ____

5. **Optional Activity**

PAL™ View and label histology slides of cardiac muscle tissue at MasteringA&P® > Study Area > Practice Anatomy Lab > Histology > Cardiovascular System > Images 18–21

Blood Flow Patterns through the Heart

The heart is a double pump; each time the two ventricles contract, blood is ejected into both the pulmonary trunk and the aorta. Deoxygenated blood ejected into the pulmonary trunk by the right ventricle enters the **pulmonary circuit** (**Figure 21-5a**). The deoxygenated blood flows to the lungs, where it gives up carbon dioxide and picks up oxygen in the pulmonary capillaries; oxygenated blood is then returned to the left atrium. Oxygenated blood pumped by the left ventricle through the aorta enters the **systemic circuit** (Figure 21-5b) and is distributed to all parts of the body before returning to the right atrium via one of the venae cavae.

A branch of the systemic circuit—the **coronary circulation**—supplies the myocardium with the oxygen and nutrients needed for muscle contraction. The right and left coronary arteries (**Figure 21-6**) branch from the ascending aorta and encircle the heart in the atrioventricular (coronary) sulcus. The **right coronary artery** branches into the **marginal artery** and the **posterior interventricular artery**. The **left coronary artery** branches into the **circumflex artery**

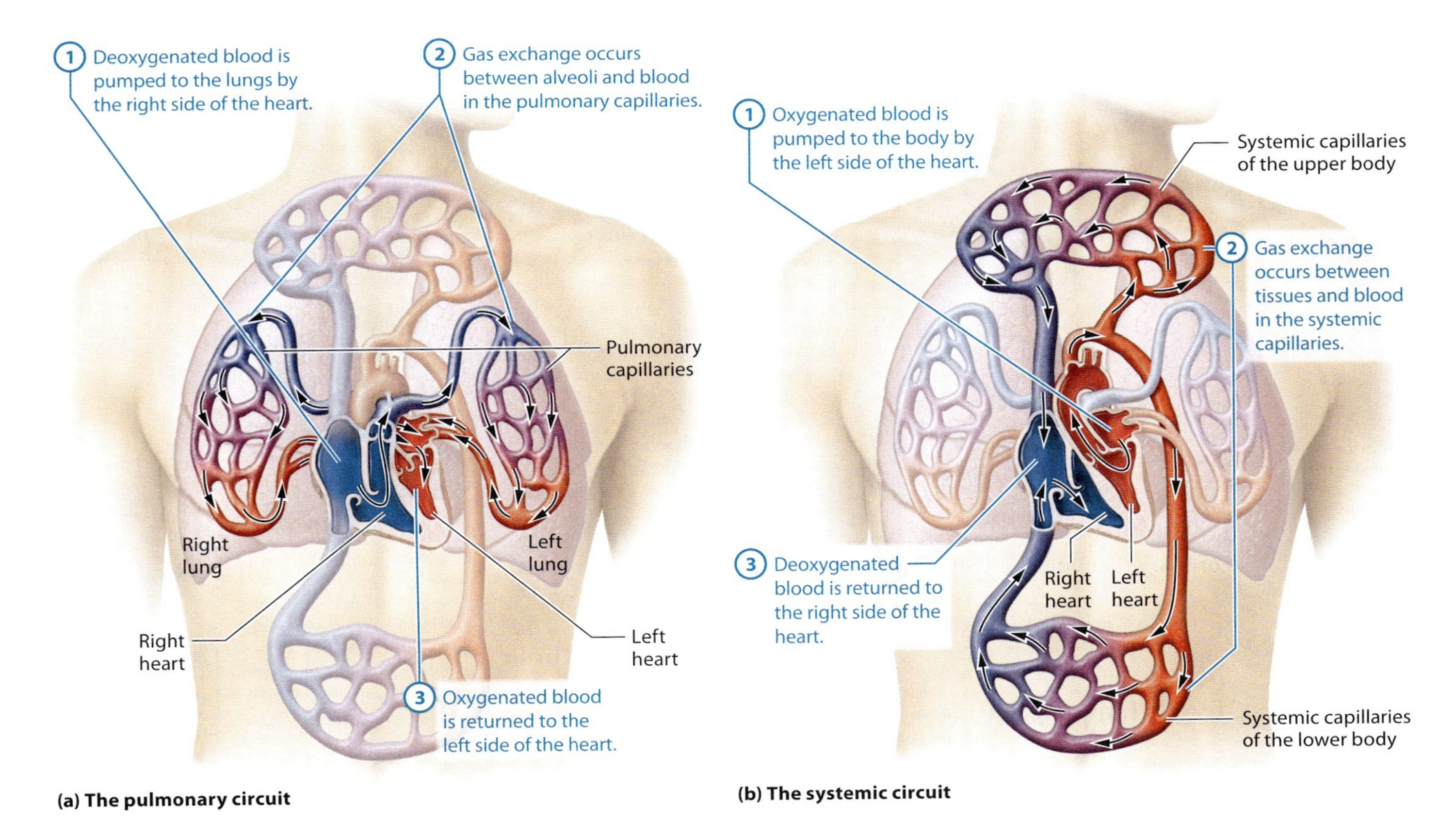

Figure 21-5 **The pulmonary and systemic circuits.**

and the **anterior interventricular artery** (clinically known as the left anterior descending branch or LAD). Arteries branch into smaller arterial vessels and eventually into capillaries, where oxygen, nutrients, and wastes are exchanged. Capillaries drain into venules, which fuse to form veins. The myocardium is drained by the **great**, **middle**, and **small cardiac veins** (**Figure 21-7**), all of which eventually empty into the **coronary sinus**, which drains into the right atrium.

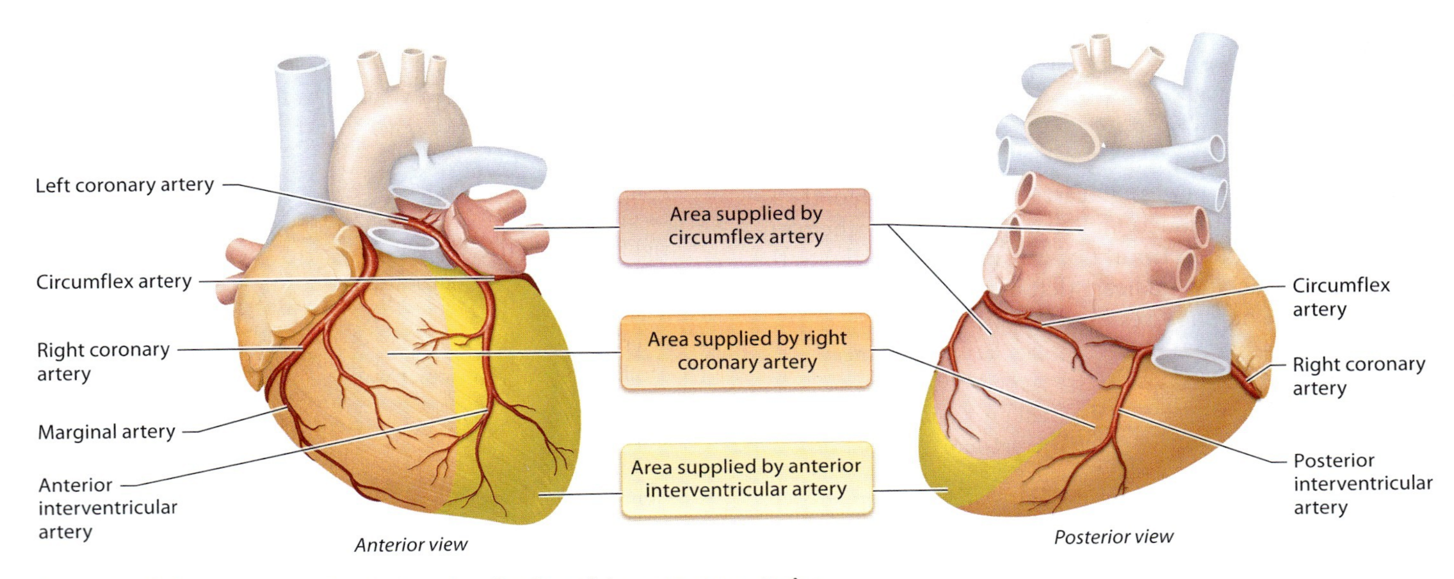

Figure 21-6 **The coronary circulation, distribution of the coronary arteries.**

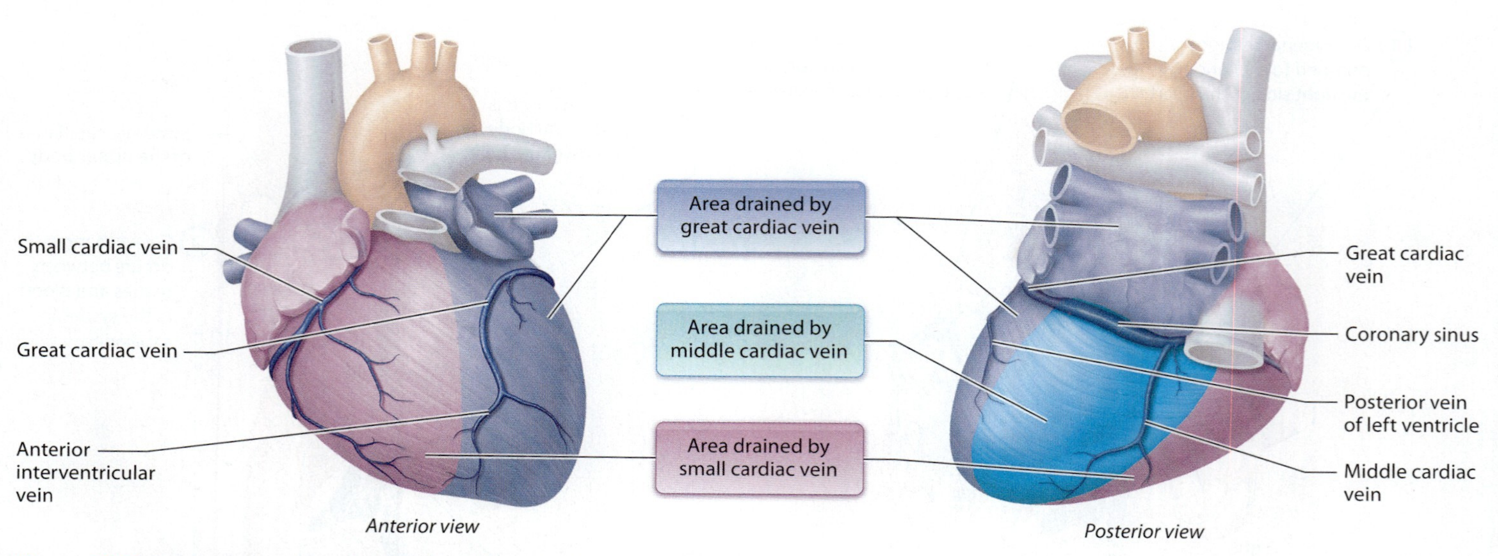

Figure 21-7 The coronary circulation, distribution of the coronary veins.

ACTIVITY 4
Tracing Circulatory Pathways

Learning Outcomes

1. Distinguish between the pulmonary and systemic circulations.
2. Trace the flow of blood through the major blood vessels in the pulmonary circuit.
3. Trace the flow of blood through the major blood vessels in the systemic circuit.
4. Trace the flow of blood through the major blood vessels in the coronary circulation.

Materials Needed

- ☐ Heart models
- ☐ Miscellaneous anatomical models and charts showing blood vessels

Instructions

A. Pulmonary Circuit

Locate each of the following heart structures and blood vessels on anatomical models and charts as you trace the pathway of a drop of blood from the heart (right ventricle) to the left lung and back to the heart (left atrium). Check off each heart structure or blood vessel as you locate it.

- ☐ right ventricle
- ☐ pulmonary trunk
- ☐ left pulmonary artery
- ☐ left lobar artery
- ☐ arteries
- ☐ arterioles
- ☐ capillary bed of lungs
- ☐ venules
- ☐ left pulmonary veins
- ☐ left atrium

B. Systemic Circuit

Locate each of the following heart structures and blood vessels on anatomical models and charts as you trace the pathway of a drop of blood from the heart (left ventricle) to the cells of the right ovary and back to the heart (right atrium). Check off each heart structure or blood vessel as you locate it.

- ☐ left ventricle
- ☐ ascending aorta
- ☐ aortic arch
- ☐ descending aorta
- ☐ thoracic aorta
- ☐ abdominal aorta
- ☐ gonadal artery
- ☐ arterioles
- ☐ capillary bed of ovary
- ☐ gonadal vein
- ☐ inferior vena cava
- ☐ right atrium

C. Coronary Circulation

Locate each of the following heart structures and blood vessels on anatomical models and charts as you trace the pathway of a drop of blood from the heart (left ventricle) to the cells of the posterior ventricular wall and back to the heart (right atrium). Check off each heart structure or blood vessel as you locate it.

- ☐ left ventricle
- ☐ ascending aorta
- ☐ right coronary artery
- ☐ posterior interventricular artery
- ☐ smaller arteries
- ☐ arterioles
- ☐ capillary bed of ventricular wall
- ☐ venules
- ☐ cardiac vein
- ☐ coronary sinus
- ☐ right atrium

Post-lab quizzes are also assignable in MasteringA&P®

POST-LAB ASSIGNMENTS

Name: ______________________ Date: ______________ Lab Section: __________

PART I. Check Your Understanding

Activity 1: Examining the Functional Anatomy of the Heart

1. Identify the components of the heart.

   a. ______________________
   b. ______________________
   c. ______________________
   d. ______________________
   e. ______________________
   f. ______________________
   g. ______________________
   h. ______________________
   i. ______________________
   j. ______________________
   k. ______________________
   l. ______________________

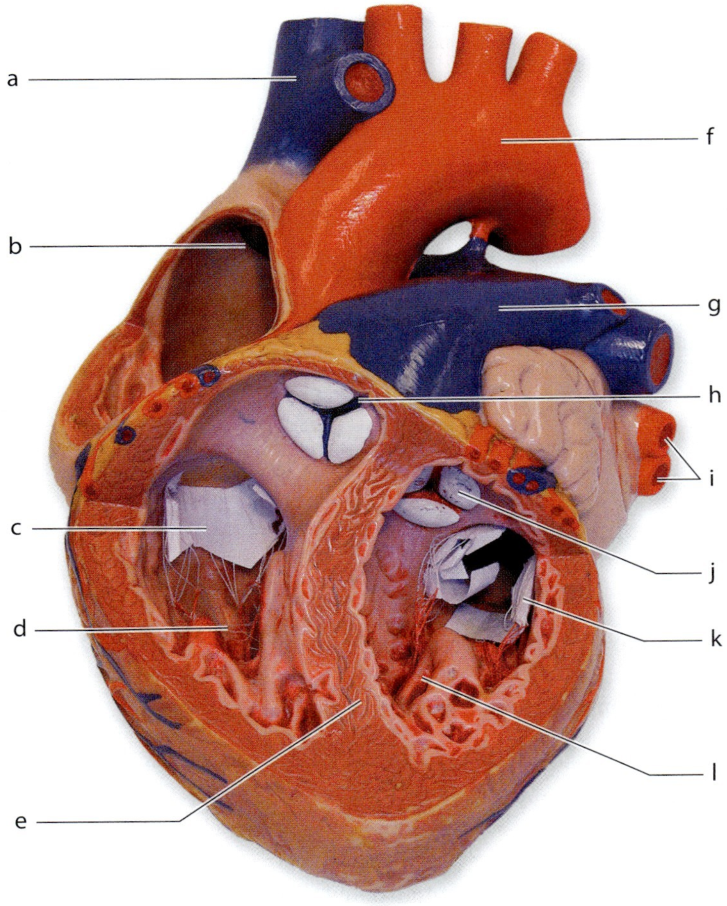

L-G12 Giant Heart, 4-part, 3B Scientific®

2. Which heart structure(s):

______________________ a. anchor atrioventricular valves to the heart wall?
______________________ b. receives oxygenated blood from the lungs?
______________________ c. separates the right and left ventricles?
______________________ d. pumps blood to the lungs?
______________________ e. receives blood from the coronary sinus?
______________________ f. prevents backflow of blood into the left ventricle?
______________________ g. is a remnant of the foramen ovale?
______________________ h. prevents backflow of blood into the right atrium

Activity 2: Dissecting a Mammalian Heart

1. Identify the components of the sheep heart.

   a. ____________________
   b. ____________________
   c. ____________________
   d. ____________________
   e. ____________________
   f. ____________________
   g. ____________________
   h. ____________________
   i. ____________________
   j. ____________________
   k. ____________________

a
b
c
d
e
f
g
h
i
j
k
Heart wall layers

Activity 3: Reviewing the Microscopic Structure of Cardiac Muscle Tissue

1. Describe the appearance of a cardiac muscle fiber. ____________________

2. What causes the striations in these cells? ____________________

3. What are two functions of an intercalated disc? ____________________

Activity 4: Tracing Circulatory Pathways

1. Trace the pathway of blood from the heart (left ventricle) to the right adrenal gland and back to the heart (right atrium). ____________________

2. List the structures of the coronary pathway serving the anterior ventricular wall in order, beginning in the left ventricle and ending in the right atrium. ____________________

3. Identify the events of the pulmonary and systemic circulations in the accompanying illustration. Color blood vessels containing oxygenated blood red, color blood vessels containing deoxygenated blood blue, and color capillary beds purple.

a. ______________________

____

b. ______________________

____

c. ______________________

____

d. ______________________

____

e. ______________________

____

f. ______________________

____

g. ______________________

____

h. ______________________

____

i. ______________________

____

Systemic capillaries

Pulmonary capillaries

Pulmonary capillaries

Systemic capillaries

PART II. Putting It All Together

A. Review Questions

Answer the following questions using your lecture notes, your textbook, and your lab notes.

1. Complete the following table.

Valve	Location	When Does the Valve Open?
Aortic valve		
Bicuspid valve		
Pulmonary valve		

2. The fetal structure ______________________ shunts blood from the right atrium to the left atrium. Its adult remnant is the ______________________.

3. The fetal structure ______________________ shunts blood from the pulmonary trunk to the aorta. Its adult remnant is the ______________________.
4. Are all cardiac muscle fibers contractile? Explain. ______________________
____
5. What is a myocardial infarction? Why is the anterior interventricular artery called the "widow-maker"? ______________________
____

B. Concept Mapping

1. Fill in the blanks to complete this concept map outlining heart anatomy.

atria	endocardium	interventricular septum	ventricles	visceral pericardium

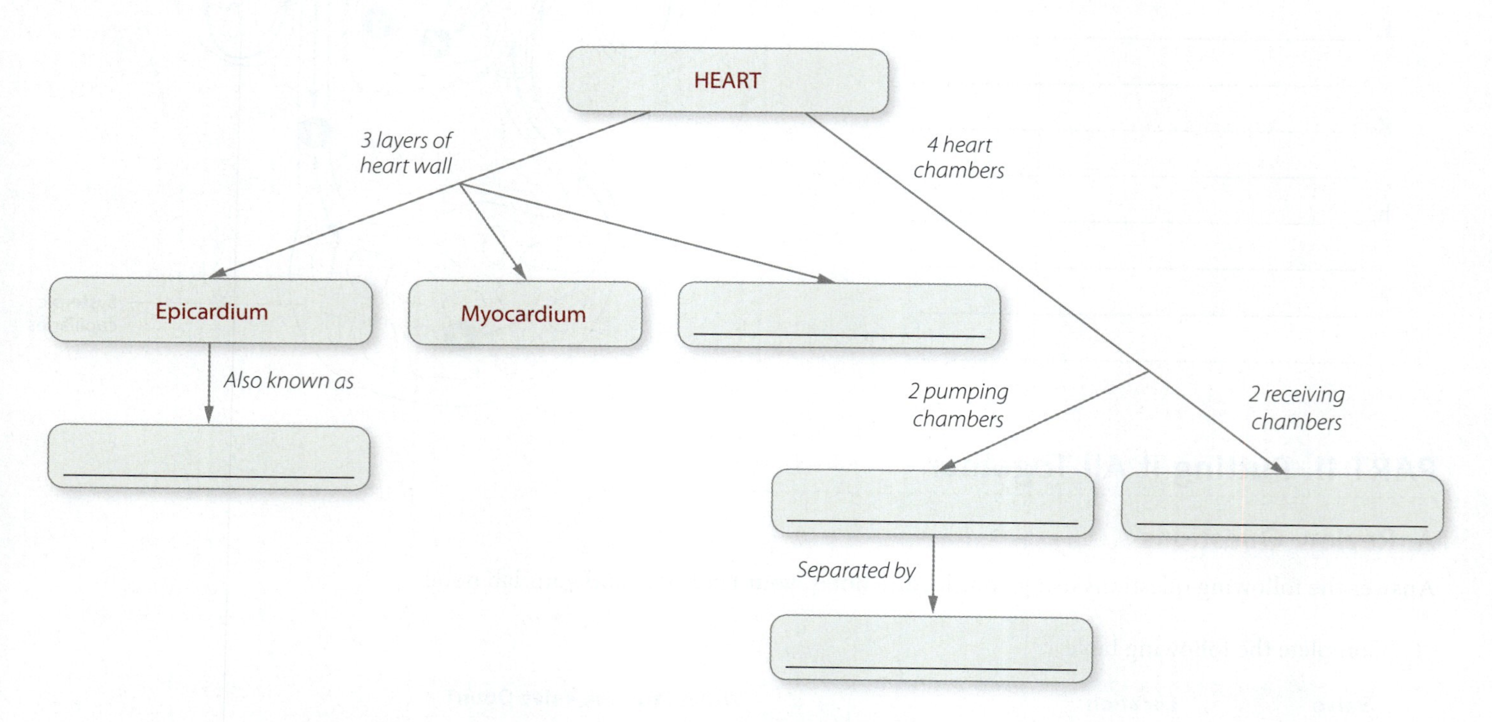

2. Construct a unit concept map to show the relationships among the following set of terms. Include all of the terms in your diagram. Your instructor may choose to assign additional terms.

aorta	atrioventricular valve	atrium	auricle
chordae tendineae	coronary sinus	desmosome	endocardium
interventricular septum	myocardium	papillary muscle	semilunar valve
superior vena cava	ventricle	visceral pericardium	

22 Physiology of the Heart

UNIT OUTLINE

The heart is an incredible organ beating an average of 100,000 times every day. Unlike skeletal muscle, the heart does not require extrinsic nervous stimulation in order to contract. Instead, a network of noncontractile cardiac muscle fibers is responsible for initiating and distributing electrical impulses through the heart. In this unit, you will examine the components of this cardiac conduction system and then explore the relationship between the electrical activity of the heart and the cardiac cycle, the events that occur during a single heartbeat.

THINK ABOUT IT *In addition to the intrinsic cardiac conduction system, which initiates and distributes the electrical impulses that control cardiac activity, many other factors affect the rate at which the heart contracts. The most important of these effects are exerted by the autonomic nervous system (ANS).*

What are the two divisions of the ANS and how does each division affect heart rate?

____

Name several other factors that influence heart rate: ______________________

____

PRE-LAB ASSIGNMENTS

Pre-lab quizzes are also assignable in MasteringA&P®

To maximize learning, BEFORE your lab period carefully read this entire lab unit and complete these pre-lab assignments using your textbook, lecture notes, and prior knowledge.

PRE-LAB Activity 1: Recording and Interpreting an Electrocardiogram

1. Use the list of terms to label the accompanying illustration of the cardiac conduction system. Check off each term as you label it.

- ☐ AV node
- ☐ AV bundle
- ☐ right bundle branch
- ☐ left bundle branch
- ☐ terminal branches
- ☐ SA node

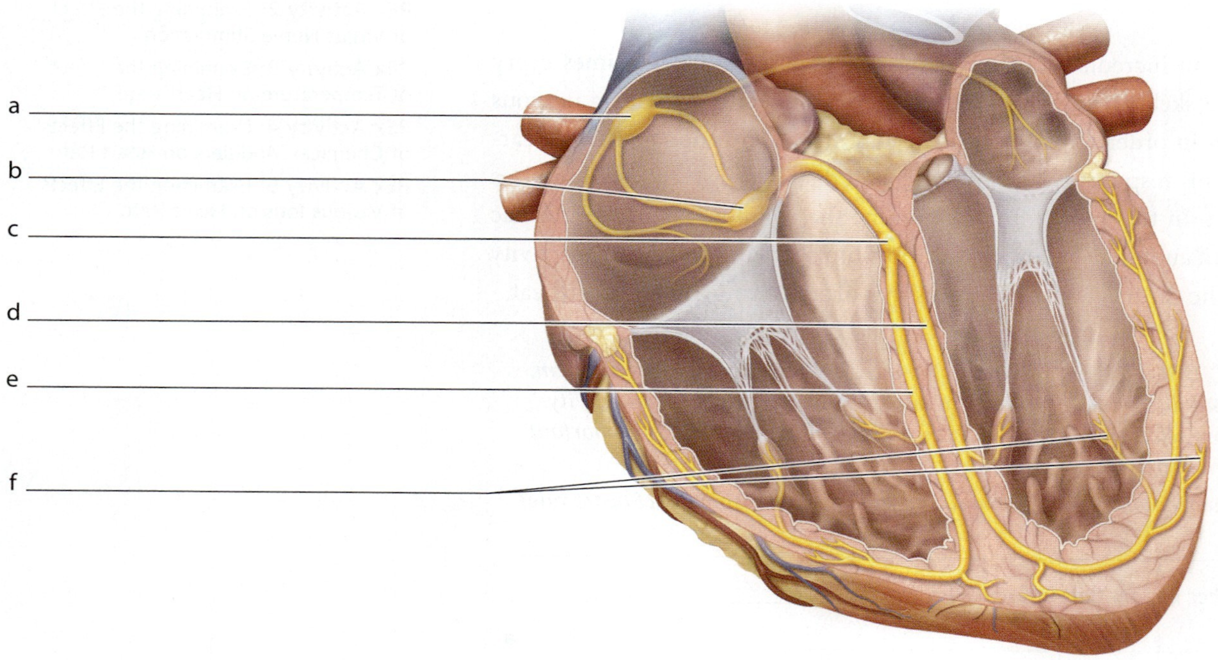

a ______________________

b ______________________

c ______________________

d ______________________

e ______________________

f ______________________

2. Which component of the cardiac conduction system is known as the pacemaker?

____

3. Name the three deflection waves of a typical ECG: ______________________

____

PRE-LAB Activity 2: Auscultating Heart Sounds

1. What causes the heart sounds heard with a stethoscope? ______________________

____

PRE-LAB Activity 3: Electrocardiography in a Human Subject Using BIOPAC®

1. How many electrodes are attached to the forearm to record the ECG?
   a. 1
   b. 2
   c. 3
   d. 4

2. The tracing of recorded electrical activity of the heart obtained by using electrodes on the skin is called an ________________________.
   a. electroencephalogram
   b. electrocardiogram
   c. electromyogram
   d. None of these is correct.

Cardiac Physiology

Here we consider two aspects of cardiac physiology: the electrical activity in the heart and heart sounds.

Electrical Activity in the Heart

During each cardiac cycle—the events that occur during one heartbeat—electrical activity of the heart stimulates contraction of the cardiac muscle. This activity is intrinsic; the heart is able to beat without neural or hormonal stimulation. This intrinsic stimulation is caused by specialized, noncontractile, autorhythmic cardiac muscle fibers that make up the **cardiac conduction system** of the heart. These autorhythmic cells both initiate action potentials that lead to cardiac muscle contraction and provide a pathway for conducting these action potentials to all cardiac muscle fibers. While the intrinsic conduction system initiates and distributes the electrical impulses that stimulate cardiac muscle contraction, both the nervous and endocrine systems regulate cardiac activity.

The cardiac conduction system (**Figure 22-1**) consists of three groups of pacemaker cells that function together to stimulate and coordinate the heart's contractions. Cells of the **sinoatrial (SA) node**, located in the posterior wall of the right atrium just inferior to the opening of the superior vena

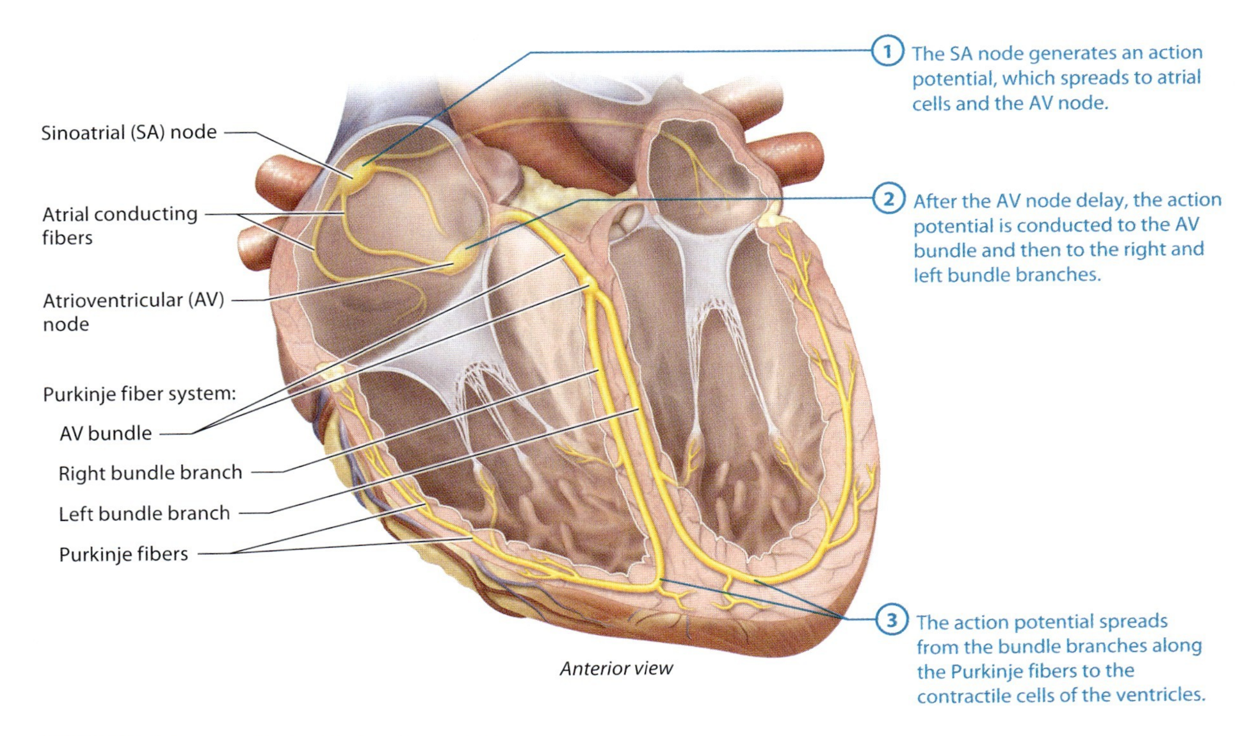

Figure 22-1 **The intrinsic conduction system.**

cava, are considered the pacemaker of the heart because the SA node generates electrical impulses (action potentials). Cells of the **atrioventricular (AV) node**, located in the floor of the right atrium anterior to the opening of the coronary sinus, receive action potentials from the SA node via the atrial conducting pathways. The third group—the **Purkinje fiber system**—consists of three components. Cells of the **atrioventricular (AV) bundle**, located in the interventricular septum, are the only electrical connection between the atria and the ventricles; the **right and left bundle branches** extend through the interventricular septum toward the apex of the heart; and the **terminal branches** conduct action potentials from the apex of the heart upward through the remainder of the ventricular myocardium.

The electrical activity that occurs in the heart spreads throughout the rest of the body; it can be detected by electrodes placed on the skin and can be recorded with an instrument called an **electrocardiograph**. A graphic recording of these electrical changes is called an **electrocardiogram** (ECG or EKG). An ECG is a record of voltage over time. It is also important to note that an ECG records electrical events occurring in the cardiac muscle as a whole, not just in the cells of the cardiac conduction system.

Figure 22-2 illustrates a normal electrocardiogram. The basic components of an ECG are a straight baseline called the **isoelectric line** and three deflection waves: the P wave, the QRS complex, and the T wave. The **P wave** is a small upward deflection that represents atrial depolarization that spreads from the SA node just prior to atrial contraction. The **QRS complex** begins as a downward deflection (Q), continues as a large upward deflection (R), and ends as a downward deflection (S). This complex represents ventricular depolarization spreading from the AV node, AV bundle, bundle branches, and terminal branches just prior to ventricular contraction. Finally, the **T wave** is an upward deflection representing ventricular repolarization; it occurs just before ventricular relaxation. Note that atrial repolarization does occur, but it is not observed in the ECG because it occurs at the same time that the ventricles depolarize and is, therefore, obscured by the QRS complex. In addition to the deflection waves, the ECG also contains intervals and segments between the waves. A **segment** is defined as a period between the end of one wave and the start of the next wave, whereas an **interval** includes one segment and one or more waves. The duration and boundaries of each deflection wave, interval, and segment of the ECG are summarized in **Table 22-1**.

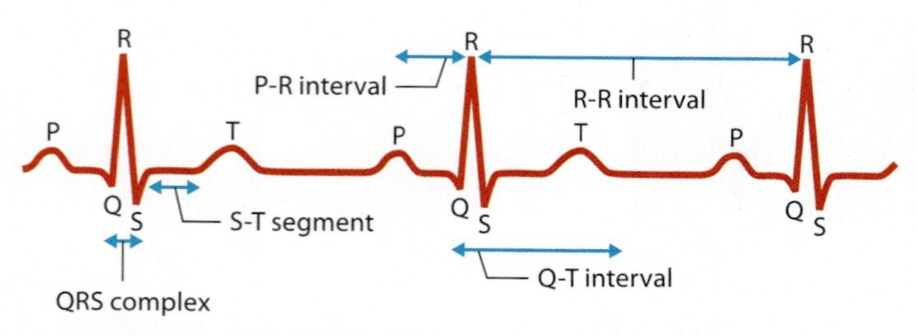

Figure 22-2 **A normal electrocardiogram.**

Abnormalities of the ECG are used to diagnose heart problems. A heart rate below 60 beats/minute (bpm) is called **bradycardia**; a heart rate over 100 bpm is called **tachycardia**. Neither of these conditions is considered pathological; however, prolonged periods of tachycardia can develop into **ventricular fibrillation**—rapid uncoordinated contractions of the heart that render it useless as a pump. The sizes of the waves can provide important clinical information. For example, larger P waves may indicate enlargement of an atrium, larger Q waves may indicate a myocardial infarction (heart attack), and larger R waves may indicate enlargement of a ventricle.

Heart Sounds

The process of listening to heart sounds with a stethoscope is called **auscultation**. The heart sounds ("lubb/dupp") are associated with the events of the cardiac cycle. During one cardiac cycle, the two atria contract at the same time and then relax while the two ventricles contract simultaneously. A cardiac cycle typically lasts 0.7–0.8 second, corresponding to an average heart rate of 75–85 beats per minute (bpm). The first sound heard during the cardiac cycle is "lubb," which is a result of blood turbulence caused by the closing of the bicuspid and tricuspid valves. The second heart sound ("dupp") is a result of blood turbulence caused by the closing of the pulmonary and aortic valves.

CLINICAL CONNECTION

Heart Murmur

Heart murmurs are abnormal blowing, whooshing, or rasping sounds made by blood flowing through the heart valves; they can be heard through a stethoscope. One type of murmur resulting from mitral valve prolapse occurs when the mitral (bicuspid) valve does not close tightly and blood flows back into the left atrium (a process called regurgitation). Another type of murmur resulting from mitral valve stenosis or aortic valve stenosis occurs when blood flows through valves that have been narrowed due to scarring caused by infections such as rheumatic fever. If left untreated, stenosis can overwork the heart and lead to heart failure.

Table 22-1 Components of the ECG

ECG Tracing	Component	Duration (in seconds)	Description
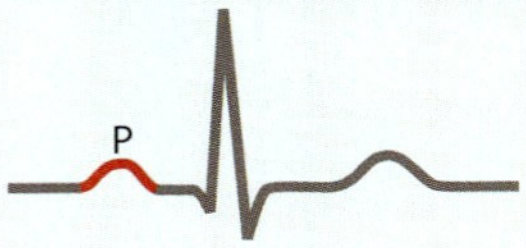 **P wave**	P wave	0.06–0.11	Start of P wave to return to isoelectric line; represents atrial depolarization
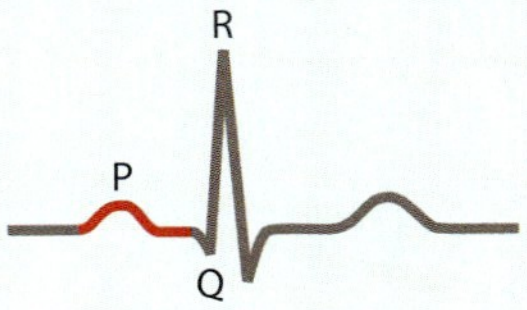 **P-R interval**	P-R interval	0.12–0.20	Start of P wave to start of the Q deflection; represents time required for electrical impulse to travel from SA node through the AV node
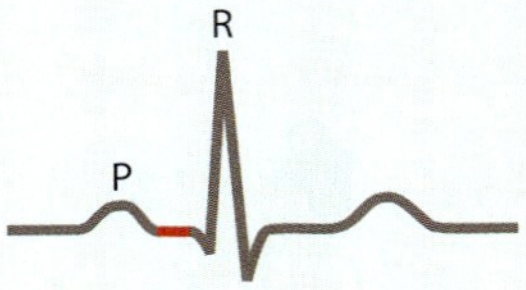 **P-R segment**	P-R segment	0.08	End of P wave to start of Q deflection; represents time required for electrical impulse to travel from the AV node to the ventricles
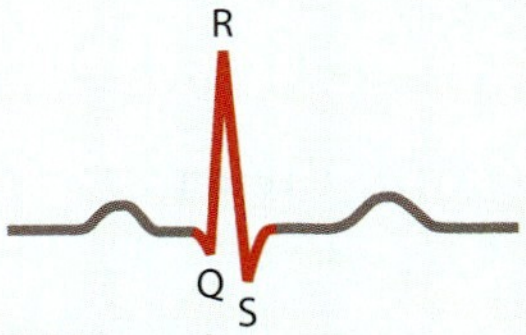 **QRS complex**	QRS complex	<0.12	Start of Q deflection to return of S deflection to isoelectric line; represents ventricular depolarization
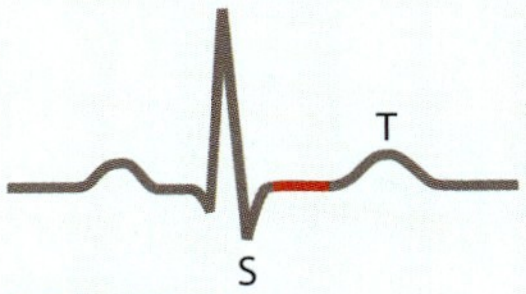 **S-T segment**	S-T segment	0.12	End of S deflection to start of T wave; represents duration of ventricular contraction
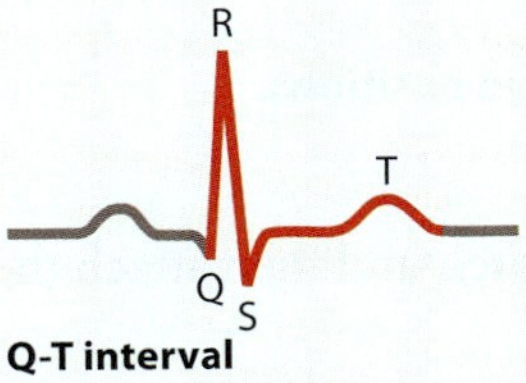 **Q-T interval**	Q-T interval	0.32–0.42	Start of Q deflection to return to isoelectric line; represents time from ventricular depolarization to the end of ventricular repolarization, including ventricular contraction
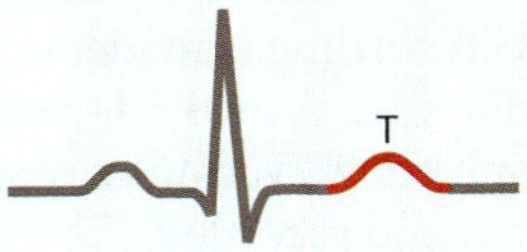 **T wave**	T wave	0.16	Start of T deflection to return to isoelectric line; represents repolarization of the ventricular fibers

ACTIVITY 1

Recording and Interpreting an Electrocardiogram

Learning Outcomes

1. Identify the components of the cardiac conduction system and explain the function of each.
2. Sketch the three deflection waves of an electrocardiogram and explain the physiological basis of each.
3. Record an electrocardiogram.
4. Compute heart rate using an electrocardiogram.
5. Compare a resting electrocardiogram and an electrocardiogram following vigorous exercise.

Materials Needed

- ☐ Alcohol swabs
- ☐ Electrodes
- ☐ Electrode gel
- ☐ Electrocardiograph (or alternative ECG recording device)
- ☐ Millimeter ruler

Instructions

To record an ECG, electrodes are placed on the body surface at sites where it is easiest to pick up the electrical changes originating in the heart. Various combinations of electrodes (leads) are activated. In a clinical setting, 12 leads are typically used, but in this activity you will use three standard limb leads (left wrist, right wrist, and left ankle). These three leads are called bipolar leads because each measures the potential difference between a positive electrode and a negative electrode.

Figure 22-3 illustrates the standard limb leads for electrocardiograms. This arrangement of leads has become known as Einthoven's triangle after the Dutch physiologist Willem Einthoven, who received a Nobel prize for his work in electrophysiology. Clinically, it is assumed that the heart lies in the center of Einthoven's triangle and that the electrical recordings are made at the vertices (corners) of the triangle. However, in practice, the electrodes are connected to each wrist and to the left ankle and they are considered to be connected (note the dashed lines in Figure 22-3) to the vertices of the triangle. Lead I connects the right wrist and the left wrist and records electrical activity spreading across the heart in a horizontal direction. Lead II connects the right wrist and the left ankle and lead III connects the left wrist and the left ankle. These two leads (leads II and III) both record electrical activity spreading from the base of the heart to its apex.

1. Record your lab partner's ECG by conducting the following steps:

   a. Use an alcohol wipe to clean the skin where the electrodes will be placed on the anterior surface of each wrist and the medial side of each ankle of the subject. (The electrode on the right ankle serves as an electrical ground; it does not measure electrical activity.)

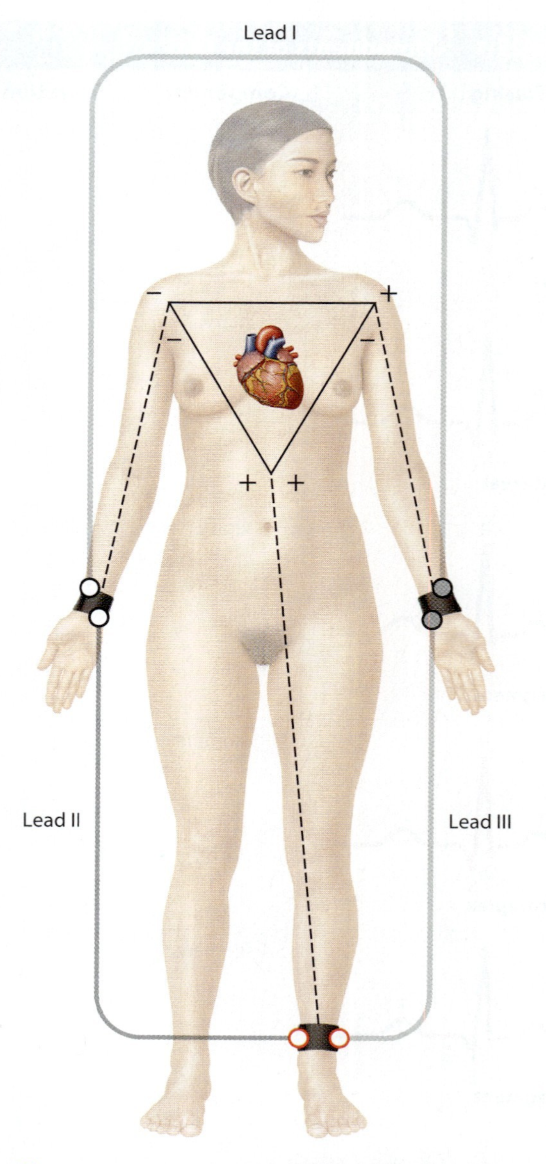

Figure 22-3 **Standard limb lead positions.**

   b. Apply electrode gel to those sites, and then attach the electrodes firmly.

   c. Connect cables to all three electrodes and to the electrocardiograph. When an ECG recording is made, only two electrodes are used at a time. A lead selector switch on the electrocardiograph allows various combinations of leads to be activated and recorded.

   Note: Follow any specific directions given by your instructor for setting up the instrument and recording an ECG with the equipment available in your lab.

   d. Turn on the electrocardiograph and set the paper speed to 2.5 cm/sec (25 mm/sec).

   e. Turn the lead selector switch to lead I (right wrist, left wrist) and record the ECG for 2 minutes. Stop the recording and mark it "Lead I."

f. Turn the lead selector switch to lead II (right wrist, left ankle) and record the ECG for 2 minutes. Stop the recording and mark it "Lead II."

g. Turn the lead selector switch to lead III (left wrist, left ankle) and record the ECG for 2 minutes. Stop the recording and mark it "Lead III."

h. Remove electrodes and clean the skin with an alcohol wipe.

i. Instruct the subject to exercise vigorously for 3 minutes.

j. Repeat steps a–g.

k. Remove electrodes and clean the skin with an alcohol wipe.

2. Interpret ECG results.

a. Each student should take a representative segment of one of the resting ECG recordings and label the segment with the name of the subject and the lead used. Identify and label the P wave, QRS complex, and T wave.

b. Compute the heart rate from the resting ECG. Recall that the ECG is recorded on paper moving at a speed of 25 mm/sec. Therefore, each millimeter of paper represents a time interval of 0.04 sec, and a distance of 5 mm (1 large square on standard ECG paper) is equivalent to 0.20 sec. Now, obtain a millimeter ruler and measure the distance from the beginning of one QRS complex to the beginning of the next QRS complex. Then, use the following equation to calculate the time that elapses during one complete heartbeat:

_____ mm/beat × 0.04 sec/mm = _____ sec/beat

Next, calculate the heart rate (in beats/min) by using the number you just calculated for seconds per beat in the following equation:

$$\frac{60 \text{ sec/min}}{\_\_\_\_\_ \text{ sec/beat}} = \_\_\_\_\_ \text{ beat/min}$$

c. Measure the P-R interval, the QRS complex, and the Q-T interval on the baseline ECG and then calculate the duration of each. Record your data in the following table.

Resting	Distance (in mm)	Duration (in seconds)
P-R interval		
QRS complex		
Q-T interval		

Are these calculated values within normal limits?

____

d. Calculate the heart rate using the ECG conducted following vigorous exercise:

_____ mm/beat × 0.04 sec/mm = _____ sec/beat

Next, calculate the heart rate (in beats/min) by using the number you just calculated for seconds per beat in the following equation:

$$\frac{60 \text{ sec/min}}{\_\_\_\_\_ \text{ sec/beat}} = \_\_\_\_\_ \text{ beat/min}$$

e. Measure the P-R interval, the QRS complex, and the Q-T interval on the ECG following vigorous exercise and then calculate the duration of each. Record your data in the following table.

After Exercise	Distance (in mm)	Duration (in seconds)
P-R interval		
QRS complex		
Q-T interval		

f. Describe the differences that you observed between the "resting" ECG and the "after exercise" ECG.

____

____

____

____

g. Explain the physiological basis of each difference that you observed between the "resting" ECG and the "after exercise" ECG. _______________________

____

____

____

____

ACTIVITY 2
Auscultating Heart Sounds

Learning Outcome

1. Auscultate heart sounds.

Materials Needed

- ☐ Stethoscope
- ☐ Alcohol wipes

Instructions

1. Obtain a stethoscope and alcohol wipes.

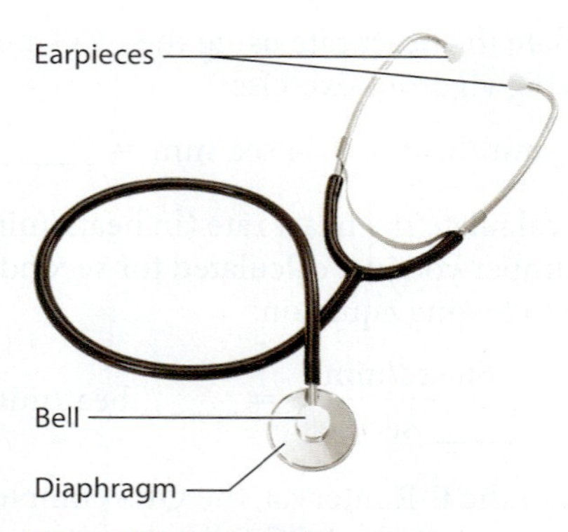

Figure 22-4 A stethoscope.

2. Clean the earpieces and the diaphragm of the stethoscope with an alcohol wipe and allow them to dry. **Figure 22-4** identifies the parts of a typical stethoscope.
3. Place the earpieces in your ears (pointed forward) and gently tap the diaphragm of the stethoscope to make sure you can hear sounds.
4. To hear the first heart sound (lubb), auscultate the mitral valve. **Figure 22-5** shows the auscultation sites. Press the diaphragm of the stethoscope on your partner's chest over the fifth intercostal space near the apex of the heart and listen to the sound.

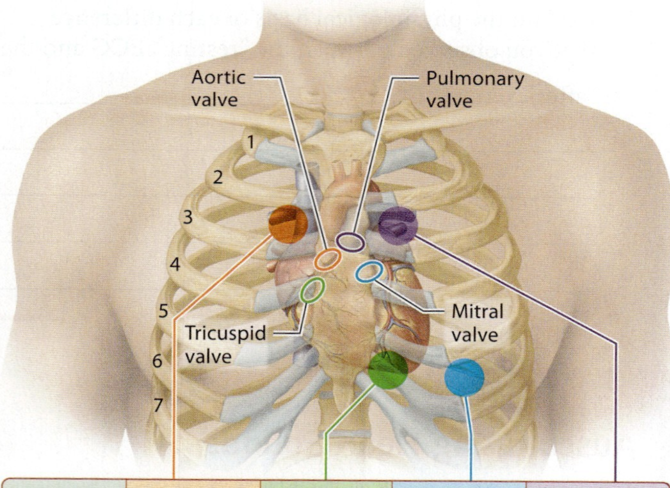

AREA	Aortic	Tricuspid	Mitral	Pulmonary
LOCATION OF SOUND	Second intercostal space, right sternal border	Fifth intercostal space, left sternal border	Fifth intercostal space, mid-clavicular line	Second intercostal space, left sternal border
TIMING OF SOUND	Aortic valve is heard here during S2.	Tricuspid valve is heard here during S1.	Mitral valve is heard here during S1.	Pulmonary valve is heard here during S2.

Figure 22-5 Heart valve locations and auscultation sites.

5. Move the stethoscope diaphragm to the second intercostal space just to the right of the sternum as shown in Figure 22-5 and listen for the closing of the aortic valve, which is associated with the second heart sound (dupp).

How did the two heart sounds differ? ______________________

____

____

ACTIVITY 3
Electrocardiography in a Human Subject Using BIOPAC®

Learning Outcomes

1. Become familiar with the electrocardiograph as a primary tool for evaluating electrical events within the heart.
2. Read an electrocardiogram (ECG) and correlate electrical events as displayed on the ECG with the mechanical events of the cardiac cycle.
3. Observe ECG rate and rhythm changes associated with changes in body position and breathing.

Materials Needed

- ☐ Computer system (Windows 7, Vista, XP, Mac OS X 10.5–10.7)
- ☐ Biopac Science Lab system (MP35 unit and software)
- ☐ Electrode lead set (SS2L lead set)
- ☐ Disposable vinyl electrodes (EL503), three electrodes per subject
- ☐ Biopac Electrode Gel (GEL1) and Abrasive Pad (ELPAD) or alcohol prep
- ☐ Biopac wall transformer (AC100A)
- ☐ Biopac serial cable (CBLSERA)
- ☐ Biopac USB adapter (USB1W)
- ☐ Lab table
- ☐ Chair with armrests

Instructions

A. Setup

1. Turn the computer ON.
2. Turn the MP35 unit OFF.
3. Plug the CBLSERA serial cable into the CH 2 channel of the acquisition unit.
4. Turn the MP35 unit ON.
5. Place the three EL503 electrodes on the subject as shown in **Figure 22-6**, using a small amount of gel (GEL1) on the skin where each electrode will be placed. The subject should not be in contact with any metal objects and should remove any metal wrist or ankle bracelets. Attach one electrode on the medial surface of the right leg just above the ankle and one on the medial surface of the left leg just above the ankle. Place the third electrode on the right anterior forearm at the wrist.

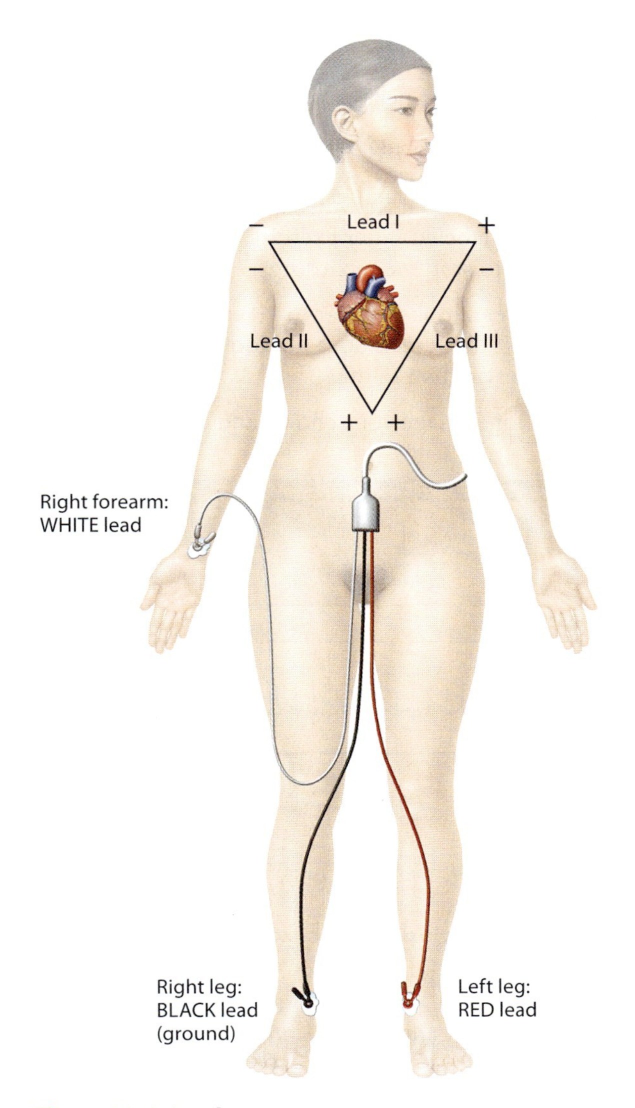

Figure 22-6 Lead setup.

6. Attach the electrode lead set (SS2L) to the electrodes as shown in Figure 22-6.
7. Have the subject lie down and relax.
8. Start the Biopac Student Lab program on the computer.
9. Choose lesson "L05-ECG.1" and click OK.
10. Type in a unique file name and click OK to end the setup section.

B. Calibration

The calibration procedure establishes the hardware's internal parameters (such as gain, offset, and scaling) and is critical for optimal performance.

1. Double check electrode connections and make sure the subject is supine, relaxed, and still.
2. Click Calibrate.
3. Wait for the calibration to stop, which will happen automatically after 8 seconds.

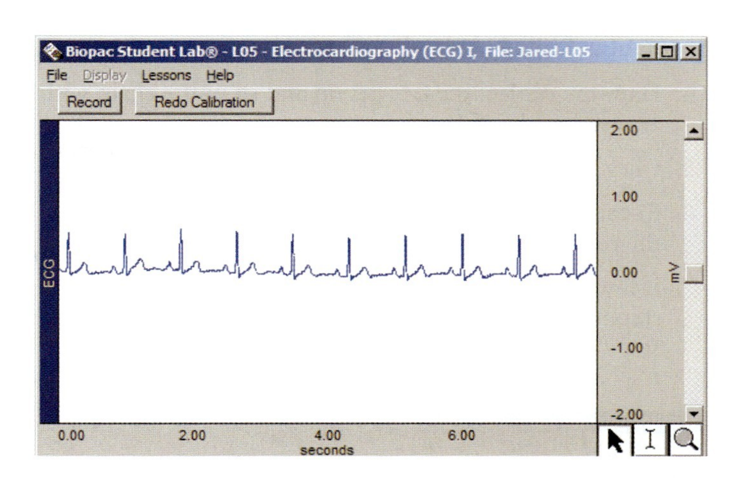

Figure 22-7 Sample calibration data.

4. Check the calibration data: If similar to **Figure 22-7**, proceed to data recording. If different from Figure 22-7, redo calibration.

C. Data Recording

Four conditions will be recorded: Supine, Seated, Breathing Deeply, and After Exercise. The subject should remain supine and relaxed while you review the lesson before beginning.

Hints for obtaining optimal data:

- ☐ The subject should be relaxed and still, and should not talk or laugh during any recording segment.
- ☐ When asked to sit up, the subject should do so in a chair, with arms relaxed.

Segment 1: Supine

1. Click Record.
2. Record for 20 seconds and then click Suspend to stop the recording.
3. Review the data on the screen. If it looks similar to **Figure 22-8**, proceed to step 4. If the data does not look similar, erase the recording by clicking Redo, and repeat steps 1 and 2.

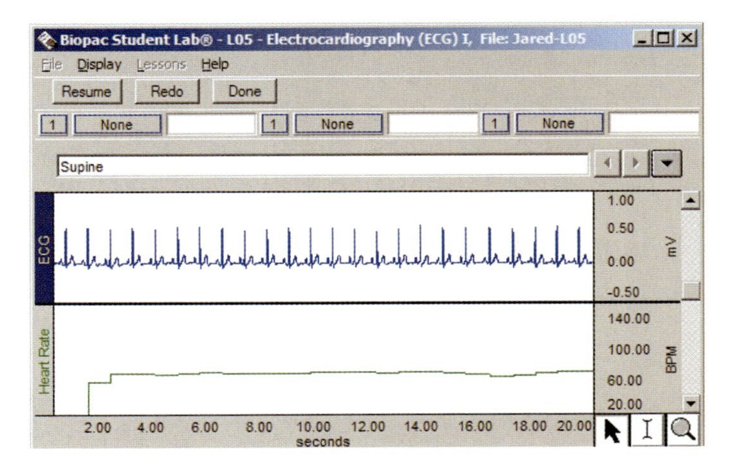

Figure 22-8 Sample data from reading in the supine position.

Segment 2: Seated

4. Have the subject quickly get up and sit in the chair, with arms relaxed and feet supported.
5. Click Resume.
6. Record for 20 seconds and then click Suspend to stop the recording.
7. If the screen looks similar to **Figure 22-9**, proceed to step 8; if it does not look similar, click Redo and repeat steps 4 through 6.

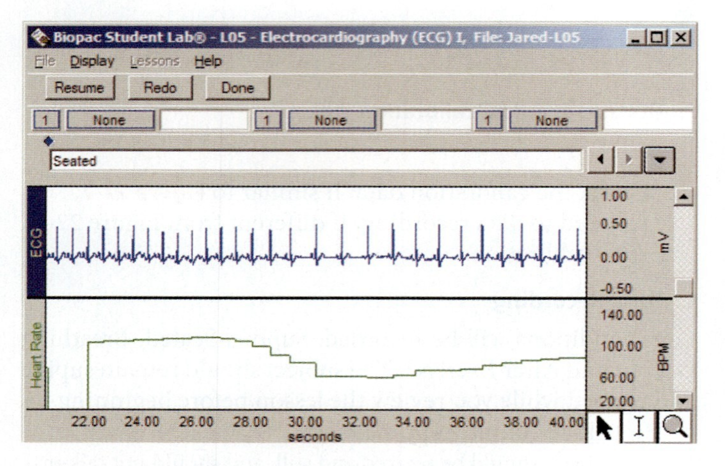

Figure 22-9 Sample data from the seated position.

Segment 3: Breathing Deeply

8. Click Resume; the recording will continue and a marker labeled "deep breathing" will be inserted.
9. Record for 20 seconds, and have the subject take five deep breaths during the 20 seconds. Insert an "inhale" marker at each inhalation and an "exhale" marker at each exhalation. Click Suspend to stop recording.
10. If the screen looks similar to **Figure 22-10**, proceed to step 11. If it does not look similar, click REDO and repeat steps 8 through 10.

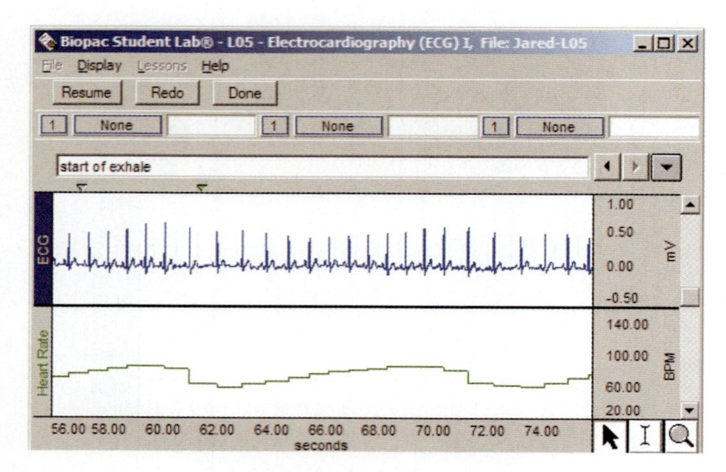

Figure 22-10 Sample data of breathing deeply.

Segment 4: After Exercise

11. Have the subject perform either pushups or jumping jacks for about 60 seconds (to elevate the heart rate).
12. Immediately after exercising, have the subject quickly sit down and click Resume, and record for 60 seconds. A marker labeled "after exercise" will be inserted. Click Suspend to stop the recording.
13. If the screen looks similar to **Figure 22-11**, proceed to step 14. If the screen does not look similar, click REDO and repeat steps 11 and 12.
14. Click Done. Make your choice and continue as directed.

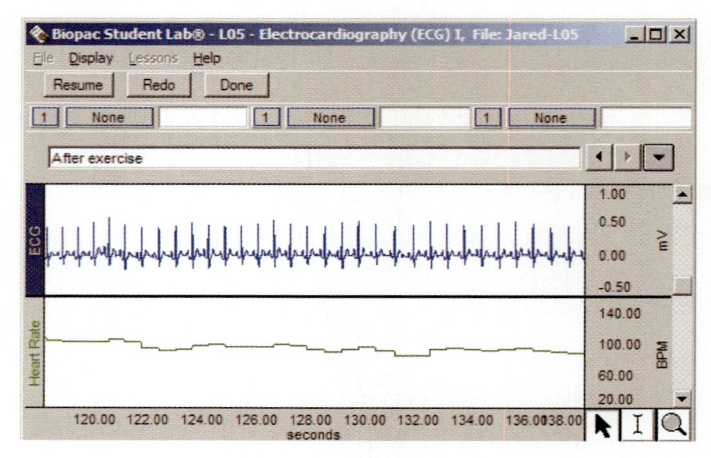

Figure 22-11 Sample data after exercise.

D. Data Analysis

In this section you will examine ECG components of cardiac cycles and measure amplitudes and durations of the ECG components. Interpreting ECGs is a skill. Do not be alarmed if your ECG is different from the examples shown or from the tables and figures.

1. Enter the Review Saved Data mode from the Lessons menu. Set up your display window for viewing four successive beats from segment 1 (Supine).
2. The measurement boxes are above the marker region in the data window. Each measurement has three sections: channel number, measurement type, and results. Set up the measurement boxes as follows:

   CH 2 ΔT (delta time; differences in time between end and beginning of any area selected by I-beam cursor)

   CH 2 BPM (beats per minute; calculates ΔT in seconds and converts this value to minutes)

   CH 2 Δ (delta amplitude, difference in amplitude between end and beginning of any area selected by I-beam cursor)

   CH 2 max (maximum amplitude; calculates maximum amplitude within any selected area, including end points)

3. Using the I-beam cursor, select and measure as precisely as possible the area from one QRS complex peak to the next QRS complex peak.
4. Take measurements at two other intervals in the current recording.
5. Use the Zoom tool to zoom in on a single cardiac cycle from segment 1 (Supine).
6. Save or print the data file.
7. Quit the program.
8. Complete the data tables (**Tables 22-2** through **22-7**) from the data analysis.

Table 22-2 Data Table for Change in Time and Beats per Minute

Measurement	From Channel	1	2	3	Mean	Range
ΔT						
BPM						

Table 22-3 Data Table for the Events of the Cardiac Cycle

ECG Component	Duration ΔT [CH 2]				Amplitude (mV) Δ [CH 2]			
	Cycle 1	Cycle 2	Cycle 3	Mean	Cycle 1	Cycle 2	Cycle 3	Mean
P wave								
P-R interval								
P-R segment								
QRS complex								
Q-T interval								
S-T segment								
T wave								

Table 22-4 Data Table for the Ventricular Readings [CH 2 ΔT]

Ventricular Readings	Cycle 1	Cycle 2	Cycle 3	Mean
Q-T				
End of T wave to subsequent R wave				

Table 22-5 Data Table for Inhalation and Exhalation

Rhythm	CH	Cycle 1	Cycle 2	Cycle 3	Mean
INHALATION					
ΔT					
BPM					
EXHALATION					
ΔT					
BPM					

Table 22-6 Data Table for Change in Time and Beats per Minute

Heart Rate	CH	Cycle 1	Cycle 2	Cycle 3	Mean
ΔT					
BPM					

Table 22-7 Data Table for Ventricular Readings [CH 2 ΔT]

Ventricular Readings	Cycle 1	Cycle 2	Cycle 3	Mean

9. Answer the following question set using the data from the previous tables:
   a. Are there changes in heart rate between conditions? What physiological mechanisms cause these changes? ______________________________

   ____

   ____

   ____

   ____

   b. Describe any differences in the cardiac cycle compared with the respiratory cycle (segment 3).

   ____

   ____

   ____

   ____

   c. What changes occurred in the duration of systole and diastole between resting and after exercise?

   ____

   ____

   ____

   ____

   d. Explain any changes in the ECG intervals and segments during exercise. ______________________________

   ____

   ____

   ____

   ____

PhysioEx EXERCISE 6
Cardiovascular Physiology

The PhysioEx 9.1 Laboratory Simulations in Physiology are easy-to-use laboratory simulations that can be used as an alternative to or as a supplement to wet lab activities in this unit. Each simulation allows you to investigate important physiological concepts, repeat labs as often as you like, and conduct experiments that are difficult to perform in a wet lab environment because of time, cost, or safety concerns.

Access the simulations in these activities at MasteringA&P® > Study Area > PhysioEx 9.1

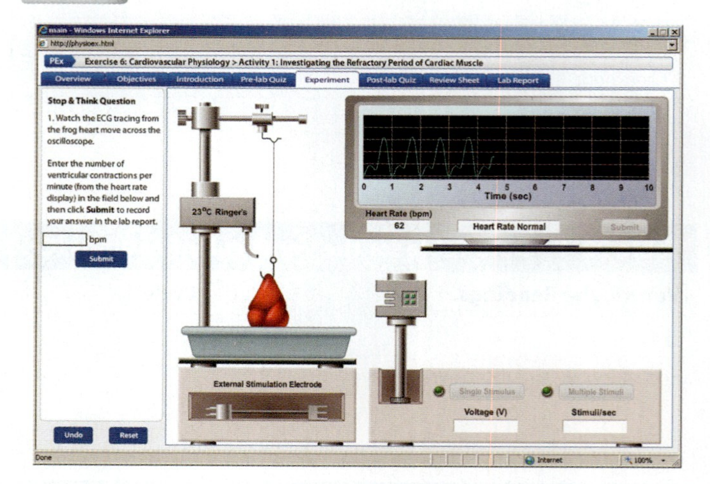

There you will find the following activities:

PEx Activity 1: Investigating the Refractory Period of Cardiac Muscle

PEx Activity 2: Examining the Effect of Vagus Nerve Stimulation

PEx Activity 3: Examining the Effect of Temperature on Heart Rate

PEx Activity 4: Examining the Effects of Chemical Modifiers on Heart Rate

PEx Activity 5: Examining the Effects of Various Ions on Heart Rate

Post-lab quizzes are also assignable in MasteringA&P®

POST-LAB ASSIGNMENTS

Name: ______________________ Date: ______________ Lab Section: ________

PART I. Check Your Understanding

Activity 1: Recording and Interpreting an Electrocardiogram

1. Which component(s) of the cardiac conduction system:

______________________ a. serves as the pacemaker?

______________________ b. is the only connection between the atria and the ventricles?

______________________ c. extend through the interventricular septum toward the apex of the heart?

______________________ d. receives action potentials from the sinoatrial node?

______________________ e. conducts action potentials from the apex of the heart to the remaining myocardium?

2. Identify the components of the electrocardiogram shown in the accompanying illustration:

a. ______________________

b. ______________________

c. ______________________

d. ______________________

e. ______________________

f. ______________________

g. ______________________

h. ______________________

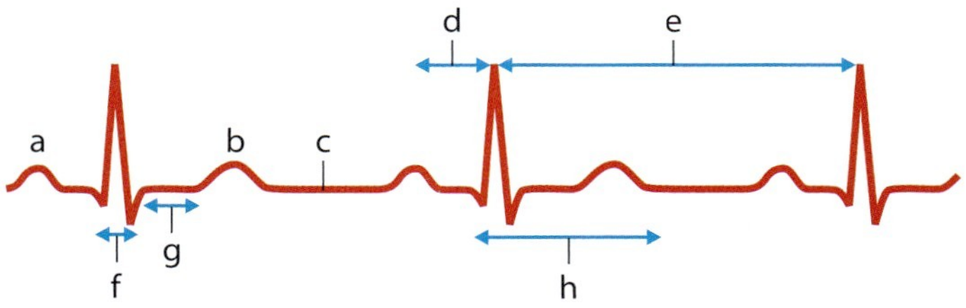

3. Which component of the electrocardiogram represents:

______________ a. depolarization of the ventricles?

______________ b. the duration of ventricular contraction?

______________ c. depolarization of the atria?

______________ d. the time required for electrical impulses to travel from the SA node through the AV node?

______________ e. repolarization of the ventricles?

______________ f. start of T deflection to return to isoelectric line?

4. State two differences between a resting electrocardiogram and an electrocardiogram following vigorous exercise. ____

____

____

Activity 2: Auscultating Heart Sounds

1. What causes the first heart sound that is heard with a stethoscope? ______________________________

____

2. What causes the second heart sound that is heard with a stethoscope? ______________________________

____

Activity 3: Electrocardiography in a Human Subject Using BIOPAC®

1. How did the subject's heart rate change when comparing seated data vs. post-exercise data? Explain the physiological basis of this difference. ______________________________

____

____

____

2. How did the Q-T intervals (which closely correspond to the duration of ventricular contraction) differ when comparing seated data vs. post-exercise data? Explain the physiological basis of this difference. ______________________________

____

____

PART II. Putting It All Together

A. Review Questions

Answer the following questions using your lecture notes, your textbook, and your lab notes.

1. How can an electrocardiogram record the heart's electrical activity when the electrodes are placed on the skin surface? ______________________________

____

2. Fibrillation is uncoordinated contraction of cardiac muscle. Which type of fibrillation is a more serious problem: atrial or ventricular? Why? ______________________________

____

____

3. Match each of the accompanying ECG tracings with its correct description:

    ______ AV heart block

    ______ Tachycardia

    ______ Atrial fibrillation

    ______ Ventricular fibrillation

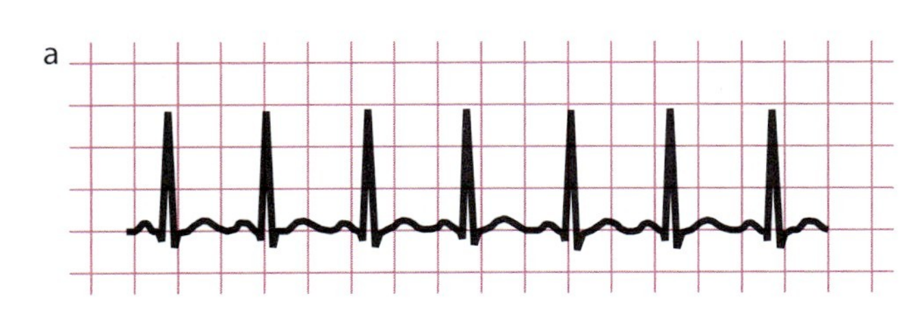

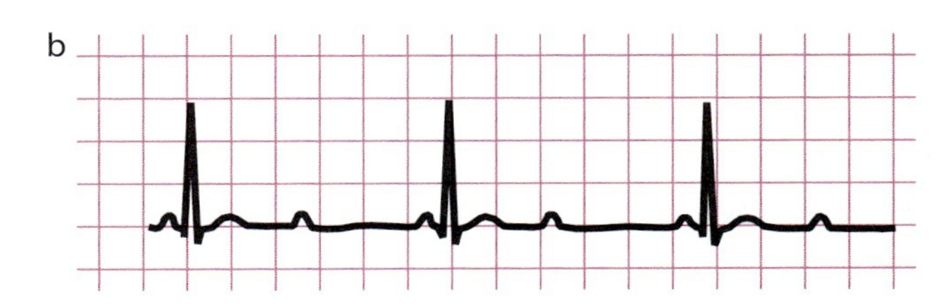

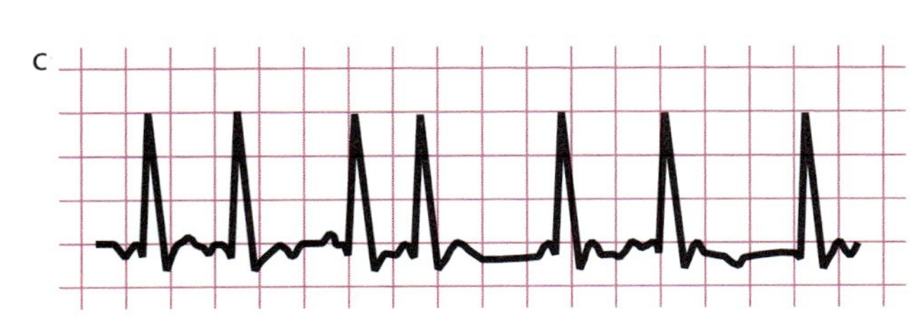

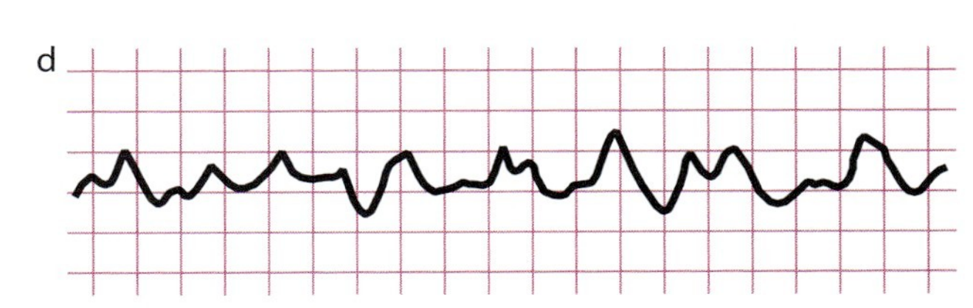

4. Pilocarpine is a parasympathomimetic drug that stimulates the release of ACh from nerve fibers of the vagus nerve. What effect does this drug have on heart rate? ______________________________

    ____

    ____

5. Beta-adrenergic blockers are used to treat high blood pressure. Explain how these drugs affect heart rate. ____

    ____

B. Concept Mapping

1. Fill in the blanks to complete this concept map outlining the electrical activity of the heart.

hormones parasympathetic fibers SA node sympathetic fibers terminal branches

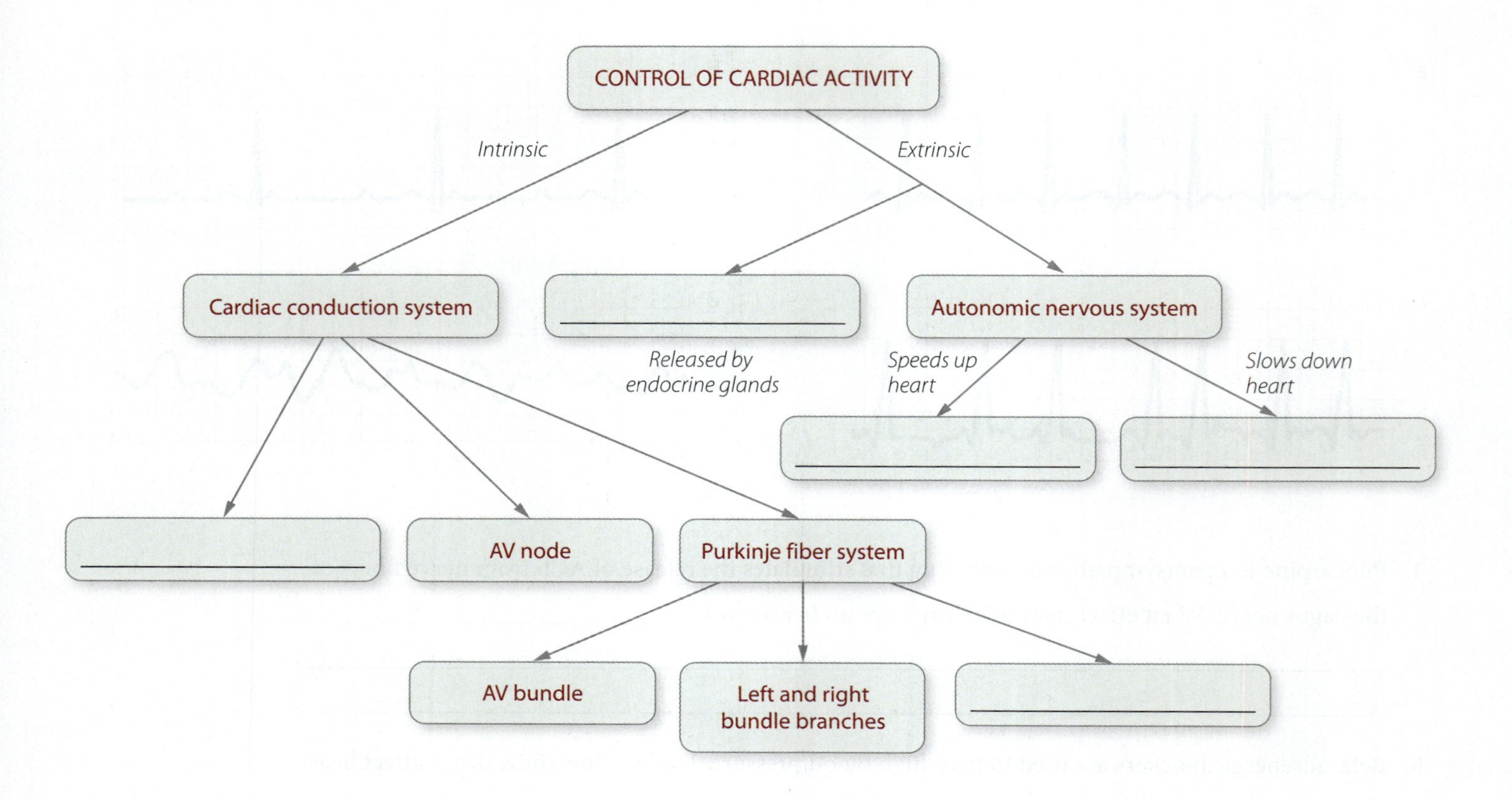

2. Construct a unit concept map to show the relationships among the following set of terms. Include all of the terms in your diagram. Your instructor may choose to assign additional terms.

AV node	diastole	electrocardiogram	exercise
hormones	P wave	parasympathetic fibers	QRS complex
right and left bundle branches		SA node	sympathetic fibers
systole	T wave	terminal branches	vagus nerve

23

Anatomy of Blood Vessels

UNIT OUTLINE

The human body is estimated to have more than 60,000 miles of blood vessels. The three major types of blood vessels are arteries, veins, and capillaries. **Arteries** carry blood away from the heart, **veins** carry blood toward the heart, and **capillaries** are the site of exchange of gases, nutrients, and wastes. As arteries carry blood away from the heart, they branch into smaller and smaller arteries that eventually lead into **arterioles**, which then feed into capillary beds throughout the body. Capillary beds are drained by **venules** that empty into progressively larger veins, which eventually empty into the venae cavae, the two great veins that return deoxygenated blood to the heart.

THINK ABOUT IT *Arteries carry blood away from the heart, so we say they "split" or "fork" as they become smaller and eventually branch into arterioles, which lead into capillary beds, where the exchange of materials occurs. From the capillaries, blood drains into venules, the smallest veins, which unite to form larger and larger veins, which eventually empty into the heart. We say that veins returning blood to the heart "merge" or "unite" as they become increasingly larger.*

Write an anatomically accurate sentence using each of the following sets of blood vessels. The first set has been completed as an example.

brachial artery, radial artery, ulnar artery

The brachial artery splits to form the radial artery and the ulnar artery.

ascending aorta, aortic arch, descending aorta

____

____

aortic arch, left common carotid artery, left subclavian artery, brachiocephalic artery

____

____

superior vena cava, right atrium, inferior vena cava

____

____

PRE-LAB ASSIGNMENTS

Pre-lab quizzes are also assignable in MasteringA&P®

To maximize learning, BEFORE your lab period carefully read this entire lab unit and complete these pre-lab assignments using your textbook, lecture notes, and prior knowledge.

PRE-LAB Activity 1: Identifying the Major Arteries That Supply the Head, Neck, Thorax, and Upper Limbs

1. Match each of the following arteries with its correct description:

_____	a. ulnar artery	1. splits to form the radial artery and the ulnar artery
_____	b. axillary artery	2. is a continuation of the subclavian artery
_____	c. external carotid artery	3. supplies the brain
_____	d. brachial artery	4. travels along the medial side of the forearm
_____	e. internal carotid artery	5. arises from the aortic arch
_____	f. radial artery	6. gives rise to the digital arteries
_____	g. superficial palmar arch	7. arises from the ascending aorta
_____	h. brachiocephalic artery	8. travels along the lateral side of the forearm
_____	i. coronary artery	9. gives rise to the occipital artery

PRE-LAB Activity 2: Identifying the Major Arteries That Supply the Abdominopelvic Organs and the Lower Limbs

1. Match each of the following arteries with its correct description:

_____	a. celiac trunk	1. gives rise to the anterior and posterior tibial arteries
_____	b. femoral artery	2. supplies the adrenal gland
_____	c. renal artery	3. supplies the kidney
_____	d. suprarenal artery	4. supplies the ovaries or testes
_____	e. popliteal artery	5. supplies the gallbladder, liver, esophagus, stomach, spleen, and duodenum
_____	f. dorsalis pedis artery	6. branches from the external iliac artery
_____	g. gonadal artery	7. supplies the ankle joint and foot
_____	h. internal iliac artery	8. perfuses the organs of the pelvic cavity
_____	i. posterior tibial artery	9. gives rise to the fibular artery

PRE-LAB Activity 3: Identifying Veins That Drain into the Venae Cavae

1. Match each of the following veins with its correct description(s). More than one correct answer is possible for each numbered item. Include all correct answers.

_____	a. external iliac vein	1. empties directly into the inferior vena cava
_____	b. brachial vein	2. drains the anterior and posterior tibial veins
_____	c. popliteal vein	3. drains the superficial palmar arch
_____	d. great saphenous vein	4. drains the radial and ulnar veins
_____	e. subclavian vein	5. empties directly into the brachiocephalic vein
_____	f. radial vein	6. fuses with the internal iliac vein to form the common iliac vein
_____	g. anterior tibial vein	7. is the longest vein in the body
_____	h. dural sinuses	8. drains the ankle and foot
_____	i. renal vein	9. empty directly into the internal jugular vein

PRE-LAB Activity 4: Examining the Histology of Arteries and Veins

1. Which blood vessel type:
   a. always transports blood away from the heart? _______________
   b. always transports blood to the heart? _______________
   c. is the site of gas, nutrient, and waste exchange? _______________
2. Which blood vessel type would you expect to have the thickest smooth muscle layer? __________
   Why? _______________

Major Systemic Arteries

The major arteries of the body are shown in **Figure 23-1**. Arteries distribute nutrient-rich oxygenated blood to all body cells. The **aorta**, the largest artery in the body, arises from the left ventricle and consists of three parts: the ascending aorta, aortic arch, and descending aorta. The **ascending aorta** gives rise to the left and right coronary arteries, which supply blood to the heart muscle. The **aortic arch** gives rise to three arteries: the **brachiocephalic artery**, the **left common carotid artery**, and the **left subclavian artery** (see Figure 23-1). The brachiocephalic artery then divides into the **right common carotid artery** and the **right subclavian artery**. The **descending aorta** is further divided into the **thoracic aorta** and the **abdominal aorta**. The abdominal aorta ends at the body of the L_4 vertebra, where it splits into the right and left **common iliac arteries**.

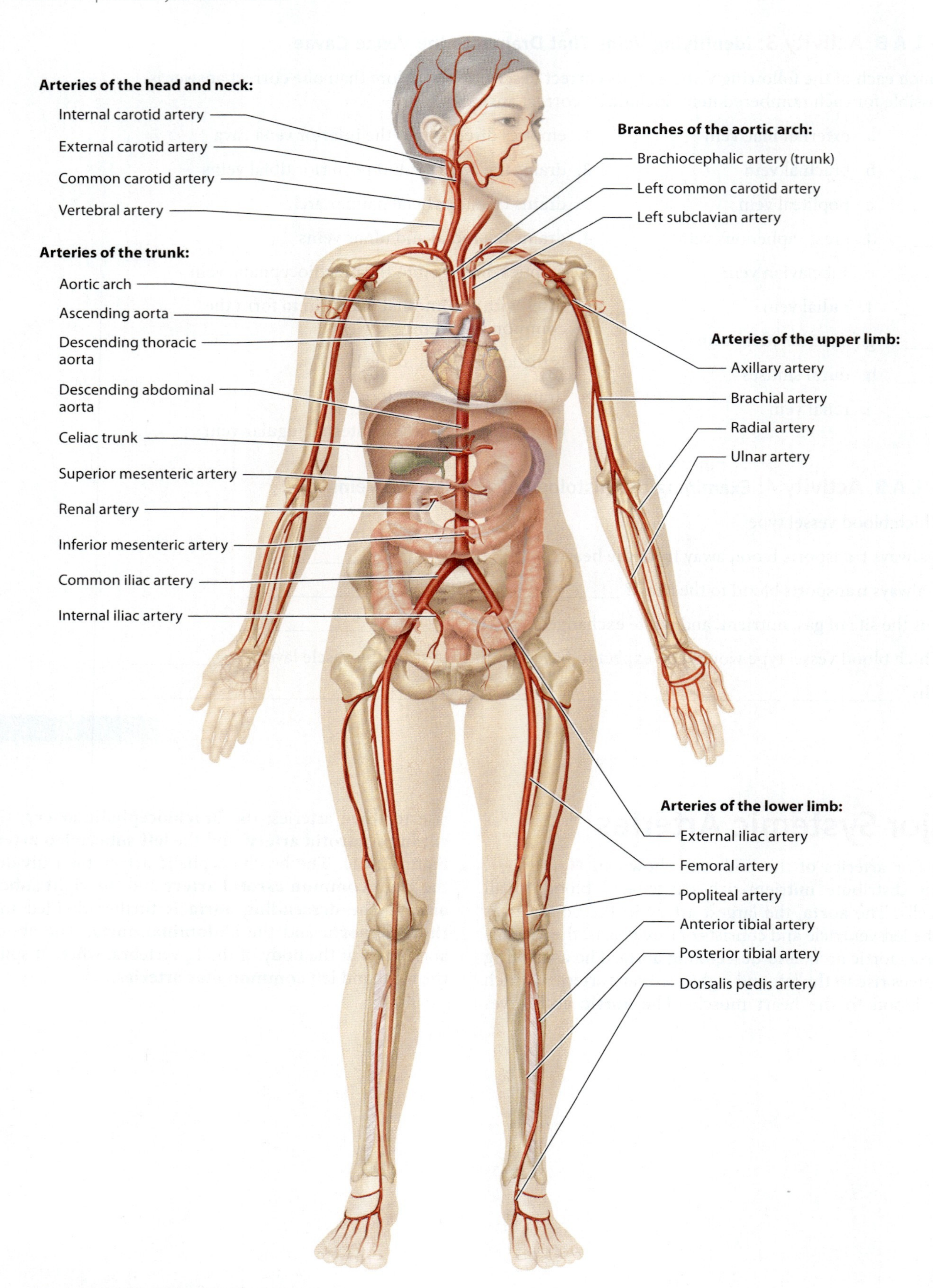
Arteries of the head and neck:
Internal carotid artery
External carotid artery
Common carotid artery
Vertebral artery
Branches of the aortic arch:
Brachiocephalic artery (trunk)
Left common carotid artery
Left subclavian artery
Arteries of the trunk:
Aortic arch
Ascending aorta
Descending thoracic aorta
Descending abdominal aorta
Celiac trunk
Superior mesenteric artery
Renal artery
Inferior mesenteric artery
Common iliac artery
Internal iliac artery
Arteries of the upper limb:
Axillary artery
Brachial artery
Radial artery
Ulnar artery
Arteries of the lower limb:
External iliac artery
Femoral artery
Popliteal artery
Anterior tibial artery
Posterior tibial artery
Dorsalis pedis artery

Figure 23-1 **Major arteries of the body.**

LabBOOST >>>

Blood Vessel Pathways Sketching blood vessel pathways is a great learning tool for understanding the relationships among blood vessels. It helps you to see how arteries branch into smaller arteries as they transport blood away from the heart, and how veins unite to form larger veins as they return blood to the heart. This Lab Boost provides an example of how you might sketch the aorta and its major branches.

The aorta, the largest artery in the body, arises from the left ventricle and has three parts: the ascending aorta, aortic arch, and descending aorta. Label the three portions of the aorta on the accompanying illustration. Then label the sites at which the left and right coronary arteries arise. Arising from the aortic arch are (in order) the brachiocephalic, left common carotid, and left subclavian arteries; label these vessels. The brachiocephalic artery splits into the right common carotid artery and the right subclavian artery; add these arteries to the illustration and label each. Next, label the two parts of the descending aorta: the thoracic aorta and the abdominal aorta.

You can use this sketch as the foundation for remembering the names and locations of all the arteries your instructor assigns. You might copy the sketch onto a poster board and then keep adding arteries and labels to the sketch. Then, make a similar sketch on another poster board showing all the veins your instructor assigns.

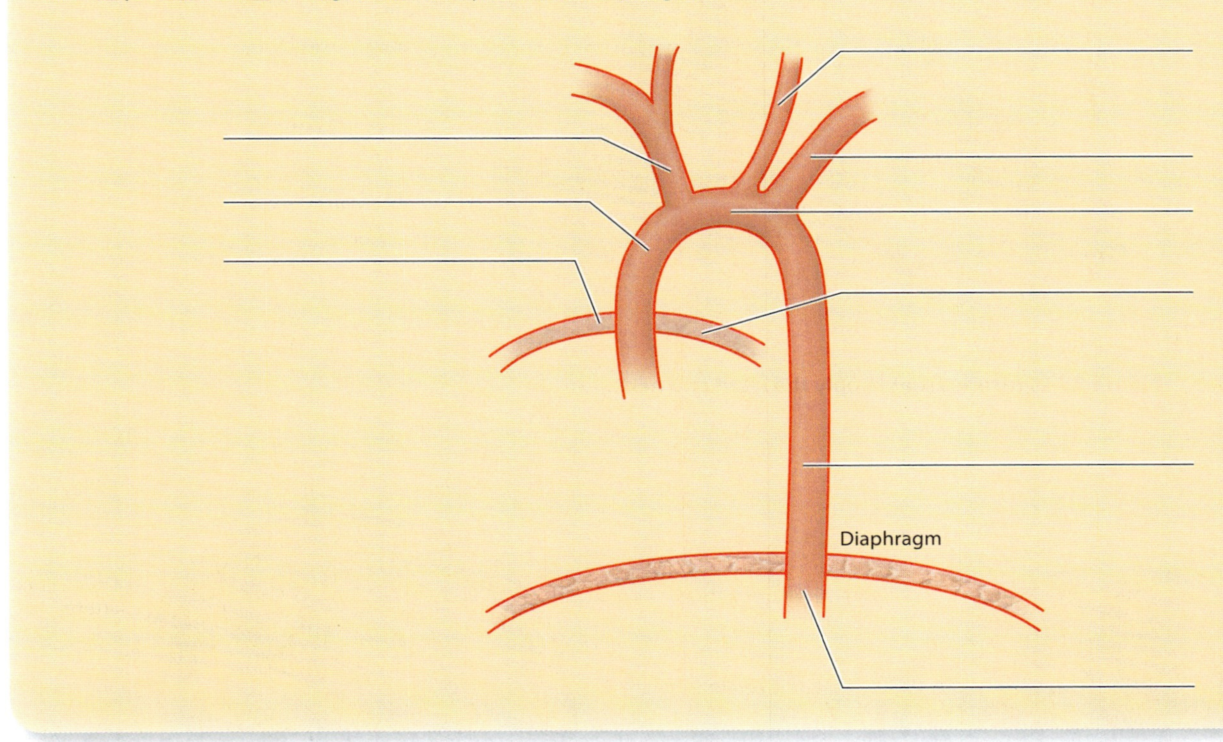

Arteries That Supply the Head and Neck

As shown in **Figure 23-2**, each of the common carotid arteries divides into an **external carotid artery** and an **internal carotid artery**. At the point where the common carotid artery splits is an enlarged sac called the carotid sinus, which contains mechanoreceptors involved in the regulation of blood pressure. The external carotid arteries supply the tissues of the face, scalp, and neck via several arterial branches. The internal carotid arteries, located deep in the superior part of the neck, enter the skull through the carotid canals of the temporal bones.

Inside the skull, the internal carotid arteries give rise to three branches: the ophthalmic, anterior cerebral, and middle cerebral arteries (**Figure 23-3**). The **ophthalmic arteries** supply the eyes and the walls of the orbit. The **anterior cerebral arteries** and **middle cerebral arteries** supply the brain with blood and will be discussed with the cerebral arterial circle in Unit 24.

The **subclavian arteries** give rise to several important branches that supply the head and neck, the first of which are the **vertebral arteries** (see Figure 23-3). As these arteries travel up toward the brain, they give off many branches that supply the spinal cord and vertebrae. After entering the skull through the foramen magnum, the vertebral arteries

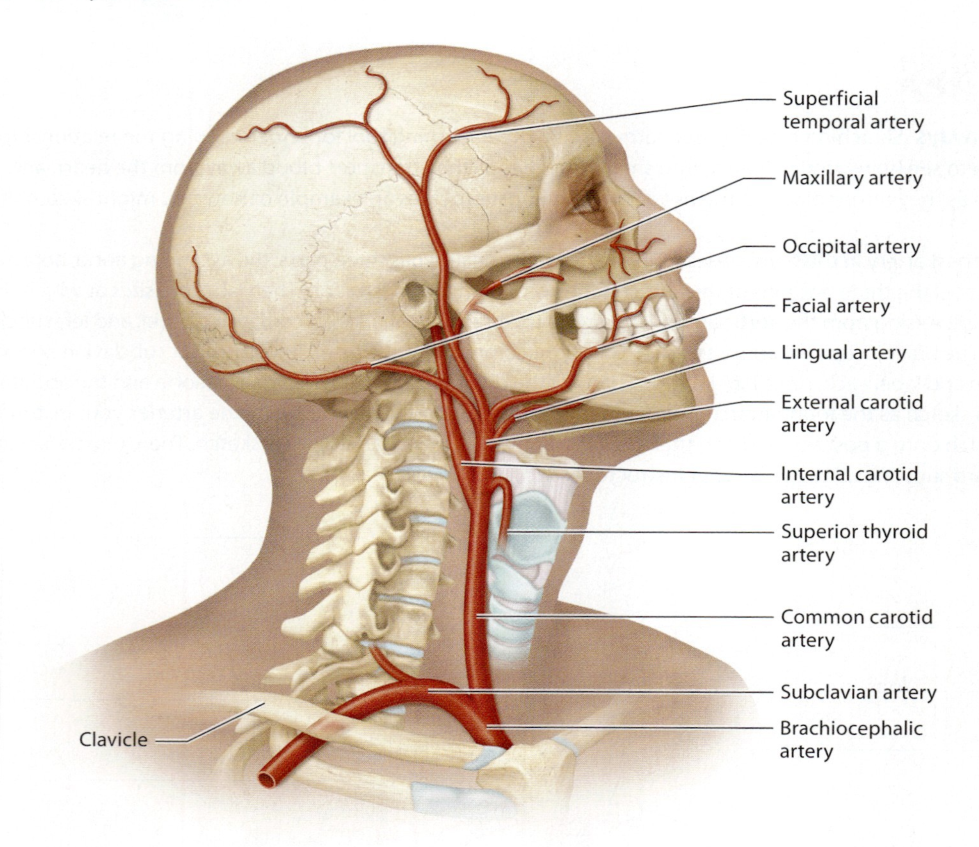

Figure 23-2 **Arteries that supply the head and neck.**

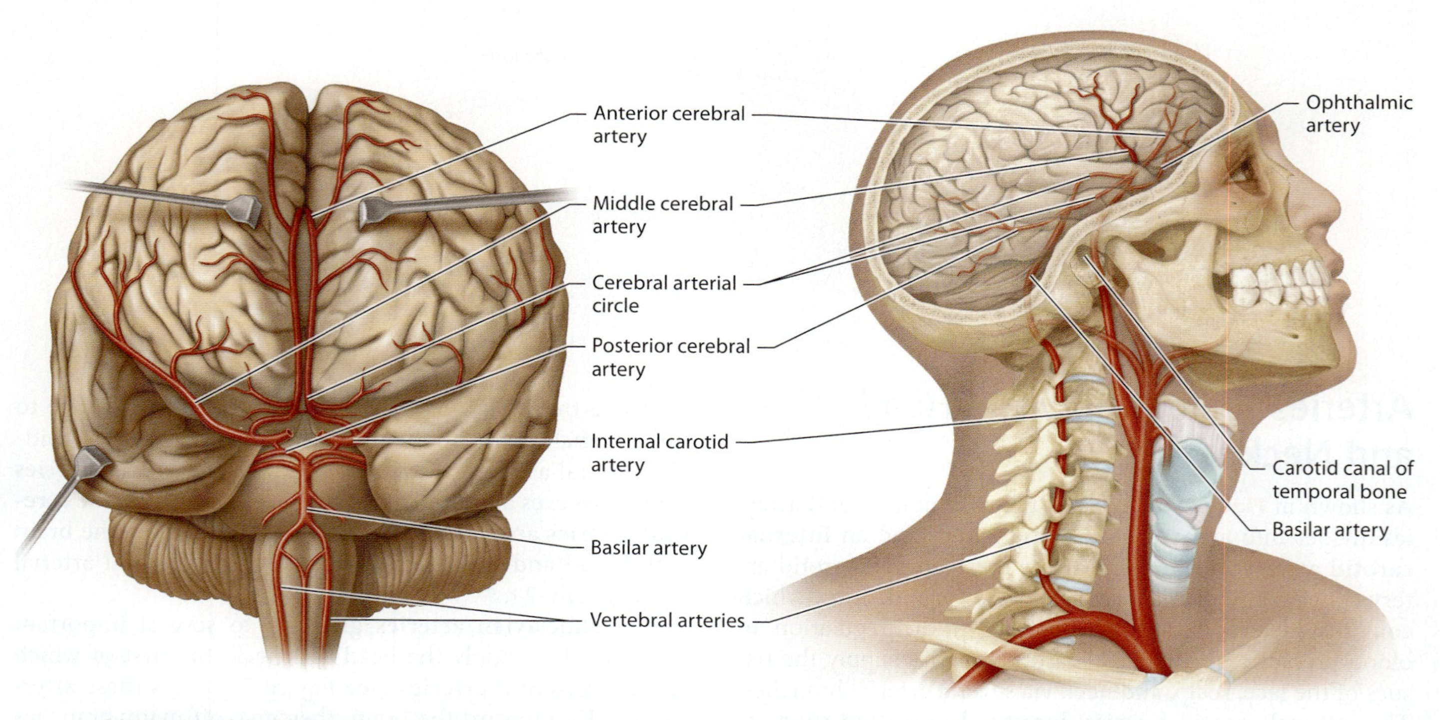

Figure 23-3 **Major arteries that supply the brain.**

fuse to form the **basilar artery**, which leads into the **cerebral arterial circle** (or circle of Willis).

Arteries That Supply the Upper Limbs

In the axilla (armpit), the subclavian artery gives rise to the **axillary artery**, which supplies the upper limbs (**Figure 23-4**). The axillary artery gives off several branches that serve the chest wall and pectoral girdle before continuing as the **brachial artery** as it enters the arm. The brachial artery gives off several branches that supply the muscles of the arm before dividing in the elbow region into its two terminal branches, the lateral **radial artery** and the medial **ulnar artery**. As the radial and ulnar arteries enter the hand, their branches anastomose to form the **deep** and **superficial palmar arches**, which give rise to the **metacarpal arteries** and the **digital arteries** (not shown) supplying the digits (fingers).

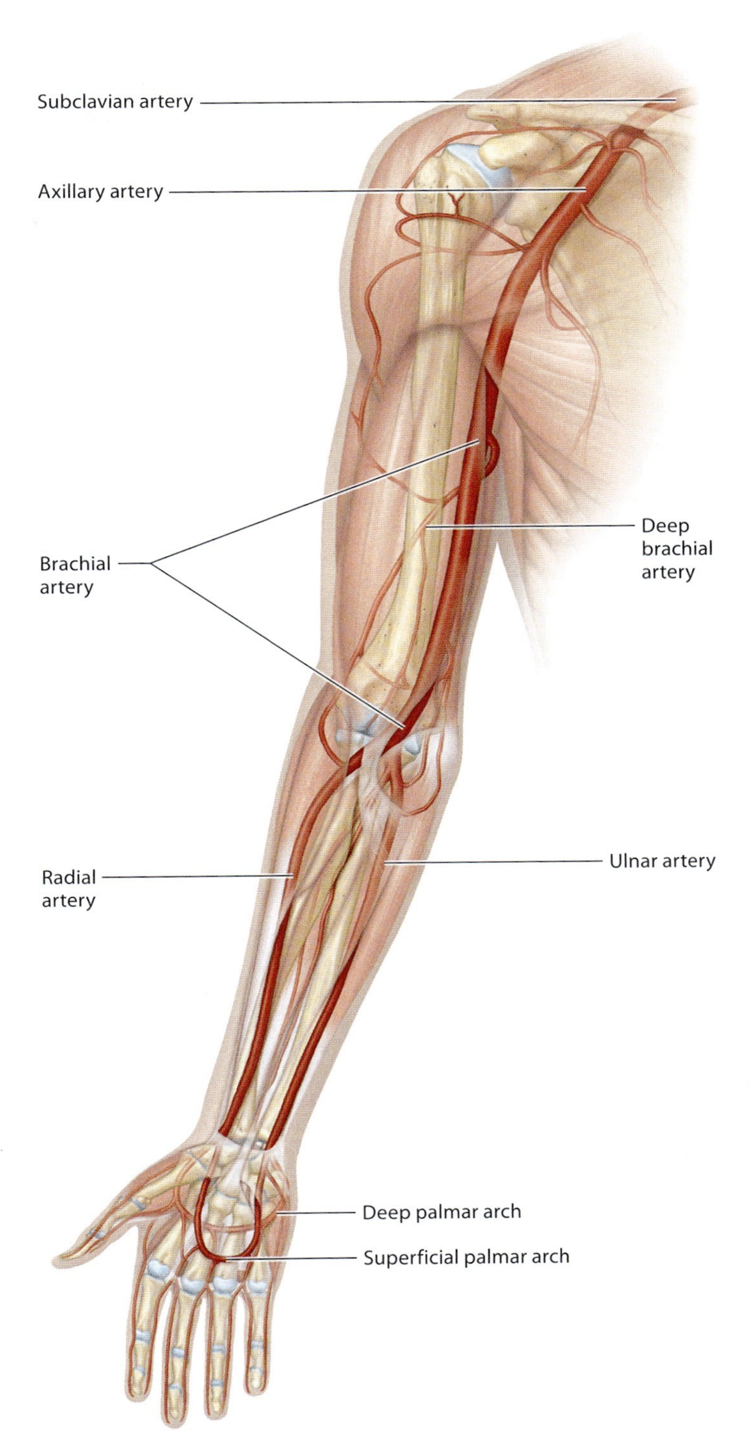

Figure 23-4 Arteries that supply the upper limbs.

Arteries That Supply the Thorax

The **thoracic aorta** gives rise to parietal (more superficial) and visceral (deeper) branches that supply the organs of the thorax (**Figure 23-5**). Parietal branches include the **posterior intercostal arteries**, which supply the intercostal muscles, spinal cord, vertebrae, and skin, and the **superior phrenic arteries**, which supply the diaphragm. Visceral branches include the **esophageal arteries**, which supply the esophagus, and the **bronchial arteries**, which supply the lungs and bronchi.

> **QUICK TIP** Do not confuse the bronchial arteries, which are vessels of the systemic circulation, with the pulmonary arteries, which function within the pulmonary circulation. The bronchial arteries supply the tissues of the lungs and bronchi with oxygen and nutrients, whereas the pulmonary arteries transport deoxygenated blood to the lungs, where the blood is then oxygenated.

Arteries That Supply the Abdomen

The abdominal aorta gives rise to arteries that supply the abdominopelvic organs (**Figure 23-6a**). The large **celiac trunk** (see Figures 23-6a and b) branches from the abdominal aorta and gives rise to arterial branches that supply the gallbladder, liver, esophagus, stomach, spleen, and duodenum. The large, unpaired **superior mesenteric artery** (see Figure 23-6a) arises below the celiac trunk and supplies almost all of the small intestine and most of the large intestine. The paired **middle suprarenal arteries**, **renal arteries**, and **gonadal arteries** supply the adrenal glands, kidneys, and gonads, respectively. The last major branch of the abdominal aorta is the unpaired **inferior mesenteric artery**, which supplies the distal portion of the large intestine from the midpoint of the transverse colon to the rectum. In addition to these visceral branches of the abdominal aorta, there are also several parietal branches (not shown): the **inferior phrenic arteries**, which supply the diaphragm; the **lumbar arteries**, which

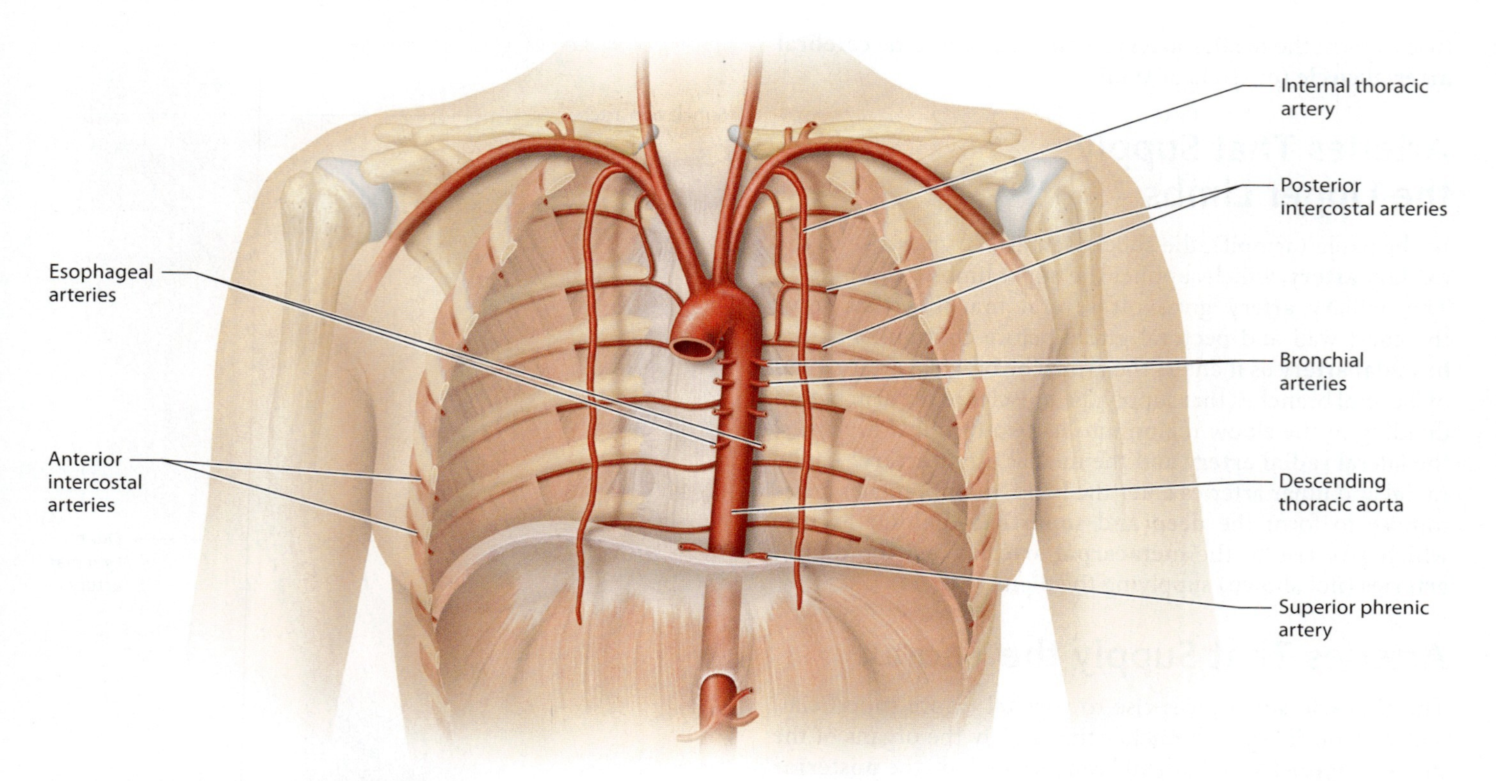

Figure 23-5 Arteries that supply the thorax.

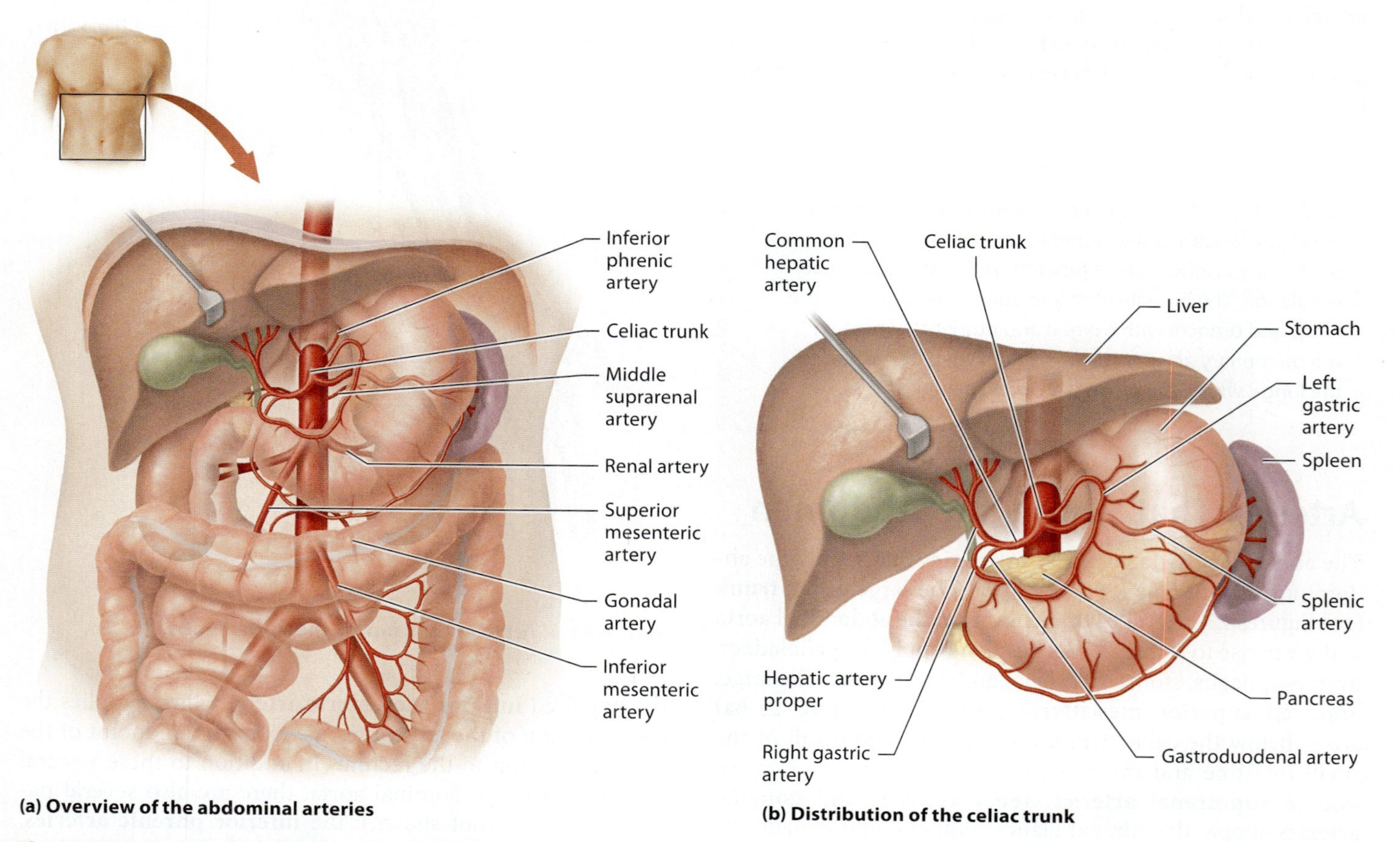

(a) Overview of the abdominal arteries

(b) Distribution of the celiac trunk

Figure 23-6 Arteries that supply the abdomen.

supply the posterior abdominal wall; and the **median sacral artery**, which supplies the sacrum and coccyx.

The abdominal aorta terminates as it splits into the right and left **common iliac arteries**, which supply the lower abdominal wall, organs of the pelvic cavity, and lower limbs (**Figure 23-7**). Each common iliac artery splits into an **internal iliac artery**, which supplies the pelvic region, and an **external iliac artery**, which supplies the lower limb.

Arteries That Supply the Lower Limbs

As the external iliac arteries enter the thigh, they become the **femoral arteries** (see Figure 23-7). As each femoral artery reaches the knee, it passes posteriorly, enters the popliteal fossa, and becomes the **popliteal artery**, which then splits into the **anterior tibial artery** and the **posterior tibial**

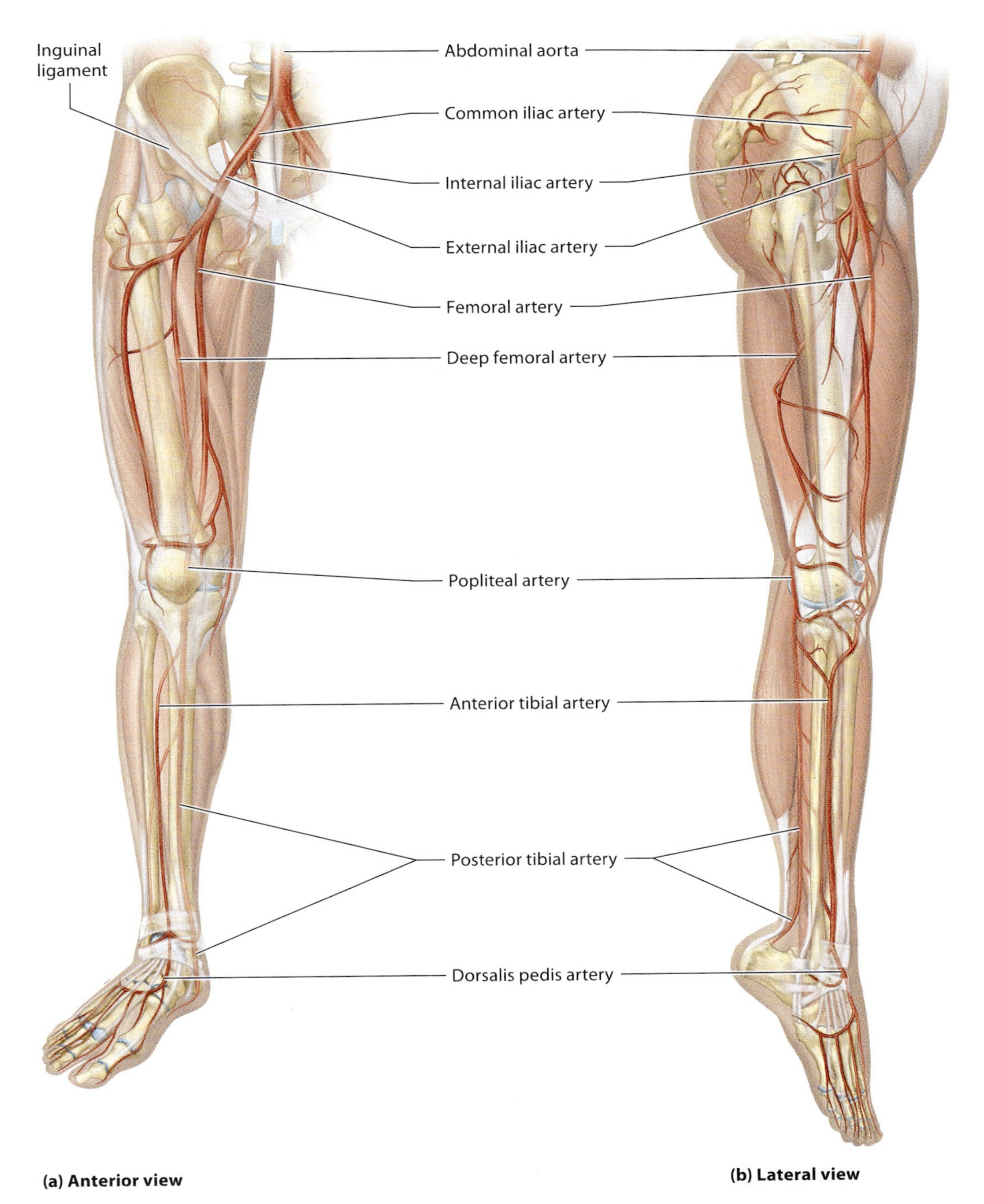

Figure 23-7 Arteries that supply the lower limbs.

artery. At the ankle the anterior tibial artery becomes the **dorsalis pedis artery**, which supplies the ankle and dorsal region of the foot. The **fibular artery** arises from the posterior tibial artery and supplies the lateral aspect of the leg.

ACTIVITY 1
Identifying the Major Arteries That Supply the Head, Neck, Thorax, and Upper Limbs

Learning Outcomes

1. Identify and describe the major arteries that supply the head and neck.
2. Identify and describe the major arteries that supply the thorax.
3. Identify and describe the major arteries that supply the upper limbs.

Materials Needed

- ☐ Anatomical charts
- ☐ Anatomical models

Instructions

CHART Locate each artery assigned by your instructor on an anatomical chart or model and then complete the following chart. Some entries have been provided as examples.

Artery	Vessel/Structure from Which It Arises	Vessel It Gives Rise To and/or Organ(s) It Supplies
Ascending aorta		
Coronary arteries		*Supply heart*
Aortic arch		
Brachiocephalic a.	*Aortic arch*	
Right common carotid a.		
Right subclavian a.		
Left common carotid a.		*Splits into left external and internal carotid a.*
Left subclavian a.		
Thoracic aorta		
Abdominal aorta		
Internal carotid a.	*Common carotid a.*	
External carotid a.		
Axillary a.		*Continues as brachial a.*
Brachial a.		
Radial a.	*Brachial a.*	
Ulnar a.		
Superficial palmar arch		*Gives rise to digital a.*
Digital a.		
Bronchial a.		

Artery	Vessel/Structure from Which It Arises	Vessel It Gives Rise To and/or Organ(s) It Supplies
Esophageal a.	*Visceral branch of thoracic aorta*	
Superior phrenic a.		
Posterior intercostal a.		*Supplies intercostal muscles*

Optional Activity

Practice labeling major arteries of the head, neck, thorax, and upper limb on human cadavers at > MasteringA&P® > Study Area > Practice Anatomy Lab > Human Cadaver: Cardiovascular System > Blood Vessels > Images 1–18 and 26–32

ACTIVITY 2

Identifying the Major Arteries That Supply the Abdominopelvic Organs and the Lower Limbs

Learning Outcomes

1. Identify and describe the major arteries that supply the abdominopelvic organs.
2. Identify and describe the major arteries that supply the lower limbs.

Materials Needed

- ☐ Anatomical charts
- ☐ Anatomical models

Instructions

CHART Locate each artery assigned by your instructor on an anatomical chart or model and then complete the following chart. Some entries have been provided as examples.

Optional Activity

Practice labeling major arteries of the abdominopelvic organs and lower limb on human cadavers at > MasteringA&P® > Study Area > Practice Anatomy Lab > Human Cadaver: Cardiovascular System > Blood Vessels > Images 10–14 and 17–19

Artery	Vessel/Structure from Which It Arises	Vessel It Gives Rise To and/or Organ(s) It Supplies
Celiac trunk	*Visceral branch of abdominal aorta*	
Middle suprarenal a.		
Renal a.		*Supplies kidneys*
Superior mesenteric a.		
Inferior mesenteric a.		
Gonadal a.		
Inferior phrenic a.	*Parietal branch of abdominal aorta*	
Median sacral a.		
Common iliac a.		*Splits to form the external and internal iliac a.*
External iliac a.		
Internal iliac a.		
Femoral a.	*External iliac a.*	

(Continued)

Artery	Vessel/Structure from Which It Arises	Vessel It Gives Rise To and/or Organ(s) It Supplies
Popliteal a.		
Anterior tibial a.		*Continues as dorsalis pedis a.*
Posterior tibial a.		
Fibular a.	*Posterior tibial a.*	
Dorsalis pedis a.		

Major Systemic Veins

The major veins of the body are illustrated in **Figure 23-8**. The superior and inferior venae cavae convey deoxygenated blood into the right atrium. The **superior vena cava**, formed by the union of the right and left brachiocephalic veins, drains the head, neck, upper extremities, and thorax. The **inferior vena cava**, formed by the union of the right and left common iliac veins, drains the abdominopelvic cavity and the lower extremities.

Venous Drainage of the Head, Face, Neck, and Brain

Figure 23-9 illustrates the major veins of the head, face, neck, and brain. The **internal jugular veins** drain blood from the brain via the **dural sinuses**, from the face via the **facial veins**, and from the eyes via the ophthalmic veins (not shown). Each internal jugular vein unites with the subclavian vein on the same side to form the **brachiocephalic vein**. The **external jugular vein** receives blood from the occipital vein (not shown), which drains blood from the superficial tissues of the head and neck and empties into the **subclavian vein**. The **vertebral veins** drain the cervical vertebrae and the spinal cord and empty into the subclavian vein, which also receives deoxygenated blood from the upper extremities.

Venous Drainage of the Upper Limb and Thorax

The **digital veins** (not shown) collect venous blood from the fingers and empty it into the superficial **palmar venous arches**, which empty into the **radial** and **ulnar veins**, which then unite to form the **brachial vein** (**Figure 23-10**). As the brachial vein enters the axilla, it continues as the **axillary vein** and then the subclavian vein. In addition to these deep veins, several superficial veins also carry blood toward the heart. The **basilic vein** travels along the medial aspect of the arm and empties into the brachial vein. The **cephalic vein** travels along the lateral aspect of the arm and empties into the axillary vein. The basilic and cephalic veins have many points of interconnection (anastomosing branches) between them. One of these anastomosing veins is the **median cubital vein**, which crosses the anterior aspect of the elbow and is a common site for drawing blood. The **median antebrachial vein** lies between the radial and ulnar veins in the forearm and ends at the elbow by draining into either the basilic vein or the cephalic vein.

A highly variable network of veins—called the **azygos system** (**Figure 23-11**)—drains most thoracic structures and the abdominal wall. The **azygos vein**, located on the right side of the vertebral column, drains the right side of the thorax and empties into the superior vena cava. Both the **hemiazygos vein**, located on the left side of the vertebral column, and the **accessory hemiazygos vein**, a superior continuation of the hemiazygos vein, drain the left side of the thorax and empty into the azygos vein.

Venous Drainage of the Abdomen

The veins that drain the abdominal region (**Figure 23-12**) include the **external iliac veins** and the **internal iliac veins**, which fuse to form the **common iliac veins**. The right and left common iliac veins fuse to form the inferior vena cava, which receives venous blood directly from the **right gonadal vein**, several pairs of lumbar veins (not shown), the right and left **renal veins**, the right suprarenal vein (not shown), and the **hepatic veins**. The left gonadal vein and the left suprarenal vein (not shown) drain into the left renal vein. The unpaired veins draining the stomach, spleen, pancreas, gallbladder, small intestine, and large intestine enter hepatic portal circulation (discussed in Unit 24) before draining into the inferior vena cava.

Venous Drainage of the Lower Limb

The veins that drain the lower limbs (**Figure 23-13**) include the **posterior tibial vein**, which drains blood from the foot and carries it superiorly toward the knee. The fibular vein (not shown) empties into the posterior tibial vein. The **anterior tibial vein**, which is a continuation of the **dorsalis pedis vein**, unites with the posterior tibial vein in the popliteal region to become the **popliteal vein**. As the popliteal vein emerges

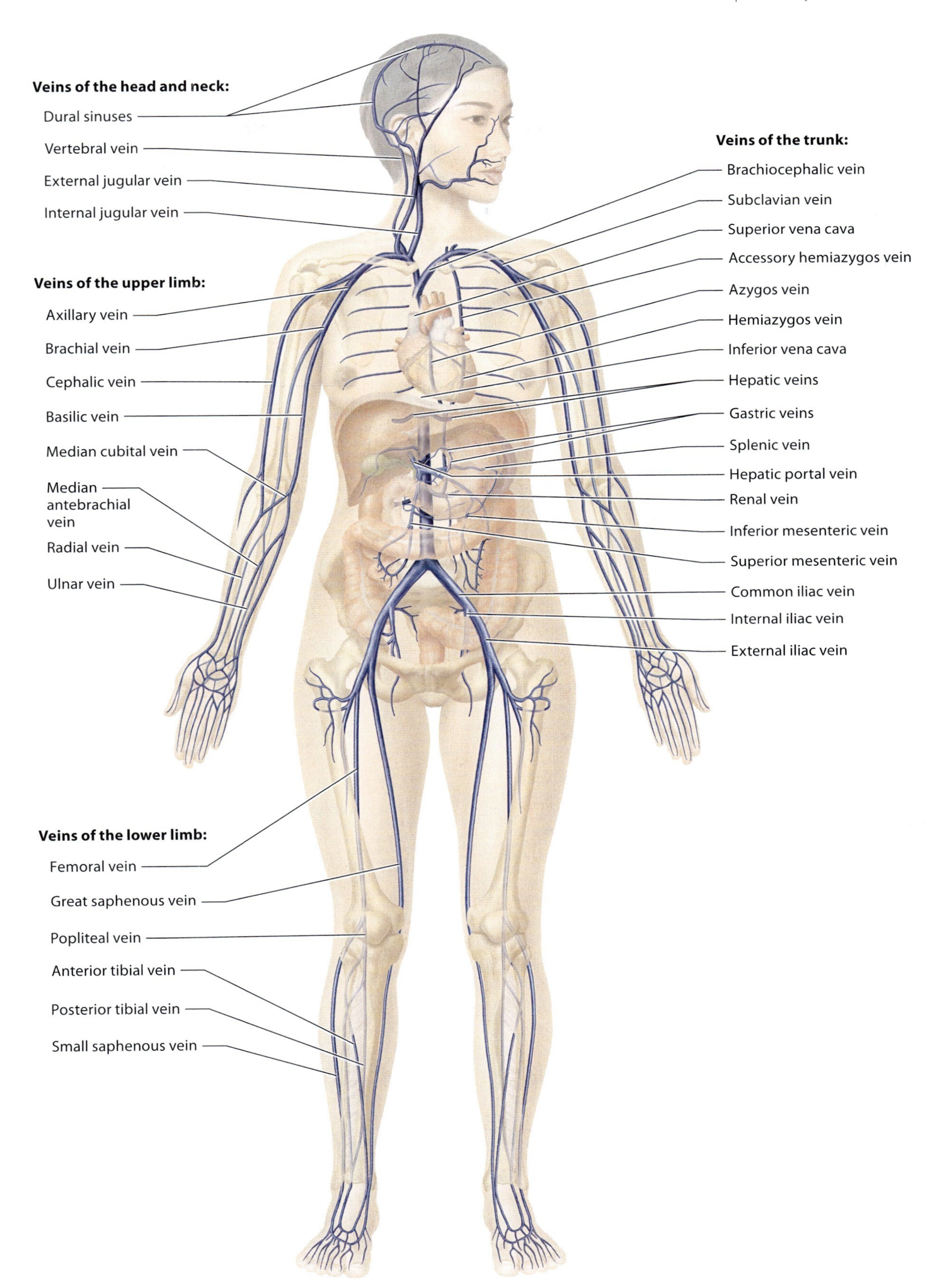

Figure 23-8 Major veins of the body.

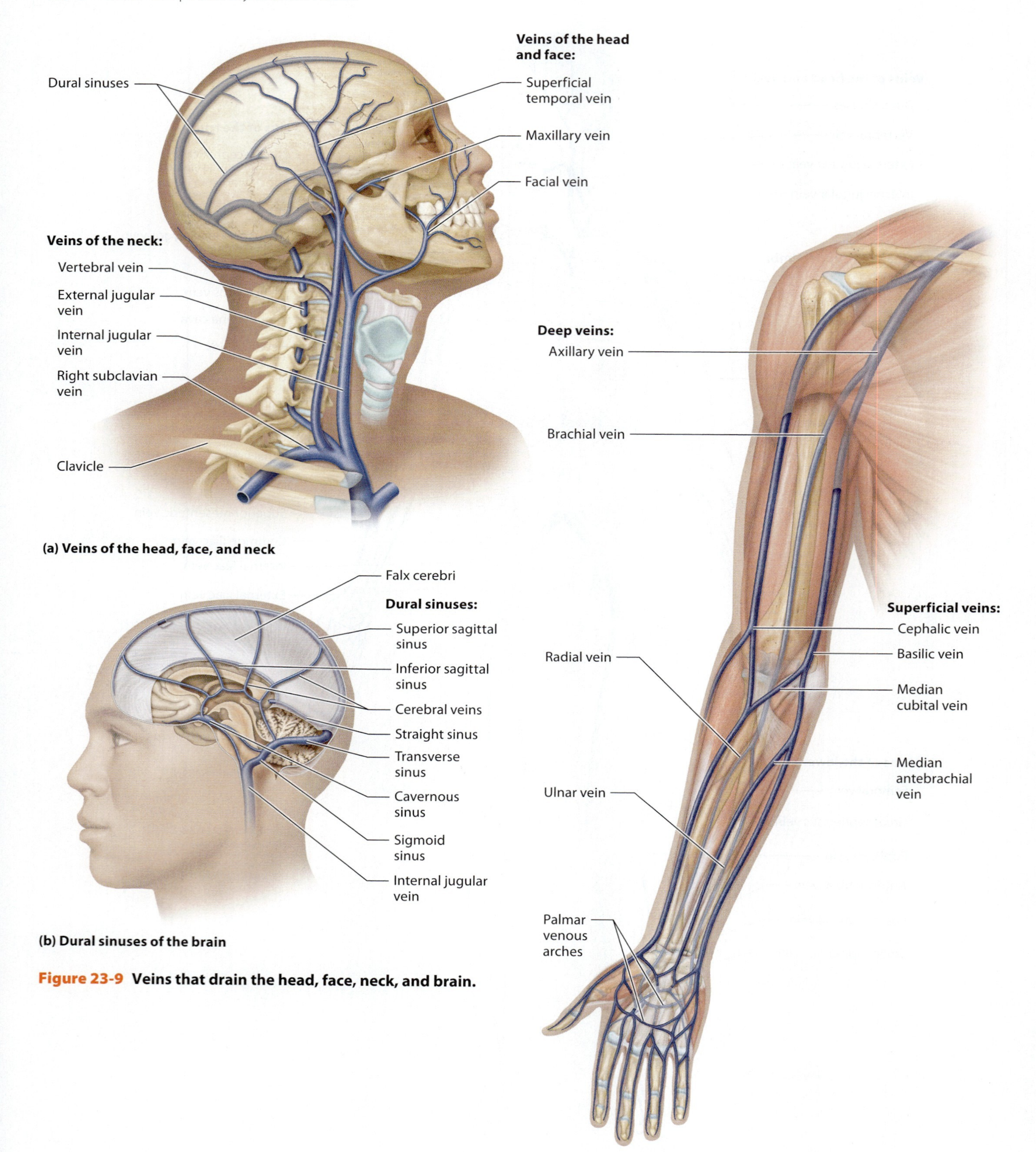

Figure 23-9 **Veins that drain the head, face, neck, and brain.**

Figure 23-10 **Veins that drain the upper limbs.**

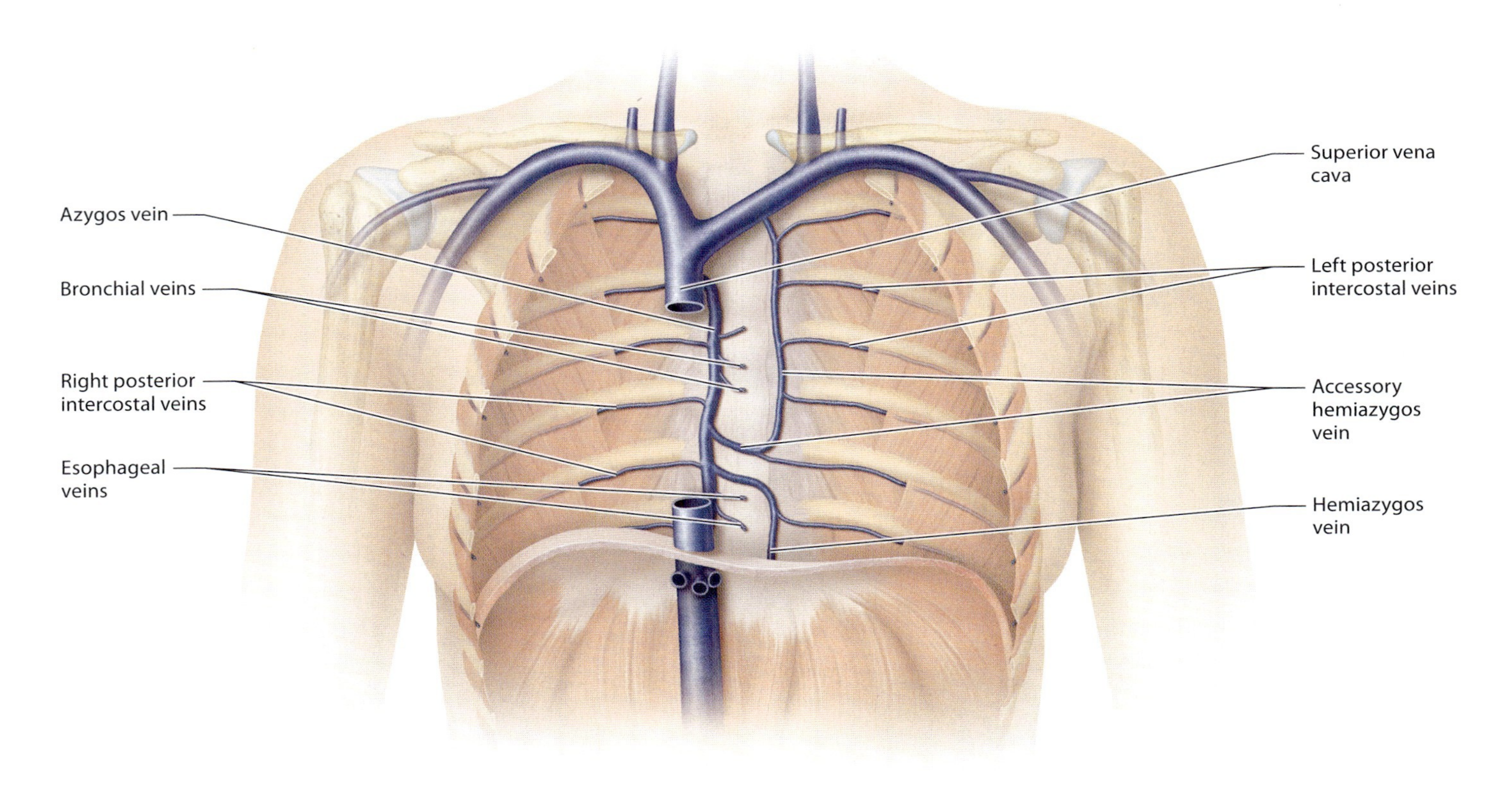

Figure 23-11 Veins that drain the thorax.

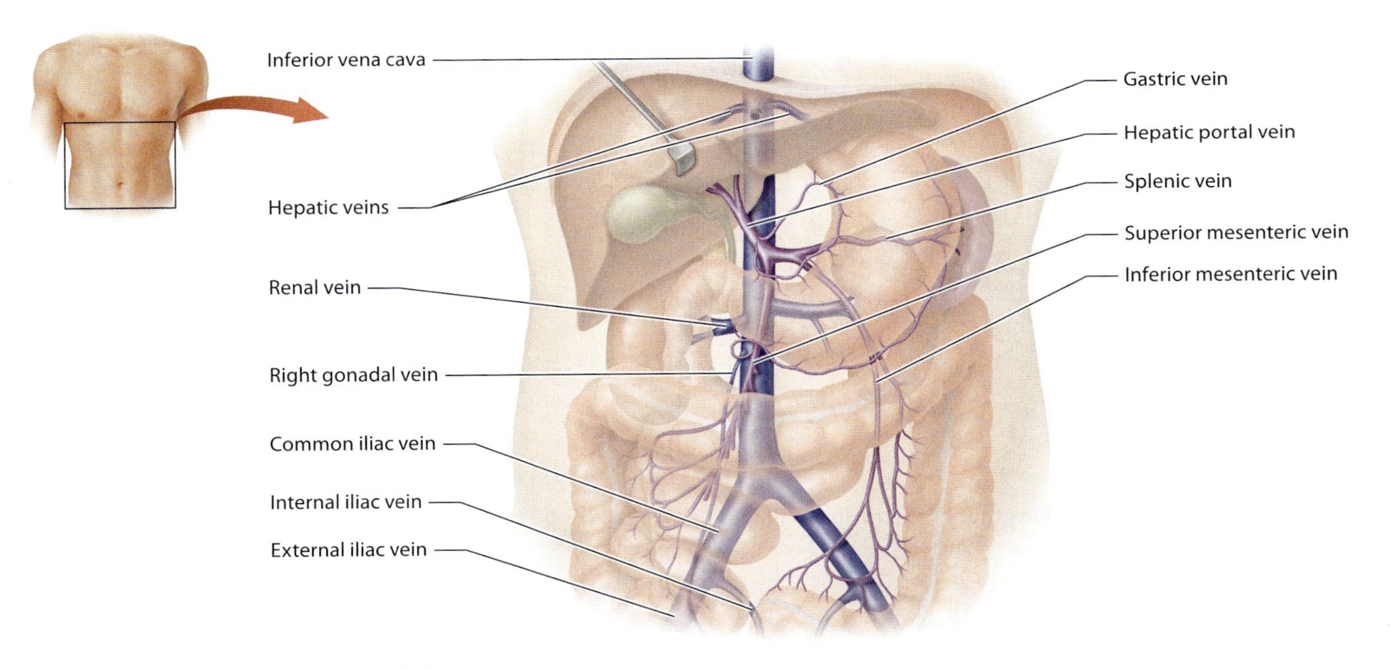

Figure 23-12 Veins that drain the abdomen.

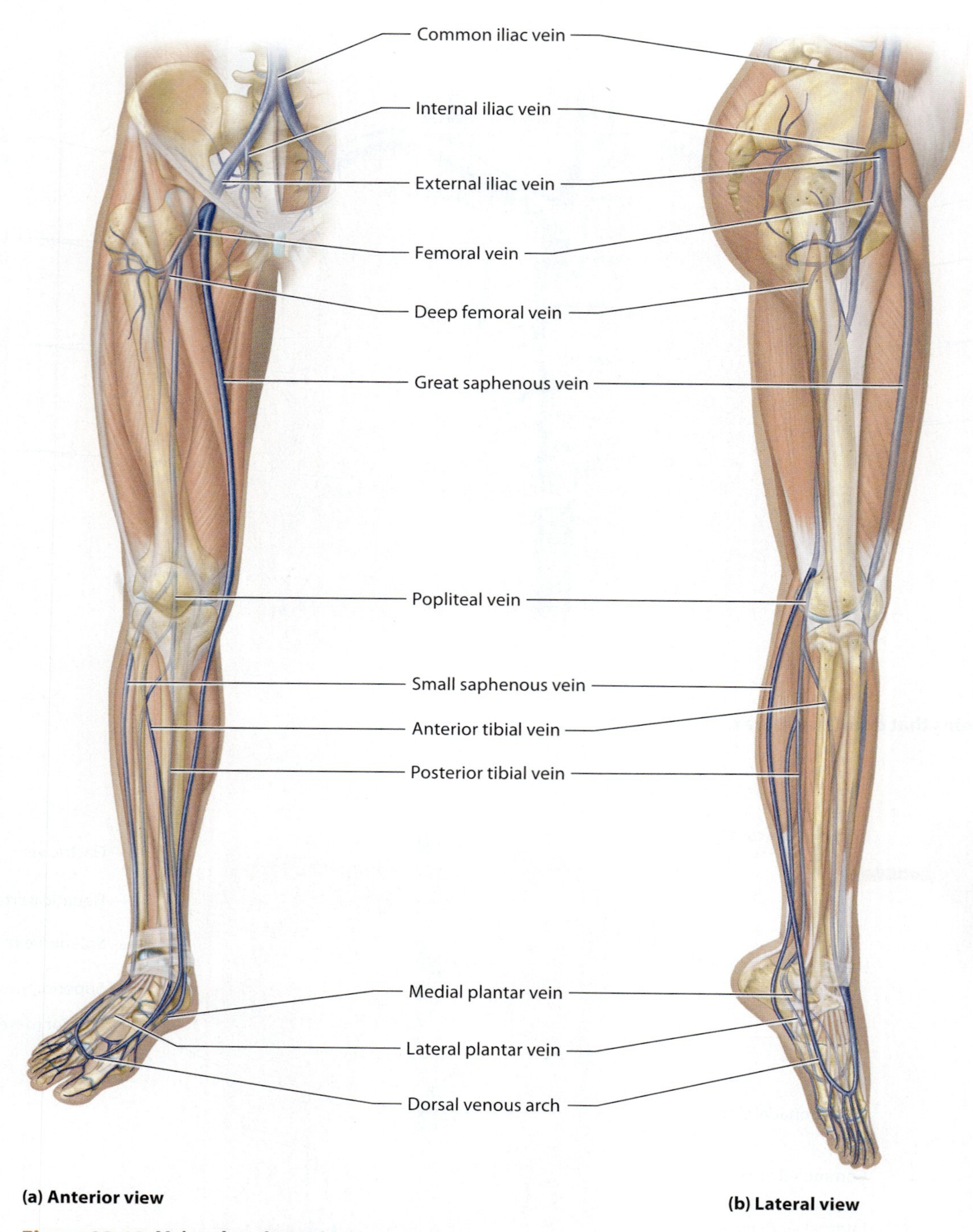

Figure 23-13 Veins that drain the lower limbs.

from the knee, it becomes the **femoral vein**. The femoral vein becomes the **external iliac vein** as it enters the pelvis. The great and small saphenous veins, arising from the dorsal venous arch, are superficial veins that anastomose frequently with each other and with the deep veins of the leg and thigh.

The **great saphenous vein**, the longest vein in the body, travels along the medial aspect of the leg and thigh and empties into the femoral vein. The **small saphenous vein** travels along the lateral aspect of the foot and through the connective tissue of the calf muscles before emptying into the popliteal vein.

ACTIVITY 3

Identifying Veins That Drain into the Venae Cavae

Learning Outcomes

1. Identify and describe the major veins that drain into the superior vena cava.
2. Identify and describe the major veins that drain into the inferior vena cava.

Materials Needed

- ☐ Anatomical charts
- ☐ Anatomical models

Instructions

CHART Locate each vein assigned by your instructor on an anatomical chart or model and then complete the following chart. Some entries have been provided as examples.

Vein	Vessel/Structure It Drains	Vessel/Structure into Which It Empties
Superior vena cava		
Brachiocephalic v.		*Superior vena cava*
Internal jugular v.		
Subclavian v.		
External jugular v.		
Dural sinuses	*Brain*	
Axillary v.		
Brachial v.		*Axillary vein*
Radial v.		
Ulnar v.		
Superficial palmar venous arch		
Digital v.		*Superficial palmar venous arch*
Cephalic v.		
Basilic v.		
Median cubital v.		
Renal v.		
Hepatic v.	*Liver*	
Common iliac v.		
External iliac v.		
Internal iliac v.		*Common iliac vein*
Femoral v.		

(*Continued*)

Vein	Vessel/Structure It Drains	Vessel/Structure into Which It Empties
Popliteal v.		
Anterior tibial v.		*Popliteal vein*
Posterior tibial v.		
Fibular v.		
Dorsalis pedis v.		

Optional Activity

Practice labeling major veins of human cadavers at > MasteringA&P® > Study Area > Practice Anatomy Lab > Human Cadaver: Cardiovascular System > Blood Vessels > Images 15–23

Microscopic Anatomy of Blood Vessels

Arteries always transport blood away from the heart and usually contain oxygenated blood. **Veins** always transport blood toward the heart and usually contain deoxygenated blood. **Capillaries** are the sites of exchange of respiratory gases, nutrients, and wastes throughout the body.

The walls of arteries and veins consist of three tunics or layers (**Figure 23-14**): the innermost tunica intima, the middle tunica media, and the outermost tunica externa. The **tunica intima** lines the **lumen** of the vessel and is composed of a simple squamous epithelium (the endothelium) and a thin layer of loose connective tissue (basal lamina). The endothelium is continuous with the endocardium of the heart and forms a smooth, slick surface that promotes blood flow. The **tunica media** is composed primarily of smooth muscle and elastic fibers. The smooth muscle is innervated by the sympathetic nervous system and plays a major role in regulating the diameter of blood vessels. The **tunica externa**, composed of dense irregular collagenous connective tissue, supports and protects the vessel. The walls of capillaries, by contrast, are one cell layer thick. They consist of an endothelium sitting on a very thin basal lamina. This thin structure facilitates the exchange of materials between the blood and body cells at the capillary beds.

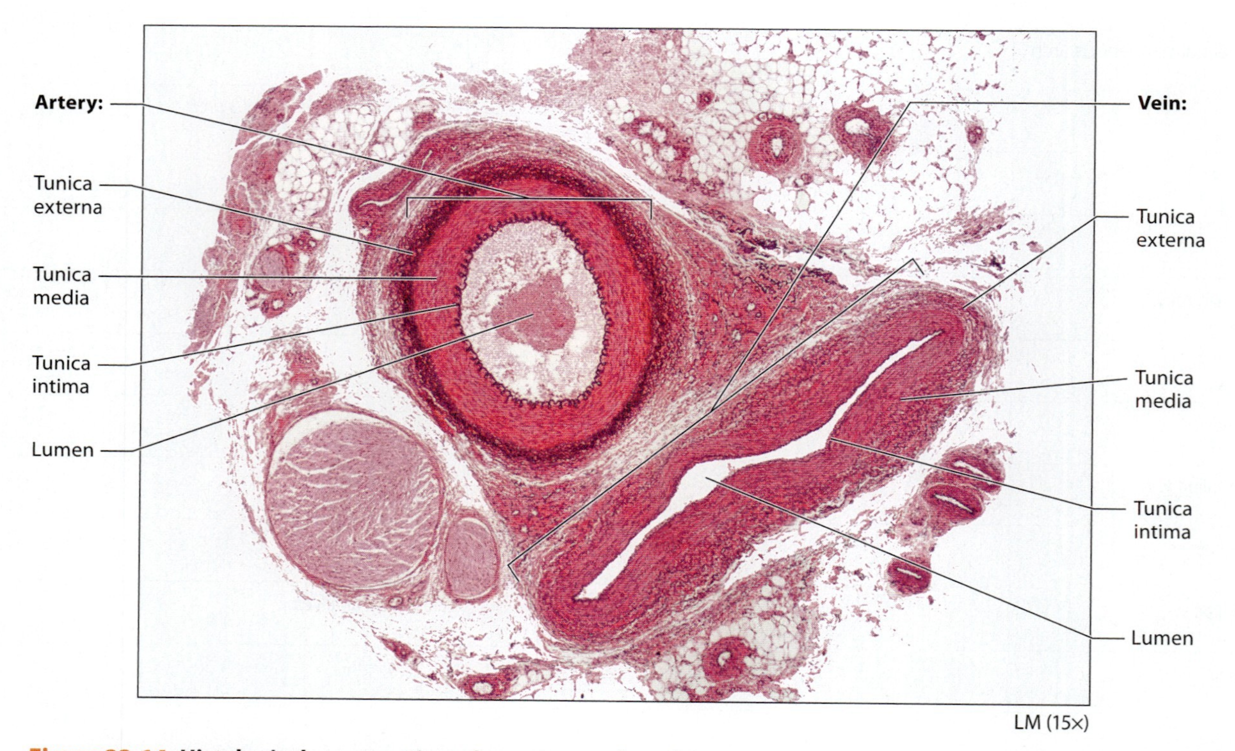

Figure 23-14 Histological cross-section of an artery and a vein.

ACTIVITY 4

Examining the Histology of Arteries and Veins

Learning Outcomes

1. Describe the structural differences among arteries, veins, and capillaries.
2. Evaluate the relationship between structure and function in arteries, veins, and capillaries.
3. Identify the three tunics of a blood vessel wall in a transverse section of an artery and a vein.

Materials Needed

- ☐ Anatomical charts
- ☐ Microscope and artery/vein slide (or photomicrograph)

Instructions

1. View a slide (or a photomicrograph) showing a transverse section of an artery and a vein under low power.
   a. Sketch the vessels and label the lumen of each.

Total magnification: ______ ×

   b. Then switch to high power (or view a photomicrograph) and sketch and label the three tunics of the walls of both vessels.

Total magnification: ______ ×

What structural differences between the artery and the vein do you see? ______________________________

____

____

____

How do these structural differences relate to the different functions of arteries and veins?

____

____

____

____

How does the structure of a capillary compare to the structure of arteries and veins?

____

____

____

____

2. **Optional Activity**

View and label histology slides of blood vessels at > MasteringA&P® > Study Area > Practice Anatomy Lab > Histology > Cardiovascular Tissue > Images 8–17

Post-lab quizzes are also assignable in MasteringA&P®

POST-LAB ASSIGNMENTS

Name: ______________________ Date: ______________ Lab Section: __________

PART I. Check Your Understanding

Activity 1: Identifying the Major Arteries That Supply the Head, Neck, Thorax, and Upper Limbs

1. Identify the arteries on the accompanying photograph of an anatomical model.

   a. ______________________

   b. ______________________

   c. ______________________

   d. ______________________

   e. ______________________

   f. ______________________

   g. ______________________

   h. ______________________

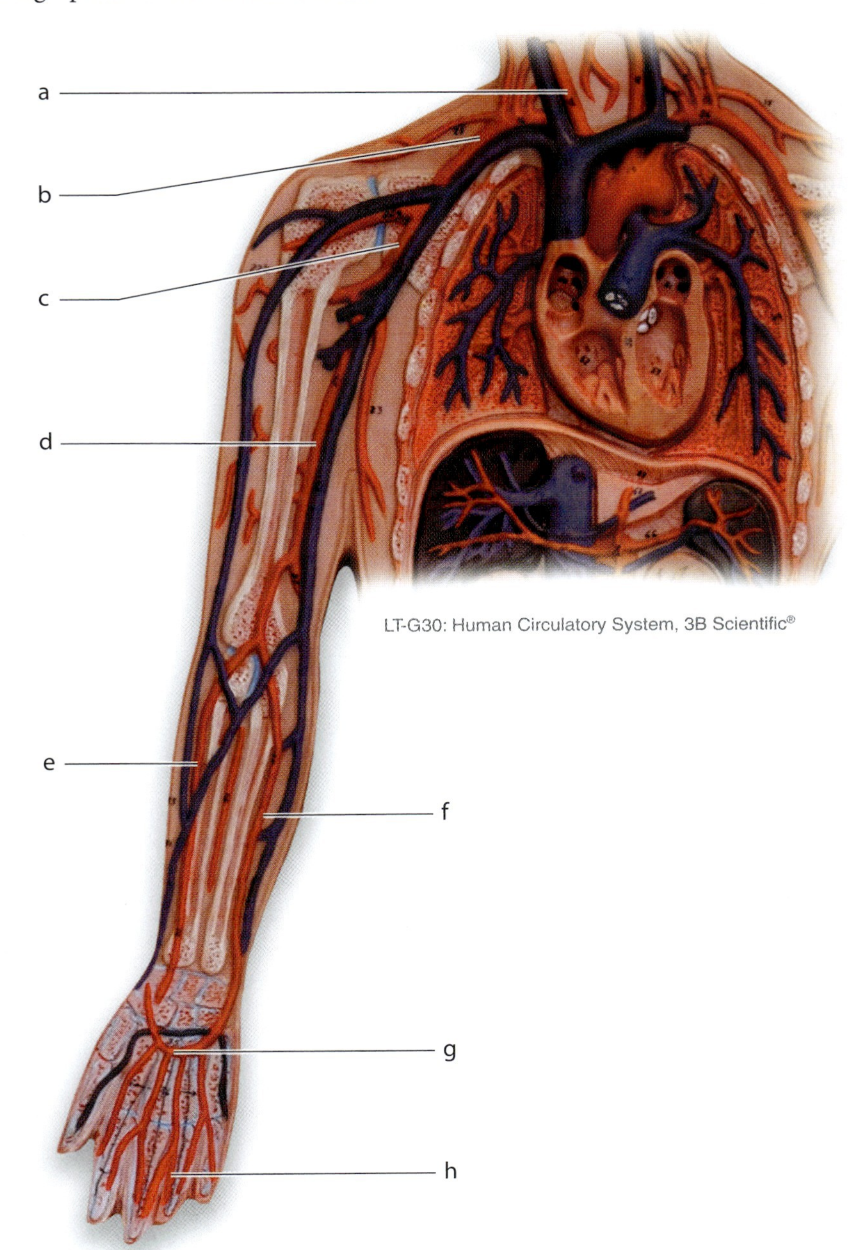

LT-G30: Human Circulatory System, 3B Scientific®

2. Which blood vessel(s):

_______________ a. supply the teeth?

_______________ b. branch from the brachial artery?

_______________ c. supply the tongue?

_______________ d. fuse to form the basilar artery?

_______________ e. is a continuation of the subclavian artery?

_______________ f. gives rise to the right common carotid artery?

_______________ g. supplies the superior diaphragm?

Activity 2: Identifying the Major Arteries That Supply the Abdominopelvic Organs and the Lower Limbs

1. Identify the arteries on the accompanying photograph of an anatomical model.

   a. _______________

   b. _______________

   c. _______________

   d. _______________

   e. _______________

   f. _______________

   g. _______________

   h. _______________

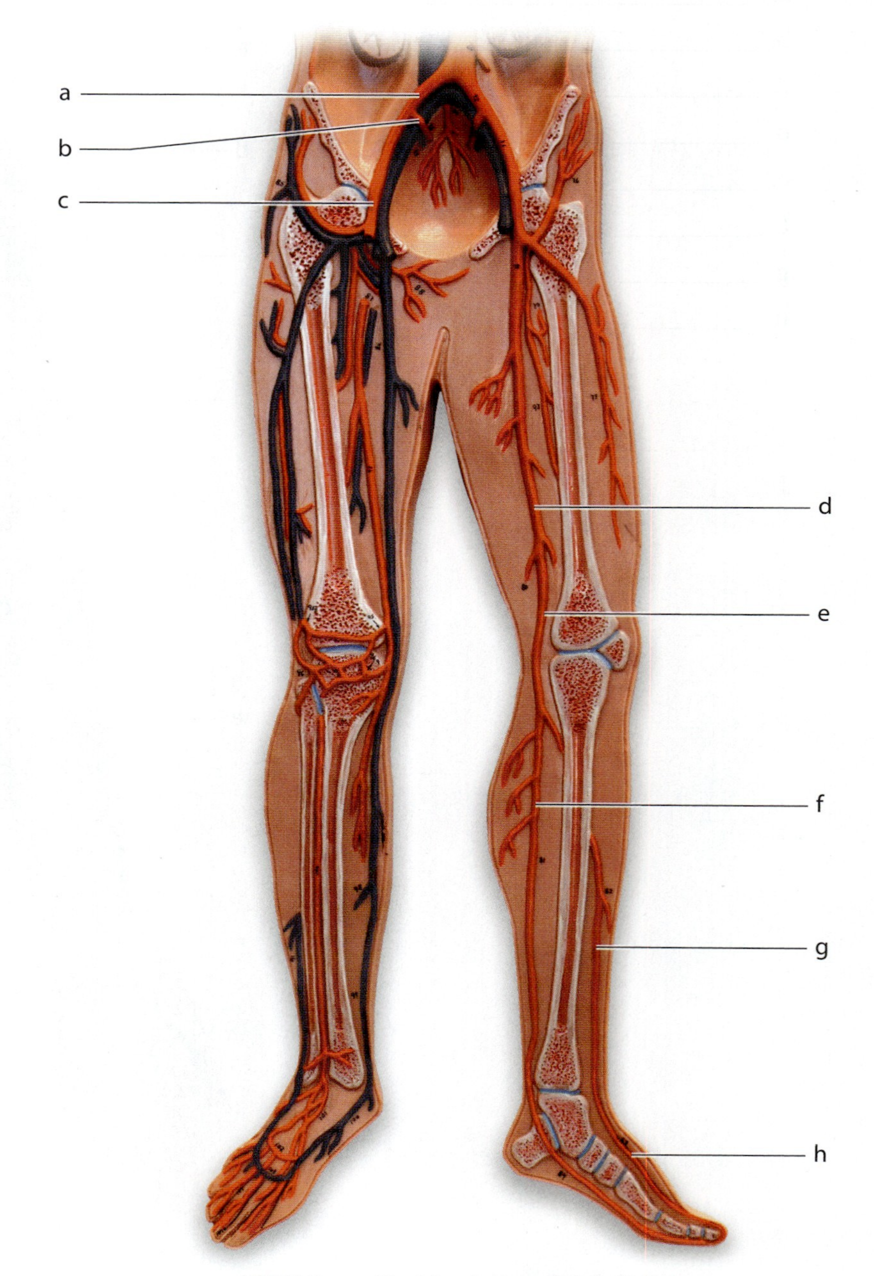

LT-G30: Human Circulatory System, 3B Scientific®

2. Which blood vessel(s):

   ______________________ a. supply the ovaries?

   ______________________ b. give rise to the femoral artery?

   ______________________ c. gives rise to the fibular artery?

   ______________________ d. splits to form the anterior tibial artery and posterior tibial artery?

   ______________________ e. supplies most of the small intestine?

   ______________________ f. supplies the coccyx?

   ______________________ g. gives rise to the dorsalis pedis artery?

   ______________________ h. supplies the pelvic organs?

   ______________________ i. passes through the adductor magnus muscle?

   ______________________ j. serves the gallbladder, liver, esophagus, stomach, and spleen?

Activity 3: Identifying Veins That Drain into the Venae Cavae

1. Identify the veins on the accompanying photograph of an anatomical model.

   a. ______________________

   b. ______________________

   c. ______________________

   d. ______________________

   e. ______________________

   f. ______________________

   g. ______________________

   h. ______________________

   i. ______________________

   j. ______________________

   k. ______________________

   l. ______________________

   m. ______________________

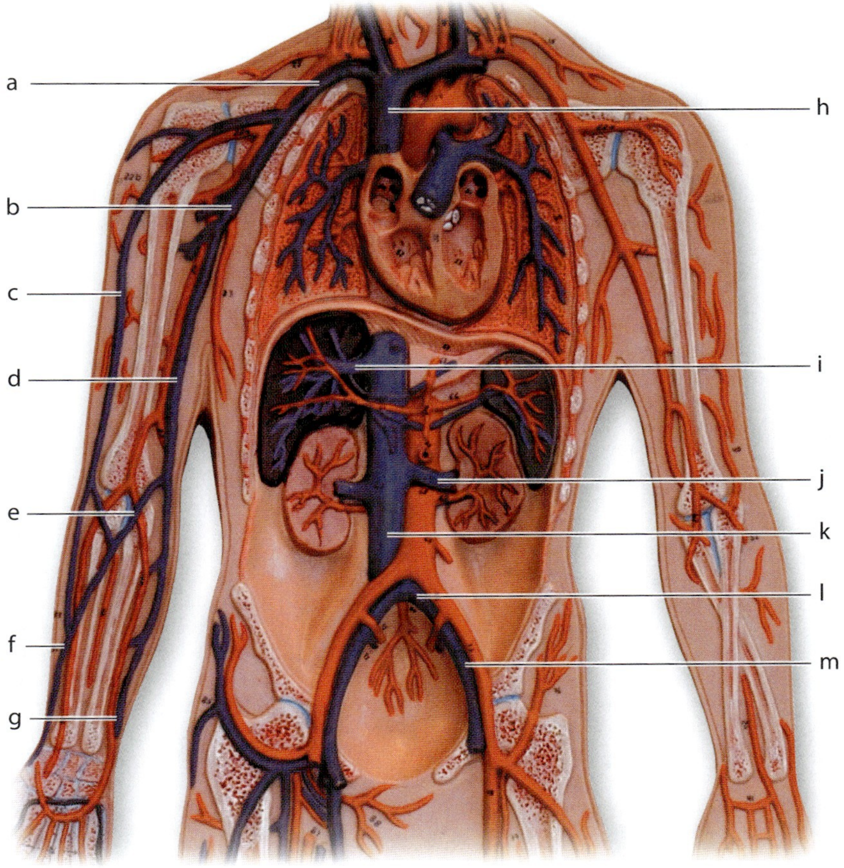

LT-G30: Human Circulatory System, 3B Scientific®

2. Which blood vessel(s):

_______________ a. drains the adrenal glands?

_______________ b. empties into the external iliac vein?

_______________ c. connects the basilic vein and the cephalic vein?

_______________ d. drains the pelvic organs?

_______________ e. drains the liver?

_______________ f. empties directly into the superior vena cava?

_______________ _______________ g. fuse to form the brachiocephalic vein?

_______________ _______________ h. fuse to form the popliteal vein?

Activity 4: Examining the Histology of Arteries and Veins

1. Compare and contrast arteries and veins with respect to both structure and function as you complete the following chart.

Characteristics Unique to Arteries	Characteristics Common to Both Arteries and Veins	Characteristics Unique to Veins

2. Identify the structures in the accompanying photomicrograph of blood vessels.

   a. ______________________

   b. ______________________

   c. ______________________

   d. ______________________

   e. ______________________

   f. ______________________

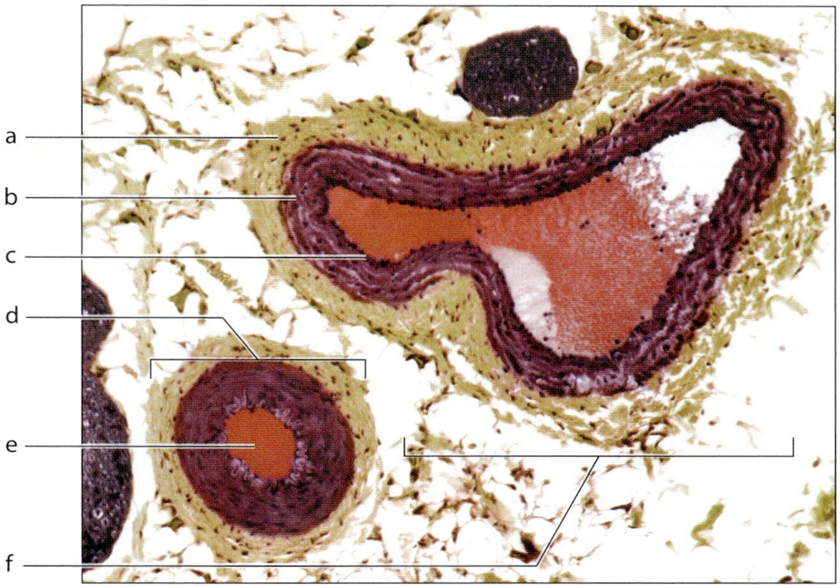

LM (13×)

PART II. Putting It All Together

A. Review Questions

Answer the following questions using your lecture notes, your textbook, and your lab notes.

1. What did you learn concerning blood vessels with the word *common* as part of their name?

   ____

2. Why are some blood vessels called "trunks"? ______________________

   ____

3. In what ways is an artery adapted for carrying blood under high pressure? ______________________

   ____

4. In what ways is a capillary adapted for the exchange of materials? ______________________

   ____

5. In what ways is a vein adapted for carrying blood under low pressure? ______________________

   ____

6. Identify the arteries on the accompanying figure.

a. ______________________

b. ______________________

c. ______________________

d. ______________________

e. ______________________

f. ______________________

g. ______________________

h. ______________________

i. ______________________

j. ______________________

k. ______________________

l. ______________________

m. ______________________

n. ______________________

o. ______________________

p. ______________________

q. ______________________

r. ______________________

s. ______________________

t. ______________________

u. ______________________

v. ______________________

w. ______________________

x. ______________________

y. ______________________

z. ______________________

aa. ______________________

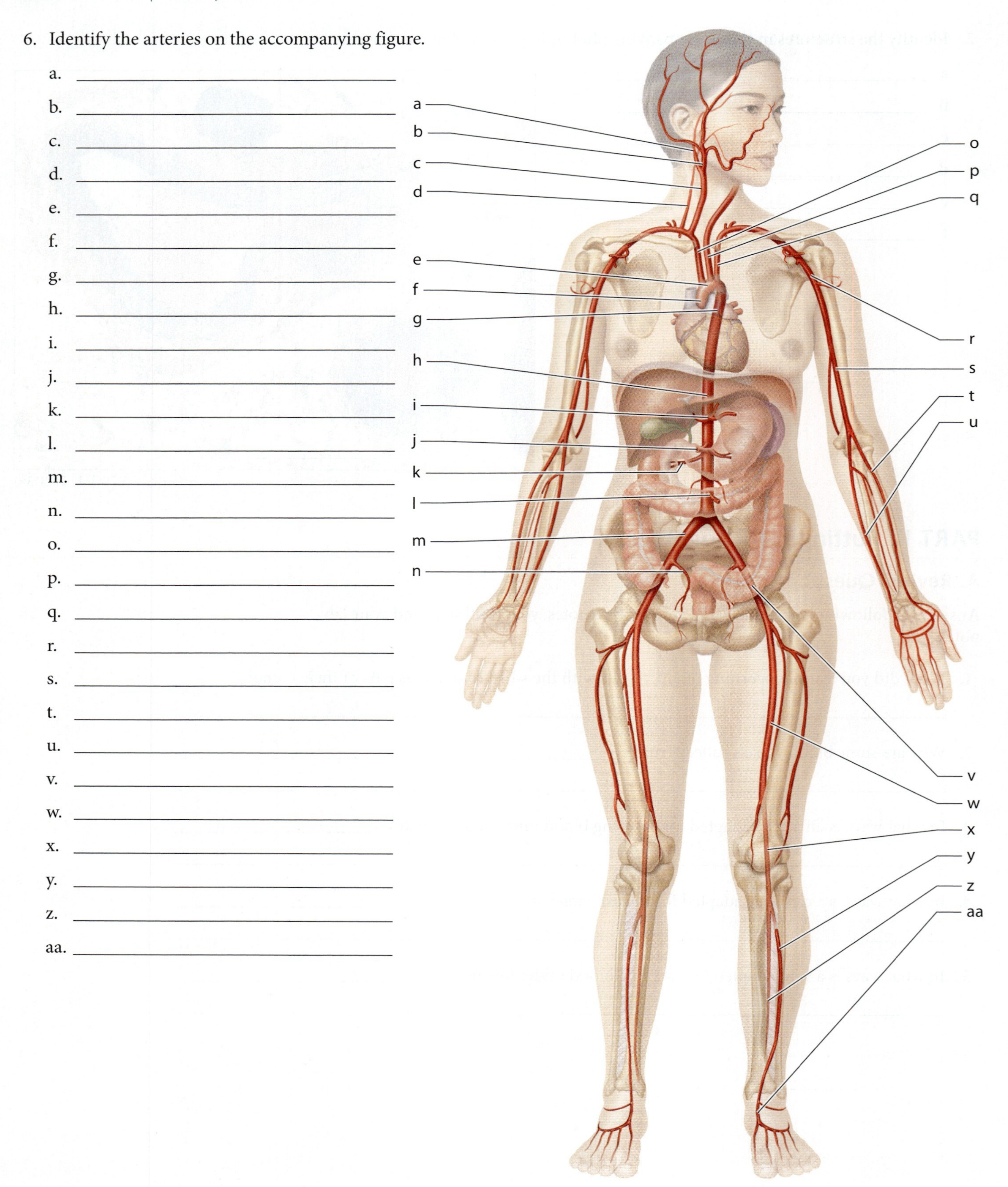

7. Identify the veins on the accompanying figure.

a. ______________________

b. ______________________

c. ______________________

d. ______________________

e. ______________________

f. ______________________

g. ______________________

h. ______________________

i. ______________________

j. ______________________

k. ______________________

l. ______________________

m. ______________________

n. ______________________

o. ______________________

p. ______________________

q. ______________________

r. ______________________

s. ______________________

t. ______________________

u. ______________________

v. ______________________

w. ______________________

x. ______________________

y. ______________________

z. ______________________

aa. ______________________

bb. ______________________

cc. ______________________

dd. ______________________

ee. ______________________

ff. ______________________

gg. ______________________

hh. ______________________

ii. ______________________

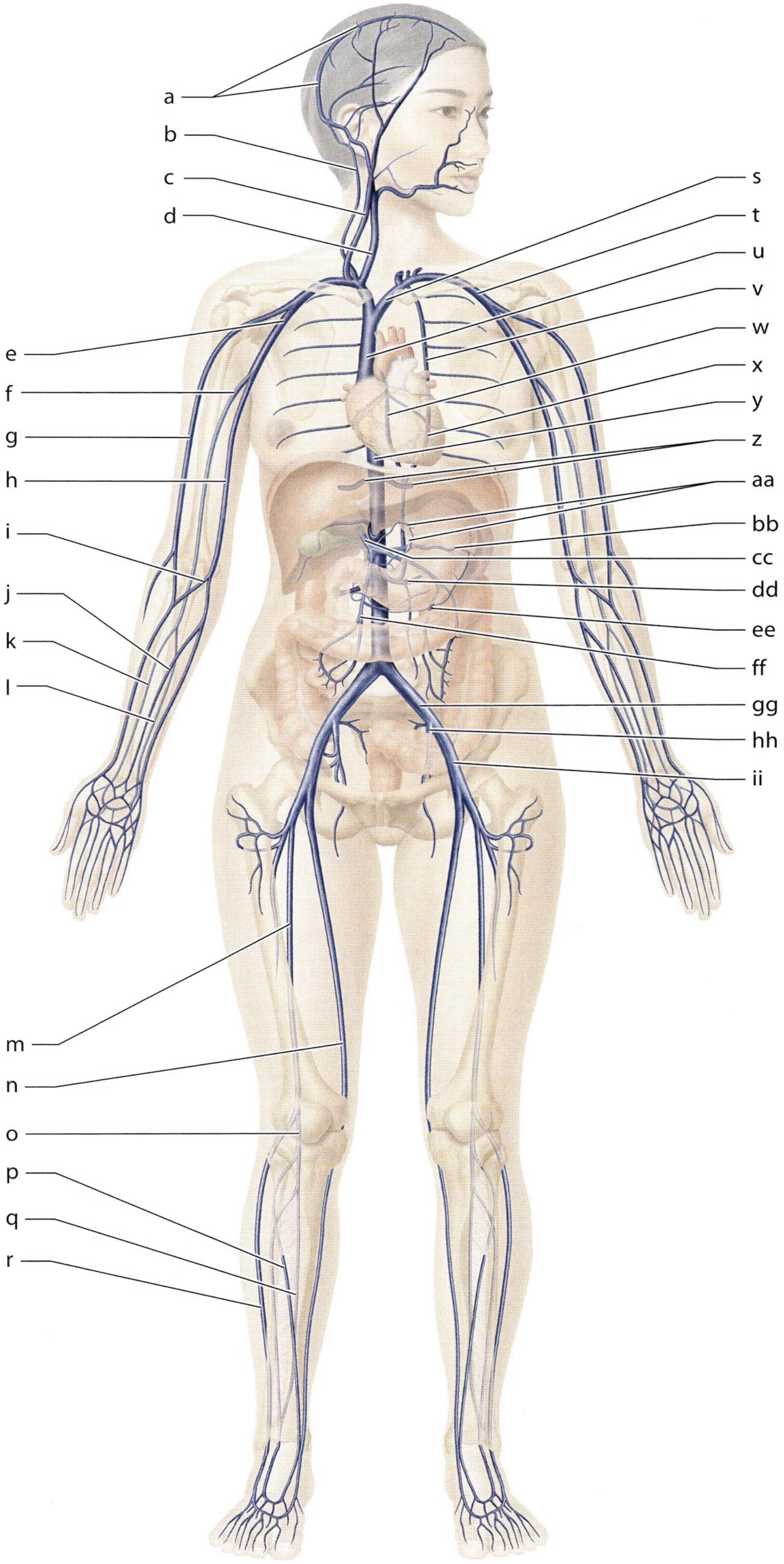

B. Concept Mapping

1. Fill in the blanks to complete this concept map outlining the structure and function of blood vessels:

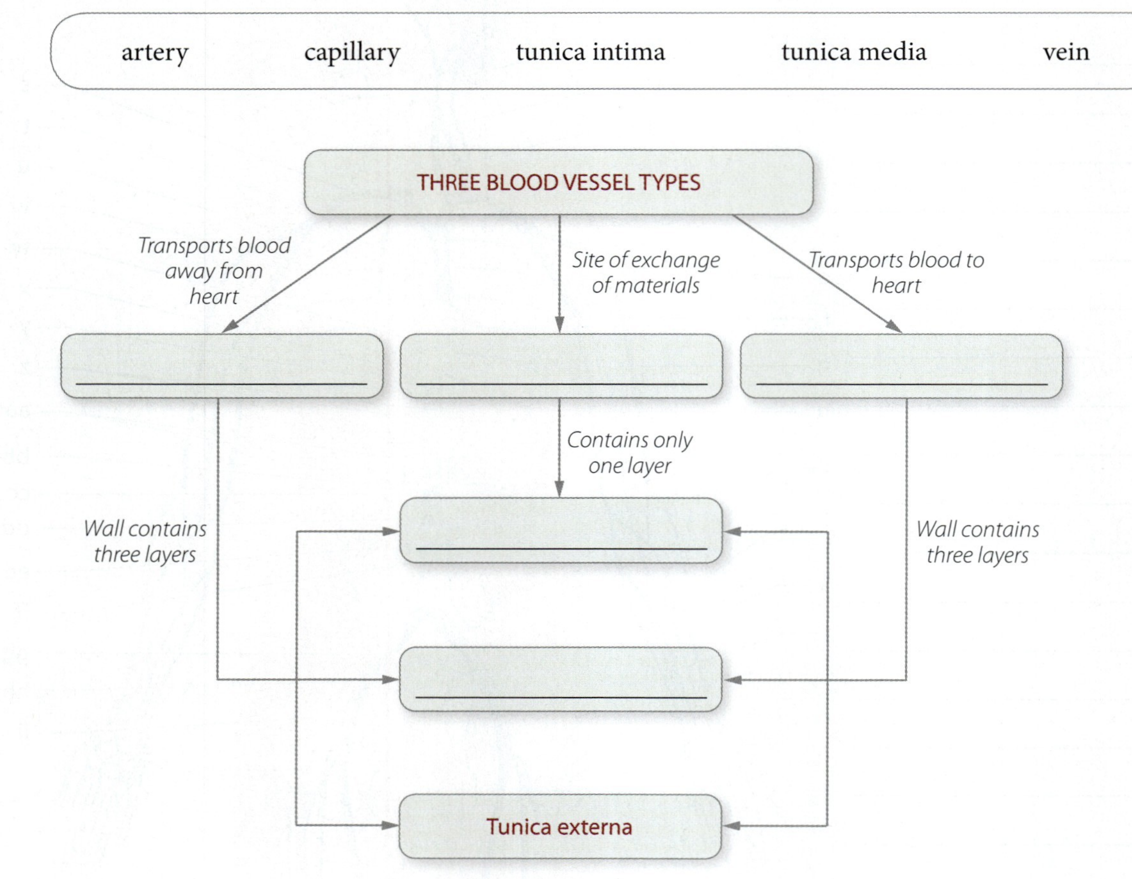

2. Construct a unit concept map to show the relationships among the following set of terms. Include all of the terms in your diagram. Your instructor may choose to assign additional terms.

aorta	artery	basal lamina	capillary
dense irregular connective tissue		elastic fibers	endothelium
lumen	simple squamous epithelium	smooth muscle	tunica intima
tunica media	valve	vein	venae cavae

24

Circulatory Pathways and the Physiology of Blood Vessels

UNIT OUTLINE

The heart is a double pump that pumps blood to all parts of the body (via the systemic circuit) and to the lungs (via the pulmonary circuit). The systemic circuit transports oxygenated blood from the left ventricle via the aorta to all parts of the body and then returns deoxygenated blood to the right atrium via the superior and inferior venae cavae. The pulmonary circuit transports deoxygenated blood from the right ventricle and through the pulmonary arteries to the lungs, where the exchange of gases occurs, and then returns oxygenated blood to the left atrium via the pulmonary veins. In this unit, you will examine several specific systemic pathways and the pulmonary circulation. You will also study the ways in which fetal circulation differs from adult circulation. As you complete the activities in this unit, you will rely heavily on the material you learned in Unit 23.

THINK ABOUT IT *You studied the coronary circulation in Unit 21. What is coronary circulation?* ______________________________

____

Is coronary circulation more similar to the systemic circuit or to the pulmonary circuit. Why? ______________________________

______________________________ ■

PRE-LAB ASSIGNMENTS

Pre-lab quizzes are also assignable in MasteringA&P®

To maximize learning, BEFORE your lab period carefully read this entire lab unit and complete these pre-lab assignments using your textbook, lecture notes, and prior knowledge.

PRE-LAB Activity 1: Tracing Blood Flow—General Systemic Pathways

1. Match each of the following blood vessels with the blood vessel from which it arises:

_____	a. superior phrenic artery	1. thoracic aorta
_____	b. ophthalmic artery	2. aortic arch
_____	c. axillary artery	3. subclavian artery
_____	d. popliteal artery	4. femoral artery
_____	e. brachiocephalic artery	5. internal carotid artery

2. Match each of the following blood vessels with a blood vessel that it drains:

_____	a. brachial vein	1. brachiocephalic vein
_____	b. external iliac vein	2. radial vein
_____	c. internal jugular vein	3. superior mesenteric vein
_____	d. superior vena cava	4. femoral vein
_____	e. hepatic portal vein	5. ophthalmic vein

PRE-LAB Activity 2: Tracing Blood Flow—Specialized Systemic Pathways

1. More than 80% of the cerebrum is supplied by the ________________ arteries.
2. The left and right vertebral arteries fuse to form the basilar artery, which leads into the ________________.
3. The superior mesenteric vein and the splenic vein fuse to form the ________________, which transports nutrient-rich blood to the liver.

PRE-LAB Activity 3: Tracing Blood Flow—Pulmonary Circulation

1. The pulmonary circuit begins in the ________________ of the heart, transports deoxygenated blood to the ________________, and ends in the ________________ of the heart.
2. State whether each of the following statements is true (T) or false (F):

   _____ a. The pulmonary trunk splits to form the right and left pulmonary arteries.

   _____ b. The pulmonary veins contain oxygenated blood.

   _____ c. Gas exchange occurs in the capillary beds of the lungs.

   _____ d. The right pulmonary artery splits into two lobar arteries.

PRE-LAB Activity 4: Tracing Blood Flow—Fetal Circulation

1. The opening in the fetal heart between the right atrium and the left atrium that allows blood to pass from the pulmonary circuit to the systemic circuit is the ________________.
2. The ductus arteriosus is a fetal structure that shunts blood from the pulmonary trunk to the ________________.

3. The umbilical cord contains two umbilical arteries that carry ________________ blood and one umbilical vein that carries ________________ blood.

PRE-LAB Activity 5: Measuring Blood Pressure and Examining the Effects of Body Position and Exercise

1. Define blood pressure. ____

____

2. What is the average blood pressure? ____

3. Is blood pressure higher in arteries or in veins? ____________________ Why? ______________

____

4. Name three factors that affect blood pressure. ____

____

Systemic Circulation

In the systemic circuit (**Figure 24-1**), oxygenated blood ejected from the left ventricle into the aorta is distributed to tissues throughout the body. Blood travels through progressively smaller arteries, then into arterioles, and eventually into capillary beds, where the exchange of respiratory gases, nutrients, and wastes occurs. Venules draining capillary beds return deoxygenated blood to progressively larger veins, which eventually drain into either the superior vena cava or the inferior vena cava. The venae cavae then return the deoxygenated blood to the right atrium of the heart.

At the capillary beds, respiratory gases (oxygen and carbon dioxide) as well as most nutrients and metabolic wastes are exchanged via filtration and diffusion. Each substance moves from an area of higher concentration to an area of lower concentration; therefore, oxygen and nutrients move out of the capillary and into the interstitial fluid, whereas carbon dioxide and other metabolic wastes move from the interstitial fluid and into the capillary.

Both hydrostatic pressure and osmotic pressure play major roles in the movement of water into and out of capillaries (**Figure 24-2**). Hydrostatic pressure, which is the result of blood pressure in the capillary, pushes water out of the capillary via filtration. Osmotic pressure, by contrast, results from the presence of plasma proteins (primarily albumin) within the capillary; it draws water into the capillary via osmosis. As shown in Figure 24-2, hydrostatic pressure is higher than osmotic pressure at the arteriole end of the capillary, whereas osmotic pressure is higher than hydrostatic pressure at the venule end of the capillary; consequently, water leaves the capillary at the arteriole end and enters the capillary at the venule end.

Blood Flow to and from the Cerebrum

The cerebral circulation is the part of the systemic circuit that provides oxygenated blood to the brain. Brain cells have a high metabolic rate; their resulting high demand for oxygen and nutrients makes a continuous blood supply to the brain crucial in protecting delicate brain tissue from damage resulting from a reduction in the delivery of oxygenated blood. To prevent damage to brain tissue from an arterial blockage, the cerebral circulation provides alternate routes for blood flow via anastomosing (interconnecting) arteries at the base of the brain by forming the **cerebral arterial circle** (or circle of Willis; **Figure 24-3b**). The cerebral arterial circle consists of the posterior cerebral arteries, the posterior communicating arteries, the anterior cerebral arteries, and the anterior communication arteries.

The brain is supplied by two pairs of arteries, the vertebral arteries and the internal carotid arteries (see Figure 24-3). The vertebral arteries arise from the subclavian arteries and pass superiorly through the transverse foramina of cervical vertebrae before entering the skull through the foramen magnum. Within the skull, the two vertebral arteries fuse to form the **basilar artery**, which travels along the ventral side of the brainstem and gives off branches to the pons, cerebellum, and inner ear. Near the superior border of the pons at the base of the cerebrum, the basilar artery divides to form the **posterior cerebral arteries**, which supply the occipital and parts of the temporal lobes. The right and left **posterior communicating arteries** connect the posterior cerebral arteries and the internal carotid arteries. The **internal carotid arteries** enter the skull through the carotid canal of the temporal bone and then divide into three parts: the **anterior cerebral arteries**, which supply the frontal and

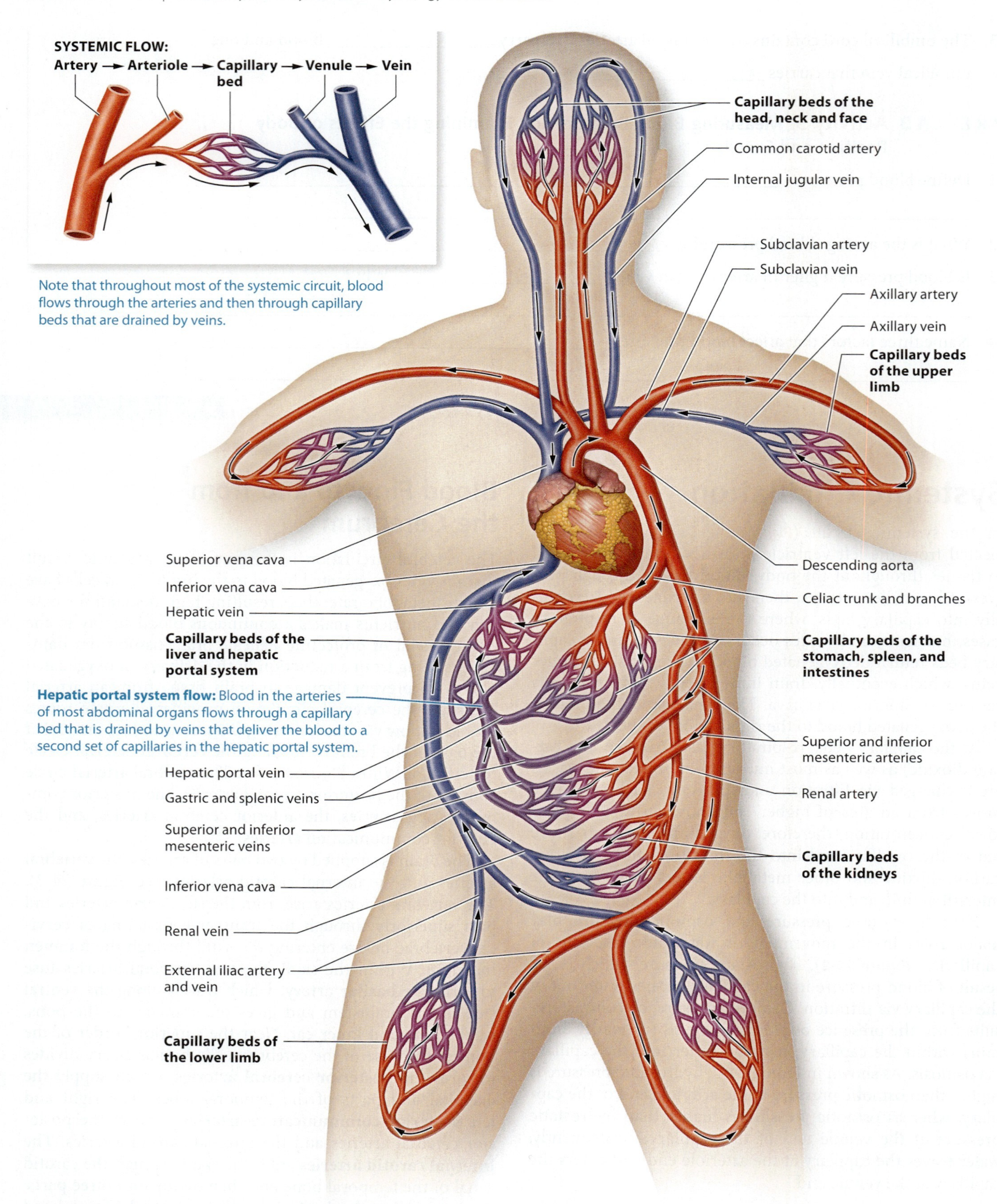

Figure 24-1 Major systemic blood flow pathways of the systemic circuit.

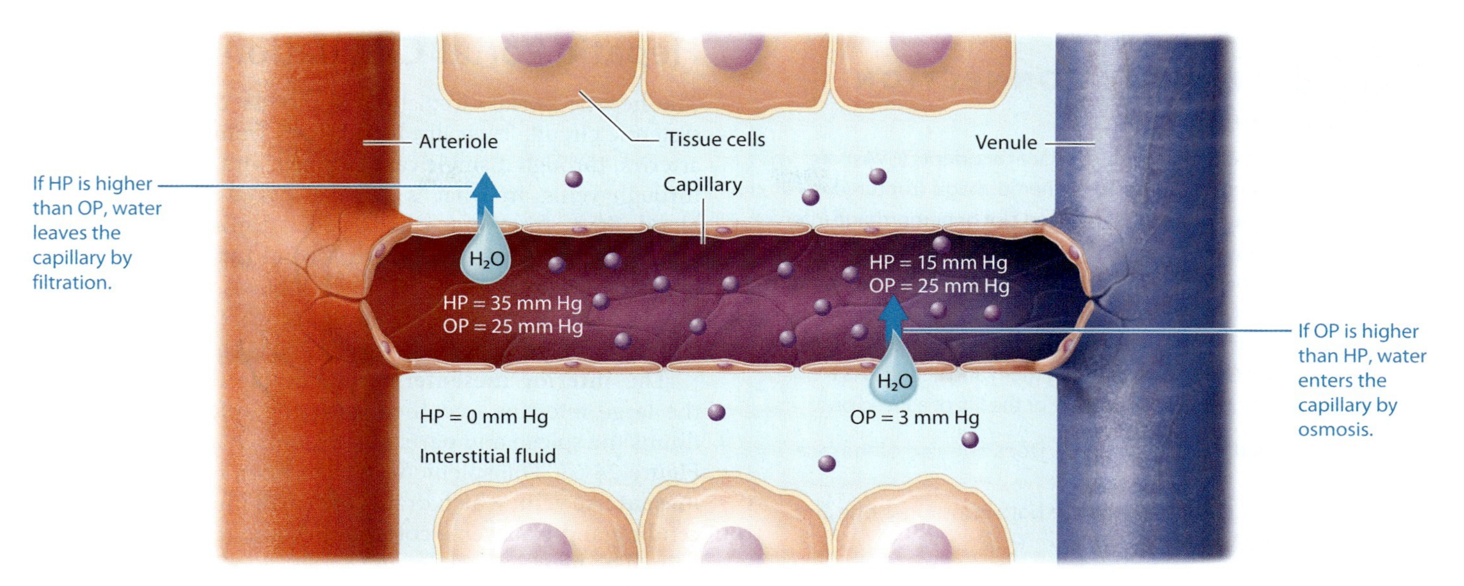

Figure 24-2 The roles of hydrostatic pressure (HP) and osmotic pressure (OP) in the exchange of materials at a capillary.

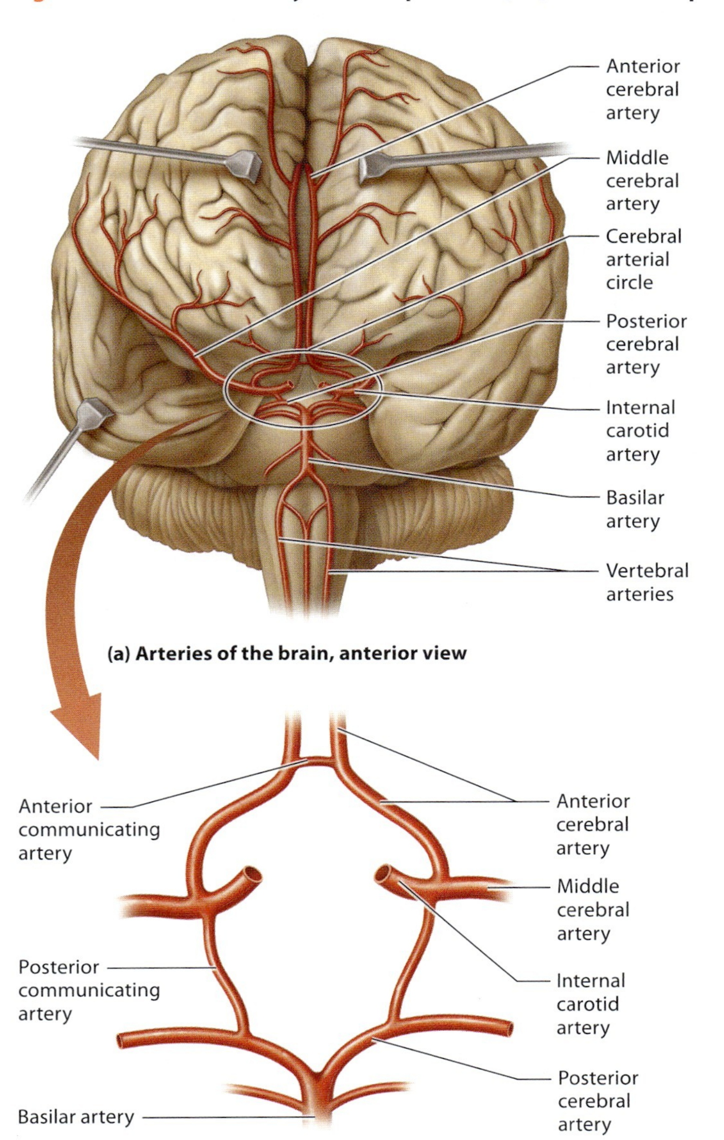

Figure 24-3 The cerebral arterial circle.

parietal lobes of the cerebrum; the **middle cerebral arteries**, which supply the midbrain and the lateral surfaces of the cerebrum; and the **ophthalmic arteries** (not shown), which supply the eyes. The right and left anterior cerebral arteries are connected by a short **anterior communicating artery** (see Figure 24-3b).

Venous blood drains into cerebral veins and then into the dural sinuses (**Figure 24-4**), which drain into the internal jugular veins, then into the brachiocephalic veins, and finally into the superior vena cava, which empties into the right atrium.

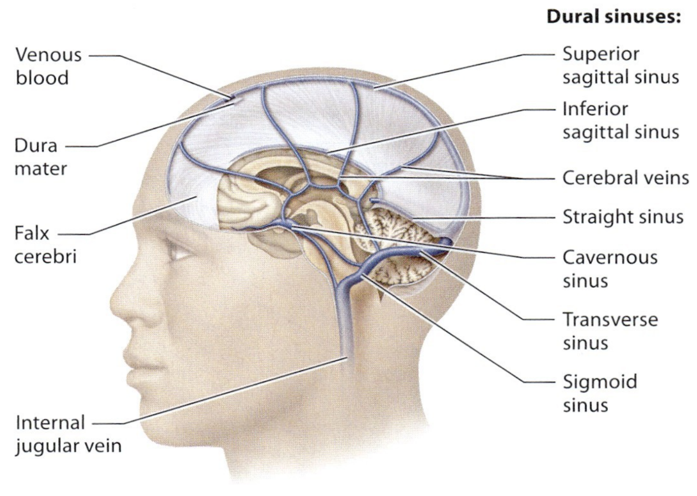

Figure 24-4 Venous drainage of the brain.

CLINICAL CONNECTION

Strokes

Strokes, or cerebrovascular accidents (CVAs), are classified as ischemic (blood clot) or hemorrhagic (ruptured blood vessel). They occur when a blood clot or hemorrhage interrupts the blood supply to a specific area of the brain. When such an interruption occurs, brain cells begin to die and brain damage results. In the United States, stroke is the fourth leading cause of death and a leading cause of serious, long-term disability. Know the warning signs of stroke. Learning them and acting **FAST** when they occur could save your life or the life of a loved one.

F = FACE—Ask the person to smile. Does one side of the face droop?
A = ARMS—Ask the person to raise both arms. Does one arm drift downward?
S = SPEECH—Ask the person to repeat a simple sentence. Is the speech slurred or strange?
T = TIME—If you observe *any* of these signs, call 911 *immediately.*

Hepatic Portal Circulation

The hepatic portal circulation is another component of the systemic circuit. Typically, blood flows from the heart within arteries, through a single capillary bed, and back to the heart through veins. In what is called a portal system, however, blood flows from one capillary bed through larger blood vessels (in this case, veins) to a second capillary bed. In the **hepatic portal system** (**Figure 24-5**), blood vessels that drain the digestive organs, spleen, and pancreas carry blood to the liver for processing before it is returned to the heart.

The **inferior mesenteric vein** drains the distal part of the large intestine and then joins the **splenic vein**, which drains the spleen and parts of the stomach and pancreas (see Figure 24-5). The splenic vein then fuses with the **superior mesenteric vein**, which drains the small intestine and the ascending and transverse colon to become the **hepatic portal vein**. The hepatic portal vein carries blood to the capillaries of the liver, where foreign substances are removed by macrophages, harmful chemicals are detoxified, and nutrients are processed for storage. The "processed" blood then returns to the inferior vena cava (IVC) via the **hepatic veins**.

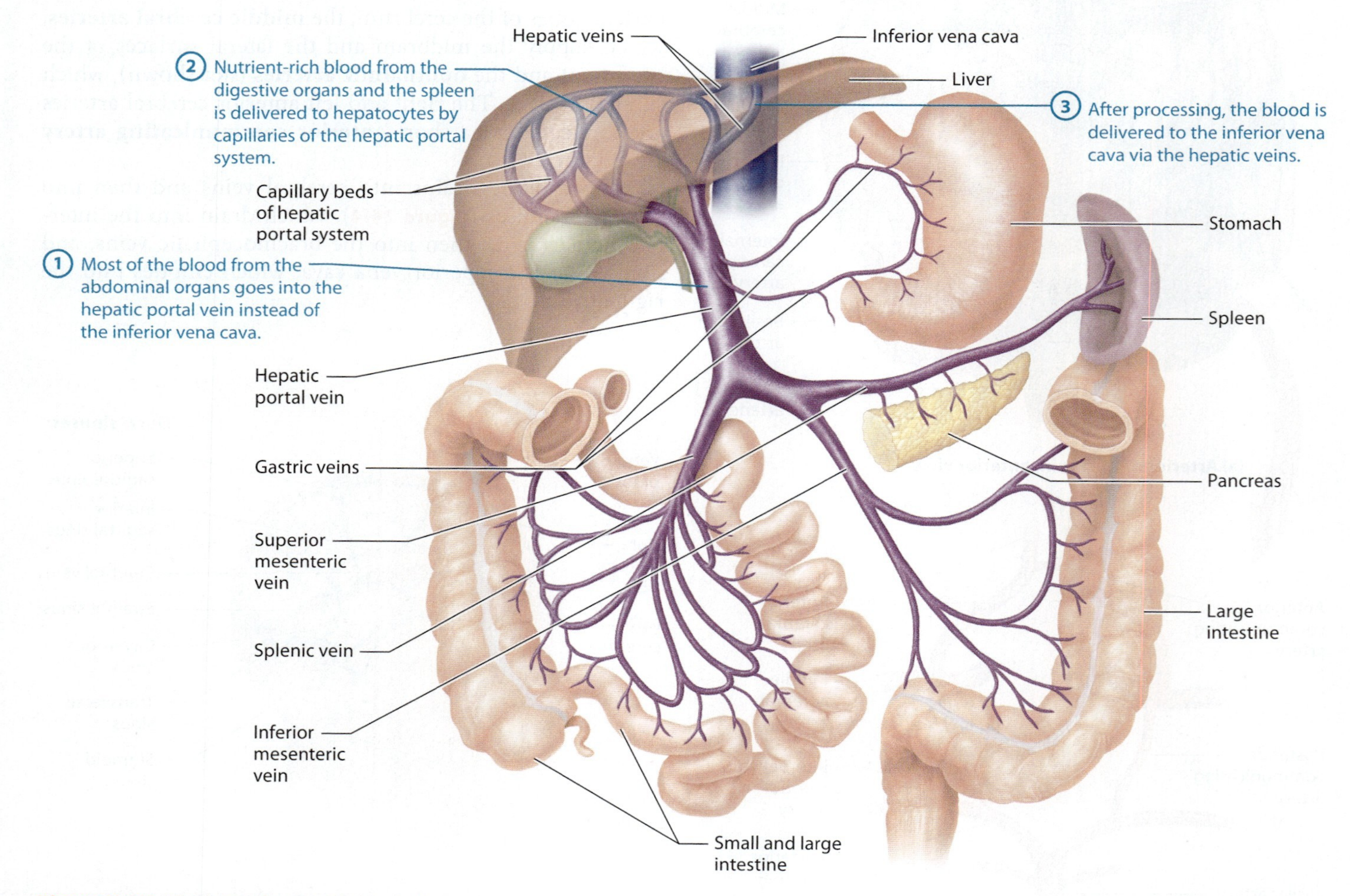

Figure 24-5 The hepatic portal circulation.

ACTIVITY 1

Tracing Blood Flow—General Systemic Pathways

Learning Outcomes

1. Trace the pathway of blood from the left ventricle to four peripheral sites (eye, forearm, abdomen, and leg) and back to the right atrium.
2. Diagram, label, and explain the exchange of materials at the capillary bed.

Materials Needed

- ☐ Anatomical charts
- ☐ Anatomical models

Instructions

1. Identify the blood vessels involved in each of the following pathways. Then, locate each blood vessel on an anatomical model or chart.

a. Trace the pathway of blood from the left ventricle to a cell in the right eye and back to the right atrium.

left ventricle → ____________ → aortic arch → ____________ → ____________ → ____________ → ____________ → ____________ → capillary bed in the right eye → ____________ → progressively larger veins → ____________ → ____________ → ____________ → ____________ → right atrium

b. Trace the pathway of blood from the left ventricle to a cell in the left flexor carpi radialis muscle and back to the right atrium.

left ventricle → ____________ → aortic arch → ____________ → ____________ → ____________ → ____________ → ____________ → capillary bed in the left flexor carpi radialis muscle → ____________ → progressively larger veins → ____________ → ____________ → ____________ → ____________ → ____________ → ____________ → right atrium

c. Trace the pathway of blood from the left ventricle to a cell in the diaphragm and back to the right atrium.

left ventricle → ____________ → aortic arch → ____________ → descending abdominal aorta → ____________ → ____________ → capillary bed in the diaphragm → ____________ → progressively larger veins → ____________ → ____________ → right atrium

d. Trace the pathway of blood from the left ventricle to a cell in the gastrocnemius muscle of the right leg and back to the right atrium.

left ventricle → ____________ → aortic arch → ____________ → descending abdominal aorta → ____________ → ____________ → ____________ → ____________ → ____________ → ____________ → capillary bed in the right gastrocnemius muscle → ____________ → progressively larger veins → ____________ → ____________ → ____________ → ____________ → ____________ → ____________ → right atrium

2. Sketch and label a diagram to explain the exchange of materials at a capillary bed.

3. "Connect" the following terms to your diagram by explaining the relationship of each term to the exchange of materials at the capillary bed. The first relationship has been explained as an example.

liver *The liver produces plasma proteins; albumin is a plasma protein largely responsible for osmotic pressure.*

alveolus ______________________________

____

____

lymph ______________________________

____

____

mitochondrion ______________________________

____

____

medulla oblongata ______________________________

____

____

ACTIVITY 2

Tracing Blood Flow—Specialized Systemic Pathways

Learning Outcomes

1. Trace the pathway of blood from the left ventricle to a cell in the frontal lobe of the cerebrum and back to the right atrium; include a description of the cerebral arterial circle.
2. Trace the pathway of blood from the left ventricle to a cell in the small intestine to the liver and back to the right atrium; include a description of the hepatic portal system.

Materials Needed

- ☐ Anatomical charts
- ☐ Anatomical models

Instructions

A. Blood Supply to the Brain, Including the Cerebral Arterial Circle

1. Make a sketch to illustrate the anatomical relationship among the following blood vessels: anterior cerebral arteries, anterior communication artery, basilar artery, internal carotid arteries, posterior cerebral arteries, posterior communicating arteries, and vertebral arteries.

2. Identify the blood vessels involved in each of the following pathways. Then, locate each blood vessel on an anatomical chart.

a. Trace the pathway of blood from the left ventricle to the basilar artery, then to the left frontal lobe of the brain, and then back to the right atrium.

left ventricle → ______________________ → aortic arch → ______________________ → ______________________ →
left basilar artery → ______________________ → ______________________ → ______________________ →
______________________ → capillary bed in the left frontal lobe → ______________________ → cerebral veins →
______________________ → ______________________ → ______________________ → ______________________ →
right atrium

b. Trace the pathway of blood from the left ventricle to the left frontal lobe of the cerebrum.

left ventricle → ______________________ → ______________________ → left common carotid artery →
______________________ → ______________________ → arterioles → left frontal lobe of the cerebrum

B. Blood Supply to and from the Small Intestine

1. Identify the blood vessels involved in each of the following pathways. Then, locate each blood vessel on an anatomical model or chart.

a. Trace the flow of blood from the left ventricle to a capillary bed of the small intestine.

left ventricle → ______________________ → ______________________ → ______________________ → descending
abdominal aorta → ______________________ → ______________________ → capillary bed of small intestine

b. Trace the flow of blood from the small intestine to the liver and then to the right atrium.

capillary bed of small intestine → ______________________ → ______________________ → ______________________ →
liver → ______________________ → ______________________ → right atrium

2. Make a sketch to illustrate the anatomical relationship among the following blood vessels of the hepatic portal system: hepatic portal veins, hepatic veins, inferior vena cava, splenic vein, and superior mesenteric vein.

What is a portal system? ______________________

____

____

What is the function of the hepatic portal system? ______

____

____

Pulmonary Circulation

The pulmonary circuit (Figure 24-6) transports blood from the heart to the lungs so that blood can pick up oxygen and release carbon dioxide. Note that the pulmonary circuit does *not* provide the cells of the lungs with oxygen and nutrients; that function is provided by the bronchial artery, a visceral branch of the thoracic aorta of the systemic circuit.

Deoxygenated blood is pumped from the **right ventricle** through the pulmonary valve into the **pulmonary trunk**,

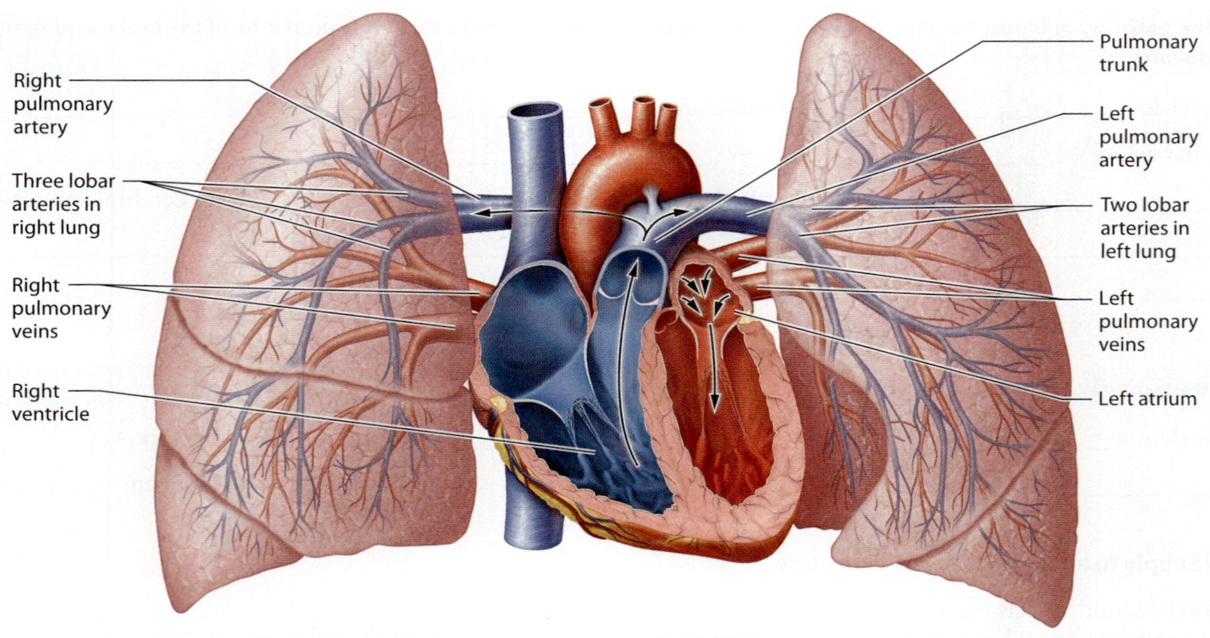

Figure 24-6 The pulmonary circulation. Note that pulmonary arteries (in blue) transport deoxygenated blood, whereas pulmonary veins (in red) transport oxygenated blood.

which splits into the right and left **pulmonary arteries** (see Figure 24-6). The right pulmonary artery enters the right lung and divides into three **lobar arteries**, each of which serves one lobe in the right lung. The left pulmonary artery enters the left lung and divides into two lobar arteries, each of which serves one lobe in the left lung. The lobar arteries branch profusely into **arterioles** that feed the **capillary beds** surrounding the alveoli (air sacs) in the lungs. Gas exchange occurs as oxygen in the alveoli enters the blood of the capillary, and carbon dioxide in the capillary enters the alveoli. The newly oxygenated blood then drains first into venules and then into the **pulmonary veins**, which transport the blood to the **left atrium**, completing the pulmonary circuit. Note that this is the only instance in adults in which veins carry oxygenated blood.

ACTIVITY 3
Tracing Blood Flow—Pulmonary Circulation

Learning Outcomes

1. Trace the pathway of blood from the right ventricle to the respiratory membrane and back to the left atrium.
2. Distinguish between a pulmonary artery and a bronchial artery.

Materials Needed

☐ Anatomical charts
☐ Anatomical models

Instructions

1. Identify the blood vessels of the pulmonary circuit. Then, locate each blood vessel on an anatomical model or chart.

Pulmonary circuit
right ventricle → ____________________ →
____________________ → ____________________ →
arterioles → ____________________ →
____________________ → ____________________ →
left atrium

2. Do the lobar arteries supply the lung tissues with oxygen? Explain. ____________________

____

____

3. By what route do the tissues of the lungs receive oxygen? ____________________

____

4. By what route is blood containing carbon dioxide released from lung tissue returned to the heart?

____

____

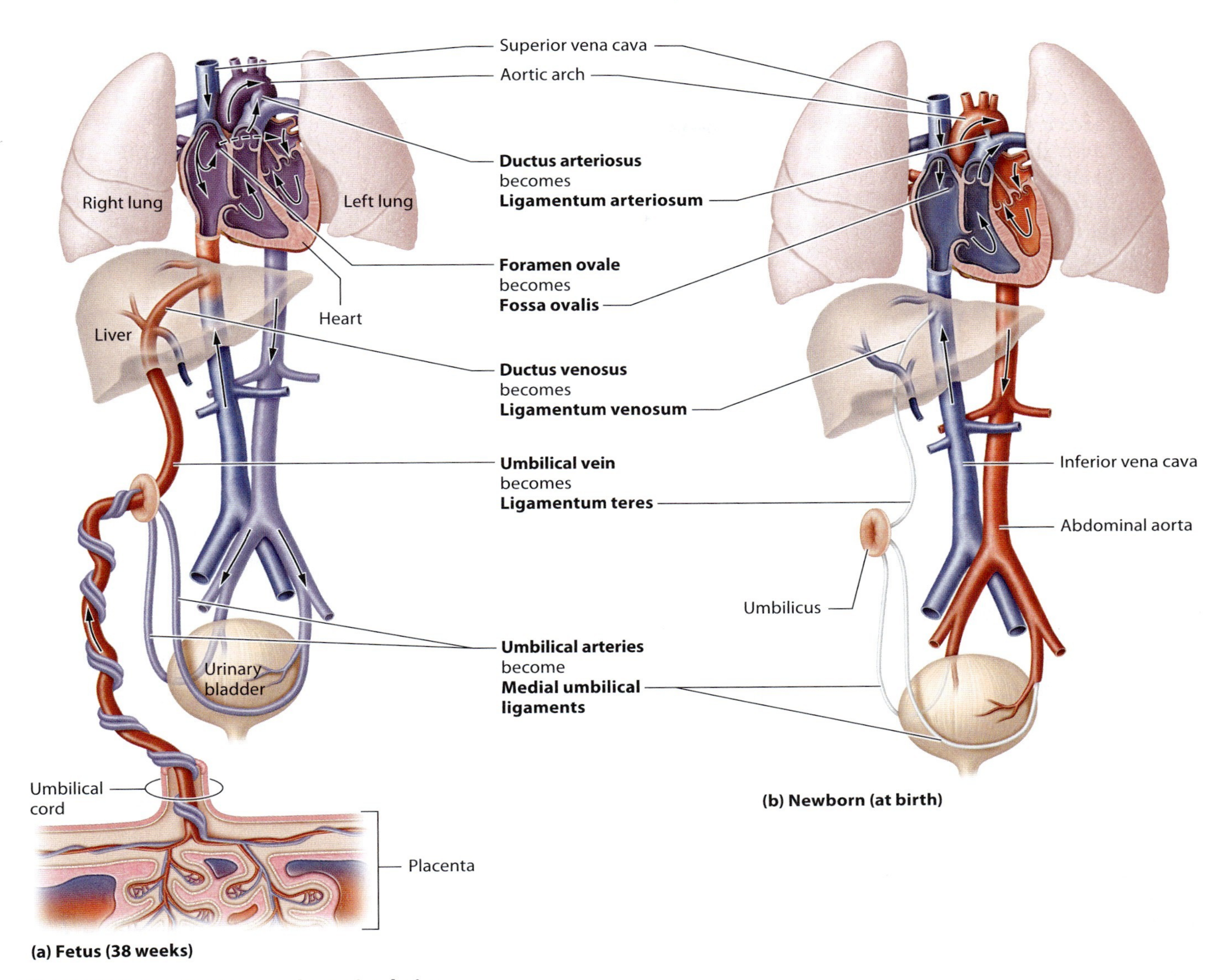

Figure 24-7 **The fetal vs. newborn circulation.**

Fetal Circulation

Because the lungs of a fetus are not functional, circulatory patterns in the fetal circulation (**Figure 24-7**) differ significantly from those in the newborn/adult circulation. The site at which materials are exchanged between fetus and mother is the **placenta**, a vascular organ composed of both fetal and maternal tissues. Maternal blood vessels bring oxygenated, nutrient-rich blood to the placenta. Fetal blood travels to and from the placenta through the **umbilical cord**, which contains one large umbilical vein and two smaller umbilical arteries. Exchange of oxygen and nutrients the fetus needs occurs in the capillaries of the placenta, and the **umbilical vein** carries the oxygen- and nutrient-rich blood to the fetus. The **umbilical arteries** carry oxygen- and nutrient-poor blood from the fetus to the placenta. As the umbilical vein travels toward the fetal heart, some blood is transported to the liver. However, most of the blood in the umbilical vein is shunted through the **ductus venosus** directly to the inferior vena cava, thereby bypassing the fetal liver. After birth, the ductus venosus collapses and becomes the **ligamentum venosum**.

Blood from the inferior vena cava enters the right atrium of the fetal heart and mixes with the deoxygenated blood entering via the superior vena cava. Blood then takes two different pathways out of the right atrium. Some blood passes through the tricuspid valve into the right ventricle, just as it does in the adult heart. Some blood, however, moves directly from the right atrium to the left atrium through the **foramen ovale** (see Figure 24-7), a hole in the interatrial septum, thereby bypassing the pulmonary circulation. Blood that enters the right ventricle is pumped into the pulmonary trunk, where it is then shunted to the systemic pathway via the **ductus arteriosus**, a blood vessel connecting the pulmonary trunk to the aortic arch, once again bypassing the lungs. Following birth, the foramen ovale becomes the **fossa ovalis**, and the ductus arteriosus collapses and becomes the **ligamentum arteriosum**. The blood vessels of the umbilical cord also collapse; the umbilical arteries become the **medial umbilical ligaments**, whereas the umbilical vein becomes the **ligamentum teres**.

ACTIVITY 4

Tracing Blood Flow—Fetal Circulation

Learning Outcomes

1. Identify the following fetal structures on an anatomical model or chart and describe the function of each: umbilical cord, umbilical artery, umbilical vein, ductus arteriosus, ductus venosus, and foramen ovale.
2. Trace the pathway of blood from the placenta to the fetal heart and back to the placenta.
3. Identify each of the following adult remnants of fetal structures on an anatomical model or chart: medial umbilical ligaments, ligamentum teres, ligamentum arteriosum, ligamentum venosum, and fossa ovalis.

Materials Needed

- ☐ Anatomical charts
- ☐ Anatomical models

Instructions

1. List the blood vessels involved in the transport of oxygenated, nutrient-rich blood in the pathway outlined below. Then, locate each blood vessel on an anatomical model or chart.

Trace the pathway of blood from the placenta to the right atrium of the fetal heart.

placenta → ______________ →

______________ → ______________ →

right atrium of the fetal heart

2. Answer the following questions and locate the structures involved in fetal circulation:

   a. Through which structure is blood shunted from the right atrium to the left atrium in the fetal heart?

   ____

   What happens to this structure after birth? ______________

   ____

   ____

   b. Through which structure is blood shunted from the pulmonary trunk to the aorta in the fetus?

   ____

   What happens to this structure after birth? ______________

   ____

   c. Why is blood shunted away from the fetal pulmonary circuit? ______________

   ____

   ____

   d. Through which structure is blood shunted away from the fetal liver?

   ____

   What happens to this structure after birth? ______________

   ____

   ____

   e. What is the function of the umbilical vein?

   ____

   What happens to this structure after birth? ______________

   ____

   ____

   f. What is the function of the umbilical arteries?

   ____

   What happens to these structures after birth? ______________

   ____

   ____

3. List the blood vessels involved in the transport of blood in the pathway outlined below. Then, locate each blood vessel on an anatomical model or chart.

Trace the pathway of blood from the left ventricle of the fetal heart to the placenta.

left ventricle of the fetal heart →

______________ → aortic arch →

______________ → ______________ →

common iliac artery → ______________ →

______________ → placenta

Blood Pressure

Blood pressure is the force exerted by blood on the walls of blood vessels. It is a result of two factors: the pumping action of the heart and the resistance to flow as the blood moves through blood vessels. Arterial blood pressure is higher than venous blood pressure because arteries are closer to the pumping action of the heart. The maximum pressure exerted on the wall of an artery is known as the systolic pressure; it occurs during ventricular contraction. The minimum pressure exerted on the wall of an artery is known as the diastolic pressure; it occurs during ventricular relaxation. Blood pressure is measured in units called millimeters of mercury (mm Hg). Average systolic pressure is 120 mm Hg; average diastolic pressure is 80 mm Hg. Average blood pressure is thus typically expressed as "120/80 mm Hg."

Several factors affect blood pressure, including peripheral resistance, cardiac output, and blood volume. Peripheral resistance is the opposition to blood flow and is a function of the amount of friction that blood encounters as it flows through blood vessels. Longer vessels have higher peripheral resistance than shorter vessels. Additionally, as blood vessels constrict (a process called vasoconstriction), vessel diameter decreases and peripheral resistance (and blood pressure) increase. Conversely, as blood vessels dilate (vasodilation), vessel diameter increases and peripheral resistance and blood pressure decrease. Increases in the thickness (viscosity) of blood and obstruction of vessels also increase peripheral resistance.

Cardiac output (CO) is a function of the stroke volume (SV) and the heart rate (HR) and is calculated as follows: CO (ml/minute) = SV (ml/beat) × HR (beats/minute). A change in cardiac output (either an increase or a decrease) causes a corresponding change in blood pressure.

Blood volume also directly affects blood pressure. As blood volume increases, blood pressure increases; when blood volume decreases, blood pressure decreases. As described in Unit 19, blood volume is regulated by hormones such as aldosterone, which is produced by the adrenal cortex, and ADH, which is produced by the hypothalamus and stored in the posterior pituitary. Both of these hormones stimulate the kidney to conserve water, thereby increasing blood volume.

ACTIVITY 5
Measuring Blood Pressure and Examining the Effects of Body Position and Exercise

Learning Outcomes

1. Demonstrate the proper technique for measuring blood pressure.
2. Distinguish between systolic blood pressure and diastolic blood pressure.
3. Evaluate the effects of body position and exercise on blood pressure.

Materials Needed

- ☐ Stethoscope
- ☐ Alcohol wipes
- ☐ Sphygmomanometer (blood pressure cuff)
- ☐ Watch

Instructions

Obtain a sphygmomanometer and a stethoscope, and clean the earpiece and diaphragm of the stethoscope. Have your lab partner sit quietly in a chair and then perform the following procedure to measure blood pressure (BP):

1. Wrap the cuff around your lab partner's right arm so that its lower edge is just superior to the antecubital region of the arm. See **Figure 24-8**.
2. Position the diaphragm of the stethoscope over the brachial artery distal to the cuff.
3. Close the valve on the cuff and squeeze the rubber bulb to inflate the cuff to a pressure of approximately 160 mm Hg.

⚠ Do not overinflate the cuff or leave the cuff inflated for more than 1 minute because doing so interrupts blood flow to the forearm and could cause fainting.

4. Turn the knob at the base of the bulb to slowly reduce the pressure in the cuff, deflating it at a rate of about 2 or 3 mm Hg per second. During this time listen to the brachial artery through the stethoscope.
5. When the cuff pressure drops below arterial pressure, blood resumes flowing through the artery. This turbulent blood flow makes characteristic sounds called Korotkoff's sounds, which can be heard through the stethoscope. Make a mental note of the pressure at which you begin to hear Korotkoff's sounds. That pressure corresponds to the systolic pressure.

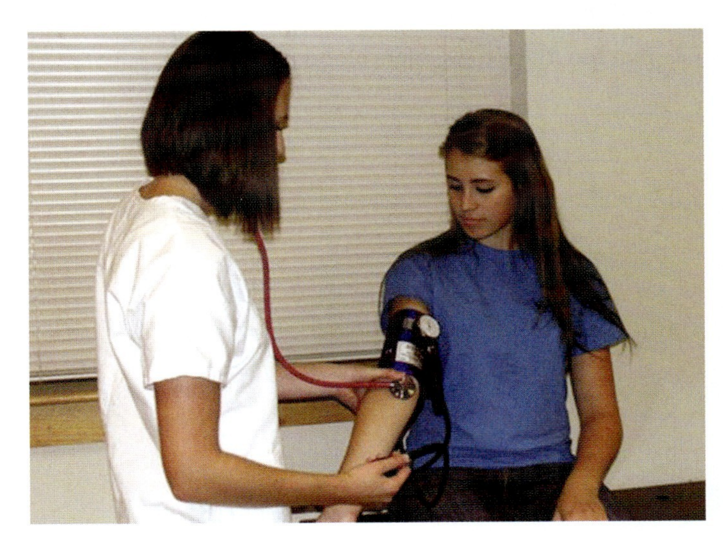

Figure 24-8 Proper positioning for taking a blood pressure reading.

Table 24-1 Blood Pressure Measurements

	BP 1	BP 2	BP 3
Sitting			
Standing			
Following exercise			

6. Continue deflating the cuff. The thumping Korotkoff's sounds will become faint and will eventually disappear. Note the pressure at which the last sound is heard; this is the diastolic pressure.
7. Release all of the pressure from the cuff, and record the results in the appropriate cell of **Table 24-1**.
8. Repeat steps 2–7 two more times and record those data.
9. Instruct your lab partner to stand for 3 minutes. Then repeat steps 2–7 three times and record your data in Table 24-1.
10. Finally, instruct your lab partner to exercise vigorously for 3 minutes. Then repeat steps 2–7 three times and record your data in Table 24-1.
11. Summarize your data concerning the effects of body position and exercise on blood pressure. ____________
____
____
____
12. Outline and describe one negative feedback mechanism that functions to regulate blood pressure. ____________
____
____
____
____
____

PhysioEx EXERCISE 5
Cardiovascular Dynamics

The PhysioEx 9.1 Laboratory Simulations in Physiology are easy-to-use laboratory simulations that can be used as an alternative to or as a supplement to wet lab activities in this unit. Each simulation allows you to investigate important physiological concepts, repeat labs as often as you like, and conduct experiments that are difficult to perform in a wet lab environment because of time, cost, or safety concerns.

Access the simulations in these activities at MasteringA&P® > Study Area > PhysioEx 9.1

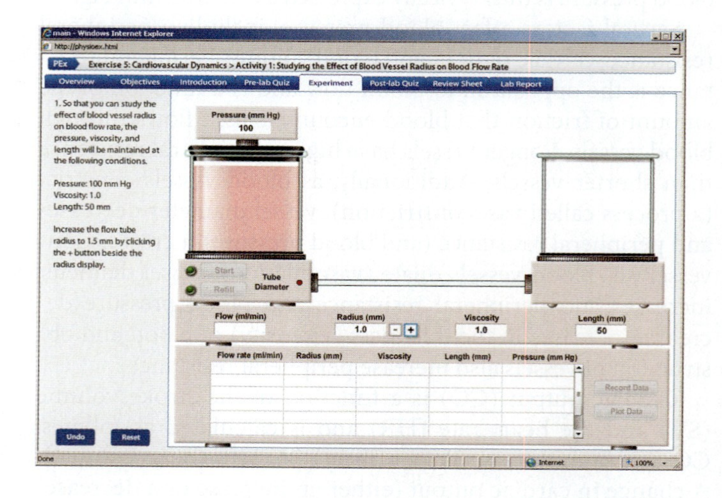

There you will find the following activities:

PEx Activity 1: Studying the Effect of Blood Vessel Radius on Blood Flow Rate

PEx Activity 2: Studying the Effect of Blood Viscosity on Blood Flow Rate

PEx Activity 3: Studying the Effect of Blood Vessel Length on Blood Flow Rate

PEx Activity 4: Studying the Effect of Blood Pressure on Blood Flow Rate

PEx Activity 5: Studying the Effect of Blood Vessel Radius on Pump Activity

PEx Activity 6: Studying the Effect of Stroke Volume on Pump Activity

Post-lab quizzes are also assignable in MasteringA&P®

POST-LAB ASSIGNMENTS

Name: ______________________ Date: ______________ Lab Section: ________

PART I. Check Your Understanding

Activity 1: Tracing Blood Flow—General Systemic Pathways

1. Trace the pathway of blood from the left ventricle to a cell in the left eye and back to the right atrium.

   left ventricle → ______________ → aortic arch → ______________ → ______________ →
   ______________ → ______________ → capillary bed of left eye → ______________ →
   ______________ → ______________ → ______________ → ______________ →
   right atrium

2. Trace the pathway of blood from the left ventricle to a cell in the right tibialis anterior muscle and back to the right atrium.

   left ventricle → ______________ → aortic arch → ______________ → ______________ →
   ______________ → ______________ → ______________ → ______________ →
   ______________ → ______________ → capillary bed of right tibialis anterior muscle →
   ______________ → ______________ → ______________ → ______________ →
   ______________ → ______________ → ______________ → right atrium

3. Oxygen, carbon dioxide, nutrients, and wastes are exchanged at the capillary bed via ______________.

4. At the arteriole end of the capillary, ______________ pressure exceeds ______________ pressure; at the venule end, ______________ pressure exceeds ______________ pressure. ______________ pressure pushes water out of the capillary bed at the arteriole end, whereas ______________ pressure draws fluids into the capillary bed at the venule end.

Activity 2: Tracing Blood Flow—Specialized Systemic Pathways

1. Branching from the subclavian arteries, the left and right ______________ arteries extend upward and fuse to form the ______________ artery, which then splits into the left and right ______________ arteries. These two arteries are connected to the anterior cerebral arteries by the ______________ arteries. The left and right anterior cerebral arteries are connected by the ______________ artery. The ______________ arteries branch from the common carotid arteries, extend continued toward the brain, and eventually divide into

Continued

three branches: the ______________________ arteries, which supply the eyes; the ______________________ arteries, which supply the midbrain and the lateral surfaces of the cerebrum; and the ______________________ arteries, which supply the frontal and parietal lobes of the cerebrum.

2. Identify the components of the cerebral arterial circle and associated blood vessels in the accompanying anterior view. Place an asterisk (*) next to the names of blood vessels that are part of the cerebral arterial circle.

   a. ______________________
   b. ______________________
   c. ______________________
   d. ______________________
   e. ______________________
   f. ______________________
   g. ______________________

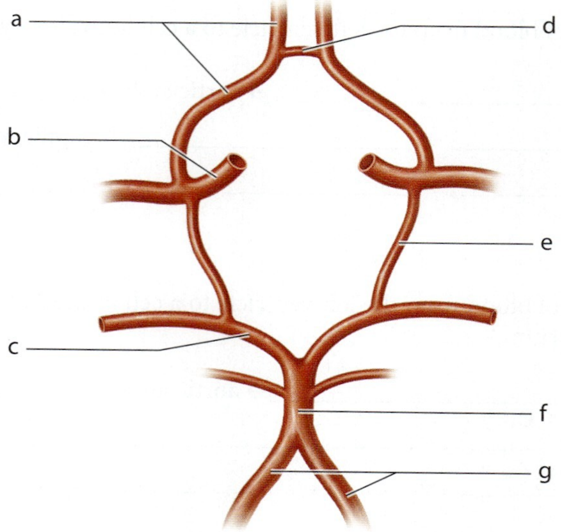

3. After nutrient-rich blood leaves a capillary bed in the small intestine, it enters venules before entering the ______________________, which fuses with the splenic vein to form the ______________________, which transports blood to the liver for processing. The ______________________ then transport blood away from the liver and empty into the ______________________.

Activity 3: Tracing Blood Flow—Pulmonary Circulation

1. Identify the structures in the accompanying figure that make up the pulmonary circulation:

   a. ______________________
   b. ______________________
   c. ______________________
   d. ______________________
   e. ______________________
   f. ______________________
   g. ______________________
   h. ______________________

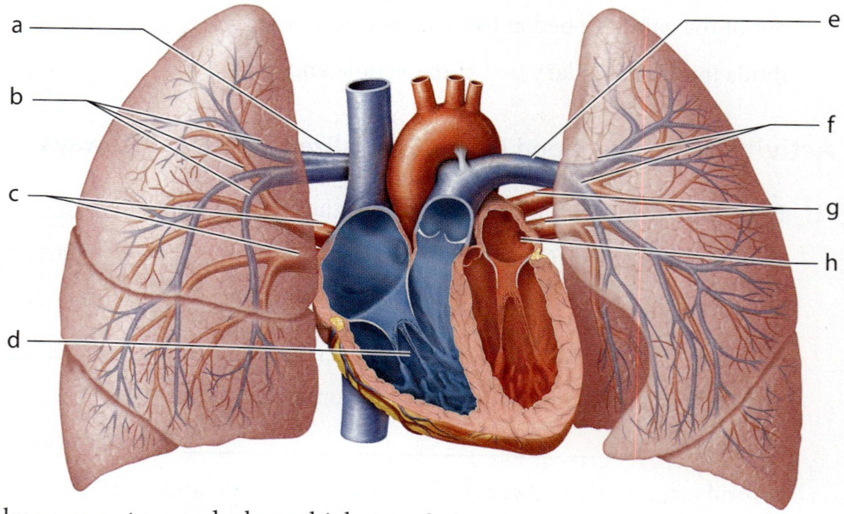

2. What is the functional difference between a pulmonary artery and a bronchial artery?

____

____

Activity 4: Tracing Blood Flow—Fetal Circulation

1. Complete the following chart:

Fetal Structure	Function	Adult Remnant
Umbilical vein		
Umbilical artery		
Ductus venosus		
Foramen ovale		
Ductus arteriosus		

Activity 5: Measuring Blood Pressure and Examining the Effects of Body Position and Exercise

1. What are Korotkoff's sounds? ______________________________

____

2. Distinguish between diastole and systole. ______________________________

____

3. Describe the effects of body position and exercise on blood pressure. ______________________________

____

____

PART II. Putting It All Together

A. Review Questions

Answer the following questions using your lecture notes, your textbook, and your lab notes.

1. What is the functional difference between a pericardial artery and a coronary artery? ______________________________

____

____

2. Describe two different pathways of blood from the left ventricle to the esophagus. ______________________________

____

____

3. In the hepatic portal system, where are the two capillary beds located? ______________________________

____

4. Name three veins that carry oxygen-rich or nutrient-rich blood. ______________________________

____

5. Describe the effect of severe malnutrition or starvation on the exchange of materials at the capillary bed.

____

____

____

6. Describe the role of the autonomic nervous system in returning blood pressure to its normal range following an increase in blood pressure. ______________________

____

____

____

B. Concept Mapping

1. Fill in the blanks to complete this concept map outlining a blood vessel's pathway.

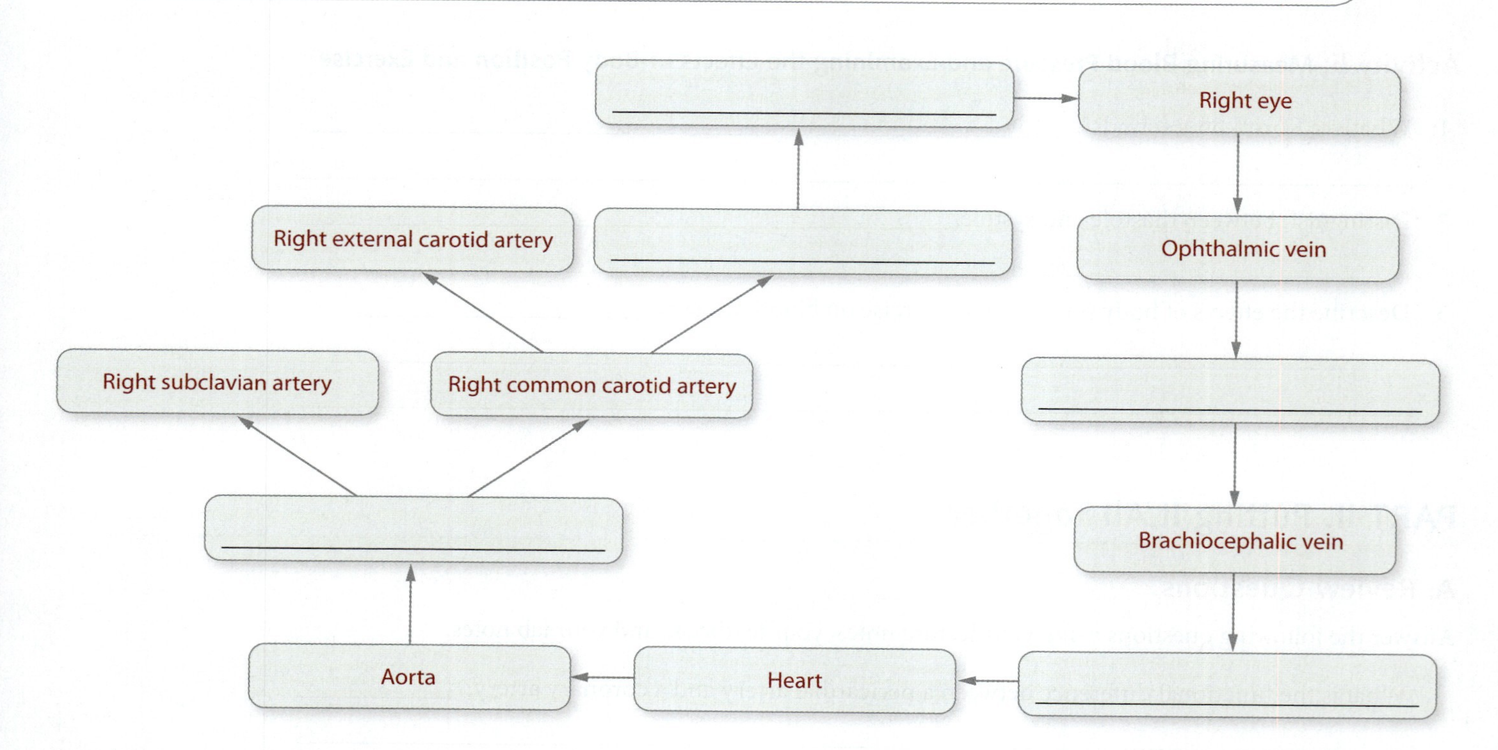

2. Construct a unit concept map to show the relationships among the following set of terms. Include all of the terms in your diagram. Your instructor may choose to assign additional terms.

axillary v.	azygos v.	brachiocephalic a.	bronchial a.	celiac trunk
femoral a.	internal carotid a.	internal jugular v.	ophthalmic a.	posterior tibial v.
radial a.	renal v.	subclavian v.	superficial palmar arch	superior vena cava

25

The Lymphatic System

UNIT OUTLINE

The lymphatic system performs three main functions: It returns excess tissue fluid to the bloodstream, absorbs fats and fat-soluble vitamins, and works with the leukocytes of the cardiovascular system to defend the body against pathogens. In this unit, you will learn how lymph is formed and transported to the bloodstream. You will also study the role of the lymphatic system in various defense mechanisms. Later, when you study the digestive system, you will learn about the role of the lymphatic system in fat absorption.

THINK ABOUT IT *List the five types of leukocytes and briefly describe the function of each in immunity:*

____

____

____

____

____

PRE-LAB ASSIGNMENTS

Pre-lab quizzes are also assignable in MasteringA&P®

To maximize learning, BEFORE your lab period carefully read this entire lab unit and complete these pre-lab assignments using your textbook, lecture notes, and prior knowledge.

PRE-LAB Activity 1: Exploring the Organs of the Lymphatic System

1. Use the list of terms to label the accompanying illustration of lymphoid organs; check off each term as you label it.

   - ☐ lymph nodes
   - ☐ lymphatic vessels
   - ☐ spleen
   - ☐ tonsils
   - ☐ thymus

a ______

b ______

c ______

d ______

e ______

2. Match each of the following lymphoid organs with its correct description.

_____	a. lymphatic vessels	1. largest lymphatic organ
_____	b. tonsils	2. transport lymph
_____	c. lymph nodes	3. filter lymph
_____	d. spleen	4. atrophies with age
_____	e. thymus	5. located in the pharynx

PRE-LAB Activity 2: Examining the Histology of a Lymph Node, a Tonsil, and the Spleen

1. Which lymphoid organ:

   ______________ a. contains red pulp and white pulp?

   ______________ b. contains more afferent vessels than efferent vessels?

   ______________ c. lacks a capsule?

   ______________ d. contains crypts?

   ______________ e. contains an outer cortex and an inner medulla?

PRE-LAB Activity 3: Tracing the Flow of Lymph through the Body

1. Interstitial fluid that moves into a lymphatic capillary is called ______________________.
2. Rank the following lymphatic vessels from smallest (1) to largest (4):

   _____ lymph duct

   _____ lymphatic capillary

   _____ lymph trunk

   _____ lymph collecting vessel

PRE-LAB Activity 4: Using a Pregnancy Test to Demonstrate Antigen–Antibody Reactions

1. Which hormone is detected by a pregnancy test? ______________________
2. Use the list of terms to label the accompanying schematic representation of the reactants in an enzyme-linked immunosorbent assay (ELISA). Check off each term as you label it.

   ☐ secondary antibody

   ☐ antigen

   ☐ enzyme substrate

   ☐ primary antibody

a ______________________ hCG in urine

b ______________________ binds to antigen

c ______________________ binds to antibody

d ______________________ produces color change

Functional Anatomy of the Lymphatic System

The lymphatic system (**Figure 25-1**) includes the lymphoid organs—lymphatic vessels, lymph nodes, the spleen, and the thymus—as well as mucosa-associated lymphoid tissues (MALT) in the body's mucous membranes. The largest collections of MALT are the tonsils, Peyer's patches (associated with the small intestine), and the appendix.

The Lymphoid Organs

When materials are exchanged at a blood capillary bed, more fluid leaves the arteriole end of the capillary than is returned to the venule end; this excess fluid becomes part of the interstitial fluid bathing the cells and is picked up by microscopic, blind-ended vessels called **lymphatic capillaries**. Like blood capillaries, lymphatic capillaries are thin-walled structures composed of simple squamous epithelial tissue (endothelium). **Figure 25-2** illustrates the close relationship between a blood capillary bed and the lymphatic capillaries that surround it. Once interstitial fluid enters a lymphatic capillary, it is called **lymph**.

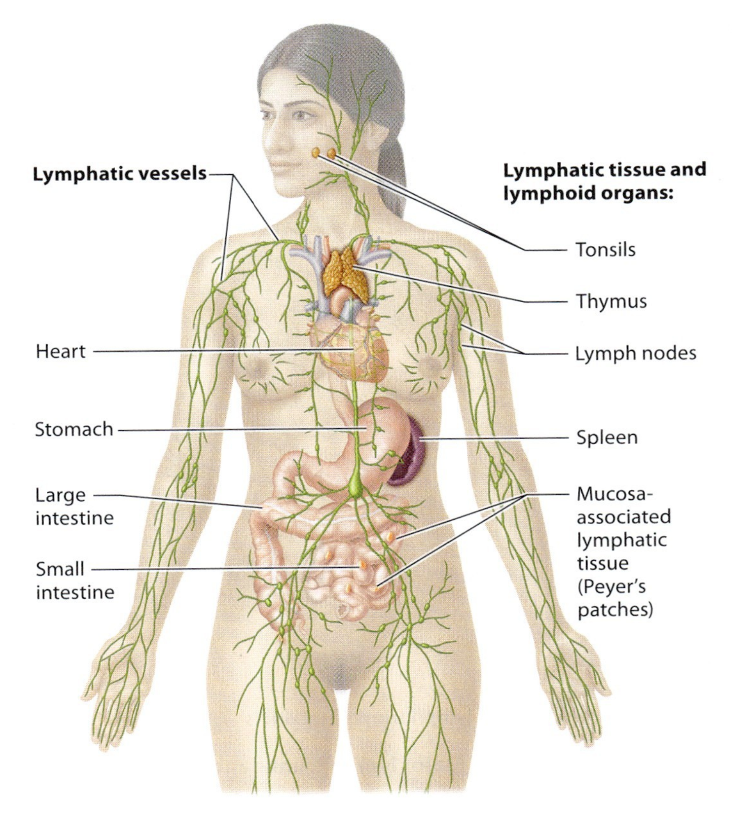

Figure 25-1 **The structures of the lymphatic system.**

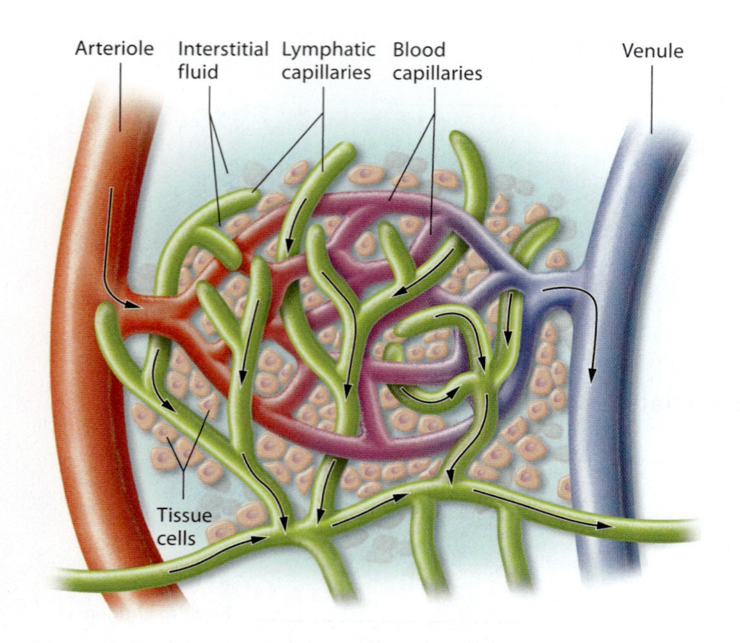

Figure 25-2 The anatomical relationship between blood capillaries and lymph capillaries.

Lymphatic vessels of increasing size transport lymph from the site of its formation until it is returned to the venous circulation; valve-like flaps prevent the backflow of lymph. Lymphatic capillaries drain into larger **lymph collecting vessels**, which drain into **lymph trunks**, which then drain into one of two **lymph ducts** (**Figure 25-3**). The **right lymphatic duct** drains the right side of the head and neck, the right shoulder, and the right upper limb; the **thoracic duct** provides lymph drainage for the rest of the body. Note that whereas the cardiovascular system transports blood in two directions (to and from the heart), the lymphatic system transports lymph in one direction only—toward the heart.

Located along the pathways of lymph collecting vessels (and clustered in the cervical, axillary, and inguinal regions) are hundreds of **lymph nodes**, which filter the lymph and rid it of pathogens and foreign debris. Each lymph node (**Figure 25-4**) is a bean-shaped structure surrounded by a connective tissue **capsule**. Extensions of the capsule, called **trabeculae**, divide the lymph node into compartments. Histologically, the lymph node has two distinct regions: a cortex and a medulla. The superficial

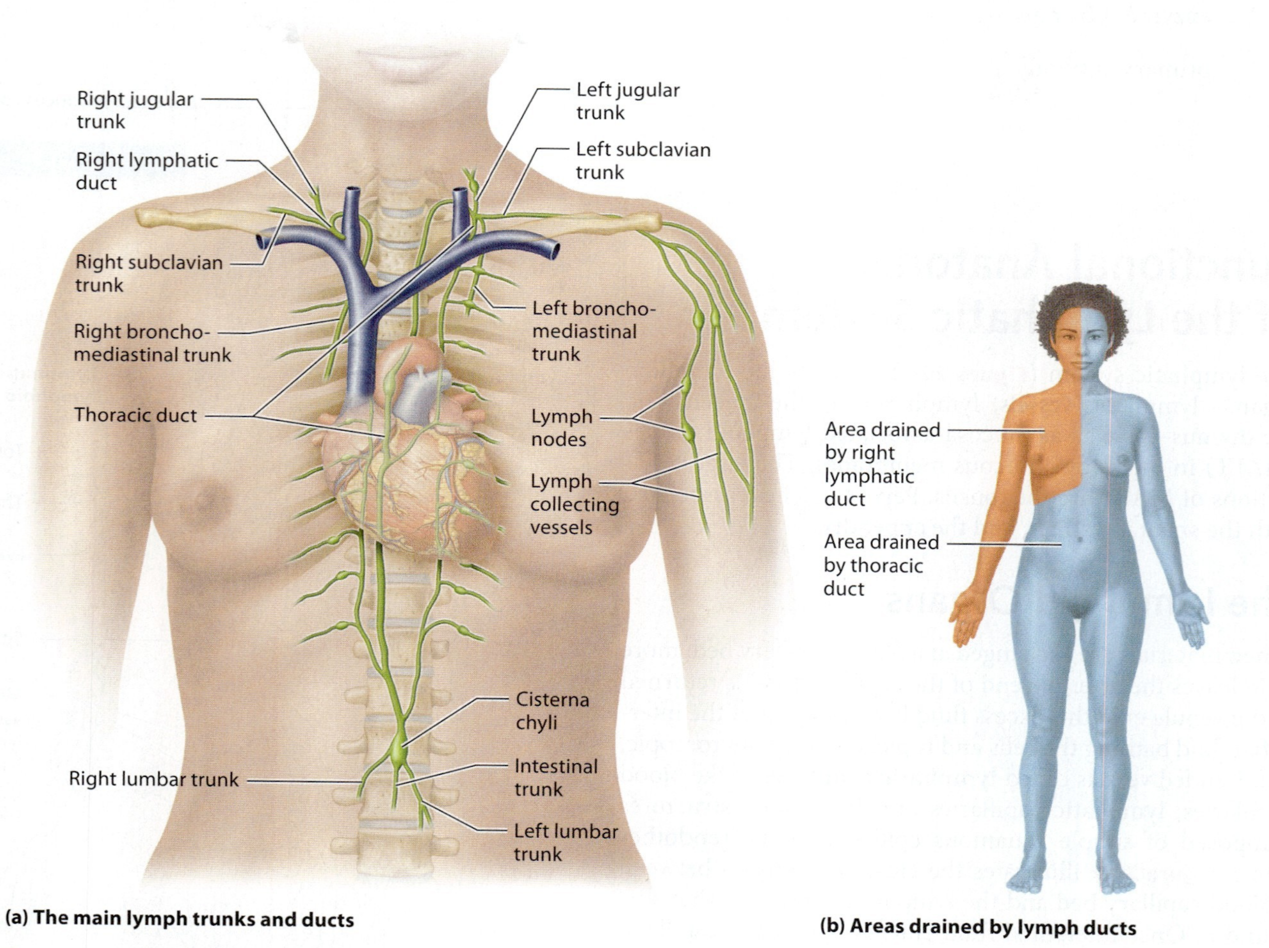

(a) The main lymph trunks and ducts

(b) Areas drained by lymph ducts

Figure 25-3 Major lymph trunks and ducts.

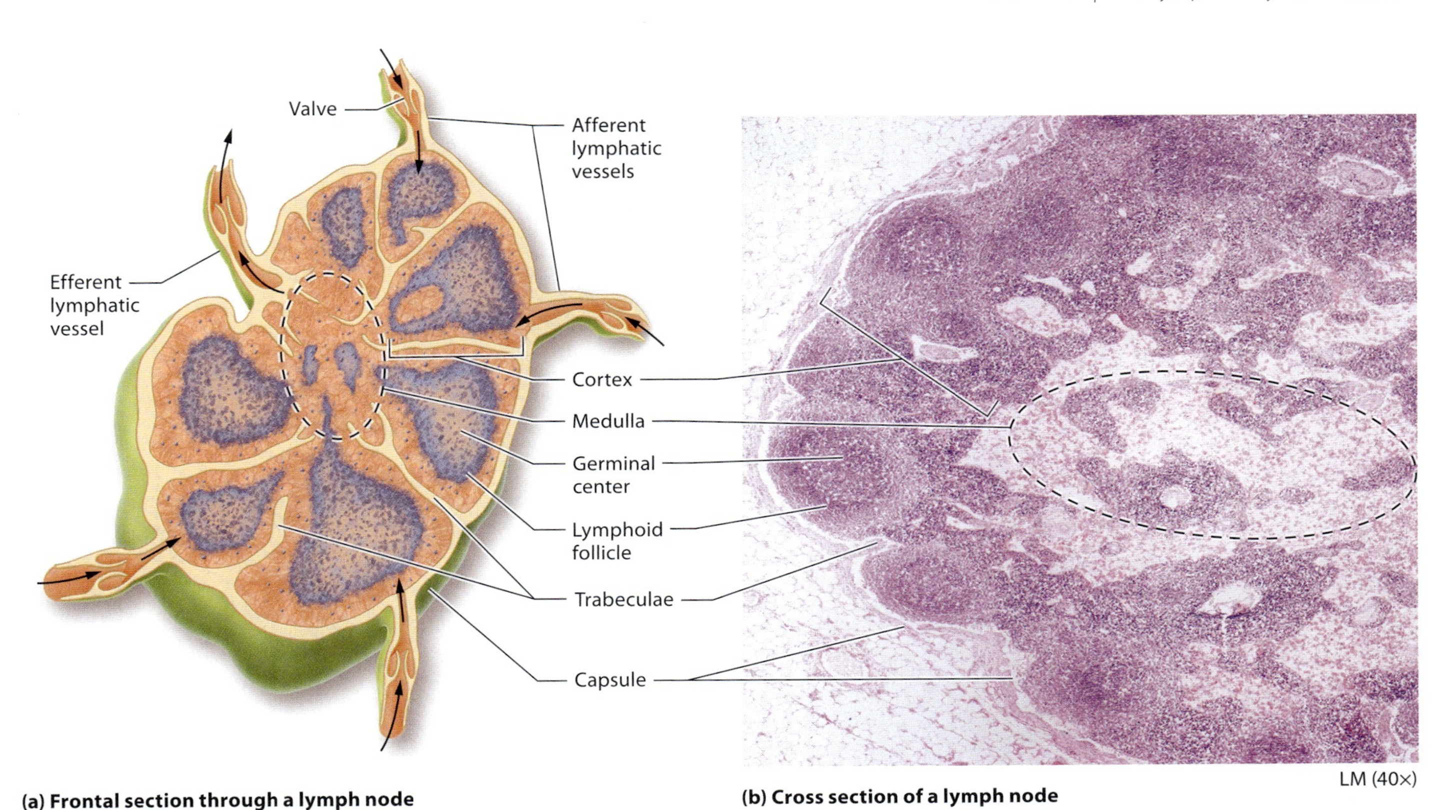

(a) Frontal section through a lymph node

(b) Cross section of a lymph node

Figure 25-4 A frontal section of a lymph node.

cortex is filled with lymphoid follicles (also called nodules) surrounding germinal centers packed with lymphocytes called B cells. Deeper parts of the cortex contain lymphocytes called T cells, which move continuously between the blood and lymph. The **medulla** is filled with medullary cords containing both B cells and T cells. Lymph enters the lymph node through several **afferent lymphatic vessels** and flows through a series of sinuses (spaces) before exiting via fewer **efferent lymphatic vessels**. This anatomical arrangement slows the flow of lymph through a lymph node, giving macrophages and lymphocytes time to destroy pathogens and launch an immune response.

The largest lymphoid organ, the **spleen** (**Figure 25-5**), is in the left hypochondriac region posterior and lateral to the stomach. The spleen filters blood, serves as a blood reservoir, and functions in hematopoiesis in the fetus. Histologically, the spleen consists of two components, red pulp and white pulp, surrounded by a connective tissue capsule. The red pulp contains splenic cords (composed of reticular connective tissue) that separate blood-filled sinuses (splenic sinusoids) where worn-out red blood cells and pathogens are phagocytized. The white pulp contains mostly lymphocytes and appears as "islands" in a "sea" of red pulp.

The bilobed thymus (**Figure 25-6**), partially overlying the heart in the mediastinum, plays an important role in the maturation of T cells. Hormones released by the thymus (thymosins) stimulate this maturation process during which T cells become able to defend the body against pathogens. The thymus is largest relative to body size in a 2-year-old child. It reaches its largest absolute size at puberty and then begins to atrophy. By old age the thymus is largely composed of adipose tissue.

Mucosa-Associated Lymphoid Tissue

Mucosa-associated lymphoid tissues (MALT) are lymphoid nodules (clusters of predominantly B cells) located in the oral, respiratory, and digestive mucosae.

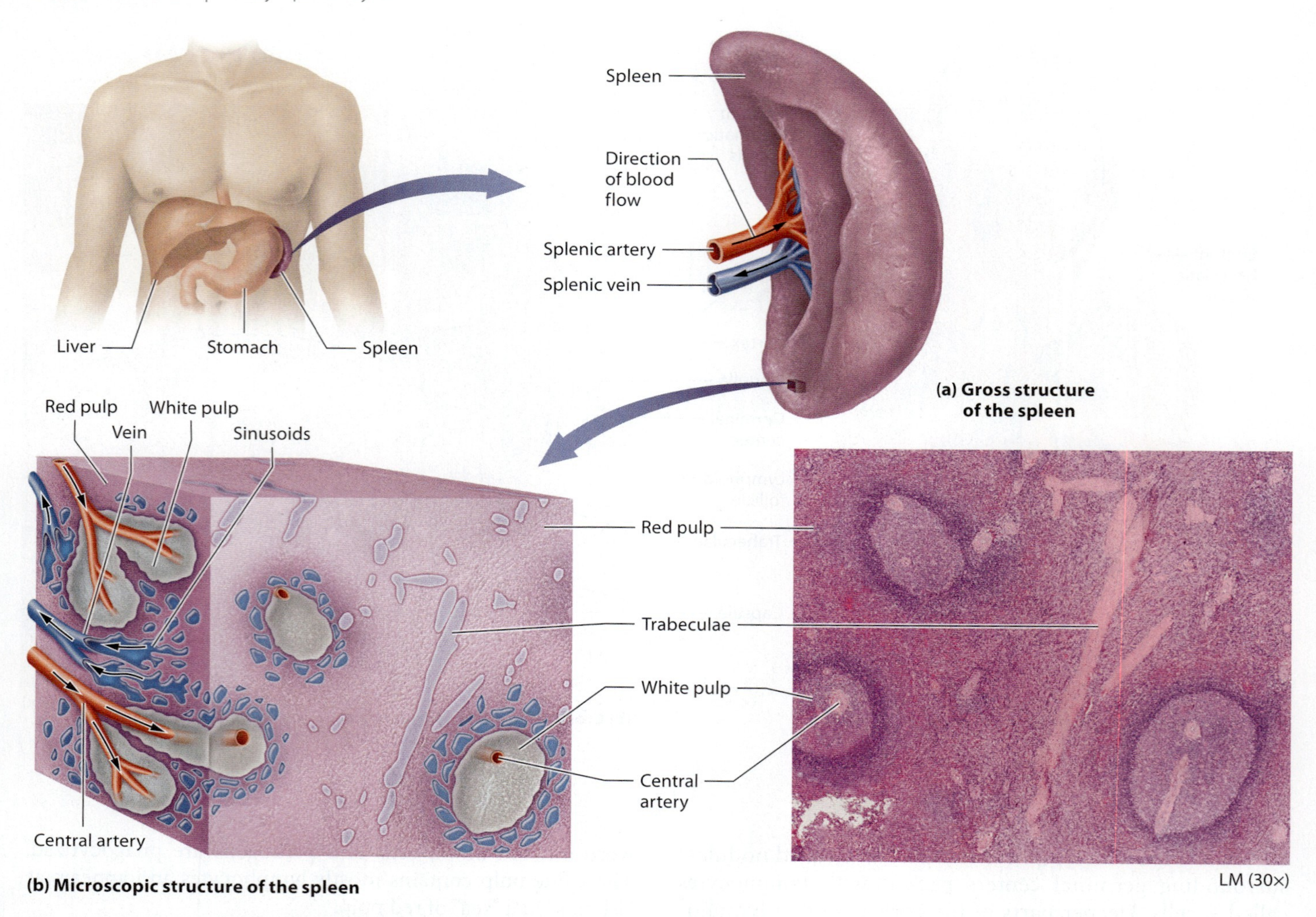

(b) Microscopic structure of the spleen

Figure 25-5 The structure of the spleen.

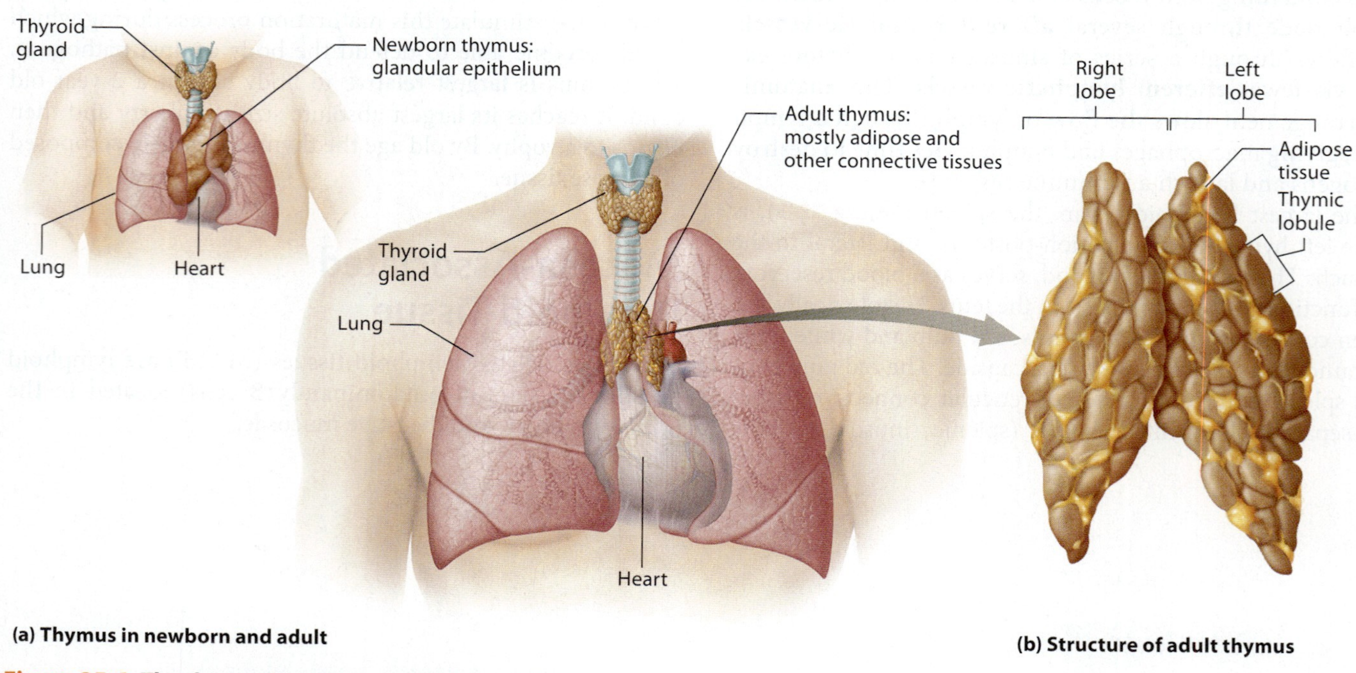

(a) Thymus in newborn and adult

(b) Structure of adult thymus

Figure 25-6 The thymus in a newborn and an adult.

The **tonsils** are rings of lymphoid tissue surrounding the entrance to the pharynx (**Figure 25-7a**). The single **pharyngeal tonsil**, in the posterior wall of the nasopharynx, is often called the adenoids (referring to the right and left halves of the tonsil) when enlarged. The paired **palatine tonsils** hang from the posterior arches of the oral cavity, and the paired **lingual tonsils** are located at the base of the tongue. Histologically, each tonsil contains follicles with germinal centers filled with B cells (**Figure 25-7b**). Tonsils are not fully encapsulated; instead, the surface epithelium invaginates to form tonsillar crypts that trap bacteria and debris. MALT is also associated with the intestines (**Figure 25-8**). **Peyer's patches** are lymphoid nodules in the wall of the ileum of the small intestine. The **appendix** is a wormlike appendage containing lymphoid nodules that extends from the wall of the cecum (the first part of the large intestine).

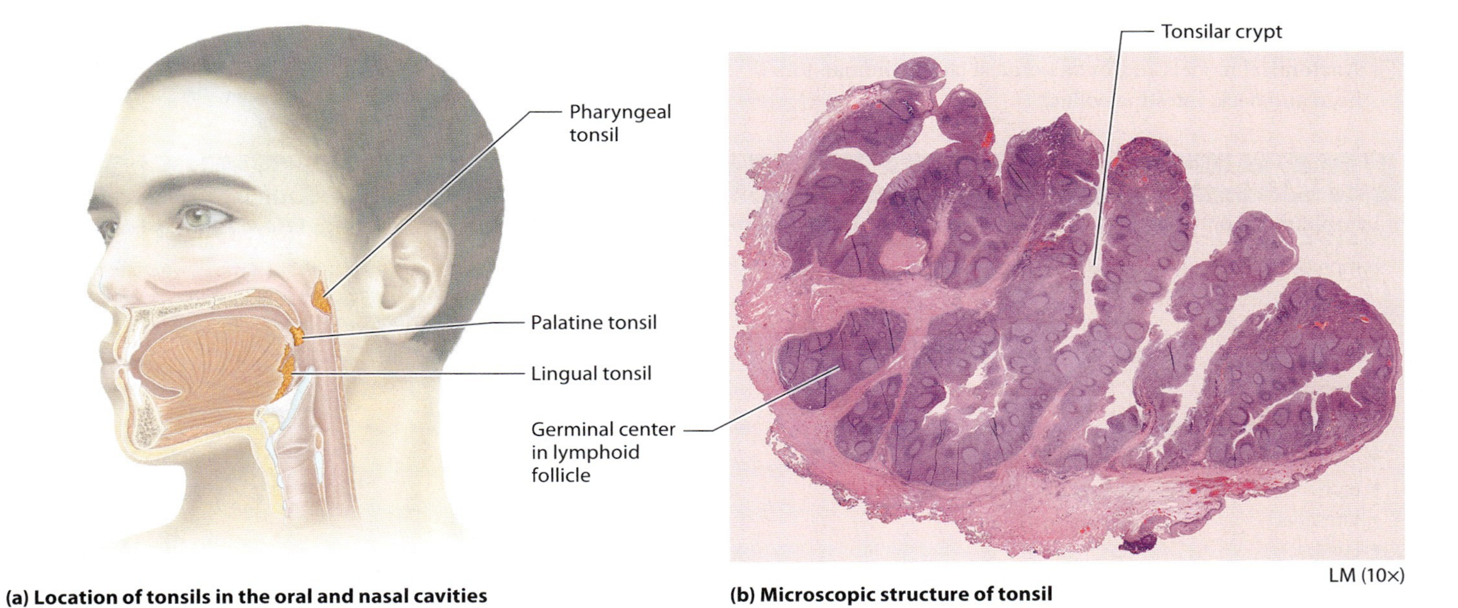

(a) Location of tonsils in the oral and nasal cavities

(b) Microscopic structure of tonsil

Figure 25-7 Tonsils, a type of mucosa-associated lymphoid tissue (MALT).

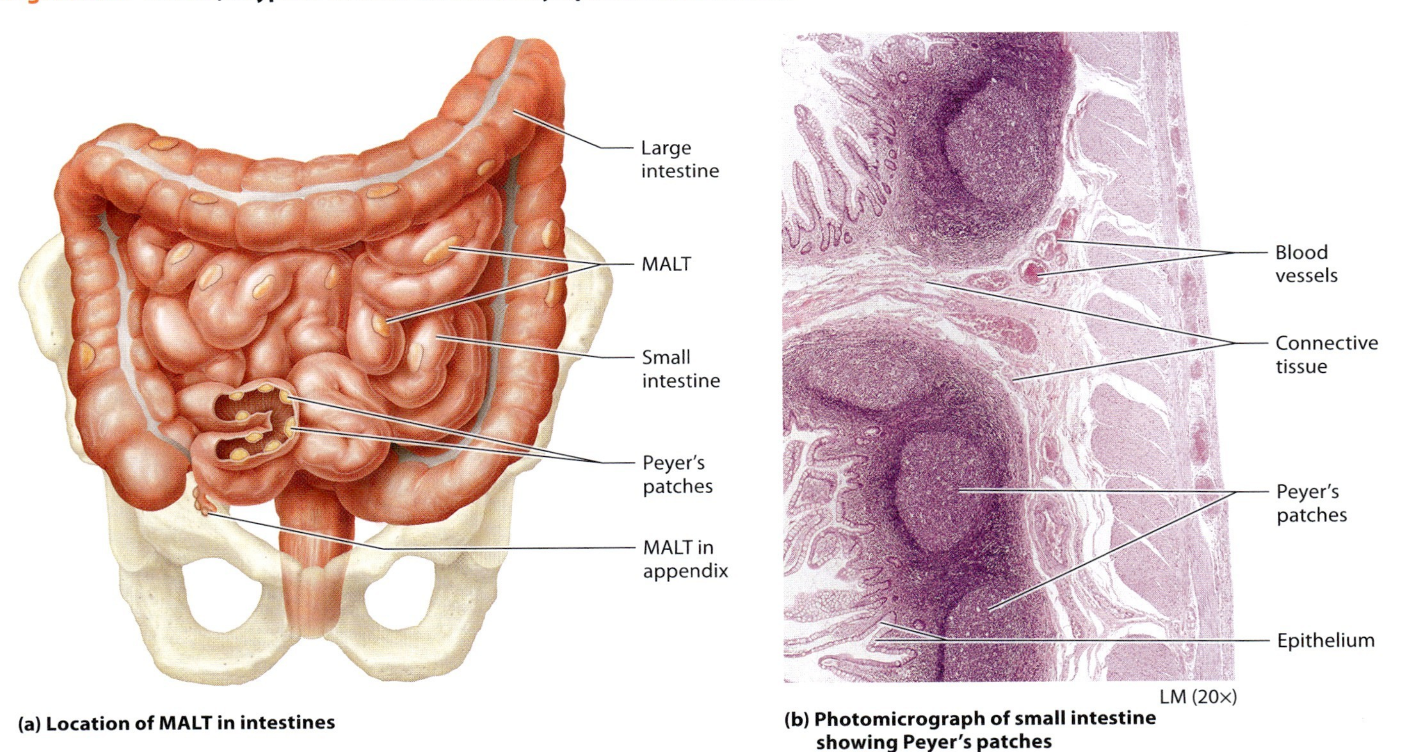

(a) Location of MALT in intestines

(b) Photomicrograph of small intestine showing Peyer's patches

Figure 25-8 The location of MALT in the intestines.

ACTIVITY 1

Exploring the Organs of the Lymphatic System

Learning Outcomes

1. Identify the major organs of the lymphatic system.
2. Describe the functions of the lymphatic system.

Materials Needed

☐ Anatomical models and/or anatomical charts (human torso, head and neck, intestinal villus)

Instructions

CHART Locate each of the lymphatic structures in the following Making Connections chart on an anatomical model or chart. Then review its description and/or function(s). Finally, make "connections" to things you have already learned in lectures, assigned readings, and lab. Part of the chart has been filled out for you.

Optional Activity

Practice labeling lymphatic organs on human cadavers at > MasteringA&P® > Study Area > Practice Anatomy Lab > Human Cadaver > Lymphatic System

Making Connections: The Lymphatic System

Lymphatic Structure	Description (Structure and/or Function)	Connections to Things I Have Already Learned
Lymph collecting vessels		*Like veins, contain three tunics and valves; lymph transport dependent on milking action of skeletal muscles and pressure changes within thorax.*
Lymph trunks • Jugular • Subclavian • Bronchomediastinal • Intestinal • Lumbar		
Lymph ducts • Thoracic duct • Right lymphatic duct		*Thoracic duct empties into left subclavian vein; right lymphatic duct empties into right subclavian vein (both at junction of subclavian and internal jugular veins).*
Lymph nodes • Cervical • Axillary • Inguinal	*Small, kidney-bean shaped organs associated with lymph collecting vessels; clean lymph; biological filter; clusters found in neck, armpit, and groin.*	
Spleen	*Largest lymphoid organ; located posterior and lateral to stomach in left hypochondriac region; filters blood, launches immune response; red pulp: many RBCs; white pulp: many lymphocytes.*	
Thymus		*Also classified as an endocrine organ; secretes thymosins, which regulate development of white blood cells.*
Mucosa-associated lymphoid tissue • Tonsils • Pharyngeal • Palatine • Lingual • Peyer's patches • Appendix		

ACTIVITY 2

Examining the Histology of a Lymph Node, a Tonsil, and the Spleen

Learning Outcomes

1. Sketch, label, and describe the histological features in a microscopic section of a lymph node.
2. Sketch, label, and describe the histological features in a microscopic section of a tonsil.
3. Sketch, label, and describe the histological features in a microscopic section of the spleen.

Materials Needed

☐ Microscope and prepared slides (or photomicrographs) of sections of a lymph node, tonsil, and spleen

Instructions

1. Observe a histological slide (or a photomicrograph) of a lymph node using the scanning-power objective.
   a. Identify the connective tissue capsule and the trabeculae. Distinguish between the cortex and the medulla. In the cortex, identify the germinal centers.
   Which white blood cell type is most prominent in the germinal centers, and what is its function?
   ____
   ____
   b. Identify the medullary cords.
   Which cell types are most prominent in the medullary cords, and what is their function?
   ____
   ____
   c. Sketch a histological view of a lymph node and label the following structures: capsule, trabeculae, cortex, lymphoid nodule, germinal center, medulla, medullary cord, medullary sinus.

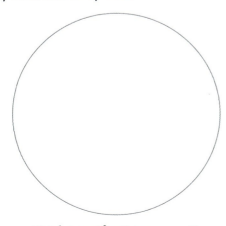

Total magnification: _____ ×

   d. Lymph enters a lymph node via afferent vessels, circulates through lymph sinuses, and then exits via efferent vessels.
   What is the physiological significance of each lymph node having fewer efferent vessels than afferent vessels?
   ____
   ____

2. View a histological slide (or a photomicrograph) of a tonsil using the scanning-power objective.
   a. Identify the surface epithelium and identify the tonsillar crypts.
   What is the purpose of the tonsillar crypts?
   ____
   b. Note the numerous lymphoid follicles containing germinal centers. Sketch a histological view of a tonsil and label the following structures: tonsillar crypt, lymphoid follicle, germinal center.

Total magnification: _____ ×

3. View a histological slide (or a photomicrograph) of the spleen using the scanning-power objective.
   a. Identify the capsule. Distinguish between the red pulp and the white pulp.
   Even though white pulp is not white in color, why is this tissue called white pulp? ______________________________
   ____
   How do the functions of red pulp and white pulp relate to their structures? ______________________________
   ____
   ____
   ____
   ____

b. Sketch a histological view of the spleen and label the following structures: capsule, red pulp, white pulp.

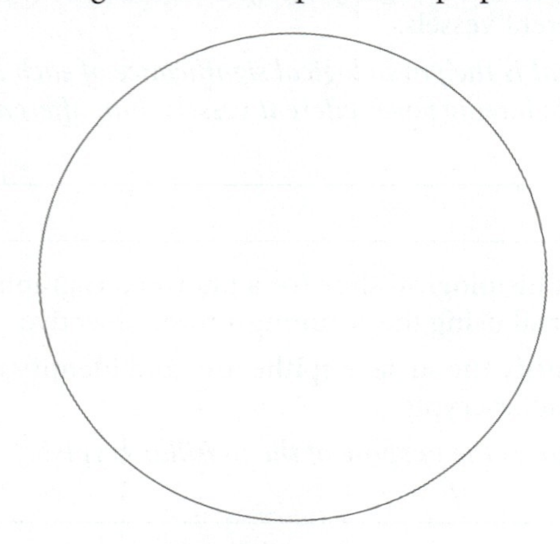

Total magnification: _____ ×

4. **Optional Activity**

View and label histology slides of the lymphatic system at > MasteringA&P® > Study Area > Practice Anatomy Lab > Histology > Lymphatic System

ACTIVITY 3

Tracing the Flow of Lymph through the Body

Learning Outcomes

1. Describe the formation and composition of lymph.
2. Trace the pathway of lymph from the site where it is produced until it enters the venous circulation.

Materials Needed

- ☐ Anatomical models (torso, head and neck, intestinal villi)
- ☐ Anatomical charts showing lymphatic vessels, lymph trunks, and lymph ducts
- ☐ Whiteboard or laminated poster board
- ☐ Dry erase markers

Instructions

1. Based on what you have learned about the exchange of materials at the capillary bed, sketch and label a blood capillary and its associated structures on a whiteboard: arteriole end of the capillary, venule end of the capillary, body cells, and lymphatic capillary.

2. Using your sketch, explain the role that hydrostatic pressure and osmotic pressure have in the exchange of materials at a capillary bed.

   Why does excess tissue fluid accumulate around body cells?

   ____

   ____

   What happens to excess tissue fluid? ______________________________

   ____

   When excess tissue fluid enters a lymphatic capillary, it becomes ________________. *Lymphatic capillaries fuse to become larger lymphatic* ________________.

3. Identify inguinal lymph nodes along lymphatic collecting vessels on a torso model or anatomical chart.

   What is the purpose of the lymph nodes? ______________________________

   ____

4. Identify the major lymph trunks on an anatomical chart. Use your textbook to determine which area of the body is drained by each of the following trunks:

   a. Jugular trunks ______________________________

   b. Subclavian trunks ______________________________

   c. Bronchomediastinal trunks ______________________________

   ____

   d. Intestinal trunk ______________________________

   ____

   e. Lumbar trunks ______________________________

   ____

5. Identify the right lymphatic duct and the thoracic duct on an anatomical chart. Which areas of the body are drained by the:

   a. right lymphatic duct? ______________________________

   ____

   b. thoracic duct? ______________________________

   ____

6. The right lymphatic duct empties into venous circulation at the junction between the right internal jugular vein and the right subclavian vein, whereas the thoracic duct empties into the blood at the junction between the left internal jugular vein and the left subclavian vein.

   a. Identify each of these blood vessels on a torso model and note the junction between them on each side of the body.

   b. Identify the two junctions of pairs of veins at which the right lymphatic and thoracic ducts empty into the venous circulation, and identify the junctions of those veins on a torso model.

   c. The thoracic duct begins at the level of vertebra L_2 as an enlarged sac called the cisterna chyli.

   Which lymph trunks unite to form the cisterna chyli?

   ____

d. The thoracic duct ascends from the cisterna chyli along the left side of the vertebral column next to the aorta and pierces the diaphragm.
From which three lymph trunks does the thoracic duct receive lymph at the level of the clavicle? ____________

____

7. Fill in the blanks to trace the pathway of lymph:

Lymph enters a ______________________ at a capillary bed. It then drains into larger lymph ______________________. Positioned along these structures are lymph ______________________ that function to ______________________. Lymph collecting vessels unite to form lymph ______________________, which then fuse to form one of two lymph ______________________. The ______________________ duct drains most of the body, whereas the ______________________ duct drains the right side of the head and neck, right shoulder, and right upper extremity. Lymph ducts return lymph to the bloodstream at the junction of the __________________________ and the __________________________ on each side of the body.

An Overview of Immunity

The immune response consists of innate (nonspecific) immunity and adaptive (specific) immunity. You are born with innate immunity; you need not be exposed to any particular foreign protein (antigen) before innate responses operate. Adaptive immunity, by contrast, requires an exposure to a specific antigen.

The body's immune response can be seen to consist of several lines of defense against threats of cellular injury (Figure 25-9). The **first line of defense** is the body's surface barriers, such as intact skin and mucosae. Also in the first line of defense are secretions these barriers produce, such as pathogen-trapping mucus produced by mucous cells in the respiratory epithelia, and hydrochloric acid (HCl) and enzymes produced by specialized stomach cells.

If the first line of defense is breached, several different mechanisms in the **second line of defense**—phagocytosis, the inflammatory response, protective proteins, and fever—come into play. These mechanisms are also elements of innate immunity. In **phagocytosis**, pathogens are engulfed by phagocytes (macrophages and neutrophils). During the **inflammatory response**, chemicals released from damaged cells attract phagocytes, cause vasodilation (leading to redness and warmth), and produce increased permeability of capillaries (leading to swelling and pain). Once activated, protective proteins such as the approximately 20 plasma proteins of the **complement system** enhance both phagocytosis and the inflammatory response and cause the lysis of foreign cells. When exposed to pathogens, certain leukocytes produce chemicals known as endogenous pyrogens (literally, "fire makers within"), which reach the hypothalamus through the bloodstream and there reset the body's "thermostat," causing the rise in body temperature we know as **fever**.

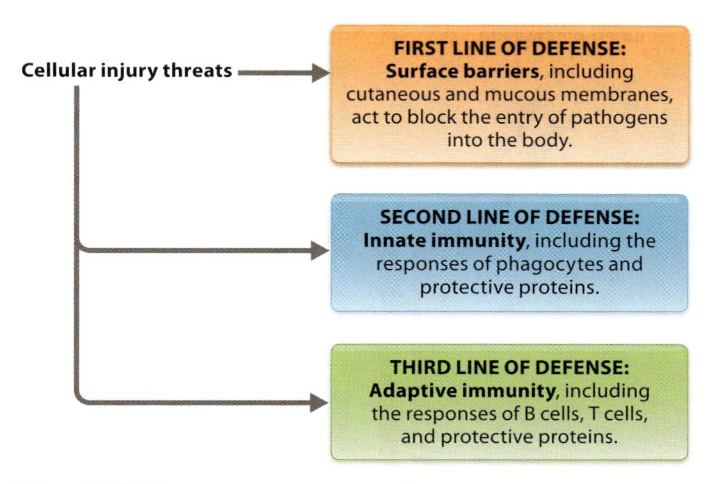

Figure 25-9 The body's three lines of defense.

The **third line of defense**—adaptive (specific) immunity—involves the responses of B cells and T cells and the proteins they produce. Both B cells and T cells are produced in bone marrow, and B cells also mature there; T cells must migrate to the thymus to mature. In those sites B cells and T cells gain immunocompetence, the ability to launch a specific immune response. Immunocompetent B cells and T cells then enter the bloodstream and travel to lymph nodes, the spleen, and tonsils, where they become activated upon encountering specific antigens on foreign cells. The plasma membranes of most of our body cells contain a variety of protein molecules including a group of glycoproteins called major histocompatibility complex (MHC) proteins. These antigens, that are not foreign to you, are called self antigens and normally do not trigger an immune response. When an immune response is directed against self antigens, autoimmune disorders such as type 1 diabetes, rheumatoid arthritis, and multiple sclerosis can result.

There are two types of adaptive immunity: cell-mediated immunity involving T cells and antibody-mediated (or humoral) immunity involving B cells. In **cell-mediated immunity**, T cells differentiate into four populations of cells: helper T cells, which aid other immune cells; cytotoxic (killer) T cells, which directly attack and destroy foreign cells; memory T cells, which "remember" the specific antigen involved in case it is encountered again; and regulatory T cells, which inhibit the immune response. In **antibody-mediated immunity**, B cells become activated, undergo a process called clonal selection, and then differentiate into populations of either plasma cells, which secrete antibodies,

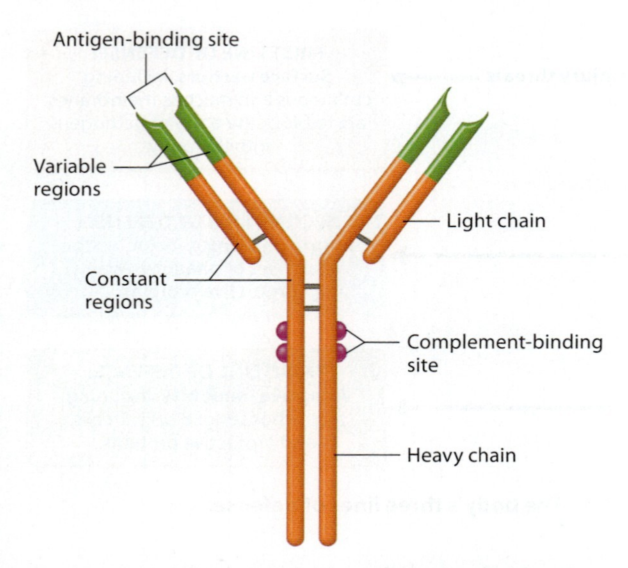

Figure 25-10 The basic structure of an antibody.

or memory B cells. Antibodies both bind antigens into clumps, which facilitates phagocytosis, and activate the complement system of the second line of defense.

All antibodies have one or more Y-shaped structures consisting of four polypeptide chains: two large heavy chains and two small light chains (**Figure 25-10**). Each chain has a constant region and a variable region, the latter of which contains an antigen-binding site with a unique shape that binds to a specific antigen in a way often likened to the fit of a lock and key. For each type of antigen that enters the body, there is a specific unique antibody to neutralize it. Binding of an antibody to an antigen immobilizes the antigen until it can be phagocytized or inactivated by the complement system. The antibody–antigen reaction is commonly used as a diagnostic tool, as you saw in Unit 20 when you identified blood types.

ACTIVITY 4

Using a Pregnancy Test to Demonstrate Antigen–Antibody Reactions

Learning Outcomes

1. Perform an enzyme-linked immunosorbent assay (ELISA) to detect the presence of human chorionic gonadotropin (hCG) in a urine sample.
2. Explain the roles of the primary and secondary antibodies and the enzyme substrate in an ELISA.
3. Perform a typical over-the-counter pregnancy test and explain how it works.

Materials Needed

- ☐ Simulated urine samples (one per student)
- ☐ Positive control (purified hCG)
- ☐ Negative control (urine from a nonpregnant woman)
- ☐ Primary antibody (PA)
- ☐ Horseradish peroxidase–labeled secondary antibody (SA)
- ☐ Enzyme substrate
- ☐ Wash buffer
- ☐ Plastic disposable transfer pipet
- ☐ 50 μl fixed-volume micropipet and disposable pipet tips
- ☐ 12-well ELISA microplate strip
- ☐ Sharpie marker
- ☐ Stack of 30 paper towels
- ☐ Watch or stopwatch
- ☐ Two pregnancy test strips
- ☐ Simulated urine – sample A
- ☐ Simulated urine – sample B

Instructions

A. Enzyme-Linked Immunosorbent Assay (ELISA)

1. Work in pairs to perform the procedure outlined in **Figure 25-11**.
2. Record results (+ or −) in the following table.

Well	1	2	3	4	5	6	7	8	9	10	11	12
Result												

3. Work with your lab group to answer the following question set:
   a. To which molecule does the PA in this procedure bind? ____________________
   b. To which molecule does the SA in this procedure bind? ____________________
   c. To which molecule is the enzyme bound? ________
   ____
   d. On which molecule does the HRP act? ________
   e. What is the purpose of the enzyme substrate in this procedure? ____________________
   ____
   f. Why are the wash steps so critical to this procedure?
   ____
   ____

B. Pregnancy Test

Over-the-counter pregnancy tests are based on the same principles as the ELISA procedure. The wick area of the dipstick is coated with anti-hCG antibodies that are labeled with a pink compound. When the dipstick is dipped in urine, any hCG present in the urine will bind with the anti-hCG antibodies. The pink antibody–antigen complexes migrate up the strip by capillary action. When the pink complexes reach the test zone (which contains fixed, unlabeled anti-hCG antibodies), the pink complexes bind and concentrate there, forming a pink stripe. In addition, the strip has a control zone containing

(1) Obtain a simulated urine sample from your instructor.

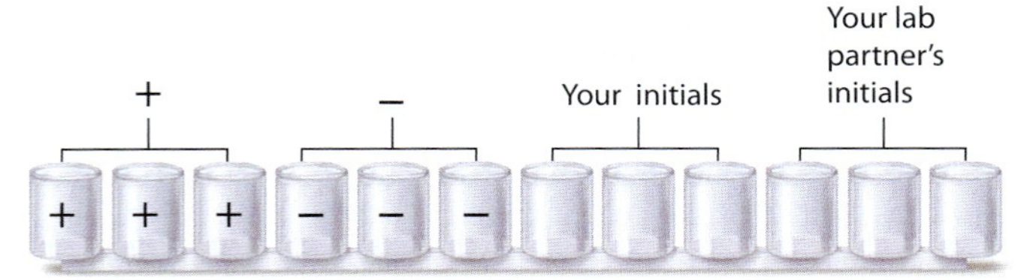

(2) Label a 12-well ELISA microplate strip as shown.

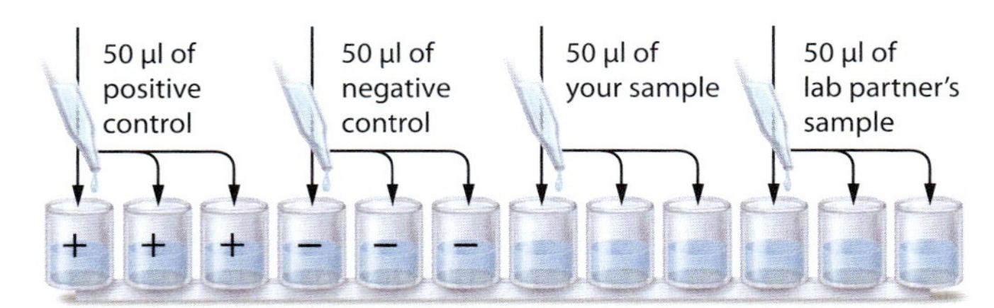

(3) Use a clean pipet to transfer samples into the wells as shown.

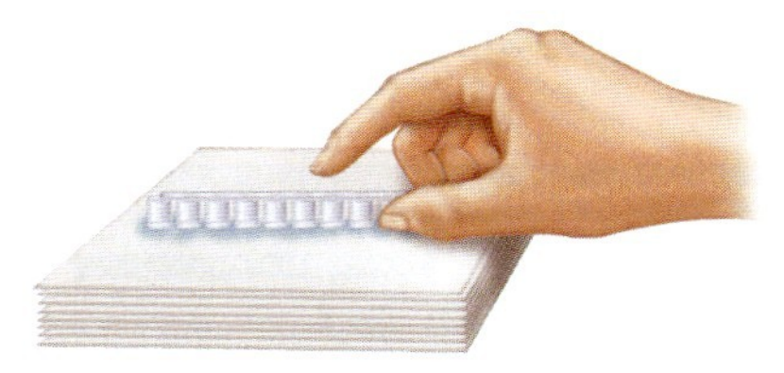

(4) Wait five minutes for proteins in the samples to bind to the wells, then turn the microplate upside down onto a stack of paper towels and gently tap three times. Discard the top paper towel.

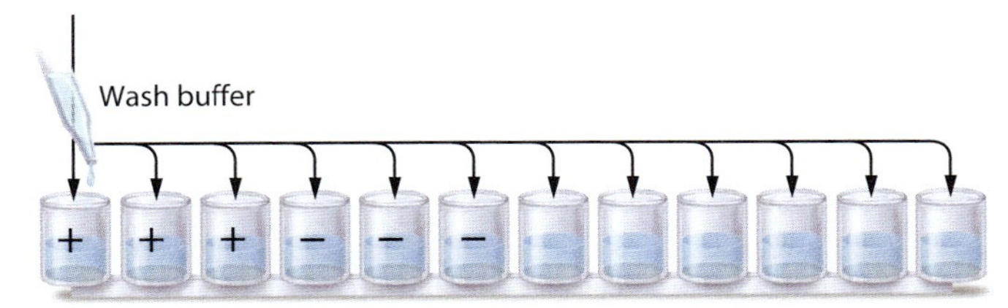

(5) Use a plastic pipet to carefully fill each well with wash buffer. Be careful not to spill any buffer from one well into neighboring wells.

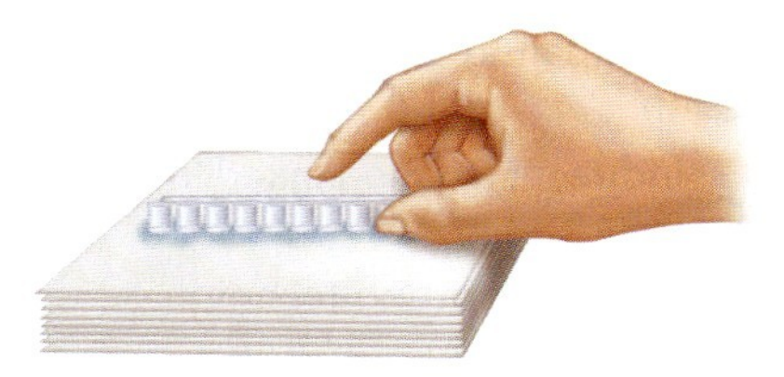

(6) Turn the microplate upside down onto stack a of paper towels and gently tap three times. Discard the top paper towel.

(7) Repeat wash steps 5 and 6.

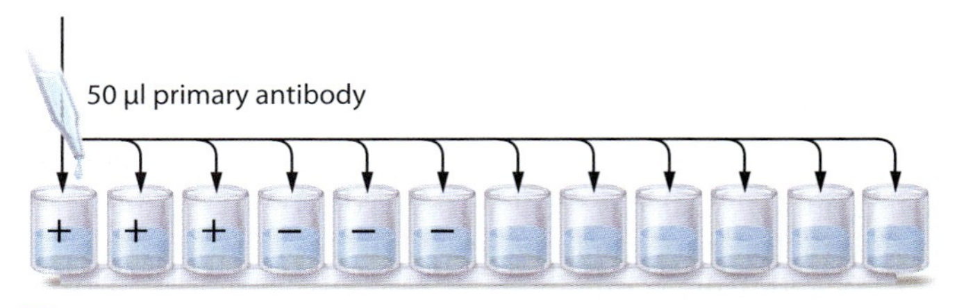

(8) Use a clean pipet tip to transfer 50 µl of primary antibody (PA) into all 12 wells. The PA is an antibody that binds to hCG (the antigen in this test).

(9) Wait 5 minutes to allow the antibodies to bind to their targets (the antigens). Then wash any unbound PA out of the well by repeating wash steps 5 and 6 twice.

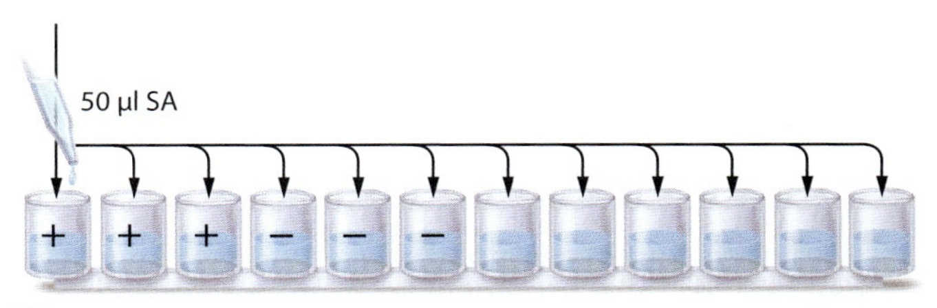

(10) Use a clean pipet tip to transfer 50 µl of horseradish peroxidase-labeled secondary antibody (SA) into all 12 wells. The SA recognizes and binds to the PA. Horseradish peroxidase (HRP) is an enzyme that oxidizes a color-producing enzyme substrate.

(11) Wait 5 minutes to allow the SAs to bind to their targets (the PAs). Then wash any unbound SA out of the well by repeating wash steps 5 and 6 *three times*.

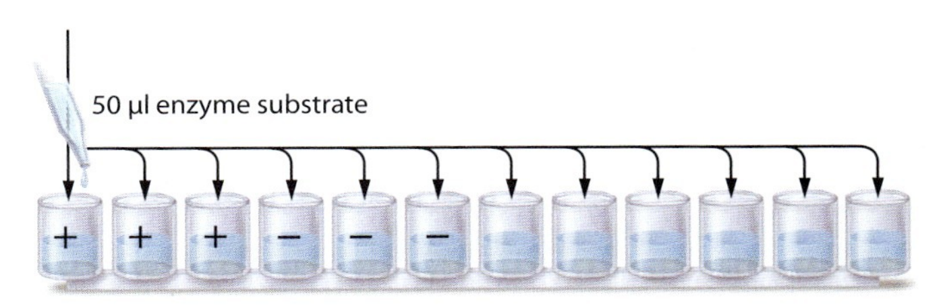

(12) Use a clean pipet tip to transfer 50 µl of enzyme substrate into all 12 wells. If HRP is present, it will oxidize the enzyme substrate.

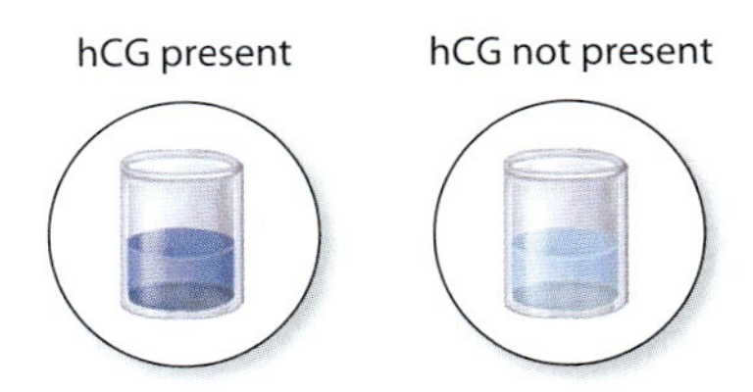

(13) Wait 5 minutes. If HRP is present (meaning that the hCG antigen is present in the sample), then the solution in the wells will turn blue. If the hCG antigen is not present, the solution will remain colorless.

Figure 25-11 ELISA procedure.

fixed, unlabeled secondary antibodies that bind any unbound pink complexes (present in both positive and negative samples); this results in the formation of a second pink stripe. As a result, every valid pregnancy test shows this second pink stripe, but only positive test results have two pink stripes.

Follow these instructions to perform a pregnancy test on a simulated urine sample:

1. Go to the demonstration area. Dip a new pregnancy test strip into simulated urine sample A and analyze the results based on the instructions provided.

   Does urine sample A test negative or positive for hCG?

   ____

2. Dip a new pregnancy test strip into simulated urine sample B and analyze the results.

   Does urine sample B test negative or positive for hCG?

   ____

3. Work with your lab group to answer the following question set:

   a. Label the components of a pregnancy test strip in the accompanying illustration. Note that although the pregnancy test strips you used for this procedure might not look exactly like the strip illustrated, the mechanism by which they work is very similar.

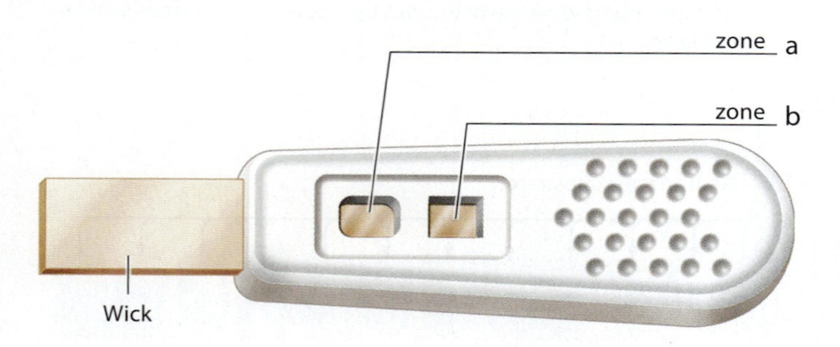

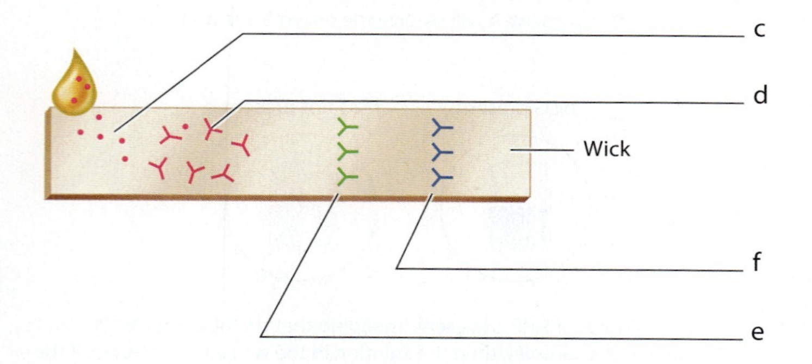

   b. What is the purpose of the control band on the pregnancy test strip? ______________________________

   ____

   ____

PhysioEx Exercise 12
Serological Testing

The PhysioEx 9.1 Laboratory Simulations in Physiology are easy-to-use laboratory simulations that can be used as an alternative to or as a supplement to wet lab activities in this unit. Each simulation allows you to investigate important physiological concepts, repeat labs as often as you like, and conduct experiments that are difficult to perform in a wet lab environment because of time, cost, or safety concerns.

Access the simulations in these activities at MasteringA&P® > Study Area > PhysioEx 9.1

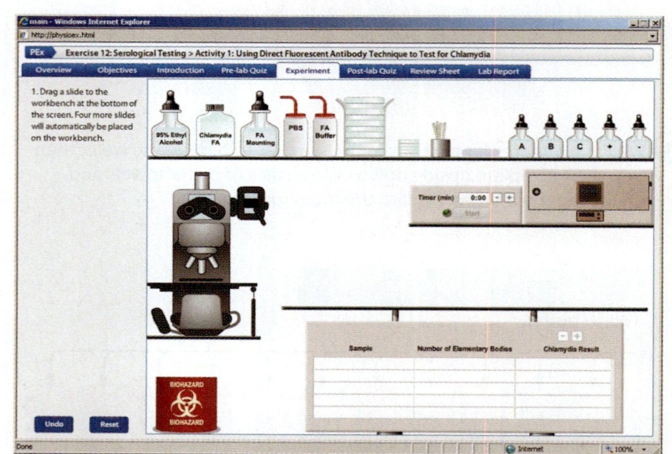

There you will find the following activities:

PEx Activity 1: Using Direct Fluorescent Antibody Technique to Test for Chlamydia

PEx Activity 2: Comparing Samples with Ouchterlony Double Diffusion

PEx Activity 3: Indirect Enzyme-Linked Immunosorbent Assay (ELISA)

PEx Activity 4: Western Blotting Technique

Post-lab quizzes are also assignable in MasteringA&P®

POST-LAB ASSIGNMENTS

Name: ______________________ Date: ______________ Lab Section: ________

PART I. Check Your Understanding

Activity 1: Exploring the Organs of the Lymphatic System

1. A lymph node filters ______________; the spleen filters ______________.
2. Lymph nodes are clustered in which three regions of the body? ______________

   ____
3. What happens to the thymus as individuals age? ______________
4. Name the unpaired tonsil located in the nasopharynx: ______________
5. Name the lymphoid tissue that extends from the cecum: ______________
6. What is MALT? ______________

Activity 2: Examining the Histology of a Lymph Node, a Tonsil, and the Spleen

1. Identify the organ in the accompanying photomicrograph and name the indicated structures.

   Organ: ______________

   a. ______________

   b. ______________

   c. ______________

   d. ______________

   e. ______________

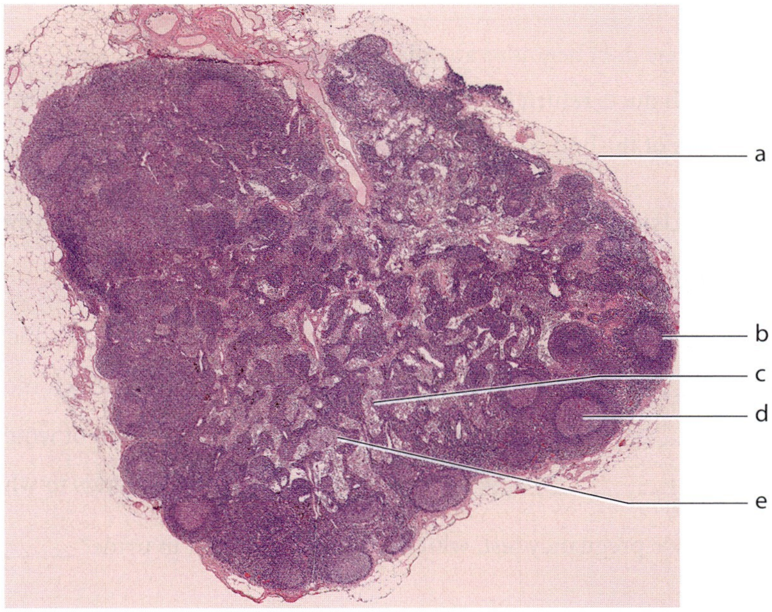

a

b

c

d

e

LM (20×)

2. How are a lymph node and a tonsil different with respect to structure?

   ____

   ____

3. Identify the organ in the accompanying photomicrograph and name the indicated structures:

   Organ: ______________________

   a. ______________________

   b. ______________________

   c. ______________________

a
b
c

LM (100×)

Activity 3: Tracing the Flow of Lymph through the Body

1. The smallest of the lymphatic vessels, ______________________, are blind-ended sacs that take in excess tissue fluid at capillary beds and transport it to larger lymph ______________________.
2. Name the four paired lymph trunks: ______________________
   ____
3. The thoracic duct begins as an enlarged sac called the ______________________.
4. The lymph ducts return lymph to the venous circulation at the junction of which two veins on either side of the body? ______________________

Activity 4: Using a Pregnancy Test to Demonstrate Antigen–Antibody Reactions

1. In an ELISA,

   ______________ a. the primary antibody binds to which molecule?

   ______________ b. the enzyme is attached to which molecule?

   ______________ c. the final step is the addition of which molecule?

   ______________ d. the secondary antibody binds to which molecule?

2. In a positive pregnancy test, which antigen is detected in urine? ______________________

PART II. Putting It All Together

A. Review Questions

Answer the following questions using your lecture notes, your textbook, and your lab notes.

1. To which two organ systems does the thymus belong? ______________________

2. After Brian suffered a ruptured spleen in an automobile crash, a surgeon removed it to stop internal bleeding. Why can Brian survive without his spleen? ____

____

3. The third line of defense is ________________ immunity, which consists of two different mechanisms: ________________-mediated immunity carried out by B cells and ________________-mediated immunity carried out by T cells.

4. Lacteals are specialized capillaries found in the small intestine that absorb dietary fat. Fats enter the lacteal as lipoprotein complexes called chylomicrons, which are then transported via lymphatic vessels to the bloodstream. Trace the pathway of a chylomicron from the point at which it enters a lacteal until it reaches a cell in the rectus femoris muscle.

lacteal → lymphatic ________________ vessel → lymph ________________ → ________________ → left subclavian vein → ________________ vein → superior vena cava → ________________ → ________________ → ________________ trunk → ________________ artery → lungs → ________________ vein → ________________ → ________________ → ascending aorta → ________________ → thoracic aorta → ________________ aorta → ________________ artery → ________________ artery → ________________ artery → arteriole → ________________ → cell in the rectus femoris

5. Compare and contrast the following lymphoid organs by completing each of the following charts:

Characteristics Unique to a Lymph Node	**Characteristics Shared by Both a Lymph Node and a Tonsil**	**Characteristics Unique to a Tonsil**

Characteristics Unique to the Thymus	**Characteristics Shared by Both the Thymus and Spleen**	**Characteristics Unique to the Spleen**

B. Concept Mapping

1. Fill in the blanks to complete this concept map outlining components of the lymphatic system.

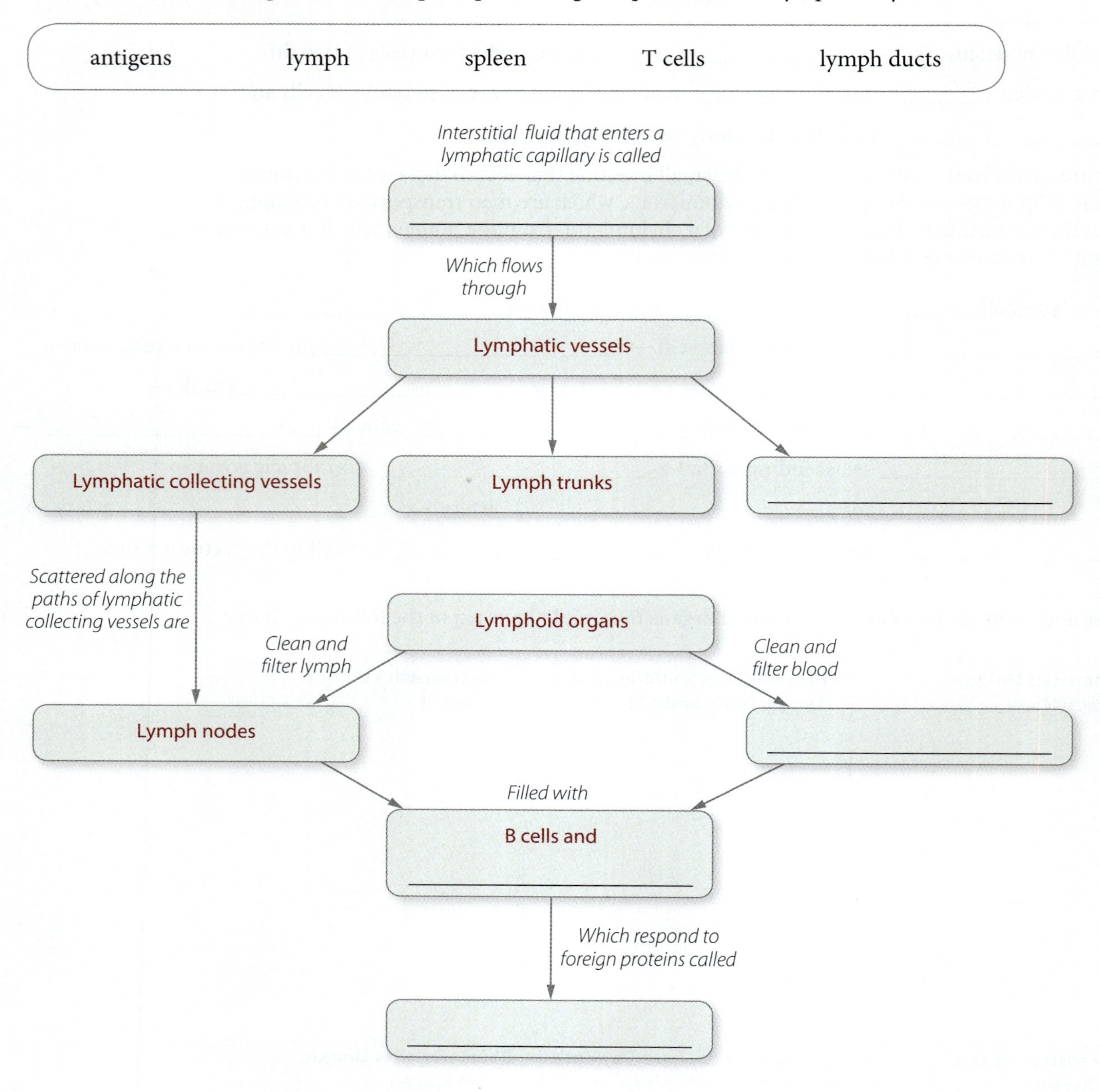

2. Construct a unit concept map to show the relationships among the following set of terms. Include all of the terms in your map. Your instructor may choose to assign additional terms.

antibody	antigens	appendix	B cells	cell-mediated immunity
complement	inflammation	lymph	lymph ducts	lymphocyte
macrophages	Peyer's patches	spleen	T cells	tonsil

26

Anatomy of the Respiratory System

UNIT OUTLINE

The respiratory system provides the body with oxygen and rids it of carbon dioxide. It consists of two parts: the upper respiratory tract (nasal cavity, pharynx, and larynx) and the lower respiratory tract (trachea, bronchial tree, and lungs). In this unit you will study the anatomy of these structures; in Unit 27 you will explore respiratory physiology.

THINK ABOUT IT *Name a respiratory organ in which each of the following tissue types is predominant. Then, describe a function of the tissue type.*

Simple squamous epithelium ____

____

Hyaline cartilage ____

____

Ciliated pseudostratified columnar epithelium ______________________________

____

Smooth muscle ____

____ ■

PRE-LAB ASSIGNMENTS

Pre-lab quizzes are also assignable in MasteringA&P®

To maximize learning, BEFORE your lab period carefully read this entire lab unit and complete these pre-lab assignments using your textbook, lecture notes, and prior knowledge.

PRE-LAB Activity 1: Exploring the Organs of the Respiratory System

1. Use the list of terms provided to label the accompanying figure showing the structures of the respiratory system. Check off each term as you label it.

☐ bronchi	☐ left lung	☐ alveoli	☐ trachea
☐ pharynx	☐ diaphragm	☐ larynx	☐ lower respiratory tract
☐ bronchiole	☐ nasal cavity	☐ right lung	☐ upper respiratory tract

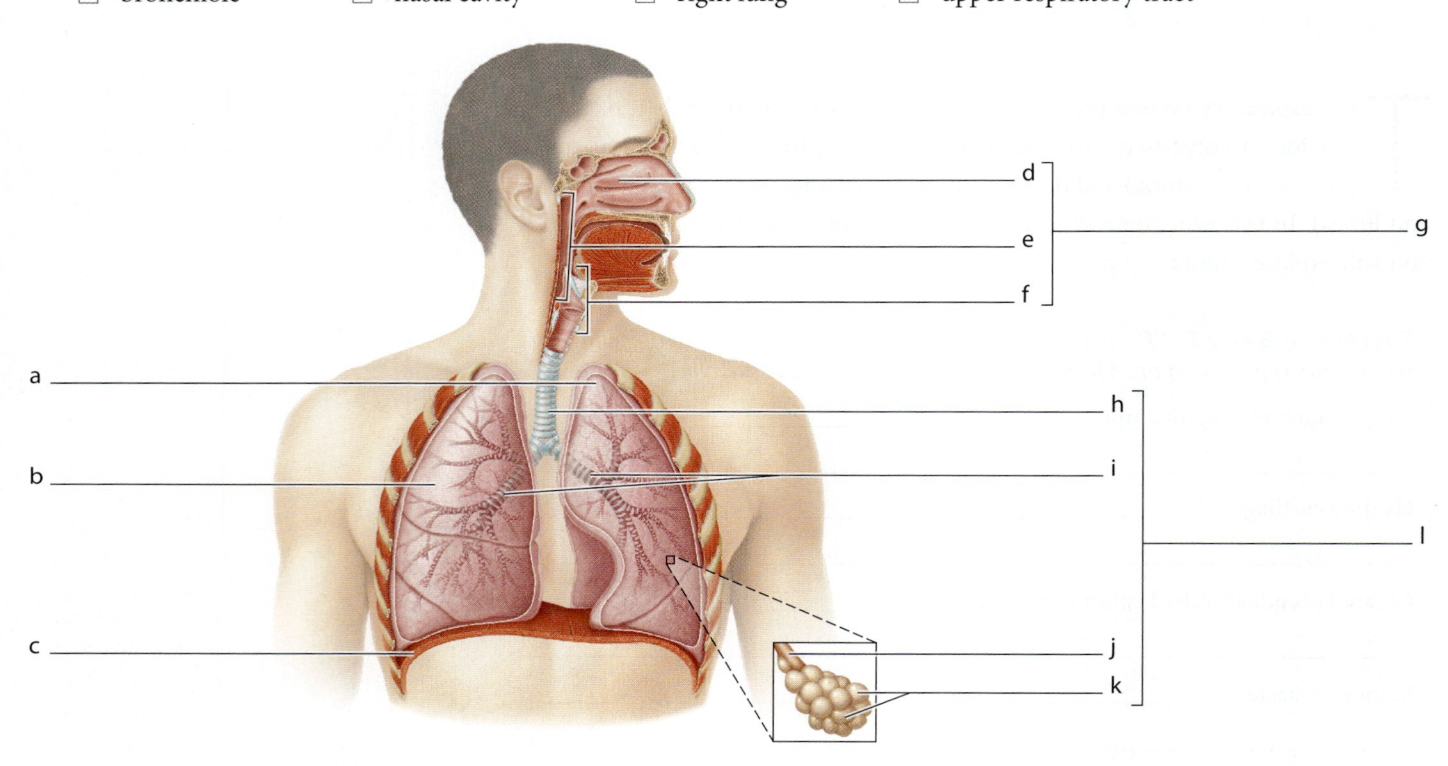

2. Match each of the following respiratory organs with its correct description.

_____	a. larynx	1. is a common passageway for respiratory and digestive systems
_____	b. nasal cavity	2. is composed of simple squamous epithelium
_____	c. bronchus	3. consists of two lobes
_____	d. pharynx	4. is composed of nine cartilaginous structures
_____	e. bronchiole	5. is innervated by the phrenic nerve
_____	f. right lung	6. leads into an alveolar duct
_____	g. alveolus	7. is divided into passageways called meatuses
_____	h. left lung	8. consists of three lobes
_____	i. diaphragm	9. branches from trachea

PRE-LAB Activity 2: Examining the Microscopic Anatomy of the Trachea and Lungs

1. Match each of the following structures with a predominant tissue type found in it.

_____	a. supporting rings of trachea	1. ciliated pseudostratified columnar epithelium
_____	b. air sacs of lungs	2. simple squamous epithelium
_____	c. inner lining of trachea	3. smooth muscle
_____	d. submucosa of trachea	4. areolar connective tissue
_____	e. arterial wall	5. hyaline cartilage

2. How is the structure of the alveolus adapted to its function? ______________________________

____

3. How is the structure of the trachea adapted to its function? ______________________________

____

PRE-LAB Activity 3: Examining a Sheep Pluck

1. What is a sheep pluck? ____

____

2. How will you inflate the lungs of the sheep pluck? ______________________________

____

Functional Anatomy of the Respiratory System

We will divide our study of the functional anatomy of the respiratory system into two parts: the upper respiratory tract and the lower respiratory tract.

The Upper Respiratory Tract

The upper respiratory tract (**Figure 26-1**) consists of the nasal cavity, the pharynx, and the larynx. Air typically enters the nasal cavity through the **anterior nares** (nostrils). The nasal cavity (**Figure 26-2**) is divided into right and left sides by the **nasal septum**, a structure composed of both bone and cartilage (not visible in Figure 26-2). Three bony shelves—the **superior**, **middle**, and **inferior nasal conchae**—project from the lateral walls of the mucosa-lined nasal cavity and help to increase air turbulence, which aids in the warming, moistening, and filtering of incoming air.

The nasal cavity is separated from the oral cavity by the anterior **hard palate** and the posterior **soft palate**. When the soft palate is elevated during swallowing, an extension of it, called the **uvula**, helps to prevents food and liquids from entering the nasal cavity by closing off the **posterior nares**, the two posterior openings of the nasal cavity that lead to the pharynx (see Figure 26-2). **Paranasal sinuses**—air-filled cavities in the frontal, ethmoid, sphenoid, and maxillary bones that are continuous with the nasal cavity and are lined with mucous membranes—surround the nasal cavity. These sinuses reduce the weight of the skull, function as resonance chambers for speech, and warm and moisten incoming air.

The pharynx is divided into three regions: the **nasopharynx**, which begins at the posterior nares and ends at the soft palate; the **oropharynx**, which begins at the soft palate and ends at the superior margin of the epiglottis; and the **laryngopharynx**, which begins at the superior margin of the epiglottis and ends at the cricoid cartilage of the larynx, where it becomes continuous with the esophagus

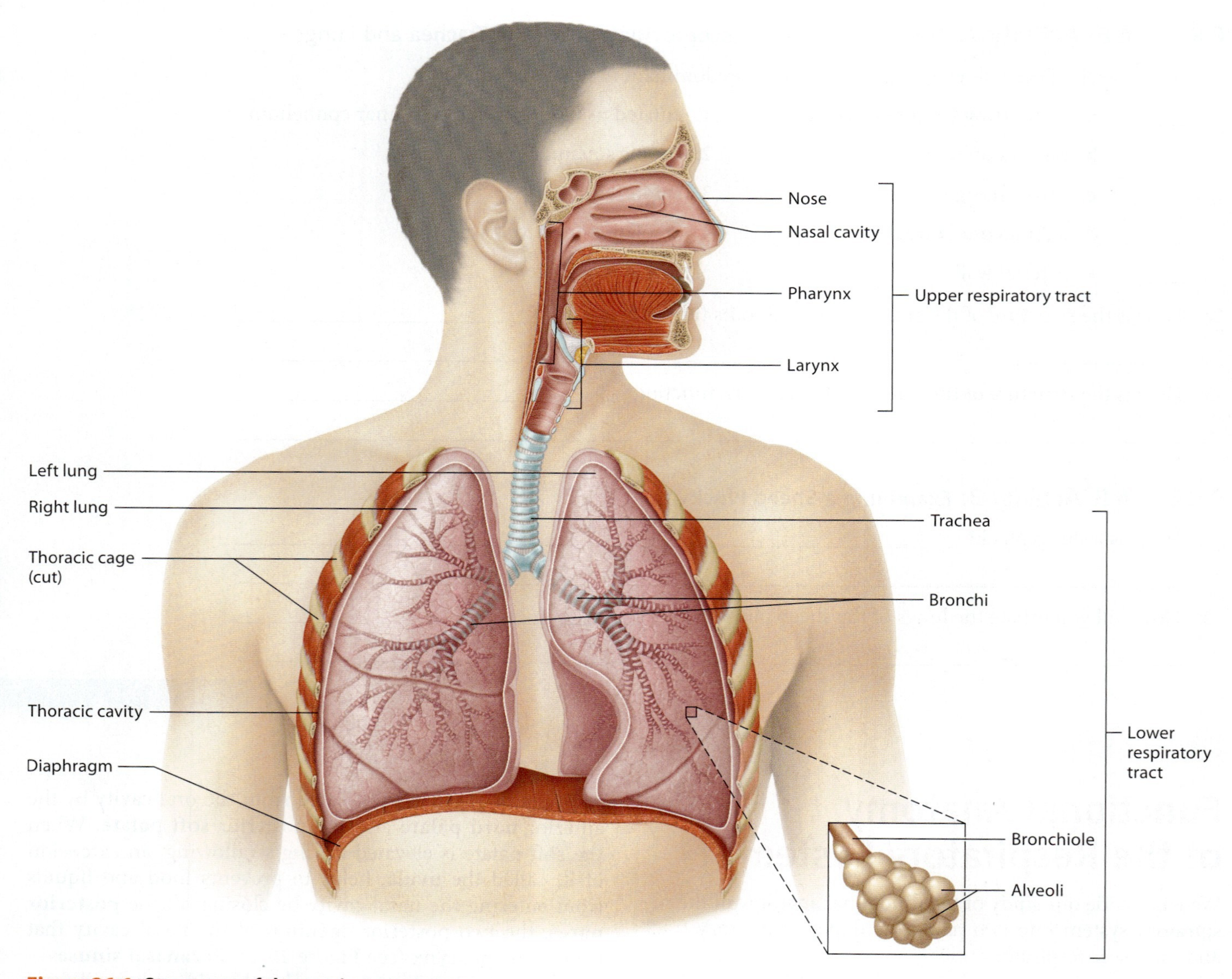

Figure 26-1 Structures of the respiratory system.

(see Figure 26-2). The pharynx conducts air to the larynx and food and fluids to the esophagus.

The **larynx**, which is located inferior to the laryngopharynx, consists of three paired cartilages and three single cartilages held together by laryngeal ligaments (**Figure 26-3**). All of these cartilaginous structures except the epiglottis are composed of hyaline cartilage. The three smaller, paired cartilages—the **arytenoid**, **cuneiform**, and **corniculate cartilages**—are located in the posterior and lateral walls of the larynx. The three single cartilages form the body of the larynx. The largest—the **thyroid cartilage**—contains an anterior laryngeal prominence commonly referred to as the Adam's apple. The **cricoid cartilage**, located inferior to the thyroid cartilage, is ring shaped; it is the only cartilage in the larynx or trachea that forms a complete ring. During swallowing, the **epiglottis**, a flap-like structure composed of elastic cartilage, forms a lid over the **glottis**, the slit-like opening that allows air into the larynx. The epiglottis is often referred to as the "guardian of the airways." Lateral to the glottis are two pairs of folds commonly called the "vocal cords." The superior **vestibular folds**, known as the false vocal cords, prevent foreign materials from entering the glottis; the inferior **vocal folds**, known as the true vocal cords, vibrate to produce speech.

The Lower Respiratory Tract

The lower respiratory tract (see Figure 26-1) consists of the trachea (commonly called the windpipe) and its branches, and the lungs. The **trachea** (**Figure 26-4**) is a tubular structure that conducts air from the larynx to the bronchi. The trachea is supported by C-shaped **cartilaginous rings** that keep the airway open. The mucosa of the trachea is lined with

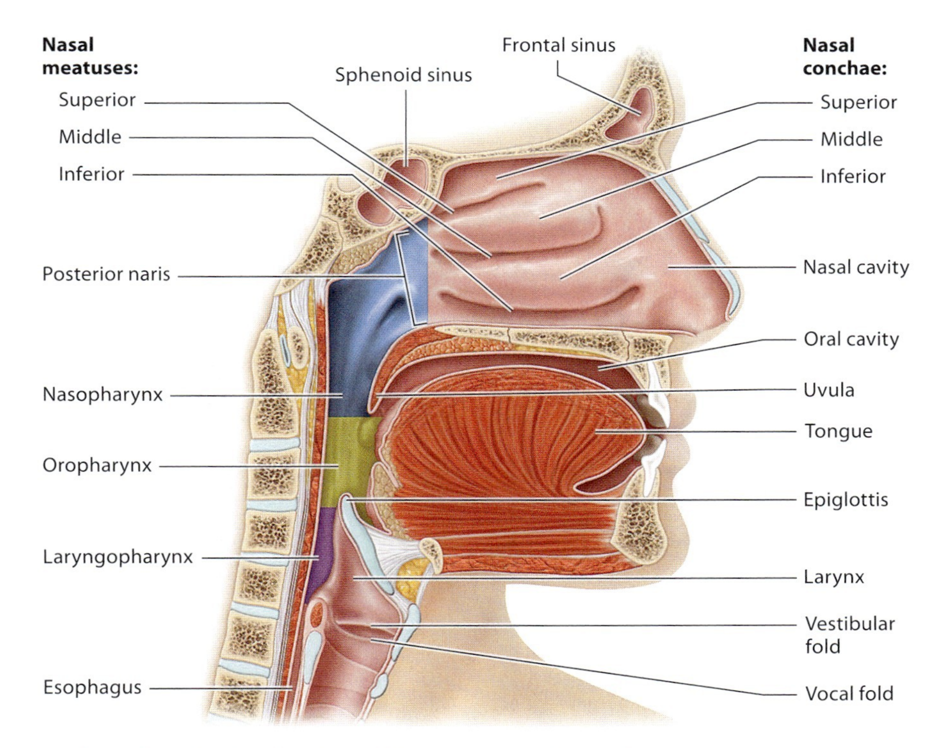

(a) Illustration showing structures of the upper respiratory system

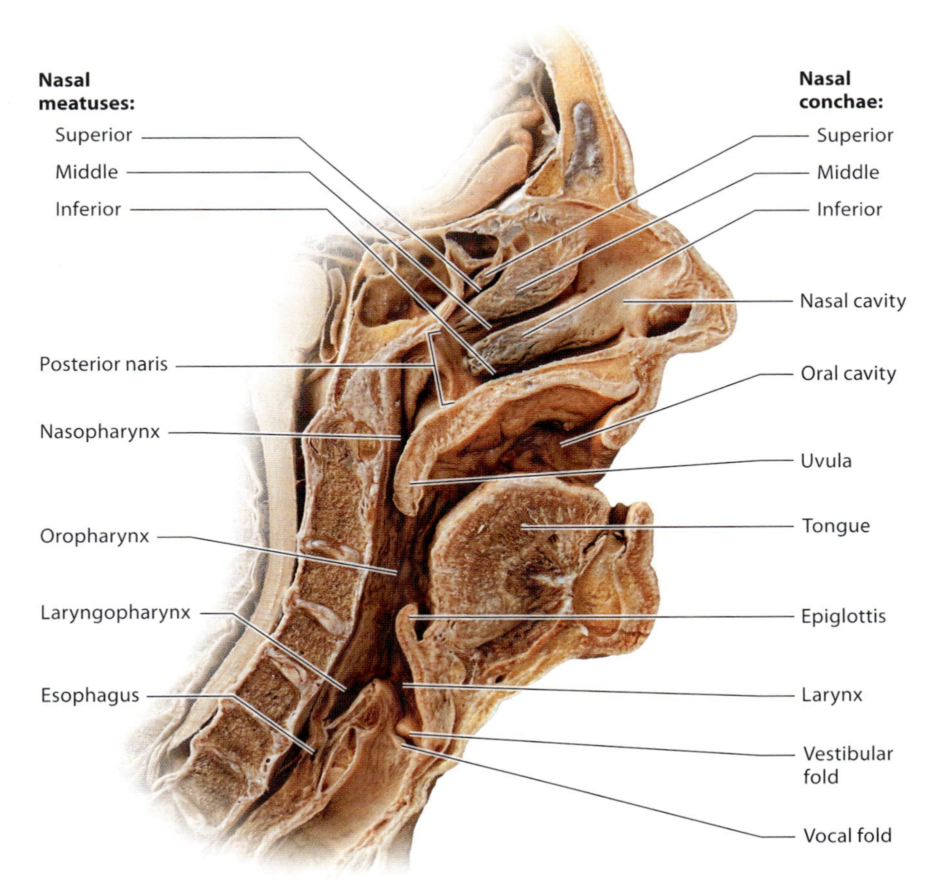

(b) Photo showing structures of the upper respiratory system

Figure 26-2 A sagittal section of the nasal cavity and the upper respiratory tract.

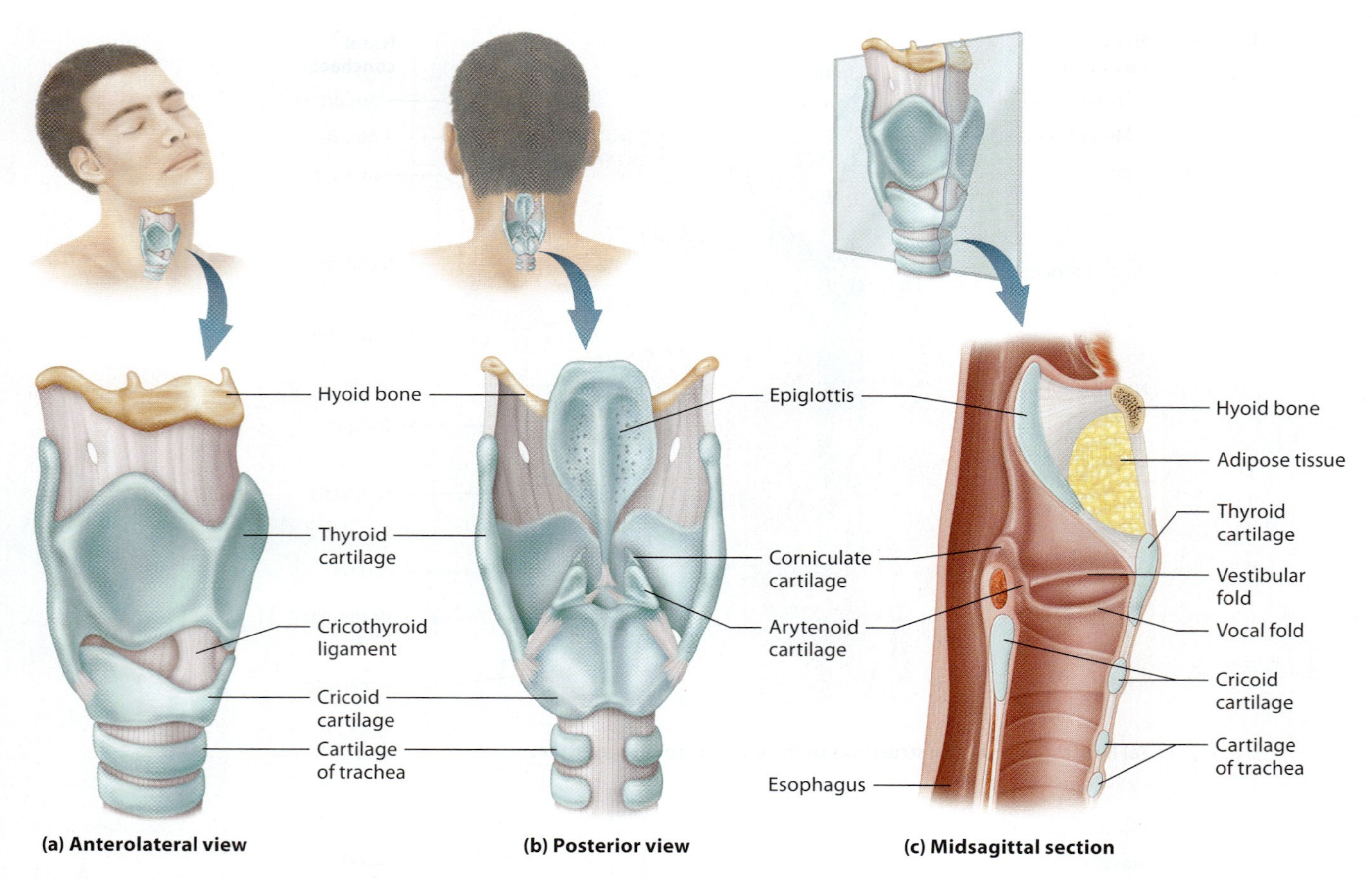

Figure 26-3 **The larynx.**

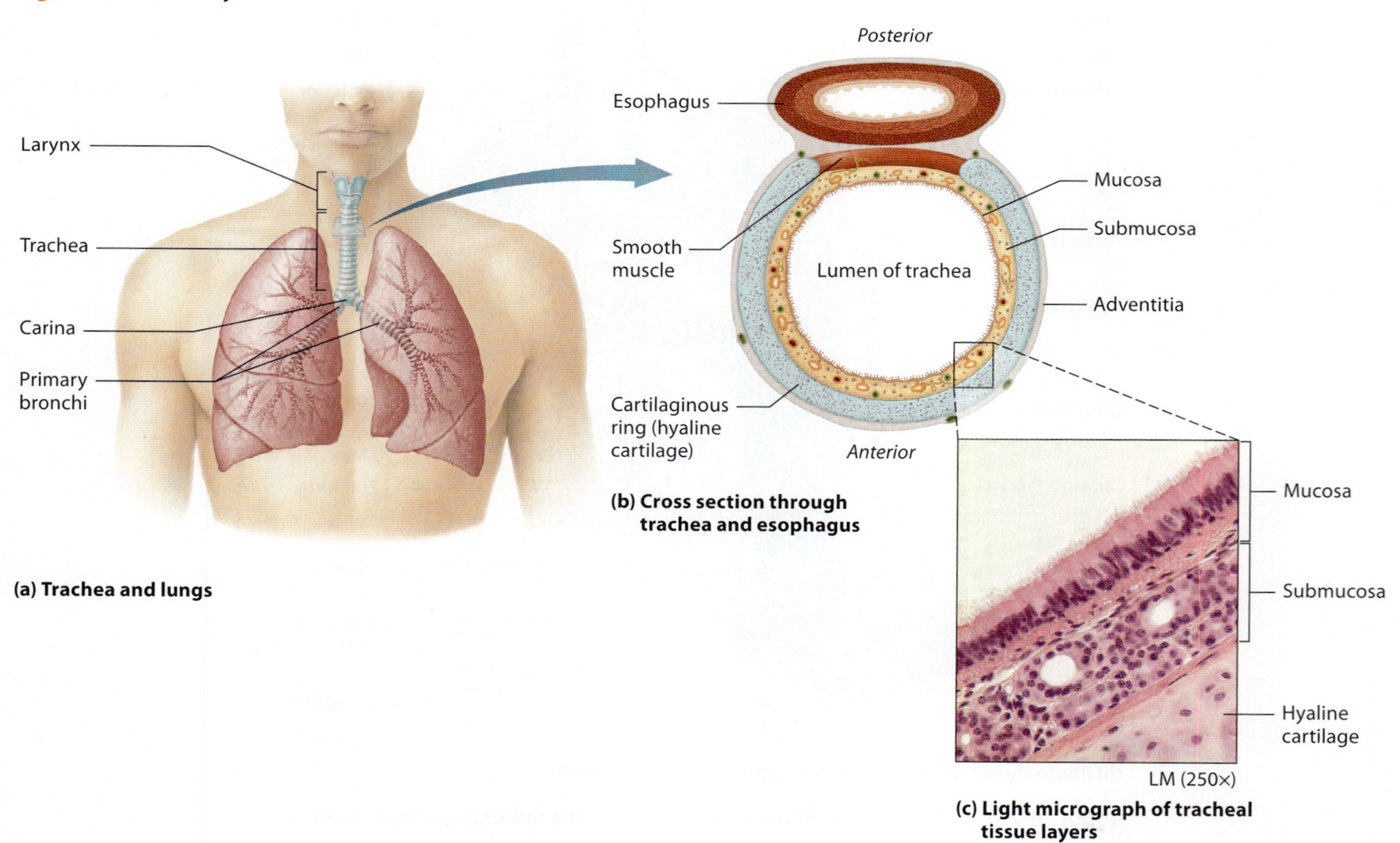

Figure 26-4 **The trachea.**

ciliated pseudostratified columnar epithelium containing many mucus-secreting goblet cells. The mucus traps dust, bacteria, and other foreign matter, and the cilia propel the mucus upward to the throat so that it can be coughed up or swallowed. At its inferior end the trachea branches at a ridge, called the carina, into the left and right **primary bronchi**, which enter the lungs.

As the primary bronchi enter the lungs, they branch into **secondary bronchi** and then into **tertiary bronchi**, forming a structure known as the bronchial tree (**Figure 26-5**). The tertiary bronchi branch repeatedly into smaller and smaller bronchi that eventually branch into **bronchioles**. Bronchioles further branch into **terminal bronchioles**, which branch into **respiratory bronchioles**.

The point at which the terminal bronchioles branch into respiratory bronchioles marks the border between the **conducting zone**, which consists of the respiratory passageways from the nasal cavity to the terminal bronchioles, and the **respiratory zone**, which also includes the alveolar ducts and alveolar sacs (**Figure 26-6**). The conducting zone is so named because it conducts air to and from the respiratory zone.

Respiratory bronchioles divide into **alveolar ducts** that lead into clusters of alveoli called **alveolar sacs.** Each alveolar sac consists of individual **alveoli** (air sacs) and their overlying pulmonary capillaries. Together the alveoli and pulmonary capillaries make up the **respiratory membrane**, where gas exchange takes place.

QUICK TIP Notice in Figure 26-6 that the alveolar duct, alveolar sacs, and alveoli look a lot like a bunch of grapes. In this comparison, think of the alveolar duct as the stem, the alveolar sac as a cluster of grapes, and the alveoli as individual grapes.

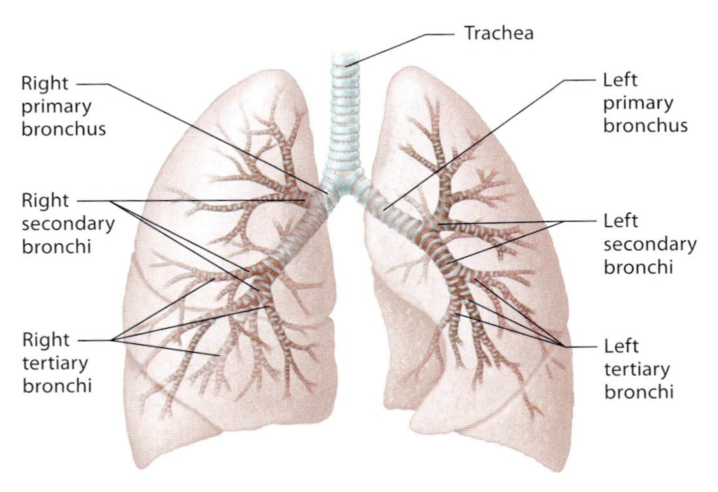

Figure 26-5 **The bronchial tree.**

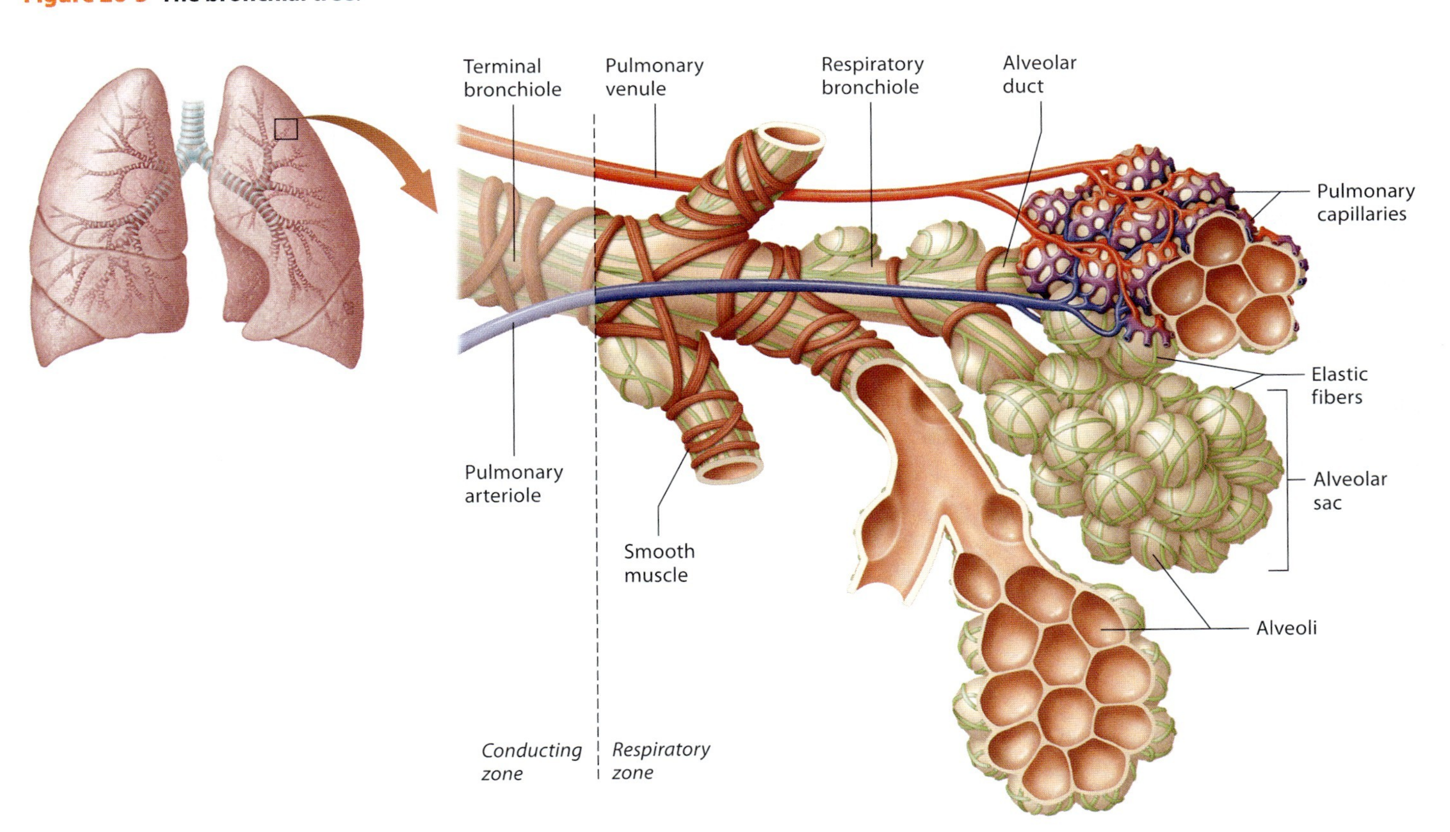

Figure 26-6 **Structures of the respiratory zone.**

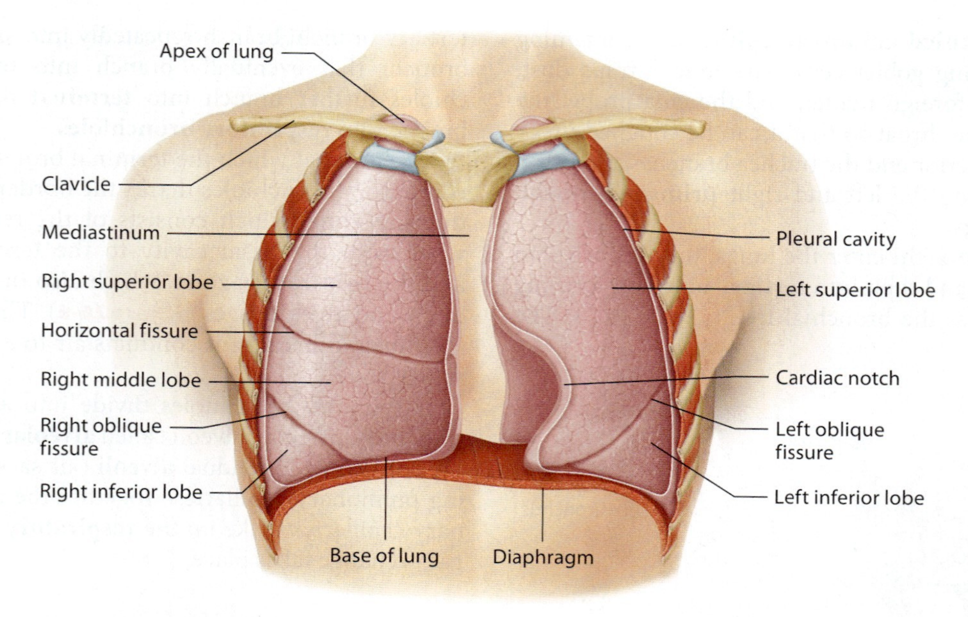

Figure 26-7 The lungs.

As the bronchial tree branches from the primary bronchi to the respiratory bronchioles, the amount of cartilage in the respiratory tubes decreases, and the amount of smooth muscle increases. The epithelium lining the respiratory tubes also changes: from pseudostratified columnar epithelium in the primary bronchi, to columnar epithelium in the smaller bronchi, to cuboidal epithelium in the bronchioles, to simple squamous epithelium in the alveoli.

The **lungs**, which are located within the thoracic cavity, are paired cone-shaped organs (**Figure 26-7**) that are surrounded by a double-layered serous membrane. The outer layer, the **parietal pleura**, is attached to the wall of the thoracic cavity and the diaphragm. The inner layer, the **visceral pleura**, covers the surface of each lung. Between these two layers is the **pleural cavity**, which is filled with a serous fluid. The right lung is divided into three lobes (the **superior**, **middle**, and **inferior lobes**) that are separated by an **oblique fissure** and a **horizontal fissure**. The left lung is divided into two lobes (the **superior** and **inferior lobes**) separated by an **oblique fissure**. The left lung also has an indention called the **cardiac notch** where the apex of the heart sits.

ACTIVITY 1
Exploring the Organs of the Respiratory System

Learning Outcomes

1. Identify the organs of the respiratory tract, and describe the functions of each organ.
2. For each respiratory organ, describe the structural adaptations that enable it to perform its functions.

Materials Needed

- ☐ Torso model or other anatomical models
- ☐ Anatomical charts

Instructions

CHART As you complete the following Making Connections chart, locate each respiratory structure on an anatomical model or chart and identify its anatomical features. Then review its description and/or function(s). Finally, make some "connections" to things you've already learned in your lectures, in your assigned reading, and in your lab. Some portions of the chart have been filled in as examples.

Optional Activity

Practice labeling respiratory organs on human cadavers at MasteringA&P® > Study Area > Practice Anatomy Lab > Human Cadaver > Respiratory System

Making Connections: Organs of the Respiratory System

Respiratory Structure	Description (Structure and/or Function)	Connections to Things I Have Already Learned
Nasal cavity • Anterior nares • Nasal septum • Nasal conchae • Posterior nares • Hard palate • Soft palate		*Nasal conchae with meatuses cause air turbulence to warm and moisten air more efficiently; olfactory nerve fibers of the olfactory nerve (CN I) pass through the olfactory foramina of the cribriform plate of the ethmoid bone.*
Paranasal sinuses • Maxillary • Sphenoid • Ethmoid • Frontal		
Pharynx • Nasopharynx • Oropharynx • Laryngopharynx		
Larynx • Epiglottis • Thyroid cartilage • Cricoid cartilage • Vestibular folds • Vocal folds	*Passageway for air; contains vocal folds, which produce sound; epiglottis prevents food and fluids from entering airways.*	
Trachea • Hyaline cartilage rings • Carina		
Bronchi • Primary • Secondary • Tertiary		*Bronchitis = inflammation of the mucous membrane lining the bronchial passages.*
Bronchiole • Terminal bronchiole • Respiratory bronchiole		

(*Continued*)

Making Connections: Organs of the Respiratory System ***(Continued)***

Respiratory Structure	Description (Structure and/or Function)	Connections to Things I Have Already Learned

Lungs
• Right lung
  • Lobes
  • Horizontal fissure
  • Oblique fissure
• Left lung
  • Lobes
  • Oblique fissure
  • Cardiac notch
	Contain alveoli; site of gas exchange.	

Blood supply • Pulmonary artery • Pulmonary capillaries • Pulmonary vein • Bronchial artery		*Pulmonary arteries branch from pulmonary trunk, which arises from RV of heart; pulmonary capillaries wrap around each alveolus to form respiratory membrane; pulmonary vein returns oxygenated blood to LA of heart; bronchial artery arises from thoracic aorta.*
Breathing muscles • Diaphragm • External intercostals • Internal intercostals		*Diaphragm is innervated by phrenic nerve; intercostal muscles are innervated by intercostal nerves, which arise from ventral rami of T_2–T_{12}.*

ACTIVITY 2

Examining the Microscopic Anatomy of the Trachea and Lungs

Learning Outcomes

1. Observe a histological section of the trachea; then identify the following structures and describe the function of each: pseudostratified columnar epithelium, cilia, goblet cell, and hyaline cartilage.
2. Observe a histological section of the lung; then identify the following structures and describe the function of each: bronchiole, alveolar duct, alveolar sac, and alveolus.

Materials Needed

☐ Microscope and slides (or photomicrographs) of trachea and lung

Instructions

1. Observe a histological slide (or a photomicrograph) of the trachea under low power. Note the distinct histological pattern that we see in many hollow organs: an innermost mucosa, a middle submucosa, and an outermost adventitia. Sketch and label these three layers of the trachea.

Total magnification: _____ ×

2. Switch to the high-power objective (or view a higher-power photomicrograph) and observe the tissues found in the mucosa and the submucosa.

   Which tissue type is predominant in the mucosa?

   ____

   Identify a goblet cell and state its function. ______________

   ____

   Identify cilia and state their function. ______________

   ____

   The epithelial tissue is separated from underlying connective tissue by which structure? ______________

   ____

   Which tissue types are prominent in the submucosa?

   ____

   ____

   What is the function of the cartilaginous tissue? ________

   ____

3. Observe a histological slide (or a photomicrograph) of the lung under low power. Identify the following structures: bronchiole, alveolar duct, alveolar sac, alveolus.

4. Switch to the high-power objective (or view a photomicrograph provided by your instructor).

   Distinguish between an alveolus and an alveolar sac.

   ____

   ____

5. Label each of the following structures on **Figure 26-8**: alveolar ducts, alveolar sac, alveolus, artery, bronchiole

6. Now, observe the lung slide using the oil-immersion objective (or view a photomicrograph provided by your instructor). Identify an alveolus.

   What is the function of an alveolus? ______________

   ____

   Which tissue type comprises the alveolar wall?

   ____

7. Note the more prominent, flattened cells found in the alveolar wall. These cells are type I alveolar cells, which serve as the sites of gas exchange in the air sac. Type I alveolar cells are also responsible for the synthesis of the enzyme ACE (angiotensin-converting-enzyme). Recall that ACE is an important enzyme in the renin-angiotensin-aldosterone pathway, which regulates blood pressure. Also located in the alveolar wall are type II alveolar cells, which produce surfactant.

   What is the function of surfactant? ______________

   ____

   Finally, note the red blood cells located within capillaries surrounding the alveoli.

8. **Optional Activity**

   PAL™ View and label histology slides of the respiratory system at MasteringA&P® > Study Area > Practice Anatomy Lab > Histology > Respiratory System

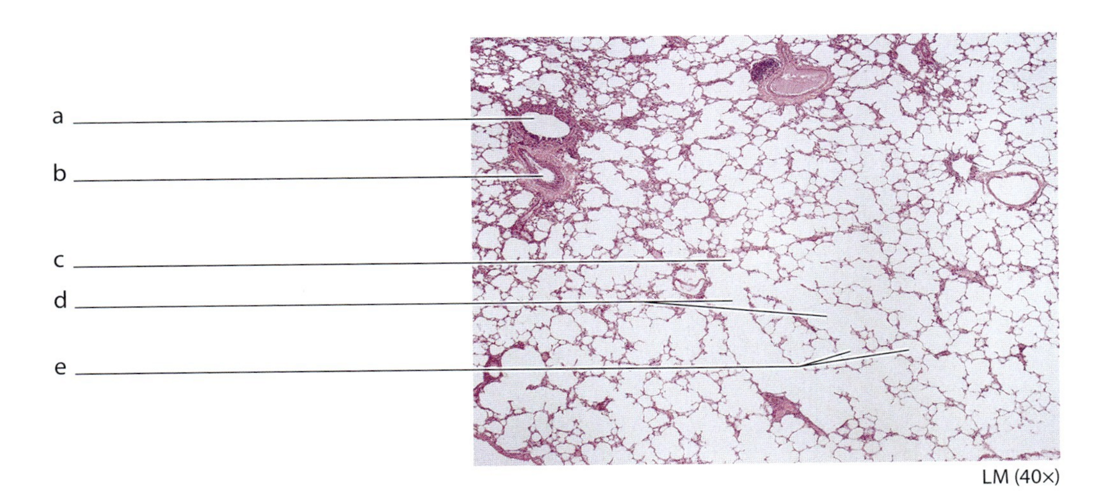

Figure 26-8 A cross-section of the lung.

ACTIVITY 3

Examining a Sheep Pluck

Learning Outcomes

1. Identify the larynx, trachea, lungs, heart, aorta, pulmonary artery, pulmonary vein, and venae cavae on a sheep pluck.
2. Demonstrate lung inflation in a sheep pluck.

Materials Needed

- ☐ Fresh or preserved sheep pluck
- ☐ Source of compressed air and an air hose
- ☐ Dissecting tray
- ☐ Disposable gloves

Instructions

A sheep pluck includes the larynx, trachea, lungs, heart, and parts of the major blood vessels (aorta, pulmonary artery, pulmonary vein, and venae cavae) located in the mediastinum (middle portion of thoracic cavity).

1. Put on disposable gloves and obtain a dissecting tray and a sheep pluck.
2. CHART Identify the organs of the sheep pluck. In the chart below, describe how the structure of each organ is related to its function. The first row has been filled in for you.
3. Feel the cartilaginous rings of the trachea. Are these rings complete or C-shaped? ____________________

   Why is this important? ____________________

   ____
4. Identify the carina, the right primary bronchus, and the left primary bronchus.
5. Identify the visceral pleura. What is the function of this serous membrane? ____________________

   ____
6. Demonstrate lung function by inserting a hose from an air compressor into the trachea and forcing air into the lungs. What happens? ____________________
7. Stop forcing air into the lungs. What happens?

   ____

Organ	Structural Characteristic	Function
Larynx	*Hyaline cartilage abundant*	*Forms supporting framework*
Trachea		
Lungs		
Heart		
Blood vessel		

Post-lab quizzes are also assignable in MasteringA&P®

POST-LAB ASSIGNMENTS

Name: ______________________ Date: ______________ Lab Section: __________

PART I. Check Your Understanding

Activity 1: Exploring the Organs of the Respiratory System

1. Label specific structures of the upper respiratory tract:

   a. ______________________
   b. ______________________
   c. ______________________
   d. ______________________
   e. ______________________
   f. ______________________
   g. ______________________
   h. ______________________
   i. ______________________
   j. ______________________

2. Label specific structures of the lower respiratory tract:

   a. ______________________
   b. ______________________
   c. ______________________
   d. ______________________
   e. ______________________
   f. ______________________

3. Fill in the blank with the appropriate term(s):

   ______________________ a. A facial bone that contains sinuses
   ______________________ b. The most superior part of the pharynx
   ______________________ c. Blood vessel that carries deoxygenated blood to lungs
   ______________________ d. Prime mover of inspiration
   ______________________ e. Major supporting cartilage of the larynx

4. State the functions of the nasal cavities and then briefly describe two specific ways in which nasal cavities are structurally adapted to their functions. ______________________________

____

____

Activity 2: Examining the Microscopic Anatomy of the Trachea and Lungs

1. Identify the following structures on the accompanying photomicrograph:

   a. ____________________

   b. ____________________

   c. ____________________

   d. ____________________

   e. ____________________

   f. ____________________

   g. ____________________

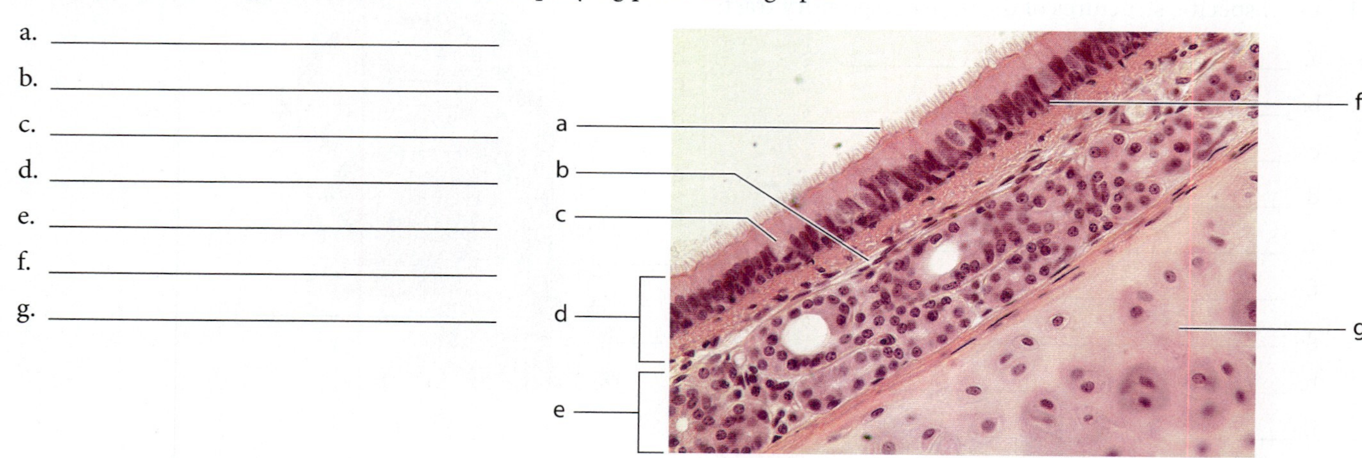

LM (400×)

2. Briefly describe the function of each of the following cell types:

   a. Type I alveolar cell ______________________________

   b. Type II alveolar cell ______________________________

Activity 3: Examining a Sheep Pluck

1. Describe the texture of the sheep trachea. ______________________________

   Describe the texture of the sheep lung. ______________________________

2. How do sheep lungs differ from human lungs? ______________________________

   ____

3. Trace the pathway of the compressed air as you forced it into the trachea. ____________

   ____

PART II. Putting It All Together

A. Review Questions

Answer the following questions using your lecture notes, your textbook, and your lab notes.

1. Describe how each of the following pairs of structures are similar and how they are different:

   a. Pharynx and larynx ______________________________

   ____

   b. Hyaline cartilage and elastic cartilage ______________________________

   ____

   c. Goblet cell and macrophage ______________________________

   ____

2. List the components of the upper respiratory system: ______________________________

   ____

3. List the components of the lower respiratory system: ______________________________

   ____

4. Where is the pharyngeal tonsil located, and what is its function? ______________________________

   ____

5. Trace the pathway of a molecule of oxygen as it enters the body via the anterior nares and travels to an alveolus.

   Anterior nares → ______________ → ______________ →

   nasopharynx → ______________ → ______________ →

   ______________ → ______________ → primary bronchi →

   ______________ → ______________ → ______________ →

   terminal bronchioles → ______________ → ______________ →

   ______________ → alveolus

B. Concept Mapping

1. Fill in the blanks to complete this concept map outlining the anatomy of the respiratory system.

alveolar duct	capillary	macrophages	surfactant	type I alveolar cell

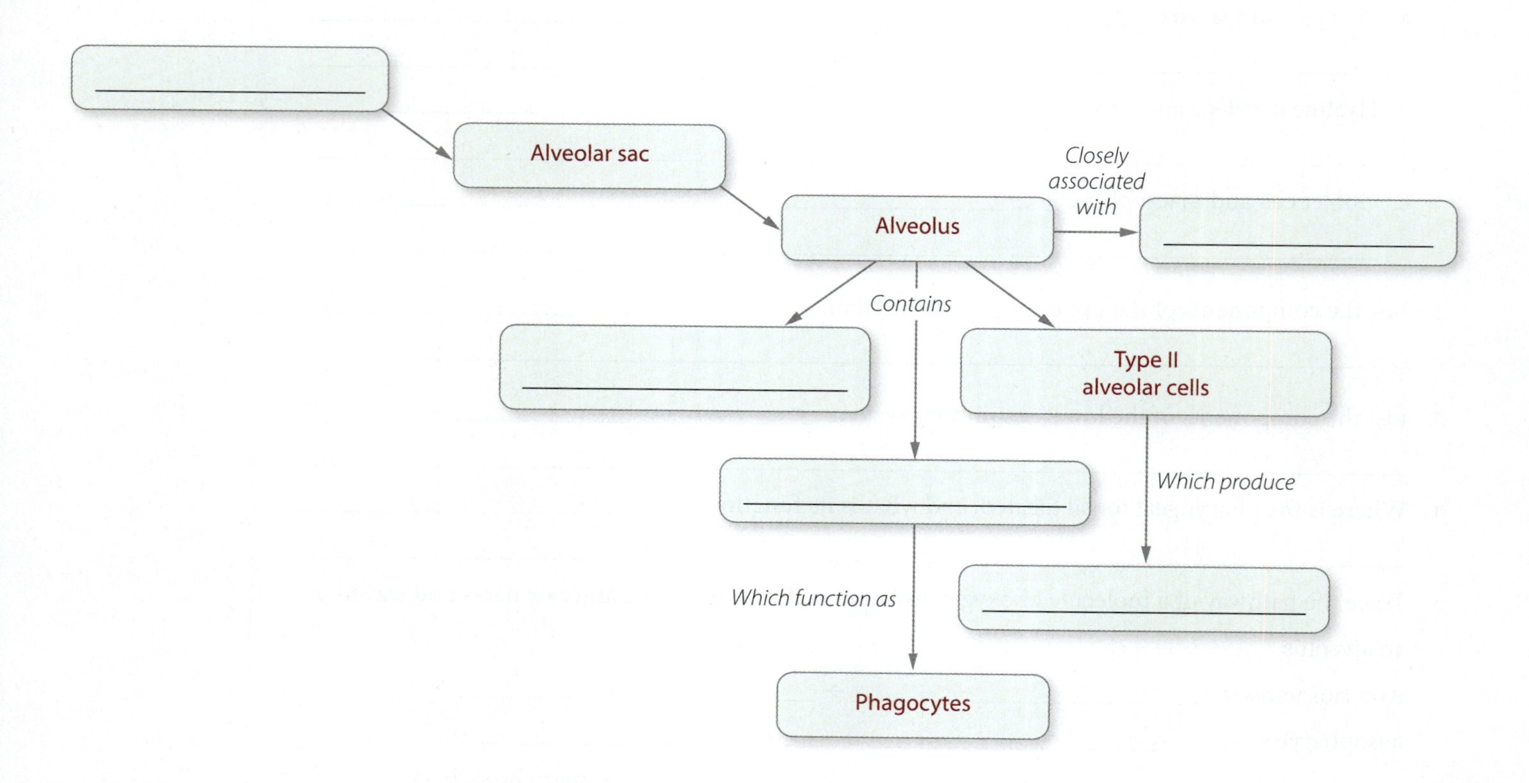

2. Construct a unit concept map to show the relationships among the following set of terms. Include all of the terms in your diagram. Your instructor may choose to assign additional terms.

alveolar duct	basement membrane	bronchiole	capillary	cartilaginous rings
cilia	cricoid cartilage	epiglottis	goblet cell	macrophages
oropharynx	superior conchae	surfactant	type I alveolar cell	vocal cords

27

Physiology of the Respiratory System

UNIT OUTLINE

In the previous unit we explored the anatomy of the respiratory system. Here we will examine the various functions of the respiratory system. First we consider the various processes involved in the mechanics of breathing; next we explore the various quantities of air that flow into or out of the lungs during a given breath. Finally, we briefly investigate the processes by which the body controls breathing.

THINK ABOUT IT *Describe the role of the organ system listed in the maintenance of homeostasis for each of the following respiratory events:*

The cardiovascular system in the exchange of oxygen and carbon dioxide in the lungs

___

___

The muscular system in inhalation and exhalation ___

___

___

The cardiovascular system in the transport of gases throughout the body ___

___

___

The lymphatic system in the exchange of gases between the blood and cells of the tissues ___

___

___

PRE-LAB ASSIGNMENTS

Pre-lab quizzes are also assignable in MasteringA&P®

To maximize learning, BEFORE your lab period carefully read this entire lab unit and complete these pre-lab assignments using your textbook, lecture notes, and prior knowledge.

PRE-LAB Activity 1: Analyzing the Model Lung and Pulmonary Ventilation

1. What will you use in this activity as a model lung?

   ____

2. The prime mover of inspiration is the:
   a. sternocleidomastoid muscle.
   b. diaphragm.
   c. internal intercostal muscles.
   d. pectoralis major muscle.
   e. transverse abdominis muscle.

3. Normal (quiet) inspiration is (an active/a passive) process. Circle the correct answer.

4. Normal (quiet) expiration is (an active/a passive) process. Circle the correct answer.

PRE-LAB Activity 2: Measuring Respiratory Volumes in a Human Subject Using BIOPAC®

1. Tidal volume is the measurement of:
   a. the volume of air that can be maximally inhaled at the end of tidal inspiration.
   b. the volume of air inspired or expired during a single breath.
   c. the volume of air that can be maximally exhaled at the end of tidal expiration.
   d. the volume of gas remaining in the lungs at the end of maximal expiration.

2. Expiratory reserve volume is the measurement of
   a. the volume of air that can be maximally inhaled at the end of tidal inspiration.
   b. the volume of air inspired or expired during a single breath.
   c. the volume of air that can be maximally exhaled at the end of tidal expiration.
   d. the volume of gas remaining in the lungs at the end of maximal expiration.

PRE-LAB Activity 3: Determining Respiratory Volumes and Capacities at Rest and Following Exercise

1. Match each of the following lung volumes with its correct description:

   _____ a. tidal volume
   _____ b. inspiratory reserve volume
   _____ c. expiratory reserve volume
   _____ d. residual volume

   1. amount of air exchanged in a normal breath
   2. volume of air remaining in the lungs after a forced expiration
   3. maximum volume of air that can be forcibly inspired after a tidal inspiration
   4. maximum volume of air that can be forcibly expired after a tidal expiration

2. Match each of the following lung capacities with its correct description:

   _____ a. vital capacity
   _____ b. inspiratory lung capacity
   _____ c. total lung capacity
   _____ d. functional residual capacity

   1. TV + IRV + ERV + RV
   2. TV + IRV
   3. TV + ERV + IRV
   4. RV + ERV

PRE-LAB Activity 4: Investigating the Control of Breathing

1. Important respiratory control centers are located in the:
   a. midbrain.
   b. choroid plexus.
   c. cerebellum.
   d. thalamus.
   e. medulla oblongata.
2. As blood CO_2 levels rise, blood H^+ levels (rise/fall) and blood pH levels (rise/fall). Circle the correct answers.
3. As blood CO_2 levels fall, blood H^+ levels (rise/fall) and blood pH levels (rise/fall). Circle the correct answers.
4. In this activity you will investigate the effects of several breathing patterns on breathing rate, ________________, and exhaled CO_2 levels.

Mechanics of Breathing

The overall function of the respiratory system is providing the body with oxygen and ridding it of carbon dioxide. Multiple processes are involved in the delivery of oxygen to body tissues and the removal of carbon dioxide from body tissues (**Figure 27-1**):

1. **Pulmonary ventilation**, or breathing, is the movement of air into and out of the lungs.
2. **External respiration** is the exchange of gases between the air and the blood. It occurs in the alveoli of the lungs.
3. **Gas transport** is a function of the blood (cardiovascular system), and it is the process by which oxygen is transported from the lungs to the tissues, and carbon dioxide is transported from the tissues to the lungs.
4. **Internal respiration** is the exchange of gases between the blood and cells of the tissues. It occurs at capillary beds throughout the body.

Additionally, in the process called **cellular respiration**, individual cells break down glucose in the presence of oxygen to produce water and energy in the form of ATP; carbon dioxide is released as a waste product. This process occurs in the mitochondria of cells. Note that three of these processes include the word *respiration*, so used alone "respiration" can refer to distinctly different processes.

During pulmonary ventilation, air is moved in and out between the atmosphere and the lungs. Pulmonary ventilation is driven by gradients between the air pressure in the atmosphere (atmospheric pressure) and the air pressure in the lungs (intrapulmonary pressure). These pressure gradients result when the actions of the respiratory muscles change the volume of the thoracic cavity and the lungs. Boyle's law explains the relationship between pressure and volume: The pressure of a gas in a closed container is inversely proportional to the volume of the container. Therefore, when lung volume increases during inspiration, intrapulmonary pressure decreases below that of atmospheric pressure, and air flows down the pressure gradient and into the lungs (**Figure 27-2**). By contrast, when lung volume decreases during expiration, intrapulmonary pressure increases above that of atmospheric pressure, and air flows down the pressure gradient and out of the lungs.

Several skeletal muscles change the size of the thoracic cavity during pulmonary ventilation. Normal (quiet) inspiration is an active process requiring the contraction of the diaphragm and external intercostal muscles. During forced inspiration, additional muscles (the scalene, sternocleidomastoid, pectoralis major, pectoralis minor, and serratus anterior muscles) pull the ribs superiorly and laterally to further increase thoracic volume. Normal (quiet) expiration, however, is a passive process initiated by so-called elastic recoil during the relaxation of the inspiratory muscles (see Figure 27-2). During forced expiration, contraction of the internal intercostal and abdominal muscles further reduces the volume of the lungs, thereby further increasing alveolar pressure and forcing additional air out of the lungs.

QUICK TIP You can remember which set of intercostal muscles is involved in quiet inspiration and in forced expiration by remembering that in both cases "I" and "E" go together:

For quiet **I**nspiration, the **E**xternal intercostals work with the diaphragm.

For forced **E**xpiration, the **I**nternal intercostals work with the abdominal muscles.

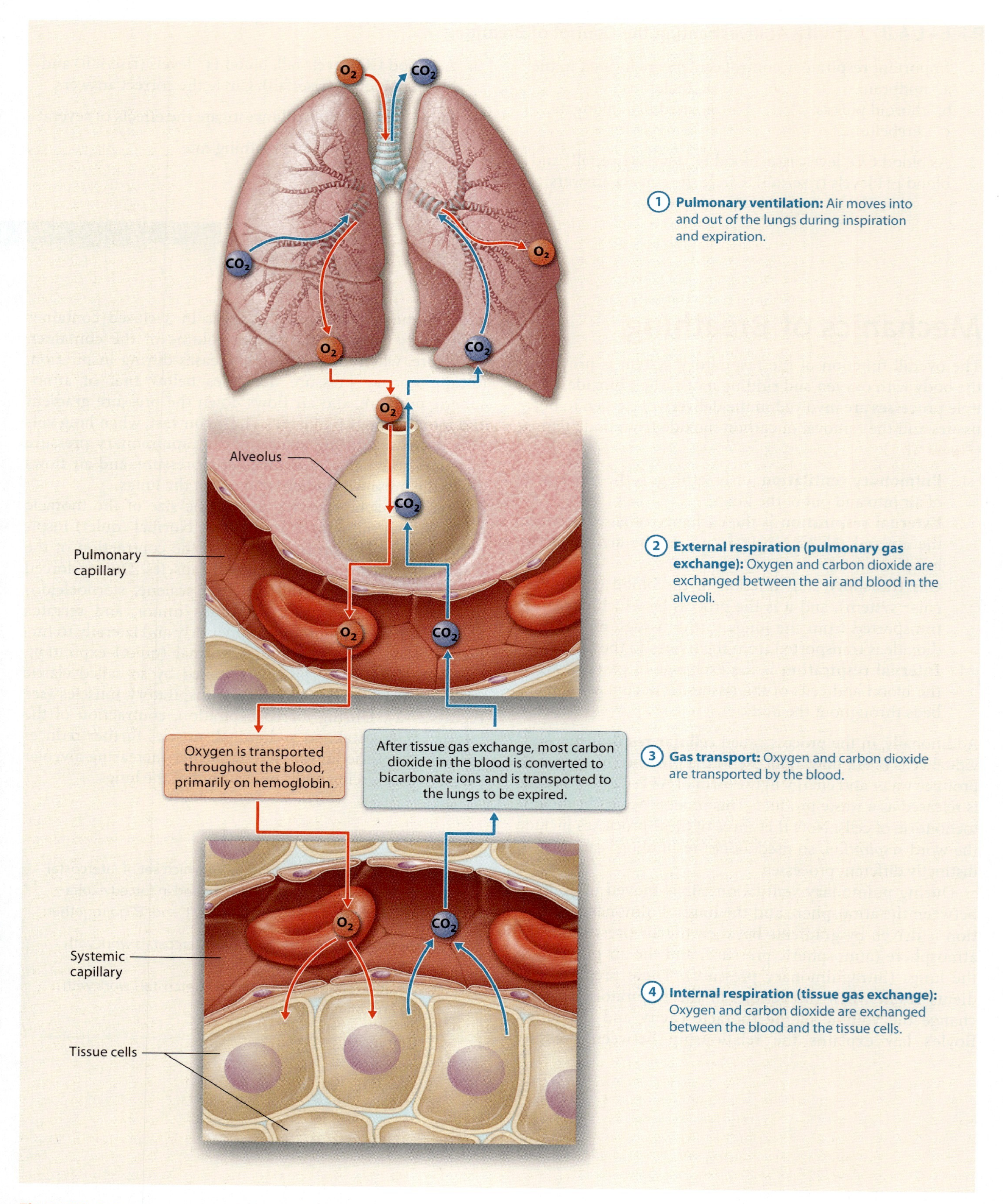

Figure 27-1 Four processes of the respiratory system.

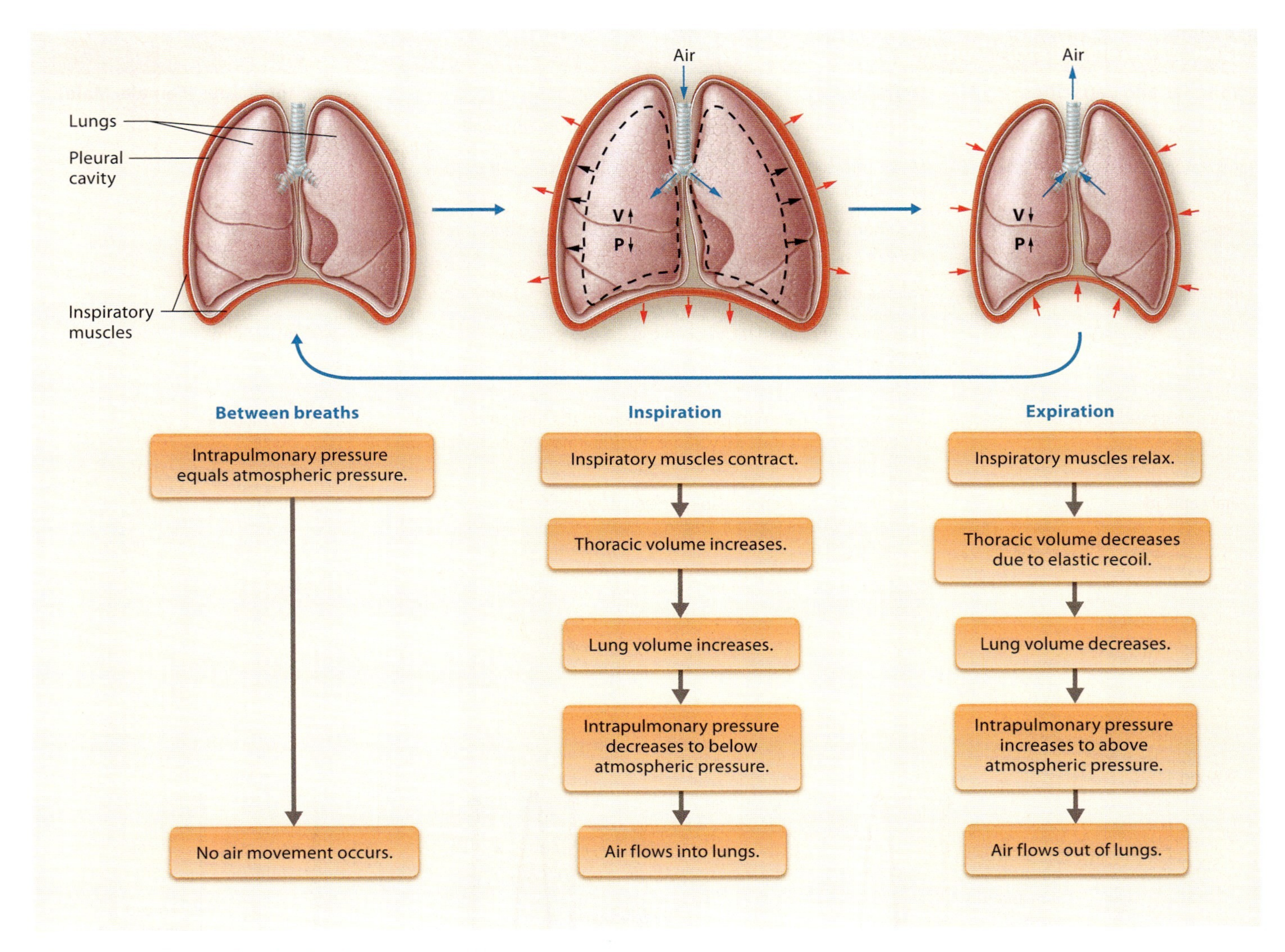

Figure 27-2 The mechanics of pulmonary ventilation.

Respiratory Volumes and Capacities

Respiratory volumes are the amounts of air that flow into or out of the lungs during a specific pulmonary event such as normal breathing. Respiratory capacities, by contrast, are the sum of two or more respiratory volumes. Thus, respiratory volumes can be measured, whereas respiratory capacities are calculated. **Table 27-1** defines respiratory volumes and capacities and gives average values for males and females; **Figure 27-3** illustrates the relationships among respiratory volumes and respiratory capacities.

Table 27-1 Respiratory Volumes and Capacities

Volumes and Capacities	Definition	Average Values (Female, Male)
Tidal volume (TV)	Volume of air exchanged during each normal (quiet) breath	500 ml
Inspiratory reserve volume (IRV)	Maximum volume of air that can be forcibly inspired after a tidal inspiration	1900 ml, 3100 ml
Expiratory reserve volume (ERV)	Maximum volume of air that can be forcibly expired after a tidal expiration	700 ml, 1200 ml
Residual volume (RV)	Volume of air that remains in the lungs after a forced expiration	1100 ml, 1200 ml
Inspiratory capacity	Total amount of air that can be inspired; equals tidal volume plus inspiratory reserve volume: (TV + IRV)	2400 ml, 3600 ml
Functional residual capacity	Total amount of air that normally remains in the lungs after a tidal expiration; equals residual volume plus expiratory reserve volume: (RV + ERV)	1800 ml, 2400 ml
Vital capacity	Total amount of exchangeable air; equals sum of tidal volume, expiratory reserve volume, and inspiratory reserve volume: (TV + ERV + IRV)	3100 ml, 4800 ml
Total lung capacity	Total amount of exchangeable and nonexchangeable air; equals sum of all of the pulmonary volumes: (TV + IRV + ERV + RV)	4200 ml, 6000 ml

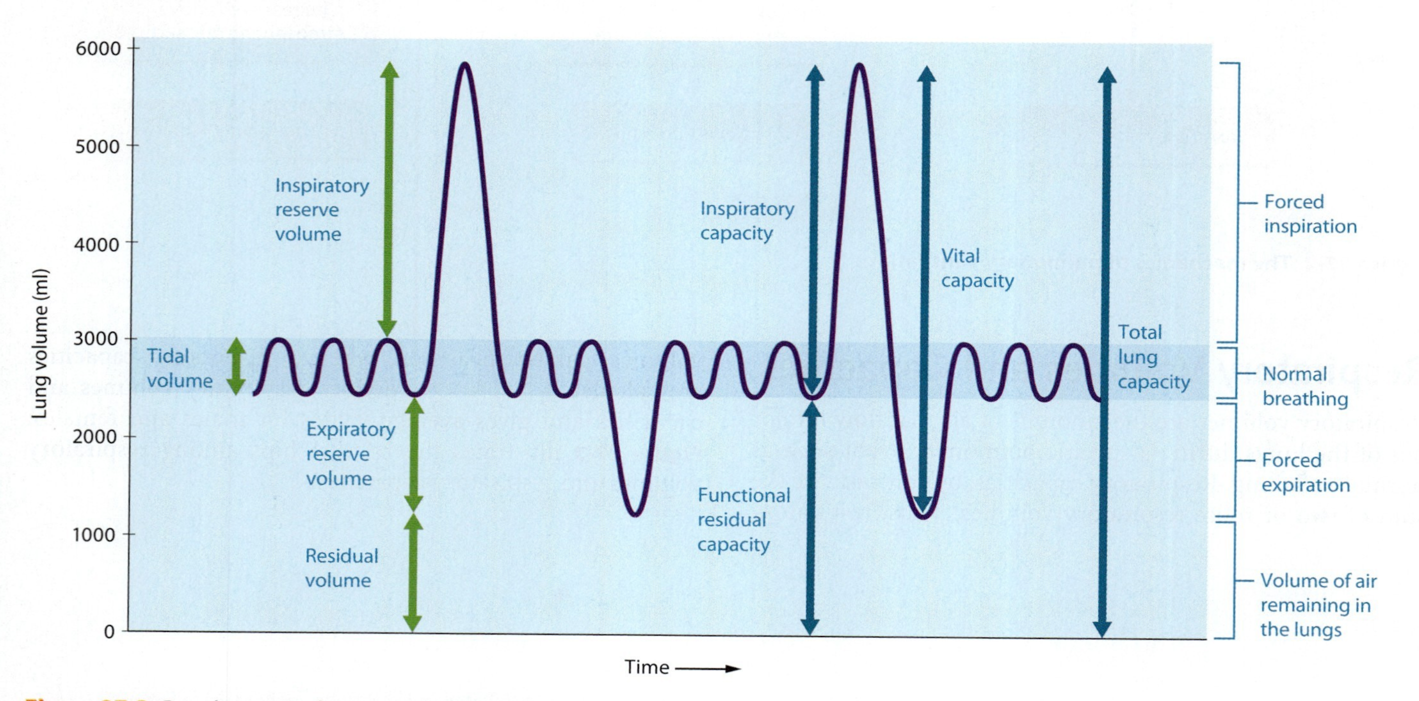

Figure 27-3 Respiratory volumes and capacities.

ACTIVITY 1

Analyzing the Model Lung and Pulmonary Ventilation

Learning Outcomes

1. Simulate pulmonary ventilation using a bell jar lung model.
2. Explain the relationship between pressure and volume as it relates to pulmonary ventilation.
3. Identify the skeletal muscles that play a role in pulmonary ventilation.
4. If required by your instructor, review the origin, insertion, innervation, and actions of the skeletal muscles that function in pulmonary ventilation.

Materials Needed

- ☐ Bell jar lung model
- ☐ Anatomical muscle models

Instructions

In this activity you will use a bell jar model of the lungs to study the principles involved in pulmonary ventilation. A bell jar model (**Figure 27-4**) consists of a heavy plastic container called a bell jar (representing the wall of pleural cavity), the interior of the bell jar (the pleural cavity), a rubber membrane (the diaphragm), two balloons (the lungs), and a Y-shaped tube (the trachea and bronchi).

1. Label the respiratory structures represented by the components of the bell jar model in Figure 27-4.

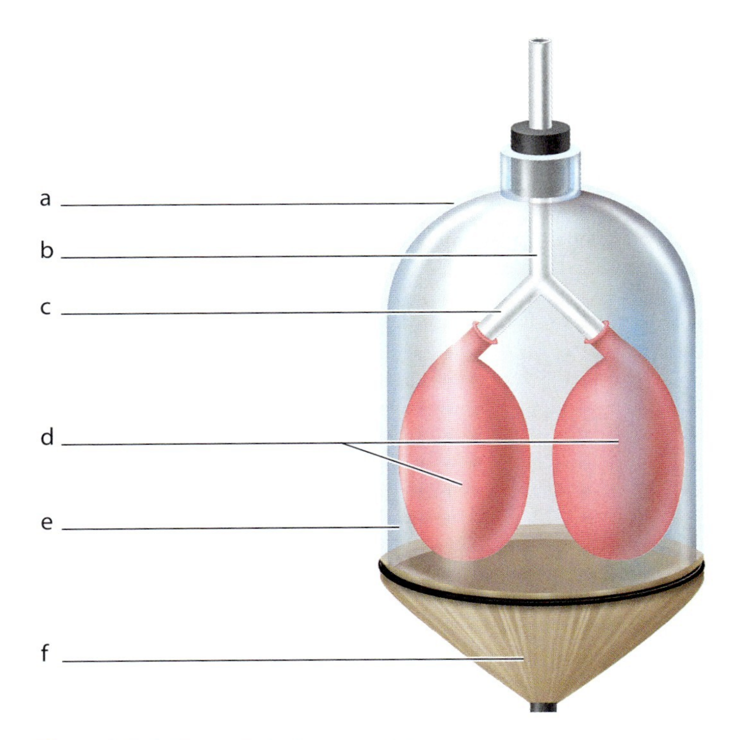

Figure 27-4 The bell jar lung model.

2. Pull the "diaphragm" downward.
   a. What happened to the "lungs"? ____________
   b. What happened to the volume of the "pleural cavity"? ____________
   c. What happened to the pressure in the "pleural cavity"? ____________
   d. Why did air enter the "lungs"? ____________

3. Push the "diaphragm" upward.
   a. What happened to the "lungs"? ____________
   b. What happened to the volume of the "pleural cavity"? ____________
   c. What happened to the pressure in the "pleural cavity"? ____________
   d. Why did air leave the "lungs"? ____________

4. Simulate a pneumothorax (a condition in which air enters the pleural cavity): Inflate the "lungs" by pulling the "diaphragm" downward, and then loosen the rubber stopper and allow some air into the bottle. What happens to the "lungs"? Why? ____________

ACTIVITY 2

Measuring Respiratory Volumes in a Human Subject Using BIOPAC®

Learning Outcomes

1. Observe experimentally, record, and/or calculate selected pulmonary volumes and capacities.
2. Compare the observed values of volume and capacity with average values.
3. Compare the normal values of pulmonary volumes and capacities of subjects differing in gender, age, weight, and height.

Materials Needed

- ☐ Computer system (Windows 7, Vista, XP, Mac OS X 10.5–10.7)
- ☐ Biopac Science Lab system (MP 35 unit and software)
- ☐ Biopac Airflow Transducer with removable, cleanable head (SS11LA)
- ☐ Biopac Bacteriological Filter (AFT1): one per subject; plus, if using calibration syringe, one dedicated to syringe

- ☐ Biopac Disposable Mouthpiece (AFT2)
- ☐ Biopac Nose Clip (AFT3)
- ☐ Biopac Calibration Syringe: 0.6-Liter (AFT6 or AFT6A+AFT11A) *or* 2-Liter (AFT26)
- ☐ Biopac wall transformer (AC100A)
- ☐ Biopac serial cable (CBLSERA)
- ☐ Biopac USB adapter (USB1W)

Instructions

In this activity, you will be using the Biopac airflow transducer to measure respiratory volumes. To begin, connect the airflow transducer to channel 1 on the BSL unit. Open Biopac Lesson 12.

A. Calibration

1. Hold the syringe parallel to the ground with the plunger pulled out.
2. Connect a filter and calibration syringe to the airflow transducer such that the cord coming out of the handle is on the left side when holding the syringe, as shown in **Figure 27-5**.
3. First read the alert box; then click Calibrate and then click OK.
4. Read all the on-screen prompts! When ready to perform the second stage of the calibration, click Yes.
   You will be simulating breathing by completing a series of five cycles: pushing the plunger in (this should take 1 second), waiting 2 seconds, pulling the plunger out (this should take 1 second), waiting 2 seconds, and repeating.
5. When you have completed five cycles, click End calibration.
6. If your data looks like **Figure 27-6**, continue on to recording the data.

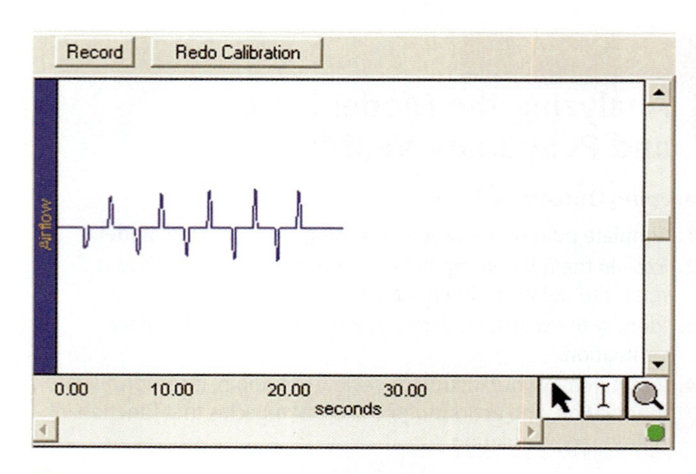

Figure 27-6 Sample calibration data.

B. Recording the Data

1. Always make sure that the air filter is on and you are breathing through the proper side labeled "Inlet." Keep the airflow transducer upright at all times. Use a nose clip during recording. Do not look at the screen while you are the subject.
2. When prepared, the subject will complete the following sequence with the lips closed tightly around the mouthpiece and a nose clip on the nose.
   a. Take three normal breaths.
   b. Inhale as much air as possible, then exhale and return to normal breathing.
   c. Take three normal breaths.
   d. Exhale as much as possible, then inhale and return to normal breathing.
   e. Take three normal breaths.

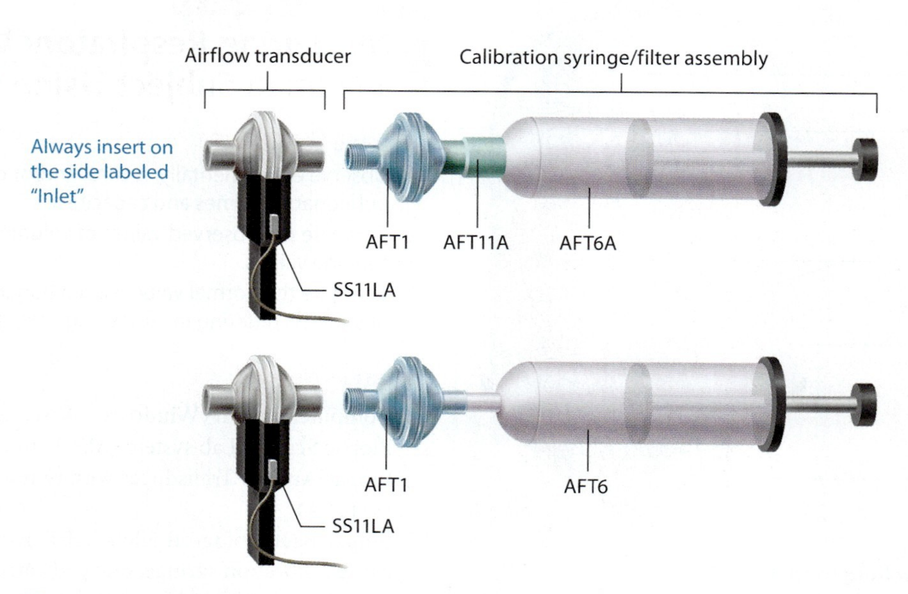

Figure 27-5 Calibration setup.

3. When the subject is ready, click Record as the subject begins the ventilation sequence.
4. When the subject finishes the last exhalation at the end of the sequence, click Stop.
5. Click Done and proceed to Analyze current data file.
6. If necessary, click Find Volume button.
7. Change channel measurement values to max, min, and p-p all on channel 2 (Volume).

C. Volume Calculations

1. Now explore the recording using the I-beam tool. **Figure 27-7** explains how to interpret a computer-generated spirogram.

   TV: For normal resting breathing, determine max and min values for a wave. Record the difference as TV in **Table 27-2**.

   IRV: Determine the max for the normal breathing cycle just before the forced inhalation and the max for forced inhalation. Record the difference as IRV in Table 27-2.

   ERV: Determine the min for the normal breathing cycle just before the forced exhalation and the min for forced exhalation. Record the difference as ERV in Table 27-2.

   VC: TV + IRV + ERV

   MRV: TV × Respiratory Rate
2. The P-P measurement can be used to obtain VC (**Figure 27-8**).
3. The delta measurement can be used to obtain IRV, ERV, and other measurements (**Figure 27-9**).

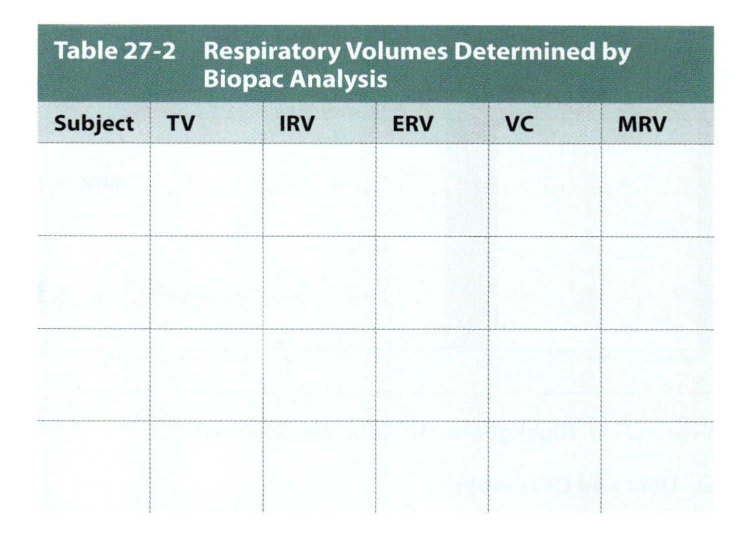

Table 27-2 Respiratory Volumes Determined by Biopac Analysis

Subject	TV	IRV	ERV	VC	MRV

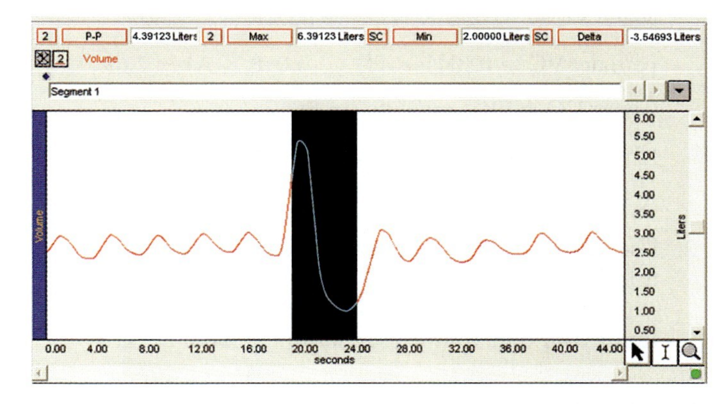

Figure 27-8 **Highlighting the P-P measurement.**

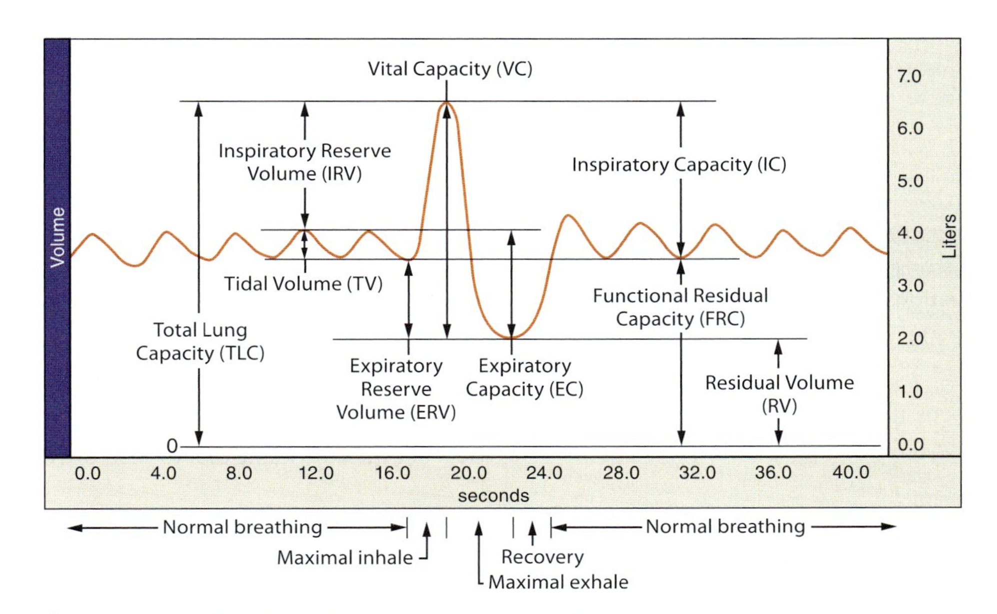

Figure 27-7 **Explanation of a computer-generated spirogram.**

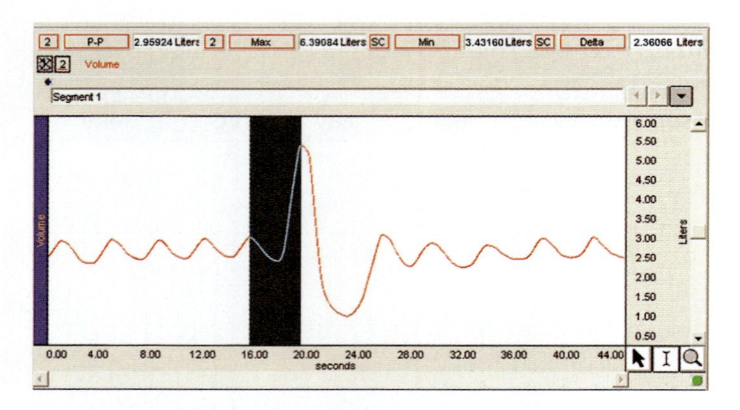

Figure 27-9 **Highlighting the delta measurement.**

D. Data and Calculations

1. Use the equations below to calculate your predicted vital capacity (VC): _____ liters

   Male: VC = (0.052 × H) − (0.022 × A) − 3.60

   Female: VC = (0.041 × H) − (0.018 × A) − 2.69

   where VC = vital capacity (in liters)

   H = height (in centimeters)

   A = age (in years).

2. Use the data in **Table 27-3** and the equations in the middle column of **Table 27-4** to calculate inspiratory, expiratory, functional residual, and total lung capacities. Enter your data in the last column of Table 27-4.

Table 27-3 Respiratory Volume and Measurements

Type of Volume	Measurement (liters)
Tidal volume (TV)	
Inspiratory reserve volume (IRV)	
Expiratory reserve volume (ERV)	
Vital capacity (VC)	

Residual volume (RV) used: _____ liters (Default is 1.2 liters.)

Table 27-4 Calculated Respiratory Values

Capacity	Equations	Your Calculations
Inspiratory capacity (IC)	IC = TV + IRV	
Expiratory capacity (EC)	EC = TV + ERV	
Functional residual capacity (FRC)	FRC = ERV + RV	
Total lung capacity (TLC)	TLC = IRV + TV + ERV + RV	

3. Compare the subject's lung volumes with the average volumes presented earlier in this exercise.

	Average	Subject
TV	500 ml	_____ ml
IRV	3300 ml	_____ ml
ERV	1000 ml	_____ ml

4. What is the subject's observed vital capacity as a percentage of the predicted vital capacity for her or his gender, age, and height?

   Liters observed _____ × 100 = _____ %

   Liters predicted: _________

ACTIVITY 3

Determining Respiratory Volumes and Capacities at Rest and Following Exercise

Learning Outcomes

1. Explain the relationships among respiratory volumes and capacities, including tidal volume, inspiratory reserve volume, expiratory reserve volume, residual volume, and vital capacity.
2. Demonstrate the proper use of a dry, handheld spirometer.
3. Measure/calculate the following respiratory volumes and capacities: tidal volume, expiratory reserve volume, vital capacity, inspiratory reserve volume, inspiratory capacity, and residual volume.

Materials Needed

- ☐ Handheld dry spirometer
- ☐ Disposable mouthpieces
- ☐ Alcohol wipes

Instructions

Respiratory volumes and capacities can be measured or calculated to assess respiratory system health. In this activity you will use an instrument called a dry, handheld spirometer (**Figure 27-10**). If you must use some other type of spirometer, your instructor will provide instructions for its use.

A. Preparing for the Activity

1. Wipe the nozzle of the spirometer with an alcohol wipe and place a new disposable mouthpiece on the spirometer tube.
2. Set the adjustable dial to zero by rotating it.

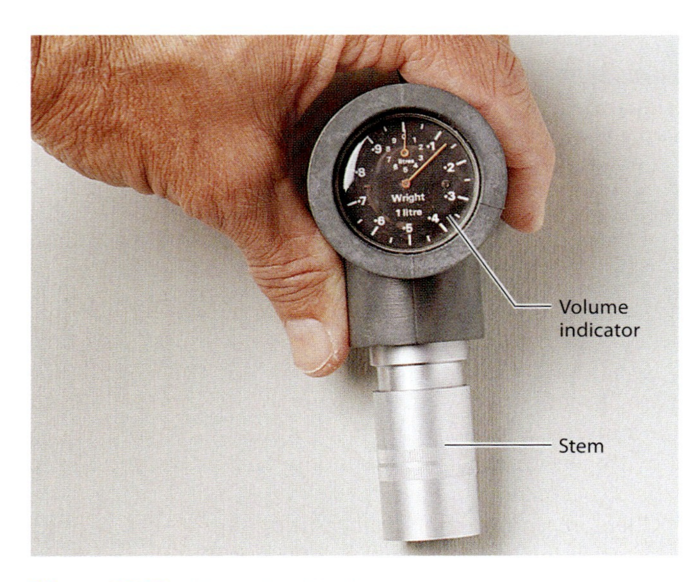

Figure 27-10 **A dry, handheld spirometer.**

3. Read the following instructions for using the spirometer:
   a. To obtain the most accurate reading possible, use your thumb and index finger to pinch your nostrils closed to prevent air from leaking out of your nose.
   b. When blowing into the spirometer, stand erect, and always hold it in a horizontal position with the dial facing upward.
   c. This type of spirometer does *not* measure inspiratory volumes, so remember to only **exhale** into the spirometer.

B. Conducting the Activity

1. Measuring Lung Volumes and Capacities at Rest:
   a. *Tidal Volume (TV) at Rest:* Set the dial on the spirometer to 0 and sit quietly for 1 minute. Inhale normally, then place the mouthpiece of the spirometer between your lips and (with nostrils pinched closed) exhale normally. Repeat this process two more times—resetting the dial on the spirometer to 0 each time—and then enter each of the values for tidal volume in **Table 27-5**. Add the values for the three exhalations and divide the total by 3 to obtain the average TV, and record it in Table 27-5.
   b. *Expiratory Reserve Volume (ERV) at Rest:* Inhale and exhale normally three times. Inhale normally again, and then place the spirometer between your lips and forcibly exhale as much air as possible. Record the value for ERV in Table 27-5. Repeat this process two times, resetting the dial on the spirometer to 0 each time. Add the values obtained for the three trials and divide by 3 to obtain the average ERV, and record it in Table 27-5.

Table 27-5 Experimental Results: Measured Volumes

At Rest	Trial 1	Trial 2	Trial 3	Average
Tidal volume (TV) in ml				
Expiratory reserve volume (ERV) in ml				
Vital capacity (VC) in ml				
Following Exercise	**Trial 1**	**Trial 2**	**Trial 3**	**Average**
Tidal volume (TV) in ml				
Expiratory reserve volume (ERV) in ml				
Vital capacity (VC) in ml				

   c. *Vital Capacity (VC) at Rest:* Inhale and exhale normally three times. Then, inhale as much air as you possibly can; quickly insert the spirometer between your lips and exhale as forcibly as you can. Record your results in Table 27-5. Repeat this process two times, resetting the dial on the spirometer to 0 each time. Add the values obtained for the three trials and divide by 3 to obtain the average VC, and record it in Table 27-5.

2. Measuring Lung Volumes and Capacities following Exercise:

   Before making the measurements, run in place for 4–5 minutes.

   a. *Tidal Volume (TV) following Exercise:* Follow the instructions for TV at rest, except do not sit quietly first and make sure that you reset the dial on the spirometer to 0 between measurements.
   b. *Expiratory Reserve Volume (ERV) following Exercise:* Follow the instructions for ERV at rest, except do not sit quietly first.
   c. *Vital Capacity (VC) following Exercise:* Follow the instructions for VC at rest, except do not sit quietly first.

C. Calculating Respiratory Capacities

1. Using the average values for TV, ERV, and VC recorded in Table 27-5, calculate the values for inspiratory reserve volume (IRV) and inspiratory capacity (IC) at rest and following exercise. Record the calculated values in **Table 27-6**.

Table 27-6 Experimental Results: Calculated Volumes and Capacities

	Inspiratory Reserve Volume (IRV)	Inspiratory Capacity (IC)	Residual Volume (RV)	Functional Residual Capacity (FRC)	Total Lung Capacity (TLC)
At rest					
Following exercise					

2. Calculate residual volume (RV). Note that RV—the amount of air remaining in the lungs after a maximum exhalation—cannot be determined experimentally. This remaining air is very important to lung function because it allows gas exchange to occur continuously, even between breaths. Calculate RV using the following equation:

$$RV = (VC) \times (\text{an age factor})$$

For individuals between the ages of 16 and 34, the factor is 0.250; for ages 35–49, the factor is 0.305; and for ages 50–69, the factor is 0.445.

3. Using the calculated value for RV, calculate your functional residual capacity (FRC) and total lung capacity (TLC), and record those values in Table 27-6.

D. Analyzing the Data

1. Briefly summarize the effect of exercise on pulmonary volumes and capacities. ______________________

____

____

Control of Breathing

Breathing is controlled by respiratory centers in the pons and medulla oblongata. Both rate and depth of breathing are affected by a wide variety of factors, including carbon dioxide (CO_2) levels in the blood.

Most CO_2 is transported in the blood as bicarbonate ion. When CO_2 diffuses into the blood, most of it enters a red blood cell and combines with water to form carbonic acid, a reaction that is catalyzed by the enzyme carbonic anhydrase. Carbonic acid then dissociates into hydrogen ions and bicarbonate ions. Note that this reaction also occurs in the plasma but it occurs at a much slower rate due to the absence of carbonic anhydrase. The chemical equation for the formation of hydrogen and bicarbonate ions from carbon dioxide and water is as follows:

$$CO_2 + H_2O \rightleftharpoons H_2CO_3 \rightleftharpoons H^+ + HCO_3^-$$

If, for example, CO_2 levels are elevated, then more carbonic acid will be formed and will dissociate into more hydrogen ions and bicarbonate ions. Conversely, if CO_2 levels drop, then more hydrogen ions and bicarbonate ions will form carbonic acid, which will then dissociate into carbon dioxide and water.

Recall that pH is determined by the concentration of hydrogen ions in a solution. As hydrogen ion levels in the blood increase, blood pH decreases (becomes more acidic); as hydrogen ion levels in the blood decrease, blood pH increases (becomes more basic).

Chemoreceptors in the brain, aortic arch, and carotid arteries detect rising hydrogen ion levels and send afferent messages to the respiratory control centers in the medulla oblongata and pons. These afferent messages stimulate increases in both breathing rate and depth of breathing, thereby enabling the lungs to exhale more CO_2. As a result, CO_2 levels, blood pH, breathing rate, and depth of breathing all return to normal.

ACTIVITY 4
Investigating the Control of Breathing

Learning Outcomes

1. Demonstrate and explain the reaction that occurs when exhaled carbon dioxide reacts with water.
2. Investigate the effects of hyperventilation and rebreathing on breathing rate, depth of breathing, and exhaled CO_2 levels.

Materials Needed

- ☐ 250-ml beaker
- ☐ 50-ml beaker
- ☐ Graduated cylinder
- ☐ Deionized water
- ☐ 5 ml of 0.05M NaOH
- ☐ Phenol red
- ☐ Plastic straw
- ☐ pH meter
- ☐ Small brown paper bag
- ☐ Safety goggles

Instructions

In this activity you will explore various factors that affect the rate and depth of breathing. Before you begin this investigation, you will first perform a quick demonstration to visualize the reaction that occurs when exhaled CO_2 reacts with water.

A. Demonstration

1. ⚠ Put on safety goggles.
2. Fill a 250-ml beaker with 100 ml of deionized water.
3. Add to the beaker 5 ml of 0.05M NaOH and five drops of phenol red, a pH indicator that turns yellow in acidic solutions.
4. Use a plastic straw to gently exhale air into the solution for 10 seconds.

Describe what you observe happening in the beaker.

___

___

Explain the physiological basis of your observation.

___

___

___

B. Investigation

1. Predict the effects of hyperventilation and rebreathing on breathing rate, depth of breathing, and CO_2 levels of exhaled air. Record your predictions in the following table.

Breathing Variation	Breathing Rate (↑, ↓, Same)	Depth of Breathing (↑, ↓, Same)	CO_2 Levels of Exhaled Air (↑, ↓, Same)
Hyperventilation			
Rebreathing			

2. Designate one member of your lab group to be the subject, one member to be the timer, and one member to collect and record the data.
3. Normal Breathing:
   a. Pour 25 ml of deionized water into the graduated cylinder and then into the 50-ml beaker.
   b. Use a pH meter to measure the pH of the water. Record this value as the control pH for normal breathing in **Table 27-7**.
   c. Instruct the subject to sit quietly and read silently for 3 minutes to ensure that the breathing rate is normal.

   ⚠ **If at any time during this activity the subject begins to feel dizzy, stop immediately to prevent the subject from fainting.**

   d. While the subject is distracted, count the number of breaths taken in 1 minute and record this value as the breathing rate for normal breathing in Table 27-7.
   e. Observe the subject's depth of breathing and record your observations in Table 27-7.
   f. Now instruct the subject to take a deep breath and to use a straw to exhale air for 10 seconds into the 25 ml of deionized water. Repeat this procedure two more times.
   g. Measure the pH of the water and record this value (experimental pH of water) in Table 27-7.
   h. Discard the water and rinse the beaker with deionized water.
4. Hyperventilation: Follow the instructions for normal breathing with the following exceptions: For instruction b, record this value in Table 27-7 as the control pH for hyperventilation. For instruction c, instruct the subject to sit quietly for 3 minutes and then to breathe rapidly and deeply for 1 minute. For instruction d, record this value as the breathing rate for hyperventilation in Table 27-7.
5. Rebreathing: Follow the instructions for normal breathing with the following exceptions: For instruction b, record this value in Table 27-7 as the control pH for rebreathing. For instruction c, instruct the subject to sit quietly for 3 minutes and then to breathe deeply for 2 minutes into a small brown paper bag held tightly over the nose and mouth. For instruction d, record this value as the breathing rate for rebreathing in Table 27-7.
6. Extend the Activity: If time allows, you can investigate the effects of breath holding and exercise on breathing rate, depth of breathing, and exhaled CO_2 levels. In both cases, follow the general sequence of instructions a through h. For instruction b, record these values as the control pHs for breath holding and exercise, respectively. For instruction c for both cases, have the subject

Table 27-7 Experimental Results

Breathing Pattern	Control pH of Water	Breathing Rate	Depth of Breathing	Experimental pH of Water
Normal breathing				
Hyperventilation				
Rebreathing				
Optional: Breath holding				
Optional: Exercise				

sit for 3 minutes; then for breath holding have the subject hold the breath for as long as possible, and for exercise have the subject jog in place for 3–5 minutes. For instruction d for both cases, record your results in the appropriate parts of Table 27-7.

C. Data Analysis

1. Which of your predictions matched your experimental results? Which did not? Explain. ______________________________
____
____
____

2. Explain the relationship between blood CO_2 levels and blood pH. ______________________________
____
____
____

PhysioEx EXERCISE 7
Respiratory System Mechanics

The PhysioEx 9.1 Laboratory Simulations in Physiology are easy-to-use laboratory simulations that can be used as an alternative to or as a supplement to wet lab activities in this unit. Each simulation allows you to investigate important physiological concepts, repeat labs as often as you like, and conduct experiments that are difficult to perform in a wet lab environment because of time, cost, or safety concerns.

PEx
Access the simulations in these activities at
MasteringA&P® > Study Area > PhysioEx 9.1

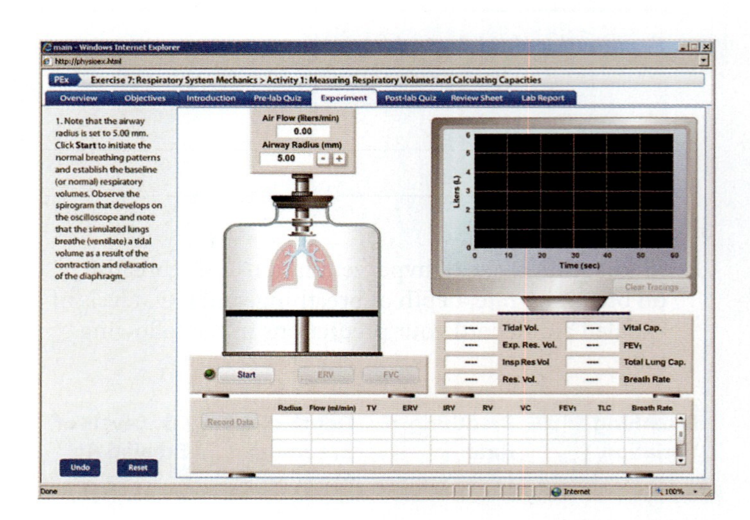

There you will find the following activities:

PEx Activity 1: Measuring Respiratory Volumes and Calculating Capacities

PEx Activity 2: Comparative Spirometry

PEx Activity 3: Effect of Surfactant and Intrapleural Pressure on Respiration

POST-LAB ASSIGNMENTS

Name: ______________________________ Date: ______________ Lab Section: ________

PART I. Check Your Understanding

Activity 1: Analyzing the Model Lung and Pulmonary Ventilation

1. Name the component of the bell jar lung model that corresponds with each of the following respiratory structures:
   a. Lungs ____________________
   b. Diaphragm ____________________
   c. Pleural cavity ____________________
   d. Trachea ____________________
   e. Bronchi ____________________
2. The pressure of a gas in a closed container is ____________________ proportional to the volume of the container.
3. What is a pneumothorax and what does it cause? ____________________

____

Activity 2: Measuring Respiratory Volumes in a Human Subject Using BIOPAC®

1. Why does the predicted vital capacity vary with height? ____________________

____

2. Explain how age and gender might affect the lung capacity. ____________________

____

____

3. How would the volume measurements change if the data were collected after vigorous exercise?

____

____

4. What is the difference between volume measurements and capacities? ____________________

____

Activity 3: Determining Respiratory Volumes and Capacities at Rest and following Exercise

1. Respiratory volumes can be (measured/calculated), whereas respiratory capacities are (measured/calculated). Circle the correct answers.

2. Identify the various lung volumes and capacities in the following diagram:

   a. ______________________ e. ______________________

   b. ______________________ f. ______________________

   c. ______________________ g. ______________________

   d. ______________________ h. ______________________

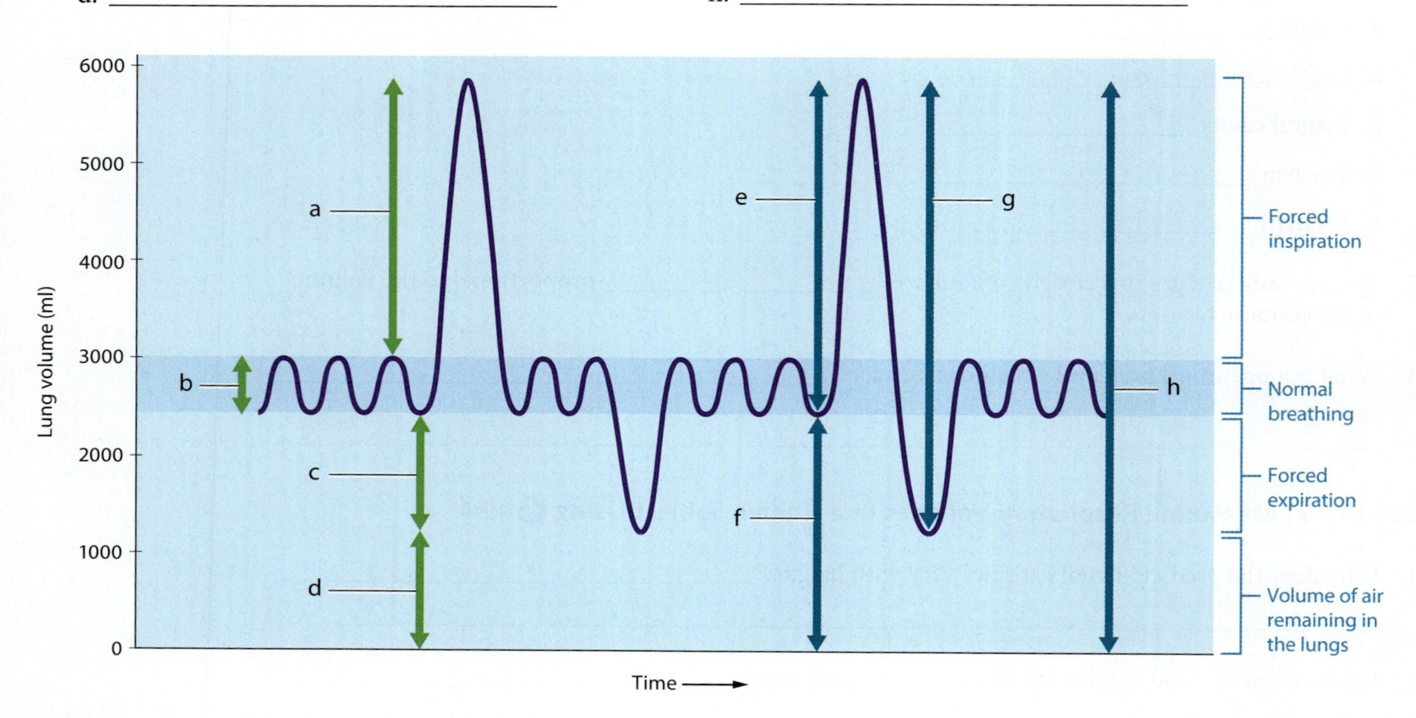

3. Describe the effects of exercise on each of the following quantities:

   a. ERV: ______________________

   b. IRV: ______________________

   c. TV: ______________________

Activity 4: Investigating the Control of Breathing

1. Explain the relationship between blood CO_2 levels and blood pH. ______________________

____

____

2. Based on your experimental results, complete the following chart:

Breathing Pattern	Did Breathing Rate Increase or Decrease?	Did the Amount of Exhaled CO_2 Increase or Decrease? How Do You Know?
Hyperventilation		
Rebreathing		

PART II. Putting It All Together

A. Review Questions

Answer the following questions using your lecture notes, your textbook, and your lab notes.

1. List and briefly describe the five processes involved in providing body cells with oxygen and ridding body cells of carbon dioxide:

____

____

____

____

____

2. How is most oxygen transported in the blood?
   a. It is dissolved in plasma.
   b. It is transported as bicarbonate ions.
   c. It is bound to hemoglobin.
   d. It attaches to the red blood cell's membrane.
   e. None of these statements is correct.

3. The following equations can be used to predict vital capacity in males and females:

   Males: $VC = (0.052 \times H) - (0.022 \times A) - 3.60$

   Females: $VC = (0.041 \times H) - (0.018 \times A) - 2.69$

   where VC = vital capacity in liters, H = height in cm, and A = age in years.

   a. Brian is a 38-year-old male who is 186 cm tall. Calculate his predicted vital capacity: __________

   b. If Brian's actual measured vital capacity is 2.80 L, suggest some possible reasons for the discrepancy between the predicted value and the actual value. ______________________________

   ____

   ____

4. What are chemoreceptors, and how are they involved in the regulation of respiratory rate?

____

____

____

B. Concept Mapping

1. Fill in the blanks to complete this concept map outlining the basic physiological concepts of respiration. Some terms are used more than once.

carbon dioxide hydrogen ion hyperventilation pH

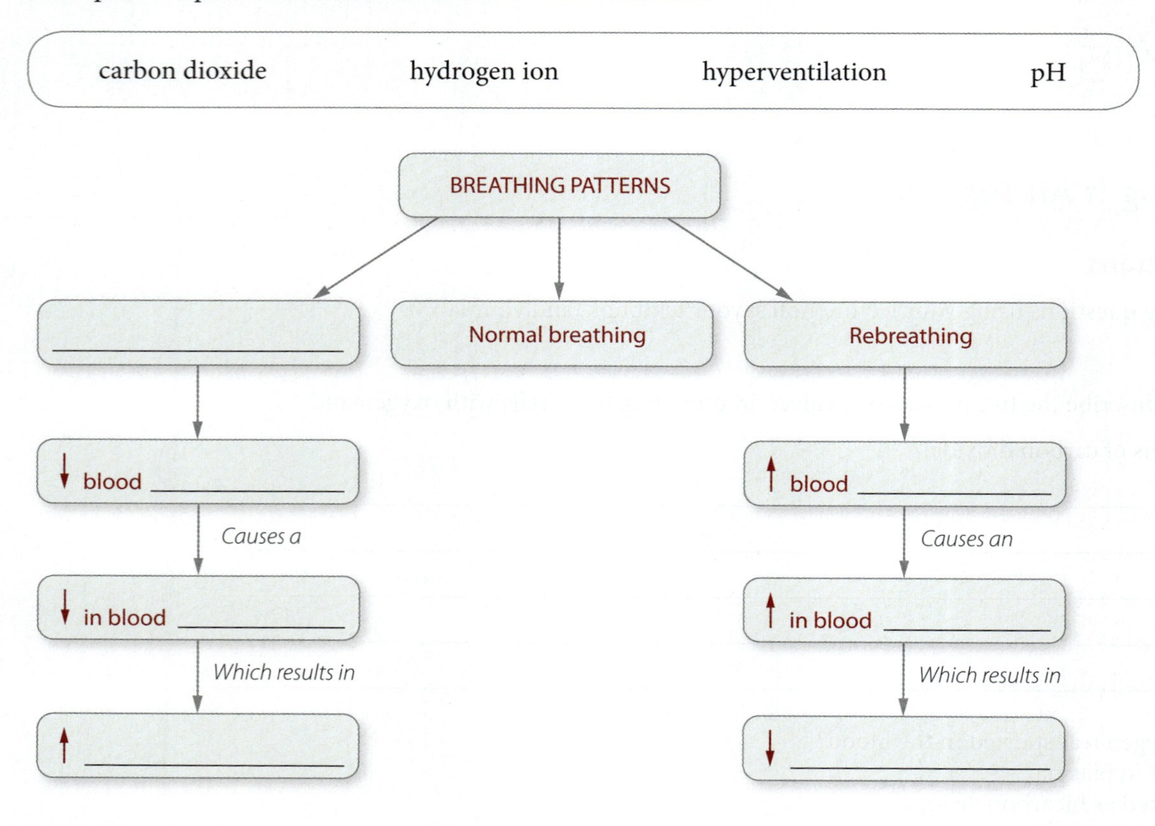

2. Construct a unit concept map to show the relationships among the following set of terms. Include all of the terms in your diagram. Your instructor may choose to assign additional terms.

bicarbonate ion	carbon dioxide	diaphragm	
expiratory reserve volume	external respiration	hydrogen ion	
hyperventilation	inspiratory reserve volume	oxygen	pH
pleural cavity	rebreathing	surfactant	tidal volume	ventilation

28 Anatomy of the Digestive System

UNIT OUTLINE

The food we eat contains important nutrients our bodies need to survive. However, the body cannot directly absorb the nutrients in the food we ingest. The digestive system must first physically and chemically break the macromolecules in food into their building blocks, which are then absorbed into the bloodstream.

The digestive system is composed of the alimentary canal and various accessory organs. The **alimentary canal**, also called the gastrointestinal tract, is a muscular tube that winds through the body from mouth to anus. The organs of the alimentary canal are the oral cavity (mouth), pharynx (throat), esophagus, stomach, small intestine, and large intestine. The **accessory organs** of the digestive system—the teeth, tongue, salivary glands, pancreas, liver, and gallbladder—assist the organs of the alimentary canal in physically and chemically breaking down food.

THINK ABOUT IT *List several cellular-level physiological events that depend on the energy (ATP) that is generated from the food you eat.*

____

____

____

____

PRE-LAB ASSIGNMENTS

Pre-lab quizzes are also assignable in MasteringA&P®

To maximize learning, BEFORE your lab period carefully read this entire lab unit and complete these pre-lab assignments using your textbook, lecture notes, and prior knowledge.

PRE-LAB Activity 1: Exploring the Organs of the Alimentary Canal

1. Use the list of terms to label the structures of the alimentary canal in the accompanying illustration. Check off each term as you label it.

   - ☐ esophagus
   - ☐ large intestine
   - ☐ oral cavity
   - ☐ pharynx
   - ☐ small intestine
   - ☐ stomach

   Use the second list of terms below to label the accessory organs of the digestive system (refer to Activity 2). Check off each term as you label it.

   - ☐ gallbladder
   - ☐ liver
   - ☐ pancreas
   - ☐ salivary glands
   - ☐ teeth
   - ☐ tongue

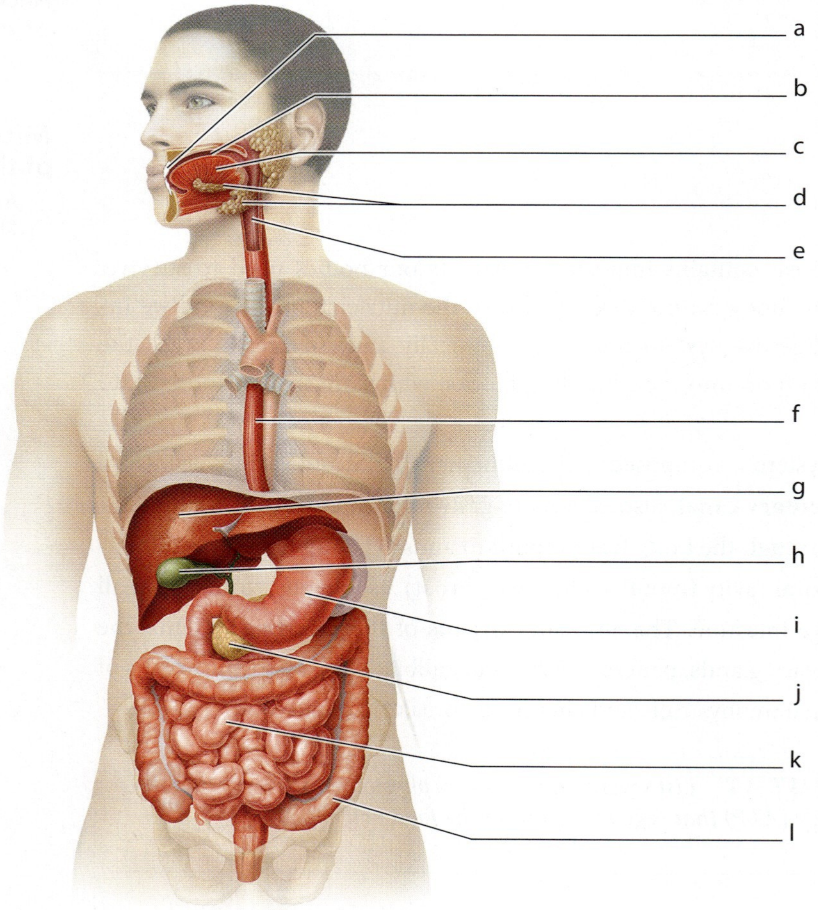

2. Match each of the following descriptions with the correct digestive structure:

_____	a. absorbs water and consolidates waste	1. esophagus
_____	b. site of most digestion and absorption	2. large intestine
_____	c. transports food	3. oral cavity
_____	d. where chemical digestion of carbohydrates begins	4. stomach
_____	e. pummels ingested food into chyme	5. small intestine
_____	f. commonly called the "throat"	6. pharynx

PRE-LAB Activity 2: Exploring the Accessory Organs of the Digestive System

1. Match each of the following descriptions with its correct accessory organ:

_____	a. produces bile	1. gallbladder
_____	b. is both an endocrine organ and an exocrine organ	2. liver
_____	c. is composed primarily of skeletal muscle	3. pancreas
_____	d. stores bile	4. salivary gland
_____	e. produces secretions that moisten food	5. tooth
_____	f. is in an alveolar margin of the mandible and maxilla	6. tongue

PRE-LAB Activity 3: Examining the Histology of Selected Digestive Organs

1. Use the list of terms to label the accompanying illustration of the alimentary canal wall. Check off each term as you label it.

☐ mucosa ☐ muscularis externa ☐ serosa ☐ submucosa

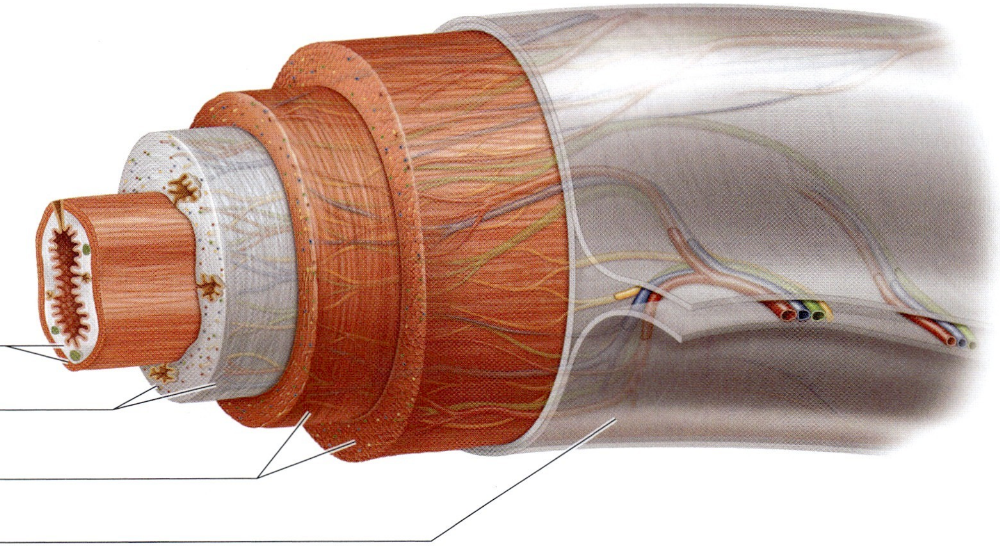

a ____________________

b ____________________

c ____________________

d ____________________

2. Match each of the following cell types with its location:

_____	a. chief cell	1. liver
_____	b. brush border cell	2. pancreas
_____	c. acinar cell	3. small intestine
_____	d. hepatocyte	4. stomach
_____	e. parietal cell	
_____	f. duodenal cell	

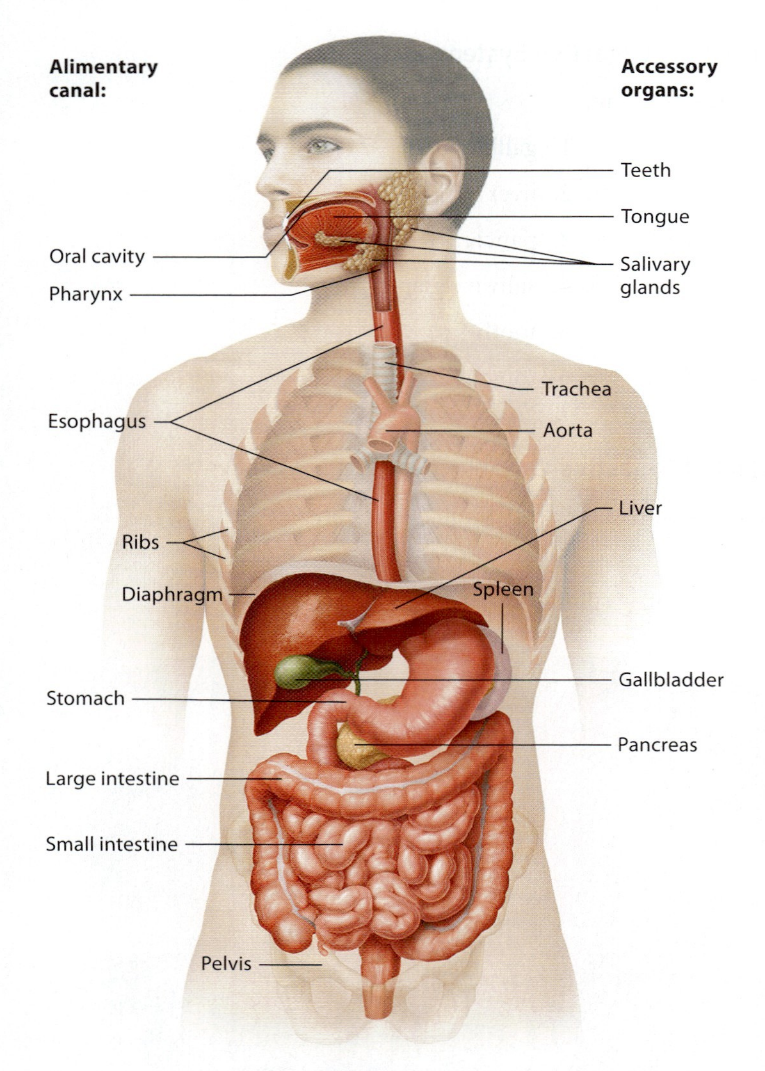

Figure 28-1 **Structures of the digestive system.**

Organs of the Alimentary Canal

The structures of the digestive system—the alimentary canal and accessory digestive organs—are illustrated in **Figure 28-1**. In this section we examine the organs that constitute the alimentary canal, which is approximately 9 meters (20 feet) long in a cadaver, but shorter in a living person due to its muscle tone.

The **oral cavity** (**Figure 28-2**) receives food and is the site at which the mechanical digestion (such as chewing, or *mastication*) and chemical digestion of food begin. Three accessory organs associated with the mouth—the salivary glands, tongue, and teeth—have important roles in digestion and will be discussed later in this unit. The anterior opening of the oral cavity is surrounded by the lips, and its lateral walls are formed by the cheeks. The space that separates the lips and cheeks from the teeth and gums is called the

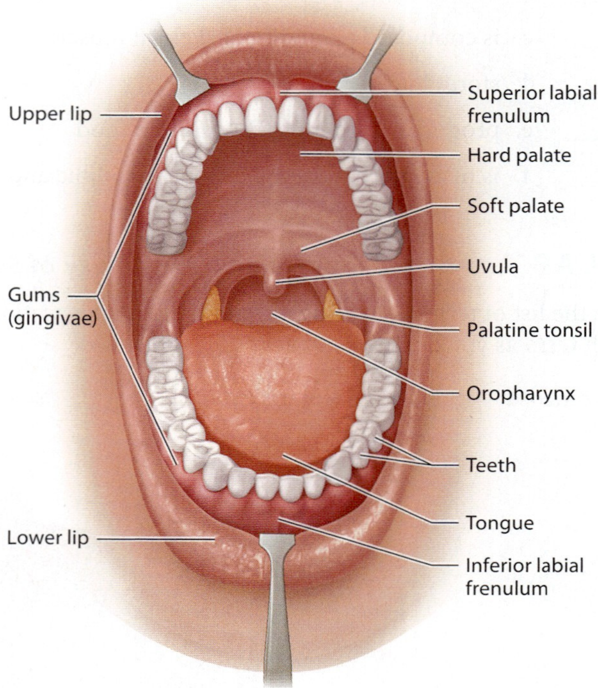

(a) Anterior view of the oral cavity

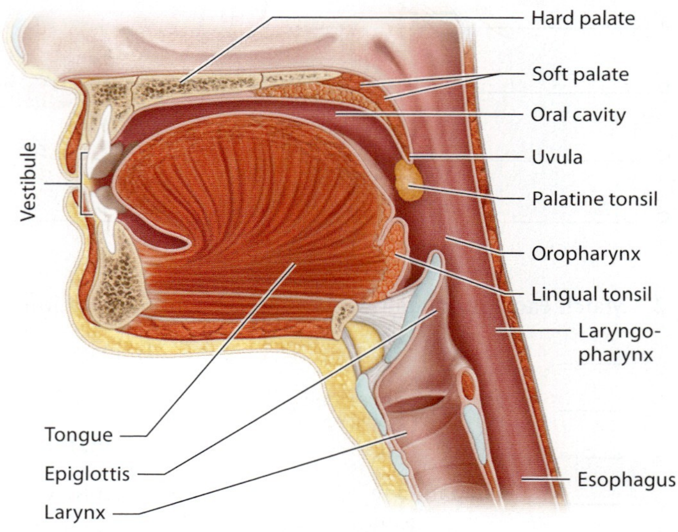

(b) Oral cavity and pharynx, sagittal section

Figure 28-2 **Anatomy of the oral cavity and pharynx.**

vestibule. The **inferior labial frenulum** and the **superior labial frenulum** are median folds that join the lower and upper lips, respectively, to the gums.

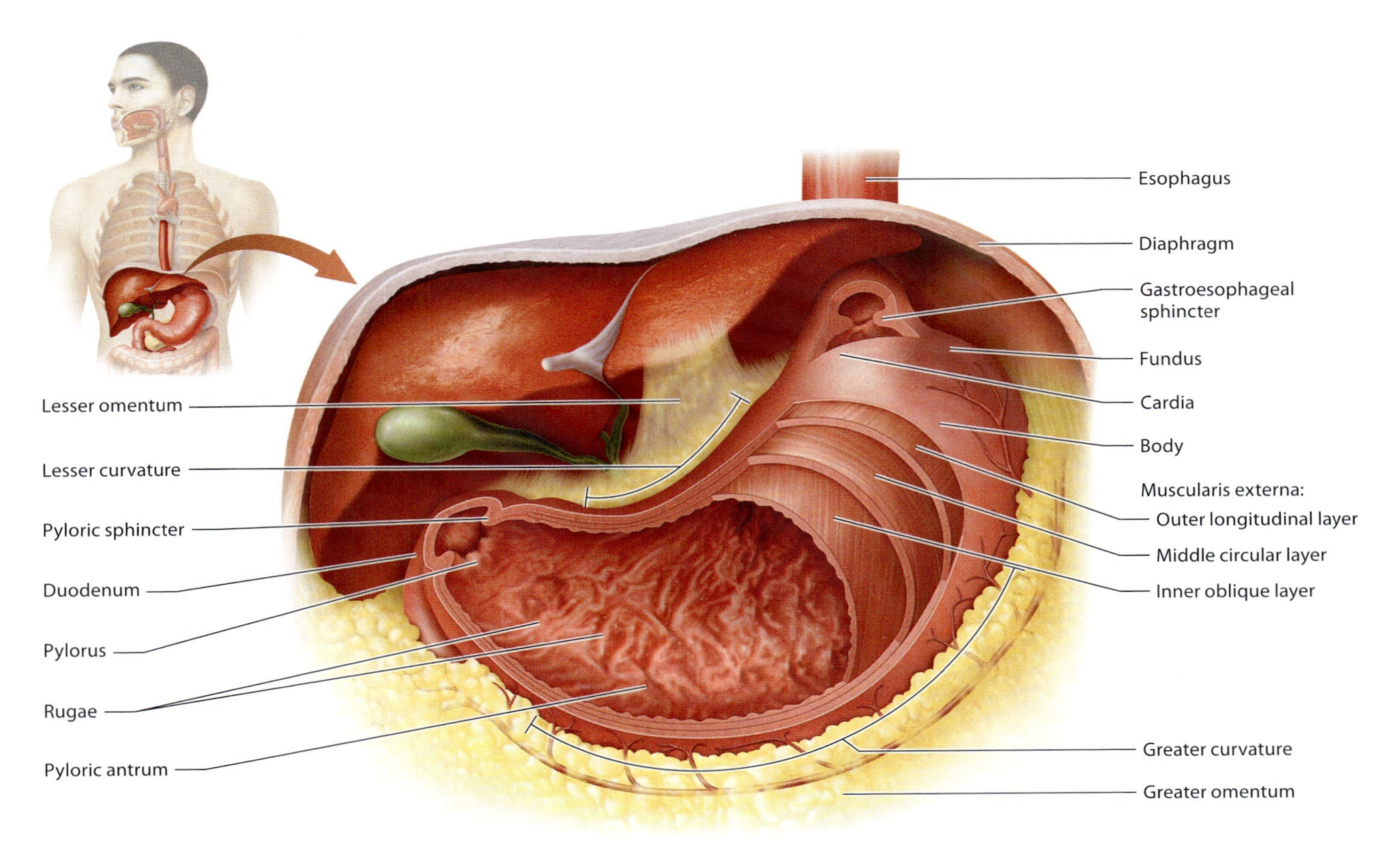

Figure 28-3 Anatomy of the stomach.

The roof of the oral cavity is formed by the palate, which is composed of the anterior **hard palate**, a rigid surface underlain with bone against which the tongue pushes food during chewing, and the posterior **soft palate**, which is primarily skeletal muscle that rises during swallowing to prevent food from entering the respiratory tract. The **uvula** is a finger-like extension of the soft palate that projects downward.

Posteriorly, the oral cavity is continuous with the **pharynx** (see Figure 28-2b), which is divided into the nasopharynx, oropharynx, and laryngopharynx. The nasopharynx, which is located behind the nasal cavity, is a passageway for air only; it is closed off by the soft palate during swallowing. The **oropharynx**, located behind the oral cavity, and the **laryngopharynx**, extending from the epiglottis to the base of the larynx, serve as passageways for food, liquids, and air. Skeletal muscles in the pharyngeal wall initiate rhythmic, wavelike contractions (called peristalsis) that propel food inferiorly along the alimentary canal. Associated with the oropharynx are the palatine tonsils and the lingual tonsil.

The **esophagus** (see Figure 28-1) is a collapsible muscular tube, approximately 25 cm (10 in.) long, that extends through the neck and mediastinum of the thorax. Its only function is as a passageway for food—it has no digestive or absorptive function. As food moves through the laryngopharynx, the epiglottis closes off the larynx, routing food into the esophagus. The esophagus passes through the diaphragm via an opening called the esophageal hiatus and then joins the stomach at the **gastroesophageal sphincter** (also known as the cardiac sphincter), which is a slight thickening of the smooth muscle that controls the passage of food into the stomach.

The **stomach** (**Figure 28-3**) is a J-shaped organ, about 15 to 25 cm (6 to 10 in.) long, located largely in the epigastric region of the abdominopelvic cavity. The muscularis externa of the stomach is made of an outer longitudinal layer, a middle circular layer, and an inner oblique layer. It is divided into four regions: the dome-shaped *fundus*, the *cardia*, the *body*, and the funnel-shaped pyloric region. The pyloric region is further subdivided into the wide, superior *pyloric antrum* and the narrow, inferior *pylorus*. The **pyloric sphincter** connects the pylorus to the duodenum of the small intestine and controls the movement of partially digested food mixed with gastric juices (a slurry mixture called **chyme**) into the duodenum. When the stomach is empty, its inner lining is thrown into folds called **rugae**.

The concave medial surface of the stomach is called the **lesser curvature**; the convex lateral surface is called the

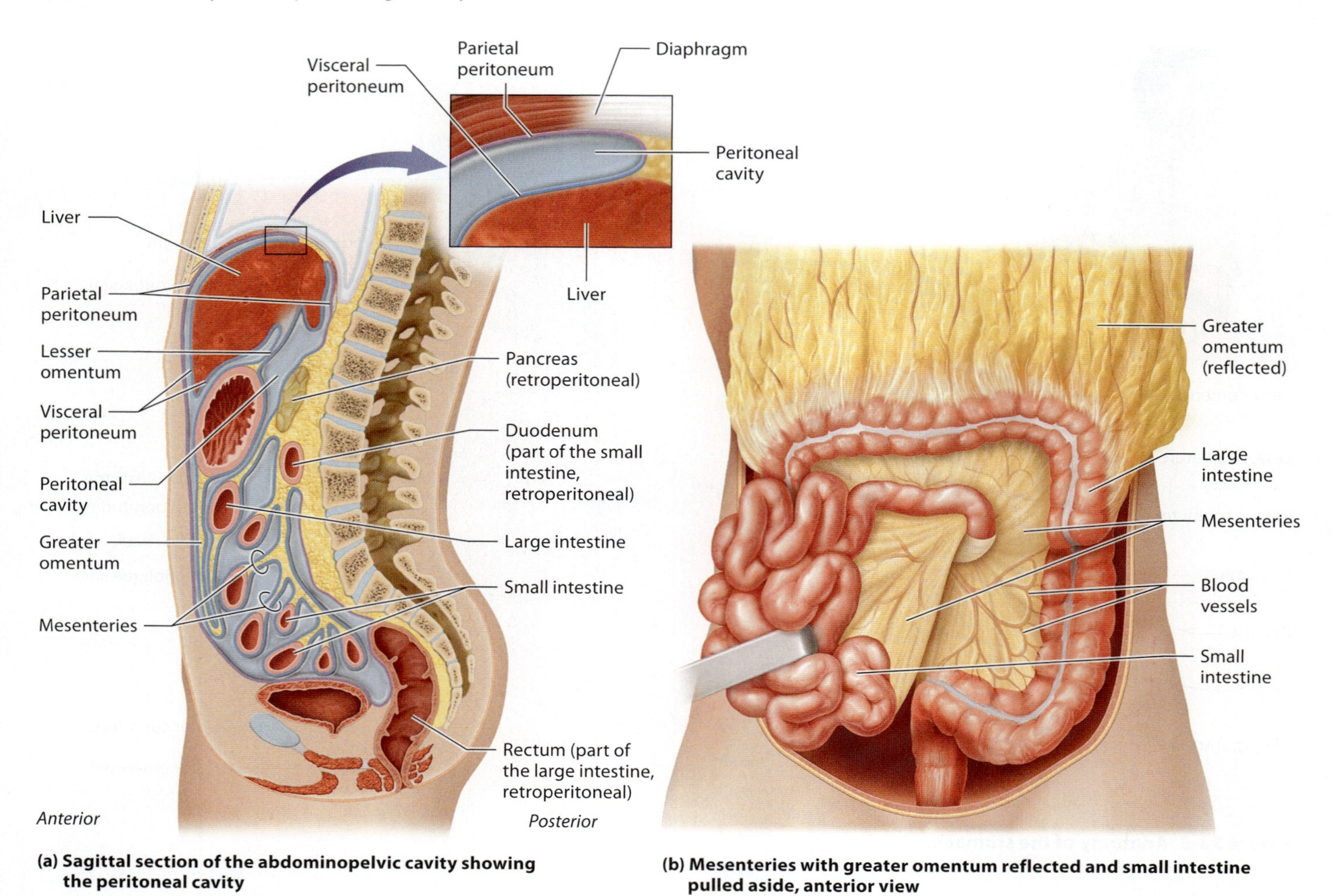

(a) Sagittal section of the abdominopelvic cavity showing the peritoneal cavity

(b) Mesenteries with greater omentum reflected and small intestine pulled aside, anterior view

Figure 28-4 The peritoneum.

greater curvature. Extending from these curvatures are two mesenteries, called omenta, that attach the stomach to other digestive organs and to the body wall. The **lesser omentum** extends from the liver to the lesser curvature; the **greater omentum** extends from the greater curvature to cover the small intestine and wrap around the spleen and transverse colon.

The stomach and most of the organs of the digestive system are located within the abdominopelvic cavity, which is lined with the **peritoneum**. As you learned in Unit 1, the peritoneum is a double-layered serous membrane consisting of the **visceral peritoneum** and the **parietal peritoneum** (**Figure 28-4a**). The visceral peritoneum, also called the serosa, lines the external surface of the abdominal organs and is continuous with the parietal peritoneum, which lines the body wall. Between the visceral and parietal layers of the peritoneal membranes is the peritoneal cavity filled with a slippery fluid produced by the serous membranes. A double sheet of peritoneum, the fan-shaped **mesentery**, suspends the small intestine in the abdominal cavity (see Figure 28-4b). Some mesenteries are given specific names such as the omenta of the stomach. The mesenteries support and stabilize the position of abdominal organs, store fat, and provide an access route for blood vessels, lymphatics, and nerves to and from the digestive tract.

The **small intestine**—the site of most chemical digestion and nutrient absorption—consists of three divisions: the duodenum, jejunum, and ileum (**Figure 28-5**). The **duodenum** is the shortest region, measuring about 25 cm (10 in.) in length, whereas the **jejunum** and **ileum** measure 2.5 m (8 ft) and 3.6 m (12 ft), respectively. The jejunum and the ileum are suspended by the fan-shaped mesentery. Three accessory structures—the liver, gallbladder, and pancreas—add secretions through a duct that enters the duodenum.

The **large intestine** (**Figure 28-6**) is about 1.5 m (5 ft) long and extends from the ileocecal valve, which connects the end of the ileum to the start of the large intestine, to the anus. The large intestine consists of four major divisions: the cecum, colon, rectum, and anal canal. The **cecum** receives contents from the ileum of the small intestine (see Figure 28-6a); extending from the cecum is the worm-like **vermiform appendix**, composed of lymphatic tissue. The

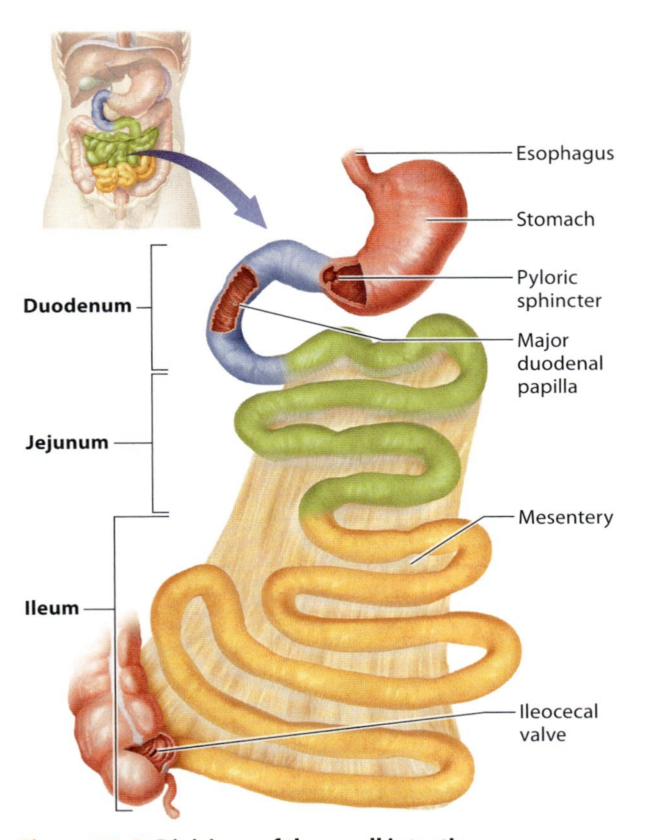

Figure 28-5 Divisions of the small intestine.

colon is further subdivided into four portions: the *ascending colon*, *transverse colon*, *descending colon*, and *sigmoid colon*. The **hepatic flexure** (right colic flexure) is the bend in the colon between the ascending colon and the transverse colon. The **splenic flexure** (left colic flexure) is the bend in the colon between the transverse colon and the descending colon. The large intestine reabsorbs water from the undigested material and forms feces.

The **rectum** begins at the level of the third sacral vertebra and extends inferiorly just anterior to the sacrum. The **anal canal**, the last division of the large intestine, lies external to the abdominopelvic cavity, penetrating the levator ani muscle and terminating at the **anus**, the opening to the exterior of the body (see Figure 28-6b). The anal canal has two sphincters: an **internal anal sphincter** composed of smooth muscle, and an **external anal sphincter** composed of skeletal muscle.

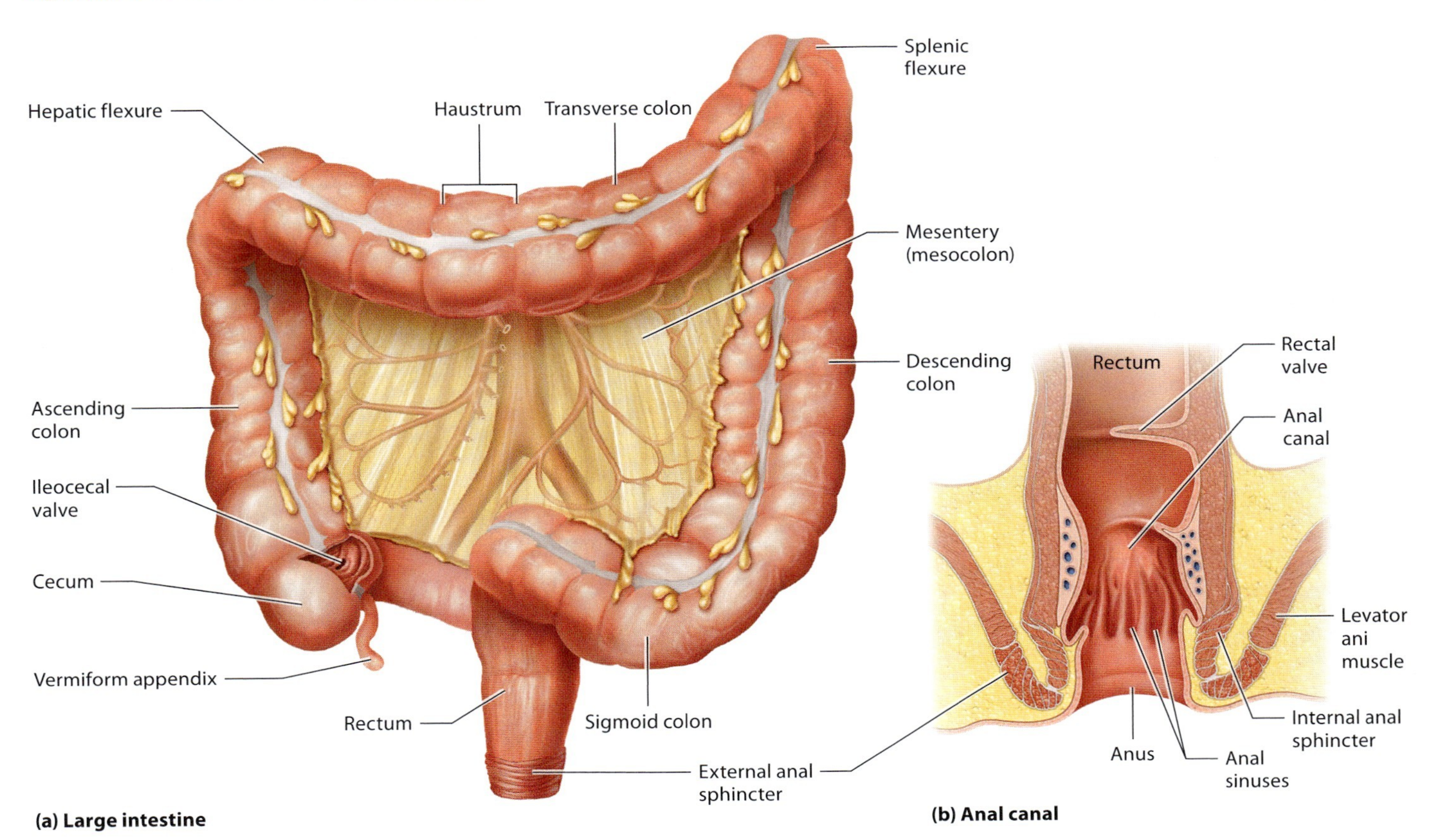

Figure 28-6 Anatomy of the large intestine.

ACTIVITY 1

Exploring the Organs of the Alimentary Canal

Learning Outcomes

1. Identify the organs of the alimentary canal on an anatomical model.
2. Describe the structure and function of each organ of the alimentary canal.

Materials Needed

- ☐ Anatomical models of the digestive organs including a midsagittal model of the head
- ☐ Anatomical charts

Instructions

1. CHART As you complete the following Making Connections chart, locate each alimentary canal structure on an anatomical model or chart and identify its anatomical features. Then write a description of each structure and some of its functions. Finally, make some "connections" to things you've already learned in your lectures, assigned reading, and lab. Some portions of the chart have been filled in as examples.
2. Answer the following questions concerning the organs of the alimentary canal:
   a. Identify the finger-like extension of the soft palate, the uvula.

      What is the function of the uvula? ____________

      ____

   b. Use your finger to feel your hard palate.

      Describe and explain the functional significance of what you feel. ____________

      ____

      ____

Making Connections: Organs of the Alimentary Canal		
Structure	**Description (Structure and/or Function)**	**Connections to Things I Have Already Learned**
Oral cavity	*Site of ingestion of food, mastication; contains muscular tongue; teeth chew, salivary glands produce saliva.*	
Pharynx		
Esophagus		*Is supplied by both the esophageal artery (visceral branch of the thoracic aorta) and a branch of the celiac trunk.*
Stomach		
Small intestine	*Most digestion and absorption take place here; three divisions: duodenum, jejunum, and ilium.*	
Large intestine		

c. Obtain a midsagittal model of the head.

Identify and name the three regions of the pharynx, and describe the location of each: ____________

____

____

____

____

In your answer above, place an * next to the pharyngeal regions that have a digestive function.

d. Identify the esophagus on an anatomical model and describe its location with respect to the trachea.

____

e. Identify the following structures on a stomach model:

- Four regions of the stomach (fundus, cardia, body, pyloric region)
- Rugae
- Lesser curvature and greater curvature

  Which structure extends from the liver to the lesser curvature? ____________

  Which structure extends from the greater curvature to cover the small intestine? ____________

- Gastroesophageal sphincter

  What is the function of this sphincter muscle?

  ____

  ____

- Pyloric sphincter

  What is the function of this sphincter muscle?

  ____

  ____

f. Identify the three divisions of the small intestine on an anatomical model.

List them in order from shortest to longest. ____________

____

g. Locate the ileocecal valve.

This valve is located between which two structures?

____

____

h. Locate the four major divisions of the large intestine on an anatomical model.

Identify the worm-like structure that extends from the cecum. Describe it and its function. ____________

____

i. Identify the four portions of the colon on an anatomical model and locate the hepatic and splenic flexures.

Describe the location of the hepatic and splenic flexures.

____

____

____

3. **Optional Activity**

PAL™ Practice labeling organs of the alimentary canal on human cadavers at > MasteringA&P® > Study Area > Practice Anatomy Lab > Human Cadaver > Digestive System

Accessory Organs of the Digestive System

Accessory organs of the digestive system include the teeth, tongue, salivary glands, liver, gallbladder, and pancreas.

The **teeth**, which sit in sockets (alveoli) of the mandible and maxilla, are classified into groups based on their shape and corresponding function: **incisors** cut food, **canines** tear food, and **premolars** and **molars** crush and grind food (**Figure 28-7a**). Humans have two sets of teeth, or dentitions: a **primary dentition** consisting of **deciduous teeth**, and a **secondary dentition** consisting of **permanent teeth** (**Figure 28-7b**).

Each tooth consists of two major regions: the crown and the root (**Figure 28-8**). The **crown** is the part of the tooth seen above the gum (gingiva). The **neck** of a tooth is the constricted region where the crown and root meet at the gum line. The **root** is the portion of the tooth embedded in the alveolar socket of the maxilla or mandible; it is held in place by the **periodontal ligament**.

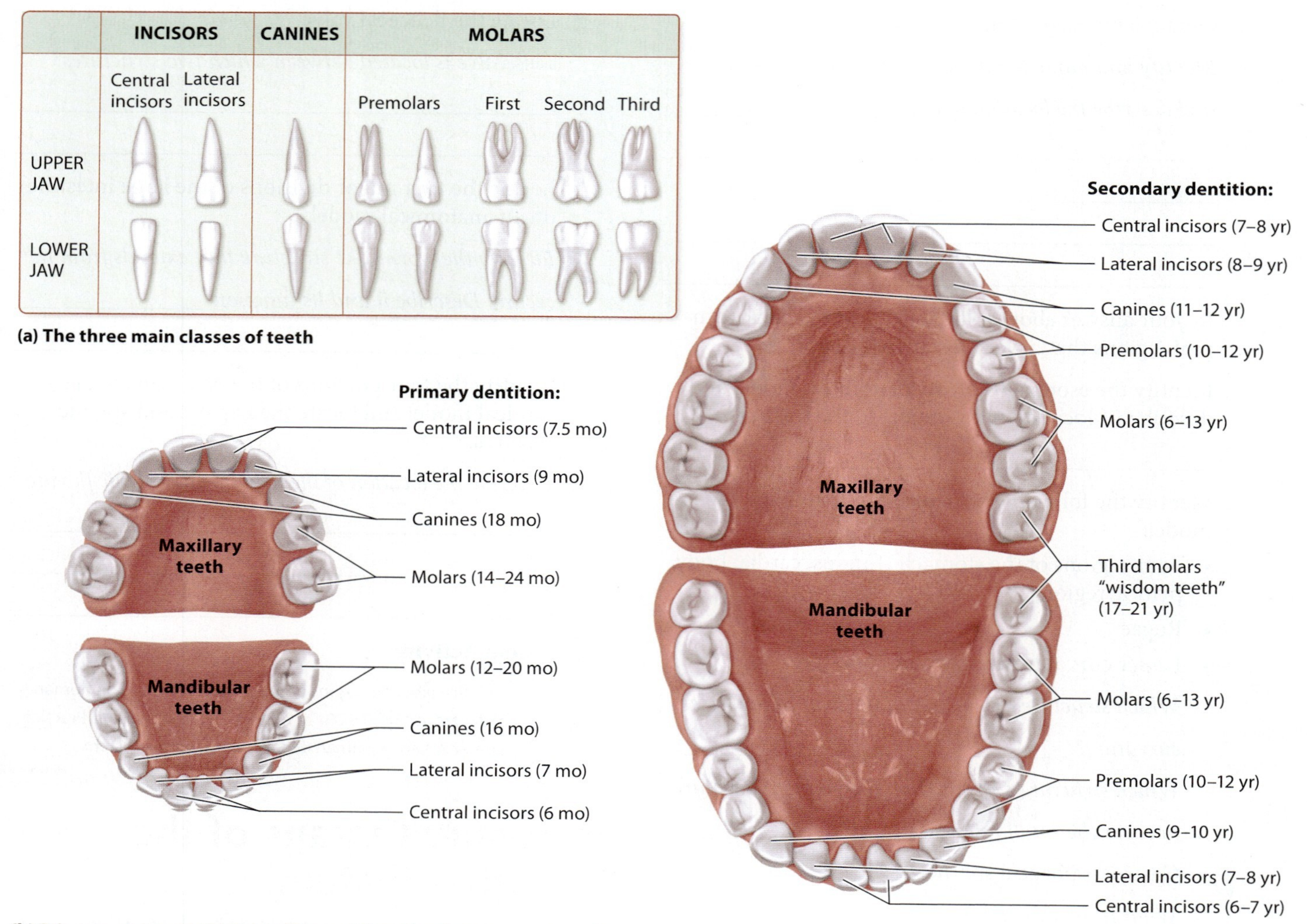

Figure 28-7 Types of teeth and the primary and secondary dentitions.

Most of a tooth is composed of **dentin**, a protein-rich, bone-like connective tissue. **Enamel**, a ceramic-like material that covers and protects the dentin of a tooth's crown, is the hardest substance in the body. **Cementum** covers the dentin of the root and attaches the tooth to the periodontal ligament. Deep to the dentin is a central **pulp cavity** filled with connective tissue, called **pulp**, that contains abundant blood vessels, lymphatics, and nerves. Pulp supplies the cells of the tooth with oxygen and nutrients and enables tooth sensation. As the pulp cavity extends into the distal root, it becomes the **root canal**. The blood vessels and nerves that supply the pulp pass through an opening called the apical foramen (not shown).

The **tongue** (see Figure 28-2), which is composed of skeletal muscles, aids the teeth in chewing (mastication) and participates in mixing the food with saliva. It also initiates the process of swallowing (deglutition) by pushing the ball of food (bolus) posteriorly into the oropharynx. The superior surface of the tongue contains **filiform papillae**, peg-like projections

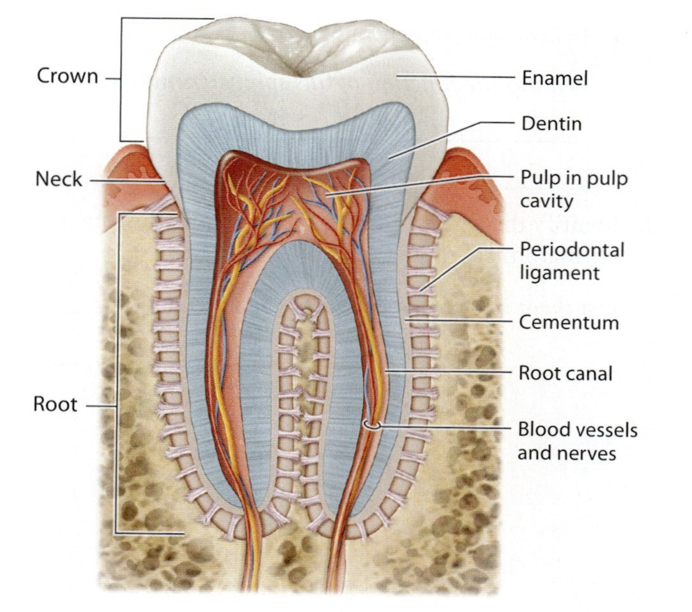

Figure 28-8 Anatomy of a tooth.

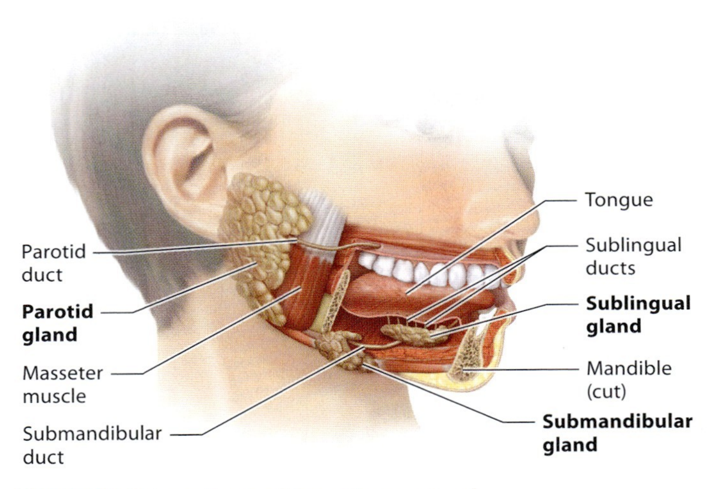

Figure 28-9 **Locations of the salivary glands.**

of the mucosa that provide a rough surface that aids in the physical breakdown of food. These papillae lack taste buds.

Intrinsic tongue muscles are not attached to bone and allow the tongue to change shape as needed for speech and swallowing. Extrinsic muscles, which extend to the tongue from their points of origin on bones of the skull or soft palate, alter the tongue's position by protruding it, retracting it, or moving it from side to side. The lingual frenulum anchors the tongue to the floor of the oral cavity.

Three major pairs of **salivary glands** (**Figure 28-9**), located in proximity to the oral cavity, produce saliva containing water, mucus, salivary amylase (a carbohydrate-digesting enzyme), and protective proteins such as antibodies, lysozyme, and defensins. The large **parotid glands** are located anterior to the ear and superficial to the masseter muscle; the parotid duct transports saliva to the oral cavity. The **submandibular glands** are located just medial to the mandible, and the small **sublingual glands** are located anteriorly in the floor of the oral cavity.

The **liver** (**Figure 28-10**), the largest gland in the body, is a reddish-brown, wedge-shaped structure weighing about 1.4 kg (3 lb) in an average adult. The liver consists of four primary **lobes**: the large left and right lobes (see Figure 28-10a and b) and the small, posterior caudate and quadrate lobes (see Figure 28-10b). The **falciform ligament** separates the right and left lobes anteriorly (see Figure 28-10a) and also suspends the liver from the diaphragm and the anterior abdominal wall. Running along the inferior edge of the falciform ligament is the round ligament (ligamentum teres), which is a remnant of the umbilical vein. The hepatic portal vein and hepatic artery enter the liver at the **porta hepatis** (see Figure 28-10b).

The liver performs a wide variety of important functions. Of greatest importance in our discussion of the digestive system is the production of bile, which plays an important role in the digestion of fats. The bile produced by the liver is stored in the **gallbladder**, a small, thin-walled green muscular sac that is about 10 cm (4 in.) long (see Figure 28-10).

The **pancreas**, located posterior to the stomach, consists of a head, body, and tail (**Figure 28-11**). The head is the broad portion of the pancreas surrounded by the C-shaped duodenum. The body is the organ's middle portion, and the tail extends laterally toward the spleen. The **main pancreatic duct** transports digestive juices to the duodenum.

Figure 28-12 illustrates the anatomical relationships among the liver, gallbladder, and pancreas, as well as the major ducts that transport bile and pancreatic juices to the

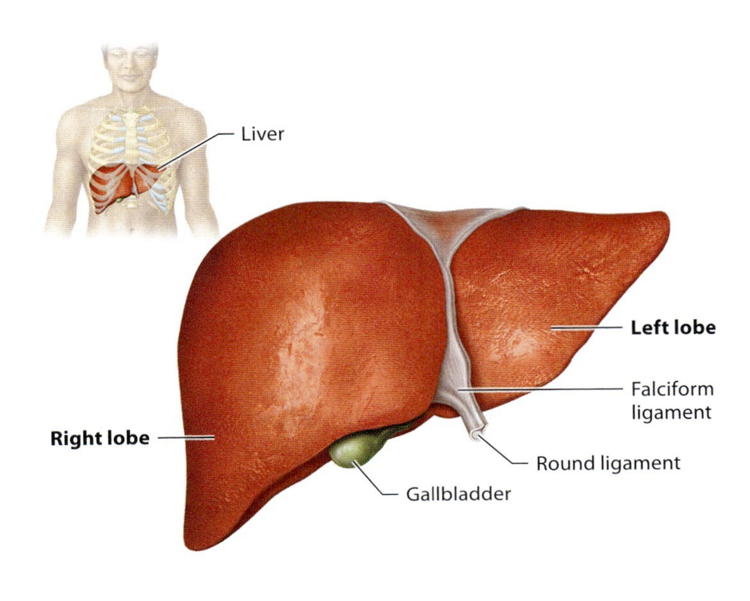

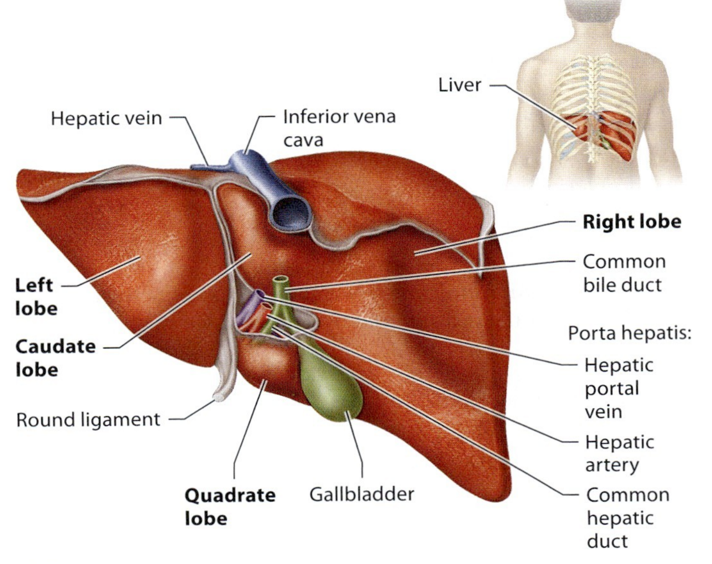

Figure 28-10 **Anatomy of the liver.**

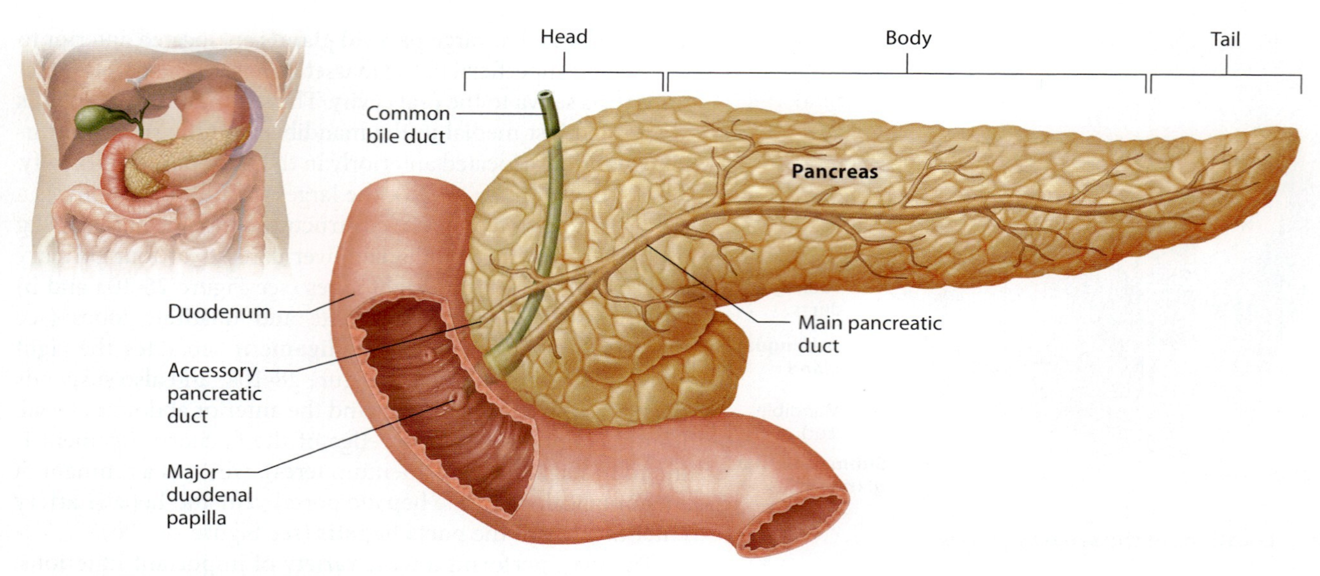

Figure 28-11 Anatomy of the pancreas.

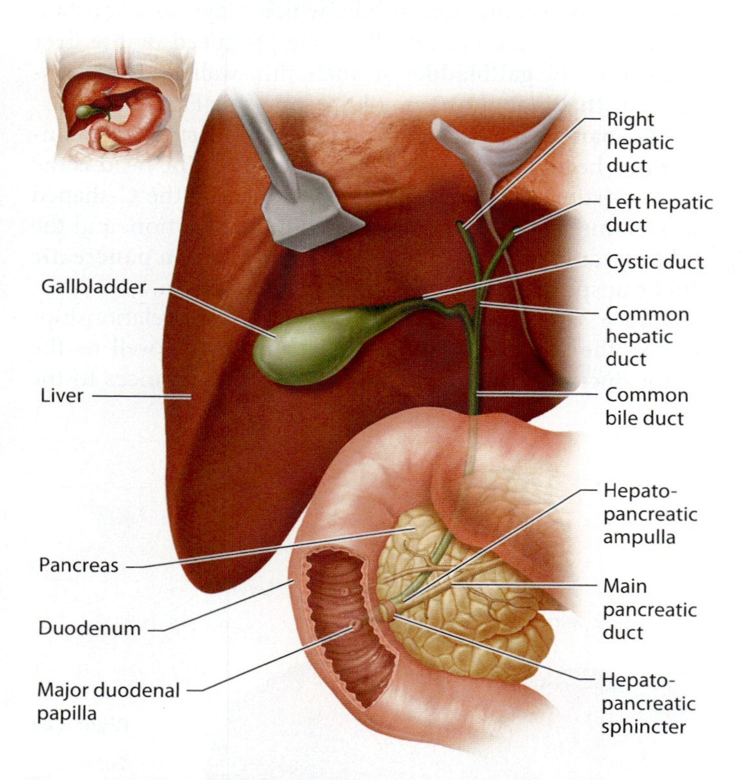

Figure 28-12 Anatomical relationships among the liver, gallbladder, and pancreas, and their respective ducts.

duodenum. Bile exits the liver through the **right** and **left hepatic ducts**, which fuse to form the **common hepatic duct**. Bile is released from the gallbladder via the **cystic duct**, which fuses with the **common hepatic duct** to form the **common bile duct**, which then empties into the duodenum. The **main pancreatic duct** typically fuses with the bile duct just as it empties into the duodenum. The point at which the main pancreatic duct and the bile duct unite in the duodenal wall is the **hepatopancreatic ampulla**, which opens into the duodenum via the **major duodenal papilla**. The **hepatopancreatic sphincter**, a valve composed of smooth muscle, controls the entry of pancreatic juice and bile into the duodenum.

ACTIVITY 2 Exploring the Accessory Organs of the Digestive System

Learning Outcomes

1. Identify the accessory organs of the digestive system on an anatomical model, and describe the structure and function of each.
2. Identify the different types of teeth, and describe how the structure of each is related to its function.
3. Trace the pathway of bile from the time it is produced until it reaches the duodenum.

Materials Needed

- ☐ Torso model
- ☐ Midsagittal head model
- ☐ Skull
- ☐ Anatomical charts

Instructions

1. CHART As you complete the following Making Connections chart, locate each accessory organ of the digestive system on an anatomical model or chart and identify its anatomical features. Then write a description of each structure and some of its functions. Finally, make some "connections" to things you've already learned in your lectures, assigned reading, and lab. Some portions of the chart have been filled in as examples.

Making Connections: Accessory Organs of the Digestive System

Structure	Description (Structure and/or Function)	Connections to Things I Have Already Learned
Salivary glands	*Produce saliva containing amylase (an enzyme that begins the chemical digestion of carbohydrates).*	
Teeth		*Sit in sockets (alveoli) located in mandible and maxilla; innervated by branches of trigeminal nerve (CN V); supplied by branches of the maxillary artery, which branches from the external carotid artery.*
Tongue	*Is composed of skeletal muscles; helps mix food with saliva; contains taste buds.*	
Liver		
Gallbladder	*Stores bile.*	
Pancreas		*Has both endocrine and exocrine components; endocrine pancreas releases insulin directly into the bloodstream; exocrine pancreas releases digestive enzymes into duodenum via pancreatic duct.*

2. Identify the tongue on the midsagittal head model.

   Describe three ways in which the tongue is structurally adapted to its functions. ______________________

   ____

   ____

3. Identify incisors, canines, premolars, and molars on a skull.

   Describe how the following tooth types are adapted to their functions:

   Incisor ______________________

   ____

   Canine ______________________

   ____

   Premolars and molars ______________________

   ____

4. Distinguish between the crown and the root of a tooth on an anatomical model or chart.

   Identify the following tooth structures, and describe the function of each:

   Enamel ______________________

   ____

   Dentin ______________________

   ____

   Periodontal ligament ______________________

   ____

   Cementum ______________________

   ____

   Pulp cavity ______________________

   ____

   Root canal ______________________

   ____

5. Locate the three pairs of salivary glands on an anatomical model.

   a. Name each pair of glands and describe its location.

   ____

   ____

   ____

   ____

Are these glands classified as endocrine glands or exocrine glands? ______________ *Explain.*

____

____

b. Locate the parotid duct.

What is the function of this duct? ______________

____

6. Identify the liver on an anatomical model.
   a. Use your lecture notes and/or text to briefly describe five functions of the liver. ______________

   ____

   ____

   ____

   ____

   ____

   b. Identify the right and left the lobes in an anterior view.

   What structure separates these two lobes and anchors the liver to the anterior body wall? ______________

   ____

   c. Identify the caudate and quadrate lobes in an inferior view.

   The gallbladder is located in a depression within which lobe? ______________

   d. Identify the porta hepatis.

   Which structures enter or exit the liver at the porta hepatis? ______________

   ____

7. Identify the common hepatic duct, the gallbladder, and the cystic duct on an anatomical model.

   The cystic duct fuses with the common hepatic duct to form the ______________.

   Name the structures through which bile flows from the point at which it is produced until it reaches the duodenum.

   ____

   ____

   ____

8. Identify the head, body, and tail of the pancreas on an anatomical model.

   Identify the pancreatic duct, and describe its appearance.

   ____

   ____

   What is the function of the pancreatic duct? ______________

   ____

9. **Optional Activity**

Practice labeling accessory digestive organs on human cadavers at > MasteringA&P® > Study Area > Practice Anatomy Lab > Human Cadaver > Digestive System

Microscopic Anatomy of the Digestive System

We have seen that the alimentary canal is a continuous hollow tube starting at the oral cavity and ending at the anus. When magnified, the wall of most alimentary canal organs can be seen to consist of four layers (or tunics): from innermost to outermost, the mucosa, submucosa, muscularis externa, and serosa (**Figure 28-13**).

The **mucosa** (or lining) consists of epithelial tissue separated from an underlying lamina propria by a basement membrane. This layer also contains a thin layer of smooth muscle called the muscularis mucosae. The surface of the mucosa is generally covered with a layer of mucus that protects the underlying epithelium from the effects of digestive enzymes and HCl. The **submucosa**, which surrounds the mucosa, is a layer of connective tissue containing abundant blood vessels, lymphatic vessels, nerves, and glands. The **muscularis externa**, which surrounds the submucosa, typically contains an outer longitudinal layer and an inner circular layer of smooth muscle tissue. These layers contract alternately to produce rhythmic peristaltic contractions, which advance the digestive contents along the alimentary canal. In most alimentary canal organs, the outermost **serosa** is composed of a single layer of simple squamous epithelium (mesothelium) sitting on top of a layer of areolar connective tissue.

The anatomical organization illustrated in Figure 28-13 is typical for most of the alimentary canal, but there are notable exceptions, many of which will be addressed as you study the microscopic anatomy of selected digestive system organs. Next we will take a look at the histology of selected organs of both the alimentary canal and the accessory organs of the digestive tract.

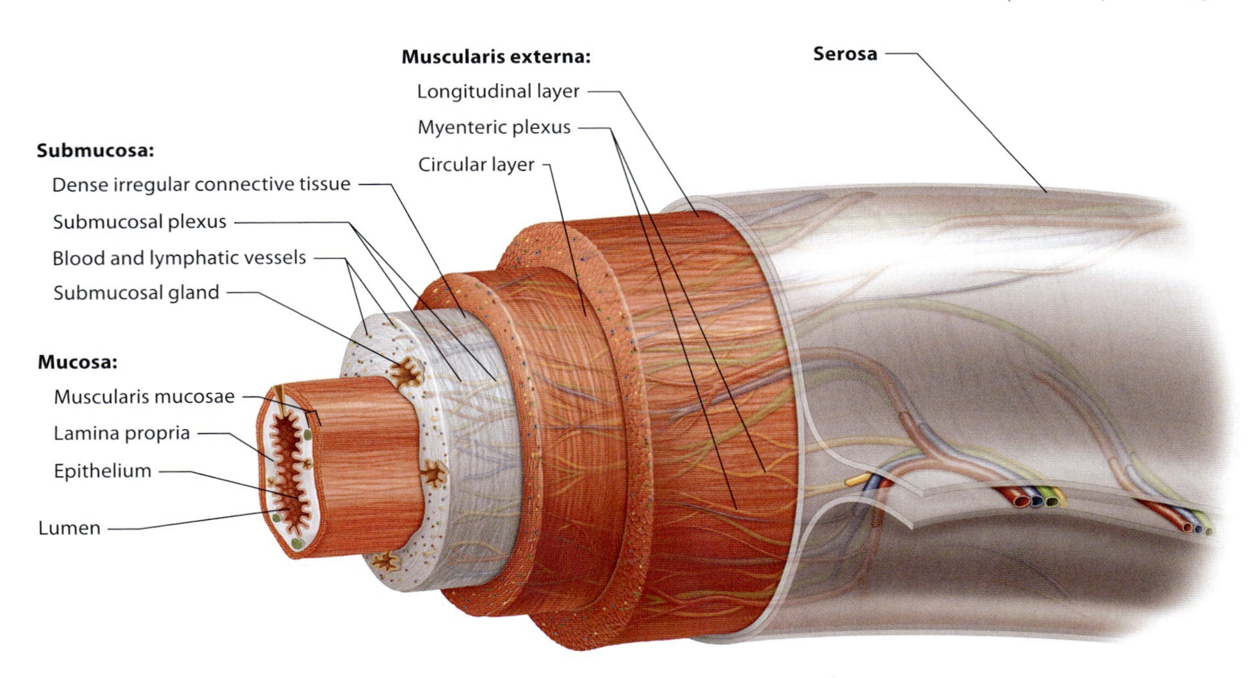

Figure 28-13 **Layers of the wall of the alimentary canal.**

Histology of the Esophagus

The histological features of the esophagus are shown in **Figure 28-14**. The esophageal mucosa contains a moist, stratified squamous epithelium separated from an underlying lamina propria by a basement membrane. The submucosa contains abundant mucus-secreting esophageal glands that secrete onto the luminal surface of the esophageal wall a slippery mucus that aids the transport of food toward the stomach. The muscularis externa of the esophagus is unique: The superior one-third contains skeletal muscle, the middle third contains a combination of smooth and skeletal muscle (see Figure 28-14a), and the inferior one-third contains smooth muscle only (see Figure 28-14b). The outermost layer of the esophageal wall is not serosa, but instead a fibrous adventitia composed of connective tissue that fuses with the connective tissue of adjacent structures.

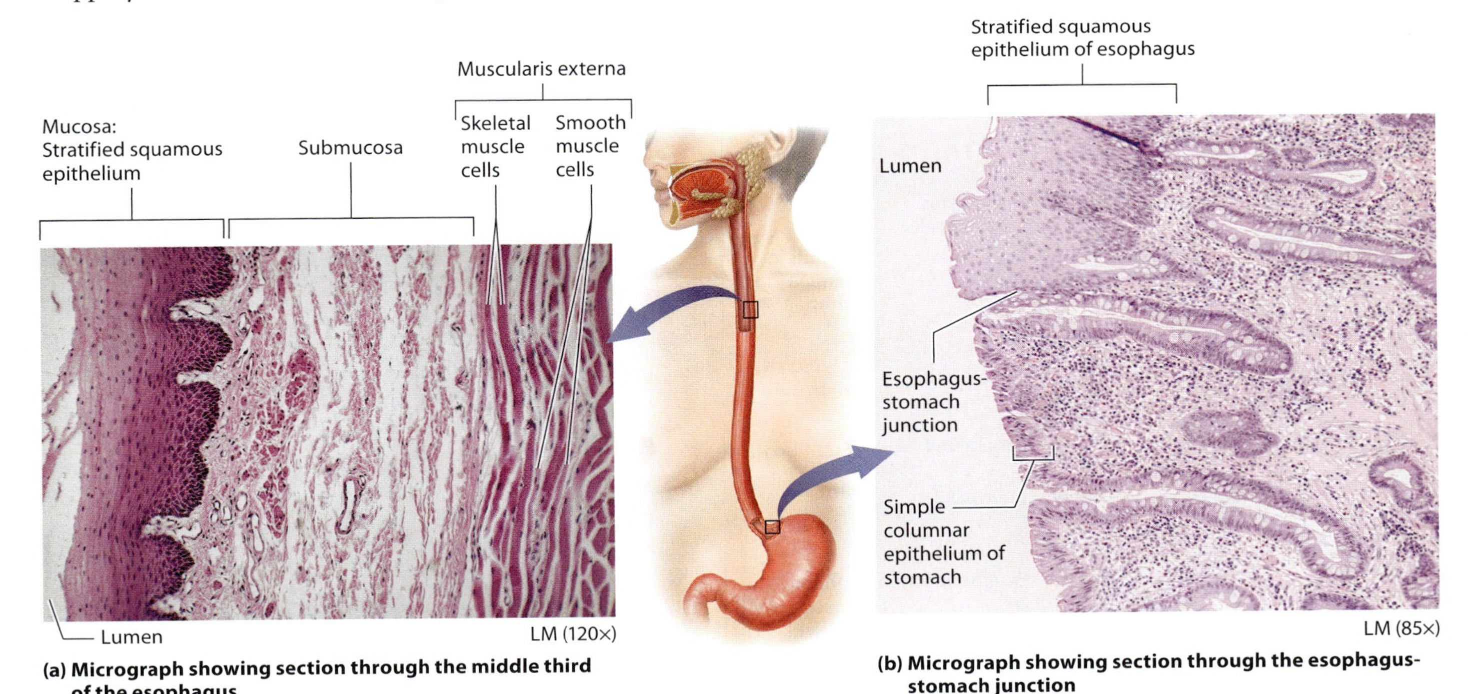

(a) Micrograph showing section through the middle third of the esophagus

(b) Micrograph showing section through the esophagus-stomach junction

Figure 28-14 **Microscopic anatomy of the esophagus.**

Histology of the Stomach

The histologic features of the stomach are shown in **Figure 28-15**. The stomach mucosa consists of a simple columnar epithelium characterized by many indentations called **gastric pits**, which lead into structures called gastric glands that produce gastric juice (see Figure 28-15a). The epithelial cells located between the gastric pits are primarily mucous neck cells, which secrete a thick alkaline mucus that protects the cells of the stomach wall from the effects of digestive enzymes and HCl.

Gastric glands (see Figure 28-15b) are composed of several types of specialized cells. **Mucous neck cells** produce mucus; **parietal cells** produce HCl and intrinsic factor. HCl produces an acidic environment that activates the protein-digesting enzyme called pepsin, whereas intrinsic factor aids in the absorption of vitamin B_{12}. **Chief cells** produce pepsinogen, an inactive

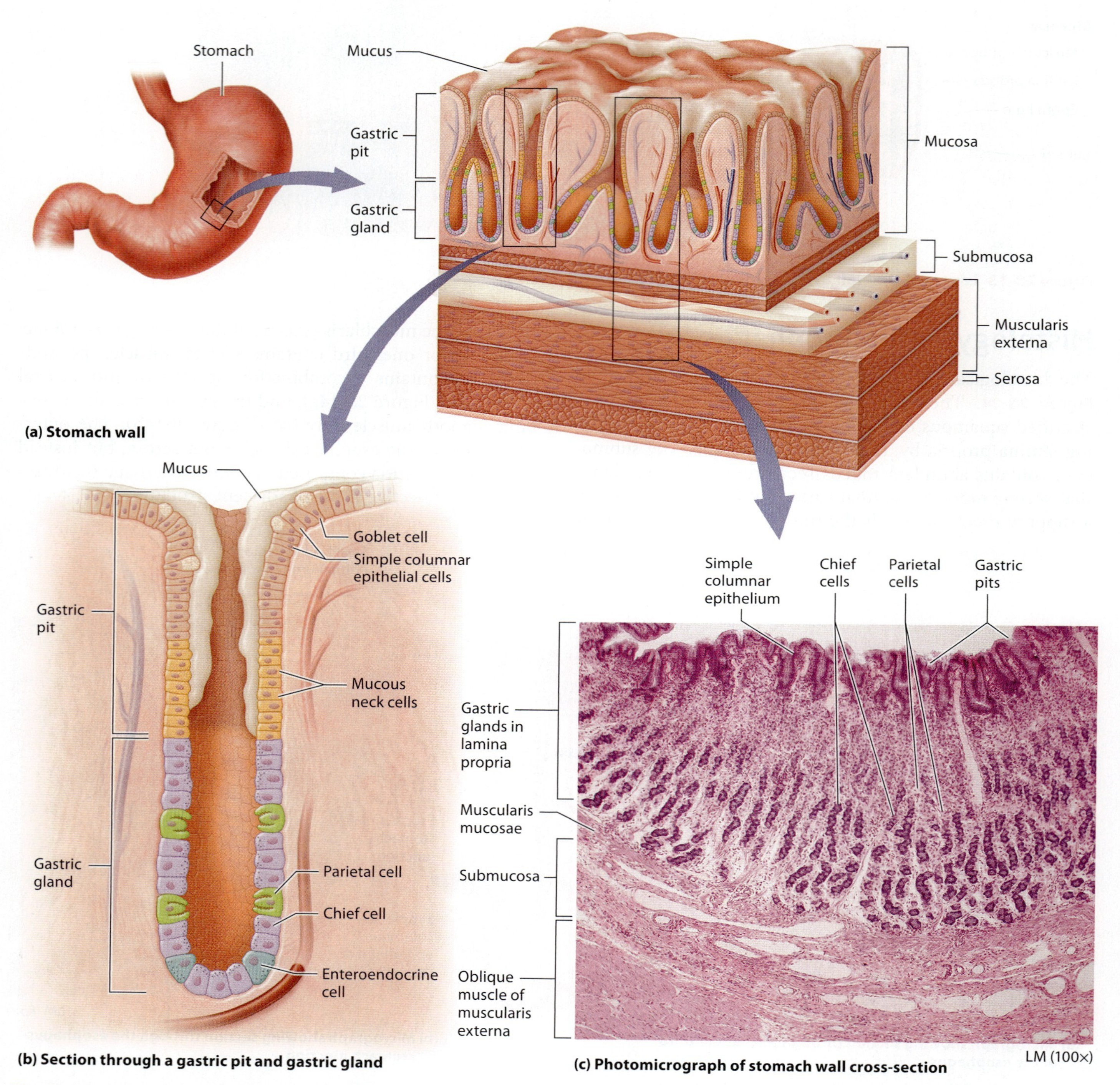

Figure 28-15 Microscopic anatomy of the stomach.

precursor of pepsin. When pepsinogen comes into contact with HCl, it is converted into pepsin, a protein-digesting enzyme. **Enteroendocrine cells** produce the hormones gastrin, which stimulates secretory activity of gastric glands, and somatostatin, which inhibits the release of gastric juices.

Histology of the Small Intestine

Figure 28-16 shows the microscopic structure of the small intestine, where the nutrients in food are absorbed. Several microscopic features produce the maximum absorptive surface possible by increasing the surface area of the small intestine. **Circular folds** (plicae circulares) are deep folds of the mucosa and submucosa. Finger-like projections of the mucosa called **villi** are lined with a simple columnar epithelium that surrounds a capillary bed and a specialized lymphatic capillary called a lacteal. The columnar epithelial cells have cytoplasmic extensions called **microvilli**, which also increase the surface area for absorption. Between the villi are indentations that lead into tubular glands called **intestinal crypts**; their cells secrete intestinal juice, a watery mixture of mucus, hormones, and various antimicrobial substances.

The mucosa is also characterized by lymphoid structures called **Peyer's patches** (not shown), which prevent bacteria from entering the bloodstream. The submucosa of the small intestine contains areolar connective tissue; in the duodenum, it is rich in mucus-secreting **duodenal glands** (**Figure 28-17**). The muscularis externa of the small intestine is typical, with inner circular and outer longitudinal layers of muscle. Most of the small intestine is covered by serosa; the exception is the duodenum, which is mostly retroperitoneal and covered by adventitia.

Histology of the Large Intestine

The large intestine absorbs water and consolidates fecal material. Because neither digestion nor significant absorption of nutrients occurs there, the microscopic anatomy of the

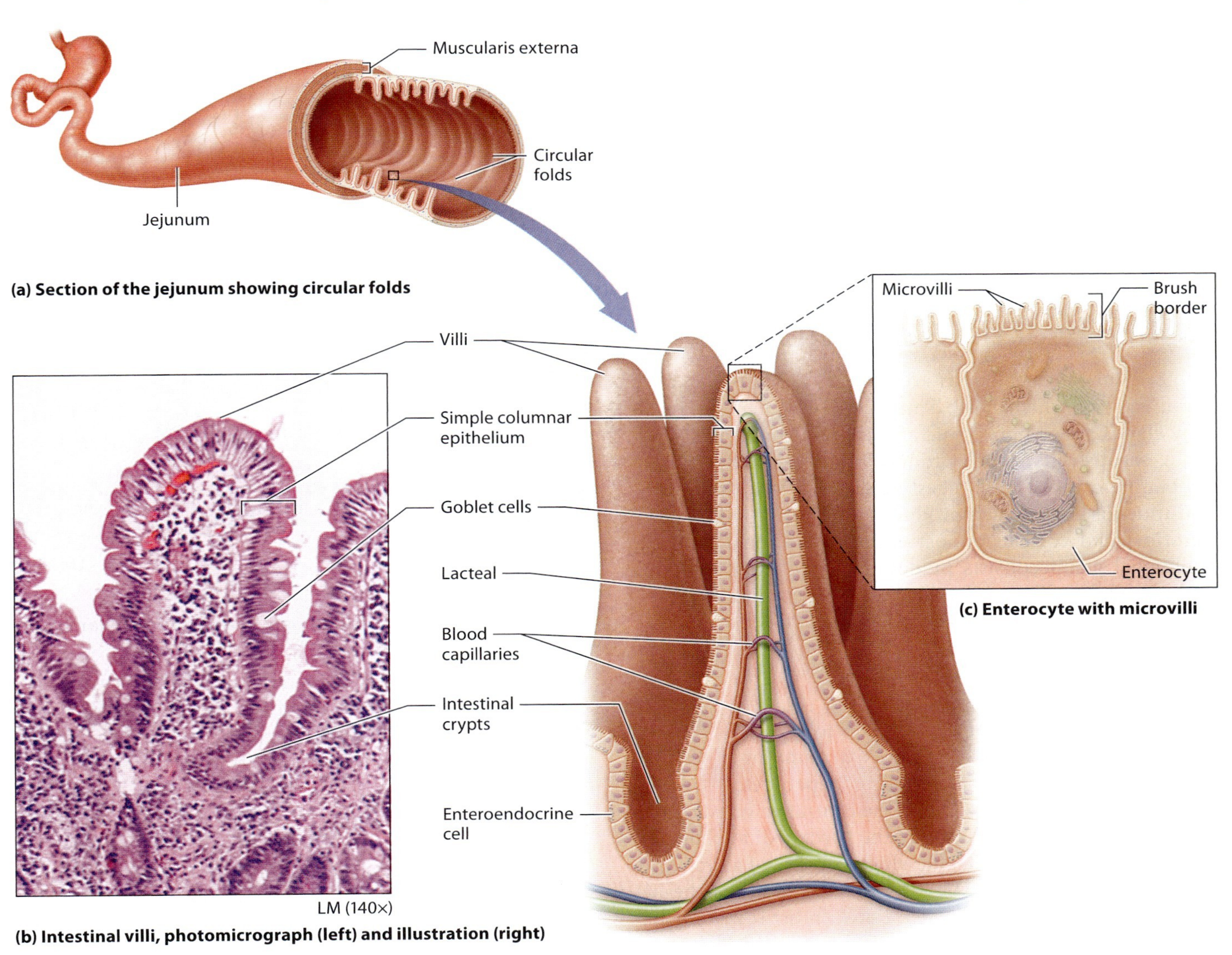

Figure 28-16 Microscopic anatomy of the small intestine.

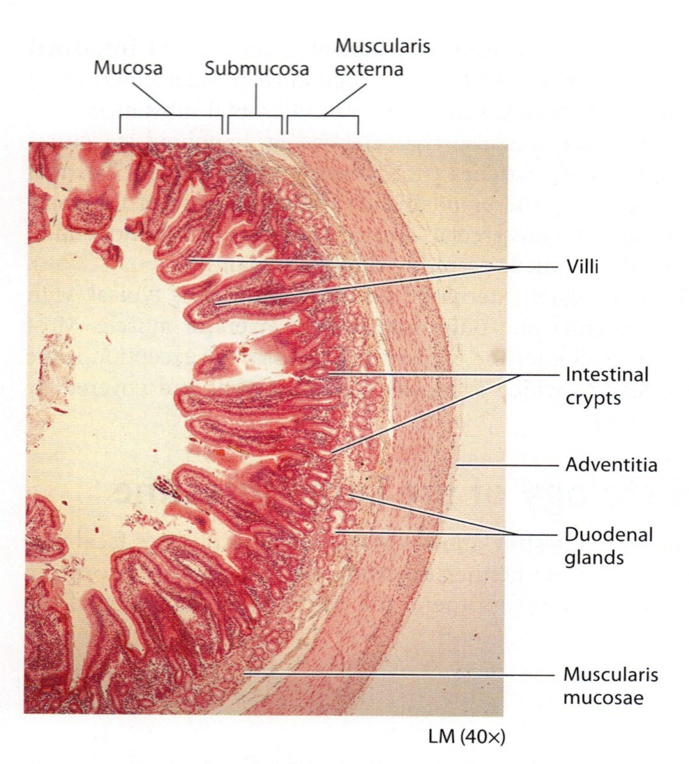

Figure 28-17 Photomicrograph of the duodenum in cross-section.

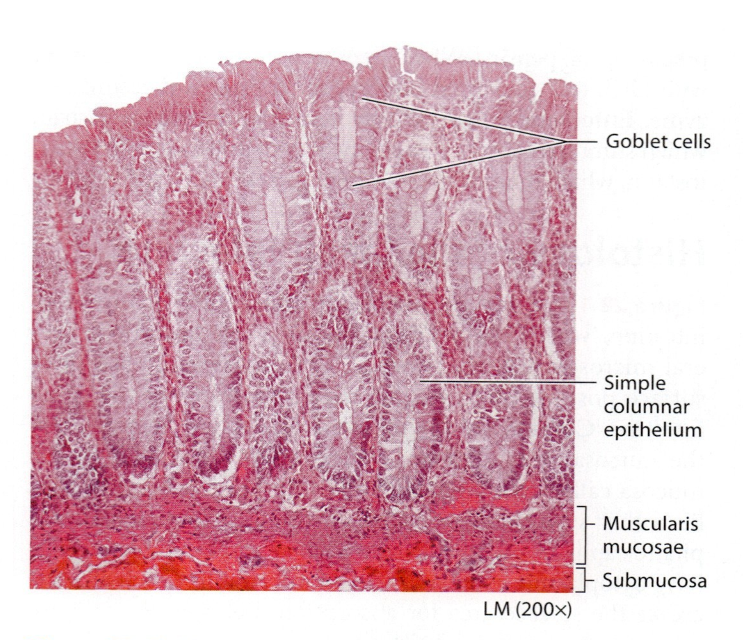

Figure 28-18 Microscopic anatomy of the large intestine.

wall of the large intestine (**Figure 28-18**) differs from that of the small intestine in important ways: The large intestine lacks circular folds, villi, and microvilli. Additionally, mucus-secreting goblet cells are abundant.

Histology of the Salivary Glands

The cells of salivary glands are arranged in clusters called acini, each of which has a small central lumen. The salivary glands are composed of various proportions of two cell types: **serous cells**, which produce saliva, a watery solution containing carbohydrate-digesting enzymes and antibodies, and **mucous cells**, which produce a lubricating mucus. The parotid gland consists almost entirely of serous acini, the sublingual gland consists predominantly of mucous acini, and the submandibular gland (**Figure 28-19**) contains a mixture of serous and mucous acini.

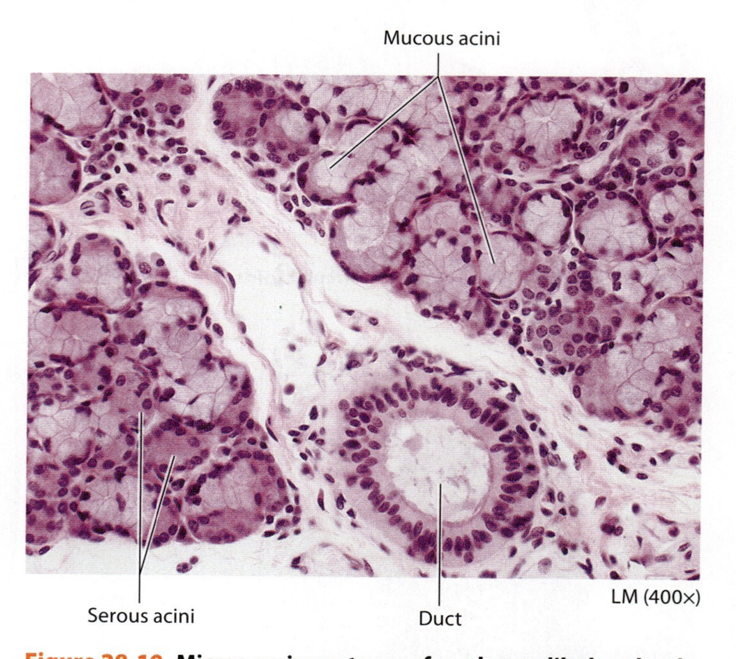

Figure 28-19 Microscopic anatomy of a submandibular gland.

Histology of the Liver

The functional unit of the liver is the liver lobule (**Figure 28-20**), a six-sided (hexagonal) structure surrounding a **central vein**. Liver cells—the bile-secreting **hepatocytes**—are arranged in rows that extend from the central vein, much like the spokes of a wheel. At each of the corners of a lobule is a structure called a **portal triad**, which consists of a branch of the hepatic artery (a hepatic arteriole), a branch of the hepatic portal vein, and a bile duct. Between the rows of hepatocytes are **hepatic sinusoids**; blood flowing through them is filtered by phagocytic cells before draining into the central vein, the hepatic vein, and eventually the inferior vena cava. Once hepatocytes have secreted bile, it travels through tiny canals called **bile canaliculi** that run between adjacent hepatocytes and then drains into branches of a bile duct.

Histology of the Pancreas

The pancreas is both an endocrine organ and an exocrine organ. We studied the endocrine pancreas in Unit 19. Here we turn our attention to the exocrine pancreas. Pancreatic acini—clusters of acinar tissue surrounding a central lumen (**Figure 28-21**)—consist of **acinar cells**, which produce

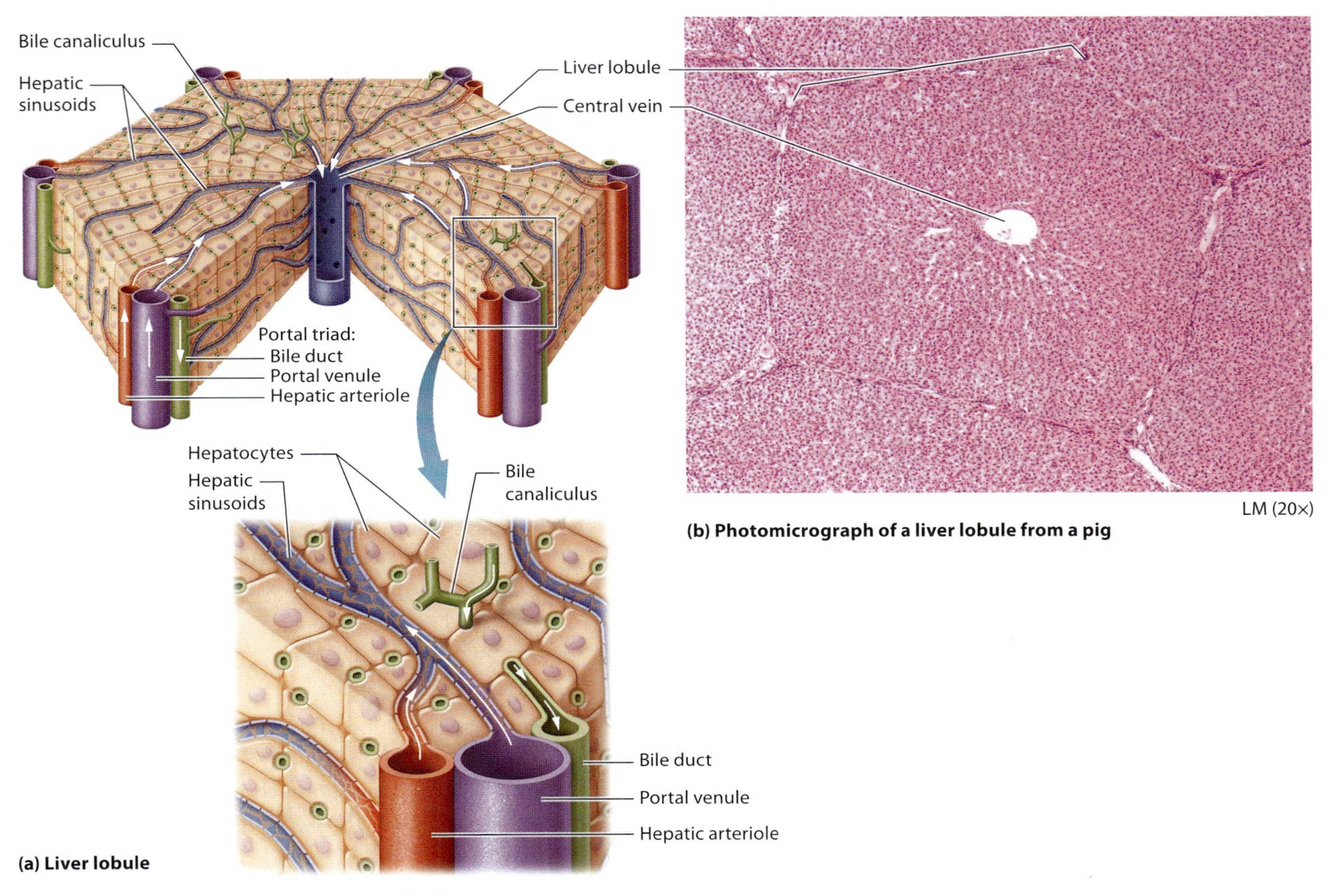

Figure 28-20 **Microscopic anatomy of the liver.**

pancreatic juice. This fluid is transported through a network of ducts to the pancreatic duct, which extends the entire length of the pancreas and empties into the duodenum.

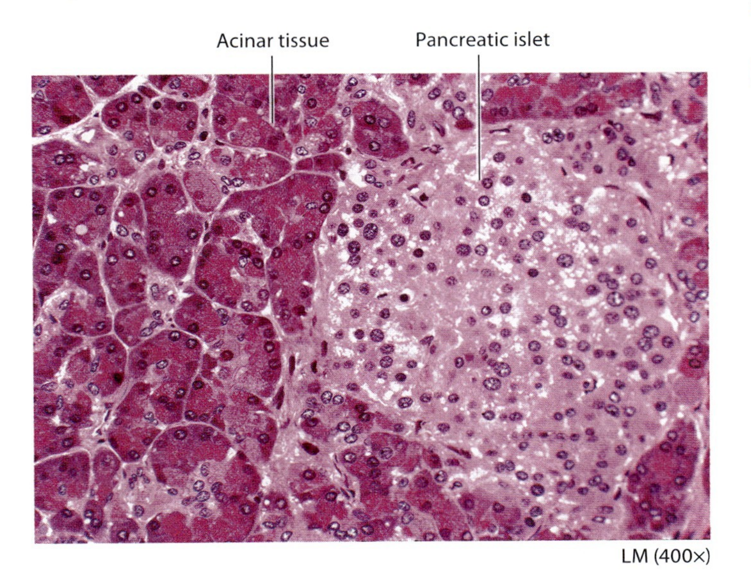

Figure 28-21 **Microscopic anatomy of the pancreas.**

ACTIVITY 3

Examining the Histology of Selected Digestive Organs

Learning Outcomes

1. Identify the layers of the alimentary canal wall on an anatomical chart, and describe the histological makeup of each layer.
2. Predict the function of each layer of the alimentary canal based on its structure.
3. Sketch, label, and describe the histological features in a microscopic section of esophagus, stomach, small intestine, and large intestine.
4. Sketch, label, and describe the histological features in a microscopic section of salivary gland, liver, and pancreas.

Materials Needed

- ☐ Anatomical charts for digestive system
- ☐ Anatomical models of digestive organs
- ☐ Microscope and prepared slides (or photomicrographs) of sections of the esophagus, stomach, small intestine (duodenum and ileum), large intestine, salivary gland, liver, and pancreas

Instructions

1. Identify the mucosa, submucosa, muscularis externa, and serosa of the alimentary canal on an anatomical chart. Then list the specific tissue types found in each layer and predict the function of each layer based on its structure.

   a. Mucosa ______________________________

   ____

   ____

   ____

   b. Submucosa ______________________________

   ____

   ____

   ____

   c. Muscularis externa ______________________________

   ____

   ____

   ____

   d. Serosa ______________________________

   ____

   ____

   ____

2. Examine a cross-section of esophagus (near its junction with the stomach) under low power.

   a. Identify the four layers of the esophageal wall and list them from innermost to outermost. ______________

   ____

   ____

   b. Switch to the high-power objective and examine the mucosa and submucosa more closely.

   The mucosa consists of which specific tissue type?

   ____

   c. Identify the numerous glands in the submucosa.

   What is the function of these glands? ______________

   ____

   d. Sketch the mucosa and the submucosa. Label the epithelium, basement membrane, lamina propria, and submucosal gland.

Total magnification: _____ ×

   e. Locate the muscularis externa.

   How does the muscularis externa of the inferior one-third of the esophagus differ from that of the superior one-third? ______________________________

   ____

   ____

3. Examine a cross-section of the stomach under low power.

   a. Identify the mucosa, submucosa, muscularis externa, and serosa. Sketch a section of the stomach wall and label the four layers.

Total magnification: _____ ×

   b. Switch to the high-power objective and examine the mucosa.

   Which specific epithelium type is found in the mucosa?

   ____

   c. Locate a gastric gland and differentiate between a chief cell and a mucous neck cell.

How do these two cell types differ histologically?

___

___

___

d. Now examine the muscularis externa more closely. Try to identify the three smooth muscle layers.

Which smooth muscle layer is missing from the alimentary canal wall of other organs?

___

What effect does this extra layer of smooth muscle have on the churning ability of the stomach?

___

___

___

4. Examine a cross-section of the duodenum under low power.

a. Identify the mucosa, submucosa, muscularis externa, and serosa. Switch to high power and sketch the mucosa. Label the epithelium, basement membrane, lamina propria, and villus.

Total magnification: _____ ×

Which specific type of epithelium is found in the mucosa?

___

What structural characteristic of the epithelial cells increases their surface area? ___

___

Name two major functions of these epithelial cells.

___

___

Name two structures located in the center of a villus and state the function of each. ___

___

___

___

b. Now observe a cross-section of the ileum using the low-power objective. Then, switch to the high-power objective and carefully examine the mucosa.

Note any differences you see between the mucosa of the ileum and the mucosa of the duodenum.

___

___

c. Find the submucosa and identify the large, spherical Peyer's patches.

What is the function of these structures? ___

___

___

5. Examine a cross-section of the large intestine under low power.

a. Identify the mucosa, submucosa, muscularis, and serosa. Sketch the large intestine and label each of the layers of its wall.

Total magnification: _____ ×

b. Note the many unstained cells in the epithelium of the mucosa.

What are these cells called? ___

c. Label these cells on your sketch.

What is the function of these cells? ___

___

d. The large intestine lacks villi and microvilli.

How does the absence of these structures correlate with the function of the large intestine? ________________

____

____

6. Examine a cross-section through a submandibular gland.

   a. Identify the darkly staining serous cells.

   What is the function of these cells? ________________

   ____

   b. Identify the much lighter mucous cells.

   What is the function of these cells? ________________

   ____

   c. Sketch a histological view of a submandibular salivary gland and label serous cells, mucous cells, and a duct.

Total magnification: _____ ×

Why are salivary glands classified as exocrine glands?

____

____

Which type of enzyme is secreted by the salivary glands?

____

7. Examine a section of nonhuman mammalian liver using the low-power objective. In humans, the lobules are not well defined. However, in pigs and other mammals, each lobule is surrounded by a connective tissue septum that makes them easier to distinguish histologically. Identify the many lobules.

   a. Sketch two lobules and label the hepatocytes, sinusoids, and a central vein.

Total magnification: _____ ×

   b. Identify the portal area at one of the corners of a lobule. Circle this area on your sketch.

   This area contains a portal triad consisting of which three structures? ________________

   ____

8. Examine a slide of the pancreas under low power. Identify the darkly staining acinar cells and the "islands" of lighter staining pancreatic islets.

   a. Switch to the high-power objective. Sketch a section of the pancreas and label the acinar cells and pancreatic islets.

Total magnification: _____ ×

What is produced by the acinar cells? ________________

Into what structures are these secretory products released? ________________

9. **Optional Activity**

View and label histology slides of the digestive system at > MasteringA&P® > Study Area > Practice Anatomy Lab > Histology > Digestive System

POST-LAB ASSIGNMENTS

Name: ______________________ Date: ______________ Lab Section: ________

PART I. Check Your Understanding

Activity 1: Exploring the Organs of the Alimentary Canal

1. Identify the labeled structures:

   a. ______________________
   b. ______________________
   c. ______________________
   d. ______________________
   e. ______________________
   f. ______________________
   g. ______________________
   h. ______________________
   i. ______________________
   j. ______________________
   k. ______________________
   l. ______________________

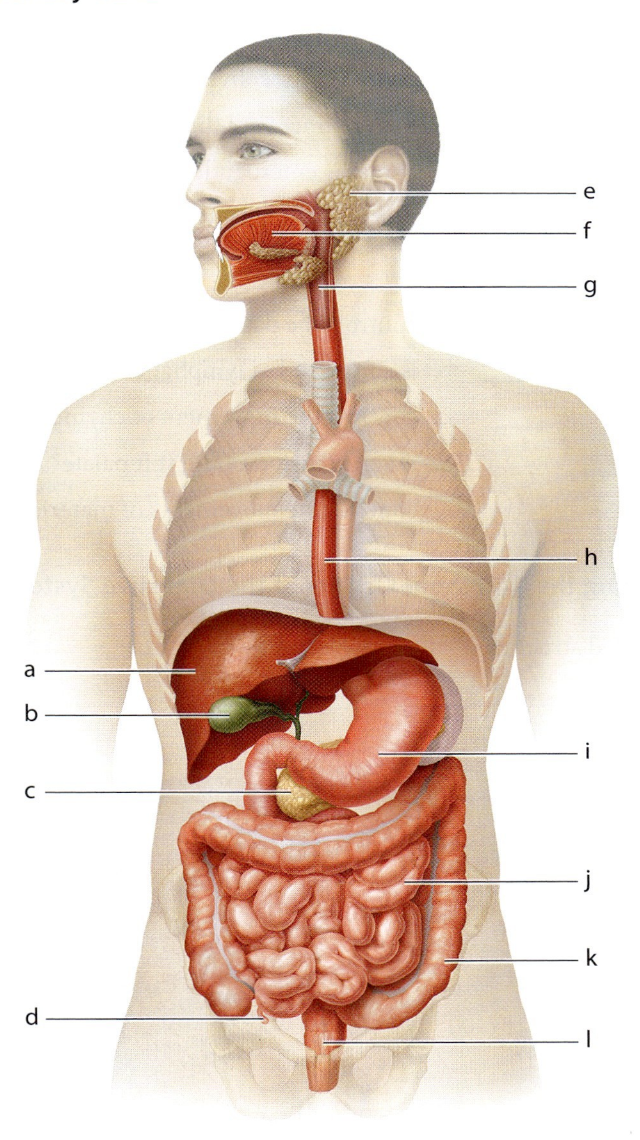

2. Name the alimentary canal organ that:

_______________ a. contains three layers of smooth muscle in the muscularis externa.

_______________ b. carries out most of the digestion and virtually all of the absorption in the alimentary canal.

_______________ c. contains pockets called haustra.

_______________ d. initiates the chemical digestion of proteins.

_______________ e. is separated from the small intestine by the pyloric sphincter.

_______________ f. initiates the chemical digestion of carbohydrates.

_______________ g. contains three divisions, one of which lacks a digestive function.

_______________ h. has an inner wall thrown into circular folds.

_______________ i. contains enteroendocrine cells that produce gastrin and somatostatin.

_______________ j. functions in mastication.

3. Which structure:

_______________ a. controls the passage of food from the esophagus to the stomach?

_______________ b. is composed of lymphatic tissue and extends from the cecum?

_______________ c. controls the movement of chyme into the duodenum?

_______________ d. is an extension of the soft palate?

_______________ e. controls the movements of materials from the small intestine to the large intestine?

Activity 2: Exploring the Accessory Organs of the Digestive System

1. Identify the labeled structures:

a. _______________

b. _______________

c. _______________

d. _______________

e. _______________

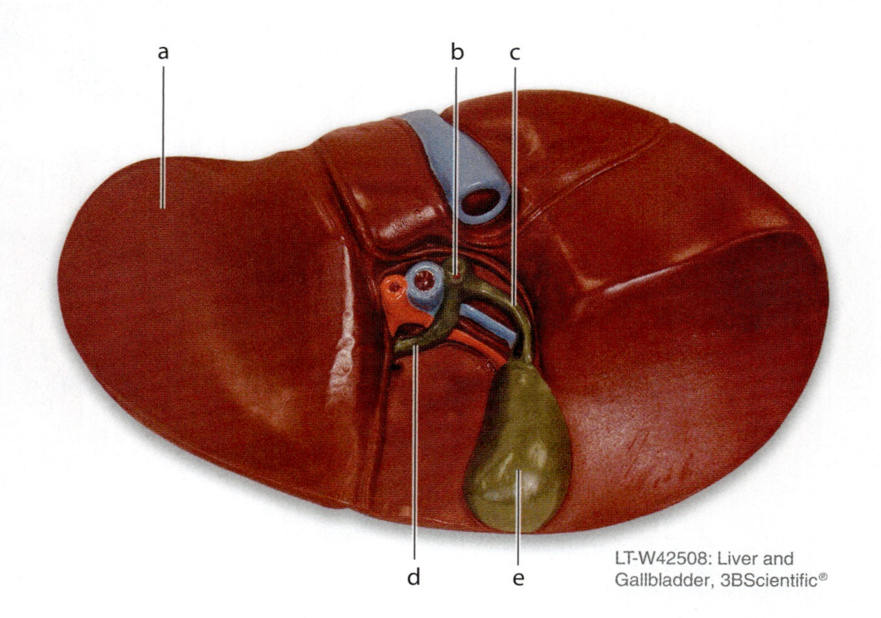

LT-W42508: Liver and Gallbladder, 3BScientific®

2. Name the accessory organ(s) of the digestive system that:

   ______________________ a. stores glucose as glycogen.

   ______________________ b. produce amylase.

   ______________________ c. contain cementum.

   ______________________ d. contains numerous filiform papillae.

   ______________________ e. initiates deglutition.

   ______________________ f. secretes enzymes that digest carbohydrates, proteins, lipids, and nucleic acids.

   ______________________ g. produces bile.

3. Which tooth type(s) is(are) adapted for:

   ______________________ a. tearing and piercing?

   ______________________ b. grinding and crushing?

   ______________________ c. biting?

4. Bile is produced by liver cells, or ______________________, and enters small canals called ______________________. It then leaves the liver through many smaller ______________________ that eventually fuse to form the large ______________________. That duct fuses with the ______________________, which transports bile from the gallbladder to form the ______________________, which then transports bile to the duodenum.

Activity 3: Examining the Histology of Selected Digestive Organs

1. Name the predominant specific tissue type in each of the following layers of the alimentary canal wall:

   a. Mucosa: ______________________

   b. Muscularis: ______________________

   c. Submucosa: ______________________

   d. Serosa: ______________________

2. Which layer of the alimentary canal wall:

   ______________________ a. functions in movements of the digestive tract?

   ______________________ b. carries away absorbed nutrients?

   ______________________ c. secretes enzymes and absorbs nutrients?

   ______________________ d. is the visceral peritoneum?

3. In what way is the anatomy of the duodenal mucosa structurally adapted for the secretion of enzymes and the absorption of nutrients? ______________________

   ____

4. From which digestive organ was this tissue taken? ______________________

   Identify the labeled structures in the photomicrograph:

   a. ______________________

   b. ______________________

   c. ______________________

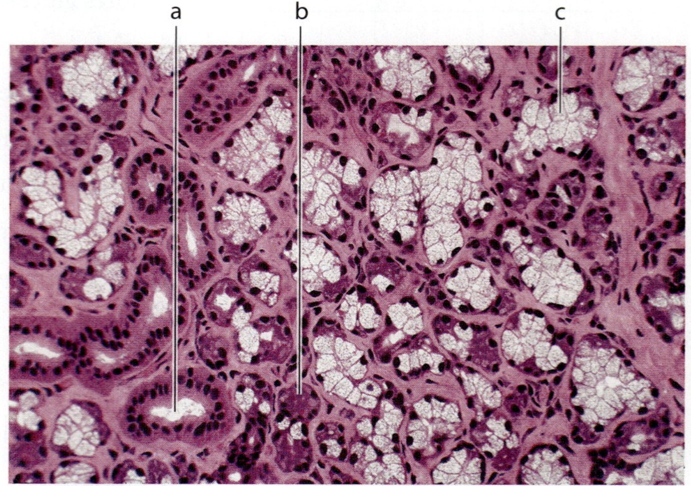

LM (400×)

5. From which digestive organ was this tissue taken? ______________________

   Identify the labeled structures in the photomicrograph:

   a. ______________________ (layer)

   b. ______________________ (layer)

   c. ______________________ (layer)

   d. ______________________ (layer)

   e. ______________________

   f. ______________________

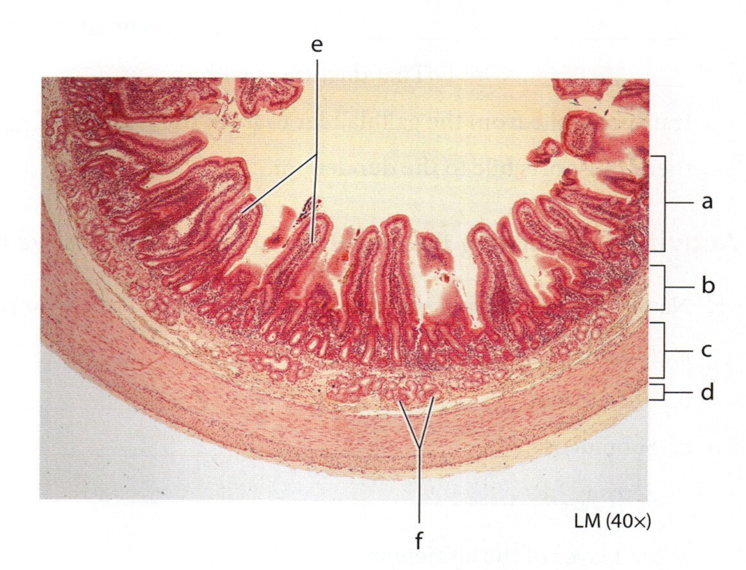

LM (40×)

PART II. Putting It All Together

A. Review Questions

Answer the following questions using your lecture notes, your textbook, and your lab notes.

1. Identify the specific part of the alimentary canal or the digestive system accessory organ in which each of the following structures is found:

   ____________________ a. palatine tonsils
   ____________________ b. vermiform appendix
   ____________________ c. microvilli
   ____________________ d. circular folds
   ____________________ e. enteroendocrine cells
   ____________________ f. Peyer's patches
   ____________________ g. portal triad
   ____________________ h. sinusoids
   ____________________ i. falciform ligament

2. Explain why acid reflux into the esophagus damages the esophageal wall and causes pain, whereas gastric juice does not typically irritate the stomach lining. ____________________
   ____
   ____

3. Predict the effect of the removal of the gallbladder on the digestion of fats. ____________________
   ____

4. Jesse, a 7-week-old baby, is experiencing frequent projectile vomiting. An MRI reveals a thickening of the pyloric sphincter and a full stomach, even though it has been hours since he was last fed. What problem is the thickened pyloric sphincter causing? ____________________
   ____

5. How are the epithelial linings of the esophagus, stomach, and small intestine structurally adapted for their specific functions? ____________________
   ____
   ____
   ____

B. Concept Mapping

1. Fill in the blanks to complete this concept map outlining important digestive structures.

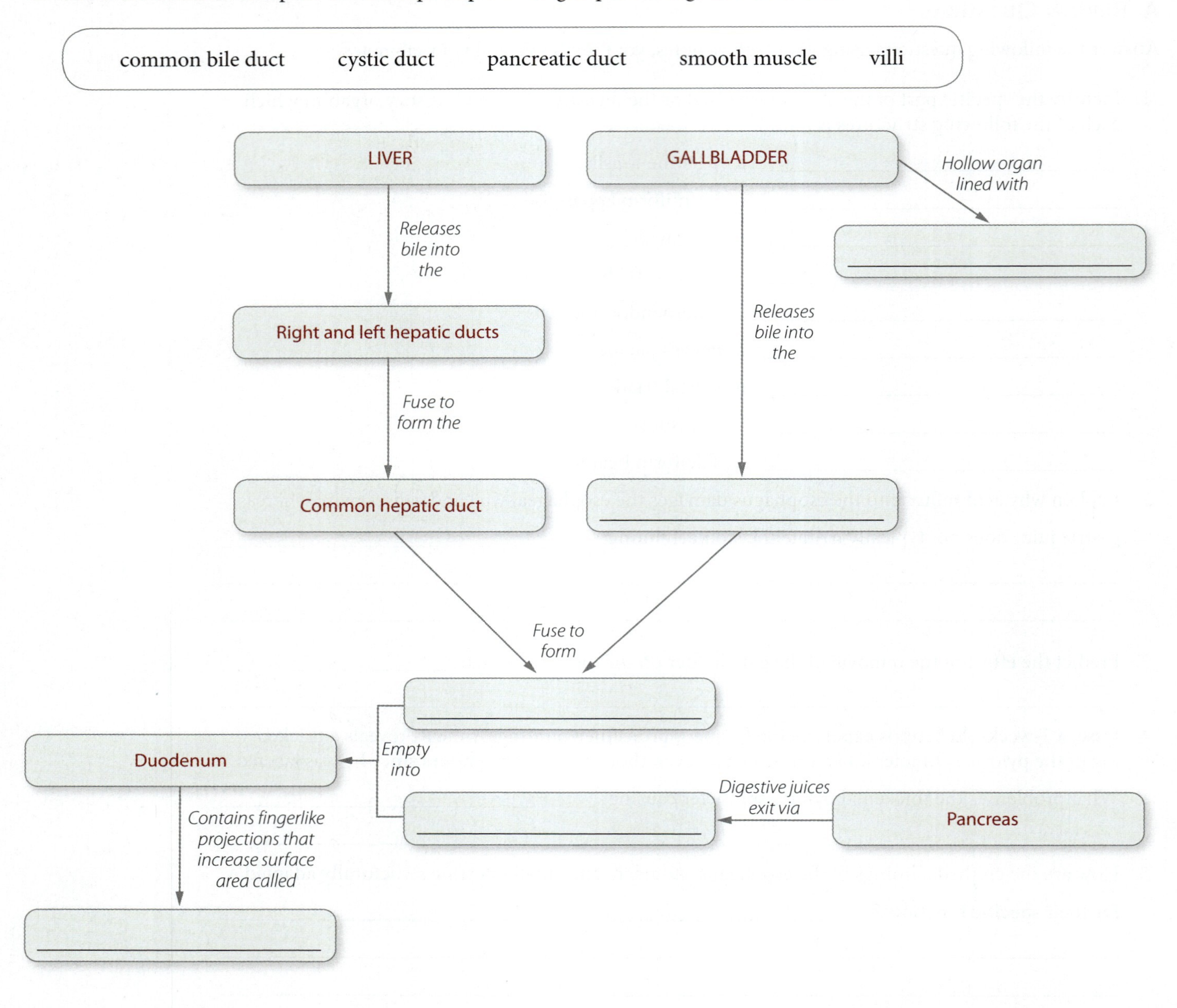

2. Construct a unit concept map to show the relationships among the following set of terms. Include all of the terms in your diagram. Your instructor may choose to assign additional terms.

appendix	bile duct	cystic duct	gastroesophageal sphincter	ileocecal valve
mucosa	oropharynx	pancreatic duct	peristalsis	peritoneum
plicae circularis	rugae	smooth muscle	splenic flexure	villi

29

Physiology of the Digestive System

UNIT OUTLINE

As we saw in Unit 28, the structures of the digestive system can be placed into two groups: organs of the alimentary canal (oral cavity, pharynx, esophagus, stomach, small intestine, and large intestine) and accessory digestive organs (teeth, tongue, salivary glands, liver, gallbladder, and pancreas). We also learned that the digestive system takes in food, breaks it down into its nutrient building blocks, absorbs the nutrients, and then rids the body of indigestible waste products. In this unit we will focus on the various physiological events that occur during digestion.

THINK ABOUT IT *Which types of food require the least amount and the most amount of chewing? (Hint: What happens in your mouth that begins the digestive process?)*

____

____

____ ■

PRE-LAB ASSIGNMENTS

Pre-lab quizzes are also assignable in MasteringA&P®

To maximize learning, BEFORE your lab period carefully read this entire lab unit and complete these pre-lab assignments using your textbook, lecture notes, and prior knowledge.

PRE-LAB Activity 1: Analyzing Amylase Activity

1. Benedict's reagent tests for the presence of ________________________.
2. Lugol's iodine solution tests for the presence of ________________________.
3. In the mouth, salivary amylase hydrolyzes ________________________ into oligosaccharides.
4. In this activity, you will test the effect of ________________________ on amylase activity.

PRE-LAB Activity 2: Analyzing Pepsin Activity

1. The building blocks of proteins are ________________________.
2. Which reagent will be used to test for the products of protein digestion?

   ____
3. The chemical digestion of proteins begins in the ________________________.
4. Inactive pepsinogen is converted to pepsin by ________________________.

PRE-LAB Activity 3: Analyzing Lipase Activity

1. What is litmus cream? ____
2. The building blocks of triglycerides (neutral fats) are ________________________ and

   ________________________.
3. Before lipids are hydrolyzed by lipase, they are emulsified by ________________________.
4. Most lipase is secreted by which accessory organ of the digestive tract? ________________________

PRE-LAB Activity 4: Tracing Digestive Pathways

1. Name three sources of amylase. ____
2. Glucose and amino acids are absorbed by active transport into the ________________________ of a villus.
3. Fats enter specialized lymphatic capillaries associated with the villi of the small intestine called

   ________________________ and are transported to venous circulation by lymphatic vessels.

The Mechanical and Chemical Events of Digestion

Figure 29-1 summarizes the processes involved in the mechanical and chemical digestion of food so that nutrients can be absorbed by the body. **Mechanical digestion** is a physical process that increases the surface area of ingested food and prepares it for chemical digestion by enzymes. The chewing of food in the mouth and the churning of food in the stomach are examples of the mechanical breakdown of food. **Chemical digestion**, by contrast, involves the action of hydrolytic enzymes that chemically break macromolecules (carbohydrates, proteins, lipids, and nucleic acids) into their building blocks. **Table 29-1** summarizes the types of macromolecules, their building blocks, and examples of enzymes that break down each type of macromolecule.

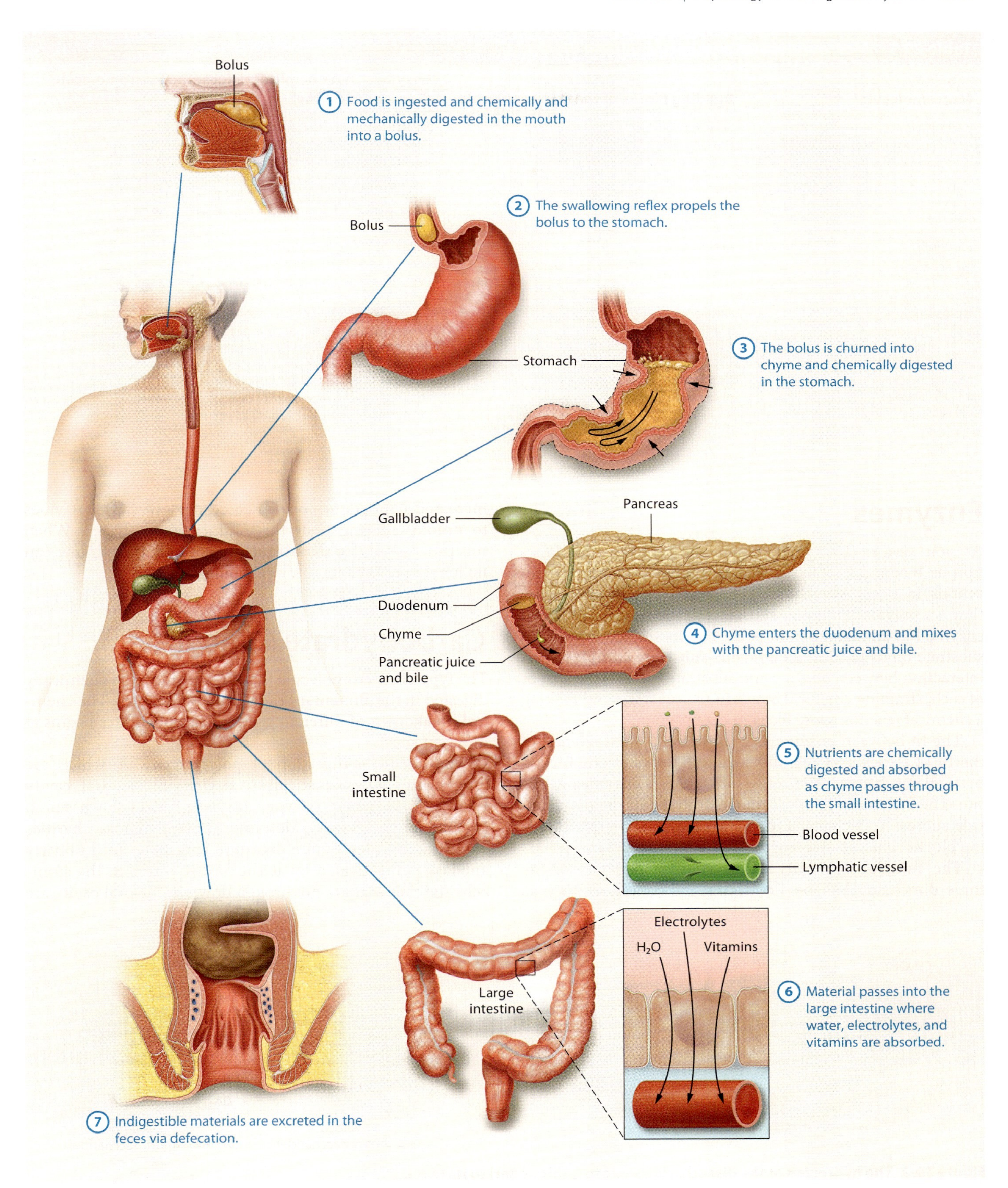

Figure 29-1 The processes involved in the mechanical and chemical digestion of food.

Table 29-1 Macromolecules and Their Building Blocks

Macromolecule	Building Blocks of the Macromolecule	Enzymes That Chemically Digest the Macromolecule (Source of Enzyme)
Carbohydrates (disaccharides, oligosaccharides, and polysaccharides)	Simple sugars (monosaccharides)	Salivary amylase (salivary gland) Pancreatic amylase (acinar cells of pancreas) Intestinal amylase (small intestine) Sucrase (brush border cells of small intestine) Maltase (brush border cells of small intestine) Lactase (brush border cells of small intestine)
Proteins	Amino acids	Pepsin (chief cells of stomach) Trypsin and chymotrypsin (pancreas) Peptidases (brush border cells of small intestine)
Lipids (triglycerides)	Fatty acids and glycerol	Gastric lipase (chief cells of stomach) Pancreatic lipase (acinar cells of pancreas) Intestinal lipase (brush border cells of small intestine)
Nucleic acids	Nitrogenous base, pentose sugar, phosphate ions	Nucleases (acinar cells of pancreas) Nucleosidases (brush border cells of small intestine) Phosphatases (brush border cells of small intestine)

Enzymes

As you saw in Unit 3, enzymes are proteins that function as biological catalysts; that is, they allow chemical reactions to occur very quickly. Enzymes exhibit specificity; they act only on particular substances called substrates. Each enzyme has an **active site**, a pocket on its surface to which a substrate binds to form an **enzyme–substrate complex**. This interaction between enzyme and substrate changes the shapes of each, straining chemical bonds in the substrate and making a chemical reaction more likely to occur.

The fit between enzyme and substrate is often likened to the fit of a lock and key. Just as a key fits only one lock, a particular substrate "fits" only into a given enzyme's active site. The enzymatic breakdown (hydrolysis) of the disaccharide sucrose (table sugar) to its two monosaccharide building blocks (glucose and fructose) is depicted in **Figure 29-2**.

The functional activity of an enzyme depends on its three-dimensional shape. Environmental conditions such as increased temperature or decreased pH can cause enzymes to unfold, altering their three-dimensional shapes. When this process—called **denaturation**—occurs, the enzyme can no longer perform its role as a biological catalyst.

Carbohydrate Digestion

The type of macromolecule that first begins to be chemically digested in the alimentary canal is carbohydrates. The chemical breakdown of different types of carbohydrates begins at different sites.

The chemical digestion of polysaccharides, which are long chains of glucose units, begins in the oral cavity (**Figure 29-3**), where **salivary amylase** breaks down starch (a kind of polysaccharide) into shorter oligosaccharides and disaccharides. Starch digestion continues until salivary amylase is inactivated by acidic gastric juices in the stomach. Starches that are not broken down in the oral cavity are

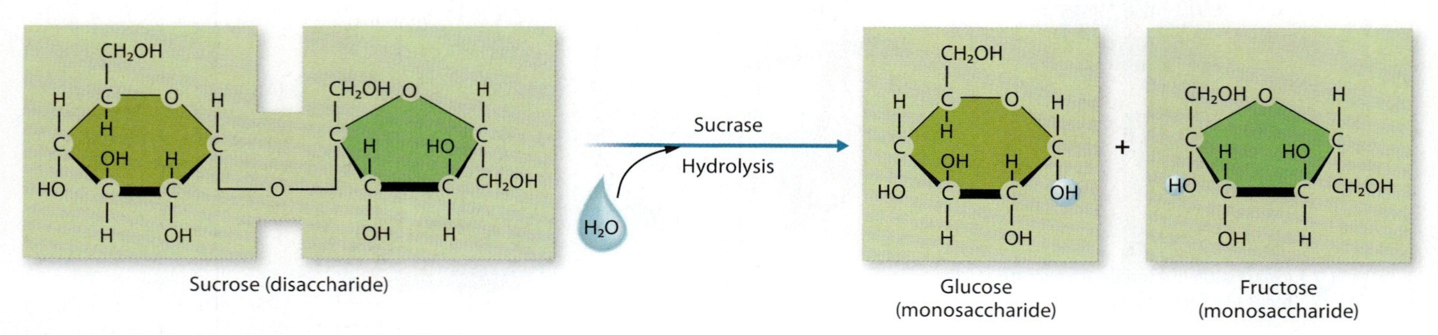

Figure 29-2 The hydrolysis of the disaccharide sucrose (table sugar) to its two monosaccharide building blocks (glucose and fructose).

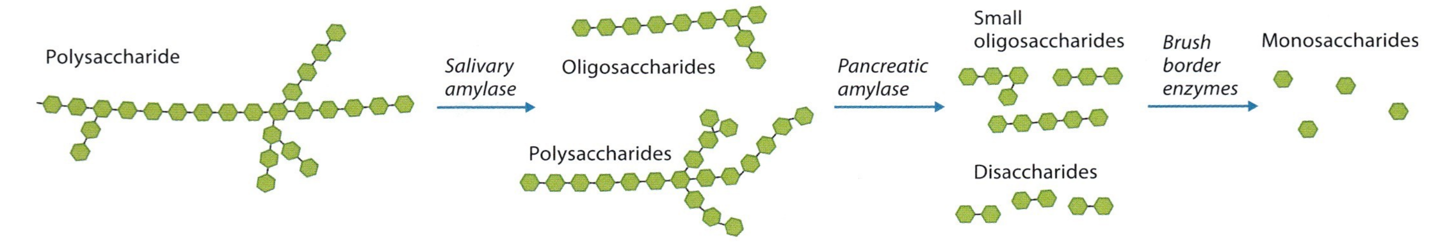

Figure 29-3 Steps in the chemical digestion of polysaccharides.

broken down in the small intestine by **intestinal amylase** and **pancreatic amylase**. Within 10 minutes of entering the small intestine, most starch has been converted to various disaccharides and oligosaccharides, primarily maltose. Intestinal brush border cells produce a variety of enzymes that break disaccharides and oligosaccharides into their monosaccharide building blocks.

Ingested disaccharides (sucrose, maltose, and lactose) travel to the small intestine before their chemical digestion begins. There, intestinal brush border cells produce the enzymes sucrase, maltase, and lactase, which hydrolyze sucrose, maltose, and lactose, respectively, into their constituent monosaccharide building blocks.

ACTIVITY 1

Analyzing Amylase Activity

Learning Outcomes

1. Distinguish between starch and maltose.
2. Name the three main sources of amylase in the body.
3. Demonstrate and explain the role of amylase in carbohydrate digestion.
4. Determine the effect of temperature on amylase activity.

Materials Needed

- ☐ Eighteen test tubes
- ☐ Two test tube racks
- ☐ Test tube clamp
- ☐ Starch solution
- ☐ Amylase solution
- ☐ Maltose solution
- ☐ Benedict's reagent
- ☐ Lugol's iodine (IKI) solution
- ☐ Parafilm (six small squares to cover the test tubes)
- ☐ Wax pencil
- ☐ Dropper, pipettor, or micropipettor (to measure in milliliters)
- ☐ 37°C water bath (to incubate five tubes per group)
- ☐ Boiling water bath (to boil one tube per group)

Instructions

1. Obtain six test tubes and label them "1C" through "6C" (C for carbohydrate) with a wax pencil.
2. Prepare the test tubes as indicated in **Table 29-2**. *Note:* Boil amylase for 5 minutes before adding the starch solution to tube 5C.
3. Cover each tube with a small square of Parafilm, and then shake each tube vigorously to mix the contents.
4. Place tubes 1C–5C in a test tube rack and incubate for 30 minutes in a 37°C water bath.

   Shake the rack gently every 15 minutes to ensure that the solutions stay thoroughly mixed. Place tube 6C in a second test tube rack and leave it at room temperature (24°C) for 30 minutes.

Table 29-2 Digestion of Starch by Amylase

Tube #	**1C**	**2C**	**3C**	**4C**	**5C**	**6C**
Contents of tube	5 ml H_2O, 5 ml starch	5 ml H_2O, 5 ml amylase	5 ml H_2O, 5 ml maltose	5 ml starch, 5 ml amylase	5 ml starch, 5 ml boiled amylase	5 ml starch, 5 ml amylase
Incubation temperature	37°C	37°C	37°C	37°C	37°C	24°C
Tube #	**1CB**	**2CB**	**3CB**	**4CB**	**5CB**	**6CB**
Benedict's test result (++ / + / −)						
Tube #	**1CL**	**2CL**	**3CL**	**4CL**	**5CL**	**6CL**
Lugol's solution test result (+ / −)						
Did carbohydrate digestion occur? Why or why not?						

5. Test the contents of each tube for the presence of sugars using the following procedure:
   a. Label six additional test tubes 1CB through 6CB (B for Benedict's reagent).
   b. Transfer 1 ml of the solution in tube 1C to tube 1CB.
   c. Add 1 ml of Benedict's reagent to tube 1CB.
   d. Using a test tube clamp, place tube 1CB in a boiling water bath for 2 minutes.
   e. Note the color of the solution in tube 1CB. Blue indicates a negative test (−) for sugar, greenish-yellow indicates that some sugar is present (+), and orangish-red indicates that a large amount of sugar is present (++).
   f. Record your data for tube 1CB in Table 29-2.
   g. Repeat steps b–f for tubes 2CB through 6CB.
6. Test the contents of each tube for the presence of starch using the following procedure.
   a. Label six additional test tubes 1CL through 6CL (L for Lugol's iodine, or IKI solution).
   b. Transfer 1 ml of the solution in tube 1C to tube 1CL.
   c. Add 3 drops of Lugol's iodine solution to tube 1CL. The solution will turn blue-black if starch is present, a positive result (+). No color change (solution stays yellow-orange) indicates that no starch is present, a negative result (−).
   d. Record your data for tube 1CL in Table 29-2.
   e. Repeat steps b–d for tubes 2CL through 6CL, recording the data, and then fill in the bottom row of the table.
7. Answer the following questions based on your data:
   a. Which tubes served as controls in this experiment? Explain. ______________________________
   ____
   ____
   b. What happened to the amylase when it was boiled?
   ____
   c. Based on your data, what is the optimal temperature for amylase activity? ______________
   d. What is the biological significance of temperature in these experiments? ______________________________
   ____

Protein Digestion

Chemical digestion of proteins (**Figure 29-4**) begins in the stomach. The gastric mucosa contains cells that secrete both hydrochloric acid (HCl) and pepsinogen. Pepsinogen reacts with the HCl to become its active form, pepsin. Pepsin hydrolyzes proteins into large polypeptides, which then enter the small intestine. The pancreatic enzymes trypsin and chymotrypsin chemically break down the large polypeptides to smaller polypeptides and peptides. Finally, brush border enzymes such as aminopeptidase and dipeptidase cleave the peptides to release the final products: small oligopeptides and amino acids.

ACTIVITY 2
Analyzing Pepsin Activity

Learning Outcomes

1. Distinguish among a protein, a polypeptide, and an amino acid.
2. Demonstrate and explain the role of pepsin in the digestion of proteins.

Materials Needed

- ☐ Five test tubes
- ☐ Test tube rack
- ☐ Graduated cylinder or pipette
- ☐ Protein solution
- ☐ Pepsinogen
- ☐ 0.5 M HCl
- ☐ 0.1 M NaOH
- ☐ Biuret reagent
- ☐ Parafilm (five small squares to cover the test tubes)
- ☐ Wax pencil
- ☐ Dropper, pipettor, or micropipettor (to measure in milliliters)
- ☐ 37°C water bath (to hold five tubes per group)

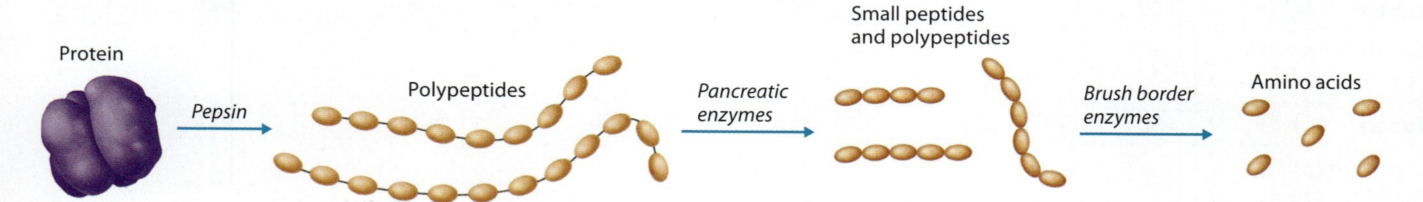

Figure 29-4 Steps in the chemical digestion of proteins.

Table 29-3 Digestion of Proteins by Pepsin

Tube #	1P	2P	3P	4P	5P
Contents of tube	10 ml H_2O, 5 ml egg white	10 ml H_2O, 5 ml pepsinogen	5 ml H_2O, 5 ml egg white, 5 ml pepsinogen	5 ml egg white, 5 ml pepsinogen, 5 ml 0.5 M HCl	5 ml egg white, 5 ml pepsinogen, 5 ml 0.1 M NaOH
Biuret reagent result: color of solution					
Did protein digestion occur? Why or why not?					

Instructions

1. Obtain five test tubes and label them "1P" through "5P" (P for protein) with a wax pencil. Then prepare the tubes as indicated in **Table 29-3**.
2. Cover each tube with a small square of Parafilm, and then shake each tube vigorously to mix the contents.
3. Place the tubes in a test tube rack and incubate for 60 minutes in a 37°C water bath.
   Shake the rack gently every 15 minutes to ensure that the solutions stay thoroughly mixed.
4. Test the contents of tube 1P for the presence of protein using the following procedure:
   a. Add five drops of Biuret reagent to tube 1P.
   b. Note the color of the solution in the tube. In the presence of undigested proteins, this reagent turns dark purple-blue; in the presence of amino acids, it turns lavender-pink. Record your data in Table 29-3 and then fill in the bottom row of the table.
   c. Repeat steps a–b for tubes 2P through 5P.
5. Answer the following questions based on your data:
   a. What was the role of HCL in this activity?

   ____

   b. How was protein digestion detected in this activity?

   ____

   ____

Lipid Digestion

Lipids include triglycerides, steroids, and phospholipds, in addition to other lipoid substances. Triglycerides, also called neutral fats, are the most abundant dietary lipids. Lipids are difficult to chemically digest because they are hydrophobic and thus tend to aggregate in large spherical droplets within the aqueous fluid in the digestive tract. The formation of droplets greatly reduces the surface area of lipid on which the water-soluble **lipases** (fat-digesting enzymes) can act, because the enzymes can contact lipid molecules only on the outer surface of the droplets.

For the digestion of triglycerides (**Figure 29-5**) to occur efficiently, large fat droplets in the stomach must first be separated by stomach churning and gastric lipases into smaller fat globules. Most digestion of triglycerides occurs in the small intestine. When chyme entering the duodenum contains fat, intestinal cells there release the hormone **cholecystokinin (CCK)**, which stimulates the gallbladder to contract and release **bile** (which contains bile salts). When bile arrives in the small intestine, the bile salts emulsify the fat droplets into still smaller fat droplets, enabling pancreatic lipase to hydrolyze fats into their building blocks: free fatty acids and glycerol. Bile salts remain associated with the digested fats to form small spheres of lipids called **micelles**.

The substrate in the activity that follows is litmus cream, a mixture of fresh cream and litmus powder. Basic or neutral solutions containing litmus are blue; in the presence of acid, litmus turns pink. The chemical breakdown of fats yields fatty acids, so if fat digestion occurs, litmus cream turns pink.

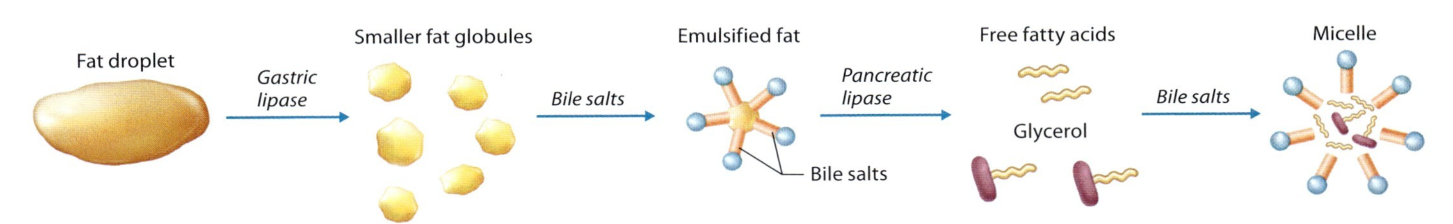

Figure 29-5 Steps in the mechanical and chemical processes involved in the digestion of triglycerides (neutral fats).

ACTIVITY 3
Analyzing Lipase Activity

Learning Outcomes

1. Distinguish among fats, fatty acids, and glycerol.
2. Demonstrate and explain the role of bile salts in lipid digestion.
3. Demonstrate and explain the role of pancreatic lipase in lipid digestion.
4. Examine the effect of temperature on lipase activity.

Materials Needed

- ☐ Six test tubes
- ☐ Test tube rack
- ☐ Litmus cream
- ☐ Bile salts (sodium taurocholate)
- ☐ Lipase
- ☐ Parafilm (six small squares to cover the test tubes)
- ☐ Wax pencil
- ☐ Dropper, pipettor, or micropipettor (to measure in milliliters)
- ☐ 37°C water bath (to hold four tubes per group)
- ☐ Ice water bath (to hold two tubes per group)

Instructions

1. Obtain six test tubes and label them "1L" through "6L" (L for lipid) with a wax pencil.
2. Prepare the tubes as indicated in **Table 29-4**.
3. Cover each tube with a small square of Parafilm, and then shake each tube vigorously to mix the contents.
4. Incubate tubes 1L through 4L in a 37°C water bath, and tubes 5L and 6L in an ice water bath, both for 60 minutes.
5. Record the color of tubes 1L through 6L in Table 29-4, and then fill in the bottom row of the table.
6. Answer the following questions based on your data:
   a. What is the role of bile in fat digestion?

   ____

   b. What is the role of lipase in fat digestion?

   ____

Table 29-4 Digestion of Lipids by Pancreatic Lipase

Tube #	1L	2L	3L	4L	5L	6L
Contents of tube	3 ml H_2O, 3 ml lipase	3 ml H_2O, 3 ml litmus cream	3 ml litmus cream, 3 ml lipase	3 ml litmus cream, 3 ml lipase, bile salts	3 ml litmus cream, 3 ml lipase	3 ml litmus cream, 3 ml lipase, bile salts
Incubation temperature	37°C	37°C	37°C	37°C	0°C	0°C
Color of contents following incubation						
Did fat digestion occur? Why or why not?						

Nutrient Absorption and Transport

So far, we have discussed the basic mechanical and chemical processes related to digestion. However, before they can be absorbed, nutrients must be digested into the individual building blocks of each macromolecule—simple sugars, amino acids, fatty acids and glycerol, and the phosphates, pentose sugars, and nitrogenous bases of nucleotides. The digestive system must first physically and chemically break the macromolecules in food into their building blocks, which are then absorbed into the bloodstream.

In the activity that follows, you will examine the pathways by which food is mechanically and chemically digested into its nutrient building blocks. As you trace the pathways for the digestion of carbohydrates, proteins, and lipids, you will review the structure of both the alimentary canal organs and the accessory digestive organs on anatomical models, revisit the functions of those organs, and answer a series of questions. Throughout you will also tie in information that you have already learned to gain a deeper understanding of how organ systems work together to maintain homeostasis.

ACTIVITY 4
Tracing Digestive Pathways

Learning Outcomes

1. Outline the fate of a starch molecule from its entry into the oral cavity until glucose molecules reach the liver. Identify all structures involved on anatomical models and in histological specimens.
2. Outline the fate of a protein from its entry into the oral cavity until amino acids reach the liver. Identify all structures involved on anatomical models and in histological specimens.

3. Outline the fate of a fat from its entry into the oral cavity until it reaches the bloodstream. Identify all structures involved on anatomical models and in histological specimens.
4. Trace the pathway of bile from the site of its production until its entry into the duodenum. Identify all structures involved on anatomical models.

Materials Needed

- ☐ Anatomical models
- ☐ Microscope and slides (or photomicrographs) of the pancreas, stomach, and small intestine

Instructions

A. Carbohydrate Digestion and Absorption

1. What happens to food when it enters the oral cavity?

 ____

 ____

2. Name two skeletal muscles that elevate the jaw during mastication. ______________________________

 Identify these two muscles on an anatomical model.

3. Name the two regions of the pharynx that have a digestive function. ______________________________

 ____

 Identify these two regions of the pharynx on an anatomical model.

4. Wavelike, rhythmic contractions called ______________________ propel food through the pharynx and into the esophagus.

5. Identify the esophagus on an anatomical model, and describe its location using at least three directional terms.

 ____

 ____

6. As a bolus enters the stomach, it passes through a circular muscle called the ______________________.

 Identify this structure on an anatomical model.

7. In what way is the stomach adapted for extra churning ability? ______________________________

 ____

8. Examine a cross-section of the stomach wall at low power. Identify the layers of the muscularis externa and then list them here. ______________________________

 ____

9. Partially digested food mixed with gastric juices is called ______________________, which enters the small intestine via the ______________________, which controls stomach emptying.

10. Name the three divisions of the small intestine.

 ____

 Identify each of these divisions of the small intestine on an anatomical model.

11. Digestive enzymes reach the duodenum from the ______________ via the ______________ duct.

 Identify these two structures on an anatomical model.

12. The brush border enzymes ______________, ______________, and ______________ digest oligosaccharides into their monosaccharide building blocks.

13. What are brush border cells? ______________________________

 ____

14. Examine a cross-section of the duodenum, and identify the villi and the microvilli. How are these two structures different? ______________________________

 ____

 ____

15. Glucose molecules enter a capillary bed at the villus and are transported via the ______________________ vein, which fuses with the ______________________ vein to become the ______________________ vein, which transports the glucose to the liver.

 Identify these three veins on an anatomical model.

B. Protein Digestion and Absorption

1. Accessory organs in the oral cavity involved in mechanical digestion include ______________ and ______________.

2. Identify the secretory product(s) of each of the following cells:
 a. chief cell ______________________________
 b. parietal cell ______________________________
 c. enteroendocrine cell ______________________________
 d. mucous cell ______________________________

3. The protein-digesting enzyme pepsin is secreted in the stomach as a precursor molecule called ______________, which is converted to pepsin in the presence of ______________. Pepsin hydrolyzes proteins into ______________ polypeptides. The

partially digested food mixed with gastric juices leaves the stomach as ________________.

4. Two protein-digesting pancreatic enzymes, ________________ and ________________, hydrolyze large polypeptides into ________________ polypeptides.

5. Next, the brush border enzymes, ________________, break the smaller polypeptides into their building blocks, ________________, which enter a ________________ bed in the ________________.

6. Amino acids are transported via the superior mesenteric vein to the ________________ vein to the ________________.

C. Lipid Digestion and Absorption

1. Ingested fats, like carbohydrates and proteins, undergo mechanical digestion in the oral cavity. Mastication and mixing of the food forms a ________________, which again undergoes further mechanical digestion in the ________________.

2. Significant chemical digestion of fats does not occur until ________________-containing fat enters the duodenum, triggering the release of ________________ from intestinal cells. This hormone targets ________________ in the gallbladder, causing it to contract and release stored ________________, which is produced by the ________________.

3. Identify the liver and the gallbladder on an anatomical model and describe the location of the gallbladder using three directional terms. ________________ ________________

4. Bile exits the gallbladder via the ________________, which fuses with the ________________ to form the ________________. Identify these ducts on an anatomical model.

5. The ________________ duct empties into the ________________. Bile breaks large particles of fat into smaller fat droplets in a process called ________________.

6. After large fat particles are broken into smaller fat droplets, the enzyme ________________ hydrolyzes each fat droplet into its building blocks: ________________ and ________________.

PhysioEx EXERCISE 8
Chemical and Physical Processes of Digestion

The PhysioEx 9.1 Laboratory Simulations in Physiology are easy-to-use laboratory simulations that can be used as an alternative to or as a supplement to wet lab activities in this unit. Each simulation allows you to investigate important physiological concepts, repeat labs as often as you like, and conduct experiments that are difficult to perform in a wet lab environment because of time, cost, or safety concerns.

Access the simulations in these activities at MasteringA&P® > Study Area > PhysioEx 9.1

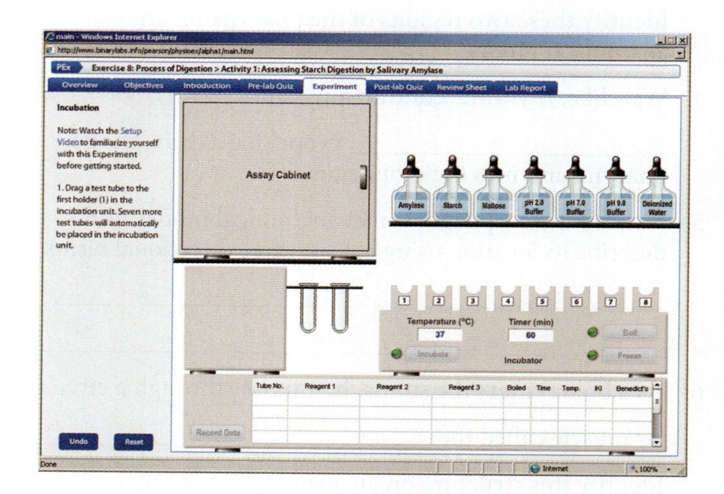

There you will find the following activities:

PEx Activity 1: Assessing Starch Digestion by Salivary Amylase

PEx Activity 2: Exploring Amylase Substrate Specificity

PEx Activity 3: Assessing Pepsin Digestion of Protein

PEx Activity 4: Assessing Lipase Digestion of Fat

POST-LAB ASSIGNMENTS

Name: ____________________ Date: ____________ Lab Section: ________

PART I. Check Your Understanding

Activity 1: Analyzing Amylase Activity

1. Name three specific sources of amylase: ____________________
2. The substrate for amylase is ____________________.
3. Why were the test tubes and their contents incubated at 37°C? ____________________
____
4. How was the amylase denatured? ____________________

Activity 2: Analyzing Pepsin Activity

1. The building blocks of proteins are ____________________.
2. The chemical digestion of proteins begins in the ____________________.
3. Chief cells release the precursor molecule ____________________, which is converted into active ____________________ by ____________________.
4. The optimum pH for pepsin is ____________________.

Activity 3: Analyzing Lipase Activity

1. What are the building blocks of triglycerides (neutral fats)? ____________________
2. Did fat digestion occur in tube 2L? Why or why not? ____________________
____
3. Did fat digestion occur in tube 4L? Why or why not? ____________________
____
4. Did fat digestion occur in tube 6L? Why or why not? ____________________
____
5. What effect did freezing have on lipase? ____________________
____

Activity 4: Tracing Digestive Pathways

1. Amylase converts starch into ____________________, which is then hydrolyzed to yield ____________________.
2. What is the optimum temperature for amylase? ____________ Why? ____________
____

3. Proteins are chemically broken down into large polypeptides by ______________________________,
large polypeptides are broken down into small polypeptides by ______________________________
and ______________________________, and small polypeptides are broken down into amino acids by
______________________________ enzymes in the cells lining the villus.
4. Amino acids enter the capillary at the villus through ______________________________ and are
transported to the superior mesenteric vein, which fuses with the ______________________________
vein to form the ______________________________ vein, which transports blood to the liver.
5. Large lipid globules in the stomach must be separated by ______________________________ and
______________________________ into smaller lipid globules.
6. When fatty chyme enters the duodenum, the release of the hormone ______________________________
triggers smooth muscle contraction in the ______________________________, causing the
release of bile. Bile travels first through the ______________________________ duct, then into the
______________________________ duct before entering the duodenum.

PART II. Putting It All Together

A. Review Questions

Answer the following questions using your lecture notes, your textbook, and your lab notes:

1. Complete the following chart concerning enzymatic activity:

	Amylase	Pepsin	Lipase
Source(s)			
Site of action			
Optimum pH			
Substrate			
Products			

2. Use the list of terms provided to complete the sentences in the illustration on the next page, which outlines the events of digestive physiology. Check off each term as you label it.

☐ absorption ☐ chyme ☐ electrolytes ☐ peristalsis
☐ bile ☐ defecation ☐ feces ☐ vitamins
☐ bolus ☐ deglutition ☐ mastication ☐ water
☐ brush border ☐ digestion ☐ pancreatic

Bolus

Chewing is called ______________________.

Bolus

Food mixes with saliva to form ______________________.

Swallowing is called ______________________.

Stomach

Rhythmic contractions of smooth muscle are called ______________________.

Bolus mixes with gastric juice to form ______________________,

Gallbladder

Pancreas

Duodenum

Chyme

Pancreatic juice and bile

which then enters the duodenum and mixes with ______________________ and ______________________ juice.

Small intestine

Most ______________________ and ______________________ occurs across the ______________________ surface of the cells of the small intestine.

Large intestine

Electrolytes

H_2O

Vitamins

In large intestine ______________________, ______________________, and ______________________ are absorbed into blood stream.

Indigestible wastes are excreted as ______________________ via ______________________.

3. Describe three structural adaptations of the small intestine that increase its surface area:

____

____

____

4. You ate a chicken sandwich for lunch. Trace the chemical digestion of the chicken (protein) from its entry into your mouth until the building blocks enter the bloodstream. Include the sites of digestion and the enzymes involved.

____

____

____

____

____

B. Concept Mapping

1. Fill in the blanks to complete this concept map outlining the chemical digestion of sucrose.

brush border enzymes	hydrolysis	monosaccharides	sucrose	villi

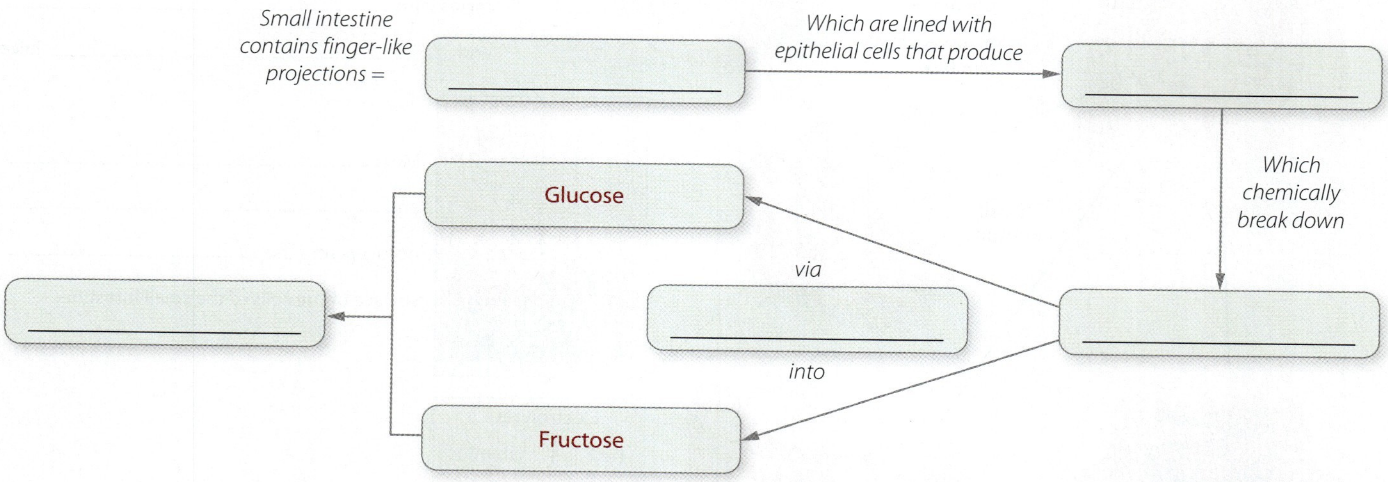

Small intestine contains finger-like projections = ____________ → *Which are lined with epithelial cells that produce* ____________ → *Which chemically break down* ____________ → *via* ____________ *into* **Glucose** and **Fructose** → ____________

2. Construct a unit concept map to show the relationships among the following set of terms. Include all of the terms in your diagram. Your instructor may choose to assign additional terms.

amino acid	brush border enzymes	chylomicron	cystic duct	fatty acid
hepatic portal vein	hydrolysis	internal jugular vein	lipase	mastication
monosaccharides	precentral gyrus	sucrose	trigeminal nerve	villi

30

Anatomy of the Urinary System

UNIT OUTLINE

The urinary system—which consists of the paired **kidneys**, two **ureters**, the **urinary bladder**, and the **urethra**—maintains blood volume and blood pressure, removes nitrogenous waste products from the blood, and regulates electrolyte, acid-base, and water balance. Within each kidney are a multitude of functional units called **nephrons**. Nephrons filter blood to form **filtrate**, reabsorb certain substances from the filtrate back into the blood, and secrete other substances into the filtrate, ultimately producing **urine**. After the urine's concentration has been fixed, it flows out of the kidneys, into the ureters and to the bladder, where it is stored. The urine is then excreted out of the body through the urethra.

In this unit you will examine the gross and microscopic anatomy of urinary system structures, and then in the next unit you will explore glomerular filtration, tubular reabsorption, and tubular secretion, the processes involved in urine formation. You will use anatomical models, a dissection of a mammalian kidney, and histological specimens to explore the anatomical features of the kidney and how those features enable it to perform its various functions.

THINK ABOUT IT *List three specific tissue types found in the organs of the urinary system and describe where each is found.*

____

____

____

____

____

____

PRE-LAB ASSIGNMENTS

Pre-lab quizzes are also assignable in MasteringA&P®

To maximize learning, BEFORE your lab period carefully read this entire lab unit and complete these pre-lab assignments using your textbook, lecture notes, and prior knowledge.

PRE-LAB Activity 1: Exploring the Organs of the Urinary System

1. Label the kidney and its associated structures:

- ☐ renal pelvis
- ☐ ureter
- ☐ renal artery
- ☐ renal medulla
- ☐ renal capsule
- ☐ renal cortex
- ☐ renal vein

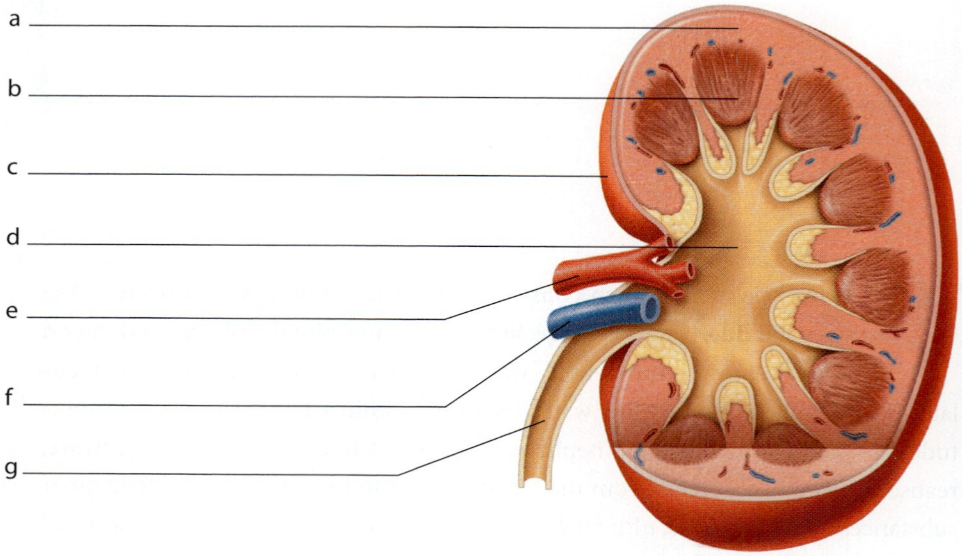

a ______

b ______

c ______

d ______

e ______

f ______

g ______

2. Match each of the following renal structures with its correct description:

_____	1. supplies the kidney with blood	a. nephron
_____	2. transports urine from the kidney to the urinary bladder	b. renal cortex
_____	3. connective tissue wrapping that surrounds kidney	c. ureter
_____	4. outer region of kidney	d. renal medulla
_____	5. houses renal pyramids	e. renal artery
_____	6. kidney region continuous with the ureter	f. renal capsule
_____	7. microscopic, functional unit of kidney	g. renal pelvis
_____	8. drains blood from the kidney	h. renal vein

PRE-LAB Activity 2: Dissecting a Mammalian Kidney

1. In which plane of section (frontal, sagittal, or transverse) will you cut the kidney? ______
   This cut will divide the kidney into ______ and ______ parts.
2. What is the hilum? ______ Which three structures enter/exit the kidney at the hilum? ______
   ______

PRE-LAB Activity 3: Examining the Microscopic Anatomy of the Kidney, Ureter, and Urinary Bladder

1. The renal corpuscle consists of the ______________________________ and the ______________________________.
2. The glomerular capsule consists of an inner ______________________________ layer and an outer ______________________________ layer. Between these two layers is the ______________________________ space.
3. A major tissue type in the urinary bladder is:
   a. moist stratified squamous epithelium.
   b. skeletal muscle.
   c. transitional epithelium.
   d. elastic cartilage.
   e. dense regular connective tissue.

Organs of the Urinary System

The organs of the urinary system (kidneys, ureters, urinary bladder, and urethra) are shown in Figure 30-1. The kidneys are retroperitoneal (located posterior to the parietal peritoneum) organs and are surrounded by three connective tissue wrappings: the **renal capsule**, the **adipose capsule**, and the **renal fascia** (Figure 30-2).

Each kidney has three distinct regions: the renal cortex, the renal medulla, and the renal pelvis (Figure 30-3a). The **renal cortex** is the most superficial kidney region. The **renal medulla**, which is deep to the cortex, is divided into cone-shaped structures called **renal pyramids** that are separated by **renal columns**. The **base** of each pyramid faces the renal cortex, and the **papilla** (apex) points toward the **renal pelvis**.

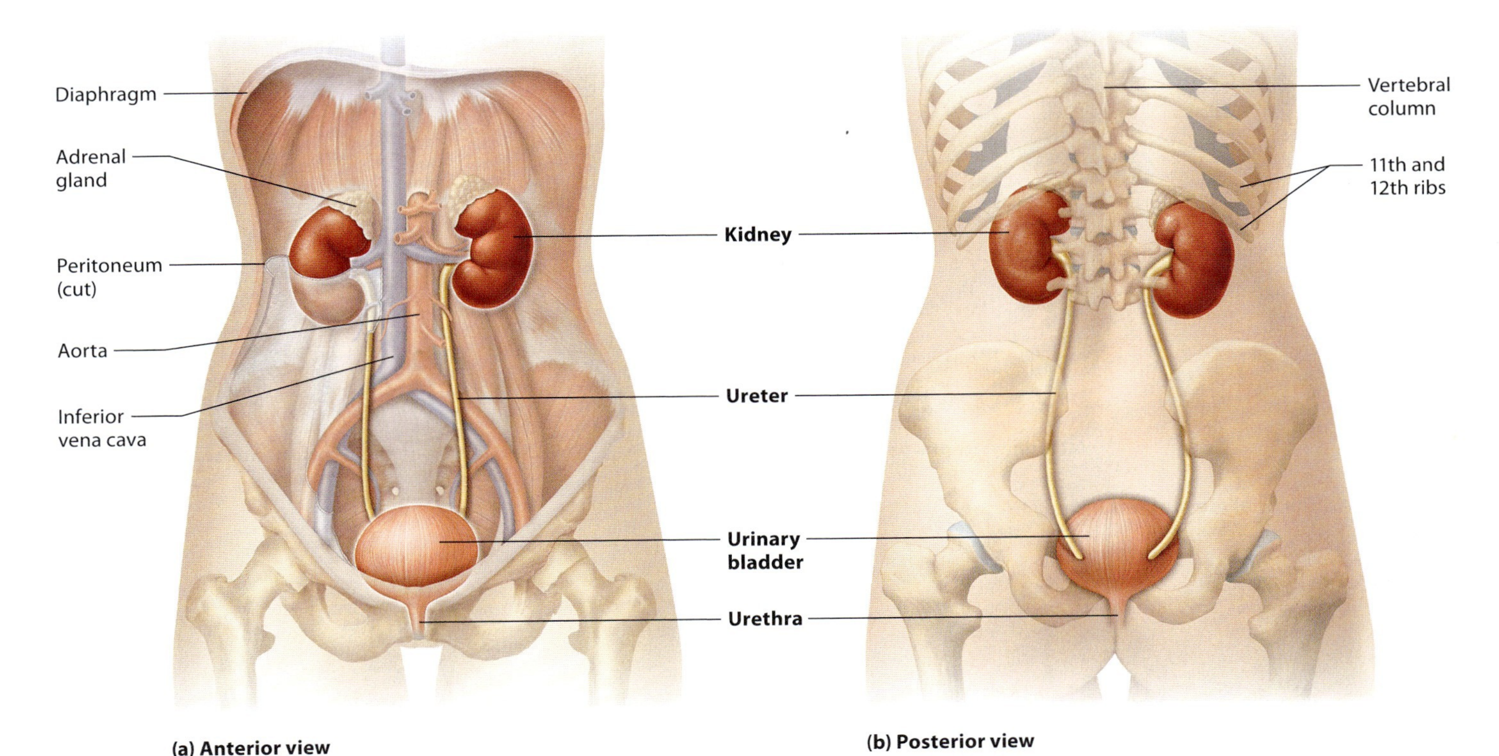

Figure 30-1 Organs of the female urinary system.

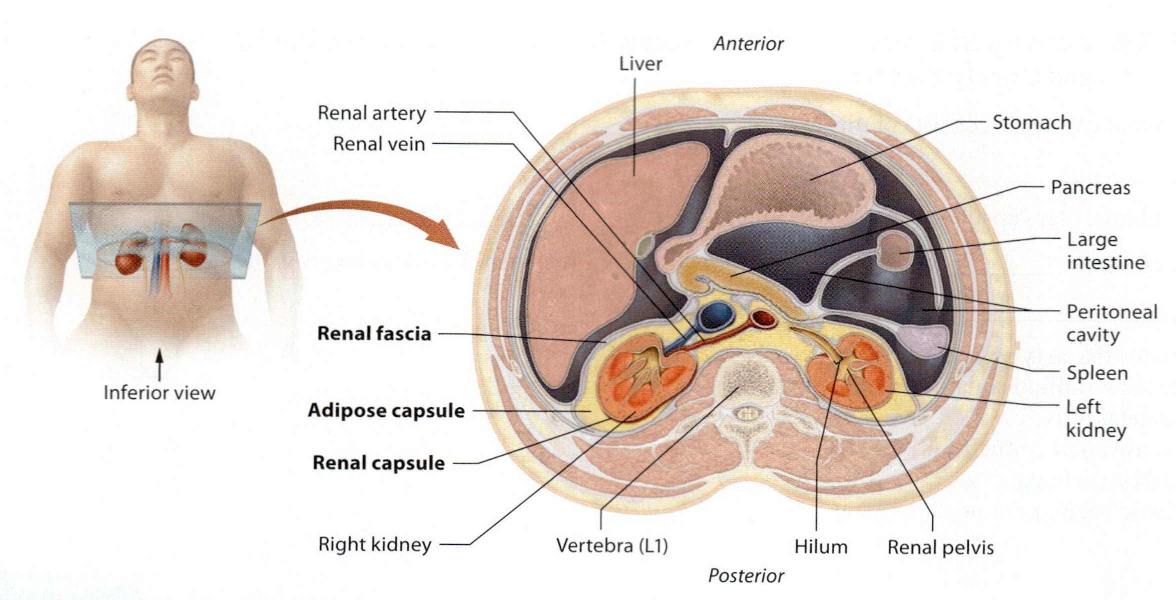

Figure 30-2 Retroperitoneal position and connective tissue wrappings of the kidneys.

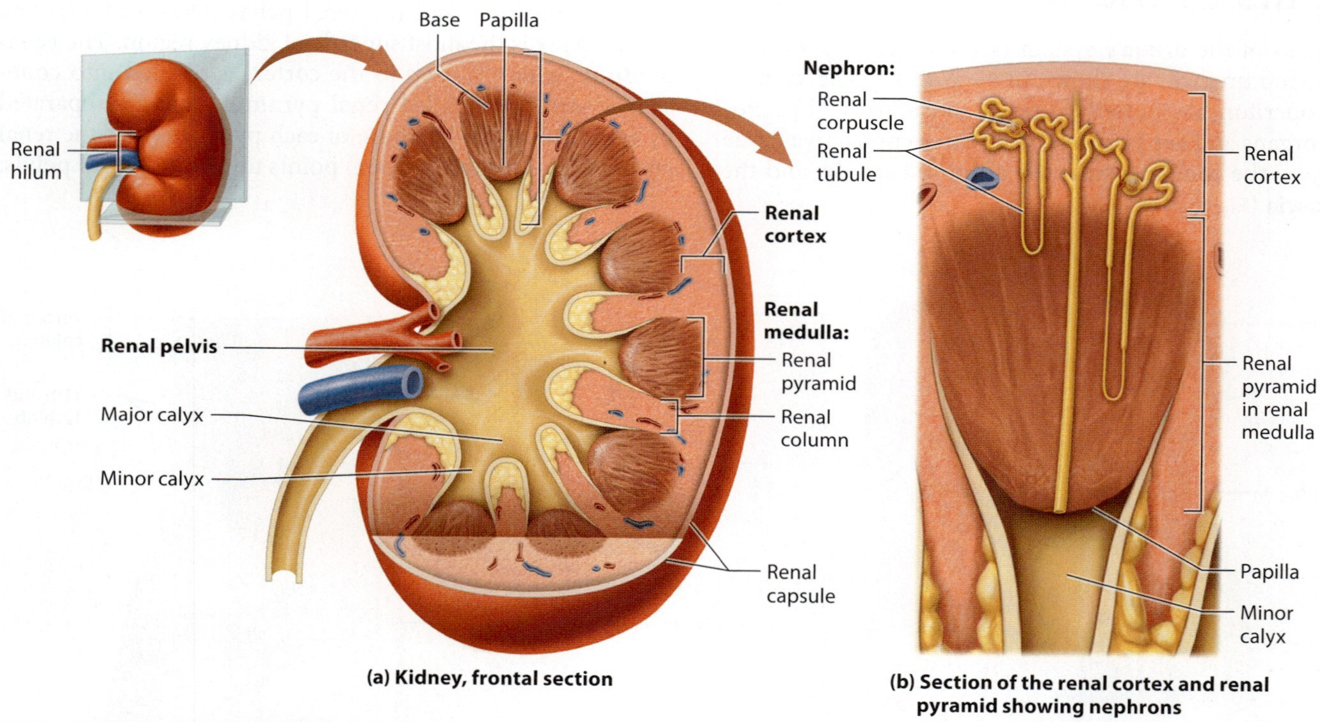

Figure 30-3 Coronal view of left kidney.

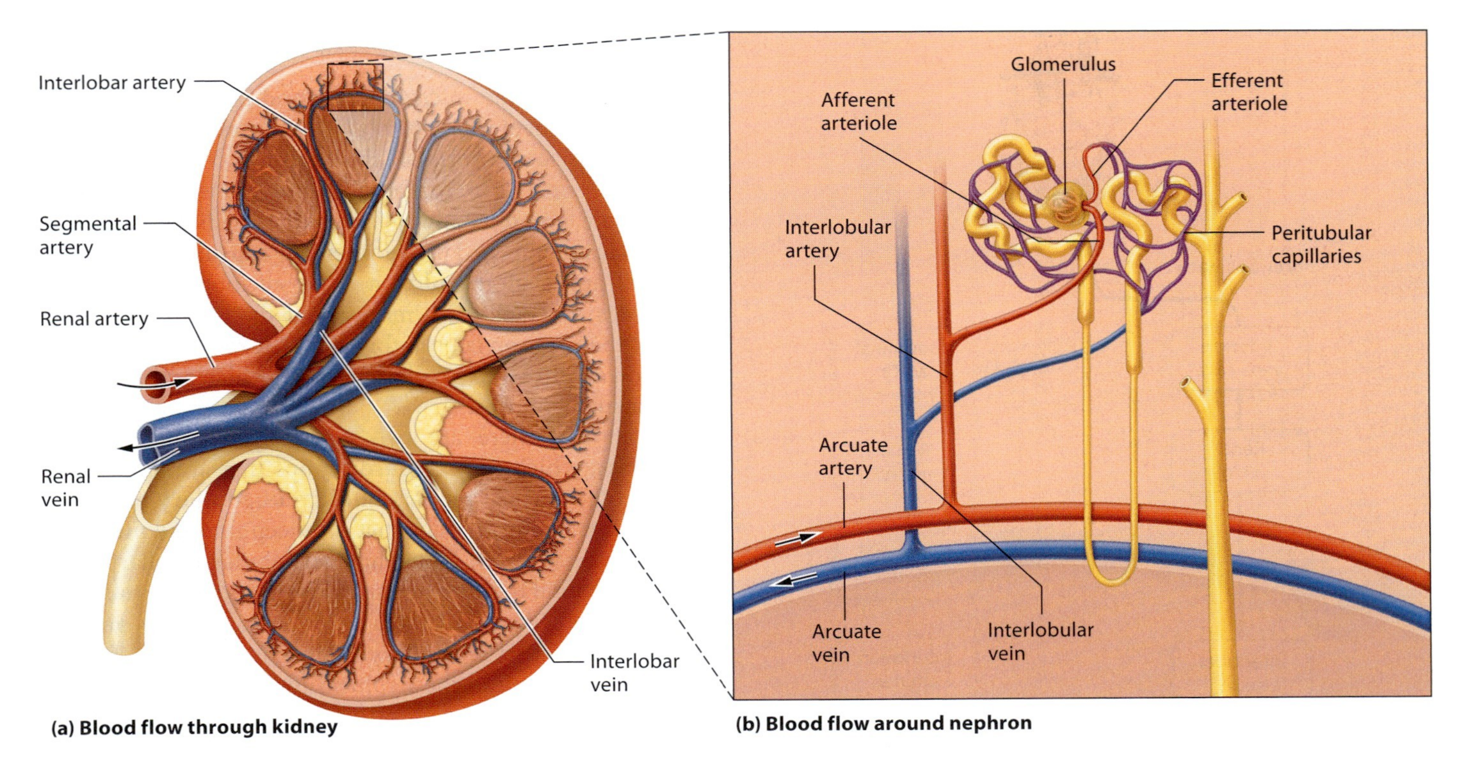

Figure 30-4 Locations of blood vessels that supply the kidney.

Both the renal cortex and the renal medulla contain **nephrons** (Figure 30-3b), the functional units of the kidney. Each kidney contains over 1 million of these microscopic blood-processing units, which filter the blood and produce urine.

Several **minor calyces** collect urine as it drains continuously from the papillae of each renal pyramid. Urine then drains from the minor calyces into two or three **major calyces** and then into the renal pelvis, which is continuous with the ureter. These structures are surrounded by smooth muscle tissue that contracts and propels urine toward the ureter.

Whereas the kidneys filter blood and produce urine, the other three organs of the urinary system transport and store the urine. The **ureters** transport urine from the kidneys to the **urinary bladder**, a muscular storage sac located posterior to the pubic symphysis. The **urethra** conveys urine out of the body.

Blood Supply to the Kidneys

The locations of the blood vessels that supply the kidney are shown in Figure 30-4; a diagram of the route of blood flow through the kidney is illustrated in Figure 30-5. Each of the two **renal arteries**, which branch from the abdominal aorta, supplies a kidney with blood and branches into progressively smaller arteries (segmental arteries, interlobar arteries, arcuate arteries, and interlobular arteries), each of which eventually leads into an **afferent arteriole** and then into a **glomerulus**. Each glomerulus is drained by an **efferent arteriole**, which leads into either a **peritubular capillary** in cortical nephrons (those that are largely confined to the renal cortex) or into a **vasa recta** in juxtamedullary nephrons (those that have components that dip down into the renal medulla). Blood then returns to the **renal vein** via the interlobular veins, the arcuate veins, and the interlobar veins. Note that there are no segmental veins.

THINK ABOUT IT Describe the path of a molecule of oxygen from the point at which it enters an alveolus in the lungs until it reaches a cell in the nephron loop of a juxtaglomerular nephron.

____

____

____

____

____

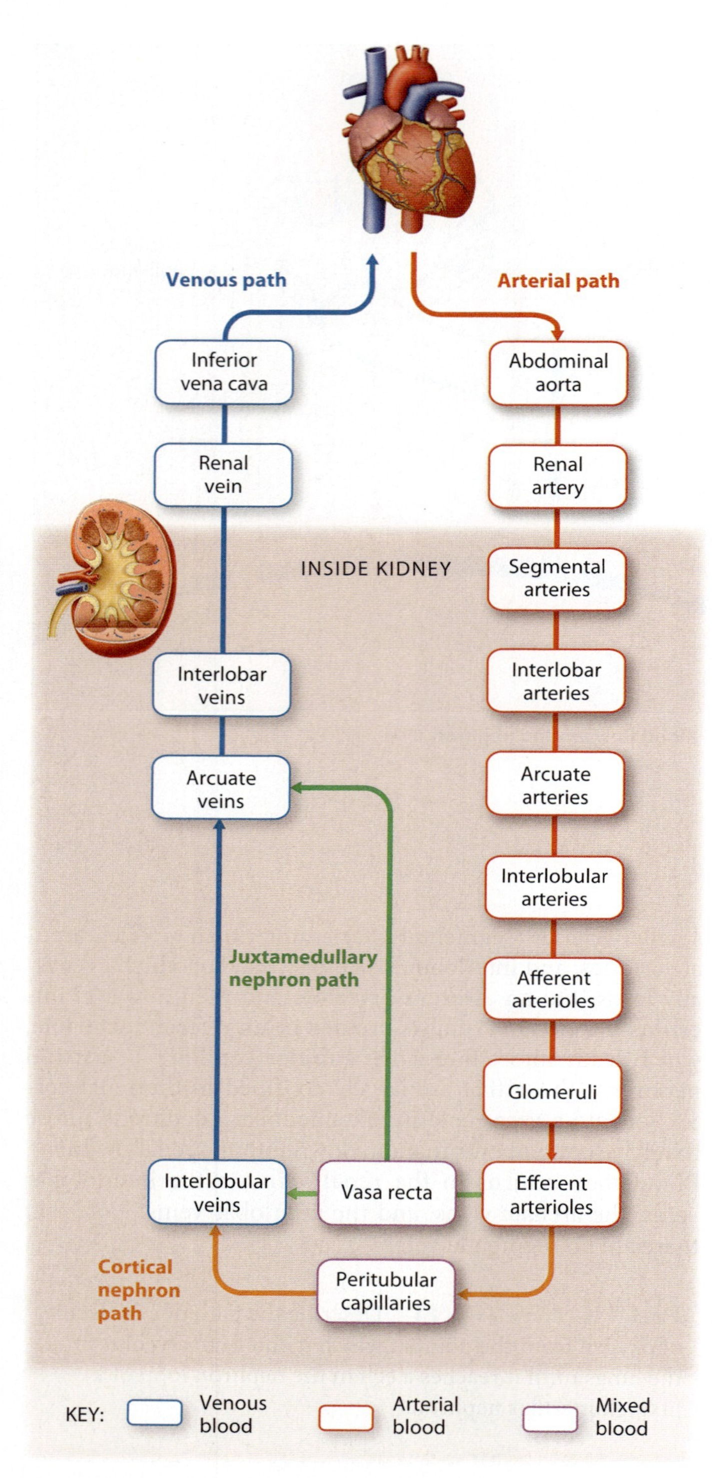

Figure 30-5 Schematic diagram of the path of blood flow through the kidney.

The Nephron

The nephron is the microscopic blood-filtering unit of the kidney. Each nephron consists of two main components: a renal corpuscle and various tubules collectively called a renal tubule (**Figure 30-6**). Each **renal tubule** consists of a **proximal tubule**, a **nephron loop**, and a **distal tubule**. Each **renal corpuscle** is composed of a **glomerular capsule** (or Bowman's capsule) and a **glomerulus**, a group of looping fenestrated capillaries. You will learn more about the microscopic anatomy of the nephron later in this unit. The renal tubule of each nephron empties into a **collecting duct**, which receives filtrate from many nephrons.

Recall that there are two types of nephrons: **cortical nephrons**, which have nephron loops that are largely confined to the cortical region, and **juxtamedullary nephrons**, which have nephron loops that dip deep into the medullary region (**Figure 30-7**). Note that peritubular capillaries surround proximal tubules, nephron loops, and distal tubules of cortical nephrons in addition to proximal tubules and distal tubules of juxtamedullary nephrons, whereas the vasa recta surrounds only the nephron loops of juxtamedullary nephrons.

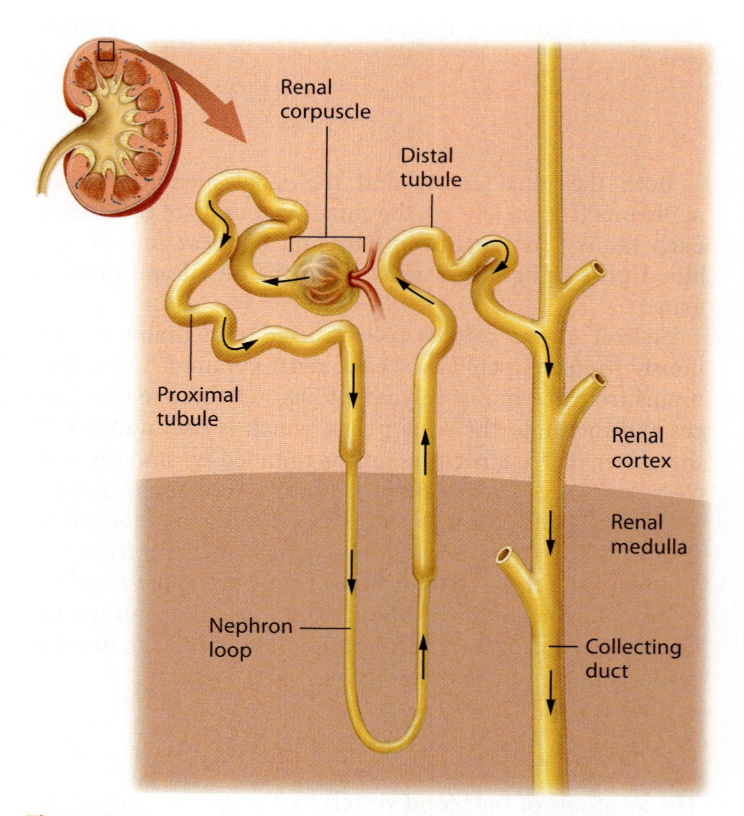

Figure 30-6 A generalized nephron and its collecting duct.

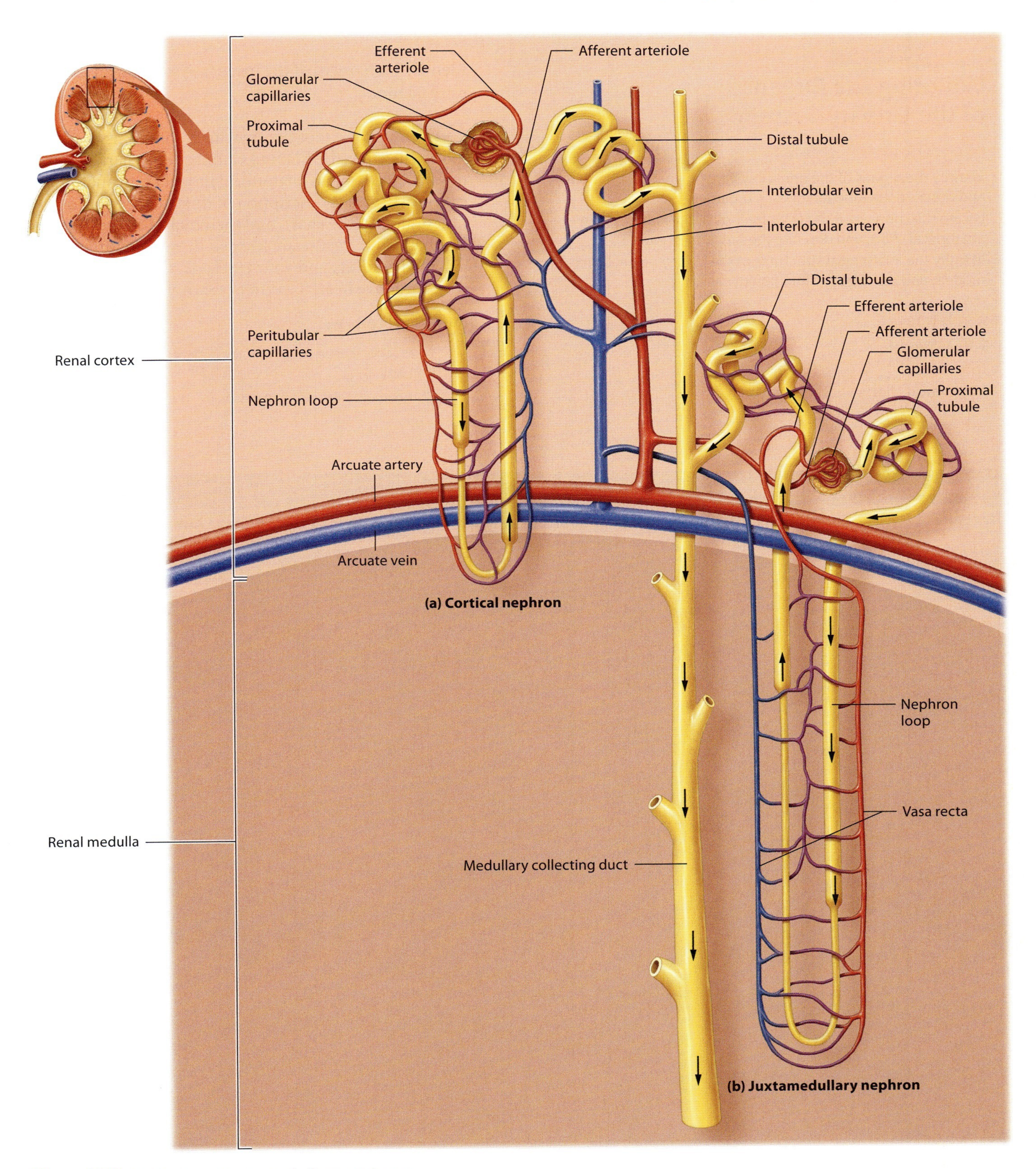

Figure 30-7 **Cortical and juxtamedullary nephrons.**

ACTIVITY 1

Exploring the Organs of the Urinary System

Learning Outcomes

1. Identify the organs of the urinary system on anatomical models, and describe the function of each organ.
2. Locate the connective tissue wrappings of the kidney on an anatomical chart, and describe the function of each.
3. Identify the anatomical structures of the kidney on anatomical models, and describe the function of each.

Materials Needed

- ☐ Kidney model
- ☐ Torso model
- ☐ Anatomical charts

Instructions

1. Identify each organ of the urinary system on an anatomical model.
2. CHART For each of the structural components of a kidney listed in the Making Connections chart, locate the structure (on an anatomical model or an anatomical chart), write a description of its structure and/or function(s) in the chart, and then make "connections" to what you have already learned in your lectures, your assigned reading, and your textbook. Some portions of the chart have been filled in as examples.
3. **Optional Activity**

Practice labeling organs of the urinary system on human cadavers at > MasteringA&P® > Study Area > Practice Anatomy Lab > Human Cadaver > Urinary System

Making Connections: The Kidney

Structure	Description (Structure and/or Function)	Connections to Things I Have Already Learned
Renal capsule		*Other organs such as a lymph node and the spleen also have capsules.*
Renal cortex		
Renal medulla	*Region of the kidney deep to the cortex*	
• Renal column		
• Renal pyramid		*Papilla (apex) means pointed; papilla points inward toward minor calyx*
Renal pelvis		
Minor calyx	*Small cavity that receives urine from the collecting duct*	
Major calyx		
Renal corpuscle	*Consists of glomerulus and glomerular capsule—filters blood*	

Structure	Description (Structure and/or Function)	Connections to Things I Have Already Learned
• Glomerular capsule		
• Glomerulus		*Glomerulus = ball; adrenal cortex has layer called zona glomerulosa (produces aldosterone)*
Renal tubule		
• Proximal tubule		*Composed of simple cuboidal epithelium*
• Nephron loop		
• Distal tubule	*Receives filtrate from the nephron loop; filtrate then travels to the collecting duct*	
Collecting duct		
Renal artery		
Afferent arteriole	*Transports blood to the glomerulus*	
Efferent arteriole		
Peritubular capillaries		*Peri = around*
Vasa recta	*Capillary associated with the nephron loop of juxtamedullary nephron*	
Renal vein		*Empties into the inferior vena cava*
Ureter		
Urinary bladder		*Inner lining composed of transitional epithelium; hollow organ—contains smooth muscle layer*
Urethra	*Transports urine out of the body in females; transports urine and sperm out of the body in males*	

ACTIVITY 2

Dissecting a Mammalian Kidney

Learning Outcomes

1. Dissect a mammalian kidney.
2. Identify the major structures of the mammalian kidney, and describe the function of each.

Materials Needed

- ☐ Fresh or preserved mammalian kidney
- ☐ Dissection tray and dissecting instruments
- ☐ Safety glasses and gloves

Instructions

1. Put on safety glasses and gloves.
2. Obtain a mammalian kidney, a dissection tray, and dissecting instruments.
3. Refer to the preserved sheep kidney in **Figure 30-8** as you complete the following steps to dissect the kidney.
4. Carefully remove any adipose tissue from the surface of the kidney, and identify the fibrous renal capsule.

   What is the function of the renal capsule?

   ____

   ____

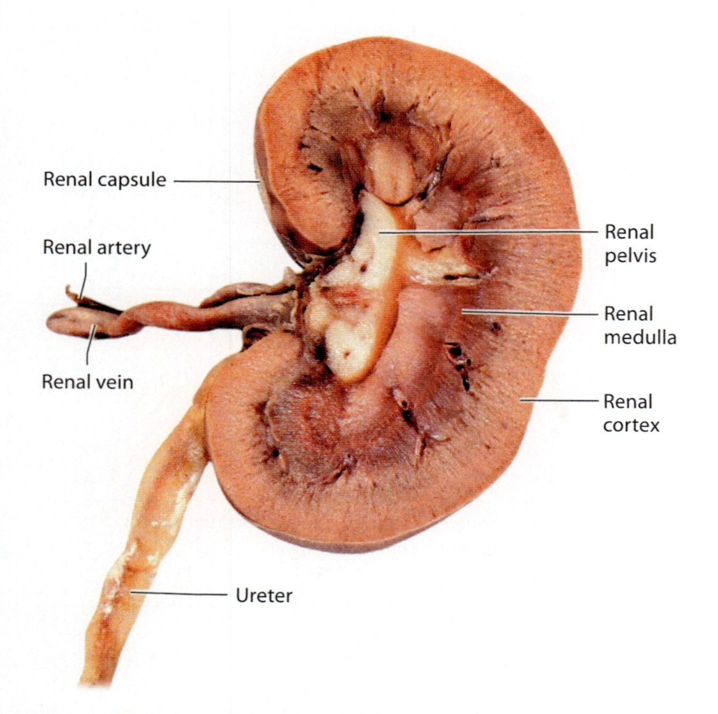

Figure 30-8 Sheep kidney, frontal section.

5. Identify the hilum (indented region) of the kidney and then locate the renal artery (it has the most muscular wall), the renal vein (it has a collapsed, thin wall), and the ureter (it has the thickest wall but is smaller in diameter than either the artery or the vein).

   State the function of each of these structures, and describe one way in which each is structurally adapted to its function.

   renal artery: ____

   ____

   renal vein: ____

   ____

   ureter: ____

   ____

6. Use a scalpel to make a frontal section of the kidney. Then make a sketch of the section and label each structure listed below. After you have labeled all of the structures in the sketch, discuss with your lab group a description and the function of the structure.

☐ renal cortex	☐ major calyx
☐ renal medulla	☐ renal pelvis
☐ renal column	☐ ureter
☐ renal pyramid	☐ renal artery
☐ minor calyx	☐ renal vein

7. Follow the instructions given by your instructor for disposing of the specimen and cleaning up.
8. **Optional Activity**

Practice labeling a sheep kidney at MasteringA&P® > Study Area > Practice Anatomy Lab > Cat > Urinary System > images 1 thru 3

Microscopic Anatomy of the Urinary Organs

The Kidney

Recall that the microscopic, functional unit of the kidney is the nephron (see Figure 30-6), which consists of a renal corpuscle and a renal tubule. The renal corpuscle (**Figure 30-9**) consists of a glomerulus and a glomerular capsule. The inner visceral layer of the glomerular capsule covers the glomerulus, and the outer parietal layer is separated from the inner visceral layer by the capsular space. Both layers of the glomerular capsule are composed of simple squamous epithelium. The epithelium of the visceral layer is composed of **podocytes** with foot-like processes called **pedicels** that interdigitate to form **filtration slits**. The glomerulus is a tuft of specialized capillaries lined with simple squamous epithelium containing pores called **fenestrations**. Together, the glomerulus and the visceral layer of the glomerular capsule form the **filtration membrane**. Blood pressure drives water and dissolved substances through both the fenestrations of the glomerulus and the filtration slits of the visceral layer of the glomerular capsule, and the resulting **filtrate** collects in the capsular space between the visceral and parietal layers of the capsule.

A histological view of a longitudinal section of the kidney at low power is shown in **Figure 30-10a**. Note the many round structures scattered throughout the cortical region. These structures are renal corpuscles and they function in glomerular filtration. When this same slide is viewed at high power (Figure 30-10b), you can identify the components of the renal corpuscle: parietal layer of glomerular capsule, glomerular space, visceral layer of glomerular capsule, and glomerulus. When using the oil-immersion lens (Figure 30-10c), you can distinguish between the proximal and distal tubules as well as the collecting ducts, structures into which the filtrate drains from distal tubules.

Ureter, Urinary Bladder, and Urethra

The **ureters** (**Figure 30-11**), are slender tubes approximately 25 cm (10 inches) long that arise as a continuation of the renal pelvis and transport urine from the kidney to the urinary bladder. The ureter wall (see Figure 30-11b) has three layers: a mucosa, a muscularis, and an adventitia. The innermost **mucosa** is composed of transitional epithelium and its underlying basal lamina (connective tissue). The middle **muscularis**, composed primarily of smooth muscle tissue, contracts rhythmically to propel urine toward the urinary bladder. The outermost **adventitia** is composed of fibrous connective tissue that supports the ureters.

The urinary bladder (see Figure 30-11) is a hollow, muscular organ that temporarily stores urine. Like the ureters, the wall of the urinary bladder (see Figure 30-11c) consists of three layers: a **mucosa** that consists of transitional epithelium and the underlying basal lamina, a thick middle smooth muscle layer called the **detrusor muscle**, and the adventia (not shown).

The urethra extends from the urinary bladder to the external urethral orifice. It drains urine from the urinary bladder and conveys it out of the body. In the female the urethra is approximately 4 cm (1.5 in) long, whereas in the male, it is approximately 20 cm (8 in) long. The urethra also has an inner mucosa, a middle muscularis, and an outer adventitia. Near the urinary bladder, the urethral mucosa consists of transitional epithelium, but it changes to pseudostratified columnar epithelium and then to a protective moist stratified squamous epithelium as it nears the external urethral orifice.

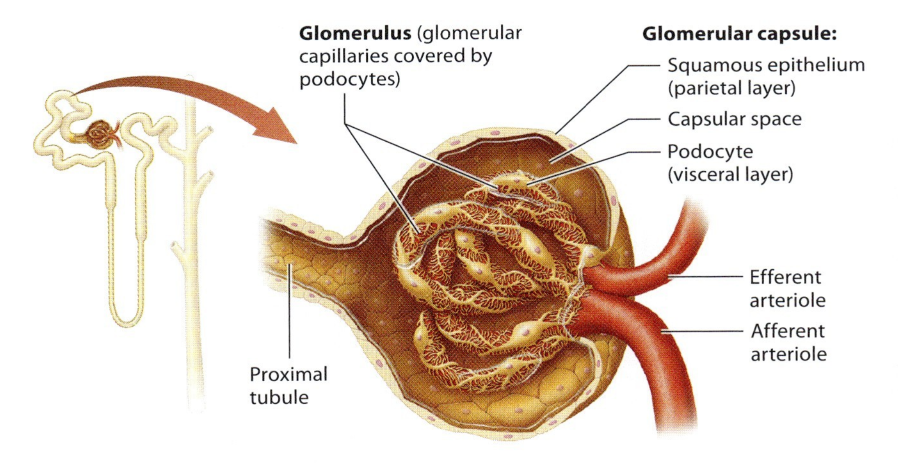

(a) The renal corpuscle

Podocyte
Filtration slits
Capillary
Pedicels
SEM (92,000×)

(b) SEM of capillary surrounded by podocytes

Figure 30-9 **The renal corpuscle.**

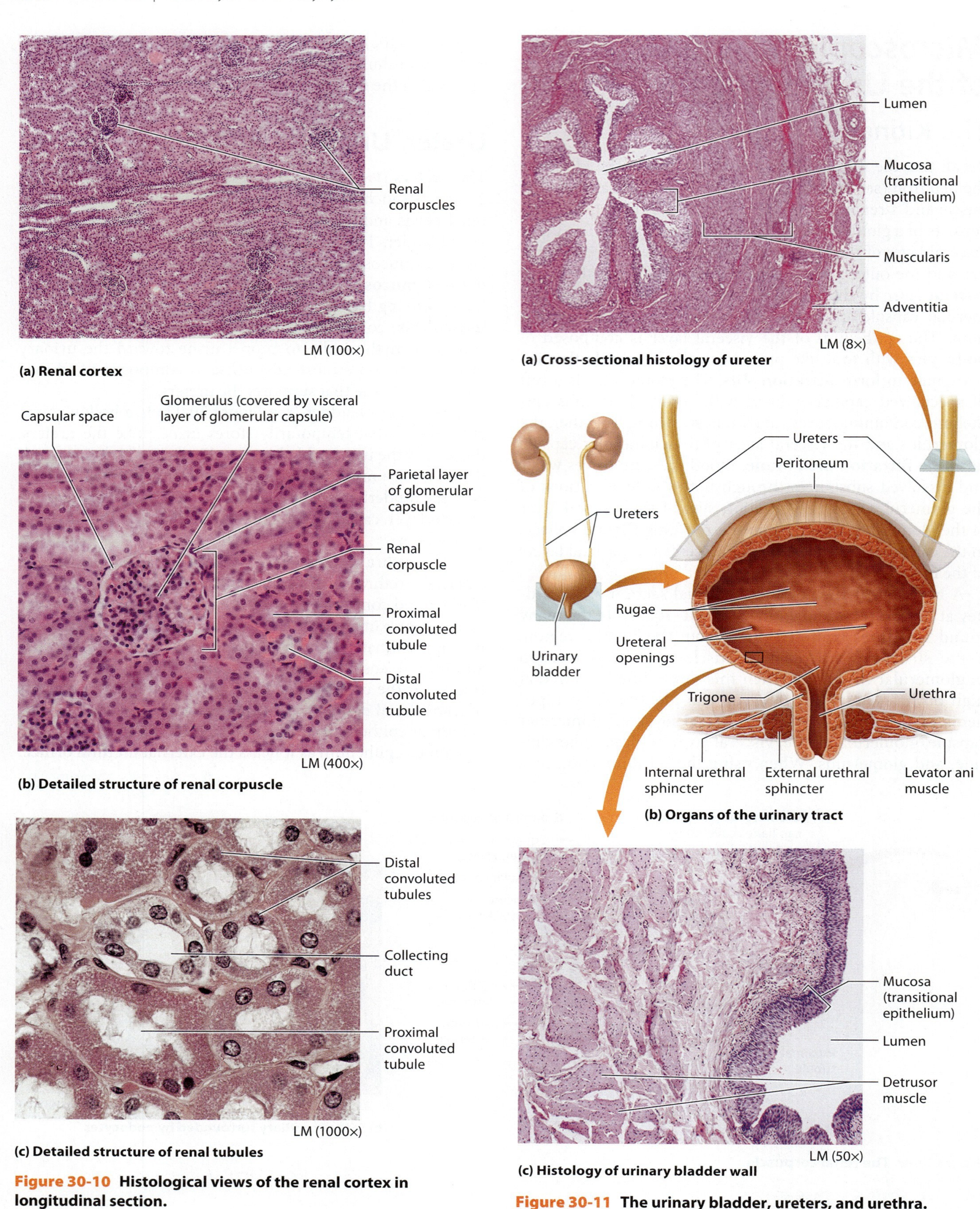

Figure 30-10 Histological views of the renal cortex in longitudinal section.

Figure 30-11 The urinary bladder, ureters, and urethra.

ACTIVITY 3

Examining the Microscopic Anatomy of the Kidney, Ureter, and Urinary Bladder

Learning Outcomes

1. Sketch, label, and describe the histological features in a microscopic section of the kidney.
2. Identify the components of the renal corpuscle (glomerulus, fenestrations, visceral layer of glomerular capsule, podocytes, pedicels, filtration slits, parietal layer of glomerular capsule, capsular space) on an anatomical model.
3. Sketch, label, and describe the histological features in a microscopic section of the ureter and the urinary bladder.

Materials Needed

- ☐ Microscope and prepared slides of sections of the kidney, ureter, and urinary bladder (Or digital photomicrographs of sections of the kidney, ureter, and urinary bladder)
- ☐ Kidney model showing enlarged renal corpuscle

Instructions

1. Examine a histological view of a longitudinal section of the kidney at low power.
   a. Identify a renal corpuscle and state its function.

   ____

   b. Now, examine this same slide at high power. Make a sketch of a longitudinal section of a kidney as seen at high power, and label the following components of the renal corpuscle: glomerulus, visceral layer of glomerular capsule, parietal layer of glomerular capsule, and capsular space.

Total magnification:____×

   c. Next, examine this same slide using the oil immersion lens. Adjacent to the glomerulus are proximal tubules and distal tubules, both composed of simple cuboidal epithelium. The apical margins of cells of the proximal tubule appear "fuzzy" due to the presence of abundant microvilli. The cells of the distal tubule do not appear "fuzzy" because they have sparse microvilli.

   What is the function of these microvilli?

   ____

   You might also be able to identify a collecting duct.

   If a collecting duct is visible, write a brief description of how you can distinguish histologically between renal tubules and collecting ducts. ______________________________

   ____

   ____

   d. Now that you have observed the renal corpuscle histologically, you can study its microscopic structure in more detail using an anatomical model of a renal corpuscle and Figure 30-9. Identify both the parietal layer and the visceral layer of the glomerular capsule as well as the glomerular space. Then, identify the podocytes that make up the visceral layer. Identify the pedicels and note how they interdigitate.

   The interdigitating pedicels form structures called

   ______________________________.

   e. Identify the glomerulus and its fenestrations on the model.

   *Which structure transports blood into the glomerulus?*______________________________

   Identify this structure on the kidney model.

   *Which structure transports blood away from the glomerulus?*______________________________

   Identify this structure on the kidney model.

   How are these two structures different?

   ____

   ____

   Predict a physiological consequence of this difference.

   ____

   ____

   Based on what you have learned about the microscopic anatomy of the renal corpuscle, how is it structurally adapted to function as an efficient filtration membrane?

   ____

   ____

LabBOOST >>>

Anatomy of the Renal Corpuscle

Understanding the anatomy of the renal corpuscle can be confusing. Here is a trick to help you learn the anatomy of the visceral layer of the glomerular capsule. Draw or tape a "nucleus" to the back of each of your hands. Your hands represent podocytes. Now, wiggle your fingers. Your fingers represent pedicels which are foot-like processes of the podocytes. Bring your fingers together so that they interdigitate (palms facing you). Note the slit-like openings between your fingers. These openings represent filtration slits. This visceral layer of the glomerular capsule overlies the glomerulus and its fenestrations to form the renal corpuscle.

2. Examine a cross section of the ureter at low power.
   a. Identify its three layers and name a predominant tissue type in each layer:

      mucosa ______________________________

      muscularis ______________________________

      adventitia ______________________________

   b. Make a sketch of a cross section of the ureter at low power, and label the following structures: lumen, transitional epithelium, muscularis (inner longitudinal layer and outer circular layer), adventitia.

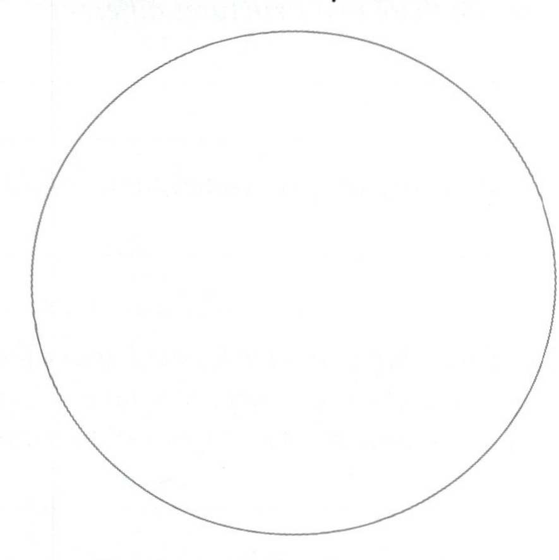

Total magnification:____×

3. Examine a cross-section of a urinary bladder at low power.
   a. Identify two layers of the bladder wall and name a predominant tissue type in each layer:

      mucosa ______________________________

      muscularis ______________________________

      The outer fibrous adventitia, the third layer of the bladder wall, might also be visible on your slide.

   b. Now, examine this same section at high power. Here we can see that the transitional epithelium is separated from the underlying basal lamina by a basement membrane. Note as well the cells of the transitional epithelium.

      Which epithelial cell shapes are visible in the transitional epithelium?

      ____

      What unique ability do these cells of the transitional epithelium possess?

      ____

      Why are the apical cells sometimes flat and sometimes dome shaped?

      ____

   c. Make a sketch of a cross-section of the urinary bladder at high power, and label the following structures: lumen, transitional epithelium, basement membrane, and basal lamina.

Total magnification:____×

4. **Optional Activity**

View and label histology slides of the urinary system at MasteringA&P® > Study Area > Practice Anatomy Lab > Histology > Urinary System

Post-lab quizzes are also assignable in MasteringA&P®

POST-LAB ASSIGNMENTS

Name: ______________________ Date: ______________ Lab Section: __________

PART I. Check Your Understanding

Activity 1: Exploring the Organs of the Urinary System

1. Label the kidney and its associated structures:

   a. ______________________

   b. ______________________

   c. ______________________

   d. ______________________

   e. ______________________

   f. ______________________

   g. ______________________

   h. ______________________

   i. ______________________

   j. ______________________

   k. ______________________

   l. ______________________

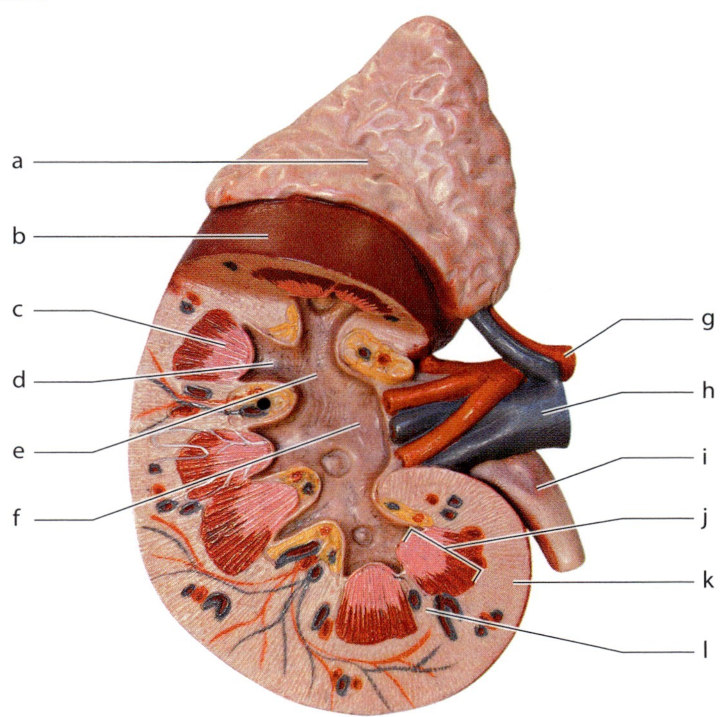

LT-K12: Kidney with adrenal gland, 2-part, 3B Scientific©

2. Which urinary organ:

   a. produces urine? ______________________

   b. transports urine out of the body? ______________________

   c. stores urine? ______________________

   d. transports urine from the kidney to the bladder? ______________________

Activity 2: Dissecting a Mammalian Kidney

1. Which connective tissue wrapping:

   ____________________ a. protects the kidney from infection?

   ____________________ b. cushions the kidney?

2. Which kidney structure:

   ____________________ a. houses the renal corpuscles?

   ____________________ b. is continuous with the ureter?

   ____________________ c. separates renal pyramids?

   ____________________ d. collects urine as it drains from a papilla?

Activity 3: Examining the Microscopic Anatomy of the Kidney, Ureter, and Urinary Bladder

1. Label the structures indicated in the following microscopic view of a kidney:

   a. ______________________________

   b. ______________________________

   c. ______________________________

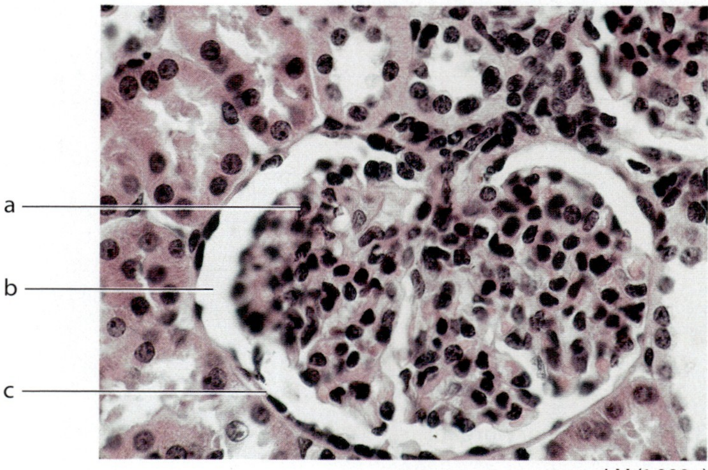

LM (1,000×)

2. The inner visceral layer of the glomerular capsule is composed of cells called ______________________. These cells have foot-like processes called ______________________ that interdigitate to form ______________________. The visceral layer of the glomerular capsule lies on top of the glomerulus, which is studded with pores called ______________________. Collectively, the inner visceral layer of the glomerular capsule and the glomerulus comprise the ______________________ membrane.

3. Label the structures indicated in the following microscopic view of the urinary bladder:

   a. ______________________

   b. ______________________

   c. ______________________

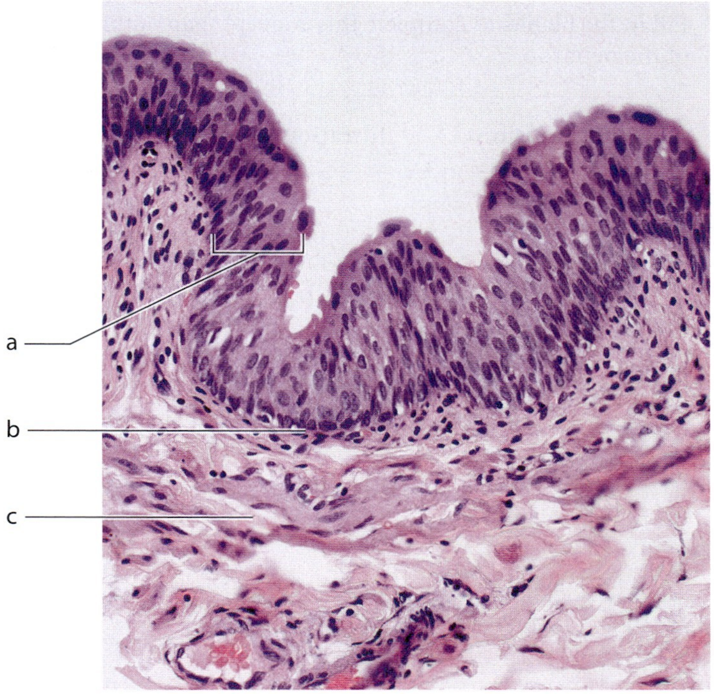

LM (400×)

PART II. Putting It All Together

A. Review Questions

Answer the following questions using your lecture notes, your textbook, and your lab notes:

1. List the connective tissue wrappings surrounding the kidney from the deepest to the most superficial wrapping:

   ____

2. Trace the pathway of blood from the left ventricle to the glomerulus. ______________________

   ____

   ____

3. What is the physiological consequence of the afferent arteriole being larger in diameter than the efferent arteriole? ______________________

   ____

4. How is the specialized epithelium of the urinary bladder related to its function? ______________________

   ____

   ____

B. Concept Mapping

1. Fill in the blanks to complete this concept map outlining the structure and function of the nephron.

fenestrations	filtration	glomerulus	podocytes	renal tubule

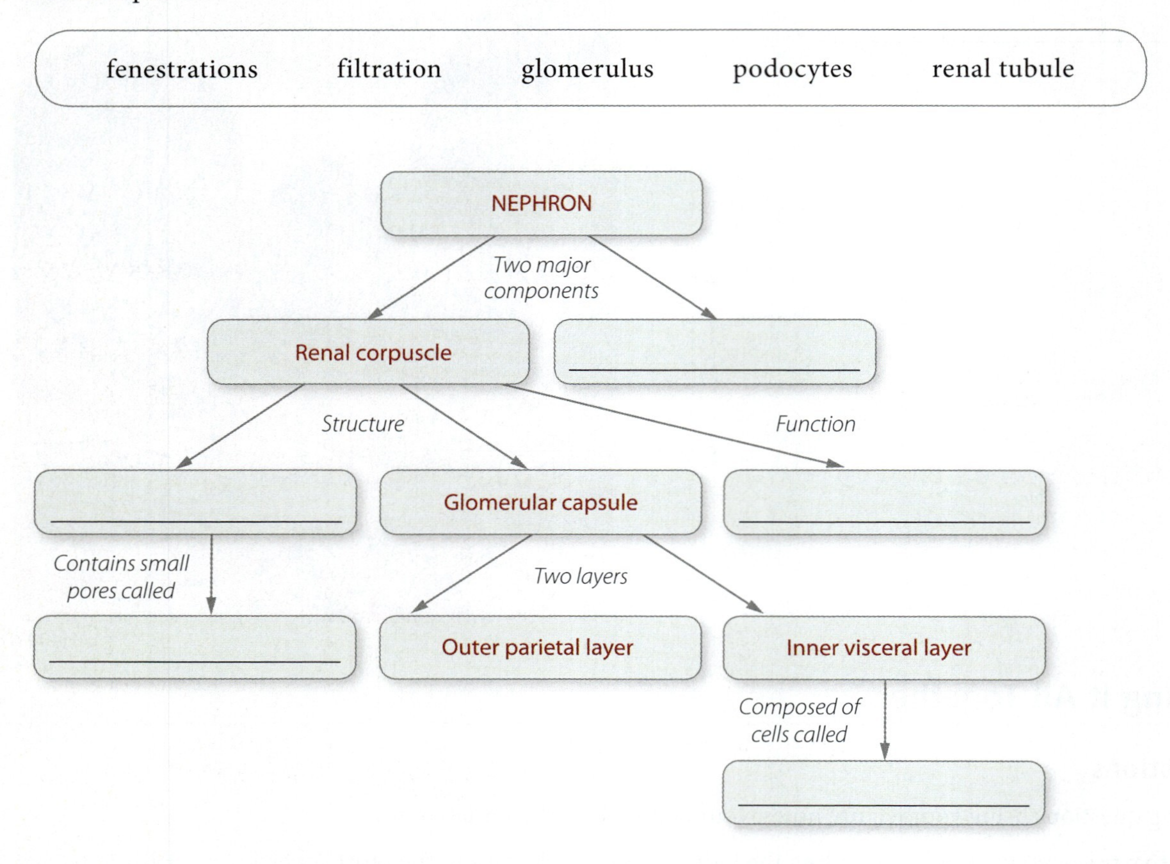

2. Construct a unit concept map to show the relationships among the following set of terms. Include all of the terms in your diagram. Your instructor may choose to assign additional terms.

collecting duct	fenestrations	filtration	glomerular capsule	glomerulus
major calyx	microvilli	nephron	podocytes	renal capsule
renal corpuscle	renal pyramid	renal tubule	urethra	urinary bladder

31

Physiology of the Urinary System

UNIT OUTLINE

In Unit 30, you examined the gross and microscopic anatomy of the urinary organs—kidneys, ureters, urinary bladder, and urethra. In this unit, you will explore **glomerular filtration**, **tubular reabsorption**, and **tubular secretion**, the processes involved in urine formation. You will engage in a variety of learning activities—building a "model" of a renal corpuscle, simulating the events that occur during the production and concentration of urine, and analyzing urine samples—each of which will further your understanding of renal physiology and the role the kidneys play in the maintenance of homeostasis.

THINK ABOUT IT *Describe several physiological links among the integumentary, digestive, urinary, and skeletal systems.*

____

____

____

____

____

PRE-LAB ASSIGNMENTS

Pre-lab quizzes are also assignable in MasteringA&P®

To maximize learning, BEFORE your lab period carefully read this entire lab unit and complete these pre-lab assignments using your textbook, lecture notes, and prior knowledge.

PRE-LAB Activity 1: Demonstrating the Function of the Filtration Membrane

1. During filtration, ________________________ pressure forces water and solutes small enough to pass through the filtration membrane into the capsular space.

2. In this activity, what does the dialysis tubing simulate? ____

____

3. How will you test for the presence of substances in the "filtrate" in this activity? ________________

____

4. What criterion determines whether or not a substance is able to pass through the filtration membrane?

____

PRE-LAB Activity 2: Simulating the Events of Urine Production (Glomerular Filtration, Selective Reabsorption, Tubular Secretion) and Urine Concentration

1. The kidneys:
   a. excrete nitrogenous wastes.
   b. regulate acid-base balance.
   c. produce renin.
   d. maintain water balance.
   e. All of these answers are correct.

2. The functional unit of the kidney is the:
   a. nephron.
   b. renal pyramid.
   c. lobe.
   d. renal corpuscle.
   e. vasa recta.

3. Which blood vessel structure dips deep into the medullary region of the kidney?
   a. peritubular capillary
   b. afferent arteriole
   c. vasa recta
   d. efferent arteriole
   e. more than one of these answers is correct.

4. Which of the following events is the first step of urine formation?
   a. tubular secretion
   b. glomerular filtration
   c. tubular reabsorption
   d. urine concentration
   e. facilitated diffusion

5. Label the components of a nephron and its blood supply.

- ☐ proximal tubule
- ☐ nephron loop
- ☐ renal corpuscle
- ☐ efferent arteriole
- ☐ afferent arteriole
- ☐ glomerular capsule
- ☐ distal tubule

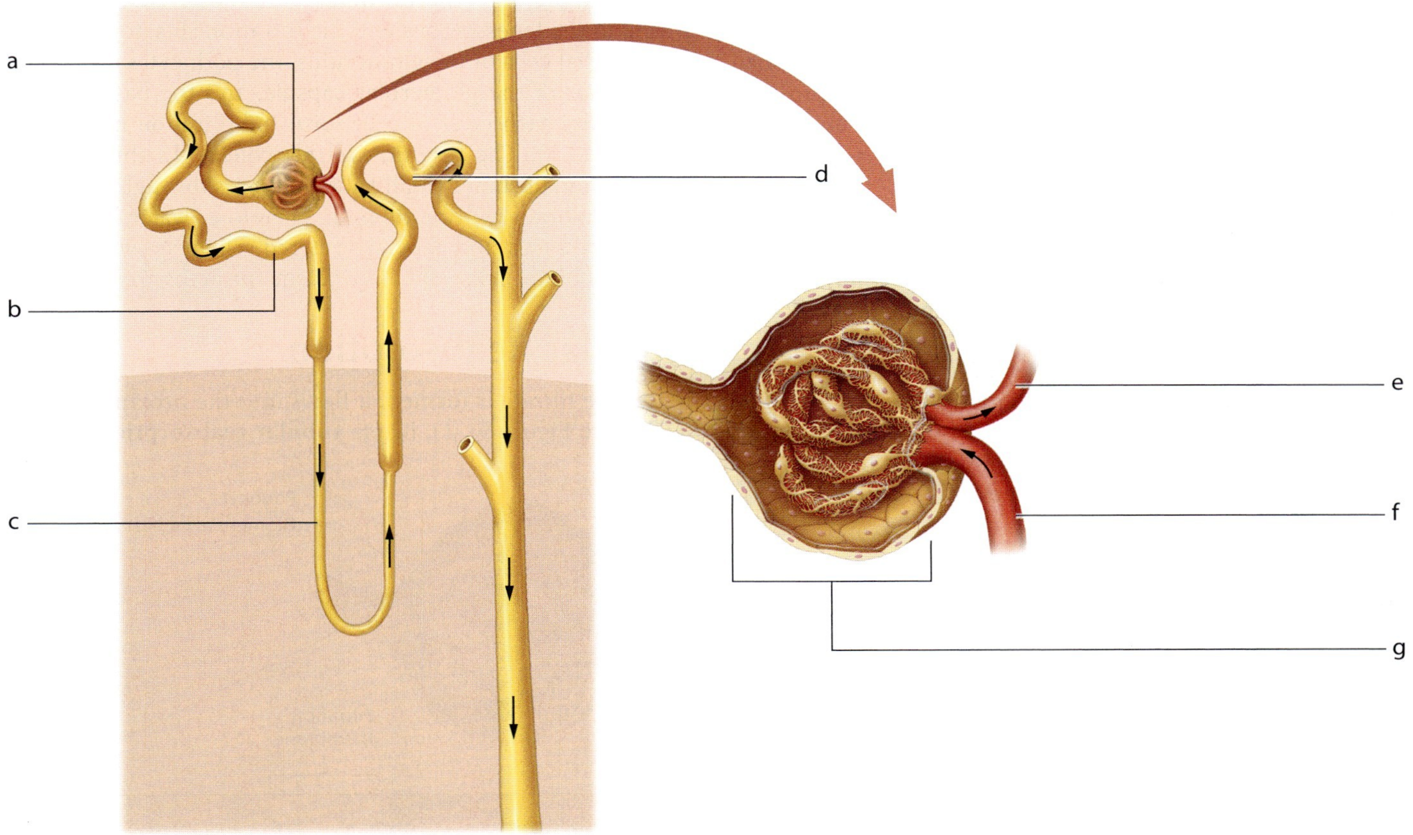

a
b
c
d
e
f
g

PRE-LAB Activity 3: Using the Results of a Urinalysis to Make Clinical Connections

1. Which of the following is normally found in urine?
   a. Plasma proteins
   b. Erythrocytes
   c. Urea
   d. Glucose
   e. All of these are normally found in urine.

2. How will you be analyzing urine in this activity?

   ____

   ____

Urinary System Physiology

During the processes of urine formation and urine concentration, the kidneys remove metabolic wastes from the blood and work to maintain homeostasis by regulating blood volume, blood pressure, ion balance, and blood pH. The formation of urine involves three physiological events: glomerular filtration, tubular reabsorption, and tubular secretion. The adjustment of urine concentration occurs during the final stages of filtrate processing as water is reabsorbed from the filtrate into the bloodstream. **Figure 31-1** provides an overview of the events of urine formation and urine concentration.

Glomerular Filtration

During **glomerular filtration** (pressure filtration), blood pressure drives the movement of water and dissolved substances through the renal corpuscle, which functions as a mechanical filter (much like a coffee filter). Recall from your study of kidney anatomy in the previous unit (see Figure 30-9) that the glomerular capsule consists of two layers: an inner visceral layer and an outer parietal layer, both of which are composed of simple squamous epithelium (see Figure 30-9).

The specialized cells of the visceral glomerular capsule are called podocytes and the foot-like processes of these cells are called pedicels. These pedicels interdigitate to form filtration slits. The tuft of specialized capillaries associated with a glomerular capsule is called a glomerulus and it is composed of simple squamous epithelium (endothelium) containing pores called fenestrations.

Blood pressure drives water and dissolved substances through the fenestrations and the filtration slits; together the endothelial cells of the glomerulus, the podocytes of the visceral layer of the glomerular capsule, and the basal lamina underlying each layer of cells form the filtration membrane (**Figure 31-2**). The resulting filtrate collects in the capsular space. Any substance small enough to pass through the filtration membrane will be filtered. Therefore, water, nutrients, waste products, and ions are filtered, whereas white blood cells, red blood cells, and large plasma proteins are not.

Tubular Reabsorption

After the filtrate is formed, it flows into the proximal tubule (PT) (see Figure 31-1), where **tubular reabsorption** begins.

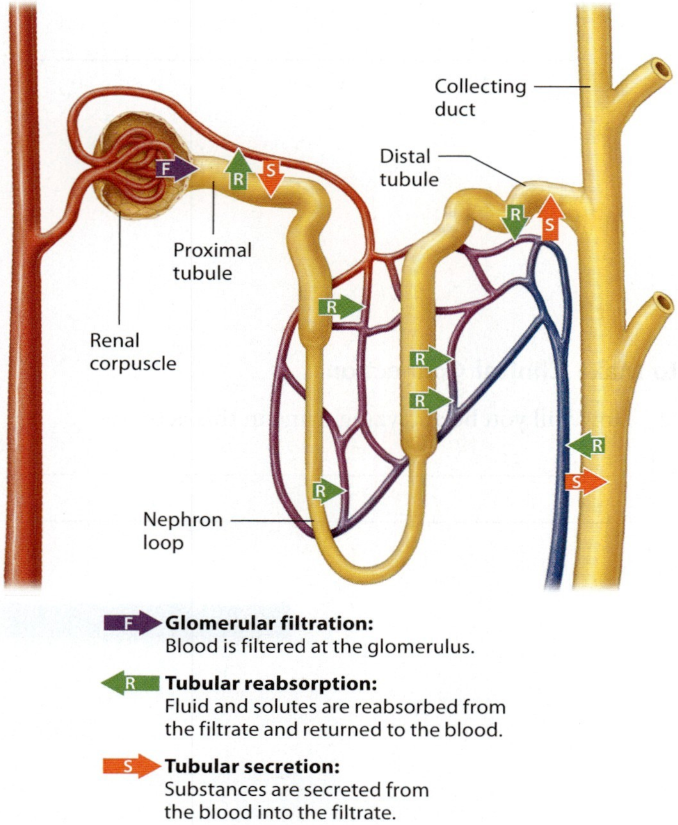

Collecting duct
Distal tubule
Proximal tubule
Renal corpuscle
Nephron loop

F **Glomerular filtration:** Blood is filtered at the glomerulus.

R **Tubular reabsorption:** Fluid and solutes are reabsorbed from the filtrate and returned to the blood.

S **Tubular secretion:** Substances are secreted from the blood into the filtrate.

Figure 31-1 An overview of the physiological processes involved in urine formation.

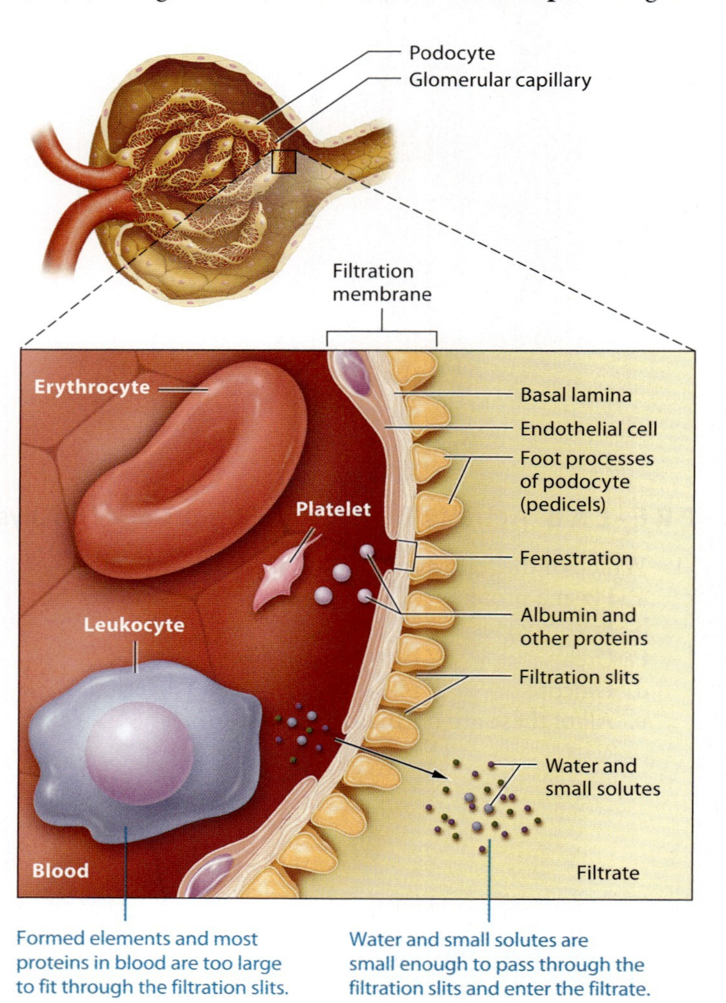

Figure 31-2 The filtration membrane.

Approximately 99% of the glomerular filtrate is reabsorbed by the renal tubules; therefore, only about 1% of the filtrate is excreted as urine. During tubular reabsorption, the renal tubule reabsorbs useful solutes from the glomerular filtrate and returns them to the blood. About 75% of tubular reabsorption occurs in the PT; the remaining 25% occurs along the rest of the renal tubule (nephron loop and distal tubule) and the collecting duct. Reabsorbed substances include ions, urea, water, and nutrients such as glucose and amino acids. As you learned in the previous unit, the cells of the PT are highly adapted for reabsorption because they contain numerous microvilli, which provide increased surface area for absorption; these cells also contain abundant mitochondria, which produce the ATP needed to drive energy-requiring transport processes.

Tubular Secretion

Tubular secretion (see Figure 31-1) is the process by which unwanted substances are transported from the blood and into the renal tubule. This process removes waste products such as urea, uric acid, and creatinine; drugs such as penicillin, aspirin, and morphine; and excess ions such as H^+, K^+, and HCO_3^-. Tubular secretion occurs primarily at the proximal and distal tubules and involves active transport mechanisms.

Urine Concentration

The final step before urine is excreted involves adjusting the concentration of the urine through the actions of what is referred to as a countercurrent mechanism that occurs in juxtamedullary nephrons. "Countercurrent" in this context means that a given fluid (in this case, either tubular filtrate or blood) is flowing in opposite directions through adjacent parts of a single tube. This circumstance occurs because both tubular fluid in the nephron loop and blood in the vasa recta first descend deep into the medulla, make a hairpin turn, and then ascend toward the cortex (see Figure 30-7).

The countercurrent mechanism establishes and maintains an osmotic gradient in the interstitial fluid in the medullary region, and it is that gradient that enables the kidneys to conserve water by producing concentrated urine. The mechanism has two components: a countercurrent multiplier and a countercurrent exchanger (**Figure 31-3**). The **countercurrent multiplier** is the nephron loop, and it establishes the osmotic gradient; the **countercurrent exchanger** is the vasa recta, and it maintains the osmotic gradient set up by the nephron loop.

The descending limb of the nephron loop is relatively impermeable to solutes and freely permeable to water; the ascending limb is impermeable to water and actively transports sodium ions into the interstitial fluid. Chloride ions then follow passively by diffusion. As a result, as the filtrate travels down the descending loop, water is drawn from the renal tubule into the interstitial fluid due to the high osmolality (the number of solute particles dissolved in 1 kilogram of water) of the medullary region (see Figure 31-3a).

Another factor contributing to the high osmolality of the interstitial fluid is the permeability of the distal portion of the collecting duct to urea. As urea moves out of the collecting duct and into the interstitial fluid, it increases the osmolality of the fluid bathing the nephrons.

The vasa recta serves as the countercurrent exchanger by maintaining the osmotic gradient set up by the nephron

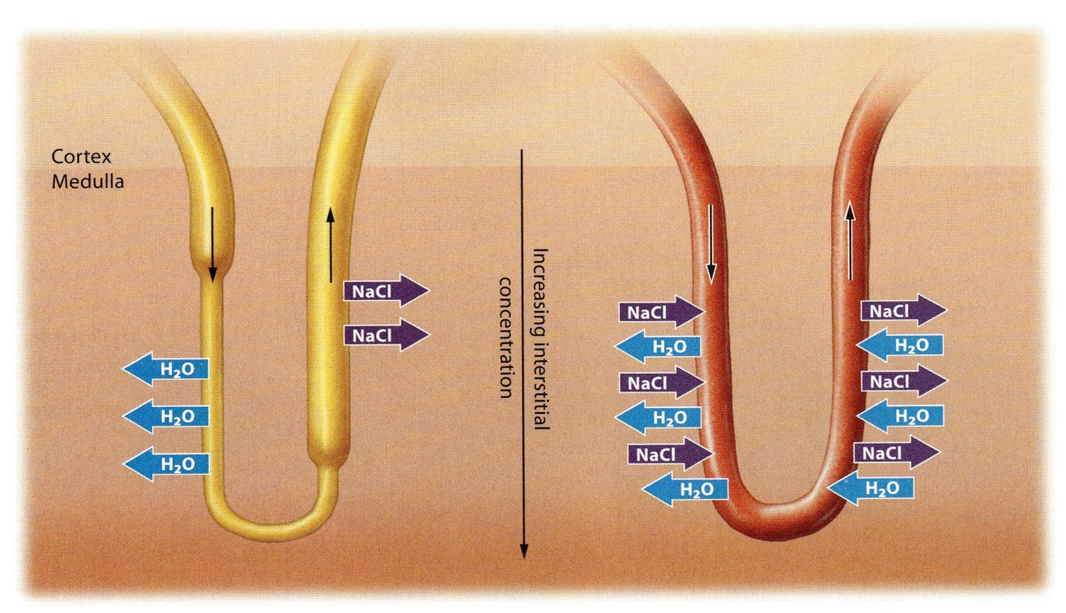

(a) Countercurrent multiplier: The nephron loop establishes the interstitial concentration gradient in the medulla.

(b) Countercurrent exchange: The vasa recta maintains the interstitial concentration gradient in the medulla.

Figure 31-3 The countercurrent mechanism.

loop. The vasa recta is freely permeable to water, sodium ions, and chloride ions; therefore, as blood flows deep into the medulla, it loses water and gains Na^+ and Cl^- (salt) (see Figure 31-3b). As blood in the vasa recta ascends and approaches the cortex, it gains water and loses salt. As a result, the vasa recta can deliver oxygen and nutrients to cells deep within the renal medulla without interfering with the interstitial concentration gradient, which is essential for producing concentrated urine.

THINK ABOUT IT What would the physiological consequences be if the blood within the vasa recta flowed from the cortex to the medulla and then left the kidney without flowing back toward the cortex?

____

____

____

____

____

The Role of the Juxtaglomerular Apparatus in Blood Pressure Regulation

A structure associated with the nephron—the **juxtaglomerular apparatus (JGA)**, which is located adjacent to the efferent and afferent arterioles at the transition point between the ascending limb of the nephron loop and the distal tubule (**Figure 31-4**)—plays an important role in regulating blood pressure. The JGA consists of two groups of cells: the juxtaglomerular cells of the afferent arteriole and the macula densa cells of the distal tubule. When blood pressure in the renal corpuscle drops, the cells of the JGA release the enzyme **renin**. A series of subsequent reactions in the blood plasma results in the production of **angiotensin II**, which has three major effects: (1) It directly stimulates blood vessels to constrict; (2) it stimulates the posterior pituitary to release **antidiuretic hormone (ADH)**, which stimulates the reabsorption of water by the kidneys; and (3) it stimulates the adrenal cortex to release **aldosterone**, which induces the kidneys to reabsorb sodium (and thus water via the resulting osmosis). The constriction of blood vessels raises blood pressure directly; the reabsorption of water resulting from the action of ADH and aldosterone increases blood volume, which also increases blood pressure. When blood pressure rises to within its homeostatic range, the JGA stops releasing renin.

ACTIVITY 1
Demonstrating the Function of the Filtration Membrane

Learning Outcomes

1. Perform a demonstration to learn about the process of glomerular filtration.
2. Describe the relationship between structure and function with respect to the filtration membrane.

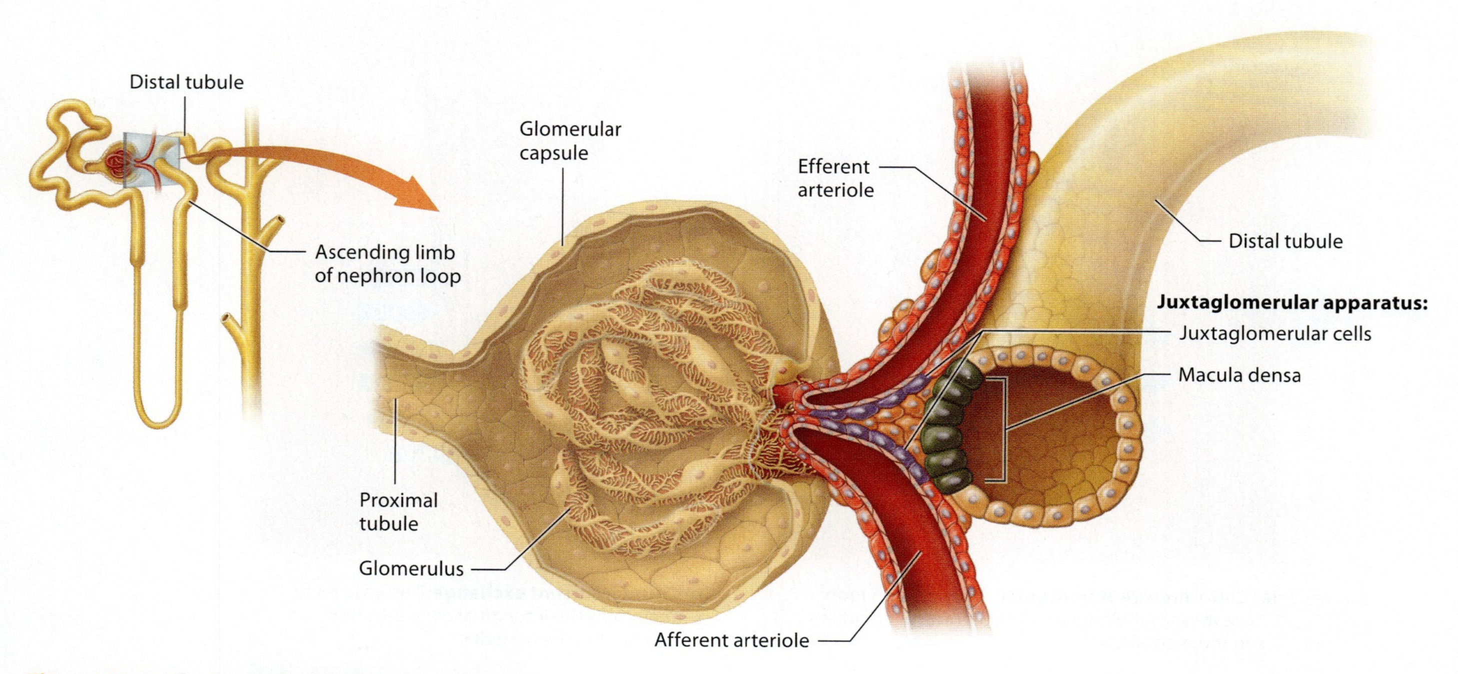

Figure 31-4 The juxtaglomerular apparatus.

Materials Needed

- ☐ Dialysis tubing
- ☐ String
- ☐ Scissors
- ☐ Simulated blood
- ☐ Beaker containing 200 ml of deionized water
- ☐ Urinalysis strips
- ☐ Safety glasses and gloves

Instructions

In this activity you will build a model using simulated animal blood and dialysis tubing in order to learn about the process of glomerular filtration. The dialysis tubing you will be using, like a filtration membrane, is a semipermeable membrane.

What determines the ability of a substance to pass through the filtration membrane?

____

The substances that are able to pass out of the "blood" through the "filtration membrane" (the dialysis tubing) will enter the "filtrate" (water), whereas those that are unable to pass through the "filtration membrane" will remain in the "blood." You will use urinalysis test strips to test the filtrate for the presence of erythrocytes, leukocytes, protein, and glucose.

1. Put on safety glasses and gloves.
2. Cut a 4-inch piece of dialysis tubing, and securely tie off one end of the tubing with a piece of string.
3. Wet the other end of the tubing and rub it between your fingers to open it.
4. Fill the tubing about half-full with simulated blood, and securely tie off the open end of the tubing with another piece of string.
5. Place the tied-up "tube" in a beaker containing 200 ml of deionized water, and leave it there for 30 minutes.
6. Predict which substances will be able to pass through the "filtration membrane" and provide a rationale for your prediction.

   ____

   ____

7. After 30 minutes, remove the "tube" from the beaker.
8. Dip a urinalysis strip into the water and then remove it quickly. Follow the directions on the side of the urinalysis strip bottle to determine and record the results for the following substances:

Substance	Result
Erythrocytes	
Leukocytes	
Protein	
Glucose	

9. Answer the following questions as you analyze and interpret your results.
   a. Which kidney structure is represented by the dialysis tubing? ______________________________

   ____

   b. Which kidney "fluid" is represented by the water?

   ____

   c. Which substance(s) is(are) "filtered"? ______________

   ____

   Why? ______________________________

   d. In what way does glomerular filtration in a real kidney differ from the process that occurred in this model?

   ____

   ____

ACTIVITY 2 Simulating the Events of Urine Production and Urine Concentration

Learning Outcomes

1. Diagram and label the structural components of a nephron.
2. Simulate the process of urine formation.
3. Compare and contrast the renal corpuscle and the renal tubule with respect to structure and function.
4. Describe the processes of glomerular filtration, tubular reabsorption, and tubular secretion.
5. Distinguish between the roles of the countercurrent multiplier and the countercurrent exchanger in urine concentration.
6. Describe the role of the juxtaglomerular apparatus in regulating blood volume and blood pressure.

Materials Needed

- ☐ Beads (1 large red, 1 large white, 2 large purple, 2 medium green, 6 small orange, 6 small yellow, 15 small blue, 15 small black, and 15 small brown)
- ☐ Pipe cleaner or string
- ☐ Two petri dishes
- ☐ Laminated simplified nephron diagram poster (or posterboard/whiteboard on which to draw the simplified nephron diagram)
- ☐ Water-soluble markers (black, brown, red, blue, orange, purple, yellow, and green)

Instructions

1. For the four parts of this activity you will use the image of a simplified nephron in **Figure 31-5** to help you visualize the movements of substances into and out of the nephron. Complete the following instructions to label the laminated poster:
   a. Use a black marker to label the following structures: the glomerular capsule (write "GC"), the proximal tubule ("PT"), the nephron loop ("NL"), the distal tubule ("DT"), and the collecting duct ("CD").
   b. Draw a brown dotted line through the nephron to separate those parts of the nephron that are in the cortex from those that are in the medulla. Label the cortex and medulla.
   c. Label the glomerulus ("G"), afferent arteriole ("AA"), efferent arteriole ("EA"), peritubular capillary ("PC"), and vasa recta ("VR") in red. Remember that peritubular capillaries actually arise from efferent arterioles of cortical nephrons, whereas the vasa recta arises from efferent arterioles of juxtamedullary nephrons. These nephrons are so near each other that each is supplied by both peritubular capillaries and the vasa recta.

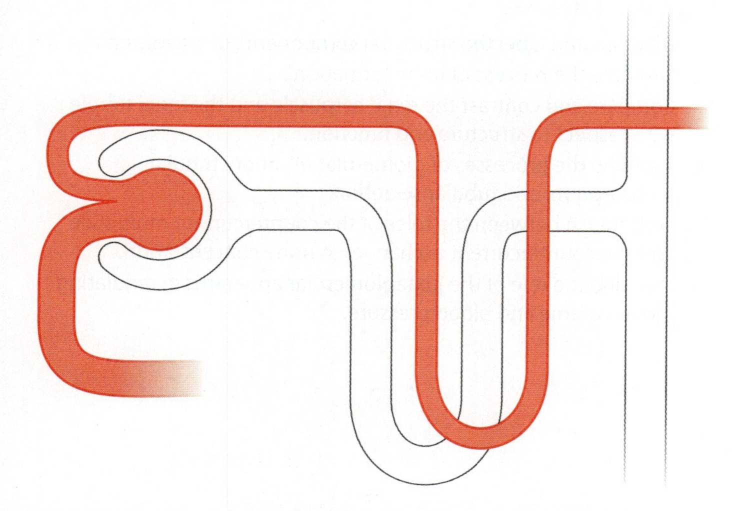

Figure 31-5 A simplified schematic view of a nephron and its blood supply for the laminated poster used in Activity 2.

   d. Draw a green box around the renal corpuscle.
      Which process occurs here?
      ____
   e. Draw an orange box around the renal tubule.
      Which processes occur here?
      ____
      ____
   f. Draw small black and brown circles in the medullary interstitium to represent sodium ions (Na^+) and chloride ions (Cl^-), respectively. Indicate the presence of an interstitial concentration gradient by increasing the number of "ions" you draw as you move from the cortex to the medulla.
      What is the function of the interstitial concentration gradient in the medullary region?
      ____
      ____
      ____
   g. Finally, label the inside of one half of a petri dish "blood" and another half "filtrate." The different beads represent different substances.
      - 1 large red bead = red blood cell
      - 1 large white bead = white blood cell
      - 2 large purple beads = plasma proteins
      - 1 medium green bead = creatinine
      - 1 medium green bead = penicillin
      - 6 small orange beads = organic nutrients
      - 6 small yellow beads = urea molecules
      - 15 small blue beads = water molecules
      - 15 small black beads = sodium ions
      - 15 small brown beads = chloride ions

Part A. Glomerular Filtration in the Renal Corpuscle

1. In this part of the activity, you will demonstrate the movement of substances from the blood of the glomerulus into the glomerular space of the glomerular capsule.
   a. Using a piece of pipe cleaner or string, connect a plasma protein and a penicillin to indicate that the two are typically bound together.
   b. Fill the petri dish containing "blood" with the following substances: 1 red blood cell, 1 white blood cell, 1 creatinine, 1 penicillin bound to a plasma protein, 1 plasma protein, 6 organic nutrients, 6 urea molecules, 15 water molecules, 15 sodium ions, and 15 chloride ions.
   c. Move the dish representing "blood" and various solutes into the afferent arteriole.
   d. Move the "blood" dish containing solutes into the renal corpuscle.

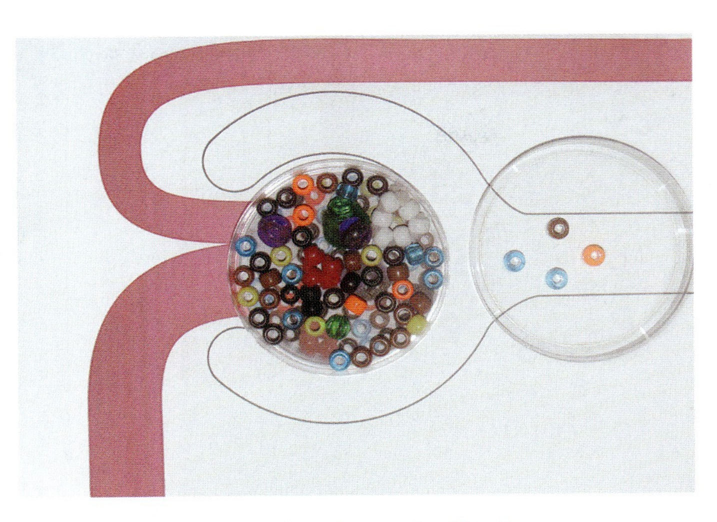

Figure 31-6 Demonstrating glomerular filtration.

e. Demonstrate glomerular filtration: For substances that are freely filtered, move all the corresponding beads into the "filtrate" dish in the glomerular space, as shown in Figure 31-6.

Which substances are freely filtered? Why?

____

____

For substances that are not in the filtrate under normal circumstances, leave the corresponding beads in the "blood" dish within the glomerulus.

Which substances are not filtered at all? Why?

____

____

In what ways is the renal corpuscle structurally adapted for the function of filtration?

____

____

____

Part B. Tubular Reabsorption into the Peritubular Capillary

1. In this part of the activity you will move the "blood" from the glomerulus into the efferent arteriole and then into the peritubular capillaries surrounding the proximal tubule. You will move the "filtrate" from the glomerular capsule into the proximal tubule.

   a. For substances that are maximally reabsorbed (99–100%) in the PT, move all the corresponding beads into the "blood" within the peritubular capillary.

   Which substances are maximally reabsorbed?

   ____

   b. For the substances that are 60–80% reabsorbed in the PT, move nine corresponding beads into the "blood" within the peritubular capillary.

   Which substances are 60–80% reabsorbed?

   ____

   c. For the substance that is 50% reabsorbed in the PT, move three corresponding beads into the "blood" within the peritubular capillary.

   Which substance is 50% reabsorbed?

   ____

   d. For the substance that is not reabsorbed, leave the corresponding bead in the "filtrate."

   Which substance is not reabsorbed?

   ____

   Why is the amount of solute that the renal tubule can absorb limited? Why would you expect to find glucose in the urine of a person with diabetes?

   ____

   ____

   ____

Part C. Tubular Secretion from the Peritubular Capillary

1. In this part of the activity you will demonstrate tubular secretion from the peritubular capillary into the renal tubule.

   a. Detach the penicillin from the plasma protein and move it into the "filtrate" in the renal tubule.

   Why was penicillin not filtered at the renal corpuscle?

   ____

   ____

   What is creatinine? Is creatinine secreted by the nephron?

   ____

   ____

   Why is it sometimes used to estimate the glomerular filtration rate (GFR)?

   ____

   ____

   b. Move the three urea beads back into the "filtrate" in the renal tubule.

   What is urea, and where is it produced?

   ____

   ____

c. Move two beads representing ions back into the "filtrate" in the renal tubule.

Name three types of ions that are secreted into the renal tubule.

____

How does the filtrate in the renal tubule differ from the urine that will eventually leave the collecting duct and drain into the renal pelvis?

____

____

Part D. Water Conservation

1. In this part of the activity you will first demonstrate the role of the countercurrent multiplier (nephron loop).

   a. Move the "filtrate" in the proximal tubule down the descending nephron loop.

   b. As the "filtrate" travels down the descending limb, move two water beads out of the nephron loop and into the interstitium, as shown in **Figure 31-7**.

   What is the driving force that pulls water out of the nephron?

   ____

   Why don't sodium ions and chloride ions move into the descending limb?

   ____

   c. As the "filtrate" travels up the ascending limb, move four sodium beads and four chloride beads out of the nephron loop and into the interstitium.

   Why do these ions leave the nephron?

   ____

   ____

   Why doesn't water move back into the ascending limb?

   ____

   ____

   Why is the nephron loop called the countercurrent multiplier?

   ____

   ____

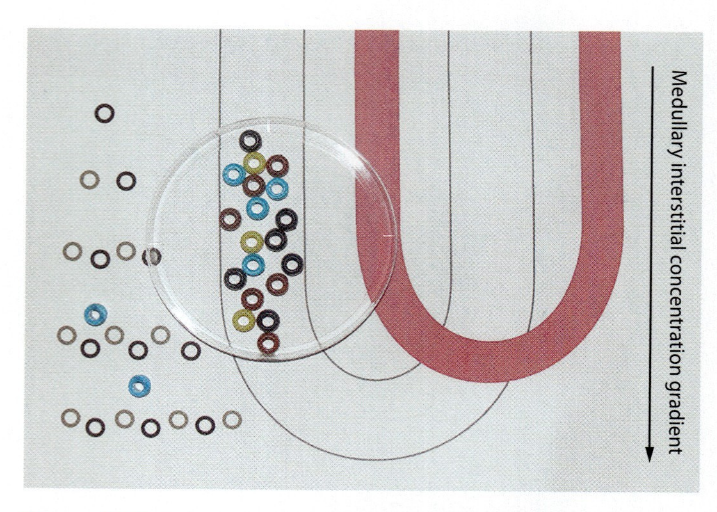

Figure 31-7 Demonstrating the countercurrent multiplier.

2. Next you will demonstrate the role of the countercurrent exchanger (vasa recta).

   a. Move the "blood" from the peritubular capillary down the descending limb of the vasa recta.

   Describe the permeability characteristics of the vasa recta.

   ____

   ____

   b. As the "blood" moves down the descending limb, demonstrate the movement of the appropriate beads from the "blood" to the interstitial fluid.

   c. As the "blood" moves up the ascending limb, demonstrate the movement of the appropriate beads from the "blood" to the interstitial fluid.

   Why is it necessary for the vasa recta to extend deep into the medullary region?

   ____

   ____

   Why is the vasa recta called the countercurrent exchanger?

   ____

   ____

   What would be the physiological consequence if the vasa recta was not part of a countercurrent mechanism?

   ____

   ____

3. Now you will demonstrate the role of the collecting duct.
   a. Move the "filtrate" from the nephron loop into the distal tubule and then into the collecting duct.
   b. As the "filtrate" moves down the collecting duct, move two water beads into the interstitial fluid of the medullary region.

      The movement of water out of the collecting duct happens only when which hormone is present?

      ____

   c. Move the "filtrate" toward the end of the collecting duct.

      What happens to some of the urea at this point?

      ____

      ____

   d. Demonstrate the movement of urea by moving two yellow beads from the "filtrate" in the collecting duct into the interstitium.

      Why is this important?

      ____

      ____

      List the contents of the urine that leaves the collecting ducts.

      ____

      ____

      Describe the path of urine from the collecting ducts to the urethra.

      ____

      ____

Lab BOOST>>>

Understanding Tonicity

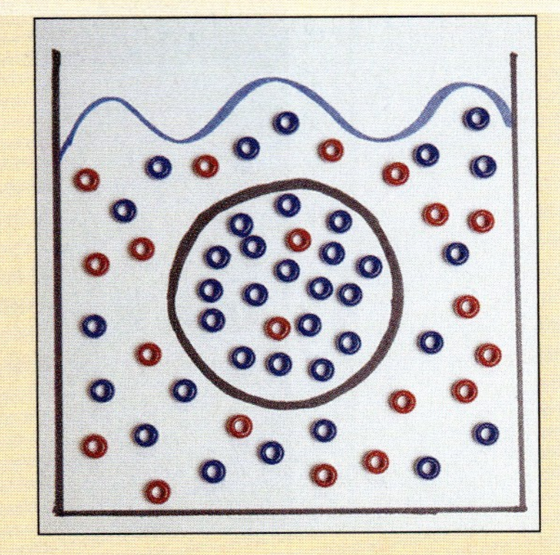

Tonicity can be a very confusing concept. To better understand the role that tonicity plays in urine concentration, use a piece of paper, pencil, blue beads (water = solvent) and red beads (ions = solute) to visually represent what happens when a cell is placed into a hypertonic environment. Remember that tonicity is a relative term used to compare the concentration of two different solutions. If Solution A is hypertonic to Solution B, then Solution A has a higher solute concentration than Solution B. On a sheet of paper, draw a large circle (3 inches in diameter) inside a beaker. "Fill" the cell with 18 water beads and 2 ion beads. Because the cell has a total of 20 beads and 2 of these beads are ions, the cell has a 1/10 or 10% solute concentration. Now "fill" the beaker with 18 water beads and 18 ion beads. The solution in the beaker has an 18/36 or 50% solute concentration. The cell membrane is impermeable to the solute; therefore, water will move by osmosis from an area of high concentration to an area of low concentration through the cell membrane. Move 10 water beads from the cell into the hypertonic solution. In the same way, water leaves the nephron loop and enters the interstitial fluid, thus allowing the kidney to reabsorb water as the urine is concentrated.

ACTIVITY 3
Using the Results of a Urinalysis to Make Clinical Connections

Learning Outcomes

1. Perform a urinalysis on urine samples and draw conclusions based on the results.

Materials Needed

- ☐ Six test tubes
- ☐ Test tube rack
- ☐ 10-ml graduated cylinder
- ☐ Wax pencil
- ☐ Student urine sample or simulated "normal" urine sample
- ☐ Simulated unknown urine samples
- ☐ Urinalysis strips
- ☐ Safety glasses and disposable gloves

Instructions

In this activity you will perform a urinalysis on a normal urine sample and on several unknown urine samples. When you have completed the urinalysis, you will use your results to match each unknown sample to a "patient." *Note:* If you are using student urine samples, follow the directions given by your instructor for obtaining these samples.

1. Put on safety glasses and gloves.
2. Obtain a test tube rack and six test tubes.
3. Use a wax pencil to label the test tubes A, B, C, D, E, and F.
4. Pour 5 ml of student urine sample (or simulated "normal" urine) "A" into test tube A.
5. Pour 5 ml each of unknown simulated urine "B" through "F" into test tubes B through F.
6. Note the color (pale yellow, medium yellow, dark yellow, or other—provide description), transparency (clear, slightly cloudy, or cloudy), and odor (characteristic or other—provide description) of each sample A–F. Record your data in the table on the next page.

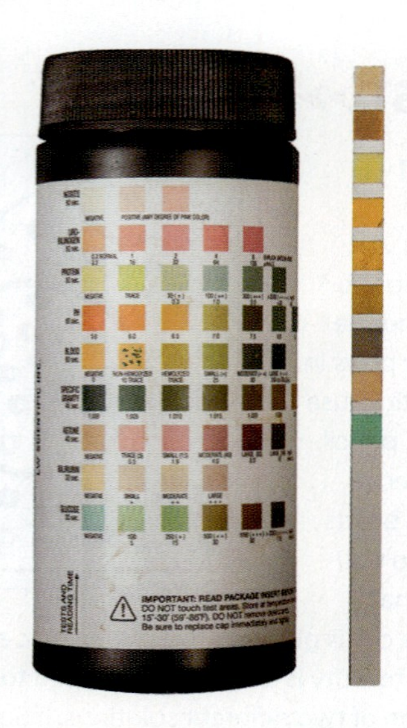

Figure 31-8 Urinalysis strip.

7. Dip one urinalysis strip into test tube A and remove it quickly. Follow the directions on the side of the urinalysis strip bottle to determine the results (**Figure 31-8**), and then record the data in the table below.
8. Repeat step 7 for test tubes B–F.
9. Use your textbook and work together as a lab group to answer the following questions:
   a. Describe the normal color of urine.
   ____
   b. What causes this normal color of urine?
   ____
   c. Describe the normal odor of urine.
   ____
   d. What causes urine to develop an ammonia-like odor when left standing?
   ____
   e. What is the pH range of normal urine? __________
   f. What might the presence of urinary nitrites indicate?
   ____
   g. Ketone bodies are intermediate products of the metabolism of ______________________.
   h. The presences of ketone bodies in excessive amounts is called ______________________ and is expected during times of starvation or while consuming diets very low in ______________.
   i. What is bilirubin?
   ____

	Sample A	Sample B	Sample C	Sample D	Sample E	Sample F
Color						
Transparency						
Odor						
Glucose						
Protein						
pH						
Leukocytes						
Nitrites						
Ketones						
Bilirubin						
Blood						
Urobilinogen						
Specific gravity						

j. Bilirubinuria (the presence of bilirubin in urine) is abnormal and usually indicates a pathology of which organ?

____

k. What is urobilinogen?

____

l. What is specific gravity?

____

m. What might a very low specific gravity indicate?

____

10. Match each of the unknown urine samples (B–F) to one of the following "patients," and briefly explain your reasoning for each match:

Patient #1: Eli is a healthy 8-year-old boy who is seeing the doctor for his annual physical exam.

____

____

____

Patient #2: Brigitte is a 19-year-old college student who recently lost 10 pounds. Additionally, she is experiencing extreme thirst and frequent urination on a regular basis.

____

____

____

Patient #3: Frank, a 60-year-old lawyer, is experiencing extreme thirst and frequent urination. Blood tests reveal that his blood glucose levels are within the normal range.

____

____

____

Patient #4: Mark, a 45-year-old chaplain, has been on an extreme low-carbohydrate diet for the past 4 months.

____

____

____

Patient #5: Sarah, a pregnant 25-year-old web developer, has been diagnosed with preeclampsia.

____

____

____

11. *(Optional)* If you have a real urine sample, then you can also analyze urine sediment microscopically. Your instructor has prepared slides for you to observe. Go to the demonstration microscope and observe these prepared slides. In the demonstration area, you will also find a picture key showing common sediments found in urine. Use this key to determine if any of these sediment types are present on each slide.

PhysioEx EXERCISE 9
Renal System Physiology

The PhysioEx 9.1 Laboratory Simulations in Physiology are easy-to-use laboratory simulations that can be used as an alternative to or as a supplement to wet lab activities in this unit. Each simulation allows you to investigate important physiological concepts, repeat labs as often as you like, and conduct experiments that are difficult to perform in a wet lab environment because of time, cost, or safety concerns.

Access the simulations in these activities at MasteringA&P® > Study Area > PhysioEx 9.1

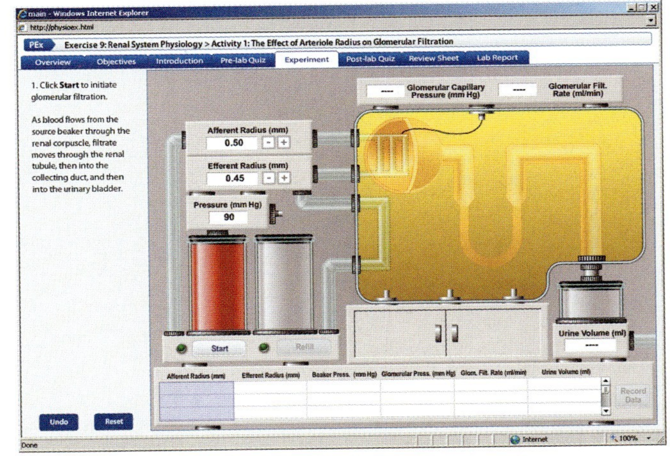

There you will find the following activities:

- **PEx Activity 1:** The Effect of Arteriole Radius on Glomerular Filtration
- **PEx Activity 2:** The Effect of Pressure on Glomerular Filtration
- **PEx Activity 3:** Renal Response to Altered Blood Pressure
- **PEx Activity 4:** Solute Gradients and Their Impact on Urine Concentration
- **PEx Activity 5:** Reabsorption of Glucose via Carrier Proteins
- **PEx Activity 6:** The Effect of Hormones on Urine Formation

PhysioEx EXERCISE 10

Acid-Base Balance

This PhysioEx 9.1 Laboratory Simulation in Physiology will help you to understand the role of both the respiratory system and the urinary system in the regulation of acid-base balance. You will rely on information that you learned in the respiratory units as well as on what you are learning in this unit as you explore an example of how organ systems interact to maintain homeostasis.

Access the simulations in these activities at MasteringA&P® > Study Area > PhysioEx 9.1

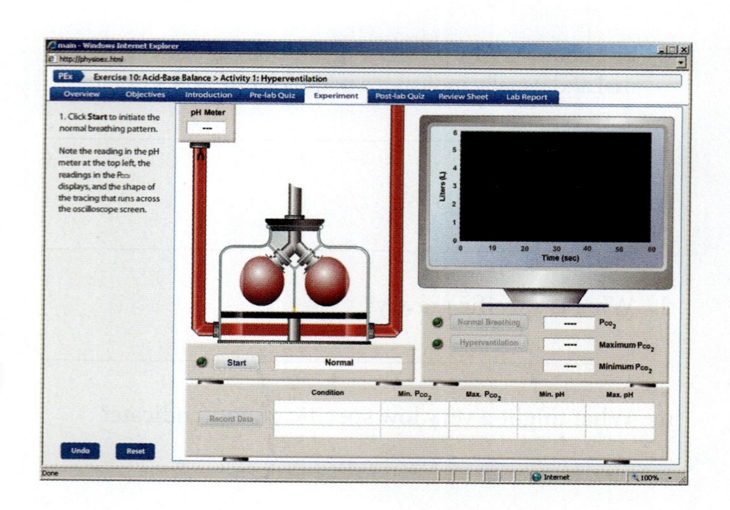

There you will find the following activities:

- **PEx Activity 1:** Hyperventilation
- **PEx Activity 2:** Rebreathing
- **PEx Activity 3:** Renal Responses to Respiratory Acidosis and Respiratory Alkalosis
- **PEx Activity 4:** Respiratory Responses to Metabolic Acidosis and Metabolic Alkalosis

Post-lab quizzes are also assignable in MasteringA&P®

POST-LAB ASSIGNMENTS

Name: ______________________________ Date: ______________ Lab Section: ________

PART I. Check Your Understanding

Activity 1: Demonstrating the Function of the Filtration Membrane

1. Is filtration an active process or a passive process? Explain.

   ____

2. In this filtration demonstration, the dialysis tubing represented the filtration membrane of the ____________________. Which two structures comprise the filtration membrane?

   ____

3. Which substance(s) was(were) able to pass through the dialysis tubing (filtration membrane) in this activity? ____

   Why? ____

4. In a real kidney, which of the following substances is able to pass through the filtration membrane? Circle all that apply.

   sodium, erythrocytes, glucose, urea, leukocytes, amino acids, immunoglobulins, potassium, fibrinogen

Activity 2: Simulating the Events of Urine Production

1. Label and color the simplified juxtaglomerular nephron diaphragm per the following directions:

   a. Label the four components of the nephron and the collecting duct.

   b. Label the afferent arteriole, the glomerulus, the efferent arteriole, and the vasa recta.

   c. Draw a purple arrow to show where the filtration of substances takes place and the direction in which filtration occurs.

   d. Draw a green arrow to show where most reabsorption of substances takes place and the direction in which reabsorption occurs.

   e. Draw orange arrows to show where most secretion of substances takes place and the direction in which secretion occurs.

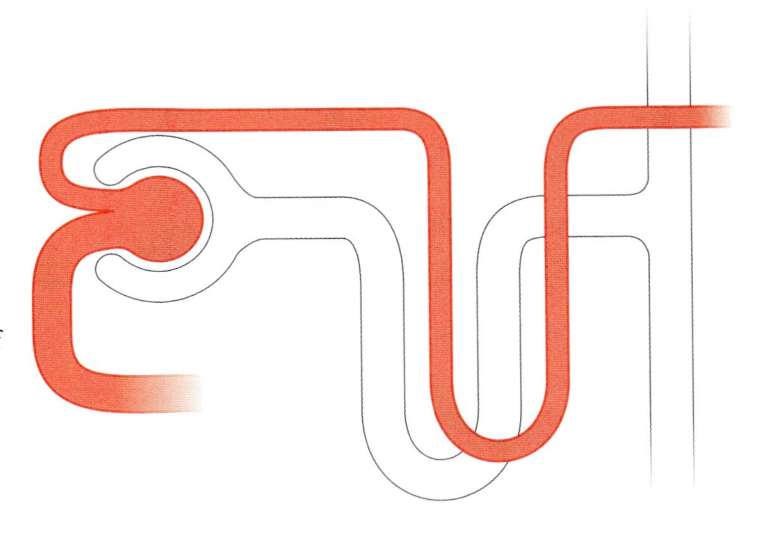

2. Match each of the following renal structures with their functions:

_____	a. renal corpuscle	1. releases renin
_____	b. vasa recta	2. contains cells with abundant microvilli
_____	c. collecting duct	3. surround cortical nephrons
_____	d. proximal tubule	4. countercurrent exchanger
_____	e. nephron loop	5. receives filtrate from many nephrons
_____	f. distal tubule	6. site of filtration
_____	g. peritubular capillaries	7. countercurrent multiplier
_____	h. juxtaglomerular apparatus	8. site of action of aldosterone

3. Which two populations of cells make up the juxtaglomerular apparatus?

____

4. Substance X is present in the urine. Does its presence prove that it was filtered at the glomerulus? Explain.

____

5. Substance Y is not present in the urine. Does its absence prove that it is neither filtered nor secreted? Explain.

____

Activity 3: Using the Results of a Urinalysis to Make Clinical Connections

1. Why is urinalysis such an important component of a medical exam?

____

____

2. Provide a possible explanation for each of the following urinalysis results:

   a. High levels of protein in the urine

   ____

   b. Low pH of the urine

   ____

   c. Glucose present in the urine

   ____

   d. Nitrites present in the urine

   ____

   e. A high specific gravity

   ____

PART II. Putting It All Together

A. Review Questions

Answer the following questions using your lecture notes, your textbook, and your lab notes:

1. Complete the following chart based on the events that occur during urine formation and concentration:

Substance	Filtered? (yes/no)	Reabsorbed? (yes/no)	Secreted? (yes/no)
Leukocytes			
Urea			
Water			
Glucose			
Penicillin			
Creatinine			
Amino acids			
Albumin			
Sodium			

2. Explain how each of the following parts of the nephron and its blood supply are structurally adapted to performing its function(s):

   a. Glomerulus ______________________________

   ____

   b. Proximal tubule ______________________________

   ____

   c. Nephron loop ______________________________

   ____

   d. Vasa recta ______________________________

   ____

3. Why are diuretics used to treat chronic high blood pressure? ______________________________

   ____

4. Do you think that the nephron loop of a desert rat would be longer or shorter than that of a human?

   ______________ Why? ______________________________

   ____

5. Complete the following chart to compare and contrast aldosterone and ADH.

	Aldosterone	ADH
What is the source of the hormone?		
How is release of the hormone controlled?		
What is the target of the hormone?		
What are the biological effects of the hormone?		

B. Concept Mapping

1. Fill in the blanks to complete this concept map outlining the structure and function of the nephron.

active transport	ATP	microvilli	osmosis	tubular reabsorption

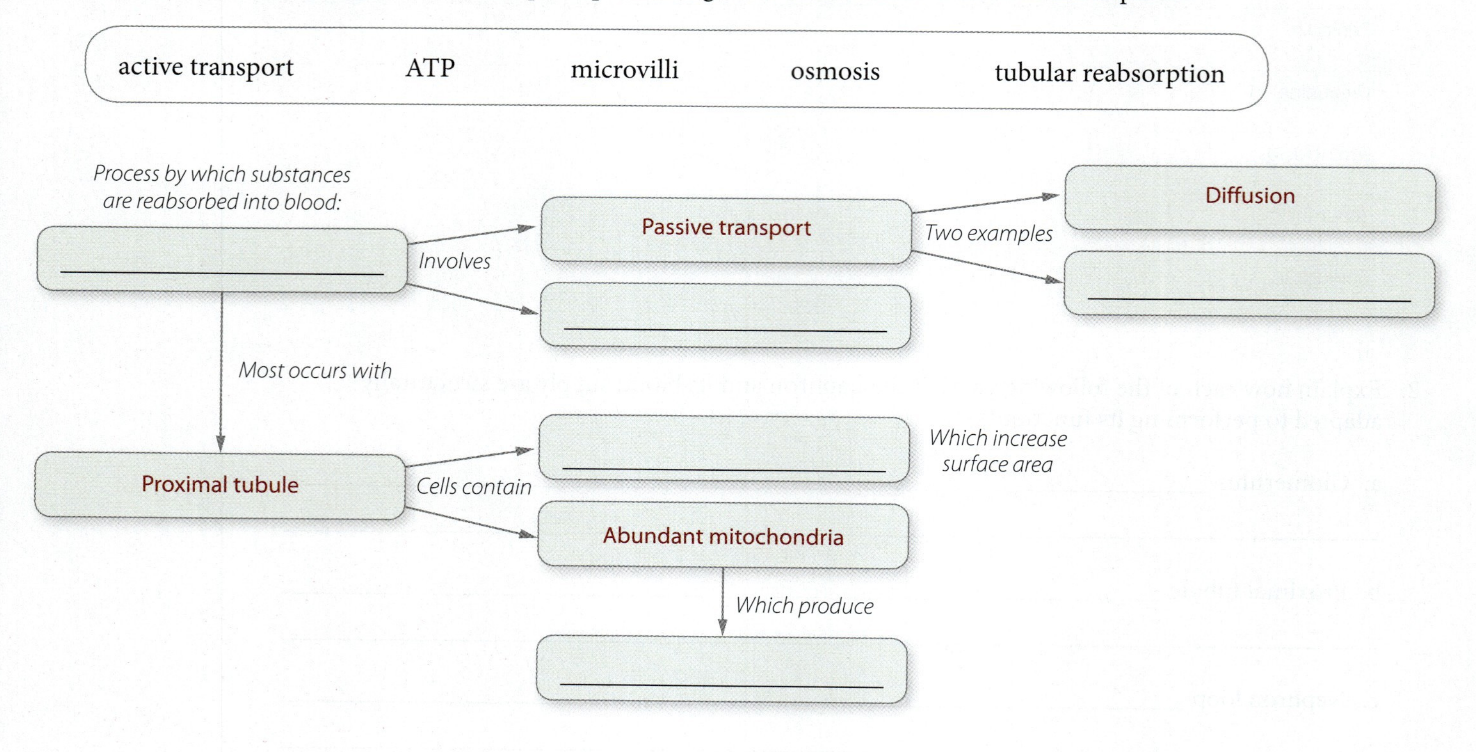

2. Construct a unit concept map to show the relationships among the following set of terms. Include all of the terms in your diagram. Your instructor may choose to assign additional terms.

active transport	ADH	angiotensin II	aquaporin	ATP
collecting duct	distal tubule	filtration	hypertonic	juxtaglomerular apparatus
microvilli	nephron loop	osmosis	tubular reabsorption	urea

32

The Reproductive System

UNIT OUTLINE

The primary function of the reproductive system is to produce offspring. The male reproductive system produces sperm and transports these gametes to the female reproductive tract so that fertilization of an ovum can take place. The female reproductive system produces ova and houses the developing embryo following fertilization. In this unit we begin with the gross anatomy of the male and female reproductive systems; we then examine the microscopic anatomy of the testis and ovary and their roles in the process of gametogenesis.

THINK ABOUT IT *Provide one specific example of how each of the following organ systems interacts with the reproductive system to carry out reproductive functions:*

Endocrine system: ______________________________

____

Cardiovascular system: ______________________________

____

Respiratory system: ______________________________

____

Urinary system: ______________________________

____

PRE-LAB ASSIGNMENTS

Pre-lab quizzes are also assignable in MasteringA&P®

To maximize learning, BEFORE your lab period carefully read this entire lab unit and complete these pre-lab assignments using your textbook, lecture notes, and prior knowledge.

PRE-LAB Activity 1: Examining Male Reproductive Anatomy

1. Use the list of terms to label the accompanying illustration of the male reproductive organs. Check off each term as you label it.

- ☐ bulbourethral gland
- ☐ ductus deferens
- ☐ epididymis
- ☐ ejaculatory duct
- ☐ penis
- ☐ prostate gland
- ☐ scrotum
- ☐ seminal vesicle
- ☐ testis
- ☐ urethra

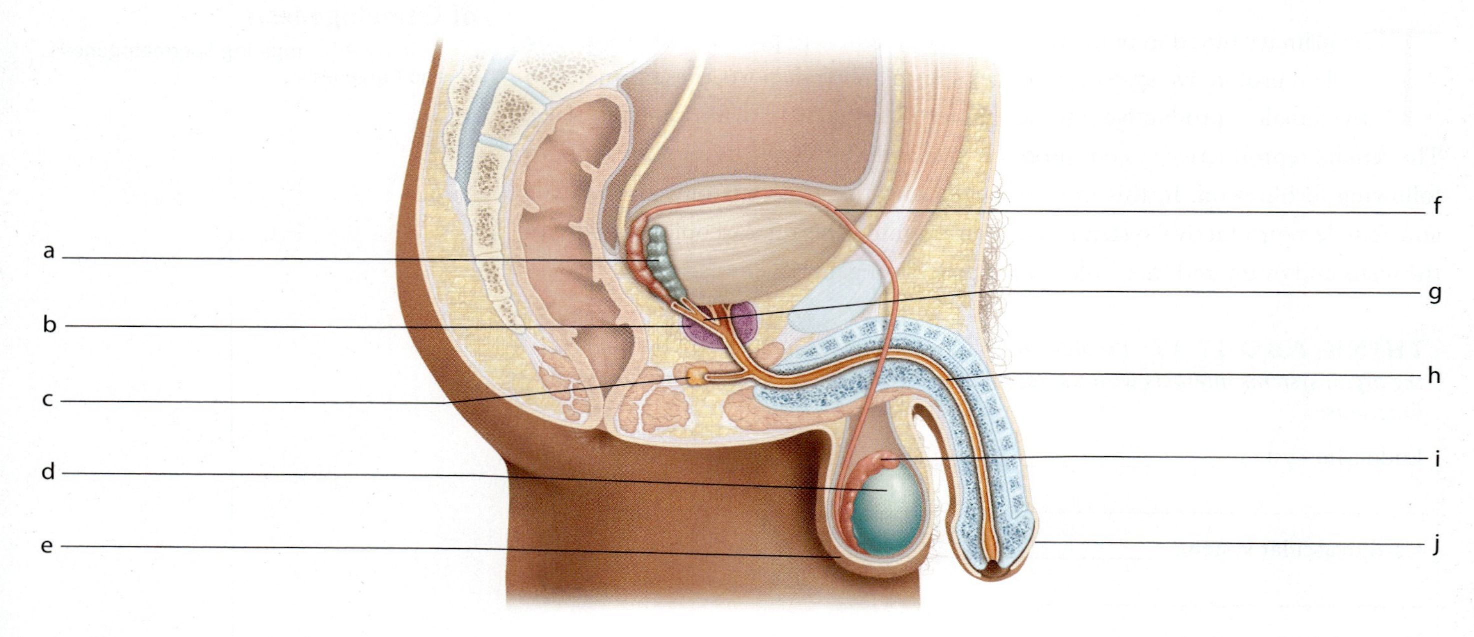

2. Match each of the following structures with its correct description:

_____	a. accessory glands	1. empties into the urethra
_____	b. testis	2. common passageway for urine and sperm
_____	c. ductus deferens	3. male copulatory organ
_____	d. epididymis	4. travels through spermatic cord
_____	e. urethra	5. produce seminal fluid
_____	f. ejaculatory duct	6. male gonad
_____	g. penis	7. site of sperm maturation

PRE-LAB Activity 2: Examining Female Reproductive Anatomy

1. Use the list of terms to label the accompanying illustration of the female reproductive organs. Check off each term as you label it.

☐ ovary ☐ vagina ☐ uterine tube ☐ uterus

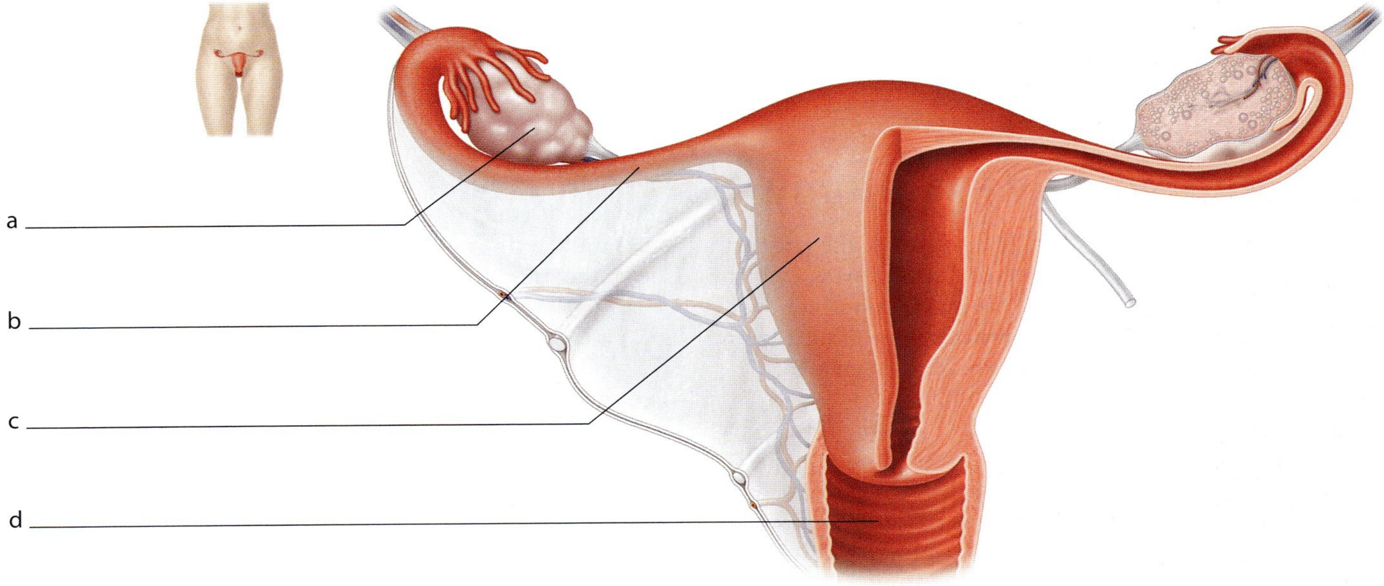

a ______

b ______

c ______

d ______

2. Match each of the following structures with its correct description:

______	a. vulva	1. site of implantation
______	b. ovary	2. birth canal and organ of copulation
______	c. uterine tube	3. site of fertilization
______	d. vagina	4. female gonad
______	e. uterus	5. external genitalia

PRE-LAB Activity 3: Modeling Meiosis

1. Match each phase of mitosis with its description:

______	a. anaphase	1. chromosomes become visible
______	b. telophase	2. chromosomes line up along the equator
______	c. prophase	3. sister chromatids separate, move toward opposite poles
______	d. metaphase	4. nuclear envelope re-forms; sister chromatids reach opposite poles

2. Division of the cytoplasm is called ______________________.

3. Do each of the following descriptions characterize mitosis, meiosis, or both mitosis and meiosis?

______________ a. Produces haploid gametes.

______________ b. Results in formation of four daughter cells.

______________ c. Crossing over of chromosomes occurs.

______________ d. Is preceded by interphase.

______________ e. Produces diploid cells.

PRE-LAB Activity 4: Comparing Spermatogenesis and Oogenesis

1. Spermatogenesis is the production of __________________ and it occurs in the __________________.
2. Oogenesis is the production of __________________ and it occurs in the __________________.
3. What is the function of meiosis? ____

____

Male Reproductive Anatomy

In the male reproductive system (**Figure 32-1**), the primary organs are the paired oval-shaped testes (male gonads), which are glands that produce sperm and the male sex hormone testosterone. Secondary reproductive structures are grouped into the duct system, accessory organs, and external genitalia.

The Testes

Each **testis** (**Figure 32-2**) is surrounded by two tunics. The outermost tunic, the **tunica vaginalis**, is a two-layered structure derived from the peritoneum. The innermost tunic, a dense connective tissue capsule called the **tunica albuginea**, invaginates to form septa that divide each testis into small compartments called **lobules**. Each lobule contains one to four tightly packed **seminiferous tubules**, the sites of spermatogenesis (the production of sperm). The seminiferous tubules unite to form the **rete testis**, which transports sperm to the epididymis, the first component of the duct system.

Duct System

The duct system includes the epididymis, ductus deferens, ejaculatory duct, and urethra (see Figure 32-1, Figure 32-4). After sperm are produced in the testis, they travel to the **epididymis**, a C-shaped structure located on the superior and posterolateral aspect of the testis (see Figure 32-2) and the site of sperm storage and maturation. During this maturation process, which takes approximately 20 days, sperm gain the ability to swim. Sperm can be stored in the epididymis for 2–3 months, after which they are phagocytized by cells of

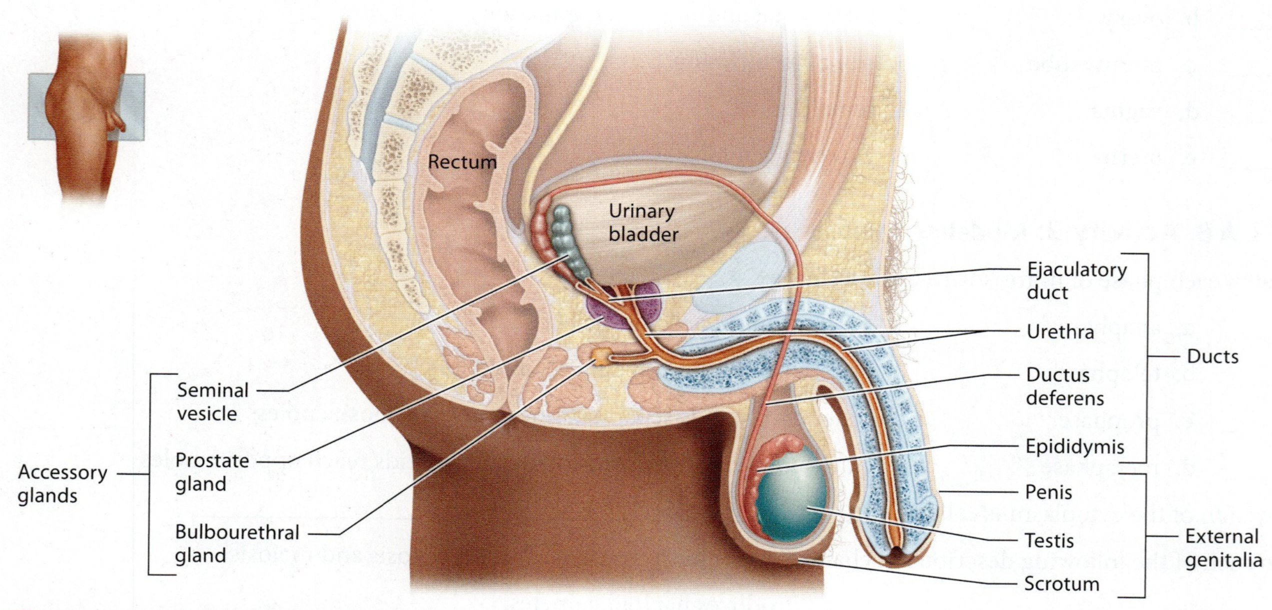

Figure 32-1 Anatomy of the male reproductive system, midsagittal section.

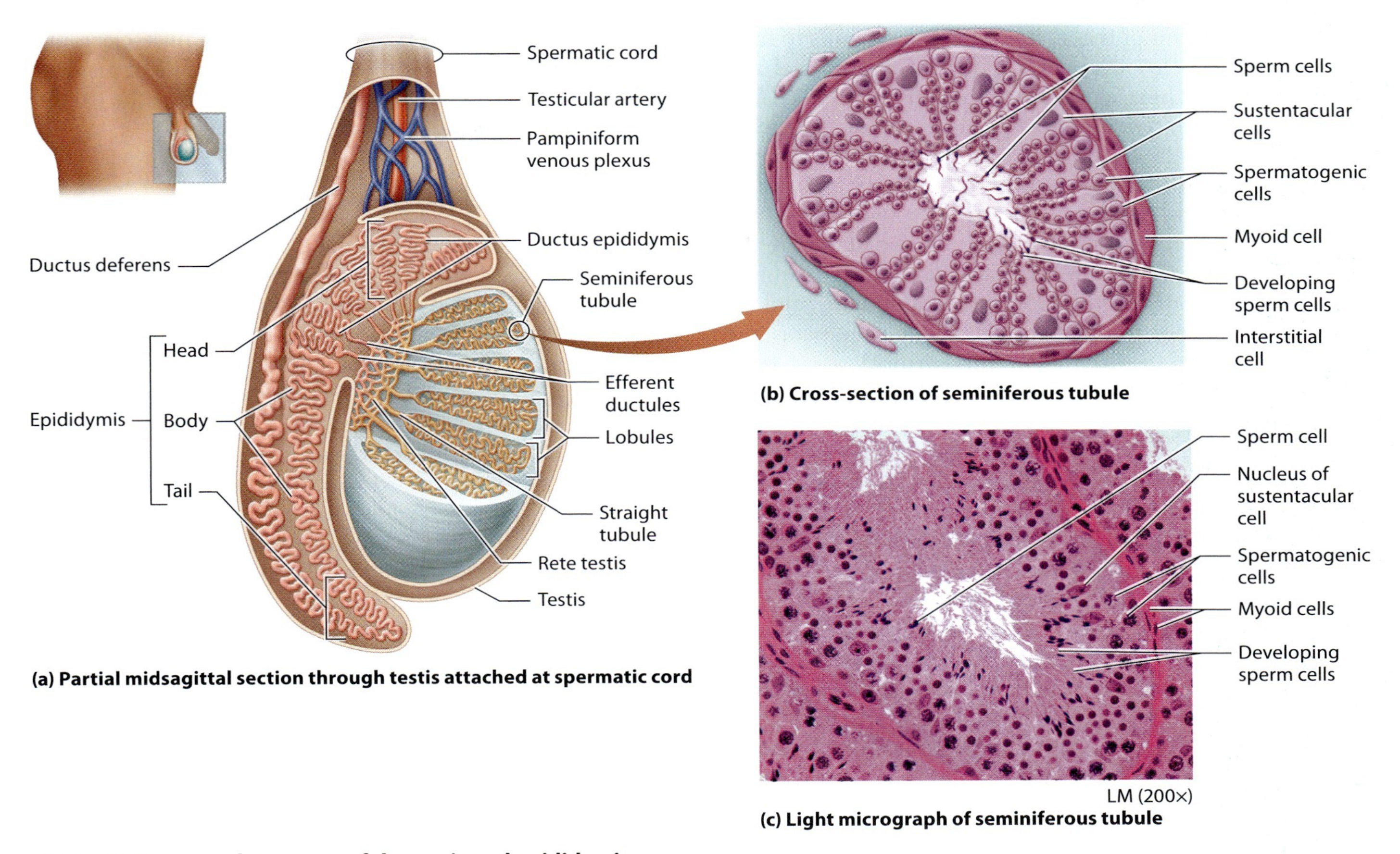

Figure 32-2 **Internal structure of the testis and epididymis.**

the epididymis. During sexual stimulation, sperm are ejaculated from the epididymis and enter the **ductus deferens** (or vas deferens). This structure, which is approximately 18 inches long, travels upward as part of the **spermatic cord** (a structure that also contains the testicular blood vessels and nerves), passes through the inguinal canal (a passageway through the anterior abdominal wall), and enters the pelvic cavity (**Figure 32-3**).

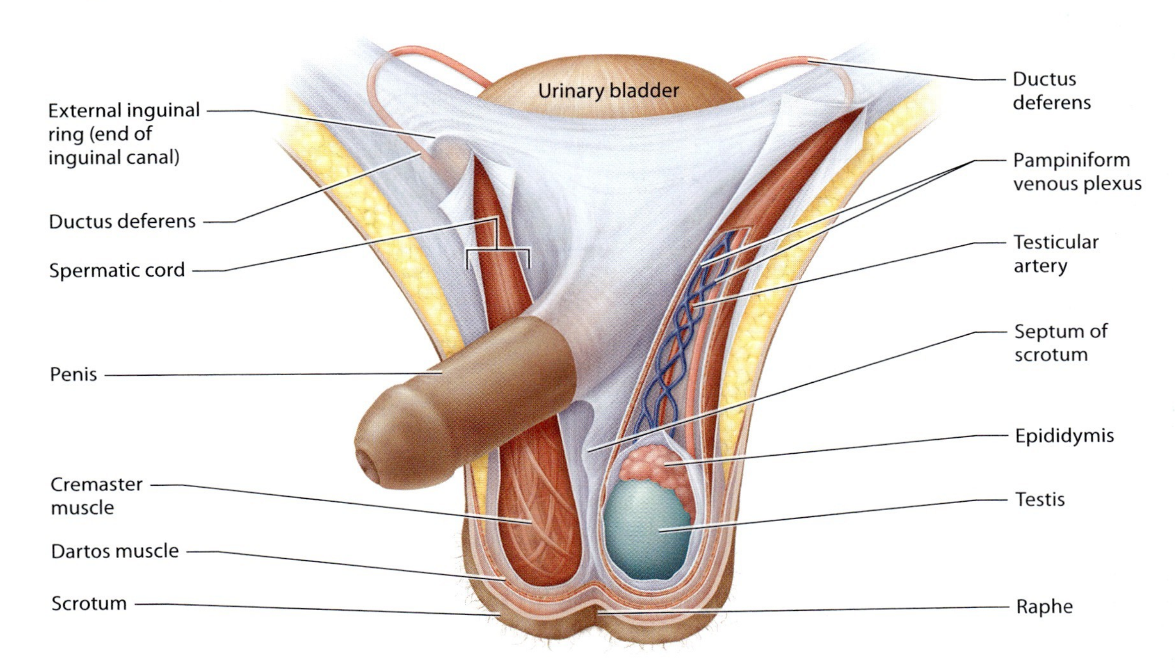

Figure 32-3 **The scrotum and spermatic cord.**

The terminal end of the ductus deferens expands to form the **ampulla**, which joins with the duct of the seminal vesicle to form the short **ejaculatory duct** (Figure 32-4). The ejaculatory duct enters the prostate gland and then empties into the **urethra**, the terminal portion of the duct system in males. The urethra belongs to both the urinary and the reproductive systems because it transports both urine and semen. It is divided into three regions: the prostatic urethra (portion surrounded by the prostate gland), the membranous urethra (located in the urogenital diaphragm), and the spongy urethra (portion that travels through the penis and opens externally at the external urethral orifice).

Accessory Glands

Two exocrine accessory glands (see Figure 32-1) contribute to seminal fluid, the fluid in which sperm are transported. The paired **seminal vesicles**, which produce approximately 70% of seminal fluid, lie near the ampulla of each ductus deferens on the posterior aspect of the urinary bladder (see Figure 32-4). The seminal vesicle's secretion provides energy for sperm. The **prostate gland**, which encircles the urethra just inferior to the urinary bladder, produces a slightly acidic, milky secretion that plays a role in activating sperm. A third type of accessory gland—the paired **bulbourethral glands**, tiny pea-shaped structures located at the base of the penis inferior to the urinary bladder—produces a thick, clear, mucous pre-ejaculate secretion that lubricates the tip of the penis and also neutralizes any traces of acidic urine in the urethra prior to ejaculation.

External Genitalia

The external genitalia of the male consist of the penis and the scrotum (see Figure 32-1). The **penis**, the organ of copulation (sexual intercourse), delivers sperm into the female reproductive tract. It consists of three erectile bodies (see Figure 32-4): the corpus spongiosum, which surrounds the spongy urethra, and the paired, dorsal corpora cavernosa. An erection results when the corpora cavernosa fill with blood. The corpus spongiosum also fills with blood, but its primary function is to keep the urethra open during an erection. The now-rigid penis facilitates the delivery of sperm into the female reproductive tract during intercourse. The distal, expanded portion of the corpus spongiosum, the **glans penis**, is covered with foreskin, the **prepuce**, in uncircumcised males. The removal of the prepuce during circumcision exposes the glans penis. The testes lie in a sac-like **scrotum** composed of skin and connective tissue and located outside of the abdominopelvic cavity. The slightly cooler temperature in the scrotum (2–3 degrees below body temperature) provides ideal conditions for the production of viable sperm.

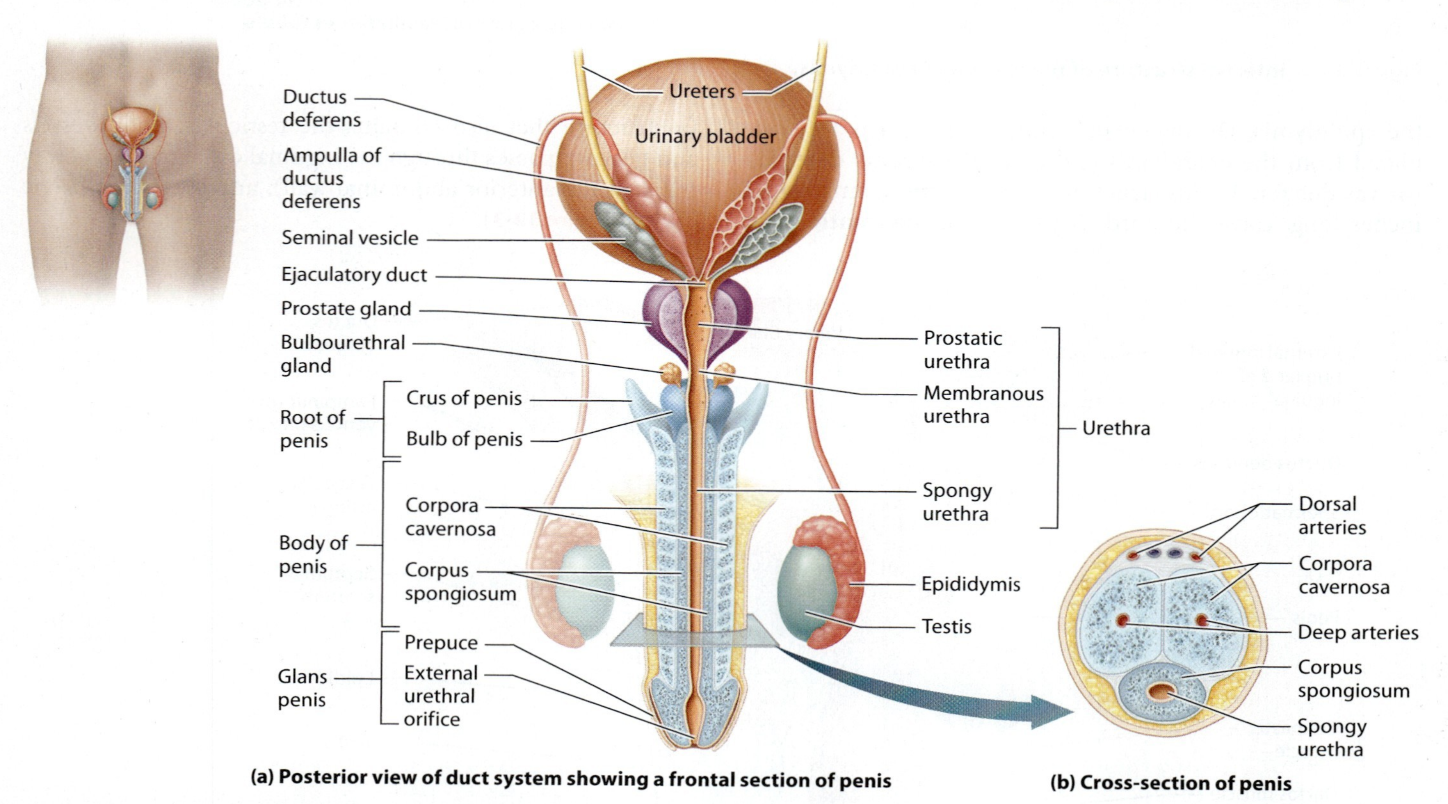

Figure 32-4 Male reproductive duct system and penis.

ACTIVITY 1

Examining Male Reproductive Anatomy

Learning Outcomes

1. Identify the organs of the male reproductive system on anatomical models.
2. Describe the function(s) of each organ.

Materials Needed

☐ Anatomical models or charts

Instructions

CHART Locate the structures in the Making Connections chart on an anatomical model or chart, and identify the anatomical features of each. Then write a description of each structure and its function(s). Finally, make some "connections" to things you have already learned in lectures, assigned readings, and lab. Some portions of the chart have been provided as examples.

Making Connections: Male Reproductive Anatomy

Reproductive Structure	Description (Structure and/or Function)	Connections to Things I Have Already Learned
Testis	*Produces sperm and testosterone.*	
Epididymis		*"Epi" means above; the epididymis sits above the testis.*
Ductus (vas) deferens		
Ejaculatory duct	*Delivers semen into the prostatic urethra.*	
Urethra • Prostatic urethra • Membranous urethra • Penile (spongy) urethra		*Belongs to both urinary and reproductive systems.*
Seminal vesicle	*Contributes ~70% of fluid to semen; secretions rich in fructose; its duct fuses with ampulla of ductus deferens to form ejaculatory duct.*	
Prostate gland		
Bulbourethral gland		*Homologous to greater vestibular glands in female*
Penis • Corpora cavernosum • Corpora spongiosum • Glans penis		
Scrotum	*Sac-like structure composed of skin and connective tissue; contains a testis.*	

Optional Activity

Practice labeling male reproductive organs on human cadavers at MasteringA&P® > Study Area > Practice Anatomy Lab > Human Cadaver > Reproductive System > Images 1—8

Female Reproductive Anatomy

In the female reproductive system (**Figure 32-5**), the primary organs are the paired almond-shaped ovaries (female gonads), which are glands that produce eggs and the female sex hormones estrogen and progesterone. Secondary female reproductive structures include internal sex organs, external sex organs, and the mammary glands.

The Ovaries

The ovaries, located in the peritoneal cavity, are held in place by several ligaments (**Figure 32-6**). The **ovarian ligaments** anchor the ovaries medially to the uterus, the **suspensory ligaments** anchor the ovaries laterally to the pelvic wall, and a part of the broad ligament called the **mesovarium** supports the ovaries posteriorly. The **broad ligament** is a fold of the peritoneum that drapes over the superior surface of the uterus and supports the uterine tubes, uterus, and vagina.

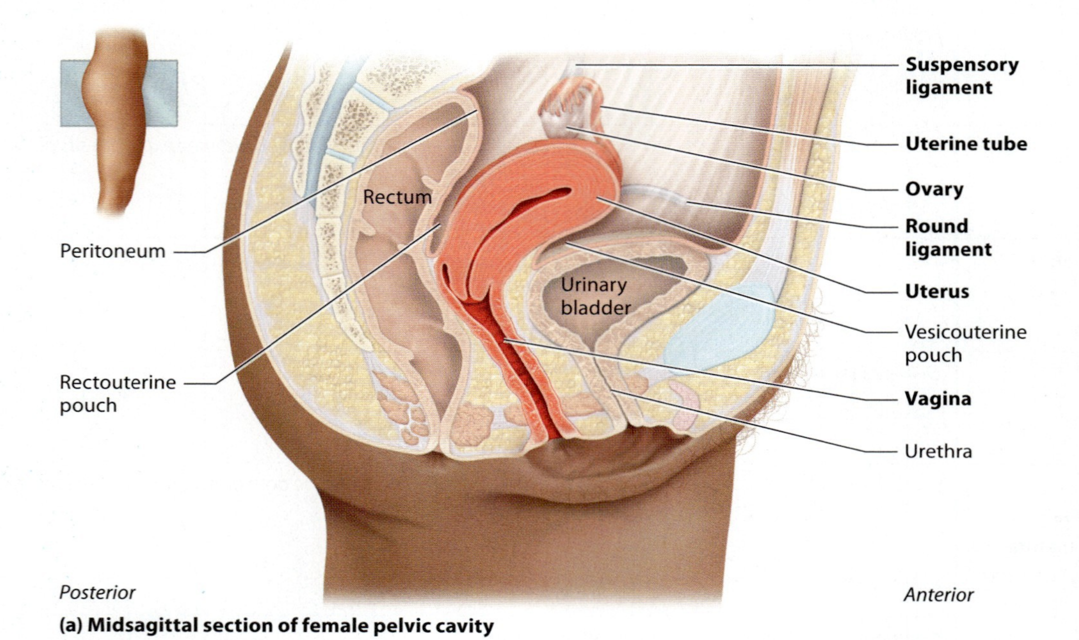

(a) Midsagittal section of female pelvic cavity

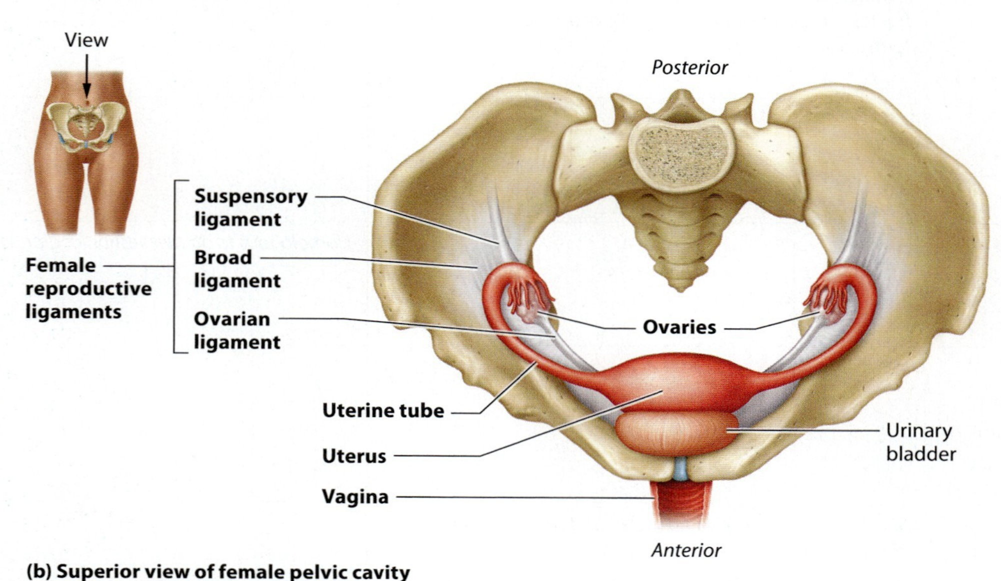

(b) Superior view of female pelvic cavity

Figure 32-5 Internal organs of the female reproductive system.

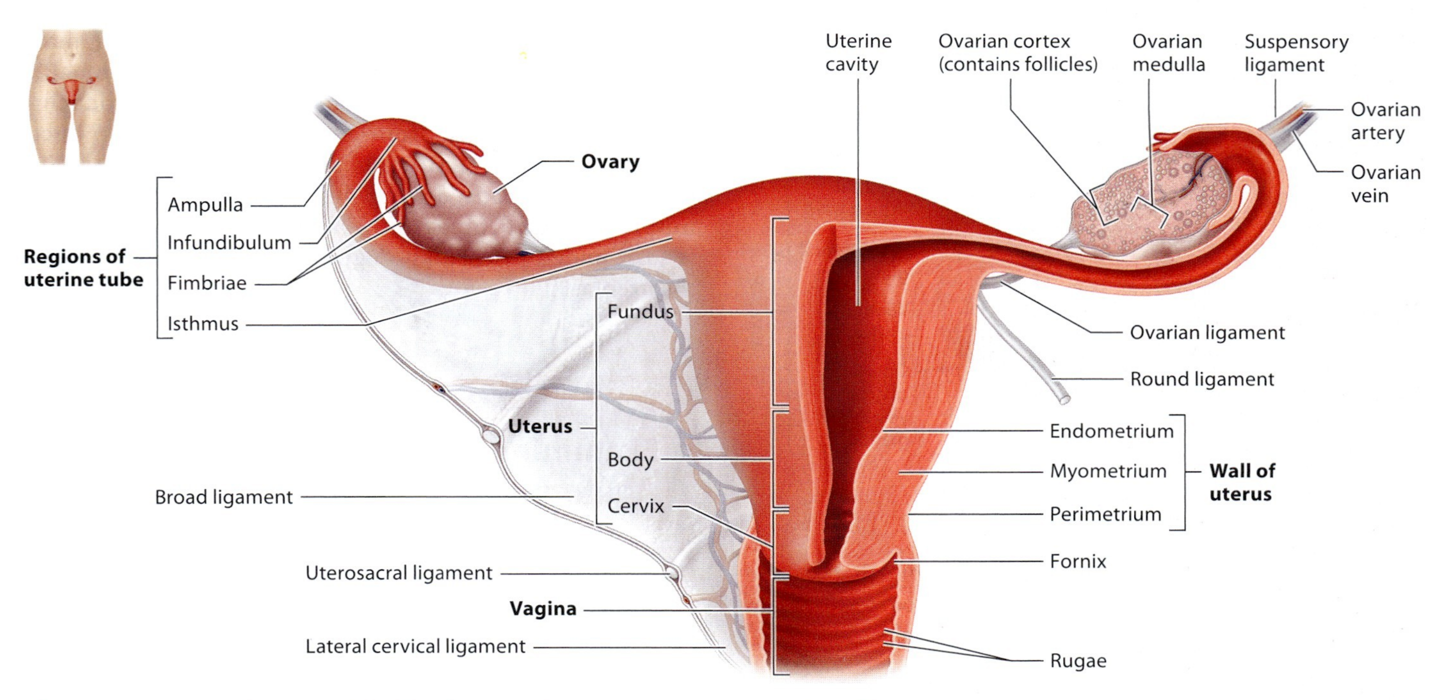

Figure 32-6 Internal organs of the female reproductive system, anterior view. The left side of the uterus, left uterine tube, left ovary, and vagina are cut to show the internal anatomy, and the broad ligament has been removed from the left side.

Like each testis, each ovary is surrounded by a fibrous tunica albuginea. External to it is a single layer of cuboidal epithelial cells known as the germinal epithelium. Internally, the ovary is divided into two regions: an outer cortex housing developing follicles, and an inner medulla consisting of blood vessels and nerves embedded in connective tissue. Each follicle consists of an oocyte (or egg) surrounded by one or more layers of supporting cells called follicle cells.

Internal Secondary Sex Organs

The **uterine tubes** (or fallopian tubes; see Figures 32-5 and 32-6) curve around the ovaries and receive oocytes released from the ovaries. Each uterine tube consists of three regions: the **infundibulum**, **ampulla**, and **isthmus** (see Figure 32-6). The distal funnel-shaped infundibulum contains finger-like extensions called **fimbriae**. When an oocyte is released from the ovary, it is actually released into the peritoneal cavity. Ciliated cells lining the fimbriae generate a current that sweeps the oocyte into the infundibulum. The oocyte then enters the ampulla, which constitutes approximately two-thirds of the tube's length and is typically the site of fertilization. The shorter isthmus is continuous with the uterus.

The **uterus**, located in the pelvic cavity anterior to the rectum and posterosuperior to the bladder (see Figure 32-5), consists of the dome-shaped fundus, the central body, and the narrow cervix (see Figure 32-6). The uterus is held in place by the **broad ligaments**, which anchor it to the lateral pelvic walls, and the **round ligaments**, which attach it to the labia majora of the external genitalia. The uterine wall consists of three layers: the **perimetrium** (or visceral peritoneum), the **myometrium** consisting of smooth muscle, and the **endometrium**, composed of simple columnar epithelium with a thick underlying lamina propria. The endometrium undergoes cyclic changes in response to ovarian hormones and is shed approximately every 28 days during menstruation. The **vagina** (see Figures 32-5 and 32-6), a thin-walled tube lying between the bladder and the rectum, extends inferiorly from the uterus to the vaginal orifice. It functions as both the female organ of copulation and the birth canal. The vaginal wall consists of three layers: an outer fibrous adventitia, a smooth muscle muscularis, and an inner mucosa consisting of a moist stratified squamous epithelium.

External Secondary Sex Organs

The external anatomy of the female (**Figure 32-7**), collectively called the vulva, consists of the mons pubis, labia majora, labia minora, clitoris, and vestibule. The **mons pubis** is a pad of adipose tissue deep to the skin overlying the pubic symphysis. The **labia majora** and **labia minora** are paired, elongated skin folds. The labia majora are homologous to the scrotum, whereas the labia minora are homologous to the ventral penis. (Homologous structures share a common embryonic origin.) The labia minora enclose a region called the **vestibule**, which contains the urethral opening, vaginal opening, and openings of the greater vestibular glands. The **clitoris**, a cylindrical erectile structure homologous to the penis, is located posterior to the mons pubis and anterior to the urethral opening. The **greater vestibular glands** are

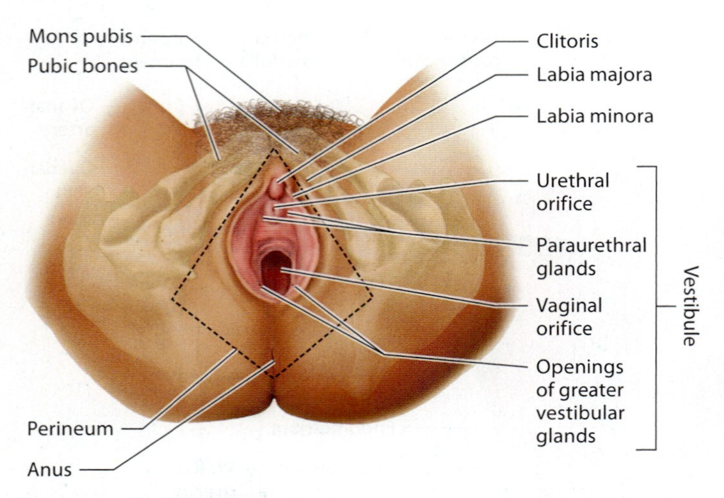

Figure 32-7 The female perineum.

pea-sized glands flanking the vaginal opening that secrete mucus to lubricate the vagina during intercourse; they are homologous to the bulbourethral glands in males.

Mammary Glands

Although present in both sexes, **mammary glands**—modified sweat glands that are derivatives of the skin and lie anterior to the pectoralis major muscles—normally function in females only to produce milk to nourish newborns. The female breast (**Figure 32-8**) consists largely of adipose tissue surrounding mammary glands. Each mammary gland consists of 15–25 **lobes**; each lobe contains smaller clusters called **lobules**, which contain milk-producing **alveoli**. Milk secreted from alveoli enters **alveolar ducts** that fuse to form **lactiferous sinuses**, which then merge to form **lactiferous ducts**. Milk is released through the **nipple** via the openings of the lactiferous ducts. The nipple is surrounded by a darkly-pigmented region of skin called the **areola**.

ACTIVITY 2
Examining Female Reproductive Anatomy

Learning Outcomes

1. Identify the organs of the female reproductive system on anatomical models.
2. Describe the function(s) of each organ.

Materials Needed

☐ Anatomical models or charts

Instructions

CHART Locate the structures in the following Making Connections chart on an anatomical model or chart, and identify the anatomical features of each. Then write a

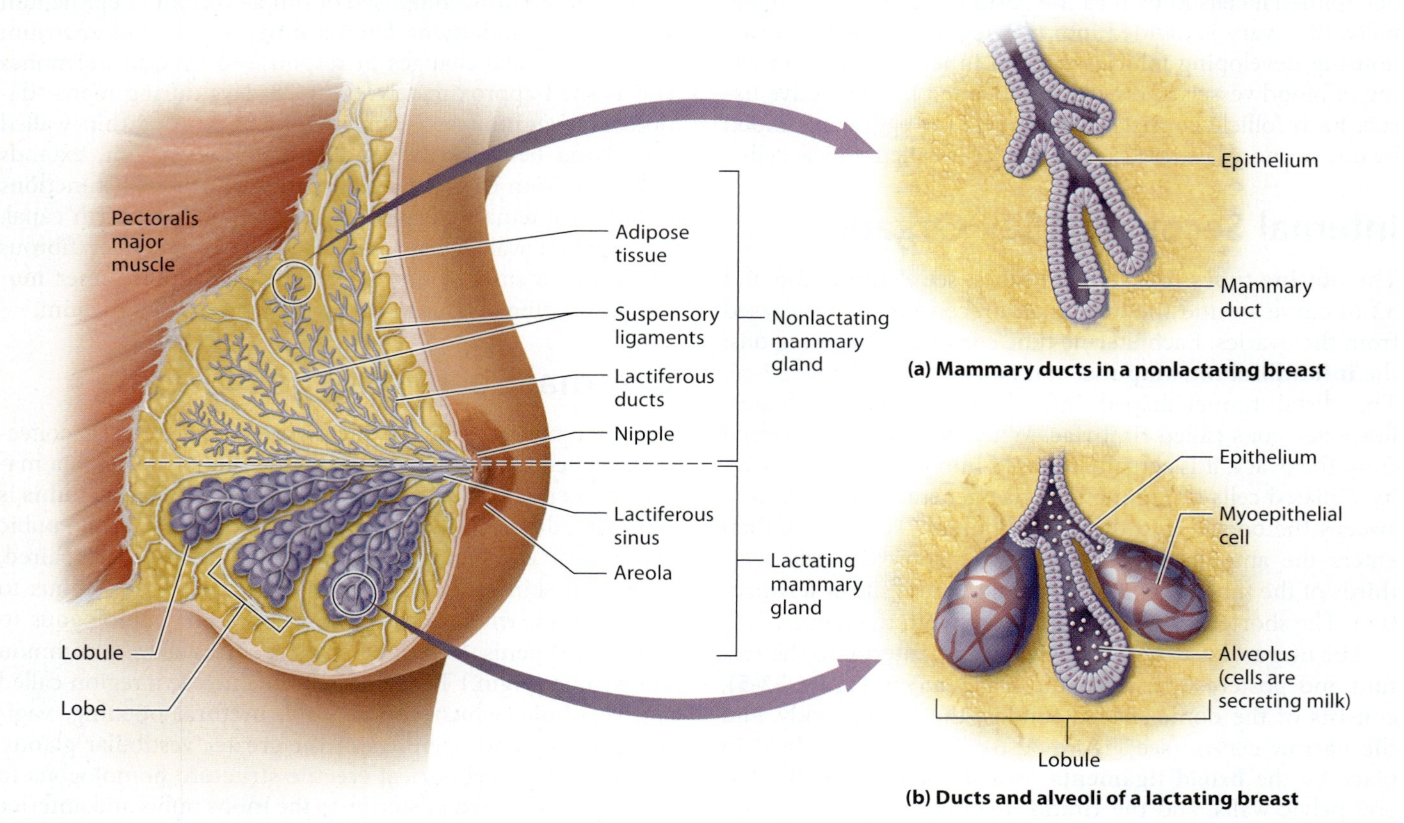

Figure 32-8 Internal anatomy of the female breast.

description of each structure and its function(s). Finally, make some "connections" to things you have already learned in lectures, assigned readings, and lab. Some portions of the chart have been provided as examples.

Optional Activity

Practice labeling female reproductive organs on human cadavers at MasteringA&P® > Study Area > Practice Anatomy Lab > Human Cadaver > Reproductive System > Images 9–11

Making Connections: Female Reproductive Anatomy

Reproductive Structure	Description (Structure and/or Function)	Connections to Things I Have Already Learned
Ovary		*Perfused by ovarian artery, a visceral branch of the abdominal aorta; contains a cortex (outer region) and a medulla (inner region) similar to kidneys, adrenal glands, and lymph nodes.*
Uterine tube • Infundibulum with fimbriae • Ampulla • Isthmus		
Uterus - regions • Fundus • Body • Cervix		*Cells are scraped from the cervix during a Pap smear.*
Uterus - layers • Perimetrium • Myometrium • Endometrium		*Myometrium is composed largely of smooth muscle, which is stimulated to contract by oxytocin during the birth process.*
Vagina	*Copulatory organ; birth canal*	
Vulva • Labia majora • Labia minora		
Vestibule • Urethral opening • Vaginal opening • Openings of the ducts of the greater vestibular glands	• *Opening through which urine flows* • *Opening to the vagina* • *Mucous secretions from greater vestibular glands pass through these openings.*	
Clitoris		
Ligaments • Ovarian ligament • Suspensory ligament • Mesovarium • Broad ligament • Round ligament		
Mammary glands		*Type of exocrine glands that release their secretory product via ducts*

BEFORE BIRTH

Cells before DNA Replication

- This is what the cell would look like if the chromatin condensed into chromosomes before the DNA replicated.

"Mother cell"
Maternal chromosomes
Paternal chromosomes

MEIOSIS I

Early Prophase I

- Chromosomes form with two sister chromatids.

Chromosomes
Centrioles

Mid- to Late Prophase I

- During synapsis, homologous chromosomes form tetrads and **crossing over** occurs.

Tetrad
Crossing over
Spindle fibers

Metaphase I

- Tetrads align randomly at equator (random orientation).

Paired homologous chromosomes
Equator
Random orientation

Anaphase I

- Random orientation in metaphase I leads to **independent assortment.**

Independent assortment

Telophase I

- Cytokinesis may follow, resulting in two genetically different haploid cells with sister chromatids still attached.

MEIOSIS II

Prophase II

- Chromosomes remain condensed.

Metaphase II

- Chromosomes line up along equator.

Anaphase II

- Sister chromatids separate.

Telophase II

- Cytokinesis follows.

Meiosis produces four genetically unique, haploid daughter cells.

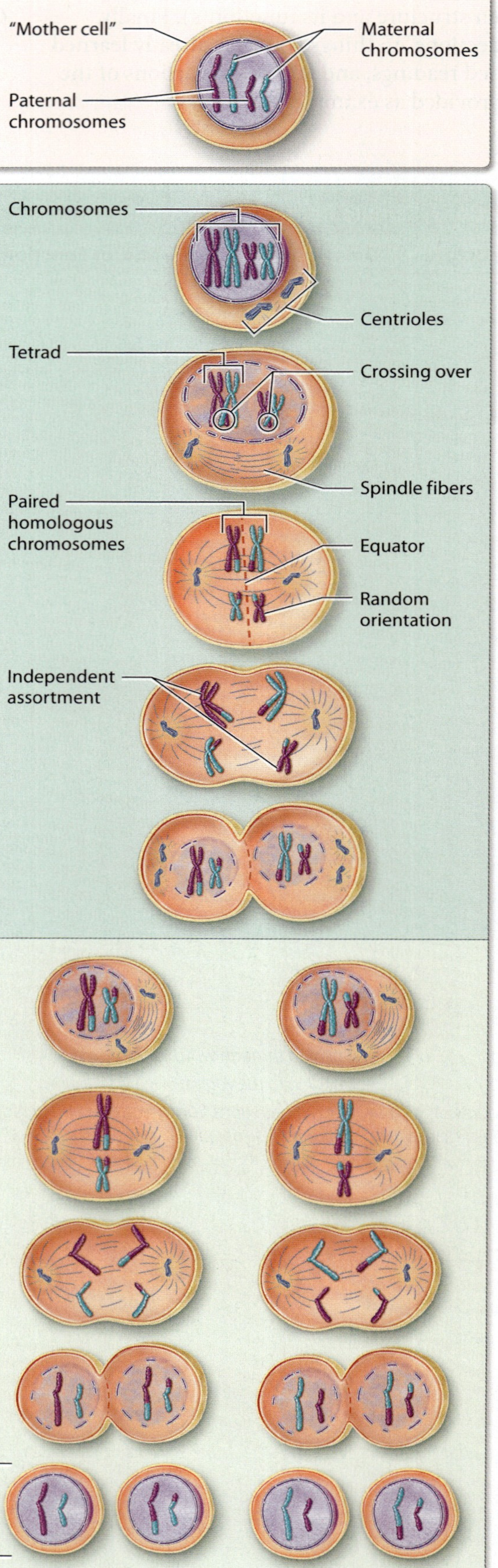

Figure 32-9 Stages of meiosis.

Meiosis

As discussed in Unit 5, most body cells divide by mitosis to produce two genetically identical diploid daughter cells. Review the stages of mitosis as illustrated and described in Figure 5-9. By contrast, gametogenesis—the formation of either sperm (spermatogenesis) or oocytes (oogenesis)—involves a process called **meiosis** (**Figure 32-9**), which produces genetically different haploid daughter cells. Like mitosis, meiosis is preceded by **interphase**, during which cells increase in size, replicate their DNA, and prepare for cell division.

Meiosis involves two cell divisions, each consisting of prophase, metaphase, anaphase, and telophase. During **prophase I**, chromosomes become visible, centrioles move to opposite poles, and the nuclear envelope breaks down. In humans, the 46 chromosomes in a diploid cell arrange themselves into 23 pairs of homologous chromosomes (called tetrads because each contains four chromatids) consisting of one maternal chromosome and one paternal chromosome. Homologous chromosomes line up so close together that they overlap in several areas, enabling a process called **crossing over** (or synapsis) in which genetic material is exchanged between maternal and paternal chromosomes. During **metaphase I** the tetrads line up independently in pairs at the equator (or metaphase plate), a process termed **independent assortment**. During **anaphase I**, the homologous chromosomes separate and move to opposite poles. During **telophase I**, the homologous chromosomes reach opposite poles, the nuclear envelopes re-form, and **cytokinesis** (division of the cytoplasm) separates the cell into two haploid cells.

Interkinesis occurs between meiosis I and meiosis II; it differs from interphase in that the genetic material is not replicated during interkinesis. Meiosis II is essentially the same as mitosis, but with one very important exception: Whereas in mitosis a diploid cell produces two diploid cells, in meiosis II a haploid cell produces two haploid cells.

During **prophase II**, chromosomes remain visible, the centrioles move to opposite poles, and the nuclear envelope breaks down. During **metaphase II**, chromosomes line up single file along the equator. During **anaphase II** the chromosomes, each composed of two sister chromatids joined by a centromere, split and move toward opposite poles. Finally, during **telophase II** the sister chromatids reach opposite poles, the nuclear envelope re-forms, and cytokinesis produces two daughter cells. In total, then, meiosis produces four genetically unique haploid cells from a single original diploid cell.

ACTIVITY 3
Modeling Meiosis

Learning Outcomes

1. Outline and explain the events of meiosis.
2. Compare and contrast meiosis and mitosis.

Materials Needed

- ☐ Set of chromosome "pop-beads" (two different colors) with magnetic centromeres
- ☐ Laminated posterboard
- ☐ Washable markers

Instructions

1. Using one color of beads for "male" chromosomes and another color for "female" chromosomes, build two homologous pairs of chromosomes to represent a cell with a diploid number ($2n$) of 4. Each chromosome should consist of two strands of beads connected by a magnetic centromere, as shown in **Figure 32-10**.
2. Place these chromosomes on the poster board, and demonstrate the events of meiosis by arranging the beads in a series of positions to represent the phases of meiosis I (include crossing over) and meiosis II. Use markers throughout the activity to draw and label structures such as the cell membrane, nuclear membrane, centrioles, spindle apparatus, equator, and cleavage furrow (when appropriate). Your instructor will check your end products for meiosis I and meiosis II to make sure that you have demonstrated the phases correctly.

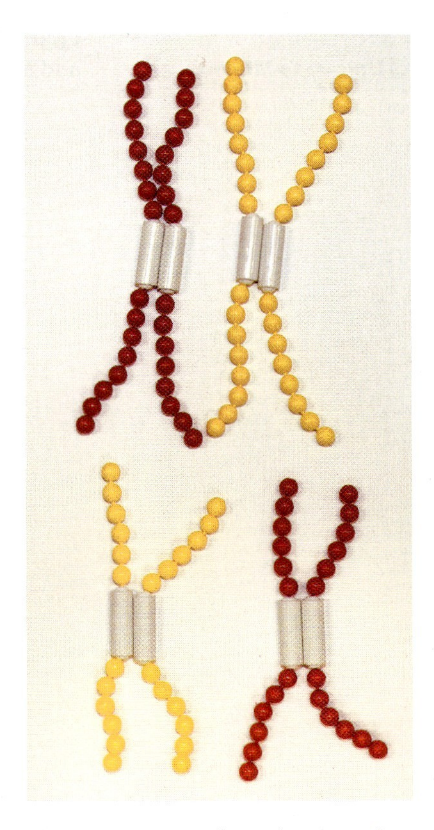

Figure 32-10 Two homologous pairs of "pop-bead" chromosomes.

3. Identify the specific meiotic phase represented in each of the following photographs, and provide a rationale for your answers.

Meiotic phase: ______________________

Rationale: ______________________

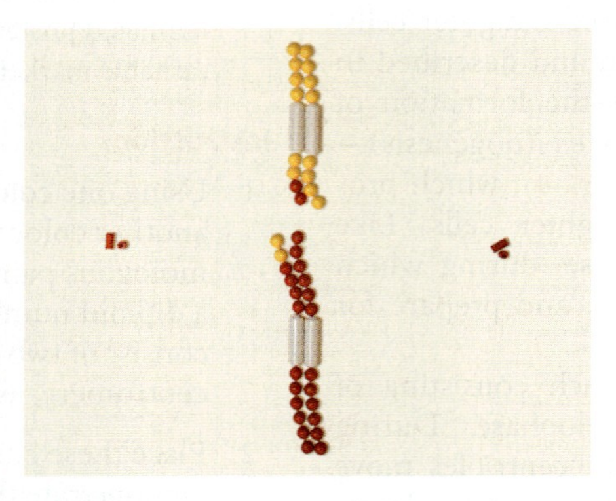

Meiotic phase: ______________________

Rationale: ______________________

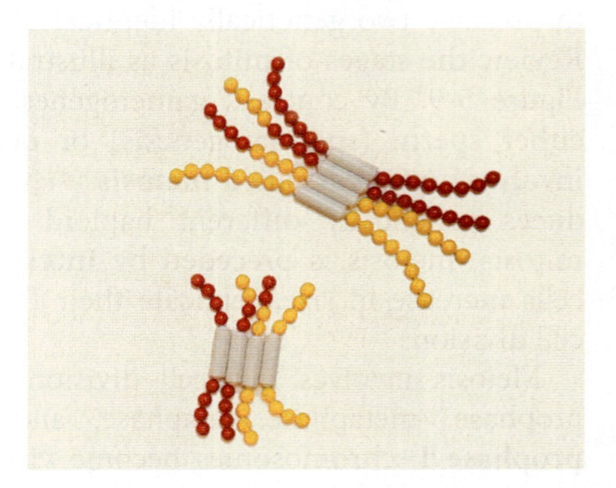

Meiotic phase: ______________________

Rationale: ______________________

4. CHART Work with your lab group to complete the following chart comparing and contrasting meiosis and mitosis.

Characteristics Unique to Mitosis	Characteristics Common to Both Mitosis and Meiosis	Characteristics Unique to Meiosis

Histology of the Gonads and the Hormonal Control of Gametogenesis

Gametogenesis is the production of haploid gametes. Next we consider the production of sperm (spermatogenesis) by the testes and of oocytes (oogenesis) by the ovaries.

Spermatogenesis

In males, **spermatogenesis** (Figure 32-11) begins at puberty and occurs within the seminiferous tubules. **Spermatogonia** adjacent to the basement membrane produce two distinct cell types: one type that remains at the basement membrane and continues to produce more of the two cell types, and a second type that becomes primary spermatocytes, each of which will eventually produce four sperm. **Primary spermatocytes** undergo meiosis I to form two haploid secondary spermatocytes. The two **secondary spermatocytes** undergo meiosis II to form four haploid **spermatids**, which then undergo a process called spermiogenesis during which the spermatids develop into sperm.

Each **sperm** (Figure 32-12) consists of a **head** containing a nucleus packed with DNA and an acrosome containing digestive enzymes that help the sperm penetrate an oocyte's cell membrane; a **midpiece**, which is packed with mitochondria that provide the energy needed for locomotion; and a **tail**, a whip-like flagellum that propels the sperm toward an

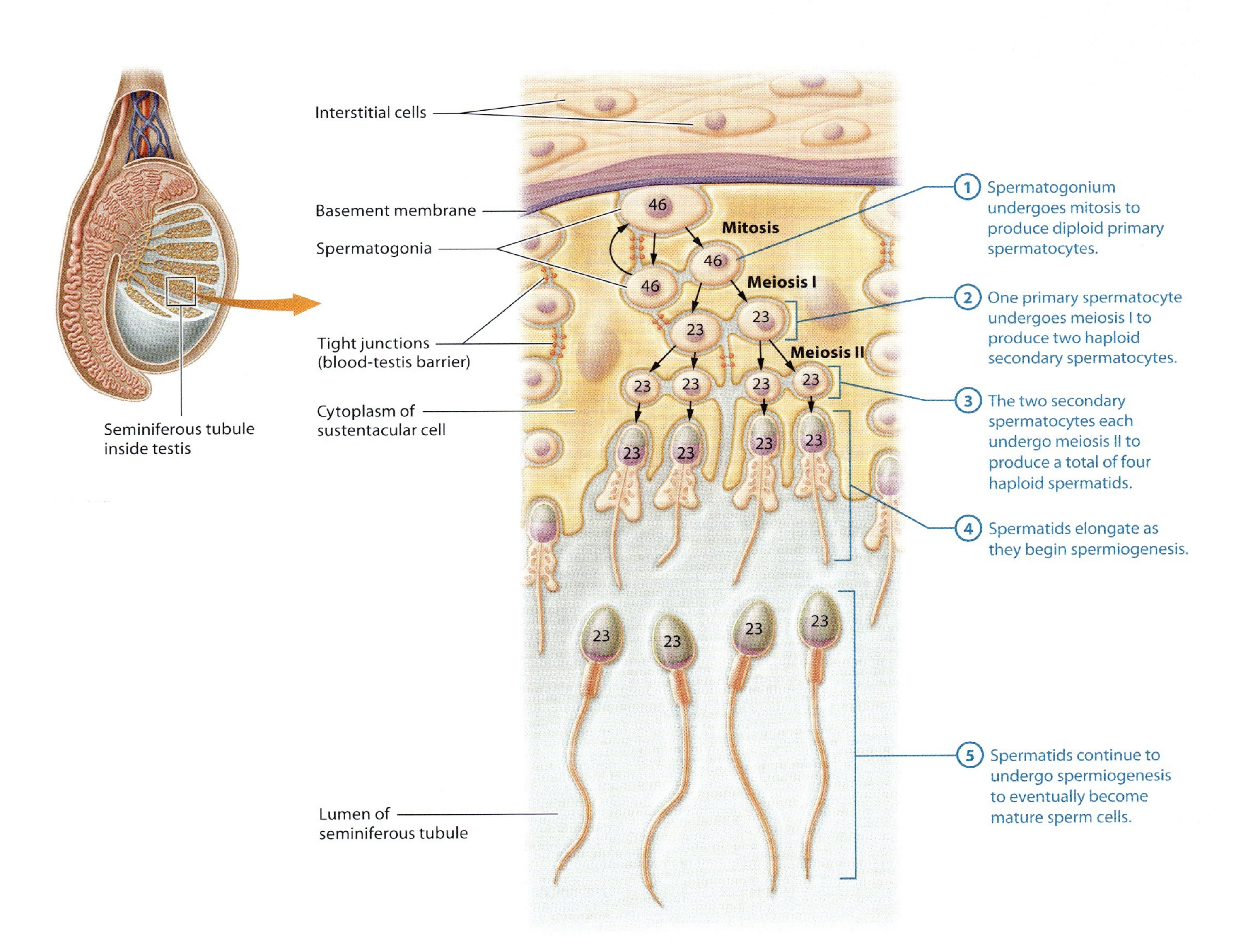

Figure 32-11 Spermatogenesis in the seminiferous tubules.

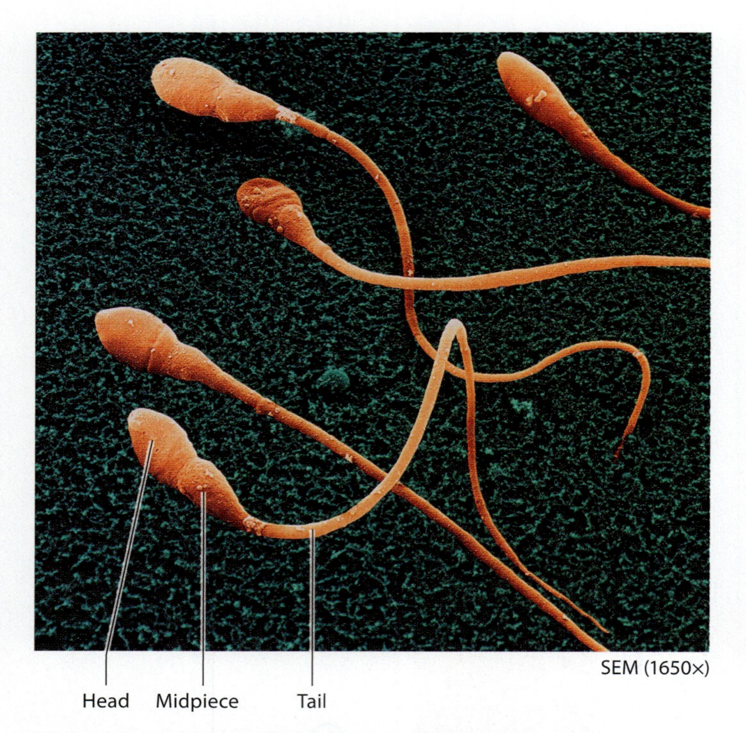

Figure 32-12 Photomicrograph of sperm.

oocyte once it has been deposited in the female's reproductive tract. Developing sperm cells are surrounded by supporting cells called **sustentacular cells**, which extend from the basement membrane to the lumen of the tubule. These cells provide developing sperm nutrients and produce testicular fluid in which sperm are transported.

Control of sperm production begins with gonadotropin-releasing hormone (GnRH), which is secreted by the hypothalamus. Once in the bloodstream, GnRH travels to the anterior pituitary gland, where it stimulates the secretion of follicle-stimulating hormone (FSH) and luteinizing hormone (LH). Both hormones target the testes: FSH indirectly stimulates the production of sperm within the seminiferous tubules by stimulating the sustentacular cells to produce androgen-binding protein (ABP), which concentrates testosterone, the hormone that stimulates spermatogenesis, in the vicinity of the spermatogenic cells. LH stimulates the secretion of testosterone from the interstitial cells. Rising testosterone levels exert a negative feedback effect on the anterior pituitary gland, inhibiting the release of both FSH and LH.

Oogenesis

In females, oogenesis (**Figure 32-13**) begins before birth. During fetal development, each **oogonium** (stem cell) undergoes rapid mitosis. The oogonia transform into **primary oocytes** and become surrounded by a layer of follicular cells, producing a **primordial follicle**. Primary oocytes enter meiosis I but are arrested, or stalled, in prophase I; meiosis I does not resume until puberty.

At birth, a female has a lifetime supply of approximately 1 million primordial follicles; only about 400,000 primordial follicles escape programmed cell death and remain at puberty. Beginning at puberty and occurring each month until menopause, several primary oocytes begin to grow, but usually only one continues in meiosis I to produce two haploid daughter cells. Interestingly, during cytokinesis the cytoplasm divides unequally, forming a smaller cell called the first **polar body** and a larger cell called the **secondary oocyte**. The secondary oocyte then enters meiosis II and is arrested in metaphase II. The first polar body may also enter meiosis II and produce two smaller polar bodies. Meiosis II is not completed until a sperm penetrates the oocyte. If sperm penetration occurs, the secondary oocyte completes meiosis II, producing one large ovum and a small second polar body. The nuclei of the sperm and ovum then fuse to complete fertilization. All polar bodies degenerate.

As in males, GnRH released by the hypothalamus stimulates the anterior pituitary to release FSH and LH. The **ovarian cycle** (**Figure 32-14**) consists of the follicular, ovulatory, and luteal phases. During the **follicular phase** (stages 1–4 in Figure 32-14), FSH stimulates follicles to grow in the ovaries and release estrogen. As the follicle grows, it becomes first a **primary follicle** containing a single layer of cuboidal epithelium and then a **secondary follicle** containing a stratified epithelium. Initially, slightly elevated estrogen levels exert a negative feedback effect on the anterior pituitary gland. However, around midcycle, as the dominant follicle and other maturing follicles grow larger and produce more estrogen, estrogen reaches a critical blood concentration and it briefly exerts a positive feedback effect on the anterior pituitary gland, causing a surge in LH that begins the **ovulatory phase** (stage 5), triggering ovulation (release of the secondary oocyte from the **vesicular**, or **Graafian, follicle**) and transforming the remaining follicle cells into a **corpus luteum**. During this **luteal phase** (stages 6 and 7), LH stimulates the release of large quantities of progesterone and some estrogen from the corpus luteum.

Estrogen and progesterone control the three phases of the **uterine cycle**: the menstrual, proliferative, and secretory phases. During the **menstrual phase**, hormone levels are low and the endometrial lining of the uterus sloughs off. This phase coincides with the earliest part of the follicular phase of the ovarian cycle. As estrogen levels begin to rise, menstruation ceases and the endometrium is repaired during the **proliferative phase**. Then, following ovulation, progesterone thickens the endometrium, preparing it for pregnancy during the **secretory phase**. If pregnancy does not occur, the corpus luteum degenerates, hormone levels decrease, and the cells of the endometrium die and slough off, beginning the cycle anew.

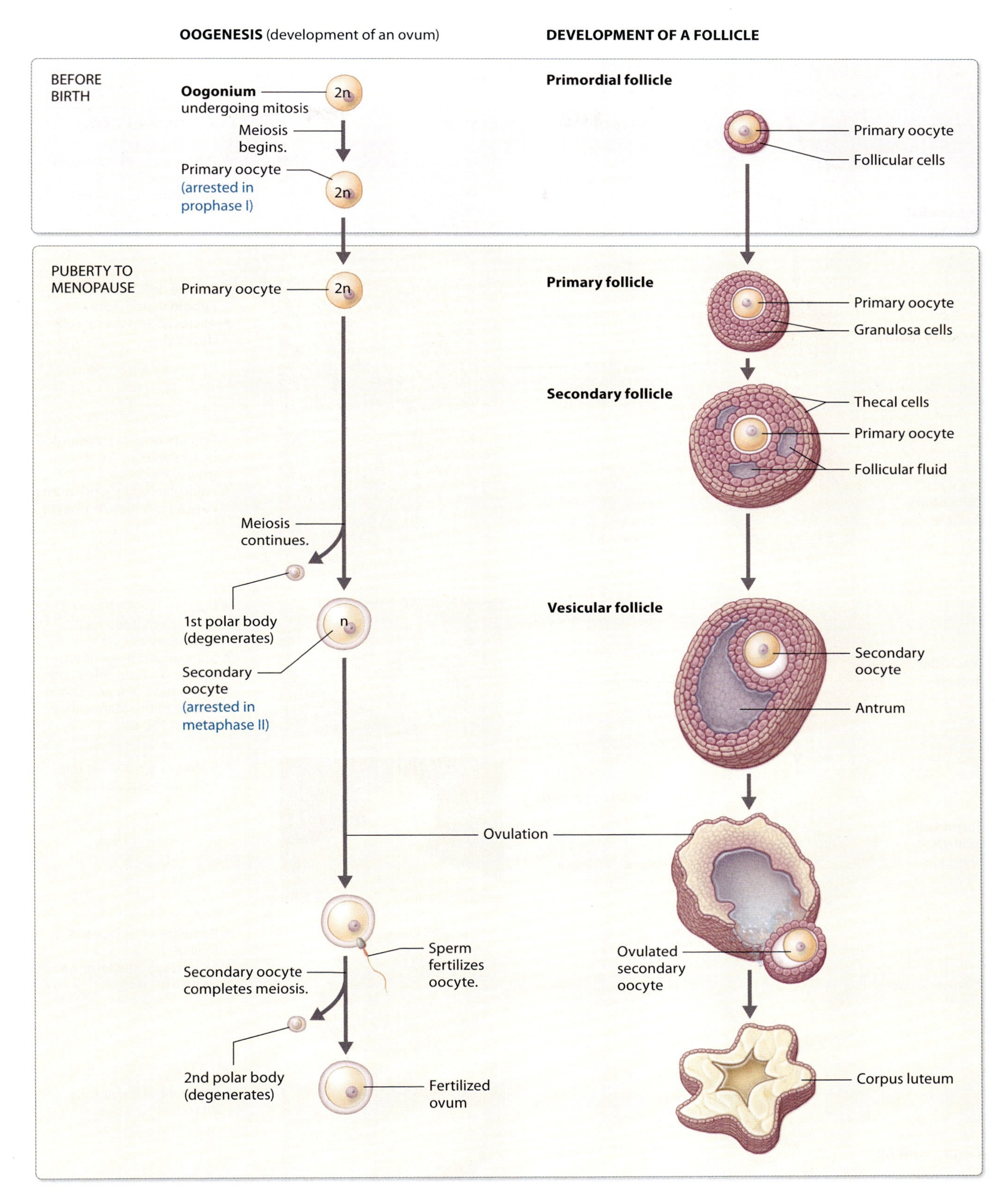

Figure 32-13 **Oogenesis and follicular development.**

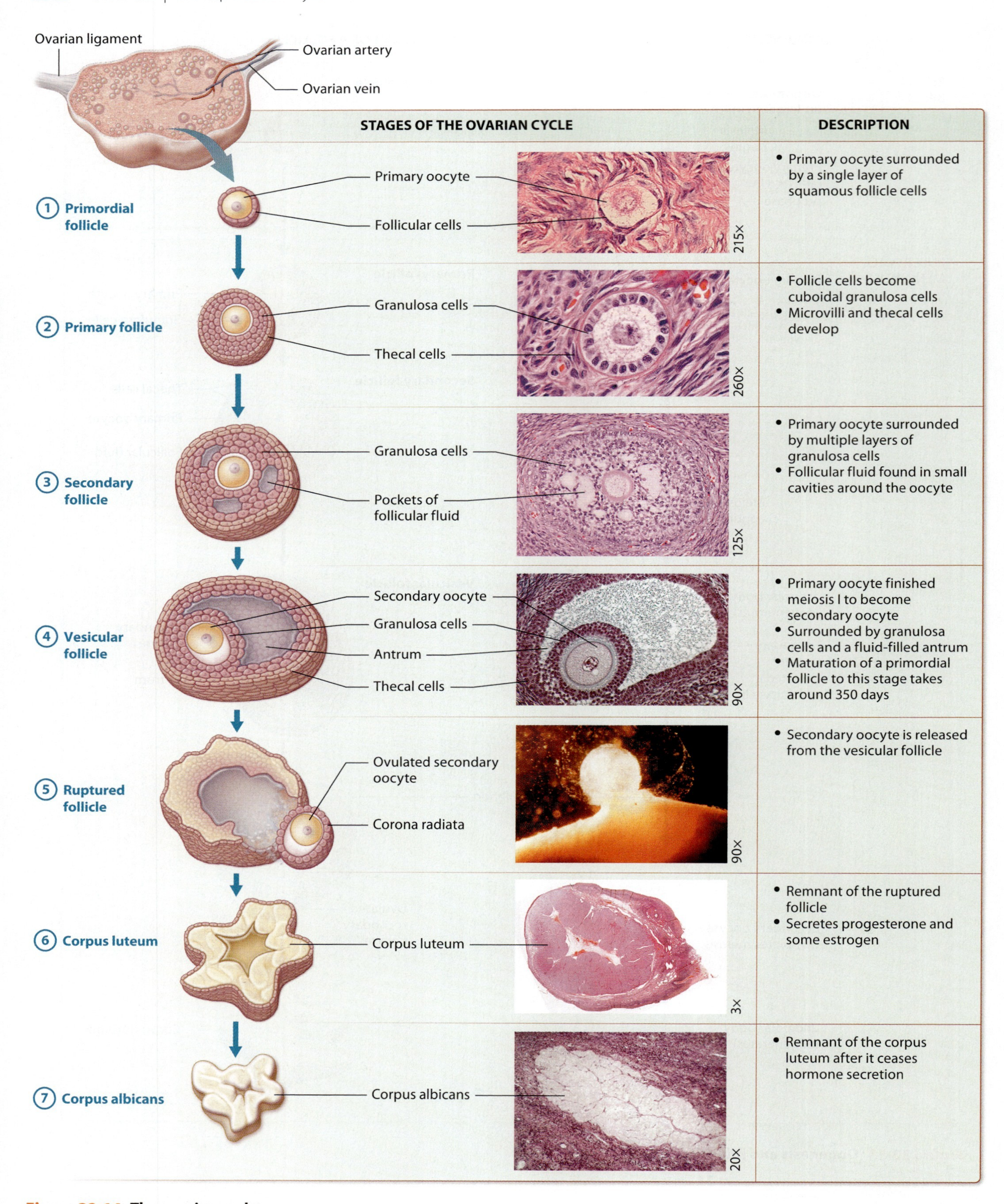

Figure 32-14 The ovarian cycle.

ACTIVITY 4

Comparing Spermatogenesis and Oogenesis

Learning Outcomes

1. Observe a histological section of the testis and identify an interstitial cell, seminiferous tubule, spermatogonium, primary spermatocyte, spermatid, sperm, and sustentacular cell.
2. Observe a histological section of the ovary and identify the germinal epithelium, cortex, medulla, primordial follicle, primary follicle (primary oocyte), secondary follicle, and vesicular follicle (secondary oocyte and antrum).
3. Compare and contrast the events of spermatogenesis and oogenesis.

Materials Needed

☐ Microscope and histological slide (or photomicrograph) of testis

☐ Microscope and histological slide (or photomicrograph) of ovary

Instructions

1. Observe a histological slide (or a photomicrograph) of the testis under low power.
   a. Identify the round seminiferous tubules. Note the interstitial cells lying between the seminiferous tubules.
   b. Switch to the high-power objective and identify the spermatogonia, with their pale-staining, centrally located nuclei. Secondary spermatocytes are often difficult to identify. Observe the cells closest to the lumen (spermatids and sperm). Sketch and label the following structures: interstitial cell, seminiferous tubule, spermatogonium, primary spermatocyte, spermatid, sperm, and sustentacular cell.

Total magnification: _____×

Which hypophyseal hormone stimulates sperm production? ______________________

Primary spermatocytes undergo ______________ *form secondary spermatocytes, which undergo* ______________ *to form spermatids.*

Compare the structure of a spermatid to that of a sperm. ______________________

____

____

Distinguish between spermatogenesis and spermiogenesis. ______________________

____

____

What is the function of interstitial cells? ______________

____

When sperm leave the lumen of the seminiferous tubule, where do they go next? ______________

What happens to the sperm there? ______________

____

2. Observe a histological slide (or a photomicrograph) of the ovary under low power.
   a. Sketch a section of the ovary and label the following structures: germinal epithelium, cortex, medulla, primordial follicle, primary follicle, secondary follicle, and vesicular follicle.

Total magnification: _____×

Which tissue type comprises the germinal epithelium, the outermost layer of the ovary? ______________

____

Describe the appearances of the ovarian cortex and medulla. ______________________________

____

____

b. Observe the cortical region, locate a primordial follicle, and describe its appearance. ______________________________

____

c. Locate a primary follicle containing one or a few layers of cuboidal cells surrounding a central developing primary oocyte. Now, locate and describe the appearance of a secondary follicle. ______________________________

____

____

Identify and describe the appearance of a vesicular follicle. ______________________________

____

____

Which hormone stimulates a follicle to grow? ________

Growing follicles release large amounts of which hormone? ______________________________

d. Locate a vesicular follicle, switch to the high-power objective, and identify the fluid-filled antrum within it. Locate the secondary oocyte, which has been pushed to one side of the follicle. Note the corona radiata, a structure consisting of follicular cells surrounding a secondary oocyte. Sketch the vesicular follicle and label the antrum, secondary oocyte, and corona radiata.

Total magnification: _____×

Which hormone stimulates ovulation of the secondary oocyte from the vesicular follicle? ______________________________

3. CHART Work with your lab group to complete the following chart comparing the events of spermatogenesis and oogenesis.

Characteristics Unique to Spermatogenesis	Characteristics Common to Spermatogenesis and Oogenesis	Characteristics Unique to Oogenesis

4. **Optional Activity**

View and label histology slides of the male and female reproductive systems at MasteringA&P® > Study Area > Practice Anatomy Lab > Histology > Reproductive System

Post-lab quizzes are also assignable in MasteringA&P®

POST-LAB ASSIGNMENTS

Name: ______________________ Date: ______________ Lab Section: __________

PART I. Check Your Understanding

Activity 1: Examining Male Reproductive Anatomy

1. Identify the organs of the male reproductive system on the accompanying photo of an anatomical model.

   a. ______________________

   b. ______________________

   c. ______________________

   d. ______________________

   e. ______________________

   f. ______________________

   g. ______________________

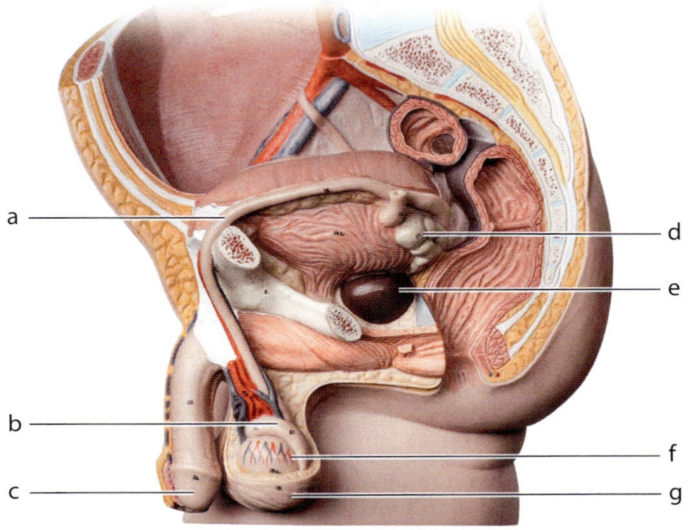

H11 Male Pelvis, 2-part, 3B Scientific®

2. Name the male reproductive structure:

   ______________________ a. where sperm maturation takes place

   ______________________ b. that produces most of the seminal fluid

   ______________________ c. where sperm and seminal fluid first mix

   ______________________ d. that transports sperm and urine

   ______________________ e. that travels through the spermatic cord

   ______________________ f. in which the testes sit

Activity 2: Examining Female Reproductive Anatomy

1. Identify the organs of the female reproductive system on the accompanying photo of an anatomical model.

   a. ______________________________

   b. ______________________________

   c. ______________________________

   d. ______________________________

   e. ______________________________

   f. ______________________________

   g. ______________________________

   h. ______________________________

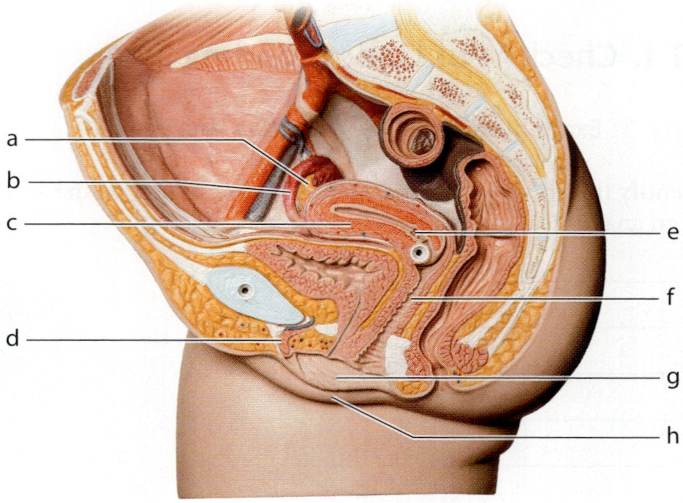

H10 Female Pelvis, 2-part, 3B Scientific®

2. Name the female reproductive structure:

   ______________________________ a. where fertilization typically occurs

   ______________________________ b. that is the birth canal

   ______________________________ c. where implantation takes place

   ______________________________ d. that anchors the ovary to the uterus

   ______________________________ e. that functions as erectile tissue

   ______________________________ f. that is the area medial to the labia minora

Activity 3: Modeling Meiosis

1. State the specific meiotic phase(s) during which:

   ______________________________ a. crossing over occurs

   ______________________________ b. individual sister chromatids move to opposite poles

   ______________________________ c. homologous chromosomes line up along the equator

   ______________________________ d. centrioles move to opposite poles

   ______________________________ e. cytokinesis occurs

   ______________________________ f. homologous chromosomes pair up

Activity 4: Comparing Spermatogenesis and Oogenesis

1. Match each of the following terms with its description:

	Term		Description
_____	a. spermatogonium	1.	undergoes meiosis II
_____	b. spermatid	2.	a stem cell
_____	c. sperm	3.	undergoes meiosis I
_____	d. primary spermatocyte	4.	a haploid gamete
_____	e. secondary spermatocyte	5.	undergoes spermiogenesis

2. Identify the structures in the accompanying illustration of a cross-section of a seminiferous tubule:

a. ______________________________

b. ______________________________

c. ______________________________

d. ______________________________

e. ______________________________

3. Match each of the following terms with its description:

	Term		Description
_____	a. vesicular follicle	1.	is formed before birth
_____	b. primordial follicle	2.	is arrested in metaphase II
_____	c. primary oocyte	3.	is arrested in prophase I
_____	d. secondary oocyte	4.	is a mature follicle

4. Identify the structures in the accompanying illustration of a cross-section of an ovary:

a. ______________________________

b. ______________________________

c. ______________________________

d. ______________________________

e. ______________________________

f. ______________________________

g. ______________________________

5. Does each of the following descriptions characterize spermatogenesis, oogenesis, or both spermatogenesis and oogenesis?

_______________ a. occurs completely in the gonad

_______________ b. is controlled by hormones

_______________ c. produces haploid gametes

_______________ d. begins at puberty

_______________ e. involves meiosis

_______________ f. involves equal division of cytoplasm

PART II. Putting It All Together

A. Review Questions

Answer the following questions using your lecture notes, your textbook, and your lab notes.

1. Identify the specific mitotic or meiotic phase shown in each of the following "pop-bead chromosome" photographs and provide a rationale for each answer. Assume a cell with a diploid number ($2n$) of 6.

Specific phase of mitosis or meiosis: _______________

Rationale: _______________

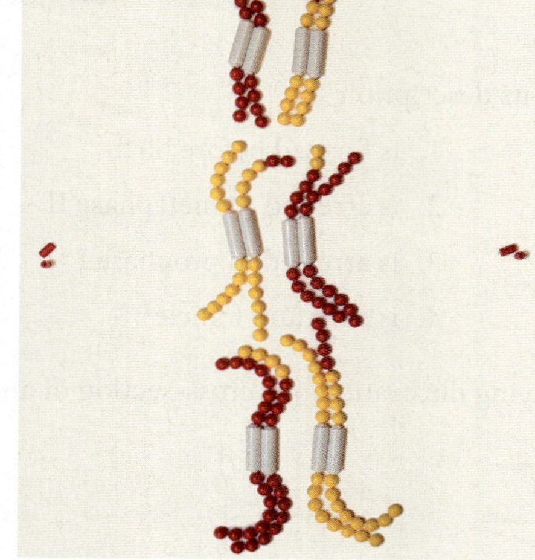

Specific phase of mitosis or meiosis: _______________

Rationale: _______________

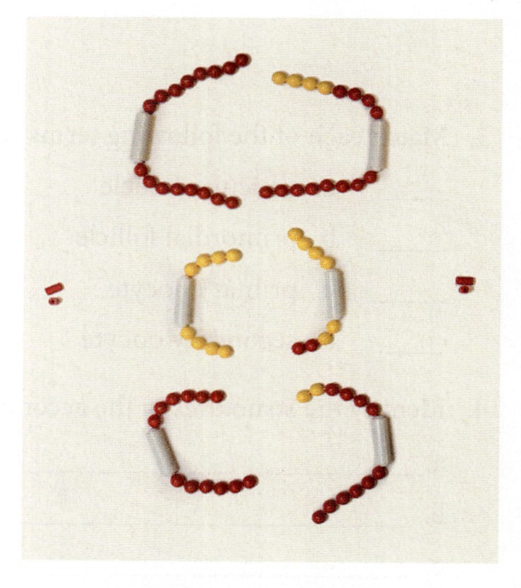

Specific phase of mitosis or meiosis: _______________

Rationale: _______________

2. Which hormone:

_______________ a. stimulates gonadotropin release?

_______________ b. triggers ovulation?

_______________ c. stimulates follicular growth?

_______________ d. stimulates testosterone release?

_______________ e. begins repair of the endometrium?

_______________ f. is known as the pregnancy hormone?

_______________ g. stimulates the corpus luteum?

3. Trace the pathway of a sperm from its site of production until it fertilizes an oocyte.

____

____

____

4. Based on your understanding of the hormonal control of spermatogenesis, how might the abuse of anabolic steroids lead to sterility? _______________

____

____

____

5. Based on your understanding of the hormonal control of oogenesis, how does the birth control pill prevent pregnancy? _______________

____

____

B. Concept Mapping

1. Fill in the blanks to complete this concept map outlining male reproductive anatomy.

ductus deferens   prostate gland   seminal vesicles   testes   urethra

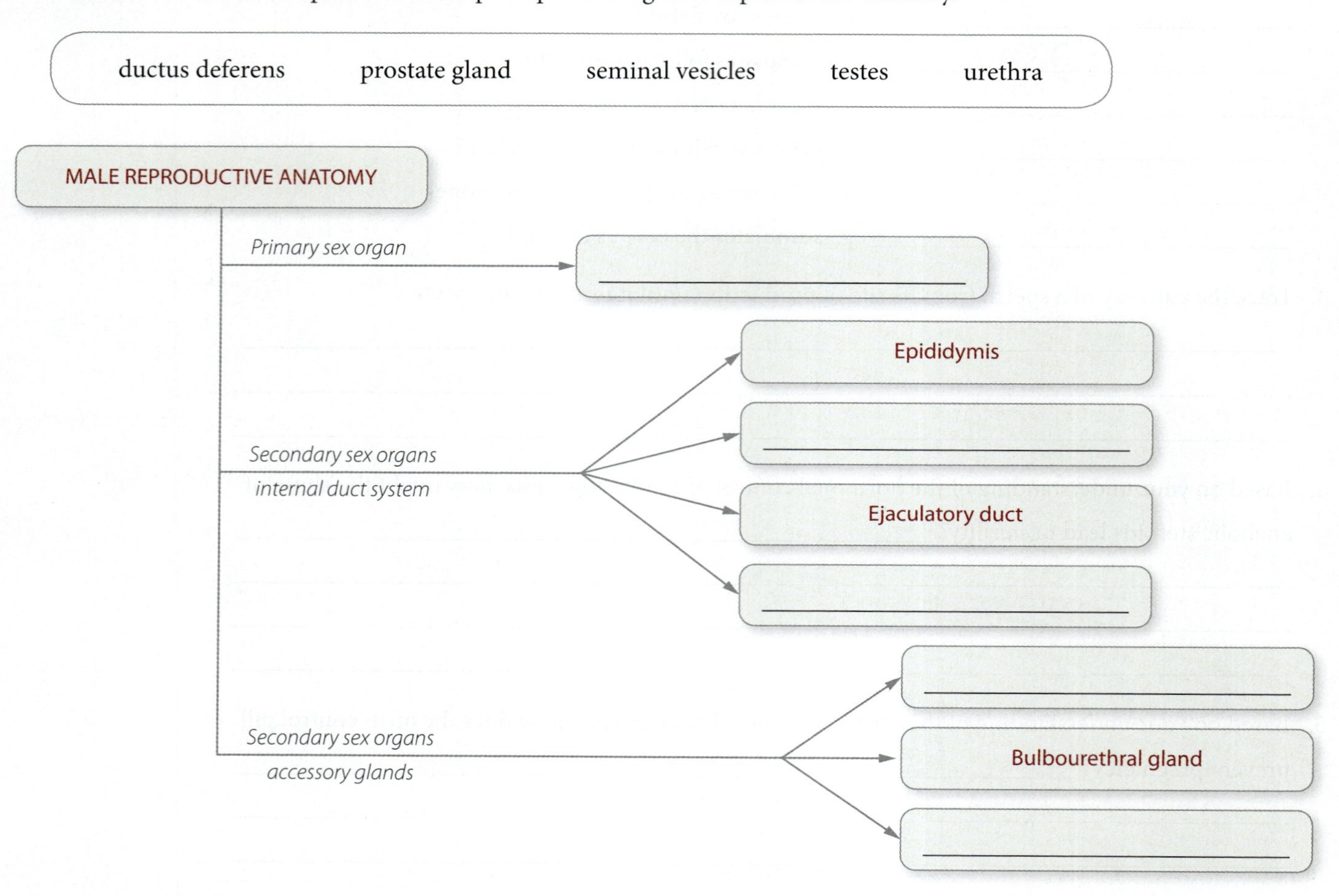

2. Construct a unit concept map to show the relationships among the following set of terms. Include all of the terms in your diagram. Your instructor may choose to assign additional terms.

bulbourethral gland   ductus deferens   epididymis   follicle   hypothalamus

interstitial cell   ovary   pituitary gland   prostate gland   seminal vesicles

testes   urethra   uterine tube   uterus   vagina

33 Embryonic Development and Heredity

UNIT OUTLINE

In Unit 32 we learned that the testes produce sperm and the ovaries produce secondary oocytes, and that in males the ducts of the reproductive system transfer semen (the mixture of sperm and fluids from accessory glands) through the penis and into the female reproductive tract.

We begin this unit by focusing on the fates of those sperm and secondary oocytes. If a sperm cell fuses with a secondary oocyte—a process called **fertilization**—the resulting cell, called a **zygote**, has the potential to develop into a new individual possessing traits passed on from the mother and father. **Embryology** is the study of the changes in form and function of a developing organism during the **prenatal period**, the time from fertilization through approximately 38 weeks of development in the womb. In this unit we explore the stages of human development during the 38-week prenatal period and examine the structures that support the developing embryo: the extraembryonic membranes, the placenta, and the umbilical cord.

We will also explore how traits are passed on from the parents to the offspring in the genetic material, or **DNA (deoxyribonucleic acid)**, that constitutes an individual's genes. The science that studies how genes are transmitted is **genetics**; the term **heredity** is used to describe the transmission of genes and genetic characteristics from parent to child. We will learn some specific terms used in the field of genetics and the examples covered in this unit will also enable us to determine the chances that a child will inherit a specific trait if it is controlled by a single gene.

THINK ABOUT IT *List some examples of specific traits (such as dimples or eye color) that are passed from parent to child.*

____

____

____

____

PRE-LAB ASSIGNMENTS

Pre-lab quizzes are also assignable in MasteringA&P®

To maximize learning, BEFORE your lab period carefully read this entire lab unit and complete these pre-lab assignments using your textbook, lecture notes, and prior knowledge.

PRE-LAB Activity 1: Exploring Fertilization and the Stages of Prenatal Development

1. Which of the following prenatal periods lasts from week 3 to week 8 of development?
   a. pre-embryonic period
   b. embryonic period
   c. fetal period
   d. gestational period
2. The ____________________ is the specific structure of the early embryo that implants in the endometrium of the uterus.
3. The primary germ layer of the early embryo that becomes the majority of the nervous system is the ____________________, whereas the germ layer that becomes the majority of the bones and muscles is the ____________________.

PRE-LAB Activity 2: Examining the Placenta

1. The ____________________ is a temporary organ that is the site of exchange of oxygen, nutrients, and wastes between the mother and fetus, and the ____________________ connects the placenta to the embryo.
2. Use the list of terms provided to label the accompanying illustration of the placenta. Check off each term as you label it.

☐ yolk sac
☐ chorion
☐ amnion
☐ allantois
☐ umbilical cord

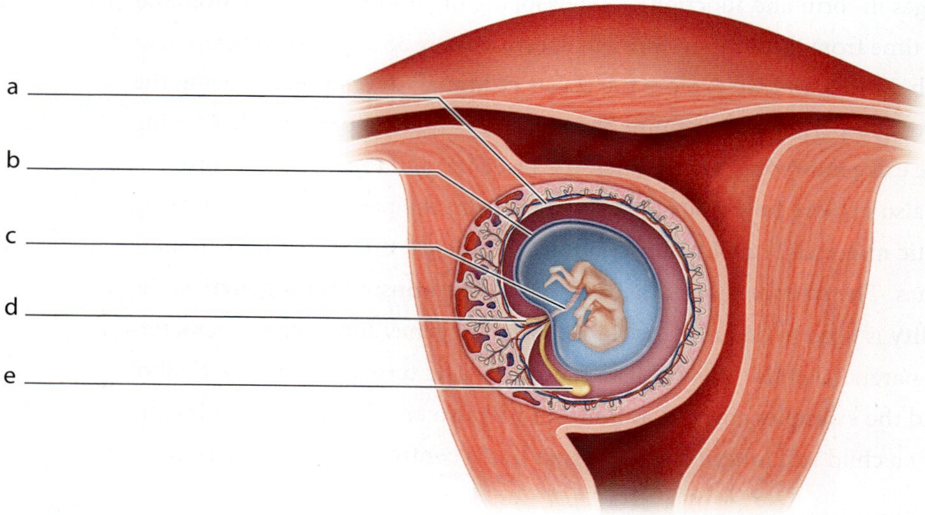

a ____________________
b ____________________
c ____________________
d ____________________
e ____________________

PRE-LAB Activity 3: Learning the Language of Genetics

1. ____________________ is the transmission of genes and genetic characteristics from parent to child; ____________________ is the science that studies how genes are transmitted.

2. Match each of the following definitions with its appropriate term.

	Definition	Term
_____	a. describes an allele that is masked	1. chromosome
_____	b. refers to two alleles that are different	2. alleles
_____	c. DNA that is folded and coiled	3. dominant
_____	d. a display of an individual's chromosome pairs	4. phenotype
_____	e. describes an allele that can mask another allele	5. homologous
_____	f. the expression of an individual's genotype	6. homozygous
_____	g. variants of a specific gene	7. karyotype
_____	h. the genetic makeup of an individual	8. heterozygous
_____	i. refers to alleles that code for the same trait	9. genotype
_____	j. an individual's entire set of DNA	10. gene
_____	k. twenty-three pairs of ___________ chromosomes	11. recessive
_____	l. segment of DNA that codes for a specific protein	12. genome

PRE-LAB Activity 4: Exploring Dominant-Recessive Inheritance

1. ____________________ inheritance refers to the interaction of alleles that are strictly dominant or recessive.

2. Name three human traits that are at least partially controlled by a dominant gene: _______________ ____

PRE-LAB Activity 5: Exploring Other Patterns of Inheritance

1. ______________________________ is a condition in which heterozygous individuals exhibit a phenotype in between those of the homozygous individuals.

2. In a phenomenon known as ______________________, some alleles are equally dominant and therefore are both expressed.

3. ______________________ are traits that are specifically expressed on the X or Y chromosomes.

4. The most common type of color blindness is __________________________.

Fertilization and Prenatal Development

Human development begins when a sperm cell fertilizes a secondary oocyte to form a zygote. For fertilization to occur, sperm must first reach the secondary oocyte. Once in the female reproductive tract, each sperm cell undergoes a series of changes—a process called *capacitation*—that makes it fully motile and modifies its plasma membrane so it can fuse with the plasma membrane of the secondary oocyte.

Typically, multiple sperm penetrate the layer of cells that surround the secondary oocyte—the **corona radiata**—and produce enzymes that penetrate the **zona pellucida**, a layer that covers the oocyte's plasma membrane. The first sperm to fuse with the oocyte's plasma membrane triggers a reaction that modifies the zona pellucida and prevents other sperm from entering. When this occurs, the oocyte finishes meiosis II, expels a polar body, and becomes an ovum. The nuclei of the sperm cell and ovum—called the **male pronuclei** and **female pronuclei**—come together, and the chromosomes of the pronuclei combine to restore the diploid number and produce a zygote. This first cell of the new individual will undergo many changes during the prenatal period leading up to birth. Prenatal development consists of three periods: the **pre-embryonic**, characterized by the production of a multicellular blastocyst that implants in the uterus; the **embryonic**, during which rudimentary organs develop within an embryo; and the **fetal**, during which a fetus develops to full term at 38 weeks (**Figure 33-1**).

Pre-Embryonic Period

During the pre-embryonic period—the 2-week period between fertilization and implantation (**Figure 33-2**)—the zygote begins to divide by **cleavage**, a series of rapid mitotic divisions that produce small, genetically identical cells called blastomeres. Fertilization and cleavage occur in the uterine tube as the developing embryo moves toward the uterus. Divisions occur quickly, so the number of cells increases

	PRE-EMBRYONIC PERIOD	EMBRYONIC PERIOD	FETAL PERIOD
Developmental stage	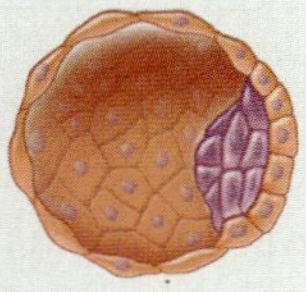 Blastocyst	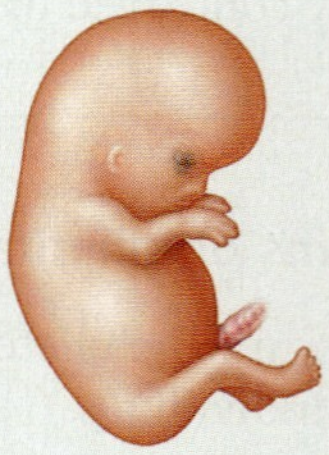 Embryo	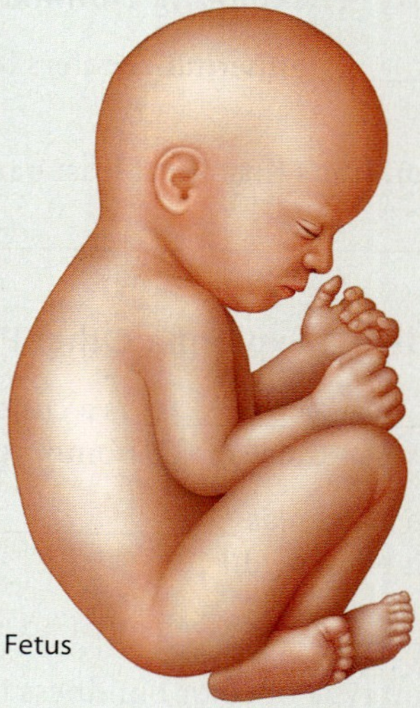 Fetus
Events	**Weeks 1 and 2:** • Zygote divides mitotically many times to produce a multicellular blastocyst that implants in the uterus.	**Weeks 3 through 8:** • Blastocyst grows, folds, and forms rudimentary organ systems. • It is now called an embryo.	**Weeks 9 through 38 (until birth):** • Embryo is now called a fetus. • It grows larger and develops until its organ systems can function without assistance from the mother.

Figure 33-1 **Fertilization and the periods of prenatal development.**

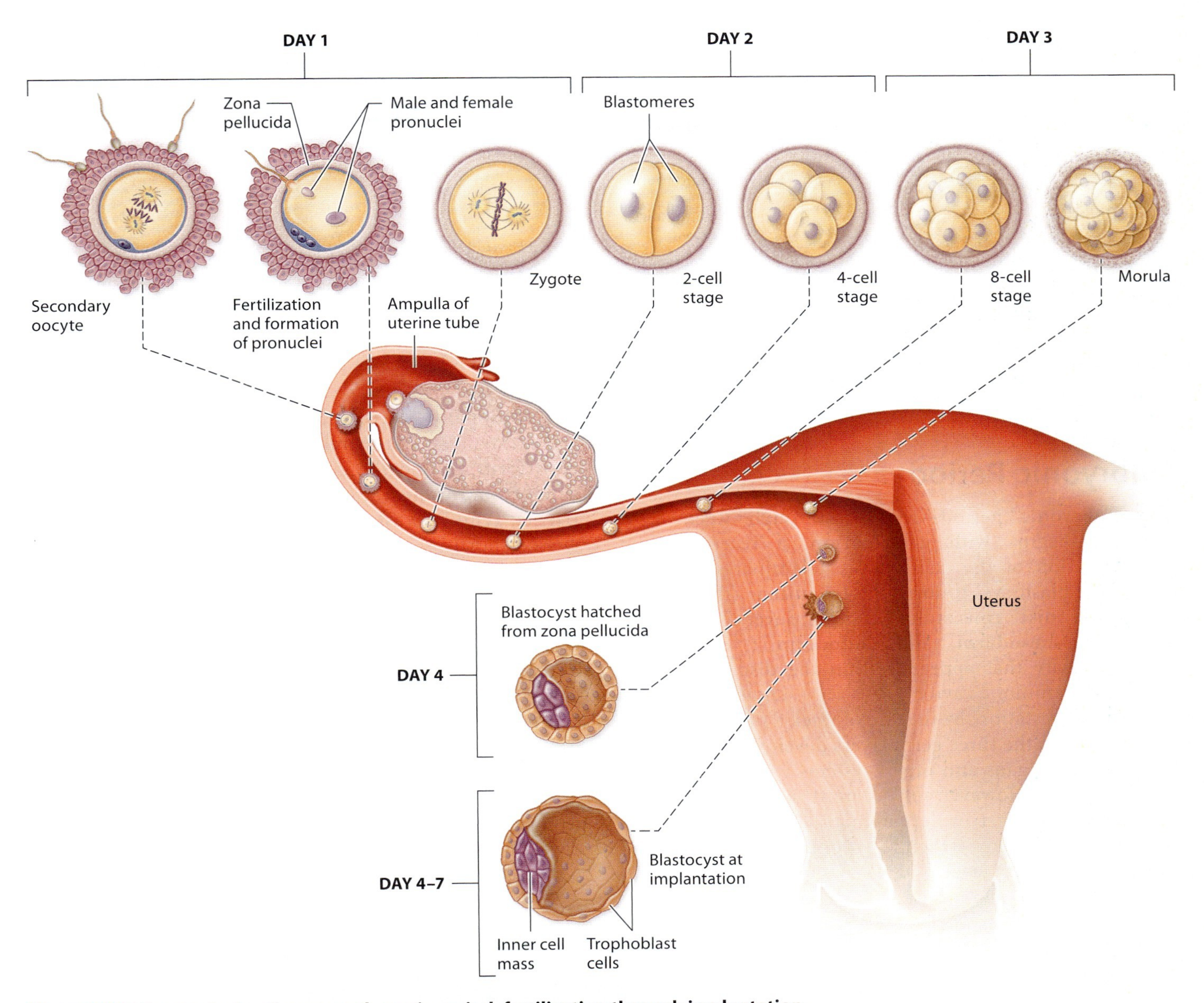

Figure 33-2 **Events during the pre-embryonic period, fertilization through implantation.**

while cell size decreases. By day 3, the embryo contains 16 solidly packed cells and is called a **morula**. On day 4 or 5, the embryo—now called a **blastocyst**—consists of a cell mass surrounding an internal fluid-filled cavity.

During the last week of the pre-embryonic period, the blastocyst arrives in the uterus and **implants**, or attaches to the endometrium of the uterus. The blastocyst has two distinct cell populations: large, flattened cells called **trophoblast cells**, and a cluster of rounded cells called the **inner cell mass**. The trophoblast cells will form the embryo's portion of the placenta, whereas the inner cell mass will form the **embryo proper**—the body of the developing offspring. The first structure of the embryo proper to develop is the flat **bilaminar** (two-layer) **embryonic disc**.

CLINICAL CONNECTION

In an **ectopic pregnancy**, a fertilized ovum implants and grows in a location other than the endometrium of the uterus. Only 1% to 2% of all pregnancies in the United States are ectopic, and almost all ectopic pregnancies are called "tubal pregnancies" because they occur in a uterine tube. However, ectopic pregnancies are doomed to fail because only the uterus is able to expand and sustain a growing embryo. If a tubal pregnancy does not terminate on its own, medical intervention is necessary to remove the embryo. Without intervention, the uterine tube will rupture, resulting in serious bleeding and possibly even death of the mother.

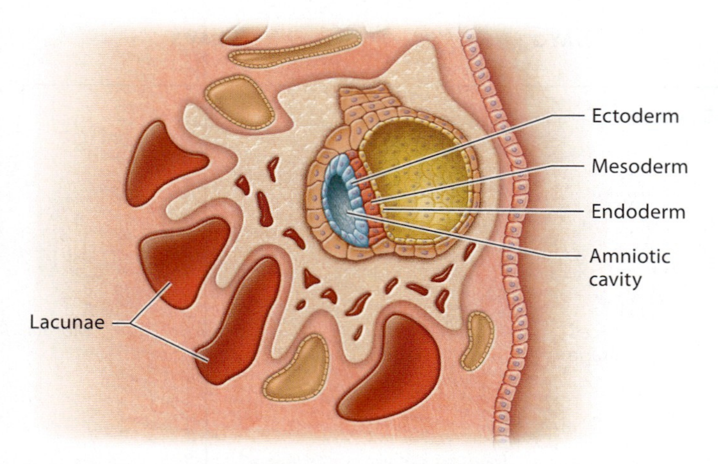

Figure 33-3 Embryo at day 16, early in the embryonic period.

Layer	Major Structures
Ectoderm	Epidermis (including accessory structures), majority of nervous system structures, and sense organs
Endoderm	Epithelium lining structures of most body systems and of some glands
Mesoderm	Dermis, bones, most muscles, most organs (except their linings), most connective tissues and membranes

Table 33-1 Primary Germ Layer Origins of Major Body Structures

Embryonic Period

The embryonic period—week 3 through week 8 of gestation—begins with **gastrulation**, a rearrangement and migration of the cells of the bilaminar embryonic disc of the blastocyst to form the **trilaminar** (three-layer) **embryonic disc** consisting of germ layers designated **ectoderm**, **mesoderm**, and **endoderm** (**Figure 33-3**). A small cavity that appears within the disc enlarges to become the amniotic cavity. **Figure 33-4a** shows an embryo at week 4 of gestation; by the end of week 8 of development (**Figure 33-4b**), the three germ layers develop into all of the body's major organs and organ systems via the process of **organogenesis**. However, for the most part organs are not functional until later in development. **Table 33-1** summarizes the major body structures produced by each of the three primary germ layers.

Additionally, during the 6-week embryonic period, the extraembryonic membranes, placenta, and umbilical cord form around the embryo. These structures, which support and protect the embryo, deliver oxygen and nutrients, remove wastes, and produce hormones, are discussed shortly.

Fetal Period

The fetal period lasts from week 9 through week 38 of gestation, or until birth. The embryo—now called a **fetus**—grows larger and continues to develop until the organ systems are functional without assistance from the mother. During this period, growth is very rapid (**Figure 33-5**) and is characterized by the maturation of tissues and organs. Fetal size, measured as crown-to-rump length (CRL), increases dramatically. In addition to an increase in size, several important developmental events occur during each month of the fetal period (**Table 33-2**).

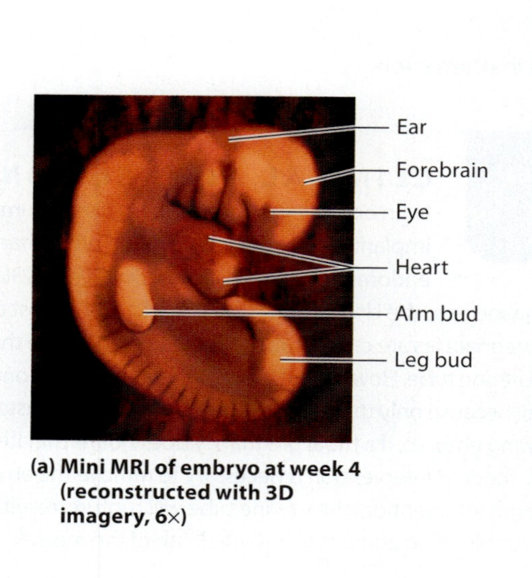

(a) Mini MRI of embryo at week 4 (reconstructed with 3D imagery, 6×)

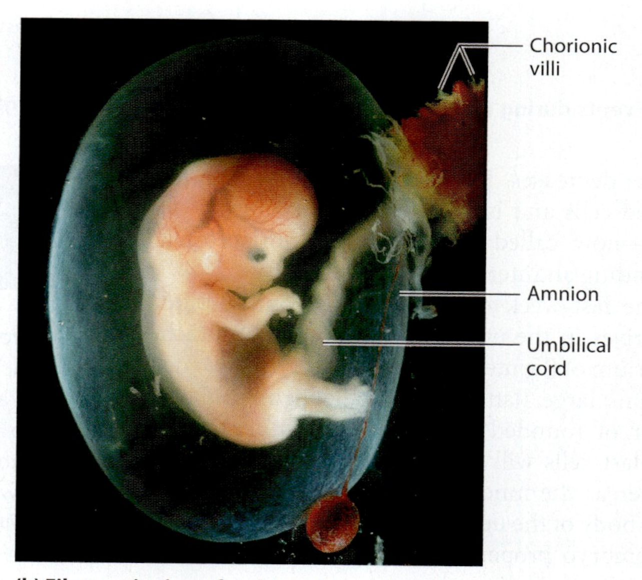

(b) Fiber optic view of embryo at week 8 (1.5×)

Figure 33-4 Development during the embryonic period.

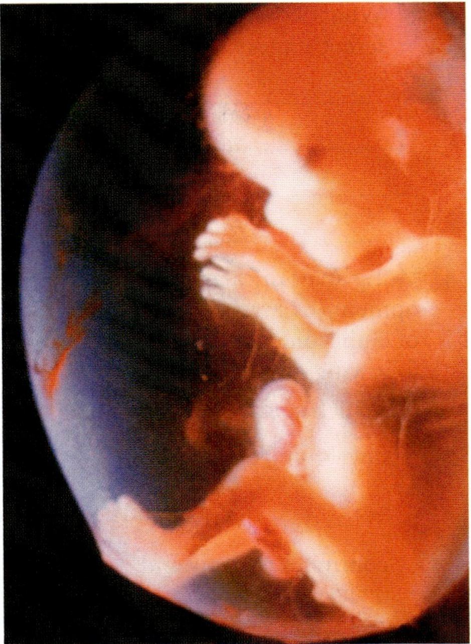

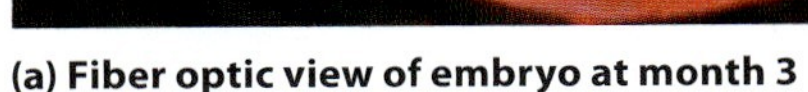

(a) Fiber optic view of embryo at month 3

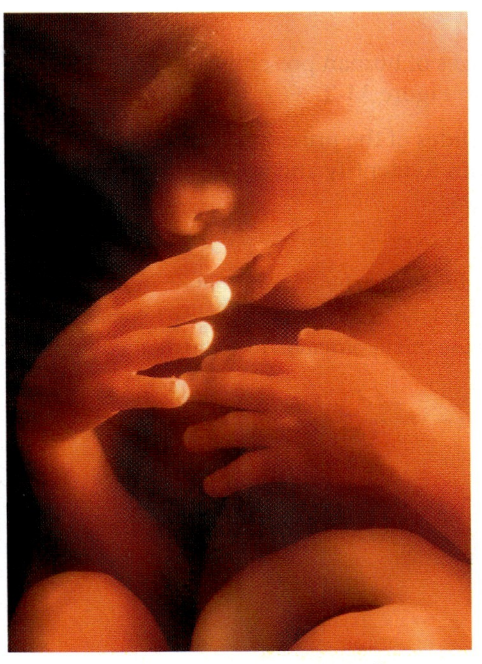

(b) Fiber optic view of embryo at month 5

Figure 33-5 **Development during the fetal period.**

Table 33-2 Major Developments during the Fetal Period of Prenatal Development

Month	Developments	Approximate Size of Fetus (CRL)
3	Ossification begins in most bones; eyes are developed but eyelids are closed; upper limbs lengthen; external reproductive organs are distinguishable.	9 cm (3.5 in.)
4	Skeleton continues to ossify; lower limbs lengthen; joints are forming; nipples and hair are present; digestive glands are forming; kidneys are formed; heartbeat can be heard with a stethoscope; fetus will startle and turn away from loud noises and bright lights.	14 cm (5.5 in.)
5	Growth of limbs slows; skeletal muscles contract; hair grows on the head; lanugo (downy hair) and vernix caseosa (a cheese-like secretion) cover the skin; brown fat forms; mothers can feel movements, a phenomenon called *quickening*.	19 cm (7.5 in.)
6	Fetus gains significant weight; eyebrows and eyelashes appear; eyelids are partially open; skin is wrinkled and translucent; lungs begin to produce surfactant.	23 cm (9.1 in.)
7	Fat is deposited in subcutaneous tissue; eyelids are completely open; in males, testes begin to descend through the inguinal canal and into the scrotum.	28 cm (11.0 in.)
8 and 9	Neurons form networks; organs specialize and grow; blood cells form in bone marrow; skin is less wrinkled; lanugo is shed; digestive and respiratory systems complete development during the ninth month; in males, testes complete descent into the scrotum.	36 cm (14.0 in.)

ACTIVITY 1
Exploring Fertilization and the Stages of Prenatal Development

Learning Outcomes

1. Identify each of the stages of the pre-embryonic period.
2. Identify the structures of the blastocyst.
3. Identify the structures present in developing embryos.
4. Describe growth during the fetal period.

Materials Needed

- ☐ Compound microscope
- ☐ Prepared slides (or photomicrographs) of embryonic development of the sea urchin or a similar animal model
- ☐ Anatomical charts and/or models of fertilization
- ☐ Anatomical charts and/or models of human embryonic development
- ☐ Anatomical charts and/or models of human fetal development

Instructions

1. Review the stages of human pre-embryonic development in Figure 33-2.
2. Use prepared slides of the pre-embryonic development of the sea urchin to identify and sketch the zygote and blastocyst stages.

> **QUICK TIP** When sketching, pay attention to relative sizes. Be sure to record the magnification you use to make your sketches and use this information to compare the zygote and blastocyst.

Zygote

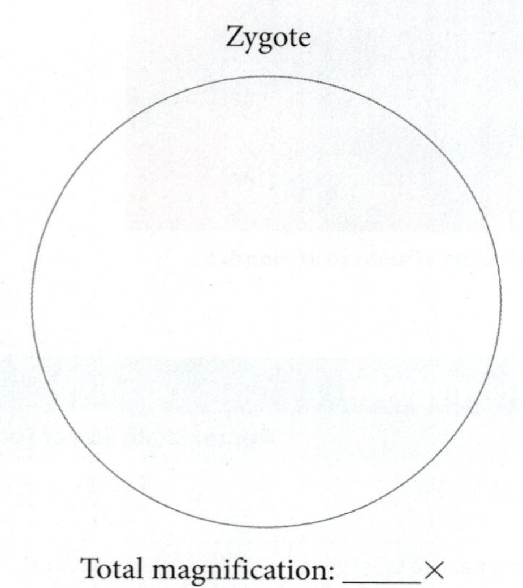

Total magnification: _____×

Blastocyst

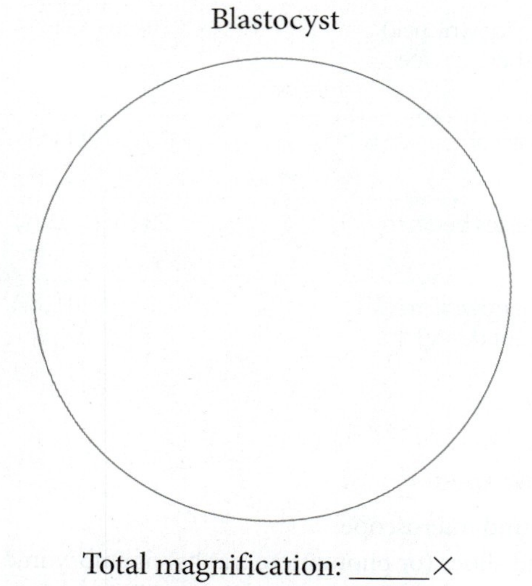

Total magnification: _____×

3. Answer the following questions:
   a. How long does it take the early embryo to travel through the uterine tube and develop into a blastocyst? ______________________________
   b. What do you notice about the relative sizes of the different stages? Why do you think this is the case?
   ____
   ____
   ____
   ____
4. Using anatomical charts and/or models of human development (or Figures 33-2 and 33-3), identify the stages of embryonic and fetal development and answer the following questions.
   a. Which specific part of the blastocyst becomes the developing embryo? ______________________________
   b. What does the trophoblast become as the embryo continues to grow and develop? ______________________________
   ____
   c. Indicate which body structures are produced from each of the primary germ layers:
   Ectoderm ______________________________
   ____
   ____
   Mesoderm ______________________________
   ____
   ____
   Endoderm ______________________________
   ____
   ____
   d. What is the most notable difference between embryos at week 4 and at week 8? ______________________________
   ____
   e. What is the most notable difference between the embryo at week 8 and the fetus at week 12?
   ____
   ____
   f. What is the general difference between embryonic development and fetal development? ______________________________
   ____
   ____
   ____

Functions of the Placenta during Pregnancy

From the clinical perspective, the period of prenatal development is called the gestation period, which the mother experiences as pregnancy. **Pregnancy** involves all the anatomical and physiological events and changes within a woman during the gestation period, which is approximately 280 days, or 40 weeks. The gestation period is longer than the prenatal period because it includes the 2 weeks after the end of the mother's last menstrual cycle—a date that women can usually recall. The gestation period is subdivided into three trimesters, which differ slightly from the subdivisions of prenatal development:

- *First trimester* (months 1–3 of pregnancy): All pre-embryonic and embryonic development is completed and fetal development begins. The rudiments of all the major organ systems are present.
- *Second trimester* (months 4–6 of pregnancy): The fetus continues to grow and develop; pregnancy usually becomes obvious as the uterus and abdomen expand.
- *Third trimester* (months 7–9 of pregnancy): The fetus grows rapidly and gains a significant amount of weight.

We saw in the previous section that the blastocyst arrives in the uterus during the last week of the pre-embryonic period and implants in the endometrium. More specifically, the trophoblast of the blastocyst invades the endometrium of the uterus by secreting digestive enzymes. Around day 12, the walls of maternal blood vessels are digested, and maternal blood pools around the developing embryo in structures called **lacunae** (see Figure 33-3). By day 16, the blastocyst moves farther into the lining and is covered over by the regrowth of maternal epithelial cells on the surface.

Together, the bilaminar embryonic disc and trophoblast produce the extraembryonic membranes, which first appear during the second week of development and continue to develop during the embryonic and fetal periods. The **extraembryonic membranes** include the yolk sac, amnion, allantois, and chorion (**Figure 33-6a**). The **yolk sac** subsequently forms part of the digestive tract and is the source of the first blood cells and blood vessels. The **amnion** secretes amniotic fluid, which protects the embryo from trauma and desiccation, helps maintain a constant temperature, allows symmetrical development, and allows freedom of movement, which is important for muscle development and for preventing body parts from adhering to each other. The **allantois** forms the base for the umbilical cord, which links the embryo to the placenta; eventually the allantois becomes part of the urinary bladder. The **chorion**, the outermost extraembryonic membrane, encloses all the other membranes and the embryo. Initially, outgrowths called **chorionic villi** form around the entire surface of the chorion. As pregnancy progresses, these outgrowths continue to grow in one area only, where they blend with the stratum functionalis of the endometrium and eventually form the principal embryonic part of the placenta (**Figure 33-6b**).

Placentation is the process of forming the disc-shaped placenta, which is the temporary organ responsible for exchange of oxygen, nutrients, and wastes between the mother and embryo. The placenta, which forms from both the uterine lining and cells of the developing embryo, also produces hormones that support the pregnancy. Once an embryo has implanted in the uterus and placentation has begun, the stratum functionalis is known as the decidua. The region of the endometrium that lies beneath the embryo becomes the **decidua basalis** (see Figure 33-6b); the region that surrounds the uterine cavity forms the **decidua capsularis** (see Figure 33-6a).

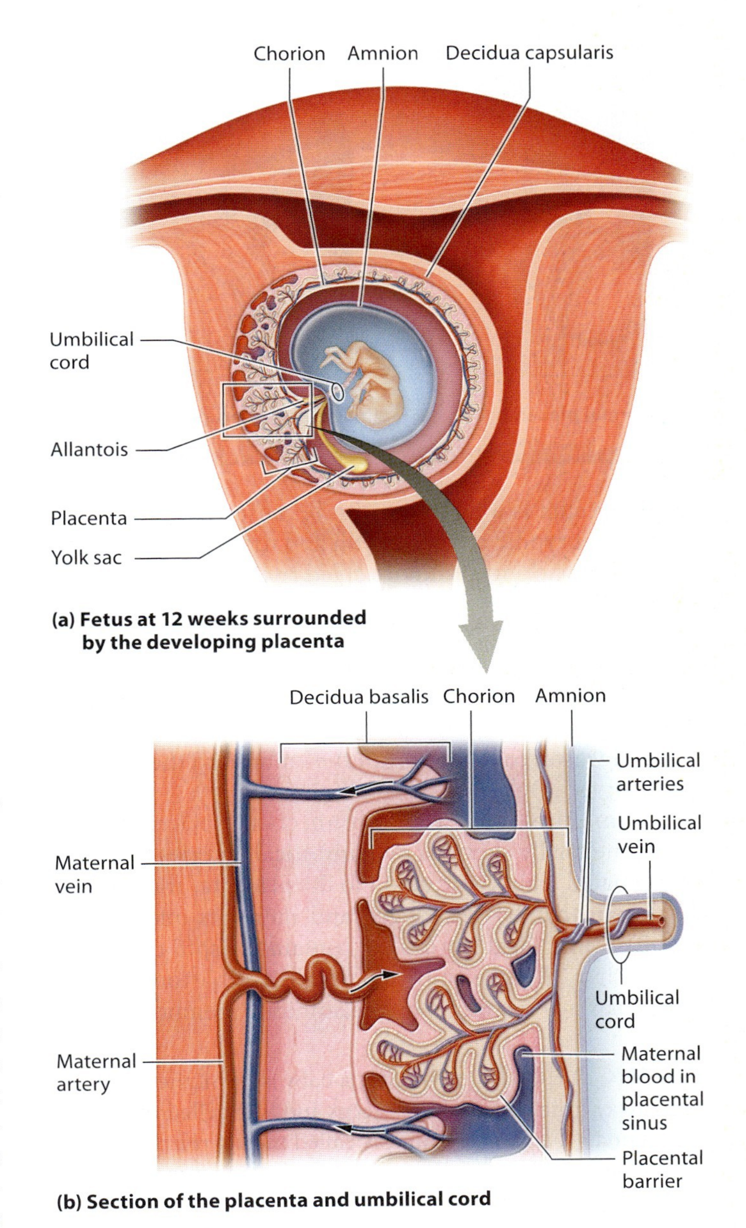

(a) Fetus at 12 weeks surrounded by the developing placenta

(b) Section of the placenta and umbilical cord

Figure 33-6 Placentation.

CLINICAL CONNECTION

Sometimes the placenta partially or completely separates from the mother's uterus before delivery of the newborn, causing **placental abruption**, the most common cause of bleeding after 20 weeks of gestation. The suddenness of placental abruption can put both fetus and mother in jeopardy by depriving the former of oxygen and nutrients and causing heavy blood loss in the latter. Bleeding through the vagina—called overt or external bleeding—occurs 80% of the time; sometimes blood pools behind the placenta, a condition known as concealed or internal placental abruption. Treatment depends on the seriousness of the abruption and the stage of pregnancy.

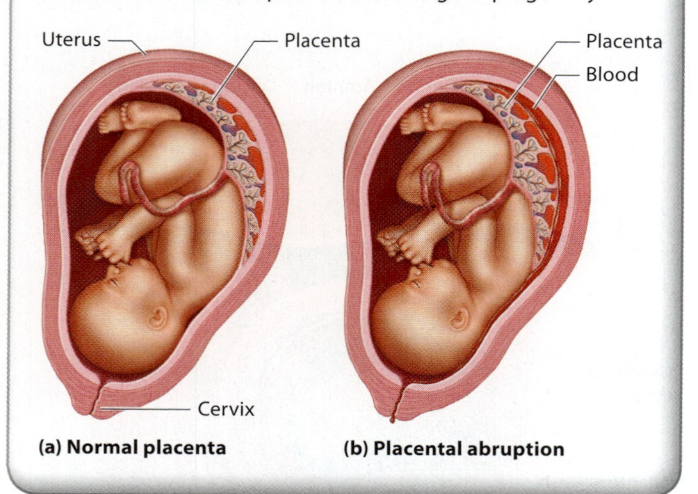

(a) Normal placenta **(b) Placental abruption**

The placenta is a unique organ because it develops from both the embryonic chorionic villi and the maternal decidua basalis. The chorionic villi are filled with embryonic blood and are surrounded by maternal blood in the **placental sinus**; however, the blood supplies do not mix because they are separated by the **placental barrier**. The proximity of the maternal and embryonic blood supplies enables oxygen and nutrients to diffuse from the maternal blood into the embryo, and wastes to diffuse from the embryo to the maternal blood (for removal).

The **umbilical cord**, which connects the placenta to the embryo, normally contains two **umbilical arteries** that carry deoxygenated embryonic blood to the placenta, and one **umbilical vein** that carries oxygen and nutrients into the embryo. The umbilical cord also contains a soft connective tissue with a jelly-like consistency called Wharton's jelly; its external surface is covered by the amniotic membrane.

THINK ABOUT IT After a baby is born, the placenta pulls away from the mother's uterus and is expelled through her vagina. (It is then commonly called the *afterbirth*.) What might happen during childbirth if the placenta was too near the cervix of the uterus? ______________________________

____

ACTIVITY 2
Examining the Placenta

Learning Outcomes

1. Identify the extraembryonic membranes and the function of each.
2. Examine the gross and microscopic structures of the placenta and umbilical cord.

Materials Needed

- ☐ Anatomical charts and/or models of a human placenta
- ☐ Preserved specimen of an animal uterus (if available)
- ☐ Model of a pregnant human torso (if available)
- ☐ Prepared slides of human placenta

Instructions

1. Using anatomical charts and/or models of the developing placenta (or Figure 33-6), identify the structures of the developing placenta and answer the following questions.
   a. What is the location of each extraembryonic membrane relative to the embryo?

      Chorion ______________________________

      Amnion ______________________________

      Allantois ______________________________

      Yolk sac ______________________________

   b. What is the primary function of the placenta? ______________________________

   c. Which part of the placenta is formed from the embryo? ______________________________

   d. Which part of the placenta is formed from maternal tissues? ______________________________

   e. What is the basic function of the umbilical cord? ______________________________

   f. What is the specific function of the umbilical vein? ______________________________

g. What is the specific function of the two umbilical arteries? ____________________________

____

2. Observe a fresh or preserved animal fetus and placenta, if available. Identify the following structures and then write a brief description of each.

   a. Placenta ____________________________

   ____

   b. Amnion (amniotic sac) ____________________

   ____

   c. Umbilical cord ________________________

   ____

3. Observe a model of a pregnant human torso, if available. The placenta is located in which region of the uterus?

   ____

4. Observe a prepared slide of a section of human placenta. Then sketch and label as many structures as possible.

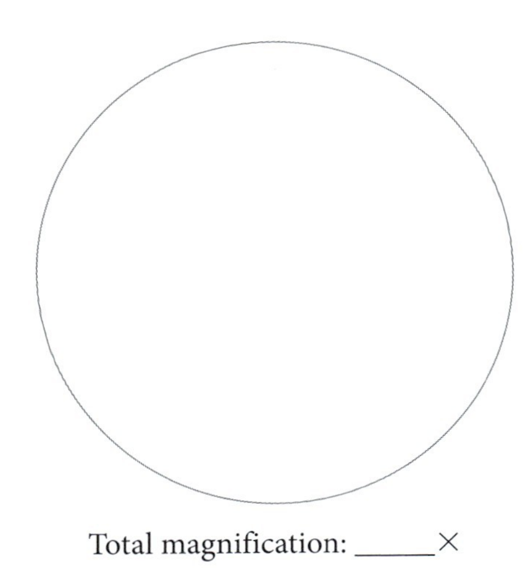

Total magnification: ______×

The Language of Genetics

All body traits, such as eye color and bone structure, are encoded in segments of DNA called **genes**, each of which codes for a specific protein. Because most body structures are proteins, genes effectively provide the blueprint for all of an individual's traits.

In a cell preparing to divide, the DNA is folded and coiled into **chromosomes**. Recall from Unit 5 that in humans, all nucleated somatic (body) cells have 46 chromosomes arranged in 23 pairs of **homologous chromosomes**; these cells are **diploid** (designated 2*n*). The individual's father

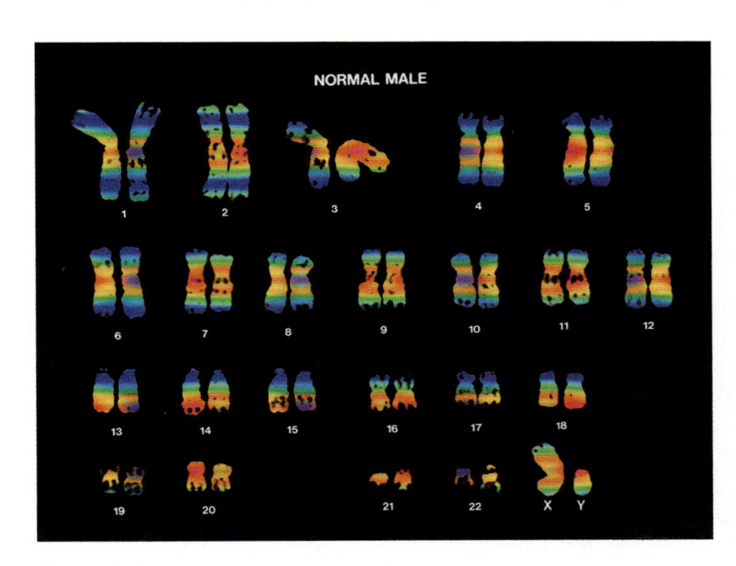

Figure 33-7 Human male karyotype with 23 pairs of chromosomes.

contributed one set of chromosomes (23 paternal chromosomes), and 23 maternal chromosomes came from the individual's mother. Twenty-two of the 23 pairs are called **autosomes**, and the members of each pair are similar in appearance; the other pair—the X and Y chromosomes, called the **sex chromosomes**—determine the sex of the individual. Females have two X chromosomes (designated XX), whereas males have one X and one Y (XY).

Collectively, an individual's entire set of DNA is termed his or her **genome**. Figure 33-7 shows a sample **karyotype**, a display of all 23 chromosome pairs of an individual arranged by size and structure. The only pair that has chromosomes of different sizes is the pair of sex chromosomes in a male—the XY.

Homologous chromosomes have pairs of genes that code for the same trait at the same location, or *locus*, on those chromosomes. However, the pairs of genes may have different variants, or **alleles**, that code for a particular trait. For example, one gene for eye color may have an allele that codes for brown eyes, whereas the other gene for eye color may have an allele that codes for blue eyes. If two alleles code for the same trait (such as brown eyes), they are said to be **homozygous** for that trait. If two alleles are different (such as one for brown eyes and one for blue eyes), they are **heterozygous**.

Some alleles mask or suppress the expression of another allele of the same gene. An allele that masks another allele is called **dominant**; an allele that is masked is called **recessive**. Dominant alleles are usually represented by capital letters, whereas recessive alleles are usually represented by lowercase letters. Freckles, for example, are at least partially the result of a dominant trait whose allele is designated "*F*"; lack of freckles is at least partially the result of a recessive trait whose allele is designated "*f*."

Individuals with the genetic makeup *FF* are **homozygous dominant** for freckles, individuals with the genetic makeup

Ff are **heterozygous**, and individuals with the genetic makeup *ff* are **homozygous recessive**. As you can conclude, only individuals who are homozygous recessive, *ff*, have no freckles, even though individuals who are heterozygous, *Ff*, possess an allele for no freckles. Not all alleles are dominant or recessive, and typically many factors influence the inheritance of a trait.

The combinations of alleles within an individual are referred to as **genotypes**. *FF*, *Ff*, and *ff* are examples of genotypes. The expression of an individual's genotype is referred to as **phenotype**. The phenotype of genotypes *FF* and *Ff* is freckles; the phenotype of genotype *ff* is no freckles. Individuals with a heterozygous genotype display the dominant phenotype, even though the recessive allele remains in the individual's genotype. Heterozygous individuals who reproduce may pass on the recessive form of the trait, which will become apparent only in the phenotype of homozygous recessive individuals in the next or subsequent generations.

A **Punnett square** is a simple diagram for illustrating the possible genotypes resulting from the mating of a man and a woman. **Figure 33-8** depicts the possible genotypes of offspring from a man and woman who are both heterozygous for freckles, or *Ff*. Recall that when sperm cells and oocytes are formed, each receives only one allele of each gene because each gamete carries half of the individual's chromosomes (that is, 23 chromosomes). Thus any given sperm cell or oocyte will have either an *F* allele or an *f* allele. When sperm and oocyte fuse during fertilization, the genotype of the offspring is the combination of alleles in those gametes.

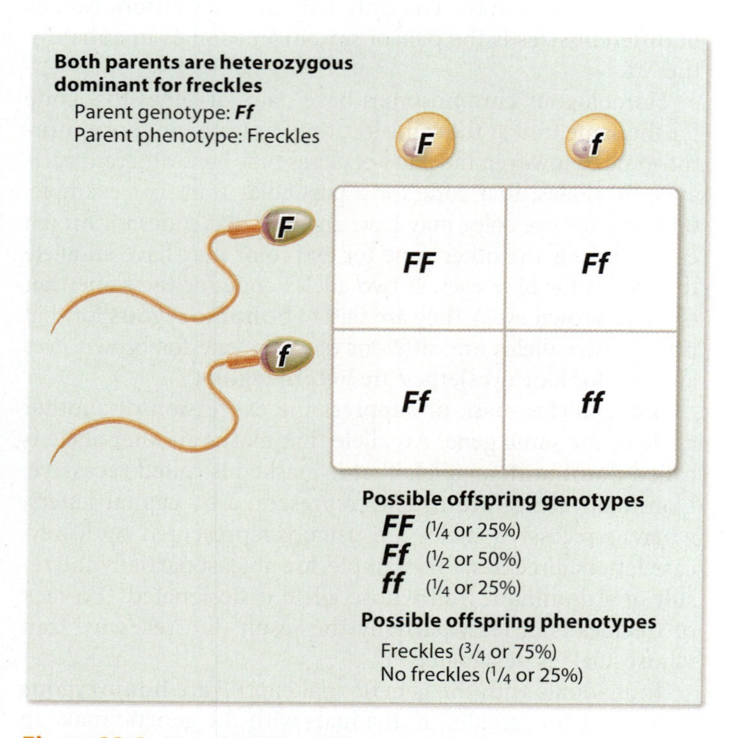

Figure 33-8 **Sample Punnett square for a monohybrid cross.**

The Punnett square in Figure 33-8 shows that when both parents are heterozygous for freckles (*Ff*), three offspring genotypes are possible: *FF*, *Ff*, and *ff*. The probabilities (percentages) for each genotype are *FF*: ¼ (25%); *Ff*: 2/4 or ½ (50%); and *ff*: ¼ (25%). The probabilities (percentages) of phenotypes resulting from this mating are freckles: ¾ (75%) and no freckles: ¼ (25%). Note that because each new fertilization event involves a different sperm and oocyte, the probabilities remain the same for each subsequent child. Put another way, the genotypes (and phenotypes) of the offspring of one mating have no bearing on the genotypes (and phenotypes) of any subsequent mating. Because the Punnett square in Figure 33-8 involves only a single trait, the mating shown is an example of a **monohybrid cross**.

ACTIVITY 3
Learning the Language of Genetics

Learning Outcomes

1. Practice using the common terms used in the field of genetics by writing a short phrase to define each term.
2. Set up monohybrid crosses using Punnett squares.

Materials Needed

None

Instructions

1. Write down a phrase that explains the meaning of each term. Two phrases for the first term have been provided as examples.

DNA *the genetic material; the blueprint for all of an individual's traits*

Gene ______________________

Allele ______________________

Chromosomes ______________________

Dominant ______________________

Recessive ______________________

Homologous chromosomes ______________________

Homozygous dominant ______________________

Heterozygous ______________________________

Homozygous recessive ______________________________

Genotype ______________________________

Phenotype ______________________________

Genome ______________________________

Punnett square ______________________________

Monohybrid cross ______________________________

Autosomes ______________________________

Sex chromosomes ______________________________

Karyotype ______________________________

2. Practice setting up monohybrid crosses in Punnett squares for the following matings. Additionally, list the genotypes and phenotypes of all possible offspring and include their percentages.
   a. Determine the possible sexes of the offspring in the mating of a father and a mother. Remember that males have X and Y chromosomes and females have two X chromosomes.

Parent genotypes: ______________________________

Offspring genotypes: ______________________________

Offspring phenotypes: ______________________________

QUICK TIP Accurately determining possible offspring genotypes and phenotypes requires that you set up the Punnett square correctly. Remember that each sperm and oocyte receives only one parental allele; thus a parent who is, for example, *Bb* for hair color produces both *B* gametes and *b* gametes. In determining what could happen when different sperm and oocytes fuse, you are in effect simulating multiple fertilizations.

   b. Determine the possible offspring of a father who is heterozygous for freckles and a mother who is homozygous recessive for freckles. Assume that F = freckles and f = no freckles.

Parent genotypes: ______________________________

Offspring genotypes: ______________________________

Offspring phenotypes: ______________________________

Patterns of Inheritance

While some human phenotypes can be linked to a single gene pair, most inherited traits are determined by multiple alleles or by the interaction of several genes. Other influences, such as environment, may also affect gene expression and ultimately phenotypic expression. We now focus on some of the most common patterns of inheritance and some examples of each.

Dominant-Recessive Inheritance

Dominant-recessive inheritance refers to the interaction of alleles that are strictly dominant or recessive. In this type of inheritance, the dominant allele is always expressed in the phenotype, whether the individual is homozygous or heterozygous for the associated trait. Recessive alleles are expressed only if two of them are present—that is, when no dominant allele is present to mask the recessive trait. While few human traits follow strict dominant-recessive inheritance, easily observed examples typically used to describe this type of inheritance include freckles, dimples, and free earlobes; the corresponding recessive traits are no freckles, no dimples, and attached earlobes.

CLINICAL CONNECTION

Phenylketonuria (PKU) is a rare recessive genetic disorder involving gene mutations that render the liver enzyme phenylalanine hydroxylase nonfunctional. Normally, this enzyme converts the amino acid phenylalanine to the amino acid tyrosine; in its absence, phenylalanine accumulates and is converted into phenylketone, which causes a variety of health problems, including intellectual disabilities and stunted growth. (This disease is called phenylketonuria because phenylketone is detected in the urine.) The only treatment for the disorder is to limit the amount of phenylalanine in the diet. Babies in the United States and many other countries are screened for phenylketonuria soon after birth because detecting it early can help prevent serious health problems.

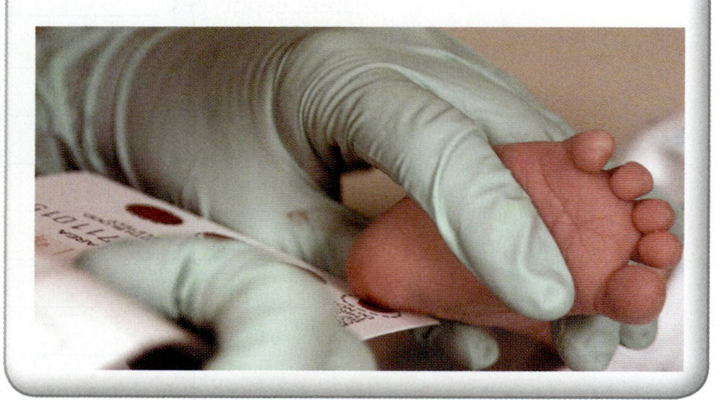

ACTIVITY 4

Exploring Dominant-Recessive Inheritance

Learning Outcomes

1. Determine the possible genotypes of classmates with specific phenotypic traits associated with a single gene pair.
2. Simulate monohybrid crosses with some commonly used examples of dominant-recessive traits.

Materials Needed

- ☐ PTC (phenylthiocarbamide) taste strips
- ☐ Sodium benzoate taste strips

Instructions

1. Assume that each of the examples in the following list (and/or in **Figure 33-9**), is controlled by dominant-recessive inheritance. Record your phenotype in the table on the next page. Then record your possible genotypes.
   - Free earlobes (*E*) are dominant; attached earlobes (*e*) are recessive.
   - Freckles (*F*) are dominant; no freckles (*f*) is recessive.
   - Dimples (*D*) in one or both cheeks are dominant; no dimples (*d*) is recessive.
   - A widow's peak (*W*)—a downward V-shaped hairline at the center of the forehead—is dominant; a straight hairline at the center of the forehead (*w*) is recessive.
   - Astigmatism (*A*)—an irregular shape of the cornea or lens that causes blurred vision—is dominant; normal vision (*a*) is recessive.
   - Ability to taste the bitterness of the harmless chemical PTC (phenylthiocarbamide) is dominant (*P*);

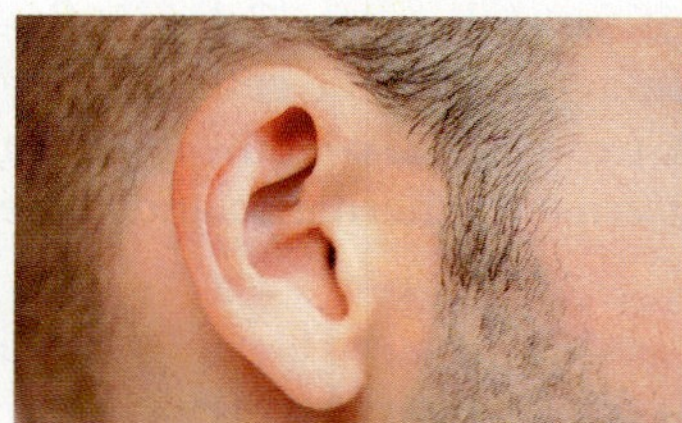

Free earlobes (*E*)

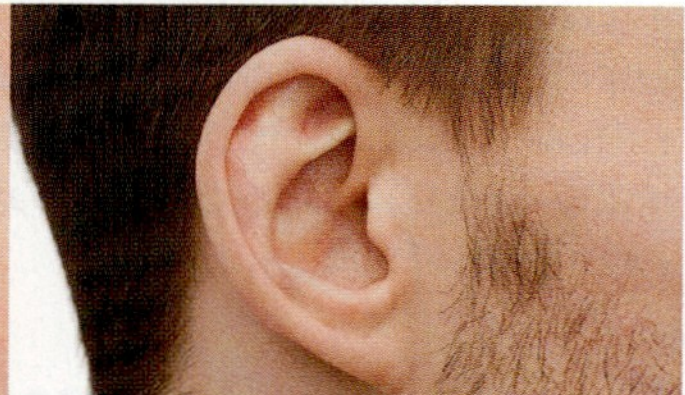

Attached earlobes (*e*)

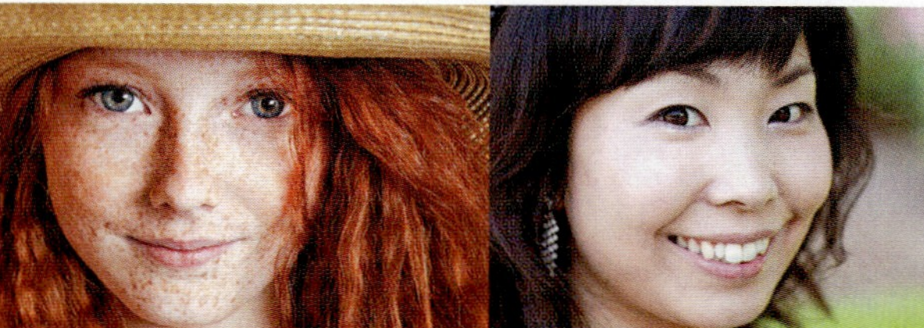

Freckles (*F*) No freckles (*f*)

Dimples (*D*)

No dimples (*d*)

Widow's peak (*W*) No widow's peak (*w*)

Figure 33-9 Some examples of human phenotypes.

Trait	Your Phenotype	Your Possible Genotypes	Percentage of Dominant Phenotypes in the Class	Percentage of Recessive Phenotypes in the Class
Free earlobes (*E*, *e*)				
Freckles (*F*, *f*)				
Dimples (*D*, *d*)				
Widow's peak (*W*, *w*)				
Astigmatism (*A*, *a*)				
Ability to taste PTC (*P*, *p*)				
Ability to taste sodium benzoate (*S*, *s*)				

inability to taste its bitterness (*p*) is recessive. Obtain and chew a PTC taste strip. If it tastes bitter (you will know!), then you have the dominant phenotype.

- Ability to taste sodium benzoate (*S*), whether salty, bitter, or sweet, is dominant; inability to taste this chemical (*s*) is recessive. Obtain and chew a sodium benzoate taste strip. If it tastes salty, bitter, or sweet, then you have the dominant phenotype.

2. Poll everyone in the class for their results, calculate the percentages of dominant and recessive phenotypes in the class, and record those data in the table below.

3. Now, simulate some dominant-recessive monohybrid crosses using three of the examples in the previous chart. After choosing the genotypes of both parents, list the possible genotypes and phenotypes of the offspring, including the percentages of the genotypes and phenotypes.

a.

Parent genotypes: ____________

____

____

Offspring genotypes: ____________

____

____

Offspring phenotypes: ____________

____

____

b.

Parent genotypes: ____________

____

____

Offspring genotypes: ____________

____

____

Offspring phenotypes: ____________

____

____

Incomplete Dominance

In **incomplete dominance**, heterozygous individuals exhibit a phenotype in between those of the homozygous individuals. A simple example involves flower color in carnations, in which the genotype *RR* is expressed as red flowers, and the genotype *rr* produces white flowers. The genotype *Rr* does not result in red flowers, as it would if dominant-recessive inheritance were operating, but instead produces pink flowers.

An example of incomplete dominance in humans is sickle cell trait, which occurs in individuals who are heterozygous for the sickling gene (*Ss*). Individuals with the genotype *SS* produce normal hemoglobin, whereas *ss* individuals produce abnormal hemoglobin that causes red blood cells to take on a sickle (or crescent) shape, resulting in a condition called sickle cell anemia. Individuals with sickle cell trait (*Ss*) have both normal and sickled RBCs (**Figure 33-10**). In most cases such individuals are healthy, but they can undergo sickle cell crises, particularly if they experience low blood oxygen levels.

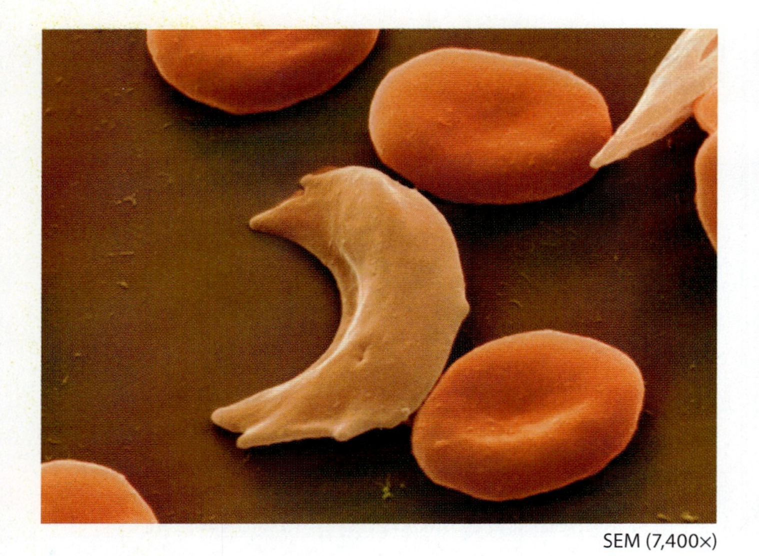
SEM (7,400×)

Figure 33-10 Red blood cells of an individual with sickle cell trait, an example of incomplete dominance.

CLINICAL CONNECTION

Because being homozygous recessive for the sickling gene can have serious consequences, we might expect that the recessive allele *s* would be selected against and would become rare. However, this allele persists, especially in African populations, because it protects against the malaria parasite, which infects and lives within red blood cells. The malaria parasite is less likely to infect sickled RBCs, and infected individuals tend to have milder symptoms than do individuals with normal RBCs because they carry fewer parasites. This advantageous situation is an example of **heterozygote superiority** or **heterozygote advantage**.

Codominance in Multiple-Allele Traits

As previously mentioned, relatively few human traits actually follow a strict dominant-recessive inheritance pattern. Many involve the interaction of **multiple alleles**. One example of this type of inheritance is the **ABO blood types** in humans. Recall from Unit 20 that antigens on red blood cells determine the ABO blood type of each individual. The presence or absence of these antigens is an inherited trait. In particular, ABO blood types exhibit a phenomenon known as **codominance**, in which two alleles are equally dominant and are therefore both expressed. If the inheritance of flower color described for incomplete dominance were instead governed by codominance, a given flower would have both red petals and white petals.

As noted in the following chart, the ABO alleles are designated *I* for the presence of either A or B antigen, and *i* for the absence of antigen. I^A and I^B are the codominant alleles; both are expressed, producing type AB blood. The homozygous recessive *ii*, which produces red blood cells that lack both A and B antigens, results in type O blood.

Genotype	Phenotype
I^AI^A	Type A blood
I^Ai	Type A blood
I^BI^B	Type B blood
I^Bi	Type B blood
I^AI^B	Type AB blood
ii	Type O blood

Sex-Linked Inheritance

Sex-linked traits are traits that are specifically expressed on the X or Y (sex) chromosomes. Even though X chromosomes are much larger and have many more genes than Y chromosomes (**Figure 33-11**), the Y chromosome has dominant alleles that specify that any individual with this chromosome is male. This dominance explains why *XX* individuals are female, and *XY* individuals are male. All oocytes contain an X chromosome, so female gametes do not determine the sex of an offspring. However, because sperm cells can contain either an X or a Y chromosome, male gametes determine the sex of an offspring.

The X chromosome includes genes that code for traits other than those that make an individual a female. If a deleterious trait on an X chromosome is inherited by a male, he will likely express that trait because his Y chromosome does not have a matching allele. If a female has only one X chromosome with a deleterious trait, she is considered a **carrier** for the trait.

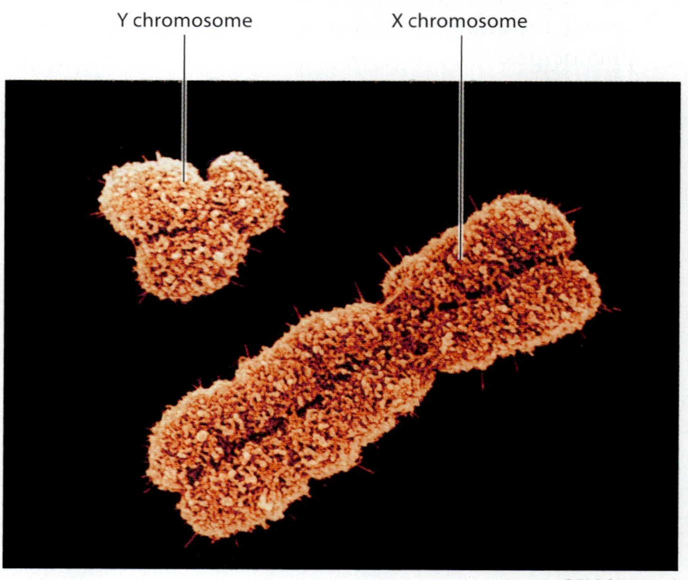

SEM (9,000×)

Figure 33-11 Human sex chromosomes.

Sex-linked disorders may be dominant or recessive, but sex-linked dominant disorders are rare. Examples of recessive sex-linked traits include some forms of baldness and certain forms of hemophilia. In addition, a common sex-linked trait is color blindness, the most common form of which is **red-green color blindness**. Individuals with this condition see red and green as the same color (either red or green). The gene for red-green color blindness is recessive (*c*); normal color vision is dominant (*C*). Possible genotypes and phenotypes of red-green color blindness are:

Genotype	Phenotype
X^CX^C	Normal female
X^CX^c	Normal female who is a carrier
X^cX^c	Red-green color-blind female
X^CY	Normal male
X^cY	Red-green color-blind male

THINK ABOUT IT Sex-linked (specifically X-linked) inheritance involves only traits passed from mothers to their sons and daughters and from fathers to their daughters. Why do sons not inherit X-linked traits from their fathers?

____

____

____

ACTIVITY 5

Exploring Other Patterns of Inheritance

Learning Outcomes

1. Determine the possible offspring of parents heterozygous for a trait that exhibits incomplete dominance.
2. Determine the possible offspring of parents with various blood types and use this information to determine paternity.
3. Determine the possible offspring of parents with a specific sex-linked trait.
4. Test individuals for red-green color blindness to determine the percentage of individuals in the class with this trait.

Materials Needed

None

Instructions

1. Set up and simulate monohybrid crosses for the following examples (use the genotypes previously listed in this unit) and answer the associated questions.

a. Both mother and father are heterozygous for the sickling gene.

Parent genotypes: ______________________________

Offspring genotypes: ______________________________

Offspring phenotypes: ______________________________

b. The mother is homozygous dominant for blood type A; the father is heterozygous for blood type B.

Parent genotypes: ______________________________

Offspring genotypes: ______________________________

Offspring phenotypes: ______________________________

c. The mother is homozygous recessive for blood type O; the father is heterozygous for blood type A.

Parent genotypes: ______________________________

Offspring genotypes: ______________________________

Offspring phenotypes: ______________________________

i. If a mother has type A blood and her child has type O blood, what possible blood types could the father have? ______________________________

____

THINK ABOUT IT You can use hypothetical monohybrid crosses to figure out the answer to the preceding question; set up crosses using the parents' possible genotypes in the following Punnett squares. (Hint: If a child has type O blood, what is his or her only possible genotype?)

____

ii. If a man has type AB blood, could he have a child with type O blood? Explain your answer. ________

____

____

d. The mother is a normal female who is a carrier for red-green color blindness; the father is a normal male.

Parent genotypes: ____________

____

____

Offspring genotypes: ____________

____

____

Offspring phenotypes: ____________

____

____

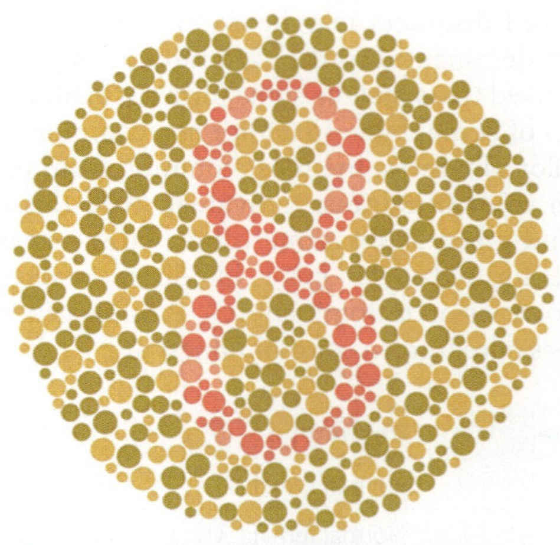

Figure 33-12 Ishihara chart for testing red-green color blindness.

i. Why is red-green color blindness more common in males? ____________________

____

____

2. Look at the **Ishihara chart** for red-green color blindness (**Figure 33-12**). Individuals with normal color vision will be able to read the number embedded in the image, but individuals with red-green color blindness will only see randomly arranged colored discs. If you see the number in the chart, you are not red-green color blind. Write your result and your sex on a slip of paper and then have someone compile the data for the entire class on a chalk board or a white board.

   a. How many individuals in the class exhibit red-green color blindness? ____________________

   b. How many of these individuals are male? ________

   c. How many of these individuals are female? ________

   d. What are the percentages of color blind individuals by sex? ________________ % males

   ________________ % females

   e. How many individuals in the class could pass the gene for red-green color blindness to their offspring?

   ____________ Why? ____________________

   ____

   ____

Post-lab quizzes are also assignable in MasteringA&P®

POST-LAB ASSIGNMENTS

Name: ______________________ Date: ______________ Lab Section: ________

PART I. Check Your Understanding

Activity 1: Exploring Fertilization and the Stages of Prenatal Development

1. Fill in the blanks with the appropriate term:
   a. Fusion of a sperm cell and a secondary oocyte to form a zygote is termed ________________.
   b. A series of rapid mitotic divisions that produce small, genetically identical cells called blastomeres is called ________________.
   c. The embryonic period begins with ________________, which is a rearrangement and migration of the cells of the bilaminar embryonic disc to form the trilaminar embryonic disc consisting of the ectoderm, mesoderm, and endoderm.
   d. During the ________________ period of prenatal development, growth is very rapid and is characterized by the maturation of tissues and organs.

2. List two well-developed organs or structures that are visible in a 4-week-old embryo.

   ____

3. List two developments that occur in month 7 of fetal development and two that occur in months 8 and 9 of fetal development. ____

   ____

   ____

Activity 2: Examining the Placenta

1. Which extraembryonic membrane ultimately becomes part of the urinary bladder? ____________

2. Which extraembryonic membrane is ultimately responsible for allowing symmetrical development and freedom of movement of the embryo? ________________________________

3. In addition to providing the embryo with oxygen and nutrients and removing wastes, the placenta also ____.

Activity 3: Learning the Language of Genetics

1. Fill in the blanks with the appropriate term:
   a. A segment of DNA that codes for a specific protein is a(an) ________________; the entire genetic makeup of an individual is the ________________.
   b. If two alleles code for the same trait, they are said to be ________________ for that trait, but if two alleles are different, they are said to be ________________.

c. An allele that has the ability to mask another is called ______________________; an allele that is masked is called ____________________.

d. The genetic makeup of an individual is the ____________________; the expression of an individual's genotype is the ____________________.

2. Why does a Punnett Square depicting a monohybrid cross only have one letter from each parent in the first blocks? ____

____

Activity 4: Exploring Dominant-Recessive Inheritance

1. List three human traits that are at least partially controlled by strict dominant-recessive inheritance.

____

For each of these traits, list the recessive trait.

____

2. Using a monohybrid cross, determine the possible phenotypes of offspring from a mother who is heterozygous for the dominant trait of polydactyly (extra fingers and/or toes) and a father who is homozygous dominant for the same trait. Calculate the percentage of each phenotype.

Activity 5: Exploring Other Patterns of Inheritance

1. Which of the following phenotypes is *not* possible in a child born to a mother who is heterozygous for the sickling gene and a father who is homozygous recessive for the sickling gene?
   a. sickle cell anemia
   b. sickle cell trait
   c. normal hemoglobin

2. Which of the following blood types is *not* possible in a child born to a mother who has type AB blood and a father who has type A blood?
   a. type A
   b. type B
   c. type AB
   d. type O

3. What is the percent chance that a male born to a mother who is a carrier for the red-green color blindness gene and a father who has normal color vision will exhibit red-green color blindness?
   a. 100%
   b. 75%
   c. 50%
   d. 25%
   e. 0%

Exploring the Muscular System of the Cat

DISSECTION OUTLINE

Materials Needed

- ☐ Safety glasses
- ☐ Gloves
- ☐ Lab coat or other protective clothing
- ☐ Preserved cat, either unskinned or skinned
- ☐ Mounted cat skeleton
- ☐ Dissecting tray
- ☐ Dissection tools
- ☐ Cheese cloth and moistening solution
- ☐ Two plastic storage bags
- ☐ Rubber bands
- ☐ Specimen label with string or wire twist tie
- ☐ Spray bottle with a solution of soap and water for cleanup

ACTIVITY 1

Skinning the Cat

Learning Outcomes

1. Learn proper techniques for dissection and for the handling and storage of preserved specimens.
2. Observe the structure of skin and the underlying fascia—epidermis, dermis, and hypodermis.
3. Remove skin and superficial fascia in preparation for dissection and identification of the skeletal musculature.

Instructions

1. Gloves, protective eyewear, and a lab coat or other protective clothing should always be worn when working with preserved specimens. Clear your lab bench so that only the supplies you need (lab textbook, dissecting tools) are out.
2. Place the cat ventral side down on the dissecting tray.
3. **Figure D1-1** shows the locations and order of incisions needed to skin the cat. Begin with a small incision at the base of the skull using either scissors or a scalpel. Once you have an opening, slide a blunt probe down the mid-dorsal line in the fascial plane between the skin and the musculature. Continue to cut with scissors down the mid-dorsal line from the neck to the tail, always keeping the probe just deep to the scissors between the skin and the muscle to protect the underlying muscles from accidental cuts.
4. Next, use scissors to cut laterally at the top of the neck around the head, down the upper limbs and around the wrist, obliquely down the lower limbs and around the ankles, and circularly around the genital/anal region (see Figure D1-1). Always keep the probe beneath the region you are cutting.
5. Use forceps to pick up one edge of the cut skin along the mid-dorsal incision and then use a blunt probe to separate the skin from the underlying musculature. As you remove the skin, remove as much of the superficial fascia as possible to simplify muscle dissection. Once you get this started, your hands will be your best tool: pull the skin with your hand, and scrape along the body wall with your fingers to peel as much fat and superficial fascia off with the skin as possible.
6. ⚠ Take note of the following cautions:
   a. Use care on the lateral thoracic wall and in the axillary area to distinguish between the large cutaneous maximus muscle (which can be removed with the skin) and the latissimus dorsi (which you wish to preserve).
   b. On the ventral surface of the neck where the platysma (a cutaneous muscle) covers the neck muscles and merges with sternocleidomastoid, use the external jugular veins and the transverse jugular vein as guidelines; all tissue superficial to these vessels can be removed.

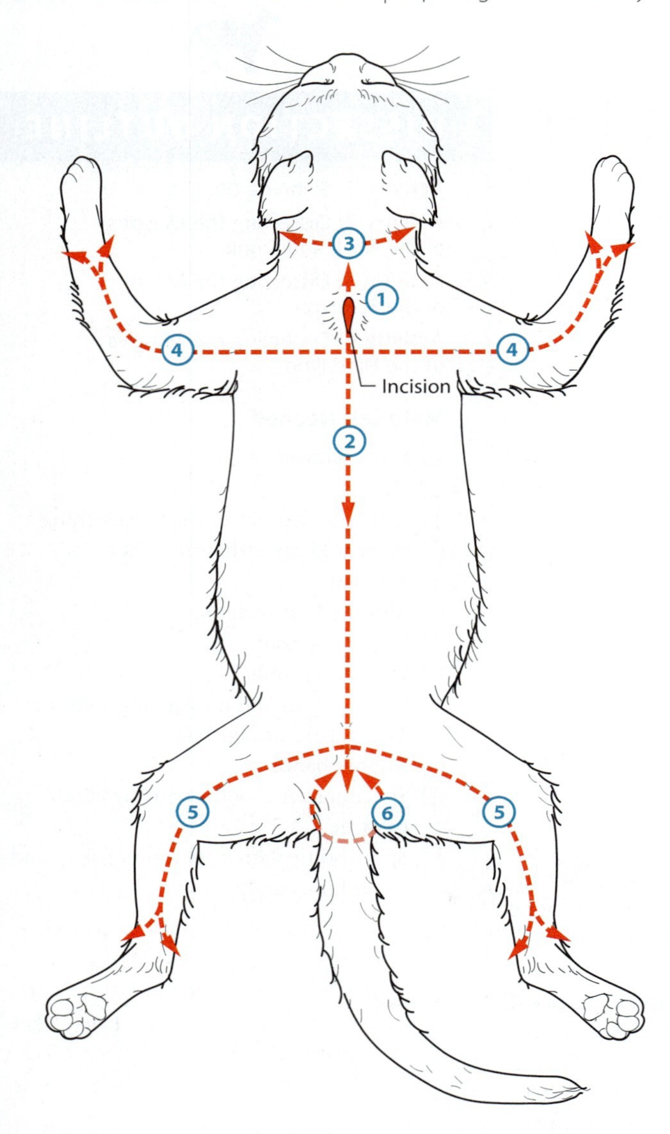

Figure D1-1 Incisions for skinning a cat. Dorsal view; numbers indicate sequence.

c. In each forearm, try to keep the superficially located cephalic vein (injected blue), the accompanying radial nerve, and the extremely thin, strap-like brachioradialis muscle with the cat, and not with the skin.

d. If dissecting a male cat, be careful not to cut or remove the spermatic cords, which extend from the body wall into the scrotum on the cat's ventral surface.

e. If dissecting a female cat, remove any mammary tissue that may be present with the skin.

7. If you can remove the skin in one piece (a challenge!!), you can use it to wrap around the specimen for storage. Observe the structure of the skin as you are removing it.

Which layer is the thickest? ______________________

Which layer is the toughest? ______________________

Describe the variations in skin thickness in different regions of the cat (back, neck, stomach, limbs).

____

____

8. You may wish to remove the tail just to get it out of the way. If so, cut the tail off close to the body using scissors or bone cutters.
9. Once the cat is fully skinned, you are ready to begin muscle dissection and identification.
10. To prepare your specimen for storage, first wrap it in cheesecloth dampened with moistening fluid. If the skin was removed in one piece, cover the specimen with it instead and place rubber bands around the trunk to hold the skin on. Double bag the specimen and close the bags with rubber bands. Attach a specimen label bearing your name to the outer bag. Store the wrapped cat as directed by your instructor.
11. Ensure that all discarded material is disposed of properly; never put such materials in regular trash or down a drain. Discard all cat parts (fascia and fur, if not used for storing the cat) in proper hazardous waste collection containers, and discard fluids into a liquid hazardous waste container.
12. Clean up your workspace. Wash your dissecting tray and dissecting tools and return them to their storage areas. Wash down your lab bench with a solution of soap and water. Discard your gloves and then thoroughly wash your hands.

ACTIVITY 2

Dissecting the Muscles of the Neck and Trunk

Learning Outcomes

1. Use proper dissection techniques to locate and isolate the muscles of the neck and trunk.
2. Identify the described muscles of the neck and trunk in the dissected cat.
3. Compare the muscles of the cat with the homologous muscles in the human. Recognize the similarities and differences in form, location, and function of homologous muscles in these two mammals.

Muscle Dissection Techniques

Your skinned cat will not look exactly like the photographs of dissected muscles shown later in this activity. In life, the individual muscles are joined by loose connective tissue, which holds or binds together all of our body structures. Muscle dissection is the process of finding muscle boundaries—white lines of connective tissue between adjacent muscles—and separating muscles from each other along these boundaries (**Figure D1-2**).

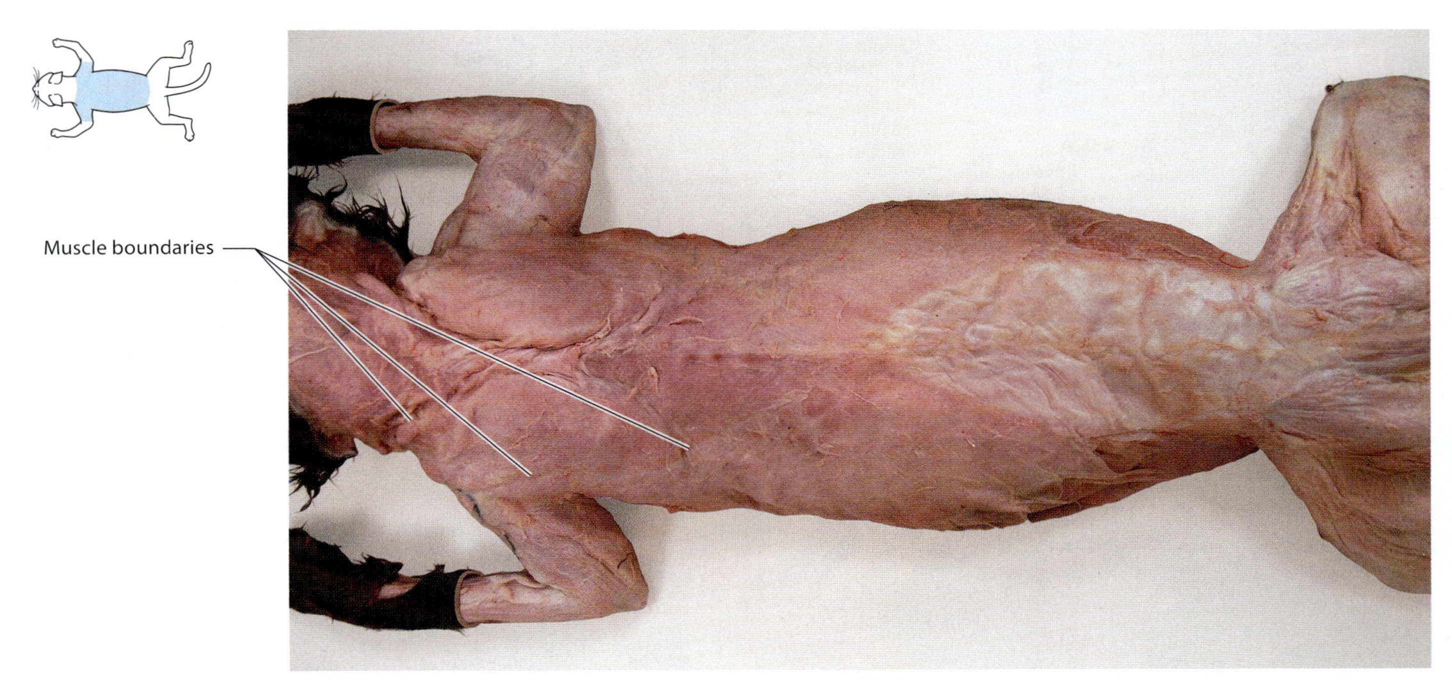

Figure D1-2 Identifying muscle boundaries. Freshly skinned, but undissected cat. White connective tissue lines indicate areas of muscle boundaries to be separated when dissecting and identifying muscles.

Care must be used to separate only the connective tissue between the muscles, and not the muscle tissue itself. You also do not want to separate a muscle from either of its skeletal attachments (origin or insertion), but instead to clear the free edges (that is, the edges not attached to the skeleton at origin and insertion). A few techniques that are used in dissection are described below.

- **Blunt dissection** refers to using the blunt probe and forceps to tease away the connective tissue between adjacent structures. Muscles separate easily at these natural boundaries. This is the technique you will most commonly use throughout this activity.
- **Reverse scissor technique** can be useful where connective tissues are particularly fibrous. To use this technique, insert the tip of closed scissors into the connective tissue boundary and then open the scissors to separate adjacent muscles. By not using the sharp (blade) edge of the scissors, you limit unintended cuts. Be careful to neither cut muscles nor "make new muscles" by creating boundaries where none exist.
- **Transection** is the cutting of superficial structures to view deeper structures. Unless otherwise specified in the instructions, use the following steps to transect a muscle:
    a. Separate the free edges of the muscle from the surrounding connective tissue.
    b. Slide a blunt probe beneath the midregion of the muscle so that the probe is perpendicular to the direction of the muscle fibers. Be certain that the only structure superficial to the probe is the muscle you are transecting.
    c. Cut the muscle with scissors along the line of the probe. A clean cut will allow you to reconnect the cut edges for easy review of superficial structures.
- To **clean** refers to the use of a blunt probe and forceps to remove fat and loose connective tissue surrounding a structure of interest to clearly expose that structure. It is important to use care while cleaning so that you remove only the fat and loose fascia, not the structures you are attempting to expose. This can be tedious, but when done well, results in beautiful, clean exposures of anatomical structures.

Instructions

A. Dissecting the Superficial Muscles of the Trunk—Dorsal Side

1. Put on safety glasses, gloves, and a lab coat, and then place the cat dorsal side up on the dissecting tray. Two large muscles form the superficial muscles on the dorsal surface of the cat: the *trapezius group* and the *latissimus dorsi*. In humans, the trapezius is a trapezoid-shaped muscle covering the dorsum of the neck and the back, medial to the scapulae. In the cat, clear muscle boundaries are visible between the three portions of the trapezius, and thus they are identified as three separate muscles.
2. Remove any remaining superficial fascia from the dorsal surface of the cat that was not removed with the skin and closely examine the dorsal surface to view the muscle boundaries between the three **trapezius** muscles: the clavotrapezius, acromiotrapezius, and spinotrapezius. Identify these muscles and use blunt dissection to separate them from each other (**Figure D1-3**).

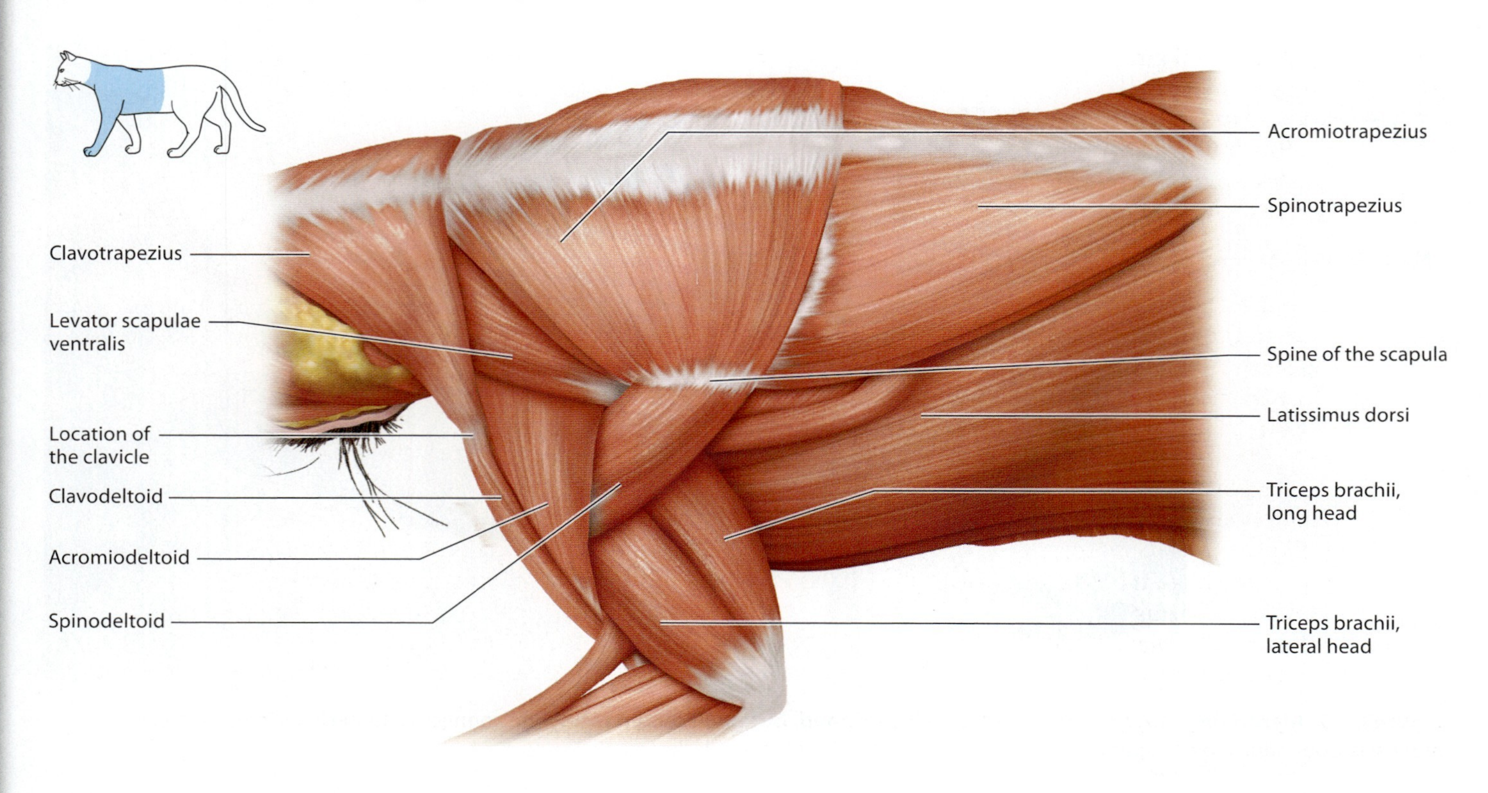

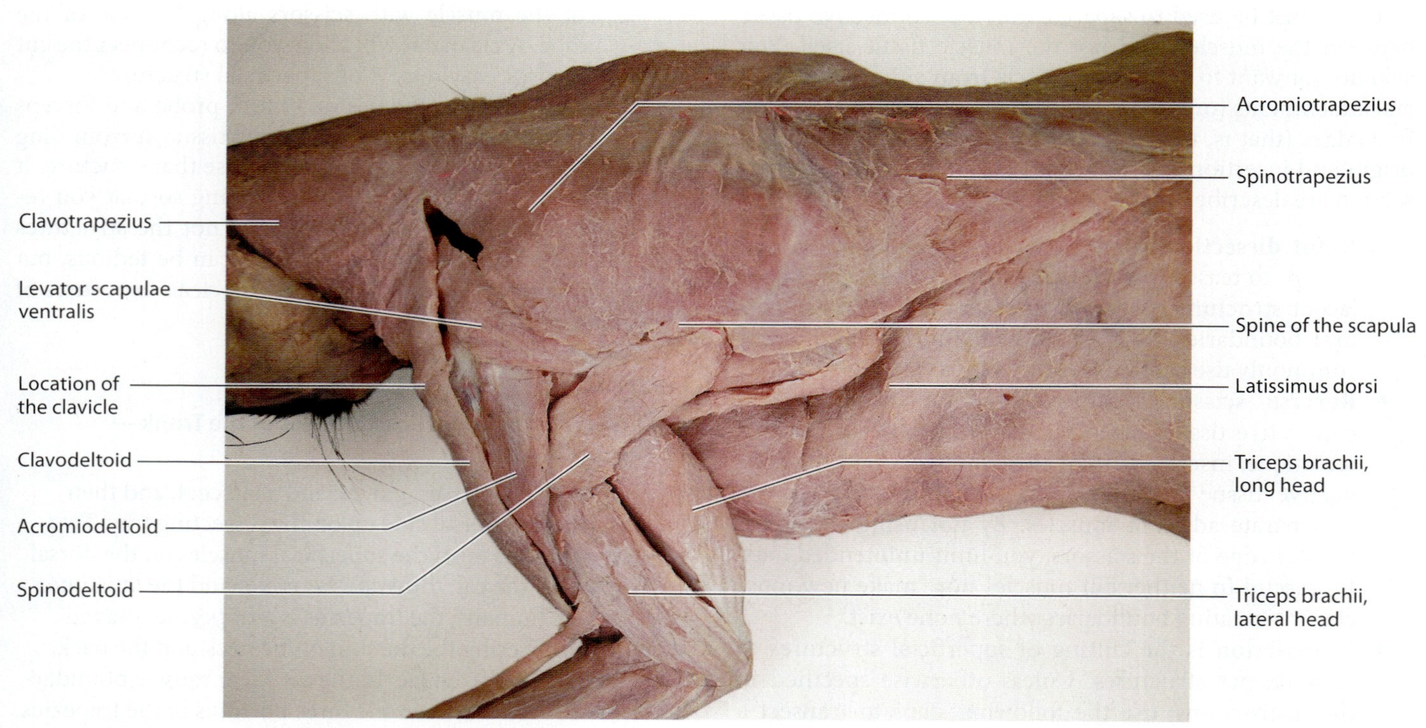

Figure D1-3 Superficial muscles of the trunk, dorsolateral view.

a. The **clavotrapezius**, the most cranial of the trapezius group, is the most superficial muscle covering the dorsum of the neck. It originates from the nuchal line and the connective tissue over the axis, and it extends laterally to insert on the clavicle. In cats, the clavicle is a tiny bone. Review its location and structure on the cat skeleton. Look carefully to see the slight constriction of the clavotrapezius at the base of the neck on the ventral side, indicating the location of the clavicle (see Figure D1-3). The clavotrapezius pulls the pectoral girdle cranially and also extends and rotates the neck.

b. Just caudal to the clavotrapezius is the **acromiotrapezius**, a square-shaped muscle that originates from the spinous processes of cervical vertebrae C_2–C_7 and thoracic vertebrae T_1–T_4 and inserts on the spine of the scapula (see Figure D1-3) and on the metacromion process, a small process extending from the spine of the scapula (refer to the cat skeleton). The acromiotrapezius adducts and stabilizes the scapula.

c. The caudal-most of the trapezius muscles, the **spinotrapezius**, is a triangular muscle that originates from the spinous processes of thoracic vertebrae T_4–T_{12} and inserts onto the spine of the scapula. It stabilizes and adducts the scapula and pulls it caudally.

d. Compare the three trapezius muscles of the cat with the human trapezius.

Which portion of the human trapezius is homologous to the clavotrapezius of the cat? ____________________

____

What are its actions? ____________________

Which portion of the trapezius in humans depresses the scapulae? ____________________

3. Using a blunt probe, separate the lateral edge of the spinotrapezius from the **latissimus dorsi** located caudal and deep to the spinotrapezius on the dorsal surface (see Figure D1-3). Use reverse scissor technique carefully to separate the lateral border of latissimus dorsi from the ventral musculature. The latissimus dorsi originates from the spinous processes of the thoracic and lumbar vertebrae and from the lumbodorsal fascia covering the lumbar vertebrae. It crosses the shoulder medial to the humerus and inserts on the proximal humerus. The latissimus dorsi is a prime mover in extension and adduction of the arm in both cats and humans.
4. Identify the **levator scapulae ventralis**, a thin strap-like muscle located in the triangular space between clavotrapezius and acromiotrapezius (see Figure D1-3). To isolate and view this muscle, use forceps and a blunt probe to carefully clean and remove the fat and fascia in the triangular area formed by the caudal border of the clavotrapezius and the cranial border of the acromiotrapezius. The levator scapulae ventralis runs obliquely, deep to the clavotrapezius, from the base of the skull and the transverse process of the atlas, and it inserts on the spine of the scapula. There is no homologue of this muscle in humans. The levator scapulae ventralis in the cat helps in moving the scapula cranially.
5. As with the trapezius muscles, there are three separate **deltoid** muscles in the cat and only a single deltoid muscle in humans. The muscles of the deltoid group originate on the pectoral girdle (clavicle or scapula) and insert on the upper limb. Look for the muscle boundaries between the three deltoid muscles. Dissect these boundaries with a blunt probe to isolate each of the three deltoid muscles (see Figure D1-3).

a. The **clavodeltoid**, also called the *clavobrachialis*, originates from the clavicle at the point where the clavotrapezius inserts. This muscle covers the cranial surface of the humerus and inserts on the proximal ulna. It flexes the arm and forearm.

b. The **acromiodeltoid** is a small triangular muscle just caudal to the clavodeltoid. Originating from the acromion of the scapula and inserting on the proximal humerus, the acromiodeltoid abducts the arm. This muscle is easily identifiable in the cat because the cephalic vein crosses it superficially.

c. Just caudal to the acromiodeltoid is the **spinodeltoid**, which originates from the spine of the scapula, runs obliquely, and inserts on the proximal humerus. This muscle extends the arm. Using blunt dissection technique, separate the inferior border of the spinodeltoid from the triceps muscles, which cover the lateral and caudal surface of the arm.

d. Compare the deltoid group of the cat with the human deltoid muscle.

Which portion of the human deltoid muscle is homologous to the clavodeltoid? ____________________

What is the action of this portion of the deltoid in humans? ____________________

Which portion is homologous to the acromiodeltoid?

____

What is the action of this portion of the deltoid in humans? ____________________

Which portion is homologous to the spinodeltoid?

____

What is the action of this portion of the deltoid in humans? ____________________

B. Dissecting the Deep Muscles of the Trunk—Dorsal Side

1. Slide a blunt probe beneath all three trapezius muscles near the mid-dorsal line. Transect the trapezius muscles along the sagittal plane just lateral to their origin from the spinous processes of the cervical and thoracic vertebrae. Be careful as you transect the acromiotrapezius, because this muscle is just a thin connective tissue aponeurosis in this region and you do not want to cut the deeper muscles.
2. Transect the latissimus dorsi midway through the muscle, cutting perpendicular to the muscle fibers.

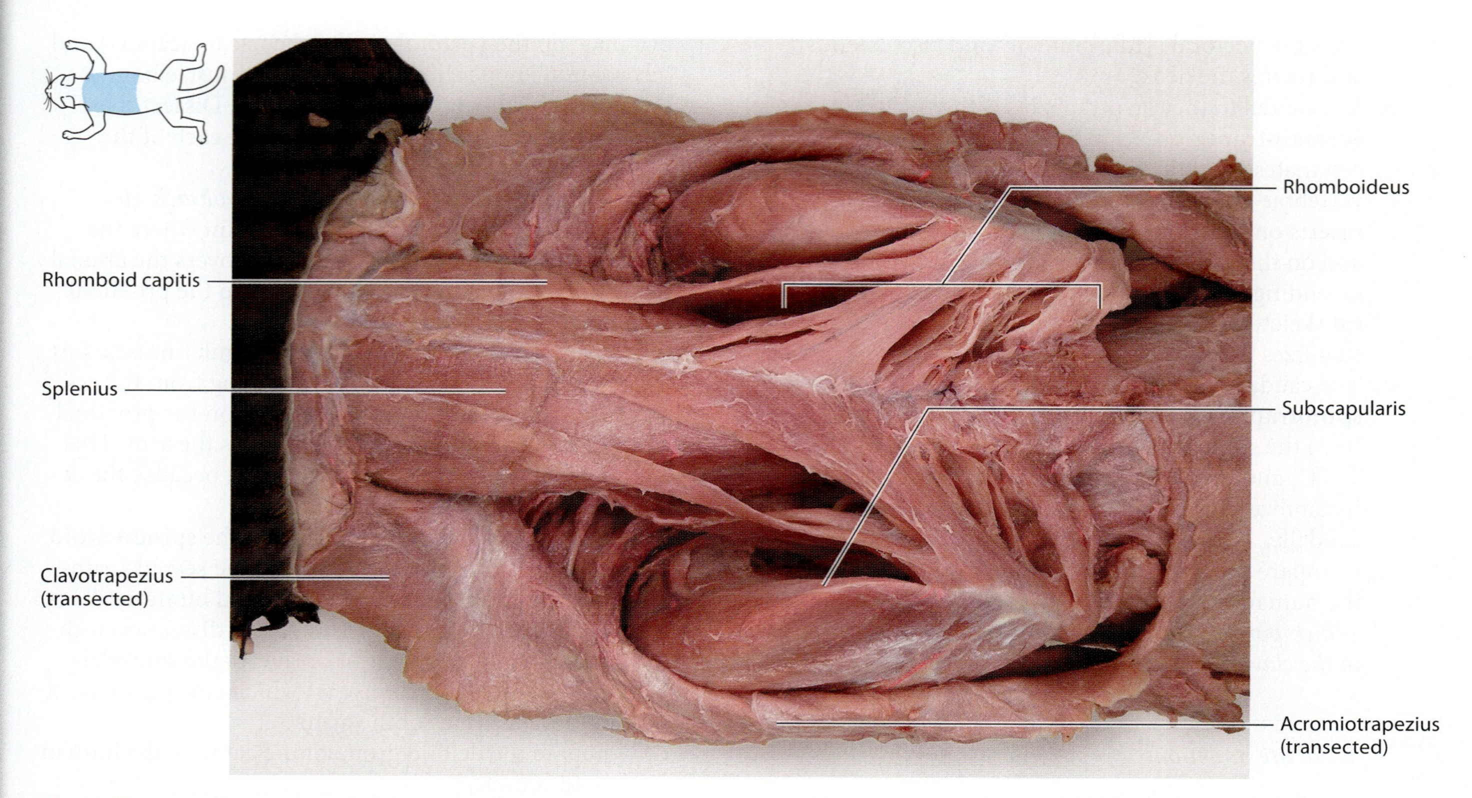

Figure D1-4 Deep muscles of the trunk, dorsal view.

3. Reflect the muscles to expose the deep muscles of the trunk and scapula. To view the deep muscles, carefully adduct the upper limbs to pull the scapulae away from the body wall. Identify the following deep muscles of the trunk and scapula (**Figure D1-4**):
   a. The **rhomboideus** lies deep to the acromiotrapezius. The rhomboideus appears as many individual slips of muscle that originate from the spinous processes of the cervical and thoracic vertebrae and insert onto the medial border of the scapula. This muscle is homologous to the rhomboideus major and rhomboideus minor in humans.
   b. The **rhomboid capitis**, a thin strip of muscle deep to the clavotrapezius, originates from the nuchal line and inserts with the other rhomboids on the vertebral border of the scapula. There is no homologue of this muscle in humans.
   c. The **splenius** is a large muscle covering the dorsal and lateral surfaces of the neck, deep to the clavotrapezius and rhomboideus capitis. The splenius originates from mid-dorsal cervical fascia and inserts onto the nuchal line. This muscle acts bilaterally in head extension and unilaterally in lateral rotation of the head (as in shaking the head "no"). It is homologous to the splenius capitis and splenius cervicis in humans.
4. Palpate the scapula to feel the spine of the scapula on its lateral surface. View the cat skeleton to see the location of the spine of the scapula. The four **rotator cuff muscles**—the supraspinatus, infraspinatus, teres minor, and subscapularis—originate from the scapula and insert on the proximal humerus surrounding the humeral head (**Figure D1-5**; see also Figure D1-4). In both cats and humans these muscles stabilize the shoulder joint. Identify the rotator cuff muscles originating from the scapular fossae:
   a. Located cranial to the spine of the scapula, the **supraspinatus** originates from the supraspinous fossa and inserts on the greater tubercle of the humerus.
   b. Located caudal to the spine of the scapula, the **infraspinatus** originates from and fills the infraspinous fossa. It is partially covered by the spinodeltoid. As with the supraspinatus, it inserts on the greater tubercle of the humerus.
   c. Deep to the spinodeltoid and arising from the caudal border of the scapula is the extremely small **teres minor**. This muscle inserts into the greater tubercle of the humerus. You will not expose this muscle in the cat.
   d. The **subscapularis** (see Figure D1-4) is located on the deep surface of the scapula, originating in the subscapular fossa. It inserts into the lesser tubercle of the humerus.
   e. In humans, in addition to stabilizing the shoulder joint, the four rotator cuff muscles also produce movement of the shoulder joint. List the action of

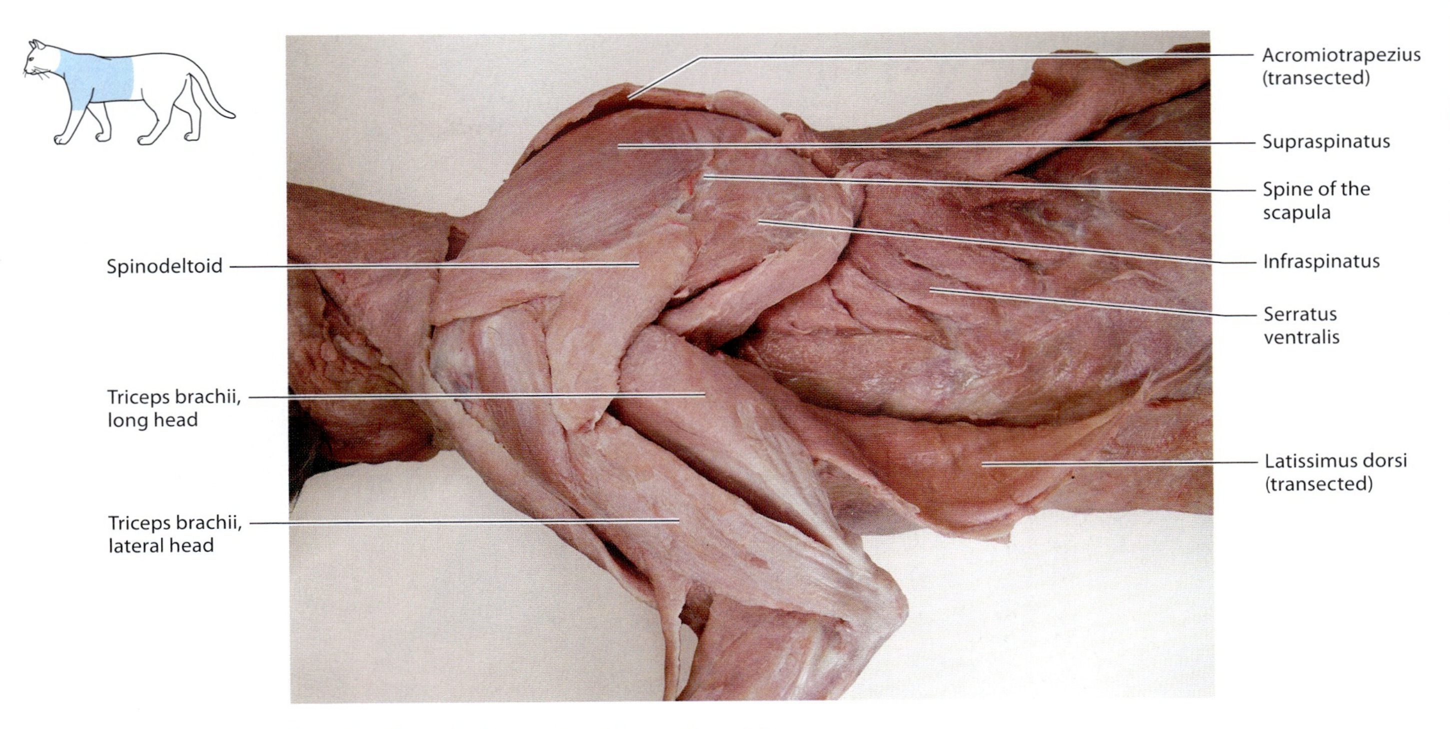

Figure D1-5 Rotator cuff muscles, lateral view of shoulder and brachium.

each of the rotator cuff muscles in humans in the following blanks:

Supraspinatus ______________________________

Infraspinatus ______________________________

Teres minor ______________________________

Subscapularis ______________________________

C. Dissecting the Muscles of the Head, Neck, and Trunk—Ventral View

1. Flip the cat onto its back to view its ventral surface. If you did not remove the cutaneous muscle of the neck (platysma) during skinning, do so now. Any tissue superficial to the external jugular and transverse jugular veins can be removed. Clean away the loose connective tissues from the ventral neck and the inferior surface of the mandible to expose, isolate, and identify the muscles of the head and neck (**Figure D1-6**) discussed next.
   a. The **sternomastoid** forms a V that frames the neck. It originates from the manubrium of the sternum, passes deep to the external jugular veins, and inserts laterally on the mastoid process of the temporal bone.
   b. Carefully clean the lateral edge of the sternomastoid and the cranial edge of the trapezius to expose the deep-lying **cleidomastoid**. Originating from the clavicle and inserting on the mastoid, this small strap-like muscle acts with the sternomastoid to rotate the head laterally. In humans the sternomastoid and the cleidomastoid muscles are fused to form the sternocleidomastoid.
   c. The **sternohyoid** and **sternothyroid** muscles both extend longitudinally on the ventral surface of the trachea from the manubrium of the sternum to the larynx. Be extremely careful when separating these muscles, because they are quite thin and tear very easily. The sternohyoid is located medially and inserts on the hyoid bone; the sternothyroid is lateral and slightly deep to the sternohyoid and inserts on the thyroid cartilage.
   d. The **digastric muscle** originates from the mastoid process of the temporal bone and inserts on the medial surface of the body of the mandible. It runs along the inferior edge of the body of the mandible and is the prime mover in depressing the mandible (opening the jaw).
   e. Separate the muscles along the medial boundary of the digastric muscle to see the **mylohyoid** more clearly. The mylohyoid is a thin sheet of muscle that extends transversely from the inferior border of the mandible and joins at the midventral line, with the mylohyoid muscle from the opposite side. It elevates the floor of the mouth.
   f. The **masseter** muscle is a large muscle located on the lateral cheek. It originates from the zygomatic arch and inserts on the angle of the mandible; it is a prime mover in elevating (closing) the jaw.

   Which of the muscles just dissected move the larynx during swallowing?

   ____

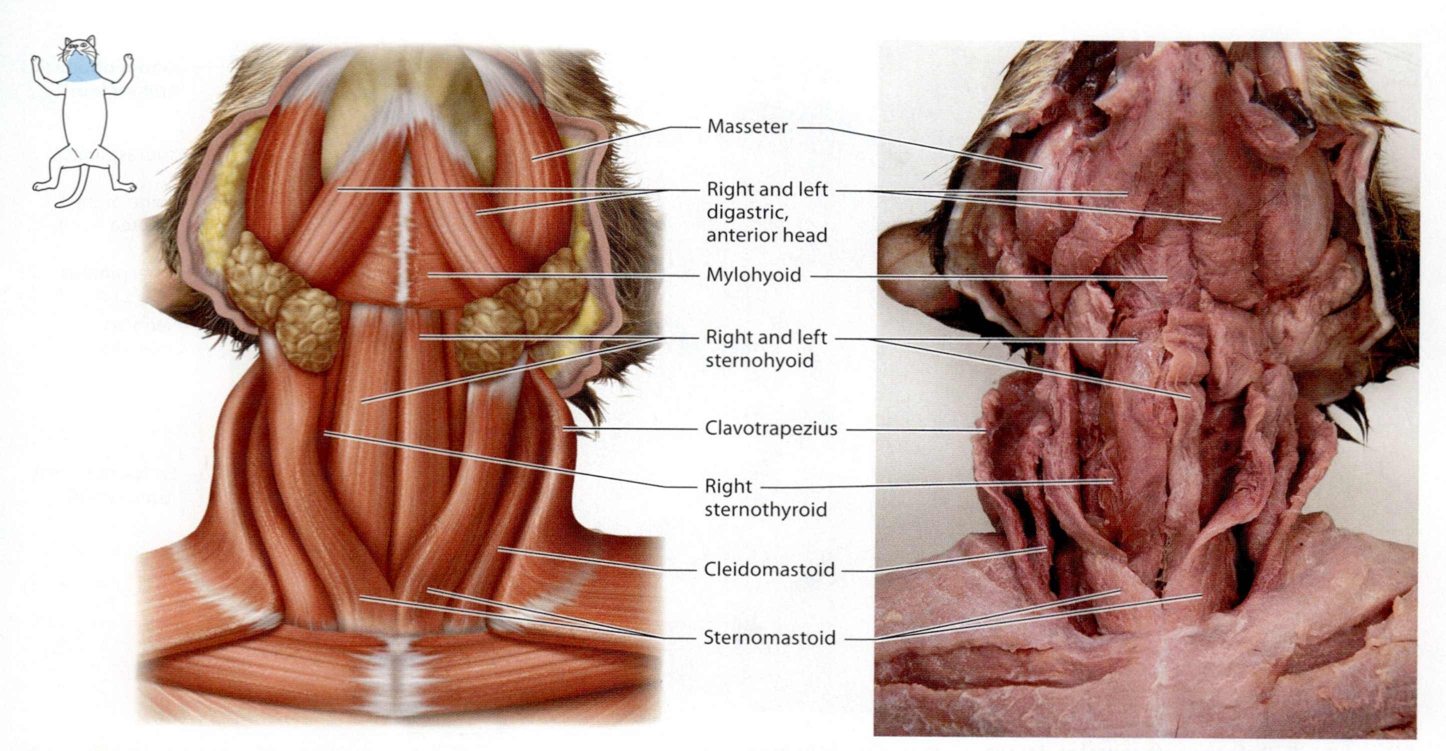

Figure D1-6 Muscles of the head and neck, ventral view.

2. Now that you have some experience with dissection, examine the muscles of the **ventral thorax**. Look carefully at the muscles to view the boundaries between the four pectoralis muscles in cats (**Figure D1-7**). (Humans have only two pectoral muscles.) In the cat, these four muscles arise from the sternum and insert on the upper limb. In many places these muscles are closely bound together. Use the reverse scissor technique to carefully separate these four muscles at their natural boundaries. The four pectoralis muscles in the cat, listed from superficial to deep, are as follows:
   a. The **pectoantebrachialis** is a thin superficial muscle that runs transversely from the manubrium of the sternum to the forearm. It has no homologue in humans.
   b. Just deep to the pectoantebrachialis is the triangular **pectoralis major**. It originates from the manubrium and body of the sternum and inserts onto the proximal humerus. In the cat, the pectoralis major is smaller than the deeper pectoralis minor.
   c. The **pectoralis minor** is the largest of the pectoralis muscles in the cat. Arising from the body of the sternum, its fibers run obliquely, caudal, and deep to the pectoralis major, and insert onto the proximal humerus.
   d. The **xiphihumeralis** is a strip of muscle originating from the xiphoid process of the sternum and inserting into the proximal humerus deep to the pectoralis minor muscle. It has no homologue in humans.

   Describe the differences between the pectoralis major and pectoralis minor in the cat as compared to the human. Consider differences in the form of each muscle as well as their skeletal attachments.

   ___

   ___

   ___

3. Push the pectoralis muscles medially to expose the deeper muscles of the thoracic wall. Use a blunt probe to clear off any fat and loose connective tissues to fully expose the deep muscles of the thorax (see Figure D1-7).
   a. The large fan-shaped **serratus ventralis** muscle lies deep to the pectoralis muscles on the lateral thoracic wall. The serratus ventralis originates from the lateral surface of the first nine ribs, producing a serrated appearance (hence its name), and it inserts on the vertebral border of the scapula. The serratus ventralis (the serratus anterior in humans) helps to stabilize and laterally rotate the scapula.
   b. Superficial to the serratus ventralis are three strips of muscle—the scalenes, the longest and most obvious of which is the **middle scalene**. It originates from the ventral surface of the ribs, extends cranially, and inserts on the transverse process of the cervical vertebrae. It flexes the neck. In humans, the scalenes are accessory respiratory muscles that help to elevate the ribs in individuals in respiratory distress.
4. With a blunt probe and forceps, clear away any fat and fascia covering the muscles of the abdominal wall (**Figure D1-8**). Once cleared, the distinct muscle fibers of

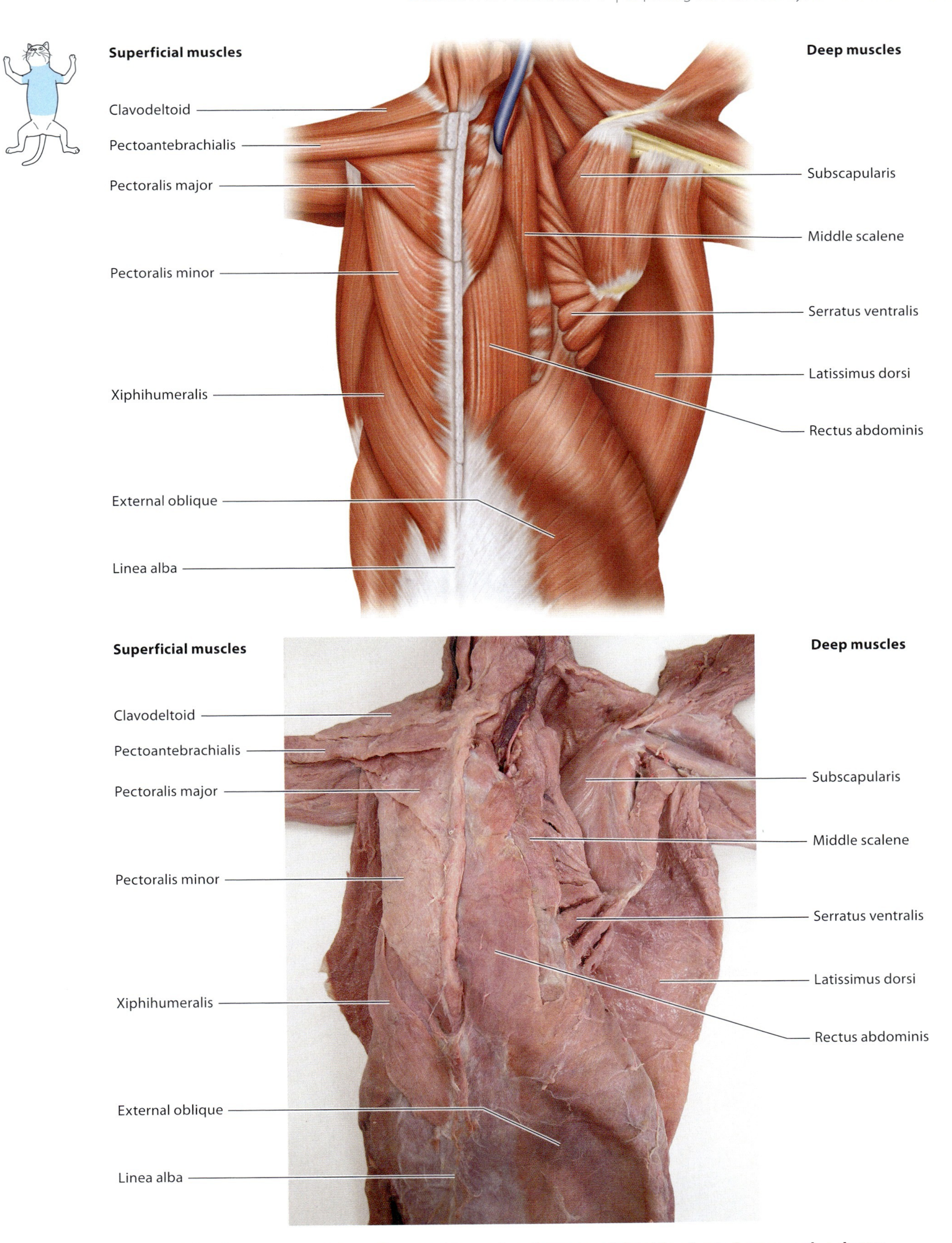

Figure D1-7 Muscles of the thorax, ventral view. Superficial muscles shown on right side of cat, deep muscles shown on left side of cat.

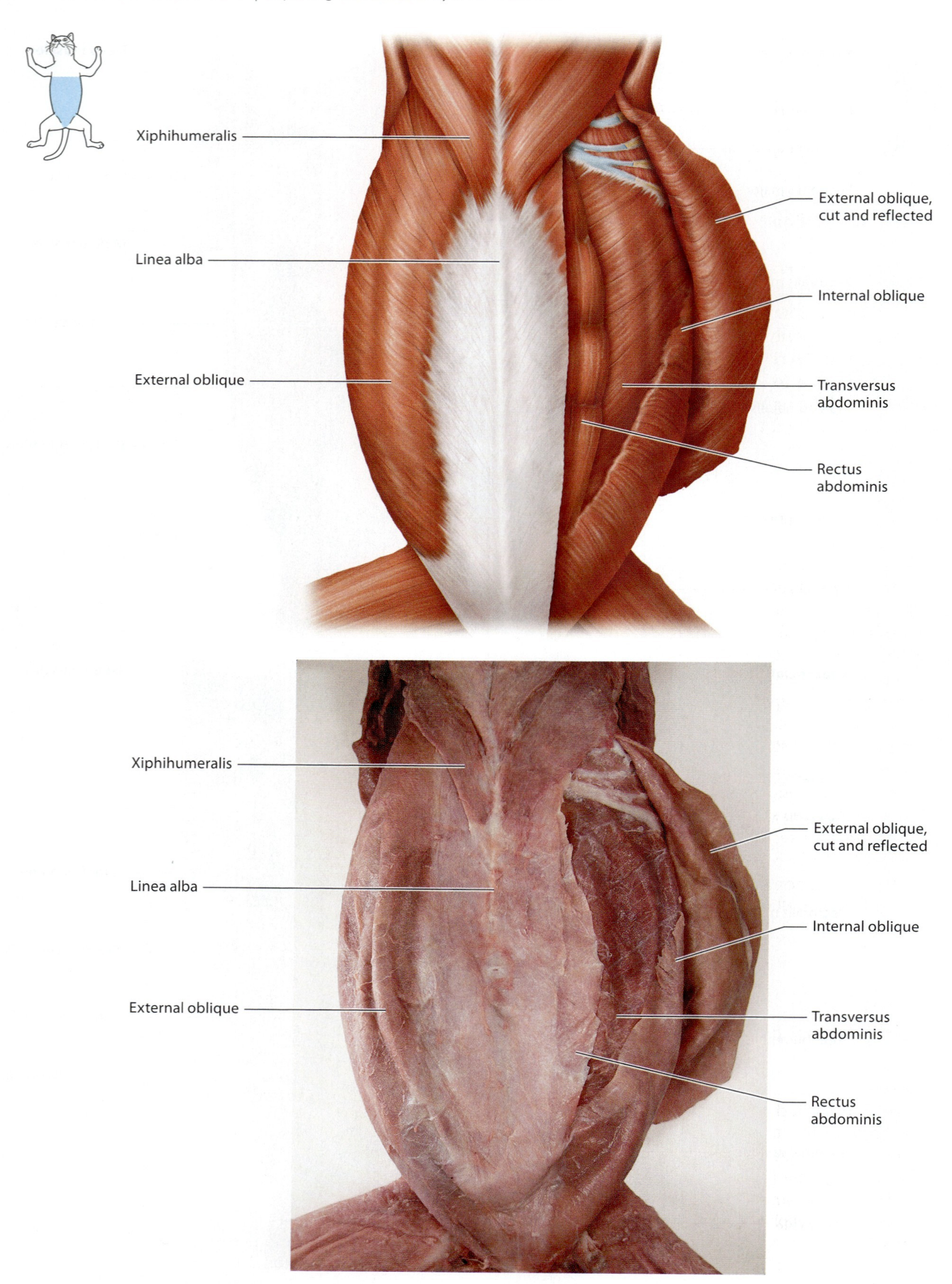

Figure D1-8 Muscles of the abdominal wall, ventral view.

the abdominal musculature become visible. Of the four abdominal muscles, three originate dorsally and wrap ventrally to insert in the midventral connective tissue raphe, the **linea alba**; the fourth lies lateral to the linea alba extending longitudinally from the pubis to the sternum. The abdominal muscles flex and laterally rotate the trunk. To view the deeper abdominal muscles, you will cut small windows in the superficial muscles.

a. The **external oblique** is the outermost layer of the abdominal muscles. It originates from the lower ribs and the fascia covering the lower back and inserts via a broad aponeurosis into the linea alba. The fibers of the external oblique run in a caudomedial direction.
b. Use scissors carefully to cut a small opening into the extremely thin lateral portion of the external oblique. Use a blunt probe to separate the external oblique from the underlying **internal oblique**. Continue the cut longitudinally; then make two transverse cuts to create a flap so that you can peel back the external oblique to view the internal oblique. It arises from the fascia on the lower back and from the iliac crest and inserts into the linea alba via a broad aponeurosis. The muscle fibers of the internal oblique run at a right angle to those of the external oblique—that is, in a craniomedial direction—and end laterally, so much of the muscle is tendinous on the ventral surface.
c. Use scissors carefully to cut longitudinally through the tendon of the internal oblique at the point where the muscle fibers end, exposing the **transversus abdominis**, the deepest layer of the abdominal muscles.
d. Finally, view the longitudinal **rectus abdominis** running from the sternum to the pelvis, deep to the insertion of the external oblique and superficial to the transversus abdominis. Tendinous intersections visible on the deep surface of the muscle divide this muscle into four sections. In humans this forms the "six pack" of body builders.

5. Follow the instructions for storage, waste disposal, and cleanup as described in Dissection 1, Activity 1, steps 10–12.

ACTIVITY 3
Dissecting the Muscles of the Forelimb

Learning Outcomes

1. Use proper dissection techniques to locate and isolate the forelimb muscles of the cat.
2. Identify the described forelimb muscles in the dissected cat.
3. Group the muscles of the forelimb into fascial compartments: anterior arm, posterior arm, anterior forearm, and posterior forearm. Use these groupings to aid in learning muscle action and innervation.
4. Compare the muscles of the cat with the homologous muscles in the human. Recognize the similarities and differences in form, location, and function of homologous muscles in these two mammals.

Instructions

A. Dissecting the Muscles of the Posterior Arm and Forearm

1. Put on safety glasses, gloves, and a lab coat, and then place the cat on its right side on the dissecting tray. You will be dissecting the left forelimb. Clean the caudal edge of the spinodeltoid to separate it from the triceps brachii. The muscles of the **posterior arm**, the **triceps brachii**, cover the caudal and lateral surface of the humerus. All heads of the triceps brachii insert on the olecranon of the ulna and extend the forearm (**Figure D1-9**).
   a. Identify the three heads of the triceps brachii: **triceps brachii, long head**, covers the caudal surface of the humerus; **triceps brachii, lateral head**, covers the lateral side of the humerus and is located cranial to the long head; **triceps brachii, medial head** (not shown in Figure D1-9), is quite small and lies deep to the lateral head. Carefully separate the connective tissue between the heads of the triceps brachii. To view the medial head, transect and reflect the lateral head.
   b. Considered a fourth head of the triceps brachii, the **anconeus** (not shown in Figure D1-9) is a small triangular muscle distal to the medial head of the triceps brachii. It covers the distal humerus and inserts on the olecranon of the ulna. Identify this muscle.
2. Identify the thin **brachioradialis** located superficially, passing with the cephalic vein and radial nerve down the forearm. Although this muscle crosses the elbow anteriorly and functions in flexion, it develops from the posterior forearm muscles.
3. The remaining muscles of the **posterior forearm** are the digital and carpal extensor muscles. These muscles all originate on the lateral epicondyle of the humerus and insert in the forefoot. To expose these muscles, cut through the deep fascia covering the forearm. Separate the muscles with a blunt probe at their natural boundaries. Identify the extensor muscles from the radial to the ulnar side of the forearm in Figure D1-9: **extensor carpi radialis longus**, **extensor carpi radialis brevis**, **extensor digitorum communis**, **extensor digitorum lateralis**, and **extensor carpi ulnaris**.

What are the insertion and action of the triceps brachii?

Insertion ______________ *Action* ______________

What is the origin of the extensor muscles in the posterior forearm? ______________

____

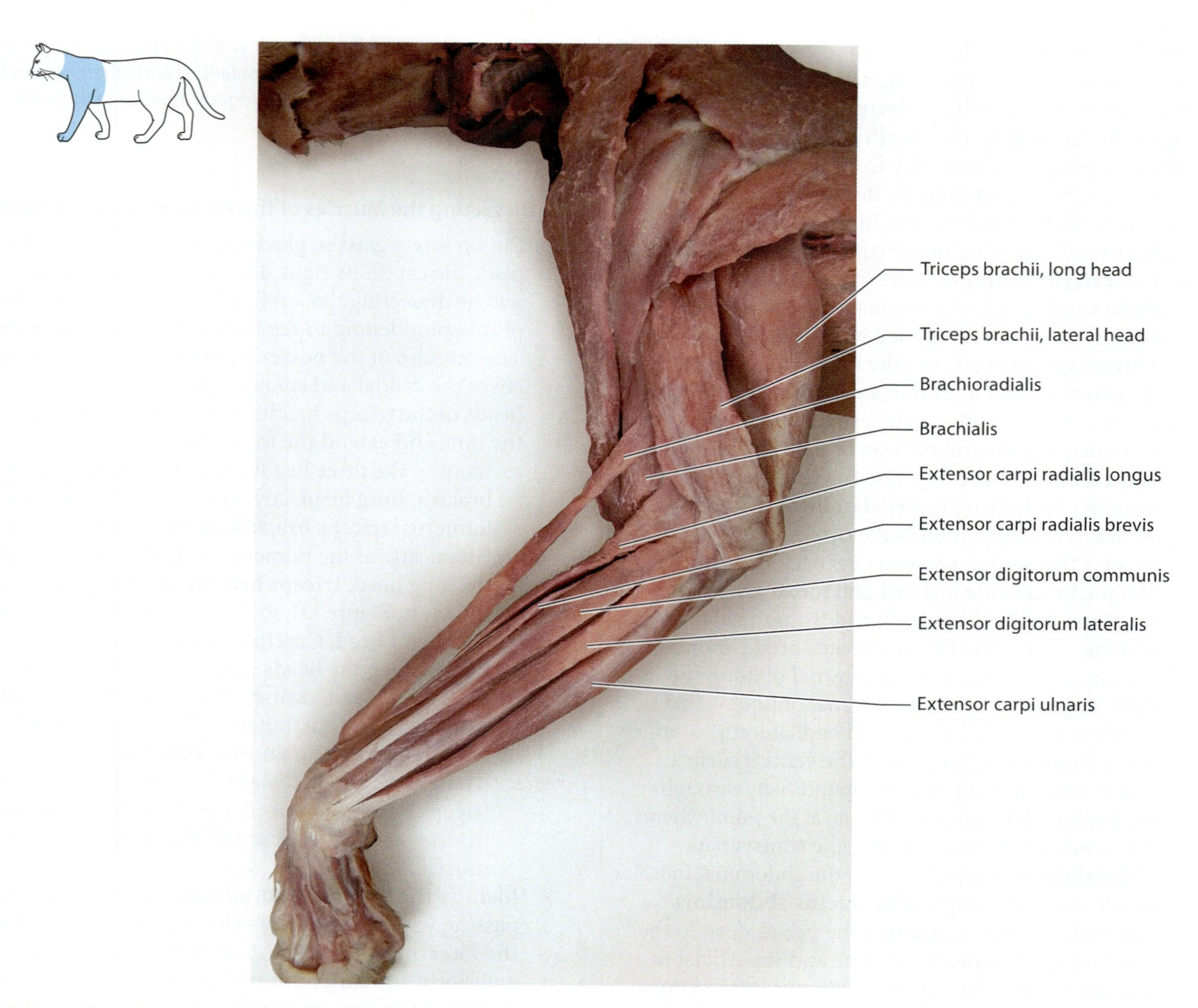

Figure D1-9 Muscles of the posterior arm and forearm.

B. Dissecting the Muscles of the Anterior Arm and Forearm

1. The muscles of the **anterior arm** flex the shoulder and/or forearm. On the lateral surface of the arm, identify the **brachialis** muscle just cranial to the lateral head of the triceps brachii (see Figure D1-9). The brachialis originates from the humeral shaft and inserts on the ulna.
2. Flip the cat to view the ventral surface. To expose the remaining muscles of the anterior arm, the pectoralis muscles must be cut and reflected. Dissect on the same limb that you used for the posterior muscles. Slide a blunt probe deep to all four pectoralis muscles. Identify the thick cluster of vessels and nerves deep to these muscles in the axillary region. Avoid cutting these structures as you transect the pectoralis muscles near their insertion on the humerus (**Figure D1-10**).
   a. Identify the **epitrochlearis** muscle (not shown in Figure D1-10), a thin sheet of muscle that originates near the insertion of the latissimus dorsi and covers the medial surface of the arm. This muscle is not present in humans. Cut and reflect the epitrochlearis.
   b. Identify the **biceps brachii** muscle deep to the pectoralis muscles. Cats have only a single head of biceps brachii. The biceps brachii and the brachialis are the prime movers for forearm flexion.
   c. At the proximal humerus, near the humeral head, identify the small **coracobrachialis** muscle. This muscle is much larger in humans. It functions in flexion of the shoulder.
3. The muscles of the **anterior forearm** all originate from the medial epicondyle of the humerus and act on the

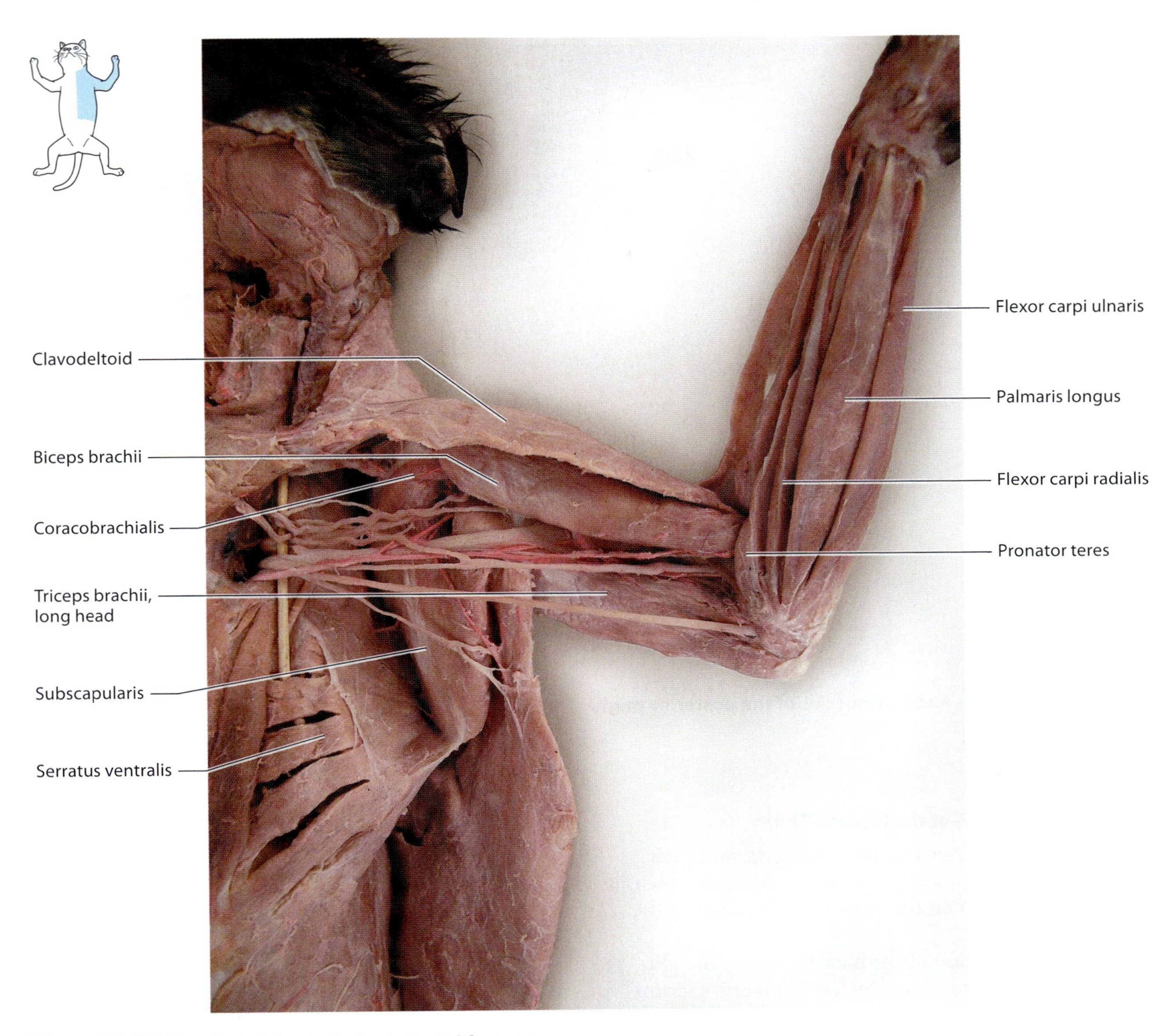

Figure D1-10 **Muscles of the anterior arm and forearm.**

forearm, wrist, and/or digits. Cut through the deep fascia on the ventral forearm. Use the blunt probe to separate and identify the superficial muscles of the forearm from ulnar to radial side (see Figure D1-10): **flexor carpi ulnaris**, **palmaris longus**, **flexor carpi radialis**, and **pronator teres**.

What is the origin of the flexor muscles in the anterior forearm? ______________________________

Identify the muscle that lies deep to epitrochlearis.

____

4. Follow the instructions for storage, waste disposal, and cleanup as described in Dissection 1, Activity 1, steps 10–12.

ACTIVITY 4 Dissecting the Muscles of the Hind Limb

Learning Outcomes

1. Use proper dissection techniques to locate and isolate the hind limb muscles of the cat.
2. Identify the described hind limb muscles in the dissected cat.
3. Group the muscles of the hind limb into fascial compartments: anterior thigh, posterior thigh, medial thigh, anterior leg, posterior leg, and lateral leg. Use these groupings to aid in learning muscle action and innervation.
4. Compare the muscles of the cat with the homologous muscles in the human. Recognize the similarities and differences in form, location, and function of homologous muscles in these two mammals.

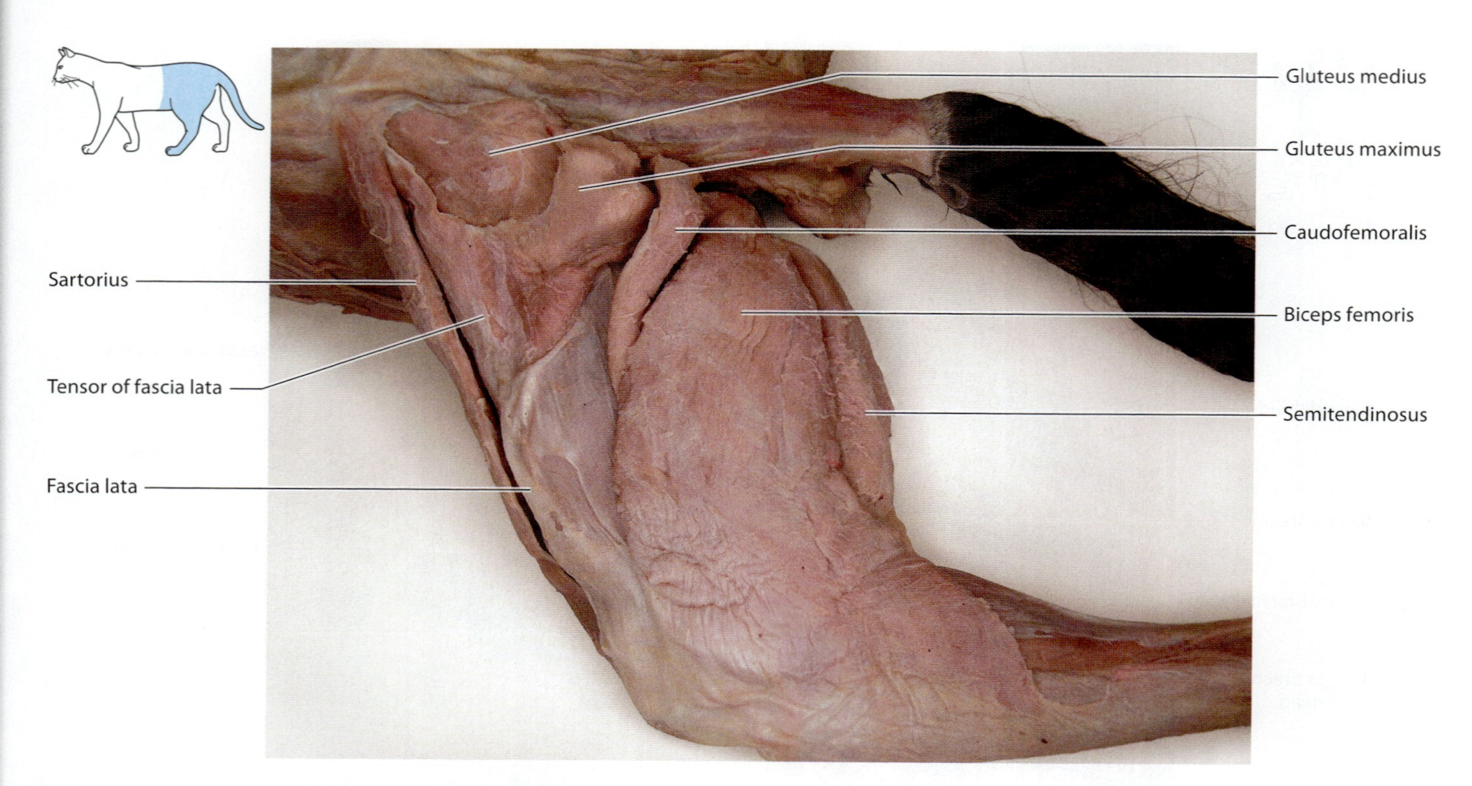

Figure D1-11 The gluteals and the muscles of the posterior thigh.

Instructions

A. Dissecting the Muscles of the Hip and Thigh

1. Put on safety glasses, gloves, and a lab coat, and then place your cat on the dissecting tray ventral side down. Dissection will begin on the dorsolateral surface of the hind limb.
2. Clean the caudal region of the back to expose the gluteal muscles, which originate on the ilium and sacrum and insert on the proximal femur. The dense connective tissue that covers the gluteus medius can be carefully removed with scissors to expose the muscle fibers of this most cranial gluteal muscle. Use blunt dissection to separate and identify the remaining muscles arising off the pelvic girdle (**Figure D1-11**).
   a. The **gluteus medius** originates from the lateral surface of the ilium and inserts on the greater trochanter of the femur. In the cat, the gluteus medius is the largest of the gluteal muscles. It extends and abducts the thigh.
   b. The extremely small **gluteus maximus** lies caudal and superficial to the gluteus medius. It arises from the last sacral and first caudal vertebrae and inserts on the greater trochanter. In the cat its primary function is abduction of the thigh. In humans, the gluteus maximus is a prime mover of thigh extension, especially from the flexed position as in climbing stairs.
   c. The **caudofemoralis** is a thin muscle located caudal to the gluteus maximus. It originates from the caudal vertebrae, becomes tendinous, and inserts with the biceps femoris on the patella.
   d. The **tensor of fascia lata** is a triangular muscle located slightly ventral to the gluteus medius. It originates on the connective tissue fascia covering the gluteus medius and inserts into a broad tendon, the **fascia lata**, that covers the lateral surface of the thigh. The fascia lata is called the *iliotibial tract* in humans.

   Why is the small, superficial gluteal muscle of the cat named gluteus maximus? ______________________

3. The muscles of the thigh in mammals are divided into three fascial compartments: the posterior thigh, anterior thigh, and medial thigh compartments. We begin the dissection of the muscles of the thigh in the **posterior compartment**.
4. On the same limb on which you exposed the gluteal muscles, separate and identify the muscles in the **posterior thigh**, the **hamstring muscles** (see Figure D1-11). This group is composed of three muscles that all originate on the ischial tuberosity and insert on the tibia. They extend the thigh and flex the leg.
   a. The **biceps femoris** is the large muscle that covers the lateral surface of the thigh and inserts laterally on the patella and the proximal tibia. Separate its cranial and caudal borders. Visualize the large sciatic nerve located deep to the biceps femoris. Being careful not to cut the underlying nerve, transect the biceps

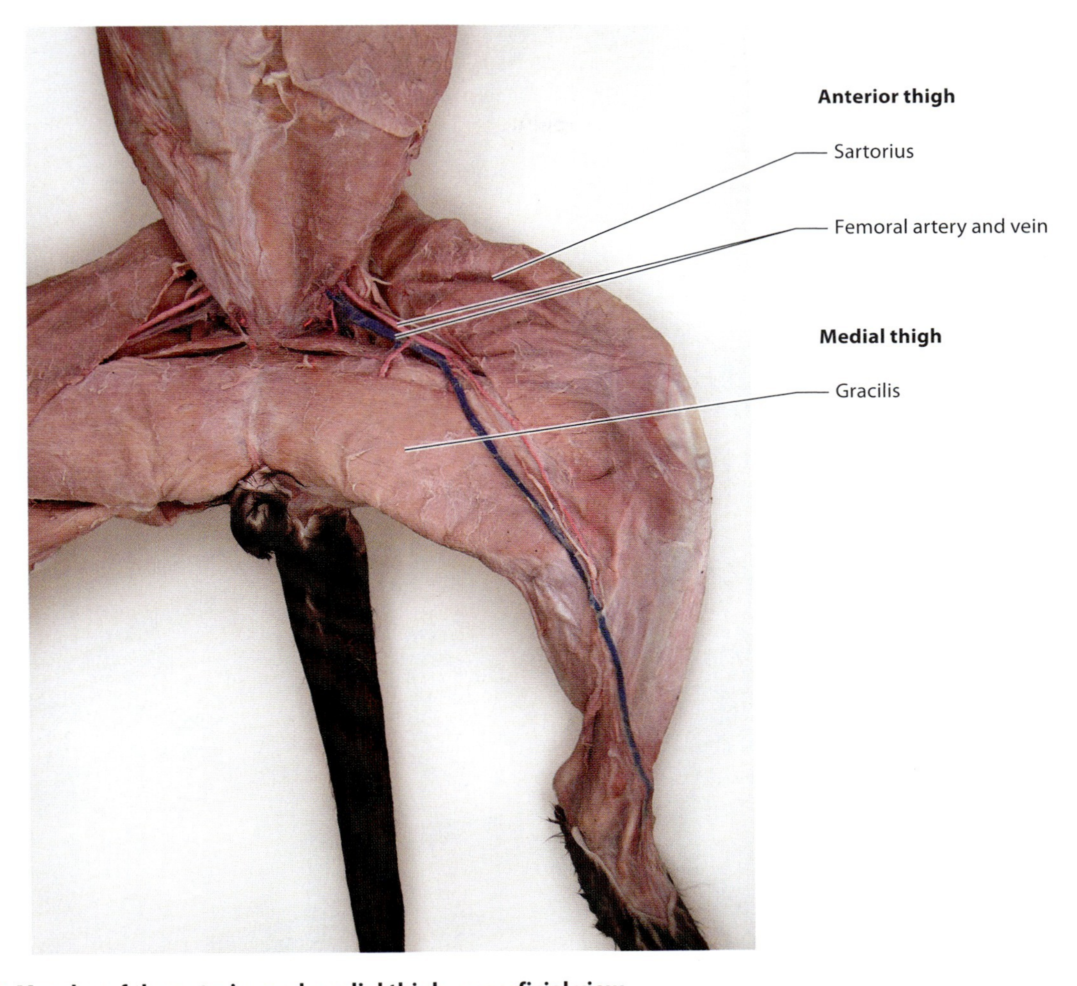

Figure D1-12 Muscles of the anterior and medial thigh, superficial view.

femoris. Reflect the biceps femoris to identify the remaining two hamstring muscles.

b. The **semimembranosus** (see Figure D2-3) is a large muscle located deep to the biceps femoris. It inserts medially on the distal femur and tibia.

c. The **semitendinosus** is a strap-like muscle posterior to the biceps femoris. It inserts medially on the tibia.

Which of the three hamstring muscles inserts on the lateral side of the leg? ______________________

5. Flip the cat over to view its ventral surface. Work on the same limb on which you dissected the posterior muscles. With a blunt probe, clean away fat and fascia to expose the superficial muscles of the anterior and medial thigh. If working with a male cat, be careful not to damage the spermatic cords, cord-like structures extending between the ventral body wall and the scrotum. Note the **femoral artery** and the **femoral vein** entering the thigh (**Figure D1-12**). These vessels mark a boundary between the muscles of the **anterior thigh**, located lateral to the femoral vessels, and the muscles of the **medial thigh**, located medial to the femoral vessels.

6. The dissection continues with the muscles of the anterior thigh. The **sartorius** is the most superficial muscle of the anterior thigh. This wide, flat muscle extends from the ilium and inserts on the medial tibia and patella. In humans the sartorius is much thinner and more obliquely oriented.

7. Transect and reflect the sartorius to view the four quadriceps muscles (**Figure D1-13**), collectively called the quadriceps femoris. All these muscles insert on the patella and the tibial tuberosity and extend the leg.
   a. The **vastus medialis** is the most medial head of the quadriceps femoris. It is located just lateral to the femoral vessels. It originates on the shaft of the femur.
   b. The **rectus femoris** is located lateral to the vastus medialis. It originates on the anterior inferior iliac spine and crosses both the hip and the knee; thus it both flexes the thigh and extends the leg.
   c. The **vastus lateralis** is lateral to the rectus femoris. Be careful not to confuse the tensor of fascia lata (identified on the dorsolateral surface) with the vastus lateralis. The tensor of fascia lata and the fascia lata cover

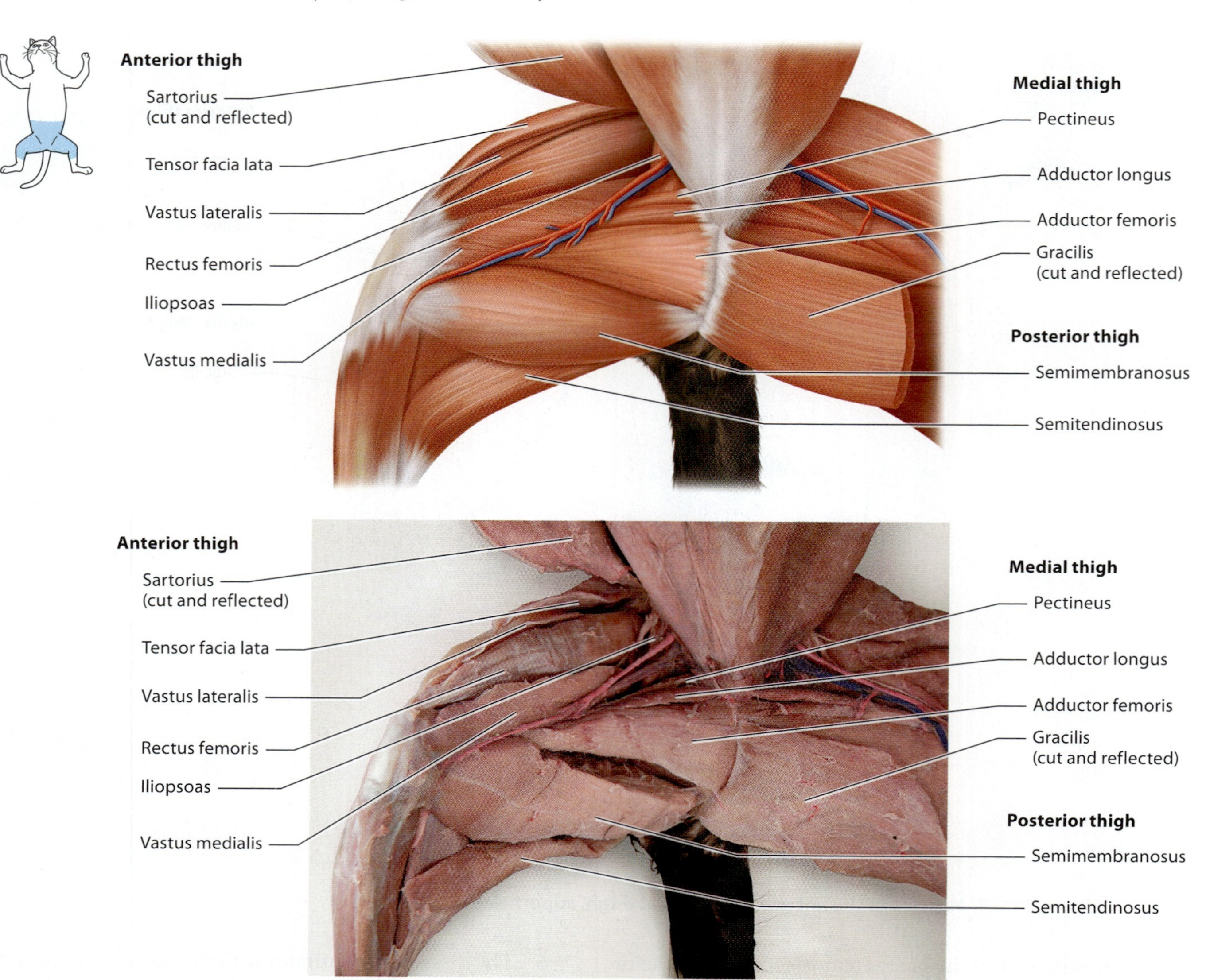

Figure D1-13 **Muscles of the anterior and medial thigh, deep view.**

the vastus lateralis. Find the boundary between these two muscles and identify the vastus lateralis.

d. The fourth head of the quadriceps femoris, the **vastus intermedius** (not shown in Figure D1-13), lies deep to the rectus femoris. To view the vastus intermedius, either transect the rectus femoris or push it aside.

Which muscle of the quadriceps femoris crosses two joints, the hip and the knee? ____________________

Which group of muscles acts as an antagonist to the quadriceps femoris? ____________________

8. The final muscle of the anterior thigh, the **iliopsoas**, is viewable only at its insertion in this dissection. It originates within the abdomen on the transverse processes of the lumbar vertebrae and on the iliac fossa. It inserts in the proximal thigh on the lesser trochanter of the femur. Carefully probe lateral to the femoral artery in the proximal thigh to see its small insertion. Don't let this view fool you; the iliopsoas has a large origin and is a powerful muscle that flexes the thigh.

9. Examine the muscles of the **medial thigh** located medial to the femoral vessels (see Figures D1-12 and D1-13). All of these muscles originate on the pubis and adduct the thigh; thus they are commonly called the *adductor muscles*.
   a. The **gracilis** is a wide superficial muscle of the medial thigh. It originates on the pubic symphysis and inserts by a broad aponeurosis on the medial tibia (see Figure D1-12). Clean its free borders and transect and reflect it to identify the remaining adductors.
   b. Deep to the gracilis is the **adductor femoris**, a large triangular muscle originating from the pubis and inserting on the femur (see Figure D1-13). This muscle is homologous to the adductor magnus and the adductor brevis in humans.
   c. The small **adductor longus** is cranial to the adductor femoris. It originates on the pubis and inserts on the femur.

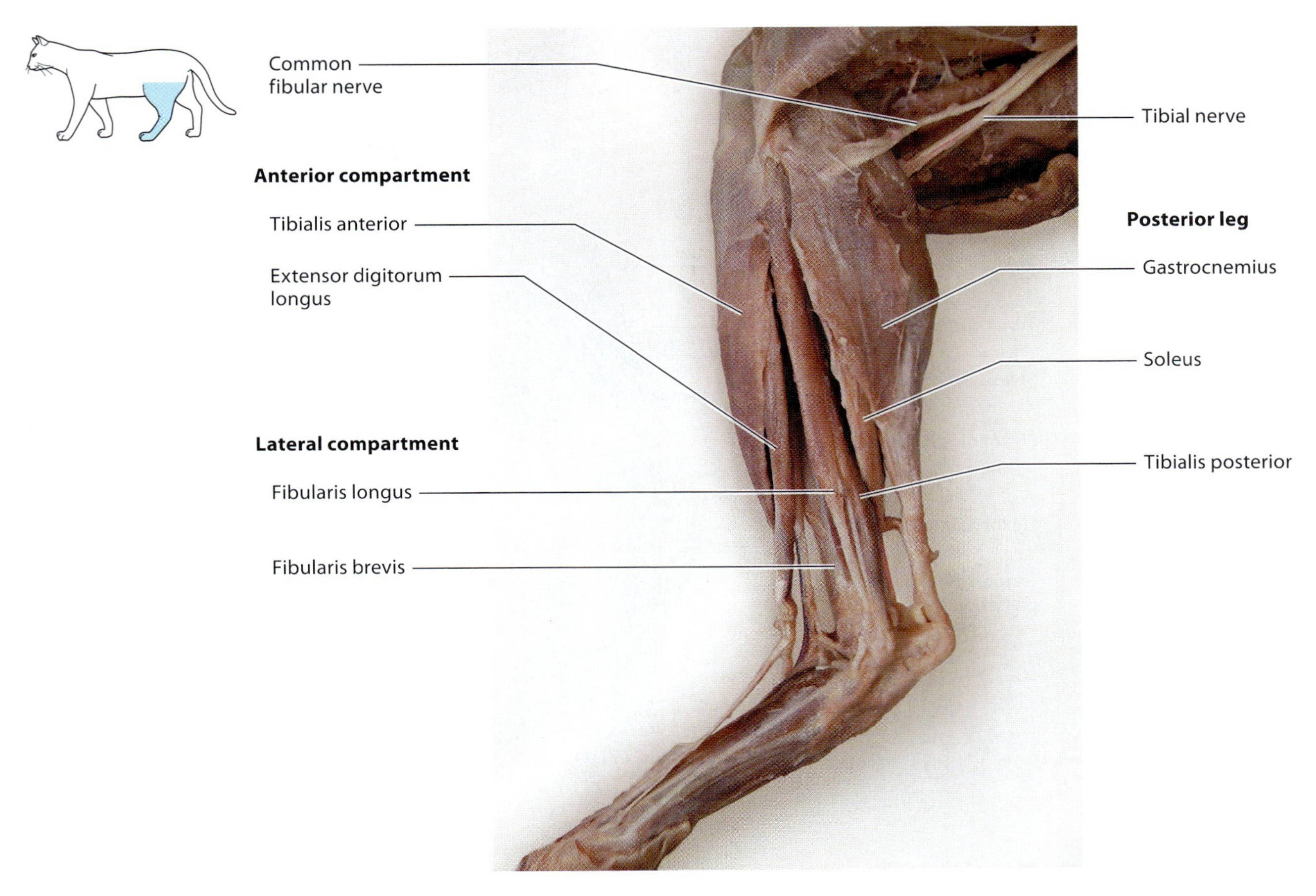

Figure D1-14 Muscles of the leg, posterolateral view.

d. The **pectineus** is an extremely small muscle cranial to the adductor longus and adjacent to the femoral vessels on their medial side. It also originates on the pubis and inserts on the proximal femur. This muscle tears very easily, so you will need to probe carefully (and deeply) to expose the pectineus.

From which bone of the pelvic girdle do the muscles of the medial compartment originate?

____

On which bone do most of these muscles insert?

____

10. Caudal to the adductor femoris, the insertions of the semimembranosus and semitendinosus can be viewed on the medial thigh and leg. Recognize the continuity of these muscles with those you identified in the dorsal view of the posterior thigh.

The muscles of the anterior and medial compartments are very similar in cats and humans. Describe any differences you have observed. _______________

____

B. Dissecting the Muscles of the Leg

1. The muscles of the leg in mammals are divided into three fascial compartments: the posterior leg, anterior leg, and lateral leg compartments. We begin the dissection with the muscles of the **posterior leg** (**Figure D1-14**). Turn your cat so that it is facing ventral side down. Reflect the cut edge of the biceps femoris and clean the popliteal fossa behind the knee to expose the muscles of the posterior leg. In general, the muscles of the posterior leg plantar flex the foot and/or flex the digits. Separate these muscles along their natural boundaries.
   a. Together the two heads of the gastrocnemius and the soleus constitute the triceps surae. The **gastrocnemius** (see Figure D1-14) arises from the medial and lateral epicondyles of the femur. The two heads merge and insert via the thick calcaneal tendon on the calcaneus. Deep to the gastrocnemius, and most easily viewed on the lateral side, is the **soleus**. It arises on the fibula and also inserts via the calcaneal tendon on the calcaneus. Both the gastrocnemius and the soleus plantar flex the foot. The gastrocnemius also flexes the leg.
   b. Flip the cat over to view its ventral surface. On the medial side of the leg, locate the **plantaris** between

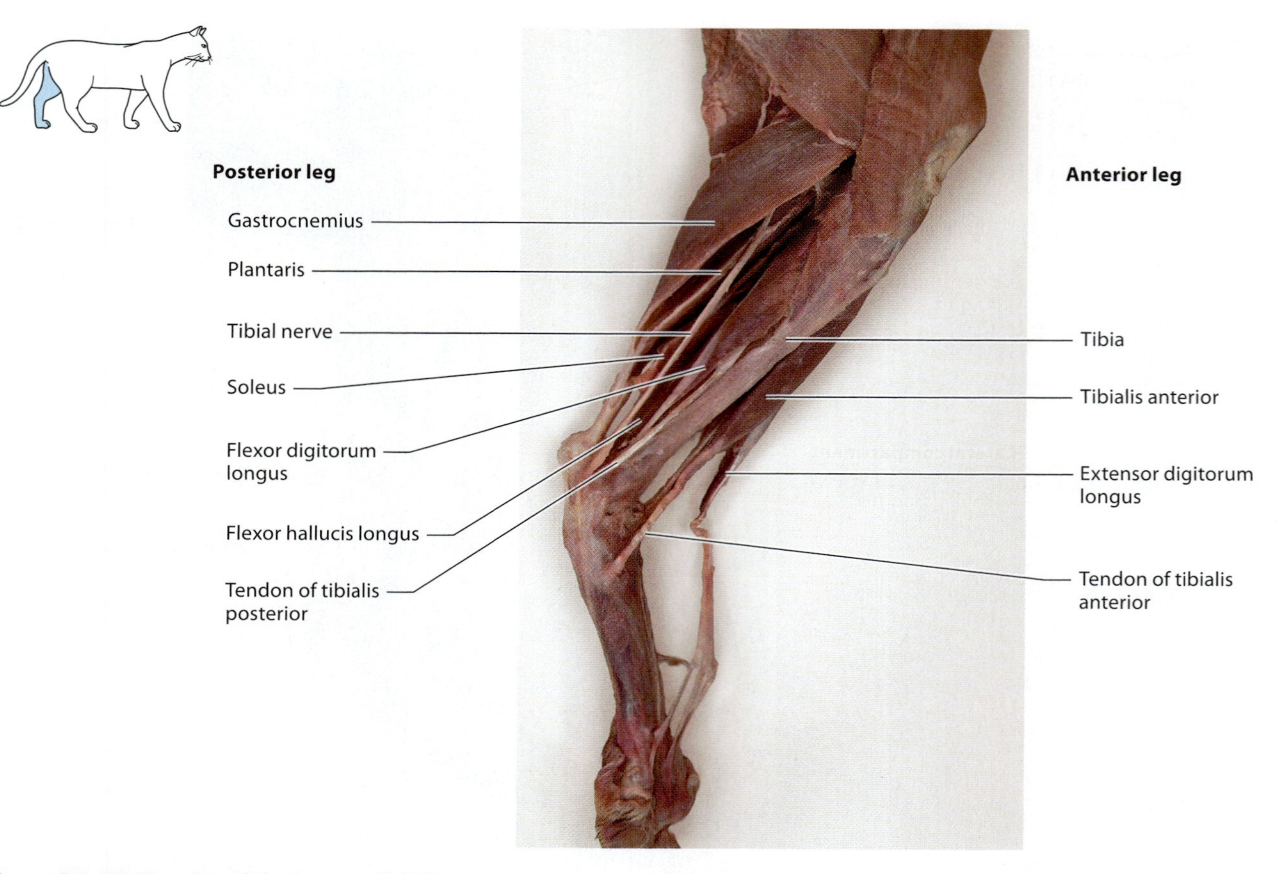

Figure D1-15 Muscles of the leg, medial view.

the gastrocnemius and the soleus (**Figure D1-15**). It arises from the lateral side of the femur and the patella, passes around the calcaneus, and splits into four tendons that insert on the middle phalanges. In the cat the plantaris flexes the digits; in humans it is extremely small and inserts on the calcaneus.

c. The **flexor digitorum longus** is located along the tibia on the medial side of the leg. It originates on the proximal tibia and fibula and inserts via four tendons on the distal phalanx of digits II–V.

d. Lateral to the flexor digitorum longus, the tendon of **flexor hallucis longus** can be viewed. The flexor hallucis longus, which originates on the tibia and with the tendons of the flexor digitorum longus, aids in flexing the digits.

e. The **tibialis posterior** (see Figure D1-14) is located deep to the flexor digitorum longus. Its tendon can be viewed medially passing around the medial malleolus of the tibia and inserting on the tarsals. The tibialis posterior aids in plantar flexion.

2. In the **lateral compartment of the leg**, lateral to the soleus, are the **fibularis muscles** (see Figure D1-14). These muscles originate on the fibula and insert on the metatarsals. They evert the foot.

3. Identify the muscles of the **anterior leg**, which function in dorsiflexion and extension of the digits.

a. The **tibialis anterior** covers the lateral surface of the tibia (see Figure D1-15). It arises from the proximal tibia and fibula. As it approaches the ankle, its tendon crosses medially to insert on the base of the first metatarsal. In addition to dorsiflexion, the tibialis anterior inverts the foot.

b. The **extensor digitorum longus** (see Figure D1-14) lies lateral to the tibialis anterior. It originates on the lateral epicondyle of the femur; it divides into four tendons that cover the dorsum of the foot and inserts on the middle and distal phalanges. As implied by its name, this muscle extends the digits.

Name the muscles that insert on the calcaneus via the calcaneal tendon. ____________________

Name a muscle that is a prime mover for dorsiflexion.

____

In which compartment of the leg is this muscle located?

____

Name the primary action or actions of the muscles in the posterior leg. ____________________

4. Follow the instructions for storage, waste disposal, and cleanup as described in Dissection 1, Activity 1, steps 10–12.

5. **Optional Activity**

Practice labeling muscles on cats at MasteringA&P® > Study Area > Practice Anatomy Lab > Cat > Muscular System

Review Questions

1. Which muscles of the trunk look significantly different in the cat as compared to the human?

____

____

2. Which muscles or muscle groups of the trunk and abdomen are quite similar in cats and humans?

____

____

3. Match the muscles listed below with their actions. Some actions may apply to more than one muscle.

_____ a. sternomastoid	1. medially rotates the arm
_____ b. masseter	2. adducts the arm
_____ c. rectus abdominis	3. extends the neck
_____ d. subscapularis	4. adducts the scapula
_____ e. acromiodeltoid	5. laterally rotates the head
_____ f. clavotrapezius	6. adducts the arm
_____ g. pectoralis major	7. laterally rotates the scapula
_____ h. latissimus dorsi	8. elevates the jaw
_____ i. rhomboideus	9. extends the arm
_____ j. serratus ventralis	10. abducts the arm
	11. flexes the trunk

4. CHART List the muscles of the forelimb you have dissected in their proper fascial compartment in the following chart. Then list the primary actions of the muscles in each compartment.

Fascial Compartment	Muscles Located in This Compartment	Primary Action(s) of the Muscles in This Compartment
Anterior arm		
Posterior arm		
Anterior forearm		
Posterior forearm		

5. Match the muscles listed below with their actions. Some actions may apply to more than one muscle.

_____ a. brachioradialis	1. flexes the forearm
_____ b. extensor carpi ulnaris	2. adducts the forefoot
_____ c. flexor carpi radialis	3. abducts the forefoot
_____ d. coracobrachialis	4. flexes the arm
_____ e. brachialis	5. extends the forearm
_____ f. triceps brachii	6. flexes the forefoot
_____ g. palmaris longus	

6. Describe how the following muscles differ between cats and humans.

a. Biceps brachii ______________________________

b. Coracobrachialis ______________________________

c. Brachioradialis ______________________________

d. Epitrochlearis ______________________________

7. CHART List the muscles of the hind limb you have dissected in their proper fascial compartment in the following chart. Then list the primary actions of the muscles in each compartment.

Fascial Compartment	Muscles Located in This Compartment	Primary Action(s) of the Muscles in This Compartment
Anterior thigh		
Posterior thigh		
Medial thigh		
Anterior leg		
Posterior leg		
Lateral leg		

8. In general, muscles in opposite compartments acts as antagonists; muscles in the same compartment act as synergists.

a. Which muscles act as an antagonist to the muscles in the medial thigh?

____

b. Name the muscles that act as synergists to the gastrocnemius.

____

c. Name the muscles that act as antagonists to the semitendinosus.

____

Exploring the Spinal Nerves of the Cat

DISSECTION OUTLINE

Activity 1: Dissecting the Nerves from the Brachial Plexus

Activity 2: Dissecting the Nerves from the Lumbosacral Plexus

Materials Needed

- ☐ Safety glasses
- ☐ Gloves
- ☐ Lab coat or other protective clothing
- ☐ Dissecting tray
- ☐ Dissection tools
- ☐ Skinned preserved cat
- ☐ Spray bottle with a solution of soap and water for cleanup

ACTIVITY 1
Dissecting the Nerves from the Brachial Plexus

Learning Outcomes

1. Locate and identify the three cords and four terminal nerves that arise from the brachial plexus and innervate the muscles of the forelimb.
2. Identify the muscular compartment and the individual muscles innervated by each nerve identified.

Instructions

1. Put on safety glasses, gloves, and a lab coat, and then place your cat on the dissecting tray ventral side up. If you haven't done so already, transect the pectoralis muscles, being careful not to damage the large neurovascular bundle deep to these muscles. Reflect the cut pectoralis muscles and carefully clean the neurovascular bundle, separating the arteries (colored red), veins (blue), and nerves (white). This cluster of nerves is the brachial plexus (**Figure D2-1**). Once separated, the nerves of the brachial plexus can be identified.
2. The three most superficial nerves arise from the **anterior divisions of the brachial plexus**. These nerves, and the cords from which they arise, form an M within the brachium. Look for this to aid in identifying the nerves to the anterior compartment muscles.
   a. The **musculocutaneous nerve** is the most lateral nerve off the brachial plexus. It arises from the lateral cord of the plexus and innervates the muscles of the anterior arm: the biceps brachii, the brachialis, and the coracobrachialis. Follow this nerve and observe it entering the biceps brachii.
   b. The **median nerve** is formed from branches off both the lateral and medial cords. It is the middle portion of the M. Follow the median nerve as it travels with the brachial artery through the arm. It passes through a small hole above the medial epicondyle of the humerus—the supracondylar foramen—and continues into the forearm to innervate most of the muscles of the anterior compartment of the forearm. In the forearm, the median nerve runs with the radial artery deep to the forearm muscles.
   c. The **ulnar nerve** is the most medial nerve off the brachial plexus. It forms from the medial cord and passes through the arm, wraps around the medial epicondyle, and continues into the forearm on the ulnar side. Trace the ulnar nerve through the arm and forearm. In the forearm it innervates the flexor carpi ulnaris and the medial portion of the flexor digitorum profundus. In the forearm, the ulnar nerve runs with the ulnar artery deep to the forearm muscles. It continues into the forefoot to innervate many of the intrinsic muscles of the forefoot.

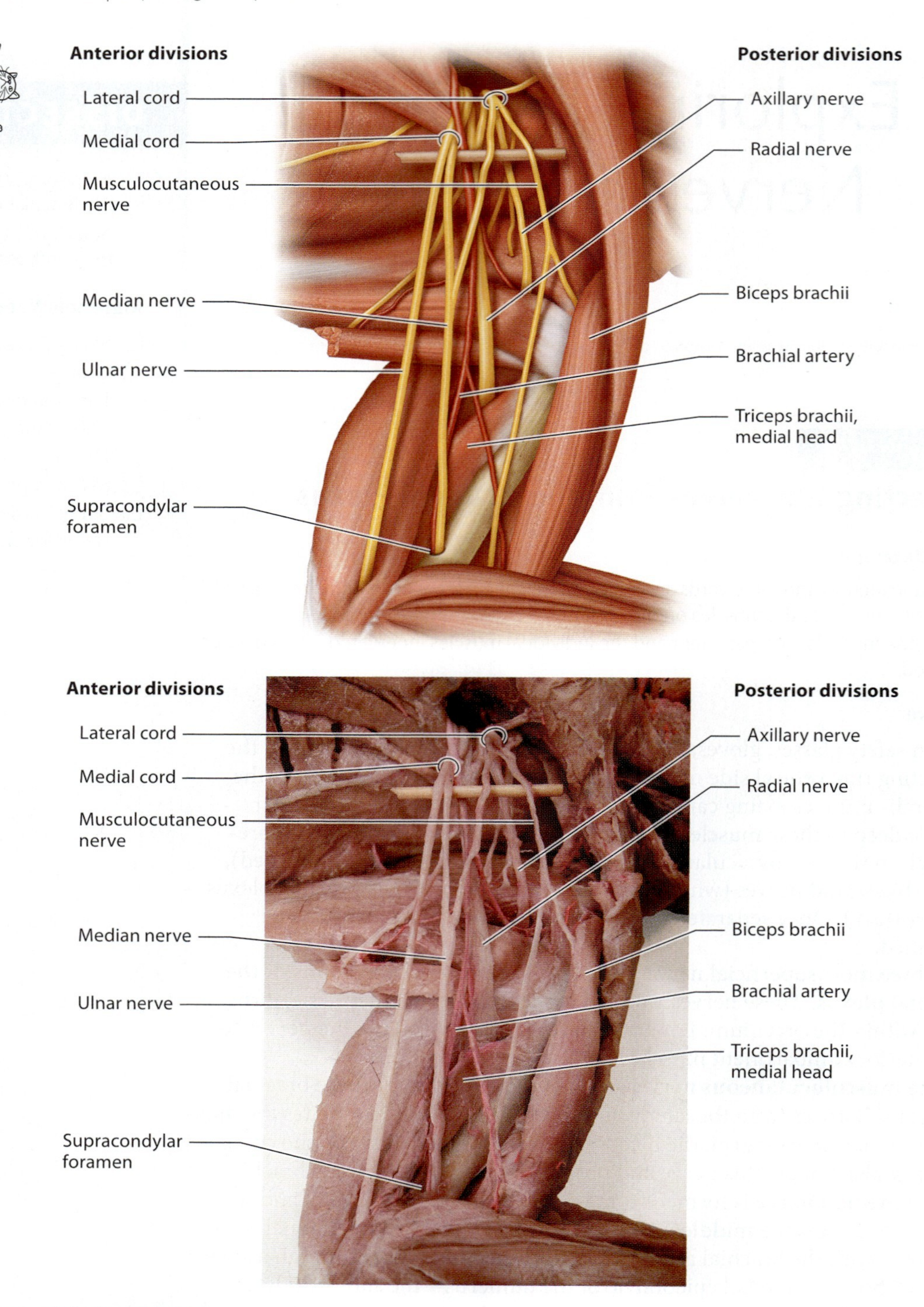

Figure D2-1 Nerves of the brachial plexus.

3. Deep to the anterior division nerves and the brachial artery and vein, identify the large **radial nerve**. The radial nerve, which is formed from the **posterior divisions of the brachial plexus**, innervates the triceps brachii and the muscles of the posterior forearm. Follow the course of the radial nerve as it passes through the triceps brachii to the posterior arm. It branches near the distal humerus to innervate the muscles of the posterior forearm. A small cutaneous branch continues superficially toward the hand, traveling with the brachioradialis muscle.

   Name the nerve that innervates ALL of the extensor muscles of the arm and forearm. ______________________

   ____

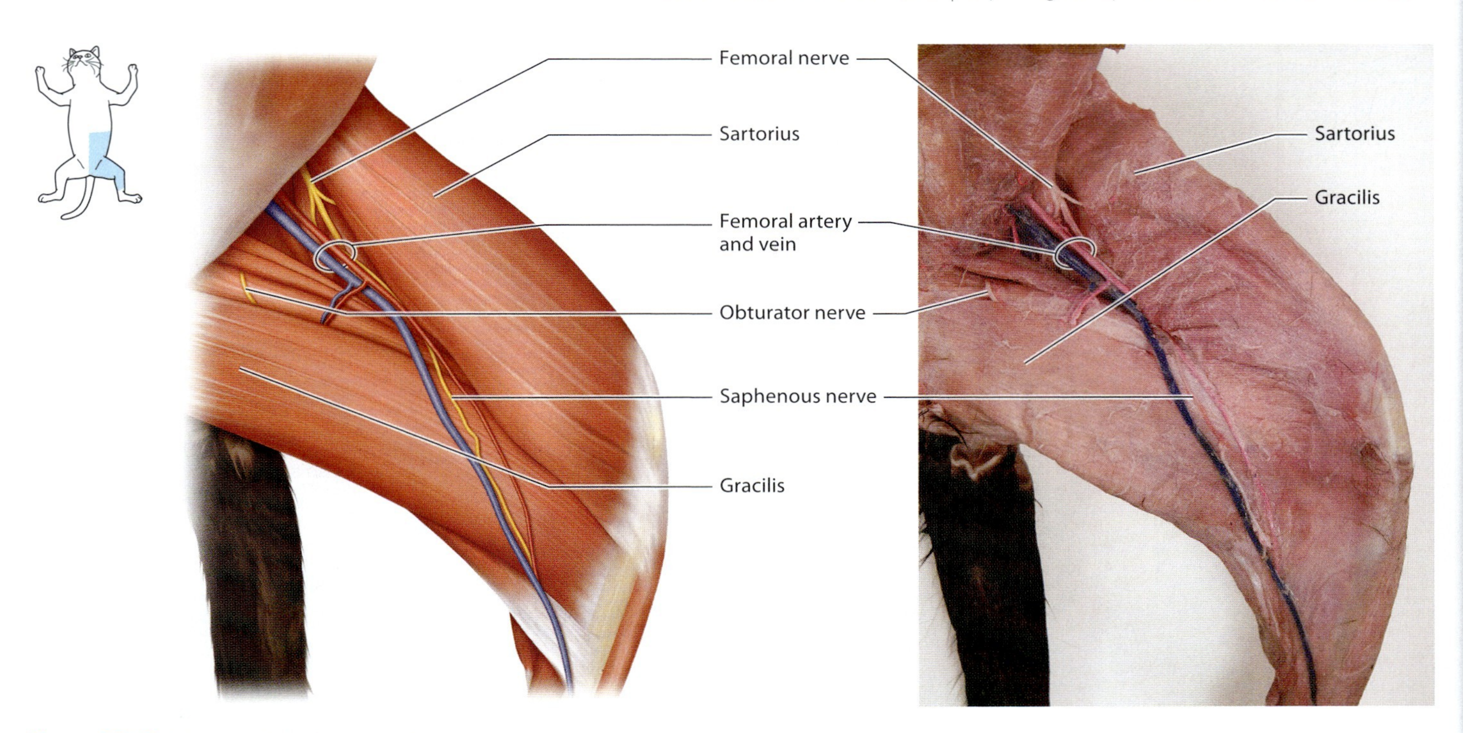

Figure D2-2 Nerves of the lumbar plexus.

Name the nerve that innervates the flexor muscles located in the anterior arm. ______________________

____

Name the nerve in the anterior forearm that runs along with the radial artery. ______________________

____

4. Follow the instructions for storage, waste disposal, and cleanup as described in Dissection 1, Activity 1, steps 10–12.

ACTIVITY 2
Dissecting the Nerves from the Lumbosacral Plexus

Learning Outcomes

1. Locate and identify the terminal nerves of the lumbar and sacral plexuses that innervate the muscles of the hind limb.
2. Identify the muscular compartment and the individual muscles innervated by each nerve identified.

Instructions

1. Put on safety glasses, gloves, and a lab coat, and then place your cat ventral side up on the dissecting tray. Reidentify the femoral artery and vein in the proximal thigh, where they enter the hind limb. Identify the **femoral nerve**, located just lateral to the femoral artery and vein, as it enters the hind limb (**Figure D2-2**). The femoral nerve innervates all of the muscles of the anterior thigh. With a blunt probe, clean the nerve and view its branches into these muscles.
2. The **saphenous nerve** is a cutaneous branch of the femoral nerve. Identify this thin, thread-like nerve as it follows the femoral vessels down the thigh between the sartorius and gracilis muscles and into the leg. The saphenous nerve innervates the skin of the medial leg.
3. The **obturator nerve** originates from the **lumbar plexus**, passes through the obturator foramen of the pelvic girdle, and enters the muscles of the medial thigh. Locate this nerve superficially as it emerges from the adductor femoris muscle and enters the deep surface of the gracilis.
4. Turn the cat over to view its dorsal side. Being careful not to cut the **sciatic nerve** located deep to the biceps femoris, transect the biceps femoris muscle if you have not done so already. The sciatic nerve is the largest nerve of the body. Formed from the **sacral plexus**, it passes through the greater sciatic notch of the ilium to enter the hind limb. Muscular branches of the sciatic nerve in the proximal thigh innervate the hamstring muscles of the posterior thigh (**Figure D2-3**).
5. With a blunt probe, clean and follow the sciatic nerve distally. Near the popliteal fossa the sciatic nerve splits into the tibial nerve and the common fibular nerve. The **tibial nerve** passes between the two heads of the gastrocnemius and innervates the muscles of the posterior

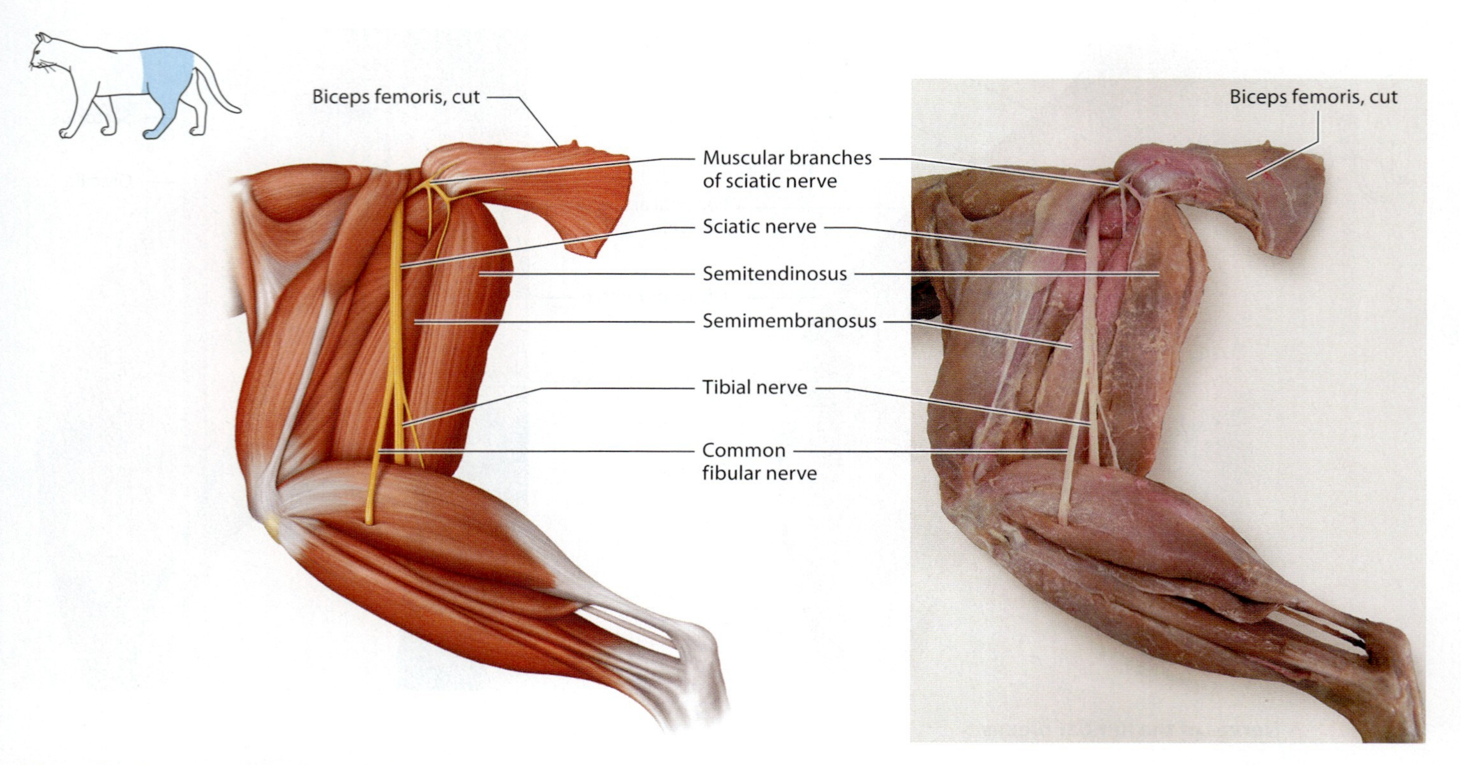

Figure D2-3 **Nerves of the sacral plexus.**

leg. The **common fibular nerve** travels laterally to innervate the muscles of the lateral and anterior leg.

Name the nerve that innervates the quadriceps femoris muscles. ____________________

Name the nerve that innervates the hamstring muscles.

____

Name the nerve that innervates the triceps surae muscles.

____

Name the nerve that innervates the adductor muscles in the thigh. ____________________

6. Follow the instructions for storage, waste disposal, and cleanup as described in Dissection 1, Activity 1, steps 10–12.
7. **Optional Activity**

Practice labeling nervous system structures on cats at MasteringA&P® > Study Area > Practice Anatomy Lab > Cat > Nervous System

Review Questions

1. CHART For each of the nerves in the following chart, list the muscular compartment in which it is located and the specific muscles it innervates.

Nerve	Muscular Compartment	Muscles Innervated
Musculocutaneous nerve		
Radial nerve		
Median nerve		
Ulnar nerve		

2. A midshaft fracture of the humerus can damage the radial nerve. What muscular deficits would occur if the radial nerve were injured at this location? ______________________________
____
____

3. Name the cord(s) that form the median nerve. ______________________________
____

4. Name the cord(s) that form the musculocutaneous nerve. ______________________________
____

5. Name the cord(s) that form the radial nerve. ______________________________
____

6. Review the muscular compartments of the hind limb and the muscles located in each compartment (see Dissection 1, Activity 4). A single nerve innervates all the muscles in one compartment. Identify the nerve that innervates the muscles in the following compartments:

   ______________________ a. Anterior thigh

   ______________________ b. Medial thigh

   ______________________ c. Posterior thigh

   ______________________ d. Posterior leg

   ______________________ e. Anterior and lateral leg

7. A herniated disc at L_5 (lumbar vertebra 5) can compress the nerve root of spinal nerve L_5. Spinal nerve L_5 contributes to the sacral plexus and the sciatic nerve. Describe the types of deficits or signs and symptoms to be expected in an individual with this condition. Include both sensory and motor deficits. ____

   ____

   ____

8. Injury to the common fibular nerve can result in foot drop, an inability to dorsiflex the foot. Individuals with this condition have difficulty clearing their foot through on the swing phase of walking. What muscles are affected by this injury? ______________________

   ____

   ____

9. Name the nerve that passes through each of the following locations:

   ______________________ a. Obturator foramen

   ______________________ b. Greater sciatic notch

   ______________________ c. Femoral triangle

   ______________________ d. Popliteal fossa

Exploring the Respiratory System of the Cat

DISSECTION OUTLINE

Activity 1: Dissecting the Organs of the Ventral Body Cavity

Activity 2: Dissecting the Structures of the Respiratory System

Materials Needed

- ☐ Safety glasses
- ☐ Gloves
- ☐ Lab coat or other protective clothing
- ☐ Dissecting tray
- ☐ Dissection tools
- ☐ Bone clippers
- ☐ Preserved cat, either unskinned or skinned
- ☐ Mounted cat skeleton
- ☐ Spray bottle with a solution of soap and water for cleanup

ACTIVITY 1

Dissecting the Organs of the Ventral Body Cavity

Learning Outcome

1. Perform the procedure for opening the thoracic and abdominal cavities, and identify the major body organs in those cavities.

Instructions

A. Opening the Thoracic Cavity

1. Put on safety glasses, gloves, and a lab coat, and then place your cat on the dissecting tray ventral side up. Use **Figure D3-1** as a guide for opening the thoracic cavity. As always, place your blunt probe deep to the area you are cutting to avoid damaging deeper structures. Use scissors or a scalpel to make a single longitudinal cut just off the midventral line (cut 1) from the neck to the xiphoid process of the sternum, cutting through the pectoralis muscles and the costal cartilages.
2. On the superior side of the diaphragm, cut the lateral body wall obliquely, following the course of the diaphragm (cut 2).
3. Note the mediastinal pleura, a thin layer of serous membrane extending from the mediastinum (the midventral portion of the thorax) to the ventral body wall. This translucent structure resembles plastic wrap. With a blunt probe, clean the vessels running through this membrane. If needed, use scissors to cut these vessels close to the sternum, leaving long ends. You will identify these vessels later in the dissection.
4. To fully open the thoracic cavity it is useful to clip through the ribs along the dorsal body wall. Reach into the thoracic cavity, pushing the lungs medially with your fingers. With bone clippers, follow each rib dorsally and snip through it lateral to its attachment to the thoracic vertebrae. Do not cut through the entire dorsal body wall; just snip each rib until you hear it break. Snip the ribs on both sides of the body. Now you can readily identify the major organs located in the thoracic cavity (**Figure D3-2**).
   a. Identify the **right** and **left lungs**, which fill most of the thoracic cavity.
   b. The space medial to the lungs, the mediastinum, contains the **heart** within the **pericardium**, as well as the roots of the great vessels that enter and leave the heart, including the **superior vena cava** and the **aorta**.
   c. Cranial to the pericardium, in the superior mediastinum, identify the **thymus gland** covering the vessels entering the heart and often extending into the neck. The thymus is an indistinct glandular structure that may be brown and fatty. It is larger in young individuals and is replaced by fatty tissue in older individuals.

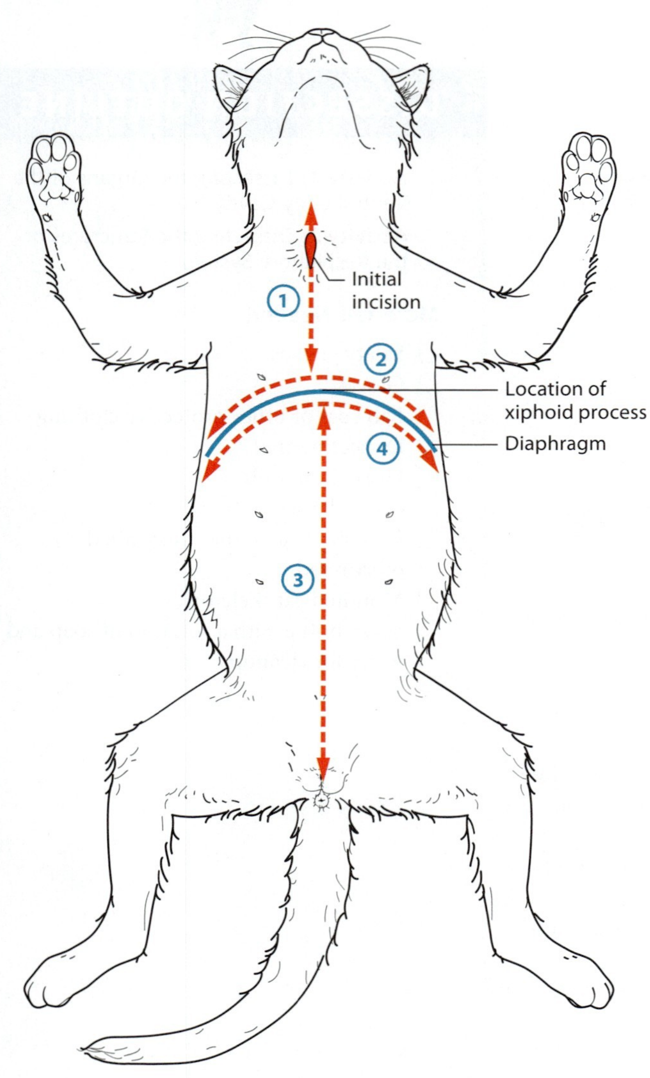

Figure D3-1 Procedure for opening the thoracic and abdominal cavities.

B. Opening the Abdominal Cavity

1. With Figure D3-1 as a guide, use scissors to cut longitudinally down the midventral abdomen at the linea alba from the inferior side of the diaphragm to the pubis (cut 3). Then cut laterally on the inferior surface of the diaphragm to open up the abdominal cavity (cut 4). Identify the major organs of the abdominal cavity (see Figure D3-2).
   a. Caudal to the diaphragm identify the large brown **liver**. On the left side, identify the dark colored **spleen** located laterally. Lift the left lobes of the liver to identify the **stomach** located cranial and slightly medial to the spleen.
   b. Extending from the stomach is a fat-filled membrane, the greater omentum, that covers the **small intestine** like an apron. The small intestine fills much of the abdominal cavity. Without removing the greater omentum, carefully free it from the coils of the intestine and lift it cranially.
   c. Lift the small intestine to identify the **large intestine** extending caudally along the dorsal body wall within the abdominal cavity.

ACTIVITY 2
Dissecting the Structures of the Respiratory System

Learning Outcomes

1. Identify the regions and organs of the respiratory system in the cat.
2. Identify the components of the larynx in the cat.
3. Identify the endocrine, cardiovascular, and nervous system structures associated with or adjacent to the respiratory structures.

Instructions

1. Put on safety glasses, gloves, and a lab coat, and then place your cat on the dissecting tray ventral side up. Examine the nostrils or **external nares**, which conduct air into the nasal cavity.
2. Open the cat's mouth to examine the **oral cavity**. Using scissors or a scalpel, cut through the soft tissue at the corners of the mouth. Use bone cutters to continue the cut through the mandible, and open the oral cavity to view the pharynx, or throat region. The **pharynx** is divided into three regions from cranial to caudal: the nasopharynx, located dorsal to the nasal cavity; the oropharynx, dorsal to the oral cavity; and the laryngopharynx, into which the esophagus and trachea open. Note the continuity of the three regions of the pharynx.
3. On the ventral neck, carefully isolate and remove the sternohyoid and sternothyroid muscles to expose the trachea and larynx. Locate the small bean-shaped **thyroid glands** located laterally at the cranial part of the trachea and medial to the common carotid arteries (**Figure D3-3**). Be careful not to remove these small endocrine glands as you clean the trachea and larynx.

   How does the thyroid gland in humans differ in appearance and location from that in the cat? ______________

   ____
4. The **larynx** is cranial to and continuous with the trachea.
   a. Identify the large, shield-shaped **thyroid cartilage**. Identify the **cricoid cartilage** located caudal to the thyroid cartilage. Both cartilages are covered with skeletal muscles that move the laryngeal cartilages. With forceps and a blunt probe, clear away these muscles to view the cartilages in their entirety.
   b. Palpate the area cranial to the thyroid cartilage to locate the small **hyoid bone**, which feels like a fish

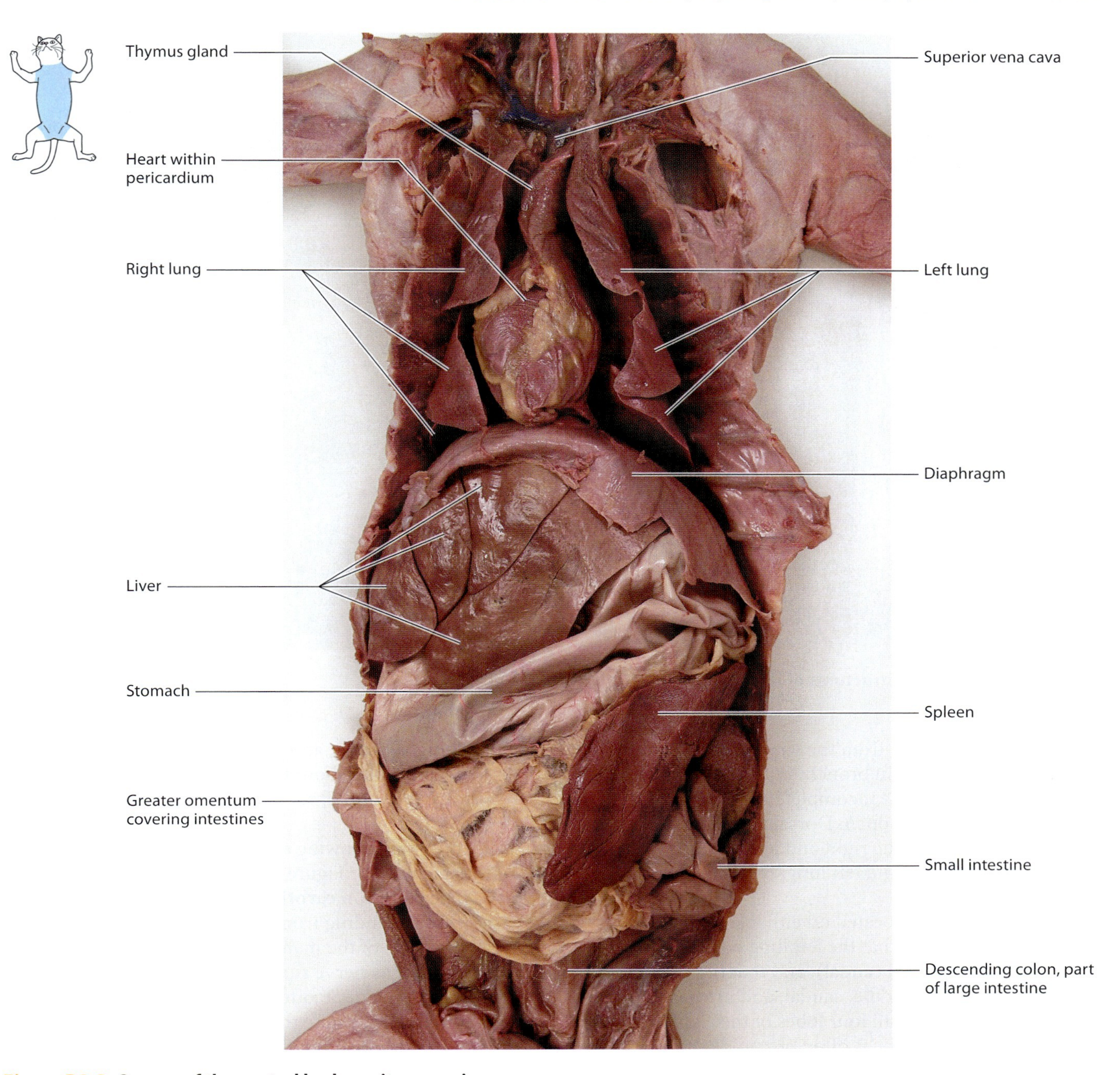

Figure D3-2 Organs of the ventral body cavity, overview.

bone. Expose it, if desired, by removing the surrounding soft tissue using forceps and a blunt probe.

c. Make a longitudinal cut through the thyroid and cricoid cartilages on the ventral surface to expose the vocal cords. The **true vocal cords** attach to the thyroid cartilage and surround the opening of the glottis. The **false vocal cords**, or vestibular folds, are slightly cranial and lateral to the true vocal cords. The true vocal cords and false vocal cords are not shown in Figure D3-3.

d. Identify the **epiglottis**, a triangular-shaped cartilage attached to the base of the tongue. It extends ventrally, cranial to the thyroid cartilage. During swallowing, the larynx moves cranially and the epiglottis flops caudally to cover the glottis, directing food into the esophagus. In this way, the epiglottis protects the airway during eating.

Identify the laryngeal cartilages described below:

True vocal folds attach to ______________________

Is a complete ring ______________________

Is composed of elastic cartilage ______________________

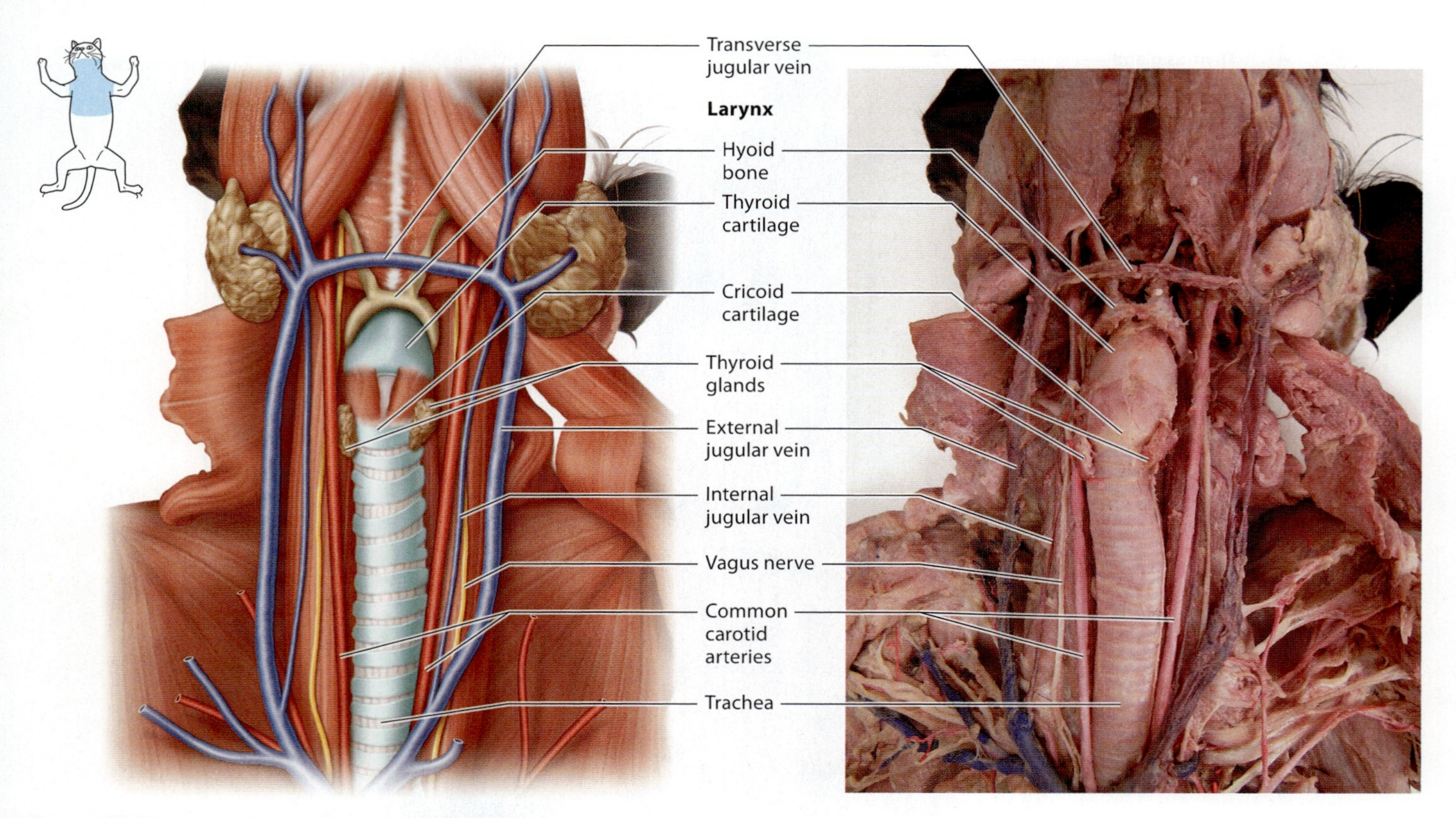

Figure D3-3 Respiratory structures of the neck.

5. The **trachea** extends from the larynx caudally to the branching of the main bronchi into the lungs. Palpate the trachea to feel the incomplete cartilaginous rings that hold the airway open. Dorsal to the heart, the trachea branches into the right and left bronchi, which enter into the right and left lung, respectively, at the root of the lung.
6. Examine the **lungs** located on either side of the **heart** (**Figure D3-4**). Note that the left lung has three lobes (cranial, middle, and caudal), and that the right lung has four (cranial, middle, caudal, and mediastinal). Be certain you observe all four lobes of the right lung; the small mediastinal lobe is often obscured; it is dorsal to the inferior vena cava.

   In humans, the right lung has ____________ lobes, and the left lung has ____________ lobes.

7. Identify the **diaphragm**. This dome-shaped skeletal muscle forms the caudal boundary of the thorax and is an important respiratory muscle. It originates on the sternum, costal cartilages, vertebrae, and the dorsal body wall, and inserts at the top of the dome into a tendon in the center of the muscle.
8. Identify the **phrenic nerve**, the somatic motor nerve that innervates the diaphragm. It originates from spinal segments C_3, C_4, and C_5, travels through the thorax on the lateral side of the pericardium, crosses the root of the lung ventrally, and terminates on the diaphragm. View and isolate both the right and left phrenic nerves; the right phrenic nerve runs along the inferior vena cava.
9. Identify the following structures in the neck, adjacent to the trachea (see Figure D3-3):
   a. The **common carotid arteries** lie lateral to the trachea, extending from the brachiocephalic trunk to the angle of the mandible.
   b. The **vagus nerve** runs with the common carotid artery in the neck, often adhering to its lateral surface. Carefully separate this nerve from the artery and follow it caudally to the root of the lung. There it branches extensively, sending parasympathetic fibers to the heart and lungs. The vagus nerve then continues caudally into the abdomen as two branches along the esophagus.
   c. The **external jugular veins** run lateral to the common carotid arteries. These large vessels drain blood from the head and neck and empty into the brachiocephalic vein.

   Differentiate between the phrenic nerve and the vagus nerve with respect to location, area of innervation, and type of innervation (somatic or visceral).

   ____

   ____

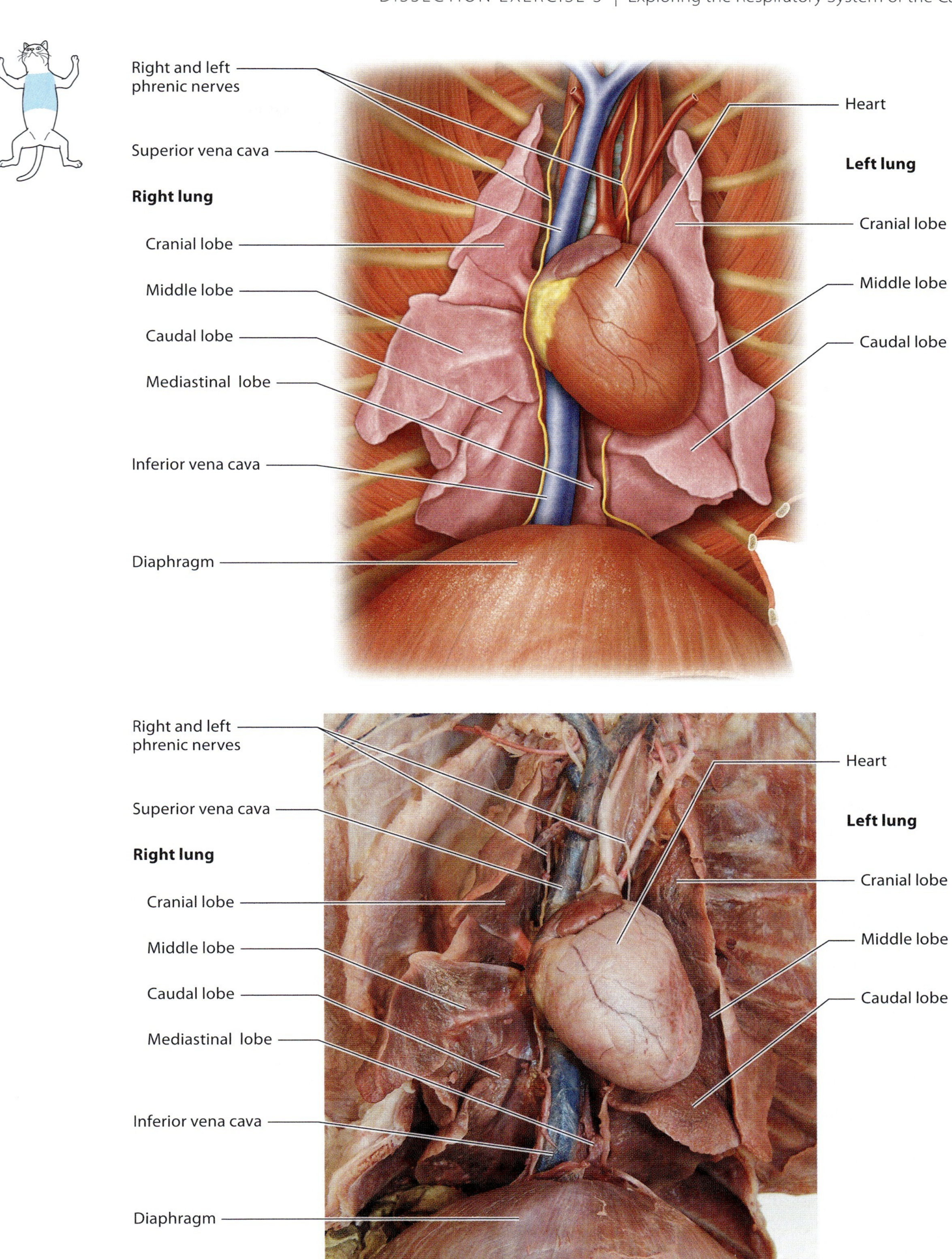

Figure D3-4 Respiratory structures of the thorax.

10. Follow the instructions for storage, waste disposal, and cleanup as described in Dissection 1, Activity 1, steps 10–12.

11. **Optional Activity**

Practice labeling respiratory structures on cats at MasteringA&P® > Study Area > Practice Anatomy Lab > Cat > Respiratory System

Review Questions

1. Name the structure that separates the thoracic cavity from the abdominal cavity. ______________

____

2. Name the organs that lie deep to the ribs. ____

3. Name the organ or organs located in the upper left abdominal quadrant of the cat. ______________

____

4. Which portion of the intestine is continuous with the stomach? ______________________________

5. What is the name of the region medial to the lungs? ______________________________________

6. Distinguish between the thymus gland and the thyroid gland in terms of structure, location, and function. ____

____

7. Number the structures of the respiratory system in the following list in the order in which air passes through them during inhalation.

_______ a. external nares	_______ e. glottis	_______ i. cricoid cartilage
_______ b. nasopharynx	_______ f. nasal cavity	_______ j. main bronchi
_______ c. laryngopharynx	_______ g. thyroid cartilage	_______ k. trachea
_______ d. oropharynx	_______ h. larynx	_______ l. lungs

8. Match each of the following structures with its description.

_____ a. thyroid cartilage	1. lies lateral to the pericardium at the root of the lung
_____ b. epiglottis	2. lies adjacent to the common carotid artery
_____ c. vagus nerve	3. located cranial to the trachea; contains the vocal cords
_____ d. phrenic nerve	4. largest cartilage of the larynx
_____ e. pharynx	5. located dorsal to the nasal and oral cavities; the throat
_____ f. larynx	6. covers the airway during swallowing

9. Draw the larynx in anterior view. Illustrate and label all of the laryngeal structures you dissected and identified in Activity 2, step 4.

Exploring the Digestive System of the Cat

DISSECTION OUTLINE

Activity 1: Identifying the Organs of the Alimentary Canal

Activity 2: Identifying the Accessory Organs of the Digestive Tract

Materials Needed

- ☐ Safety glasses
- ☐ Gloves
- ☐ Lab coat or other protective clothing
- ☐ Dissecting tray
- ☐ Dissection tools
- ☐ Bone cutters
- ☐ Preserved cat, either unskinned or skinned
- ☐ Spray bottle with a solution of soap and water for cleanup

ACTIVITY 1

Identifying the Organs of the Alimentary Canal

Learning Outcomes

1. Identify the digestive system structures located in the head, neck, and thorax.
2. Identify the organs of the alimentary canal and the specific portions or regions of each organ.

Instructions

1. Put on safety glasses, gloves, and a lab coat, and then place your cat ventral side up on the dissecting tray. If not already completed, cut open the mouth as described in Dissection 3, Activity 2, step 2. Examine the **oral cavity** (**Figure D4-1**).
   a. The roof of the oral cavity is formed by the **hard palate** cranially and the **soft palate** caudally. Run a blunt probe along the roof of the oral cavity to locate the boundary between the hard and soft palates.
   b. Identify the **tongue**. Note the extensive number of long papillae, which give the tongue its rough feel and aid the cat in grooming.
   c. On the inferior side of the tongue identify the **lingual frenulum**, the connective tissue "cord" that secures the tongue to the floor of the oral cavity.
   d. Examine the **teeth** and observe the dental formula for the adult domestic cat:

$$\frac{\text{I3, C1, PM3, M1}}{\text{I3, C1, PM2, M1}}$$

   Note if your specimen varies from this expected formula.

   What is the dental formula for humans? ______________________________

   The premolars in cats have a sharp, slicing edge. This specialization indicates that cats are adapted to eating which type of food (nuts, seeds, grasses, meat, leaves, bones, etc.)? ______________________________

   In humans, the premolars are small and the molars are modified into crushing surfaces. The human dentition is modified for eating which types of food?

   ____

2. Examine the **pharynx**. Using a blunt probe, demonstrate the continuity of the nasopharynx, oropharynx, and laryngopharynx.

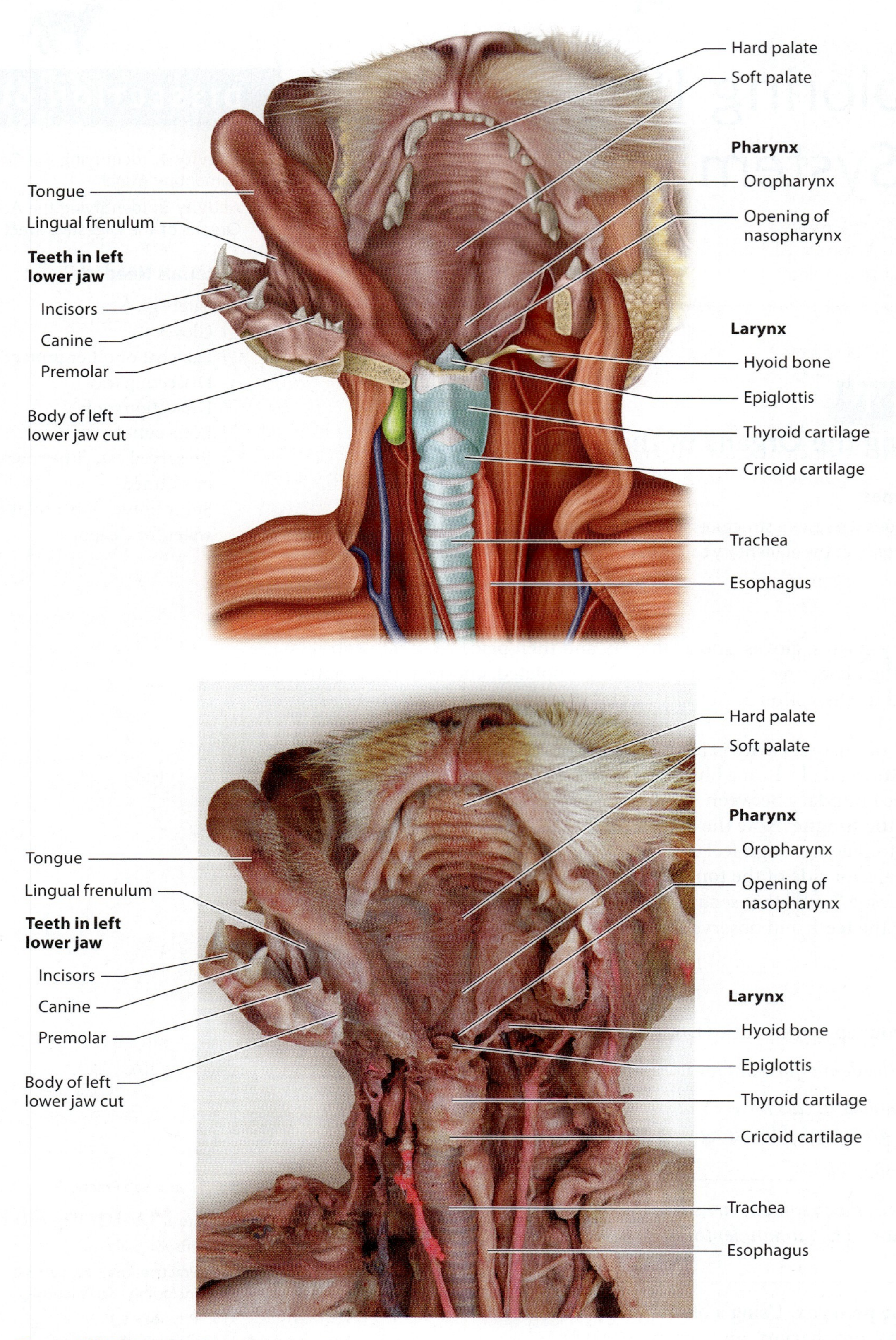

Figure D4-1 Oral cavity, internal view.

3. As food is swallowed, the pharyngeal constrictor muscles push the bolus of food from the pharynx into the **esophagus**. Pass a blunt probe into the esophagus from the laryngopharynx. Identify the esophagus both in the neck dorsal to the trachea and in the thorax if the thoracic cavity is opened.
4. If you haven't done so already, open the abdominal cavity using the instructions in Dissection 3, Activity 1, step 1. Identify the esophagus where it passes through the diaphragm and enters the stomach. Locate the **stomach** in the upper-left quadrant of the abdomen (**Figure D4-2**).
   a. Identify the serous membranes associated with the stomach: the **greater omentum**, the large apron-like drape that extends from the greater curvature of the stomach, and the **lesser omentum** (not shown in Figure D4-2), which extends between the lesser curvature of the stomach and the liver.
   b. The location of the **pyloric sphincter**, a smooth muscle between the stomach and the duodenum, is indicated by a constriction at the end of the pyloric region of the stomach.
   c. With scissors, make a longitudinal cut in the wall of the stomach to examine its internal surface. Note the longitudinal folds of the mucosa layer, the **rugae**, which increase the surface area of the stomach and allow it to stretch considerably to accommodate a meal.
5. The **small intestine** consists of three regions: the duodenum, jejunum, and ileum. The boundaries between these regions are not apparent in gross dissection; they are distinguished histologically. Begin at the stomach and hold the proximal portion of the small intestine with your fingers. Slide your fingers along the length of the small intestine, following it to its termination at the beginning of the large intestine.
   a. The **duodenum** is the proximal portion of the small intestine and is approximately 12–18 cm (5–7 in.) long. In the cat, the duodenum is located in the right side of the abdomen. The **jejunum**, the middle section of the small intestine, forms approximately two-fifths of its length. The last part of the small intestine, the **ileum**, meets with the large intestine at the **ileocecal junction**.
   b. The **mesentery** is a thin, translucent serous membrane that attaches the coils of the small intestine to the dorsal body wall. Spread out the coils of the small intestine. In specimens in which the hepatic portal vessels are injected (triple-injected specimens), the mesentery looks like a stained glass window filled with red (arteries) and yellow (veins). If the hepatic portal vessels are not injected in your specimen (double-injected specimens), you will only see the red arteries traveling within the mesentery.
6. The small intestine ends at the ileocecal junction, a T-shaped connection in which the base of the T is the ileum, and the top of the T is the **large intestine**.
   a. The **cecum** is a small pouch at the beginning of the large intestine. With scissors, make a small longitudinal cut on the side of the cecum opposite the ileocecal junction (Figure D4-2). Open the cecum and view the **ileocecal valve**, a muscular sphincter at the ileocecal junction. In humans, the appendix extends off of the cecum. Cats do not have an appendix.
   b. The large intestine continues as the **colon**. The three parts of the colon—the ascending, transverse, and descending colon—are difficult to distinguish in the cat.
   c. Identify the **mesocolon**, a serous membrane that attaches the colon to the dorsal body wall.
   d. The **rectum**, the terminal portion of the large intestine, is located in the pelvic cavity. The rectum is not viewable in this dissection but will be exposed in a later dissection.
   e. The **anus** is the external opening of the alimentary canal. Examine it on the external surface of the cat.

   What differences do you observe between the GI tract of cats and the GI tract of humans?

   ____

   ____

7. Follow the instructions for storage, waste disposal, and cleanup as described in Dissection 1, Activity 1, steps 10–12.

ACTIVITY 2

Identifying the Accessory Organs of the Digestive Tract

Learning Outcomes

1. Identify the accessory organs of digestion, the secretion each produces, and the duct(s) that drain each.
2. Identify the spleen.
3. Identify the mesenteries of the abdominal cavity and the organs to which each attaches.

Instructions

1. Put on safety glasses, gloves, and a lab coat, and then place the cat on its lateral side on the dissecting tray. To remove the skin from the lateral cheek, push a blunt probe beneath the skin of the cheek from the angle of the mandible dorsally toward the ear. With a scalpel cut through this thick layer of skin. Using the forceps, a blunt probe, and reverse scissor technique if needed, peel the skin of the cheek dorsally, freeing it from the underlying fascia.

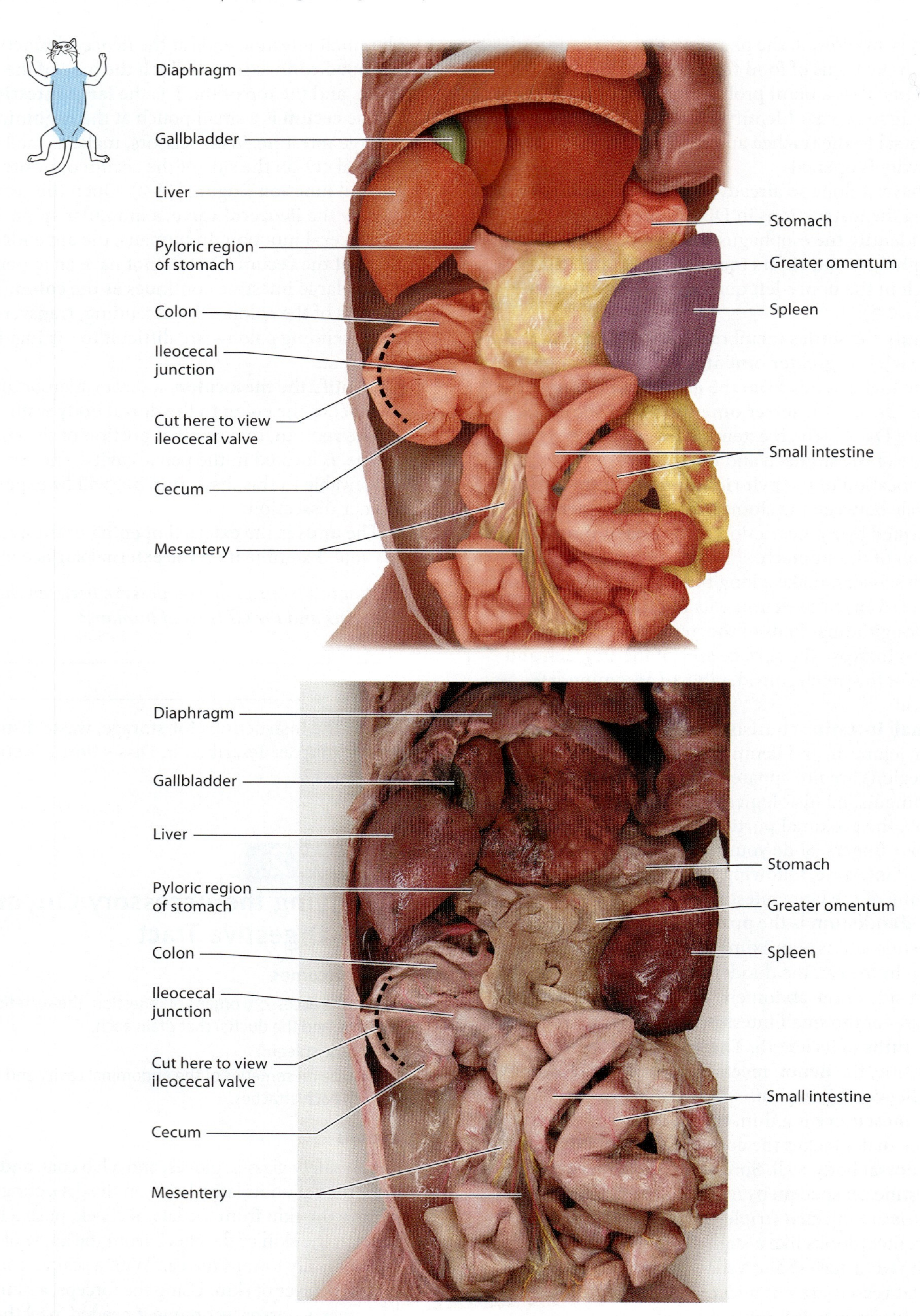

Figure D4-2 Digestive organs of the abdominal cavity.

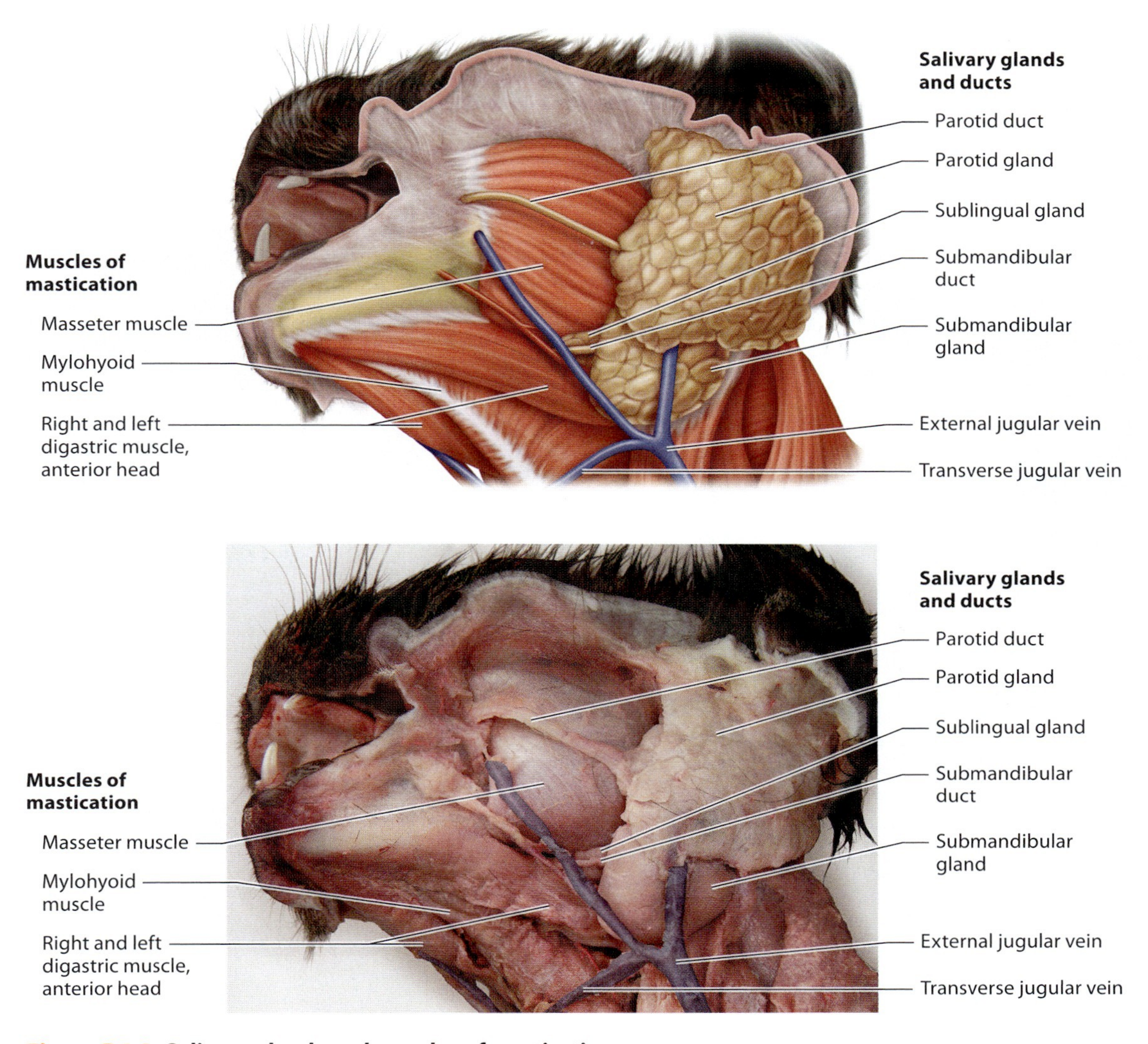

Figure D4-3 Salivary glands and muscles of mastication.

Be careful not to remove the glandular tissue just ventral to the ear.

2. On the ventral surface of the jaw, with the forceps and a blunt probe, clean and expose the large external jugular veins running along the lateral neck. Using the blunt probe and forceps, identify and remove the two lymph nodes (which resemble glands) that surround each external jugular vein so you can view the **salivary glands** (**Figure D4-3**).
   a. Identify the large **parotid gland** located just inferior to the ear. This is the largest of the three salivary glands. The **parotid duct**, which drains this gland, extends from the deep surface of the gland, crosses over the masseter muscle on the cheek, and opens into the lateral wall of the oral cavity. Identify this duct superficial to the masseter muscle.
   b. Ventral to the parotid gland is a smaller, circular salivary gland, the **submandibular gland**. With a blunt probe clean away the surrounding fascia to view this gland clearly. The submandibular duct exits the deep surface of the gland and runs cranially along the angle of the mandible.
   c. Cranial to the submandibular gland, lying along the submandibular duct, is the small, triangular **sublingual gland**. The sublingual duct runs with the submandibular duct but is too small to see in this dissection.

   Which salivary gland is largest in both cats and humans?

   ____

3. Reposition the cat so you are viewing its ventral surface. Return to the abdominal cavity and identify the **liver**, the large, dark-colored organ just inferior to the diaphragm (see Figure D4-2). The liver is divided into right and left lobes, each of which is subdivided into medial and lateral lobes. A small caudate lobe is located dorsally, close to the right kidney.
   a. On the cranial surface of the liver identify the **falciform ligament**, a serous membrane that extends

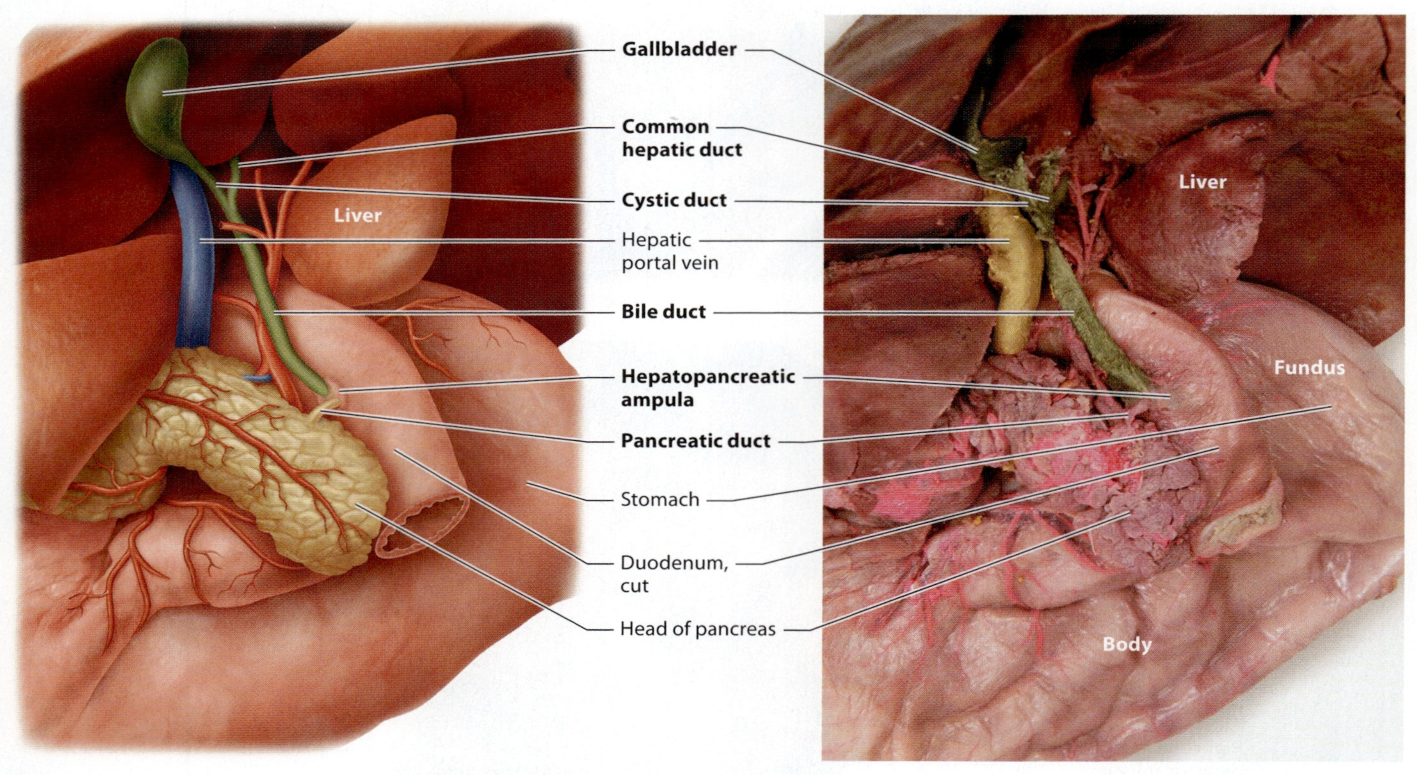

Figure D4-4 **Ducts of the liver, gallbladder, and pancreas.**

from the ventral body wall and the caudal surface of the diaphragm to the cranial surface of the liver.

b. Identify the **ligamentum teres**, a slight thickening of the ventral edge of the falciform ligament. The ligamentum teres is a remnant of the umbilical vein.

c. Locate again the **lesser omentum**, a serous membrane that extends from the liver to the lesser curvature of the stomach and to the duodenum. The hepatic artery, hepatic portal vein, and the bile ducts run through the lesser omentum.

The liver produces bile, a digestive secretion that emulsifies fat. Bile is secreted from the liver and carried via ducts to the gallbladder, where it is stored.

4. Identify the **gallbladder**, a small green, balloon-like structure that is imbedded in the right lobe of the liver (**Figure D4-4**). When chyme enters the duodenum from the stomach, bile is released from the gallbladder through ducts that enter the duodenum.

What digestive secretion does the liver produce?

____

On which nutrient molecule does this secretion act?

____

What organ stores this secretion? _______________

5. The ducts that carry bile from the liver and gallbladder are imbedded within the lesser omentum. These ducts are commonly green or dark in color, and they tear very easily, so use caution as you dissect. With a blunt probe, carefully tease away the lesser omentum to view the **bile duct** as it approaches the duodenum (see Figure D4-4). Follow the bile duct cranially. It forks into the **cystic duct**, which drains the gallbladder, and the **common hepatic duct**, which drains the right and left lobes of the liver.

6. The **pancreas** is an elongated, glandular structure that secretes into the duodenum. Lift the greater omentum cranially to locate the body of the pancreas imbedded in the greater omentum near the greater curvature of the stomach. The head of the pancreas lies along the curve of the pylorus and duodenum. With a blunt probe, tease away the tissue in the head of the pancreas to expose

the thin **pancreatic duct**, which drains the pancreas. The pancreatic duct merges with the bile duct to form the **hepatopancreatic ampulla**, which empties into the duodenum. This junction is difficult to view.

7. Although not part of the digestive system, reidentify the **spleen** in the upper-left quadrant of the abdomen, lateral to the stomach (see Figure D4-2). It is part of the lymphatic/immune system. The spleen removes damaged or aged blood cells from the blood, activates the immune response to foreign antigens, and stores platelets.
8. Follow the instructions for storage, waste disposal, and cleanup as described in Dissection 1, Activity 1, steps 10–12.
9. **Optional Activity**

Practice labeling digestive structures on cats at MasteringA&P® > Study Area > Practice Anatomy Lab > Cat > Digestive System

Review Questions

1. Identify the serous membrane(s) in the cat that attach to each of the following digestive system organs:

   a. Stomach ______________________________

   ____

   b. Small intestine ______________________________

   c. Large intestine ______________________________

   d. Liver ______________________________

   ____

2. CHART In the following chart, list the accessory digestive glands you have identified in this lab. Then name the duct that drains each gland and the region of the alimentary canal into which each duct empties.

Gland	Duct	Region of Alimentary Canal That Receives Secretion

3. Match the following organs with the functions described.

_____ a. stomach	1. removes aged or damaged blood cells
_____ b. small intestine	2. forms the bony roof of the oral cavity
_____ c. pancreas	3. produces bile
_____ d. liver	4. produces pepsin, which digests proteins
_____ e. gallbladder	5. empties into the cardia of the stomach
_____ f. colon	6. stores bile
_____ g. salivary gland	7. secretes into the duodenum enzymes that act on carbohydrates, proteins, and fats
_____ h. spleen	8. secretes enzymes that act on carbohydrates
_____ i. pyloric sphincter	9. is a gateway between the small and large intestines
_____ j. ileocecal valve	10. absorbs nutrients
_____ k. hard palate	11. absorbs water
_____ l. esophagus	12. is a gateway between the stomach and duodenum

5 Exploring the Cardiovascular System of the Cat

DISSECTION OUTLINE

Activity 1: Identifying Arteries
Activity 2: Identifying Veins

Materials Needed

- ☐ Safety glasses
- ☐ Gloves
- ☐ Lab coat or other protective clothing
- ☐ Dissecting tray
- ☐ Dissection tools
- ☐ Dye-injected, preserved cat, either unskinned or skinned
- ☐ Spray bottle with a solution of soap and water for cleanup

ACTIVITY 1

Identifying Arteries

Learning Outcomes

1. Identify the major arteries of the thorax, head and neck, and forelimb, and indicate the region supplied by each identified vessel.
2. Identify the major arteries branching off the abdominal aorta, and trace each to the organ(s) it supplies.
3. Identify the major arteries of the hind limb and the region supplied by each identified vessel.

Instructions

A. Arteries of the Thorax

1. Put on safety glasses, gloves, and a lab coat, and then place your cat ventral side up on the dissecting tray. If you haven't done so already, open the thoracic and abdominal cavities following the instructions in Dissection 3, Activity 1. Reidentify the thymus in the superior mediastinum and on the ventral surface of the pericardium (Dissection 3, Activity 1). Holding the thymus with forceps, use a blunt probe to carefully tease away the thymus and fatty tissue from the pericardium. Reidentify and preserve the thin thread-like phrenic nerve closely adhered to the pericardium along the root of the lung. The phrenic nerve innervates the diaphragm and was examined in Dissection 3, Activity 2.
2. Using scissors, cut open the pericardium to expose the heart; do not cut too deeply or you will damage the heart. The fibrous connective tissue of the pericardium is tough, but relatively thin. Once the pericardium is opened, use a blunt probe and forceps to clear away the fat and connective tissue surrounding the heart and the blood vessels running to and from the heart (**Figure D5-1**). Clean the vessels so you can follow their course. Identify the large (blue) superior vena cava emptying into the right atrium. This vessel will be studied again during Activity 2 of this dissection.
3. Identify the **pulmonary trunk** exiting from the right ventricle. The pulmonary trunk bifurcates adjacent to the aortic arch to form the **right** and **left pulmonary arteries**. These arteries carry poorly oxygenated blood (and thus are injected with blue latex) from the heart to the lungs to be oxygenated.
4. The **aorta** exits the left ventricle of the heart as the **ascending aorta**. This vessel curves to the left just cranial to the heart to become the **aortic arch** and then continues as the **descending aorta** through the thorax and into the abdomen. The aorta and all of the systemic arteries you will be identifying carry highly oxygenated blood and thus are injected with red latex.
5. The arteries that supply the myocardium of the heart, the **right** and **left coronary arteries**, branch off the ascending aorta just cranial to the aortic semilunar valve and run in the coronary sulcus. Identify branches of these

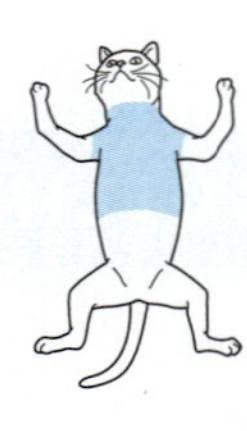

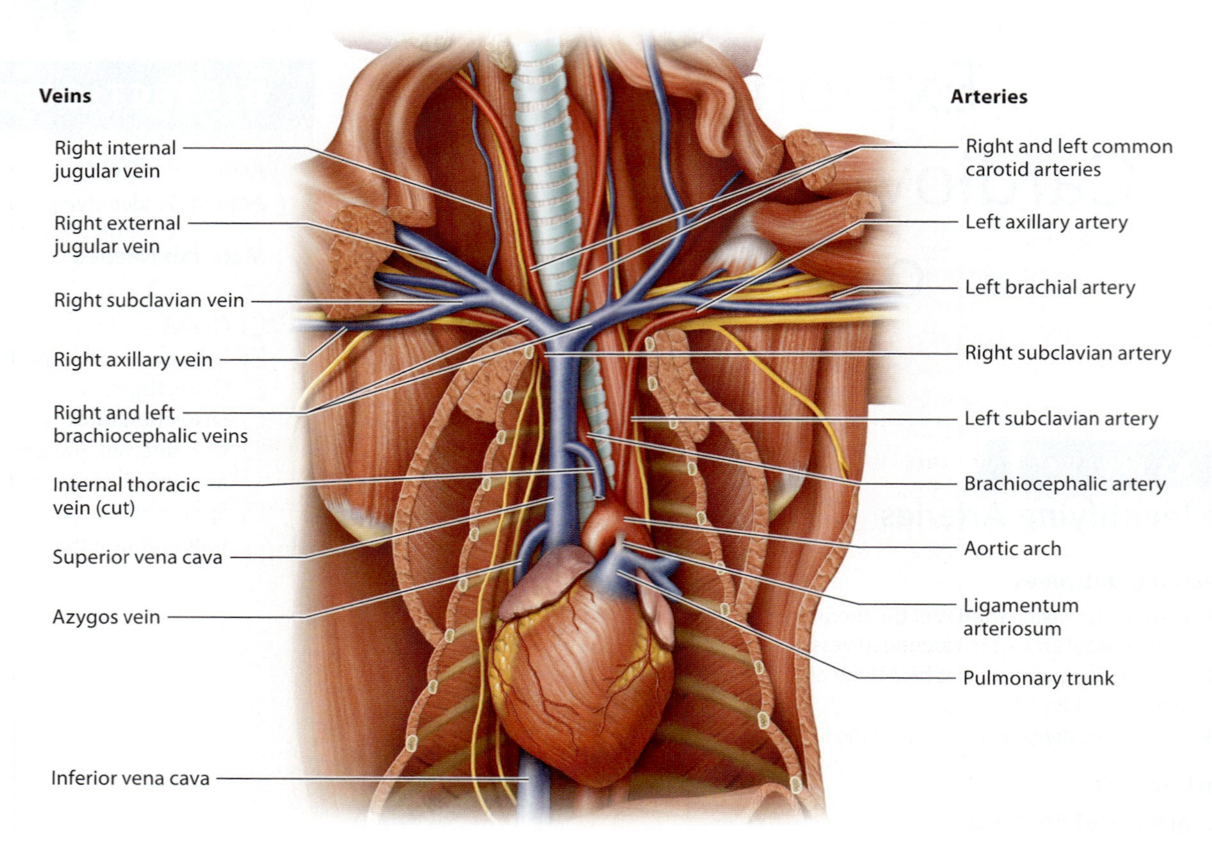

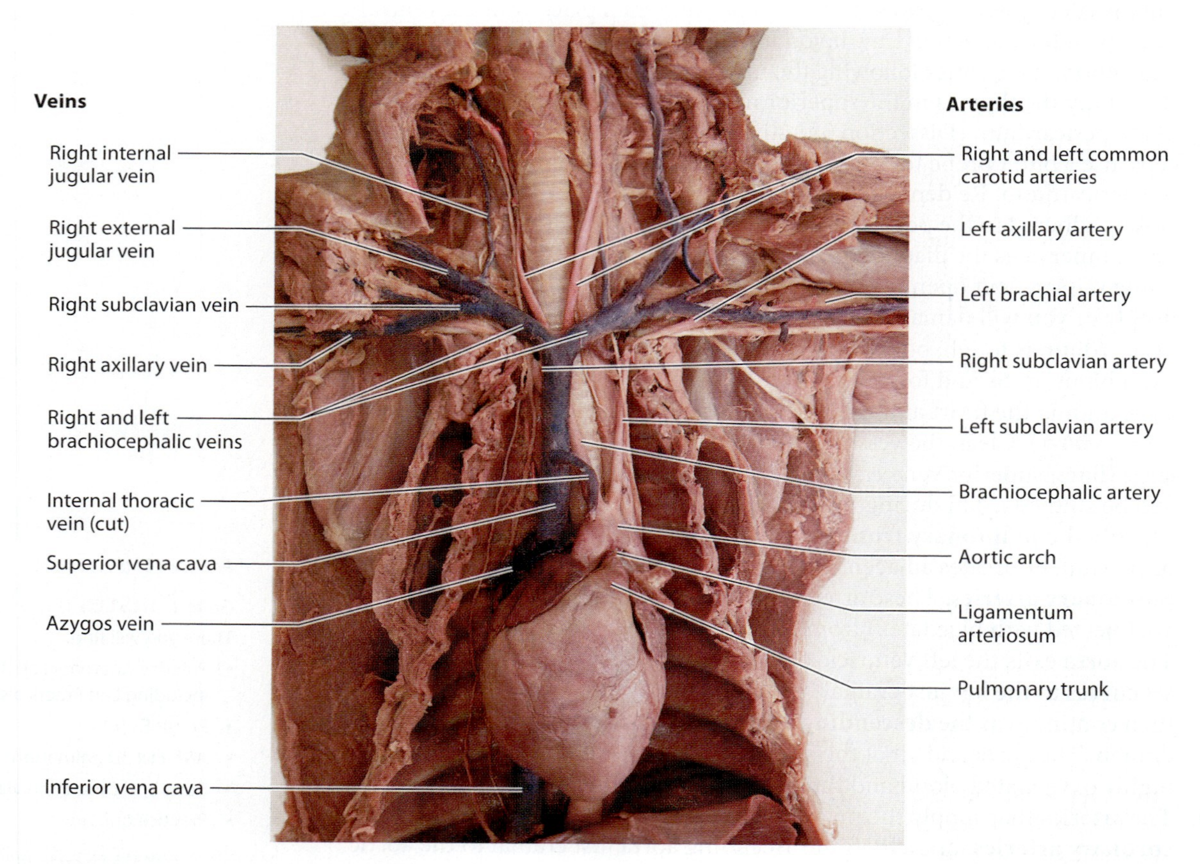

Figure D5-1 Arteries and veins of the thorax and forelimb.

vessels on the anterior surface of the heart in the coronary and interventricular sulci.

6. Return to the aortic arch to identify the large vessels that supply the head, neck, thorax, and upper limbs. Clean and follow the vessels to identify their branches. You may transect the superior vena cava to view these vessels more clearly.
   a. Identify the two large vessels that branch off the aortic arch. The **brachiocephalic artery** branches off first and carries blood to the right upper limb and thoracic wall and to the head and neck. The **left subclavian artery** branches off next and supplies the left upper limb and thoracic wall.
   b. Three vessels branch off the brachiocephalic artery: the **left common carotid artery**, the **right common carotid artery**, and the **right subclavian artery**. Follow the common carotid arteries as they extend cranially on either side of the trachea. Near the angle of the mandible, the common carotid arteries bifurcate into the **external** and **internal carotid arteries**. The external carotid artery supplies the external structures of the head; the internal carotid artery passes through the carotid canal, enters the skull, and supplies blood to the brain.

   How does the pattern of branching off the aortic arch in the cat differ from the pattern typical in humans?

   ____

   ____

   ____

7. Return to the subclavian arteries; clean them and locate their branches. It is easiest to identify branches off the left subclavian artery because it is longer; however, the same branches can be located off the right subclavian artery.
   a. Identify the **vertebral artery**, which branches off the subclavian artery and dives deep toward the vertebrae. It continues cranially, passing through the transverse foramina of the cervical vertebrae, enters the skull through the foramen magnum, and supplies the brain. Three other vessels branch off the subclavian artery and supply the thoracic body wall and neck: *the internal thoracic artery, costocervical trunk*, and *thyrocervical trunk*.

B. Arteries of the Forelimb

1. Follow the subclavian artery as it passes through the thoracic body wall and enters the forelimb. Once in the axilla, this vessel is called the **axillary artery**. Three branches arise from each axillary artery: the *ventral thoracic* and *lateral thoracic arteries* serve the pectoralis muscles, and the *subscapular artery* serves the scapular muscles.
2. After the subscapular artery branches off the axillary artery, the vessel continues into the arm as the **brachial artery**. The brachial artery runs along the humerus, traveling with the median nerve through the supracondylar foramen into the forearm. Deep to the forearm flexor muscles in the proximal forearm, the brachial artery branches into the **radial artery**, which runs along the radial (lateral) side of the forearm, and the **ulnar artery**, which runs along the ulnar (medial) side of the forearm.

   Name the nerve that travels with each of the following:

   Radial artery ______________________________

   Ulnar artery ______________________________

   Brachial artery ______________________________

3. Return to the thoracic aorta. Identify the numerous paired **intercostal arteries** that branch off the thoracic aorta, continue in the costal groove of each rib, and supply the thoracic body wall.

C. Arteries Supplying the Abdominal Organs

1. The aorta continues caudally, passing through the diaphragm (**Figure D5-2**). Move all of the digestive viscera to the right side of the abdomen. With the blunt probe, probe through the parietal peritoneum dorsally to the left of the midline just caudal to the diaphragm to identify the aorta as it passes through the diaphragm. The arteries that branch off the abdominal aorta can be divided into two groups: midline vessels that supply the organs of the digestive system, and paired vessels that supply the paired abdominal organs and the abdominal wall. To expose these vessels, you must probe through the parietal peritoneum and the mesenteries that attach the abdominal organs to the body wall. Clean and follow each vessel from the abdominal aorta to the area it supplies. The abdominal arteries are described next from cranial to caudal.
2. As the aorta passes through the diaphragm, the short, thick **celiac trunk** branches off the abdominal aorta. This midline vessel sends branches to the liver (*hepatic artery*), the stomach (*left gastric artery*), and the spleen (*splenic artery*).

   What mesentery does this vessel travel through?

   ____

3. The **superior mesenteric artery**, also a midline vessel, branches off the aorta just caudal to the celiac trunk. Branches of this vessel supply the small intestine (the *pancreatoduodenal artery* and the numerous *intestinal branches*) and the proximal part of the large intestine (the *ileocolic artery* and the *middle colic* artery).

   What mesentery does this vessel travel through?

   ____

4. The **adrenolumbar arteries** supply the adrenal glands, the lumbar body wall, and the diaphragm. Being careful not to damage the adrenal glands (small pea-sized glands located cranial and medial to the kidneys and adjacent to the aorta), carefully clean the abdominal

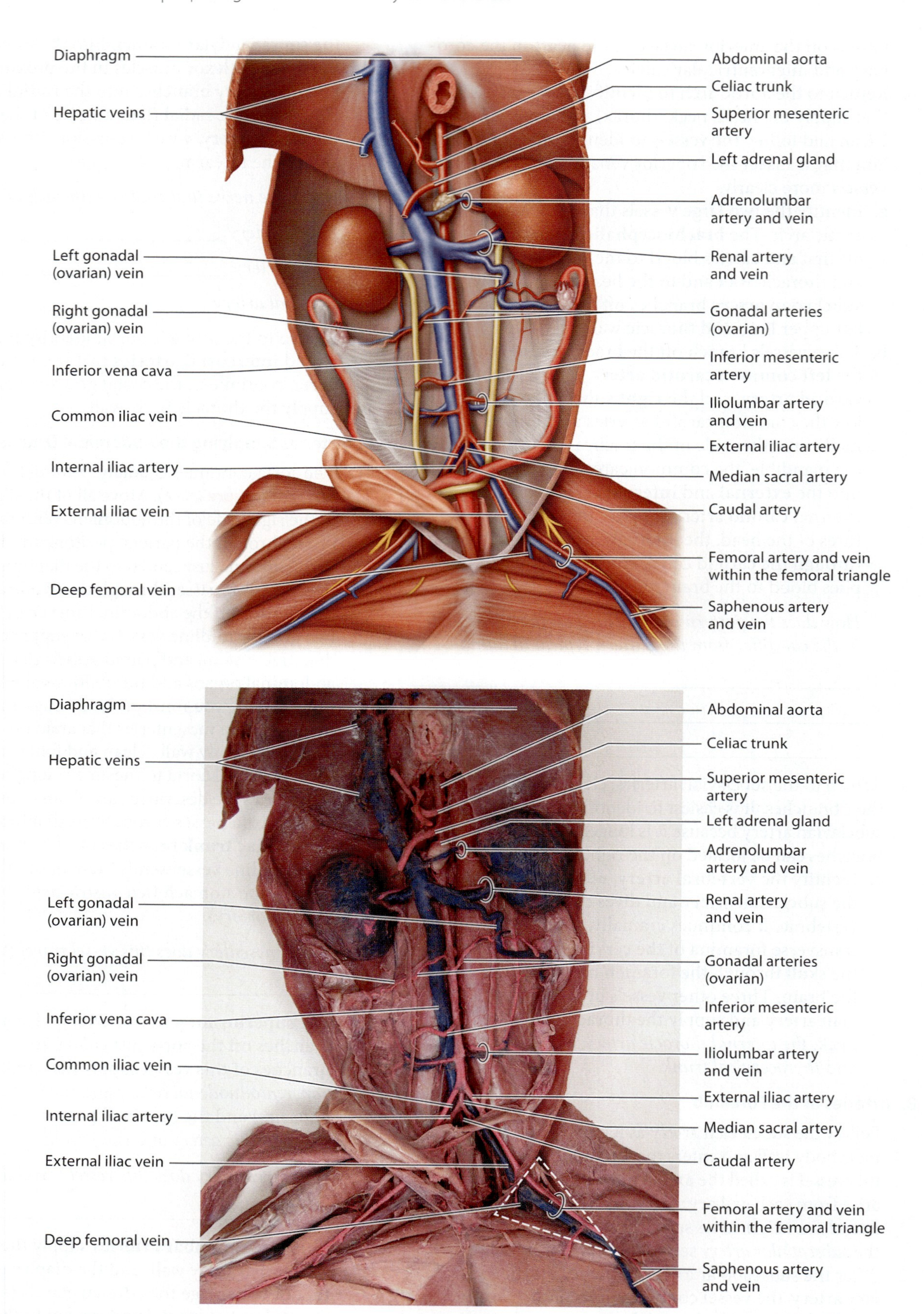

Figure D5-2 Arteries and veins of the abdomen and hind limb (excluding the hepatic portal veins).

aorta cranial to the renal arteries to expose these small, paired arteries.

5. The **renal arteries** are large, paired arteries that supply the kidneys.
6. The **gonadal arteries** are extremely thin (be careful you do not destroy them), paired arteries that branch off the abdominal aorta just caudal to the renal arteries. These vessels supply the testes in males (**testicular arteries**) and the ovaries in females (**ovarian arteries**). The testicular arteries pass through the inguinal canal in the anterior abdominal wall and continue caudally to the testes in the scrotum.
7. The **inferior mesenteric artery** is a midline arterial branch of the aorta that supplies the descending colon and rectum.

   What mesentery does this vessel travel through?

   ____

8. The paired **iliolumbar arteries** branch off the aorta, cross the deep surface of the ilium, and supply the lumbar body wall.
9. At its caudal extent, the aorta branches into the paired **external iliac arteries**. These arteries pass through the abdominal wall and enter the lower limb as the **femoral arteries**.
10. Caudal to the branching of the external iliac arteries, the aorta continues as a small vessel, the **median sacral artery**, from which the **internal iliac arteries** branch. The internal iliac arteries supply the structures in the pelvic cavity. The **caudal artery** continues into the tail.

    How does the arterial branching of the iliac arteries differ between cats and humans? ______________________________

    ____

D. Arteries of the Hind Limb

1. Blood supply to the hind limb begins with the femoral artery entering the thigh (see Figure D5-2). Identify the **femoral artery** in the femoral triangle, which is bounded by the sartorius laterally, the adductor longus medially, and the ventral body wall cranially. The femoral artery continues with the saphenous nerve along the ventral surface of the thigh toward the knee.

   In the femoral triangle, what nerve runs adjacent to the femoral artery? ______________________________

2. Just proximal to the knee the femoral artery branches: the **saphenous artery** continues superficially into the leg, whereas a deeper branch, the **popliteal artery**, passes between the vastus medialis and semimembranosus muscles.
3. In the leg, the **sural artery** branches medially off the popliteal artery. The vessel then bifurcates as the **anterior** and **posterior tibial arteries** to supply the foot.

ACTIVITY 2
Identifying Veins

Learning Outcomes

1. Identify the major veins of the thorax, head and neck, and forelimb.
2. Identify the major veins in the abdomen.
3. Describe the different routes of venous return from the paired abdominal organs versus the digestive tract organs.
4. Identify the major veins of the hind limb.

Instructions

A. Veins of the Neck, Thorax, and Forelimb

1. Begin this dissection in the thorax by reidentifying the **superior vena cava (SVC)** as it enters the right atrium returning blood from the body regions above the diaphragm (see Figure D5-1). This vessel is the most ventral vessel adjacent to the heart and is colored blue. Clean this vessel with a blunt probe to view its tributaries.
2. Identify the **internal thoracic (mammary) veins** merging together and entering into the ventral surface of the SVC. These vessels receive blood from the anterior thoracic body wall.
3. Lift the right lung to view the **azygos vein**, which runs along the right dorsal thoracic wall. Run a blunt probe longitudinally along the right dorsal body wall near the vertebrae to expose it. The intercostal veins from both the right and left sides drain into the azygos vein, which then joins with the SVC just as it enters the right atrium.
4. The **right** and **left brachiocephalic veins**, which receive blood from the forelimb and head, drain into the SVC.
5. Locate the **right vertebral vein** (not illustrated), which typically empties into the right brachiocephalic vein near its junction with the SVC.
6. The **inferior vena cava (IVC)** is on the right side of the thorax, inferior to the heart. It continues cranially from the abdomen, passes through the central tendon of the diaphragm, travels through the right side of the thorax, and enters the right atrium. The IVC drains blood from all regions below the diaphragm.
7. Probe in the root of the lung to view the **pulmonary veins**. These vessels, colored red if injected, carry oxygenated blood from the lungs to the left atrium of the heart. Do not confuse these vessels with the pulmonary arteries, colored blue if injected, which branch off the pulmonary trunk and carry poorly oxygenated blood to the lungs.
8. Return to the right and left brachiocephalic veins. Clean these vessels to observe their tributaries. Each brachiocephalic vein is formed from the junction of a **subclavian vein**, which drains the upper limb, and an **external jugular vein**, which drains the head and neck.

Identify these vessels. The **internal jugular vein**, carrying blood from the brain, empties into the external jugular vein near its junction with the brachiocephalic vein. Locate the internal jugular vein as it travels with the common carotid artery and the vagus nerve through the neck.

9. Follow the left subclavian vein out of the thoracic wall and into the axilla. If the veins on the left side were destroyed during earlier dissections, expose these vessels on the right side of the cat. Once the vein has left the thorax, it is called the **axillary vein**. The large **subscapular vein** emerges from the boundary between the subscapularis and teres major muscles and joins the axillary vein. Distal to this junction, the vein continues into the arm as the **brachial vein**.
10. In the forearm, the brachial vein bifurcates into the **radial** and **ulnar veins**, traveling parallel to the radial and ulnar arteries.
11. Identify the large superficial vein of the forelimb, the **cephalic vein**, located superficial to the acromiodeltoid muscle in the shoulder and continuing along the lateral arm and into the forearm with the cutaneous branch of the radial nerve.

Name the vessels that empty directly into the SVC.

____

____

____

Both the cephalic and brachial veins drain blood from the upper limb. How do these two vessels differ?

____

____

B. Veins of the Paired Abdominal Organs

1. Open the abdominal wall and move the organs of the digestive system to the right side. Identify the **inferior vena cava (IVC)** along the dorsal body wall slightly to the right of the abdominal aorta (see Figure D5-2). The blood from the hind limbs and from the paired abdominal organs drains into the IVC. In most cases venous return from the paired abdominal organs parallels the arterial supply. The veins draining the abdomen are described here from cranial to caudal. To expose these vessels, carefully clean along the IVC with a blunt probe. Then follow each vessel toward the region it drains.
2. The **hepatic veins** collect blood from the liver and deliver it to the IVC. These veins are imbedded within the liver and will not be exposed in this dissection.
3. The **adrenolumbar veins** drain the lumbar body wall and the adrenal glands, which are small, pea-sized glands that lie medial and slightly cranial to the kidney. Each adrenolumbar vein crosses over the adrenal gland and empties into the IVC superior to the renal veins. It is not uncommon to observe the left adrenolumbar vein emptying into the left renal vein.
4. The **right** and **left renal veins** are large veins that drain the right and left kidney, respectively.
5. The **right gonadal vein** empties into the IVC. The **left gonadal vein** empties into the left renal vein. These vessels are very thin and may or may not be injected. Clean carefully to expose these veins. Note that in males, the gonadal veins (**testicular veins**) extend from the scrotum, pass through the anterior abdominal wall, and then run along the dorsolateral abdominal wall toward the IVC.
6. The **iliolumbar veins**, which lie alongside the iliolumbar arteries, drain into the IVC.
7. The IVC is formed from the junction of the **common iliac veins**. The **internal iliac vein** branches off the common iliac vein. The vessel then continues toward the hind limb as the **external iliac vein**. With a blunt probe, clean along the pelvic brim to expose these vessels.
8. The **deep femoral vein**, returning blood from the muscles of the medial thigh, drains into the medial side of the external iliac vein just before the external iliac vein leaves the abdominal cavity.

Which vessel draining a paired abdominal organ does not empty directly into the IVC? ______________

C. Veins of the Hind Limb

1. The external iliac vein passes through the body wall and into the hind limb as the **femoral vein** (see Figure D5-2), which lies adjacent to the femoral artery and femoral nerve in the anterior thigh. The femoral vein travels with the femoral artery through the thigh, forming a boundary between the muscles of the anterior and medial compartments.
2. Near the knee, the **saphenous vein** enters the medial side of the femoral vein. This superficial vessel extends down the medial leg into the foot. It may have been removed during skinning.
3. At the knee, the femoral vein dives deep, traveling between the vastus medialis and semimembranosus muscles. It continues as the **popliteal vein** to the back of the knee.
4. The **anterior** and **posterior tibial veins**, which drain the leg and foot, drain into the popliteal vein.

The femoral vein empties into what vessel? ______________

D. Venous Drainage of the Digestive Tract

1. Identify the veins that drain the organs of the digestive tract—the **hepatic portal system** (**Figure D5-3**). These veins do not empty directly into the IVC but instead carry blood from the digestive organs to the liver for processing before it is returned to the IVC. If you are working with a triple-injected specimen, the vessels of the hepatic portal system are colored yellow. If you are working with a double-injected specimen, identifying

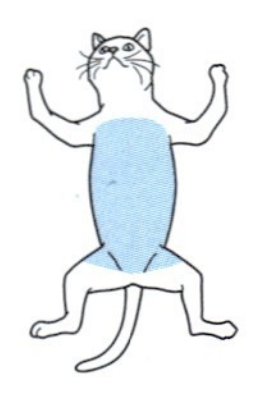

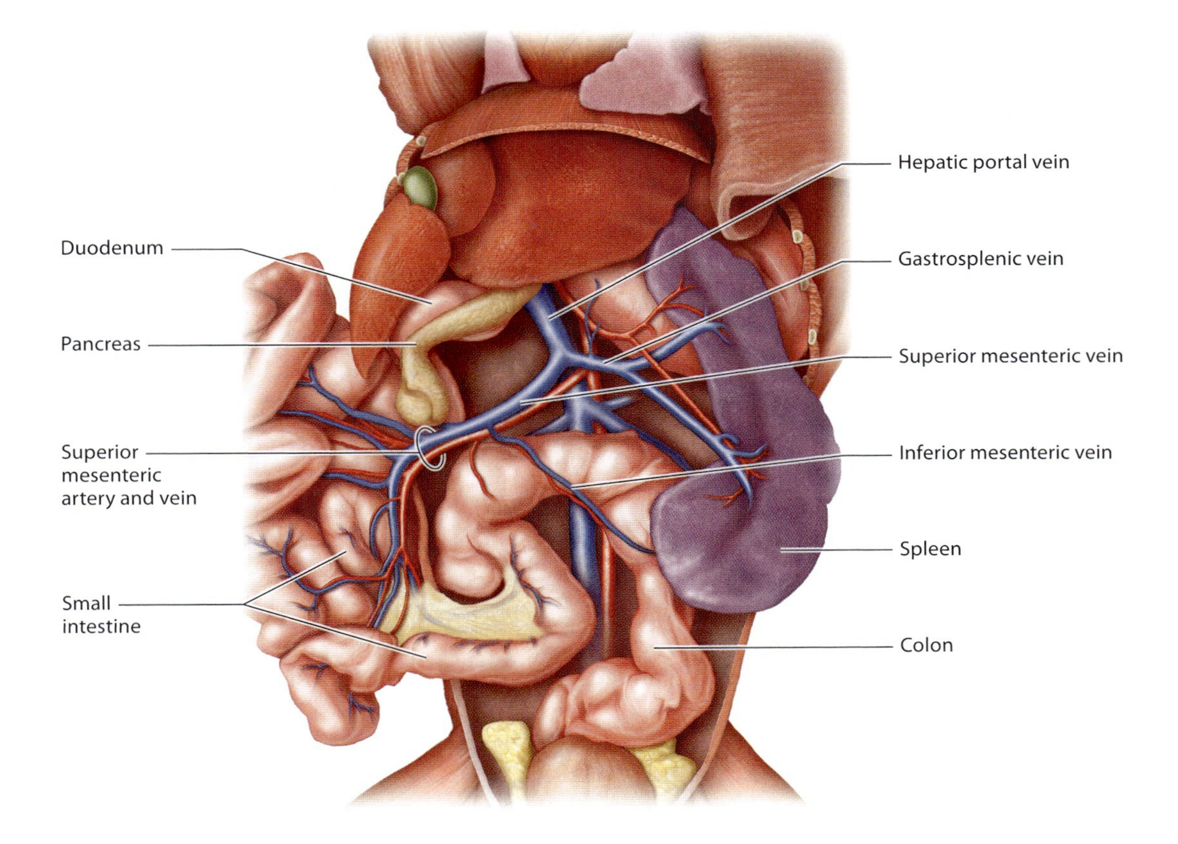

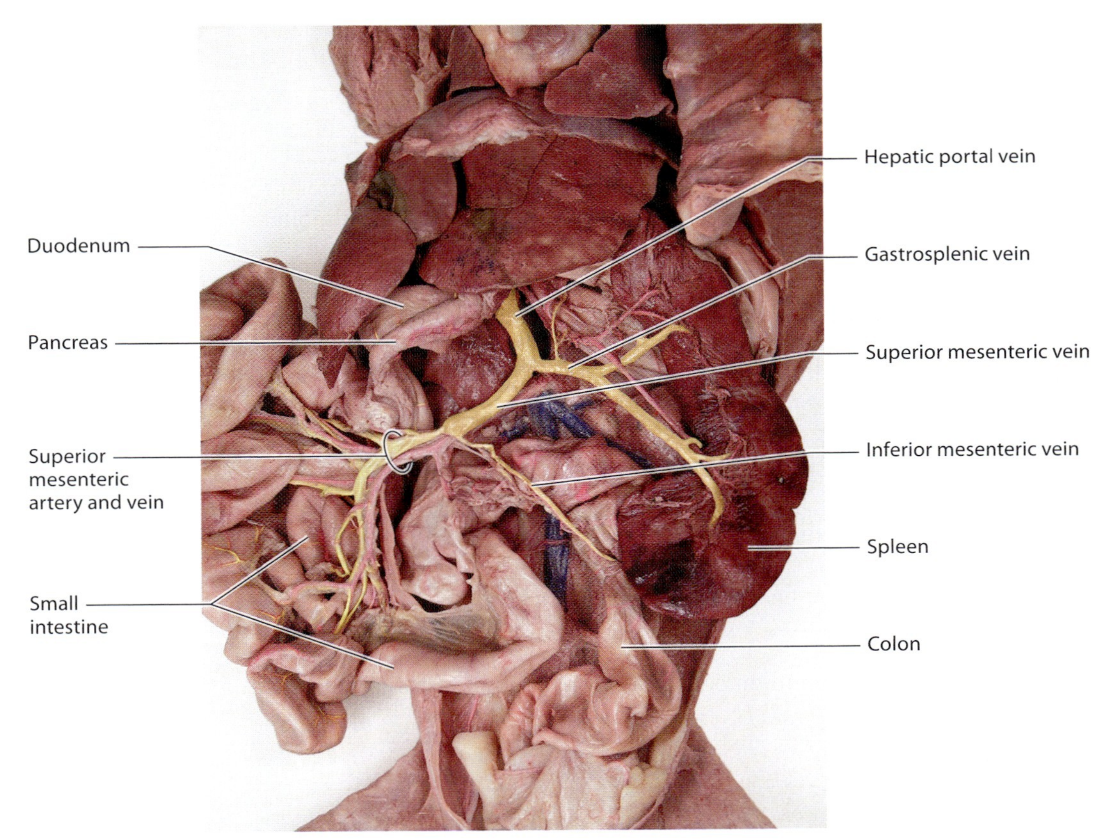

Figure D5-3 The hepatic portal circulation.

these vessels will be difficult. Check with your instructor for guidance.

2. Position the small intestine on the cat's right side to view the superior mesenteric artery and its branches into the intestines. Identify the **superior mesenteric vein** following a similar course. Clean the superior mesenteric vein both cranially and caudally.
3. Caudally, find the junction of the **inferior mesenteric vein**, which drains blood from the distal portions of the large intestine. Cranially the **gastrosplenic vein** joins with the superior mesenteric vein. Flip the stomach cranially to expose and view this vessel. The gastrosplenic vein drains blood from the spleen and stomach.
4. Cranial to the junction of the gastrosplenic vein, the vessel continues as the **hepatic portal vein**, delivering blood to the liver.
5. The **pancreaticoduodenal vein** drains the proximal portion of the small intestine and empties into the hepatic portal vein from the right side as it approaches the liver. Clean the hepatic portal vein to expose this tributary.
6. Blood delivered to the liver via the hepatic portal vein is processed by the liver and then collected by the hepatic veins, which, as previously noted, empty into the IVC.
7. Follow the instructions for storage, waste disposal, and cleanup as described in Dissection 1, Activity 1, steps 10–12.
8. **Optional Activity**

Practice labeling cardiovascular structures on cats at MasteringA&P® > Study Area > Practice Anatomy Lab > Cat > Cardiovascular System

Review Questions

1. Name the three midline vessels that branch off the abdominal aorta and the organs that each supplies.

____

____

____

2. Name the arteries that supply blood to the brain, and the artery from which each vessel branches.

____

____

3. The pulmonary arteries are typically injected with blue latex; the pulmonary veins (when injected) contain red latex. Explain what these colors indicate, and why the pulmonary vessels are so colored.

____

____

4. Fill in the vessels of the pathway that blood takes in the cat to reach the listed artery.

   a. left ventricle → ____________ → ____________ → ____________ → ____________ → right brachial a.

   b. left ventricle → ____________ → ____________ → ____________ → left internal carotid a.

5. Fill in the vessels in the pathway of venous return in the cat from the listed organ.

   a. left kidney → ____________ → ____________ → right atrium

   b. stomach → ____________ → ____________ → liver sinusoids → ____________ → ____________ → right atrium

6. Most vessels are paired (found on both sides of the body). List the veins identified in this dissection that are *not* paired. ____

____

____

Exploring the Urinary System of the Cat

DISSECTION OUTLINE

Activity 1: Identifying the Organs of the Urinary System

Materials Needed

- ☐ Safety glasses
- ☐ Gloves
- ☐ Lab coat or other protective clothing
- ☐ Dissecting tray
- ☐ Dissection tools
- ☐ Preserved cat, either unskinned or skinned
- ☐ Mounted cat skeleton
- ☐ Spray bottle with a solution of soap and water for cleanup

ACTIVITY 1

Identifying the Organs of the Urinary System

Learning Outcomes

1. Identify the organs of the urinary system in male and female cats.
2. Identify the internal structures of the kidney, and trace the pathway of urine from its production in the kidney to its elimination from the body.

Instructions

1. Put on safety glasses, gloves, and a lab coat, and then place your cat ventral side up on the dissecting tray. If you haven't done so already, cut open the abdominal cavity following the instructions in Dissection 3, Activity 1. To expose the urinary structures in the pelvis, the pubic symphysis must be transected. If working with a male cat, carefully expose the spermatic cords extending from the ventral abdominal wall to the testes. Be careful not to cut these important reproductive system structures as you transect the pubic symphysis. Place a blunt probe deep to the pubic symphysis to protect the underlying organs. Palpate the pubis to locate the pubic tubercles. The pubic symphysis is located midline between the pubic tubercles. Use a sharp scalpel to slice longitudinally through the pubic symphysis. Once the pubic symphysis is cut, bend the thighs back to "crack" open the pelvic cavity for better visibility.
2. Open the abdominal cavity. If necessary, push the abdominal organs to the right side of the cat to expose the dorsal region of the abdominal cavity. The two **kidneys**, located cranially in the abdominal cavity, lie along the dorsal body wall in a retroperitoneal position—behind the parietal peritoneum—and are surrounded by a fat capsule. With a blunt probe, clear away the peritoneum and the surrounding fat to expose the kidneys (**Figure D6-1**). Reidentify the **renal artery** (branching off the **abdominal aorta**) and **renal vein** (draining into the **inferior vena cava**) as they emerge from the hilum of the kidney on its medial surface.
   a. Identify the **renal capsule**, the connective tissue forming the outermost layer of the kidney.
   b. With a scalpel, make a frontal section through one kidney to identify its internal structures. Notice the outer region, the **renal cortex**, which appears granular in texture. The darker inner region, the **renal medulla**, contains the renal pyramids, pyramid-shaped structures with a striped texture.

   In which region of the kidney does filtration occur, the renal cortex or the renal medulla?

   ____

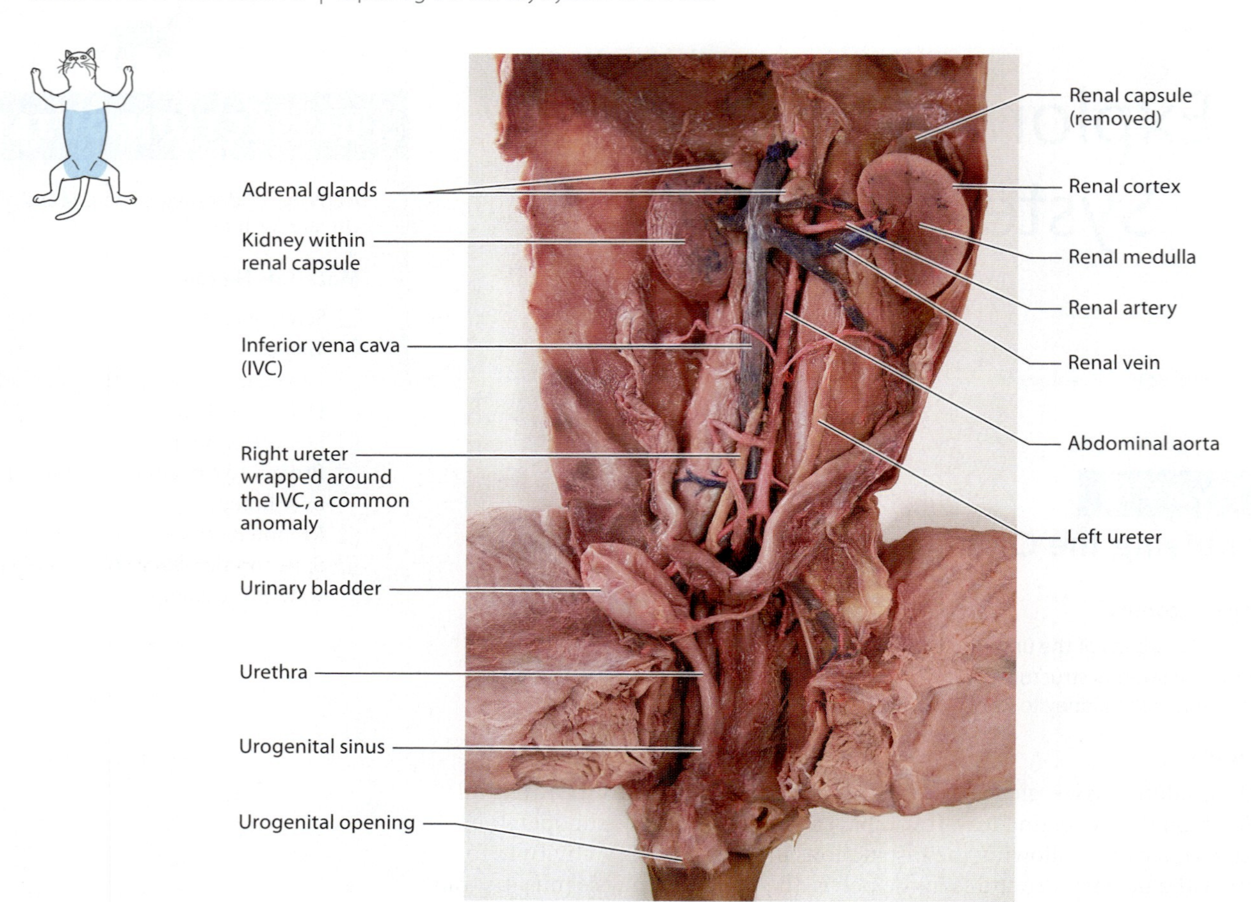

Figure D6-1 Urinary system in the female cat.

Which portions of the nephron are located in the renal medulla?

____

____

____

3. Cranial and medial to the kidneys, adjacent to the abdominal aorta, lie the **adrenal glands**. In humans, these glands sit directly on top of each kidney.
4. Identify and trace the **ureters**, muscular tubes that carry urine from the kidney to the urinary bladder. Be careful of the gonadal vessels as you expose the ureter. The ureters enter the posterolateral wall of the urinary bladder. In males, the ductus deferens (which carries sperm from the testis) loops over the ureter near its entry into the urinary bladder (**Figure D6-2**).
5. In the ventral caudal region of the abdominal cavity, identify the **urinary bladder**, a smooth muscle sac that stores urine. In a preserved cat the urinary bladder may resemble a deflated balloon or a thick muscular ball, depending on how distended it is.
6. The **urethra** extends caudally from the urinary bladder. It drains the bladder, conducting urine out of the body.
   a. In males, the urethra passes through the **prostate**, a pea-sized gland located midway down the urethra's length. The urethra continues through the urogenital diaphragm as the membranous urethra and enters the **penis** as the spongy urethra (see Figure D6-2).
   b. In female cats, the urethra merges with the vagina to form the **urogenital sinus**, which opens externally at the **urogenital opening** to allow the excretion of urine (see Figure D6-1).

   How do the external openings of the urinary and reproductive tracts differ between human females and female cats?

   ____

   ____

   ____

7. Follow the instructions for storage, waste disposal, and cleanup as described in Dissection 1, Activity 1, steps 10–12.
8. **Optional Activity**

Practice labeling urinary structures on cats at MasteringA&P® > Study Area > Practice Anatomy Lab > Cat > Urinary System

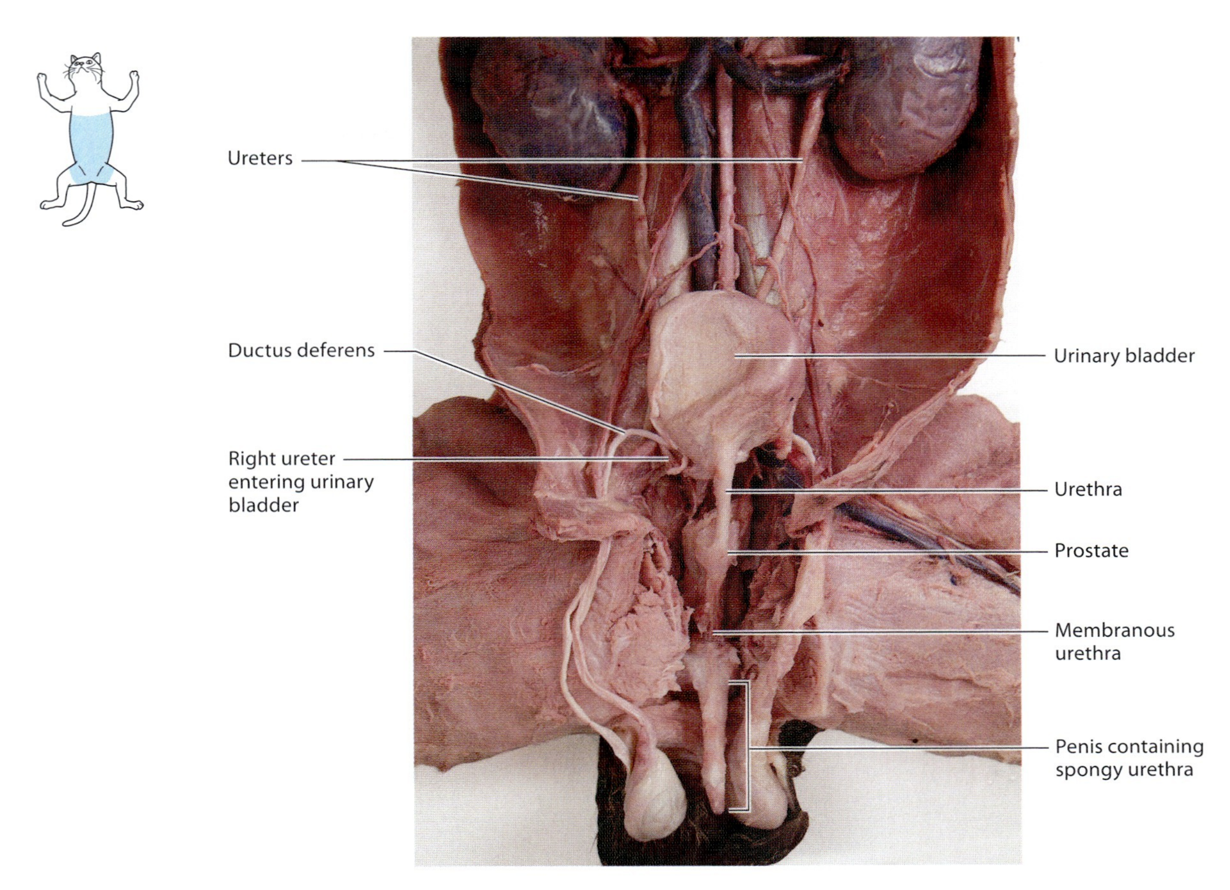

Figure D6-2 **Pathway of ureters and urethra in the male cat.**

Review Questions

1. Number the structures of the urinary system in order from the production of urine (1) to the elimination of urine (5).

   _______ a. urethra

   _______ b. kidney

   _______ c. ureter

   _______ d. urogenital opening

   _______ e. urinary bladder

2. If you can only view the urinary bladder and ureters, what information can you use to determine if a specimen is a male or a female? ____

   ____

   ____

3. In males, which portion of the urethra passes through the urogenital diaphragm?

   ____

4. In female cats, the urethra empties into the ____.

Exploring the Reproductive System of the Cat

DISSECTION OUTLINE

Activity 1: Identifying the Reproductive Structures of the Female Cat

Activity 2: Identifying the Reproductive Structures of the Male Cat

Materials Needed

- ☐ Safety glasses
- ☐ Gloves
- ☐ Lab coat or other protective clothing
- ☐ Dissecting tray
- ☐ Dissection tools
- ☐ Bone cutters
- ☐ Preserved cat, either unskinned or skinned
- ☐ Mounted cat skeleton
- ☐ Spray bottle with a solution of soap and water for cleanup

ACTIVITY 1

Identifying the Reproductive Structures of the Female Cat

Learning Outcomes

1. Identify the organs of the reproductive system in female cats.
2. Identify the peritoneal folds that support the female reproductive organs.
3. Trace the route through the female reproductive tract from oocyte production in the ovary to fertilization, development, and delivery of offspring.

Instructions

1. Put on safety glasses, gloves, and a lab coat, and then place your cat ventral side up on the dissecting tray. In female cats, the reproductive structures are located in the abdominopelvic cavity. If you haven't done so already, cut open the abdominal cavity following the instructions in Dissection 3, Activity 1. Open the abdominal wall and push aside the digestive system organs to view and identify the structures of the female reproductive system. Along the dorsolateral abdominal wall, locate the **ovaries**, small pea-sized structures, and the **uterine horns**, tubular structures extending caudally from the ovaries (**Figure D7-1**). If the pubic symphysis has not been cut open, follow the instructions in the urinary system dissection (Dissection 6, Step 1).
2. The reproductive structures develop behind the parietal peritoneum. In females, folds of the peritoneum, called ligaments, help support the ovaries, uterine tubes, and uterus.
   a. The **broad ligament** attaches the uterine horn, uterine tube, and ovary to the dorsolateral body wall. Identify this thin serous membrane extending from the uterine horn to the lateral body wall.
   b. The **round ligament** extends perpendicular to the broad ligament from the midregion of the uterine horn to the anterior body. It passes through the inguinal canal and attaches to the labia majora in the vulva. Observe the pathway of the round ligament from the uterine horn to the anterior body wall.
3. Reidentify the **ovary**, the primary reproductive organ in females. In the cat the ovaries lie along the lateral abdominal wall caudal to the kidney. Reidentify the vessels that supply and drain the ovary, the **ovarian artery** and the **ovarian vein**. Note again the asymmetry of venous return: the right ovarian vein drains into the inferior vena cava (IVC); the left ovarian vein empties into the left renal vein.
4. Identify the **ovarian ligament**, a short connective tissue cord that attaches the ovary to the cranial end of the uterine horn.

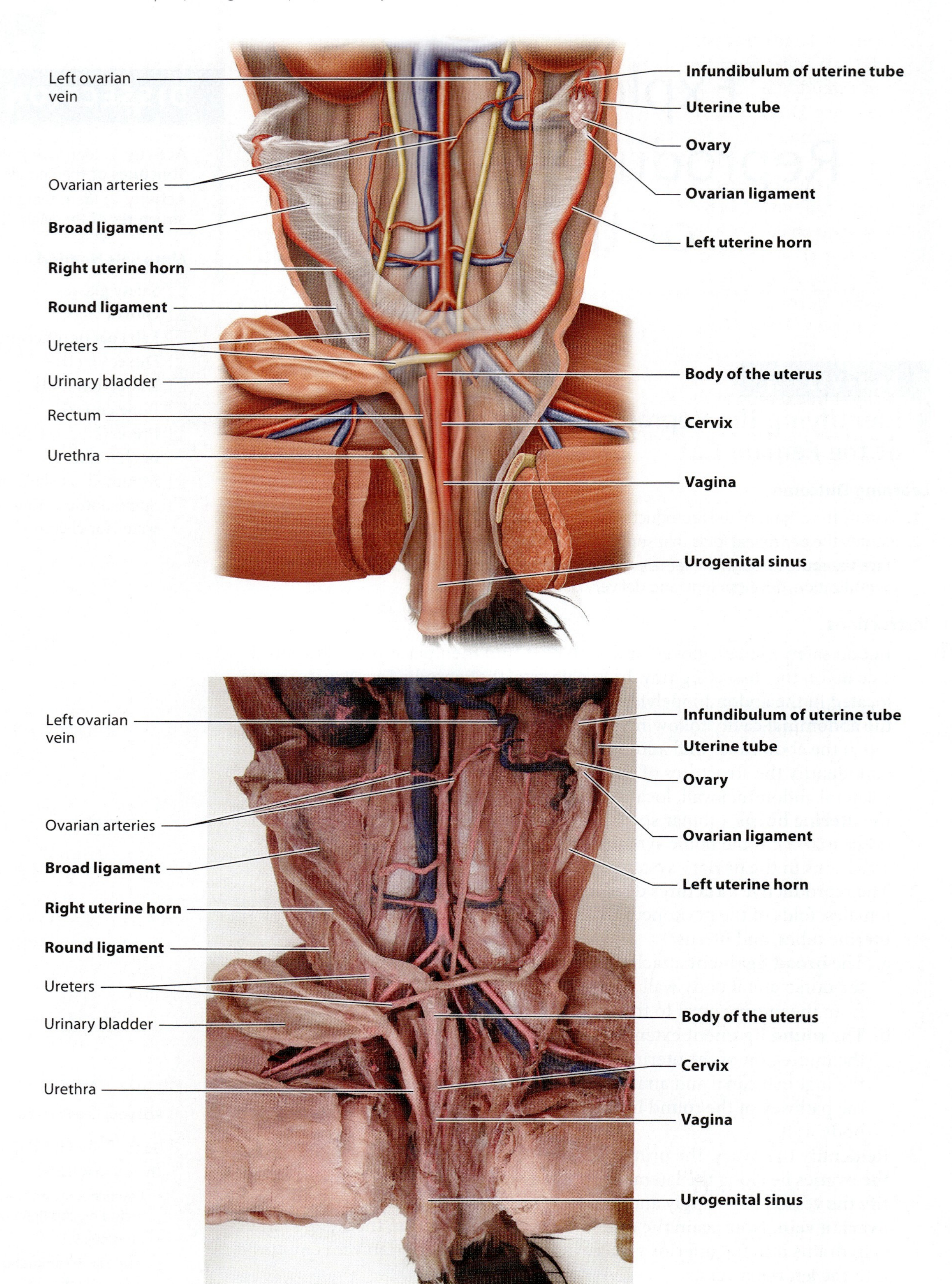

Figure D7-1 **Reproductive organs of the female cat.**

5. Identify the **uterine tube**, an extremely short, coiled tube that arises from the cranial end of the uterine horn and extends cranially and medially, wrapping around the ovary. Do not confuse the tiny uterine tube with the long uterine horn lying along the lateral body wall. The distal expansion of the uterine tube, the **infundibulum**, receives the oocyte upon ovulation.
6. Unlike humans, cats have a bipartite uterus. This Y-shaped structure is composed of two **uterine horns** that fuse caudally, forming the **body of the uterus**. Offspring develop in the uterine horns. Reidentify the uterine horns along the dorsolateral abdominal wall. Identify the body of the uterus dorsal to the urinary bladder. The cervix, the most caudal region of the uterus, opens to the vagina. The cervix may be visible as a slight thickening at the base of the body of the uterus.

   How does the uterine structure in human females differ from that observed here in the female cat?

   ____

   ____

7. The **vagina** is dorsal to the urethra. It merges with the urethra to open into the **urogenital sinus**, a common passageway for the urinary and reproductive systems. The urogenital sinus opens externally at the **urogenital opening**.
8. Externally, identify the **vulva** and the **labia majora** surrounding the urogenital opening.
9. Follow the instructions for storage, waste disposal, and cleanup as described in Dissection 1, Activity 1, steps 10–12.
10. **Optional Activity**

Practice labeling reproductive structures on female cats at MasteringA&P® > Study Area > Practice Anatomy Lab > Cat > Reproductive System > Images 5–9

ACTIVITY 2

Identifying the Reproductive Structures of the Male Cat

Learning Outcomes

1. Identify the organs of the reproductive system in male cats.
2. Identify the peritoneal coverings of the testis.
3. Trace the pathway of sperm from production in the testis to ejaculation from the penis.

Instructions

1. Put on safety glasses, gloves, and a lab coat, and then place your cat ventral side up on the dissecting tray. If you haven't done so already, cut open the abdominal cavity following the instructions in Dissection 3, Activity 1. If the pubic symphysis has not been cut open, follow the instructions in the urinary system dissection (Dissection 6, Step 1).
2. Carefully expose the **spermatic cords** extending from the ventral abdominal wall to the testes (**Figure D7-2**). The spermatic cords are just lateral to the pubic symphysis. Follow the spermatic cords to the testes, located within their skin covering, the **scrotum**. Remove a testis from the scrotum by cutting the lateral wall of the scrotum with scissors and using a blunt probe to separate the testis from the scrotal sac.
3. Follow the spermatic cord cranially from the testis through the **inguinal canal**, the passageway through the musculature of the anterior abdominal wall. Identify the **superficial inguinal ring**, an opening in the external oblique muscle through which the spermatic cord passes, and the **deep inguinal ring**, an opening internally through the transversus abdominis. Identify the individual components of the spermatic cord within the abdominal cavity: the **ductus deferens**, the **testicular artery**, and the **testicular vein**. Note the asymmetry of drainage of the testicular vein: the right testicular vein drains into the IVC; the left empties into the left renal vein.
4. The **testes** are the primary reproductive organ in males. The testes develop in the abdomen and descend into the scrotum during *in utero* development, carrying with them the nerves, vessels, and ducts that supply them. As they pass through the abdominal body wall, these structures are covered with a connective tissue wrapping and together are called the spermatic cord.
   a. A two-layered membranous sac derived from the peritoneum, the **tunica vaginalis**, covers the anterior surface of each testis. Use scissors to carefully slit the outer layer of the tunica vaginalis to fully expose the testis and epididymis. The inner layer of the tunica vaginalis is a shiny peritoneal covering over the testis.
5. The **epididymis** is a comma-shaped structure covering the dorsal and lateral surfaces of the testis. Sperm are moved from the seminiferous tubules in the testis into the epididymis where they mature, gaining motility and the ability to fertilize an egg.
6. Locate the "squiggly" **ductus deferens** as it arises from the caudal surface of the epididymis and ascends within the spermatic cord. Follow it cranially as it travels through the inguinal canal into the abdominal cavity. In the abdomen, the ductus deferens travels medially, looping over the **ureter**, and then continues caudally, dorsal to the urinary bladder and the urethra. The two ductus deferentes enter the urethra at the prostate gland.
7. The **prostate** is located farther caudally in cats than in humans. To locate the prostate, follow the two ductus deferens behind the **urinary bladder** and the **urethra** to the point where they merge in the midregion of the urethra. This is the location of the prostate. Pass your

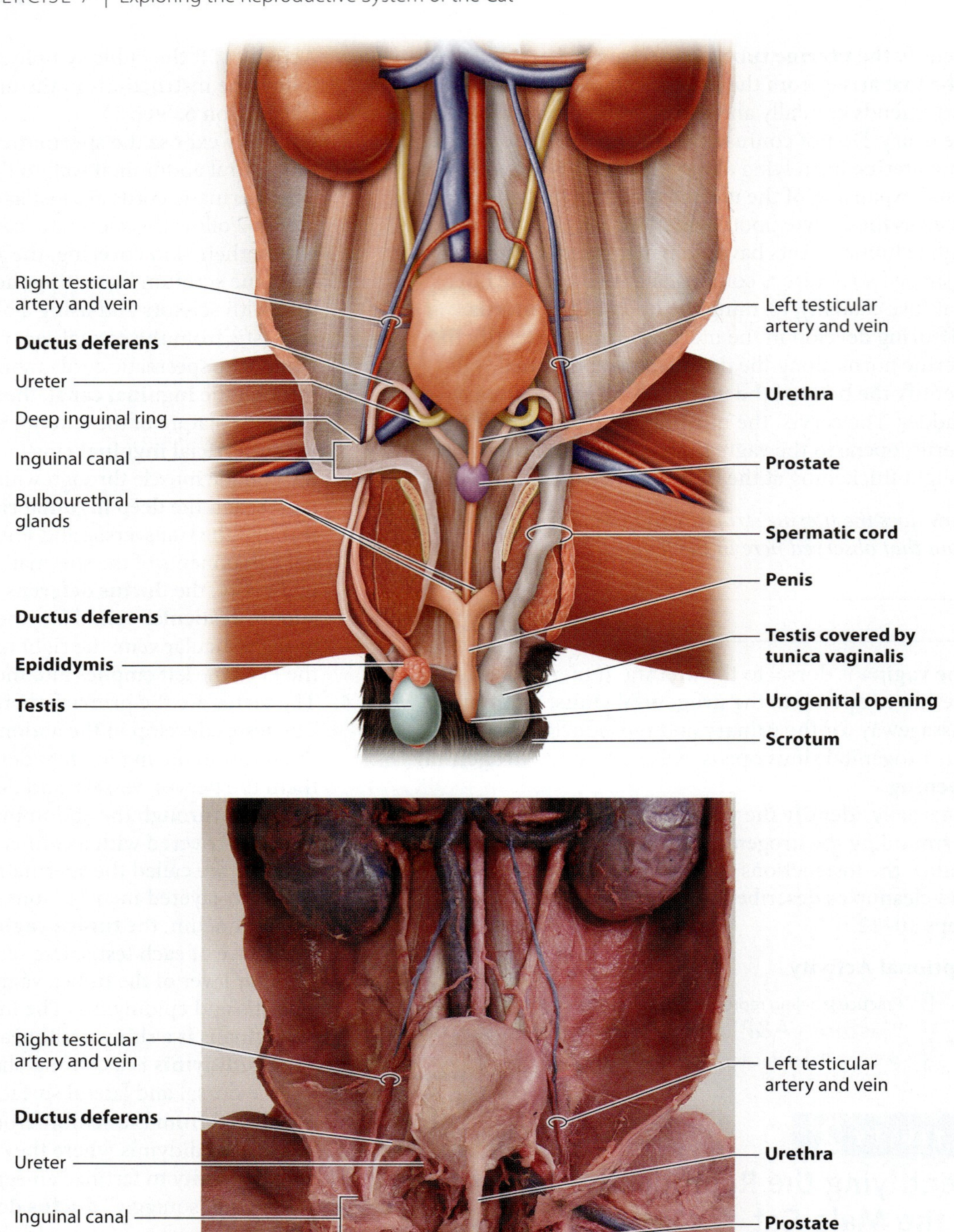

Figure D7-2 Reproductive organs of the male cat.

finger dorsal to the bladder and urethra to better view the prostate, which appears as a slight thickening.

Describe the location of the prostate relative to the urinary bladder in human males.

____

8. In males the urethra has three portions: the prostatic urethra, which passes through the prostate; the membranous urethra, which passes through the urogenital diaphragm; and the spongy urethra, which passes through the corpus spongiosum in the penis. Identify these urethral regions in the male cat.
9. Located on either side of the membranous urethra, dorsal to the root of the penis, are the **bulbourethral glands**. They are difficult to identify in the cat because they are extremely small and resemble fat.
10. Identify the **penis**, composed of three erectile tissues: the paired lateral corpora cavernosa and the midline corpus spongiosum. The spongy urethra travels through the corpus spongiosum and opens at the **urogenital opening**.
11. Follow the instructions for storage, waste disposal, and cleanup as described in Dissection 1, Activity 1, steps 10–12.
12. **Optional Activity**

Practice labeling reproductive structures on male cats at MasteringA&P® > Study Area > Practice Anatomy Lab > Cat > Reproductive System > Images 1–4

Review Questions

1. Number the ducts in the male reproductive system in the order through which sperm passes, beginning with their production in the seminiferous tubules of the testis (1) to ejaculation through the urogenital opening (7).

1 ______ a. seminiferous tubules of testis

_______ b. prostatic urethra

_______ c. ductus deferens

_______ d. membranous urethra

_______ e. epididymis

_______ f. spongy urethra

7 ______ g. urogenital opening

2. Name the two glands in the cat that contribute to semen.

____

3. Name the additional gland that contributes to semen in humans.

____

4. In which structure of the reproductive tract of the female cat do kittens develop?

____

5. Trace the path of an oocyte as it travels through the female reproductive tract, beginning with ovulation from the ovary (1) and ending at the urogenital opening (7).

1 _______ a. ovulation from the ovary
_______ b. uterine horn
_______ c. uterine body
_______ d. uterine tube
_______ e. urogenital sinus
_______ f. vagina
7 _______ g. urogenital opening

A

B

D

E

F

J

K

L

M

N

Q

T

U

V

W

X

Y

Z